PROCEEDINGS OF THE FOURTEENTH ANNUAL SYMPOSIUM ON

COMPUTATIONAL GEOMETRY

June 7-10, 1998
Minneapolis, Minnesota

Sponsored by the ACM Special Interest Groups

for Graphics and Algorithms and Computation Theory

acm PRESS

The Association for Computing Machinery
1515 Broadway
New York New York 10036

ACM ISBN: 0-89791-973-4

Additional copies may be ordered prepaid from:

ACM Order Department
PO Box 11414
Church Street Station
New York, NY 10286

Phone: 1-800-342-6626
(US and Canada)
+1-212-626-0500
(all other countries)
Fax: +1-212-944-1318
E-mail: acmpubs@acm.org

ACM Order Number: 429980
Printed in the USA

Foreword

The papers in this volume were presented at the *Fourteenth Annual Symposium on Computational Geometry*, held June 7-10, 1998, in Minneapolis, Minnesota. The conference included an applied track, a theoretical track, and a video review. In addition, there were a number of invited speakers on a variety of topics related to computational geometry, and a panel discussion on "The Theory/Application Interface." These talks were organized by Nina Amenta and Shang-Hua Teng, with funding provided by the Geometry Center, a National Science Foundation Science and Technology Center at the University of Minnesota.

The symposium was sponsored by two Special Interest Groups of the Association for Computing Machinery: SIGGRAPH, for computer graphics, and SIGACT, for algorithms and computation theory. Some support for students was provided by Lucent Technologies.

The symposium continues the two-track format begun at the 1997 meeting, with an applied track and a theory track, each with its own program committee, and with separate sessions. Of the 53 submissions to the applied track, 19 were accepted, and of the 57 submitted to the theory track, 25 were accepted.

These papers generally present preliminary results of ongoing research. It is expected that most papers will eventually appear in more polished form in refereed journals.

The program committees are grateful to the people who helped referee the papers; these referees are listed on page v. Many referees gave quite substantial and specific comments and suggestions, and many of these were forwarded to the authors.

Each video in the Seventh Video Review of Computational Geometry is accompanied by a short description; these papers are introduced on page 402.

The program committees wish to thank all authors who submitted abstracts in response to the call for papers.

Conference Chair: Ravi Janardan (Univ. Minnesota)

Program Committees:

Applied track:

Marshall Bern (Xerox Parc)
Jean-Daniel Boissonnat (INRIA)
Fred Bookstein (Univ. Mich)
Tamal Dey (I.I.T. Kharagpur)
Gershon Elber (Technion)
Steve Fortune (Bell Labs)
Randy Franklin (RPI)
Ken Goldberg (Berkeley)
Leo Guibas (Stanford Univ.)
Christoph Hoffmann (Purdue Univ.)
Dinesh Manocha (UNC)
Nick Patrikalakis (MIT)
Jarek Rossignac, chair (GVU/Georgia Tech)
Francois Sillion (Imag)
Steven Skiena (SUNY at Stony Brook)

Theoretical track:

Chanderjit Bajaj (Univ. Texas, Austin)
Bernard Chazelle (Princeton Univ.)
Ken Clarkson, chair (Bell Labs)
John Hershberger (Mentor Graphics)
Jiří Matoušek (Charles Univ.)
Mark Overmars (Utrecht Univ.)
János Pach (New York, Budapest)
Micha Sharir (Tel Aviv Univ.)
Michiel Smid (Univ. Magdeburg)
Steve Vavasis (Cornell Univ.)
Mariette Yvinec (CNRS, I3S)

Symposium Sponsored by:
SIGGRAPH, the ACM Special Interest Group for Computer Graphics, and SIGACT, the ACM Special Interest Group for Algorithms and Computation Theory

With support from:
The Geometry Center
Lucent Technologies

Referees

Pankaj Agarwal
Steve Alpern
Nina Amenta
Nur Arad
Estie Arkin
Lou Billera
Hervé Brönnimann
Paul Callahan
Frederic Cazals
J-M Chassery
Danny Chen
Lawrence Cowsar
Tim Culver
Mark de Berg
Matthew Dickerson
Fredo Durand
Alon Efrat
Ioannis Emiris
Amos Fiat
Anath Fischer
Zoltán Füredi
Michael Garland
Danny Halperin
Sariel Har-Peled
Tiko Kameda
John Keyser
Rolf Klein
Jon Kleinberg
Jan Kratochvil
Shankar Krishnan
Hans-Peter Lenhof
John Leonard
Christos Levcopoulos
Ming C. Lin
Yarden Livnat
Takashi Maekawa
Anil Maheshwari
Victor Milenkovic
Joe Mitchell
Frederic Noel
Ricky Pollack
Stephane Riviere
Günter Rote
Takis Sakkalis
Stefan Schirra
Sven Schuierer
Raimund Seidel
Guoling Shen
Evan Sherbrooke
Michiel Smid
Jack Snoeyink
Myung Soo Kim
Torsten Suel
Ayellet Tal
Marek Teichmann
Monique Teillaud
Seth Teller
George Turkiyyah
Pavel Valtr
Frank van der Stappen
Chris Van Wyk
Amitabh Varshney
Stephen Vavasis
Remco Veltkamp
Gordon Wilfong
Chee Yap
Mariette Yvinec

Fourteenth Annual ACM Symposium on Computational Geometry

June 7 – 10, 1998
Minneapolis, Minnesota, USA

Table of Contents

Sunday, June 7

Session 1, applied track
Chaired by Jarek Rossignac, GVU/Georgia Tech

Session 2, theoretical track
Chaired by Michiel Smid, Univ. Magdeburg

Monday, June 8

Session 3, applied and theoretical track
Chaired by Steve Fortune, Bell Labs

Session 8, theoretical track
Chaired by Mariette Yvinec, CNRS, I3S

Session 9, theoretical track
Chaired by John Hershberger, Mentor Graphics

Wednesday, June 10

Session 10, applied track
Chaired by Tamal Dey, I.I.T. Kharagpur

Session 11, theoretical track
Chaired by Mark Overmars, Utrecht Univ.

Video Review
Chaired by Dan Halperin, Tel Aviv Univ.

Rotational Polygon Containment and Minimum Enclosure

Victor J. Milenkovic
University of Miami
Department of Mathematics and Computer Science*

Abstract

An algorithm and implementation is given for *rotational polygon containment*: given polygons $P_1, P_2, P_3, \ldots, P_k$ and a container polygon C, find rotations and translations for the k polygons that place them into the container without overlapping. A version of the algorithm and implementation also solves *rotational minimum enclosure*: given a class $\mathcal{C}$ of container polygons, find a container $C \in \mathcal{C}$ of minimum area for which containment has a solution. Minimum enclosure algorithms are given for the following classes: 1) rectangles of fixed width, 2) scaled copies of a fixed convex polygon, 3) arbitrary rectangles. Containment and minimum enclosure are NP-hard (even in the purely translational case). The minimum enclosure is approximate: it bounds the the minimum area between $(1-\epsilon)A$ and A. Experiments are done to determine the largest practical value of k for both containment and minimum enclosure. Important applications for these algorithm to industrial problems are discussed. The paper also gives practical algorithms and numerical techniques for robustly calculating polygon set intersection, Minkowski sum, and *range intersection*: the intersection of a polygon with itself as it rotates through a range of angles.

1 Introduction

A number of industries generate new parts by cutting them from stock material: cloth, leather (hides), sheet metal, glass, *etc.* These industries need to generate dense non-overlapping layouts of polygonal shapes. Because fabric has a grain, apparel layouts usually permit only a finite set of orientations. Stripes, plaids, or other patterns on the fabric can further limit the allowed orientations and translations. Nevertheless, apparel manufacturers often allow small rotations (called "tilting"), typically not more than 3 degrees, but still large enough to make a significant difference in cloth utilization. Glass and sheet metal (and sometimes leather) have no grain (or stripes or plaids), and therefore layout on these materials allows arbitrary orientations for the parts.

The problem these manufacturers need to solve is either *containment* or *minimum enclosure*.

Containment Given polygons P_1, P_2, ..., P_k and a fixed container C, place the polygons into C without overlapping. As in the case of overlap minimization, P_0 denotes $\overline{C}$, in which case the goal becomes to find a non-overlapping layout of P_0, P_1, ..., P_k.

Minimum Enclosure Find a non-overlapping layout of P_1, P_2, ..., P_k that minimizes some measure of the container. The most common version in the textile industry is *strip packing*: minimize the length of a rectangular container R of fixed width. It is also useful to minimize the area of R.

The values of k one sees is often in the dozens or even hundreds. Unfortunately, even the translational versions of containment and minimum enclosure are NP-hard, and therefore one has to expect the running time of a containment or minimum enclosure algorithm to be exponential in k. In the apparel industry, no layout software has yet replaced a human. All hope is not lost: one has to use a heuristic or meta-heuristic of which there are many for the layout problem.

In the translational case, we have demonstrated [8] that a containment *algorithm* for modest k ($k = 5$ or even perhaps $k = 2$) is an excellent *tool* for the creation of containment *heuristics* for larger values of k, even exceeding expert human performance in some case. This work is based on packing one column at a time, but it is only one among many ways a "large k" heuristic can use a "small k" algorithm. For instance, subsets of the polygons might be packed together tightly and then the union treated as a single polygon.

This paper examines the following question: for what values of k is it possible to practically solve rotational containment and minimum enclosure problems *algorithmically*? In other words, for what value of k is it necessary to switch to heuristics and give up on performance guarantees? The two main issues are running time and numerical robustness.

1.1 New Results.

This paper presents new algorithms and robust implementations for rotational polygon containment and minimum enclosure and implementations. As far as we know, there are

*http://www.cs.miami.edu/~vjm, vjm@cs.miami.edu. This research was funded by the Textile/Clothing Technology Corporation from funds awarded by the Alfred P. Sloan Foundation, by NSF grants CCR-91-157993 and NSF-CCR-97-12401, and by a subcontract of a National Textile Center grant to Auburn University, Department of Consumer Affairs.

SCG 98 Minneapolis Minnesota USA
Copyright ACM 1998 0-89791-973-4/98/ 6...$5.00

no other *algorithms* for $k > 2$ or implementations for $k \geq 2$ for multiple non-convex polygons. Minimum enclosure algorithms are given for the following classes: 1) rectangles of fixed width, 2) scaled copies of a fixed convex polygon, 3) arbitrary rectangles. The minimum enclosure is approximate: it bounds the the minimum area between $(1-\epsilon)A$ and A. Experiments are done to determine the largest practical value of k for both containment and minimum enclosure.

The property that distinguishes an *algorithm* from a *heuristic* is the ability to say "no". The key to detecting infeasibility in our algorithms is the *range intersection* $P(\alpha, \beta)$, of a polygon P: the intersection of P with itself as it rotates through the ranges of angles from α to β about the origin. The new containment/enclosure algorithms integrate the range intersection into the framework we developed originally for solving translational containment/minimum enclosure [28, 33].

This paper also shows that the range intersection has linear complexity, although it may have circular arcs in its boundary. It gives practical algorithms for computing the range intersection and for approximating it by a polygon in a manner that is suitable for the containment/enclosure algorithms.

Like our earlier translational algorithms, our new containment/enclosure algorithms are very numerically demanding. They apply polygon set intersection and Minkowski sum over and over to the same collection of polygons. *Cascading*, the repeated use of the output as the next input, cannot be implemented using "pure" exact arithmetic because the number of bits required to represent coordinates can grow exponentially [31]. Cascading also quickly crashes any naive rounded arithmetic implementation. This paper presents a new hybrid technique *nearest pair rounding* for implementing polygon set operations in floating point arithmetic. Currently, nearest pair rounding must use some exact arithmetic to ensure correctness, but we have created an version that uses no exact arithmetic at all and which, so far, has been numerically robust, even under the cascading "torture test".

1.2 Related Work.

In the literature, there are essentially three approaches to layout of non-convex polygons: 1) heuristics and meta-heuristics (neural nets, simulated annealing, genetic algorithms), 2) "classical" computational/combinatorial geometry, 3) a combination of computational geometry and mathematical programming (which we use here).

Heuristics and meta-heuristics are limited in theory because they cannot say "no" if there is no solution. According to our sources in the apparel industry, available layout software is limited in practice to falling about 5% behind humans in cloth utilization. There are a number of surveys of packing/nesting heuristics [11, 7, 38, 10, 36, 12]. Recent work has used simulated annealing [21], boundary matching [22], grouping of polygons into sub-rectangles [13], genetic algorithms [5], database driven layout [23, 24], or a hybrid approach [18, 19, 37, 34]. Heuristic approaches have been tailored to sheet metal [35] and leather [20], both of which permit rotations.

Chazelle [6] introduced the single-polygon containment problem: place m-gon P into n-gon container Q. For convex Q, the running time bound is $O(mn^2)$, and for non-convex P and Q, $O(m^3n^3(m+n)\log(m+n))$. Avnaim *et al.* [4, 3] improve this to $O(m^3n^3\log(m+n))$. Most recent work deals with finding the largest copy of a convex P that can be placed, which is equivalent to finding the minimum (scaled) enclosure. For convex Q the best running time is $O(mn^2\log n)$ [1], and for non-convex Q, $O(m^2n^2)$ [2]. Grinde and Cavalier use an extensive case analysis and linear programming to place a single convex P [15] or convex P_1 and P_2 [16]. The running time of the first algorithm appears to be $O(m^2n^3)$, and it is not clear what the running time of the second is, but for one of its cases they are able to use parametric programming and find a solution by solving $O(m^4n^4)$ linear programs.

For multi-polygon *translational* layout, we have had considerable theoretical and practical success using a combination of computational geometry and mathematical programming (CG/MP) [9, 29, 28]. These algorithms can practically solve containment for up to ten polygons and minimum enclosure for at least five. We have also proved a number of theoretical running time bounds including $O((m^2+mn)^{2k}\log n)$ for placing k non-convex m-gons into a non-convex n-gon and $O(m^{4k-4}\log m))$ for placing them into a minimum area, fixed orientation rectangle [27] (this paper also surveys other results in multi-polygon translational layout).

1.3 Relation of New Algorithm to Previous Work

Avnaim and Boissonnat gave a formula for placing two polygons in a container. This formula works for both the translational and the rotational case. Our previous translational algorithm generalizes this work to $k > 2$ polygons through the use of a branch and bound paradigm. It is possible to generalize our translational algorithm to handle the rotational case in the same manner. One might call this the *true generalization* of our translational algorithm.

The algorithm presented here is *not* the true generalization. The true generalization would require robust set operations on subsets of $\mathbf{R}^2 \times S^1$. These subsets would either represent either a) valid configurations (translation plus rotation) of a single polygon to place it in the container or b) the set of valid (non-overlapping) relative positions of one polygon with respect to another. Unfortunately, the configuration sets are three dimensions with curved surface boundary. As yet, there are no numerically robust ways to implement the necessary set operations. As we have previously stated, the branch and bound paradigm cascades these set operations in a manner that is extremely stressful numerically.

The algorithm presented here only requires set operations on polygons in two dimensions, which we know how to do robustly. It is likely that the true generalization would run faster than the current algorithm–if and when it could be implemented robustly. The algorithm presented here represents a practical tradeoff of running time for numerical robustness.

1.4 Outline.

Section 2 gives the algorithms for rotational containment and minimum enclosure using the range intersection. Section 3 analyzes the range intersection and gives algorithms for constructing it and useful polygonal approximations to it. Section 4 gives the nearest pair rounding algorithm. Finally, Section 5 gives results.

2 Containment/Enclosure Algorithms

The rotational containment/enclosure algorithm uses a "branch and bound" approach which generalizes our previous practical translational containment algorithm [28, 33]. However, this is also essentially the approach used by many mathematical programming algorithms.

The input to the containment algorithm is a set of polygons $P_1, P_2, P_3, \ldots, P_k$ and a container polygon C. In the case of minimum enclosure, the container is replaced by a set $\mathcal{C}$ of containers. A *configuration* is an assignment of translations $t_1, t_2, \ldots, t_k$ and rotations $\theta_1, \theta_2, \ldots, \theta_k$ to the polygons. To simplify the notation, it is convenient to replace the container by an additional polygon $P_0 = \overline{C}$ which is the complement of the container and to fix its translation t_0 and rotation θ_0 at $(0,0)$ and 0, respectively.

In the case of containment, the goal is to find all configurations (or at least one configuration) with zero overlap among the polygons. Since the complement of the container is P_0, this means that $P_1, P_2, \ldots, P_k$ lie inside the container. In what follows these configurations are called the *solutions*. In the case of minimum enclosure, the goal is to find a non-overlapping layout which fits in the minimum area container $C \in \mathcal{C}$.

2.1 Abstract Algorithm.

We use the same language as our translational work. A *hypothesis* is a set of constraints on the configuration. *Restriction* adds to this set of constraints in a way that is guaranteed not to eliminate any solutions. *Evaluation* attempts to find a solution within the hypothesis (that satisfies its constraints). *Subdivision* splits a hypothesis H into two sub-hypotheses H' and H'' such that all solutions in H reside in (satisfy all the constraints of) either H' or H''.

The containment algorithm first generates a root hypothesis: a set of constraints which all solutions must satisfy. It restricts this hypothesis and evaluates it. If it cannot either a) restrict it to the empty set or b) find a solution, it subdivides the hypothesis and recurses on the two subhypotheses. A hypotheses for which evaluation is successful is called a *solution hypothesis*. If only one solution is desired, the algorithm can stop when it finds the first one. Otherwise, it generates a set of solution hypotheses. Note that we only find one solution for each solution hypothesis, not all. In the translational case, it is possible to find all solutions [27], but we have not done this for the rotational case.

The minimum enclosure algorithm acts the same as the containment algorithm, except that every time it finds a solution hypothesis, it establishes a new upper bound on the area of the minimum container. It uses this upper bound as a constraint to restrict the previously discovered solution hypotheses, and it recurses on these. Its output is a solution hypothesis with minimum area (or possibly several solution hypothesis whose solutions have the same area). To establish that the area is indeed minimal, it is necessary to run the algorithm on its output hypotheses with a slightly smaller upper bound. We first diminish the upper bound by 1%. If no new solution is found, then we roll back and try diminishing the upper bound by 0.01%, and so forth. By these means, the upper bound can be established to any degree of numerical accuracy. In theory, it might be possible to apply symbolic perturbation to the solution to establish that there is no solution for any smaller area. In practice, this is not possible since much of our algorithm is numerical.

2.2 Root Hypothesis.

The hypothesis for the containment algorithm takes the following form,

$$\alpha_i \leq \theta_i \leq \beta_i, \qquad 0 \leq i \leq k, \tag{1}$$

$$t_j - t_i \in U_{ij}, \qquad 0 \leq i < j \leq k, \tag{2}$$

where α_i and β_i are bounds on the angle and U_{ij} is a (usually non-convex) planar region.

In the case of translational containment [28, 33] all solutions must satisfy,

$$t_j - t_i \in \overline{P_i \oplus P_j}, 0 \leq i < j \leq k, \tag{3}$$

where $\oplus$ is the Minkowski sum,

$$A \oplus B = \{a + b \,|\, a \in A \text{ and } b \in B\}.$$

Hence, we set $U_{ij} = \overline{P_i \oplus P_j}$ in the root hypothesis. However, this does not work for rotational containment.

Define the *range intersection*

$$P(\alpha, \beta) = \bigcap_{\alpha \leq \theta \leq \beta} P(\theta), \tag{4}$$

where $P(\theta)$ is P rotated by angle θ.[1] The range intersection is the intersection of all copies of P as it is rotated from α to β. The following lemma provides a way of constructing a root hypothesis for rotational containment.

Lemma 2.1 *There is a solution to rotational containment for $\alpha_i \leq \theta_i \leq \beta_i$, $0 \leq i \leq k$ only if (but not if, alas) there is a solution to translational containment for $P(\alpha_i, \beta_i)$, $0 \leq i \leq k$.*

Proof. If $\alpha_i \leq \theta_i \leq \beta_i$, then $P(\alpha_i, \beta_i) \subseteq P(\theta_i)$. Given the solution to rotational containment, just use the same translations. □

In the case of rotational containment, we set

$$\alpha_i = 0, \beta_i = 2\pi, 0 \leq i \leq k,$$

in the root hypothesis, and set

$$U_{ij} = \overline{P_i(0, 2\pi) \oplus P_j(0, 2\pi)}, 0 \leq i < j \leq k.$$

By Equation 2.1 and Equation 3, all solutions to containment will satisfy the constraint Equations 1 and 2 for this hypothesis.

In the case of minimum enclosure, if the set $\mathcal{C}$ of containers is either 1) a set of rectangles with fixed width, 2) a set of similar (scaled) copies of a convex polygon, then we set C equal to the element whose size (length or scale) equals a known upper bound. (For sufficiently large size, it is trivial to find a solution.) We postpone the discussion of more general sets, such as 3) *all rectangles*, until Section 2.6.

2.3 Restriction.

In our previous work on translational containment, we established two types of restriction: *geometric restriction* [7] and *linear programming restriction* [27]. These restrictions were both derived from Equation 2, and therefore they apply in the case of rotational containment as well.

[1] All rotations are about the origin. It is assumed that all polygons contain the origin.

Repeatedly, for all triples h, i, j, geometric restriction replaces U_{ij} by a subset,[2] $U_{ij} \cap (U_{ih} \oplus U_{hj})$. It stops when no polygon diminishes in area by more than a threshold fraction. This repetition (cascading) can lead to numerical difficulties which we deal with using nearest pair rounding (Section 4). Note: for $k = 2$, geometric restriction is an exact algorithm. In this special case, it becomes essentially the same as Avnaim and Boissonnat's exact formula for this case [4, 3]. If either geometric restriction or linear programming restriction generates the empty set, then there is no solution in the current hypothesis.

2.4 Evaluation.

We have previously developed a practical algorithm for *rotational overlap minimization* [32]: find translations and rotations that minimizes the sum of the overlaps among $P_0, P_1, P_2, \ldots, P_k$. This algorithm can also perform *rotational compaction*: given a non-overlapping layout of P_1, P_2, ..., P_k inside a rectangular container C, find a layout which minimizes (local minimum) of the container while keeping the width fixed. In the full paper we show how to generalize compaction to minimize the area of a scaled convex enclosure or even the area of an arbitrary rectangular container. (The latter problem is non-linear.)

Evaluation for the containment algorithm simply consists of running rotational overlap minimization with a set of additional constraints derived from the current hypothesis. Since compaction is based on linear programming, the additional constraints must be convex. The additional constraints are a) the angle constraints, Equation 1, and b) the "relaxed" translation constraints, Equation 2 with each U_{ij} replaced by its convex hull.

For the minimum enclosure algorithm, if the algorithm finds a solution hypothesis, it compacts the non-overlapping layout into a local minimum area container. This establishes an upper bound on the area of the container which is also a local minimum.

2.5 Subdivision.

Subdivision has two cases based on the output of evaluation. If the translations satisfy Equation 2, then the output is a non-overlapping layout of the range intersections $P_i(\alpha_i, \beta_i)$ (Lemma 2.1). It is necessary in this case to subdivide an angle range. If the output does not satisfy Equation 2, then the algorithms use essentially the same subdivision algorithm as the translational containment algorithm. We explain the difference in the full paper.

To subdivide an angle range, the algorithm selects the polygon P_i whose total overlap with the other polygons is maximal at its current translation and rotation (the output of evaluation). If θ_i is closer to the midpoint of $[\alpha_i, \beta_i]$ than the endpoints, then the algorithm divides the range at θ_i. Otherwise, it divides the range at its midpoint. Let the two new ranges be $[\alpha'_i, \beta'_i]$ and $[\alpha''_i, \beta''_i]$. To generate H', it generates the root hypothesis for $[\alpha_i, \beta_i]$ replaced by $[\alpha'_i, \beta'_i]$ and adds these constraints to H. Generating H'' is analogous.

2.6 Minimum Area Rectangle Enclosure.

The preceding sections have described the whole of the containment and minimum enclosure algorithms for the case of a rectangle with fixed width and variable length or the case of scaled copies of a fixed convex polygon. The case of minimum area rectangle with arbitrary length and width requires a more sophisticated root hypothesis. We summarize the idea here. It is necessary to add an additional "container" polygon P_{k+1} where $U_{0,k+1}$ is the region $xy \leq A$. This enforces that that container an arbitrary rectangle with area no greater than A. To keep everything polygonal, we approximate the hyperbola $xy = A$ by a polygon.

[2] Replacing U_{ij} by a subset is equivalent to adding more constraints since the previous constraints must still hold.

3 Range Intersection

Section 2.2 gave a definition for the *range intersection* $P(\alpha, \beta)$ of a polygon P: the intersection of all copies of P as it is rotated from angle α to angle β about the origin. The range intersection is an "arc-gon": a planar region bounded by line segments and circular arcs. In particular, these circular arcs are concentric with the origin.

The set intersection, convex hull, and even Minkowski sum of arc-gons is an arc-gon, and therefore it is possible in principle to implement all the operations used in Section 2 on arc-gons. However, generalizing the overlap minimization/compaction algorithm to arc-gons would require a switch from linear program to quadratic or convex programming. Also, it is difficult to handle the intersection of circular arcs in a numerically robust fashion, which would make it difficult to implement the intersection and Minkowski sum of arc-gons.

This section establishes that the complexity of the range intersection $P(\alpha, \beta)$ is linear in the complexity of P. It gives a way to approximate arc-gons by polygons in a way that is suitable for the containment/enclosure algorithms. In particular, the approximation is a) a subset b) whose error is proportional to $(\alpha - \beta)^2$ and c) whose complexity is also linear. Finally, it gives the "practical" implementation we use to construct approximate range intersections.

3.1 Theoretical Complexity.

Lemma 3.1 *The range intersection $P(\alpha, \beta)$ has complexity* $O(|P|)$.

Proof. Construct a *concentric trapezoidalization* of P which is analog to the standard trapezoidalization of P. From each vertex v of P extend an arc (centered at the origin) in each direction (clockwise or counter-clockwise) that goes into the interior of P. Terminate the arc when it hits the boundary of P. This procedure cuts P into a linear number of *concentric trapezoids*: regions bounded by two line segments and two concentric circular arcs.

The intersection $P(\alpha, \beta)$ is a rotated copy of $P(0, \gamma)$, where $\gamma = \beta - \alpha$. The set $P(0, \gamma)$ is equivalent to the set of counter-clockwise endpoints of arcs of angle γ which lie inside P. It is clear that each such arc lies in a single concentric trapezoids. Therefore, the range intersection of each concentric trapezoids can be computed independently in constant time. It is also an concentric trapezoids. The range intersections of the concentric trapezoids meet only at their boundaries and in a nice fashion. It is easy to show that the complexity of the boundary of their union is also linear: a linear number of arcs and line segments. □

3.2 Polygonal Approximations.

In order to compute a polygonal approximation to $P(\alpha, \beta)$, it is necessary to perform a *radial trapezoidalization.* From each vertex v of $P(\alpha, \beta)$, extend a line towards the origin

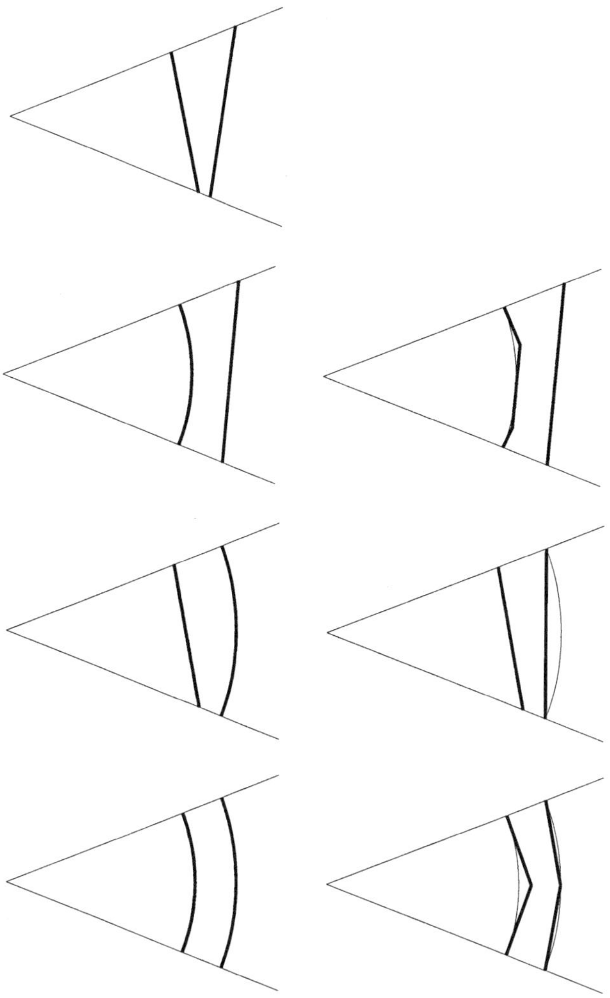

Figure 1: Radial trapezoids and their inner polygonal approximations.

and/or away from the origin, in each direction that goes into the interior of $P(\alpha, \beta)$. Terminate a line when it hits the boundary.

Figure 1 shows the four types of radial trapezoids and how to form inner polygonal approximations. The most complex case has an inner arc and outer line segment. We approximate the inner arc by three tangent line segments: two of them tangent at the endpoints and the middle one parallel to the outer bounding line segment. In the case of two arcs, the polygonal approximations can intersect. In this case, we lose the middle part of the arc, and the topology changes as illustrated in Figure 2.

Lemma 3.2 *Let $B(r)$ be the ball of radius r. Let $Q(\alpha, \beta)$ be the polygonal approximation to the range intersection $P(\alpha, \beta)$. For $|\gamma|$ sufficiently small ($\gamma = \beta - \alpha$),*

$$Q(\alpha, \beta) \subseteq P(\alpha, \beta) \subseteq Q(\alpha, \beta) \oplus B(\mathrm{O}(\gamma^2)).$$

Proof. Clearly the approximation is a subset. It can be shown that $P(\alpha, \beta)$ can have no arc in its boundary longer than $|\gamma|$. Basic trigonometry shows that a chordal approximation to an outer arc has error at most $1 - \cos(\gamma/2) \approx \gamma^2/4$ times the radius. A tangential approximation to an inner arc has error at most $1/\cos(\gamma/2) - 1 \approx \gamma^2/4$ times its radius.

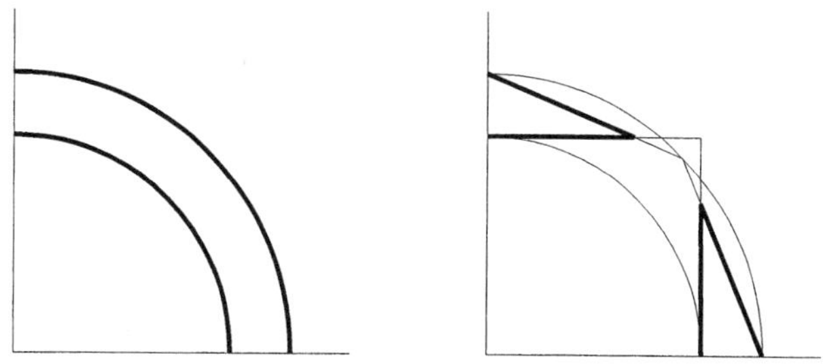

Figure 2: Polygonal approximation of two-arc trapezoid can have different topology.

The difference between the inner and the outer radius is at least the thickness of the thinnest concentric trapezoid of the original trapezoidalization of P. This value is independent of γ. Therefore, for sufficiently small γ, the topological change of Figure 2 does not occur. Each individual radial trapezoid approximation satisfies the condition of the lemma, and therefore their union does too. □

Lemma 3.2 is exactly the property we need to ensure that we can substitute the polygonal approximation to the range intersection for the actual range intersection in the containment/enclosure algorithms. As the containment algorithm subdivides the angle ranges, the accuracy of the approximation increases with the square of the angle range. Thus we expect to have to divide a given angle range no more than b times to obtain accuracy at least $\epsilon = 2^{-b}$.

3.3 Practical Algorithm.

In the full paper we describe the "practical algorithm" we use in lieu of constructing a full circular trapezoidalization. We decompose the complement $\overline{P}$ into *angularly monotonic* regions: each ray out of the origin intersects the region in a single line segment, a points, or the empty set. This implies that a monotone component M is bounded by two functions of angle:

$$M = \{(r, \theta) \mid f(\theta) \leq r \leq g(\theta)\},$$

in polar coordinates. We compute the minimum of $f()$ over a angle "window" of size α as the window rotates about the origin. Similarly, we compute the maximum of $g()$ over a sliding window of angle α. We compute an inner polygonal approximation to the minimum curve and an outer polygonal approximation to the maximum curve. The two curves bound an approximation to the range union $M(0, \gamma)$ of the component M. Finally, we take the complement of the union of the range unions.

4 Nearest Pair Rounding

In previous work, [26, 25, 30, 31], we give methods for *geometric rounding* of line segment arrangements: reducing the precision of the coordinates while maintaining a consistent topology. Geometric rounding is absolutely necessary in cases of *cascading*: the output of one operation repeatedly becomes the input to another. This section summarizes a new geometric rounding technique: *nearest pair rounding.* We use this technique in our current implementations of polygon intersection and Minkowski sum.

Like our previous method, *shortest path rounding*, nearest pair rounding can round to the non-uniform lattice of points with floating point coordinates. Other rounding methods [14, 17], can only round to the integer lattice. Nearest pair rounding is much simpler than shortest path rounding, but it may have larger numerical error. We also discuss an experimental method in which we do not use exact arithmetic at all. So far this has worked well, even under the containment algorithm "torture test".

4.1 Exact Rounding Algorithm.

Nearest pair rounding repeats a rounding step until no more rounding is necessary. Each step "rounds together" the nearest pair of non-incident features: vertex/vertex or vertex/edge. If after rounding, any pair of edges are intersecting, a new vertex is created as near as possible to their actual intersection, and rounding resumes. Rounding stops when every pair of non-incident features is farther apart than some threshold and when no two edges are intersecting.

In the vertex/vertex case, the rounding step creates new vertex at the midpoint of the two vertices and moves both vertices to that new vertex along with an edges that have them as an endpoint. In the vertex/edge case, a new vertex is created halfway from the vertex to the edge. The original vertex is moved to this new vertex, along with edges that have it as an endpoint. The original edge is broken into two edges each which have the new vertex as one of their endpoints. The rounding step does not consider a vertex/edge pair unless the vertex forms an angle less than 45 degrees with either endpoint.

Theorem 4.1 *Nearest pair rounding converges without introducing an new intersections.*

Proof. The key to nearest path rounding is that the perturbation it introduces is at most 3/4 of the minimum distance between non-incident features. Therefore, applying the rounding step cannot cause other non-incident features to "collide". Because we ignore vertex/edge pairs unless the angles with the endpoints are less than 45 degrees, the measured minimum feature separation may be $\sqrt{2}$ times the actual minimum. However, rounding move features to a midpoint, and therefore the actual perturbation is at most $\sqrt{2}/2 < 3/4$ times the actual minimum feature separation.

Consider a partial order on sets of line segments, ordered on: 1) number of edge/edge intersections, 2) number of vertices, $3, 4, 5, \ldots$) length of longest edge, length of second longest edge, and so forth. Adding an intersection point decreases (1). Rounding two vertices together decreases (2) without increasing (1). Because of the 45 degree rule, breaking an edge forms two shorter edges without increasing (1) or (2). □

4.2 Experimental Rounded Arithmetic Rounding Algorithm.

In practice, we use floating point arithmetic for all operations and apply nearest pair rounding without using exact arithmetic to measure distances. Because the rounding step has $1 - 3/4 = 25\%$ "leeway" for error, we expect it to be numerically stable. So far, it has been stable even after hundreds of cascaded operations.

5 Results

All algorithms were implemented in C++. Overlap minimization code: 9100 lines. Polygon set operations: 2700 lines. Containment/Enclosure algorithms: 1200 lines. Overlap minimization uses CPLEX 4.0 by CPLEX Optimization, Inc. Experiments were run on a DEC Alpha 3000/700 which has roughly the same speed as a 266MhZ Intel Pentium II.

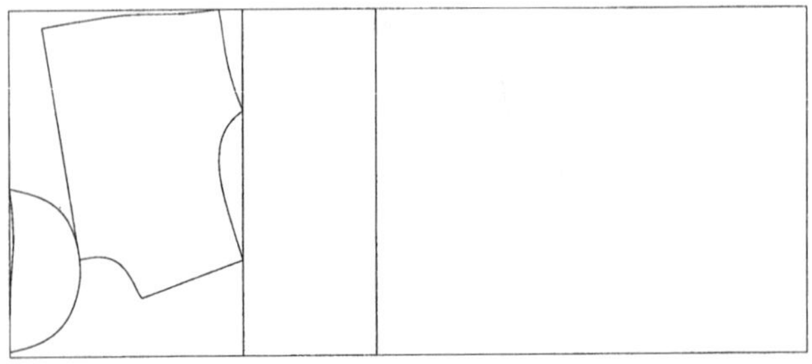

Figure 3: Target length: 2519. Output length after containment and leftward compaction: 1749.750.

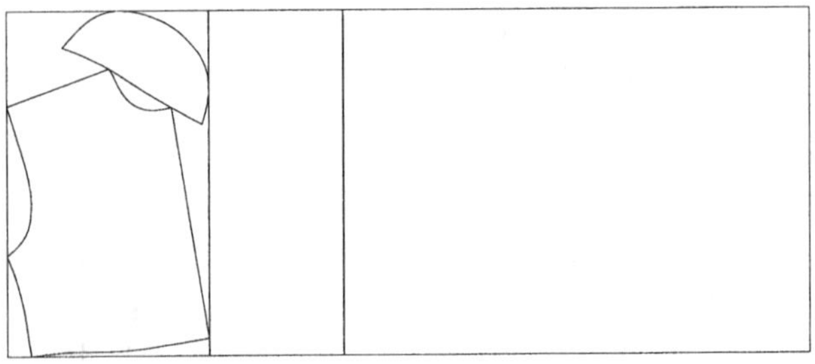

Figure 4: Target length: 1749. Output length after containment and leftward compaction: 1505.643.

5.1 Two Polygons

For the two-polygon example, we alternated containment with leftward compaction. Each positive example in Figures 3 through 5 ran in under a minute. There are still so many checks, prints, and dumps in the program, that running times under a minute are not very accurate.

Finally, in about 18 minutes, the containment algorithm evaluated 191 hypotheses and verified that there was no solution shorter than 1386.

5.2 Three Polygons

We ran the algorithm on industrial parts with 71, 65, and 64 vertices. The good news is that it appeared to find the minimum length enclosure quickly as shown in Figures 6 though 8.

The bad news is that we challenged the system with a third input length of 2173 (1% less than previous output) and it took more than 5 hours to evaluate 811 hypotheses and verify that there was no solution with this length. We did not even try running it with a target closer to the previous output, as we did in the two-polygon example.

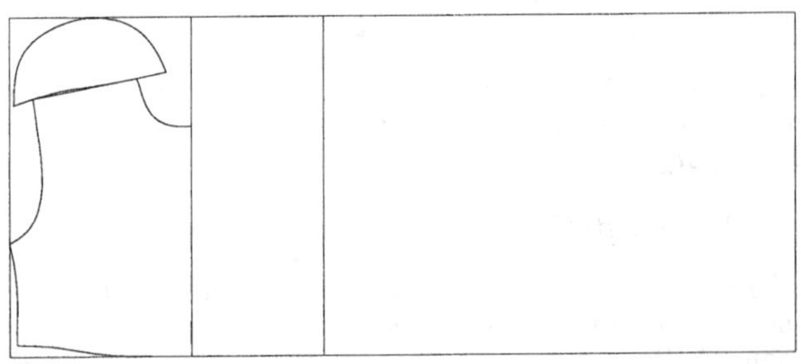

Figure 5: Target length: 1505. Output length after containment and leftward compaction: 1386.546.

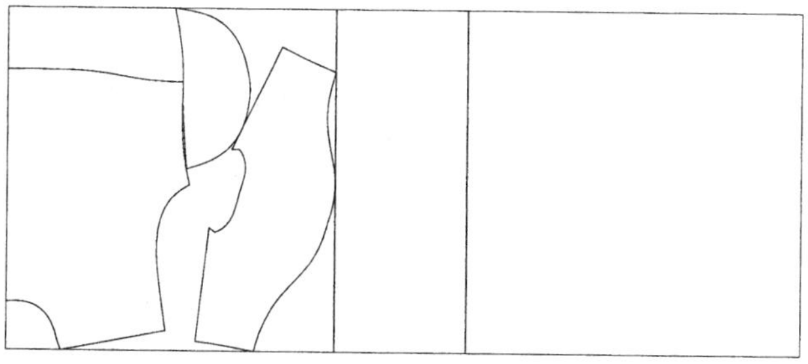

Figure 6: Hand placed and compacted leftwards. Length: 2496.542.

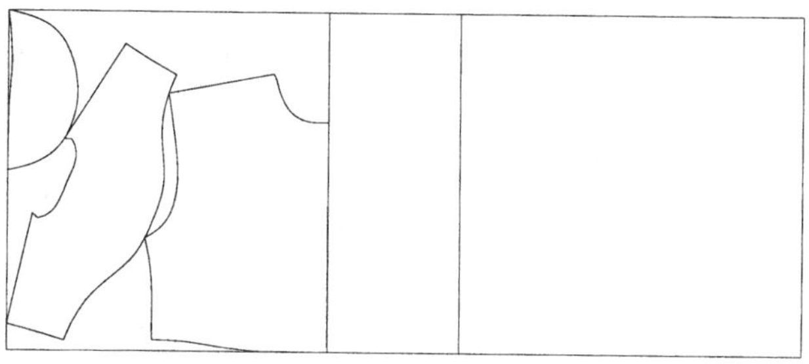

Figure 7: Target length: 2496. Output length after containment and leftward compaction: 2429.647. Time: 42 seconds.

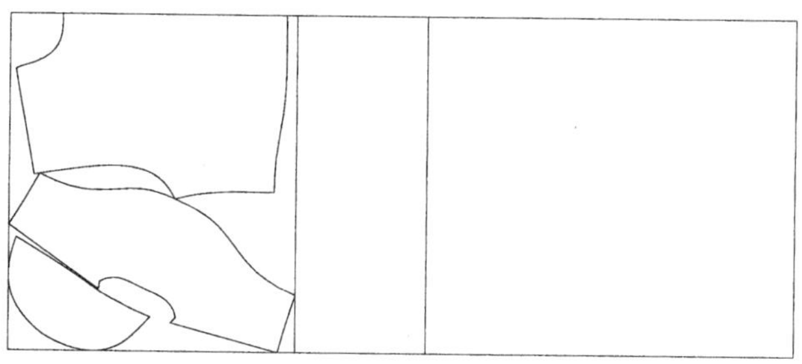

Figure 8: Target length: 2429. Output length after containment and leftward compaction: 2196.433. Time: 96 seconds.

6 Conclusion

At present, the range intersection method appears to be practical for two-polygon minimum enclosure. For three polygons, it appears to generate a minimum quickly but require a very long time to verify it.

Some modifications might speed up the algorithm. For instance, it can search the hypothesis space for each individual polygon and then each pair before considering three at time. That might speed up the three-polygon example somewhat because it appeared to sometimes make a bad choice for on polygon's angle and then spend much time investigating all hypotheses with this angle.

Another conclusion is that it is worth pursuing the "true generalization" of translational containment described in Section 1.3. This will involve performing complex set operations on curved three dimensional objects. It will probably be difficult to make these operations robust, but the current results demonstrate that these algorithms would be worth the effort. For instance, in the special case of two polygons, the true generalization would be essentially the same as Avnaim and Boissonnat's formula and thus solve this case exactly. Our experience with translational work would indicate that it could solve at least the three-polygon case and possibly the four-polygon case also in a reasonable amount of time.

Finally, we should not forget the first paragraph in which it was mentioned that apparel manufacturers only allow small tilts. All experiments in this paper were run with a full 360 degrees of rotation allowed. With only small rotations allowed, the range intersection algorithm should be practical for much larger values of k.

References

[1] P. Agarwal, N. Amenta, and M. Sharir. Largest placement of one convex polygon inside another. In *Proceedings of the Second Workshop on Algorithmic Foundations of Robotics*, July 1996.

[2] P.K. Agarwal, B. Aronov, and M. Sharir. Motion planning for a convex polygon in a polygonal environment. In preparation., 1992.

[3] F. Avnaim. *Placement et déplacement de formes rigides ou articulées.* PhD thesis, Université de Franche-Comté, France, 1989.

[4] F. Avnaim and J. Boissonnat. Polygon placement under translation and rotation. In *Proceedings of the 5th Annual Symposium on Theoretical Aspects of Computer Science; Lecture Notes in Computer Science (294)*, pages 322–333. Springer-Verlag, 1988.

[5] C. Bounsaythip and S. Maouche. Irregular shape nesting and placing with evolutionary approach. In *1997 IEEE International Conference on Systems, Man, and Cybernetics. Computational Cybernetics and Simulation (Cat. No.97CH36088-5*, page 5. IEEE; New York, NY, USA, 12-15 October 1997.

[6] B. Chazelle. The Polygon Containment Problem. In Preparata, editor, *Advances in Computing Research, Volume 1: Computational Geometry*, pages 1–33. JAI Press, Inc., Greenwich, Connecticut, 1983.

[7] K. Daniels. *Containment algorithms for nonconvex polygons with applications to layout.* Ph.D. thesis, Harvard University, Cambridge, MA, 1995.

[8] K. Daniels and V. Milenkovic. Column-based strip packing using ordered and compliant containment. In *Proc. 1st ACM Workshop on Appl. Comput. Geom.*, pages 33–38, 1996.

[9] K. Daniels and V. J. Milenkovic. Multiple Translational Containment, Part I: An Approximate Algorithm. *Algorithmica*, 19:148–182, 1997.

[10] K. A. Dowsland and W. B. Dowsland. Packing Problems. *European Journal of Operational Research*, 56:2 – 14, 1992.

[11] K.A. Dowsland and W.B. Dowsland. Solution approaches to irregular nesting problems. *European Journal of Operational Research*, 84(3):506–21, August 1995.

[12] H. Dyckhoff. A typology of cutting and packing problems. *European Journal of Operations Research*, 44:145–159, 1990.

[13] R. M. S. Abd El-Aal. A new technique for nesting irregular shapes based on rectangular modules. *Current Advances in Mechanical Design and Production 1996*, 6:533–540, 1996.

[14] D. H. Greene and F. F. Yao. Finite-resolution computational geometry. In *Proc. 27th Annu. IEEE Sympos. Found. Comput. Sci.*, pages 143–152, 1986.

[15] R. Grinde and T. Cavalier. Containment of a Single Polygon Using Mathematical Programming. Technical Report IMSE Working Paper 92-164, The Pennsylvania State University, Department of Industrial and Management Systems Engineering, 1993.

[16] R. B. Grinde and T. M. Cavalier. A new algorithm for the two-polygon containment problem. *Computers and Operations Research*, page to appear, 1996.

[17] Leonidas Guibas and David Marimont. Rounding arrangements dynamically. In *Proc. 11th Annu. ACM Sympos. Comput. Geom.*, pages 190–199, 1995.

[18] G. C. Han and S. J. Na. Two-stage approach for nesting in two-dimensional cutting problems using neural network and simulated annealing. *Journal of Engineering Manufacture*, 210(B6):509–19, 1996.

[19] R. Heckmann and T. Lengauer. Computing upper and lower bounds on textile nesting problems. *LECTURE NOTES IN COMPUTER SCIENCE*, 0302-9743(1136):392–405, 1996.

[20] Khoi Hoang. Aspects in automatic nesting of irregular shapes. In *Proc. SPIE - Int. Soc. Opt. Eng. (USA), Proceedings of the SPIE - The International Society for Optical Engineering, vol.2589*, pages 234–41. SPIE-Int. Soc. Opt. Eng, 23-26 October 1995.

[21] P. Jain, P. Fenyes, and R. Richter. Optimal blank nesting using simulated annealing. *Journal of mechanical design*, 114(1):160–165, March 1992.

[22] H. J. Lamousin, W. N. Waggenspack, and G. T. Dobson. Nesting of complex 2-d parts within irregular boundaries. *Journal of Mechanical Design*, 118(4):615, 1996.

[23] Z. Li. *Compaction algorithms for nonconvex polygons and their applications.* Ph.D. thesis, Harvard University, Cambridge, MA, 1994.

[24] Z. Li and V. Milenkovic. Compaction and separation algorithms for nonconvex polygons and their applications. *European Journal of Operations Research*, 84:539–561, 1995.

[25] V. Milenkovic. Double precision geometry: a general technique for calculating line and segment intersections using rounded arithmetic. In *Proc. 30th Annu. IEEE Sympos. Found. Comput. Sci.*, pages 500–505, 1989.

[26] V. Milenkovic. Rounding face lattices in the plane. In *Abstracts 1st Canad. Conf. Comput. Geom.*, page 12, 1989.

[27] V. Milenkovic. Translational polygon containment and minimal enclosure using linear programming based restriction. In *Proc. 28th Annu. ACM Sympos. Theory Comput. (STOC 96)*, pages 109–118, 1996.

[28] V. Milenkovic and K. Daniels. Translational polygon containment and minimal enclosure using geometric algorithms and mathematical programming. Technical Report 25-95, Center for Research in Computing Technology, Division of Applied Sciences, Harvard University, Cambridge, MA, 1995.

[29] V. J. Milenkovic. Multiple Translational Containment, Part II: Exact Algorithms. *Algorithmica*, 19:183–218, 1997.

[30] Victor J. Milenkovic. Practical methods for set operations on polygons using exact arithmetic. In *Proc. 7th Canad. Conf. Comput. Geom.*, pages 55–60, 1995.

[31] Victor J. Milenkovic. Shortest path rounding. *Algorithmica*, page (submitted), 1997.

[32] Victor J. Milenkovic. Rotational polygon overlap minimization. In *Proceedings of the 13th Annual ACM Symposium on Computational Geometry.* ACM, 1997 (Journal version accepted with revisions by *Computational Geometry: Theory and Applications*).

[33] V.J. Milenkovic and K. Daniels. Translational Polygon Containment and Minimal Enclosure using Mathematical Programming. *International Transactions in Operational Research (accepted with revisions)*, 1997.

[34] Y.K.D.V. Prasad. A set of heuristic algorithms for optimal nesting of two- dimensional irregularly shaped sheet-metal blanks. *Computers in Industry*, 24(1):55–70, May 1994.

[35] Y.K.D.V. Prasad, S. Somasundaram, and K.P. Rao. A sliding algorithm for optimal nesting of arbitrarily shaped sheet metal blanks. *International Journal of Production Research*, 33(6):1505–20, June 1995.

[36] P. E. Sweeney and E. R. Paternoster. Cutting and Packing Problems: A Categorized, Application-Oriented Research Bibliography. *Journal of the Operational Research Society*, 43(7):691–706, 1992.

[37] J. Y. Wang, D. Y. Liu, E. W. Lee, and T. H. Koh. An algorithm for nesting patterns in apparel. In *COMPUTER INTEGRATED MANUFACTURING - INTERNATIONAL CONFERENCE- WORLD SCIENTIFIC 1995; VOL 1*, pages 377–384. World Scientific; 1995, 1995.

[38] P.F. Whelan and B.G. Batchelor. Automated packing systems: review of industrial implementations. In *Proceedings of the SPIE - The International Society for Optical Engineering, vol.2064*, pages 358–69. SPIE, September 1993.

A General Framework for Assembly Planning: The Motion Space Approach

Dan Halperin*
Tel Aviv University

Jean-Claude Latombe†
Stanford University

Randall H. Wilson‡
Eastman Kodak Company

Abstract

Assembly planning is the problem of finding a sequence of motions to assemble a product from its parts. We present a general framework for finding assembly motions based on the concept of *motion space*. Assembly motions are parameterized such that each point in motion space represents a mating motion that is independent of the moving part set. For each motion we derive blocking relations that explicitly state which parts collide with other parts; each subassembly (rigid subset of parts) that does not collide with the rest of the assembly can easily be derived from the blocking relations. Motion space is partitioned into an arrangement of cells such that the blocking relations are fixed within each cell.

In the first part of the paper we give background material, present the motion space approach and describe applications of the approach to assembly motions of several useful types, including one-step translations, multi-step translations, and infinitesimal rigid motions. Several efficiency improvements over the basic algorithms are described. The resulting algorithms have been implemented and tested extensively on complex assemblies and we describe some of the experimental results. Most of the results described in this part have been published separately before and we integrate them here under the unifying concept of motion space.

In the second part (the last section) of the paper we survey major open problems in this domain. We discuss four topics where some progress has recently been achieved but which are in general still largely open: assembly planning with compound motions, with toleranced parts, and with additional practical constraints on assembly plans, and optimizing cost measures of assembly sequences.

1 Introduction

Assembly planning is the problem of finding and sequencing the motions that put the initially separated parts of an assembly together to form the assembled product. This is a problem of considerable importance. Its solution would allow CAD systems to provide better feedback to designers, in order to help them create products that are more cost-effective to manufacture. Assembly sequences are also useful to select assembly tools and equipment, and to layout manufacturing facilities.

In this paper, we present a general framework for assembly planning based on the concept of motion space. This framework is similar in spirit to the configuration space framework used for path planning. But, while in path planning one represents any possible placement of the moving object as a point in a configuration space, here we represent each possible motion of a subassembly as a point in a motion space. For each motion in the motion space, certain parts collide with others. These pairs of parts constitute constraints on which subassemblies may perform the motion. Then the idea is to precompute a decomposition of motion space into regions such that the constraints among the parts in the subassembly are the same for all the motions in the same region. We represent the constraints over a region in the form of a graph called the *directional blocking graph*, or DBG. We call the collection of all regions in motion space together with their associated DBGs the *non-directional blocking graph*, or NDBG, of the assembly. Once the NDBG has been computed, it can be queried to generate one assembly sequence, or several, or all. Throughout this paper, we present several techniques to efficiently compute and query NDBGs for a variety of motion families.

Most of the techniques presented here have already been described in a series of recent papers and reports, many of which we have authored or co-authored. But the connections among these techniques and the underlying general framework were only partially noticed. The present paper outlines the motion space framework in its full generality and re-presents (without all the details) previous techniques within this framework. It provides foundations for future research in assembly planning and, along the way, it identifies several problems that remain open.

*Department of Computer Science, Tel Aviv University, Tel Aviv 69978, Israel. halperin@math.tau.ac.il. Work on this paper by D. Halperin has been supported in part by an Alon Fellowship, by ESPRIT IV LTR Project No. 21957 (CGAL), by the USA-Israel Binational Science Foundation, by The Israel Science Foundation founded by the Israel Academy of Sciences and Humanities, and by the Hermann Minkowski – Minerva Center for Geometry at Tel Aviv University.

†Department of Computer Science, Stanford University, Stanford, CA 94305, USA. latombe@cs.stanford.edu. Work on this paper by J.C. Latombe has been supported by Army MURI grant DAAH04-96-1-007.

‡Eastman Kodak Company, Albuquerque, NM 87112-1881, USA. rwilson@kodak.com.

SCG 98 Minneapolis Minnesota USA

The paper is organized as follows. Section 2 defines the assembly planning problem in more formal terms and introduces the assembly-by-disassembly approach that is used in the motion space framework. It references some of the previous work and surveys related work. Section 3 presents the general principles of the motion space approach. Section 4 exemplifies the motion space approach for what we call one-step and multi-step motions, presenting polynomial-time algorithms to compute and use NDBGs for many useful families of motions, describing improvements of the efficiency of the basic algorithms, and surveying implementations of these algorithms. In Section 5 we present what we consider to be major challenges in the area of assembly planning. We concentrate on four topics where some results have already been obtained but which are still largely open: assembly planning with compound motions, assembly of toleranced parts, planning with additional constraints, and optimizing cost measures of assembly sequences.

2 Assembly Planning

2.1 Definitions

A **workspace** is a subset of the two or three-dimensional physical space modeled by the Euclidean space $\Re^k$ ($k = 2$ or 3). A **body** is a rigid physical object modeled as a compact manifold with boundary[1] in $\Re^k$. We say that two bodies **overlap** if their interiors intersect. We say that they **touch** each other if they intersect without overlapping.

An **assembly** is a collection of bodies (also called **parts**) in some given relative placements in the workspace, such that no two bodies overlap. A **subassembly** is a subset of the bodies composing an assembly A in their relative placements in A. In most practical cases, each part of an assembly is in contact with a few other parts so that the union of all the parts forms a connected subset of the workspace. However, except when we indicate otherwise, we do not make such an assumption. The motion space approach described in this paper applies to both the case where there are contacts among parts of an assembly and the case where there are no such contacts.

We say that two subassemblies are **separated** if they are arbitrarily far apart from one another. Given an assembly A, an **assembly operation** is a motion that merges s ($s \geq 2$) pairwise separated subassemblies of A into a new subassembly of A. During this motion, each subassembly moves as a single body and no overlapping between bodies is allowed. The parameter s is called the **number of hands** of the operation. We call the reverse of an assembly operation a **partitioning** operation.

An **assembly sequence** is a total ordering on assembly operations that merges the separated parts composing an assembly into this assembly. The maximum, over all the operations in the sequence, of the number of hands required by an operation is called the number of hands of the sequence. A **two-handed** assembly sequence is one that requires two hands only.

It has been shown in [16] that every assembly of convex polygons in the plane admits a two-handed assembly sequence of translations; but, in the worst-case, s hands are necessary and sufficient for assemblies of s star-shaped polygons/polyhedra [35]. In the rest of this paper, *we will only consider two-handed assembly sequences.*

[1] A subset M of $\Re^k$ is a manifold with boundary if every point of M has a neighborhood V such that $V \cap M$ is homeomorphic to either an open ball of $\Re^k$ or a closed half-space of $\Re^k$.

With the above definitions, an assembly sequence only generates subassemblies of the final assembly. Such a sequence is said to be **monotone** [47]. A more general assembly sequence is one in which some operation brings a body to an intermediate placement (relative to other bodies), before another operation transfers it to its final placement. Such a sequence is called **non-monotone.** In the rest of this paper, *we will only consider monotone assembly sequences.*

An **assembly sequence planner** (or **assembly planner**) is an algorithm that takes the description of an assembly as input and computes the assembly sequences for this assembly.

A classical strategy for assembly planning is called **assembly-by-disassembly.** It consists of partitioning, first the input assembly A into subassemblies and then, recursively, the generated subassemblies that are not individual parts. This strategy is implemented by two key procedures: `partition` and `disassemble`. The procedure `partition` takes the description of an assembly S as input and generates two subassemblies S_1 and S_2 ($S_1 \cup S_2 = S$), along with a path p such that moving S_1 along p separates it from S_2. Whenever such subassemblies and direction do not exist, the procedure returns failure. The procedure `disassemble` applies `partition` to the given assembly A and, recursively, to the generated subassemblies. The motion space approach described in this paper provides general and efficient means for designing `partition`.

Note that the above definitions ignore the constraints introduced by the mechanical system manipulating the parts; in other words, they assume that the parts of an assembly can move and maintain stability without any external help. In Section 5.3 we discuss the addition of other constraints.

Finally, it is convenient to introduce the following categories of motions. Let us say that a motion step translates a body along a single direction t by some distance ℓ, while rotating it at constant rate about an axis a that is fixed relative to the body. A **one-step motion** consists of a single motion step in which ℓ is arbitrarily large. If the rotation rate is null (i.e., the body has fixed orientation), the motion is a **one-step translation.** A **multi-step motion** is the concatenation of several motion steps, in which the last step has arbitrarily large ℓ. An **infinitesimal motion** consists of a single motion step in which ℓ is arbitrarily small.

2.2 Previous and Related Work

The complexity of assembly sequencing has been investigated in several papers. When arbitrary sequences are allowed, assembly sequencing is PSPACE-hard, both for assemblies of polyhedra each with a constant number of vertices [35] and for assemblies of polygons [47]. When only two-handed monotone sequences are permitted, deciding if an assembly can be partitioned into two subassemblies by an arbitrary motion is NP-complete [44]. The problem remains NP-complete when motions are restricted to multi-step translations and other variations [27].

The assembly-by-disassembly strategy prevails throughout the literature on assembly planning [2, 3, 9, 21, 23, 29, 48]. It is motivated by the observation that a product in its assembled state exhibits far more constraints on its components than in its disassembled state. These constraints drastically reduce the range of assembly motions which a planner must consider. Moreover, when all parts are rigid, there is a bijection between assembly and disassembly sequences, as far as geometric constraints are concerned. (This bijection would also be broken when physics, e.g., gravity, and motion control uncertainty are taken into account.)

In many planners (e.g. [3, 23]), the procedure `partition` is based on generate-and-test: given S, enumerate all candidate partitions $\{S_1, S_2\}$ of S, until a path p is found that separates S_2 from S_1. The partitions are usually generated from cutsets of the assembly connection graph, which in many cases lowers the number of partitions to consider and ensures S_1 and S_2 are both connected. Most often p is a simple path (e.g., a translation) that is inferred from contacts between parts [22, 48]. But even with the cutset approach the number of candidate partitions is exponential in the number of parts in S, while the number of feasible partitions is usually much smaller. Several effective planners were nevertheless built along this principle. Some reduced time complexity by only allowing partitioning operations in which one subassembly is a single part [21, 29, 47]. Others incorporated heuristics, for example, to drastically reduce the number of separating motions considered by the planner at each partitioning step [9, 21]. Some were quite general and hence limited to assemblies with small part count [3]. The NDBG was introduced to avoid this combinatorial trap [42, 45].

Arkin, Connelly, and Mitchell [2] use the concept of a monotone path among obstacles to derive a polynomial-time algorithm for partitioning an assembly of polygons in the plane with a one-step translation. Their algorithm for this special case is more efficient than a straightforward application of the NDBG approach, but does not generalize to other motions.

If we restrict ourselves to moving one part at a time and allow only a single direction d of translation, we obtain a very simple assembly sequence (and hence desirable, whenever possible). Such a sequence, when one exists, defines a depth order of the parts in the assembly along the direction d. Depth orders were extensively studied, mainly in relation to computer graphics [6]. A depth order for an assembly of polyhedral parts with a total of N vertices can be computed in $O(N^{4/3+\epsilon})$, for any $\epsilon > 0$ [8]. If we are allowed to move the parts along k different directions (still one part at a time and each move is a single translation to infinity) then a more elaborate algorithm needs to be used [1] which runs in $O(kN^{4/3+\epsilon})$ time. The latter algorithm has been also extended to deal with rotational motions along a small fixed set of axes, and with a slightly higher running time. For details see [1].

The motion space approach that we present below is primarily concerned with determining one or several (or all) assembly sequences of a type, if these exist. More involved questions arise when one attempts to optimize certain cost measures of an assembly sequence—see Section 5.4.

The representation of a motion planning problem in a motion space is not completely new. In motion planning with uncertainty in control and sensing, the notion of a nondirectional preimage is also based on a decomposition of a motion space [10, 32]. The part-orienting technique proposed in [11] makes use of a similar decomposition.

3 The Motion Space Approach

A widespread representation tool used in robot motion planning [30, 33] is **configuration space**. The configuration space of a given robot is the space of parametric representations of the robot configurations: every point in the configuration space defines a unique placement of the robot in the physical space. The dimension of the configuration space is the number of **degrees of freedom** of the robot. It is finite for robots made of rigid components, but infinite when components are flexible. Each obstacle B in the physical space maps into a configuration-space obstacle, or **C-obstacle**, CB in configuration space: every point in CB corresponds to a configuration of the robot where it intersects the obstacle B.

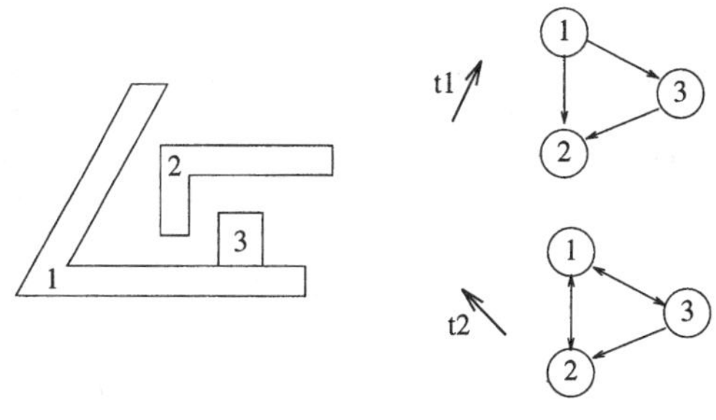

Figure 1: Examples of directional blocking graphs

Using a similar approach for assembly planning, we define the **motion space** (or **M-space**) to be the space of parametric representations of all allowable motions for partitioning operations: every point in M-space uniquely defines a path of the subassemblies moved by an operation. Hence, the dimension of the motion space is the minimal number of parameters required to define a path with a fixed starting point (the initial position of the corresponding subassembly). If no restriction is placed on assembly motions, the dimension is infinite. If, for instance, motions are restricted to be one-step translations in 3-D, the motion space has dimension 2, corresponding to the two parameters describing the direction of a translation.

Note that the motion space must be parameterized in such a way that the representation of a motion is independent of the subassembly that will eventually be moved away from the rest of the assembly. Usually, this is accomplished by attaching coordinate frames to the parts such that all the coordinate frames coincide with a universal frame $\mathcal{U}$ when the parts are in their assembled configurations. This allows us to describe a motion as a displacement from the universal frame, independently of the parts that perform it.

The analog of a C-obstacle in motion space is the M-space region, or **M-region**: for every ordered pair of parts P_i and P_j in an assembly we define their M-region $\mathcal{P}_{ij}$ to be the collection of points p in motion space such that if we move P_i along the path that p represents, P_i will overlap with P_j at some point. For each path p in $\mathcal{P}_{ij}$ we say that P_j **blocks** P_i.

Given an assembly A made of n parts $P_1, \ldots, P_n$, we associate a directed graph $G(p)$ with every point p of motion space. The nodes of $G(p)$ are the n parts $P_1, \ldots, P_n$ composing the assembly, and each ordered pair (P_i, P_j) such that $p \in \mathcal{P}_{ij}$ induces an arc of $G(p)$ directed from P_i to P_j. We call this graph the **directional blocking graph**[2] (or DBG) of A for path p. See Figure 1 for an illustration of DBG's for one-step translation in 2-D.

Let $\partial\mathcal{P}_{ij}$ denote the set of all paths p such that if P_i moves along p it will eventually touch P_j (i.e., the boundaries of P_i and P_j will intersect) without overlapping it. In most cases $\partial\mathcal{P}_{ij}$ and the boundary of $\mathcal{P}_{ij}$ are identical; in some cases, however, $\partial\mathcal{P}_{ij}$ is a superset of $\mathcal{P}_{ij}$'s bound-

[2] In the earlier papers on the NDBG, we only considered paths that were directions of motion, hence the term "directional blocking graph." We retain this terminology for consistency with the existing work.

ary [30]. The sets $\partial\mathcal{P}_{ij}$ for all $i,j \in [1,n], i \neq j$ decompose the motion space into an arrangement of cells [18] such that the DBG of A remains fixed over each cell. The arcs of the DBG on any cell c in this arrangement correspond exactly to the M-regions that contain c. We call the arrangement of cells thus defined, along with the DBG of each cell, the **non-directional blocking graph** (or NDBG) of A.

Consider a DBG $G(p)$ of A. We can partition A into two subassemblies A_1 and A_2 by moving A_1 along p if and only if there exists no arc in $G(p)$ connecting a part of A_1 to a part of A_2. Hence, A can be partitioned by a motion along p if and only if $G(p)$ is not strongly connected. If $G(p)$ is not strongly connected, then one strong component must have no outgoing arcs, and is therefore removable along p. A straightforward computation on the strong components of $G(p)$ yields all possible partitions of A for path p if desired [42]. Notice also that the NDBG of any subassembly S of A is obtained by restricting every DBG to the parts of S; it does not have to be recomputed from scratch, it is computed once, before running `disassemble`. We define `partition` as follows:

procedure `partition`(S);
  for every cell c in the NDBG of S **do:**
    if the DBG associated with c is not strongly connected
      then return c and a feasible partition of S;
  return failure;

The dimension of M-space is independent of the number of parts in A. But it grows with the complexity of the allowable motions. As will become apparent in the following sections, the motion space approach to assembly planning yields efficient algorithms only when the allowable motions can be described with relatively few parameters.

4 Applications, Efficiency Improvements, and Implementations

In this section we review applications of the motion space approach, together with techniques to speed up the running time of the basic algorithms in certain instances. We then survey software implementations of the approach and experiments carried out with them.

4.1 One-Step motions

4.1.1 One-Step Translations in 2-D and 3-D

Let A be a planar assembly made of n polygonal parts $P_1, \ldots, P_n$, which we want to assemble with one-step translations only. Each possible path is defined by a single angular parameter representing the direction of the translation. Therefore, the motion space is represented by the unit circle $\mathcal{S}^1$.

The M-region $\mathcal{P}_{ij}$ of translations along which P_i is blocked by P_j is an arc of $\mathcal{S}^1$. The endpoints of all the arcs $\mathcal{P}_{ij}$ decompose $\mathcal{S}^1$ into an arrangement of cells (open arcs and points) of complexity $O(n^2)$, such that the DBG of A remains fixed over each cell of this arrangement. The circular sequence of cells and their DBGs form the NDBG of A.

Assume that there are no "tight" insertions in A, i.e., no insertion like a peg in a hole that leaves a single translation along which a part is not blocked by another. Each arc $\mathcal{P}_{ij}$ can be constructed by extending the two extreme rays from the origin O of the coordinate system embedded in the plane containing the assembly and tangent to $P_j \ominus P_i = \{p_j - p_i | p_i \in P_i, p_j \in P_j\}$, the Minkowski sum of P_j and

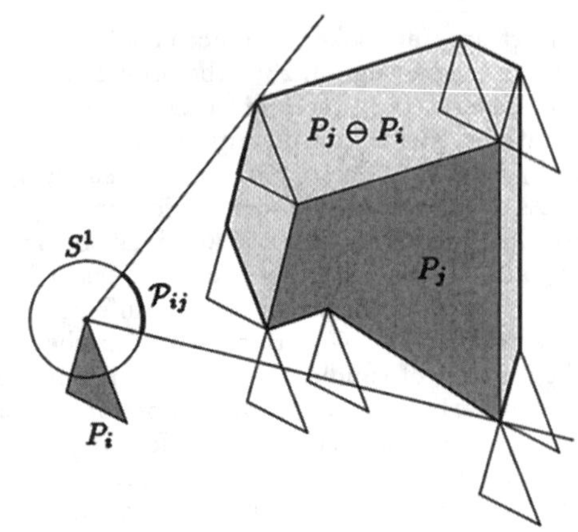

Figure 2: Construction of the M-region $\mathcal{P}_{ij}$

$-P_i$. See Figure 2, where the polygon in bold contour is $P_j \ominus P_i$. If P_i and P_j touch each other, then O lies in the contour of $P_j \ominus P_i$. If we allowed P_i to be tightly inserted into P_j, we would have to be more careful, since the set of positions where P_i touches P_j would then be a superset of the boundary of $P_j \ominus P_i$ [30]. In the general case where P_i and P_j are non-convex polygons with q_i and q_j edges, respectively, the computation of $\mathcal{P}_{ij}$ can easily be done in total time $O(q_i q_j)$.

Let q be the maximal number of edges in a single part of A. The $O(n^2)$ arcs $\mathcal{P}_{ij}$ are computed in time $O((nq)^2)$. They determine $O(n^2)$ points in $\mathcal{S}^1$ that are sorted in time $O(n^2 \log n)$. The DBG in any cell can be obtained in time $O(n^2)$. However, as we scan all the cells in $\mathcal{S}^1$, the DBG undergoes $O(n^2)$ changes that can be handled each in constant time. Thus, once a DBG has been computed, all other DBGs can be computed in total time $O(n^2)$ by scanning the sequence of cells in $\mathcal{S}^1$ and, for each cell, modifying the DBG constructed for the previous cell [42]. The complete NDBG takes time $O(n^2(\log n + q^2))$ to compute. Computing the strong components of a DBG of A takes time $O(n^2)$. Hence, `partition` runs in time $O(n^4)$ and `disassemble` generates an assembly sequence in time $O(n^5)$.

The above techniques easily generalize to polyhedral assemblies. Then the motion space is the unit sphere $\mathcal{S}^2$ in $\Re^3$, and each region $\mathcal{P}_{ij}$ is obtained by "projecting" $P_j \ominus P_i$ onto $\mathcal{S}^2$, for pairs of polyhedral parts P_i and P_j. For further details see [42].

4.1.2 Infinitesimal Motions in 3-D

The direction of a one-step motion is given by a unit vector in 6-D. Hence, the set of all possible directions of motion make up the unit 5-D sphere $\mathcal{S}^5$. Computing the M-regions requires projecting 6-D regions of complex shapes onto $\mathcal{S}^5$. This computation is feasible in principle (as long as all parts of A are described as semi-algebraic sets), but it would be much too expensive to be practical.[3] This leads us to consider infinitesimal motions, instead. The directions of infinitesimal motion still span $\mathcal{S}^5$, but computing the M-regions is much easier.

Considering infinitesimal motions requires us to adapt some of our early definitions since no infinitesimal motion can truly separate two subassemblies. For the rest of this

[3]However, this computation should be more reasonable in a 2-D workspace.

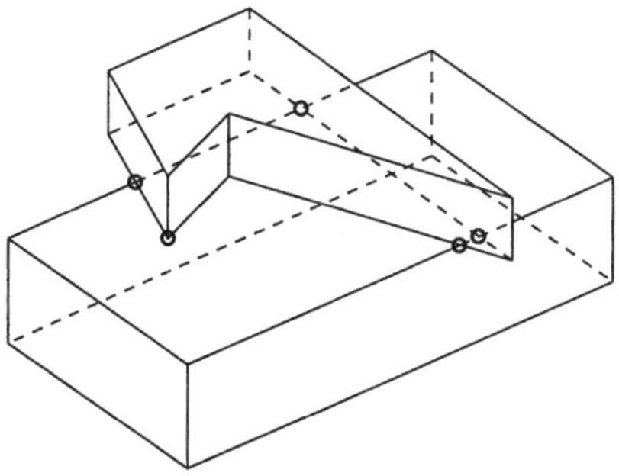

Figure 3: The face-face contact is equivalent to a set of point-plane contacts at the vertices V_k of the convex hull of the contact area (circled).

section, we will say that an infinitesimal motion separates two subassemblies if it displaces one relative to the other, by an arbitrarily small amount, without overlapping of their interiors. Clearly, the existence of a separating infinitesimal motion is a necessary condition for the existence of an arbitrarily complex extended motion that would move the two subassemblies far apart, but it is not a sufficient condition. Note also that considering infinitesimal motions makes sense only in the case where the assembly forms a connected body. In an assembly with no contacts among parts, any assembly sequence with infinitesimal motions would be valid.

As mentioned above, to ensure that motion space is parameterized independently of which parts follow a path, we attach a Cartesian frame to each part in A, such that all the frames coincide in the assembled configuration. We call the initial placement of the frames the universal frame and denote it by $\mathcal{U}$. An infinitesimal motion of any part P_i is described by a 6-D vector $dX = (dx, dy, dz, d\alpha, d\beta, d\gamma)$. The components dx, dy, and dz give the translation of the frame of P_i with respect to the frame $\mathcal{U}$. The components $d\alpha$, $d\beta$, and $d\gamma$ are the infinitesimal angles by which the frame of P_i rotates about the axes of $\mathcal{U}$.

Let V be a vertex of P_i. The motion described by dX causes V to undergo a translation $J_V dX$, where J_V is a constant 3×6 Jacobian matrix: each column of J_V gives the translation that V experiences due to a unit motion of P_i in the corresponding parameter of dX. Assume that P_i and P_j are in contact such that the vertex V of P_i is contained in the face F of P_j. Let n_F be the outgoing normal vector to F. The motion dX causes V to penetrate F when $n_F J_V dX < 0$, to break the contact with F when $n_F J_V dX > 0$, and to slide in F when $n_F J_V dX = 0$. The set of motions dX of P_i that are blocked by the contact between V and F are those satisfying $n_F J_V dX < 0$.

Now let P_i and P_j be two parts in contact such that a face F_i of P_i and a face F_j of P_j intersect in a piece of planar surface (see Figure 3). The set of motions dX allowed by this contact is the intersection of all the closed half-spaces $n_{F_j} J_{V_k} dX \geq 0$ computed for the vertices V_k of the convex hull of the intersection of F_i and F_j [20, 46]. For each vertex V_k of the convex hull of the intersection of two parts, the equation $n_F J_{V_k} dX = 0$ defines a 5-D hyperplane in the 6-D space of infinitesimal motions, which partitions $\mathcal{S}^5$ into two open half-spheres and a great circle. The set of all such hyperplanes, determined by the vertices of the convex hulls of the planar contacts, induces an arrangement of cells of dimensions $0, 1, \ldots, 5$ on $\mathcal{S}^5$. The DBG is fixed over each such region.

From this point on applying the general scheme is fairly straightforward (see [46] for details). In [17] we show that for the purpose of infinitesimal partitioning it is not necessary to build the entire NDBG, and this observation yields big savings in computation time; see also Section 4.3.2 below.

4.2 The Case of Multi-Step Motions

We present two ways to cope with multi-step motions. First, in Section 4.2.1 we describe a structure called the *interference diagram*. This structure explicitly represents the overlapping relations between the parts in the assembly, for all possible relative positions of the parts. The interference diagram is an intermediate structure from which one can compute the NDBG. It provides a conceptually general method for constructing NDBGs for multi-step motions (in fact, for motions of any type), but the resulting algorithm may not be very efficient. In Section 4.2.2 we propose an efficient direct M-space algorithm that handles the specific case of two-translations in the plane.

4.2.1 The Interference Diagram

Let A be a planar assembly of n polygons $P_1, \ldots, P_n$. We wish to solve the partitioning problem for A when the allowable motions are multi-step translations.

We select an arbitrary placement of A and we place a coordinate system of origin O in the plane containing A. We then describe each part of A as a set of points in this coordinate system. With each part P_i we associate a point O_i such that when P_i is at its placement in A, O_i coincides with O. Note that O_i does not have to be a point of P_i, but its position relative to P_i's points is fixed. Since we restrict part motions to be translations, the position of O_i always determines the placement of P_i in the plane.

For every pair of parts (P_i, P_j) we define $P_i/P_j = P_j \ominus P_i$. Assuming no tight insertion of P_i in P_j, the interior of the region P_i/P_j is the set of all positions of O_i where P_i overlaps P_j. Note that P_j/P_i is simply P_i/P_j rotated by π radians. The boundaries of all the regions P_i/P_j define a polygonal arrangement in $\Re^2$, which we call the **interference diagram** $I(A)$ of A. We label each cell c of this arrangement with the list of all regions P_i/P_j that contain c. All cells are bounded, except one which surrounds all others and have an empty label; we call this cell the **external** cell of $I(A)$. We call the cell that contains O the **central** cell of $I(A)$.

Let us now consider the M-space of all multi-step translations. We are only interested in paths that partition A into two subassemblies. Each such path p must be a curve[4] connecting O to the external cell of $I(A)$. The DBG $G(p)$ for this path can be constructed as follows. Initialize the set of arcs in $G(p)$ to the empty set. Then, for every cell c of $I(A)$ traversed by p, for every P_i/P_j in the label of c, add (P_i, P_j) to the set of arcs of $G(p)$.

This construct yields a decomposition of the M-space into an arrangement of path cells: each path cell is defined by a sequence $(c_1, \ldots, c_r)$ of cells of $I(A)$, such that c_1 is the central cell, c_r is the external cell, and c_k and c_{k+1} are adjacent for all $k = 1, \ldots, r-1$. The DBG remains fixed over each path cell.

This construction is very general, with the interference diagram used as an intermediate structure toward the NDBG. The same construction can be made for polyhedral assemblies. More generally, it can also be extended to any assembly whose parts are modeled as semi-algebraic sets, for

[4] We can consider this curve to be a polygonal line without loss of generality.

general multi-step motions. However, even for multi-step translations in 2-D, the resulting NDBG has impractically large size. This leads us to investigate the case of two-step translations in 2-D in a different way.

4.2.2 Two-Step Translations in 2-D

Let A be a planar assembly of n polygons $P_1, \ldots, P_n$. We wish to solve the partitioning problem for A when the allowable motions are two-step translations.

As in the construction of the interference diagram we place a coordinate system of origin O in the plane and we describe each part P_i as a point set in this system. A partitioning two-step translation consists of a line segment s connecting O to some point (x, y) and a ray r originating at this point. We denote the direction of r (i.e., the angle it makes with the positive x-axis) by θ. Since each possible partitioning path can be represented by the parameters (x, y, θ), the motion space in this case is three-dimensional.

We consider every pair (P_i, P_j) of parts and compute the boundaries of the M-region $\mathcal{P}_{ij}$. We first compute the boundary patches induced by the first translation s, and then the boundary patches induced by the second translation r to infinity. The union of these two sets of patches constitutes the boundary of $\mathcal{P}_{ij}$.

Constructing the M-regions precisely may be a rather difficult task. Therefore, we simplify this step by computing a family $\mathcal{S}$ of surface patches (here called surfaces, for short) whose union contains all the boundaries of all the M-regions. Note that by this simplification we do not give up the completeness of the assembly planner. $\mathcal{S}$ induces an arrangement that decomposes M-space into cells of dimensions $0, 1, 2$, and 3, such that the DBG is fixed over each cell. The resulting NDBG is a refinement of the NDBG that would have been computed if we had only computed the boundaries of the M-regions.

In [19] we give a detailed account of how to compute the family $\mathcal{S}$; we omit the details here for lack of space. The next step is to construct the arrangement induced by $\mathcal{S}$. For a three-dimensional motion space, this can be done by a fairly simple algorithm [7] whose output enables us to traverse the arrangement cell by adjacent cell. In [19] we also extend this approach to handle paths consisting of three or more translations in the plane.

4.3 Improved Efficiency

This section describes two improvements to the efficiency of the motion space approach that apply in certain cases. The first efficiency gain is purely combinatorial: it notes that successive DBGs vary only slightly, and amortizes some shared costs of testing their connectivity. The second gain is geometric: it reduces the number of DBGs that must be considered and finds them without computing the entire arrangement of M-regions, based on a property of that arrangement for certain types of motions.

4.3.1 Amortizing Strong Connectivity Tests

In some representations of NDBG, the DBGs are most efficiently computed as one DBG G_0 for a first cell in the NDBG and a list of arc insertions and deletions (stepping into and out of M-regions) that generate all other DGBs $G_i, i > 0$. Khanna *et al* [28] derive a more efficient algorithm to test all the DBGs for strong connectivity that is applicable in cases where the number of arcs in a blocking graph can be $\Omega(n^2)$, such as in one-step translation NDBGs. They group the graphs into **phases**, preprocess the common subgraph for each phase, then compute the strong connectivity of the full graphs.

Let a phase consist of some number K of arc operations in sequence, and assume that we are testing the first phase $G_0, \ldots, G_{K-1}$.[5] The total number of nodes involved in the K arc operations within the phase cannot exceed $2K$; call these the **active** nodes. Let E be the set of all possible arcs between the active nodes. Each graph G_i can now be divided into a common subgraph $G_i \setminus E = G_0 \setminus E$ and an active subgraph $G_i \cap E$. Preprocess the common subgraph $G_0 \setminus E$ for the phase by computing its transitive closure and storing the result in a subgraph H restricted to active nodes. Khanna *et al* [28] prove that G_i is strongly connected if and only if both $G_0 \cup E$ and $H_i = H \cup (G_i \cap E)$ are strongly connected.

The graph H_i has at most $2K$ nodes, so testing its strong connectivity requires $O(K^2)$ time. The preprocessing for each phase is dominated by computing the transitive closure of $G_0 \setminus E$. Let $T(N)$ denote the time complexity of computing transitive closure on an N-node graph. Then the running time for the phase is $O(n^2 + T(n) + K^3)$. This expression is minimized when $K = T(n)^{1/3}$.

However, observe that testing the sequence of graphs H_i for strong connectivity is equivalent to our original problem. This leads to a recursive approach with further improved running time. Using fast matrix multiplication to compute transitive closure, [28] achieve an amortized cost of $O(T(n)/n)$ or $O(n^{1.376})$ per graph tested.

4.3.2 Maximally Covered Cells

Let $\Gamma(A)$ be an NDBG for an assembly A. Suppose G_1 and G_2 are two DBGs of $\Gamma(A)$ such that $E_1 \subseteq E_2$ for their respective arc sets E_1 and E_2. Then it suffices to test G_1 for strong connectivity and we do not need to test G_2 for strong connectivity as well. This follows from the monotonicity of strong connectivity: if G_1 is strongly connected then so is G_2, and if G_2 is *not* strongly connected then neither is G_1.

This simple observation implies that if we can identify subset containment between DBGs we could spare strong connectivity tests. Moreover, if we could predict such containment relationship without even constructing the DBG for the 'containing' set (E_2 in the example above), then the saving becomes more meaningful; this is because asymptotically, the time to test the DBG for strong connectivity is subsumed by the time to construct the DBG.

In M-space we let $\mathcal{Q}_i$ denote the complement of an M-region $\mathcal{P}_i$, and call it the Q-region. Thus for an ordered pair of parts P_i, P_j, the corresponding Q-region denotes motions that do *not* cause P_i to overlap P_j. The key concept of the improvement is that of a **maximally covered cell**. Informally, a cell is maximally covered if it is covered by (or contained in) more Q-regions than any of its immediate neighbors. Being maximally covered implies that the set of DBG arcs of the cell is locally minimal. In [17] we show that it suffices to test the DBGs of maximally covered cells for strong connectivity in order to solve the partitioning problem. Note that if some Q-regions touch one another without overlapping, the maximally covered cell may not be full-dimensional.

The discussion of maximally covered cells so far applies to any NDBG. Although we do not know how to exploit it in arbitrary NDBGs, we have been able to make good use of it

[5] We will also use the notation G_i somewhat loosely to mean the set of arcs of graph G_i.

for NDBGs induced by infinitesimal motions. We have shown the following result in [17]:

Given an assembly A of n polyhedral parts, the NDBG $\Gamma(A)$ of the assembly for infinitesimal motions, where any direction of motion is defined by d parameters, has at most $O(K^d)$ maximally covered cells, where K is the number of ordered pairs of parts in contact in the assembly. A sample direction in each maximally covered cell can be computed in total time $O(K^{d-1}N)$, where N denotes the number of "equivalent" point-plane contacts (Section 4.1.2) in the assembly.

For infinitesimal motions in 3-D, the dimension of the motion space is 5. The maximally covered cells approach leads to significant savings over the best previously known algorithm [46] to solve this partitioning problem. In addition, the algorithm is based on linear programming techniques and hence simpler to implement robustly than the arrangement computations required by most NDBGs. In particular, we have implemented an algorithm based on this approach and have been able to solve the partitioning problem for sizable instances; see more details in Section 4.4.

4.4 Implementations

Several forms of NDBG have been implemented and tested. Some of these were part of experimental assembly planning systems, while others were stand-alone implementations. In this section we give brief descriptions and pointers to those implementations and corresponding experiments.

The NDBG was initially implemented in the assembly planner GRASP, described in [42, 45]. GRASP implemented an infinitesimal translation NDBG in 3-D for polyhedral parts. Its implementation was simple but inefficient: it created and analyzed a DBG for the direction of translation given by the intersection of every pair of contact planes in the target assembly. This was supplemented by DBGs for additional directions of motion identified from feature-based part contacts, such as cylinder–cylinder and threaded contacts. The resulting one-step translations were then checked for collisions. GRASP explicitly compared NDBG-based partitioning against the cutset approach [23] for assemblies of up to 42 parts. The NDBG was somewhat faster in most tests; however, one assembly constituted a worst case for the cutset approach, which not surprisingly took a very long time to partition that assembly.

Wilson and Schweikard [40] describe experiments with an extended 3-D translational NDBG for assemblies of polyhedral parts. They calculate the M-region for a pair of parts by triangulating the faces of both polyhedra, computing the M-region for every pair of triangles, and taking their union. They use a standard line sweep [37] to construct the 2-D arrangement of regions and corresponding blocking graphs. The implementation was tested on hand-generated and random examples.

Romney *et al* [39] reimplement and refine the methods of [42] and apply them to determine various measures of assembly complexity. They pay particular attention to issues affecting average-case performance. Their program, STAAT, explicitly constructs an NDBG for infinitesimal translations on the unit sphere, resulting in a much smaller set of translation directions to analyze than GRASP analyzes. STAAT not only generates assembly sequences, but also analyzes the NDBG directly to detect certain assembly properties, such as whether the product is a stack assembly (i.e., all parts can be assembled from a single direction). One of the distinguishing features of STAAT is an efficient implementation of the decomposition of $\mathcal{S}^2$ induced by infinitesimal translations in 3-D [12].

The observation regarding maximally covered cells (see Section 4.3.2) has been exploited in an implementation of an algorithm for polyhedral assembly partitioning under infinitesimal rigid motions (translation and rotation) [17]. This implementation uses efficient linear programming to access the maximally covered cells. The program has been successfully run on elaborate examples, and experimental results are reported in [17]. In Figure 4 we display a model aircraft engine with twelve parts inducing 1,112 point-plane contacts that has been successfully handled by this program.

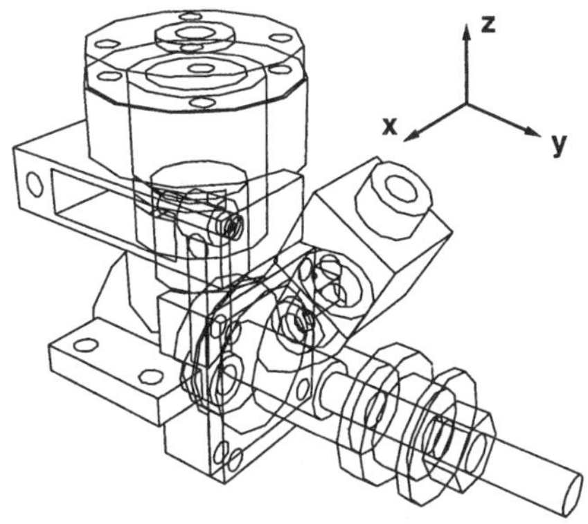

Figure 4: Assembled engine with 12 parts inducing $K = 24$ ordered pairs of parts in contact and $N = 1,112$ point plane contacts [17]: the program finds infinitesimal partitionings under translation and rotation. Previous solutions required time $O(N^5K^2)$, while in the novel approach the required time is $O(K^4N + K^6 \log N + N^{2+\varepsilon})$, for any $\varepsilon > 0$.

The special NDBG mentioned below in Section 5.2 to deal with toleranced parts was implemented in an assembly planner described in [31]. This planner is written in C++. It was tested on several polygonal assemblies of small size. Experiments with this planner show that tolerances can have profound impact on the set of feasible assembly sequences.

The Archimedes computer-aided assembly planning system [25, 26] was designed to bring NDBG-based assembly sequence generation to practical use. Its NDBG implementation is very similar to GRASP's (above), but it includes NDBG extensions to handle additional constraints (such as grasping—see Section 5.3), including a library of over 100 common hand and robotic tools. Despite the inefficient NDBG implementation, planning time is dominated by other computations, including contact finding, collision detection, and determining feasible tool placements. Archimedes can input CAD data from several commercial mechanical design systems, and has been applied to over 20 products at this time. The most complex is the guidance section of the Hughes AIM-9X air-to-air missile shown in Figure 5, consisting of 472 parts described by over 55Mb of CAD data.

A method which concurrently plans an assembly sequence and generates a fixture that stabilizes all intermediate subassemblies (see [38] and Section 5.3 below) is implemented in a system called Atlas 3 written in C. It has been successfully tested on many planar assemblies having 5 to 11 parts [38].

Figure 5: Guidance section of the Hughes AIM-9X air-to-air missile. (Figure used by permission of Hughes Aircraft and the U.S. Naval Air Systems Command.)

5 Open Problems

In this section we focus on four problems in assembly planning that, in spite of progress achieved in recent years, are still largely open.

5.1 Compound Motions

In most existing implementations of assembly planners the allowable motions of the parts in each assembly operation are rather limited (e.g., each is a single translation). Two reasons can be pointed out for this limitation: (i) simple motions are favorable in manufacturing, (ii) the algorithms for generating the sequence become too complicated if the allowable motions are not restricted.

However, there are occasions where more complex motions are required. In particular when dealing with product maintenance, namely with service and repair. For example, in aircraft engines certain parts need to be serviced every several dozens of flight hours. It is required that these parts can be accessed and removed without an overhaul of the engine.

Many new products are densely packed with a large number of parts of complex and diverse shapes. Answering maintainability questions for such products can be a very difficult task—it often calls for planning a single operation which may be extremely complicated. Motion planning techniques have already been applied to solve problems of this nature [5]. However, recent experiments have shown that current motion planning technology can offer only partial solution to these problems. Even the most advanced practical motion planners take enormous time to solve maintainability problems. Furthermore, their running time varies greatly and unpredictably. Note that if we need to remove a single part, then this is a variant of the motion planning problem where the obstacle environment is cluttered. If we need to find a removable subset along a complicated path, then this is the partitioning problem which is even harder.

Can the motion space approach be of help when compound assembly motions are necessary? We are currently examining a randomized variant of the motion space approach [4]. We aim to solve the assembly partitioning problem for polygonal parts with multi-translation paths, with about five-to-ten segments in a path. The algorithm for solving the problem as described in Section 4 would require the construction of an arrangement in ten (or more) dimensional space—far beyond what current geometric software technology offers. In [4] instead of constructing the full arrangement, we sample vertices, namely 0-dimensional cells, of this arrangement. This is in contrast with prevailing randomized motion planners that sample with very high probability points inside d-dimensional cells in d-dimensional arrangements. The reason to sample vertices rather than d-dimensional cells is that in tight assemblies a solution to the assembly planning problem may only lie in vertices.

While preliminary experimental results with sampling vertices are encouraging, they are still rather limited. The maintenance problem mentioned above may require dozens of steps in a single operation. Also, it is desirable to extend the approach to three-dimensional space and to compound motions involving rotations.

In summary, assembly planning raises natural problems with many degrees of freedom that call for the development of new and more advanced machinery. Unlike typical motion planning problems, assembly planning problems often require handling "tight" situations, difficult for existing motion planners.

5.2 Toleranced Assemblies

Tolerancing addresses the fact that manufacturing is inherently imprecise and produces parts of variable shapes. Tolerancing languages provide designers with symbolic means to specify acceptable variations. Given an assembly made of toleranced parts, one would like to compute assembly sequences that are valid for all possible instances of parts manufactured according to the specified tolerances. The NDBG concept can be adapted to solve this problem.

Consider an assembly A made of n polygonal parts $P_1, \ldots, P_n$. Each part edge e is supported by a line of fixed orientation located anywhere in a narrow strip called the **tolerance zone** of e. The tolerance zones are narrow enough so that all instances of a given part have the same topology. The placement of the parts in A is uniquely defined by spatial relations. Each relation fixes the relative placement of two parts P_i and P_j. It consists of elementary relations between edges and vertices of P_i and P_j, such as: vertex v_i of P_i is at some distance from edge e_j of P_j.

Let the **relation graph** of A be defined as follows: its nodes are the parts $P_1, \ldots, P_n$; each pair (P_i, P_j) is a non-directed link if and only if a spatial relation has been specified between P_i and P_j. This graph is assumed to be connected and without cycles.

Let us allow one-step translations only, so that the M-space is $\mathcal{S}^1$. Consider any two parts P_i and P_j. Due to the toleranced geometry of the parts in A, the arc of $\mathcal{S}^1$ representing the translations along which P_i is blocked by P_j is not fixed over multiple instances of A. We define $\mathcal{P}_{ij}$ as the *union* of all such arcs, when the parts in A span all their possible geometries. Using the various $\mathcal{P}_{ij}$'s, we create the NDBG of A exactly as in Section 4.1.1. Any partitioning of A that can be computed from this NDBG is guaranteed to be feasible for any geometry of the parts in A.

The novel issue here is the computation of $\mathcal{P}_{ij}$. In [31] we describe an algorithm that performs this computation in time $O(r_{ij}(q_i q_j + \log r_{ij}))$, where q_i and q_j are the number of edges of P_i and P_j, respectively, and r_{ij} is the length of the path connecting P_i and P_j in the relation graph of A. Hence, if q is the maximal number of edges in a part of A and r is the length of the longest path in the relation graph, the NDBG of A is obtained in time $O(n^2 r(q^2 + \log r))$. The algorithm generalizes to polyhedral assemblies; in that case, the NDBG is computed in time $O((nr)^2 q^4 \log(nq))$.

The algorithms in [31] assume a relatively simple tolerancing language. In particular, it allows no tolerancing on the orientation of an edge/face of a polygonal/polyhedral part. Extending the algorithms effectively to handle such angular tolerances, as well as developing strategies for different tolerancing languages, remain open problems.

5.3 Additional Constraints

The motion space approach presented here addresses the motions of the parts of an assembly, but many other constraints apply to the assembly process [24]. For example, fixtures are required to support unstable subassemblies; human hands or robot grippers must manipulate the parts; and various tools must be applied to fasten, move, or test parts [43]. These are active research areas by themselves, but the application to assembly raises important additional issues. A human-interactive approach to various assembly constraints that integrates well with the motion space approach is presented in [26], but we will focus here on computational solutions. We describe some characteristic open problems in fixturing and grasping for assembly.

Fixturing An assembly fixture must locate and support all partial assemblies in a chosen assembly sequence, but cannot interfere with any of the assembly operations. The sequence and fixture can be very dependent on each other. Hence solving the assembly fixturing problem efficiently requires considering fixture design and assembly sequencing together.

An initial step in this direction is described in [38]. The method models a fixture as a collection of frictionless points called **fixels** (for "fixture elements"). An initial fixture is created by distributing fixels at random on the boundary of the assembly. As assembly operations are generated, fixels that interfere with the part motions are removed. The remaining fixels must be sufficient to guarantee stability of all partial assemblies that appear in the sequence. Linear programming techniques test for assembly stability.

The approach described in [38] is limited in several ways. A single fixture must be used for the entire assembly sequence, and the set of fixels specified may not be physically realizable. It only allows adding a single part at a time to the assembly. Finally, a worst-case exponential-time algorithm searches the space of possible sequences and fixtures. Improvements in these areas will require careful representation and integration of the various geometric constraints.

Grasping Grasping and gripper design for single parts has been well studied (see e.g. [15, 34, 36, 41]). When applied in an assembly context, two important open problems arise: grasping sets of parts, and grasping sequences of parts. The former problem is closely related to fixture design for a static assembly. For the second problem, consider a single low-volume assembly robot that must assemble an entire product or a subset of a product. The robot can have several grippers available, each with a different set of parts it can grasp. However, grippers add to the cost of the cell and gripper changes are slow. A gripper can place a part if it can grasp it stably and neither the gripper nor the part collide with already-placed parts during the mating operation. Most industrial grippers are quite simple, such as parallel jaws of varying size and travel distance, configurations of suction cups of different sizes, and chucks that hold vertical cylindrical features of a certain diameter range.

The problem is to choose an assembly sequence and design a set of grippers so as to minimize the number of grippers, the number of gripper changes, or a given weighted sum of the two. A more complex problem is to choose a sequence to assemble the product on a given number of sequential robot workcells, so as to minimize the maximum assembly time taken by any workcell and/or the total number of grippers in the workcells. Solutions to these problems could significantly impact the economics of low-volume robotic assembly.

5.4 Cost Measures of Assembly Sequences

The motion space approach is primarily concerned with determining one or several (or all) assembly sequences of a type, if these exist. More involved questions arise when one attempts to optimize certain cost measures of an assembly sequence. For example, one may wish to minimize the number of directions of extended motion (one-step translation) used in an assembly sequence. Recently it has been shown that several variants of cost measure optimization in assembly planning are hard, even in their approximate versions [13, 14]. We cite here one hardness result from [14], for a problem that at first glance seems very simple. We are give n pairwise interior-disjoint unit disks in the plane. One disk is marked as the *key* disk that we need to access, and we are allowed to remove disks in one-step translational motions. The goal is to minimize the number of the disks removed before the key disk can be removed. It is shown that obtaining a $2^{\log^{1-\gamma} n}$-approximation to this problem is hard for any $\gamma > 0$. We refer the reader to [13, 14] for full technical details on this and other hardness results in this area. In general the problem of efficient computation of cost measures is open.

References

[1] P. Agarwal, M. de Berg, D. Halperin, and M. Sharir. Efficient generation of k-directional assembly sequences. In *Proc. 7th ACM-SIAM Symp. on Discrete Algorithms*, pages 122–131, 1996.

[2] E. M. Arkin, R. Connelly, and J. S. B. Mitchell. On monotone paths among obstacles, with applications to planning assemblies. In *Proc. 5th ACM Symp. on Computational Geometry*, pages 334–343, 1989.

[3] D. F. Baldwin, T. E. Abell, M.-C. M. Lui, T. L. De Fazio, and D. E. Whitney. An integrated computer aid for generating and evaluating assembly sequences for mechanical products. *IEEE Trans. on Robotics and Automation*, 7(1):78–94, 1991.

[4] N. Birnbaum and D. Halperin. A randomized approach to assembly planning along paths. Manuscript, 1998.

[5] H. Chang and T. Li. Assembly maintainability study with motion planning. In *Proc. IEEE Int. Conf. on Rob. and Autom.*, pages 1012–1019, 1995.

[6] M. de Berg. *Ray Shooting, Depth Orders and Hidden Surface Removal*, volume 703 of *Lecture Notes in Computer Science*. Springer-Verlag, Berlin, Germany, 1993.

[7] M. de Berg, L. J. Guibas, and D. Halperin. Vertical decompositions for triangles in 3-space. *Discrete Comput. Geom.*, 15:35–61, 1996.

[8] M. de Berg, M. Overmars, and O. Schwarzkopf. Computing and verifying depth orders. *SIAM Journal on Computing*, 23:437–446, 1994.

[9] A. Delchambre and P. Gaspart. KBAP: An industrial prototype of knowledge-based assembly planner. In *Proc. IEEE Intl. Conf. on Robotics and Automation*, pages 2404–2409, 1992.

[10] B. R. Donald. The complexity of planar compliant motion planning under uncertainty. *Algorithmica*, 5:353–382, 1990.

[11] K. Y. Goldberg. Orienting polygonal parts without sensors. *Algorithmica*, 10:201–225, 1993.

[12] M. Goldwasser. An implementation for maintaining arrangements of polygons. In *Proc. 11th Annu. ACM Sympos. Comput. Geom.*, pages C32–C33, 1995.

[13] M. Goldwasser, J.-C. Latombe, and R. Motwani. Complexity measures for assembly sequences. In *Proc. IEEE Intl. Conf. on Robotics and Automation*, pages 1581–1587, 1996.

[14] M. Goldwasser and R. Motwani. Complexity measures for assembly sequences. *International Journal of Computational Geometry and Applications*. To appear.

[15] R. Grupen, T. Henderson, and I. McCammon. A survey of general-purpose manipulation. *Intl. J. of Robotics Research*, 8(1):38–62, 1989.

[16] L. Guibas and F. F. Yao. On translating a set of rectangles. *Advances in Computing Research*, 1:235–260, 1983.

[17] L. J. Guibas, D. Halperin, H. Hirukawa, J.-C. Latombe, and R. H. Wilson. A simple and efficient procedure for polyhedral assembly partitioning under infinitesimal motions. In *Proc. IEEE Intl. Conf. on Robotics and Automation*, pages 2553–2560, 1995.

[18] D. Halperin. Arrangements. In J. E. Goodman and J. O'Rourke, editors, *Handbook of Discrete and Computational Geometry*, chapter 21, pages 389–412. CRC Press LLC, 1997.

[19] D. Halperin and R. H. Wilson. Assembly partitioning along simple paths: the case of multiple translations. *Advanced Robotics*, 11:127–145, 1997.

[20] H. Hirukawa, T. Matsui, and K. Takase. Automatic determination of possible velocity and applicable force of frictionless objects in contact from a geometric model. *IEEE Trans. on Robotics and Automation*, 10(3):309–322, 1994.

[21] R. L. Hoffman. Automated assembly in a CSG domain. In *Proc. IEEE Intl. Conf. on Robotics and Automation*, pages 210–215, 1989.

[22] L. S. Homem de Mello. *Task Sequence Planning for Robotic Assembly*. PhD thesis, Carnegie-Mellon Univ., 1989.

[23] L. S. Homem de Mello and A. C. Sanderson. A correct and complete algorithm for the generation of mechanical assembly sequences. *IEEE Trans. on Robotics and Automation*, 7(2):228–240, 1991.

[24] R. E. Jones and R. H. Wilson. A survey of constraints in assembly planning. In *Proc. IEEE Intl. Conf. on Robotics and Automation*, pages 1525–32, 1996.

[25] R. E. Jones and R. H. Wilson. An interactive assembly planning system. In *Video Proc. IEEE Intl. Conf. on Robotics and Automation*, 1997.

[26] R. E. Jones, R. H. Wilson, and T. L. Calton. Constraint-based interactive assembly planning. In *Proc. IEEE Intl. Conf. on Robotics and Automation*, pages 913–920, 1997.

[27] L. Kavraki and M. N. Kolountzakis. Partitioning a planar assembly into two connected parts is NP-complete. *Information Processing Letters*, 55(3):159–165, 1995.

[28] S. Khanna, R. Motwani, and R. H. Wilson. On certificates and lookahead in dynamic graph problems. In *Proc. 7th ACM-SIAM Symp. on Discrete Algorithms*, pages 222-231, 1996.

[29] H. Ko and K. Lee. Automatic assembling procedure generation from mating conditions. *Computer Aided Design*, 19(1):3–10, 1987.

[30] J.-C. Latombe. *Robot Motion Planning*. Kluwer, 1991.

[31] J.-C. Latombe, R. H. Wilson, and F. Cazals. Assembly sequencing with toleranced parts. *Computer-Aided Design*, 29(2):159–174, Feb. 1997.

[32] A. Lazanas and J. C. Latombe. Landmark-based robot navigation. *Algorithmica*, 13:472–501, 1995.

[33] T. Lozano-Pérez. Spatial planning: A configuration space approach. *IEEE Transactions on Computers*, C-32(2):108–120, 1983.

[34] B. Mishra, J. T. Schwartz, and M. Sharir. On the existence and synthesis of multifinger positive grips. *Algorithmica*, 2:541–558, 1987.

[35] B. K. Natarajan. On planning assemblies. In *Proc. 4th ACM Symp. on Computational Geometry*, pages 299–308, 1988.

[36] J. Ponce, S. Sullivan, A. Sudsand, J.-D. Boissonnat, and J.-P. Merlet. On computing four-finger equilibrium and force-closure grasps of polyhedral objects. *Intl. J. of Robotics Research*, 1997. To appear.

[37] F. P. Preparata and M. I. Shamos. *Computational Geometry: An Introduction*. Springer-Verlag, 1985.

[38] B. Romney. *On the Concurrent Design of Assembly Sequences and Fixtures*. PhD thesis, Dept. of Computer Science, Stanford Univ., 1997.

[39] B. Romney, C. Godard, M. Goldwasser, and G. Ramkumar. An efficient system for geometric assembly sequence generation and evaluation. In *Proc. ASME Intl. Computers in Engineering Conf.*, pages 699–712, 1995.

[40] A. Schweikard and R. H. Wilson. Assembly sequences for polyhedra. *Algorithmica*, 13(6):539–552, 1995.

[41] C. Wentink, A. F. van der Stappen, and M. Overmars. Algorithms for fixture design. In J.-P. Laumond and M. Overmars, editors, *Algorithms for Robotic Motion and Manipulation: 1996 WAFR*, pages 321–346. A. K. Peters, 1997.

[42] R. H. Wilson. *On Geometric Assembly Planning*. PhD thesis, Stanford Univ., March 1992. Stanford Technical Report STAN-CS-92-1416.

[43] R. H. Wilson. Geometric reasoning about assembly tools. Technical Report SAND95-2423, Sandia National Labs, 1996. To appear in *Artificial Intelligence*.

[44] R. H. Wilson, L. Kavraki, T. Lozano-Perez, and J.-C. Latombe. Two-handed assembly sequencing. *Intl. J. of Robotics Research*, 14(4):335–350, 1995.

[45] R. H. Wilson and J.-C. Latombe. Geometric reasoning about mechanical assembly. *Artificial Intelligence*, 71(2):371–396, 1994.

[46] R. H. Wilson and T. Matsui. Partitioning an assembly for infinitesimal motions in translation and rotation. In *Proc. Intl. Conf. on Intelligent Robots and Systems*, pages 1311–1318, 1992.

[47] J. D. Wolter. *On the Automatic Generation of Plans for Mechanical Assembly*. PhD thesis, Univ. of Michigan, 1988.

[48] T. C. Woo. Automatic disassembly and total ordering in three dimensions. In *Conf. on Intelligent and Integrated Manufacturing Analysis and Synthesis*, pages 291–303, 1987.

Multi-criteria geometric optimization problems in Layered Manufacturing

Jayanth Majhi* Ravi Janardan* Michiel Smid† Jörg Schwerdt†

Abstract

In Layered Manufacturing, the choice of the build direction for the model influences several design criteria, including the number of layers, the volume and contact-area of the support structures, and the surface finish. These, in turn, impact the throughput and cost of the process. In this paper, efficient geometric algorithms are given to reconcile two or more of these criteria simultaneously, under three formulations of multi-criteria optimization: Finding a build direction which (i) optimizes the criteria sequentially, (ii) optimizes their weighted sum, or (iii) allows the criteria to meet designer-prescribed thresholds. While the algorithms involving "support volume" or "contact area" apply only to convex models, the solutions for "surface finish" and "number of layers" are applicable to any polyhedral model. Some of the latter algorithms have also been implemented and tested on real-world models obtained from industry.

The geometric techniques used include construction and searching of certain arrangements on the unit-sphere, 3-dimensional convex hulls, Voronoi diagrams, point location, and hierarchical representations. Additionally, solutions are also provided, for the first time, for the *constrained* versions of two fundamental geometric problems, namely polyhedron width and largest empty disk on the unit-sphere.

1 Introduction

Layered Manufacturing (LM) is an exciting new technology which enables complex 3D parts to be built directly from their CAD models, as a stack of 2D layers. It is revolutionizing the field of CAD/CAM because it allows the designer to create rapidly a physical version of the CAD model (literally on the desktop) and to "feel and touch" it, thereby detecting and correcting flaws in the model early on in the design cycle. Moreover, it opens up the possibility of directly building functional parts composed of multiple materials—something that is not possible via conventional manufacturing methods.

Figure 1 depicts a well-known LM process called *StereoLithography* [12]. The input is a surface triangulation of the CAD model in a format called STL. The STL model is oriented suitably, sliced by horizontal planes, and built vertically, as follows: The laser traces out the contour of each layer on the surface of the photocurable liquid and cross-hatches the interior, causing it to harden and attach to the previous layer. The platform is then lowered by an amount equal to the layer thickness and the next layer is built similarly atop the previous one. Simultaneously, *support structures* are also generated to prop up portions of the current layer that overhang the previous layer. For this the CAD model is analyzed beforehand and a description of the supports is generated and merged into the STL file. After it is built, the part is postprocessed to remove the supports and to improve the finish, which has a stair-stepped appearance due to the non-zero layer thickness. (See Figure 2, which shows the cross-section of a four-sided "cylinder" that is normal to the paper. Facets that are not parallel to the build direction **d** will experience stair-stepping.)

1.1 Geometric issues in Layered Manufacturing

A key process-planning step in LM is choosing the model's orientation, or, equivalently, the *build direction*. This choice impacts, among other things, the number of layers, the surface finish, the quantity of supports, and the area of contact between supports and part. These, in turn, affect the speed, throughput, and cost of the process. For example, if the "cylinder" in Figure 2 is built in the indicated direction **d**, then it will have a stair-stepped finish and will require supports under facet $\overline{14}$ (along its entire length normal to the paper). If it is built normal to the paper, no supports are needed and there is no stair-stepping except possibly on the top and bottom facets (where we mean "top"/"bottom" w.r.t. the build direction); however, the number of layers will be very large. Presently, the build direction is chosen by the human operator, based on experience, so that the number of layers and the quantity of supports is "small" and the surface finish is "good". However, this approach is both time-consuming and prone to errors and there is an urgent need for computer algorithms that can automate this step. Indeed, the problems that we consider here are motivated, in part, by discussions with engineers at Stratasys, Inc.—a

*Department of Computer Science and Engineering, University of Minnesota, Minneapolis, MN 55455, U.S.A. E-mail: {majhi,janardan}@cs.umn.edu . Research supported in part by NSF grant CCR-9712226.

†Fakultät für Informatik, Otto-von-Guericke-Universität Magdeburg, D-39106 Magdeburg, Germany. E-mail: {michiel,schwerdt}@isg.cs.uni-magdeburg.de .

SCG 98 Minneapolis Minnesota USA
Copyright ACM 1998 0-89791-973-4/98/ 6...$5.00

leading Minnesota-based manufacturer of LM machines.

As illustrated above, a single build direction will, in general, not optimize multiple design criteria simultaneously. Thus it is important to find a build direction which reconciles multiple criteria in a meaningful way. In this paper we formulate and solve several such multi-criteria optimization problems.

1.2 Some design criteria for LM

Throughout the paper, we denote by $\mathcal{P}$ the polyhedron that we wish to build and by n the number of vertices in $\mathcal{P}$.[1] We let $\mathbf{d}$ denote the build direction. The criteria of interest include:

STAIR-STEP ERROR (CRITERION $\mathcal{C}_{str}$): The degree of stair-stepping on a facet, f, depends on the angle, $\theta_f(\mathbf{d})$, between the facet normal and $\mathbf{d}$, and it can be mitigated by a suitable choice of $\mathbf{d}$. In [3], the notion of an *error-triangle* for a facet is introduced as a way of quantifying stair-stepping (Figure 2). For a direction $\mathbf{d}$, we define the *stair-step error*, $\mathcal{C}_{str}(\mathbf{d})$, as the maximum height of the error triangle, taken over all facets.

VOLUME OF SUPPORTS (CRITERION $\mathcal{C}_{vol}$): The quantity of supports used affects the build time and the cost of raw materials. If $\mathcal{P}$ is convex, then the support volume, $\mathcal{C}_{vol}(\mathbf{d})$, for direction $\mathbf{d}$ is the volume of the region enclosed between $\mathcal{P}$ and the platform, i.e., the volume of the polyhedron which is bounded from below by the platform and from above by the *back facets* of $\mathcal{P}$ w.r.t. $\mathbf{d}$, i.e., the facets whose outward normals make an angle greater than $\pi/2$ radians with $\mathbf{d}$. If $\mathcal{P}$ is non-convex, then the supports for some facets may actually be attached to other facets instead of to the platform, and hence the structure of the supports is more complex.

CONTACT-AREA OF SUPPORTS (CRITERION $\mathcal{C}_{area}$): For direction $\mathbf{d}$, we define $\mathcal{C}_{area}(\mathbf{d})$ as the area of $\mathcal{P}$'s surface that is in contact with the supports. This area affects the post-processing time, since the supports that "stick" to $\mathcal{P}$ must be removed. If $\mathcal{P}$ is convex, then $\mathcal{C}_{area}(\mathbf{d})$ is the total area of the back facets. If $\mathcal{P}$ is non-convex, then $\mathcal{C}_{area}(\mathbf{d})$ is the total surface area that is in contact with the supports. (It includes the areas of all back facets and portions of the other facets.)

NUMBER OF LAYERS (CRITERION $\mathcal{C}_{wid}$): As layer thickness is often measured in thousandths of an inch, the number of slices can run into the thousands if the part is oriented along its longest dimension. This can increase the build time considerably. If the layer thickness is fixed, then for a given build direction $\mathbf{d}$, the number of layers is proportional to the *width of* $\mathcal{P}$ *in direction* $\mathbf{d}$, i.e., the smallest distance between two parallel planes that are normal to $\mathbf{d}$ and which enclose $\mathcal{P}$. We denote this by $\mathcal{C}_{wid}(\mathbf{d})$.

In previous work [14], we show how each of the first three criteria can be optimized, when considered independently. Specifically, we show how to find a $\mathbf{d}$ which minimizes $\mathcal{C}_{str}(\mathbf{d})$ in time $O(n \log n)$, for any simple polyhedron $\mathcal{P}$, and how to find a $\mathbf{d}$ which minimizes $\mathcal{C}_{vol}(\mathbf{d})$ or $\mathcal{C}_{area}(\mathbf{d})$ in $O(n^2)$ time for convex $\mathcal{P}$. Finally, a $\mathbf{d}$ minimizing $\mathcal{C}_{wid}(\mathbf{d})$ can be computed in $O(n^2)$ time [11].[2]

[1] Note that we may confine ourselves to polyhedral models, since the industry-standard STL representation is polyhedral, even if the original CAD model is not.

[2] A faster (randomized) algorithm, which runs in roughly $O(n^{1.5})$ time is known [1] but it is not well-suited for our purposes here.

1.3 Formulation of multi-criteria optimization problems in LM

We investigate our problem under three formulations of multi-criteria optimization. We describe these below in the context of two generic criteria, $\mathcal{C}_1$ and $\mathcal{C}_2$, but we emphasize that our methods extend easily to more than two criteria.

THRESHOLD FORMULATION: Find a build direction $\mathbf{d}$ such that $\mathcal{C}_1(\mathbf{d}) \leq \rho_1$ and $\mathcal{C}_2(\mathbf{d}) \leq \rho_2$, where $\rho_1 \geq 0$ and $\rho_2 \geq 0$ are designer-specified real numbers.

For instance, if $\mathcal{C}_1$ and $\mathcal{C}_2$ correspond to support volume and contact-area, respectively, then the designer could insist that the support volume be a fraction (e.g., 50%) of the part's volume and the contact-area be a fraction (e.g., 25%) of the part's surface area. Then ρ_1 and ρ_2 would be the appropriate fraction times the part's volume or surface area, respectively.

WEIGHTED FORMULATION: Find a build direction such that $w_1\mathcal{C}_1 + w_2\mathcal{C}_2$ is minimum, where w_1 and w_2 are designer-specified real numbers reflecting the relative importance of the two criteria.

SEQUENTIAL FORMULATION: Among all build directions minimizing $\mathcal{C}_1$, find one minimizing $\mathcal{C}_2$.

1.4 Contributions of this paper

Our work makes three main contributions:

(1) A SUITE OF ALGORITHMS: We have developed a set of (over a dozen) algorithms that provably optimize different combinations of the above-mentioned criteria under the three formulations. To our knowledge, ours are the first such results for this important problem. All our algorithms run in $O(n^2)$ time, with the exception of one algorithm which runs in $O(n \log n)$ time and one which runs in $O(n^3)$ time. These solutions employ our earlier algorithms for single-criterion optimization [14] as building blocks. Moreover, the algorithms are modular in the sense that solutions for different combinations of criteria can be derived quickly by combining the appropriate single-criterion algorithms.

(2) IMPLEMENTATION AND TESTING ON REAL DATA: We have implemented two of our algorithms and tested them on a variety of real-world STL files obtained from industry (see, for instance, Figures 5 and 6). These implementations are valid for any polyhedral model, even one with holes. The implementations are somewhat simplified versions of the asymptotically efficient algorithms described in the paper. Nevertheless, they run quite fast, taking, for instance, under 9 seconds on the model in Figures 5 and 6. (See also Tables 1 and 2.)

(3) NEW RESULTS ON CONSTRAINED GEOMETRIC PROBLEMS: In the course of solving the multi-criteria problems, we also provide—for the first time—algorithms for the constrained versions of two fundamental geometric problems, namely, computing the *width* of a polyhedron in 3-space and computing the *largest empty disk* on the unit-sphere, $\mathbf{S}^2$. The problems are constrained in the sense that the search space (in the former case, the set of directions realizing the width, and, in the latter case, the center of the largest empty disk) is restricted to a polygonal region on $\mathbf{S}^2$. Previously, only the unconstrained width problem and the planar version of the largest empty disk problem had been investigated [11, 16].

We establish new geometric characterizations for these constrained problems, which should be of independent interest.

We remark that our results also demonstrate the applicability and power of a wide range of advanced techniques from computational geometry in an applied setting. The geometric techniques that we bring into play include algorithms for constructing and searching various types of arrangements on the unit-sphere, 3-dimensional convex hulls, Voronoi diagrams, point location, and hierarchical representations of polyhedra. Additionally, our solutions also employ methods from continuous optimization such as Lagrange Multipliers.

A limitation of our work is that, at the present time, the solutions involving "support volume" or "contact area" as a criterion are applicable only to convex polyhedra. We are investigating these problems for simple polyhedra. We note that our solutions for "stair-step error" and "number of layers" are applicable to any polyhedron, even one with holes—and these are the ones that we have implemented.

1.5 Prior work

While there has been a fair amount of work done on single-criterion optimization for LM (e.g., [2, 10, 14]), the only work that we have come across for multi-criteria optimization in LM is [6], which considers two criteria—accuracy and build time—under the sequential formulation. (Reference [6] also mentions two technical reports [13, 17] that appear to consider related problems.) The solution proposed in [6] is not optimal since it restricts the search space to the set of facet normals, which are then checked exhaustively. (By contrast, our solutions here are all provably optimal.)

In the remainder of this paper, we review briefly the earlier results on single-criterion optimization (Section 2), describe our new results for constrained geometric problems (Section 3), and show how to combine these results to solve the multi-criteria problems (Section 4). A more detailed discussion of our results can be found in the full paper [15].

2 Single-criterion optimization—a quick review of results

2.1 Minimizing stair-step error

Given a simple polyhedron $\mathcal{P}$, we wish to find a direction $\mathbf{d}$ which minimizes the maximum error-triangle height, $\mathcal{C}_{str}(\mathbf{d})$. Let $t_f(\mathbf{d})$ denote the error triangle for facet $f \in \mathcal{P}$ and let $h_f(\mathbf{d})$ denote the height of $t_f(\mathbf{d})$. If $\mathbf{n}_f$ is the outward unit-normal to f, it is easy to see that $h_f(\mathbf{d}) = L\cos\theta_f(\mathbf{d})$, where L is the layer thickness and $\theta_f(\mathbf{d})$ is the smaller of two angles, one between $\mathbf{d}$ and $\mathbf{n}_f$ and the other between $\mathbf{d}$ and $-\mathbf{n}_f$. (See Figure 3.) We seek a direction $\mathbf{d}$ such that $\mathcal{C}_{str}(\mathbf{d}) = \max_f L\cos\theta_f(\mathbf{d})$ is minimized, i.e., $\min_f \theta_f(\mathbf{d})$ is maximized.

Consider the set $S = \{\mathbf{n}_f \cap \mathbf{S}^2, -\mathbf{n}_f \cap \mathbf{S}^2 | f \text{ is a facet of } \mathcal{P}\}$ of $O(n)$ *sites* on $\mathbf{S}^2$. Our problem is to find a point $\mathbf{d}$ on $\mathbf{S}^2$ such that the minimum angle between it and the sites is maximized. Let us define a *disk* on $\mathbf{S}^2$, with *center* $\mathbf{d}$ and *radius* θ, as the set of all points on $\mathbf{S}^2$ that are at a distance of at most θ from $\mathbf{d}$, as measured along the surface of $\mathbf{S}^2$. Clearly, our problem is equivalent to finding the center $\mathbf{d}$ of the largest-radius disk whose interior does not contain any site, i.e., a *largest empty disk.* It can be shown that the desired center is a vertex of *Vor(S)*—the Voronoi diagram of S on $\mathbf{S}^2$. (Equivalently, the desired center is the normal from the origin to a facet of the convex hull of S [14].) Thus our problem can be solved in $O(n\log n)$ time.

Let R be a face of *Vor(S)* and let $\mathbf{n}_R$ be the site of S associated with it (i.e., any point in R's interior is closer to $\mathbf{n}_R$ than to any other site). Then, for any $\mathbf{d} \in R$, the largest empty disk with its center at $\mathbf{d}$ contains $\mathbf{n}_R$ on its boundary. Hence, if $\theta_R(\mathbf{d})$ is the angle between $\mathbf{n}_R$ and $\mathbf{d}$, then the stair-step error for any $\mathbf{d} \in R$ is given by $L\cos\theta_R(\mathbf{d}) = L(\mathbf{n}_R \cdot \mathbf{d})$, which is a linear function of the form $A_R x + B_R y + C_R z$, where $\mathbf{d} = x\mathbf{i} + y\mathbf{j} + z\mathbf{k}$ and A_R, B_R, and C_R are constants depending only on $\mathbf{n}_R$ and L. We will use this fact in Section 4.2.

2.2 Minimizing contact-area of supports

For convex $\mathcal{P}$, a facet $f \in \mathcal{P}$ needs support for direction $\mathbf{d}$ iff it is a back facet. Thus, all directions for which f needs support can be represented on $\mathbf{S}^2$ by an open hemisphere, h_f, with pole $-\mathbf{n}_f$. Let $\mathcal{A}_{area}$ be the arrangement of the great circles bounding all h_f's. If R is any face of $\mathcal{A}_{area}$, then for all $\mathbf{d} \in R$, the same facets of $\mathcal{P}$ require support, and so the total contact area, $\mathcal{C}_{area}(\mathbf{d})$, for all such $\mathbf{d}$ is a constant $\mathcal{C}_{area}(R)$. Thus we need to find a face, R, of $\mathcal{A}_{area}$ for which $\mathcal{C}_{area}(R)$ is minimum. In [14], we show how to do this in $O(n^2)$ time using a planar (topological) sweep algorithm [9].

2.3 Minimizing volume of supports

For convex $\mathcal{P}$, the combinatorial structure of the supports for a direction $\mathbf{d}$ is determined by the back facets and by the vertex of $\mathcal{P}$ that is farthest away in direction $-\mathbf{d}$. (This is the vertex touching the platform when $\mathcal{P}$ is built along $\mathbf{d}$; we call it the *extreme vertex* of $\mathcal{P}$ w.r.t. $\mathbf{d}$.)

Let $\mathcal{A}_{vol}$ be an arrangement on $\mathbf{S}^2$ with the property that, for any face $R \in \mathcal{A}_{vol}$, the back facets and the extreme vertex are the same for any $\mathbf{d} \in R$. Thus, the structure of the supports (but not their volume) is invariant within R. $\mathcal{A}_{vol}$ can be computed as the overlay of two arrangements of great arcs, $\mathcal{A}_{area}$ (Section 2.2) and $\mathcal{A}_{ext}$ [4], whose faces represent directions for which the back facets and extreme vertex, respectively, are invariant.

For any $\mathbf{d} = x\mathbf{i} + y\mathbf{j} + z\mathbf{k}$ in R, the support volume is of the form $Ax^2 + By^2 + Cz^2 + Dxy + Eyz + Fxz$. In [14], we find the best $\mathbf{d}$ within R by using the Lagrange Multipliers method, subject to the constraints $x^2 + y^2 + z^2 = 1$ and $a_i x + b_i y + c_i z \geq 0$ for each great arc r_i bounding R. (Here A–F and a_i–c_i are constants.) The volume formula is updated incrementally and efficiently by visiting the faces in a depth-first order. The algorithm runs in $O(n^2)$ time.

2.4 Minimizing the number of layers

Let $\mathcal{P}$ be a simple polyhedron and let $CH(\mathcal{P})$ be its convex hull. Vertices u and v of $CH(\mathcal{P})$ are an *antipodal vertex-vertex pair (VV pair)* if $\mathcal{P}$ can be contained between parallel planes passing through u and v. *Antipodal vertex-edge (VE), vertex-facet (VF), and edge-edge (EE) pairs* are defined similarly. In [11], it is shown that a direction $\mathbf{d}$ minimizing the width $\mathcal{C}_{wid}(\mathbf{d})$ must be perpendicular to the parallel planes for a VF or an EE pair. The VF and EE antipodal pairs can be determined as follows [5]:

The upper and lower hulls of $CH(\mathcal{P})$ are mapped to two subdivisions, $\mathcal{A}_u$ and $\mathcal{A}_l$, on $\mathbf{S}^2$. The vertices of these subdivisions are are in 1–1 correspondence with the facet normals of $CH(\mathcal{P})$, and two vertices are joined by a great arc if the corresponding facets share an edge. The faces of the subdivisions are in 1–1 correspondence with the vertices of

$CH(\mathcal{P})$. The VF pairs are obtained from $\mathcal{A}_u$ and $\mathcal{A}_l$ by locating each vertex of $\mathcal{A}_u$ in a face of $\mathcal{A}_l$, and vice versa. The EE pairs are obtained by computing the overlay, $\mathcal{A}_{wid}$, of $\mathcal{A}_u$ and $\mathcal{A}_l$ and taking the edge intersections. The running time is $O(n^2)$.

For later reference, we note the following properties of $\mathcal{A}_{wid}$: The vertices correspond to directions that are normal to the parallel supporting planes of antipodal VF or EE pairs. The interior of any edge (resp., face) corresponds to directions that are normal to the parallel supporting planes of antipodal VE (resp., VV) pairs. Consequently, we have the following observation: Let E be any *element* of $\mathcal{A}_{wid}$, i.e., a vertex, edge interior, or face interior, and let f_1 and f_2 be any of the following *features* of $CH(\mathcal{P})$: two vertices, two edges, a vertex and a facet, or a vertex and an edge. Then, for any $\mathbf{d} \in E$, the same two features determine $\mathcal{C}_{wid}(\mathbf{d})$. Moreover, let $\mathbf{u_1}$ and $\mathbf{u_2}$ be the position vectors of any points on f_1 and f_2 respectively. Then for any $\mathbf{d} \in E$, $\mathcal{C}_{wid}(\mathbf{d}) = (\mathbf{u_1} - \mathbf{u_2}) \cdot \mathbf{d}$ (assuming $\mathbf{u_1}$ is above $\mathbf{u_2}$ w.r.t. $\mathbf{d}$). Thus, $\mathcal{C}_{wid}(\mathbf{d})$ is a linear function of the form $A_E x + B_E y + C_E z$, where A_E–C_E depend only on E.

3 Solving geometric problems in a constrained domain

We now discuss how to solve the width, stair-step error, and support volume problems when $\mathbf{d}$ is restricted to a polygonal region on $\mathbf{S}^2$. These results will be used as subroutines in our multi-criteria algorithms in Section 4. We begin with a utility lemma:

Lemma 3.1 *(i) Let $\mathcal{A}$ and $\mathcal{B}$ be arrangements of $O(n)$ great arcs each on $\mathbf{S}^2$. Then the overlay, $\mathcal{C}$, of $\mathcal{A}$ and $\mathcal{B}$ can be computed in $O(n^2)$ time.*

(ii) Let R be a convex polygon on $\mathbf{S}^2$ composed of great and/or small arcs and let $\mathcal{A}$ be an arrangement of $O(n)$ great arcs on $\mathbf{S}^2$. Then the overlay, $\mathcal{A}'$, of R and $\mathcal{A}$, restricted to the region R, has size $O(n^2 + |R|)$ and can be computed in $O(n^2 + n \log |R| + |R|)$ time.

Proof The proof of part (i) is easy; we sketch the proof for part (ii). $\mathcal{A}'$ has three types of vertices: (i) vertices of $\mathcal{A}$ inside/on R, (ii) vertices of R, and (iii) intersections between arcs of $\mathcal{A}$ and arcs of R. There are $O(n^2)$, $O(|R|)$, and $O(n)$ vertices of types (i)–(iii), respectively. Therefore, $\mathcal{A}'$ has size $O(n^2 + |R|)$. It can be computed as follows:

We start with the polygon R, and add the great circles, G_a, corresponding to the arcs, a, of $\mathcal{A}$ one at a time. We compute the intersection (if any) of the current G_a and R in $O(\log |R|)$ time using binary search on R (type (iii) vertices). We then walk along G_a through the part of the current arrangement that is in the interior of R, from one intersection point to the other, and compute the intersection of G_a with previously added great circles (type (i) vertices). We retain only those intersections that are also in a. By the zone theorem [7], applied to G_a and the great circles, this takes $O(n)$ time per G_a. The time bound follows. ∎

3.1 The constrained width problem

The problem here is to minimize $\mathcal{C}_{wid}(\mathbf{d})$ when $\mathbf{d}$ is restricted to lie within a convex polygonal region, R, on $\mathbf{S}^2$, bounded by great arcs. Note that the antipodal VF or EE pair which minimizes the width of $\mathcal{P}$ in the unconstrained case may not yield a direction lying in R. Indeed the minimum width may now be determined by some other type of antipodal pair, as the following lemma shows.

Lemma 3.2 *Let $\mathcal{P}$ be a simple polyhedron and let R be a convex polygon on $\mathbf{S}^2$. A width-minimizing direction, $\mathbf{d}$, lying in R satisfies one of the following:*

(i) *it is perpendicular to the parallel supporting planes associated with an antipodal VF or EE pair of $CH(\mathcal{P})$ and lies in the interior of R, or in the interior of an edge of R, or coincides with a vertex of R,*

(ii) *it is perpendicular to the parallel supporting planes associated with an antipodal VE pair of $CH(\mathcal{P})$ and lies in the interior of an edge of R or coincides with a vertex of R,*

(iii) *it is perpendicular to the parallel supporting planes associated with an antipodal VV pair of $CH(\mathcal{P})$ and coincides with a vertex of R.*

Proof Any direction in R must lie either in the interior of R, or in the interior of an edge of R, or must coincide with a vertex of R. Moreover, any such direction must be perpendicular to the parallel supporting planes associated with an antipodal VF, or VE, or EE, or VV pair of $CH(\mathcal{P})$. The only possibilities for $\mathbf{d}$, other than those given by the lemma, are that (a) it lies in the interior of R and is perpendicular to the parallel supporting planes of an antipodal VE or VV pair, or (b) it lies in the interior of an edge of R and is perpendicular to the parallel supporting planes of an antipodal VV pair.

Consider such a direction $\mathbf{d}$. Suppose that case (a) holds and suppose that the antipodal pair in question is a VE pair. Thus, one of the parallel supporting planes, h_1, contains an edge, e, of $CH(\mathcal{P})$, and the other plane, h_2, contains a vertex, u, of $CH(\mathcal{P})$. Direction $\mathbf{d}$ is perpendicular to h_1 and h_2 and is in R's interior; w.l.o.g. assume $\mathbf{d}$ coincides with the normal of h_1 that is directed outwards from the region enclosed by h_1 and h_2. If we rotate h_1 and h_2 infinitesimally about e in either direction, such that they remain parallel and contain e and u, then the outward normal of h_1 moves along a great arc and remains in R's interior. As shown in [11], one of the two rotations will yield a width less than $\mathcal{C}_{wid}(\mathbf{d})$—a contradiction. A similar discussion applies for a VV pair in case (a) or case (b). (Note that the above argument does not apply to parts (ii) and (iii) of the lemma because the width-reducing rotation may yield a direction outside R.) ∎

From Lemma 3.2 and the properties of $\mathcal{A}_{wid}$ in Section 2.4, it follows that the optimal direction for the constrained minimum width problem is **(i)** a vertex of R, or **(ii)** a vertex of $\mathcal{A}_{wid}$ lying in the interior of R or in the interior of an edge of R, or **(iii)** the intersection of the interior of an edge of $\mathcal{A}_{wid}$ with the interior of an edge of R.

Theorem 3.1 *A width-minimizing direction for a simple n-vertex polyhedron, where the directions are restricted to a convex polygon R on $\mathbf{S}^2$ composed of great arcs, can be found in $O(n^2 + n \log |R| + |R|)$ time.*

Proof The number of candidate directions of types (i)–(iii) listed above are, respectively, $O(|R|)$, $O(n^2)$, and $O(n)$. These points can be identified by constructing the overlay of R and $\mathcal{A}_{wid}$ in $O(n^2 + n \log |R| + |R|)$ time (Lemma 3.1(ii)), which implies the same time bound for the constrained width problem. ∎

Remark 3.1 The bounds in Theorem 3.1 can be shown to hold even if R is composed of both great and small arcs [15].

3.2 The constrained largest empty disk problem on $\mathbf{S}^2$

Let S be a set of n sites on $\mathbf{S}^2$ and let R be a convex polygon on $\mathbf{S}^2$ which is bounded by great and/or small arcs. We wish to find a largest disk, D, on $\mathbf{S}^2$ which contains no site in its interior and whose center lies in R. The unconstrained problem can be solved in $O(n \log n)$ time using the Voronoi diagram, *Vor(S)* [16]. The following lemma characterizes the optimum center for the constrained problem.

Lemma 3.3 *The center,* $\mathbf{d}$*, of the largest empty disk,* D*, on* $\mathbf{S}^2$*, where* $\mathbf{d}$ *is restricted to a convex polygon* R *composed of great and/or small arcs, satisfies one of the following:*

(i) *it coincides with a vertex of Vor(S) lying in the interior or on the boundary of* R*,*

(ii) *it is the intersection of an edge of Vor(S) with the boundary of* R*,*

(iii) *it coincides with a vertex of* R*,*

(iv) *it lies in the interior of an edge,* $r \in R$ *and in the interior of a face,* $F \in Vor(S)$*, and is the farthest point on* $r \cap F$ *from the site* p_F *associated with* F*.*

Proof Assume, for a contradiction, that $\mathbf{d}$ does not satisfy conditions (i)–(iv) of the lemma. Then the only possibilities for $\mathbf{d}$ are: (a) it is in the interior of a face, F, of *Vor(S)* and in the interior of R, or (b) in the interior of a face, F, of *Vor(S)* and in the interior of an edge, r, of R and it is not the farthest point in the interior of $r \cap F$ from the site associated with F, or (c) it is in the interior of an edge, e, of *Vor(S)* and in the interior of R.

Consider case (a). Let p_F be the site of S associated with F. Thus, p_F is the only site on the boundary of D. Since $\mathbf{d}$ is in R's interior, there is a point $\mathbf{d}' \in R$, on the great circle through p_F and $\mathbf{d}$, such that $\mathbf{d}'$ is further away from p_F than is $\mathbf{d}$. This implies that $\mathbf{d}'$ is the pole of an empty cap which is larger than D—a contradiction.

Next, consider case (b). Let p_F be the site associated with F. Then, there is another point $\mathbf{d}'$ in the interior of $r \cap F$ such that its distance from p_F is larger than the distance between $\mathbf{d}$ and p_F. Therefore, $\mathbf{d}'$ yields a larger empty cap than D.

Finally, consider case (c). Let p_1 and p_2 be the sites whose Voronoi faces share e. The great arc through p_1 and p_2 is perpendicular to (the great circle through) e. Moreover, p_1 and p_2 are the only sites on D's boundary. Clearly, there is a point $\mathbf{d}'$ on e, which is on one side of $\mathbf{d}$ or the other, such that $\mathbf{d}'$ is further away from p_1 and p_2 than is $\mathbf{d}$. Again, $\mathbf{d}'$ yields a larger empty cap than D. ∎

This leads to the following result for finding a constrained largest empty disk.

Theorem 3.2 *For* n *sites on* $\mathbf{S}^2$*, a largest empty disk whose center is constrained to lie in a convex polygonal region* R *composed of great and/or small arcs can be found in time* $O(n \log n + n \log |R| + |R| \log n)$*.*

Proof There are $O(n)$, $O(n)$, and $O(|R|)$ candidate directions of types (i)–(iii) in Lemma 3.3. These can be identified in $O(n \log |R|)$ time, via binary search on R. The corresponding empty disks can be found in time $O(n \log n + |R| \log n)$ using *Vor(S)*.

For type (iv) points, we divide the boundary of R into $O(n + |R|)$ segments r, each of which lies completely within a face F of *Vor(S)*, with associated site p_F. It can be shown [15] that the point on r that is farthest from p_F is an endpoint of r or the intersection between r and the plane passing through p_F, the origin, and the center of the small circle containing r. Thus, the type (iv) points can be found in $O(n + |R|)$ total time. ∎

3.3 The constrained minimum support volume problem

We wish to find a $\mathbf{d}$ which minimizes the volume of supports required by a convex polyhedron $\mathcal{P}$ when $\mathbf{d}$ is restricted to lie in a convex polygon R whose boundary consists of great arcs and/or small arcs. The idea is to use the support volume minimization algorithm given in [14] on only those faces of $\mathcal{A}_{vol}$ (or portions thereof) that are also in R. Towards this end, we compute the overlay of R and $\mathcal{A}_{vol}$ in $O(n^2 + n \log |R| + |R|)$ time (Lemma 3.1(ii)). The arrangement of interest to us is the part of $\mathcal{A}_{vol}$ which is inside R. This has size $O(n^2 + |R|)$. Thus, the support volume minimization algorithm runs in $O(n^2 + |R|)$ time [14].

Theorem 3.3 *A support volume minimizing direction for a convex* n*-vertex polyhedron, where the directions are constrained to lie inside a polygonal region* R *on* $\mathbf{S}^2$ *composed of great and/or small arcs, can be found in* $O(n^2 + n \log |R| + |R|)$ *time.* ∎

4 Computing a build direction to reconcile multiple criteria

In this section we combine the results established in previous sections to solve several multi-criteria optimization problems arising in LM. For brevity, we will describe only a subset of our results here, selected to illustrate the variety of techniques employed. Throughout, the polyhedron $\mathcal{P}$ is assumed to be convex if one of the criteria is "support volume" or "contact-area"; otherwise, $\mathcal{P}$ can be non-convex.

4.1 Threshold formulation

Recall that our problem is to find a direction $\mathbf{d}$ such that $\mathcal{C}_1(\mathbf{d}) \leq \rho_1$ and $\mathcal{C}_2(\mathbf{d}) \leq \rho_2$.

STAIR-STEP ERROR AND WIDTH: Suppose that $\mathcal{C}_1 = \mathcal{C}_{str}$ and $\mathcal{C}_2 = \mathcal{C}_{wid}$. We compute the directions $\mathbf{d}$ for which $\mathcal{C}_{str}(\mathbf{d}) \leq \rho_1$ as follows: We require that $\mathcal{C}_{str}(\mathbf{d}) = \max_f L \cos \theta_f(\mathbf{d})$ be at most ρ_1. That is, $L \cos \theta_f(\mathbf{d}) \leq \rho_1$ for all f, i.e., $\theta_f(\mathbf{d}) \geq \cos^{-1}(\rho_1/L) =: \theta$ for all f. For each f, the directions $\mathbf{d}$ such that $\theta_f(\mathbf{d}) \geq \theta$ can be represented by the complement of the union of two open disks of radius θ and centers $\mathbf{n}_f$ and $-\mathbf{n}_f$. This complement, which we call a *band*, consists of two parallel small circles at a distance of $\pi/2 - \theta$ on either side of the great circle defined by $\mathbf{n}_f$ (Figure 4). Thus, the desired set of directions $\mathbf{d}$ for which $\mathcal{C}_{str}(\mathbf{d}) \leq \rho_1$ is the intersection of such bands for all facets $f \in \mathcal{P}$. This gives two antipodal convex polygons, R and R', on $\mathbf{S}^2$, whose boundaries are composed of small arcs. Each polygon has size $O(n)$.

We now solve the constrained minimum width problem for R and R' using Theorem 3.1 and check if the resulting width meets the threshold ρ_2. Since $|R| = |R'| = O(n)$, the total time is $O(n^2)$ in both cases.

A similar approach works if $\mathcal{C}_2$ is $\mathcal{C}_{vol}$ except that we solve the constrained minimum support volume problem in R and R' (Theorem 3.3).

Theorem 4.1 *For an* n*-vertex simple polyhedron* $\mathcal{P}$*, the threshold formulation for the pair* $(\mathcal{C}_{str}, \mathcal{C}_{wid})$ *can be solved in* $O(n^2)$ *time. If* $\mathcal{P}$ *is convex, then the problem for the pair* $(\mathcal{C}_{str}, \mathcal{C}_{vol})$ *can also be solved in* $O(n^2)$ *time.* ∎

SUPPORT CONTACT-AREA AND WIDTH: Suppose that $C_1 = C_{area}$ and $C_2 = C_{wid}$. We construct $\mathcal{A}_{area}$ and compute $C_{area}(R)$ for each face R in $O(n^2)$ time (Section 2.2). For each face R such that $C_{area}(R) \leq \rho_1$, we compute the direction $\mathbf{d}_R \in R$ which minimizes C_2. If $C_2(\mathbf{d}_R) \leq \rho_2$, then $\mathbf{d}_R$ satisfies both thresholds; if $C_2(\mathbf{d}_R) > \rho_2$ for all R such that $C_{area}(R) \leq \rho_1$, then no direction exists which satisfies both thresholds.

Minimizing C_2 over R is merely the constrained minimum width problem (Theorem 3.1). A straightforward approach is to apply this theorem to each candidate face R in turn. However, there is a more efficient approach based on the observation that the candidate faces all come from the same arrangement, i.e., $\mathcal{A}_{area}$.

We compute the overlay, $\mathcal{A}$, of $\mathcal{A}_{area}$ and $\mathcal{A}_{wid}$. $\mathcal{A}$ has size $O(n^2)$ and can be computed in $O(n^2)$ time (Lemma 3.1(i)). $\mathcal{A}$ has three types of vertices: (i) vertices of $\mathcal{A}_{area}$, (ii) vertices of $\mathcal{A}_{wid}$, and (iii) intersections between arcs of $\mathcal{A}_{area}$ and $\mathcal{A}_{wid}$. This coupled with Lemma 3.2 implies that the directions $\mathbf{d}_R$ of interest, taken over all candidate faces, R, of $\mathcal{A}_{area}$, are the vertices of $\mathcal{A}$ that are in the interior or on the boundary of R. We can identify these vertices as follows:

Any region $R \in \mathcal{A}_{area}$ is divided into one or more regions R' in $\mathcal{A}$. Thus the candidate directions are boundary vertices of R'. We visit all the faces of $\mathcal{A}$ in the order given by a depth-first search (dfs) of the dual graph of $\mathcal{A}$. As we visit each face, $R' \in \mathcal{A}$, we update the value of $C_{area}(R')$ and also the formula for $C_{wid}(R')$. We can update these quantities incrementally, in $O(1)$ time, using the corresponding quantities from the previous face visited in the dfs. For each face R' such that $C_{area}(R') \leq \rho_1$, we also compute $C_{wid}(\mathbf{v})$ for each vertex, $\mathbf{v}$, on the boundary of R'. If $C_{wid}(\mathbf{v}) \leq \rho_2$, then $\mathbf{v}$ satisfies both thresholds; if $C_{wid}(\mathbf{v}) > \rho_2$ for all such $\mathbf{v}$, then no direction exists which satisfies both thresholds.

A similar approach works if $C_2 = C_{str}$ or $C_2 = C_{vol}$.

Theorem 4.2 *For an n-vertex convex polyhedron $\mathcal{P}$, the threshold formulation for pairs (C_{area}, C_{wid}), (C_{area}, C_{str}), and (C_{area}, C_{vol}) can be solved in $O(n^2)$ time.* ∎

WIDTH AND SUPPORT VOLUME: Let $C_1 = C_{wid}$ and $C_2 = C_{vol}$. The idea is to identify a (possibly disconnected) region R on $\mathbf{S}^2$ representing directions $\mathbf{d}$ such that $C_{wid}(\mathbf{d}) \leq \rho_1$. Within each component of R we find the minimum support volume, using Theorem 3.3, and check the threshold ρ_2.

We form R as follows: Let E be any element (vertex, edge interior, or face interior) of $\mathcal{A}_{wid}$. From Section 2.4, we know that for any $\mathbf{d} \in E$, $C_{wid}(\mathbf{d})$ can be written as $A_E x + B_E y + C_E z$. We want the subset of E for which $A_E x + B_E y + C_E z \leq \rho_1$. Let h_E be the halfspace $A_E x + B_E y + C_E z \leq \rho_1$. Thus we want the subset E' of E contained in $h_E \cap \mathbf{S}^2$. For each E, we can compute E' in $O(|E|)$ time. E' is either empty, a point, an edge, or a portion of a face of $\mathcal{A}_{wid}$, in which case it consists of great arcs and one small arc, or it consists of a face of $\mathcal{A}_{wid}$ with a hole (composed of a small circle). The total time to compute all the E''s is $O(n^2)$. The desired region R is the union of the E''s and it has size $O(n^2)$. Using Theorem 3.3 and Remark 3.1, which carry through even though we now have a set of convex polygons and convex polygons with holes, we can compute a direction within R that minimizes the support volume in $O(n^2 + n\sum_R \log|E'| + \sum_R |E'|) = O(n^3)$ time. (Since R can have $O(n^2)$ components E, each of size $O(1)$, the term $n\sum_R \log|E|$ is $O(n^3)$.)

Theorem 4.3 *For an n-vertex convex polyhedron, the threshold formulation for the pair (C_{wid}, C_{vol}) can be solved in $O(n^3)$ time.* ∎

4.2 Weighted formulation

Recall that here we wish to find a build direction such that $w_1 C_1 + w_2 C_2$ is minimum.

STAIR-STEP ERROR AND WIDTH: Suppose that $C_1 = C_{str}$ and $C_2 = C_{wid}$. We construct $Vor(S)$ and $\mathcal{A}_{wid}$, and compute their overlay, $\mathcal{A}$, in $O(n^2)$ time. Then we know (from Sections 2.1 and 2.4) that within a face, R, of $\mathcal{A}$, the formula for the stair-step error is of the form $C_{str} = A_R x + B_R y + C_R z$ and the formula for the width is of the form $C_{wid} = A'_R x + B'_R y + C'_R z$. Therefore, within R, we need to minimize $w_1(A_R x + B_R y + C_R z) + w_2(A'_R x + B'_R y + C'_R z)$, which is again a linear function. This problem is similar to the optimization problem described in Section 2.3 (except that the objective function is now linear), and it can be solved in $O(n^2)$ total time. The formula can also be updated in a total time of $O(n^2)$ as we visit all the faces of $\mathcal{A}$ in a depth-first order.

We can use this approach to solve for other criteria combinations also. Note that C_{area} is a constant within a face of $\mathcal{A}_{area}$, while C_{vol} is a quadratic function within a face of $\mathcal{A}_{vol}$ (Section 2.3).

Theorem 4.4 *For an n-vertex simple polyhedron $\mathcal{P}$, the weighted formulation for pair (C_{str}, C_{wid}) can be solved in $O(n^2)$ time. If $\mathcal{P}$ is convex, then the weighted formulation for the other pairs can also be solved in $O(n^2)$ time.* ∎

4.3 Sequential formulation

Here we wish to find among all directions that minimize C_1, the one that minimizes C_2.

STAIR-STEP ERROR AND WIDTH: Consider the case $C_1 = C_{str}$ and $C_2 = C_{wid}$. Recall from Section 2.1 that the directions that minimize C_{str} must coincide with a vertex of $Vor(S)$. There could be $O(n)$ such directions. Among these directions, we need to find a $\mathbf{d}$ minimizing $C_{wid}(\mathbf{d})$. We do this by computing and preprocessing $CH(\mathcal{P})$ into the hierarchical data structure described in [8], so that given any query direction the extreme vertices in that direction, and hence the width of $\mathcal{P}$ in that direction, can be found in $O(\log n)$ time. Thus, we have:

Theorem 4.5 *For an n-vertex simple polyhedron $\mathcal{P}$, the sequential formulation for pair (C_{str}, C_{wid}) can be solved in $O(n \log n)$ time.* ∎

STAIR-STEP ERROR AND SUPPORT VOLUME: Next, let $C_1 = C_{str}$ and $C_2 = C_{vol}$. We construct the arrangement $\mathcal{A}_{vol}$, and compute and store the formula for the volume of supports, C_{vol}, for each face, R, in $\mathcal{A}_{vol}$ (Section 2.3). Then for each candidate direction, $\mathbf{d}$ we do a point location in $\mathcal{A}_{vol}$ to determine the face R and compute $C_{vol}(\mathbf{d})$. We then find the overall minimum. All this can be done in $O(n^2)$ time ($O(n \log n)$ time to find the candidate directions, $O(n^2)$ time to compute $\mathcal{A}_{vol}$ and the formulae for C_{vol} within each face of $\mathcal{A}_{vol}$, and $O(n \log n)$ time for the $O(n)$ point location queries). Clearly, the same approach can be used to solve the problem if $C_2 = C_{area}$.

Theorem 4.6 *For an n-vertex convex polyhedron $\mathcal{P}$, the sequential formulation for pairs (C_{str}, C_{vol}) and (C_{str}, C_{area}) can be solved in $O(n^2)$ time.* ∎

WIDTH AND SUPPORT VOLUME: We could use the above approach for $(\mathcal{C}_{str}, \mathcal{C}_{vol})$ to also handle the pair $(\mathcal{C}_{wid}, \mathcal{C}_{vol})$. However, this will require $O(n^2 \log n)$ time as there could be $O(n^2)$ candidate directions (since each vertex of $\mathcal{A}_{wid}$ is a candidate direction). Fortunately, we can solve this problem in $O(n^2)$ time using a different approach. Let $\mathcal{A}$ be the overlay of $\mathcal{A}_{wid}$ and $\mathcal{A}_{vol}$. Note that the minimum width can occur only at vertices of $\mathcal{A}$ that are also vertices of $\mathcal{A}_{wid}$. We visit all the faces of $\mathcal{A}$ in the order given by a dfs of the dual graph of $\mathcal{A}$. As we visit each face, R, we update the width formula $\mathcal{C}_{wid}(R)$, for R, calculate $\mathcal{C}_{wid}$ at each vertex of R, and keep track of the minimum value $\mathcal{C}^*_{wid}$, of $\mathcal{C}_{wid}$. Next, we perform a second dfs of the dual of $\mathcal{A}$, this time keeping track of the volume formula, $\mathcal{C}_{vol}(R)$, for R. At each vertex, $\mathbf{v}$ of R, if $\mathcal{C}_{wid}(\mathbf{v}) = \mathcal{C}^*_{wid}$, then we calculate $\mathcal{C}_{vol}(\mathbf{v})$. We finally output that $\mathbf{v}$ for which $\mathcal{C}_{wid}(\mathbf{v}) = \mathcal{C}^*_{wid}$ and $\mathcal{C}_{vol}(\mathbf{v})$ is the minimum. All this can be done in $O(n^2)$ time. A similar approach can be used if $\mathcal{C}_2 = \mathcal{C}_{area}$ or $\mathcal{C}_2 = \mathcal{C}_{str}$, with the latter solution holding for any simple polyhedron. Therefore, we have:

Theorem 4.7 *For an n-vertex simple polyhedron $\mathcal{P}$, the sequential formulation for pair $(\mathcal{C}_{wid}, \mathcal{C}_{str})$ can be solved in $O(n^2)$ time. If $\mathcal{P}$ is convex, then the problem for pairs $(\mathcal{C}_{wid}, \mathcal{C}_{vol})$ and $(\mathcal{C}_{wid}, \mathcal{C}_{area})$ can be solved in $O(n^2)$ time.* ∎

5 Experimental results

We have implemented simplified versions of our algorithm for the pairs $(\mathcal{C}_{str}, \mathcal{C}_{wid})$ and $(\mathcal{C}_{wid}, \mathcal{C}_{str})$, under the sequential formulation. We chose the criteria $\mathcal{C}_{str}$ and $\mathcal{C}_{wid}$ since our algorithms for these are applicable to simple polyhedra. We have tested our implementation on real-world STL models obtained from industry. Both implementations are written in C, run on an SGI 100MHz, R4000 processor machine, and do convex hull computations using the `qhull` routine (`www.geom.umn.edu/software/download/qhull.html`). We ran `qhull` with the specification that whenever the cosine of the angle between the normals of two neighboring facets of the convex hull was greater than 0.99, then the facets should be treated as co-planar and merged.

5.1 $(\mathcal{C}_{str}, \mathcal{C}_{wid})$ under sequential formulation

In this version, we have not implemented the hierarchical data structure of [8] to find the extreme vertices. Instead, we do this in time linear in the size of $CH(\mathcal{P})$, which is usually much smaller than that of $\mathcal{P}$. To compute $\mathcal{C}_{str}$, we compute $CH(S)$ and find the facet(s) closest to the origin; the normal from the origin to each such facet yields the center of a largest empty disk. Figure 5 shows the optimal orientation found by our algorithm for one of our test parts—a speedometer component for an automobile, containing 16,720 facets. The value of $\mathcal{C}_{str}$ is $0.968 \times L$ and that of $\mathcal{C}_{wid}$ is 3.33, where L is the layer thickness. By comparison, the width of this part (ignoring the stair-stepping criterion) is 1.89—about half as large as found above. For this part, the algorithm took about 9 seconds, excluding the time for graphical output.

Table 1 gives the results for all our test parts. (The sizes of the polyhedron and the convex hulls in the table are the number of facets; also the $\mathcal{C}_{str}$ values should be multiplied by L.) The test parts, chosen so as to encompass a variety of geometries, included (i) a box cover with several sloping internal surfaces and one hole, (ii) a cylinder, (iii) a rectangular base for a part with numerous ridges, (iv) a mushroom-like part composed of a hemisphere and a cylinder, with four holes and four protrusions, and (v) the speedometer component (rows 1–5 of the table, respectively). One observation worth making is that for parts whose facet normals were distributed fairly uniformly over $\mathbf{S}^2$, the value of $\mathcal{C}_{str}$ approached the maximum of $1.00 \times L$. This is to be expected since the largest empty disk then becomes quite small.

$(\mathcal{C}_{str}, \mathcal{C}_{wid})$					
$\|\mathcal{P}\|$	$\|CH(\mathcal{P})\|$	$\|CH(S)\|$	*time (s)*	$\mathcal{C}_{str}$	$\mathcal{C}_{wid}$
906	13	1088	0.58	0.885	11.21
3600	37	110	1.14	0.706	2.81
5487	16	1448	1.97	0.822	5.22
9530	179	24494	10.17	0.958	3.14
16720	67	17180	8.76	0.968	3.33

Table 1: *Results for the pair $(\mathcal{C}_{str}, \mathcal{C}_{wid})$, under the sequential formulation*

$(\mathcal{C}_{wid}, \mathcal{C}_{str})$				
$\|\mathcal{P}\|$	$\|CH(\mathcal{P})\|$	*time (s)*	$\mathcal{C}_{wid}$	$\mathcal{C}_{str}$
906	13	2.61	4.78	1.000
3600	37	3.30	1.98	0.997
5487	16	4.19	1.18	1.000
9530	179	5.77	2.48	1.000
16720	67	7.27	1.89	0.998

Table 2: *Results for the pair $(\mathcal{C}_{wid}, \mathcal{C}_{str})$, under the sequential formulation*

5.2 $(\mathcal{C}_{wid}, \mathcal{C}_{str})$ under sequential formulation

We first compute the VF pairs by finding for each facet of $CH(\mathcal{P})$, the vertex farthest away from the plane through the facet. Next we compute $\mathcal{A}_u$ and $\mathcal{A}_l$ directly on $\mathbf{S}^2$ and find all EE pairs by computing edge intersections between $\mathcal{A}_u$ and $\mathcal{A}_l$. For each width-minimizing direction thus found, we compute $\mathcal{C}_{str}$ by checking each point in S. Figure 6 shows the optimal orientation found by our algorithm for the speedometer part. The value of $\mathcal{C}_{wid}$ is 1.89 and that of $\mathcal{C}_{str}$ is $0.998 \times L$, while the minimum stair-step error (ignoring the width criterion) is $0.967 \times L$. For this part, the algorithm took about 7 seconds, excluding the time for graphical output. Table 2 gives the results for the same set of test parts as in Table 1. In each case, the width-minimizing direction is realized by a VF-pair of $CH(\mathcal{P})$ and the corresponding facet of $CH(\mathcal{P})$ is parallel (or nearly parallel) to a facet of $\mathcal{P}$, thereby yielding a value for $\mathcal{C}_{str}$ that is equal to (or close to) $1.00 \times L$.

6 Conclusion

We have presented efficient algorithms for several multicriteria optimization problems in Layered Manufacturing, under different formulations. These algorithms employ solutions to the single-criterion optimization problems as building blocks. We have also implemented and tested simplified versions of some of our algorithms on actual STL data. We are currently working on alternative implementations of our algorithms that mirror more closely the solutions described in Section 4.3. We note that although we have described our results in the context of two criteria, they can be extended in a straightforward way to handle three or more criteria. In general, the running times are asymptotically the same as for two criteria, since the algorithms involve overlaying three or more arrangements of total size $O(n^2)$.

Acknowledgement

We thank the software development group at Stratasys, Inc. for providing us with test data and for stimulating discussions.

References

[1] P. K. Agarwal and M. Sharir. Efficient randomized algorithms for some geometric optimization problems. *Discrete Comput. Geom.*, vol. 16, pages 317–337, 1996.

[2] S. Allen and D. Dutta. Determination and evaluation of support structures in layered manufacturing. *Journal of Design and Manufacturing*, 5:153–162, 1995.

[3] M. Bablani and A. Bagchi. Quantification of errors in rapid prototyping processes and determination of preferred orientation of parts. In *Transactions of the 23rd North American Manufacturing Research Conference*, 1995.

[4] P. Bose, M. van Kreveld, and G. Toussaint. Filling polyhedral molds. In *Proc. 3rd Workshop Algorithms Data Structures*, volume 709 of *Lecture Notes in Computer Science*, pages 210–221. Springer-Verlag, 1993.

[5] B. Chazelle, H. Edelsbrunner, L. Guibas and M. Sharir. Diameter, width, closest line pair and parametric searching. *Discrete Comput. Geom.* vol. 10, pages 183–196, 1993.

[6] W. Cheng, J.Y.H. Fu, A.Y.C. Nee, Y.S. Wong, H.T. Loh and T. Miyazawa. Multi-objective optimization of part-building orientation in stereolithography. *Rapid Prototyping Journal*, vol. 1, no. 4, pages 12–23, 1995.

[7] B. Chazelle, L. J. Guibas and D. T. Lee. The power of geometric duality. *BIT*, vol. 25, pages 76–90, 1985.

[8] D. P. Dobkin and D. G. Kirkpatrick. A linear algorithm for determining the separation of convex polyhedra. *J. Algorithms*, vol. 6, pages 381–392, 1985.

[9] H. Edelsbrunner and L. Guibas. Topologically sweeping an arrangement. *Journal of Computer and System Sciences*, vol 38, pages 165–194, 1989.

[10] D. Frank and G. Fadel. Preferred direction of build for rapid prototyping processes. In *Proceedings of the 5th International Conference on Rapid Prototyping*, pages 191–200, 1994.

[11] M.E. Houle and G.T. Toussaint. Computing the width of a set. In *IEEE Transactions on Pattern Analysis and Machine Intelligence*, vol. 10, no. 5, pages 761–765, 1988.

[12] P. Jacobs. *Rapid Prototyping & Manufacturing: Fundamentals of StereoLithography*. McGraw-Hill, 1992.

[13] J. Kim, K. Lee and J. Park, Determination of optimal orientation in stereolithographic rapid prototyping. *Tech. Report*, Dept. of Mechanical Design and Production Engineering, Seoul National University, July 1994.

[14] J. Majhi, R. Janardan, M. Smid and P. Gupta. On some geometric optimization problems in layered manufacturing. In *Proc. 5th Workshop on Algorithms and Data Structures*, vol. 1272 of Lecture Notes in Computer Science, pages 136–149, Springer-Verlag, 1997.

[15] J. Majhi, R. Janardan, M. Smid, and J. Schwerdt. Multi-criteria geometric optimization problems in Layered Manufacturing *Tech. Report TR-97-055*, Dept. of Computer Science and Engineering, University of Minnesota, Nov. 1997.

[16] F.P. Preparata and M.I. Shamos. *Computational Geometry–An Introduction*. Springer-Verlag, 1988.

[17] P. Sreeram and D. Dutta. Determination of optimal orientation based on variable slicing thickness in layered manufacturing. *Tech. Report UMEAM-94-14*, University of Michigan, 1994.

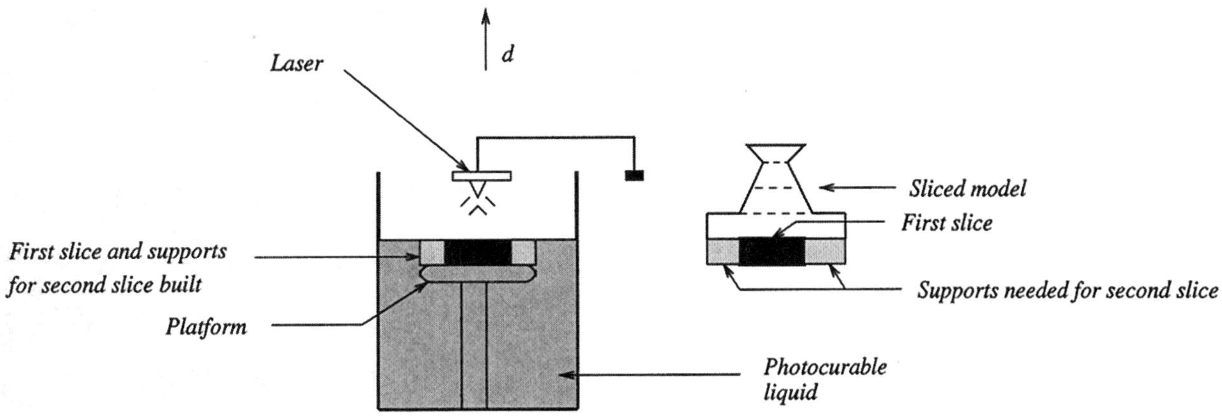

Figure 1: *The stereolithograpy apparatus.*

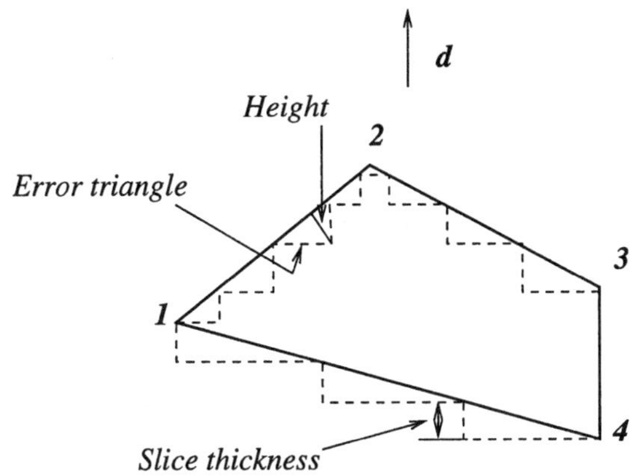

Figure 2: *Illustrating stair-stepping. The figure shows the cross-section of a a 4-sided "cylinder". The cylinder is normal to the paper and has a uniform cross-section along its entire length. Facets that are not parallel to the build direction will have a stair-stepped finish.*

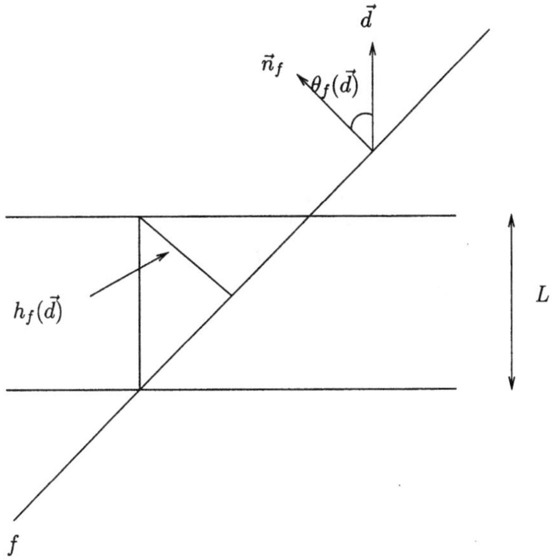

Figure 3: *The error triangle height for a facet f.*

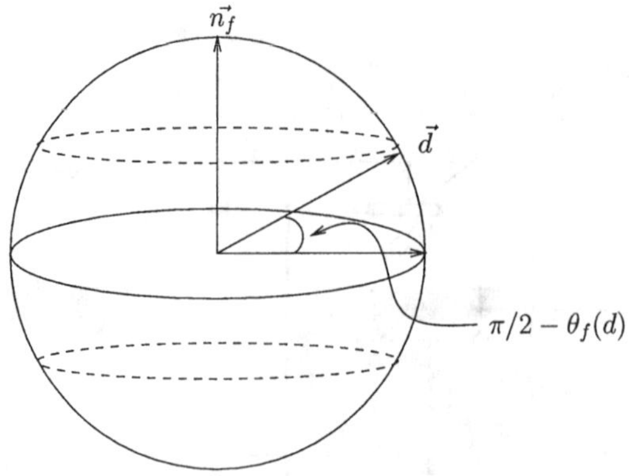

Figure 4: *The band corresponding to a facet f.*

Figure 5: *Optimal orientation found by our algorithm for the pair ($\mathcal{C}_{str}, \mathcal{C}_{wid}$), under the sequential formulation. The part has been oriented so that the build direction is vertical. The part shown is a speedometer component for an automobile, with about 16,720 facets (of which only a third are displayed for clarity).*

Figure 6: *Optimal orientation found by our algorithm for the pair ($\mathcal{C}_{wid}, \mathcal{C}_{str}$), under the sequential formulation. The part has been oriented so that the build direction is vertical.*

Design and Analysis of Planar Shape Deformation *

Siu-Wing Cheng[†] Herbert Edelsbrunner[‡] Ping Fu[§] and Ka-Po Lam[†]

Abstract

Shape deformation refers to the continuous change of one geometric object to another. We develop a software tool for planning, analyzing, and visualizing deformations between two shapes in $\mathbb{R}^2$. The deformation is generated automatically without any user intervention or specification of feature correspondences. A unique property of the tool is the explicit availability of the two-dimensional shape space, which can be used for designing the deformation either automatically by following constraints and objectives or manually by drawing deformation paths.

1 Introduction

This paper describes a method for the deformation of one geometric shape in the plane to another and a software tool that implements it. In computer graphics this operation is commonly referred to as *morphing*. In that area the focus is on the creation of image sequences that *display* the morph [8, 9, 10]. In contrast, this paper constructs the deformation as a 1-dimensional family of genuine geometric objects.

Rationale. This paper follows the general ideas about shape representation and deformation laid out in [2, 3]. It focuses on the special case of *two* shapes in *two* dimensions. Even in this simplest of all interesting cases there is a frightening amount of freedom and an abundant variety of possible deformations. It has been suggested we follow the general trend of making up an evaluation function and compute the optimal deformation defined by that function. The authors of this paper believe that this strategy fails to rationalize the process as it only shifts the irrational decisions one level higher. Instead, we follow a path of deformation design that permits efficient quantitative analysis. The deformations satisfy basic requirements, such as economy in shape variation, local smoothness, and geometric integrity. Other than that the method employs natural geometric operations, such as lifting, taking convex hulls, slicing, and projecting. All constructions are global and mathematically defined. The full description of these constructions and higher-dimensional interpretations is beyond the scope of this paper and we refer the interested reader to [2].

The criticism of lacking an optimality criterion and the above response apply only to the automatic reconstruction of deformations. Section 4 explains how the system permits the specification of evaluation functions and the optimization within a 2-dimensional space of shapes.

Summary of results. The software tool described in this paper constructs and deforms 2-dimensional shapes and it rationalizes the process through analysis and visualization. Some of the features of that tool are now listed.

- The source and target shapes are regular sets in $\mathbb{R}^2$: each set is the closure of its interior. The sets can have arbitrarily many components and holes. The chosen data structure places no principle limitations on the kind of shapes although it is better in representing some types than others. The appearance of the shapes for small data sets is similar to the blobby models introduced by Blinn [1].
- The deformation can be computed fully automatically, which means that also the topology of the

*This research is partially supported by RGC CERG HKUST 190/95E and by NSF CCR-96-19542.

[†]Department of Computer Science, Hong Kong University of Science and Technology, Kowloon, Hong Kong.

[‡]Department of Computer Science, University of Illinois at Urbana-Champaign, and Raindrop Geomagic, Inc.

[§]Raindrop Geomagic, Inc., and National Computational Science Alliance, University of Illinois at Urbana-Champaign.

shape is changed without user intervention. See figure 1 for an example where a '0' with one hole is deformed into a '1' without hole.

- Each intermediate shape is determined by a time parameter, $t \in [0,1]$, and a size parameter, $\alpha^2 \in \mathbb{R}$. In the automatic and unconstrained deformation the size is a function of t and the initial and final size values. Deformation paths satisfying shape constraints can be designed in the 2-dimensional shape space of time and size.

- The deformation is monitored by keeping track of quantitative information, such as the area, boundary length, number of components, and number of holes. Constraints guiding the automatic design of a deformation can be formulated in terms of these quantities.

The tool provides VCR functionalities for the interactive visualization, such as play, reverse, step forward, step backward, etc.

Ramifications. The widespread use of image morphing in the movie and advertisement industries justifies the related research, maybe also the one reported in this paper. The particular properties and features of the method described in this paper suggests additional applications.

- The shape is well defined at every moment during the deformation. If we stack up all 2-dimensional shapes we get a pipe-like 3-dimensional shape that smoothly starts, branches, joins, and ends. Such pipes are good models of blood vessels. Another example are complicated exhaust systems, and the constrained deformation option can be used to maintain, say, a constant cross-section area.

- The automatic deformation between two given shapes is canonical. The kind and amount of change can be used to define a distance measure between shapes useful for example in automatic pattern recognition. It might be worth trying out this idea in the classification of printed and possibly even handwritten Chinese characters.

- The shape space embedding the deformations is explicitly given as a 2-dimensional strip of pairs $(t, \alpha^2) \in [0,1] \times \mathbb{R}$. If the source and target shapes are significantly different then this space is an index into a rich variety of shapes. This suggests applications in shape compression and the design of shape databases.

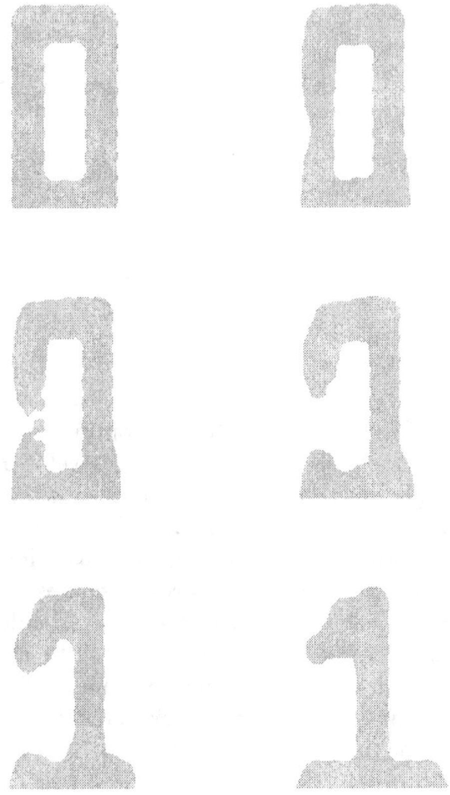

Figure 1: Read the pictures from left to right and top to bottom as a deformation from '0' to '1'.

Outline. Section 2 reviews the two basic ingredients to our deformation tool: the shape representation of [3] and the deformation technique of [2]. Section 3 explains the various geometric concepts used in the construction and visualized by the software tool. Section 4 discusses the structure of the shape space and how it is exploited in the design of special deformation paths. Section 5 concludes the paper.

2 Shape and Deformation

The full details of the shape representation and the shape deformation methods can be found in [3] and in [2]. This section explains the minimum of both formalisms and relies on illustrations to provide an intuition for the ideas. Every shape is represented by two data structures: a complex that captures structure and connectivity and a set with smooth boundary that serves as the appearance of the shape. We explain first the two data structures and second the deformation method.

Structure and connectivity. We begin by specifying the input data, which is a set of weighted points or disks, and the Voronoi decomposition of the plane this set defines. Each disk in the input set $B = \{b_1, b_2, \ldots, b_n\}$ is specified by its center and its radius: $b_i = (z_i, \varrho_i)$ with $z_i \in \mathbb{R}^2$ and ${\varrho_i}^2 \in \mathbb{R}$. We permit negative ${\varrho_i}^2$ which corresponds to an imaginary radius, ϱ_i, and an imaginary disk, b_i. The *(weighted) distance* of a point $x \in \mathbb{R}^2$ from b_i is $\pi_i(x) = \|x - z_i\|^2 - {\varrho_i}^2$, and the *(weighted) Voronoi cell* of b_i is the set of points no closer to any other disk:

$$V_i \quad = \quad \{x \in \mathbb{R}^2 \mid \pi_i(x) \leq \pi_j(x), \text{ for all } j\},$$

see figure 2. The Voronoi cells decompose the union of

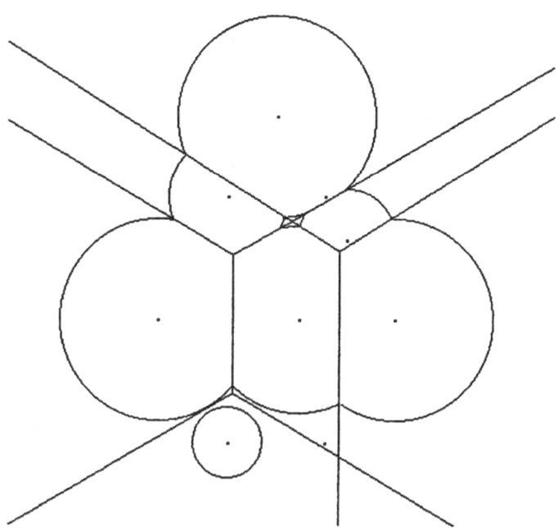

Figure 2: The Voronoi cells of 9 disks, 7 with positive and 2 with imaginary radius. Two of the centers do not lie in their Voronoi cells.

disks into convex cells: $\bigcup B = \bigcup_{i=1}^{n}(V_i \cap b_i)$. Any two of these cells are either disjoint or they meet along a portion of their boundaries.

The *complex* data structure is denoted as Dsx B and records the overlap between the cells $R_i = V_i \cap b_i$. Specifically, z_i is a vertex in Dsx B iff $R_i \neq \emptyset$ and $z_i z_j$ is an edge iff $R_i \cap R_j \neq \emptyset$. Furthermore, $\text{conv}\,\{z_i \mid i \in I\}$ is a 2-dimensional cell, for I a set of three or more disk indices, iff $\bigcap_{i \in I} R_i \neq \emptyset$ and I is maximal with this property. For B in general position all cells in Dsx B are triangles. However, we cannot reasonably make this simplifying assumption because the deformation method described shortly systematically creates degenerate disk sets, such as the one shown in figure 2. Even if we have general position at times $t = 0$ and $t = 1$ there are parallelograms in the intermediate complexes, see figure 3.

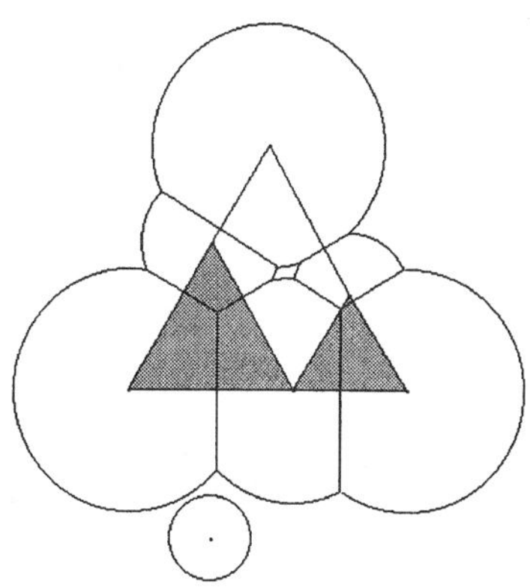

Figure 3: The Voronoi decomposition and complex of the union of disks in figure 2. The complex contains only two cells, namely the shaded triangles. The white parallelogram would belong to the complex, along with other cells, if the disks were larger.

Shape appearance. The body is defined as a union of an infinite family of disks, and the skin is its boundary, which is the envelope of an infinite family of circles. To define these families observe that $b_i = {\pi_i}^{-1}(-\infty, 0]$ and its bounding circle is ${\pi_i}^{-1}(0)$. The weighted distance functions span the usual vector space defined by function addition and scaling. Let $\Pi = \{\pi_i \mid 1 \leq i \leq n\}$ be the collection of weighted distance functions. The affine hull of Π is the family of combinations $\sum \gamma_i \cdot \pi_i$, where the γ_i add up to 1. The convex hull is the subfamily defined by non-negative scaling factors:

$$\text{conv}\,\Pi \quad = \quad \{\sum_{i=1}^{n} \gamma_i \cdot \pi_i \mid \sum_{i=1}^{n} \gamma_i = 1, \gamma_i \geq 0 \text{ for all } i\}.$$

The convex hull is an infinite family of paraboloid functions and we construct the skin and body by shrinking each $\pi \in \text{conv}\,\Pi$ using a parameter $0 \leq s \leq 1$. For $\pi(x) = \|x - z\|^2 - \varrho^2$ define the shrunken function as

$$\pi^s(x) \quad = \quad \frac{1}{s} \cdot \|x - z\|^2 - \varrho^2.$$

The zero-set of π^s is the circle with center z and radius $\sqrt{s}\varrho$. The envelope and union of the infinite family of shrunken disks is now defined with the help of the pointwise minimum function, $\Gamma_s(x) = \min\{\pi^s(x) \mid \pi \in$

conv Π}:

$$\begin{aligned}\text{skn}^s B &= \Gamma_s^{-1}(0),\\ \text{bdy}^s B &= \Gamma_s^{-1}(-\infty, 0].\end{aligned}$$

They are referred to as the *s-skin* and the *s-body* of B, see figure 4.

Figure 4: The skins for $s = 0.1, 0.3, 0.5, 0.7, 0.9$ of the same set of disks as in figures 2 and 3. Observe that the complex is contained in the five bodies and they are all connected the same way, with two components of which one has a hole and the other does not.

Shape deformation. Let B and C be two finite sets of disks defining the initial and the final shapes of the deformation. Intermediate shapes are constructed by interpolating between the two sets. It is convenient to project a cross-section of the vector space of weighted distance functions onto the set of disks. Formally, if b and c are disks with weighted distance functions π and φ and $\gamma_1 + \gamma_2 = 1$ then $a = \gamma_1 \cdot b + \gamma_2 \cdot c$ is the disk with weighted distance function $\gamma_1 \cdot \pi + \gamma_2 \cdot \varphi$. With this introduction define

$$\begin{aligned}A_t &= (1-t) \cdot B + t \cdot C\\ &= \{(1-t) \cdot b + t \cdot c \mid b \in B, c \in C\}\end{aligned}$$

for every $t \in [0, 1]$. Clearly $A_0 = B$ and $A_1 = C$. For a given value of s, we thus have a one-parameter family of shapes,

$$\mathbb{X}_t = \text{bdy}^s A_t,$$

that deforms the initial shape, $\mathbb{X}_0 = \text{bdy}^s B$, into the final shape, $\mathbb{X}_1 = \text{bdy}^s C$, see figure 5.

Observe the automatic change of topology that happens synchronously for the skin and the complex. In figure 5 there is a change in topology between all contiguous snapshots except for the last three. Even though the seventh and the eight snapshot have the same topology, the latter is obtained from the former by two changes, one filling a hole and one opening a hole. The second, third, fourth, and sixth snapshots are taken right at the time of a topology change when two components or two portions of the same component reach each other in a point and locally separate the complement.

3 Visualization of Concepts

The display panel of the software tool can be used to visualize all geometric concepts mentioned in section 2 and more.

Basic structures. Let B be a finite set of disks as before. The *(weighted) Voronoi complex*, denoted as Vor B, is the collection of 2-dimensional Voronoi cells together with common intersections among them. The *(weighted) Delaunay complex*, denoted as Del B, records the overlap among 2-dimensional Voronoi cells, see the rightmost example in figure 6.

Next we discuss a nested sequence of subcomplexes of the Delaunay complex. Let $\alpha^2 \in \mathbb{R}$ and define $b_i(\alpha)$ as the disk with center z_i and radius $\sqrt{\varrho_i{}^2 + \alpha^2}$. We have $b_i(0) = b_i$, and if $\varrho_i{}^2 = 0$ then the radius of $b_i(\alpha)$ is α. Let $B(\alpha) = \{b_i(\alpha) \mid b_i \in B\}$ and observe that the Voronoi complex of $B(\alpha)$ is the same as that of B. In other words, the Voronoi complex and the Delaunay complex do not vary with α^2. The cells of the convex decomposition of $\bigcup B(\alpha)$ grow with α:

$$R_i(\alpha_1) = V_i \cap b_i(\alpha_1) \subseteq V_i \cap b_i(\alpha_2) = R_i(\alpha_2)$$

if $\alpha_1 \leq \alpha_2$. We call $\text{Dsx}_\alpha B = \text{Dsx}\, B(\alpha)$ the *α-complex* of B. The collection of such complexes is nested and the last one is the Delaunay complex:

$$\emptyset = \text{Dsx}_{\alpha_0} B \subseteq \text{Dsx}_{\alpha_1} B \subseteq \text{Dsx}_{\alpha_2} B = \text{Del}\, B,$$

where $\alpha_0^2 \leq \alpha_1^2 \leq \alpha_2^2$, α_0^2 is sufficiently small, and α_2^2 is sufficiently large, see figure 6.

Decomposed skin and body. Similar to the union of disk we can decompose the skin and the body into simple pieces. The complex that produces the decomposition consists of 2-dimensional Voronoi cells, 2-dimensional Delaunay cells, and rectangles that are

Cartesian products of matching Voronoi and Delaunay edges. All these cells are shrunk so they fit together to form a decomposition of $\mathbb{R}^2$. To describe this complex let $0 \leq s \leq 1$ be arbitrary but fixed. Let I be a subset of indices with

$$\nu_I = \bigcap_{i \in I} V_i \in \operatorname{Vor} B,$$
$$\delta_I = \operatorname{conv}\{z_i \mid i \in I\} \in \operatorname{Del} B.$$

The dimension of ν_I is $k \in \{0,1,2\}$ and that of δ_I is $2-k$. The corresponding (2-dimensional) cell in the complex is

$$\mu_I{}^s = s \cdot \nu_I + (1-s) \cdot \delta_I,$$

where addition and scaling refer to the operations in the vector space of points in $\mathbb{R}^2$. The *s-mixed complex*, denoted as $\operatorname{Mix}^s B$, consist of all cells $\mu_I{}^s$ and all non-empty common intersections among them, see figure 7.

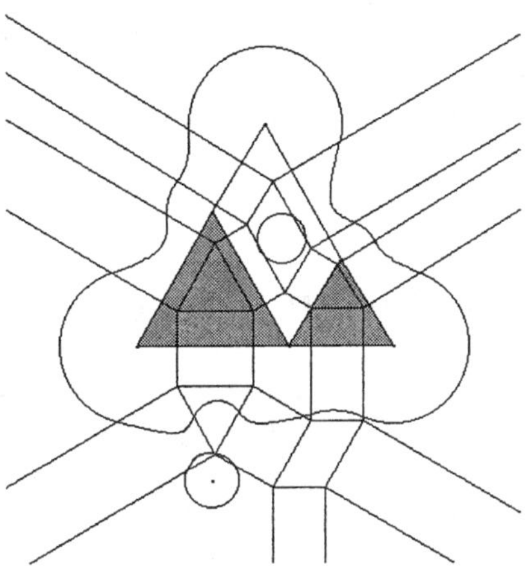

Figure 7: The 0.5-skin of the disk set in figure 2 is decomposed into circle and hyperbola arcs by the 0.5-mixed complex.

Observe that $\operatorname{Mix}^0 B = \operatorname{Del} B$ and $\operatorname{Mix}^1 B = \operatorname{Vor} B$. For all other values of s, $\operatorname{Mix}^s B$ contains a shrunken copy of every Voronoi and Delaunay cell. The copy of V_i intersects $\operatorname{skn}^s B$ in a collection of arcs that all belong to a common circle, namely the zero-set of $\pi_i{}^s$. Similarly, the copy of a Delaunay cell intersects the skin in a collection of arcs that all belong to a common circle, but this circle is not the boundary of any shrunken disk in B. This can be seen from the fact that the body lies outside every circle of the latter type. The remaining 2-dimensional cells in $\operatorname{Mix}^s B$ are rectangles, and each rectangle intersects the skin in a collection of arcs that belong to a common hyperbola. A more detailed description of the skin decomposition including formulas for all circles and hyperbolas can be found in [3].

Complementarity. We develop an explicit expression of the symmetry between shape and complement by introducing another set of disks. Call two disks $b_i = (z_i, \varrho_i)$ and $b_j^\perp = (y_j, \rho_j)$ *orthogonal* if

$$\|z_i - y_j\|^2 = \varrho_i{}^2 + \rho_j{}^2,$$

and *further than orthogonal* if the square distance between the centers exceeds the sum of square radii. In case of real radii the bounding circles of two orthogonal disks meet at right angles. We construct a new set of disks, $B^\perp = \{b_1^\perp, b_2^\perp, \ldots, b_m^\perp\}$, that cover the part of $\mathbb{R}^2$ outside $\bigcup B$ and overlap the disks in B orthogonally or less. Specifically, for each vertex y_j of $\operatorname{Vor} B$ choose the radius ρ_j so $b_j^\perp = (y_j, \rho_j)$ is orthogonal to all b_i with $y_j \in V_i$. By definition of Voronoi complex, $b_j^\perp$ is further than orthogonal from all other disks in B. For completeness, we stipulate a Voronoi vertex at infinity in the direction of each unbounded Voronoi edge and define the corresponding disk as a half-plane. It is not difficult to see that $\operatorname{Vor} B = \operatorname{Del} B^\perp$ and $\operatorname{Del} B = \operatorname{Vor} B^\perp$, and therefore $(B^\perp)^\perp = B$. Two additional noteworthy properties are:

$$\operatorname{bdy}^s B \cup \operatorname{bdy}^{1-s} B^\perp = \mathbb{R}^2,$$
$$\operatorname{bdy}^s B \cap \operatorname{bdy}^{1-s} B^\perp = \operatorname{skn}^s B = \operatorname{skn}^{1-s} B^\perp.$$

Figure 8 illustrates the relationship between B and $B^\perp$ by showing three skins of B, for $s = 0, 0.5, 1$. The 0-body of B is the closed complement of a union of disks, namely the complement of the interior of $\bigcup B^\perp$. The 0.5-skin shrinks away from the boundaries of $\bigcup B$ and of $\bigcup B^\perp$ and lies between $\operatorname{skn}^0 B = \operatorname{skn}^1 B^\perp$ and $\operatorname{skn}^1 B = \operatorname{skn}^0 B^\perp$.

Deformation and topology change. We return to the deformation constructed by interpolating disk sets: $A_t = (1-t) \cdot B + t \cdot C$, see section 2. Let $0 < s < 1$ be fixed and let t vary inside $[0,1]$ and generate a 1-parameter family of shapes, $\mathbb{X}_t = \operatorname{bdy}^s A_t$. We focus on the local picture of a change in topology as t varies continuously. Each change occurs at a particular moment in time and a particular point in the plane. Let this point be z and distinguish three cases depending on the type of cell in the mixed complex that contains z.

CASE $I = \{i\}$. Point z lies inside the shrunken Voronoi cell $\mu_I{}^s$. The cell specifies the portion of the circle defined by b_i that belongs to the skin. This circle is

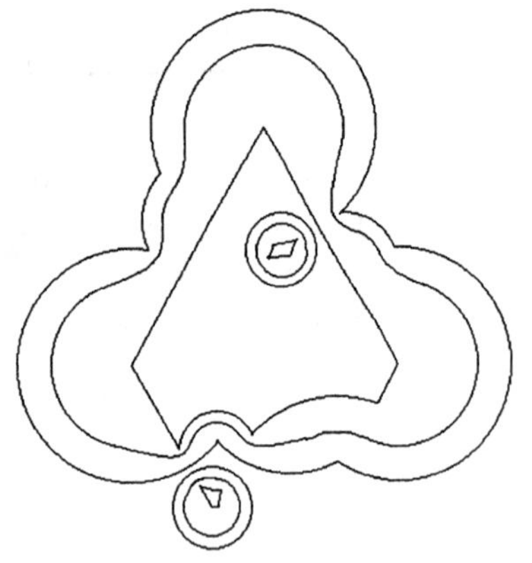

Figure 8: The boundary of $\bigcup B$, the boundary of $\bigcup B^{\perp}$, and the common 0.5-skin of B and $B^{\perp}$ separating the two boundaries.

the zero-set of ${\pi_i}^s$. In the non-degenerate case the radius of this circle is either imaginary or positive real. The topology change happens at the transition from the former case to the latter or vice versa. This transition either creates a new component by growing a disk from z or it removes a component by shrinking a disk to z.

CASE $I = \{i, j\}$. Point z lies inside the rectangle ${\mu_I}^s$. The rectangle specifies the portion of the hyperbola defined by b_i and b_j that belongs to the skin. The hyperbola consists of two branches living in diagonally opposite quadrants defined by the two asymptotic lines, see figure 9. In the non-degenerate case the two branches either locally separate or locally sandwich the body. The topology change happens at the transition when the hyperbola equals the two asymptotic lines and flips from on pair of opposite quadrants into the other. The transition either creates a bridge by locally joining two portions of the body at z, or it removes a bridge by locally separating two portions at z.

CASE $I = \{i, j, k\}$. Point z lies inside the shrunken Delaunay triangle ${\mu_I}^s$. In degenerate configurations I may contain more than three indices and ${\mu_I}^s$ may have more than three sides. The triangle specifies the portion of the circle defined by $b_\ell^{\perp}$ orthogonal to b_i, b_j, b_k that belongs to the skin. Observe the symmetry to card $I = 1$. The non-degenerate cases are again when the radius of the circle is either positive real or imaginary. The transition either removes a hole by shrinking a disk to z or it creates a hole by growing a disk from z.

Asymmetry of deformation. For a fixed $t \in [0, 1]$, $A_t = (1-t) \cdot B + t \cdot C$ is a set of disk and we consider its Voronoi and Delaunay complexes, see figure 10. Combinatorially, the Delaunay complex changes only twice, from $t = 0$ to $t \in (0, 1)$ to $t = 1$. Geometrically, it moves all the time as the cells gradually change shape. Interestingly, the situation is different for the Voronoi complex, and this in spite of the symmetry between the two complexes noted above. The Voronoi complexes of the A_t are the same for all $t \in (0, 1)$. Specifically, each such Voronoi complex is the overlay of Vor A_0 and Vor A_1, see again figure 10. The asymmetry is explained by the asymmetry of the deformation. An alternative definition of deformation uses the shapes defined by the complementary disk sets:

$$D_t^{\perp} = (1-t) \cdot B^{\perp} + t \cdot C^{\perp}.$$

$D_t^{\perp}$ is different from $A_t^{\perp}$ except at the beginning, at $t = 0$, and at the end, at $t = 1$. To see that the two are indeed different, in general, note that $(A_t^{\perp})^{\perp} = A_t$. So when $A_t^{\perp}$ changes as t increases, the Voronoi complex changes and the Delaunay complex is invariant. On the other hand, when $D_t^{\perp}$ changes as t increases, the Delaunay complex changes and the Voronoi complex is invariant.

4 Deformation Design

The control panel of the software tool contains the VCR buttons that drive the deformation through the selection and alteration of time and size. The explicit representation of the 2-dimensional shape space permits the rational design of deformations.

Shape space. Let the two disk sets B and C define the initial and the final shape of a deformation problem. We have an intermediate shape defined for each value $t \in [0, 1]$ of the time parameter and each value $\alpha^2 \in \mathbb{R}$ of the size parameter. Specifically, the shape is defined by the set of disks $A_t(\alpha)$, where $A_t = (1-t) \cdot B + t \cdot C$. We use (t, α^2) as a coordinate pair to specify shapes in

$$\Omega = [0, 1] \times \mathbb{R},$$

which we refer to as *shape* or *state space.* Ω is a 2-dimensional strip that can be decomposed into regions within which the shape variation is insignificant. As an example consider the decomposition into shapes whose complexes are the same: $\mathbb{X}_1 = \text{bdy} A_{t_1}(\alpha_1)$ and $\mathbb{X}_2 = \text{bdy} A_{t_2}(\alpha_2)$ are *similar* if $\text{Dsx}_{\alpha_1} A_{t_1} = \text{Dsx}_{\alpha_2} A_{t_2}$.

Since the moments in time the two shapes exist are in general different this needs some clarification. Although the Delaunay complex moves with time, its combinatorial structure is the same during the entire open time interval. We therefore have an unambiguous notion of *sameness* for Delaunay cells over time and it makes sense to refer to a Delaunay cell δ_I without specifying the time, which can be anywhere in $(0,1)$. Two complexes at different moments in time are the *same* if they contain the same cells. Similarity among shapes is therefore well defined and it is an equivalent relation.

Curve arrangement. We gain insight into the decomposition of Ω by considering a single Delaunay cell, δ_I. For each time $t \in (0,1)$ let $f_I(t)$ be the smallest real number so $\delta_I(t)$ belongs to $\text{Dsx}_\alpha A_t$ iff $\alpha^2 \geq f_I(t)$. For each 2-dimensional δ_I the discriminating function is linear:

$$f_I(t) \quad = \quad a_1 \cdot t + a_0.$$

For Delaunay edges and vertices the description of f_I is made complicated by the somewhat technical distinction between attached and unattached cells, see e.g. [5]. A cell δ_I is *attached* at time t if for every $\alpha^2 \in \mathbb{R}$ the presence of $\delta_I(t)$ in $\text{Dsx}_\alpha A_t$ implies the presence of a higher-dimensional cell, δ_J, that contains δ_I as a face. Since δ_J cannot be in the complex without its faces this implies $f_I(t) = f_J(t)$ for this value of t. The interesting case is when $\delta_I(t)$ is unattached. For example if δ_I is 2-dimensional then it is unattached during the entire open time interval. In general, δ_I is unattached during a single and possibly empty time interval. In that time interval the discriminating function is at most quadratic:

$$f_I(t) \quad = \quad a_2 \cdot t^2 + a_1 \cdot t + a_0,$$

with $a_2 \leq 0$. We picture Ω and its decomposition by drawing time from left to right and size from bottom to top. The graphs of the discriminating functions are lines that pass through the strip from left to right and upside-down parabolas that begin and end on the strip boundary or on graphs of other discriminating functions, see figure 11. The graphs decompose Ω into regions of similar shapes. It can however happen that two similar shapes correspond to points in different regions. In this case there is no deformation representable in Ω that transforms one shape to the other and uses only similar shapes.

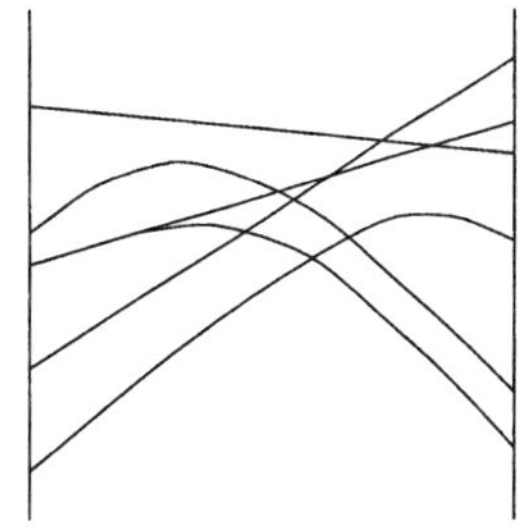

Figure 11: The decomposition of Ω into regions of similar shapes. All curves are pieces of lines and parabolas.

Designing in shape space. The initial shape corresponds to the point $P_0 = (0,0)$ in Ω and the final shape corresponds to $P_1 = (1,0)$. Every continuous deformation between the two shapes that can be represented in our framework corresponds to a path $p : [0,1] \to \Omega$ with $p(0) = P_0$ and $p(1) = P_1$. For example the basic deformation described in section 2 corresponds to the straight line segment connecting P_0 with P_1, which is the image of the path $p(u) = (u,0)$. It may be that more complicated paths give better deformations and we describe how the decomposition of Ω is used to find such paths.

An important first step in the rational design of deformations is the computation of ***signatures*** that are functions from Ω to $\mathbb{R}$. An example is the area signature that maps each (t,α^2) to the area of the corresponding body. Strictly speaking that area also depends on the value of the parameter s and we have a signature for each $0 \leq s \leq 1$. Another example is the connectivity of shapes captured by two signatures, $\beta_0, \beta_1 : \Omega \to \mathbb{R}$, counting components and holes. β_0 and β_1 are independent of s and can be computed directly from the complex, which is homotopy equivalent to the body of the shape [3].

Let $\mathcal{G}$ be the dual graph of the decomposition of Ω created by the arrangement of lines and parabolas described above. Let $\beta_0(\nu)$ be the number of components and $\beta_1(\nu)$ be the number of holes of any one shape represented by the region or node ν in $\mathcal{G}$. Let μ_0 and μ_1 be the nodes whose regions contain P_0 and P_1 in their boundaries. A non-degenerate path, $p : [0,1] \to \Omega$, traces out a sequence of regions and thus translates into a discrete path in $\mathcal{G}$: $\mu_0 = \nu_0, \nu_1, \ldots, \nu_\ell = \mu_1$. Suppose we are interested in a deformation that goes through as few topological changes as possible. The number of such changes that occur along the path p is a sum of absolute differences:

$$\sum_{i=1}^{\ell} |\beta_0(\nu_i) - \beta_0(\nu_{i-1})| + |\beta_1(\nu_i) - \beta_1(\nu_{i-1})|.$$

Think of the term inside the sum as the length of the

arc connecting nodes ν_{i-1} and ν_i. The problem of finding a path in Ω with minimum topological changes thus reduces to finding a shortest path in $\mathcal{G}$, which can be solved by standard graph algorithms. An improvement of this approach based on the monotonicity of minimizing paths is described in [7].

5 Discussion

This paper describes a method for the automatic deformation of shapes in $\mathbb{R}^2$ and a software tool that implements it.

Algorithms. The basic algorithmic problem raised by the deformation method is the construction of the 1-parameter family of Delaunay complexes $\text{Del}\, A_t$, for $t \in [0,1]$. As explained in [2], this problem reduces to computing a 4-dimensional convex hull, or equivalently a 3-dimensional (weighted) Delaunay complex. In the latter interpretation the third coordinate is time and $\text{Del}\, A_t$ is the cross-section at $x_3 = t$. $\text{Dsx}\, A_t$ is a subcomplex of $\text{Del}\, A_t$ and the body and skin of the shape are computed from that subcomplex. Our software tool uses the 3-dimensional Delaunay complex algorithm mentioned in [5] and selects subcomplexes as explained in the same reference.

A second interesting algorithmic problem is the construction of the line and parabola arrangement that decomposes the shape space. We implement the incremental algorithm of [4]. The decomposition is typically too fine to be of much use within the user interface. We plan to add a coarse version of the decomposition to the control panel for visualization purposes and to support the manual design of deformation paths. Such coarsenings can be construction with the snap-rounding method described in [6].

Extensions. There are means to impose structure on the shape space other than the arrangement of lines and parabolas described in section 4. Take for example the area signature, $a : \Omega \rightarrow \mathbb{R}$, say for $s = 0.5$. It is continuous and increases with growing α^2. Each level set, $a^{-1}(c)$, is therefore a monotone path connecting a point on the left with a point on the right boundary of Ω. If P_0 and P_1 are these two points then $a^{-1}(c)$ describes an area preserving deformation between the two corresponding shapes.

Any application to image morphing will almost surely require a way to carry color and pattern along with the shape through the deformation. Maybe one of the background structures, such as the mixed complex that decomposes the body and moves with it, can be used for that purpose.

Applications in data compression and databases for shapes will find the restriction to *two* base shapes unreasonably limiting. As explained in [2], the mathematical framework for deformation readily extends to any fixed number of base shapes. However, all algorithmic problems get significantly harder as the dimension of the shape space increases.

References

[1] J. Blinn. A generalization of algebraic surface drawing. *ACM Trans. Graphics* **2** (1980), 235–256.

[2] H.-L. Cheng, H. Edelsbrunner and P. Fu. Shape space from deformation. Report rgi-tech-011, Raindrop Geomagic, Champaign, Illinois, 1998.

[3] H. Edelsbrunner. Deformable smooth surface design. *Discrete Comput. Geom.*, to appear.

[4] H. Edelsbrunner, L. J. Guibas, J. Pach, R. Pollack, R. Seidel and M. Sharir. Arrangements of curves in the plane-topology, combinatorics and algorithms. *Theoret. Comput. Sci.* **92** (1992), 319–336.

[5] H. Edelsbrunner and E. P. Mücke. Three-dimensional alpha shapes. *ACM Trans. Graphics* **13** (1994), 43–72.

[6] L. J. Guibas and D. Marimont. Rounding arrangements dynamically. *In* "Proc. 11th Ann. Sympos. Comput. Geom. 1995", 190-199.

[7] K. P. Lam. Two-dimensional geometric morphing. Master thesis, Dept. Comput. Sci., Hong Kong University of Science and Technology, 1996.

[8] S. Y. Lee, K. Y. Chwa, J. Hahn and S. Y. Shin. Image morphing using deformation techniques. *J. Visualization Comput. Animation* **7** (1996), 3–23.

[9] S. M. Seitz and C. R. Dyer. View morphing. *Computer Graphics* **50** (1996), 21–30.

[10] G. Wolberg. Recent advances in image morphing. *In* "Proc. Computer Graphics Internat. 1996", 64–71.

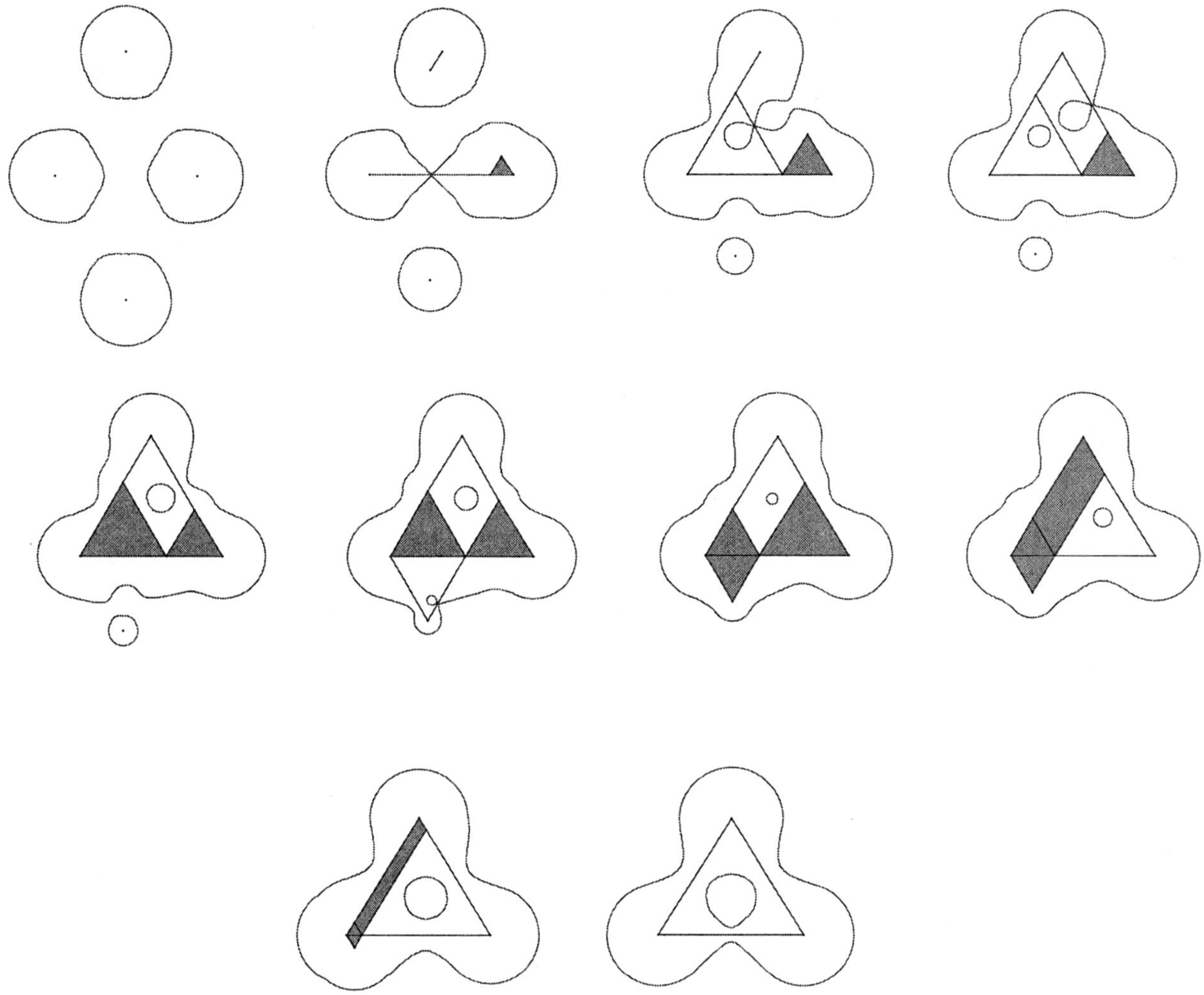

Figure 5: Ten snapshots of a deformation with complex and skin displayed. The parameter s for all shapes is 0.5. The set of disks that generates the complex and the skin in the fifth snapshot is the same as in figures 2, 3, 4.

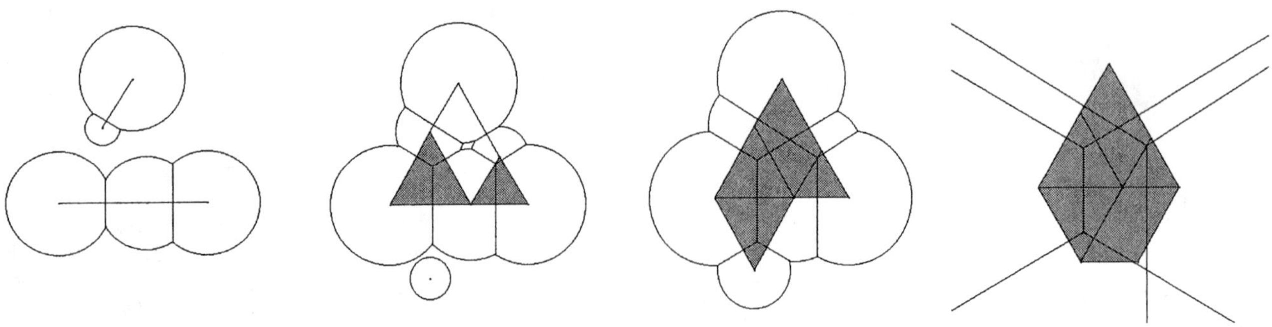

Figure 6: Decomposed disk union and α-complex for four values of α. In the second example we have $\alpha = 0$ and therefore the same complex as in figure 3. In the fourth example we have α sufficiently large so that the α-complex is the Delaunay complex.

Figure 9: From left to right the sequence illustrates the creation of a bridge through flipping a hyperbola. From right to left it illustrates the removal of that bridge.

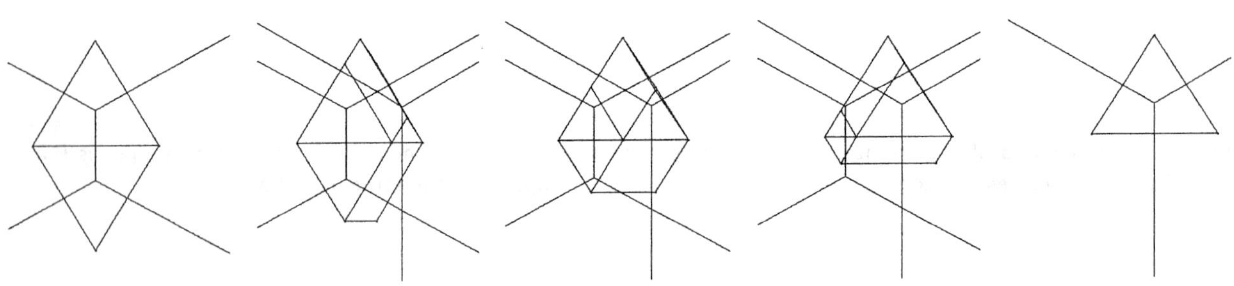

Figure 10: The Vornonoi and Delaunay complexes at times $t = 0, 0.25, 0.5, 0.75, 1$ during the deformation.

Surface Reconstruction by Voronoi Filtering

Nina Amenta* Marshall Bern†

Abstract

We give a simple combinatorial algorithm that computes a piecewise-linear approximation of a smooth surface from a finite set of sample points. The algorithm uses Voronoi vertices to remove triangles from the Delaunay triangulation. We prove the algorithm correct by showing that for densely sampled surfaces, where density depends on "local feature size", the output is topologically valid and convergent (both pointwise and in surface normals) to the original surface. We describe an implementation of the algorithm and show example outputs.

1 Introduction

The problem of reconstructing a surface from scattered sample points arises in many applications such as computer graphics, medical imaging, and cartography. In this paper we consider the specific reconstruction problem in which the input is a set of sample points S drawn from a smooth two-dimensional manifold F embedded in three dimensions, and the desired output is a triangular mesh with vertex set equal to S that faithfully represents F. We give a "provably correct" combinatorial algorithm for this problem. That is, we give a condition on the input sample points, such that if the condition is met the algorithm gives guaranteed results: a triangular mesh of the same topology as the surface F, with position and surface normals within a small error tolerance. The algorithm relies on the well-known constructions of the Delaunay triangulation and the Voronoi diagram.

This paper is an extension of previous work by Amenta, Bern, and Eppstein [1] on reconstructing curves in two dimensions. Our previous work defined a planar graph on the sample points called the "crust". The crust is the set of edges in the Delaunay triangulation of the sample points that can be enclosed by circles empty not only of sample points, but also of Voronoi vertices. The crust comes with a guarantee: if the curve is "well-sampled", then the crust contains exactly the edges between sample points adjacent on the curve. Our notion of well-sampled, which involves the medial axis of the curve, is sensitive to the local geometry. Hence our algorithm, unlike other algorithms for this problem, allows highly nonuniform sampling, dense in detailed areas yet sparse in featureless areas. Any provably correct algorithm must impose some sampling density requirement, similar to the Nyquist limit in spectral analysis.

The extension to three dimensions in this paper requires both new algorithmic ideas and new proof techniques. Most notably the algorithm uses only a subset of the Voronoi vertices to remove Delaunay triangles. The algorithm picks only two Voronoi vertices—called *poles*—per sample point: the farthest vertices of the point's cell on each side of the surface. With this modification, the straightforward generalization of our two-dimensional algorithm now works. Delaunay triangles with circumspheres empty of poles give a piecewise-linear surface pointwise convergent to F. The poles, however, also enable further filtering on the basis of triangle normals. Adding this filtering gives a piecewise-linear surface that converges to F both pointwise and in surface normals (and hence in area). We believe that poles may be applicable to other algorithms as well, perhaps whenever one wishes to estimate a surface normal or tangent plane.

This paper is organized as follows. Section 2 describes previous work on surface reconstruction. Section 3 gives our algorithm. Section 4 states our theoretical guarantees, and Section 5 sketches their proofs. Section 6 shows some example outputs.

2 Previous Work

Previous work on the reconstruction problem has been mostly heuristic. Only recently have researchers started publishing algorithms for the two-dimensional problem with provable properties.

Hoppe et al. [20, 21, 22] brought the reconstruction problem to the attention of the computer graphics community. Their algorithm computes an approximating surface—not interpolating but close. The algorithm estimates a tangent plane at each sample using the k nearest neighbors, and uses the distance to the plane of the closest sample point as a signed distance function. The zero set of this function is then contoured by a continuous piecewise-linear surface using the marching cubes algorithm. A similar algorithm by Curless and Levoy [13] is tuned for data samples collected by a laser range scanner, but could be applied to the general

*Computer Sciences, University of Texas, Austin, TX 78712. Work performed in part while at Xerox PARC, partially supported by NSF grant CCR-9404113.

†Xerox Palo Alto Research Center, 3333 Coyote Hill Rd., Palo Alto, CA 94304

SCG 98 Minneapolis Minnesota USA

reconstruction problem. Their algorithm sums anisotropically weighted contributions from the samples to compute a signed distance function, which is then discretized on voxels to eliminate the marching cubes step. These two algorithms are quite successful in practice, but have no provable guarantees. Indeed there exist arbitrarily dense sets of samples, for example ones with almost collinear nearest neighbor sets, for which the algorithm of Hoppe et al. would fail.

The most famous computational geometry construction for associating a polyhedral shape with an unorganized set of points is the α-shape of Edelsbrunner et al. [15, 16]. Like our reconstructed surface, the α-shape is a subcomplex of the Delaunay triangulation. A Delaunay simplex (edge, face, etc.) belongs to the α-shape of S if its circumsphere has radius at most α. The major drawback of using α-shapes for surface reconstruction is that the optimal value of α depends on the sampling density, which often varies over different parts of the surface. For uniformly sampled surfaces, however, α-shapes are workable. Bernardini et al. [8] follow α-shape-based reconstruction with a clean-up phase to resolve sharp dihedral angles. Edelsbrunner and Raindrop Geomagic [14] are continuing to develop α-shape-based reconstruction along with proprietary extensions.

An early algorithm due to Boissonnat [10] is related to ours. He proposed a "sculpting" heuristic for selecting a subset of Delaunay tetrahedra to represent the interior of an object. The heuristic is motivated by the observation that "typical" Delaunay tetrahedra have circumspheres approximating maximal empty balls centered at points of the medial axis; our algorithm relies on this same observation. Boissonnat's algorithm, however, overlooks the fact that even dense sample sets can give Delaunay tetrahedra with circumspheres that are arbitrarily far from the medial axis; indeed it is this second observation which motivates our definition of poles. Goldak, Yu, Knight and Dong [19] made a similar oversight, asserting incorrectly that the Voronoi diagram vertices asymptotically approach the medial axis as the sampling density goes to infinity.

Finally, for the two-dimensional problem there are a few recent algorithms with provable guarantees. Figueiredo and Miranda Gomes [18] prove that the Euclidean minimum spanning tree can be used to reconstruct uniformly sampled curves in the plane. Bernardini and Bajaj [7] prove that α-shapes also reconstruct uniformly sampled curves in the plane. Attali [3] gives yet another provably correct reconstruction algorithm for uniformly sampled curves in the plane, using a graph in which edges are defined by empty regions between vertices. Our previous paper showed that both the crust and the β-skeleton [23] (another empty-region planar graph) correctly reconstruct curves even with nonuniform sampling. Our two-dimensional results [1] are in this way strictly stronger than those of the other authors.

3 Description of the Algorithm

We start by describing the algorithm of Amenta et al. [1] for the problem of reconstructing curves in $\mathbb{R}^2$. Let F be a smooth (twice differentiable) curve embedded in $\mathbb{R}^2$, and S be a set of sample points from F. Let V denote the vertices of the Voronoi diagram of S. The *crust* of S contains exactly the edges of the Delaunay triangulation of $S \cup V$ with both endpoints from S. Saying this another way, the crust contains exactly those Delaunay edges around which it is possible to draw a circle empty of Voronoi vertices. In our earlier paper, we proved that if S is a sufficiently dense sample, this simple algorithm constructs a polygonal approximation of F (Theorem 1 in Section 4 below).

1. Compute the Voronoi diagram of the sample points S.
2. For each sample point s:
   (a) If s does not lie on the convex hull of S, let p^+ be the vertex of $Vor(s)$ farthest from s.
   (b) If s does lie on the convex hull of S, let p^+ be a point at "infinite distance" outside the convex hull with the direction of sp^+ equal to the average of the outward normals of hull faces meeting at s.
   (c) Pick the vertex p^- of $Vor(s)$ farthest from s with negative projection onto sp^+.
3. Let P denote all poles p^+ and p^-, except those p^+'s at infinite distance. Compute the Delaunay triangulation of $S \cup P$.
4. (Voronoi Filtering) Keep only those triangles in which all three vertices are sample points.
5. (Filtering by Normal) Remove each triangle T for which the normal to T and the vector to the p^+ pole at a vertex of T form too large an angle (greater than θ for the largest-angle vertex of T, greater than 2.2θ for the other vertices of T).
6. (Trimming) Orient triangles and poles (inside and outside) consistently, and extract a piecewise-linear manifold without sharp dihedral angles.

Figure 1. The surface reconstruction algorithm.

The straightforward generalization of this algorithm fails for the task of reconstructing a smooth two-dimensional manifold embedded in three dimensions. The problem is that vertices of the Voronoi diagram may fall very close to the surface, thereby punching holes in the crust. For example, the Voronoi center of a sliver can lie arbitrarily close to the surface F. A *sliver* is a tetrahedron with bad aspect ratio yet a reasonably small circumradius to shortest edge ratio, such as the tetrahedron formed by four nearly equally spaced vertices around the equator of a sphere.

The fix is to consider only the *poles*. The poles of a sample point s are the vertices of the Voronoi cell $Vor(s)$ farthest from s, one on each side of the surface. Since the algorithm does not know the surface, only the sample points, it chooses the poles by the method given in step 2 in Figure 1. Lemma 2, parts (a) and (c), in Section 5 imply that that this method is correct for well-sampled surfaces. Voronoi cells of sample points on the convex hull of S are unbounded in the outwards direction; for such a point the outwards pole simply represents a direction in which the cell is unbounded. Such a pole is used only to help find the sample point's opposite pole and is not included in the eventual Delaunay triangulation. Denoting the poles by P, we define the *crust* of S to be the triangles of the Delaunay triangulation of $S \cup P$, all of whose vertices are members of S.

Steps 1–4 compute the crust (sometimes called the *raw crust* to distinguish it from the more finished versions). The crust has a relatively weak theoretical guarantee: it is pointwise convergent to F as the sampling density increases. Steps 5 and 6 are "postprocessing" steps that produce an output with a stronger guarantee: convergence both pointwise and in surface normals, and topological equivalence.

Step 5 removes triangles based on the directions of their surface normals. Let T be a triangle of the crust and let s be its vertex of maximum angle. Step 5 removes T if the angle between the normal to T and the vector from any one of T's vertices to its first-chosen pole is too large. The definition of "too large" depends on which vertex of T is under consideration: for the vertex with largest angle, too large means greater than an input parameter θ, and for the other two vertices it means greater than 2.2θ. Angles are unsigned angles in the range $[0, \pi/2]$. As stated in Theorem 4, the choice of θ is connected with the sampling density. If the user of our algorithm does not have an estimate of the sampling density (the parameter r in Definition 3 below), then the user can slowly decrease θ, backing off when holes start to appear in the surface, similar to choosing a surface from the spectrum of α-shapes [16].

Step 6 ensures that the reconstructed surface has the topology of the original surface; before this final step, the computed surface will resemble the original surface geometrically, but may have some extra triangles enclosing small bubbles and pockets. The problem once again is slivers: all four faces of a flat sliver may make it past steps 4 and 5.

Step 6 first orients all triangles. Start with any sample point s on the convex hull of S. Call the direction to p^+ at s the *outside* and the direction to p^- the *inside*. Pick any triangle T incident to s, and define the outside side of T to be the one visible from points on the sp^+ ray. Orient the poles of the other vertices of T to agree with this assignment. Orient each triangle sharing a vertex with T so that they agree on the orientations of their shared poles, and continue by breadth-first search until all poles and triangles have been oriented.

Define a *sharp* edge to be an edge for which the cyclic order of triangle sides alternates between outside and inside. (As we show in the proof of Theorem 4, a sharp edge indeed has all triangles within a small wedge. Notice that an edge bounding only a single triangle is necessarily a sharp edge.) Step 6 trims off pockets by greedily removing triangles with *sharp* edges. Now the remaining triangles form a "quilted" surface, in which each edge borders at least two triangles, with consistent orientations. Finally, Step 6 extracts the outside of this quilted surface by a breadth-first search on triangles.

4 Theoretical Guarantees

What sets our algorithm apart from previous algorithms are its theoretical guarantees. We start with some definitions. Figure 2 gives an example of the medial axis in $\mathbb{R}^2$; we allow F to have more than one connected component.

Definition 1. *The* **medial axis** *of a manifold F embedded in $\mathbb{R}^d$ is the closure of the set of points in $\mathbb{R}^d$ with more than one nearest neighbor on F.*

Definition 2. *The* **local feature size** *$LFS(p)$ at a point p on F is the Euclidean distance from p to (the nearest point of) the medial axis.*

Definition 3. *Set $S \subset F$ is an* **r-sample** *of F if no point p on F is farther than $r \cdot LFS(p)$ from a point of S.*

Notice that the notion of r-sample does not assume any global—or even local—uniformity. Further notice that to prove an algorithm correct, we must place some condition on the set of sample points S, or else the original surface could be any surface passing through S. Our paper on curve reconstruction [1] proved the following theoretical guarantee.

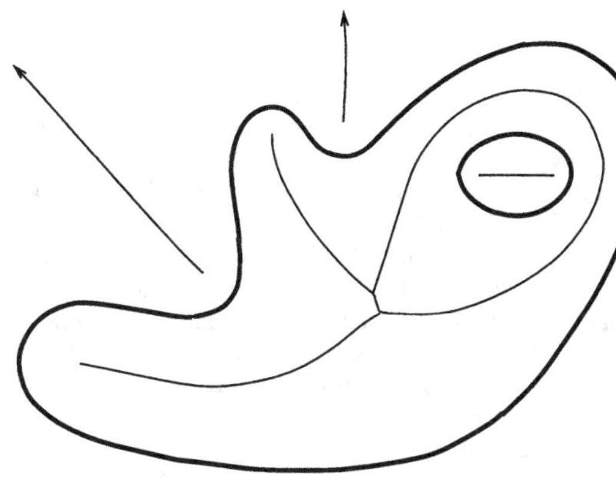

Figure 2. The medial axis of a smooth curve.

Theorem 1 (Amenta et al. [1]). *If S is an r-sample of a curve in $\mathbb{R}^2$ for $r \leq .40$, then the crust includes all the edges between pairs of sample points adjacent along F. If S is an r-sample for $r \leq .25$, then the crust includes exactly those edges.*

To state our results for the three-dimensional problem, we must define a generalization of adjacency. Consider the Voronoi diagram of the sample points S. This Voronoi diagram induces a cell decomposition on surface F called the *restricted Voronoi diagram*: the boundaries of the cells on F are simply the intersections of F with the three-dimensional Voronoi cell boundaries. We call a triangle with vertices from S a *good triangle* if it is dual to a vertex of the restricted Voronoi diagram; good triangles are necessarily Delaunay triangles. Our first three-dimensional result shows that good triangles deserve their name. To our knowledge, our proof of this result is the first proof that the three-dimensional Delaunay triangulation of a sufficiently dense set of samples contains a piecewise-linear surface homeomorphic to F.

Theorem 2. *If S is an r-sample of F for $r \leq .1$, then the good triangles form a polyhedron homeomorphic to F.*

Our next theorem states the theoretical guarantees for the three-dimensional (raw) crust.

Theorem 3. *(a) If S is an r-sample for $r \leq .1$, then the crust includes all the good triangles. (b) If S is an r-sample for $r \leq .06$, then the crust lies within a fattened surface formed by placing a ball of radius $5r LFS(q)$ around each point $q \in F$.*

Step 5 adds another guarantee: convergence in surface normals. This step is indeed necessary for this guarantee, as the raw crust sometimes includes small skinny triangles with deviant surface normals. For example, the insides of the sausages shown on the left in Figure 11 have a sort of "washboard" texture due to small deviant triangles lining the inside curves. Finally, Step 6 adds the guarantee of topological equivalence.

Theorem 4. *Assume S is an r-sample and set $\theta = 3r$. (a) Let T be a triangle of the θ-crust and t a point on T. The angle between the normal to T and the normal to F at the point $p \in F$ closest to t measures $O(r)$ radians. (b) For sufficiently small r, the trimmed θ-crust is homeomorphic to F.*

5 Proofs

In this section we sketch proofs of the theoretical guarantees. We start with some terminology. At each point $p \in F$, there are two tangent *medial balls* centered at points of the medial axis. The vectors from p to the centers of its medial balls are normal to F, and F does not intersect the interiors of the medial balls. Since $LFS(p)$ is at most the radius of the smaller medial ball, F is also confined between the two tangent balls of radius $LFS(p)$. We call a maximal empty ball centered at a Voronoi vertex a *Voronoi ball*, and the Voronoi ball centered at a pole a *polar ball*.

Our first lemma is rather basic: Lipschitz conditions for the $LFS(p)$ function and for the direction of surface normals (a function from F to the two-dimensional sphere). We use $d(p,q)$ to denote the Euclidean distance from p to q. Angles are measured in radians.

Lemma 1. *(a) For any two points p and q on F, $|LFS(p) - LFS(q)| \leq d(p,q)$. (b) For any two points p and q on F with $d(p,q) \leq \rho \min\{LFS(p), LFS(q)\}$, where $\rho < 1/3$, the angle between the normals to F at p and q is at most $d(p,q)/((1-3\rho)LFS(p))$.*

Proof: For part (a), $LFS(p) \geq LFS(q) - d(p,q)$, since the ball of radius $LFS(q)$ around q contains the ball of radius $LFS(q) - d(p,q)$ around p and contains no point of the medial axis. Similarly, $LFS(p) - d(p,q) \leq LFS(q)$.

For part (b), let us parametrize the line segment pq by length. Let $p(t)$ denote the point on pq with parameter value t and let $f(t)$ denote the nearest point to $p(t)$ on the surface F. In other words, $f(t)$ is the point at which an expanding sphere centered at $p(t)$ first touches F. Point $f(t)$ is unique, because otherwise $p(t)$ would be a point of the medial axis, contradicting $d(p,q) \leq \rho LFS(p)$.

Let $n(t)$ denote the unit normal to F at $f(t)$, and $|n'(t)|$ the magnitude of the derivative with respect to t, that is, the rate at which the normal turns as t grows. The change in normal between p and q is at most $\int_{pq} |n'(t)| dt$, which is at most $d(p,q) \max_t |n'(t)|$.

There are tangent balls of radius $LFS(f(t))$ on either side of F at $f(t)$, so $|n'(t)|$ is at most the rate at which the normal turns on one of these tangent balls. Referring to Figure 3, we see that

$$dt \geq (LFS(f(t)) - d(f(t),p(t))) \cdot \sin\theta.$$

Now $\sin\theta$ approaches θ as θ goes to zero, so

$$|n'(t)| = \lim_{\theta \to 0} \theta/dt \leq 1/(LFS(f(t)) - d(f(t),p(t))).$$

We have that

$$d(f(t),p(t)) \leq d(p(t),p) \leq \rho LFS(p)$$

and

$$d(f(t),p) \leq d(f(t),p(t)) + d(p(t),p) \leq 2\rho LFS(p),$$

so by Lemma 1, $LFS(f(t)) \geq (1-2\rho)LFS(p)$. Altogether we obtain $\max_t |n'(t)| \leq 1/((1-3\rho)LFS(p))$, which yields the lemma. ∎

We next show that the cells of the Voronoi diagram of S are long and skinny, with long direction nearly normal to the surface F. Together, Lemma 2(a) and (c) below show that the vector from a sample point to a pole gives a good approximation to the surface normal. This observation may

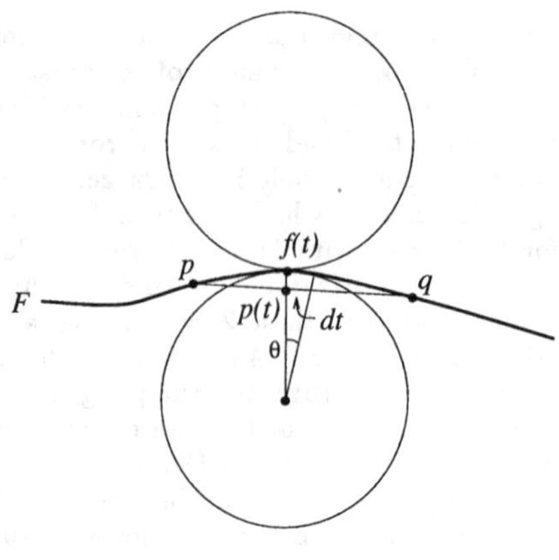

Figure 3. Bounding $|n'(t)|$ in terms of the radius $LFS(f(t))$ and $d(f(t),p(t))$.

have wider applicability than to our own surface reconstruction algorithm; for example, the Voronoi diagram and the poles could be used to obtain provably reliable estimates of tangent planes in the algorithm of Hoppe et al.

Lemma 2. *(a) On either side of F, the distance from s to its pole is at least $LFS(s)$. (b) The intersection of $Vor(s)$ and F is contained in a ball of radius $\frac{r}{1-r}LFS(s)$. (c) Let v be a vertex of $Vor(s)$ such that $d(v,s) \geq LFS(s)$. The angle at s between the vector to v and the normal to the surface is at most $2\arcsin\frac{r}{2-2r}$.*

Proof: (a) On either side of F at s, the tangent ball of radius $LFS(s)$ is empty, since it is contained in a medial ball at s. The center c of such a ball lies within $Vor(s)$. So either $Vor(s)$ has a vertex at distance at least $LFS(s)$, or the pole on that side lies at infinity.

(b) Let $p \in Vor(s) \cap F$. Since s is the closest sample point to p, $d(p,s) \leq rLFS(p) \leq r(LFS(s) + d(p,s))$ by Lemma 1(a). So $d(p,s) \leq \frac{r}{1-r}LFS(s)$.

(c) Let B_v be the Voronoi ball centered on v. Let B_m be the medial ball touching s on the same side of the surface F, and let m be its center. Let ϕ be the angle between the segments sv and sm, that is, the angle referred to in the lemma. Let B_p be the ball of radius $LFS(s)$, tangent to F at s, but lying on the opposite side of F from B_m; let p be the center of B_p. The surface F passes between B_m and B_p at s, and does not intersect the interior of either of them, as shown in Figure 4.

Since p and v lie on opposite sides of F, line segment pv must intersect F at least once. Let q be the intersection point closest to p. No sample point can lie in either B_p or B_v, so the nearest sample point to q must be s. Since B_p and B_v each have radius at least $LFS(s)$, $d(q,s) \geq 2\sin(\phi/2)LFS(s)$. Since S is an r-sample, $d(q,s)$ must be less than $\frac{r}{1-r}LFS(s)$. Combining the last two inequalities, $2\sin(\phi/2) \leq \frac{r}{1-r}$, or $\phi \leq 2\arcsin\frac{r}{2-2r}$. ∎

Lemma 3. *Let T be a good triangle and s its vertex with largest angle. (a) The angle between the normal to T and the normal to F at s is at most $\arcsin(\sqrt{3}r/(1-r))$. (b) The angle between the normal to T and the normal to F at any other vertex of T is at most $2r/(1-7r) + \arcsin(\sqrt{3}r/(1-r))$.*

Proof: For part (a), let C be the circumcircle of T and let ρ_C be its radius. Consider the balls of radius $LFS(s)$

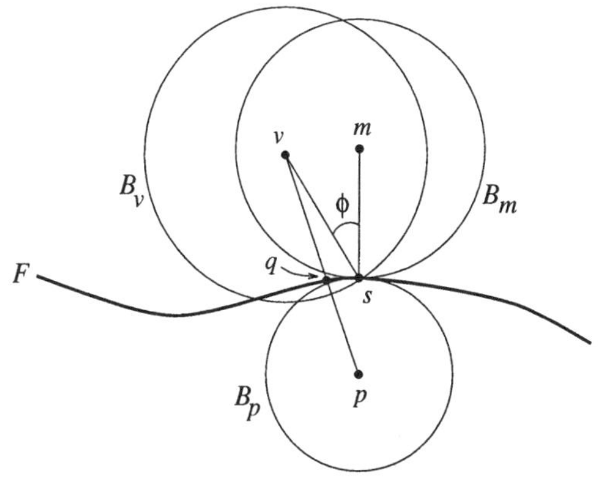

Figure 4. The vector from s to a distant Voronoi vertex such as a pole must be nearly normal to the surface. B_p, B_m and B_v all have radius at least $LFS(s)$.

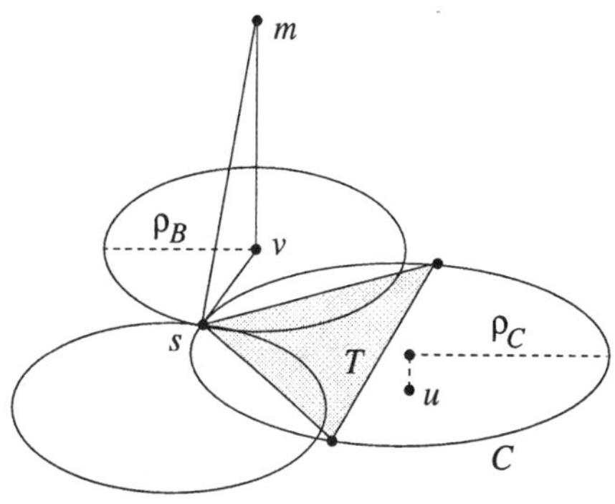

Figure 5. Bounding the angle between the normal to the triangle and the normal to the surface at s.

tangent to F at s on either side of F. These balls intersect the plane of T in two disks of the same radius, which we shall denote ρ_B. Since the balls are empty of sample points, the disks cannot contain the other two vertices of T. The other two vertices are at most distance $\sqrt{3}\rho_C$ away from s, which in turn implies that $\rho_B \leq \sqrt{3}\rho_C$.

We can rewrite these radii in terms of $LFS(s)$. Let u denote the restricted Voronoi diagram vertex dual to T. Since u lies on the line through the center of C normal to the plane of C, $\rho_C \leq d(u, s)$. By Lemma 2(b), $d(u, s) \leq \frac{r}{1-r} LFS(s)$, so altogether $\rho_B \leq \sqrt{3}r LFS(s)/(1-r)$.

Now to find the angle between the normal to T and the normal to F at s, we consider one of the tangent balls B at s. Let m denote the center of B and v denote the center of the disk of radius ρ_B that is the intersection of B with the plane of T, as shown in Figure 5. The segment sm is normal to F at s and the segment mv is normal to T, so the angle we would like to bound is $\angle smv$. The triangle smv is right, with hypotenuse of length $LFS(s)$ and leg opposite $\angle smv$ of length $\rho_B \leq \sqrt{3}r LFS(s)/(1-r)$. Hence $\angle smv$ measures at most $\arcsin(\sqrt{3}r/(1-r))$.

For part (b), let s' be one of the other vertices of T. Since T is a good triangle, s and s' are neighbors in the restricted Voronoi diagram. Let p be a point on the boundary of both restricted Voronoi diagram cells. Then

$$d(p, s) \leq r LFS(p) \leq \frac{r}{1-r} \min\{LFS(s), LFS(s')\}.$$

So $d(s, s') \leq \frac{2r}{1-r} \min\{LFS(s), LFS(s')\}$. By Lemma 1(b), the angle between the normals to F at s and s' is at most $2r/(1-7r)$. ∎

We are now ready to sketch a proof of Theorem 2: the good triangles form a polyhedron homeomorphic to F. The proof uses a result of Edelsbrunner and Shah [17].

Proof of Theorem 2: It suffices [17] to show that S has the following *closed-ball property*: the closure of each k-dimensional face, $1 \leq k \leq 3$, of the Voronoi diagram of S intersects F in either the empty set or in a closed $(k-1)$-dimensional topological ball.

Let s be a sample point and $Vor(s)$ its Voronoi cell. Let the direction of the normal to F at s be vertical. Lemma 2(b) shows that $Vor(s) \cap F$ is small, fitting inside a ball B of radius $\frac{r}{1-r} LFS(s)$. Lemma 1(b) then shows that F is nearly horizontal throughout $Vor(s) \cap F$, more precisely, the normal to F is nowhere farther than $r/(1-3r) \leq .15$ radians from vertical.

First consider an edge e of $Vor(s)$, that is, the case $k = 1$. If e has nonempty intersection with F, then e is normal to the good triangle T dual to its intersection point. By Lemma 3(b), e must be within $2r/(1-7r) + \arcsin(\sqrt{3}r/(1-r))$ radians from the normal to F at s. For $r \leq .1$, this expression is less than .9, so e is within .9 radians from vertical, and consequently can intersect F only once within B.

Next consider a face f of $Vor(s)$, that is, the case $k = 2$. Face f is contained in the perpendicular bisector of s and another sample point s'. If f intersects F, then some side of f must pierce F, and since such an edge can form an angle no greater than .9 radians with vertical, f itself lies within .9 radians of vertical. Now consider a single connected component C of $f \cap F$. We use the curved segment C to divide $\mathbb{R}^3$ into two pieces. Let H be the set of points p in $\mathbb{R}^3 \setminus C$ such that the line segment from p to its closest point on C forms an angle smaller than .2 radians with horizontal. Set H is thus a union of wedges with vertices on C. We assert that all points of $F \cap B_s$ lie either in C or in H. If F crossed the boundary of H other than at C, then there must be a point of F with normal more than .2 radians from the vertical. Similarly, we assert that all points of f lie in either C or $\mathbb{R}^3 \setminus H$. Face f lies in a plane within .9 radians of vertical, and within a strip on this plane bounded by lines within .9 radians of vertical. Any point within this strip can be connected to C by a line within .9 radians of vertical. Altogether, we can conclude that C is the only connected component of $f \cap F$, and of course is a topological 1-ball.

Finally we consider $Vor(s)$ itself, the case $k = 3$. Consider any connected component C of the intersection of F and the Voronoi cell. We mimic the argument from the case $k = 2$, again dividing $\mathbb{R}^3$ into two pieces using the angle formed by a shortest line segment to C. Again H will be a union of wedges with vertices on C. Except at C itself, $F \cap B_s$ must lie inside the wedges (closer to horizontal), whereas $Vor(s)$ must lie outside the wedges. ∎

Next we give a proof of Theorem 3(a): the crust contains all the good triangles. The intuition behind this proof is that restricted Voronoi cells are small and poles are far away, so that the ball centered at a vertex u of the restricted Voronoi diagram, passing through the three sample points whose cells meet at u, must be empty of poles.

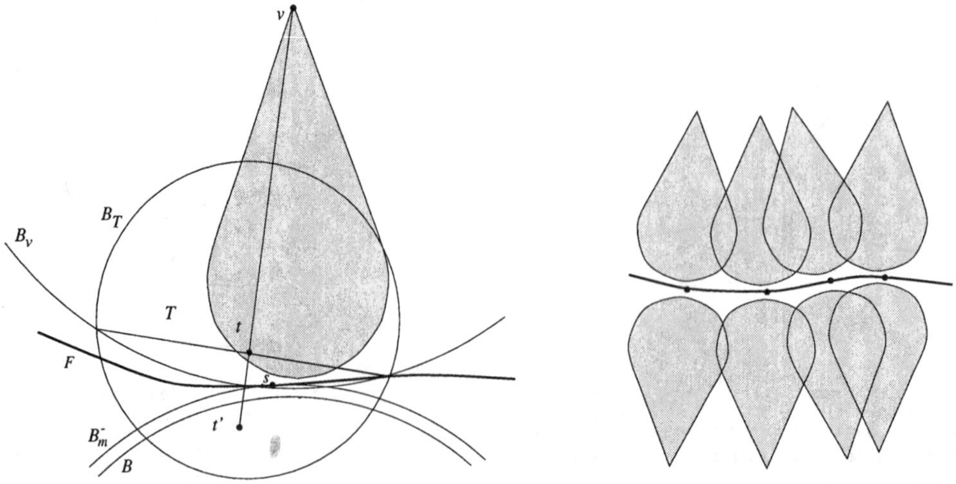

Figure 6. (a) The Delaunay ball B_T of a triangle intersecting the spindle must contain a big patch of surface F. (b) Spindles of sample points fuse so that all triangles must lie close to F.

Proof of Theorem 3(a): Let T be a triangle dual to a vertex u of the restricted Voronoi diagram. Consider the ball B_u centered on u with boundary passing through the vertices of T. Since T is a Delaunay triangle, B_u contains no point of S in its interior. Since S is an r-sample of F for $r < 1$, the radius of B_u is less than $rLFS(u)$. By the definition of LFS, even the larger ball B'_u of radius $LFS(u)$ centered on u cannot contain a point of the medial axis.

Now assume that B_u contains a pole v of a sample point s. We will show that under this assumption the polar ball B_v must be contained in B'_u, and that B_v must contain a point of the medial axis, thereby giving a contradiction. Let B_m be the medial ball with center m, tangent to F at s on the same side of F as v; this ball has radius at least $LFS(s)$. By Lemma 2(c), $\angle msv$ measures at most $2\arcsin(r/2)$, which is less than .12 for $r \leq .1$. An easy calculation shows that B_v must contain the medial axis point m.

Since v lies in B_u, the radius of B_v is no greater than the distance from v to the nearest vertex of T, which is $2rLFS(u)$ since S is an r-sample. Since $d(u,v) \leq rLFS(u)$, ball B_v lies entirely within B'_u since $3rLFS(u) \leq LFS(u)$. ∎

We now move on to the proof of Theorem 3(b). Let s be a sample point and v a pole of s. We shall define a forbidden region inside polar ball B_v, which cannot be penetrated by large crust triangles.

Let B_m^+ be the ball of radius $LFS(s)$ tangent to F at s, on the same side of F as v, and let B_m^- be the tangent ball of radius $LFS(s)$ on the opposite side of F from v. Surface F must lie between these two balls, since these balls are contained in medial balls at s. Let B be the ball concentric with B_m^- with radius $(1-r)LFS(s)$, as shown in Figure 6(a). Notice that Lemma 2(a) shows that the radius of B_v is at least that of B.

Definition 4. *The* **reflection** *of a point t through B_v is the point t' along ray vt such that line segment tt' is divided into equal halves by the boundary of B_v. The* **spindle** *of s is* $\{ t \in B_v \mid$ segment tt' intersects $B \}$, *that is, all points in B_v whose reflection lies in or beyond B.*

The spindle is shaded in Figure 6(a). Our plan is to confine large crust triangles between the union of spindles on each side of F as shown in Figure 6(b). (Small crust triangles lie within the fattened surface simply due to their size.) We start by proving two lemmas about spindles: they are indeed forbidden regions, and they have relatively "flat" bottoms, meaning that their width does not shrink with r.

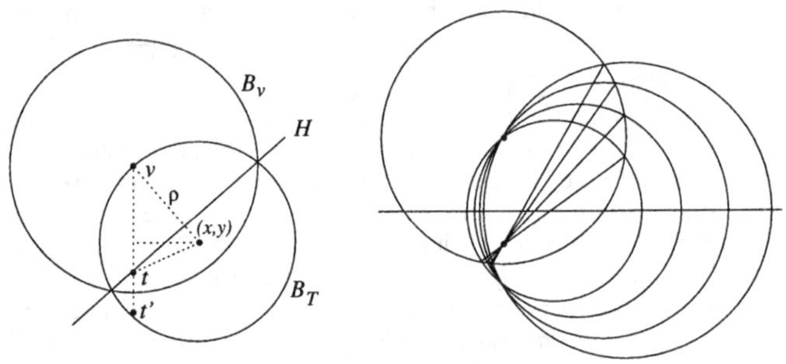

Figure 7. (a) B_T must contain reflection point t'. (b) The family of possible B_T circles.

Lemma 4. *Assume $r \leq .06$. No crust triangle T whose Delaunay ball B_T has radius greater than $5rLFS(s)$ can penetrate the spindle of s.*

Proof: Assume t is a point inside B_v on a crust triangle T with Delaunay ball B_T. We first assert that B_T contains the reflection point t'. Let H be the plane containing the intersection of the boundaries of B_v and B_T. Since the vertices of T lie on B_T outside B_v, T must be contained in the closed halfspace bounded by H not containing v. We may assume that t lies on H, since this is the worst case for our assertion.

Now consider any plane containing line vt. Balls B_v and B_T intersect this plane in circles and plane H intersects in a line containing the mutual chord of these circles, as shown in Figure 7(a). We may assume that ball B_T passes through v, for this situation is again the worst case.

Assume w.l.o.g. that the cross-section of B_v is the unit circle with center $v = (0,1)$. Let $t = (0, y_t)$. Denote the center and radius of B_T's cross-section by (x,y) and ρ. Since t lies along the mutual chord, it has equal "power distance" to $(0,1)$ and (x,y):

$$(1-y_t)^2 - 1 = x^2 + (y-y_t)^2 - \rho^2.$$

Substituting $(1-y)^2$ for $\rho^2 - x^2$, we obtain

$$y_t^2 - 2y_t = (y-y_t)^2 - (1-y)^2,$$

which simplifies to $y = (1-2y_t)/(2-2y_t)$. Thus the centers of all possible B_T circles lie on the same horizontal line, as shown in Figure 7(b).

Any B_T passes through the reflection of $(0,1)$ across the horizontal line, the point $(0,(1-2y_t)/(1-y_t)-1)$. For any value of $y_t < 1$, $(1-2y_t)/(1-y_t)-1 < -y_t$, so B_T contains $t' = (0,-y_t)$.

Thus if the original point t lies within the spindle of s, then B_T must intersect B. Since F is confined between B_m^- and B_m^+, this statement implies that B_T intersects F. In fact, a short calculation using the assumption that the radius of B_T is greater than $5rLFS(s)$ reveals that B_T must intersect F in a patch large enough that it must contain a sample point in its interior, a contradiction to B_T being a Delaunay ball. ∎

The next lemma shows that spindles have flat bottoms. In this lemma we assume that B and B_v have equal radius. It is not hard to confirm that this assumption is worst case: a larger B_v just gives a larger, flatter spindle.

Lemma 5. *Assume that B and B_v are unit balls, and that the distance between them is at most $\delta \leq .06$. Let t be a point outside B and outside the spindle induced by B in B_v. Let p be the closest point on B to t. If $|\angle omp|$, the measure of $\angle omp$ in radians, is less than .20, then $d(t,p) \leq \delta + |\angle omp|$.*

Proof: Assume v has coordinates $(0,1)$. The worst case for the lemma occurs when δ assumes its maximum value, as larger δ means a higher and narrower spindle, thereby maximizing $d(t,p)$ relative to $\delta + |\angle omp|$. So assume m has coordinates $(0,-1.06)$.

Draw the .20-radian ray with origin m and the .32-radian ray with origin v as shown in Figure 8. The rays intersect at a point x with coordinates about $(.259,.218)$. By computing the distances to the boundaries of B_v and B along ray vx, we can confirm that x lies inside the spindle. Thus the boundary of the spindle lies below x on the .20-radian ray with origin m. Assume t and p are as stated in the lemma, and $|\angle omp| = .20$. The distance from x to m is less than 1.252, so $d(t,p) - \delta \leq .192 \leq |\angle omp|$. Since $d(t,p)$ increases ever more rapidly as $|\angle omp|$ increases, this inequality also applies to points t and p such that $|\angle omp| < .20$ as well. ∎

We are now in a position to finish the proof of the theorem: all crust triangles lie within the fattened surface formed by placing a ball of radius $5rLFS(q)$ around each point $q \in F$.

Proof of Theorem 3(b): Let B_T be the Delaunay ball of the crust triangle containing point t. Let s be the sample point nearest t. If B_T has radius less than $5rLFS(s)$, then there is nothing to prove, since s itself could be the q of the theorem.

So assume B_T has radius at least $5rLFS(s)$. Let B_v, B_m^-, and B be the polar ball of s, the opposite medial axis ball, and the concentric ball with radius reduced by $rLFS(s)$ as in Figure 9. Let o and o' be the points of lune $B_m^- \cap B_v$ closest to the centers of B_m^- and B_v, respectively. Surface F could pass through the point o', and if it did, s would necessarily

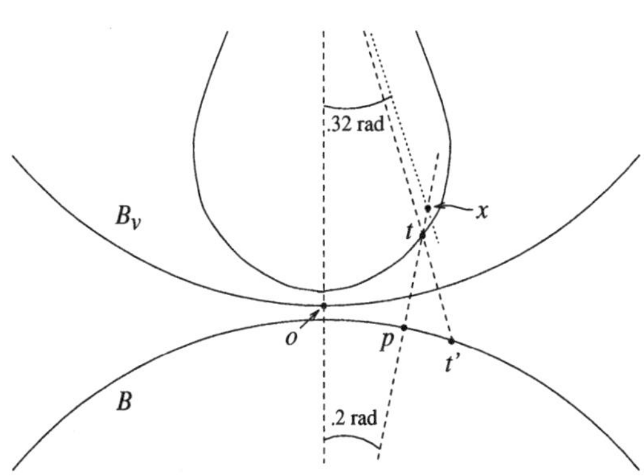

Figure 8. The spindle curves gradually, so t must be close to B.

be the closest sample point to o', since B_m^- and B_v are both empty. Hence by Lemma 2(b), $d(s,o') \leq rLFS(s)/(1-r)$. Since B_v has radius at least that of B_m^-, $d(s,o) \leq d(s,o')$.

Let p and p' be the closest points to t on B and B_m^-, respectively, and let q be the point of F on line pt closest to t. Hence $d(t,q) \leq d(p,t)$. By an argument analogous to that used for o', $d(s,p') \leq rLFS(s)/(1-r)$, and so by the triangle inequality, $d(o,p') \leq 2rLFS(s)/(1-r)$. So $\angle omp' \leq 2\arcsin(r/(1-r))$, which for $r \leq .06$, is less than .20 radians. The set-up satisfies the hypotheses of Lemma 5, only with radii scaled by $(1-r)LFS(s)$.

By Lemma 4, t must lie between the spindle and B_m. Applying Lemma 5,

$$d(t,p) \leq rLFS(s) + |\angle omp|(1-r)LFS(s).$$

We now use the fact that $|\angle omp| \leq 2\arcsin(r/(1-r)) \leq 3r$, to obtain

$$d(t,p) \leq rLFS(s) + 3r(1-r)LFS(s) \leq 4rLFS(s).$$

Finally, $d(s,q) \leq rLFS(s)/(1-r)$, so

$$LFS(q) \geq (1-2r)LFS(s)/(1-r),$$

and hence $5rLFS(q) \geq d(t,p) \geq d(t,q)$. ∎

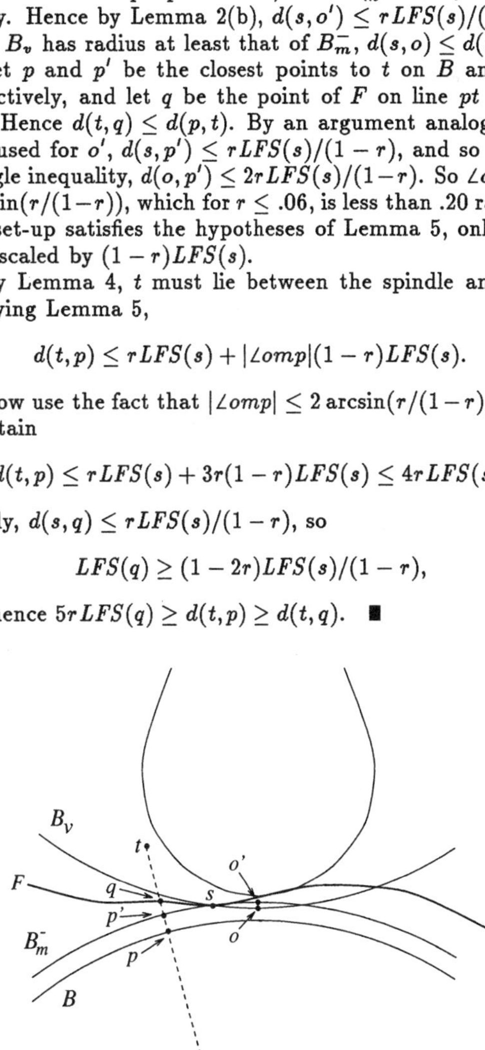

Figure 9. Crust point t must be near surface point q.

Let T be a triangle of the θ-crust, with $\theta = 3r$, and let t be a point on T. Theorem 4, part (a), states that for the angle between the normal to T and the normal to F at the point $p \in F$ closest to t measures $O(r)$ radians. Part (b) states that for sufficiently small r, the trimmed θ-crust is homeomorphic to F.

Proof of Theorem 4: We first prove that the (untrimmed) θ-crust contains all the good triangles, so that we do not lose the guarantees of the raw crust. Theorem 3(a) shows that the crust contains all the good triangles, so we need only show that each good triangle passes the filtering-by-normal step. Let T be a good triangle and s its vertex of maximum angle. By Lemma 3(a), the angle between the normal to T and the normal to F at s measures at most $\arcsin(\sqrt{3}r/(1-r))$ radians. By Lemma 2(c), the angle between the pole vector at s and the normal to F at s measures at most $2\arcsin(r/2)$. Combining these two bounds, the angle between the normal to T and either pole vector at s must be less than $3r = \theta$. Similarly, Lemmas 3(b) and 2(c) combine to show that the angle between the normal to T and the pole angle at any other vertex of T is at most $2\arcsin(r/2) + 2r/(1-7r) + \arcsin(\sqrt{3}r/(1-r))$ radians, which is less than $6.6r = 2.2\theta$.

We next prove that at each sample point s, the normals to incident θ-crust triangles do not deviate by more than $O(r)$ radians from the normal to F. This statement follows from the fact that Step 5 of the algorithm removes each triangle around s whose normal forms an angle larger than $6.6r$ with the vector to the pole. By Lemma 2(c), the pole vector deviates from the normal to F by at most $2\arcsin(r/2)$, which is less than $1.1r$ for $r \leq .06$.

Now let t be any point on a θ-crust triangle T, and let p be the closest point on F to T. By Theorem 3(b), $d(t,p) \leq 5rLFS(p)$. We next prove that the normal to T does not deviate by more than $O(r)$ radians from the normal to F at p. If t is less than distance $LFS(s)/4$ from a vertex s of T, then combining the bounds from Lemmas 1(b) and 3(b) establishes the result. If t is not close to any vertex of T, then we use an argument related to our spindle argument above. Let B_T be the Delaunay ball of T, and s and s' be two vertices of T at least $LFS(s)/3$ apart. Since B_T is fairly large with respect to both $LFS(s)$ and $LFS(s')$, the vectors from s and s' to the center v of B_T deviate only slightly (linearly in r) from the normals to F at s and s'. Since these surface normals in turn deviate only slightly (again linearly in r) from the normal to T, as r shrinks the vectors from s and s' to v become nearly parallel, By picking r sufficiently small, we force the boundary of B_T to be very close to t and parallel to T. Now consider the medial balls, which have radius at least $LFS(p)$, on either side of F at p. As in the proof of Lemma 4, F is confined between these medial balls. If the normal to F at p were not nearly parallel to the normal to T, then F would be forced to penetrate B_T in a large patch (a constant times $LFS(p)$). F cannot avoid B_T since its curvature is bounded by Lemma 1(b). Thus we can conclude that Theorem 4(a) holds.

For part (b), we must show that the trimming operation (Step 6) produces a set of triangles with the same topology as the good triangles. Let s be a sample point, and assume the normal to F at s is vertical. Step 5 ensures that for $r \leq .06$, all triangles around s remaining after Step 5 have normals within .5 radians of vertical. By Lemma 2(c), the vector from s to one of its poles is within .2 radians of vertical. Since the sum of .5 and .2 is bounded below $\pi/2$, the vertex-to-triangle breadth-first-search in Step 6 orients triangles consistently: the orientations do not depend on the actual search order, and at each vertex they agree with an orientation of F.

Sharp edges are exactly those edges at which a walk on the triangles can cross from an outside side to an inside side without piercing a triangle. After all triangles with sharp edges have been removed, all walks along the remaining triangles must run along either only inside or only outside sides.

Consider the mapping that takes each point of space to its closest point on F. We claim that this mapping is a homeomorphism. By Theorem 4(a) each triangle is nearly parallel to F, so the map is one-to-one on each triangle. Because the triangles are consistently oriented, points on two different triangles cannot map to the same point on F. ∎

Figure 10. A reconstructed minimal surface along with the poles of sample points. The crust contains exactly the original triangles. (Sample points courtesy of Hugues Hoppe)

6 Implementation and Examples

Manolis Kamvysselis, an undergraduate from MIT, implemented steps 1–4 of the crust algorithm during a summer at Xerox PARC. We used Clarkson's *Hull* program [12] for Delaunay triangulation, and *Geomview* [24] to visualize and print the results. We used vertices from pre-existing polyhedral models as inputs, in order to compare our results with "ground truth". A companion paper [2] reports on our experimental findings.

The only tricky part of the implementation was the handling of degeneracies and near degeneracies. Our test examples, many of which started from approximately gridded sample points, included numerous quadruples of points supporting slivers. Kamvysselis incorporated an explicit tolerance parameter ϵ; the circumcenter of quadruples within ϵ of cocircularity was computed by simply computing the circumcenter of a subset of three. This "hack" did not affect the overall algorithm, as these centers were never poles. Running time was only a little more than the time for two three-dimensional Delaunay triangulations. Notice that the Delaunay triangulation in step 3 involves at most three times the original number of vertices.

Figure 10 shows an especially advantageous example for our algorithm, a well-spaced point set on a smooth surface. Even though our algorithm is not designed for surfaces with boundary, it achieves perfect reconstruction on this example. Of course, the trimming step should not be used in reconstructing a surface with boundary.

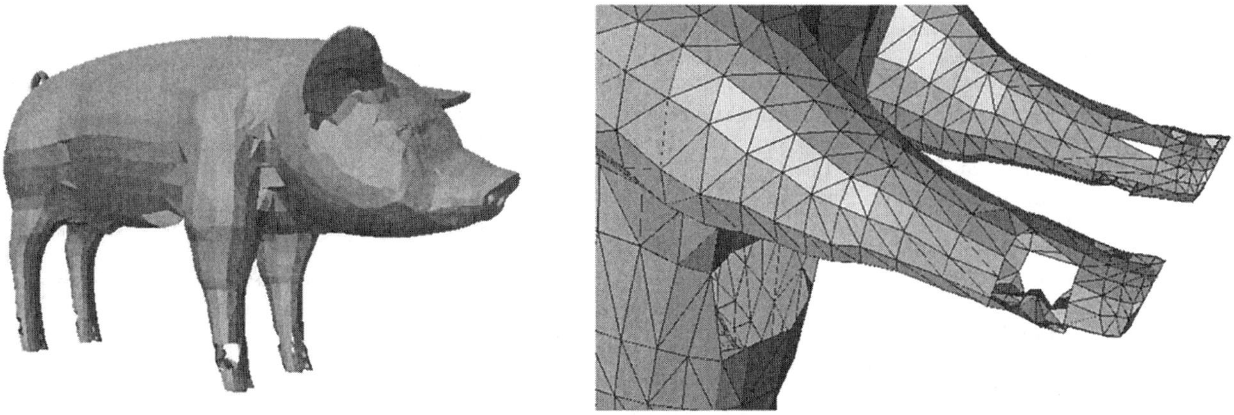

Figure 12. (a) The pig sample set contains 3511 points. (b) A close-up of the front feet shows an effect of undersampling. (Sample points courtesy of Tim Baker)

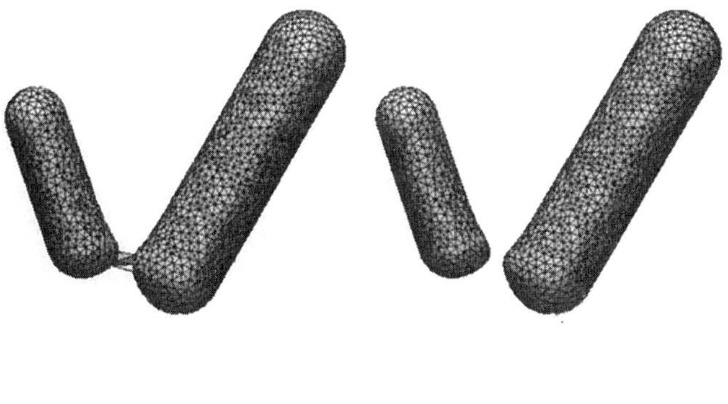

Figure 11. The raw crust contains some extra triangles linking the sausages; this defect is corrected by step 5. (Sample points courtesy of Paul Heckbert)

Figure 11 shows an effect of undersampling. (We say we have *undersampled* if the sample set is not an r-sample for a sufficiently small r.) In this example, the raw crust contains all the good triangles, along with some extra triangles. The extra triangles turn separated sausages into link sausages, and as mentioned above roughen the inside surfaces of the sausages. Both of these defects are corrected by step 5, filtering by normals. Figure 12 shows another effect of undersampling: missing triangles around the chest and hooves. Some sample points are not "opposed" by samples on the other side of these roughly cylindrical surfaces; hence Voronoi cells extend too far and poles filter out some good triangles. An r-sample for a sufficiently small r would be very dense near the hooves, which include some rather sharp corners.

7 Conclusions and Future Work

In this paper we have given an algorithm for reconstructing an interpolating surface from sample points in three dimensions. The algorithm is simple enough to analyze, easy enough to implement, and practical enough for actual use.

Our previous paper [1] gave two provably good algorithms for reconstructing curves in two dimensions, one using Voronoi filtering as in this paper, and the other using the β-skeleton. It is interesting to ask whether the β-skeleton can be generalized to the problem of surface reconstruction. (We know that the most straightforward generalization of the β-skeleton does not work.)

Another interesting question concerns the generalization of Voronoi filtering to higher dimensions. *Manifold learning* is the problem of reconstructing a smooth k-dimensional manifold embedded in $\mathbb{R}^d$. This problem arises in modeling unknown dynamical systems from experimental observations [11]. Even if Voronoi filtering can be generalized to this problem, its running time for the important case in which $k \ll d$ would not be competitive with algorithms that compute triangulations only in k-dimensional subspaces [11], rather than in $\mathbb{R}^d$.

Along with the two theoretical open questions outlined above, there are several quite practical directions for further research on our algorithms. What is the empirical maximum value of r for which our algorithm gives reliable results? We believe that the value of $r \leq .06$ in Theorem 3 is much smaller than necessary. Is the crust useful in simplification and compression of polyhedra? Can the crust be extended to inputs with creases and corners, such as machine parts? Can the crust be modified for the problem of reconstruction from cross-sections, in which the input is more structured than scattered points?

Acknowledgements

We would like to thank Alan Cline, Bob Connelly, David Goldberg, David Eppstein, and Manolis Kamvysselis for helpful suggestions, and Ken Clarkson and The Geometry Center for making their software available.

References

[1] N. Amenta, M. Bern, and D. Eppstein. The crust and the β-skeleton: combinatorial curve reconstruction. To appear in *Graphical Models and Image Processing*.

[2] N. Amenta, M. Bern, and M. Kamvysselis. A new Voronoi-based surface reconstruction algorithm. To appear in *Siggraph* 1998.

[3] D. Attali. r-Regular shape reconstruction from unorganized points. In *Proc. 13th ACM Symp. Computational Geometry*, 1997, 248–253.

[4] G. Barequet. Piecewise-linear interpolation between polygonal slices. In *Proc. 10th ACM Symp. Computational Geometry*, 1994, 93–102.

[5] C. Bajaj, F. Bernardini, and G. Xu. Automatic reconstruction of surfaces and scalar fields from 3D scans. *Proc. SIGGRAPH '95*, 1995, 109–118.

[6] M. Bern, H. Edelsbrunner, D. Eppstein, S. Mitchell, and T.-S. Tan. Edge insertion for optimal triangulations. *Disc. and Comp. Geometry* 10 (1992) 47–65.

[7] F. Bernardini and C. Bajaj. Sampling and reconstructing manifolds using α-shapes, *9th Canadian Conference on Computational Geometry*, 1997, 193–198.

[8] F. Bernardini, C. Bajaj, J. Chen and D. Schikore. Automatic reconstruction of 3D CAD models from digital scans. Technical report CSD-97-012, Purdue University (1997).

[9] F. Bernardini, C. Bajaj, J. Chen, D. Schikore. A triangulation-based object reconstruction method. *Proc. 13th ACM Symp. Computational Geometry*, 1997, 481–484.

[10] J-D. Boissonnat. Geometric structures for three-dimensional shape reconstruction. *ACM Trans. Graphics* 3 (1984) 266–286.

[11] C. Bregler and S. M. Omohundro. Nonlinear manifold learning for visual speech recognition. *Proc. 5th International Conf. on Computer Vision*, 1995, 494–499.

[12] K. Clarkson. *Hull*: a program for convex hulls. http://cm.bell-labs.com/netlib/voronoi/hull.html.

[13] B. Curless and M. Levoy. A volumetric method for building complex models from range images. *Proc. SIGGRAPH '96*, 1996, 303–312.

[14] H. Edelsbrunner. Surface reconstruction by wrapping finite sets in space. Tech. Rept. 96-001, Raindrop Geomagic, Inc., 1996.

[15] H. Edelsbrunner, D.G. Kirkpatrick, and R. Seidel. On the shape of a set of points in the plane. *IEEE Trans. on Information Theory* 29 (1983), 551-559.

[16] H. Edelsbrunner and E. P. Mücke. Three-dimensional alpha shapes. *ACM Trans. Graphics* 13 (1994) 43–72.

[17] H. Edelsbrunner and N. Shah. Triangulating topological spaces. *Proc. 10th ACM Symp. Computational Geometry*, 1994, 285–292.

[18] L. H. de Figueiredo and J. de Miranda Gomes. Computational morphology of curves. *Visual Computer* 11 (1995) 105–112.

[19] J. Goldak, X. Yu, A. Knight, and L. Dong. Constructing discrete medial axis of 3-D objects. *Int. J. Computational Geometry and its Applications* 1 (1991) 327–339.

[20] H. Hoppe. *Surface Reconstruction from Unorganized Points*. Ph.D. Thesis, Computer Science and Engineering, U. of Washington, 1994. http://www.research.microsoft.com /research/graphics/hoppe/thesis/thesis.html

[21] H. Hoppe, T. DeRose, T. Duchamp, J. McDonald, and W. Stuetzle. Surface reconstruction from unorganized points. *Proc. SIGGRAPH '92*, 1992, 71–78.

[22] H. Hoppe, T. DeRose, T. Duchamp, H. Jin, J. McDonald, and W. Stuetzle. Piecewise smooth surface reconstruction. *Proc. SIGGRAPH '94*, 1994, 19–26.

[23] D. G. Kirkpatrick, J. D. and Radke. A framework for computational morphology. *Computational Geometry* (G. Toussaint, ed.), Elsevier, pp. 217-248.

[24] S. Levy, T. Munzner, and M. Phillips. Geomview. http://www.geom.umn.edu/software/download/ geomview.html

[25] S. Lodha. Scattered Data Techniques for Surfaces. To appear in *Geometry Detection, Estimation and Synthesis for Scientific Visualization*, Academic Press.

[26] S. Mann, C. Loop, M. Lounsbery, D. Meyers, J. Painter, T. DeRose, and K. Sloan. A survey of parametric scattered data fitting using triangular interpolants. *Curve and Surface Design*, H. Hagen, ed., SIAM, 1992, 145–172.

[27] R. C. Veltkamp. *Closed object boundaries from scattered points*. LNCS Vol. 885, Springer, 1994.

Cross-ratios and angles determine a polygon

Jack Snoeyink*
UBC Dept. of Comp. Sci.
Vancouver, BC, Canada
snoeyink@cs.ubc.ca

Abstract

We prove that a unique simple polygon is determined, up to similarity, by the interior angles at its vertices and the cross-ratios of diagonals of any given triangulation. (The cross-ratio of a diagonal is the product of the ratio of edge lengths for the two adjacent triangles.) This establishes a conjecture of Driscoll and Vavasis, and shows the correctness of a key step of their algorithm for computing Schwarz-Christoffel transformations mapping a disk to a polygon.

1 Introduction

Computing meshes for numerical solution of problems on domains with irregular geometry poses interesting challenges because it interweaves continuous mathematics with discrete geometry and then seeks an implementation on an even more discrete computer. One excellent example is a new algorithm by Driscoll and Vavasis for computing a conformal mapping of a unit disk to a simple plane polygon [2]. We describe this algorithm in some detail in Section 2. The correctness of one of its steps depends on a purely geometric conjecture [2, Conjecture 1]:

> Let P be an n-vertex, triangulated, simple polygon. Then P is uniquely determined up to similarity transform by the following data: the sequence of all interior angles of P at its vertices, and the list of $n-3$ absolute values of cross-ratios of the quadrilaterals determined by the triangulation of P.

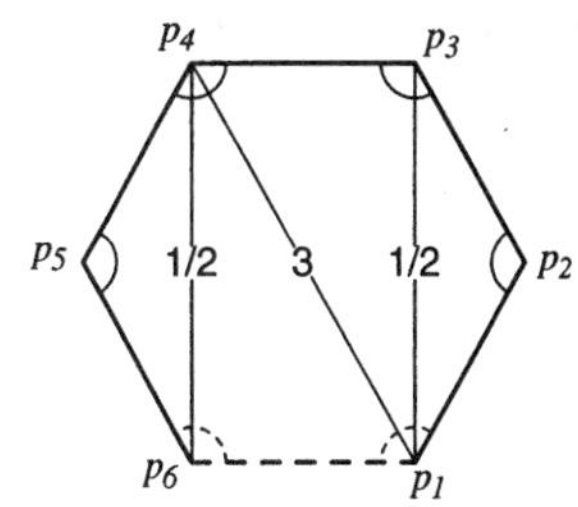

Figure 1: Angles and cross-ratios

Pairs of adjacent triangles, which share a *diagonal edge* of the triangulation, form the above-mentioned quadrilaterals. The *cross-ratio* of a quadrilateral is the product of the lengths of two sides clockwise from the endpoints of the diagonal divided by the product of the lengths of the two sides counterclockwise. In the triangulated hexagon of Figure 1, diagonal edges are labeled with their cross-ratios.

In this paper, we establish their conjecture by establishing properties of *partially-specified polygons*, which we define in Section 4. We employ some observations on the ratios of triangles that we make in Section 3.

---

*Supported in part by an NSERC Research Grant. Portions of this work were performed while on leave at the Geometry Center at the Johns Hopkins University, and in project Prisme at INRIA Sophia-Antipolis.

SCG 98 Minneapolis Minnesota USA
Copyright ACM 1998 0-89791-973-4/98/ 6...$5.00

2 The Schwarz-Christoffel transform and the CRDT algorithm

If we think of a simple polygon $P = \{p_1, p_2, \ldots, p_n\}$ as a simply-connected, open subset of the complex plane and let D be the unit disk, then the Riemann Mapping Theorem [3] says that there is an analytic function $f: D \to P$ mapping the unit disk to the polygon. If the point $f(0)$ and direction of $f'(0)$ are specified, then f is unique. With this mapping, some problems on P (e.g., Laplace's equation) can be solved on the simpler domain D. Other problems that require a mesh on P can begin with a mesh on D; conformal mapping preserves angles in the mesh.

All mappings $f: D \to P$, from the unit disk D to an n-gon P, can be written using the Schwarz-Christoffel formula

$$f(w) = p + q \int_0^w \prod_{1 \le j \le n} \left(1 - \frac{\omega}{w_j}\right)^{\beta_j} d\omega.$$

In this formula, the complex numbers p and q translate, rotate, and scale the image, the scalars $\beta_j = \psi_j/\pi - 1$, where ψ_j is the interior angle at polygon vertex p_j, and the unit-length complex numbers w_j are the *prevertices*, the points on the boundary of the disk that map to the vertices of the polygon. The integration is a complex contour integral in the disk. The S-C formula has $n+4$ parameters. Eliminating similarity transformations of the polygon and the freedom to choose $f(0)$ and the direction of its derivative leaves $n-3$ parameters unspecified. These are determined by numerically solving a difficult nonlinear system [1, 2, 4, 5].

This direct approach suffers from *crowding*—skinny regions of P cause the angle between the prevertices to decrease exponentially in the aspect ratio. This causes problems for numerical techniques; many bits of precision can be lost when determining the angle between two prevertices.

Driscoll and Vavasis [2] have developed an clever combination of numerical and geometric ideas to minimize the impact of crowding. They use a constrained Delaunay triangulation of P, with new vertices introduced on the boundary so that triangles in P have good aspect ratio. There is a corresponding triangulation of prevertices in D (join prevertices iff an edge joins their vertices in P); Driscoll and Vavasis reformulate the nonlinear system to compute the cross-ratios of this triangulation. These $n-3$ cross-ratios are three parameters too few to specify the n prevertices completely. They note that when an application is concerned with accuracy of computations inside a particular triangle, it can specify the three prevertices of this triangle to be spread out around the disk, then determine the rest of the prevertices from the cross-ratios in linear time. Crowding increases further from the specified triangle, but usually the importance of a vertex decreases with distance as well.

As an illustration of their technique, they produced the right side of Figure 2, which shows a mapping from a rectangle to a "maze" polygon. Solid lines in the rectangle at $i/5$, for $1 \le i \le 4$, and dotted lines at 10^{-2}, 10^{-4}, ..., 10^{-16} map to their images at right. Driscoll and Vavasis [2] report that "all computations were performed in double precision."

While the experimental evidence supports the correctness and efficiency of their approach, it does depend on some conjectures. The first, cited in Section 1 and established in this paper, relates to their iteration to find the cross-ratios of the triangulation of prevertices. Their iteration uses the S-C transform to determine cross-ratios of P from the guessed cross-ratios of prevertices; the question is whether, having determined the cross-ratios of polygon P, the algorithm has determined P. There

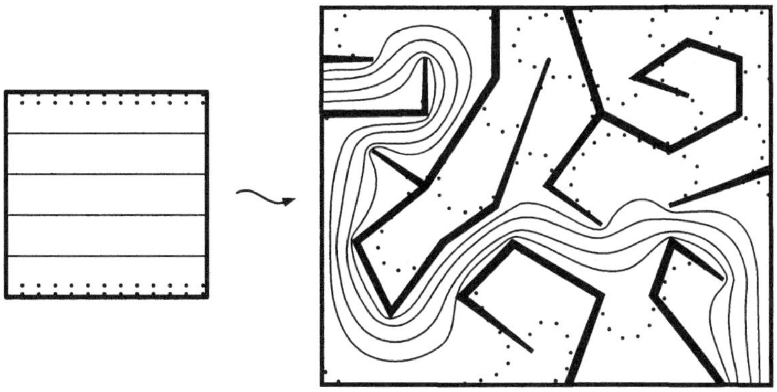

Figure 2: Conformal mapping from rectangle to maze. (From Driscoll and Vavasis [2])

are other interesting conjectures as to the relationship between the magnitudes of cross-ratios in P and in D, which we cannot yet address.

3 Observations on the ratios of triangles

To begin, we establish several sufficient conditions on sides and angles of a triangle for its ratio to increase. We use the notation illustrated in Figure 3: Points p and q are fixed so that the segment $\overline{pq}$ is horizontal from left to right. Points r_1 and r_2 are chosen above the line $\overleftrightarrow{pq}$, and the angles of triangle pqr_i measure α_i, β_i, and γ_i at p, q, and r_i, respectively. Let a_i and b_i denote the lengths of qr_i and pr_i, respectively. The *ratio* of the triangle pqr_i, will thus be a_i/b_i.

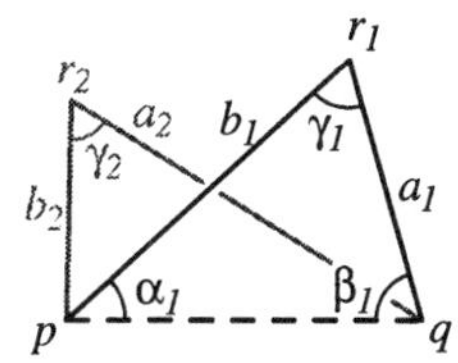

Figure 3: Triangle notation

If we fix $\gamma_1 = \gamma_2$, then the triangle is determined when α_i is chosen. Thus, we establish the next condition by a continuous argument and drop the subscripts.

Lemma 1 *Using the above notation, when angle γ is fixed, the ratio $r(\alpha) = a/b$ is a strictly-increasing function of α, for $\alpha \in [0, \pi - \gamma]$.*

Proof: By the law of sines, the ratio

$$r(\alpha) = \frac{a}{b} = \frac{\sin\alpha}{\sin(\alpha+\gamma)} = \cos\gamma - \sin\gamma \cdot \cot(\alpha+\gamma).$$

This is an invertible function for $0 < \alpha \leq \pi - \gamma$, and its derivative $\sin\gamma/\sin^2(\alpha+\gamma)$ is positive for $0 < \gamma < \pi$, which we must have for pqr to be a triangle. ∎

By the way, all of our conditions are established by algebraic arguments, which is a good way to build character. To develop intuition, it is sometimes more pleasant to observe the conditions geometrically by considering the interaction of four families of constraint curves that cover the plane, which are illustrated in Figure 4. We use these to illustrate the algebraic proofs.

Observation 2 *Let pqr be a triangle with p and q fixed and r to the left of $\overline{pq}$. The following families are the loci of points r with the corresponding parameter of triangle pqr fixed:*

A. points with fixed angle $\alpha = \angle rpq$ lie on the lines through p,

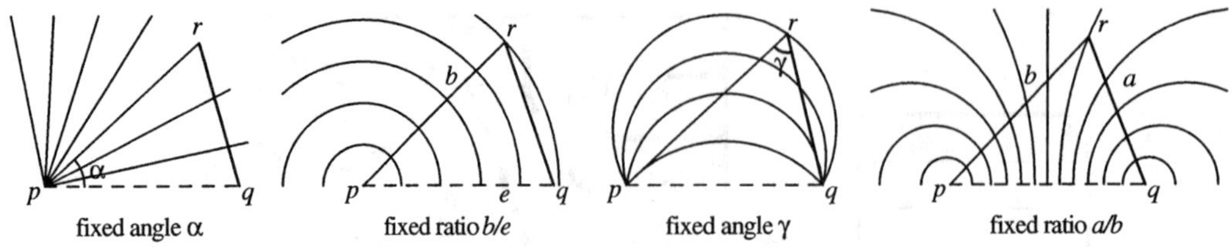

Figure 4: Four families of constraint curves for triangle endpoint r.

B. points with fixed ratio $b/e = rp/pq$ *lie on the circles centered at* p,

C. points with fixed angle $\gamma = \angle qrp$ *lie on circles through* p *and* q, *and*

D. points with fixed ratio $a/b = qr/rp$ *lie on circles orthogonal to those of C.*

These curves intersect nicely; a pair of curves chosen from A and B, A and C, B and D, or from C and D will have only one intersection above the x axis. Families A and B and families C and D are orthogonal—any two of their curves intersect at right angles.

A geometric argument for Lemma 1 observes that with angle γ fixed, the point r moves along a circular arc of family C as α_i increases, see Figure 5. As r moves, it crosses each curve of family D, and the ratio increases over $(0, \infty)$.

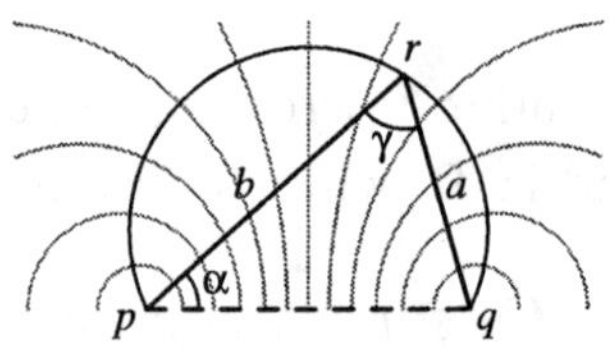

Figure 5: Fixed γ

Lemma 3 *Using the triangle notation of Figure 3, the ratios* $a_1/b_1 < a_2/b_2$ *if* $r_1 \neq r_2$ *and any of the four following conditions hold:*

(1). if $a_1 \leq a_2$ *and* $b_1 \geq b_2$,

(2). if $b_1 \geq b_2$, $\alpha_1 \leq \alpha_2$, *and* $\gamma_1 \geq \gamma_2$,

(3). if $a_1 \leq a_2$, $\beta_1 \geq \beta_2$, *and* $\gamma_1 \leq \gamma_2$, *or*

(4). if $a_1 \geq a_2$, $b_1 \leq b_2$, $\alpha_1 \leq \alpha_2$, *and* $\beta_1 \geq \beta_2$.

Proof: Condition (1) is the easiest. With the assumption that $r_1 \neq r_2$, one of the inequalities $a_1 \leq a_2$ or $b_1 \geq b_2$ must be strict, so when condition (1) holds, $a_1/b_1 < a_2/b_2$.

For the remaining conditions, it helps to express ratios of sides using the law of sines and trigonometric identities. For example,

$$\frac{a_i}{b_i} = \frac{\sin \alpha_i}{\sin \beta_i} = \frac{\sin(\gamma_i + \beta_i)}{\sin \beta_i} = \cos \gamma_i + \cot \beta_i \sin \gamma_i = \cos \gamma_i + \cos \beta_i \Big(\frac{e}{b_i}\Big).$$

Note that all angles will be in $(0, \pi)$, so sines are positive. Cosine and cotangent are decreasing functions over this angle range, crossing zero at angle $\pi/2$.

When condition (2) holds we consider two cases, depending on whether angle β_1 is obtuse.

(2), β_1 obtuse: Assume, for the sake of deriving a contradiction, that $a_1/b_1 \geq a_2/b_2$, or equivalently

$$\cos \gamma_1 + \cot \beta_1 \sin \gamma_1 \geq \cos \gamma_2 + \cot \beta_2 \sin \gamma_2.$$

Since $\beta_1 > \pi/2$, the value $-\cot\beta_1$ is positive. Also, $\pi/2 > \gamma_1 \geq \gamma_2$, so $\cos\gamma_1 \leq \cos\gamma_2$ and $\sin\gamma_1 \geq \sin\gamma_2$. Combining these inequalities, we observe that $-\cot\beta_1 \leq -\cot\beta_2$; We conclude that $\pi_2 < \beta_1 \leq \beta_2$ and $\alpha_1 \leq \alpha_2 < \pi/2$.

But then, $\sin\alpha_1 \leq \sin\alpha_2$ and $\sin\beta_1 \geq \sin\beta_2$, and, if $r_1 \neq r_2$, one of these inequalities is strict. Thus we derive a contradiction,

$$\frac{a_1}{b_1} = \frac{\sin\alpha_1}{\sin\beta_1} < \frac{\sin\alpha_2}{\sin\beta_2} = \frac{a_2}{b_2}.$$

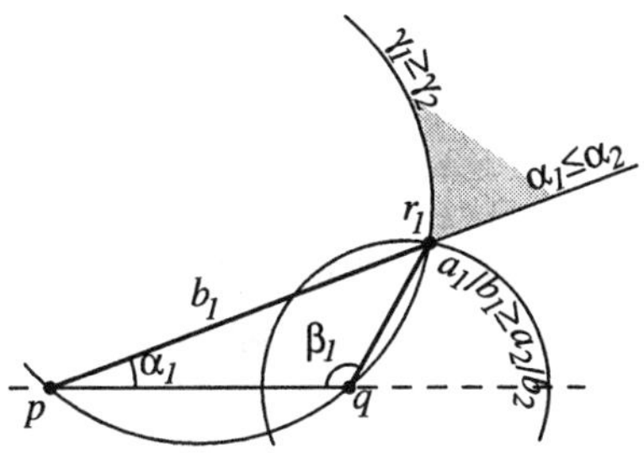

Figure 6: (2), β_1 obtuse

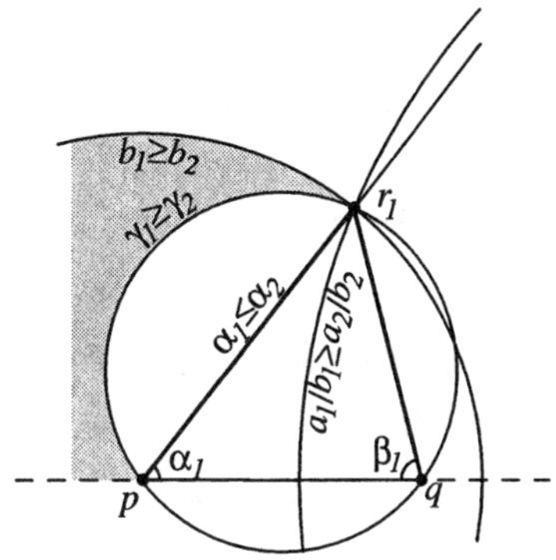

Figure 7: (2), β_1 not obtuse

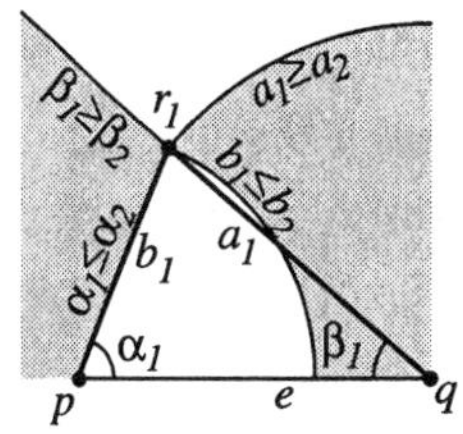

Figure 8: (4), angle & distance constraints

(2), β_1 not obtuse: Assume, for the sake of deriving a contradiction, that $a_1/b_1 \geq a_2/b_2$, The condition $\alpha_1 \leq \alpha_2$ implies $\cos\alpha_1 \geq \cos\alpha_2$. By the condition $b_1 \geq b_2$, we know

$$\cos\alpha_1 + \cos\beta_1\Big(\frac{a_1}{b_1}\Big) = \frac{e}{b_1} \leq \frac{e}{b_2} = \cos\alpha_2 + \cos\beta_2\Big(\frac{a_2}{b_2}\Big).$$

Since $\cos\beta_1$ is non-negative, we can combine the above inequalities to derive that $0 \leq \cos\beta_1 \leq \cos\beta_2$. But since $\cos\gamma_1 \leq \cos\gamma_2$ and $e/b_1 \leq e/b_2$, with one of these inequalities strict if $r_1 \neq r_2$, we again derive a contradiction:

$$\frac{a_1}{b_1} = \cos\gamma_1 + \cos\beta_1\Big(\frac{e}{b_1}\Big) < \cos\gamma_2 + \cos\beta_2\Big(\frac{e}{b_2}\Big) = \frac{a_2}{b_2}.$$

(3): We reduce condition (3) to condition (2). Reflect the triangle about a vertical line, which exchanges 'a's with 'b's, and 'α's with 'β's. Then exchange '1's with '2's, and apply the proof for condition (2). When we return the conclusion to the original notation, we obtain that $b_2/a_2 < b_1/a_1$. Taking reciprocals gives the desired result.

(4): The final part of the lemma is vacuously true—condition (4) can never be satisfied.

By the law of sines $b_i = e\sin\beta_i/\sin\gamma_i = e\cos\alpha_i + a_i\cos\gamma_i$. Since $\cos\alpha_1 \geq \cos\alpha_2$ and $a_1 \geq a_2$, in order to have $b_1 \leq b_2$ (and $r_1 \neq r_2$), we must either have $0 \leq \cos\gamma_1 < \cos\gamma_2$ or $\cos\gamma_1 < 0$. Similarly, $a_i = e\cos\beta_i + b_i\cos\gamma_i$, $\cos\beta_1 \leq \cos\beta_2$, and $b_1 \leq b_2$. In order to have $a_1 \geq a_2$ (and $r_1 \neq r_2$), we must either have $0 \leq \cos\gamma_2 < \cos\gamma_1$ or $\cos\gamma_2 < 0$. Because the inequalities with non-negative cosines are incompatable, we must have that both γ_1 and γ_2 are obtuse angles.

We can now observe a contradiction: angles α_i and β_i are not obtuse, so $\sin\alpha_1 \leq \sin\alpha_2$ and $\sin\beta_1 \geq \sin\beta_2$. By the law of sines, $a_i/b_i = \sin\alpha_i/\sin\beta_i$, but

$$\frac{a_1}{b_1} \leq \frac{a_2}{b_2} \qquad \text{and} \qquad \frac{\sin\alpha_1}{\sin\beta_1} \geq \frac{\sin\alpha_2}{\sin\beta_2}.$$

Only if $r_1 = r_2$ can these inequalities be satisfied simultaneously. ∎

4 A proof using partial polygons

The remainder of this paper proves that the interior angles and cross-ratios of a triangulated polygon P are sufficient to determine P up to similarity. The proof uses *partial polygons,* for which all cross-ratios and all angles except the first and last are specified. In notation, let $P = p_1p_2\ldots p_n$ be a simple polygon in the Euclidean plane. Assume that we are given:

- $n-2$ angles $\psi_2, \psi_3, \ldots, \psi_{n-1}$, where $0 < \psi_i \leq 2\pi$ measures the internal angle at p_i.
- $n-3$ cross-ratios for the quadrilaterals formed by the pairs of triangles adjacent to the $n-3$ diagonals in the triangulation of P.

Note that angles ψ_1 and ψ_n at vertices p_1 and p_n are not specified. We call ψ_n the *variable angle* of partial polygon P. If we set $\psi_n = \theta$, we can determine ψ_1 since the sum of interior angles $\sum_{1\leq i\leq n}\psi_i = (n-2)\pi$.

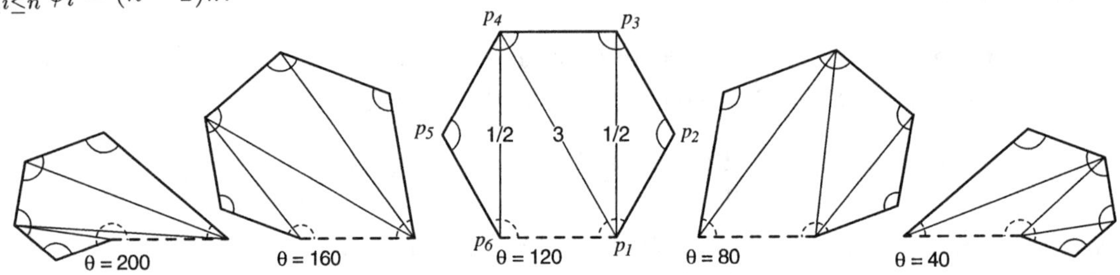

Figure 9: Realizations of a partial polygon P for five different angles θ. In each, the angles at p_2 through p_5 are 120°, and cross-ratios are 1/2, 3, and 1/2 as labeled in the middle hexagon.

A *realization of partial polygon* P is an embedding that respects the specified angles and cross-ratios and has positive, finite area for all triangles; Figure 9 shows five realizations of a triangulated hexagon for different values of θ. No global monotonicity properties for realizations of P seem to be evident as θ varies—e.g., the area and sides of triangle $p_1p_2p_3$ shrink and grow again as θ decreases in Figure 9.

We call $\overline{p_np_1}$ the *unspecified edge* of partial polygon P. It is one side of a triangle in the triangulation of P; let a and b be the lengths of the other two sides in counter-clockwise order, and define the *ratio of* P to be a/b. We prove by induction that the ratio of P is a partial function of θ that is strictly-increasing over its domain.

Theorem 4 *Let P be a partial polygon with variable angle θ. For each θ, the realization of P, if it exists, is unique, and the ratio of P is a strictly-increasing partial function $r(\theta)$.*

Proof: We prove this by induction on the number of triangles in P.

Base: In the base case, partial polygon P is a single triangle $p_1p_2p_3$. We can choose the positions of p_1 and p_3 as in Figure 10 to fix the similarity transformation. Since angle ψ_2 at p_2 is specified, Lemma 1 applies to show that the ratio is a strictly-increasing function of the variable angle θ at p_1.

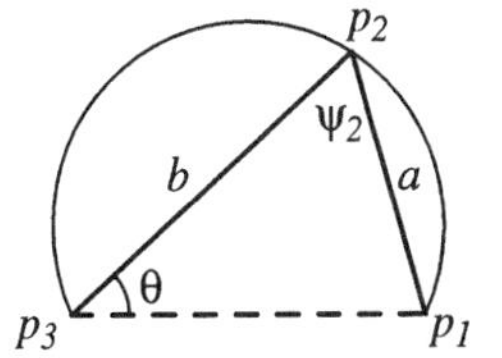

Figure 10: Base case

In the induction step, we are given a partial polygon P, having $n > 3$ vertices, $n-2$ triangles, and variable angle θ and unspecified edge e, which we fix as horizontal. Removing the triangle incident on the unspecified edge leaves a pair of partial polygons or, when this triangle uses vertex p_{n-1} or p_2, a single partial polygon. We handle these cases separately.

Single using p_{n-1}: When triangle $p_1p_{n-1}p_n$ is adjacent to the unspecified edge e of P, then removing this triangle leaves a partial polygon P' from p_1 to p_{n-1} to which we may apply the induction hypothesis.

We define notation as in Figure 11: Let ϕ be the variable angle for P', and let a be the length of the unspecified edge of P'. Let $\gamma = \psi_{n-1} - \phi$. Let b be the length of $\overline{p_{n-1}p_n}$, which is the third side of the removed triangle, and let β be the angle opposite the side of length b. Let κ be the cross-ratio specified for diagonal $\overline{p_{n-1}p_2}$. Define the *cross-ratio function* $s(\phi)$ to be the ratio for P' divided by κ. By the induction hypothesis, $s(\phi)$ is a strictly-increasing partial function.

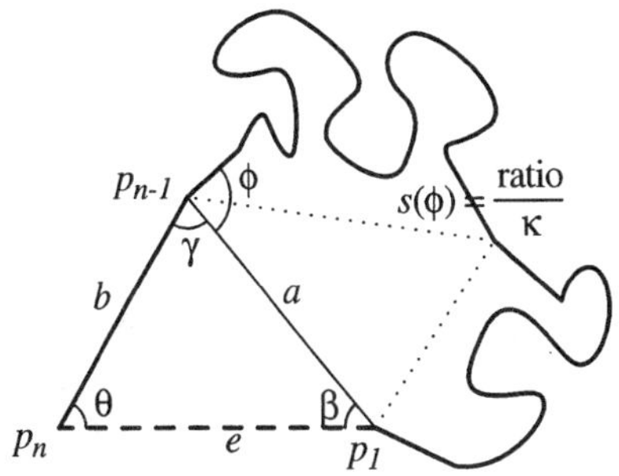

Figure 11: Single partial polygon

We want to prove that the ratio $r(\theta) = a/b$ is a strictly-increasing partial function. First, we note that a given angle θ determines at most one realization of P. As ϕ varies from 0 to ψ_{n-1}, the function $s(\phi)$ increases. At the same time, the angle $\gamma = \psi_{n-1} - \phi$ decreases; for a fixed θ, Lemma 1 says that the ratio e/b also decreases. Because any realization of P has $e/b = s(\phi)$, the realization for any fixed θ is unique.

Now, to show that $r(\theta) = a/b$ is strictly increasing, we compare the ratios for two realizations of P with two different angles, $\theta_1 < \theta_2$. We consider two subcases, depending on whether the corresponding angles $\gamma_1 \geq \gamma_2$ or $\gamma_1 < \gamma_2$. (As with the lemmas of the previous section, we can assume that e remains fixed, and subscript other lengths and angles in the two realizations.)

If $\gamma_1 \geq \gamma_2$, then $\phi_1 \leq \phi_2$, so the ratio $e/b_1 = s(\phi_1) \leq s(\phi_2) = e/b_2$. Since angles $\theta_1 < \theta_2$ correspond to the αs Lemma 3, we have the conditions of part (2) and can conclude that the ratio increases: $r(\theta_1) = a_1/b_1 < a_2/b_2 = r(\theta_2)$.

On the other hand, if $\gamma_1 \leq \gamma_2$, then $\phi_1 \geq \phi_2$, so the ratio $e/b_1 = s(\phi_1) \geq s(\phi_2) = e/b_2$. Since the angles of the triangle sum to π, we know that $\beta_1 \geq \beta_2$. Thus, when $a_1 \leq a_2$ we can apply Lemma 3(3), and when $a_1 \geq a_2$ we apply Lemma 3(4); in either case, the ratio increases: $r(\theta_1) = a_1/b_1 < a_2/b_2 = r(\theta_2)$.

This completes the argument for the single case using p_{n-1}.

Single using p_2: When triangle $p_1p_2p_n$ is adjacent to the unspecified edge of P, we can reduce this case to the previous. Reflect P about a line, taking the reciprocals of all specified cross-ratios, to obtain P_R. Apply the argument of the previous case to show that there is a strictly-increasing function of angle θ_R for the ratio $r_R(\theta_R)$ of P_R. Then there is a strictly-increasing function of angle θ for the ratio $r(\theta)$ of P, namely

$$r(\theta) = \frac{1}{r_R\Big((n-2)\pi - (\sum_{2\le i\le n-1} \psi_i) - \theta\Big)}.$$

Uniqueness is preserved under reflection.

Pair using p_j: When triangle $p_1p_jp_n$ is adjacent to the unspecified edge e of P, for some $2 < j < n-1$, we can label the sides $a = \overline{p_1p_j}$ and $b = \overline{p_jp_n}$, and angles α, β, and $\gamma = \pi - \alpha - \beta$ opposite the sides a, b, and e. We will fix the similarity by choosing edge e to be horizontal as in Figure 12.

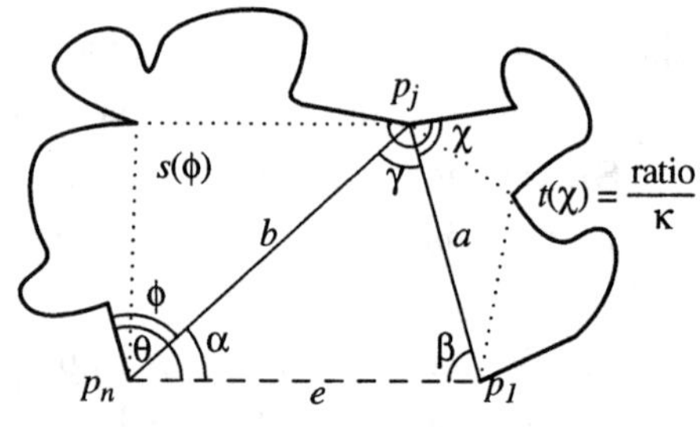

Figure 12: Pair of partial polygons

Removing triangle abe leaves two partial polygons: one from p_j to p_n with variable angle $\phi = \theta - \alpha$ and cross-ratio function $s(\phi)$, and another from p_1 to p_j with variable angle χ and cross-ratio function $t(\chi)$. (Because the angle at p_j is specified as ψ_j, and the angles of a k-gon sum to $(k-2)\pi$, we know that $\chi+\gamma-\phi$ is a constant, namely $(n-j-1)\pi - \sum_{j\le k<n} \psi_k$. Similarly, $\theta + \beta - \chi$ is a constant.) We can apply the induction hypothesis to both partial polygons, so the cross-ratio functions $s(\phi)$ and $t(\chi)$ are strictly-increasing partial functions of their angles.

We begin with a different decomposition of P, however, to show that the realization of P for a given θ is unique. Break P into two partial polygons at edge a. If we fix θ, then the induction hypothesis applies to the polygon from $p_j, \ldots, p_n, p_1$: the ratio b/e is a strictly-increasing partial function of variable angle β, and its reciprocal e/b is decreasing. With $\theta + \beta - \chi$ a constant and θ fixed, however, angle χ and ratio function $t(\chi)$ are also strictly-increasing (partial) functions of β, and $t(\chi) = e/b$. Equality holds at most once for any specified θ.

Now, to show that $r(\theta) = a/b$ is strictly increasing, we compare the ratios for two realizations of P with two different angles, $\theta_1 < \theta_2$, and consider four subcases depending on the comparisons of corresponding angles ϕ_1 with ϕ_2 and χ_1 with χ_2. (As before, we assume that e remains fixed, and subscript other lengths and angles in the two realizations.)

Pair, $\phi_1 < \phi_2$ and $\chi_1 \le \chi_2$: Since $s(\phi_1) = a_1/e < s(\phi_2) = a_2/e$ and $t(\chi_1) = e/b_1 \le t(\chi_2) = e/b_2$, the ratio $r(\theta_1) = a_1/b_1 < a_2/b_2 = r(\theta_2)$ increases in this case. Note that this is Lemma 3(1), but it is easier to observe directly than by paging back to the lemma.

Pair, $\phi_1 \ge \phi_2$ and $\chi_1 \le \chi_2$: Here, $t(\chi_1) = e/b_1 \le t(\chi_2) = e/b_2$, so $b_1 \ge b_2$. Since $\alpha_i = \theta_i - \phi_i$, angles $\alpha_1 < \alpha_2$, and since $\gamma_i + \chi_i - \phi_i$ is a constant, $\gamma_1 \ge \gamma_2$. This satisfies the conditions for Lemma 3(2), so $r(\theta_1) = a_1/b_1 < a_2/b_2 = r(\theta_2)$ in this case, also.

Pair, $\phi_1 < \phi_2$ and $\chi_1 > \chi_2$: Here, $s(\phi_1) = a_1/e < s(\phi_2) = a_2/e$. The angles $\gamma_1 < \gamma_2$, since $\gamma_i + \chi_i - \phi_i$ is a constant, and $\beta_1 > \beta_2$, since $\beta_i + \theta_i - \chi_i$ is a constant. This satisfies the conditions for Lemma 3(3), so $r(\theta_1) = a_1/b_1 < a_2/b_2 = r(\theta_2)$.

Pair, $\phi_1 \geq \phi_2$ and $\chi_1 > \chi_2$: Now, $s(\phi_1) = a_1/e \geq s(\phi_2) = a_2/e$ and $t(\chi_1) = e/b_1 \geq t(\chi_2) = e/b_2$. Angles $\alpha_1 < \alpha_2$ and $\beta_1 > \beta_2$, since $\alpha_i = \theta_i - \phi_i$ and $\beta_i + \theta_i - \chi_i$ is a constant. This satisfies the conditions for Lemma 3(4), so $r(\theta_1) = a_1/b_1 < a_2/b_2 = r(\theta_2)$—vacuously, in fact.

This completes the pair case; the theorem is established by mathematical induction. ∎

Because any fully-specified polygon can be made into a partial polygon by omitting the first and last internal angles, we have an immediate corollary.

Corollary 5 *A triangulated n-gon P is uniquely determined, up to similarity, by its sequence of internal angles and the cross-ratios of its diagonals.*

5 Conclusions

We have proved by induction that angles and cross-ratios determine a unique simple polygon. It might be nice to extract from the proof an algorithm to compute the realization, but because the proof reveals that the polygon's vertex coordinates are related by compositions of trigonometric functions, an exact implementation would require expensive, high-precision arithmetic. The CRDT algorithm of Driscoll and Vavasis [2] performs an iterative computation to find a Schwarz-Christoffel mapping of a disk to a polygon; our theorem shows that their algorithm does produce a mapping to the correct polygon.

Many interesting questions about the CRDT algorithm remain, including to analyze its rate of convergence and quantify the worst-case crowding. It would be good to consider other interesting combinations of continuous and combinatorial geometry that address precision questions.

Acknowledgement

Marshall Bern posed this problem at the Mount Holyoke workshop on Discrete and Computational Geometry in July 1996. Günter Rote provided encouragement and suggestions to change my initial geometric proof into algebra, and Bob Thurman provided counterbalancing suggestions and encouragement. I thank Marshall, Günter, Bob, Steve Vavasis, and Jim Varah for discussions on this problem, proof, and the CRDT algorithm.

References

[1] T. A. Driscoll. A MATLAB toolbox for Schwarz-Christoffel mapping. *ACM Transactions on Mathematical Software*, 22:168–186, 1996.

[2] T. A. Driscoll and S. A. Vavasis. Numerical conformal mapping using cross-ratios and Delaunay triangulation. *SIAM Journal of Scientific and Statistical Computing*, to appear. ftp://ftp.cs.cornell.edu/pub/vavasis/papers/crdt.ps.gz, 1996.

[3] P. Henrici. *Applied and Computational Complex Analysis*, volume 1. Wiley, 1974.

[4] L. H. Howell and L. N. Trefethen. A modified Schwarz-Christoffel transformation for elongated regions. *SIAM Journal of Scientific and Statistical Computing*, 11(5):928–949, 1990.

[5] L. N. Trefethen. Numerical computation of the Schwarz-Christoffel transformation. *SIAM Journal of Scientific and Statistical Computing*, 1(1):82–102, 1980.

Raising Roofs, Crashing Cycles, and Playing Pool:
Applications of a Data Structure for Finding Pairwise Interactions

David Eppstein* Jeff Erickson†

Abstract

The straight skeleton of a polygon is a variant of the medial axis, introduced by Aichholzer *et al.*, defined by a shrinking process in which each edge of the polygon moves inward at a fixed rate. We construct the straight skeleton of an n-gon with r reflex vertices in time $O(n^{1+\varepsilon} + n^{8/11+\varepsilon}r^{9/11+\varepsilon})$, for any fixed $\varepsilon > 0$, improving the previous best upper bound of $O(nr \log n)$. Our algorithm simulates the sequence of collisions between edges and vertices during the shrinking process, using a technique of Eppstein for maintaining extrema of binary functions to reduce the problem of finding successive interactions to two dynamic range query problems: (1) maintain a changing set of triangles in $\mathbb{R}^3$ and answer queries asking which triangle would be first hit by a query ray, and (2) maintain a changing set of rays in $\mathbb{R}^3$ and answer queries asking for the lowest intersection of any ray with a query triangle. We also exploit a novel characterization of the straight skeleton as a lower envelope of triangles in $\mathbb{R}^3$. The same time bounds apply to constructing non-self-intersecting offset curves with mitered or beveled corners, and similar methods extend to other problems of simulating collisions and other pairwise interactions among sets of moving objects.

1 Introduction

Suppose we are given the floor plan of a building, and a specification for the slope of its roof planes. How can we design a roof, meeting all the walls at a consistent height, with no dips or flat spots where rainwater can accumulate? Aichholzer *et al.* [8, 7] determined an answer: the *straight skeleton*.

The straight skeleton of a polygon P is, as its name implies, a *skeleton* (one-dimensional topological retract) of the polygon formed from straight line segments. It is closely related to the *medial axis*, another type of skeleton commonly defined as the set of points in P with more than one closest point on the boundary of P. The medial axis is a subset of the Voronoi diagram of the vertices and edges of the polygon and consists of line segments and parabolic arcs. (Voronoi edges that meet reflex vertices of the polygon are technically not part of the medial axis.) Medial axes are used in several applications, including shape recognition and reconstruction [14, 16, 52], mesh generation [31, 47, 50, 51], motion planning [45, 49], and computer-aided manufacturing [32, 33, 50]. Despite these many uses of medial axes, their curved arcs have been considered a shortcoming and have led several researchers to form piecewise-linear approximations by sampling the input [50], rectilinear Voronoi diagrams [50], or other techniques [15, 42]. The straight skeleton provides an alternate piecewise linear construction which unlike these other approaches is not sensitive to sampling rates or to the polygon's orientation.

Although, as described above, the medial axis can be defined in terms of closest points or Voronoi diagrams, there is a third definition which Aichholzer *et al.* generalized in their definition of the straight skeleton. An *offset curve* of P is defined as the set of points having some fixed distance d to P; this curve consists of straight line segments parallel to P's sides and circular arcs centered at P's reflex vertices. As d grows, the straight segments of the offset curve move at a constant speed away from the corresponding edges of P, and the circular arcs grow radially away from their centers. The breakpoints between consecutive line segments and circular arcs in the offset curve trace out edges of the Voronoi diagram of the polygon, and the medial axis can be defined as the

*Department of Information and Computer Science, University of California, Irvine, CA 92697, USA, eppstein@ics.uci.edu, http://www.ics.uci.edu/~eppstein. Research partially supported by NSF grant CCR-9258355 and by matching funds from Xerox Corporation.

†Center for Geometric Computing, Department of Computer Science, Duke University, Box 90129, Durham, NC 27708–0129, USA, jeffe@cs.duke.edu, http://www.cs.duke.edu/~jeffe. Research supported by NSF grant DMS-9627683 and by U. S. Army Research Office MURI grant DAAH04-96-1-0013.

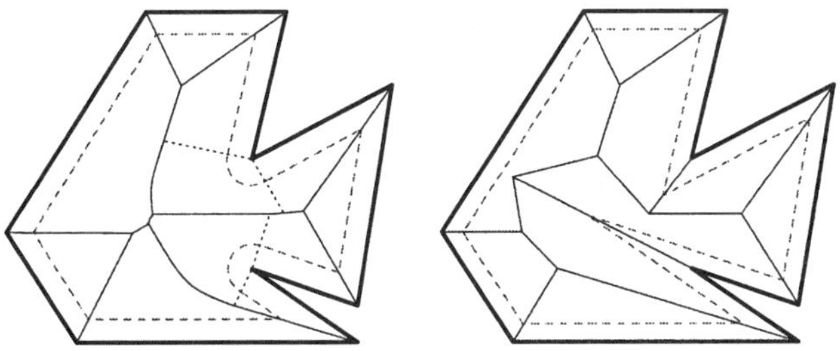

(a) (b)

Figure 1. Skeletons and offset curves of a simple polygon. (a) The medial axis (solid), the Voronoi diagram (solid and dotted), and a rounded offset curve. (b) The straight skeleton and a mitered offset polygon.

locus traced out by a subset of these breakpoints [32]. See Figure 1(a). A generalization of offset curves (available for instance in drawing programs such as Adobe Illustrator) again has straight segments paralleling the polygon edges at a fixed distance, but allows the circular arcs connecting these segments to be replaced by other types of *caps*. The form of offset curve that concerns us has *mitered caps* in which the straight segments parallel to P's edges are extended until they meet each other. Similarly to the way the medial axis can swept out by offset curves with round caps, the straight skeleton is swept out by offset polygons with mitered caps. See Figure 1(b).

The straight skeleton has several applications, including architecture, where it describes the shape of a fixed-slope roof rising over a given set of walls [8, 46], and geographic information systems, where it can be used to reconstruct terrains from a given set of rivers and coastlines [7]. The straight skeleton can be used to reconstruct the offset polygons from which it was defined, and form a consistent family of non-self-intersecting mitered-corner offset polygons (unlike those formed by, *e.g.*, Adobe Illustrator, in which crossings must be removed manually). Techniques related to straight skeletons have also been used in origami constructions [37].

Several efficient algorithms are known for constructing medial axes [38, 24, 21, 32]. Unlike the medial axis, however, the straight skeleton cannot be defined by a distance measure or as an abstract Voronoi diagram [36]; consequently, its construction is considerably more difficult. The fastest previously published algorithms both use $O(n^2 \log n)$ time in the worst case [7, 8]; a more careful analysis shows that the running time of one of these algorithms [8] is actually $O(nr \log n)$, where r is the number of reflex vertices. This can be further improved to $O(nr+n \log n) = O(n^2)$ using a quadtree-like data structure of Eppstein of size $O(nr)$ [26]. This paper describes an algorithm that constructs the straight skeleton of a polygon in time and space $O(n^{8/5+\varepsilon})$ for any fixed $\varepsilon > 0$. In fact, our algorithm can construct weighted straight skeletons (where each edge moves at a different rate) of arbitrary planar straight line graphs, within the same time and space bounds. As an immediate application, we also obtain the first subquadratic algorithm for computing mitered offset polygons.

Like earlier algorithms, our algorithm simulates the sequence of interactions between edges and vertices in the shrinking process described above. A technique of Eppstein [25] for maintaining closest pairs reduces the problem of finding the next interaction to two dynamic range query problems: (1) maintain a changing set of triangles in $\mathbb{R}^3$ and answer queries asking which triangle would be first hit by a query ray, and (2) maintain a changing set of rays in $\mathbb{R}^3$ and answer queries asking for the lowest intersection of any ray with a query triangle. Standard range-searching techniques solve these problems in sublinear time per operation, hence our subquadratic time bounds. The same method applies to other problems of simulating pairwise interactions among a set of moving objects. For example, we can trace the paths of a collection of moving billiard balls in sublinear time per collision.

We also present a novel characterization of the unweighted straight skeleton as the projection of the lower envelope of $n + O(r)$ triangles in $\mathbb{R}^3$. These triangles are not (and cannot be [8]) locally defined; nevertheless, we can exploit the structure of these triangles to obtain a "reflex-sensitive" algorithm that runs in time $O(n^{1+\varepsilon} + n^{8/11+\varepsilon} r^{9/11+\varepsilon}) = O(n^{17/11+\varepsilon})$. Practical variants of our algorithm run in time $O(n \log n + nr)$ using space $O(n + r^2)$ and in time $O(n \log n + nr + r^2 \log r)$ using space $O(n)$.

The rest of the paper is organized as follows. In Section 2, we describe Eppstein's technique for maintaining closest pairs. We describe data structures for lowest intersection queries in Section 3. Section 4 contains a description of our straight skeleton algorithms. In Section 5, we describe and solve a similar problem, constructing the *motorcycle graph* of a set of moving points. We briefly describe our algorithm for dynamically simulating moving balls in Section 6. Finally, in Section 7, we list a few open problems.

2 Maintaining Closest Pairs

Our straight skeleton algorithm is based on maintaining a set of moving offset curve features and using a data structure to find each successive interaction between these features. We formalize this approach as follows. Let R and B be sets of objects, and let $d : R \times B \to \mathbb{R}$ be an arbitrary "distance" function that can be computed in constant time. A data structure supports *minimization queries* if, for any object $r \in R$, an object $b \in B$ minimizing $d(r, b)$ can be determined quickly, and vice versa. Our algorithm uses a data structure that efficiently maintains a "closest" pair of objects $r \in R$ and

$b \in B$ minimizing $d(r, b)$ as objects are inserted and deleted, using a dynamic data structure for minimization queries as a black box.

Theorem 2.1 (Eppstein [25, 26]). *Suppose that after $P(n)$ preprocessing time, we can maintain a data structure of size $S(n)$ that supports insertions, deletions, and minimization queries, each in amortized time $T(n)$. Then after $O(P(n) + nT(n))$ preprocessing time, we can maintain the closest pair between R and B in $O(S(n))$ space, $O(T(n) \log n)$ amortized insertion time, and $O(T(n) \log^2 n)$ amortized deletion time.*

The original statement of the theorem [25] required $T(n)$ to be a worst-case bound on the query and update times for the minimization query structure, but the proof still works if $T(n)$ is only an amortized time bound.

Eppstein previously used this technique to maintain closest and furthest bichromatic pairs among a changing set of red and blue points, and for maintaining the Euclidean minimum spanning tree of a changing set of points in the plane [25]. In our application, R and B will be sets of offset curve features, and d will describe the time at which some pair of features interacts. The closest pair in this setting will give us the time of the earliest interaction.

A simplified version of this data structure for maintaining closest pairs from non-geometric point sets (for which $T(n)$ is the trivial $O(n)$ bound for sequential search), is described along with other non-geometric closest pair algorithms, applications, and experiments, in a companion paper [26]. That paper also contains the following result which we use later as part of a more practical simplification of our algorithm.

Theorem 2.2 (Eppstein [26]). *We can maintain the closest pair between two sets of objects in $O(n)$ time per insertion or deletion, using $O(n^2)$ space and preprocessing time.*

The technique consists of forming a quadtree-like structure on the distance matrix of the objects, constructed by forming pairs of members of R and B and defining the distance between pairs to be the minimum of the four distances between members of the pairs, then by continuing recursively in the resulting smaller distance matrix. The closest pair can be found as the single distance at the root of this hierarchical structure; each update causes $O(n)$ distances to be recomputed.

We can obtain sharper bounds if one of the sets is much smaller than the other. Let $r = |R|$ and $n = |B|$ and suppose $r < n$. A simple modification of Eppstein's quadtree structure maintains the closest pair between R and B in $O(n)$ time per change in R, and in $O(r + \log(n/r)) = O(r + \log n)$ time per change in B, using $O(nr)$ space and preprocessing time. We omit further details.

3 Lowest Intersection Queries

We now consider a dynamic range searching problem that arises as a subproblem in our straight skeleton algorithm: maintaining a set of rays in $\mathbb{R}^3$ so that we can quickly determine the lowest intersection between a ray and a query triangle. These *lowest intersection queries* are, in a sense, the inverse of ray shooting queries; instead of the first triangle hit by a query ray, we want the "first" ray hit by a query triangle.

Throughout this section, we describe several dynamic range searching data structures that allow continuous tradeoffs between space and query time. In the interest of brevity, we explicitly state only the query time as a function of the size s of the data structure, omitting additive terms of the form $O(\operatorname{polylog} n)$ or $O(n^\varepsilon)$. All the data structures we describe can be constructed in time $O(s^{1+\varepsilon})$ and can be modified to support insertions and deletions in time $O(s^{1+\varepsilon}/n)$, at the expense of at most an $O(n^\varepsilon)$ factor in both the preprocessing and query times [2, 5, 40].

Theorem 3.1. *Given n rays in $\mathbb{R}^3$, we can answer lowest intersection queries for triangles in time $O(n^{1+\varepsilon}/s^{1/4})$.*

Proof: We follow the same pattern used to construct ray shooting data structures for halfplanes and triangles in $\mathbb{R}^3$ [3, 4, 5]. It suffices to construct a data structure that supports queries asking whether any of the rays intersects a query quadrilateral. To answer a lowest intersection query using this data structure, we apply parametric search [43] or Chan's recent randomized reduction technique [17] to find the largest value z^* so that no ray crosses the intersection of the query triangle and the halfspace $z \le z^*$. The query algorithm we describe below can easily be executed in parallel in $O(\log n)$ time using $O(n^{1+\varepsilon}/s^{1/4})$ processors, so the additional cost of the parametric search is negligible. Chan's reduction is significantly simpler than parametric search, but the resulting randomized query algorithm is only efficient in expectation; in the worst case, the algorithm is actually slower than brute force.

To detect intersections, we construct a multi-level data structure. (We assume the reader is familiar with this standard technique for composing geometric range searching data structures. Otherwise, see [41] or [1].) The first level is a halfspace range searching data structure of Matoušek [41], which lets us (implicitly) find the rays intersecting the plane containing a query quadrilateral q, in time $O(n^{1+\varepsilon}/s^{1/3})$. This reduces the problem to detecting intersections between q and a set of *lines*.

Let $\ell_1, \ell_2, \ell_3, \ell_4$ be the lines containing the edges of q, oriented so that q lies on the same "side" of each line. A line λ intersects q if and only if it has the same relative orientation with respect to all the ℓ_i. The relative orientation of two oriented lines λ, ℓ is defined to be the orientation of any simplex $abcd$, where λ passes through the points a and b, and ℓ through c and d, in that order. Equivalently, the relative orientation is given by the inner product of the Plücker coordinates of the two lines [48].

To determine whether any line intersects q, we use a four-level data structure. Each of the first three levels is used to find the lines oriented correctly with respect to one of the ℓ_i. We use a data structure of Agarwal and Matoušek [4], which supports such queries in time $O(n^{1+\varepsilon}/s^{1/4})$. Finally, for the last level, we only need to know whether a query line ℓ_4 lies entirely above a set of lines. Chazelle *et al.* [19] describe a data structure that supports such queries in time $O(n^{1+\varepsilon}/s^{1/2})$. □

3.1 Fixed-Slope Lines and Nice Triangles

The most difficult part of answering a lowest-intersection query is finding the lines oriented positively with respect to a query line; in the remainder of this section, we refer to these as *line queries*. This is also the bottleneck in answering ray shooting queries among triangles [4, 5]. Let the *slope* of a line in $\mathbb{R}^3$ denote the tangent of its angle from the xy-plane. In some of our applications of ray shooting and lowest intersection queries, either the rays, the edges of the triangles, or both, have fixed slope. In such cases, we can speed up the relevant line queries slightly, and thus improve the time to answer ray shooting and lowest intersection queries.

A line with fixed slope has only three degrees of freedom: its intersection with the xy-plane and the slope of its projection onto the xy-plane, for example. Except for extreme cases, which we can handle with a lower-dimensional data structure, we can represent any fixed-slope line as a point in $\mathbb{R}^3$. For an *arbitrary* line ℓ, the set of fixed-slope lines that intersect ℓ forms a bounded-degree algebraic surface in $\mathbb{R}^3$. One halfspace of this surface contains the fixed-slope lines oriented positively with respect to ℓ.

Thus, preprocessing a set of fixed-slope lines for arbitrary line queries is equivalent to preprocessing a set of points in $\mathbb{R}^3$ for semialgebraic range queries. Agarwal and Matoušek [4] describe a linear-space data structure that supports such queries in time $O(n^{2/3+\varepsilon})$. Combining this with a data structure of Agarwal and Sharir [5] that supports line queries for an *arbitrary* set of lines in $O(\log n)$ time, using $O(n^{4+\varepsilon})$ space and preprocessing, we obtain the following tradeoff between space and query time. (For further details on space-time tradeoffs, see [1, 41].)

Lemma 3.2. *Given a set of n lines in $\mathbb{R}^3$ with fixed slope, we can answer arbitrary line queries in time $O(n^{8/9+\varepsilon}/s^{2/9})$.*

Conversely, preprocessing a set of arbitrary lines for fixed-slope line queries is equivalent to preprocessing a set of algebraic surfaces in $\mathbb{R}^3$ for point location queries. Chazelle *et al.* [18] describe a data structure of size $O(n^{3+\varepsilon})$ that supports such queries in $O(\log n)$ time. Agarwal and Matoušek [4] describe a linear-size data structure that supports *arbitrary* line queries in $O(n^{3/4+\varepsilon})$ time. Combining these two data structures, we achieve the following space-time tradeoff.

Lemma 3.3. *Given a set of n arbitrary lines in $\mathbb{R}^3$, we can answer fixed-slope line queries in time $O(n^{9/8+\varepsilon}/s^{3/8})$.*

Finally, by combining our two three-dimensional structures, we can support fixed-slope line queries for a set of fixed-slope lines in time $O(n^{1+\varepsilon}/s^{1/3})$. (The slope of the given lines and the slope of the query lines could be different.)

Say that a triangle is *nice* if two of its edges have fixed (but possibly different) slopes. By replacing the general line query data structure used in Theorem 3.1 with our fixed-slope data structure, we obtain the following result.

Theorem 3.4. *Given a set of n rays in $\mathbb{R}^3$, we can support lowest intersection queries for nice triangles in time $O(n^{9/8+\varepsilon}/s^{3/8})$. If all the rays have fixed slope, we can support lowest intersection queries for nice triangles in time $O(n^{1+\varepsilon}/s^{1/3})$ and for arbitrary triangles in time $O(n^{8/9+\varepsilon}/s^{2/9})$.*

We can make a similar improvement for the standard ray shooting problem. Agarwal and Matoušek [4] and Agarwal and Sharir [5] describe data structures for ray shooting among triangles. The bottleneck in both of their data structures is answering line queries. By substituting our data structure into theirs, we obtain a faster algorithm for ray shooting among nice triangles, or with fixed-slope rays, or both.

Theorem 3.5. *Given a set of n triangles in $\mathbb{R}^3$, we can support fixed-slope ray shooting queries in time $O(n^{9/8+\varepsilon}/s^{3/8})$. If all the triangles are nice, we can support arbitrary ray shooting queries in time $O(n^{8/9+\varepsilon}/s^{2/9})$ and fixed-slope ray shooting queries in time $O(n^{1+\varepsilon}/s^{1/3})$.*

4 Straight Skeletons (Raising Roofs)

We now describe our subquadratic straight skeleton algorithms.

The definition of straight skeleton generalizes easily to disconnected polygons, polygons with holes, and even arbitrary planar straight-line graphs [7]. Although for simplicity we explicitly consider only simple polygons, all of the results in this section apply to these more general cases as well.

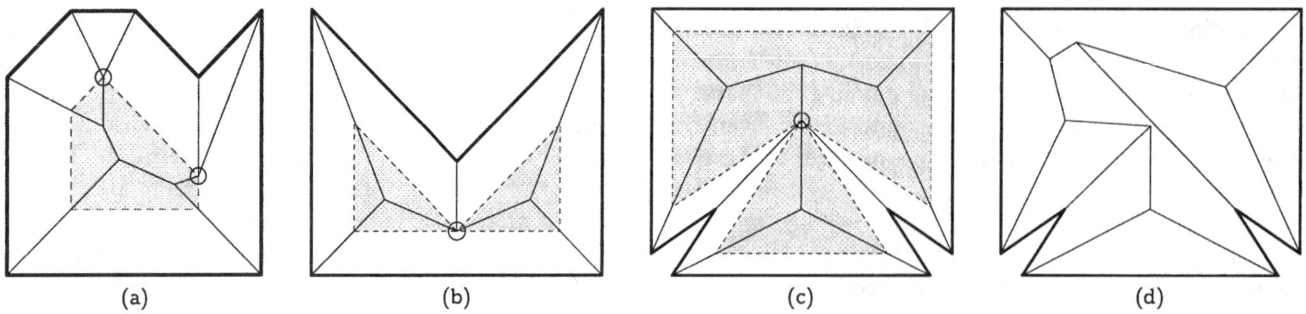

Figure 2. (a) Two simultaneous edge events. (b) A split event. (c) A vertex event. (d) Perturbing the polygon to eliminate the vertex event produces a radically different skeleton.

4.1 Events

The straight skeleton of a polygon is constructed by a shrinking process in which each edge moves inwards at a fixed rate. Assuming the polygon is in general position, it undergoes only two types of combinatorial changes as it shrinks. An *edge event* occurs when an edge collapses down to a point; if its neighboring edges still have nonzero length, they become adjacent. See Figure 2(a). Note that one of the endpoints of the disappearing edge can be a reflex vertex. A *split event* occurs when a reflex vertex collides with and splits an edge; the edges adjacent to the reflex vertex are now adjacent to the two parts of the split edge. Each split event divides a component of the shrinking polygon into two smaller components. See Figure 2(b). Each event introduces a node of degree three into the evolving straight skeleton.

In degenerate cases, the straight skeleton can have vertices of degree higher than three, introduced by simultaneous events at the same location. Most of the time, we can handle these events one at a time using standard perturbation techniques, replacing the high-degree node with several nodes of degree three. The only exception occurs when two or more reflex vertices (and nothing else) reach the same point simultaneously. We call this a *vertex event*. See Figure 2(c). Unlike edge or split events, a vertex event can introduce a new reflex vertex into the shrinking polygon, although the total number of reflex vertices always decreases. Any perturbation of the polygon that removes a vertex event radically changes the structure of the polygon's straight skeleton; consequently, we must handle vertex events directly. See Figure 2(d).

4.2 A Subquadratic Algorithm

Like earlier algorithms [8, 7], our basic approach is to simulate the sequence of edge, split, and vertex events that define the skeleton. We view time as a third spatial dimension, so that the shrinking process becomes an upward sweep of the *roof* of the polygon with a horizontal plane. The key observation is that although we do not know the entire sequence of events until we are finished, at any time we can efficiently compute the *next* event using only local information.

Each edge of the polygon defines a (possibly unbounded) triangle in $\mathbb{R}^3$, whose other two edges are defined by the initial paths of the two endpoints; similarly, each reflex vertex defines a ray in $\mathbb{R}^3$. If the first event is an edge event, it happens when the sweep plane reaches the top of a triangle. If the first event is a split or vertex event, it happens when a ray hits a triangle.

We store the triangles and rays in a data structure that maintains the first event of each type. To update the data structure at each event, we perform a constant number of insertions and deletions. At each edge event, we delete three triangles (defined by the collapsed edge and its two neighbors), possibly delete one ray (defined by a reflex endpoint of the collapsed edge), and insert two triangles (defined by the newly adjacent edges). At each split event, we delete one ray (the reflex vertex), delete three triangles (the edge being split and the two edges adjacent to the reflex vertex), and insert four triangles (the two pairs of adjacent edges defined by the split). At a vertex event involving k reflex vertices, we delete $2k$ triangles and k rays, and insert $2k$ triangles and possibly one ray. Over the entire event sequence, we perform $O(n)$ insertions and deletions.

We maintain potential edge events by storing the triangles in a simple priority queue, where the priority of a triangle is the t-coordinate of its top vertex. The overall time spent maintaining the priority queue is $O(n \log n)$.

To find the next split or vertex event, we use a data structure that that maintains the lowest intersection between a set of rays and a set of triangles in $\mathbb{R}^3$. Theorem 2.1 reduces this to two range-searching problems: (1) maintain a set of triangles and answer ray shooting queries, and (2) maintain a set of rays and answer lowest-intersection queries. To answer ray-shooting queries, we use a data structure of Agarwal and Matoušek [4], and to answer lowest-intersection queries, we use Theorem 3.1. Both data structures support queries in time $O(n^{1+\varepsilon}/s^{1/4})$ and updates in time $O(s^{1+\varepsilon}/n)$; balancing the query and update times, we obtain data structures of size $s = O(n^{8/5+\varepsilon})$ that sup-

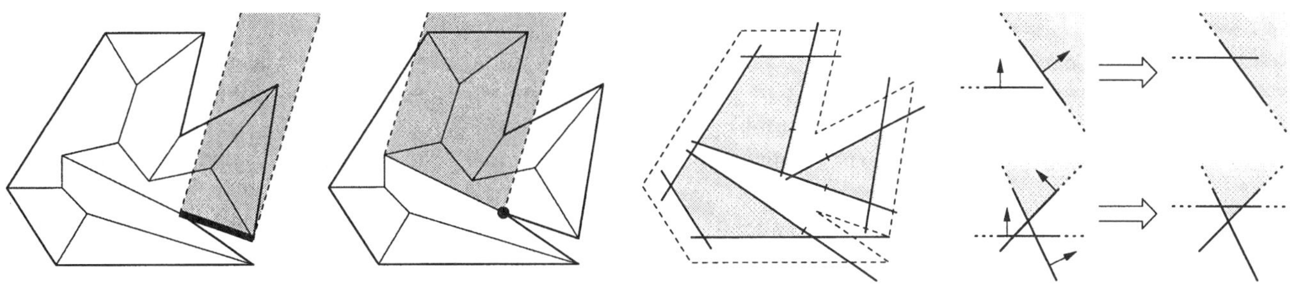

(a) (b) (c) (d)

Figure 3. (a) Top view of an edge slab. (b) Top view of a reflex slab. (c) A cross section of the slabs, after two split events. (d) Two impossible transitions; see Lemma 4.2.

ports queries or updates in time $O(n^{3/5+\varepsilon})$. By Theorem 2.1, the total time to initialize the combined data structure and perform $O(n)$ insertions and deletions is $O(n^{8/5+\varepsilon})$. Since this dominates the time to handle the edge events, we obtain the following theorem.

Theorem 4.1. *The straight skeleton of an n-gon can be constructed in time and space $O(n^{8/5+\varepsilon})$.*

Our algorithm can also be used to construct *weighted* straight skeletons, where each edge moves at a different different speed, in the same time and space. The only difference from the unweighted case is that both endpoints of a collapsing edge can be reflex vertices, in which case the edge event introduces a new reflex vertex.

4.3 A Faster Reflex-Sensitive Algorithm

We can improve our algorithm in the unweighted case. For any polygon in the xy-plane, we define two families of *slabs* in $\mathbb{R}^3$. Each slab is an semi-infinite strip, bounded on two sides by parallel rays pointing upwards from the xy-plane at a 45-degree angle and perpendicular to some edge of the polygon. Each edge e defines an *edge slab*, bounded below by e and on the sides by rays perpendicular to e. Each reflex vertex $v = e \cap e'$ defines two *reflex slabs*, bounded below by the edge of the roof induced by v, and bounded on the sides by rays perpendicular to either e or e'. See Figure 3(a) and (b).

Lemma 4.2. *The straight skeleton of a polygon is the intersection of the polygon and the vertical projection of the lower envelope of its edge slabs and reflex slabs.*

Proof: Consider the shrinking process for generating the straight skeleton, or equivalently, sweep a horizontal plane upwards through the roof. Each point in a cross-section of a slab corresponds to exactly one point on the base edge of the slab, so we can think of these base edge points as all traveling upwards along the slabs.

Each of these points starts on the boundary of the shrinking polygon; points on edge slabs appear at the beginning, and points on reflex slabs appear later. As the polygon shrinks, slab points leave the polygon boundary. See Figure 3(c). As long as every slab point stays outside the polygon after it leaves the boundary, only points on the boundary of the polygon can contribute to the lower envelope of the slabs.

There are only two ways that a point can reenter the polygon after it leaves: either a slab endpoint overtakes an edge of the polygon, or a slab overtakes a convex vertex of the polygon. See Figure 3(d). Since all the segments and their endpoints are moving at the same speed, neither of these transitions can occur. □

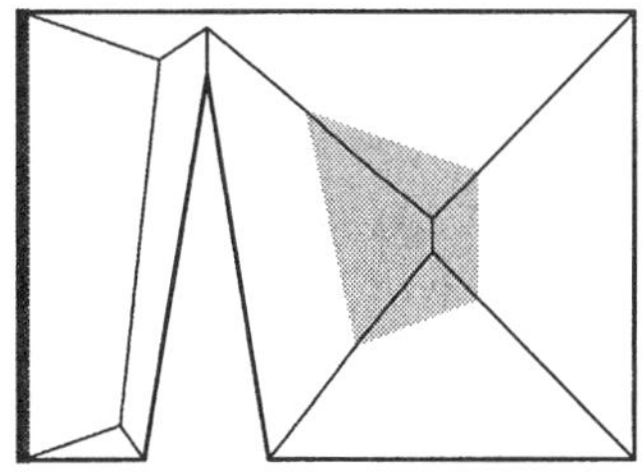

Figure 4. The roof of a weighted polygon is not the lower envelope of its slabs. The leftmost edge has higher weight than the other edges, and its edge slab cuts through the roof in the shaded region.

Aichholzer *et al.* [8] show that the roof cannot be expressed as the lower envelope of partial linear functions, each function locally defined by a small neighborhood of an edge. Our reflex slabs are not locally defined. We emphasize that Lemma 4.2 holds only for unweighted straight skeletons. Counterexamples for the weighted case are easy to construct; see Figure 4.

To exploit Lemma 4.2, we classify split events into two classes, *edge splits* and *reflex splits*, depending on which type of slab is hit by the ray. For the moment, let us assume that the polygon is in general position, so there are no vertex events. We preprocess the edge slabs into a ray shooting data structure, and for each ray, we find the first edge slab that it hits. Since two sides of each slab have slope 1, the total time to preprocess the n slabs and answer r ray shooting queries is $O(n^{1+\varepsilon} + n^{8/11+\varepsilon}r^{9/11+\varepsilon})$ by Theorem 3.5. This gives us a superset of the possible edge split events, which we then sort chronologically.

We simultaneously maintain a priority queue for the next edge event, scan the sorted list of potential edge split events, and maintain the next reflex split event using the following variant of our earlier algorithm. We maintain a data structure storing only the r rays and $2r$ reflex slabs defined by reflex vertices. Initially, we store *unbounded* reflex slabs, each with only two edges, since we do not yet know where the third edge is. At each split event of either type, we replace two unbounded slabs with the correct reflex slabs and delete one ray. If an edge event involves a reflex vertex, we update its slab and delete its ray. Theorem 2.1 reduces maintaining the next reflex split event to maintaining data structures for ray shooting and lowest intersection queries. Since one side of every unbounded slab and two sides of every reflex slab have slope 1, we can maintain the next reflex split in time $O(r^{6/11+\varepsilon})$ per split event, by Theorems 3.4 and 3.5. Since the data structure is updated r times, the total time spent maintaining this data structure is $O(r^{17/11+\varepsilon})$.

Theorem 4.3. *The straight skeleton of an n-gon with r reflex vertices can be constructed in time $O(n^{1+\varepsilon} + n^{8/11+\varepsilon}r^{9/11+\varepsilon})$.*

In degenerate cases, we must also handle vertex events. At each vertex event involving k reflex vertices, we delete k rays and replace $2k$ unbounded slabs with the correct reflex slabs. If the event creates a new reflex vertex, we perform another ray shooting query among the edge slabs, update the sorted sequence of potential edge splits, and insert a new ray into the next-reflex-split data structure. Even with vertex events, there are at most $2r - 1$ reflex vertices over the entire history of the polygon, so the asymptotic running time of our algorithm is unchanged.

4.4 Final Remarks

Although our algorithms are a significant theoretical improvement over earlier results, because of the complexity of the range-searching algorithms involved, they would be less efficient in practice. The running time of Aichholzer and Aurenhammer's algorithm derives from the number of topological changes in a triangulation (or trapezoidal decomposition) of the shrinking polygon [7]. Although there can be $\Theta(n^2)$ such changes in the worst case, "typical" inputs require much less work. Even for polygons specially constructed to make their algorithm run slowly, our algorithm would be slower for reasonable values of n.

However, if we use our modification of Eppstein's quadtree-based data structure (Theorem 2.2) to maintain the next split event, instead of Theorem 2.1, we obtain a practical algorithm for weighted skeletons that runs in time $O(n \log n + nr)$ using $O(nr)$ space. In the unweighted case, we can improve the space bound to $O(n + r^2)$ by using brute force ray-shooting to find the potential edge splits and Eppstein's quadtree to maintain reflex splits. Alternately, if we use Aichholzer and Aurenhammer's triangulation-based algorithm [7] to find reflex splits, we obtain a practical linear-space algorithm that runs in time $O(n \log n + nr + r^2 \log r)$. All of these algorithms improve the previously best time bound $O(nr \log n)$ [8].

If the input polygon is c-oriented, we can compute its straight skeleton in time $O(c^4 n \operatorname{polylog} n)$. Each of the relevant ray shooting and inverse ray shooting queries can be answered using one of $O(c^4)$ orthogonal range query data structures, since each ray has one of $O(c^2)$ orientations, and there are $O(c^2)$ parallel families of slabs. Finally, if the input polygon is convex, we can construct its straight skeleton in linear time [6, 20], since the straight skeleton is identical to the medial axis in that case.

Once we compute the straight skeleton, we can compute any desired mitered offset polygon from it in linear time. Alternatively, we can compute offset polygons directly, with the same worst-case time bound, by halting the inward sweep when we reach the desired offset. We can also compute offset polygons with *beveled* caps by adding a zero-length edge at each reflex vertex, perpendicular to the vertex's bisector, and applying the same algorithm.

5 Crashing Cycles

In this section, we consider a problem that captures the most difficult part of constructing straight skeletons: determining how the reflex vertices interact. Imagine several people riding motorcycles out in the desert. Each motorcycle is specified by an initial location and a velocity. All the bikes start moving simultaneously, and thereafter, no turning, braking, or acceleration is allowed. The bikes are extremely fragile; if any motorcycle runs over the track previously left by another bike, it immediately crashes. If two motorcycles collide, they both crash. After all the bikes either crash or escape to infinity, their tracks form a planar directed graph, which we call the *motorcycle graph.* See Figure 5. The problem is to construct this graph as quickly as possible. A similar problem was previously considered by Lisberger *et al.* [39]

If we form a non-simple polygon in which each motorcycle is replaced by a small hole in the form of a sharp isosceles triangle, the straight skeleton edges traced out by the sharp reflex vertices of these triangles will approximate the motorcycle graph. Motorcycle graphs are also a generalization of the *weighted planar partitions*

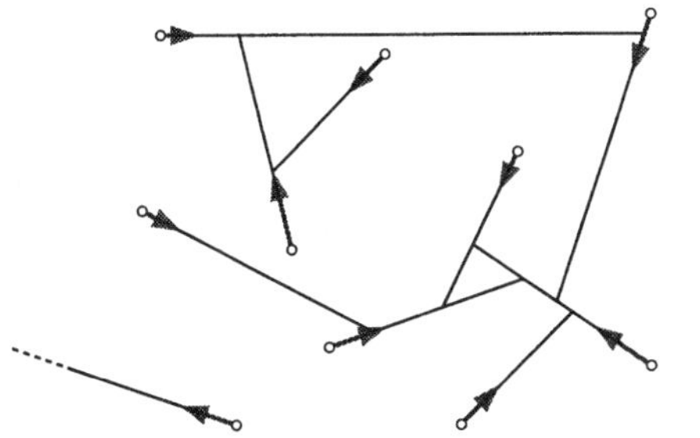

Figure 5. A motorcycle graph.

introduced by Czyzowicz *et al.* as a tool to solve certain art gallery problems [23]. Given a sequence of n non-intersecting line segments in the plane, the weighted planar partition is obtained by extending each segment, one at a time in the order presented, until each endpoint reaches another (possibly extended) segment.

Every motorcycle graph is a pseudo-forest with at most $2n$ vertices and $2n$ edges. Despite the simplicity of this graph, no near-linear algorithm is known for its construction, except in a few special cases, described briefly below. Aichholzer and Aurenhammer's straight skeleton algorithm can easily be adapted to construct motorcycle graphs in $O(n^2 \log n)$ time; the time bound can be improved to $O(n^2)$ using Eppstein's quadtree data structure [26]. Our methods yield a subquadratic algorithm.

Theorem 5.1. *The motorcycle graph of n moving points can be constructed in time and space* $O(n^{17/11+\varepsilon})$.

Proof: Our algorithm simulates the interactions between bikes and tracks as the bikes move and the tracks grow. As in our previous straight skeleton algorithms, we treat time as a third spatial dimension. A bike with initial position (x, y) and velocity (u, v) induces a ray, based at $(x, y, 0)$ and pointing in the direction $(u, v, 1)$. We classify each track as either *live* or *dead*, depending on whether the corresponding bike has crashed or not. Each track induces a vertical *curtain*, consisting of all the points directly above the corresponding ray (if the track is live) or segment (if the track is dead) in $\mathbb{R}^3$. Initially we have a set of n rays and n live curtains. Whenever a bike crashes, we delete the bike's ray and add a third side to its curtain. The next collision is always given by the lowest intersection of a bike with a curtain.

Theorem 2.1 reduces maintaining the next collision to two range searching problems: (1) maintain a changing set of curtains and answer queries asking which curtain is first hit by a query ray, and (2) maintain a changing set of rays in $\mathbb{R}^3$ and answer queries asking for the lowest intersection of any ray with a query curtain. Since all but one side of every curtain is vertical, curtains are nice triangles. We can apply Theorems 3.4 and 3.5 to solve both of these subproblems in time $O(n^{6/11+\varepsilon})$ per query or update, using a data structure of size $O(n^{17/11+\varepsilon})$. □

We can improve the performance of some of the data structures. De Berg *et al.* [13] describe a data structure that supports ray shooting queries among curtains with query and update time $O(n^{1/3+\varepsilon})$. For each dead track, it suffices to store the vertical strip consisting of points directly above *or below* each dead track, instead of three-sided curtains; this allows us to reduce the lowest-intersection query time for dead tracks to $O(n^{1/3+\varepsilon})$. Unfortunately, these improved time bounds are still dominated by the time to find the lowest intersection with a live curtain, so we do not obtain a faster algorithm overall.

There are several special cases of motorcycle graphs that we can compute more quickly. If all the bikes move at the same speed (or in a constant number of different speeds), we can build the motorcycle graph in time and space $O(n^{3/2+\varepsilon})$. If the bikes move in only a constant number of directions, we need only $O(n^{4/3+\varepsilon})$ time and space. If there are only a constant number of different velocity vectors, we can compute the graph in $O(n \operatorname{polylog} n)$ time and space using orthogonal range searching data structures. Finally, if all the velocity vectors have, for example, non-negative x-coordinates, we can use a simple sweep-line algorithm to compute the motorcycle graph in $O(n \log n)$ time. Further details will appear in the full version of the paper.

6 Playing Pool

Our techniques can be applied to any problem that calls for the efficient detection of collisions among moving objects. For example, consider the following *billiard problem*. We are given a set of unit disks in the plane representing billiard balls, each with an initial position and velocity, and are asked to compute the sequence of collisions that occur as the balls move and bounce off each other, following the standard laws of classical physics.

Basch *et al.* [12] observe that only the closest pair of balls can collide, and suggest simulating collisions using a kinetic data structure to efficiently maintain the closest pair. (See also [11, 30].) In the worst case, however, the closest pair changes $\Theta(n^2)$ times without a single collision, which makes this approach less efficient than even the most naïve algorithm that simply checks every pair of balls. Kim *et al.* [35] describe an event-driven algorithm that divides space into a uniform grid of cells and uses $O(\log n)$ time whenever a ball enters a cell, leaves a cell, or collides with another ball in the same cell; a similar approach (but for more general objects) is described by Mirtich and Canny [44]. These algorithms perform quite badly if the balls are very far apart or if there are few collisions. Several other collision-detection algorithms are known that are efficient, at least in practice, for many types of objects; see, for example, [10, 22, 27, 34]. Almost none of these have theoretical guarantees on their performance. In fact, like the algorithms in [35, 44], most perform well only in dense environments.

Our methods yield the first known collision-detection algorithm with a guaranteed sublinear time bound per collision.

Theorem 6.1. *The billiard problem described above can be solved in time* $O(n^{20/29+\varepsilon}) = O(n^{0.6897})$ *per collision using space* $O(n^{49/29+\varepsilon}) = O(n^{1.6897})$.

Proof: Theorem 2.1 reduces the billiard problem to maintaining a ray shooting data structure for a changing set of elliptical cylinders in $\mathbb{R}^3$ with circular horizontal cross-sections. We express each ray shooting query as a composition of several four- and five-dimensional semi-algebraic range queries, which we can solve using techniques of Agarwal and Matoušek [4]. We omit further details. □

Using trivial ray-shooting data structures, we also immediately obtain practical algorithms for the billiard problem that use $O(n \log^2 n)$ time per collision using linear space, or $O(n)$ time per collision using quadratic space. We can also maintain collisions between the balls and any collection of stationary line segments—the sides of a pool table, for example—with the same asymptotic space and time bounds. (Deciding whether a system of moving balls and stationary line segments ever reaches a steady state is PSPACE-complete, since a polynomial-space Turing machine can be simulated by such a system [28].)

7 Open Problems

The most obvious open problem is to improve the running times of our algorithms. The best lower bound for constructing either skeletons of planar straight-line graphs or motorcycle graphs is $\Omega(n \log n)$, by an easy reduction from sorting; no nontrivial lower bound is known for constructing straight skeletons of simple polygons. It seems especially unlikely that $O(n^{17/11})$ is the best possible time bound for constructing a motorcycle graph or a single offset polygon.

Atallah *et al.* [9] show that the construction of weighted planar partitions is P-complete. A similar technique was used by Griffeath and Moore to prove the P-completeness of certain two-dimensional cellular automata [29]. A simple modification of either proof shows that construction of motorcycle graphs is also P-complete, even when the motorcycles move only north, east, or north-east. (There is a simple $O(n \log n)$-time sequential algorithm for this special case.) Similar arguments show that constructing the straight skeleton of an arbitrary planar straight-line graph is a P-complete problem, but the proof does not extend to simple polygons. Can straight skeletons of simple polygons be constructed efficiently in parallel?

The main difficulty in constructing straight skeletons is computing the nonlocal effects of the reflex vertices. This non-locality may also prevent straight skeletons from being as useful as medial axes in applications such as mesh generation and motion planning. Aichholzer and Aurenhammer [7] observe that if the polygon contains no acute angles, then the resulting skeleton shares several properties with Voronoi diagrams. Can we construct the straight skeleton more quickly if we have a constant bound on the smallest external angle in the input polygon? If the minimum external angle is bounded, then so is the ratio between the speed of any reflex vertex and the speed of any edge. Although the straight skeleton is still not an abstract Voronoi diagram in this case, it seems likely that this property restores enough locality to the skeleton that faster divide-and-conquer or incremental techniques can be applied.

Acknowledgments The second author would like to thank Pankaj Agarwal, Lars Arge, and T. M. Murali for helpful discussions on shooting rays, crashing motorcycles, and raising roofs.

References

[1] P. K. Agarwal and J. Erickson. Geometric range searching and its relatives. Technical report CS-1997-11, Duke University, May 1997. To appear in *Discrete & Computational Geometry: Ten Years Later*, B. Chazelle, J. E. Goodman, and R. Pollack, editors, American Mathematical Society Press, 1998. ⟨http://www.cs.duke.edu/~jeffe/pubs/survey.html⟩.

[2] P. K. Agarwal and J. Matoušek. Dynamic half-space range reporting and its applications. *Algorithmica* 13:325–345, 1995.

[3] P. K. Agarwal and J. Matoušek. Ray shooting and parametric search. *SIAM J. Comput.* 22(4):794–806, 1993.

[4] P. K. Agarwal and J. Matoušek. On range searching with semialgebraic sets. *Discrete Comput. Geom.* 11:393–418, 1994.

[5] P. K. Agarwal and M. Sharir. Applications of a new space-partitioning technique. *Discrete Comput. Geom.* 9:11–38, 1993.

[6] A. Aggarwal, L. J. Guibas, J. Saxe, and P. W. Shor. A linear-time algorithm for computing the Voronoi diagram of a convex polygon. *Discrete Comput. Geom.* 4(6):591–604, 1989.

[7] O. Aichholzer and F. Aurenhammer. Straight skeletons for general polygonal figures in the plane. *Proc. 2nd Annu. Internat. Conf. Computing and Combinatorics*, pp. 117–126. Lecture Notes in Computer Science 1090, Springer, 1996.

[8] O. Aichholzer, F. Aurenhammer, D. Alberts, and B. Gärtner. A novel type of skeleton for polygons. *J. Universal Comput. Sci.* 1(12):752–761, 1995. ⟨http://www.iicm.edu/jucs_1_12/a_novel_type_of⟩.

[9] M. J. Atallah, P. Callahan, and M. T. Goodrich. P-complete geometric problems. *Internat. J. Comput. Geom. Appl.* 3:443–462, 1993.

[10] D. Baraff. Interactive simulation of solid rigid bodies. *IEEE Comput. Graph. Appl.* 15(3):63–75, May 1995.

[11] J. Basch, L. Guibas, and J. Hershberger. Data structures for mobile data. *Proc. 8th ACM-SIAM Sympos. Discrete Algorithms*, pp. 747–756, 1997.

[12] J. Basch, L. Guibas, and L. Zhang. Proximity problems on moving points. *Proc. 13th Annu. ACM Sympos. Comput. Geom.*, pp. 344–351, 1997.

[13] M. de Berg, D. Halperin, M. Overmars, J. Snoeyink, and M. van Kreveld. Efficient ray shooting and hidden surface removal. *Algorithmica* 12:30–53, 1994.

[14] H. Blum. A transformation for extracting new descriptors of shape. *Models for the Perception of Speech and Visual Form*, pp. 362–380. MIT Press, 1967.

[15] F. L. Bookstein. The line-skeleton. *Comput. Graph. Image Process.* 11:123–137, 1979.

[16] L. Calabi and W. E. Hartnett. Shape recognition, prairie fires, convex deficiencies and skeletons. *Amer. Math. Monthly* 75:335–342, 1968.

[17] T. M. Chan. Geometric applications of a randomized optimization technique. To appear in *Proc. 14th Annu. ACM Sympos. Comput. Geom.*, 1998.

[18] B. Chazelle, H. Edelsbrunner, L. Guibas, and M. Sharir. A singly-exponential stratification scheme for real semi-algebraic varieties and its applications. *Theoret. Comput. Sci.* 84:77–105, 1991.

[19] B. Chazelle, H. Edelsbrunner, L. J. Guibas, M. Sharir, and J. Stolfi. Lines in space: Combinatorics and algorithms. *Algorithmica* 15:428–447, 1996.

[20] L. P. Chew. Building Voronoi diagrams for convex polygons in linear expected time. Technical Report PCS-TR90-147, Dept. Math. Comput. Sci., Dartmouth College, 1986.

[21] F. Chin, J. Snoeyink, and C.-A. Wang. Finding the medial axis of a simple polygon in linear time. *Proc. 6th Annu. Internat. Sympos. Algorithms Comput.*, pp. 382–391. Lecture Notes Comput. Sci. 1004, Springer-Verlag, 1995.

[22] J. D. Cohen, M. C. Lin, D. Manocha, and M. K. Ponamgi. I-collide: An interactive and exact collision detection system for large-scale environments. *Proc. ACM Interactive 3D Graphics Conf.*, pp. 189–196, 1995.

[23] J. Czyzowicz, I. Rival, and J. Urrutia. Galleries, light matchings and visibility graphs. *Proc. 1st Workshop Algorithms Data Struct.*, pp. 316–324. Lecture Notes Comput. Sci. 382, Springer-Verlag, 1989.

[24] O. Devillers. Randomization yields simple $O(n \log^* n)$ algorithms for difficult $\Omega(n)$ problems. *Internat. J. Comput. Geom. Appl.* 2(1):97–111, 1992. ⟨http://www.inria.fr/prisme/biblio/search.html⟩.

[25] D. Eppstein. Dynamic Euclidean minimum spanning trees and extrema of binary functions. *Discrete Comput. Geom.* 13:111–122, 1995.

[26] D. Eppstein. Fast hierarchical clustering and other applications of dynamic closest pairs. *Proc. 9th Annu. ACM-SIAM Sympos. Discrete Algorithms*, pp. 619–628, 1998.

[27] A. Foisy, V. Hayward, and S. Aubry. The use of awareness in collision prediction. *Proc. 1990 IEEE Internat. Conf. Robotics and Automation*, pp. 338–343, 1990.

[28] E. Fredkin and T. Toffoli. Conservative logic. *Internat. J. Theoret. Phys.* 21:219–253, 1981/82. Proceedings of Conference on Physics of Computation, Dedham, Mass., 1981.

[29] D. Griffeath and C. Moore. Life Without Death is P-complete. Working Paper 97-05-044, Santa Fe Institute, 1997. To appear in *Complex Systems*. ⟨http://psoup.math.wisc.edu/java/lwodpc/lwodpc.html⟩.

[30] L. J. Guibas. Kinetic data structures: A state of the art report. To appear in *Proc. 3rd Workshop on Algorithmic Foundations of Robotics*, P. K. Agarwal, L. Kavraki, and M. Mason, editors, A. K. Peters, 1998.

[31] H. N. Gürsoy and N. M. Patrikalakis. An automatic coarse and fine surface mesh generation scheme based on medial axis transform, Part I: Algorithms. *Engineering with Computers* 8:121–137, 1992.

[32] M. Held. Voronoi diagrams and offset curves of curvilinear polygons. To appear in *Comput. Aided Design*. ⟨http://www.cosy.sbg.ac.at/~held/papers/cad96.ps.gz⟩.

[33] M. Held, G. Lukács, and L. Andor. Pocket machining based on contour-parallel tool paths generated by means of proximity maps. *Comput. Aided Design* 26(3):189–203, Mar. 1994.

[34] P. M. Hubbard. Collision detection for interactive graphics applications. *IEEE Trans. Visualization and Computer Graphics* 1(3):218–230, Sept. 1995.

[35] D. Kim, L. Guibas, and S. Shin. Fast collision detection among multiple moving spheres. *Proc. 13th Annu. ACM Sympos. Comput. Geom.*, pp. 373–375. 1997.

[36] R. Klein. *Concrete and Abstract Voronoi Diagrams*. Lecture Notes Comput. Sci. 400. Springer-Verlag, 1989.

[37] R. J. Lang. A computational algorithm for origami design. *Proc. 12th Annu. ACM Sympos. Comput. Geom.*, pp. 98–105, 1996.

[38] D. T. Lee. Medial axis transformation of a planar shape. *IEEE Trans. Pattern Anal. Mach. Intell.* PAMI-4:363–369, 1982.

[39] S. Lisberger, director. *Tron*. Walt Disney Productions, 1982. Motion picture, 96 minutes.

[40] J. Matoušek. Reporting points in halfspaces. *Comput. Geom. Theory Appl.* 2(3):169–186, 1992.

[41] J. Matoušek. Range searching with efficient hierarchical cuttings. *Discrete Comput. Geom.* 10(2):157–182, 1993.

[42] M. McAllister, D. Kirkpatrick, and J. Snoeyink. A compact piecewise-linear Voronoi diagram for convex sites in the plane. *Discrete Comput. Geom.* 15:73–105, 1996.

[43] N. Megiddo. Applying parallel computation algorithms in the design of serial algorithms. *J. ACM* 30:852–865, 1983.

[44] B. Mirtich and J. Canny. Impulse-based simulation of rigid bodies. *Symposium on Interactive 3D Graphics*. ACM Press, 1995.

[45] C. Ó'Dúnlaing and C. K. Yap. A "retraction" method for planning the motion of a disk. *J. Algorithms* 6:104–111, 1985.

[46] A. Recuaero and J. P. Gutiérrez. Sloped roofs for architectural CAD systems. *Microcomputers in Civil Engineering* 8:147–159, 1993.

[47] V. Srinivasan, L. R. Nackman, J.-M. Tang, and S. N. Meshkat. Automatic mesh generation using the symmetric axis transform of polygonal domains. *Proc. IEEE* 80(9):1485–1501, Sept. 1992.

[48] J. Stolfi. *Oriented Projective Geometry: A Framework for Geometric Computations*. Academic Press, New York, NY, 1991.

[49] J. A. Storer and J. H. Reif. Shortest paths in the plane with polygonal obstacles. *J. ACM* 41(5):982–1012, 1994.

[50] A. Sudhalkar, L. Gursoz, and F. Prinz. Box-skeletons of discrete solids. *Comput. Aided Design* 28:507–517, 1996.

[51] T. K. H. Tam and C. G. Armstrong. 2D finite element mesh generation by medial axis subdivision. *Advances in Engineering Software and Workstations* 13(5–6):313–324, Sept. 1991.

[52] P. J. Vermeer. *Medial Axis Transform to Boundary Representation Conversion*. Ph.D. thesis, CS Dept., Purdue University, West Lafayette, Indiana 47907-1398, USA, 1994.

Construction of Contour Trees in 3D in $O(n \log n)$ steps

Sergey P. Tarasov *
Computing Center of RAS
Vavilova 40
117967 Moscow GSP-1, Russia.
e-mail: sergey@ext1.ccas.ru

Michael N. Vyalyi †
Computing Center of RAS
Vavilova 40
117967 Moscow GSP-1, Russia.
e-mail: misha@vyalyi.mccme.ru

Abstract

We outline an $O(n \log n)$ algorithm for computing the contour trees for simplicial meshes with n elements in 3D. As a byproduct we describe an $O(n \log n)$ algorithm for "resolution" of singularities of piecewise-linear functions in 3D (i.e., transforming singularities into simple — Morse-like ones by subdividing the mesh).

Keywords: contour tree, Reeb graph, singularity, resolution of singularity, Morse theory.

Introduction

Interpolation technique is standard for the representation of a real-valued function in computer. In particular, a simplest approach of this kind — linear interpolation over simplicial meshes — is commonly used. In GIS design such model is known as TIN (Triangulated Irregular Network). Typical queries should quickly report mesh elements through which the contours (the level surfaces of the function involved) pass. In general, due to storage limitations, it is impossible to have a representation of all contours passing through all the vertices of the simplicial mesh, and thus, we are naturally led to the idea of some concise representation of the contours' evolution. One of possible approaches here is the *contour tree* (CT) or the topographic change graph (a special case of the *Reeb graph*). Informally, the contour tree records topological changes in the evolution of the level surfaces.

The contour tree can be constructed in a time $O(n \log n)$ for 2D triangulations [1, 2] and in $O(n^2)$ steps for meshes with n elements [1].

This paper can be regarded as a feedback to [1], where in particular was formulated the problem of subquadratic contour tree algorithms in higher dimensions. Moreover, to achieve this purpose we will use a modification of the $O(n \log n)$ two-dimensional procedure of [1].

Our result is the following. Let $\mathcal{T}$ be any triangulation of any simply connected polyhedral region in $\mathbf{R}^3$ having n simplices and $f(\cdot)$ is a piecewise-linear (PL) function on $\mathcal{T}$.

* Supported in part by RFFI grant 96-01-00662
† Supported in part by RFFI grant 97-01-00159

SCG 98 Minneapolis Minnesota USA

Theorem 1 *The CT for the pair $(\mathcal{T}, f(\cdot))$ can be constructed in $O(n \log n)$ time.*

The proof of the Theorem 1 is organized as follows. In the next Section we state formal definition of the problem. Data structures and basic subroutines of the algorithm are described in Section 2 and in Section 3, respectively. The *Main algorithm* for the CT construction is outlined in Section 4. Boundary effects are discussed separately in Section 5.

The *Main Algorithm* works properly only on functions with *simple* PL singularities that correspond to Morse (nondegenerate) singularities in the smooth case. Unlike smooth case, nonsimple singularities in general can not be removed by some small enough changes of the function $f(\cdot)$ at the vertices of $\mathcal{T}$. It is nevertheless possible to *refine* (we call this construction a *resolution*) the initial triangulation and continue $f(\cdot)$ onto it, so that the resulting pair $(\tilde{\mathcal{T}}, \tilde{f}(\cdot))$ would have only simple singularities and that, moreover, we could recover the initial CT by shrinking some edges of the corresponding CT for the new pair. The resolution is effective and does not increase the asymptotic running time of the overall algorithm. This result (perhaps of independent interest) is proved in the Appendix. It is applied at Stage 0 of the *Main Algorithm.*

Theorem 2 *The resolution for the pair $(\mathcal{T}, f(\cdot))$ can be constructed in $O(n \log n)$ steps.*

In what follows S^d and B^d denote, respectively, d-sphere and d-ball. The *total* number of all simplices of a finite simplicial complex C and the number of its connected components are denoted by $|C|$ and $b_0(C)$, respectively. ε denotes a small positive number that is assumed to be less than the minimum among different consecutive values of $f(\cdot)$ at the vertices of the mesh.

1 Main Definitions

We begin with some notation and useful definitions.

Let $\mathcal{T}$ be a finite triangulation of some bounded polyhedral region $P \in \mathbf{R}^3$. We assume that P is *simply connected,* i.e. that any circular path in P can be continuously contracted within P to a point. In our consideration P may be thought of as an embedded into $\mathbf{R}^3$ polyhedral topological ball, from which a finite system of disjoint polyhedral topological balls (cavities) is removed.

For any vertex $v \in \mathcal{T}$ its *star* $\mathrm{St}(v)$ is a union of all simplices of T incident to v, and its *link* $\mathrm{Lk}(v)$ consists of all simplices of the simplicial complex $\mathrm{St}(v)$ not incident to v. (Hence, the star of any vertex is a cone over its link.)

Performing if necessary barycentric subdivision of some simplices of $\mathcal{T}$ (see [8]), we can assume that the *link of any boundary* *vertex $v \in \mathcal{T}$ is a two-dimensional PL ball.* Therefore, the star of any $v \in \mathcal{T}$ is a 3D PL ball.

Let $f(\cdot)$ be a PL function on $\mathcal{T}$, i. e. $f(\cdot)$ is specified at the vertices of $\mathcal{T}$ and is continued onto P by linearity.

We assume throughout the paper that $f(\cdot)$ takes different values at different vertices of the triangulation. Such PL functions will be called regular.

Thus, we fix some triangulation $\mathcal{T}$ of P (given by the list of 3D simplices $\mathcal{T} = \{T_1, \ldots, T_n\}$) and specify the values of some regular PL function $f(\cdot)$ at the vertices of $\mathcal{T}$.

Denote by N_0, N_1, and N_2, respectively, the numbers of vertices, edges, and 2D faces in $\mathcal{T}$. Evidently, $N_i = O(|\mathcal{T}|) = O(n)$, $i = 0, 1, 2$, and so n measures the *input size* of our problem[1].

A *level set* of $f(\cdot)$ passing through a vertex $v \in \mathcal{T}$ is defined as follows, $\mathcal{F}_v = f^{-1}(f(v)) \subset P$.

Let V^-, respectively V^+, be the union of all simplices of the star involved on which the values of $f(\cdot)$ are strictly less, respectively greater, than $f(v)$.

Prior to starting the formal definition of CT we remind the notion of a *singularity* for PL functions.

For a smooth function $f(\cdot)$, by definition, *singularities* are the points where the gradient vanishes. The remaining points are called *regular.* Assume now that all the singularities of $f(\cdot)$ are isolated, and let's consider a small enough ball B_v centered at a particular point v. Let c be the number of the connected pieces into which the level set $\mathcal{F}_v = f^{-1}(f(v))$ breaks the B_v. It is easy to see that v is regular iff $c = 2$, and v is a local extremum iff $c = 1$. Furthermore, if $c \geq 3$, then v is called a *saddle point.* In particular, a saddle point is called *simple* if $c = 3$.

This definition of singularities can be extended almost literally to the PL case.

Note, additionally, that the PL singularities can be grouped only at the vertices of the triangulation $\mathcal{T}$. Indeed, if v is not a vertex of $\mathcal{T}$, then the level set $\mathcal{F}_v$ breaks small enough ball centered at v into exactly two parts. In particular, if $\tilde{\mathcal{T}}$ is a triangulation obtained by an arbitrary subdivision of $\mathcal{T}$ and if $f(\cdot)$ is linearly extended to $\tilde{\mathcal{T}}$, then there can be no singularities at the *additional* points of $\tilde{\mathcal{T}}$.

We summarize the discussion above by giving an explicit definition of a PL singularity slightly adapted for our 3D case[2]. Since, by our assumption, the star of any vertex in $\mathcal{T}$ is a 3D ball, the definition below can be applied to all vertices of $\mathcal{T}$ including the boundary ones.

Definition 1 *Let c be the number of connected components into which the level set $\mathcal{F}_v = f^{-1}(f(v))$ breaks* $\mathrm{St}(v)$. *(In our notation $c = b_0(V^-) + b_0(V^+) = b_0(\mathrm{St}(v) \setminus \mathcal{F}_v)$.) Then*

- *if $c = 1$ and $|V^+| = \emptyset$ (or respectively, $|V^-| = \emptyset$), then v is a local maximum (respectively, v is a local minimum);*
- *if $c = 2$, then v is regular;*
- *if $c > 2$, then v is a saddle point, a simple one iff $c = 3$.*

A formal definition of a contour (or Reeb) graph can be given for an arbitrary pair $\{P, f(\cdot)\}$. But at first, for clarity, we adapt the formal definition of the contour graph for our setting.

Contours by definition are the connected components of the level set $\mathcal{F}_v = f^{-1}(f(v)) \subset P$. The connected component of $\mathcal{F}_v$ containing the point v is called the *main component* [2] of $\mathcal{F}_v$ or simply a *contour through the vertex v.* Now delete all main components from the set P and denote the remaining set by P'. The connected components of P' are called *regions.*

Each region R is PL homeomorphic to the product of some closed oriented compact 2D-manifold with boundary $M^2_{g,b}$ (here g is the genus and b is the number of the removed boundary disks, respectively) by an open interval (open 1D-ball). Indeed, consider the sections of R by sweeping level sets $f^{-1}(t)$. By construction, there are no vertices of $\mathcal{T}$ in these sections. One can define the PL homeomorphism between the sections identifying the points along the edges of $\mathcal{T}$ in R and then extending it by linearity.

Definition 2 *A region R is* incident *to the main contour component C iff $\bar{R} \cap C \neq \emptyset$ ($\bar{R}$ denotes the closure of R).*

Definition 3 The contour graph *of a pair $(\mathcal{T}, f)$ is the graph whose nodes are the main contour components and whose edges are regions. Edge-vertex incidences are specified by definition 2.*

Furthermore, each node is labeled by the function value of its countour and each edge (corresponding to some region R) is labeled by a pair (g, b) of nonnegative integers counting the genus and the number of removed boundary disks of the corresponding manifold $M^2_{g,b}$.

Remark 1 Note that our definition of the contour graph is somewhat redundant, since the nodes corresponding to the contours through regular points of $f(\cdot)$ have valency 2, and the edges incident to one such node have identical labels. Removing all these nodes and shrinking the corresponding edges (this operation is often called *contraction* in graph theory) results in a new graph often called the *Reeb graph* or the *component graph.* The nodes of the component graph are in a one-to-one correspondence with the singularities of $f(\cdot)$. Thus, in graph-theoretic terms the contour graph is a *homeomorph* of the component graph. To account for this relation we, following [1], refer to as *supernodes* all the nodes of the contour graph corresponding to singularities of $f(\cdot)$. Similarly, sequences of edges of the contour graph connecting supernodes are called *superarcs.* From the algorithmic point of view, the removal of all regular nodes by contraction of the corresponding edges can be performed in linear time in the size of the graph involved.

Topologically, the Reeb graph is a 1D-complex resulting from indentification of all points of P having the same function value and belonging to the same connected component of the corresponding level set. In other words, if we define the equivalence relation, $\sim_f$, on P by $x \sim_f y$ if x and y lie in the same connected component of a level set $f^{-1}(a)$ for some $a \in \mathbf{R}$, then the Reeb graph is a quotient space $P/\sim_f$[3].

To our knowledge the next statement belongs to the folklore, but we sketch its proof for completeness[4].

Statement 1 *If the fundamental group of P is finite (in particular, if P is simply connected), then for any PL function $f(\cdot)$ the Reeb graph (and, hence, the contour graph) is a tree.*

Proof. At first, note that the quotient space $P/\sim_f$ is an 1D-complex with a finite number of vertices. We prove that the resulting Reeb graph is a tree by checking that its fundamental group is finite. Indeed, the quotient map $\bar{f} : P \to P/\sim_f$ induces a homomorphism of the fundamental groups of the spaces involved. Moreover, it can be checked that this homomorphism is surjective. Therefore, the fundamental group of the Reeb graph is finite, and hence, it is a tree. □

[1] Our complexity estimates don't take into account the *bit length* of the input data. We assume that any arithmetic operation takes a unit of time.

[2] Analogous definitions of PL singularities, but expressed in somewhat different terms, can be found, e.g., in [9, 7]. We are grateful to O.Musin for stimulating discussions on this subject.

[3] We note in passing that the Reeb graph is an important tool in some profound topological works, see e.g., [4]. We are grateful to Y. M. Burman for pointing the corresponding reference.

[4] The proof along the lines of computational geometry for the case when P is a PL 2D-ball can be found in [2].

This tree will be referred to as the CT. Assign orientation to the edges of the CT from lesser to greater values of $f(\cdot)$ in the incident nodes [5]. Two types of the CT nodes (necessarily corresponding to singular vertices) will be of special importance to us in the sequel. *Junctions* are the nodes of the (oriented) CT with out-degree 1 and in-degree greater than 1. *Bifurcations* are the nodes of the CT with in-degree 1 and out-degree greater than 1. Passing a junction corresponds to merging several contours into one and passing a bifurcation results in splitting one contour into several ones.

Now we formally state the Contour Tree Construction Problem (CTCP). The input for the problem consists of the triangulation $\mathcal{T}$ of some simply connected region in $\mathbf{R}^3$ and a PL function $f(\cdot)$ on it. The triangulation is represented by the list of the coordinates of its vertices $\{v_1, \ldots, v_{N_0}\}$ and the list of 3D simplices $\{(v_{i_1}^1, v_{j_1}^1, v_{k_1}^1, v_{m_1}^1), \ldots, (v_{i_n}^n, v_{j_n}^n, v_{k_n}^n, v_{m_n}^n)\}$. The function is specified by its values $f_1, \ldots, f_{N_0}$ at the vertices of $\mathcal{T}$.

The output consists of the description of the CT given by the list of its nodes $\{V_1, \ldots, V_{N_0}\}$ and edges $\{R_1, \ldots, R_{N_0-1}\}$. Vertices are labeled by the corresponding function values and each edge R_j is labeled by a pair (g_j, b_j), which specifies topological parameters (the genus and the number of removed boundary disks) of the underlying contour.

Theorem 1 states the existence of $O(n \log n)$ algorithm for CTCP. Here we informally describe the CT algorithm. A simple idea behind it is an adaptation of the standard for elementary Morse theory (see, e.g., [11]) procedure of recovering of the topological structure of the manifold by tracing topological changes of the "sweeping" level sets of the appropriate smooth functions. It is important that all topological changes of the level sets could happen only while passing through singular values. And all these changes (for a special, yet sufficiently broad, class of the so called Morse functions) are well documented.

Similar consideration can be used in the case of PL functions. Actually, to construct the CT we keep track of the evolution of contours resulting from the evolution of the "sweeping" level set[6] from $-\infty$ to $+\infty$ and in the opposite direction[7].

We remark in passing that several fragments of our text are necessarily very close to the corresponding parts from [1]. We regret all possible confusions.

2 Data structures for the CTCP algorithm

The proof of the Theorem 1 is contained in the description of the *Main Algorithm* (see Section 4) below together with complexity estimates.

Besides the initial information, the algorithm operates with two basic data structures: *a simplex pool* and *a vertex pool.*

The simplex pool represents the level set at a particular moment of a sweep. The first part of it is analogous to the corresponding data structure from [1]. It consists of a list of simplices (called *active*) each storing a pointer to the *name* of some contour and to the corresponding superarc. Note that different contours may get the same names during the *special sweep procedure* at Stage 2 of the algorithm.

Moreover, we need to keep track of the level surface itself modifying it locally while passing vertices[8]. (Incidentally, we need this part of the simplex pool for the last stages of the algorithm only.) For this purpose we support the second part of the simplex pool, which consists of a *level graph.* Namely, we store the current level set as an *embedded* graph. Recall that for any $t \in \mathbf{R}$ a level set $f^{-1}(t)$ consists of a certain number of polyhedral components $S_j(t)$, $j = 1, \ldots, q \leq n$. And for any component $S_j(t)$ its 2-cells (faces of the contour) are the intersections of this contour with simplices of the triangulation; its 1-cells (the edges of the contour) are its intersections with the 2D-faces of simplices of the triangulation; its 0-cells (the vertices of the contour) are its intersections with the edges of simplices of the triangulation. For algorithmic purposes we will consider the contours as the (dual) embedded *level graphs* $G_j, j = 1, \ldots, q$. The vertex set of the level graph is the set of 2-cells of the contour, the edge set is the set of 1-cells. The 2D-faces of the embedding correspond to 0-cells. Moreover, we add some extra *special* vertices corresponding to the removed boundary disks of the level set involved and the corresponding edges. The edge-vertex incidences and the cyclic orders on faces are defined naturally. Thus, we maintain the adjacency structure for each G_j together with the cyclic orders on the edges outcoming from the corresponding vertices of G_j.

Remark 2 Note that the degree of any vertex in G_j except for the special ones is not more than 4. Moreover, the sum of the degrees of all special vertices is bounded by the number of all active simplices. Therefore, for an arbitrary level set the sum of the numbers of vertices and edges of all level graphs $G_j, j = 1, \ldots, q \leq n$ is $O(n)$.

The vertex pool accumulates topological information resulting from passing vertices during the sweep. It consists of a list of records for each vertex of $\mathcal{T}$ including a list of simplices in the star of this vertex; a vertex *type* (regular, local extremum or saddle point) and a pair of numbers $c(v) = (c^-, c^+)$. The records are computed in the course of the first two sweeps of the *Main algorithm.* Informally, $c^{\pm}$ counts the number of *different* contours approaching v from the top (c^+) and respectively, from the bottom (c^-). We define this pair constructively and indicate the corresponding topological changes of the level set resulting from passing v. Recall that all PL-singularities are assumed to be *simple.* So we can easily list all possible cases. Let v be any vertex of the triangulation involved, then:

1. v is regular. Then $c^- = b_0(V^-) = 1, c^+ = b_0(V^+) = 1$.
2. v is a local minimum. Then $c^- = b_0(V^-) = 0, c^+ = b_0(V^+)$.

   In this case a new contour — a two-dimensional PL sphere centered at v — is formed.
3. v is a local maximum. Then $c^- = b_0(V^-) = 1$, $c^+ = b_0(V^+) = 0$.

   If v is a local maximum, then the two-dimensional PL sphere centered at v disappears.
4. v is a (simple) saddle point and $b_0(V^-) = 2$. Then, by definition, $c^+ = b_0(V^+) = 1$.

   Either two different contours C_1 and C_2 are merged to form the *connected sum* $C_1 \# C_2$ (i. e., we remove small open 2-balls B_1 and B_2 from C_1 and C_2, respectively, and then identify $C_1 \setminus B_1$ and $C_2 \setminus B_2$ along their boundary circles) or attach additional *handle* to some contour (thus increasing its *genus* by one). In the first case $c^- = 2$, otherwise, $c^- = 1$.

[5] We use such an orientation of the CT for convenience as it conforms to the standard direction of sweeping level surface in Morse theory. Note that in [1, 2] opposite orientation of the CT is adopted.

[6] Of course this approach is not new. In particular, it was used for the CT and the Reeb graph construction in [5] (for the purposes of computer graphics) and in [1, 2] and [6]. As have been pointed already, we are closely following [1] with some modifications.

[7] This is the main difference of our approach from that of [1], where to construct the CT only one sweep pass is used.

[8] At this point our data structure differs from that of [1], where only the first part of the simplex pool was supported.

5. v is a saddle point and $b_0(V^+) = 2$. By definition, we have $c^- = b_0(V^-) = 1$.

   Either we decrease the genus of some contour by one or split some contour into two. In the first case $c^+ = 1$, otherwise, $c^+ = 2$.

   Topological change resulting from passing this singularity can be modelled by removing from the main component a circular "band" homeomorphic to $S^1 \times B^1$ and then attaching a pair of balls B^2 along the boundary circles of the band. Different topology is realized depending on whether removing of the band disconnects the main component.

Thus, $c = (1, 2)$ for bifurcations and $c = (2, 1)$ for junctions.

3 Basic procedures

The *Main Algorithm* three times passes through all vertices in the order determined by the function values. During every pass the algorithm updates the simplex pool resulting from passing each vertex. The first two sweeps are similar and we refer to them as *special*. They use operations somewhat different from the final sweeping pass. After special sweeps all bifurcations and junctions are identified.

Level graphs $G_j, j = 1, \ldots, q$ and names are considered as a union-find structure allowing the following standard operations *Merge, Split, Find*, and a specific one — *Local Change*.

The *Local Change* updates the graphs of contours resulting from passing through a vertex v. We distinguish *Local Changes* for the special and the final sweeps. The former is easy. In this case the current contour C is represented by a list of already included into it active simplices. We just enlarge it by all simplices from $\mathrm{St}(v)$ that are not in the list.

The *Local Change* for the final sweep is somewhat more complicated, since now the level sets are represented by level graphs. To perform it, we compute in turn "$\pm$" intersections of all simplices from $\mathrm{St}(v)$ with the level sets $f^{-1}(f(v) \pm \varepsilon)$, respectively. The "+" and "−" intersections can be naturally arranged as dual subgraphs $G^\pm$ (G^- being a proper subgraph of the corresponding level graph G). In order to update the change of G resulting from passing v we should delete the G^- from G and properly attach the G^+. This can be done in a straightforward way for nonboundary vertices. The effects resulting from passing boundary vertices are briefly sketched in Section 5. This finishes the description of the *Local Change* for this case.

We also assume that while performing special sweep from top to bottom at any merge vertex v we label as "initial" some simplices from $\mathrm{St}(v)$ belonging to each of the two contours that are merged in v. These labels will be used later while implementing *Split* in the final sweep.

Find-operation, i. e., recovering the name of any simplex in the current level set, takes $O(1)$.

An ingenious idea proposed in [1] to beat the $O(n^2)$ bound for the 2D case is to implement merge in time proportional to the *smaller* of the two contours involved and to implement split in time linear in the size of the *smaller of the resulting contours*. The minimal of the two contours involved is chosen by application of a *tandem* search that is explained below.

Merging of the two contours C_1 and C_2 now reduces to renaming of the *smaller one*. Following this idea, we are able to implement merge in $O(\min(|C_1|, |C_2|)$ time, where, by definition, the length $|C|$ of any contour C is the number of its (active) simplices. To merge the contours on a special sweep, we use *the tandem search* of the lists of active simplices as is explained in [1] for the 2D case. For the final sweep we apply the tandem depth-first vertex scan of the corresponding level graphs G_1 and G_2. Thus, we should alternatingly make one "step" in G_1 in the depth-first search scan (starting from any vertex) of the vertices of G_1 and one "step" in G_2 in the depth-first search scan (again starting from any vertex) of the vertices of G_2, until we scan (and then rename) one of the contours. The smaller contour is renamed by resetting pointers of all its simplices to the corresponding name field of the bigger contour. Since the depth-first search for a level graph $G_j = (V, E)$ takes $O(|V| + |E|)$ time (that is proportional to the the number of active simpices of the contour involved), we obtain the required bound for the time.

A similar consideration can be applied to *Split*. If it is known in advance that a contour C_0 splits into contours C_1 and C_2 then we are able to apply the tandem search. *Split* will be used only during the final pass. We again apply the tandem depth-first scan to corresponding level graphs G_1 and G_2 obtained after *Local Change* starting the process from corresponding initial vertices that have been labeled on the previous special sweep, as was explained above. (It takes $O(|\mathrm{St}(v)|)$ time to identify the initial vertices.) Thus, we find the smaller level graph and reset the name pointer at its vertices.

Next Lemma is clear from the discussion above.

Lemma 1 Local Change *takes time* $O(|\mathrm{St}(v)|)$. Merge *and* Split *procedures take time* $O(|\mathrm{St}(v)| + m)$*, where* m *is the number of renamed simplices.*

Summing these estimates for all vertices, we obtain that the overall time complexity of all *Local Changes* per pass is $O(n)$ (namely, $\sum_{v \in \mathcal{T}} O(|\mathrm{St}(v)|) = O(\sum_{v \in \mathcal{T}} |\mathrm{St}(v)|) = O(|\mathcal{T}|) = O(n)$). Similarly, we get that the time bound for all *Merges* and *Splits* per pass is $O(n) + M$, where M is the total number of renamings during the pass.

It is worth reminding here that the main stumbling block of the described sweeping level set procedure is distinguishing *Split* or *Merge* of the current contours and the genus change. In this respect *Split* is a more subtle operation than *Merge*, since in order to distinguish *Merge* from genus change it is sufficient to compare names of the contours that have come across at the singularity involved. If the names (of the vertices of the corresponding level graphs) are different, then we should merge the contours, otherwise — implement genus change.

As already pointed out, only for the case when the Reeb graph is a tree, we do know how to bypass the analogous problem for *Split*.

4 Main algorithm

Input: the pair $(\mathcal{T}, f(\cdot))$.

Output: CT of this pair.

Stage −1. Sort vertices of $\mathcal{T}$ by values of function $f(\cdot)$. It takes $O(N_0 \log N_0) = O(n \log n)$ operations.

Stage 0. Resolution of Singularities. This stage is described in the Appendix. As an output in $O(n \log n)$ time we get a new $O(n)$ triangulation $\tilde{\mathcal{T}}$ having simple singularities only. Recall that the required CT of $(\mathcal{T}, f(\cdot))$ can be obtained in a linear time by shrinking some edges of the CT for $(\tilde{\mathcal{T}}, \tilde{f}(\cdot))$, where by $\tilde{f}(\cdot))$ we denote the modified PL function.

Stage 1. Initialization. Sort vertices of $\tilde{\mathcal{T}}$ by values of function $\tilde{f}(\cdot)$.

1. Initialize data structures. It takes $O(n)$ time.

2. Compute the numbers $b_0(V^-)$ and $b_0(V^+)$. For any vertex $v \in \mathcal{T}$ it takes $O(|\operatorname{St}(v)|)$ time to identify its type and to compute the numbers $b_0(V^-)$ and $b_0(V^+)$. Thus, as above, summing for all vertices we obtain the overall $O(n)$ complexity estimate for this step. After this step all local extrema are labeled.

Stage 2. Identification of junction and bifurcation singularities. We perform twice *a special sweeping procedure*. At first, we sweep from bottom to top and identify all junctions in the CT. We proceed as follows.

We start the procedure from the global minimum value of $\tilde{f}(\cdot)$ and sequentially sweep the whole range of $\tilde{f}(\cdot)$. For each current vertex v we check names for all simplices in $\operatorname{St}(v)$.

- If v is a local minimum, then we perform *Local Change* and use a new name for simplices of the newly arisen contour.
- If v is a saddle point and $b_0(V^-) = 2$ (or equivalently $b_0(V^+) = 1$), then we identify the names of the contours that have come across at v. If these names are different, we merge these contours by the tandem search of the corresponding lists of simplices, label v as junction, and set $c^-(v) = 2$. Otherwise, we set $c^-(v) = 1$. Then we perform *Local Change*.
- In all other cases we perform *Local Change* only.

Then we perform an analogous procedure sweeping from top to bottom to identify all bifurcations and compute the corresponding c^+ labels. Besides that, as was explained above, in each bifurcation vertex we label some simplices from its star as initial.

The correctness proof and complexity estimates for these procedures follow from Lemma 2 below.

Thus, at the end of this stage all junctions and bifurcations together with corresponding pairs (c^-, c^+) are identified and some simplices are labeled as initial.

Stage 3. Boundary CT construction. Construct in $O(n \log n)$ time the CT for the function restricted to the boundary. It will be referred to as the *boundary CT* and will be used to perform properly *Local Change* while passing boundary vertices in the last sweep. We omit the details and briefly sketch the procedure in Section 5.

Stage 4. CT construction. We make one more sweep from bottom to top.

Again the process starts from the global minimum of $\tilde{f}(\cdot)$. At any current vertex we perform *Local Change.* Furthemore, at vertices labeled on the previous stage as junctions and bifurcations the renaming of the contours involved is made by *Merge* or *Split* procedures. *Merge* is performed before *Local Change*. *Split* is performed after *Local Change.*

All these changes in the topology of contours are translated into the CT structure that is simultaneously grown during this sweep. Let's describe the details of the CT construction. Let v be the current point.

1. If v is a local minimum, then we perform *Local change*, create a new supernode and a new superarc starting from it, and assign the pointer to this superarc to all new active simplices. We set the genus label of the superarc to be zero.
2. If v is a local maximum, then we find the corresponding superarc. It should be done before local modifications. Then we perform *Local change* and disable active simplices from the $\operatorname{St}(v)$. Finally, we create a new supernode and end the current superarc at it.
3. If v is regular, we identify the corresponding superarc, insert a node corresponding to v into the superarc. Then we perform *Local change* and update the list of active simplices.
4. If v is a saddle point and $b_0(V^-) = 2$, then we perform a merge or a genus change depending on the label of v obtained on the previous stage.
   - If v is labeled as a junction then *Merge* of the corresponding level graphs is performed. It consists of a *Local change* that follows topological change of passing the saddle point and renaming using tandem depth-first scan of the level graphs involved, as was explained above. A new junction supernode is created. The superarcs of the merged contours are made incident to this supernode. A new superarc starting from this supernode is created, and we reset all pointers from all active simplices of the merged contour to its name. Genus labels are summed up to get genus label for the new superarc.
   - If v is not a junction, then we perform *Local Change.* Further, we create a new supernode p and make the current superarc e incident to it. We create a new superarc e_1 starting from p. Let g be the genus label of e. We set $g(e_1) = g + 1$.
5. If v is a saddle point and $b_0(V^-) = 1$ (or equivalently, if $b_0(V^+) = 2$), then we perform a split or a genus change depending on the label of v obtained on the previous stage.
   - If v is labeled as a bifurcation, then *Local Change* followed by *Split* are performed. A new bifurcation supernode is created. The superarc of the split contour is made incident to this supernode. Two new superarcs starting from this supernode are created, and we reset all pointers from all active simplices of the arisen contours to their names. To calculate genus labels for the new superarcs we identify the smaller of the newly arisen contours by the tandem depth-first search of the corresponding level graphs and compute its genus g_{small} in time linear in the size of the smaller of the graphs involved. The genus of the bigger contour is now $g_{big} = g - g_{small}$, where g is the genus of the splitted contour.
   - If v is not a bifurcation, then we perform *Local Change.* Further, we create a new supernode p and made the current superarc e incident to it. We create a new superarc e_1 starting from p. Let g be genus label of e. We set $g(e_1) = g - 1$.

It has been already pointed out that the $O(n \log n)$ time complexity estimate for this stage follows from [1], since while sweeping the tree we implement *Merge* and *Split* in time proportional to the smaller of the (merged or, respectively, split) contours. (Namely, it is shown in [1] how to change the tree involved preserving the complexity esimates so that no bifurcations are ancestors to junctions and then apply the tandem search trick, residing in the fact that no simplex can be renamed more than $O(\log n)$ times.)

Final Stage. We shrink in linear time some edges of CT obtained on the previous stage to get the contour tree for the initial pair $(\mathcal{T}, f(\cdot))$.

Now to finish the proof of the Theorem 1 it remains to demonstrate the following Lemma.

Lemma 2 *The special sweeping procedures of the Stage 2 correctly identify all junctions and bifurcations. Moreover, the time complexity of the Stage 2 is $O(n \log n)$.*

Proof. Since the two special sweeps are symmetrical, it is sufficient to consider the first one (from bottom to top) and prove that all junctions are identified.

This is a consequence of the following observation.

Let $\mathcal{Q}$ be any subgraph of the CT. Denote by $\mathcal{A}(\mathcal{Q}) \subset \mathcal{T}$ the set of simplices whose vertices have nonempty intersection with the vertices of $\mathcal{Q}$.

Let v be an arbitrary vertex of $\mathcal{T}$. Denote by $CT^{f(v)}$ a subgraph of the CT obtained by removing from the CT all nodes with function values larger than $f(v)$ and all the corresponding incident edges. Obviously, $CT^{f(v)}$ is a forest. Denote its connected components by $CT_1^{f(v)}, \ldots, CT_s^{f(v)}$. Note that the sets $\mathcal{A}(CT_1^{f(v)}), \ldots, \mathcal{A}(CT_s^{f(v)})$ are always *disjoint*.

Therefore, by induction, in the special sweep the *different* (i.e., bearing different names) contour components resulting from passing an arbitrary vertex v are just $\mathcal{A}(CT_1^{f(v)}), \ldots, \mathcal{A}(CT_s^{f(v)})$. Moreover, if v is a junction in the CT, then passing it in the special sweep, amounts to merging two components necessarily bearing *different* names. This implies that all junctions will be correctly identified.

The complexity of the Stage 2 can be estimated by Lemma 1. Since we are using the tandem search for renaming, we get an $O(n \log n)$ estimate for the total number of all renamings (since no simplex can be present more than $O(\log n)$ times in the smallest set to be renamed). □

5 Boundaries

In this section we briefly sketch how to manage the boundary effects.

Recall that the removed boundary disks correspond to some special vertices of the corresponding level graphs. Thus, as has been already pointed out, in order to take into account boundary effects it is sufficient to describe how to perform *Local Change* of the level graphs while sweeping boundary vertices of $\mathcal{T}$ in the last pass. Recall that we have already calculated the boundary CT, and for a particular level set the removed boundary disks correspond to some contours of the corresponding boundary CT. Loosely speaking, the boundary CT records the evolution of the disks that are removed from the sweeping level set. Thus, when we pass a boundary vertex in the final sweep, we additionally check its type (junction, bifurcation, minimum, maximum) in the boundary and update some special vertices of the corresponding level graph. For example, if the vertex involved is a bifurcation in the boundary CT, then one additional boundary disk should be removed from the level set. Thus, we should add a new special vertex to the corresponding level graph and use local analysis of its star to set necessary incidencies in the level graphs. Finally, we should update labels counting the number of the removed disks from the corresponding contour component(s). Other cases are treated analogously and are performed within the same bound of $O(|\operatorname{St}(v)|)$ operations.

Concluding remarks

We have presented an $O(n \log n)$ method for constructing CT in 3D. An interesting remaining open problem is of constructing subquadratic algorithms not only for contour trees but for general Reeb graphs in 3D.

References

[1] M. van Kreveld, R. van Ostrum, C. L. Bajai, V. Pascucci, and D. Schikore. Contour trees and small seed sets for isosurface traversal. In *Proc. 13th Annu. ACM Sympos. Comput. Geom.*, 1997, 213-220.

[2] M. de Berg and M. van Kreveld. Trekking in the Alps without freezing and getting tired. In *First Annu. European Symp. on Algorithms (ESA '93), Lect. Notes in Computer Science*, **726**, 1993, 121 - 132.

[3] G. Reeb. Sur les point singuliers d'une forme de pfaff completement integrable ou d'une function numerique. *Comptes Rendus Acad. Sciences Paris*, **222**, 1946, 847 - 849.

[4] A. Hatcher and W. Thurston. A presentation for the mapping class group of a closed orientable surface. em Topology, **19**, 1980, 221 - 237.

[5] Yu. Shinagava, T. L. Kunii, and Yu. L. Kergosien. Surface coding based on Morse Theory. *IEEE Computer Graphics and Applications*, **11**, Sept. 1991, 66 - 78.

[6] S. Tarasov and M. Vyalyi. Some PL functions on surfaces are not height functions. In *Proc. 13th Annu. ACM Sympos. Comput. Geom.*, 1997, 113 - 118.

[7] O. R. Musin. On some problems of computational geometry and topology. *Lect. Notes in Math.*, **1520**, 1992, 57 - 80.

[8] C. Rourke and B. Sanderson. Introduction to Piecewise-Linear Topology. Springer-Verlag, NY, 1972.

[9] T. F. Banchoff. Critical points and curvature for embedded polyhedra, *J. Diff. Geom.* **1**, 1967, 245 - 256.

[10] B. Dubrovin, S. Novikov, and A. Fomenko. Modern Geometry. Moscow, 1979.

[11] J. Milnor. Morse Theory. Princeton, NJ, Princeton University Press, 1963.

APPENDIX: Resolution of Singularities

Stages 2 - 4 of the *Main Algorithm* can be applied to the triangulations with simple singularities only. Here we describe an $O(n \log n)$ subdivision algorithm performing resolution of singularities for an arbitrary triangulation.

At first, we state a definition.

Definition 4 *Let $(\mathcal{T}, f(\cdot))$ and $(\tilde{\mathcal{T}}, \tilde{f}(\cdot))$ be two arbitrary triangulations given together with regular PL functions defined on them. Let G and G' be the corresponding Reeb graphs. A pair $(\tilde{\mathcal{T}}, \tilde{f}(\cdot))$ is called a* resolution *of a pair $(\mathcal{T}, f(\cdot))$ if*

1. *$\tilde{\mathcal{T}}$ is an arbitrary* refinement *of the triangulation $\mathcal{T}$ (i.e., we arbitrarily subdivide some simplices of $\mathcal{T}$ preserving the structure of a simplicial complex).*

2. *The values of $\tilde{f}(\cdot)$ are equal to the values of $f(\cdot)$ in the corresponding common vertices.*

3. *$\tilde{f}(\cdot)$ has only simple singularities on $\tilde{\mathcal{T}}$.*

4. *The Reeb graph G is a* contraction *of the Reeb graph G' (i.e., G can be obtained by shrinking some edges of G').*

We prove the following theorem.

Theorem 3 *The resolution for the pair $(\mathcal{T}, f(\cdot))$ can be constructed in $O(n \log n)$ steps.*

Proof We first informally sketch the proof. The refinement will be obtained after three intermediate steps. At the initial step, by the standard barycentric subdivision of $\mathcal{T}$, we obtain a new pair $(\mathcal{T}_0, f_0(\cdot))$, where $f_0(\cdot)$ is a specification of $f(\cdot)$ at the vertices of $\mathcal{T}_0$ so that *all singular points of $f_0(\cdot)$ are nonadjacent in $\mathcal{T}_0$*. Unfortunately, $f_0(\cdot)$ may become nonregular (i.e., some values taken by $f_0(\cdot)$ at the vertices of $\mathcal{T}_0$ may coincide). Now, at the second step, we slightly change the triangulation $\mathcal{T}_0$ to make the function $f_0(\cdot)$ regular. Thus, we get a new pair $(\mathcal{T}_1, f_1(\cdot))$, where $f_1(\cdot)$ is a regular PL function on $\mathcal{T}_1$ and all its singular points are nonajacent. Finally, we make one more subdivision of the triangulation $\mathcal{T}_1$ and additionally alter the values of $f_1(\cdot)$ at the additional points transforming all singularities into simple ones. It is worth noting that after the refinement all singular points are grouped only at the additional points (so that the old singular points are regular in the refinement!).

Now we give a more detailed description of the refinement algorithm. At first, we construct the barycentric subdivision $\mathcal{T}_0$ of $\mathcal{T}$ (see [8]) and extend the function $f(\cdot)$ by linearity. In the process, each simplex is subdivided into 24 ones, so the size grows linearly, and each additional vertex is regular. Note two consequences. Firstly, the contour tree for $\mathcal{T}$ can be obtained by contraction of all additional edges in $\mathcal{T}_0$ (see Remark 1). Secondly, all singular vertices are pairwise nonadjacent in the barycentric subdivision, as all additional vertices are regular.

The remaining problem is the violation of the regularity condition $f_0(v_1) \neq f_0(v_2)$, for different vertices v_1, v_2 of $\mathcal{T}_0$. To maintain this condition we need to perturb the function values at the additional points preserving their regularity. This can be done, for example, as follows. Sort all *different* function values at the vertices of $\mathcal{T}_0$ and find the minimal gap δ between them. Let $v_i \in \mathcal{T}_0$, $i = 1, 2, \ldots m \leq 11n$, be any additional vertex of $\mathcal{T}_0$. Let F be the minimal simplex of the initial triangulation $\mathcal{T}$ containing v_i. Shift v_i to the *nearest* in F vertex $\tilde{v}_i \in F$ having function value $f(\tilde{v}_i) = f(v_i) + i\frac{\delta}{12n}$. The required triangulation $\mathcal{T}_1$ is obtained from $\mathcal{T}_0$ by replacing all additional vertices by their correspondent shifted counterparts. Note that the minimal gap between successive function values at the vertices of $\mathcal{T}_1$ is not less than $\frac{\delta}{12n}$. Also note that in passing from $\mathcal{T}_0$ to $\mathcal{T}_1$ the relative change of any edge length is at most $\frac{\delta}{6n}$.

Thus, we can assume in what follows that singular vertices of the pair $(\mathcal{T}_1, f_1)$ are pairwise nonadjacent. We construct a refinement $\tilde{\mathcal{T}}$ of the triangulation $\mathcal{T}_1$ and make a PL extension of the function $f_1(\cdot)$ onto $\tilde{\mathcal{T}}$. The refinement is obtained by the application of a *vertex resolution scheme* to all singular nonsimple vertices of $\mathcal{T}_1$ and is described below.

The following definition will be useful.

Definition 5 *Let v be any vertex of $\mathcal{T}_1$. Label each vertex u of a 2D simplicial sphere* $\mathrm{Lk}(v)$ *(or 2D simplicial ball if v is a boundary vertex) by a $+$ or a $-$ label depending on the increase or decrease of $f(\cdot)$ along the ray vu. Labeled in this way set* $\mathrm{Lk}(v)$ *will be denoted as* $\mathrm{Lk}_v^{\pm}(f)$.

Define a graph $G_v(f)$ on the vertices of $\mathrm{Lk}_v^{\pm}(f)$ whose edges correspond to 1-simplices of $\mathrm{Lk}_v^{\pm}(f)$ with the same $\pm$-labels. The number $b_0(G_v(f))$ of its connected components equals to the number c of the connected components of the set $\mathrm{St}(v) \setminus \mathcal{F}_v$ introduced in the Definiton 1 above.

We are ready now to describe the vertex resolution scheme. Each simplex of $\mathrm{Lk}_v^{\pm}(f)$ gives an additional vertex to the refined triangulation $\tilde{\mathcal{T}}$. 0-simplices correspond to the opposite ends of the edges incident to v. The additional vertex in $\tilde{\mathcal{T}}$ is the image of this opposite end under relative to v homothety with the dilation ratio $\frac{1}{4}$. Any 1D-simplex is a line segment. The additional point is the image of the middle of this segment under homothety relative to v with the dilation ratio $\frac{1}{3}$. Finally, each 2-simplex is a triangle. Take the barycenter of this triangle and its image under homothety relative to v with the dilation ratio $\frac{1}{3}$. This point is the additional point for the 2D-face. We refer to the additional vertices as d-vertices according to the dimension d of the corresponding face.

Simplices of $\tilde{\mathcal{T}}$ are defined as follows.

Let v be any nonsimple singular point. Take an arbitrary 3D-simplex $S \subset \mathrm{St}(v)$. By construction, it contains three 0-vertices, three 1-vertices and one 2-vertex. S is partitioned into 15 simplices in $\tilde{\mathcal{T}}$. The first one is formed by v and three 0-vertices. The rest of S is now a prism with triangular bases. This prism contains a 2-vertex inside, and each "wall" of the prism contains a 1-vertex. We define the star triangulations of the "walls" using corresponding 1-vertices. Then we define the star triangulations of the prism using 2-vertices. Thus, the prism is triangulated into $1 + 3 \cdot 4 + 1 = 14$ three-dimensional simplices.

We triangulate all simplices from the stars of all nonsimple singularities of $\mathcal{T}_1$ along similar lines. The construction of $\tilde{\mathcal{T}}$ is finished.

Next we define the extension $\tilde{f}(\cdot)$ of $f_1(\cdot)$ onto $\tilde{\mathcal{T}}$. By definition $\tilde{f}(\cdot)$ takes the same values as $f_1(\cdot)$ at points of $\mathcal{T}_1$. To define the values of $\tilde{f}(\cdot)$ at the additional points near some singularity v, choose a linear function $h(\cdot)$ in $\mathbf{R}^3$ which takes different values at all vertices of $\tilde{\mathcal{T}}$ (i.e., $h(\cdot)$ is a regular linear function on $\tilde{\mathcal{T}}$). Let H be the total variation of $h(\cdot)$ on $\tilde{\mathcal{T}}$, i.e., $H = \max_{x,y \in \tilde{\mathcal{T}}}(h(x) - h(y))$. Let δ be the minimal gap between the successive values of $f_1(\cdot)$ at different vertices of $\mathcal{T}_0$. If u is an additional vertex adjacent to v, then $\tilde{f}(u)$ is defined as follows:

$$\tilde{f}(u) = f(v) + \frac{\delta}{2H} h(u).$$

Now we check some properties of the pair $(\tilde{\mathcal{T}}, \tilde{f}(\cdot))$.

1. *$\tilde{\mathcal{T}}$ and $\mathcal{T}_1$ coincide everywhere except in the neighborhoods of nonsimple singular points of $f_1(\cdot)$ on $\mathcal{T}_1$.*

2. *All vertices of $\mathcal{T}_1$ are in $\tilde{\mathcal{T}}$.*

3. *The values of $\tilde{f}(\cdot)$ at the vertices of $\mathcal{T}_1$ are equal to the corresponding values of $f_1(\cdot)$.*

4. *The value of $\tilde{f}(\cdot)$ at any additional vertex differs from the value of $f_1(\cdot)$ at the corresponding initial vertex by one half of δ or less.*

   All items above are valid by construction.

5. *Each simplex of $\mathcal{T}_1$ is subdivided at most once.*

   Recall that all singular vertices are assumed nonadjacent.

6. *$\tilde{\mathcal{T}}$ can be obtained in $O(n \log n)$ time from $\mathcal{T}_1$.*

   The algorithmic procedure is straightforward. The size grows at most 15 times[9] (each simplex is subdivided into 14+1 simplices or does not change).

7. *For a regular or simple singular vertex v from $\mathcal{T}_1$ the equality*

$$b_0(G_v(f_1)) = b_0(G_v(\tilde{f}))$$

holds.

[9] Note that total (including barycentric subdivisions) blowup is $24 \times 15 = 360$ times!

Note that $\mathrm{Lk}_v^{\pm}(\tilde{f})$ refines $\mathrm{Lk}_v^{\pm}(f_1)$. Some vertices are replaced by the clusters of new vertices. The new vertices have the same $\pm$-labels (due to the property 4) and are connected by edges. Contracting these edges leads to the initial set $\mathrm{Lk}_v^{\pm}(f_1)$. This completes the demonstration.

8. *A nonsimple singular vertex v from $\mathcal{T}_1$ becomes regular for $\tilde{f}(\cdot)$ in $\tilde{\mathcal{T}}$.*

   Indeed, $\tilde{f}(\cdot)$ is linear on $\mathrm{St}(v)$ in $\tilde{\mathcal{T}}$.

9. *An additional 0-vertex v in $\tilde{\mathcal{T}}$ is regular or simple.*

   Note that except for one edge, all edges incident to v in $\tilde{\mathcal{T}}$ start from the additional vertices. The function $\tilde{f}(\cdot)$ at these vertices is linear. So there are at most two connected components in $\mathrm{Lk}_v^{\pm}(\tilde{f})$ (relative to the edges with similar end-labels). An exceptional vertex may increase the number of connected components by 1. In this case the vertex v is simple, otherwise it is regular. The argument does not depend on whether v belongs to the boundary or not.

10. *An additional 1-vertex v in $\tilde{\mathcal{T}}$ is regular or simple.*

    If v is a nonboundary vertex, then the set $\mathrm{Lk}_v^{\pm}(\tilde{f})$ has a combinatorial type of the boundary complex of the octahedron If v is a boundary vertex, then the set $\mathrm{Lk}_v^{\pm}(\tilde{f})$ has a combinatorial type of the lateral boundary complex of a pyramid with a quadrangle as a base. In each of these two cases the following reasoning holds.

    Let us arbitrarily label the vertices of $\mathrm{Lk}_v^{\pm}(\tilde{f})$ by $\pm$-labels and delete all edges whose ends are labeled by different labels. If this deletion results in 4 connected components then some vertex q forms a separate component. So all vertices adjacent to q have the same label. Thus, they belong to the same connected component. We come to a contradiction (the remaining vertex should form 2 connected components).

11. *An additional 2-vertex v in $\tilde{\mathcal{T}}$ is regular or simple.*

    At first, note that v is a nonboundary vertex. The set $\mathrm{Lk}_v^{\pm}(\tilde{f})$ has 9 vertices. Three out of them belong to $\mathcal{T}$ (the top of the prism), next three are 0-vertices (the bottom of the prism) and the rest three are 1-vertices (each subdividing the corresponding lateral face of the prism into 4 triangles). There can be at most two connected components of the graph $G_v(\tilde{f})$ of similarly labeled edges on the top of the prism involved (it is a triangle in $\mathrm{Lk}_v^{\pm}(\tilde{f})$). The remaining vertices form at most two connected components, since $\tilde{f}(\cdot)$ is linear on them.

    Suppose that there are 4 connected components in total. Then (up to rotational symmetry and changing all signs) the $\pm$-labels of the top and the bottom are $++-$ and $--+$, respectively. Definitely, a 1-vertex on a lateral face with the cyclic order of labels $+-+-$ is not $+$-labeled (since positive components on the top and the bottom should be different). Similarly, it is not $-$-labeled. We come to a contradiction.

All vertices of $\tilde{\mathcal{T}}$ fall into one of the items 7 – 11. Thus we construct the required resolution and the Theorem follows. □

A Condition Guaranteeing the Existence of Higher-Dimensional Constrained Delaunay Triangulations

Jonathan Richard Shewchuk
School of Computer Science
Carnegie Mellon University
Pittsburgh, Pennsylvania 15213
jrs@cs.cmu.edu

Abstract

Let X be a complex of vertices and piecewise linear constraining facets embedded in E^d. Say that a simplex is *strongly Delaunay* if its vertices are in X and there exists a sphere that passes through its vertices but passes through and encloses no other vertex. Then X has a d-dimensional constrained Delaunay triangulation if each k-dimensional constraining facet in X with $k \leq d-2$ is a union of strongly Delaunay k-simplices.

This theorem is especially useful in E^3 for forming tetrahedralizations that respect specified planar facets. If the bounding segments of these facets are subdivided so that the subsegments are strongly Delaunay, then a constrained tetrahedralization exists. Hence, fewer vertices are needed than in the most common practice in the literature, wherein additional vertices are inserted in the relative interiors of facets to form a conforming (but *un*constrained) Delaunay tetrahedralization.

1 Introduction

Many applications can benefit from triangulations that have properties similar to Delaunay triangulations, but are constrained to contain specified edges or faces. For instance, Delaunay triangulations have desirable properties when used for function interpolation, but a triangulation might be required to conform to specified facets so that discontinuities can be represented. Delaunay triangulations can also serve as meshes that represent objects for rendering or the numerical solution of partial differential equations. In these cases, each triangulation is required to conform to the shape of the object being modeled.

In two dimensions, there are two popular alternatives for creating a Delaunay-like triangulation that conforms to constraints. In either case, the input is a *planar straight line graph* (PSLG) X, which is a set of vertices and segments (constraining edges) as illustrated in Figure 1 (upper left). A PSLG is required to contain both endpoints of every segment it contains, and a segment may intersect vertices and other segments only at its endpoints. A triangulation is sought that includes the vertices of X and respects the segments of X.

The first alternative is to form a *conforming Delaunay triangulation* (Figure 1, lower left). The vertices of X are augmented by additional vertices (sometimes called *Steiner points*) carefully chosen so that the Delaunay triangulation of the augmented vertex set conforms to all the segments—in other words, so that each segment is represented by a contiguous linear sequence of edges of the triangulation. Edelsbrunner and Tan [4] show that any PSLG can be triangulated with the addition of no more than $\mathcal{O}(m^2n)$ augmenting vertices, where m is the number of segments in X, and n is the number of vertices. Not every PSLG requires this many augmenting vertices, but the numbers required in practice may nonetheless seem undesirably large for some applications.

The second alternative is to form a *constrained Delaunay triangulation* (CDT) [1] (Figure 1, lower right). A CDT of X has no vertices not in X, and every segment of X is a single edge of

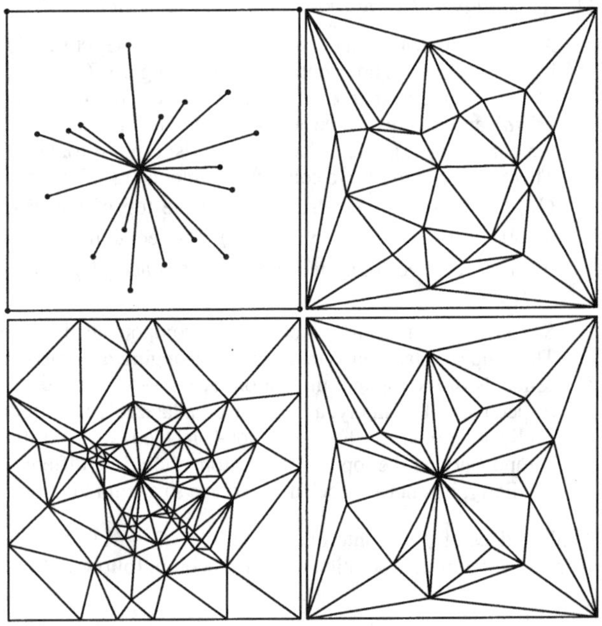

Figure 1: The Delaunay triangulation (upper right) of the vertices of a PSLG (upper left) might not respect the segments of the PSLG. These segments can be incorporated by adding vertices to obtain a conforming Delaunay triangulation (lower left), or by forgoing Delaunay triangles in favor of constrained Delaunay triangles (lower right).

Supported in part by the Advanced Research Projects Agency and Rome Laboratory, Air Force Materiel Command, USAF under agreement number F30602-96-1-0287, in part by the National Science Foundation under Grant CMS-9318163, and in part by a grant from the Intel Corporation.

SCG 98 Minneapolis Minnesota USA

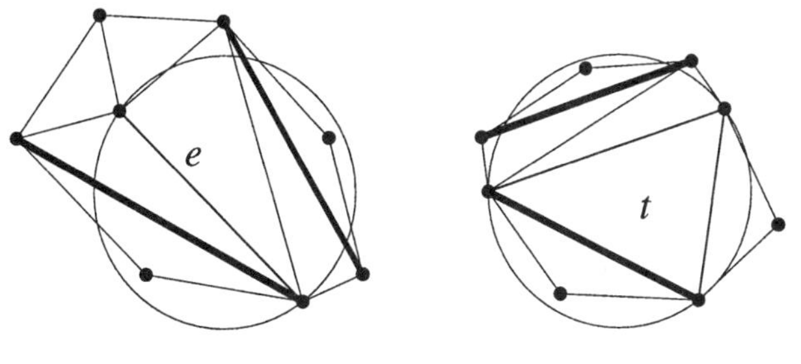

Figure 2: The edge e and triangle t are each constrained Delaunay. Bold lines represent segments.

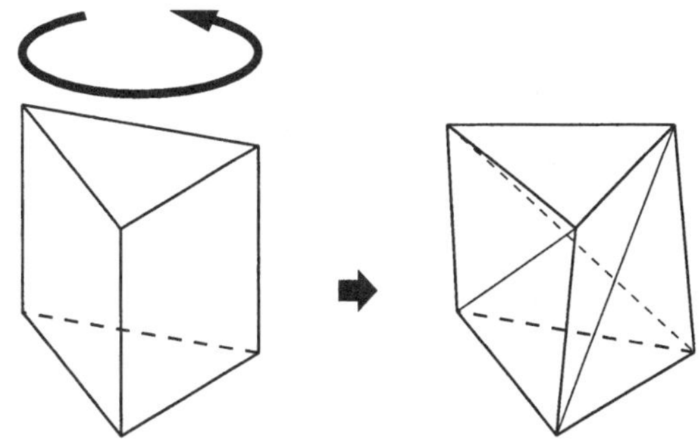

Figure 3: Schönhardt's untetrahedralizable polyhedron (right) is formed by rotating one end of a triangular prism (left), thereby creating three diagonal reflex edges.

the CDT. However, a CDT is not a Delaunay triangulation. In an ordinary Delaunay triangulation, every simplex (of any dimensionality) is *Delaunay*. A simplex is Delaunay if its vertices are in X and there exists a *circumcircle* of the simplex—a circle that passes through all its vertices—that encloses no vertex of X (although any number of vertices is permitted on the circle itself). In a CDT, this requirement is waived, and instead every simplex must either be a segment specified in X or be *constrained Delaunay*. A simplex is constrained Delaunay if it has a circumcircle that encloses no vertex of X that is *visible* from any point in the relative interior of the simplex; and furthermore, the relative interior of the simplex does not intersect any segment. Visibility is occluded only by segments of X.

Figure 2 demonstrates examples of a constrained Delaunay edge e and a constrained Delaunay triangle t. Input segments appear as bold lines. Although there is no empty circle that contains e, the depicted circumcircle of e encloses no vertex that is visible from the relative interior of e. There are two vertices inside the circle, but both are hidden behind segments. Hence, e is constrained Delaunay. Similarly, the circumcircle of t contains two vertices, but both are hidden from the interior of t by segments, so t is constrained Delaunay.

The advantage of a CDT over a conforming Delaunay triangulation is that it has no vertices other than those in X. However, a conforming Delaunay triangulation's triangles are Delaunay, but those of a CDT are not. Nevertheless, a CDT retains many of the desirable properties of Delaunay triangulations. For instance, a two-dimensional CDT maximizes the minimum angle in the triangulation, compared with all other constrained triangulations of X [7].

Unfortunately, CDTs have not been generalized to dimensions higher than two. One reason is that in three or more dimensions, there are polytopes that cannot be triangulated at all without additional vertices. Schönhardt [12] furnishes a three-dimensional example depicted in Figure 3 (right). The easiest way to envision this polyhedron is to begin with a triangular prism. Imagine grasping the prism so that one of its two triangular faces cannot move, while the opposite triangular face is rotated slightly about its center without moving out of its plane. As a result, each of the three square faces is broken along a diagonal *reflex edge* (an edge at which the polyhedron is locally nonconvex) into two triangular faces. After this transformation, the upper left corner and lower right corner of each (former) square face are separated by a reflex edge and are no longer visible to each other within the polyhedron. Any four vertices of the polyhedron include two separated by a reflex edge; thus, any tetrahedron whose vertices are vertices of the polyhedron will not lie entirely within the polyhedron. Therefore, Schönhardt's polyhedron cannot be tetrahedralized without additional vertices.

Ruppert and Seidel [11] add to the difficulty by proving that it is NP-hard to determine whether a polyhedron is tetrahedralizable. Even among polytopes that can be triangulated without additional vertices, there might not always be a triangulation that is in any reasonable sense "constrained Delaunay." Work is needed to determine what features of polytopes make them amenable to being triangulated with Delaunay-like simplices. This paper takes a first step by demonstrating that there is an easily tested condition that guarantees that a CDT exists, for a conservative definition of CDT. Furthermore, there are useful consequences to this guarantee.

The input for which a triangulation is sought is called a *piecewise linear complex* (PLC), following Miller, Talmor, Teng, Walkington, and Wang [8]. Before considering PLCs in their full generality, consider the special case in which a PLC X is a set of vertices and *constraining simplices* in E^d. Each constraining simplex is a k-dimensional simplex (henceforth, k-simplex) having $k+1$ affinely independent vertices, where $1 \leq k \leq d-1$. Each constraining simplex must appear as a face of the final triangulation. (Observe that a PSLG is a two-dimensional PLC.)

PLCs have restrictions similar to those of PSLGs. If X contains a simplex s, then X must contain every lower-dimensional face of s, including its vertices. Any two simplices of a PLC, if one is not a face of the other, may intersect only at a shared lower-dimensional face or vertex.

Several definitions are needed prior to a formal statement of the main result. Say that the visibility between two points p and q in E^d is *occluded* if there is a $(d-1)$-simplex s of X such that p and q lie on opposite sides of the hyperplane that contains s, and the line segment pq intersects s (either in the boundary or in the relative interior of s). If either p or q lies in the hyperplane containing s, then s does not occlude the visibility between them. Simplices in X of dimension less than $d-1$ do not occlude visibility. The points p and q can *see* each other if there is no occluding $(d-1)$-simplex of X.

Let s be a k-simplex (for any k) whose vertices are in X (but s is not necessarily a constraining simplex of X). Let S be a (full-dimensional) sphere in E^d; S is a *circumsphere* of s if S passes through all the vertices of s. If $k = d$, then s has a unique circumsphere; otherwise, s has infinitely many circumspheres. The simplex s is *Delaunay* if there is a circumsphere S of s such that no vertex of X lies inside S. The simplex s is *strongly Delaunay* if there is a circumsphere S of s such that no vertex of X lies inside *or on* S, except the vertices of s. Every 0-simplex is trivially strongly Delaunay.

The simplex s is *constrained Delaunay* if no constraining simplex of X intersects the interior of s unless it contains s in its entirety, and there is a circumsphere S of s such that no vertex of X inside S is visible from any point in the relative interior of s.

A PLC X is said to be *ridge-protected* if each constraining simplex in X of dimension $d-2$ or less is strongly Delaunay.

The main result of this paper is that if X is ridge-protected, and if no $d+2$ vertices of X lie on a common sphere, then the

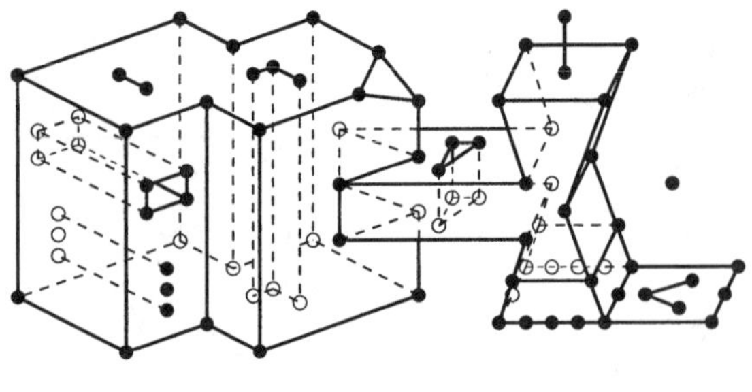

Figure 4: Any facet of a PLC may contain holes, slits, and vertices, which may be used to enforce the presence of specific faces or to support intersections with other facets.

constrained Delaunay d-simplices defined on the vertices of X collectively form a triangulation of X. This triangulation is called the *constrained Delaunay triangulation* of X. The condition that X be ridge-protected holds trivially in two dimensions (hence the success of two-dimensional CDTs), but not in three or more.

Note that it is not sufficient for each lower-dimensional constraining simplex to be Delaunay; if Schönhardt's polyhedron is specified so that all six of its vertices lie on a common sphere, then all of its edges (and its faces as well) are Delaunay, but it still does not have a tetrahedralization. It is not possible to place the vertices of Schönhardt's polyhedron so that all three of its reflex edges are strongly Delaunay (though any two may be).

If some $d + 2$ vertices of X are cospherical, the existence of a CDT can be established through perturbation arguments (not presented here), but the CDT is not necessarily unique. There may be constrained Delaunay d-simplices whose interiors are not disjoint. Some of these simplices must be omitted to yield a proper triangulation.

It is appropriate, now, to consider a more general definition of PLC. To illustrate the limitation of the definition given heretofore, consider finding a tetrahedralization of a three-dimensional cube. If each square face is represented by two triangular constraining simplices, then the definition given above requires that the diagonal edge in each face be a constraining edge of X. However, the result of this paper may be obtained even if these edges are not strongly Delaunay. Hence, a PLC may contain constraining *facets*, which are more general than constraining simplices. Each facet is a polytope of any dimension from one to $d-1$, possibly with holes and lower-dimensional facets inside it. A k-simplex s is said to be a *constraining simplex in* X if s appears in the k-dimensional CDT of some k-facet in X. If X is ridge-protected, then the CDT of a facet is just its Delaunay triangulation, because the boundary simplices of each facet are strongly Delaunay.

Figure 4 illustrates a three-dimensional PLC. As the figure illustrates, a facet may have any number of sides, may be nonconvex, and may have holes, slits, or vertices inside it. Two facets, if one is not a boundary facet of the other, may intersect only at shared lower-dimensional facets and vertices. See Miller et al. [8] for a complete list of restrictions.

The advantage of employing whole facets, rather than individual constraining simplices, is that only the lower-dimensional boundary facets of a $(d-1)$-facet need be included in the PLC (including interior boundaries, at holes and slits). The lower-dimensional faces that are introduced in the relative interior of a $(d-1)$-facet when it is triangulated do not need to be strongly Delaunay for this paper's result to hold.

This advantage does not extend to lower-dimensional facets, because a ridge-protected k-facet, for $k \leq d-2$, must be composed of strongly Delaunay k-simplices, and it is easy to show that any lower-dimensional face of a strongly Delaunay simplex is strongly Delaunay. Hence, in a ridge-protected PLC, all simplicial faces in the Delaunay triangulation of a facet are required to be strongly Delaunay, except for the faces in the triangulation of a $(d-1)$-facet that are not in the facet's boundary.

Testing whether a PLC is ridge-protected is straightforward. Form the Delaunay triangulation of the vertices of the PLC. If a constraining simplex s is missing from the triangulation, then s is not strongly Delaunay. Otherwise, s is Delaunay; testing whether s is strongly Delaunay is a local operation equivalent to determining whether the dual face of s in the corresponding Voronoi diagram is nondegenerate.

Why is it useful to know that ridge-protected PLCs have CDTs? Although a given PLC X may not be ridge-protected, it can be made ridge-protected by splitting simplices that are not strongly Delaunay into smaller simplices, with the insertion of additional vertices. The result is a new ridge-protected PLC Y, which has a CDT. The CDT of Y is not a CDT of X, because it has vertices that X lacks, but it is what I call a *conforming constrained Delaunay triangulation* (CCDT) of X: conforming because of the additional vertices, and constrained because its simplices are constrained Delaunay (rather than Delaunay).

One advantage of a CCDT over a conforming Delaunay triangulation is that the number of additional vertices needed is generally smaller. Once a PLC X has been augmented to form a ridge-protected PLC Y, the Delaunay triangulation of the vertices of Y contains all the constraining simplices of Y of dimension $d-2$ or smaller (because they are strongly Delaunay), but does not necessarily respect the $(d-1)$-simplices of Y. The result of this paper implies that the CDT of Y may be formed without adding any more vertices to Y. This idea stands in apposition to the most common method of recovering unrepresented facets in three-dimensional Delaunay-based mesh generation algorithms [14, 6, 15, 10], wherein additional vertices are inserted within the facet (for instance, where an edge of the tetrahedralization intersects the missing facet).

Because of the large disparity between the number of vertices required for two-dimensional constrained and conforming Delaunay triangulations, I suspect that asymptotically more vertices are needed (in the worst case) to produce a Delaunay tetrahedralization that conforms to a set of segments and facets in E^3 than are needed to produce a Delaunay tetrahedralization that merely conforms to the segments. Because the latter can be converted into a CCDT, the former may not be needed.

The utility of the CCDT is further buttressed by a practical algorithm for three-dimensional mesh generation that I describe elsewhere [13]. One variant of the Delaunay refinement algorithm described therein uses the constrained Delaunay property of the tetrahedra it produces to prove guaranteed bounds on the quality of the final mesh. The results of this paper are the necessary underpinnings of that algorithm.

2 Proof of Existence of the CDT

Throughout the proof, the terms "simplex" and "convex hull" refer to closed, convex sets of points; hence, they include all the points on their boundaries and in their interiors.

Let X be a ridge-protected PLC in E^d, and suppose that some subset of $d + 1$ vertices in X is affinely independent. The set of constrained Delaunay d-simplices defined on X forms a triangulation of the vertices of X if X contains no $d + 2$ vertices that lie on a common sphere. This fact may be proven in two steps: first, by showing that every point in the convex hull of X is contained in some constrained Delaunay d-simplex of X; second, by showing that any two constrained Delaunay d-simplices have disjoint interiors, and that the d-simplices are aligned on their $(d-1)$-faces. Hence, the union of the constrained Delaunay d-simplices is the convex hull of X, and the constrained Delaunay d-simplices (and

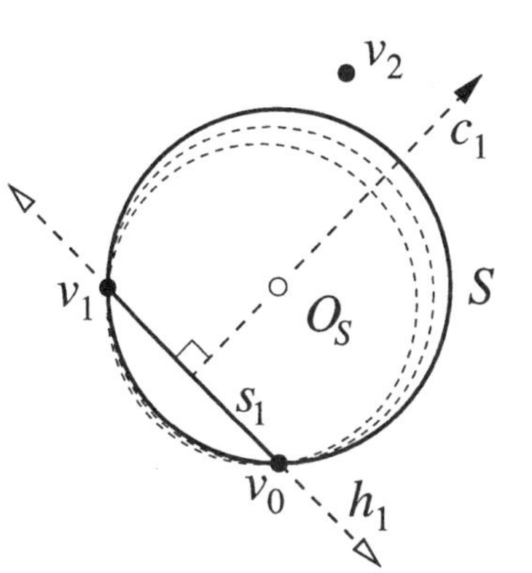

Figure 5: An empty sphere S that circumscribes s_1, expanding in search of another vertex v_2.

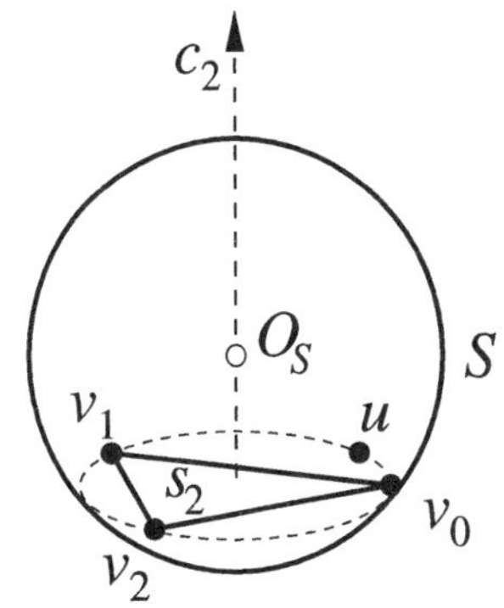

Figure 6: The point u lies on S, but is ignored because it is not affinely independent from $\{v_0, v_1, v_2\}$.

their faces) form a simplicial complex. Only the second step requires the assumption that no $d+2$ vertices are cospherical.

Both steps are based on a straightforward procedure for "growing" a simplex, vertex by vertex. The procedure is presumed to be in possession of a constrained Delaunay k-simplex s_k, whose vertices are $v_0, \ldots, v_k$, and produces a constrained Delaunay $(k+1)$-simplex s_{k+1} that possesses a face s_k and an additional vertex v_{k+1}.

Because s_k is constrained Delaunay, it has a circumsphere S that encloses no vertex visible from the interior of s_k. The growth procedure expands S like a bubble, so that its center O_S moves in a direction orthogonal to the k-dimensional hyperplane h_k that contains s_k, as illustrated in Figure 5. Because O_S is moving in a direction orthogonal to h_k, O_S remains equidistant from the vertices of s_k, and hence it is always possible to choose a radius for S that ensures that S continues to pass through all the vertices of s_k. The expansion ends when S contacts an additional vertex v_{k+1} that is visible from some point in the interior of s_k. Once it is found, the forthcoming Theorem 5 guarantees that v_{k+1} is visible from every point in s_k. The simplex s_{k+1} defined by v_{k+1} and the vertices of s_k is constrained Delaunay, a fact attested to by the circumsphere S, which can be shown (again by Theorem 5) to enclose no vertex visible from any point in the interior of s_{k+1}.

The motion of the sphere center O_S is governed by a *direction vector* c_k, which is constrained to be orthogonal to h_k, but may otherwise be specified freely. In the limit as O_S moves infinitely far away, S will approach the $(d-1)$-dimensional hyperplane that contains s_k and h_k and is orthogonal to c_k. The region enclosed by S will approach an open half-space bounded by this hyperplane. A point is said to be *above* h_k if it lies in this open half-space. Any vertex outside S that S comes in contact with while expanding must lie above h_k; the portion of S below h_k is shrinking toward the inside of S.

A special case occurs if the sphere S already contacts a vertex u not in s_k before S begins expanding. If u is affinely independent of s_k, and is visible from the interior of s_k, then it is immediately accepted as v_{k+1}. Otherwise, u is ignored. (The affinely dependent case can only occur for $k \geq 2$; for instance, when a triangle grows to become a tetrahedron, as illustrated in Figure 6, there may be ignored vertices on the triangle's planar circumcircle.)

If the procedure is to succeed, c_k must satisfy the following *visibility hypothesis*. The open half-space above h_k must contain a vertex of X that is visible from the interior of s_k. In the case of unconstrained Delaunay triangulations, this simply means that the half-space contains a vertex of X. Theorem 4, to follow, clarifies the case of constrained Delaunay triangulations by showing that if X is ridge-protected and there is a vertex above h_k, then there is a vertex above h_k that is visible from an arbitrarily chosen point in s_k. Hence, for a ridge-protected PLC, the visibility hypothesis holds in the constrained case whenever it holds in the unconstrained case.

This growth procedure is the basis for the following proof.

Theorem 1 *Let p be any point in the convex hull of X. If X is ridge-protected, then some constrained Delaunay d-simplex of X contains p.*

Proof: The proof is based on a two-stage constructive procedure similar to one used by Fortune [5] to prove the existence of unconstrained Delaunay triangulations. The first stage finds an arbitrary constrained Delaunay d-simplex, and involves $d+1$ steps, numbered zero through d. For step zero, define a sphere S whose center is an arbitrary vertex v_0 in X, and whose radius is zero. The vertex v_0 seeds the starting simplex $s_0 = \{v_0\}$.

During the remaining steps, illustrated in Figure 7, S expands according to the growth procedure described above. At step $k+1$, the direction vector c_k is chosen so that some vertex of X lies above h_k, and thus the visibility hypothesis is established. Such a choice of c_k is always possible because X contains $d+1$ affinely independent vertices, and thus some vertex is not in h_k. Theorem 4 guarantees that some vertex is visible above h_k from the interior of s_k. If there are several vertices visible, let v_{k+1} be the first such vertex contacted by the expanding sphere, so that no vertex inside the sphere is visible from the interior of s_k. Theorem 5 shows that the new simplex s_{k+1} is constrained Delaunay.

After step d, s_d is a constrained Delaunay d-simplex. If s_d contains p, the procedure is finished; otherwise, the second stage begins.

Consider the directed line segment qp, where q is an arbitrary point in the interior of s_d chosen so that qp does not intersect any simplices (defined on the vertices of X) of dimension $d-2$ or smaller, except possibly at the point p. (Such a choice of q is always possible; the set of points in s_d that *don't* satisfy this condition has measure zero.) The second stage "walks" along the segment qp, traversing a sequence of constrained Delaunay d-simplices that intersect qp (illustrated in Figure 8), until it finds one that contains p.

Wherever qp exits a d-simplex, the next d-simplex may be constructed as follows. Let s be the face through which qp exits the current d-simplex. The $(d-1)$-simplex s is either a constraining simplex in X, or is constrained Delaunay. s is the base from which another constrained Delaunay d-simplex is grown, with the direction vector chosen to be orthogonal to s and directed out of the previous d-simplex. Is the visibility hypothesis satisfied? Clearly, the point p is above s. Some vertex of X must lie above s, because if none did, the convex hull of X would not intersect the half-space

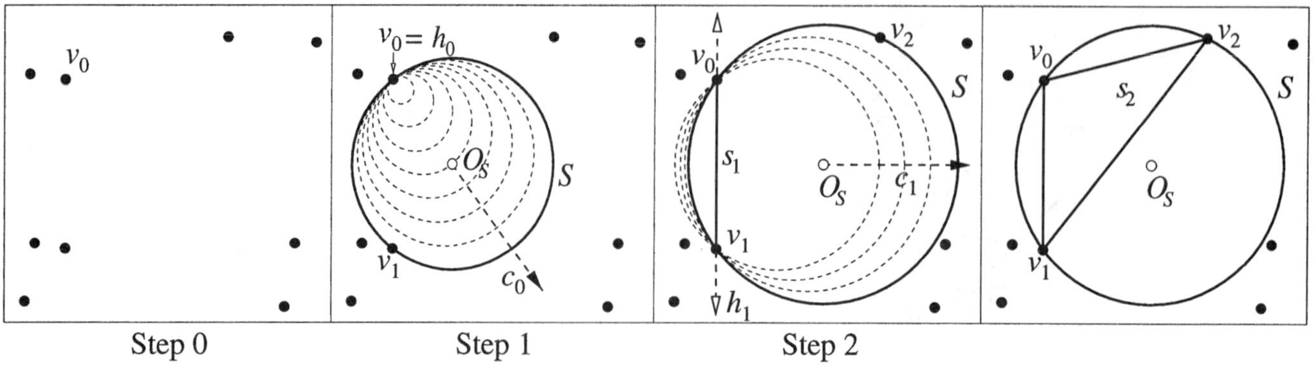

Figure 7: Growing a Delaunay simplex. If $d > 2$, c_2 will point directly out of the page.

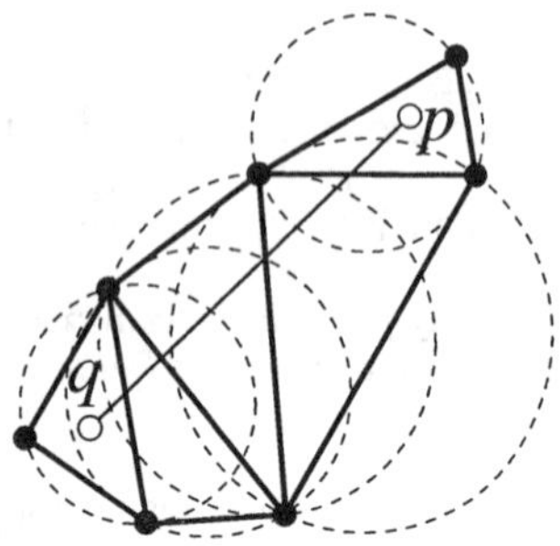

Figure 8: Walking to the d-simplex that contains p.

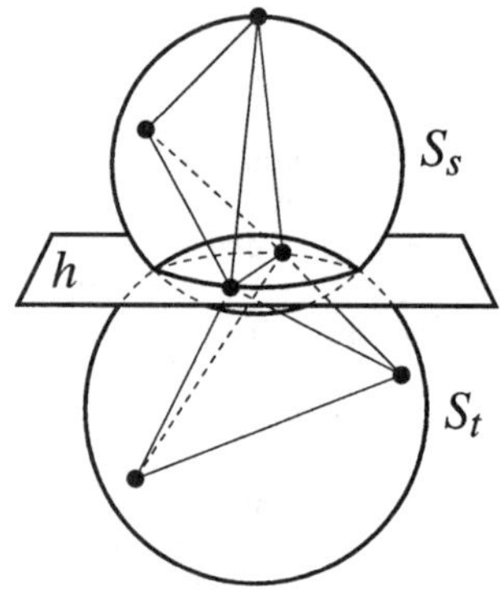

Figure 9: The constrained Delaunay d-simplices s and t intersect at a lower-dimensional shared face (in this illustration, an edge).

above s and thus would not contain p. Hence, by Theorem 4, some vertex above s is visible from the interior of s.

If s is constrained Delaunay, a new constrained Delaunay d-simplex may be formed as in the first stage. If s is a constraining $(d-1)$-simplex in X, but is not constrained Delaunay, then the growth procedure must be modified slightly to account for the fact that s does not have a circumsphere that encloses no vertex visible from s. In this case, the "expanding" sphere S begins with its center infinitely far below s (so that the "inside" of S is the open half-space below s), and the portion of S above s is initially empty but expands as usual. A different proof (Theorem 8) is needed to show that the new d-simplex is constrained Delaunay.

Because the new d-simplex is above s, it is distinct from the previous d-simplex. Because each successive d-simplex intersects a subsegment of qp having nonzero length, each successive $(d-1)$-face intersects qp closer to p, and thus no simplex is visited twice. Since only a finite number of simplices can be defined (over a finite set of vertices), the procedure must terminate; and since the procedure will not terminate unless the current d-simplex contains p, there exists a Delaunay d-simplex that contains p. ■

This procedure is recognizable as the basis for a well-known algorithm, called *gift-wrapping*, *graph traversal*, or *incremental search*, for constructing Delaunay triangulations [2]. Gift-wrapping begins by finding a single Delaunay d-simplex, which is used as a seed upon which the remaining Delaunay d-simplices crystallize one by one. Each $(d-1)$-face of a Delaunay d-simplex is used as a base from which to search for the vertex that serves as the apex of an adjacent d-simplex. Theorem 1 shows that gift-wrapping can be used to produce CDTs as well, at least for ridge-protected PLCs.

As every point in the convex hull of X is contained in a constrained Delaunay d-simplex, it remains only to show that the set of constrained Delaunay d-simplices do not occupy common volume or fail to intersect neatly. It is only here that the assumption that no $d+2$ vertices are cospherical is needed.

Theorem 2 *Suppose that no $d+2$ vertices of X lie on a common sphere. Then the constrained Delaunay d-simplices of X (and their faces) collectively form a simplicial complex.*

Proof: First, I show that constrained Delaunay d-simplices have disjoint interiors. Suppose for the sake of contradiction that some point p lies in the interior of two distinct constrained Delaunay d-simplices s and t. Because s and t are constrained Delaunay, every vertex of s and t is visible from p.

Let S_s and S_t be the circumspheres of s and t. S_s and S_t cannot be identical, because s and t each have at least one vertex not shared by the other; if S_s and S_t were the same, at least $d+2$ vertices would lie on S_s. S_s cannot enclose S_t, nor the converse, because S_s and S_t enclose no vertices visible from p. Hence, either S_s and S_t are entirely disjoint (and thus so are s and t), or their intersection is a $(d-1)$-dimensional circle or point and is contained in a $(d-1)$-dimensional hyperplane h, as Figure 9 illustrates. (If S_s and S_t intersect at a single point, h is chosen to be tangent to both spheres.) Without loss of generality, suppose h is oriented horizontally, with the center of S_s directly above the center of S_t. Because p lies in the interiors of s and t, either some vertex of s lies below h, or some vertex of t lies above h. In the former case, there is a vertex of s inside S_t that is visible from a point (p) inside t, so t is not constrained Delaunay. In the latter case, there is a vertex of t inside S_s that is visible from a point inside s, so s is not constrained Delaunay. Either case implies a contradiction, so s and t have disjoint interiors and can intersect only at their boundaries.

Recall from the proof of Theorem 1 that if s is a constrained Delaunay d-simplex, then from any $(d-1)$-face of s not in the

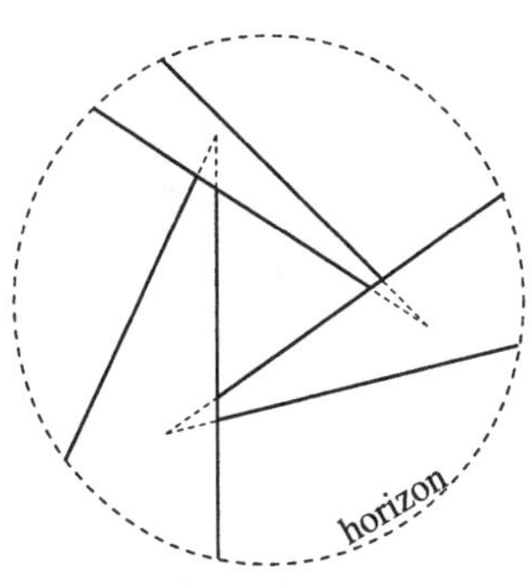

Figure 10: Spherical projection of the half-space above your vantage point.

boundary of the convex hull of X, one can find an adjoining constrained Delaunay d-simplex that shares that $(d-1)$-face. No other constrained Delaunay d-simplex may occupy the same volume. It follows that constrained Delaunay d-simplices meet neatly on their $(d-1)$-faces; and since the union of all constrained Delaunay d-simplices is the convex hull of X, the constrained Delaunay d-simplices collectively form a simplicial complex. ■

It is now straightforward to show that every constraining simplex in X appears as a face of the CDT. Simply note that from any constraining simplex in X, one may grow a constrained Delaunay d-simplex whose faces include the constraining simplex.

The remainder of this section is devoted to completing the proof of Theorem 1. The first step is to prove the visibility hypothesis. One potential difficulty is illustrated (for the three-dimensional case) in Figure 10. Imagine that you are standing on a $(d-1)$-simplex, scanning the half-space above the simplex for a vertex that can serve as the apex of a d-simplex. Looking up into the sky, you see the three illustrated $(d-1)$-facets, each of which occludes the apical vertex of another; the remaining vertices of these facets are hidden below the horizon (in the half-space below you). Hence, no vertex in the half-space is visible from your vantage point.

To prove the existence of a constrained Delaunay triangulation, one must show that this possibility is precluded if X is ridge-protected. In Figure 10, observe that the inner edges of the three facets form a cycle of overlapping simplices. The proof operates by attacking the possibility that the strongly Delaunay faces that bound the occluding $(d-1)$-facets can form such a cycle.

Let p be an arbitrary vantage point in E^d. Let s and t be any two strongly Delaunay simplices. Say that s *overlaps* t from the viewpoint p if some point of s not shared by t lies directly between p and t. In other words, there exists a point p_s of s and a point p_t of t such that $p_s \notin t$ and p_s lies between p and p_t.

Lemma 3 *From any fixed vantage point p, X contains no cycle of consecutively overlapping strongly Delaunay constraining simplices.*

Proof: For each strongly Delaunay simplex s of X, let S_s be a sphere that passes through the vertices of s, but neither passes through nor encloses any other vertex. Hence, S_s attests to the fact that s is strongly Delaunay. There may be many such spheres, but one should be chosen arbitrarily and used for all pairwise comparisons of simplices herein.

Let O_s and r_s be the center and radius of S_s, respectively. Consider the function $\Psi_p(s) = |pO_s|^2 - r_s^2$ defined over the set of strongly Delaunay simplices relative to a vantage point p. The proof stands on the fact that if s and t are strongly Delaunay simplices and s overlaps t from the viewpoint p, then $\Psi_p(s) < \Psi_p(t)$. Hence, the overlap relation among strongly Delaunay simplices defines a partial order, and no cycle of consecutively overlapping strongly Delaunay simplices is possible.

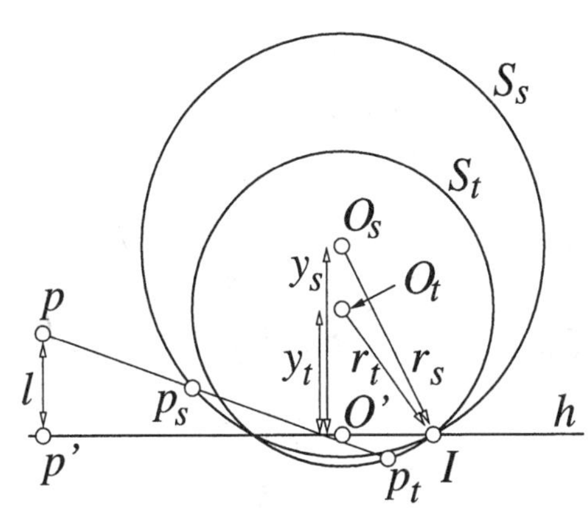

Figure 11: A two-dimensional cross-section of E^d that passes through p, O_s, and O_t. The hyperplane h is orthogonal to O_sO_t, and thus extends orthogonally out of the page. Note that p_s and p_t do not necessarily lie in this cross-section; they are depicted here as a reminder of why p must lie above h.

It remains only to show that $\Psi_p(s) < \Psi_p(t)$ if s overlaps t from the viewpoint p. Because s contains a point that t lacks (namely p_s), s must also possess a vertex that t lacks; hence, S_s (which passes through this vertex) and S_t (which does not) are distinct. If S_s and S_t intersect in a $(d-1)$-dimensional circle, let h be the $(d-1)$-dimensional hyperplane that passes through the circle of intersection (recall Figure 9). Otherwise, let h be the $(d-1)$-dimensional hyperplane tangent to S_s at the point of S_s nearest S_t.

Assume without loss of generality that h is oriented horizontally, with the center of S_s directly above the center of S_t. Observe that S_s encloses any portion of S_t above h, and S_t encloses any portion of S_s below h. Because s and t are strongly Delaunay, every vertex of s, and hence every point in s, lies in or above h; and every point in t lies in or below h. Every vertex of s or t that lies in h must belong to both s and t; hence, every point of s or t that lies in h belongs to both. Recall that there exists a point p_s of s and a point p_t of t such that $p_s \notin t$ and p_s lies between p and p_t. The point p_s must lie strictly above h, and p_t lies in or below h, so p lies strictly above h.

Consider Figure 11, which depicts the two-dimensional cross-section through E^d that passes through p and the centers O_s and O_t of the spheres S_s and S_t. Let y_s be the signed height of O_s above h; y_s is negative if O_s is below h. Similarly, let y_t be the signed height of O_t above h, and let l be the signed height of p above h. Let r_s and r_t be the radii of S_s and S_t, respectively. Let p' be the orthogonal projection of p onto h (in other words, pp' is orthogonal to h), and let O' be the orthogonal projection of O_s onto h, which coincides with the orthogonal projection of O_t onto h. Let I be any point of intersection of S_s, S_t, and h; if the spheres do not intersect, let $I = O'$. Observe that if S_s and S_t intersect, then $r_t^2 = y_t^2 + |O'I|^2$; otherwise, $r_t^2 < y_t^2$ and $|O'I| = 0$, as Figure 12 illustrates.

From these relationships and the Pythagorean Theorem, we have

$$\begin{aligned}
\Psi_p(t) - \Psi_p(s) &= |pO_t|^2 - |pO_s|^2 - r_t^2 + r_s^2 \\
&\geq [(y_t - l)^2 + |p'O'|^2] \\
&\quad -[(y_s - l)^2 + |p'O'|^2] \\
&\quad -[y_t^2 + |O'I|^2] + [y_s^2 + |O'I|^2] \\
&= 2l(y_s - y_t).
\end{aligned}$$

Because l and $y_s - y_t$ are both positive, $\Psi_p(t) > \Psi_p(s)$. ■

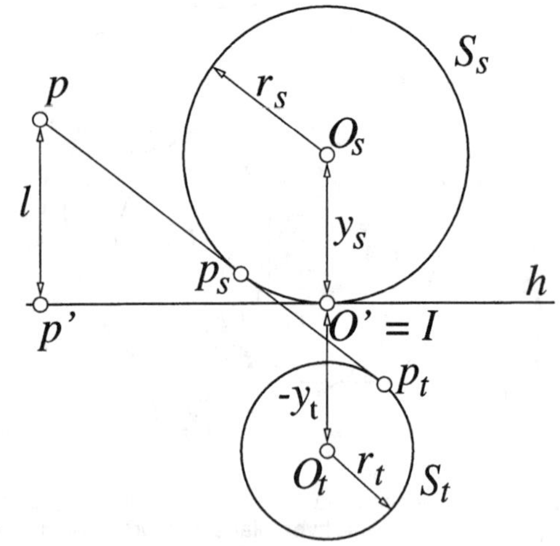

Figure 12: The case where S_s and S_t do not intersect.

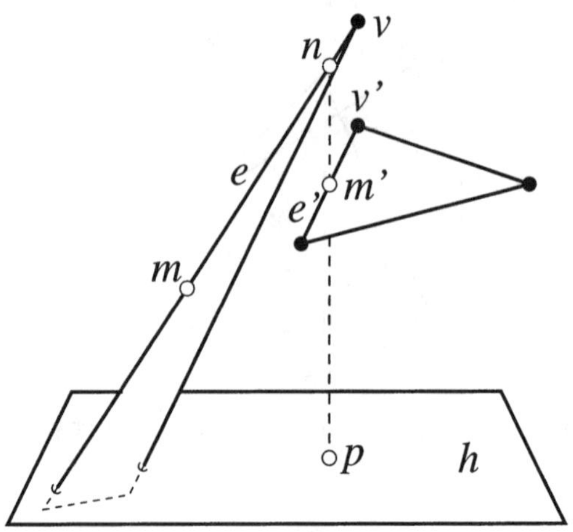

Figure 13: Some vertex above h is visible from p.

Edelsbrunner [3] presents an acyclicity theorem that is nearly identical[1] to Lemma 3. The proof given here is much simpler.

The impossibility of a cycle of overlapping strongly Delaunay simplices is the key to proving the visibility hypothesis for ridge-protected PLCs.

Theorem 4 (Visibility Hypothesis) *Let h be a $(d-1)$-dimensional hyperplane, and let p be a point in h. Suppose there is at least one vertex of X in the open half-space above h. If X is ridge-protected, then at least one vertex of X in the open half-space above h is visible from p.*

Proof: There is at least one vertex above h. Either it is visible from p and the result follows, or there must be a $(d-1)$-facet of X occluding its visibility. Hence, at least a portion of the boundary of at least one $(d-1)$-facet lies above h. It follows that some simplex e (of dimension $d-2$ or less) in the boundary of a $(d-1)$-facet is at least partly visible from p in the half-space above h. (The only possible alternative is a single facet blotting out the whole sky, which would only be possible if facets could be curved). Let m be a point in e that is above h and visible from p, as illustrated in Figure 13. Assume without loss of generality that m is in the relative interior of e by choosing e to be of as low dimension as possible; for instance, if m lies in an edge of a tetrahedron in E^5, choose e to be the edge and not the tetrahedron. If e is a vertex the result follows, so assume e is of dimension at least one.

The proof proceeds by "walking" from m toward a vertex of e, replacing e with any simplex that occludes its visibility from p, and continuing the walk on the new simplex. This process is repeated until a vertex visible from p is found.

Begin by observing that because e is a simplex that contains m, e must have at least one vertex v above h. (The other vertices of e might lie on or below h.) If v is visible from p, the result follows. Otherwise, m is visible from p but v is not.

Let n be the point nearest m on the line segment mv that is not visible from p. (In other words, n is the first occluded point encountered when walking from m to v.) The segment pn must intersect some $(d-1)$-facet of X at some point m'. (If there are several $(d-1)$-facets occluding the view from p to n, consider only the facet that intersects pn closest to p, so that m' is visible from p). Since n is the first occluded point on mv, m' must lie in the boundary of the occluding facet. Generally, m' will lie in the interior of a $(d-2)$-simplex bounding the facet, but in some cases m' may lie in a lower-dimensional boundary; let e' be the lowest-dimensional boundary simplex that contains m'. If e' is a vertex, the proof is complete, so assume e' is not a vertex.

Because e and e' lie in boundaries of facets in X, e and e' are strongly Delaunay. Clearly, e' overlaps e from the viewpoint p.

Let v' be a vertex of e' that is above h. If v' is visible from p, the proof is complete. Otherwise, find another facet-bounding simplex e'' of dimension $d-2$ or less that overlaps e' from the viewpoint p, and search for one of its vertices. Repeat the process until a vertex visible from p is found. Because X contains only a finite number of constraining simplices, and because Lemma 3 rules out the possibility of a cycle of consecutively overlapping strongly Delaunay simplices, the search must end. ■

The visibility hypothesis tells us that if the open half-space above a k-simplex s_k contains a vertex, then at least one candidate vertex u above s_k is visible from the interior of s_k. Can we find a constrained Delaunay simplex s_{k+1} by taking the convex hull of s_k and u? There are two catches. First, the candidate vertex u, although visible from at least one point inside s_k, might not be visible from *all* points inside s_k. Second, some vertex visible from inside s_{k+1} might prevent s_{k+1} from being constrained Delaunay. The following theorem shows that these problems do not arise in the case that matters: when a sphere circumscribing s_k and u encloses no vertex visible from the interior of s_k.

Theorem 5 *Let X be a ridge-protected PLC. Let s_k be a constrained Delaunay k-simplex, for some $k < d$. Let u be a vertex of X that is affinely independent of s_k and is visible from some point p in the relative interior of s_k. Suppose that there is a sphere S that passes through u and all the vertices of s_k, and encloses no vertex of X that is visible from p. Let s_{k+1} be the convex hull of s_k and u. Then no constraining facet of X intersects the relative interior of s_{k+1} unless it contains s_{k+1}. Furthermore, no vertex inside S is visible from any point in the relative interior of s_{k+1}; hence, s_{k+1} is constrained Delaunay.*

The proof requires the following two lemmata.

Lemma 6 *Let S be a sphere, and let H_S be the convex hull of all the vertices of X that lie on or inside S. Let t be a strongly Delaunay simplex. Suppose that some point of t lies between two points of H_S not in t. Then at least one vertex of t lies inside S.*

Proof: Suppose, for the sake of contradiction, that all vertices of t lie on or outside S. Because t is strongly Delaunay, there is some

[1] Edelsbrunner [3, Section 3] investigates relationships among Delaunay, rather than strongly Delaunay, simplices, and arrives at roughly the same conclusion as the present Lemma 3. I believe that his conclusion is incorrect in the case of simplices that share a common circumsphere.

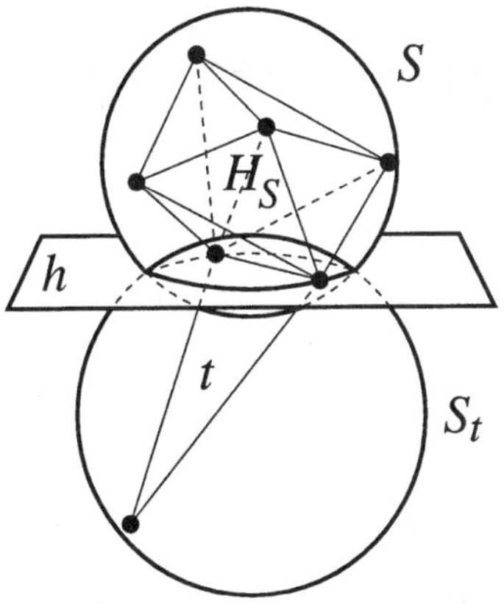

Figure 14: No point of t can lie between two points of H_S not in t.

sphere S_t that circumscribes t, but has no other vertices on or inside it. Because there are points of H_S not in t, a portion of S must lie outside S_t. By assumption, some point of t falls between two points of H_S; this point lies inside S and hence cannot be a vertex of t, so t has at least two vertices. Because these vertices are on or outside S, and S_t cannot be identical to S, a portion of S_t must lie outside S. It follows that the spheres S and S_t intersect in a $(d-1)$-dimensional circle.

Let h be the $(d-1)$-dimensional hyperplane that contains the intersection of S and S_t, as illustrated in Figure 14. Without loss of generality, suppose h is oriented horizontally with the center of S directly above the center of S_t. The vertices of t lie on S_t but not inside S, so all points of t must lie on or below h. Because t is strongly Delaunay, every vertex of H_S lies on or above h, and only those vertices of H_S that are shared with t may lie on h. Therefore, any point of H_S not in t lies strictly above h. Hence, no point of t can lie between two points of H_S not in t. The result follows by contradiction. ■

Lemma 7 *Let S be a sphere, and let H_S be the convex hull of all the vertices of X that lie on or inside S. Let q and z be two points in H_S with the property that qz intersects a strongly Delaunay simplex e that contains neither q nor z. Let m be the intersection point, and let p be any point in H_S from which m is visible. If X is ridge-protected, then there is a vertex of X inside S that is visible from p.*

Proof: See Figure 15. By assumption, q and z do not lie in e. Because the point m of e lies between two points that are in H_S but not in e, Lemma 6 implies that at least one vertex v of e lies inside S. If v is visible from p, the result follows. Otherwise, observe that m and v are both in H_S and are both in e. The point m is visible from p, but v is not.

Let n be the point nearest m on the line segment mv that cannot see p. The line segment pn must intersect the boundary of some $(d-1)$-facet of X at some point m' (if there are several $(d-1)$-facets occluding the view from p to n, consider only the facet that intersects pn closest to p). Let e' be the lowest-dimensional boundary simplex that contains m'. Because e' lies in the boundary of a facet, e' is strongly Delaunay. Clearly, e' overlaps e from the viewpoint p. Observe that p and n both lie in H_S, but they cannot lie in e'; if either of them did, the facet containing e' would not obstruct the visibility between them. Again, Lemma 6 implies that at least one vertex v' of e' lies inside S. If v' is visible from p, the result follows. Otherwise, find another simplex e'' that overlaps e' from the viewpoint p and continue.

The act of iterating in this manner yields a sequence of strongly Delaunay simplices, each overlapping the previous one. These iterations must terminate because the number of constraining simplices in X is finite and, by Lemma 3, a cycle of consecutively overlapping strongly Delaunay simplices is not possible. Hence, the procedure will eventually yield a vertex visible from p. ■

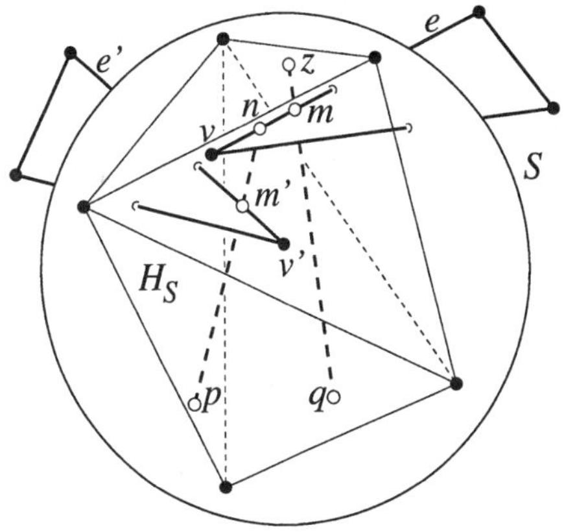

Figure 15: Because qz intersects a strongly Delaunay simplex e at m, and m is visible from p, some vertex inside S is visible from p.

Proof of Theorem 5: Consider the first claim. Suppose, for the sake of contradiction, that some j-facet f (for any $j < d$) intersects the relative interior of s_{k+1} but does not contain s_{k+1}.

Let s be a j-simplex of the Delaunay triangulation of f that intersects the relative interior of s_{k+1}. Because s_k is constrained Delaunay, it follows that if s contains any point in the interior of s_k, then s contains s_k entirely. By assumption, f does not contain s_{k+1}, so s lacks at least one vertex of s_{k+1}. Hence, s either does not contain u or does not contain any point in the relative interior of s_k.

There exists a simplex t, which is s or a face of s, that intersects the relative interior of s_{k+1} and contains neither u nor any point in the relative interior of s_k. How do we know this? If s contains u, then s does not contain any point in the relative interior of s_k; let t be the $(j-1)$-face of s that does not have u for a vertex. Because s intersects the relative interior of s_{k+1}, so must t. On the other hand, if s contains s_k, then s does not contain u; let t be any $(j-1)$-face of s that lacks one vertex of s_k and intersects the relative interior of s_{k+1}. (There must be at least one such face between u and the relative interior of $s \cap s_{k+1}$.) Otherwise, let $t = s$. In each of these cases, t contains neither u nor any point in the relative interior of s_k.

Let r be a point in the relative interior of s_k such that ru intersects t. Note that t contains neither r nor u. Let q be the point nearest p on the line segment pr such that qu intersects a constraining simplex or facet of dimension $d-2$ or less that contains neither q nor u, as Figure 16 illustrates. To see that such a choice of q exists, imagine sliding q along pr from p (which can see u) toward r, stopping when the line segment qu first intersects an appropriate simplex. If t is a constraining simplex of dimension $d-2$ or less, then q will stop sliding no later than when it reaches r. Alternatively, if t is not such a simplex, then f must be a $(d-1)$-facet (as illustrated), with t lying within it. Because p can see u, the line segment pu does not intersect f. Hence, the 2-simplex $\blacktriangle pru$ must intersect a boundary of f, so q will stop sliding no later than when qu meets this boundary.

Let m be the intersection point of qu and the constraining simplex, as illustrated. The point m is visible from p, because the relative interior of $\blacktriangle pqu$ intersects no obstructing facet (otherwise, q would have stopped earlier). Lemma 7 holds that there is a vertex of X inside S that is visible from p. This contradicts the assump-

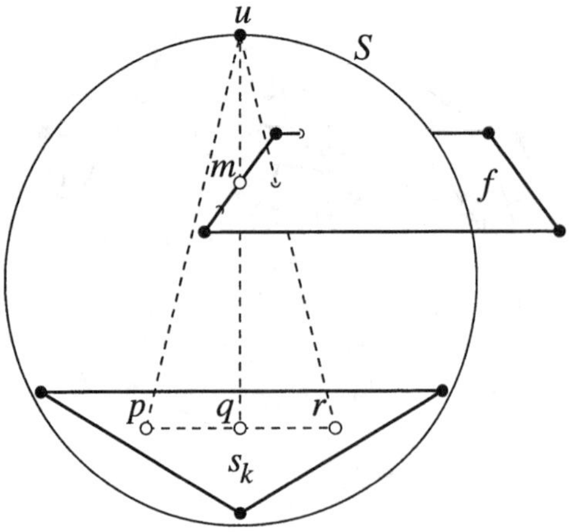

Figure 16: The circumstance depicted here, wherein p can see u but r cannot, could not occur if the vertices of f were on or outside S and the edges of f were strongly Delaunay.

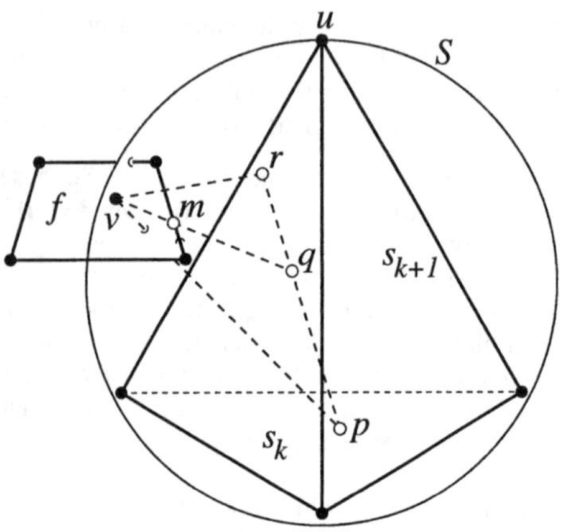

Figure 17: The circumstance depicted here, wherein p cannot see v but r can, could not occur if the vertices of f were on or outside S and the edges of f were strongly Delaunay.

tion that no vertex inside S is visible from p; hence, no constraining facet of X intersects the relative interior of s_{k+1} unless it contains s_{k+1}.

Next, consider the claim that the $(k+1)$-simplex s_{k+1} formed by s_k and u is constrained Delaunay. Suppose, for the sake of contradiction, that there is some point r in the interior of s_{k+1} that can see a vertex v of X inside S, as Figure 17 illustrates. By assumption, v is not visible from p.

Let f be the $(d-1)$-facet that obstructs the visibility of v from p; if there are several such facets, choose f to be the facet that intersects pv closest to p.

Let q be the point nearest p on the line segment pr such that qv intersects the boundary of a $(d-1)$-facet that contains neither q nor v, and the intersection point m is visible from p. To see that such a choice of q exists, imagine sliding q along pr from p (whose view of v is obstructed by f) toward r (from which v is visible). As q moves, either qv will strike the boundary of a $(d-1)$-facet lying between q and f, or the intersection of qv and f will remain visible from p until qv intersects a boundary of f (which will occur before q reaches r).

Lemma 7 holds that there is a vertex of X inside S that is visible from p. By contradiction, s_{k+1} is constrained Delaunay. ∎

Theorem 5 shows that a constrained Delaunay simplex can be "grown" into a higher-dimensional constrained Delaunay simplex. However, a separate result is needed to show that a constraining $(d-1)$-simplex of X—which might not be constrained Delaunay—can also be "grown" into a constrained Delaunay d-simplex.

Theorem 8 *Let s be a constraining $(d-1)$-simplex of X. Let u be a vertex of X that is above s and visible from some point p in the interior of s. Suppose that there is a sphere S that passes through u and all d vertices of s, and encloses no vertex of X that is above s and visible from a point in the interior of s. Let s_d be the convex hull of s and u. Then no constraining facet of X intersects the interior of s_d, and s_d is constrained Delaunay.*

Proof: The proof is similar to that of Theorem 5, with a few modifications to account for the fact that there may be vertices of X inside S below s. Consider the first claim. Suppose, for the sake of contradiction, that some facet intersects the interior of s_d.

Let p' be a point on the line segment pu that is extremely close to p. Specifically, p' is such a small distance above s that p' cannot see any vertex of X below s (because they are all occluded by s). Furthermore, for any vertex w of X above s, the line wp' passes through the interior of s. Such a choice of p' is always possible because p is in the relative interior of s. Additionally, p' is close enough to p that the convex hull of s and p' intersects no facet above s. Such a choice of p' is always possible because no facet of X can intersect the interior of s (except the facet that contains s).

Because s is a Delaunay triangle of a ridge-protected facet, no vertex of X coplanar with s is inside S. Hence, if there is a vertex inside S that is visible from p', it lies above s. Observe that p' lies in H_S and can see u.

By repeating the argument from the proof of Theorem 5, with p replaced by p', one may deduce that there is a vertex w of X inside S that is visible from p'; w must be above s. Because the line wp' passes through the relative interior of s, w is visible from some point in the relative interior of s, a contradiction. Hence, no facet intersects the interior of s_d.

Next, consider the claim that the d-simplex s_d formed from s and u is constrained Delaunay. Suppose, for the sake of contradiction, that there is some point r in the interior of s_d from which a vertex v of X inside S is visible.

If v is below s, let $p'' = p'$. If v is above s, let p'' be a point on the line segment pv that is extremely close to p, satisfying all the same conditions listed above for p' as well as the condition that p'' is in the interior of s_d. In either case, p'' lies in the interiors of s_d and H_S, and can see r but not v.

By repeating the argument from the proof of Theorem 5, with p replaced by p'', one may deduce that there is a vertex w of X inside S that is visible from p''. Because the line wp'' passes through the relative interior of s, w is visible from some point in the interior of s, a contradiction. Hence, s_d is constrained Delaunay. ∎

3 Algorithms for Constructing CDTs

Starting with one constrained Delaunay d-simplex, a naïve implementation of the foregoing gift-wrapping algorithm might construct each additional d-simplex by considering all of the vertices as candidates and testing the visibility of each one against every facet. The running time is thus $\mathcal{O}(d^4 n_s n_v n_f)$, where n_s is the number of d-simplices in the output, n_v is the number of input vertices, and n_f is the number of $(d-1)$-simplices in the triangulations of the input $(d-1)$-facets. (The factor of d^4 is the cost of testing whether a $(d-1)$-simplex obstructs the visibility between two points.) This leaves much room for improvement, which I hope will be filled by future research. For instance, the sophisticated search algorithm and analysis techniques applied by Dwyer [2] to unconstrained gift-wrapping might be generalizable to the constrained case as well.

Here, I offer a few practical suggestions without asymptotic guarantees.

For most applications, the fastest way to form the CDT of a ridge-protected PLC (albeit not in the worst case) is to use the best available algorithm to find an unconstrained Delaunay triangulation of the input vertices, then recover the $(d-1)$-facets one by one. Each $(d-1)$-facet f may be recovered by deleting the d-simplices whose interiors it intersects, then retriangulating the polytopes now left empty on each side of f. It is easy to show that all of the simplices not thus deleted are still constrained Delaunay. Since a CDT of the new configuration exists, each empty polytope can be triangulated with constrained Delaunay simplices. If these polytopes are typically small, the performance of the algorithm used to triangulate them is not critical, and gift-wrapping will suffice.

Vertices may be incrementally inserted into and deleted from a CDT just like an ordinary Delaunay triangulation, so long as the underlying PLC (which changes incrementally with the triangulation) remains ridge-protected. When a vertex is inserted, the simplices that are no longer constrained Delaunay are deleted. When a vertex is deleted, the simplices that contain it are deleted. In either case, the resulting polytopal hole is retriangulated to complete the new CDT. As with facet recovery, the existence of a CDT of the entire underlying PLC ensures that a CDT of the hole can be produced. Hence, the best approach to triangulating a PLC might be to start with a Delaunay triangulation of the vertices of the $(d-1)$-facets, then recover the $(d-1)$-facets themselves, and then finally insert the remaining vertices incrementally.

The ability to incrementally insert and delete vertices is also useful for mesh generation, especially in circumstances where the constrained Delaunay property can be used to establish provable properties of the meshing algorithm [13].

Unfortunately, subsets of $d+2$ or more cospherical vertices can cause real difficulties for gift-wrapping. A gift-wrapping algorithm may make decisions that are mutually inconsistent, and find itself unable to complete the triangulation. For an example affecting *unconstrained* Delaunay tetrahedralizations in E^3, imagine a large vertex set that includes six cospherical vertices whose surroundings have been inadvertently tetrahedralized so as to form a hollow space shaped like Schönhardt's polyhedron.

This problem can be solved by using symbolic perturbation to simulate general position, thereby ensuring that all decisions made by gift-wrapping are mutually consistent. Additionally, cospherical or nearly-cospherical vertices create the need for exact arithmetic when performing the *insphere* tests associated with Delaunay triangulation.

4 Conclusions

In their paper on two-dimensional conforming Delaunay triangulations, Edelsbrunner and Tan [4] write:

> A seemingly difficult open problem is the generalization of our polynomial bound to three dimensions. The somewhat easier version of the generalized problem considers a graph whose vertices are embedded as points in $\mathbb{R}^3$, and edges are represented by straight line segments connecting embedded vertices. More relevant, however, is the problem for the crossing-free embedding of a complex consisting of vertices, edges, *and* triangles.

The present result shifts the emphasis back to the former of these two problems. An algorithm that could create a CCDT by inserting only a modest number of additional vertices on input segments might have great practical importance.

Several other questions also deserve investigation. Are there higher-dimensional constrained Delaunay triangulation algorithms that have the same running time as optimal algorithms for unconstrained Delaunay triangulations? Do higher-dimensional CDTs have optimality properties such as minimizing the largest min-containment sphere, as higher-dimensional Delaunay triangulations do [9]? Is there a less conservative definition of "constrained Delaunay" (perhaps allowing visibility to be affected by constraining simplices of dimension less than $d-1$) that defines triangulations over a larger class of PLCs?

Acknowledgments

I thank Dafna Talmor for her many helpful comments, and in particular for suggesting an intuitive way to recognize whether a simplex is strongly Delaunay.

References

[1] L. Paul Chew. *Constrained Delaunay Triangulations.* Algorithmica **4**(1):97–108, 1989.

[2] Rex A. Dwyer. *Higher-Dimensional Voronoi Diagrams in Linear Expected Time.* Discrete & Computational Geometry **6**(4):343–367, 1991.

[3] Herbert Edelsbrunner. *An Acyclicity Theorem for Cell Complexes in d Dimension.* Combinatorica **10**(3):251–260, 1990.

[4] Herbert Edelsbrunner and Tiow Seng Tan. *An Upper Bound for Conforming Delaunay Triangulations.* Discrete & Computational Geometry **10**(2):197–213, 1993.

[5] Steven Fortune. *Voronoi Diagrams and Delaunay Triangulations.* Computing in Euclidean Geometry (Ding-Zhu Du and Frank Hwang, editors), Lecture Notes Series on Computing, volume 1, pages 193–233. World Scientific, Singapore, 1992.

[6] Carol Hazlewood. *Approximating Constrained Tetrahedrizations.* Computer Aided Geometric Design **10**:67–87, 1993.

[7] D. T. Lee and A. K. Lin. *Generalized Delaunay Triangulations for Planar Graphs.* Discrete & Computational Geometry **1**:201–217, 1986.

[8] Gary L. Miller, Dafna Talmor, Shang-Hua Teng, Noel Walkington, and Han Wang. *Control Volume Meshes using Sphere Packing: Generation, Refinement and Coarsening.* Fifth International Meshing Roundtable (Pittsburgh, Pennsylvania), pages 47–61, October 1996.

[9] V. T. Rajan. *Optimality of the Delaunay Triangulation in $\mathbb{R}^d$.* Proceedings of the Seventh Annual Symposium on Computational Geometry, pages 357–363. ACM, 1991.

[10] James Martin Ruppert. *Results on Triangulation and High Quality Mesh Generation.* Ph.D. thesis, University of California at Berkeley, Berkeley, California, 1992.

[11] Jim Ruppert and Raimund Seidel. *On the Difficulty of Triangulating Three-Dimensional Nonconvex Polyhedra.* Discrete & Computational Geometry **7**(3):227–254, 1992.

[12] E. Schönhardt. *Über die Zerlegung von Dreieckspolyedern in Tetraeder.* Mathematische Annalen **98**:309–312, 1928.

[13] Jonathan Richard Shewchuk. *Delaunay Refinement Mesh Generation.* Ph.D. thesis, School of Computer Science, Carnegie Mellon University, Pittsburgh, Pennsylvania, May 1997. Available as Technical Report CMU-CS-97-137.

[14] Nigel P. Weatherill and O. Hassan. *Efficient Three-Dimensional Grid Generation using the Delaunay Triangulation.* Proceedings of the First European Computational Fluid Dynamics Conference (Brussels, Belgium) (Ch. Hirsch, J. Périaux, and W. Kordulla, editors), pages 961–968, September 1992.

[15] Nigel P. Weatherill, O. Hassan, D. L. Marcum, and M. J. Marchant. *Grid Generation by the Delaunay Triangulation.* Von Karman Institute for Fluid Dynamics 1993–1994 Lecture Series, 1994.

Tetrahedral Mesh Generation by Delaunay Refinement

Jonathan Richard Shewchuk
School of Computer Science
Carnegie Mellon University
Pittsburgh, Pennsylvania 15213
jrs@cs.cmu.edu

Abstract

Given a complex of vertices, constraining segments, and planar straight-line constraining facets in E^3, with no input angle less than $90°$, an algorithm presented herein can generate a conforming mesh of Delaunay tetrahedra whose circumradius-to-shortest edge ratios are no greater than two. The sizes of the tetrahedra can provably grade from small to large over a relatively short distance. An implementation demonstrates that the algorithm generates excellent meshes, generally surpassing the theoretical bounds, and is effective in eliminating tetrahedra with small or large dihedral angles, although they are not all covered by the theoretical guarantee.

1 Introduction

Meshes of triangles or tetrahedra have many applications, including interpolation, rendering, and numerical methods such as the finite element method. Most such applications demand more than just a triangulation of the object or domain being rendered or simulated. To ensure accurate results, the triangles or tetrahedra must be "well-shaped," having small aspect ratios or bounds on their smallest and largest angles.

Mesh generation algorithms based on *Delaunay refinement* are effective both in theory and in practice. Delaunay refinement algorithms operate by maintaining a Delaunay or constrained Delaunay triangulation, which is refined by inserting carefully placed vertices until the mesh meets constraints on triangle quality and size.

It is difficult to trace who first used Delaunay triangulations for finite element meshing, and equally difficult to tell where the suggestion arose to use the triangulation to guide vertex creation. These ideas have been intensively studied in the engineering community since the mid-1980s; for instance, Frey [7] eliminates poorly shaped triangles from a triangulation by inserting new vertices at their circumcenters (defined in Section 2), whereas Weatherill [17] inserts new vertices at their centroids.

These ideas bore vital theoretical fruit when the problem of mesh generation began to attract interest from the computational geometry community in the early 1990s. The first provably good Delaunay refinement algorithm is due to Paul Chew [2], and takes as its input a set of vertices and segments that define the region to be meshed. By inserting additional vertices, Chew's algorithm generates a two-dimensional constrained Delaunay triangulation whose angles are bounded between $30°$ and $120°$. Dey, Bajaj, and Sugihara [5] and Chew [4] generalize Chew's algorithm to three dimensions, but only for unconstrained point set inputs. All three algorithms produce *uniform* meshes, whose triangles or tetrahedra are of roughly the same size.

Uniform meshes sometimes have many more triangles or tetrahedra than are necessary, and thus impose an excessive computational load upon the applications that make use of them. Jim Ruppert [13] and Paul Chew [3] have each proposed two-dimensional Delaunay refinement algorithms that produce meshes of well-shaped triangles whose sizes are graded, and Ruppert has furthermore proven that his algorithm produces meshes that are nicely graded in a theoretical sense described in Section 7.

These algorithms are successful because they exploit several favorable characteristics of Delaunay triangulations. Delaunay triangulations have been extensively studied, and good algorithms are available. Inserting a vertex is a local operation, and is inexpensive except in unusual cases. In two dimensions, Delaunay triangulations maximize the minimum angle, compared with all other triangulations of the same vertex set [8].

The greatest advantage of Delaunay triangulations is less obvious. The central question of any Delaunay refinement algorithm is "where should the next vertex be inserted?" As Section 3 will demonstrate, a reasonable answer is "as far from other vertices as possible." If a new vertex is inserted too close to another vertex, the resulting short edge will engender thin triangles or tetrahedra.

Because a Delaunay triangle has no vertices in its circumcircle (and a Delaunay tetrahedron has no vertices in its circumsphere), a Delaunay triangulation is an ideal search structure for finding points that are far from other vertices.

Herein, I build upon the algorithmic and analytical framework of Ruppert to design a new tetrahedral Delaunay refinement algorithm. This algorithm generates meshes whose tetrahedra have circumradius-to-shortest edge ratios (defined shortly) no greater than the bound $B = 2$. Upon relaxing B to be greater than two, one can also guarantee good grading. My algorithm is distinguished from those of Dey et al. and Chew by this guarantee of good grading (as opposed to uniform meshes), and by its ability to handle the rather general constrained inputs described in Section 4 (as opposed to unconstrained sets of vertices). The main theoretical deficiency of the algorithm is its requirement that incident input segments and facets be separated by angles of at least $90°$. (The requirements are less onerous in practice.) This deficiency can be partly ameliorated; Section 9 includes a brief summary of the steps involved.

Supported in part by the Advanced Research Projects Agency and Rome Laboratory, Air Force Materiel Command, USAF under agreement number F30602-96-1-0287, in part by the National Science Foundation under Grant CMS-9318163, and in part by a grant from the Intel Corporation.

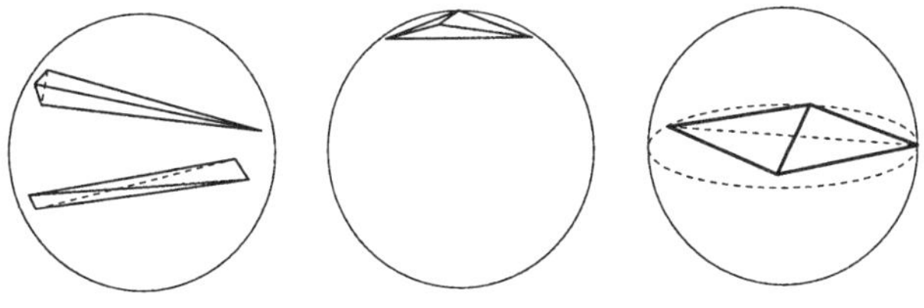

Figure 1: Tetrahedra with poor angles. Needles and wedges have edges of greatly disparate length; caps have a large solid angle; slivers have neither, but can have good circumradius-to-shortest edge ratios.

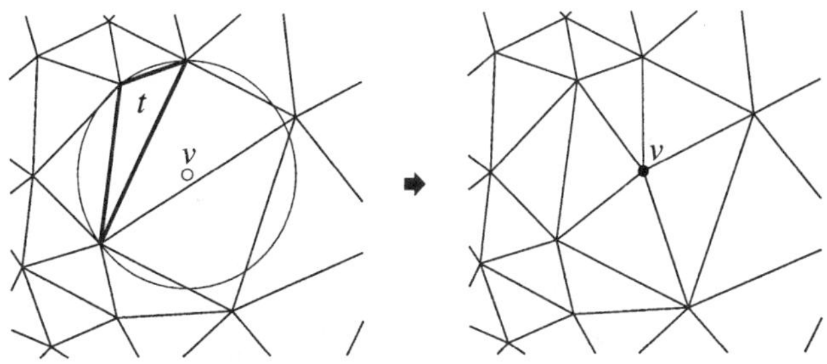

Figure 2: Any triangle whose circumradius-to-shortest edge ratio is larger than some bound B is split by inserting a vertex at its circumcenter. The Delaunay property is maintained, and the triangle is thus eliminated. Every new edge has length at least B times that of shortest edge of the poor triangle.

Two other tetrahedral mesh generation algorithms (not based on Delaunay refinement) have provable bounds. The octree-based algorithm of Mitchell and Vavasis [11, 12] is a theoretical *tour de force*, obtaining provable bounds on dihedral angles and grading, but its bounds are too weak to offer any practical reassurance (and have not been explicitly stated). An algorithm by Miller, Talmor, Teng, Walkington, and Wang [10] generates its final vertex set before triangulating it. Their algorithm has provable bounds similar to those of the algorithm described herein, but the algorithm herein has the opportunity to stop early if fewer vertices are needed than the theory suggests. As a result, the present algorithm typically uses fewer vertices by several orders of magnitude; details are provided elsewhere [14].

2 A Quality Measure for Simplices

Miller, Talmor, Teng, and Walkington [9] have pointed out that the most natural and elegant measure for analyzing Delaunay refinement algorithms is the *circumradius-to-shortest edge ratio* of a triangle or tetrahedron. The *circumsphere* of a simplex is the unique circle or sphere that passes through all its vertices. The *circumcenter* and *circumradius* of a simplex are the center and radius of its circumsphere, respectively. The quotient of a simplex's circumradius and the length of its shortest edge is the metric that is naturally optimized by Delaunay refinement algorithms. One would like this ratio to be as small as possible.

But does optimizing this metric aid practical applications? In two dimensions, the answer is yes. A triangle's circumradius-to-shortest edge ratio r/ℓ is related to its smallest angle $\theta_{\min}$ by the formula $r/\ell = 1/(2\sin\theta_{\min})$. The smaller a triangle's ratio, the larger its smallest angle. Chew's two-dimensional algorithms produce meshes whose triangles' circumradius-to-shortest edge ratios are bounded below one, and hence their angles are bounded between 30° and 120°. Ruppert's algorithm can produce meshes whose triangles' ratios are bounded below $\sqrt{2}$, and hence their angles range between 20.7° and 138.6°.

In three dimensions, however, a mesh of tetrahedra whose circumradius-to-shortest edge ratios are bounded is not entirely adequate for the needs of interpolation. As Dey et al. illustrate, most tetrahedra with poor angles have circumcircles much larger than their shortest edges, including the needle, wedge, and cap illustrated in Figure 1, but there is one type called a *sliver* or *kite* tetrahedron that does not. The canonical sliver is formed by arranging four vertices, equally spaced, around the equator of a sphere, then perturbing one of the vertices slightly off the equator. A sliver can have a circumradius-to-shortest edge ratio as low as $1/\sqrt{2}$, yet be considered awful by most other measures, because its volume and its shortest altitude can be arbitrarily close to zero, and its dihedral angles can be arbitrarily close to 0° and 180°.

Despite slivers, Delaunay refinement methods are valuable for generating three-dimensional meshes. Slivers having good circumradius-to-shortest edge ratios typically arise in small numbers in practice. As Section 8 will demonstrate, the worst slivers can often be removed by Delaunay refinement, even if there is no theoretical guarantee. Meshes with bounds on the circumradius-to-shortest edge ratios of their tetrahedra are an excellent starting point for mesh smoothing and optimization methods that remove slivers and otherwise improve the quality of an existing mesh [6]. Even if slivers are not removed, the Voronoi dual of a tetrahedralization with bounded circumradius-to-shortest edge ratios has nicely rounded cells, and is sometimes ideal for use in the control volume method [9].

3 The Key Idea Behind Delaunay Refinement

The central operation of Chew's, Ruppert's, Dey's, and my own Delaunay refinement algorithms is the insertion of a vertex at the circumcenter of a triangle or tetrahedron of poor quality. The Delaunay property is maintained, perhaps by the Bowyer/Watson algorithm for the incremental update of Delaunay triangulations [1, 16]. The poor simplex cannot survive, because its circumsphere is no longer empty. For brevity, the act of inserting a vertex at a simplex's circumcenter is called *splitting* a simplex. If poor simplices are split one by one, either all will eventually be eliminated, or the algorithm will run forever.

The main insight of all these Delaunay refinement algorithms is that Delaunay refinement is guaranteed to terminate if the notion of "poor quality" includes only simplices that have a circumradius-to-shortest edge ratio larger than some appropriate bound B. The only new edges created by the Delaunay insertion of a vertex v are edges connected to v (see Figure 2). Because v is the circumcenter of some Delaunay simplex t, and there were no vertices inside the circumsphere of t before v was inserted, no new edge can be shorter than the circumradius of t. Because t has a circumradius-to-shortest edge ratio larger than B, every new edge has length at least B times that of the shortest edge of t.

Ruppert's Delaunay refinement algorithm [13] employs a bound of $B = \sqrt{2}$, and Chew's second Delaunay refinement algorithm [3] employs a bound of $B = 1$. Chew's first Delaunay refinement algorithm [2] splits any triangle whose circumradius is greater than the length of the shortest edge in the entire mesh, thus achieving a bound of $B = 1$, but forcing all triangles to have uniform size. In the same manner, the three-dimensional algorithms of Dey et al. and Chew achieve a bound of $B = 2$. With these bounds, every new edge created is at least as long as some other edge already in the mesh. Hence, no vertex is ever inserted closer to another vertex than the length of the shortest edge in the initial triangulation. Delaunay refinement must eventually terminate, because the augmented triangulation will run out of places to put vertices.

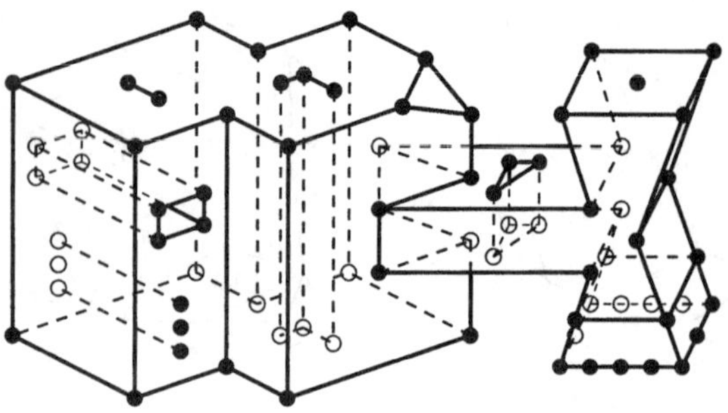

(a)

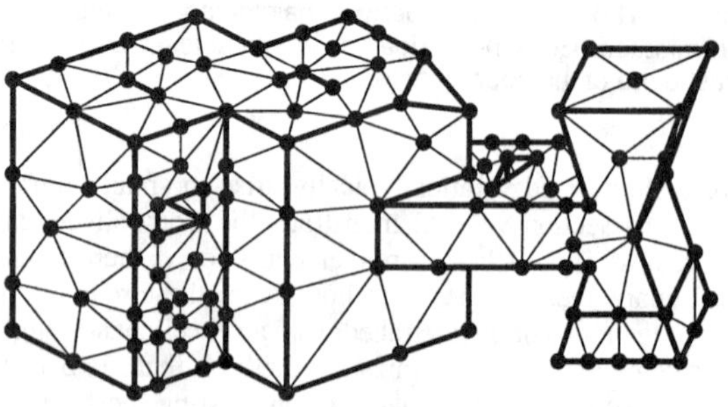

(b)

Figure 3: (a) Any facet of a PLC may contain holes, slits, and vertices; these may support intersections with other facets and segments or allow a user of the finite element method to apply boundary conditions. (b) When a PLC is tetrahedralized, each facet of the PLC is partitioned into triangular subfacets, which respect the holes, slits, and vertices.

This idea generalizes without change to higher dimensions. Unfortunately, my description of Delaunay refinement thus far has a gaping hole: mesh boundaries have not been accounted for. The flaw in the procedure presented above is that the circumcenter of a poor simplex might not lie in the mesh at all. Delaunay refinement algorithms, including those of Chew, Ruppert, and Dey et al., are distinguished primarily by how they handle boundaries.

4 Piecewise Linear Complexes

The input upon which my three-dimensional Delaunay refinement algorithm operates is called a *piecewise linear complex* (PLC). See Miller, Talmor, Teng, Walkington, and Wang [10] for a definition in E^d. In three dimensions, a PLC is a set of vertices, segments, and facets. A segment is a constraining edge that must be represented by a sequence of contiguous edges in the final mesh. A facet is a constraining planar surface that must be represented by a set of triangular faces in the final mesh. A facet can be quite complicated in shape; it is a polygon (not necessarily convex), possibly augmented by holes, slits, and vertices in its interior. Figure 3(a) demonstrates some of the possibilities.

A piecewise linear complex X is required to have the following properties. First, X contains both endpoints of each segment of X. Similarly, facets of X are *segment-bounded*: for any facet in X, every edge and vertex of the facet must be a segment or vertex of X.

Second, X is closed under intersection. For example, if two facets of X intersect at a line segment, that line segment must be a segment of X. Third, if a segment of X intersects a facet of X at a point in the segment's interior, then the segment must be entirely contained in the facet.

The process of tetrahedral mesh generation necessarily divides each segment into smaller edges called *subsegments*. (Each segment is initially represented by one subsegment, until it is subdivided.) The bold edges of the tetrahedralization illustrated in Figure 3(b) are subsegments; other edges are not. Similarly, each facet is subdivided into triangular faces called *subfacets*. All of the triangular faces visible in Figure 3(b) are subfacets, but most of the faces in the interior of the tetrahedralization are not.

Any vertex inserted into a segment or facet during Delaunay refinement remains there permanently. However, keep in mind that the edges that partition a facet into subfacets are *not* permanent, are *not* treated like subsegments, and are subject to flipping according to the Delaunay criterion.

Many approaches to tetrahedral mesh generation permanently triangulate the input facets as a separate step prior to tetrahedralizing the interior of a region. The problem with this approach is that these independent facet triangulations may not be collectively ideal for forming a good tetrahedralization. The algorithm discussed herein uses an alternative approach, wherein facet triangulations are refined in conjunction with the tetrahedralization. The tetrahedralization process is not beholden to poor decisions made earlier.

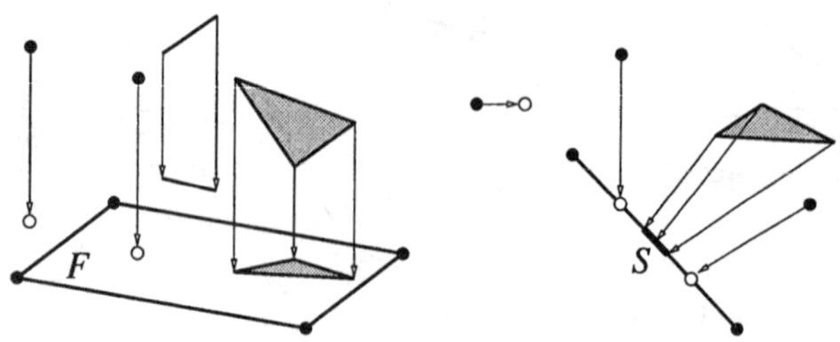

Figure 4: The orthogonal projections of points and sets of points onto facets and segments.

5 Orthogonal Projections

The following sections use the notion of the *orthogonal projection* of a geometric entity onto a line or plane. Given a facet or subfacet F and a point p, the orthogonal projection $\mathrm{proj}_F(p)$ of p onto F is the point that is coplanar with F and lies in the line that passes through p orthogonally to F, as illustrated in Figure 4. The projection exists whether or not it falls in F.

Similarly, the orthogonal projection $\mathrm{proj}_S(p)$ of p onto a segment or subsegment S is the point that is collinear with S and lies in the plane through p orthogonal to S.

Sets of points may be projected as well. If F and G are facets, then $\mathrm{proj}_F(G)$ is the set $\{\mathrm{proj}_F(p) : p \in G\}$.

6 A 3D Delaunay Refinement Algorithm

In this section, I describe a three-dimensional Delaunay refinement algorithm that produces well-graded tetrahedral meshes satisfying any circumradius-to-shortest edge ratio bound greater than two.

The algorithm takes a *facet-bounded* PLC as its input. Tetrahedralized and untetrahedralized regions of space must be separated by facets so that, in the final mesh, any triangular face not shared by two tetrahedra is a subfacet.

The first step is to form a Delaunay tetrahedralization of the input vertices. Some input segments and facets might be missing (or partly missing) from this mesh. The tetrahedralization is refined by inserting additional vertices into the mesh, using the Bowyer/Watson algorithm to maintain the Delaunay property, until all segments and facets are respected and all constraints on tetrahedron quality are met. Vertex insertion is governed by three rules.

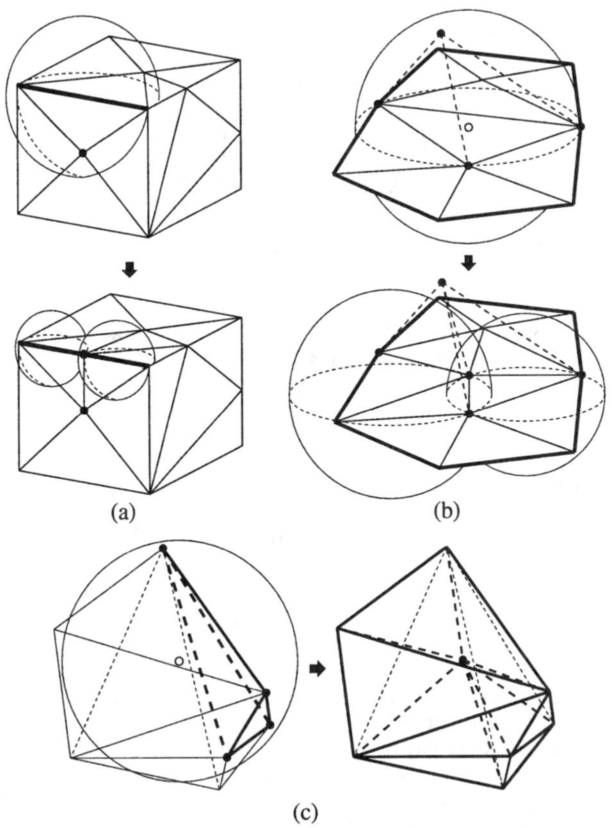

Figure 5: Three operations for three-dimensional Delaunay refinement. (a) Splitting an encroached subsegment. The original subsegment is encroached because there is a vertex in its diametral sphere. In this example, the two subsegments created by bisecting the original subsegment are not encroached. (b) Splitting an encroached subfacet. The triangular faces shown are subfacets of a larger facet, with tetrahedra (not shown) atop them. In this example, all equatorial spheres (included the two illustrated) are empty after the split. (c) Splitting a skinny tetrahedron.

- The *diametral sphere* of a subsegment is the (unique) smallest sphere that encloses the subsegment. Following Ruppert [13], a subsegment is said to be *encroached* if a vertex other than its endpoints lies inside or on its diametral sphere. A subsegment may be encroached whether or not it actually appears as an edge of the tetrahedralization. It is a property of Delaunay tetrahedralizations that if a subsegment is missing from the tetrahedralization, then it is encroached. Any encroached subsegment that arises is immediately split into two subsegments by inserting a vertex at its midpoint, as illustrated in Figure 5(a). These subsegments may or may not be encroached themselves; splitting continues until no subsegment is encroached.

- The *equatorial sphere* of a triangular subfacet is the (unique) smallest sphere that passes through the three vertices of the subfacet. A subfacet is encroached if a non-coplanar vertex lies inside or on its equatorial sphere. It is a property of Delaunay tetrahedralizations that if a subfacet does not appear in the tetrahedralization, and it is not covered by other faces that share the same equatorial sphere, then it is encroached. (The question of what subfacets should not be missing from the mesh will be considered shortly.) Each encroached subfacet is normally split by inserting a vertex at its circumcenter; see Figure 5(b). However, if the new vertex would encroach upon any subsegment, it is not inserted; instead, all the subsegments it would encroach upon are split.

- A tetrahedron is said to be *skinny* if its circumradius-to-shortest edge ratio is larger than some bound B. Each skinny tetrahedron is normally split by inserting a vertex at its circumcenter, thus eliminating the tetrahedron; see Figure 5(c). However, if the new vertex would encroach upon any subsegment or subfacet, then it is not inserted; instead, all the subsegments it would encroach upon are split. If the skinny tetrahedron is not eliminated as a result, then all the subfacets its circumcenter would encroach upon are split.

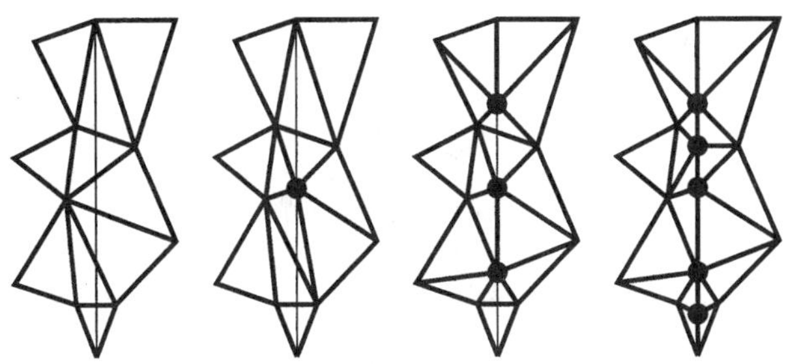

Figure 6: Missing segments are recovered through the same recursive splitting procedure used for encroached subsegments that aren't missing. In this sequence of illustrations, the thin line represents a segment missing from the triangulation. Segment recovery is illustrated here in two dimensions for clarity, but operates no differently in three.

Encroached subsegments are given priority over encroached subfacets, which have priority over skinny tetrahedra. These encroachment rules are intended to recover missing segments and facets, and to ensure that all vertex insertions are valid. Although the proof is omitted here, one can show that if there are no encroached subsegments, then each subfacet circumcenter lies in the containing facet; and if there are no encroached subfacets, then each tetrahedron circumcenter lies in the mesh.

Because encroached subsegments have priority, the algorithm begins by recovering all the segments that are missing from the initial tetrahedralization. Each missing segment is bisected by inserting a vertex into the mesh at the midpoint of the segment (more accurately, at the midpoint of the place where the segment should be). After the mesh is adjusted to maintain the Delaunay property, the two resulting subsegments may appear in the mesh. If not, the procedure is repeated recursively for each missing subsegment until the original segment is represented by a contiguous linear sequence of edges of the mesh, as illustrated (in two dimensions) in Figure 6. We are assured of eventual success because the Delaunay triangulation always connects a vertex to its nearest neighbor; once the spacing of vertices along a segment is sufficiently small, its entire length will be represented. In the engineering literature, this process is sometimes called *stitching*.

When no encroached subsegment remains, missing facets are recovered in an analogous manner. The main complication is that if a facet is missing from the mesh, it is difficult to say what its subfacets are. With segments there is no such problem; if a segment is missing from the mesh, and a vertex is inserted at its midpoint, one knows unambiguously where the two resulting subsegments are. But how may we identify subfacets that do not yet exist?

The solution is straightforward. For each facet, it is necessary to maintain a two-dimensional Delaunay triangulation of its vertices, independently from the tetrahedralization in which we hope its subfacets will eventually appear. By comparing the triangles of a facet's triangulation against the faces of the tetrahedralization, one can identify subfacets that need help in forcing their way into

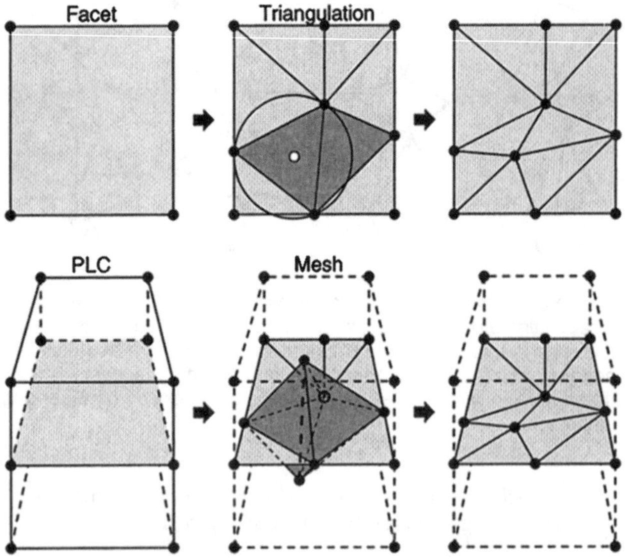

Figure 7: The top illustrations depict a rectangular facet and its triangulation. The bottom illustrations depict the facet's position as an interior boundary of a PLC, and its progress as it is inserted into the tetrahedralization. Most of the vertices and tetrahedra of the mesh are omitted for clarity. The facet triangulation and the tetrahedralization are maintained separately. Shaded triangular subfacets in the facet triangulation (top center) are missing from the tetrahedralization (bottom center). The bold dashed line (bottom center) represents a tetrahedralization edge that passes through the facet. A missing subfacet is forced into the mesh by inserting a vertex at its circumcenter (top right and bottom right). The vertex is independently inserted into both the triangulation and the tetrahedralization.

the mesh. For each triangular subfacet in a facet triangulation, look for a matching face in the tetrahedralization; if the latter is missing, insert a vertex at the circumcenter of the subfacet (subject to rejection if subsegments are encroached), as illustrated in Figure 7. The new vertex is independently inserted into both the facet triangulation and the tetrahedralization. Similarly, the midpoint of an encroached subsegment is independently inserted into the tetrahedralization and into *each* facet triangulation that contains the subsegment.

In essence, the same procedure is used to recover missing segments. However, the process of forming a one-dimensional triangulation is so simple that it passes unnoticed.

Which vertices of the tetrahedralization need to be considered in a facet triangulation? If a facet appears in a Delaunay tetrahedralization as a union of faces, then the triangulation of the facet is determined solely by the vertices of the tetrahedralization that lie in the plane of the facet. If a vertex lies near a facet, but is not coplanar with the facet, it may cause a subfacet to be missing (as in Figure 7, bottom center), but it cannot otherwise affect the shape of the triangulation. If a facet appears as a union of faces in a Delaunay tetrahedralization, then those faces form a two-dimensional Delaunay triangulation of the facet.

Furthermore, because each facet is segment-bounded, and segments are recovered (in the tetrahedralization) before facets, each facet triangulation can safely ignore vertices that lie outside the facet (coplanar though they may be). The requirements set forth in Section 4 ensure that all of the vertices and segments of a facet must be explicitly identified in the input PLC. The only additional vertices to be considered are those that were inserted in segments to help recover those segments and other facets. The algorithm maintains a list of the vertices on each segment, ready to be called upon when a facet triangulation is initially formed.

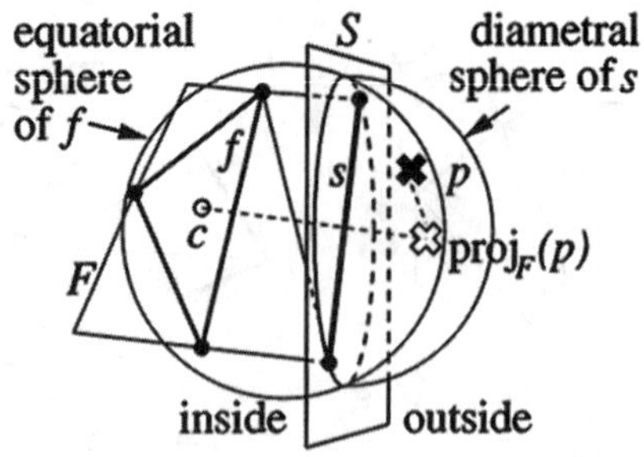

Figure 8: If a vertex p encroaches upon a Delaunay subfacet f of a facet F, but its projection into the plane containing F lies outside F, then p encroaches upon some subsegment s of F as well.

Unfortunately, if a facet's Delaunay triangulation is not unique because of cocircularity degeneracies, then the facet might be represented in the tetrahedralization by faces that do not match the independent facet triangulation. An implementation must detect these cases and correct the triangulation so that it matches the tetrahedralization.

When no encroached subsegment or subfacet remains, every input segment and facet is represented by a union of edges or faces of the mesh. The first time the mesh reaches this state, all external tetrahedra (lying in the convex hull of the input vertices, but outside the region enclosed by the facet-bounded PLC) are removed prior to splitting any skinny tetrahedra. This measure prevents problems that might arise if superfluous skinny tetrahedra are split, such as overrefinement and failure to terminate because of spurious small angles formed between the PLC and its convex hull.

One further amendment to the algorithm is necessary to obtain the best possible bound on the circumradius-to-shortest edge ratios of the tetrahedra. When several encroached subfacets exist, they should not be split in arbitrary order. If a vertex p encroaches upon a subfacet f of a facet F, but the projected point $\text{proj}_F(p)$ does not lie in f, then splitting f is not the best choice. One can show (with the following lemma) that there is some other subfacet g of F that is encroached upon by p and contains $\text{proj}_F(p)$. (The lemma assumes that there are no encroached subsegments in the mesh, as they have priority.) A better bound is achieved if the algorithm splits g first and delays the splitting of f indefinitely.

Lemma 1 (Projection Lemma) *Let f be a subfacet of the Delaunay triangulated facet F. Suppose that f is encroached upon by some vertex p, but p does not encroach upon any subsegment of F. Then $\text{proj}_F(p)$ lies in the facet F, and p encroaches upon a subfacet of F that contains $\text{proj}_F(p)$.*

Proof: First, I prove that $\text{proj}_F(p)$ lies in F. Suppose for the sake of contradiction that $\text{proj}_F(p)$ lies outside the facet F. Let c be the centroid of f; c clearly lies inside F. Because all facets are segment-bounded, the line segment connecting c to $\text{proj}_F(p)$ must intersect some subsegment s in the boundary of F. Let $\mathcal{S}$ be the plane that contains s and is orthogonal to F, as illustrated in Figure 8.

Because f is a Delaunay subfacet of F, its circumcircle (in the plane containing F) encloses no vertex of F. However, its equatorial sphere may enclose vertices—including p—and f might not appear in the tetrahedralization.

It is apparent that p and $\text{proj}_F(p)$ lie on the same side of $\mathcal{S}$, because the projection is defined orthogonally to F. Say that a point is *inside* $\mathcal{S}$ if it is on the same side of $\mathcal{S}$ as c, and *outside* $\mathcal{S}$ if it is on the same side as p and $\text{proj}_F(p)$. The circumcircle of f cannot enclose the endpoints of s, because f is Delaunay in F. Furthermore, the circumcenter of f lies in F; otherwise, a vertex of f would encroach upon some boundary segment of F. It follows that

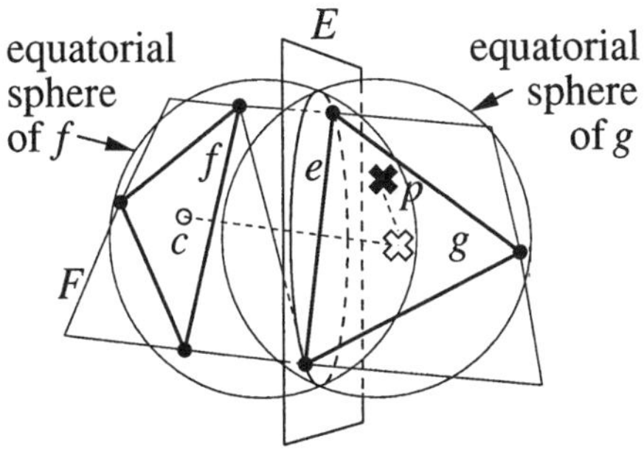

Figure 9: If a vertex p encroaches upon a subfacet f of a Delaunay triangulated facet F, but does not encroach upon any subsegment of F, then p encroaches upon the subfacet(s) g of F that contains $\text{proj}_F(p)$.

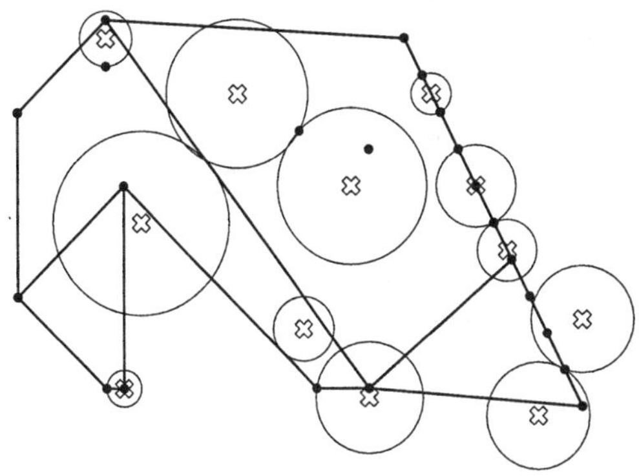

Figure 10: The radius of each disk illustrated is the local feature size of the point at its center.

the portion of f's equatorial sphere outside S lies entirely inside or on the diametral sphere of s (as the figure demonstrates). Because p is inside or on the equatorial sphere of f, p also lies inside or on the diametral sphere of s, contradicting the assumption that p encroaches upon no subsegment of F.

It follows that $\text{proj}_F(p)$ must be contained in some subfacet g of F. (The containment is not necessarily strict; $\text{proj}_F(p)$ may fall on an edge interior to F, and be contained in two subfacets.) To complete the proof of the lemma, I shall show that p encroaches upon g. If $f = g$ the result follows immediately, so assume that $f \neq g$.

Again, let c be the centroid of f. The line segment connecting c to $\text{proj}_F(p)$ must intersect some edge e of the subfacet g, as illustrated in Figure 9. Let $\mathcal{E}$ be the plane that contains e and is orthogonal to F. Say that a point is on the *g-side* if it is on the same side of $\mathcal{E}$ as g. Because the triangulation of F is Delaunay, the portion of f's equatorial sphere on the g-side lies entirely inside or on the equatorial sphere of g. The point p lies on the g-side or in $\mathcal{E}$ (because $\text{proj}_F(p)$ lies in g), and p lies inside or on the equatorial sphere of f, so it must also lie inside or on the equatorial sphere of g, and hence encroaches upon g. ■

If p encroaches upon some subfacet of F but no subfacet of F contains $\text{proj}_F(p)$, then p also encroaches upon some boundary subsegment of F. Because encroached subsegments have priority, subsegments encroached upon by p are split until none remains. The Projection Lemma guarantees that any subfacets of F that were encroached upon by p are eliminated in the process.

On the other hand, if $\text{proj}_F(p)$ lies in a subfacet g of F, and no subsegment is encroached, then splitting g is a good choice. As a result, several new subfacets will appear, at least one of which contains $\text{proj}_F(p)$; if this subfacet is encroached, then it is split as well, and so forth until the subfacet containing $\text{proj}_F(p)$ is not encroached. The Projection Lemma guarantees that any other subfacets of F that were encroached upon by p are eliminated in the process.

7 Proof of Termination and Good Grading

The proof of termination for this algorithm is related to that of Ruppert's. Ruppert's analysis requires that any two incident input segments be separated by an angle of at least 90°. Similarly, the three-dimensional analysis requires that the input PLC satisfy the following *projection condition*: if two constraints (each being a segment or facet) in X intersect, then the orthogonal projection of either one onto the other cannot intersect the interior of the other. (Here, "interior" is defined to exclude all boundaries, including isolated slits and input vertices in the interior of the facet.) For example, if two convex facets intersect along a segment, they must be separated by a dihedral angle of at least 90°. Thanks to the Projection Lemma, this rule ensures that no vertex in the interior of one segment or facet can cause the splitting of a subsegment or subfacet in an incident segment or facet, except that a vertex in the interior of a facet may encroach upon a subsegment of that facet. In theory (and more so in practice), the projection condition can be relaxed somewhat, although the analysis becomes too complicated to present here.

Ruppert's algorithm and the three-dimensional algorithm described here produce nicely graded meshes, in the sense that a small feature in one part of a mesh does not unreasonably reduce the edge lengths at other, distant parts of the mesh. To formalize this idea, Ruppert introduces a function called the *local feature size*, defined on all points.

Given a PLC X, the local feature size $\text{lfs}(p)$ at any point p is the radius of the smallest ball centered at p that intersects two nonincident features of X (where each of the two features might be a vertex, segment, or facet). Figure 10 illustrates the notion in two dimensions (with no facets) by giving examples of such balls for a variety of points.

The function $\text{lfs}(\cdot)$ is continuous and has the property that its directional derivatives (where they exist) are bounded in the range $[-1, 1]$, as the following lemma shows. This property sets a lower bound (within a constant factor to be derived shortly) on the rate at which edge lengths grade from small to large as one moves away from a small feature.

Lemma 2 (Ruppert [13]) *For any PLC X, and any two points u and v,* $\text{lfs}(v) \leq \text{lfs}(u) + |uv|$.

Proof: The ball having radius $\text{lfs}(u)$ centered at u intersects two nonincident features of X. The ball having radius $\text{lfs}(u) + |uv|$ centered at v contains the prior ball, and thus also intersects the same two features. Hence, the smallest ball centered at v that contains two nonincident features of X has radius no larger than $\text{lfs}(u) + |uv|$. ■

With each vertex v, associate an *insertion radius* r_v equal to the length of the shortest edge connected to v immediately after v is introduced into the tetrahedralization. Consider what this means in each possible circumstance. If v is an input vertex, then r_v is the Euclidean distance between v and the input vertex nearest v. If v is a vertex inserted to split a subsegment or subfacet, then r_v is the distance between v and the nearest encroaching mesh vertex. If there is no encroaching vertex in the mesh (a vertex was considered for insertion but rejected as encroaching), then r_v is the radius of the diametral/equatorial sphere of the encroached subsegment/subfacet. If v is a vertex inserted at the circumcenter of a skinny tetrahedron, then r_v is the circumradius of the tetrahedron. If a vertex is considered for insertion but rejected because of an encroachment, its insertion radius is defined the same way, even though it is never actually inserted.

Each vertex v, whether inserted or rejected, has a *parent* vertex $p(v)$, unless v is an input vertex. Intuitively, for any non-input vertex v, $p(v)$ is the vertex that is "responsible" for the insertion of v.

If v is a vertex inserted to split a subsegment or subfacet, then $p(v)$ is the encroaching vertex, whether that vertex is inserted or rejected. If there are several encroaching vertices, choose the one nearest v. If v is a vertex inserted (or rejected) at the circumcenter of a skinny tetrahedron, then $p(v)$ is the most recently inserted endpoint of the shortest edge of that tetrahedron.

Lemma 3 *Let v be a vertex of the mesh, and let $p = p(v)$ be its parent, if one exists. Then either $r_v \geq \text{lfs}(v)$, or $r_v \geq Cr_p$, where*

- *$C = B$ if v is the circumcenter of a skinny tetrahedron;*
- *$C = \frac{1}{\sqrt{2}}$ if v is the midpoint of an encroached subsegment or the circumcenter of an encroached subfacet.*

Proof: If v is an input vertex, there is another input vertex a distance of r_v from v, so $\text{lfs}(v) \leq r_v$, and the theorem holds.

If v is inserted in an encroached subsegment or subfacet, and its parent p is an input vertex or lies on another subsegment or subfacet, then there are two cases to consider. The case in which v lies in a boundary segment of a facet that contains p will be considered shortly. Otherwise, v and p must lie on nonincident features (because X satisfies the projection condition), so $\text{lfs}(v) \leq r_v$.

If v is inserted at the circumcenter of a skinny tetrahedron, then its parent $p = p(v)$ is the most recently inserted endpoint of the shortest edge of the tetrahedron; see Figure 11(a). Hence, the length of the shortest edge of the tetrahedron is at least r_p. Because the tetrahedron is skinny, its circumradius-to-shortest edge ratio exceeds B, so its circumradius is $r_v > Br_p$.

If v is inserted at the midpoint of an encroached subsegment s, and its parent p is a tetrahedron/subfacet circumcenter that was considered for insertion but rejected because it encroaches upon s, then p lies inside or on the diametral sphere of s. Because the tetrahedralization/facet triangulation is Delaunay, the circumsphere/circumcircle centered at p encloses no vertices, and in particular does not enclose the endpoints of s. Hence, $r_p \leq \sqrt{2}r_v$; see Figure 11(b) for an example where the relation is equality.

If v is inserted at the circumcenter of an encroached subfacet f, and its parent p is a tetrahedron circumcenter that was considered for insertion but rejected because it encroaches upon f, then p lies inside or on the equatorial sphere of f. Because the tetrahedralization is Delaunay, the circumsphere centered at p encloses no vertices, including the vertices of f. Recall that the algorithm splits f only if f contains $\text{proj}_f(p)$. Given these facts, one can show that $r_p \leq \sqrt{2}r_v$; see Figure 11(c) for an example where the relation is equality. ■

Lemma 3 limits how quickly the insertion radius can decrease as one traverses a sequence of descendants of a vertex. If vertices with ever-smaller insertion radii cannot be generated, then edges shorter than existing ones cannot be introduced, and Delaunay refinement is guaranteed to terminate. Figure 12 expresses this notion as a dataflow graph. Mesh vertices are divided into four classes: input vertices (which are omitted from the figure because they cannot contribute to cycles), vertices inserted into segments, vertices inserted into facet interiors, and vertices inserted at circumcenters of tetrahedra. Labeled arrows indicate how a vertex can cause the insertion of a child whose insertion radius is some factor times that of its parent. If the graph has no cycle whose product is less than one, termination is guaranteed. This goal is achieved by choosing B to be at least 2.

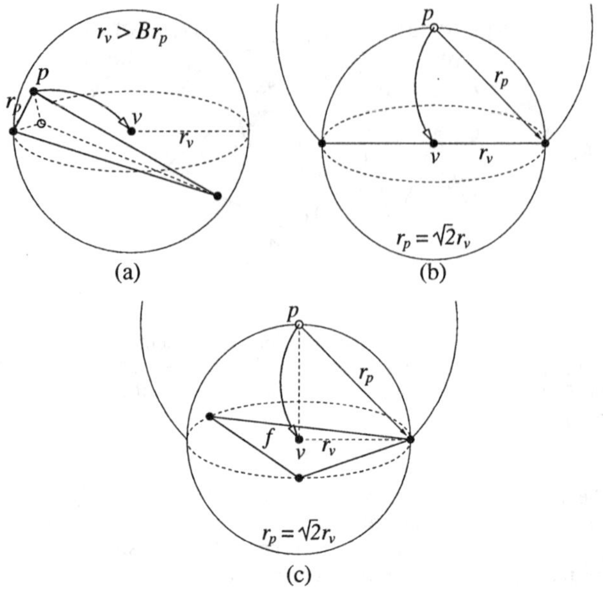

Figure 11: The relationship between the insertion radii of a child and its parent. (a) When a skinny tetrahedron is split, the child's insertion radius is at least B times larger than that of its parent. (b) When a subsegment is encroached upon by a circumcenter, the child's insertion radius may be a factor of $\sqrt{2}$ smaller than its parent's. (c) When a subfacet is encroached upon by the circumcenter of a skinny tetrahedron, the child's insertion radius may be a factor of $\sqrt{2}$ smaller than its parent's.

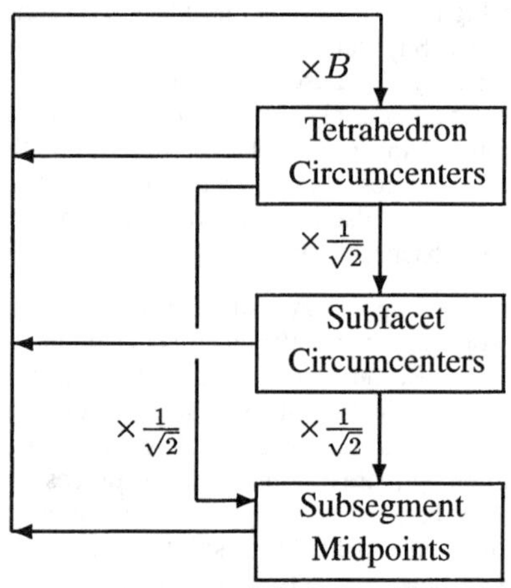

Figure 12: Dataflow diagram illustrating the worst-case relation between a vertex's insertion radius and the insertion radii of the children it begets. If no cycle has a product smaller than one, Delaunay refinement will terminate.

A guarantee of termination alone is not satisfying when a graded mesh is required. What follows is a proof that each edge of the output mesh has length proportional to the local feature sizes of its endpoints. Hence, edge lengths are determined by local considerations; features lying outside the ball that defines the local feature size of a point can only weakly influence the sizes of edges that contain that point.

Lemma 3 was concerned with the relationship between the insertion radii of a child and its parent; the next lemma is concerned with the relationship between $\frac{\text{lfs}(v)}{r_v}$ and $\frac{\text{lfs}(p)}{r_p}$. For any vertex v, define $D_v = \frac{\text{lfs}(v)}{r_v}$. Think of D_v as the one-dimensional density of vertices near v when v is inserted, weighted by the local feature size. One would like this density to be as small as possible. $D_v \leq 1$ for any input vertex, but D_v tends to be larger for a vertex inserted late.

Lemma 4 *Let v be a vertex with parent $p = p(v)$. Suppose that $r_v \geq Cr_p$ (following Lemma 3). Then $D_v \leq 1 + \frac{D_p}{C}$.*

Proof: By Lemma 2, $\text{lfs}(v) \leq \text{lfs}(p) + |vp|$. By definition, the insertion radius r_v is $|vp|$ if p is a mesh vertex, whereas if p is rejected, then $r_v \geq |vp|$. Hence, we have

$$\begin{aligned} \text{lfs}(v) &\leq \text{lfs}(p) + r_v \\ &= D_p r_p + r_v \\ &\leq \frac{D_p}{C} r_v + r_v. \end{aligned}$$

The result follows by dividing both sides by r_v. ■

Theorem 5 *Suppose the quality bound B is strictly larger than 2, and the input PLC satisfies the projection condition. Then there exist fixed constants $D_T \geq 1$, $D_F \geq 1$, and $D_S \geq 1$ such that, for any vertex v inserted (or rejected) at the circumcenter of a skinny tetrahedron, $D_v \leq D_T$; for any vertex v inserted (or rejected) at the circumcenter of an encroached subfacet, $D_v \leq D_F$; and for any vertex v inserted at the midpoint of an encroached subsegment, $D_v \leq D_S$. Hence, the insertion radius of every vertex has a lower bound proportional to its local feature size.*

Proof: Consider any non-input vertex v with parent $p = p(v)$. If p is an input vertex, then $D_p = \frac{\text{lfs}(p)}{r_p} \leq 1$ by Lemma 3. Otherwise, assume for the sake of induction that the lemma is true for p. Hence, $D_p \leq \max\{D_T, D_F, D_S\}$.

First, suppose v is inserted or considered for insertion at the circumcenter of a skinny tetrahedron. By Lemma 3, $r_v \geq Br_p$. Therefore, by Lemma 4, $D_v \leq 1 + \frac{\max\{D_T, D_F, D_S\}}{B}$. It follows that one can prove that $D_v \leq D_T$ if D_T is chosen sufficiently large that

$$1 + \frac{\max\{D_T, D_F, D_S\}}{B} \leq D_T. \tag{1}$$

Second, suppose v is inserted or considered for insertion at the circumcenter of a subfacet f. If its parent p is an input vertex or lies on a segment or facet not incident to the facet containing f, then $\text{lfs}(v) \leq r_v$, so $D_v \leq 1$ and the theorem holds. If p is the circumcenter of a skinny tetrahedron (rejected for insertion because it encroaches upon f), $r_v \geq \frac{r_p}{\sqrt{2}}$ by Lemma 3, so by Lemma 4, $D_v \leq 1 + \sqrt{2} D_T$. It follows that one can prove that $D_v \leq D_F$ if D_F is chosen so that

$$1 + \sqrt{2} D_T \leq D_F. \tag{2}$$

Third, suppose v is inserted at the midpoint of a subsegment s. If its parent p is an input vertex or lies on a segment or facet not incident to the segment containing s, then $\text{lfs}(v) \leq r_v$, and the theorem holds. If p is the circumcenter of a skinny tetrahedron or encroached subfacet (rejected for insertion because it encroaches upon s), $r_v \geq \frac{r_p}{\sqrt{2}}$ by Lemma 3, so by Lemma 4, $D_v \leq 1 + \sqrt{2}\max\{D_T, D_F\}$. It follows that one can prove that $D_v \leq D_S$ if D_S is chosen so that

$$1 + \sqrt{2}\max\{D_T, D_F\} \leq D_S. \tag{3}$$

If the quality bound B is strictly larger than 2, Inequalities 1, 2, and 3 are simultaneously satisfied by choosing

$$D_T = \frac{B + 1 + \sqrt{2}}{B - 2}, \qquad D_F = \frac{(1 + \sqrt{2})B + \sqrt{2}}{B - 2},$$

$$D_S = \frac{(3 + \sqrt{2})B}{B - 2}.$$ ■

Theorem 6 (Ruppert [13]) *For any vertex v of the output mesh, the distance to its nearest neighbor w is at least $\frac{\text{lfs}(v)}{D_S + 1}$.*

Proof: Inequality 3 indicates that D_S is larger than D_T and D_F, so Theorem 5 shows that $\frac{\text{lfs}(v)}{r_v} \leq D_S$ for any vertex v. If v was added after w, then $|vw| = r_v \geq \frac{\text{lfs}(v)}{D_S}$, and the theorem holds.

If w was added after v, apply the theorem to w, yielding

$$|vw| \geq r_w \geq \frac{\text{lfs}(w)}{D_S}.$$

By Lemma 2, $\text{lfs}(w) + |vw| \geq \text{lfs}(v)$, so

$$|vw| \geq \frac{\text{lfs}(v) - |vw|}{D_S}.$$

It follows that $|vw| \geq \frac{\text{lfs}(v)}{D_S + 1}$. ■

To provide an example, suppose $B = 2.5$. Then $D_T \doteq 9.8$, $D_F \doteq 14.9$, and $D_S \doteq 22.1$. Hence, the spacing of vertices is at worst about 23 times smaller than the local feature size. Note that as B approaches 2, the values of D_T, D_F, and D_S approach infinity.

As Figure 13 shows, the algorithm performs much better in practice. In this example, as soon as all encroached subsegments and subfacets have been eliminated (upper left), the largest circumradius-to-shortest edge ratio is already less than 2.1. The shortest edge length is 1, and $\text{lfs}_{\max} = \sqrt{5}$, so the spectre of edge lengths 23 times smaller than the local feature size has not materialized. As the quality bound B decreases, the number of tetrahedra in the final mesh increases gracefully until B drops below 1.05. With $B = 1.04$, the algorithm fails to terminate on this example.

Figure 14 offers a demonstration of the grading of a tetrahedralization generated by Delaunay refinement. A cube has been truncated at one corner, cutting off a portion whose width is one-millionth that of the cube. Although this mesh satisfies a bound on circumradius-to-shortest edge ratio of $B = 1.2$, reasonably good grading is apparent. For this bound there is no theoretical guarantee, but the worst edge is 73 times shorter than the local feature size at one of its endpoints. If a bound of $B = 2.5$ is applied, the worst edge is 9 (rather than 23) times smaller than the local feature size at one of its endpoints.

After the initial Delaunay tetrahedralization has been formed, my implementation runs in constant time per vertex insertion—in practice. The theoretical worst-case running time is surely much worse, and may be $\mathcal{O}(n^2)$ time per vertex insertion, even after amortization, in pathological cases. I do not know how to devise such a case, however. In practice, the first few vertex insertions usually banish any large vertex-free regions, after which further vertex insertions tend to be well-localized and thus execute in constant time.

8 Sliver Removal by Delaunay Refinement

The only known theoretical guarantee about Delaunay refinement's ability to remove sliver tetrahedra is provided by Chew [4], whose bound on tetrahedron quality is too weak to provide any practical assurance. Nonetheless, it is natural to wonder whether Delaunay refinement might be effective in practice. If one inserts a vertex at the circumcenter of any tetrahedron with a small dihedral angle, will the algorithm fail to terminate?

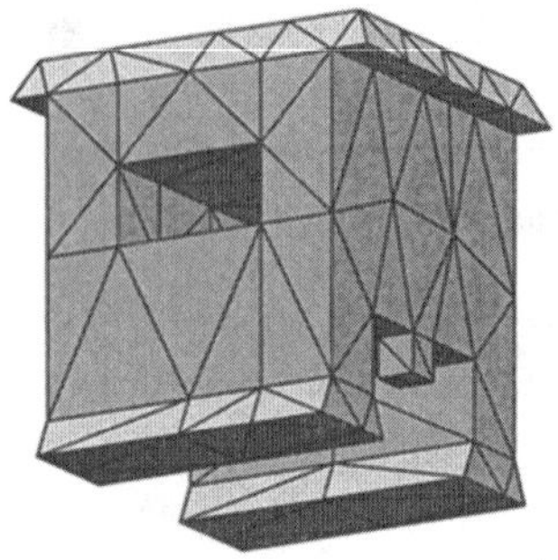

$B = 2.095$, $\theta_{\min} = 1.96°$, $\theta_{\max} = 176.02°$, $h_{\min} = 1$, 143 vertices, 346 tetrahedra.

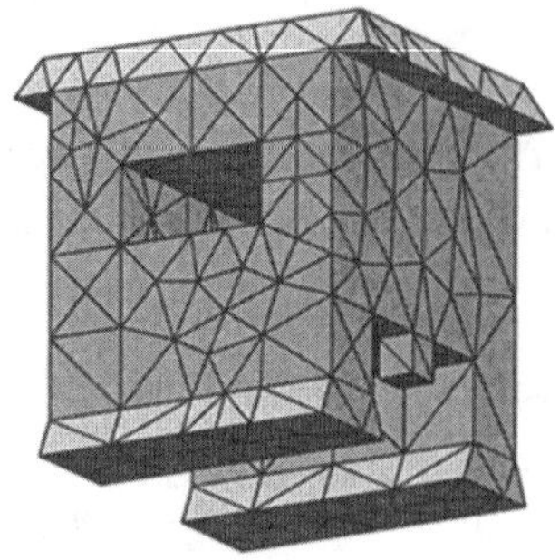

$B = 1.2$, $\theta_{\min} = 1.20°$, $\theta_{\max} = 178.01°$, $h_{\min} = 0.743$, 334 vertices, 1009 tetrahedra.

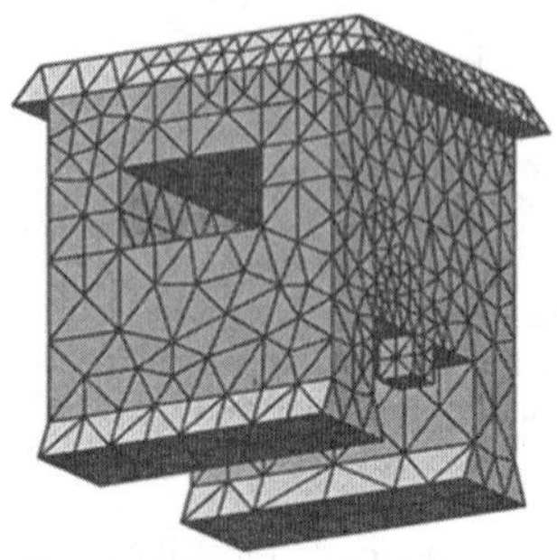

$B = 1.07$, $\theta_{\min} = 1.90°$, $\theta_{\max} = 177.11°$, $h_{\min} = 0.369$, 1397 vertices, 5596 tetrahedra.

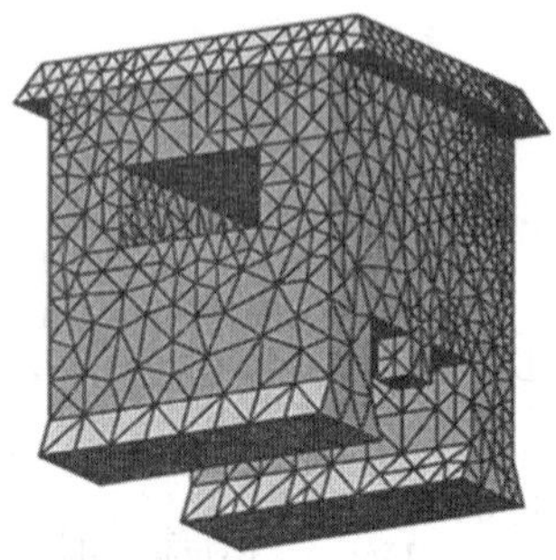

$B = 1.041$, $\theta_{\min} = 0.93°$, $\theta_{\max} = 178.40°$, $h_{\min} = 0.192$, 3144 vertices, 13969 tetrahedra.

Figure 13: Several meshes of a $10 \times 10 \times 10$ PLC generated with different bounds (B) on circumradius-to-shortest edge ratio. Below each mesh is listed the smallest dihedral angle $\theta_{\min}$, the largest dihedral angle $\theta_{\max}$, and the shortest edge length $h_{\min}$. The algorithm does not terminate on this PLC for the bound $B = 1.04$.

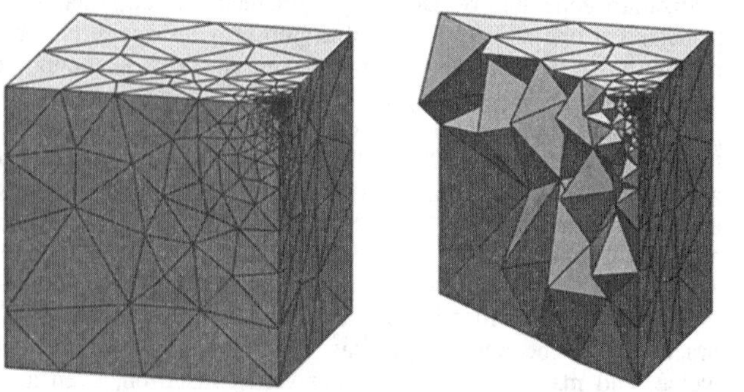

Figure 14: At left, a mesh of a truncated cube. At right, a cross-section through a diagonal of the top face.

As Figure 15 demonstrates, Delaunay refinement can succeed for useful dihedral angle bounds. Each of the meshes illustrated was generated by enforcing a lower bound $\theta_{\min}$ on dihedral angles, rather than a circumradius-to-shortest edge ratio bound. However, the implementation prioritizes poor tetrahedra according to their ratios, and thus slivers are split last. I suspect that the program generates meshes with fewer tetrahedra this way, and that the likelihood of termination is greater. Intuitively, one expects that a vertex inserted at the circumcenter of the tetrahedron with the largest ratio is more likely to eliminate more bad tetrahedra.

Both meshes illustrated have dihedral angles bounded between

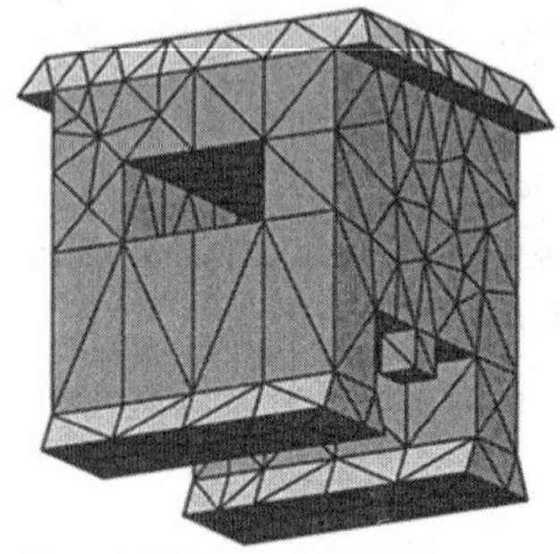

$B = 1.88$, $\theta_{\min} = 23.5°$, $\theta_{\max} = 144.8°$, $h_{\min} = 0.765$, 307 vertices, 891 tetrahedra.

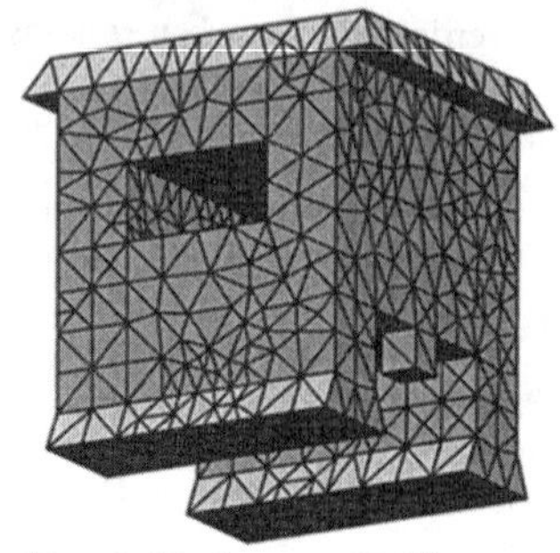

$B = 2.02$, $\theta_{\min} = 21.3°$, $\theta_{\max} = 148.8°$, $h_{\min} = 0.185$, 1761 vertices, 7383 tetrahedra.

Figure 15: Meshes created by Delaunay refinement with bounds on the smallest dihedral angle $\theta_{\min}$.

21° and 149°. The mesh on the right was generated with bounds on both tetrahedron volume and dihedral angle, so that enough tetrahedra were generated to ensure that the mesh on the left wasn't merely a fluke. (The best attainable lower bound drops by 2.2° as a result.) Experiments with very large meshes suggest that a minimum angle of 19° can be obtained reliably. See Chew [4] for some intimations as to why slivers might be eliminated so readily. One of his useful observations is that the region in which a tetrahedron's apex must lie so that the tetrahedron can simultaneously have poor dihedral angles and a good circumradius-to-shortest edge ratio is relatively small.

Unfortunately, my success in removing slivers is probably due in part to the severe restrictions on input angles I have imposed. Practitioners report that they have the most difficulty removing slivers at the boundary of a mesh, especially near small angles. Mesh improvement techniques such as optimization-based smoothing and topological transformations can likely remove some of the imperfections that cannot be removed directly by Delaunay refinement.

9 Conclusions

Delaunay refinement is an effective technique for three-dimensional mesh generation. Its theoretical guarantees on tetrahedron quality and grading make it attractive. However, these guarantees are not entirely satisfying. There is no guarantee that slivers can be eliminated. The bounds on edge length are reassuring, but might not be strong enough for all practical purposes, as the number of tetrahedra is inversely proportional to the cube of the edge length. Although Ruppert uses Theorem 6 to prove the size-optimality of the triangular meshes his algorithm produces, an analogous result is not possible in three dimensions.

Fortunately, Delaunay refinement algorithms outperform their worst-case bounds. There is a great deal of "slack" in the analysis; the relationship between the insertion radii of a child and its parent is usually looser than in the worst case. This slack accumulates as one traces a sequence of descendants from an input vertex, and makes it possible to achieve better bounds on tetrahedron quality and edge length than the theory promises. (Miller et al. [10] cannot exploit this slack, and thus require about 8,000 times as many vertices to guarantee a bound of $B = 2.5$ on the PLC in Figure 13).

The main outstanding problem in three-dimensional Delaunay refinement is the question of how best to handle small input angles. The difficulty is that if vertices lying in segments or facets can encroach upon incident subsegments and subfacets, then additional arcs appear in the dataflow diagram of Figure 12, creating cycles of ever-diminishing insertion radii. Fortunately, it is possible to modify the algorithm so that it always terminates, even when the

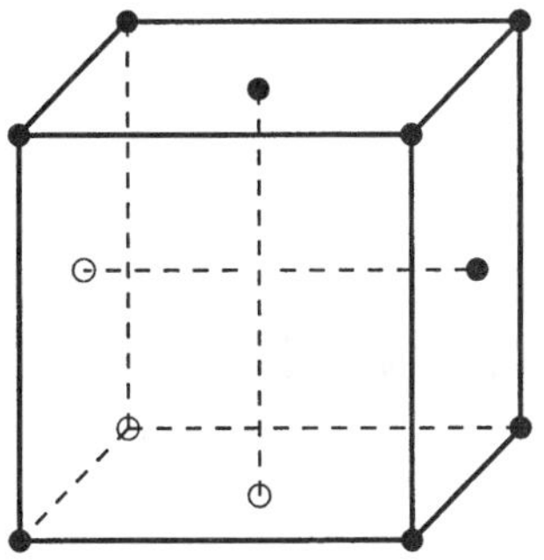

Figure 16: A counterexample demonstrating that the three-dimensional Delaunay refinement algorithm is not size-optimal.

projection condition is not satisfied. The insertion radius of each vertex can be explicitly computed. A vertex should be rejected if its insertion radius is too small compared to its parent's. As a result, some skinny tetrahedra may remain in the final mesh, but only in the vicinity of small input angles. A few other tricks are needed to ensure that all segments and facets are recovered: segments are recovered by using a technique based on concentric spherical shells suggested by Ruppert [13], then facets are recovered by forming a constrained Delaunay tetrahedralization [15].

The reason why a size-optimality result cannot be established is worth exploring. Gary Miller and Dafna Talmor (private communication) have pointed out the counterexample depicted in Figure 16. Inside this PLC, two segments pass very close to each other without intersecting. The PLC might reasonably be tetrahedralized with a few dozen tetrahedra having bounded circumradius-to-shortest edge ratios, if these tetrahedra include a sliver tetrahedron whose four vertices are the endpoints of the two internal segments. However, the best the present Delaunay refinement algorithm can promise is to fill the region with tetrahedra whose edge lengths are proportional to the distance between the two segments. Because this distance may be arbitrarily small, the algorithm is not size-optimal. However, if guaranteed bounds could be established for the dihedral angles, and not merely the circumradius-to-shortest edge ratios, then size-optimality might be proven using ideas like those with which Mitchell and Vavasis [11, 12] demonstrate the size-optimality of their octree-based algorithms.

Many theoretical improvements can be made to the core algorithm described here. The bound on circumradius-to-shortest edge ratio can be improved to 1.63 by replacing equatorial spheres with narrower protective shells called *equatorial lenses.* (This technique also requires the use of constrained Delaunay tetrahedralizations.) The bound can be further improved to 1.15 with a technique called *range-restricted segment splitting*, albeit by sacrificing good grading in theory (not in practice). With or without these improvements, a bound as small as 1 can be applied to any tetrahedron that does not intersect the interior of a segment or facet, and thus only boundary tetrahedra are subject to the weaker bound of 2, 1.63, or 1.15. The projection condition can be somewhat weakened; for example, a 60° separation between incident segments suffices. All these improvements are described in detail elsewhere [14]. The improvements in circumradius-to-shortest edge ratio, in particular, are significant and further propel the approach described here beyond the pioneering tetrahedral mesh generation algorithms of Dey et al., Chew, and Miller et al.

References

[1] Adrian Bowyer. *Computing Dirichlet Tessellations.* Computer Journal **24**(2):162–166, 1981.

[2] L. Paul Chew. *Guaranteed-Quality Triangular Meshes.* Technical Report TR-89-983, Department of Computer Science, Cornell University, 1989.

[3] ——. *Guaranteed-Quality Mesh Generation for Curved Surfaces.* Proceedings of the Ninth Annual Symposium on Computational Geometry, pages 274–280. ACM, May 1993.

[4] ——. *Guaranteed-Quality Delaunay Meshing in 3D.* Proceedings of the Thirteenth Annual Symposium on Computational Geometry, pages 391–393. ACM, June 1997.

[5] Tamal Krishna Dey, Chanderjit L. Bajaj, and Kokichi Sugihara. *On Good Triangulations in Three Dimensions.* International Journal of Computational Geometry & Applications **2**(1):75–95, 1992.

[6] Lori A. Freitag and Carl Ollivier-Gooch. *A Comparison of Tetrahedral Mesh Improvement Techniques.* Fifth International Meshing Roundtable (Pittsburgh, Pennsylvania), pages 87–100. Sandia National Laboratories, October 1996.

[7] William H. Frey. *Selective Refinement: A New Strategy for Automatic Node Placement in Graded Triangular Meshes.* International Journal for Numerical Methods in Engineering **24**(11):2183–2200, November 1987.

[8] Charles L. Lawson. *Software for C^1 Surface Interpolation.* Mathematical Software III (John R. Rice, editor), pages 161–194. Academic Press, New York, 1977.

[9] Gary L. Miller, Dafna Talmor, Shang-Hua Teng, and Noel Walkington. *A Delaunay Based Numerical Method for Three Dimensions: Generation, Formulation, and Partition.* Proceedings of the Twenty-Seventh Annual ACM Symposium on the Theory of Computing, pages 683–692, May 1995.

[10] Gary L. Miller, Dafna Talmor, Shang-Hua Teng, Noel Walkington, and Han Wang. *Control Volume Meshes using Sphere Packing: Generation, Refinement and Coarsening.* Fifth International Meshing Roundtable, pages 47–61, October 1996.

[11] Scott A. Mitchell and Stephen A. Vavasis. *Quality Mesh Generation in Three Dimensions.* Proceedings of the Eighth Annual Symposium on Computational Geometry, pages 212–221, 1992.

[12] ——. *Quality Mesh Generation in Higher Dimensions.* Submitted to SIAM Journal on Computing, February 1997.

[13] Jim Ruppert. *A Delaunay Refinement Algorithm for Quality 2-Dimensional Mesh Generation.* Journal of Algorithms **18**(3):548–585, May 1995.

[14] Jonathan Richard Shewchuk. *Delaunay Refinement Mesh Generation.* Ph.D. thesis, School of Computer Science, Carnegie Mellon University, Pittsburgh, Pennsylvania, May 1997. Available as Technical Report CMU-CS-97-137.

[15] ——. *A Condition Guaranteeing the Existence of Higher-Dimensional Constrained Delaunay Triangulations.* This proceedings, June 1998.

[16] David F. Watson. *Computing the n-dimensional Delaunay Tessellation with Application to Voronoi Polytopes.* Computer Journal **24**(2):167–172, 1981.

[17] Nigel P. Weatherill. *Delaunay Triangulation in Computational Fluid Dynamics.* Computers and Mathematics with Applications **24**(5/6):129–150, September 1992.

Implementations of the LMT Heuristic for Minimum Weight Triangulation

Ronald Beirouti* Jack Snoeyink†
Dept. of Computer Science
University of British Columbia

Abstract

No polynomial-time algorithm is known to compute the minimum weight triangulation (MWT) of a finite planar point set. In this paper we present efficient implementations of the LMT-skeleton heuristic, which identifies edges that must be, and cannot be, in an MWT. For uniformly distributed points, we can compute the exact MWT of tens of thousands of points in minutes.

These results are obtained by improving the asymptotic time and memory usage of the LMT-skeleton heuristic and of filters that identify initial candidate edges, and also by bucketing and further tuning for evenly distributed points. Further details and an implementation as a macro for the IPE drawing program are available on the web: `http://www.cs.ubc.ca/spider/snoeyink/mwt`.

1 Introduction

Given a set S of points in the Euclidean plane, a *triangulation* is a maximum set of non-crossing segments (or edges) with endpoints in S. These edges are said to *triangulate S*. The *weight of a triangulation* is the sum of the Euclidean lengths of its edges.

It is natural to ask for an algorithm to compute a minimum-weight triangulation (MWT), and applied papers in areas such as stock cutting, finite element analysis, and terrain modeling occasionally do so. The proper response from theoretical computational geometers is to advise practitioners that computing an MWT is difficult, and to suggest other triangulations with nice properties (e.g. the Delaunay triangulation [29]). As both a theoretical and practical question, however, the problem of computing an MWT is intriguing. No polynomial time algorithm is known, nor has the problem been proven NP-hard. In fact, computing an MWT is one of the few problems in Garey and Johnson's book on NP-completeness [16] whose complexity is still unknown.

We briefly review approaches to the MWT problem in Section 2. We focus on two: the diamond test of Das and Joseph [10], which shows that certain edges with endpoints in S cannot be in an MWT, and the LMT-skeleton heuristic of Dickerson, Keil, and Montague [12], which identifies edges that are *possible* and edges that are *certain* to be in an MWT. Actually, they identify edges for any *locally-minimal triangulation*; it is heuristic in the sense that, for many but not all point sets, it identifies sufficiently many edges that the MWT can be completed by dynamic programming.

We implemented several improved versions of this heuristic: Section 3.1 describes an improvement from $O(n^6)$ time to $O(n^4)$ and from $O(n^3)$ space to $O(n^2)$ that was referred to in the 5th Video Review of Computational Geometry [4]. Section 4 describes a modified diamond test that makes the algorithm run in $O(n^2 \log d + d^2 E)$ time and $O(E)$ space, where d is the maximum degree of the graph of edges that survive the pretest and $E \leq nd$ is the number of edges that survive. Section 5 describes our method for bucketing, which gives linear expected time when the point set is uniformly distributed over a square. The constants hidden by the big-O notation are small, although we omit most of the details from this extended abstract.

Until recently, MWTs were known only for special point sets having less than 400 points. Our code computes MWTs of 40,000 points in less than 2.5 minutes on a 200Mhz SGI XZ. The code

*Supported in part by a grant from FCAR Quebec.
†Supported in part by grants from NSERC Canada and Facet Decision Systems.

SCG 98 Minneapolis Minnesota USA

is available from `http://www.cs.ubc.ca/spider/snoeyink/mwt` as a user macro for Schwarzkopf's drawing system IPE [31]. This macro has already been useful for exploring the difficulty of the MWT problem: because the LMT heuristic identifies many edges that are certainly in or out of an MWT, one can begin to find structure among the edges that remain. Section 6 lists some examples.

2 History

One can identify several types of attacks on the MWT problem: direct attacks by dynamic programming, studying the weight properties of known triangulations, approximation algorithms, efficient algorithms for restricted classes of point sets, and algorithms that identify edges that must be or must not be in a MWT. Two of the later, the *diamond test* and *LMT-skeleton*, are the foci of this paper.

Dynamic programming: Gilbert [17] and Klincsek [21] independently presented a dynamic programming algorithm that computes a minimum weight triangulation of a simple polygon in $O(n^3)$ time. We note in Section 3 that when one is told a subset of the edges that contain the MWT, then the time for dynamic programming is reduced. Many others have applied dynamic programming with branch and bound techniques to the general problem. E.g, Cheng et al. [7] use dynamic programming to complete a subgraph of a minimum weight triangulation composed of k unconnected components in $O(n^{k+2})$ time. Kyoda [22] used branch and cut to compute MWTs of 100 points.

Approximations to the MWT: The Delaunay is not always an MWT, and may be longer by a factor of $\Omega(n)$ [28, 20]. The greedy triangulation may be longer by a factor of $\Omega(\sqrt{n})$ [23]. Plaisted and Hong [30] proposed a heuristic that approximates the MWT within a factor of $\Omega(\log n)$. To compute this triangulation the algorithm required $O(n^2 \log n)$ worst-case time. Lingas [27], Levcopoulos et al. [26] and Levcopoulos and Krznaric [24, 25] have an in-depth study of the MWT and its relation to greedy triangulations, which includes a modified greedy triangulation that achieves a constant factor approximation, and, for convex polygons, a $(1+\epsilon)$ approximation factor in linear time, for any fixed ϵ.

Restricted Classes of Point Sets: Many of the heuristics mentioned do compute MWTs of special point sets, but have no characterization of the point sets for which they work. Examples that do, include Anagnostou and Corneil's $O(n^{3k+1})$ time algorithm [3] for point sets that are the vertices of k nested convex polygons.

Aichholzer et al. [1] define a *light edge* in a point set S to be any segment joining two points of S that does not intersect another segment joining two other points of S. They proved that any triangulation composed entirely of light edges was a MWT and exhibited an MWT of 150 points. They have also extended the concept of light edges [2] to characterize when greedy triangulations perform well.

Subgraphs of the MWT: The β-skeleton, for $\beta \geq 1$, of a point set S is the set of edges with endpoints in S such that for each edge $\overline{ab}$ the two circles of radius $\frac{\beta}{2}|\overline{ab}|$ passing through a and b are empty of all points. Keil [19] showed that the β-skeleton is also a subgraph of the MWT when $\beta \leq \sqrt{2}$. Cheng and Xu [9] showed that the β-skeleton is still a subgraph of the MWT for $\beta \leq 1.17682$. Unfortunately this subgraph usually contains many disconnected components, and there is little room for improvement of β since Keil [19] also found a four point example for which the β-skeleton is not a subgraph of the MWT for $\beta < 1/\sin(\pi/3) < 1.154701$).

Yang, Xu, and You [33] show that edges that join mutual nearest neighbors are in the MWT. Wang, Chin, and Xu [32] propose a test that can identify which edges of the greedy triangulation are in the MWT in $O(n^3)$ time and $O(n^2)$ space.

Supergraphs of the MWT: Diamond Test: Das and Joseph [10] argued that all edges in an MWT have the following *diamond property*.

Theorem 2.1 If an edge $\overline{ab}$ is in MWT(S) then at least one of the two isosceles triangles with base $\overline{ab}$ and base angles $\pi/8$ contains no points of S (Figure 1).

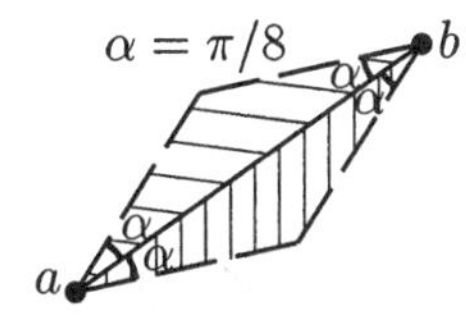

Figure 1: $\overline{ab}$ diamond test

The *diamond test* eliminates from the initial possible edge set those edges with at least one point in each of the two triangles of the diamond. For uniformly-distributed points in the plane, it is not hard show that only a linear number of edges survive the diamond test. A similar lemma, with larger constants, can be proved for points uniformly distributed in the unit square.

Lemma 2.1 The expected number of edges that pass the complete diamond test with angle θ in a uniformly distributed set of n points is less than $(3\pi/\sin\theta)n$.

With angle $\pi/8$, we expect less than $25n$ edges to survive the diamond test. Recently, Drysdale, McElfresh, and Snoeyink [14] have improved the angle bound from $\pi/8$ to $\pi/4.6$, for which $15n$ edges are expected to survive.

A straightforward implementation of the diamond test could consider each edge individually and simply check all points against the two triangles. This would take $\Theta(n^3)$ time. In sections 4 and 5 we give methods that are faster—in section 4 by weakening the property to allow a more efficient test, and in section 5 by bucketing under the assumption that few edges pass the diamond test. Bucketing has been successfully employed in the MWT and related problems by a number of people, including Dickerson, Drysdale, Aichholzer, and ourselves.

Subgraphs of the MWT: LMT-skeleton: Dickerson, Keil, and Montague [12] describe the LMT-skeleton, a new subgraph of the MWT that has been the focus of much recent effort.

Let e be an edge in a triangulation $T(S)$. Edge e is *locally-minimal* if e is on the convex hull of S, if the adjacent quadrilateral is not convex, or if $weight(e) \leq weight(d)$, where d is the diagonal of the adjacent quadrilateral. Note that, if e is not locally-minimal, then the weight of the triangulation $T(S)$ can be decreased by swapping e for d. When e is locally-minimal, the pair of triangles forming Q is called the *certificate* for e; Figure 2 gives two examples.

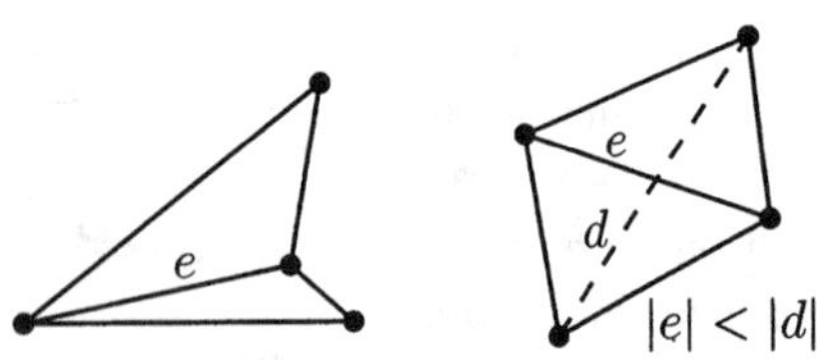

Figure 2: Certificates for edge e

A *locally-minimum triangulation* is one in which every edge is locally-minimum; clearly, a MWT will be a locally-minimum triangulation. The *locally-minimum skeleton* (LMT-skeleton) is defined as the set of edges that are in all locally-minimum triangulations of a point set.

Given a point set S, an iterative algorithm can find edges of the LMT-skeleton, and therefore of the MWT. The algorithm partitions the set of all edges joining pairs of points in S into three classes: *impossible* in an LMT, *possible* in an LMT, or *certain* in an LMT, and therefore in the LMT-skeleton.

All edges are initially labeled *possible*. The algorithm then considers each possible edge e and makes e *impossible* if e has no certificate made of possible edges. Note that making an edge impossible may invalidate previous certificates. Edge e is made *certain* if e is on the convex hull of S or e has a certificate and no possible or certain edge crosses e. All *certain* edges form a subgraph of the LMT-skeleton of S, which is a subgraph of the MWT(S). If the set of *certain* edges is connected, then the MWT can be completed by applying the dynamic programming algorithm for polygons.

The implementation of Dickerson et al. [12] differs from ours primarily in two points. First, Dickerson et al. not only store all edges that join point in S, but also the set of all *possible triangles*, which initially contains all empty triangles on S—that is, all triangles with vertices in S, that intersect no other points of S. Their certificate for an edge e is a pair of adjacent, possible triangles that form a non-convex quadrilateral or a convex quadrilateral whose other diagonal is longer than e. To find a certificate for edge e, their algorithm inspects all pairs of possible triangles adjacent to e. If no certificate for e is found, then e and all adjacent triangles are made impossible. Space is a bottleneck since, in the worst case, $\Theta(n^3)$ empty triangles must be stored. For uniformly distributed point sets, they observed that $O(n^2 \log n)$ space was expected, and were able to run their implementation on 250 points in about 12 hours on a Macintosh 8500/120.

Second, because making edges or triangles impossible may make earlier certificates invalid, Dickerson et al. apply the test in several passes. When no edge or triangle is made impossible in a pass, then the algorithm is done. For a worst-case set of n point, each pass takes at most $O(n^4)$ time, and there may be $O(n^2)$ passes for a total bound of $O(n^6)$.

As described in the next section, we base our data structure on edges, rather than triangles. Using the fact that the *possible* edges always contain a triangulation, we can search locally for possible certificates, for certificates that are invalidated when an edge is made impossible, and for possible edges that are not crossed by other possible edges and are therefore *certain.* Then, using the diamond test and bucketing to eliminate edges before we compute the LMT-skeleton, we obtain algorithms that have better worst-case, expected-case, and observed time bounds.

Others have also made modifications and improvements to the LMT-skeleton algorithm: Cheng, Katoh, and Sugai [8] weakened the test to allow each pass to run in $O(n^3 \log n)$ time and $O(n^2)$ space, and ran tests on up to 2,000 vertices. Hainz, Aichholzer, and Aurenhammer [18] incorporate bucketing techniques of [15] and other local tests to obtain an algorithm that has linear performance on uniformly-distributed point sets. They are able to compute the LMT-skeleton of up to 5,000 points in 20 minutes and have run tests on up to 15,000 vertices.

3 Basic Data Structure and Algorithms

This section presents the data structure and basic procedures used by the *LMT-skeleton* heuristic. The bottleneck of the algorithm described by Dickerson (Section 2) was the $O(n^3)$ space required to store all empty triangles. Our LMT-skeleton algorithm tries to minimize memory use by not keeping a list of empty triangles but by scanning for the empty triangles when they are needed.

```
Edge := { Point ptr dest
          Status    s
          Edge ptr  rev, next
          Edge ptr  i, j, dstart
          Edge ptr  rightPoly, leftPoly
          real      polyWeight
        }
```

Figure 3: Data for a directed edge

Our data structure is edge-based, storing origin and destination points of the edge, two radially-sorted lists of edges around each endpoint, and pointers to scan for empty triangles. We use the directed edge structure of Figure 3, which we now describe in more detail.

Between every two points a and b there are two directed edges, $\overline{ab}$ and $\overline{ba}$, where $\overline{ba} = \overline{ab}{\rightarrow}rev$ and $\overline{ab} = \overline{ba}{\rightarrow}rev$. The edge $\overline{ab}$ holds a pointer to its destination endpoint, $\overline{ab}{\rightarrow}dest = b$. The point a is then $\overline{ab}{\rightarrow}rev{\rightarrow}dest$. Variable s stores the edge's status: whether the edge is *possible*, *certain* or *impossible.*

The pointer $\overline{ab}{\rightarrow}next$ points to the next counter-clockwise edge in the radially sorted list of edges around a; pointer $\overline{ba} \rightarrow next$ points to the next counter-clockwise edge around b. The radially sorted list $\overline{ab}{\rightarrow}next$ only contains *possible* or *certain* edges; this is maintained by lazy deletion, as described in the full paper.

Pointers $\overline{ab} \rightarrow i$ and $\overline{ab} \rightarrow j$ are used to scan for empty triangles left of $\overline{ab}$ in time proportional to the degrees of a and b. The edge $\overline{ab}{\rightarrow}$ *dstart* points to the first edge ccw around b from the ray $\overrightarrow{ab}$, and is used to initialize the scan. All pairs of empty triangles can be checked for certificates, therefore, in $O(d^2)$ time. When a certificate is found for $\overline{ab}$, the pointers $\overline{ab}{\rightarrow}i$ and $\overline{ab}{\rightarrow}j$ point to the edges of the empty triangle to the left of $\overline{ab}$ and the pointers $\overline{ab}{\rightarrow}rev{\rightarrow}i$ and $\overline{ab}{\rightarrow}rev{\rightarrow}j$ point to the two edges of the other empty triangle as seen in figure 4. When a certificate becomes invalid, the scan can resume where it left off.

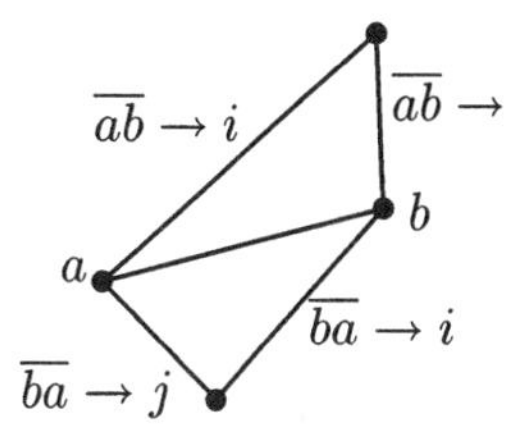

Figure 4: Pointers for $\overline{ab}$ and $\overline{ab}{\rightarrow}rev = \overline{ba}$

Finally, *rightPoly*, *leftPoly* and *polyWeight* are used to triangulate the polygonal holes remaining in the LMT-skeleton. This is done by a standard dynamic programming algorithm, with the modification that it can work efficiently with a given subset of the edges that are known to contain all the edges of an MWT, again described in the full paper.

The Sum of Square Roots: In the dynamic programming algorithm, we implicitly assume that one can determine if a sum of radicals is less than, equal to, or greater than zero. No polynomial-time for this problem is known for machine models that lack square roots as a primitive operation [5].

Summing of square roots is common in geometric optimization problems that involve computing Euclidean lengths. In the paper we assume that the dynamic programming algorithm can use square roots as a primitive and, thus, compute the exact MWT. It is notable that the LMT-skeleton algorithm does not involve comparing sums of square roots, but works with at most two radicals at a time. This allows it to be implemented exactly, even in fixed precision models of computation.

3.1 The *LMT-skeleton* Implementation

We sketch how our implementation computes an LMT-skeleton using the edge-based data structure; more detail is included in the full paper. First, make all edges *possible* and sort around each vertex to create the *next* lists. Push all edges onto a stack, indicating that certificates need to be found. (We sort the stack so the longest appear on top, since they are least likely to have certificates.)

```
Initialize data structures
Push all edges onto the stack ST
Sort ST so that longest edges are on top
while ST is not empty
  e := pop ST
  check certificate of e
  if e has no certificate
    check crossing edges
    if e has no crossing edge
      e→s := certain
    else
      e→s := impossible
      restack edges
for remaining possible edges
  check crossing edges
```

Figure 5: The algorithm for computing the *LMT-skeleton*

The top edge $\overline{ab}$ is then popped from the stack and adjacent pairs of triangles are scanned in an attempt to find a certificate for $\overline{ab}$. If no certificate is found, then $\overline{ab}$ is either a convex hull edge or is impossible—this can be determined by a local scan, since the possible edges are guaranteed to contain a triangulation. (See the full paper.)

If $\overline{ab}$ is impossible, then one must restack any edges that depended on $\overline{ab}$ for their certificates—they no longer have valid certificates. These edges can be found by scanning the neighbors of $\overline{ab}$ and looking at the i and j pointers of the scanned segments.

Finally, when the stack is empty, each edge has a certificates or has been declared *impossible*. Some edges may be *certain*—namely, those whose removal prevents us from completing a triangulation. We can identify that an edge $\overline{ab}$ is not *certain* if we can find an empty triangle $\triangle acd$ such that $\overline{ab}$ crosses $\overline{cd}$. This is another local scan of the neighbors of a.

The most time-consuming step in this algorithm is the check for certificates, which takes $O(n^4)$ time in total. If the input is a set of possible edges where the maximum degree of a vertex is d, then the algorithm takes $O(Ed^2)$ total time.

The edges that remain *possible* can be isolated into polygonal regions. If these regions are connected, then the minimum weight triangulation can be completed by dynamic programming.

3.2 Experimental Results

This algorithm was implemented in *cweb* and run on an *SGI Indy* in producing the video [4]. To correctly perform all arithmetic operations in 53-bit double precision, the input points were scaled to 20 bit positive integers. Degeneracies, such as co-linearities and equal length edges, were handled. Trials showed that for small sets of uniformly distributed random points, the algorithm computes the complete *MWT*. For 250 and 1000 points, the algorithm completed in, respectively, 25 seconds and half an hour. The bottleneck was definitely the $O(n^2)$ space required by making all edges initially possible.

4 The Diamond Test Algorithm

The LMT-skeleton implementation just described spends most of its time discarding long edges that are *impossible* in an MWT. As suggested by Dickerson et al. [12], we use the diamond test as a filter.

It is natural, in our data structure, to implement the diamond test for all edges around a point a by sorting radially and scanning to see if each diamond is occupied. Unfortunately, this takes $\Theta(n^3)$ in both worst and expected cases. In the expected case, the constant is about 1/32, so when n is larger than 250, we prefer to use a weaker test that can be combined with the sorting step to produce an algorithm that runs in expected $O(n^2 \log d)$ time, where d is the average degree of the vertices after the test has been applied. Specifically, we use Quicksort to sort the edges around point a, with the modification that it chooses a partitioning point p in order of increasing distance from a, and uses p to kill off diamonds for edges as it partitions. A few extra edges are accepted since edges are tested only against partitioning points and not against all points, but the full paper shows that expected degree remains constant for uniformly distributed points.

n	# edges pass diamond test	avg degree after test	time in secs
100	1 511	30.21	0.18
200	3 711	37.11	0.60
400	8 088	40.44	2.58
600	12 227	40.76	4.14
800	16 972	42.43	7.32
1 000	21 386	42.77	11.46
2 000	44 846	44.85	44.64
4 000	92 974	46.49	175.19
6 000	141 290	47.10	390.48
8 000	189 438	47.36	721.38
10 000	238 079	47.62	1 283.94
12 000	285 956	47.66	2 210.94
16 000	383 538	47.94	4 818.90
20 000	483 174	48.32	8 242.50
30 000	729 612	48.64	20 018.76
40 000	973 547	48.68	36 998.64

Table 1: Statistics for LMT with diamond test on uniformly distributed points

4.1 Experimental Results

The algorithm was implemented in *C++* and run on an *SGI XZ* with a 200 MHz *IP22* processor. The algorithm was run on several sets of uniformly distributed random points ranging from 100 to 40,000 points; timing data is plotted in Figure 10. For n points, about $50n$ directed edges pass the weakened diamond test.

For every point set, a connected graph was obtained, which means the algorithm was able to compute MWT(S) by triangulating the unresolved polygonal holes. Table 1 gives a summary of the times observed and the number of edges that passed the diamond test.

5 Eliminating Edges by Buckets

The diamond test defined in previous section is faster than the straightforward algorithm that tests all diamonds, but it still must look at every edge. Bucketing techniques, which have been successfully applied in greedy triangulations [11, 15], can be used to throw out clusters of edges at a time.

We apply bucketing to radially sort only those edges that pass the diamond test around each vertex; our main ideas can be summarized in two figures. First, given three points o, a and b with angle $\angle aob$ less than $\pi/4$, there is a *dead sector S* consisting of all the points p such that $\overline{op}$ fails the diamond test by having a and b in the isosceles triangles to the left and right of $\overline{op}$.

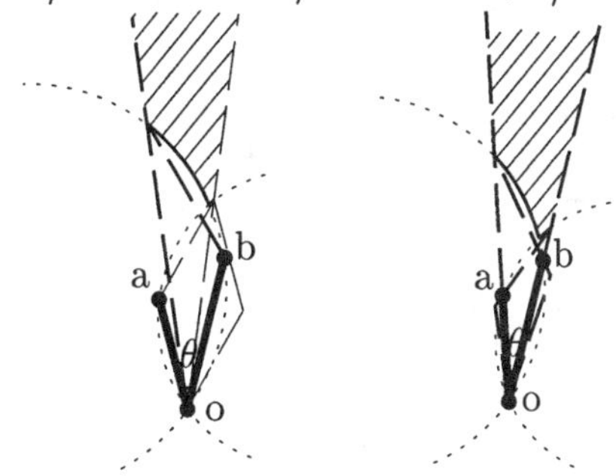

Figure 6: Sectors defined by two line segments

As shown in Figure 6, the dead sector has a complicated boundary that depends on the actual angle and lengths of $\overline{oa}$ and $\overline{ob}$. When the angle is less than $\pi/8$, the dead sector is the region between the rays $\overrightarrow{oa}$ and $\overrightarrow{ob}$, but outside the two circles that contain the points p such that angles $\angle opa = \pi/8$ and $\angle bpo = \pi/8$. When the angle is between $\pi/8$ and $\pi/4$, then the sector is bounded by rays at angles $\pi/8$ from $\overrightarrow{oa}$ and $\overrightarrow{ob}$, and again outside the circles.

To make bucketing as effective as possible, we must be able to accept or reject buckets quickly. Thus, we do not compute sector boundaries except where they intersect the line that separates a row of inspected buckets from a row of uninspected buckets. This intersection, partially illustrated in Figure 7, is obtained without trigonometric calculations from a quadratic equation.

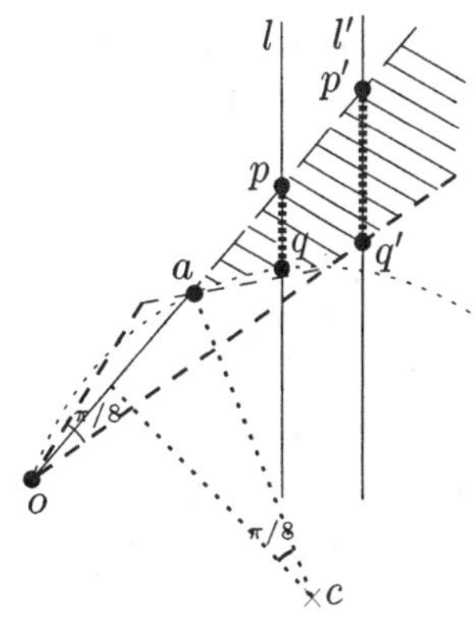

Figure 7: Sector & bucket-line intersect

n	Diamond with angle $\pi/8$					Diamond with angle $\pi/4.6$				
	edges passing	average degree	avg bkts visited	avg layers visited	time in secs	edges passing	average degree	avg bkts visited	avg layers visited	time in secs
100	1 705	34.10	13.34	6.00	0.20	1 022	20.44	12.48	6.00	0.1
200	3 759	37.59	18.81	8.96	0.59	2 320	23.20	16.88	8.37	0.3
400	8 474	42.37	21.10	10.30	1.25	5 004	25.02	18.75	9.32	0.8
600	13 342	44.47	23.19	11.29	2.16	7 592	25.31	20.43	10.19	1.1
800	18 111	45.28	25.33	12.23	3.14	10 181	25.45	21.68	10.78	1.7
1 000	22 777	45.55	27.31	13.02	4.09	13 031	26.06	22.41	11.12	2.1
2 000	47 531	47.53	30.11	14.17	10.65	27 081	27.08	23.93	11.84	4.9
4 000	98 488	49.24	31.31	14.73	21.07	55 755	27.88	25.03	12.32	10.8
6 000	149 985	50.00	31.63	14.93	33.36	83 947	27.98	25.42	12.44	17.1
8 000	201 574	50.39	32.90	15.38	45.86	112 450	28.11	25.89	12.65	23.2
10 000	254 145	50.83	32.62	15.32	59.54	141 817	28.36	26.04	12.71	30.2
12 000	306 397	51.07	32.72	15.34	75.73	171 082	28.51	26.12	12.74	37.2
16 000	411 778	51.47	33.41	15.62	106.21	229 137	28.64	26.47	12.88	52.2
20 000	513 869	51.39	33.74	15.72	138.96	285 460	28.55	26.66	12.95	65.9
25 000	642 468	51.40	33.69	15.76	178.65	356 331	28.51	26.71	12.98	89.4
30 000	770 904	51.39	34.00	15.88	224.03	427 900	28.53	26.87	13.05	104.5
35 000	901 881	51.54	34.10	15.92	255.16	500 259	28.59	26.93	13.06	123.0
40 000	1 031 140	51.56	34.25	15.97	293.27	571 224	28.56	26.99	13.10	141.0
60 000						859 795	28.66	27.16	13.15	229.3
80 000						1 148 420	28.71	27.30	13.21	330.7

Table 2: Statistics observed for the LMT heuristic with bucketing on uniformly distributed point sets, using diamonds of angle $\pi/8$ or $\pi/4.6$

5.1 Optimizations and Experimental Results

We ran our *C++* implementation on a SGI XZ with 200MHz IP22 processor on uniformly distributed point sets of 100 to 40,000 points, and obtained the results in the left half of Table 2. After some experimentation, we settled on a grid size for which we would expect about five points per bucket, namely $\lceil\sqrt{n/5}\rceil^2$ cells. We started for each point by sorting edges to all points at most two layers away. This explains why the number of buckets visited rises quickly to about 25, and then grows more slowly. The number of buckets visited is less than 25 for smaller problem sizes because of edge effects on the grid.

A significant improvement in the running time was observed in our implementation of the algorithm. For uniformly distributed point sets, the exact minimum weight triangulation of 40,000 points was found in less than 5 minutes. As we had hoped, only a constant number of layers and buckets are visited around each point. The time saved by not sorting all $n-1$ edges for each point was reflected in the observed linear behavior of the overall algorithm.

Space remains a bottleneck in our implementation. Once the data structure size outgrows main memory (somewhere between 40 and 60 thousand points on the machine we were using) our reliance on pointer-following in an edge-based data structure leads to severe thrashing in the LMT-skeleton heuristic. Some spatial partitioning could be used on top of our structure to improve the locality of reference and enable the handling of larger point sets. (Since our implementation sorts edges by decreasing length before running the LMT-skeleton algorithm, it actually destroys any spatial partitioning that might have been induced by bucketting.)

Improving the filter is also an attractive alternative. The right side of Table 2 shows the statistics when the angle of the diamond test is changed from $\pi/8$ to $\pi/4.6$, in accord with a recent proof of Drysdale et al. [14]. This change reduces running time—40,000 points now run in less than 2.5 minutes—and also allows sets of 80,000 points to be tested.

6 Structure Revealed by the LMT-skeleton

We would like to take this opportunity to mention some of the structures in point sets that have been revealed by our LMT-skeleton implementation.

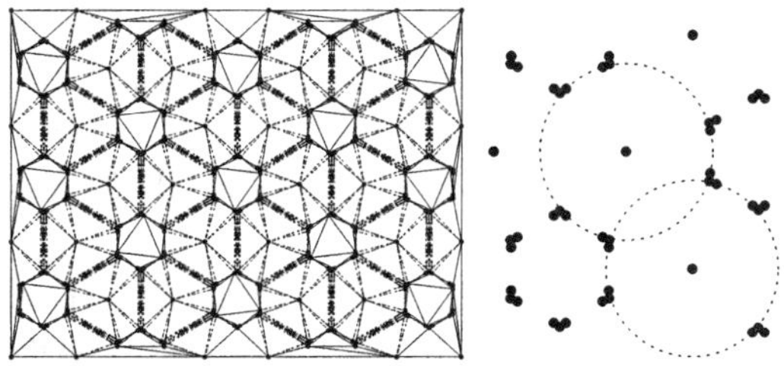

Figure 8: Tiled wheels

Keil (personal communication, 1995) suggested that sufficiently many points around a circle with one point in the center would produce a "wheel" with an isolated point. Bose, Devroye and Evans [6] show that the probability that a such a wheel occurs in a uniformly distributed point set is a (very small) constant, which means that a linear number of connected components can be expected. Even worse, Belleville, Keil, McAllister and Snoeyink [4] were able to place the wheels at the vertices of a hexagonal grid in such a way that the heuristic produces a graph with disconnected 18-gons as shown in Figure 8. With n points one can generate as many as $2n/19 - o(n)$ disconnected regions. Such a structure shows that the LMT-skeleton heuristic does not provide a polynomial time algorithm to solve the minimum weight triangulation problem. If we would like prove the MWT problem *NP*-hard

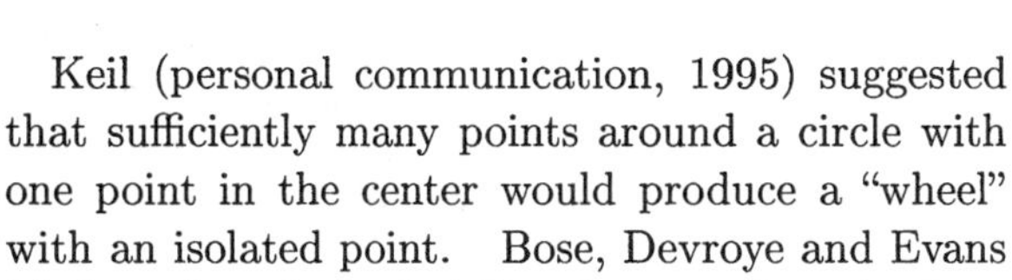

Figure 9: An example of a wire

by a reduction from satisfiability [16], we need point structures that act as variables, gates and wires. Drysdale and Snoeyink used the implementation described in this paper to construct *wires*—point sets with two different MWTs that can carry a boolean signal an arbitrary distance in the plane. The point set in Figure 9 is symmetric about its length except for the ends; the minimum weight triangulation, shown with dashed edges, is asymmetric. These dashed edges and their mirror images are both possible triangulations, and which triangulation is chosen depends on the ends. A *variable* can be any structure that admits two triangulations. One can build a variable by constructing a wire such that the two extremities connect. As yet, however, no structures have been found to represent gates or to link variables and wires together.

7 Conclusion

We have presented efficient implementations of the LMT-skeleton heuristic that involved a number of steps:

1. Convert to an edge-based data structure and use local scanning algorithms whose running time is related to vertex degrees.
2. Minimize the data structure to allow large problems to fit in main memory, but don't sacrifice pointers that preserve the local status.
3. Filter edges by an efficient test to further minimize time and storage.
4. Apply bucketing to accelerate the filter.

One aspect of this order of development caused us some surprise. Since we designed the bucketing scheme last, after the bottlenecks in the other algorithms were understood, it wound up looking quite different from what we would have designed if we had started with a bucketing scheme.

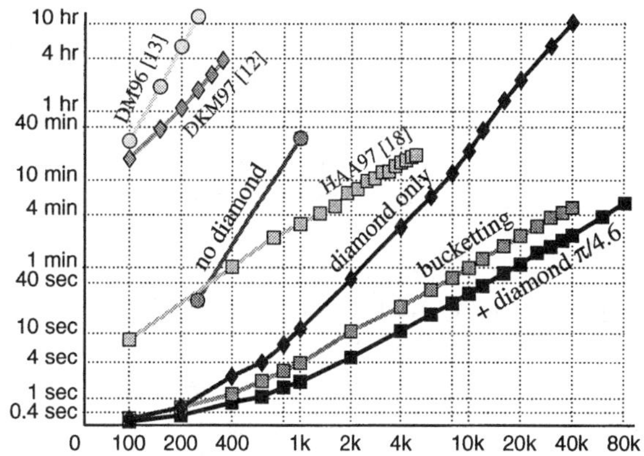

Figure 10: Log-log graph of times to execute LMT heuristics

We can summarize our experimental results in a log-log plot of Figure 10. We also include some of the reported results of Dickerson et al. [12, 13] and Hainz et al. [18]—even though these are not directly comparable due to execution on different machines, it is interesting to observe the similarities in slope for similar approaches. The diamond tests of Section 4 and of [12] improve on the basic algorithms of Section 3.1 and of [13] by improving the slope, which corresponds to improving the degree of the polynomial bound on the running time. The bucketting algorithms of Section 5 and of [18] all show linear growth rates. Memory usage, in each case, is the primary obstacle to extending the graphs further.

We have used our algorithm to compute the largest minimum weight triangulations known, assuming that the subsequent dynamic programming algorithm correctly summed radicals. This has also allowed us to investigate structures that can be built into difficult instances of the MWT—we show examples where the LMT-skeleton heuristic fails to determine an MWT. Since it is still open whether the MWT problem is *NP*-hard or whether it can be solved by a polynomial-time algorithm, we hope that making this implementation available will allow progress in resolving the complexity of MWT.

Acknowledgments

We have many people to thank for discussions, ideas, and keeping us informed. Mark Keil described the initial idea, which we used as an exercise to learn *cweb* in October 1995. Oswin Aichholzer, Siu-Wing Cheng, Matt Dickerson, and Scot Drysdale contributed valuable ideas that were tested in the code and valuable confirmation that their implementations computed the same results as ours. Franz Aurenhammer, Naoki Katoh, David Kirkpatrick, and Cao-an Wang also contributed to the discussions.

Otfried Schwarzkopf's IPE system made the UBC code a valuable tool. Patrice Belleville, Jit Bose, Luc Devroye, Matt Dickerson, Will Evans, Mark Keil, and Michael McAllister have all helped to design difficult instances while at or visiting UBC; Scot Drysdale and Jack Snoeyink designed the wire while at the DREI workshop in Princeton.

References

[1] O. Aichholzer, F. Aurenhammer, S.-W. Chen, N. Katoh, M. Taschwer, G. Rote, and Y.-F. Xu. Triangulations intersect nicely. *Discrete Comput. Geom.*, 16:339–359, 1996.

[2] O. Aichholzer, F. Aurenhammer, G. Rote, and Y.-F. Xu. Constant-level greedy triangulations approximate the MWT well. In D.-Z. Du, X.-S. Zhang, and K. Cheng, editors, *Proc. Second Internat. Symp. Operations Research and its Applications, Guilin, China, December 11–13, 1996*, volume 2 of *Lecture Notes in Operations Research*, pages 309–318, Beijing, 1996. World Publishing Corp.

[3] E. Anagnostou and D. Corneil. Polynomial time instances of the minimum weight triangulation problem. *Comput. Geom. Theory Appl.*, 3:247–259, 1993.

[4] P. Belleville, M. Keil, M. McAllister, and J. Snoeyink. On computing edges that are in all minimum-weight triangulations. In *Proc. 12th Annu. ACM Sympos. Comput. Geom.*, pages V7–V8, 1996.

[5] J. Blömer. Computing sums of radicals in polynomial time. In *Proc. 32nd Annu. IEEE Sympos. Found. Comput. Sci.*, pages 670–677, 1991.

[6] P. Bose, L. Devroye, and W. Evans. Diamonds are not a minimum weight triangulation's best friend. In *Proc. 8th Canad. Conf. Comput. Geom.*, pages 68–73, 1996.

[7] S. W. Cheng, M. J. Golin, and J. C. F. Tsang. Expected case analysis of β-skeletons with applications to the construction of minimum-weight triangulations. In *Proc. 7th Canad. Conf. Comput. Geom.*, pages 279–284, 1995.

[8] S.-W. Cheng, N. Katoh, and M. Sugai. A study of the LMT-skeleton. In T. Asano et al., editors, *Algorithms and Computation*, number 1178 in LNCS, pages 256–265. Springer-Verlag, 1996. ISAAC'96.

[9] S.-W. Cheng and Y.-F. Xu. Approaching the largest β-skeleton within a minimum weight triangulation. In *Proc. 12th Annu. ACM Sympos. Comput. Geom.*, pages 196–203, 1996.

[10] G. Das and D. Joseph. Which triangulations approximate the complete graph. In *Proc. International Symposium on Optimal Algorithms*, volume 401 of *Lecture Notes Comput. Sci.*, pages 168–192. Springer-Verlag, 1989.

[11] M. T. Dickerson, R. L. S. Drysdale, S. A. McElfresh, and E. Welzl. Fast greedy triangulation algorithms. In *Proc. 10th Annu. ACM Sympos. Comput. Geom.*, pages 211–220, 1994.

[12] M. T. Dickerson, J. M. Keil, and M. H. Montague. A large subgraph of the minimum

weight triangulation. *Disc. Comp. Geom.*, 18(3):289–304, 1997.

[13] M. T. Dickerson and M. H. Montague. A (usually?) connected subgraph of the minimum weight triangulation. In *Proc. 12th Annu. ACM Sympos. Comput. Geom.*, pages 204–213, 1996.

[14] R. L. S. Drysdale, S. McElfresh, and J. S. Snoeyink. An improved diamond property for minimum-weight triangulations. Preliminary version at CG'98, 14th Euro. Workshop on Comp. Geom., 1998.

[15] R. L. S. Drysdale, G. Rote, and O. Aichholzer. A simple linear time greedy triangulation algorithm for uniformly distributed points. Technical Report IIG-408, Institutes for Information Processing, Technische Universität Graz, Feb. 1995.

[16] M. R. Garey and D. S. Johnson. *Computers and Intractability: A Guide to the Theory of NP-Completeness.* W. H. Freeman, New York, NY, 1979.

[17] P. D. Gilbert. New results in planar triangulations. Report R-850, Coordinated Sci. Lab., Univ. Illinois, Urbana, IL, 1979.

[18] R. Hainz, O. Aichholzer, and F. Aurenhammer. New results on minimum weight triangulations and the LMT-skeleton (extended abstract). In *CG'97 13th Euro. Workshop on Comp. Geom.*, Würzburg Germany, 1997.

[19] M. Keil. Computing a subgraph of the minimum weight triangulation. *Comput. Geom. Theory Appl.*, 4:13–26, 1994.

[20] D. G. Kirkpatrick. A note on Delaunay and optimal triangulations. *Inform. Process. Lett.*, 10:127–128, 1980.

[21] G. T. Klincsek. Minimal triangulations of polygonal domains. *Discrete Math.*, 9:121–123, 1980.

[22] Y. Kyoda. A study of generating minimum weight triangulation within practical time. Master's thesis, Grad. School Info. Sci., Univ. Tokyo, Feb. 1996. `http://naomi.is.s.u-tokyo.ac.jp/papers/THESIS/MASTER/kyoda.ps.gz`.

[23] C. Levcopoulos. An $\Omega(\sqrt{n})$ lower bound for the nonoptimality of the greedy triangulation. *Inform. Process. Lett.*, 25:247–251, 1987.

[24] C. Levcopoulos and D. Krznaric. Quasi-greedy triangulations approximating the minimum weight triangulation. In *Proc. 7th ACM-SIAM Sympos. Discrete Algorithms*, pages 392–401, 1996.

[25] C. Levcopoulos and D. Krznaric. A near-optimal heuristic for the minimum weight triangulation of convex polygons. *unpublished*, 1997.

[26] C. Levcopoulos, A. Lingas, and J.-R. Sack. Heuristics for optimum binary search trees and minimum weight triangulation problems. *Theoret. Comput. Sci.*, 66(2):181–203, Aug. 1989.

[27] A. Lingas. A linear-time heuristic for minimum weight triangulation of convex polygons. In *Proc. 23rd Allerton Conf. Commun. Control Comput.*, 1985.

[28] E. L. Lloyd. On triangulations of a set of points in the plane. In *Proc. 18th Annu. IEEE Sympos. Found. Comput. Sci.*, pages 228–240, 1977.

[29] A. Okabe, B. Boots, and K. Sugihara. *Spatial Tessellations: Concepts and Applications of Voronoi Diagrams.* John Wiley & Sons, Chichester, UK, 1992.

[30] D. A. Plaisted and J. Hong. A heuristic triangulation algorithm. *J. Algorithms*, 8:405–437, 1987.

[31] O. Schwarzkopf. The extendible drawing editor Ipe. In *Proc. 11th Annu. ACM Sympos. Comput. Geom.*, pages C10–C11, 1995.

[32] C. A. Wang, F. Chin, and Y.-F. Xu. A new subgraph of the minimum weight triangulation. In T. Asano et al., editors, *Algorithms and Computation*, number 1178 in LNCS, pages 266–274. Springer-Verlag, 1996. ISAAC'96.

[33] B.-T. Yang, Y.-F. Xu, and Z.-Y. You. A chain decomposition algorithm for the proof of a property on minimum weight triangulations. In *Proc. 5th Annu. Internat. Sympos. Algorithms Comput.*, volume 834 of *Lecture Notes Comput. Sci.*, pages 423–427, Beijing, 1994. Springer-Verlag.

Improved Incremental Randomized Delaunay Triangulation.*

Olivier Devillers†

1 Introduction

The computation of the Delaunay triangulation of a set of n points in the plane is one of the classical problems in computational geometry and plenty of algorithms have been proposed to solve it.

These Delaunay algorithms can have different characteristics:

- Optimal on worst case data, i.e. $O(n \log n)$ time.
- Good complexity on random data only
- Randomized
- On-line vs off-line

In the current trade-off between algorithmic simplicity, practical efficiency and theoretical optimality, practitioners often choose the simplicity and practical efficiency taking the risk of having bad performance on some special kind of data.

Our aim is to conciliate many of the above aspects, namely to obtain an incremental algorithm using simple data structure having good practical performance on realistic input and still provable $O(n \log n)$ computation time on any data set.

Previous related work

*This work was partially supported by ESPRIT LTR 21957 (CGAL)

†INRIA, BP93, 06902 Sophia Antipolis. Olivier.Devillers@sophia.inria.fr.

SCG 98 Minneapolis Minnesota USA

Our work is strongly related to some previous algorithms for Delaunay triangulation. All these algorithms are incremental and their complexity is randomized, they use some location structure to find where the new point is inserted, and then update the triangulation.

The first idea of a randomized incremental construction for the Delaunay triangulation [BT86] uses a location structure based on the history of the Delaunay triangulation: the Delaunay tree. Point p_i is inserted at time i, and to find where point p_n fell, p_n is located in all the triangulations at times 1 to $n-1$; the location at time $i+1$ is deduced from the location at time i. This idea yields an expected $O(n \log n)$ complexity [BT93, GKS92] if the points are inserted in a random order. The drawbacks of this approach are the following: the location structure consists of the history of the construction and thus strongly depends on the insertion order, and the additional memory needed cannot be controlled. (The expected memory is proved to be $O(n)$ and is experimentally about twice the size of the final triangulation.)

Mulmuley [Mul91] proposed a location structure independent of the insertion order. The structure has $O(\log n)$ levels, each level being a random sample of the level below. At each level, the Delaunay triangulation of the points is computed, and the overlapping triangles at different levels are linked to enable location of new points. This structure has the advantage of being independent of the order of insertion, of ensuring an $O(\log^2 n)$ location time for any point, and of allowing deletions in an easier way than the Delaunay tree [DMT92]. However, the additional memory is still important and the location structure is not especially simple.

In 1996, Mücke, Saias and Zhu [MSZ96] proposed a very simple structure to handle triangulation of random points. The structure reduces to a random subset of $\sqrt[3]{n}$ points, and pointers from these points to an incident triangle in the Delaunay triangulation. A new point is located by finding the nearest neighbor in the sample by brute force, and walking in the triangulation. For evenly distributed points, the expected complexity of the algorithm is $O(n^{\frac{4}{3}})$ with a small constant, which makes it competitive with many $O(n \log n)$ algorithms. But for some data (for example points on a parabola) the complexity increases to $O(n^{\frac{5}{3}})$.

Overview

Our approach uses a structure with levels similar to Mulmuley, but with simple relations between levels. This allows better control of the memory overhead. The transition between two levels is not direct as in Mulmuley, but uses a march similar Mücke, Saias and Zhu to locate point in triangulations.

In Section 2 we present the algorithm, in Section 3 we prove that the expected complexity of constructing the Delaunay triangulation is $O(n \log n)$. The parameters of the data structure are then tuned to minimize the constant in the case of random points and are shown to yield an excellent behavior in Section 4, we pay special attention to the comparison with the method of Mücke, Saias and Zhu. Finally we give some implementation remarks and practical results in Section 5.

2 Algorithm

Let $\mathcal{S}$ be a set of n sites in the plane. The aim is to compute the Delaunay triangulation $\mathcal{DT}_{\mathcal{S}}$ of $\mathcal{S}$ and to maintain it efficiently under insertions and deletions.

2.1 The location structure

The algorithm uses a data structure composed of different levels. Level i contains the Delaunay triangulation $\mathcal{DT}_i$ of a set of sites $\mathcal{S}_i$.

The sets $\mathcal{S}_i$ forms a decreasing sequence of random subsets of $\mathcal{S}$ based on a Bernoulli sampling technique [MR95, Mul94]:

$$\mathcal{S} = \mathcal{S}_0 \supseteq \mathcal{S}_1 \supseteq \mathcal{S}_2 \supseteq \ldots \supseteq \mathcal{S}_{k-1} \supseteq \mathcal{S}_k$$

$$Prob(p \in \mathcal{S}_{i+1} \mid p \in \mathcal{S}_i) = \frac{1}{\alpha} \in]0,1[.$$

The data structure is fairly simple: it contains the points of $\mathcal{S}$ and the triangles of all the triangulations $\mathcal{DT}_i$. A point $p \in \mathcal{S}$ such that $p \in \mathcal{S}_i \subseteq \ldots \subseteq \mathcal{S}_0$ and $p \notin \mathcal{S}_{i+1}$ is said to be a *vertex of level i* and has a link to a Delaunay triangle of $\mathcal{DT}_j$ incident to p for all j for $0 \leq j \leq i$. A triangle of $\mathcal{DT}_i$ has links to its three neighbors in $\mathcal{DT}_i$ and to its three vertices.The number k of levels is not fixed; for each point random trials decide its level, and the point with highest level determines k.

2.2 Location of a query

For the location of a query q, we start at a known vertex v_{k+1} of the highest level k. Then we search for v_k, the vertex of $\mathcal{DT}_k$ nearest to q. Since v_k is also a vertex of $\mathcal{DT}_{k-1}$, we search for v_{k-1}, the nearest neighbor of q in $\mathcal{DT}_{k-1}$, starting at v_k. The search is continued descending the different levels. At each level i, the nearest vertex v_i of q in $\mathcal{DT}_i$ is determined.

At level i the search of v_i is carried out in three phases:

- First phase: from v_{i+1}, we have a link to a triangle of $\mathcal{DT}_i$ having v_{i+1} as vertex. All triangles incident to v_{i+1} are explored to find the triangle containing the segment $v_{i+1}q$.
- Second phase: all the triangles of $\mathcal{DT}_i$ intersected by $v_{i+1}q$ are visited, walking along the segment $v_{i+1}q$ up to the triangle t_i that contains q.
- Third phase: using neighborhood relationships between triangles, we will traverse few triangles of $\mathcal{DT}_i$ from t_i to find v_i. If $vv'v''$ are the three vertices of t_i, and, without loss of generality, v is closer to q than v' and v'', then v_i is either v or it lies in the disk of center q and passing through v (shaded on Figure 1a); thus the search for v_i has to be done only in the direction of the neighbors of t_i through the edges vv' and vv'' and the neighbor through the edge $v'v''$ can be ignored (the portion of the shaded disk in that direction is inside the disk through $vv'v''$ which is empty) . For each such triangle, the distance to the new vertex is computed and the algorithm maintains the closest visited vertex. For a visited triangle $ww'w''$ such that w is the nearest to q among $ww'w''$ the neighbor triangle through edge ww' (resp ww'') will be visited if angle qww' is smaller than $\frac{\pi}{2}$ (Figure 1b).

Figure 1c show the triangles visited by the different phases of the search.

2.3 Updates

Because of its simplicity, the data structure is fairly easy to update. Maintaining it dynamically provides a fully dynamic triangulation algorithm. The links between the different levels do not use any complicated data structure simply vertices know a triangle at all levels in which they appear.

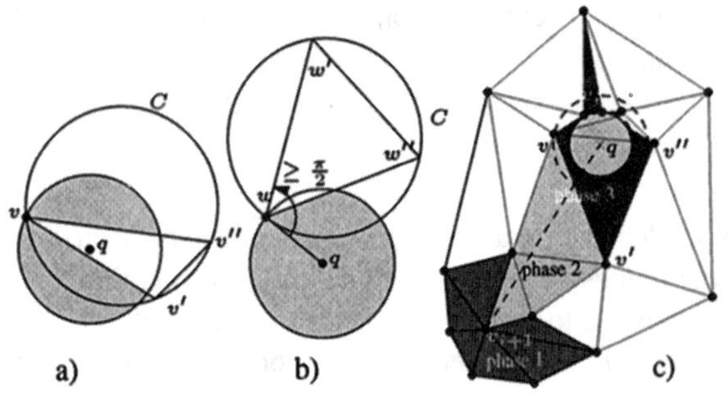

Figure 1: Search for v_i.

To delete a point from $\mathcal{S}$, just delete the corresponding vertex at all the levels where it appears, which can be done in time sensitive to d the degree of that vertex. On average $d = 6$ and thus some of the following algorithms can be used. A complicated algorithm [AGSS89] of deterministic complexity $O(d)$, a simple randomized $O(d)$ algorithm [Che86] can be used or simpler solutions of complexity $O(d \log d)$ or even $O(n^2)$ may be good in practice.

Inserting a point in $\mathcal{S}$ reduces to locating the new point at all levels, computing its level i and inserting the new vertex at all levels $j, 0 \leq j \leq i$ (which is sensitive to the degree of the new vertex once the location is done). The insertion using the standard algorithm [Law77].

3 Worst-case randomized analysis

The analysis will rely on the randomization in the construction of the random subsets $\mathcal{S}_i$ and the points of $\mathcal{S}$ are assumed to be inserted in a random order. In this section, no assumption applies to the data distribution, which can be in the worst case. As usual in theoretical computational geometry, we make only an asymptotic analysis and give rough upper bounds for the constants. In the next section, parameter α will be tuned to get a tight constant in the special case of evenly-distributed points.

Let $\mathcal{S}$ be a set of n points organized in the structure described in Section 2 and q a point to be inserted in $\mathcal{S}$. Since we have assumed a random insertion order, q is a random point of $\mathcal{S} \cup \{q\}$.

We denote $n_i = |\mathcal{S}_i|$ and $\mathcal{R}_i = \mathcal{S}_i \cup \{q\}$.

Notice that, thanks to the random insertion order, $\mathcal{R}_i$ is a random subset of size $n_i + 1$ of $\mathcal{R}_{i-1}$ and q is a random element of $\mathcal{R}_i$.

The cost of exploring all the triangles incident to v_{i+1} at the first phase of the march of level i is the degree of v_{i+1} in $\mathcal{DT}_i$. The cost of the second phase is the number of triangles intersected by segment $v_{i+1}q$. The cost of the third phase is the number of candidate vertices visited during the search of v_i from t_i.

Lemma 1 *The expected degree of v_i in $\mathcal{DT}_{i-1}$ is $O(1)$.*

Proof Let $\mathcal{NN}$ be the nearest neighbor graph of $\mathcal{R}_i$: that is, the vertices of $\mathcal{NN}$ are the points of $\mathcal{R}_i$, and $q, v \in \mathcal{R}_i$ define an edge of $\mathcal{NN}$ if and only if v is the nearest neighbor of q (denoted by $v = NN(q)$) or q is the nearest neighbor of v in $\mathcal{R}_i$. $\mathcal{NN}$ is well known to be a subgraph of $\mathcal{DT}_{\mathcal{R}_i}$, the Delaunay triangulation of $\mathcal{R}_i$, and to have maximum degree 6 [PY92].

We denote by $d^\circ_{\mathcal{DT}_{i-1}}(v)$ the degree of v in $\mathcal{DT}_{i-1}$, and by $E_{v \in \mathcal{R}_i \subset \{q\}}$ the expectation when v is chosen uniformly in $\mathcal{R}_i \subset \{q\}$. Then we have

$$E_{v \in \mathcal{R}_i \subset \{q\}} \left(d^\circ_{\mathcal{DT}_{i-1}}(v) \right)$$

$$= E_{v \in \mathcal{R}_{i-1} \subset \{q\}} \left(d^\circ_{\mathcal{DT}_{i-1}}(v) \right) < 6$$

notice that $d^\circ_{\mathcal{DT}_{i-1}}(v)$ is a random variable; result holds since $\mathcal{R}_i$ and $\mathcal{R}_{i-1} \subset \{q\}$ are random subsets of $\mathcal{R}_{i-1}$ and that the average degree of a vertex in a triangulation is less than 6.

But even if q is a random point in $\mathcal{R}_i$, the vertex v_i, the nearest neighbor of q in $\mathcal{R}_i$, is not uniformly random.

$$\begin{aligned}
& E_{q \in \mathcal{R}_i} \left(d^\circ_{\mathcal{DT}_{i-1}}(NN(q)) \right) \\
&= E \left(\frac{1}{|\mathcal{R}_i|} \sum_{q \in \mathcal{R}_i} d^\circ_{\mathcal{DT}_{i-1}}(NN(q)) \right) \\
&= \frac{1}{|\mathcal{R}_i|} E \left(\sum_{v \in \mathcal{R}_i} \sum_{q \in \{\rho; v = NN(\rho)\}} d^\circ_{\mathcal{DT}_{i-1}}(v) \right) \\
&< \frac{1}{|\mathcal{R}_i|} E \left(\sum_{v \in \mathcal{R}_i} 6 d^\circ_{\mathcal{DT}_{i-1}}(v) \right) \\
&\leq 36
\end{aligned}$$

■

Lemma 2 *Given $w \in \mathcal{R}_i$, the expected number of vertices q of $\mathcal{R}_i$ such that w belongs to the disk of center q and passing through the nearest neighbor of q in $\mathcal{R}_{i+1}$ is less than 6α.*

Proof Let $w \in \mathcal{R}_i$ and let $w = q_0, q_1, q_2 \dots q_k$ be the points of $\mathcal{R}_i$ lying in a section of angle $\frac{\pi}{3}$ having apex w sorted by increasing distance to w. Clearly, a disk of center q_l passing through q_j ($j < l$) cannot contain w and thus, if $q = q_l$, a necessary condition for w to be in the disk having as diameter the segment defined by q and the nearest neighbor of q in $\mathcal{R}_{i+1}$ is that no point of $\{q_0, \dots q_{l-1}\}$ is in the sample $\mathcal{R}_{i+1}$ which has probability $(1 - \frac{1}{\alpha})^l$.

Using six sections around w to cover the whole plane, and summing over the choice of $q \in \mathcal{R}_i$ we get the claimed result. Notice that the disk of center q and passing through the nearest neighbor of q contain the disk of diameter the line segment defined by these two points, and thus the bound apply also to that circle. ■

Lemma 3 *The expected number of edges of $\mathcal{DT}_i$ intersecting segment qv_{i+1} is $O(\alpha)$.*

Proof Let e be an edge of $\mathcal{DT}_i$ intersecting segment qv_{i+1}. If e does not exist in $\mathcal{DT}_{\mathcal{R}_i}$, it means that e is an internal edge of the region retriangulated when q is inserted in $\mathcal{DT}_i$. Since q is a random point in $\mathcal{R}_i$, the expected number of such edges is 3 since it equals the average degree of q in $\mathcal{R}_i$ minus 3.

If e exists in $\mathcal{DT}_{\mathcal{R}_i}$, one end-point w of e must belong to the disk of diameter qv_{i+1}, denoted disk$[qv_{i+1}]$, (otherwise any disk through the end-points of e must contain q or v_{i+1} and e cannot belong to $\mathcal{DT}_{\mathcal{R}_i}$).
The expected number of edges of $\mathcal{DT}_{\mathcal{R}_i}$ intersecting disk$[qv_{i+1}]$ is bounded by the sum of the degrees of the vertices in disk$[qv_{i+1}]$

$$
\begin{aligned}
&E(\#\{e \in \mathcal{DT}_{\mathcal{R}_i} \text{ having an end-point} \in \mathcal{R}_i \cap \text{disk}\,[qv_{i+1}]\}) \\
&= \frac{1}{|\mathcal{R}_i|} \sum_{q \in \mathcal{R}_i} \sum_{w \in \mathcal{R}_i \cap \text{disk}\,[qv_{i+1}]} d^{\circ}_{\mathcal{DT}_{\mathcal{R}_i}}(w) \\
&= \frac{1}{|\mathcal{R}_i|} \sum_{w \in \mathcal{R}_i} d^{\circ}_{\mathcal{DT}_{\mathcal{R}_i}}(w)\, |\{q \in \mathcal{R}_i | w \in \text{disk}\,[qv_{i+1}]\}| \\
&\leq \frac{1}{|\mathcal{R}_i|} \sum_{w \in \mathcal{R}_i} d^{\circ}_{\mathcal{DT}_{\mathcal{R}_i}}(w) 6\alpha \quad \text{using Lemma 2} \\
&\leq 36\alpha \quad \text{using the bound of 6 on the average degree of } w
\end{aligned}
$$

Notice that Lemma 2 was established for a fixed w and a random q which allows to use it inside the sum over w. Thus we get a total expected cost for the march bounded by $36\alpha + 3$. ■

Lemma 4 *The expected number of triangles of $\mathcal{DT}_i$ visited during the search for v_i from t_i is $O(\alpha)$.*

Proof All the triangles t examined in phase 3 have a vertex in the disk of center q passing through v_{i+1}. Thus we can argue similarly as in Lemma 3, denoting disk $|_c qv_{i+1}]$ the disk of center q through v_{i+1}:

$$
\begin{aligned}
&E(\#\{t \in \mathcal{DT}_{\mathcal{R}_i} \text{having an end-point} \in \text{disk}\,|_c qv_{i+1}]\}) \\
&\leq \frac{1}{|\mathcal{R}_i|} \sum_{q \in \mathcal{R}_i} \sum_{w \in \mathcal{R}_i \cap \text{disk}\,|_c qv_{i+1}]} d^{\circ}_{\mathcal{DT}_{\mathcal{R}_i}}(w) \\
&\leq \frac{1}{|\mathcal{R}_i|} \sum_{w \in \mathcal{R}_i} d^{\circ}_{\mathcal{DT}_{\mathcal{R}_i}}(w)\, |\{q \in \mathcal{R}_i | w \in \mathcal{R}_i \cap \text{disk}\,|_c qv_{i+1}]\}| \\
&\leq \frac{1}{|\mathcal{R}_i|} \sum_{w \in \mathcal{R}_i} d^{\circ}_{\mathcal{DT}_{\mathcal{R}_i}}(w) 6\alpha \quad \text{using Lemma 2} \\
&\leq 36\alpha \quad \text{using the bound on the average degree of } w
\end{aligned}
$$

■

Theorem 5 *The expected cost of inserting n^{th} point in the structure is $O(\alpha \log_\alpha n)$*

Proof By linearity of expectation, Lemmas 1, 3 and 4 prove that the expected cost at one level is $O(\alpha)$. Since the expected height of the structure is $\log_\alpha n$, we get the claimed result. (The analysis is similar to the ananlysis for skip lists [MR95].) ■

Theorem 6 *The construction of the Delaunay triangulation of a set of n points is done in expected time $O(\alpha n \log_\alpha n)$ and $O(\frac{\alpha}{\alpha-1} n)$ space. The expectation is on the randomized sampling and the order of insertion, with no assumption on the point distribution.*

Proof Easy corollary of Theorem 5. ■

4 Tuning parameters

We have proved that our structure is worst case optimal in the expected sense for any set of points. In this section, we will focus on more practical cases, and tune the algorithm to be optimal on random distribution. In that case, many events such as that a point has high degree and that it is the nearest neighbor of a random point can be considered as independent.

4.1 Phase 1

We can assume that, $d^{\circ}_{\mathcal{DT}_i}(v_{i+1}) = 6$ (and not only ≤ 36 as proved in Lemma 1). And thus if the turn around v_{i+1} is done in clockwise or counterclockwise direction depending on the position of segment $v_{i+1}q$ with respect to the starting

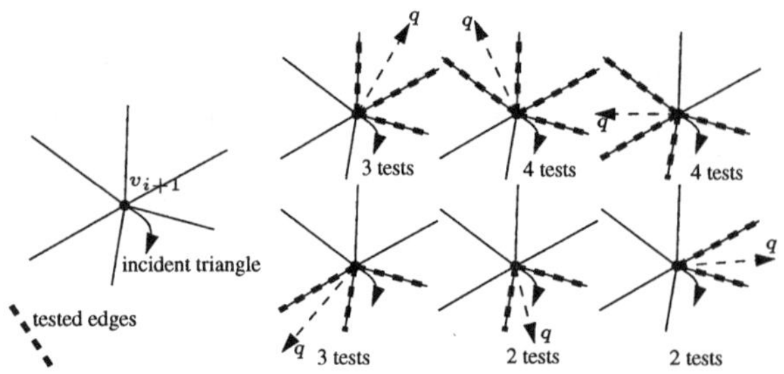

Figure 2: Number of orientation tests in phase 1

triangle, and assuming that this position is random around v_{i+1} the expected number of orientation tests is 3. Figure 2 shows the different cases to average, the edges $v_{i+1}w$ such that an orientation test $v_{i+1}wq$ is performed are indicated, for a typical degree 6 vertex in the triangulation.

4.2 Phase 2

Bose and Devroye [BD95] proved that the expected number of edges of a Delaunay triangulation of random points crossed by a line segment of length l is $O(l\sqrt{\gamma})$ where γ is the point density. Our experiments shows that the constant is 2.

The expected number of points in the disk of center q passing through v_{i+1} is $\alpha - 1$. Indeed, if the points of $\mathcal{R}_i$ are sorted by increasing distance from q, v_{i+1} is the first point in $\mathcal{R}_{i+1}$, thus the number of points in the disk is k with probability $(1-\frac{1}{\alpha})^k\frac{1}{\alpha}$, and the expected number is $\frac{1}{\alpha}\sum(1-\frac{1}{\alpha})^k = \alpha - 1$. Thus if l is the length of qv_{i+1} the density of points in $\mathcal{DT}_i$ is $\frac{\alpha}{\pi l^2}$.

Thus we conclude that the expected number of edges of $\mathcal{DT}_i$ intersecting segment qv_{i+1} is $2l\sqrt{\frac{\alpha}{\pi l^2}} = \frac{2\sqrt{\alpha}}{\sqrt{\pi}}$.

For each edge ww' crossed, two orientation tests are performed: if w is the newly examined vertex, orientations of triangles wqv_{i+1} and qww' are computed.

We have to point out, that in the orientation tests of kind wqv_{i+1}, the edge qv_{i+1} remains constant, and thus some computations do not need to be done for each test.

4.3 Phase 3

Phase 3 is more difficult to analyze precisely, but a rough bound is that the number of candidate vertices examined (with shortest distance) is less than two and that we examine less than 8 triangles in total.

In fact, we modified phase 3, instead of really searching for v_i, the nearest neighbor of q in $\mathcal{S}_i$, we just define v_i as the nearest among the three vertices of t_i. Thus this modified phase 3 reduced to three distance computations and two comparisons.

4.4 Tuning α

We will count more precisely the number of operation needed to evaluate our primitives. More exactly, we count the number of floating point operations (f.p.o.) without making diistinctions between additions, subtractions or multiplications.

The total evaluation at a given level is $3 + \frac{\sqrt{\alpha}}{\sqrt{\pi}}$ orientation tests involving qv_{i+1}, $\frac{\sqrt{\alpha}}{\sqrt{\pi}}$ other orientation tests and 3 distance computations.

Orientation tests always using points q and v_{i+1} can be done using 5 f.p.o. to initialize plus 4 f.p.o. for each test. Other orientation tests need 7 f.p.o. each, and square distance computations need 5 f.p.o. each.

Thus the total cost in terms of number of f.p.o. at level i is

$$5 + 4(3 + \frac{\sqrt{\alpha}}{\sqrt{\pi}}) + 7\frac{\sqrt{\alpha}}{\sqrt{\pi}} + 5 \cdot 3 = 32 + 6.2\sqrt{\alpha}.$$

Since the number of level is $\log_\alpha n = \frac{\log_2 n}{\log_2 \alpha}$ we get a cost of $c_0(n) = (32 + 6.2\sqrt{\alpha})\left\lceil\frac{\log_2 n}{\log_2 \alpha}\right\rceil$ which is close to its minimum ($\in [13.3\log_2 n, 14\log_2 n]$) for $\alpha \in [18, 90]$, with the minimum occuring for $\alpha \simeq 40$.

4.5 Comparison with [MSZ96]

Similar counting of f.p.o. in Mücke et al. algorithm, using a random sample of $\beta\sqrt[3]{n}$ points, produces a cost of

$$c_{MSZ}(n) = 5 + 4(3 + \frac{\frac{n}{\beta n^{\frac{1}{3}}}}{\sqrt{\pi}}) + 7\frac{\frac{n}{\beta n^{\frac{1}{3}}}}{\sqrt{\pi}} + 5\beta\sqrt[3]{n}$$

$$= 17 + \sqrt[3]{n}\left(\frac{6.2}{\sqrt{\beta}} + 5\beta\right)$$

which is close to its minimal value for $0.5 < \beta < 1$.

As shown by the comparison of the two curves in Figure 3, our method is potentially much better than [MSZ96], even for a small number of points. However, this method to analyze our approach hides the discontinuity of the cost, since the effective number of levels is necessarily an integer.

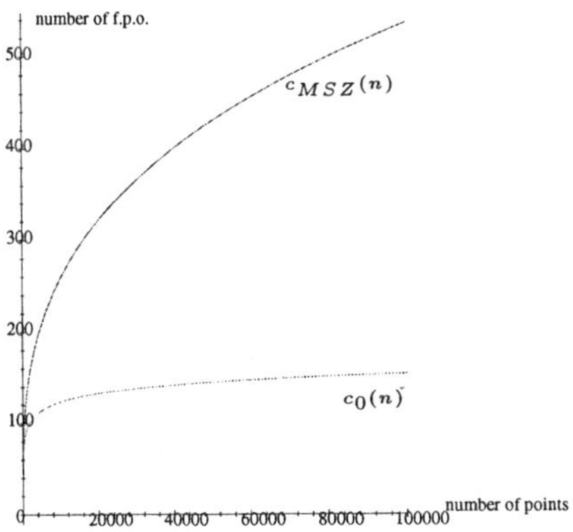

Figure 3: Comparison of number of floating point operations between $c_0(n)$ and $c_{MSZ}(n)$ for $\alpha = 40$ and $\beta = 1$.

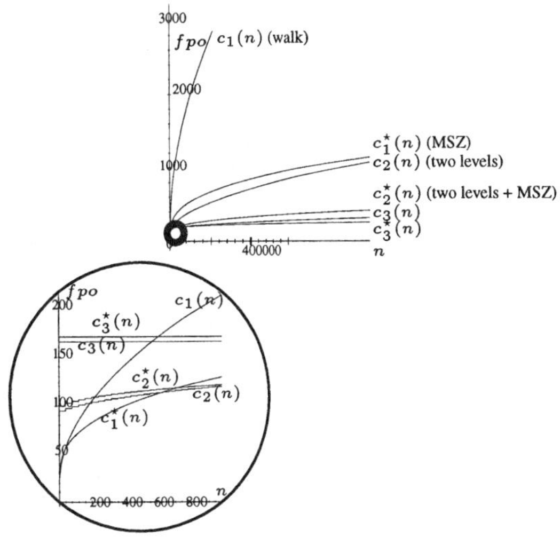

Figure 4: Comparison of number of floating point operations between $c_k(n)$ and $c_k^\star(n)$ for $\alpha = 40$.

To have a better comprehension of what happens for a small number of points, we can draw the cost of inserting a point in a structure having a fixed number of levels.

The classical walk from a random point in the structure costs

$$c_{walk}(n) = 5 + 4(3 + \frac{\sqrt{n}}{\sqrt{\pi}}) + 7\frac{\sqrt{n}}{\sqrt{\pi}} = 17 + 6.2\sqrt{n}$$

which is also the cost of inserting in our structure up to the time a second level is created.

When k levels have been created, the cost is

$$c_k(n) = c_{walk}\left(\frac{n}{\alpha^k}\right) + 15k + k \cdot c_{walk}(\alpha)$$

We can alternatively mix this multilevel approach with Mücke et al's. sampling at the first level of the structure. In that case, the cost is

$$c_k^\star(n) = c_{MSZ}\left(\frac{n}{\alpha^k}\right) + 15k + k \cdot c_{walk}(\alpha)$$

This comparison (see Figure 4) shows that [MSZ96] ($c_1^\star(n)$) becomes better than the simple march ($c_1(n)$) for $n > 40$. The two level structure ($c_2(n)$) becomes better than the single level structure ($c_1(n)$) for $n > 180$ and better than [MSZ96] ($c_1^\star(n)$) for $n > 600$. The main information is that the structure presented in that paper should be significantly better than [MSZ96] for $10000 < n$.

5 Implementation

5.1 Deletion

The above structure supports insertions and queries as explained above, but also deletions. Since there is no complicated data structure to maintain, deletions can be handled by just deleting the removed point at each level where it appears.

This can be done in output-sensitive time [Che87, AGSS89], and thus the deletion of a random point is done in expected constant time since a point appears at an expected constant number of levels and its expected degree k is also constant.

From a practical point of view, and to keep the simplicity of the algorithm, a simpler suboptimal algorithm should be preferred. It can be done in $O(k^2)$ time, for example by flipping to reduce the degree of the deleted vertex to 3, and flipping again to restore the Delaunay property. Another simple algorithm consists in finding the Delaunay triangle incident to an edge of the hole in $O(k)$ time which also yields an $O(k^2)$ time algorithm. Both algorithms are efficient in practice and needs only few micro-seconds (about 30 in a random triangulation) to delete a point once it had been localized.

5.2 Arithmetic degree

The algorithm above is designed to make a parsimonious use of high degree tests [TLP96]. More precisely, the location phase uses only orientation tests on three points in phases

1 and 2, and distance computation and angle comparisons with $\frac{\pi}{2}$ in phase 3. All these tests are degree 2 tests. Clearly, updates need to use in-circle tests which are of degree 4.

An alternative to phase 3 should have to use in-circle tests to limit the explored triangles in $\mathcal{DT}_i$ to those whose circumcircle contains q. Such variant may explore fewer triangles and be easier to analyze, but may use more degree 4 tests.

5.3 Robustness issues and degeneracies

Degeneracies are solved by handling special cases: if two points have the same coordinates, then the insertion is not done, if four points are cocircular, then the last point inserted is considered as inside the disk defined by the others.

We use exact arithmetic for 24 bits integers, and thus coordinates of our points are integers in range $[-16777216, 16777216]$ (up to a multiplication by a power of 2). Using this restricted kind of data, double precision computation is exact on degree 2 tests and almost never leads to precision problems on degree 4 predicates. Nevertheless, the exactness of all computations are verified by an arithmetic filter and exact computation is performed if needed.

5.4 Code parameters

The following parameters can be specified:

- maximal number of levels
- α the ratio between two levels
- the minimal number of points to use the higher level for point location
- the minimal number of points to use MSZ sampling at one of the higher levels
- β the constant for the size of MSZ sample.

Our default parameters are

- number of levels unlimited
- $\alpha = 30$.
- minimal size to use hierarchy is 20.
- minimal size to use MSZ is 20.
- $\beta = 1$.

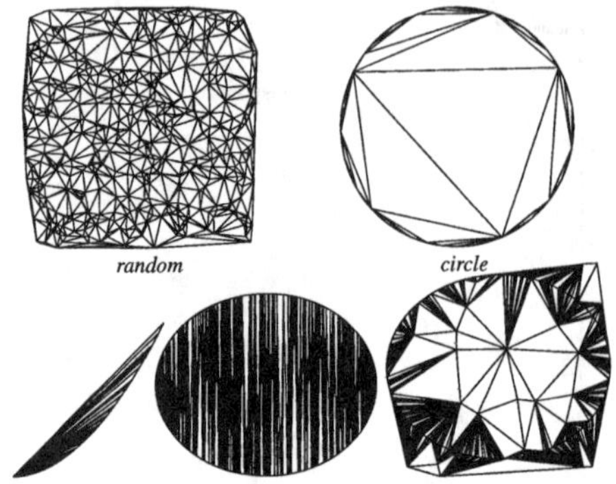

Figure 5: Data sets.

We found that the code is relatively insensitive to the parameters. For reasonable changes of these parameters, (up to a factor 2) the computation time is not greatly affected. Using these configuration parameters, our code can be used to run

- the usual walk algorithm (only one level and minimal size for MSZ=∞),
- the Mücke et al. algorithm [MSZ96] (only one level),
- the hierarchical algorithm described in this paper (minimal size for MSZ=∞),
- the mixed method suggested in Section 4.5 (default parameters above).

5.5 Experimental results

5.5.1 Data sets

We claim that our algorithm performs well on random point sets, and has acceptable worse case complexity. To illustrate this fact, we will test it with the realistic and degenerate data sets. For each kind of data, we used sets of size 5,000, 50,000 and 500,000 points. The coordinates are random on 24 bits and the constraints such that the points are on a parabola are verified, up to the rounding arithmetic errors.

- *random*: points evenly distributed in a square.
- *ellipse*: points evenly distributed on an ellipse.
- *ellipse2*: 95% points evenly distributed on an ellipse plus 5% points evenly distributed in a square.
- *circle*: points evenly distributed on a circle.
- *parabola*: points evenly distributed on a parabola,

distribution	size	walk	[MSZ96]	hierarchy	hierarchy + MSZ
random	5000	0.3	0.17	0.15	0.14
random	50000	12	3.8	2.7	2.3
random	500000	460	72	36	31
ellipse2	5000	0.53	0.34	0.21	0.20
ellipse2	50000	49	21	3.9	3.5
ellipse2	500000	930	760	57	49
ellipse	5000	2.2	0.46	0.31	0.21
ellipse	50000	187	21	3.9	3.7
ellipse	500000	long	270	54	55
parabola	5000	2.5	0.31	0.21	0.16
parabola	50000	87	5.9	3.2	3.0
parabola	500000	long	74	69	45
circle	5000	0.15	0.13	0.13	0.14
circle	50000	2.4	2.6	2.4	2.4
circle	500000	39	44	36	36

Figure 6: Running times

If the *circle* and *parabola* examples can be considered as pathological inputs, the *ellipse* and *ellipse2* examples are more realistic, Delaunay triangulation of points distributed on a curve occurs in practical applications, for example in shape reconstruction (see Figure 5).

5.5.2 Results

Following results are obtained on a Sun-Ultra1 200 MHz. The code is written in C++ and compiled with AT-T compiler with optimizing options. Time has been obtained with the `clock` command and is given in seconds. The time which is measured is just the Delaunay computation; it does not take into account the time for input or output.

Figure 6 gives the computation times for execution of the code with the different parameters described in Section 5.4. Since it is the same code, the low level primitives such as incircle tests or the walk in the triangulation are identical and it provides a fair comparison between the different methods.

The last column is always the fastest method. It is significantly better than MSZ for very large sets of random points, and the difference is even more important on data set *ellipse2* which is representative of real applications.

5.5.3 Comparison with other software

We have compared with some Delaunay softwares available on the WWW:

- `qhull` by Bradford Barber and Hannu Huhdanpaa, duality with 3D convex hull [BDH93] (available at http://www.geom.umn.edu/locate/qhull).
- `div-conquer` by Jonathan Shewchuk, divide and conquer [She96]
- `sweep` by Jonathan Shewchuk, plane sweep
- `incremental` by Jonathan Shewchuk, incremental with Mücke et al. localization.
  These three codes supports exact arithmetic on `double` (available at http://www.cs.cmu.edu/~quake/triangle.research.html).
- `Dtree` Delaunay tree structure[BT93] (time includes input)
  (available at http://www.inria.fr/prisme/logiciel/del-tree.html).
- `hierarchy` this paper, mixed with MSZ.

The execution times in seconds are in Figure 7. Our method is significantly faster than the other incremental method, especially in the ellipse cases. Our method is about 50% slower than the divide and conquer algorithm.

6 Conclusion

We proposed a new hierarchical data structure to compute the Delaunay triangulation of a set of points in the plane. It combines good worst case randomized complexity, fast behavior on real data, small memory occupation and dynamic updates (insertion and deletion of points).

Referring to Su and Drysdale [SD97] study of several techniques and our comparisons with Shewchuk implementation [She96] of some of these techniques, we have

distribution	size	qhull	sweep	div-conq	incr	Dtree	hier.
random	5000	0.65	0.21	0.11	0.29	1.4	0.14
random	50000	8.0	3.6	1.6	6.6	17	2.3
random	500000	101	53	22	150	swap	31
ellipse2	5000	0.54	0.21	0.13	0.75	1.3	0.20
ellipse2	50000	7.8	3.2	2.16	42	16	3.5
ellipse2	500000	420	46	29	2100	swap	49
ellipse	5000	0.83	0.18	0.14	2.1	1.3	0.21
ellipse	50000	57	2.8	2.4	110	14	3.7
ellipse	500000	swap	39	33	1400	swap	55
parabola	5000	3.9	0.16	0.11	2.0	1.2	0.16
parabola	50000	790	2.7	2.0	110	14	3.0
parabola	500000	swap	39	28	1800	swap	45
circle	5000	93	0.17	0.17	0.52	1.4	0.14
circle	50000	220	3.1	1.8	11	15	2.4
circle	500000	swap	22	43	240	swap	36

Figure 7: Comparisons with other softwares

shown that our implementation is competitive with other approaches on random data. Furthermore, we can prove that the performances remains good on pathological inputs. Finally, one of the main advantage of this algorithm is to allow a dynamic setting.

The main idea of our structure is to perform point location using several levels. The lowest level just consists of the triangulation, then each level contains the triangulation of a small sample of the levels below. Point location is done by marching in a triangulation to determine the nearest neighbor of the query at that level, then the march restart from that neighbor at the level below. Location at highest level is done using [MSZ96] which is efficient for small set of points.

One characteristics of the structure is that best time performance is obtained with a ratio of about three per cent between two levels, which yields to few levels (three or four typically) and a small memory occupation. The structure is simple and does not need additional features such as buckets.

Such structure can be generalized to other problems. The two main ingredients of the proofs are bounds on the maximal degree of the nearest neighbor graph and the expected degree of a random vertex in the Delaunay triangulation. The first generalizes well in higher dimension, while the second becomes an data sensitive parameter (constant for random points, $n^{\lceil (d-1)/2 \rceil}$ in the worst case). A generalization for computing the trapezoidal map can also be done.

Code

A demo version compiled for Sun Solaris and SGI is available at

http://www.inria.fr/prisme/logiciels/del-hierarchy/.

Acknowledgement

The author would like to thank Hervé Brönnimann, Jonathan Shewchuk, Jack Snoeyink and Mariette Yvinec for helpful discussions and careful reading of this paper.

References

[AGSS89] A. Aggarwal, L. J. Guibas, J. Saxe, and P. W. Shor. A linear-time algorithm for computing the Voronoi diagram of a convex polygon. *Discrete Comput. Geom.*, 4(6):591–604, 1989.

[BD95] P. Bose and L. Devroye. Intersections with random geometric objects. Technical report, School of Computer Science, McGill University, 1995. Manuscript.

[BDH93] C. B. Barber, D. P. Dobkin, and H. Huhdanpaa. The Quickhull algorithm for convex hull. Technical Report GCG53, Geometry Center, Univ. of Minnesota, July 1993.

[BT86] J.-D. Boissonnat and M. Teillaud. A hierarchical representation of objects: The Delaunay tree. In *Proc. 2nd Annu. ACM Sympos. Comput. Geom.*, pages 260–268, 1986.

[BT93] J.-D. Boissonnat and M. Teillaud. On the randomized construction of the Delaunay tree. *Theoret. Comput. Sci.*, 112:339–354, 1993.

[Che86] L. P. Chew. Building Voronoi diagrams for convex polygons in linear expected time. Technical Report PCS-TR90-147, Dept. Math. Comput. Sci., Dartmouth College, Hanover, NH, 1986.

[Che87] L. P. Chew. Constrained Delaunay triangulations. In *Proc. 3rd Annu. ACM Sympos. Comput. Geom.*, pages 215–222, 1987.

[DMT92] O. Devillers, S. Meiser, and M. Teillaud. Fully dynamic Delaunay triangulation in logarithmic expected time per operation. *Comput. Geom. Theory Appl.*, 2(2):55–80, 1992.

[GKS92] L. J. Guibas, D. E. Knuth, and M. Sharir. Randomized incremental construction of Delaunay and Voronoi diagrams. *Algorithmica*, 7:381–413, 1992.

[Law77] C. L. Lawson. Software for C^1 surface interpolation. In J. R. Rice, editor, *Math. Software III*, pages 161–194. Academic Press, New York, NY, 1977.

[MR95] R. Motwani and P. Raghavan. *Randomized Algorithms*. Cambridge University Press, New York, NY, 1995.

[MSZ96] Ernst P. Mücke, Isaac Saias, and Binhai Zhu. Fast randomized point location without preprocessing in two- and three-dimensional Delaunay triangulations. In *Proc. 12th Annu. ACM Sympos. Comput. Geom.*, pages 274–283, 1996.

[Mul91] K. Mulmuley. Randomized multidimensional search trees: Dynamic sampling. In *Proc. 7th Annu. ACM Sympos. Comput. Geom.*, pages 121–131, 1991.

[Mul94] K. Mulmuley. *Computational Geometry: An Introduction Through Randomized Algorithms*. Prentice Hall, Englewood Cliffs, NJ, 1994.

[PY92] M. S. Paterson and F. F. Yao. On nearest-neighbor graphs. In *Proc. 19th Internat. Colloq. Automata Lang. Program.*, volume 623 of *Lecture Notes Comput. Sci.*, pages 416–426. Springer-Verlag, 1992.

[SD97] P. Su and R. Drysdale. A comparison of sequential Delaunay triangulation algorithms. *Comput. Geom. Theory Appl.*, 7:361–386, 1997.

[She96] J. R. Shewchuk. Triangle: engineering a 2d quality mesh generator and Delaunay triangulator. In *First Workshop on Applied Computational Geometry*. Association for Computing Machinery, May 1996.

[TLP96] R. Tamassia, G. Liotta, and F. P. Preparata. Robust proximity queries in implicit Voronoi diagrams. In *Proc. 8th Canad. Conf. Comput. Geom.*, page 1, 1996.

Vertex-rounding a three-dimensional polyhedral subdivision

Steven Fortune
Bell Laboratories
Murray Hill, New Jersey 07974
sjf@research.bell-labs.com

Abstract

Let P be a polyhedral subdivision in $\mathbb{R}^3$ with a total of n faces. We show that there is an embedding σ of the vertices, edges, and facets of P into a subdivision Q, where every vertex coordinate of Q is an integral multiple of $2^{-\lceil \log_2 n+2 \rceil}$. For each face f of P, the Hausdorff distance in the L_∞ metric between f and $\sigma(f)$ is at most 3/2. The embedding σ preserves or collapses vertical order on faces of P. The subdivision Q has $O(n^4)$ vertices in the worst case, and can be computed in the same time.

1 Introduction

Geometric algorithms are usually described in the "real-number RAM" model of computation, where arithmetic operations on real numbers have unit cost. A programmer implementing a geometric algorithm must find some substitution for real arithmetic. The substitution of exact arithmetic on a subset of the reals, say the integers or the rationals, avoids the difficulties that can arise from naive substitution of floating-point arithmetic [4, 12, 14, 15]. The substitution is not trivial, since the required arithmetic bit-length usually exceeds the native arithmetic bit-length of most computer hardware, and some form of software arithmetic is required.

Recent research has made the use of software exact arithmetic for geometric algorithms much more attractive. A predicate on geometric data is determined by the sign of an arithmetic expression in the coordinates of the data. A promising strategy for sign-evaluation is adaptive-precision arithmetic [6, 13, 20], where the expression is evaluated to higher and higher precision until its sign is known, i.e. until the magnitude of the expression exceeds an error bound. Low precision, even floating-point, suffices most of the time, since most instances of geometric predicates are easy. In addition, for some basic predicates like the sign of a determinant, there are alternative evaluation strategies that require arithmetic with relatively low precision [1, 2, 3].

Exact arithmetic would be more useful if high-level geometric rounding algorithms were available. Virtually any geometric construction that produces new geometric data increases the bit-length of geometric coordinates. For example, suppose points are represented with homogeneous integer coordinates. The plane through three such points has coefficients whose bit-lengths are about three times the point coordinate bit-lengths; the point of intersection of three such planes has coordinate bit-length about nine times that of the original points. Thus a solid modeler, which implements boolean operations and rigid motions on polyhedra, might produce a polyhedron with high coordinate bit-length even if the original polyhedra had short coordinate bit-length. Typically an application requires only a low-precision approximation, not the exact answer. Hence there is a need for high-level rounding, which replaces a geometric structure with high bit-length coordinates with an approximating structure with short bit-length coordinates. It does not suffice to round each coordinate independently, since such rounding is a geometric perturbation, and may introduce inconsistencies between geometric and combinatorial information. Furthermore, some change in combinatorial structure is inevitable; indeed, in certain cases it is NP-hard to determine if it is possible to round to low-precision without changing combinatorial structure [19].

There are satisfactory high-level rounding algorithms for polygonal subdivisions in two dimensions. One such algorithm is snap-rounding [10]. Fix a polygonal subdivision, with arbitrary-precision coordinates. A *pixel* is a unit square in the plane centered at a point with integer coordinates; a pixel is *hot* if it contains a vertex of the subdivision. Snap-rounding replaces each vertex by the center of the pixel containing the vertex, and each edge by the polygonal chain through the centers of the hot pixels met by the edge, in the same order as met

SCG 98 Minneapolis Minnesota USA

by the edge. The snap-rounded subdivision approximates the original subdivision in the sense that each vertex and edge of the original subdivision has an image in the snap-rounded arrangement whose Hausdorff distance is at most 1/2 in the L_∞ metric. Snap-rounding may change the combinatorial structure of the subdivision, for example, vertices and edges may collapse together, but some combinatorial ordering information is preserved [10].

This paper generalizes snap-rounding to polyhedral subdivisions in three dimensions. Fix a polyhedral subdivision P with a total of n vertices, edges, and facets. We show that there is a polyhedral subdivision Q so that each vertex coordinate is an integer multiple of $1/2^{\lceil \log_2 n \rceil + 2}$. Each face f of P has an image $\sigma(f)$ in Q so that the Hausdorff distance between f and $\sigma(f)$ is at most 3/2. As with snap-rounding in two dimensions, f and $\sigma(f)$ may have different combinatorial structure: an edge may be replaced with a polygonal chain, and a facet with a triangulation. Two vertices may collapse together; the polygonal chains for two edges or the triangulations for two facets may collapse together or overlap partially, perhaps in several places. However, vertical order is preserved (or collapsed): if face f is vertically above face f' (i.e. there is a line parallel to the z-axis meeting both faces, and the intersection with f has higher z-coordinate), then $\sigma(f)$ is above (or overlaps) $\sigma(f')$. In the worst case Q has $O(n^4)$ vertices and can be computed in time $O(n^4)$.

Other work. Greene and Yao were the first to suggest a rounding scheme for polygonal subdivisions in two dimensions [8]. Hobby [11] and Greene [9] give algorithms to compute the snap-rounding of the arrangement formed by a set of intersecting edges. Guibas and Marimount [10] show how to maintain the snap-rounded arrangement of edges under insertion and deletion of edges; they also give elementary proofs of basic topological properties of snap rounding. Goodrich *et al* [7] give improved algorithms to snap-round a set of intersecting edges, in the case when there are many intersections within a pixel. Milenkovic [18] suggests a "shortest-path" geometric rounding scheme that sometimes introduces fewer bends than snap rounding.

Goodrich *et al* [7] propose a scheme for snap-rounding a set of edges in three dimensions after first adding as vertices the points of "closest encounter" between nearby edges. Milenkovic [16] sketches a scheme for rounding a polyhedral subdivision in three dimensions (in fact, any dimension). Unfortunately, both schemes have the property that rounded edges can cross (see below), which violates any notion of topological consistency.

Fortune [5] suggests a high-level rounding algorithm for polyhedra in three dimensions. His algorithm assumes that a polyhedron is presented by the equations of its face planes (and the combinatorial incidence structure of faces), not the coordinates of vertices as assumed by snap-rounding. His algorithm does not appear to extend from polyhedra to polyhedral subdivisions.

1.1 Overview

We give a brief overview of the rounding algorithm. We start by mentioning some difficulties with the three-dimensional extension of snap-rounding.

The obvious way to snap-round a vertex in three dimensions is to replace it with the center of the voxel containing it. (A *voxel* is a unit cube centered at an integer point.) However, it is less clear how to snap-round edges and facets.

Snap-rounding a set of edges in three-dimensions requires the addition of new vertices, unlike the situation in two dimensions. Consider two transverse nearby edges. Rounding the endpoints to voxel centers perturbs the edges, and hence the edges may change orientation or cross. We can attempt to prevent this by adding a vertex in the interior of each edge near the other edge; then either the two new vertices are in the same voxel and snap-round together, or they are in different voxels and the snap-rounded edges will not cross. Clearly, it might be necessary to add quadratically many vertices, if the edges form a "cross-hatch" pattern.

Snap-rounding with facets as well is more problematic. If a vertex v and a facet f are nearby, we can add a new vertex v' to f to ensure that v and f are properly separated or collapsed. However, this requires that f be triangulated, which introduces new edges. Potentially these edges are close to old edges, which could require new vertices, and it is not immediate that the process is finite. We can attempt to ensure termination by projecting nearby edges onto a facet, and then triangulating the facet compatibly with the projection. The actual rounding algorithm is a formalization of this idea.

The rounding algorithm. Orthogonally project all the edges of the subdivision P onto the xy-plane, form the arrangement, snap-round, and compute a triangulation T. Each face of P has an image within the triangulation: the image of an edge is a polygonal chain, and the image of a facet is a subtriangulation of T. The rounding of each facet f is obtained by lifting the image of f in T to three dimensions in such a way as to approximate f. By considering each cylinder over a vertex, edge, or triangle of T separately, we can ensure that the lifting preserves (or collapses) the vertical order on faces of P.

There are several technical difficulties with this algorithm. We must first ensure that there are no crossings among the polygonal chains that result from rounding the edges of P. Figure 1 indicates one way such a crossing could occur. To prevent crossings, we subdivide the edges of P by all $xy-$, $xz-$, and yz-intersection points. (If the orthogonal projections of e and e' into the xz-plane cross at a point p, and l is the line parallel to the y-axis through p, then $e \cap l$ and $e' \cap l$ are *xz-intersection points.*) Unfortunately, this subdivision is not quite suf-

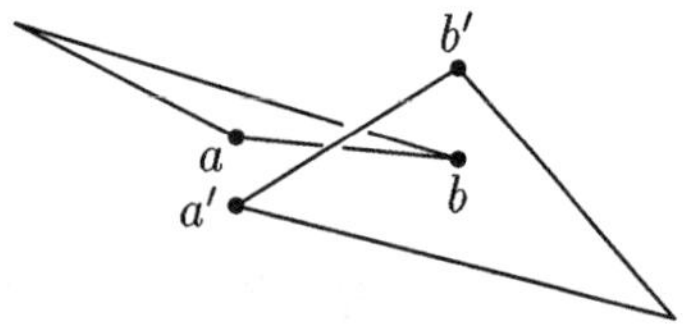

Figure 1: Vertices a and a' project to the same pixel in the xy-plane, as do b and b'. Hence in three dimensions, the snap-rounding of ab crosses the snap-rounding of $a'b'$.

ficient to prevent crossings among snap-rounded edges. In figure 2, d^* and e^* have endpoints on column facets. The xy-, xz-, and yz-projections of d^* and e^* are all disjoint, but their snap-roundings cross. Fortunately, the configuration of figure 2 is almost the only way this can happen, and we can show that there is a slight modification of snap-rounding that avoids crossings. In figure 2, the modified snap-rounding of d^* is a two-edge polygonal chain, connecting a snap-rounded endpoint of d^* to the snap-rounded endpoint of e^* on the same vertical line, and then to the other snap-rounded endpoint of d^*. We show that the distance between an edge and its modified snap-rounding increases slightly, to at most 3/2. (The configuration in figure 2 can be modified to show that the "close encounter" subdivision of Goodrich *et al* [7] does not prevent edge crossings.) Section 4 describes the modification of snap-rounding.

Let T_f be the image of facet f within triangulation T. We lift T_f to three dimensions by first lifting each edge and then the interior of each triangle. The lifting of an edge must satisfy three conditions: it must be close to f; it must not cross any other edge; and it must preserve vertical order. This last condition is a bit tricky. Consider the situation given in cross-section in figure 3, with e, e', and f orthogonal to the figure. Assume that edges e and e' project and snap-round to the same edge d in the triangulation T. Edge e is above facet f, so the lifting of edge d for facet f must be on

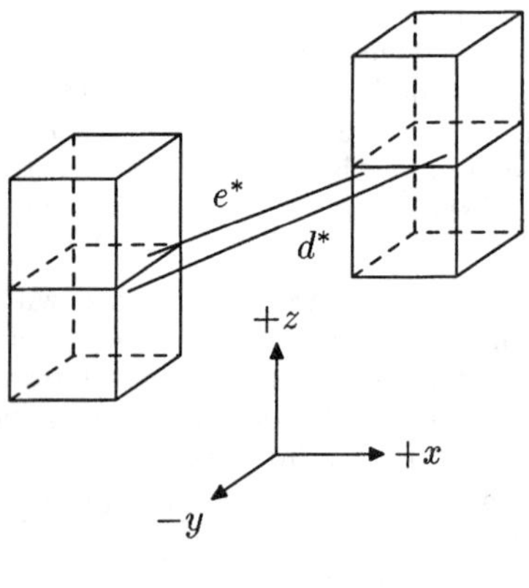

Figure 2: $\rho(d^*)$ and $\rho(e^*)$ cross, although the xy-, xz-, and yz-projections of d^* and e^* do not.

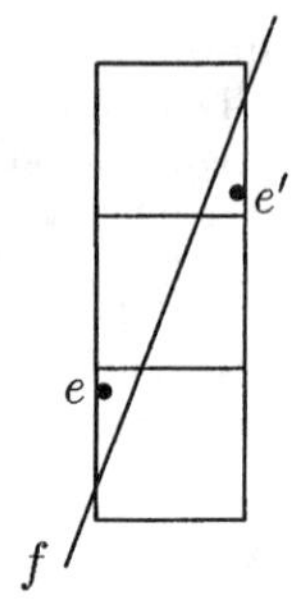

Figure 3: Side view. Edge e is above facet f and e' is below. Hence the rounding of facet f must contain the vertical interval from the rounding of e to the rounding of e'.

or below the rounding of e. Similarly, edge e' is below facet f, so the lifting of edge d for facet f must on or above the rounding of e. Hence the lifting of d for facet f must at least span the vertical interval between the rounding of e and the rounding of e'. Section 5 below describes how to lift triangulation edges.

Consider a triangle $\Delta = \Delta abc$ of T_f. If the lifted edges ab, ac, and bc for facet f are all pairwise incident, then the lifting of Δ for facet f, $l_f(\Delta)$, is simply the triangle with those edges. Unfortunately, incidence cannot be guaranteed (though, of course, both lifted edges ab and ac meet the vertical line through a, and similarly for b and c). Hence $l_f(\Delta)$ must be a triangulation of the polygon formed by the lifted edges ab, ac, and bc, and perhaps edges along the vertical lines through a, b, and c. See figure 4. It is easy to triangulate the polygon using a central vertex whose xy-projection is within triangle Δ. However, a vertical boundary edge may be shared among several different liftings. To ensure that there are no crossings among edges, each central vertex must have distinct coordinates. Since there may be n central vertices, coordinates that are integer multiples of roughly $1/n$ are necessary. More details of the lifting appear in section 6.

Naively the rounded subdivision Q has at most $O(n^3)$ faces: the triangulation T has $O(n^2)$ triangles, so for each facet f the rounding $\sigma(f)$ consists of $O(n^2)$ lifted triangles $\{l_f(\Delta)\}$. However, in the worst case each lifted triangle $l_f(\Delta)$ may consist of $O(n)$ faces, since there could be linearly many vertices on the vertical edges of its boundary. Hence Q has $O(n^4)$ faces.

2 The main theorem

For points $a, b \in \mathbb{R}^3$ and sets $A, B \subset \mathbb{R}^3$, $d(a,b)$ is the L_∞ distance between a and b (the L_∞ distance is used exclusively in this paper); $d(a,B)$ is $\inf_{b\in B} d(a,b)$; and $d(A,B)$ is $\sup_{a\in A} d(a,B)$. Note that d is symmetric for points, but not in general for sets. *Hausdorff distance* $d_H(A,B)$ is $\max(d(A,B), d(B,A))$.

The direction parallel to the z-axis is the *vertical*

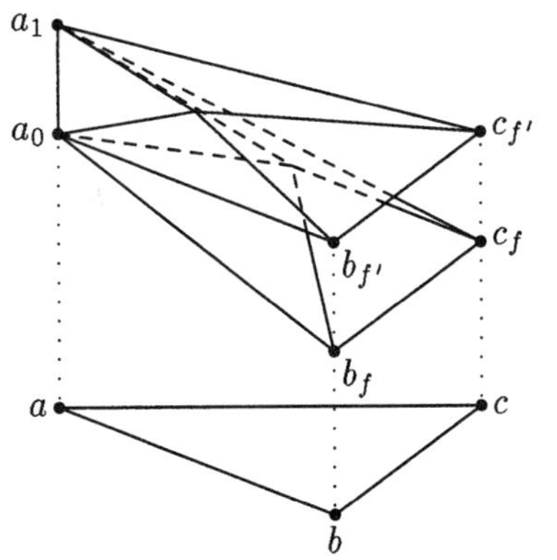

Figure 4: The liftings of triangle Δabc for facets f and f' have boundary $a_1a_0b_fc_f$ and $a_1a_0b_{f'}c_{f'}$, respectively.

direction. Two sets $A, B \subset \mathbb{R}^3$ are *vertically ordered* $A \prec B$ (read "A is below B") if there is a vertical line meeting both A and B, and for every vertical line l meeting A and B, $A \cap l$ is below $B \cap l$, i.e. the z-coordinate of every point of $A \cap l$ is less than the z-coordinate of every point in $B \cap l$. Sets A and B satisfy $A \preceq B$ if there is a vertical line meeting both, and for every vertical line meeting both, $A \cap l$ is below or intersects $B \cap l$. As is well-known, $\prec$ is not transitive in general; it is transitive among a family of sets that have the same xy-projection. If in addition, every set in the family is a *surface*, i.e. every vertical line misses the set or meets it at one point, then $\preceq$ is also transitive.

A *subdivision* P in $\mathbb{R}^3$ is a set of compact convex polyhedral *cells* so that every face of every cell is in the subdivision and so that the intersection of two cells is a face of both. Cells of dimension 0, 1, and 2 are *vertices*, *edges*, and *facets*, respectively. $|P|$ is the union of the cells of P. An *embedding* of a subdivision P into a subdivision Q is a mapping σ that maps each cell of P into a subdivision contained in Q so that if f is a face of f', then $\sigma(f) \subseteq \sigma(f')$.

To simplify notation somewhat, we extend d and $\prec$ to subdivisions. Thus for subdivisions P and Q, $P \prec Q$ means $|P| \prec |Q|$ and $d(P,Q)$ means $d(|P|,|Q|)$.

Throughout this paper we assume that subdivisions in $\mathbb{R}^3$ do not include cells of dimension 3. Furthermore, we assume that every subdivision is in general position, specifically, that no edge or facet is parallel to a coordinate axis and that no vertex has a coordinate that is an integer multiple of 1/2. The general position assumption simplifies presentation; it is not hard to remove (either explicitly or for example by an infinitesimal symbolic rigid motion).

Theorem 2.1 *Let P be a subdivision in $\mathbb{R}^3$ with a total of n cells; set $\kappa = 3/2$. There is a subdivision Q and an embedding σ of P into Q so that:*

1. *For each cell f of P, $d_H(f, \sigma(f)) < \kappa$.*
2. *Each vertex coordinate of Q is an integral multiple of $1/2^{\lceil 2+\log_2 n \rceil}$.*
3. *If cells f, f' of P satisfy $f \preceq f'$, then $\sigma(f) \preceq \sigma(f')$.*
4. *Q can be computed in time $O(n^4)$ and has $O(n^4)$ cells.*

This theorem follows from the discussion below. At a high level, the algorithm required for step (4) has three steps.

1. Subdivide the vertices and edges of P, forming a set of vertices and edges P^* (Section 4).
2. Orthogonally project P^* onto the xy-plane, snap-round, and triangulate the convex hull of the resulting subdivision. Let T be the resulting triangulation.
3. For each cell f in P, lift T_f (the image of f in T) to a subdivision $Q_f \subset \mathbb{R}^3$ (Sections 5 and 6).

3 Definitions

A *pixel* is an open unit square in the xy-plane centered at an integer point; pixel(q) is the pixel containing point q. A *voxel* is an open unit cube in $\mathbb{R}^3$ centered at an integer point; voxel(q) is the voxel containing point q. A *column* (of voxels) is all voxels whose centers have the same x- and y-coordinates; column(q) is the column containing q.

Let A be a subdivision in the xy-plane. A pixel is *hot* (with respect to A) if it contains a vertex. The *snap-rounding* (with respect to A) of an edge e of A is the polygonal chain connecting the centers of the hot pixels met by e in the same order as met by e; similarly the *snap-rounding* of a vertex of A is the center of the hot pixel containing it. A basic fact[10] is that two polygonal chains that result from snap-rounding intersect only at vertices and edges of both chains. The *snap-rounding* of A is obtained by replacing each edge and vertex of A with its snap-rounding with respect to A; it is a polygonal subdivision whose vertices are hot pixel centers, i.e. integer points, and whose edges connect integer points.

Let π_{xy} be orthogonal projection onto the xy-plane, and similarly for π_{xz} and π_{yz}. A set $A \subset \mathbb{R}^3$ is *over* a set P in the xy-plane if $\pi_{xy}(A) = P$. If A is a surface with $p \in \pi_{xy}(A)$, then A_p is the point of A over p (i.e. $\pi_{xy}(A_p) = p$). If A and B are surfaces over the same set, then max(A, B) is the pointwise maximum (viewed as functions of the xy-plane), and min(A, B) is the pointwise minimum. If A, B, C are surfaces over the same set with $A \succeq B$, then snap($C, [A, B]$) is min(A, max(B, C)). Clearly, $A \succeq$ snap($C, [A, B]$) $\succeq B$.

Suppose a set P in the xy-plane is fixed. We define symbolic sets $\top$ (top) and $\bot$ (bottom) satisfying $\bot \prec A \prec \top$ for any other set A over P. We have for example

$\min(A, \top) = A = \max(A, \bot)$; we define min and max of an empty collection to be $\top$ and $\bot$, respectively.

Two edges *cross* if they intersect at a point interior to at least one of the edges.

Proposition 3.1 *Let $T \subset \mathbb{R}^3$ be convex, $\{s_1, \ldots, s_k\}$ be a finite set of points in $\mathbb{R}^3$ with convex hull S, and let $\kappa \geq 0$. If $d(s_i, T) \leq \kappa$ for $i = 1, \ldots, k$, then $d(S, T) \leq \kappa$.*

Proof: Any point in S can be expressed as $\sum \alpha_i s_i$ with $0 \leq \alpha_i \leq 1$ and $\sum \alpha_i = 1$. For each s_i, there is a point $t_i \in T$ so that $d(s_i, t_i) \leq \kappa$. Clearly $\sum \alpha_i t_i \in T$ and $d(\sum \alpha_i s_i, \sum \alpha_i t_i)$ is the maximum absolute value of any coordinate of $\sum \alpha_i (s_i - t_i)$, which is bounded by κ since $\sum \alpha_i = 1$, $\alpha_i \geq 0$, and the absolute value of each coordinate of $s_i - t_i$ is bounded by κ. □

4 Snap-rounding edges

Define $\rho(q)$ to be the center of the voxel containing q, and extend to ρ to edges: $\rho(qq')$ is the edge $\rho(q)\rho(q')$. The mapping ρ is the obvious extension of snap-rounding to three dimensions (ignoring snapping to hot voxels, which is unimportant here). Unfortunately, ρ may cause two edges to cross. We now define a refinement P^* of the vertices and edges of P and a modification τ of ρ so that no two edges in $\tau(P^*)$ cross.

4.1 The subdivision P^*

Let e and e' be two edges of P whose xy-projections cross at a point p. An *xy-intersection point* (of P) is either point on e or e' that meets the line through p parallel to the z-axis. The definition of an $xz-$ or yz-intersection point is similar.

Subdivision P^* results from subdividing the edges of P. At any point in the process, $\hat{e}$ denotes the subdivision of edge e of P; any voxel containing a vertex is a *hot* voxel; and any column of voxels containing a hot voxel is a *hot* column. There are two steps in the subdivision:

1. Subdivide the edges of P at all xy-, xz-, and yz-intersection points of P.
2. For each edge e of P, split $\hat{e}$ by each hot column C it meets: $\hat{e}$ must meet C in a consecutive set of voxels; $\hat{e}$ is *split* by C by further subdividing $\hat{e}$ at any point in the first voxel (if $\hat{e}$ does not yet have a vertex in the first voxel) and similarly by subdividing $\hat{e}$ in the last voxel.

Splitting by hot columns has an easy consequence: for any edge e of P, the snap-rounding of $\pi_{xy}(\hat{e})$ with respect to $\pi_{xy}(P^*)$ is identical to the snap rounding of $\pi_{xy}(\hat{e})$ with respect to $\hat{e}$. Henceforth we use a superscript '*' for edges and vertices of P^*. For e^* an edge of P^*, we write $s(e^*)$ for the snap-rounding of $\pi_{xy}(e^*)$. It is immediate that if d^*, e^* are edges of P^*, then $\rho(d^*)$ crosses $\rho(e^*)$ only if $s(d^*) = s(e^*)$.

Lemma 4.1 *P^* has $O(n^3)$ vertices; there are $O(n^2)$ hot columns and $O(n^3)$ hot voxels.*

Proof: Clearly there are at most $O(n^2)$ xy-, xz-, and yz-intersection points, and $O(n)$ vertices of P. Splitting edges by hot columns adds no new hot columns, hence there are $O(n^2)$ hot columns. For each edge e of P and for each hot column, there are at most two vertices added when $\hat{e}$ is split by the column. Hence there are $O(n^3)$ vertices altogether. □

As mentioned earlier, T is a triangulation of the convex hull of $s(P^*)$. Consider the edges E^* in P^* bounding a facet f of P. The projection $\pi_{xy}(E^*)$ forms a simple cycle, but the snap-rounding $s(E^*)$ need not. However, it is not hard to see that $s(E^*)$ consists of some number of simple cycles connected by polygonal chains. Let T_f be the subtriangulation of T consisting of the vertices and edges of $s(E^*)$ plus any vertices, edges and triangles of T interior to the simple cycles in $s(E^*)$.

For v a vertex of T, e an edge of T, and Δ a triangle of T, define

$$\begin{aligned} P_e^* &= \{e^* \in P^* : s(e^*) = e\} \\ P_v^* &= \{v^* \in P^* : s(v^*) = v\} \\ F_e &= \{f \in P^* : e \in T_f\} \\ F_\Delta &= \{f \in P^* : \Delta \in T_f\} \end{aligned}$$

where v^* and e^* are vertices and edges of P^*, respectively, and f is a facet of P.

4.2 The mapping τ

Lemma 4.2 *Let e be an edge of T. If $d^*, e^* \in P_e^*$ and $\rho(d^*), \rho(e^*)$ cross, then either there is an endpoint w of $\rho(d^*)$ with $d(w, e^*) < \kappa$ or an endpoint w' of $\rho(e^*)$ with $d(w', d^*) < \kappa$.*

The intricate proof of this lemma is omitted due to lack of space.

Lemma 4.3 *Let e be an edge of T. There is a mapping τ defined on P_e^* so that*

1. *For all edges $e^* \in P_e^*$, $\tau(e^*)$ is an edge over e with endpoints among the endpoints of $\rho(P_e^*)$.*
2. *For all edges e^*, $d(\tau(e^*), e^*) < \kappa$.*
3. *$\tau(P_e^*)$ is noncrossing.*
4. *τ can be computed in time quadratic in the size of P_e^*.*

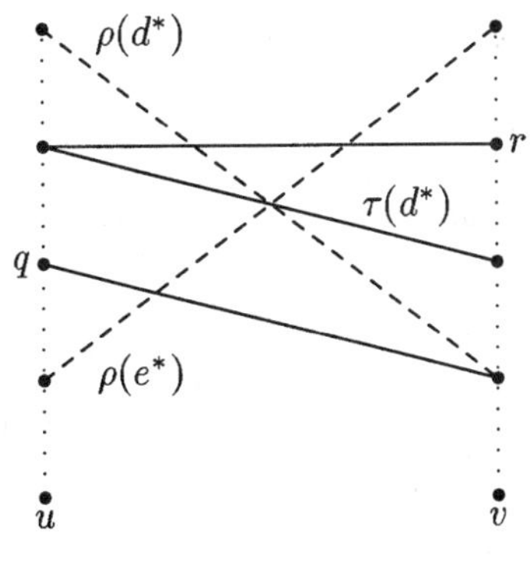

Figure 5: Definition of τ on new edge e^*.

Proof: We define τ inductively, adding edges of P_e^* one by one in arbitrary order. The addition of an edge may change the definition of τ on other edges as well; however, properties (1) through (3) of the lemma statement are maintained.

So suppose τ has been defined on a subset S of P_e^* and e^* is the next edge. If no edge of $\tau(S)$ crosses $\rho(e^*)$, then simply define $\tau(e^*) = \rho(e^*)$. Otherwise, since $\tau(S)$ is noncrossing, we can assume up to a symmetric argument that every edge $\tau(d^*)$ crossing $\rho(e^*)$ has $\tau(d^*)_u \succ \rho(e^*)_u$ and $\tau(d^*)_v \prec \rho(e^*)_v$. Let q be the highest (in $\prec$) vertex in $\tau(S)_u$ so that $d(q, e^*) < \kappa$ and let r be the lowest vertex in $\tau(S)_v$ so that $d(r, e^*) < \kappa$. If there is an edge $q'r'$ not crossing any edge in $\tau(S)$ with $q' \in \tau(S)_u$, q' between $\rho(e^*)_u$ and q, $r' \in \tau(S)_v$, and r' between $\rho(e^*)_v$ and r, define $\tau(e^*) = q'r'$; condition (2) of the lemma is easily verified. Otherwise, let S' be the subset of S crossing qr; S' must not be empty. Clearly for any $d^* \in S'$, $\tau(d^*)_u \succ q$ and $\tau(d^*)_v \prec r$. See figure 5.

We claim that for any edge $\tau(d^*) \in S'$, either $d(q, d^*) < \kappa$ or $d(r, d^*) < \kappa$. If $\rho(d^*)_u \preceq q$, then certainly $d(q, d^*) < d(\tau(d^*)_u, d^*) < \kappa$. Similarly if $\rho(d^*)_v \succeq r$, then $d(r, d^*) < \kappa$. Otherwise $\rho(d^*)_u \succ q \succeq \rho(e^*)_u$ and $\rho(d^*)_v \prec r \preceq \rho(e^*)_v$, so $\rho(d^*)$ crosses $\rho(e^*)$. See figure 5. By lemma 4.2 and the definition of q and r, either $d(\rho(e^*)_v, d^*) < \kappa$ or $d(\rho(e^*)_u, d^*) < \kappa$, so either $d(q, d^*) < \kappa$ or $d(r, d^*) < \kappa$.

Let Q be the set of edges $d^* \in S'$ so that $d(q, d^*) < \kappa$, and $R = S' \setminus Q$. Define $\tau(e^*) = qr$; for $d^* \in Q$, redefine $\tau(d^*)_u = q$; and for $d^* \in R$, redefine $\tau(d^*)_v = r$. It is easy to check that τ satisfies conditions (1) through (3). The running time is immediate. □

Henceforth we let τ be defined on all edges of P^*, by choosing a definition on P_e^* separately for each edge e of T, using lemma 4.3. Since there can be $O(n^2)$ edges e in T, and $O(n)$ edges in P_e^*, computation of τ takes time $O(n^4)$.

There is no guarantee that $d(e^*, \tau(e^*)) \leq \kappa$ nor that $\tau(e^*)$ and $\rho(e^*)$ have the same endpoints. In section 6, we guarantee both properties by in effect augmenting $\tau(e^*)$ to a polygonal chain using vertical edges connecting its endpoints to the endpoints of $\rho(e^*)$.

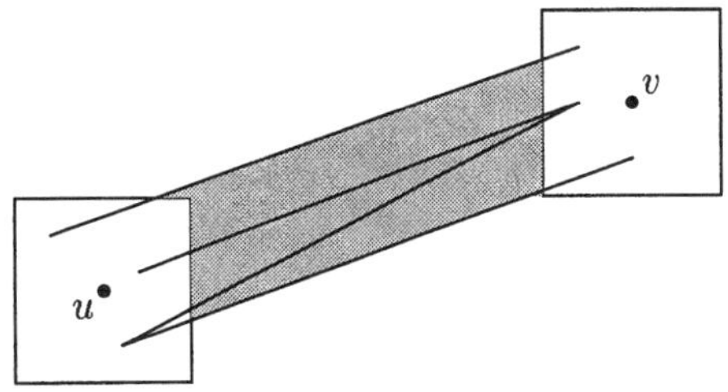

Figure 6: R_e is the shaded region plus the portion of the edges inside pixel(u) and pixel(v).

5 Lifting triangle edges

The desired embedding $\sigma(f)$ of facet f of P will eventually be obtained by lifting each vertex, edge, and triangle of T_f to three dimensions. This section handles a technically difficult case, the lifting of an edge e of a triangle Δ of T_f to the "lifted edge" $l_{f\Delta}(e)$. The lifted edge will satisfy three properties: $f \preceq f'$ implies $l_{f\Delta}(e) \preceq l_{f'\Delta}$, i.e. vertical order is preserved or collapsed; $d(l_{f\Delta}(e), f) \leq \kappa$; and no pair of lifted edges cross.

5.1 The order $\lhd$ and the snapping lemma

Let edge e of T have endpoints u and v. Define R_e to be the convex hull of $\pi_{xy}(P_e^*)$, less the interior of pixel(u) and pixel(v), unioned with $\pi_{xy}(P_e^*)$. See figure 6. Notice that there are no intersections among the boundaries of $\{\pi_{xy}(f) : f \in F_\Delta\}$ within R_e except possibly at the endpoints of edges of $\pi_{xy}(P_e^*)$. Facet $f \in F_e$ *covers* e if no edge in P_e^* bounds f; it is easy to check that $R_e \subseteq \pi_{xy}(f)$. A facet f *covers facet* f' *at* e if $\pi_{xy}(f') \cap R_e \subseteq \pi_{xy}(f) \cap R_e$. For any two facets $f, f' \in F_e$, either f covers f' at e, or f' covers f at e.

Suppose that e is an edge of triangle Δ of T. The *covering order* $\lhd$ on the facets in F_Δ is any total order so that $f \lhd f'$ implies f' covers f at e. (The order depends on both e and Δ, but to keep the notation simple we do not make this dependence explicit.) The order $\lhd$ can be described as follows. Assume that Δ lies to the left of the e, directed from endpoint u to endpoint v; direct all edges in P_e^* from pixel(u) to pixel(v). If facets $f_0, f_1 \in F_\Delta$ have bounding edges $e_0^*, e_1^* \in P_e^*$, then $f_0 \lhd f_1$ if e_0^* is to the left of e_1^*; all facets covering e appear at the end of the order $\lhd$, and are ordered arbitrarily among themselves.

For a set $S \subset \mathbb{R}^3$, let $V(S)$ be all points on all vertical lines through S. Let f_A be a facet of P, e an edge of T, and A an edge over e with endpoints u and v. Edge A *approximates* f_A *at* e if $d(A_u, f_A \cap V(R_e)) < \kappa$ and $d(A_v, f_A \cap V(R_e)) < \kappa$. Clearly, if A approximates f_A at e, then $d(A, f_A) < \kappa$. Also, if $e^* \in P_e^*$ is a boundary

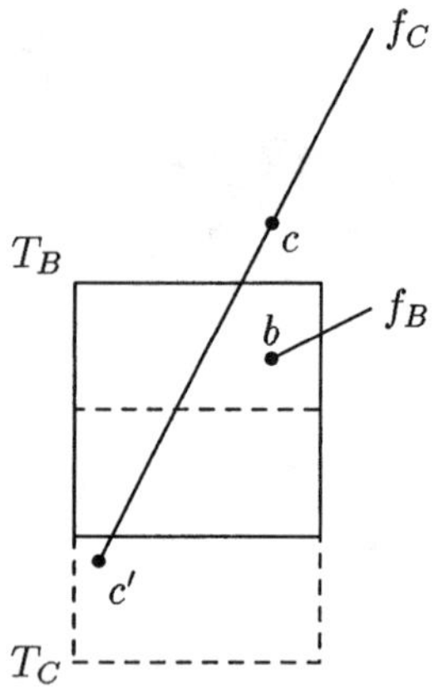

Figure 7: Proof of lemma 5.1, side view. T_B is solid square, T_C dashed.

edge of face f, then $\tau(e^*)$ approximates f at e.

Lemma 5.1 *Let edge e bound triangle Δ of T. Suppose $f_A, f_B, f_C \in F_\Delta$ with $f_A \succeq f_C \succeq f_B$, A, B, C are edges over e approximating f_A, f_B, f_C, respectively, $A \succeq B$, and f_C covers f_A and f_B. Then $\mathrm{snap}(C, [A, B])$ also approximates f_C at e.*

Proof: We claim $\max(B, C)$ approximates f_C at e; a similar result holds for min, from which the lemma follows. Let u be an endpoint of e. We show

$$d(\max(B_u, C_u), f_C \cap V(R_e)) < \kappa.$$

If $C_u \succeq B_u$, there is nothing to prove, so suppose $C_u \prec B_u$.

Let T_B and T_C be the cubes of sidelength 2κ centered at B_u and C_u, respectively, and $T = V(T_B)$ (clearly also $T = V(T_C)$). See figure 7.

Since B approximates f_B, there is a point $b \in f_B \cap V(R_e) \cap T_B$. Since $f_C \succeq f_B$ and f_C covers f_B at e, there is a point $c \in f_C$ with $c \succeq b$; clearly $c \in T$. Since C approximates f_C, there is a point $c' \in f_C \cap V(R_e) \cap T_C$. Since $f_C \cap V(R_e) \cap T$ is path-connected, there is a path in $f_C \cap V(R_e) \cap T$ from c to c'. Since c is above the bottom facet of T_B, c' is below the top facet of T_C, and $T_B \succeq T_C$, some point of the path meets T_B. Hence $d(B_u, f_C) < \kappa$. □

5.2 Default edges.

Let e be an edge of T_f with endpoints u and v and with some triangle incident. We define the *default lifting of edge e for facet f*, $c_f(e)$, which is to be used in the absence of other constraints. If e is a boundary edge of T_f, then there is a unique edge $e^* \in P_e^*$ bounding facet f, and we define $c_f(e) = \tau(e^*)$. Otherwise suppose e is an interior edge of T_f; clearly f covers e and no edge in P_e^* meets f. The definition is slightly complicated because of the requirement that e not cross any edge in $\tau(P_e^*)$.

Define $\mathrm{low}_f(e)$ to be the edge $\hat{u}\hat{v}$, where $\hat{u}$ is the center of the lowest voxel X in column(u) so that $X \cap f \cap V(R_e)$ is not empty, and similarly for $\hat{v}$. A pair of distinct edges (a^*, b^*) in P_e^* is a *bracketing pair* if $a^* \succ f$ and $f \succ b^*$, $\tau(a^*) \succeq \tau(b^*)$ and no edge $\tau(d^*)$, $d^* \in P_e^*$, lies between $\tau(a^*)$ and $\tau(b^*)$ (possibly $\tau(a^*) = \tau(b^*)$). The existence of a bracketing pair can be seen by indexing the edges of $P_e^* = \{e_0^*, \ldots, e_k^*\}$ so that $\tau(e_0^*) \succeq \tau(e_1^*) \succeq \ldots \succeq \tau(e_k^*)$. Either $f \succ e_0^*$, and the pair $(\top, e_0^*)$ suffices (with the definition $\tau(\top) = \top$); or $e_k^* \succ f$, and the pair $(e_k^*, \bot)$ suffices; or there is i so that an $e_i \succ f$ and $f \succ e_{i+1}$, and the pair (e_{i+1}, e_i) suffices. Define

$$c_f(e) = \mathrm{snap}(\mathrm{low}_f(e), [\tau(a^*), \tau(b^*)]),$$

choosing bracketing pair (a^*, b^*) so that $(\tau(a^*), \tau(b^*))$ is minimal in $\prec$ among bracketing pairs. Set $C(e) = \{c_f(e) : f \in F_\Delta\}$.

Lemma 5.2 *Let f be a facet of P and e an edge of T_f.*

1. *$c_f(e)$ approximates f at e.*
2. *If f, f' cover e and $f \preceq f'$, then $c_f(e) \preceq c_{f'}(e)$.*
3. *$\tau(P_e^*) \cup C(e)$ is noncrossing.*

Proof:

1. If e is a bounding edge of T_f, then the claim is immediate. Otherwise

$$c_f(e) = \mathrm{snap}(\mathrm{low}_f(e), [\tau(a^*), \tau(b^*)]),$$

for some bracketing pair (a^*, b^*). Let a^* and b^* be incident to faces f_a and f_b, respectively. It is easy to check that $\mathrm{low}_f(e)$ approximates f. Clearly $\tau(a^*)$ approximates f_a, $\tau(b^*)$ approximates f_b, and f covers f_a and f_b (since e is not a bounding edge of T_f). Part (1) is thus immediate from lemma 5.1.

2. We have $R_e \subseteq \pi_{xy}(f)$, $R_e \subseteq \pi_{xy}(f')$, and $f \prec f'$, so $\mathrm{low}_f(e) \preceq \mathrm{low}_{f'}(e)$. Let a_f^*, b_f^* and $a_{f'}^*, b_{f'}^*$ be bracketing pairs for f and f', respectively. We claim $\tau(a_f^*) \preceq \tau(a_{f'}^*)$ and $\tau(b_f^*) \preceq \tau(b_{f'}^*)$, from which $c_f(e) \preceq c_{f'}(e)$ follows easily. Clearly $\tau(a_f^*) \succeq \tau(b_f^*)$. It cannot be that $\tau(b_f^*) \succ \tau(a_{f'})$, for $a_{f'} \succ f' \succeq f$ and f would have a bracketing pair below (a_f^*, b_f^*), contradicting minimality. No edge of $\tau(P_e^*)$ lies between $\tau(a_f^*)$ and $\tau(b_f^*)$, so it must be that $\tau(a_{f'}^*) \succeq \tau(a_f^*)$. Similarly $\tau(b_{f'}^*) \succeq \tau(b_f^*)$.

3. Clearly no edge $c_f(e)$ crosses an edge of $\tau(P_e^*)$. Also clearly, if f and f' have distinct bracketing pairs, then $c_f(e)$ and $c_{f'}(e)$ do not cross. If f and f' have the same bracketing pair, then $c_f(e)$ and $c_{f'}(e)$ do not cross because $\mathrm{low}_f(e)$ and $\mathrm{low}_{f'}(e)$ do not cross. □

5.3 Lifting triangle edges.

Let e be an edge of triangle Δ of T. For facets $f \in F_\Delta$ in the order $\lhd$, simultaneously and inductively define

$a_{f\Delta}(e)$ and $b_{f\Delta}(e)$ (the *constraints from above and below, respectively*) and $l_{f\Delta}(e)$ (the *lifting of edge e of* Δ *in* f), as follows:

$$\begin{aligned} a_{f\Delta}(e) &= \min\{l_{f'\Delta}(e) : f' \lhd f \text{ and } f' \succ f\} \\ b_{f\Delta}(e) &= \max\{l_{f'\Delta}(e) : f' \lhd f \text{ and } f' \prec f\} \\ l_{f\Delta}(e) &= \text{snap}(c_f(e), [a_{f\Delta}(e), b_{f\Delta}(e)]). \end{aligned}$$

We have $a_{f\Delta}(e) \succeq b_{f\Delta}(e)$ by lemma 5.3(1) below.

Lemma 5.3 *Let e be an edge of triangle* Δ *of* T *with edge e, and let* $f, f' \in F_\Delta$.

1. $a_{f\Delta}(e) \succeq b_{f\Delta}(e)$.
2. *If* $f \preceq f'$, *then* $l_{f\Delta}(e) \preceq l_{f'\Delta}(e)$.
3. $l_{f\Delta}(e)$ *approximates* f *at* e.
4. $l_{f\Delta}(e) \in \tau(P_e^*) \cup C(e)$.

Proof: 1. and 2. We prove both simultaneously by induction on $\lhd$. If $a_{f\Delta}(e) = \top$ or $b_{f\Delta}(e) = \bot$, $a_{f\Delta}(e) \succeq b_{f\Delta}(e)$ is immediate. Otherwise $a_{f\Delta}(e) = l_{f_0\Delta}(e)$ and $b_{f\Delta}(e) = l_{f_1\Delta}(e)$ for some facets $f_0 \succeq f \succeq f_1$, so by induction hypothesis $a_{f\Delta}(e) \succeq b_{f\Delta}(e)$. For (2), suppose $f \preceq f'$; without loss of generality assume $f' \lhd f$. Then by definition $l_{f'\Delta}(e) \succeq a_{f\Delta}(e) \succeq l_{f\Delta}(e)$.

3. Since $l_{f\Delta}(e)$ is defined in the order $\lhd$, the claim follows from an easy induction using lemma 5.1.

4. By lemma 5.2, $\tau(P_e^*) \cup C(e)$ is noncrossing, so the "snap" in the definition of $l_{f\Delta}(e)$ results in an element of $\tau(P_e^*) \cup C(e)$. □

6 The subdivision Q

In this section we define the subdivision Q and the embedding σ of P into Q required by theorem 2.1. For technical reasons σ is defined on P^* as well.

We first define a "vertical carrier" over each vertex and edge of T. For v a vertex of T, let $VC(v)$ be the vertical chain of edges through $\rho(P_v^*)$, i.e. all edges connecting two vertices of $\rho(P_v^*)$ that are adjacent in vertical order. Let e be an edge of T with endpoints u and v. Consider the edges $\tau(P_e^*) \cup \{l_{f\Delta}(e) : \Delta \text{ incident to } e, f \in F_\Delta\}$; they are noncrossing, by lemmas 5.1(3) and 5.3(4). Split each edge at its midpoint. These edges together with $VC(u)$ and $VC(v)$ form a planar graph (in the plane through $VC(u)$ and $VC(u)$); let $VC(e)$ be any triangulation of this graph.

For v^* a vertex of P^*, define $\sigma(v^*) = \rho(v^*)$. Let $e^* \in P_e^*$, where edge e in T has endpoints u and v. Define $\sigma(e^*)$ to be the subdivision consisting of $\tau(e^*)$, the subchain of $VC(u)$ connecting $\tau(e^*)_u$ to $\rho(e^*)_u$ and the subchain of $VC(v)$ connecting $\tau(e^*)_v$ to $\rho(e^*)_v$. Extend σ to edges e of P:

$$\sigma(e) = \bigcup_{e^* \in P^*,\ e^* \subseteq e} \sigma(e^*)$$

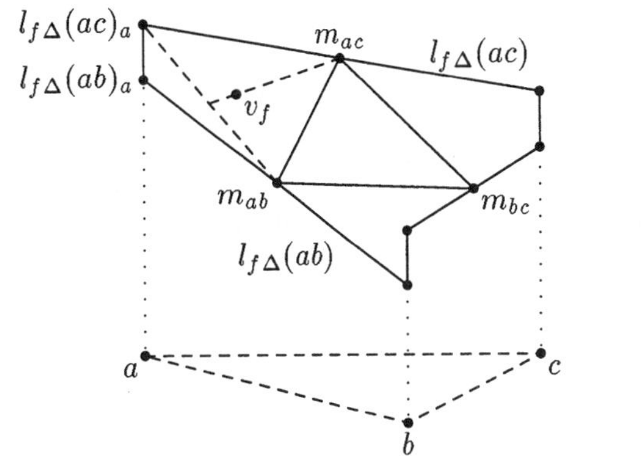

Figure 8: Construction of $l_f(\Delta)$ for triangle Δ of T_f with vertices a, b, c.

Clearly $\sigma(e)$ is a subdivision. We extend σ to facets in section 6.2.

Lemma 6.1 *If* w, w' *are vertices or edges of* P *and* $w \preceq w'$, *then* $\sigma(w) \preceq \sigma(w')$.

Proof: The claim is immediate for two vertices. Suppose w is a vertex and w' an edge; the symmetric case is similar. Then $\sigma(w')$ contains a vertical chain of edges from the center of the first voxel in column(w) met by w' to the center of the last such voxel. Since $w \prec w'$, $\rho(w)$ is below or on the chain, and $\sigma(w) \preceq \sigma(w')$. The case of two edges is similar. □

6.1 Lifting triangles

Let Δ be a triangle of T_f with vertices a, b, c. Consider the edges $l_{f\Delta}(ab)$, $l_{f\Delta}(ac)$, $l_{f\Delta}(bc)$. There is no guarantee that these edges are pairwise incident (of course both $l_{f\Delta}(ab)$ and $l_{f\Delta}(ac)$ are incident to vertices over a, and similarly for the other pairs). We form a (three-dimensional) polygon from $l_{f\Delta}(ab)$, $l_{f\Delta}(ac)$, $l_{f\Delta}(bc)$ by adding the vertical subchain of $L(a)$ connecting $l_{f\Delta}(ab)_a$ to $l_{f\Delta}(ac)_a$ (if they are not equal) and similarly for the b and c endpoints. The *lifting of* Δ *for facet* f, $l_f(\Delta)$, is a triangulation of this polygon, described as follows.

Split edges $l_{f\Delta}(ab)$, $l_{f\Delta}(ac)$, $l_{f\Delta}(bc)$ at their respective midpoints m_{ab}, m_{ac}, m_{bc}, and add the three edges connecting midpoints. This forms a triangle $m_{ab}m_{ac}m_{bc}$ and three polygons, where e.g. the *a-polygon (of f)* consists of edge $m_{ab}m_{ac}$, the two subedges of $l_{f\Delta}(ab)$ and $l_{f\Delta}(ac)$ with endpoints over a, and possibly a vertical chain over a. See figure 8.

For points $p, q \in \mathbb{R}^3$ and $\alpha \in \mathbb{R}$, let $\alpha[p, q]$ be the point $(1-\alpha)p + \alpha q$, i.e. the point a fraction α of the way from p to q.

The *a-index* of f is the number of distinct pairs $(l_{f'\Delta}(ab), l_{f'\Delta}(ac))$, where $f' \succeq f$. Let $\alpha_f = i/2^{\lceil \log_2 n \rceil}$, where i is the a-index of f; clearly $0 < \alpha_f < 1$. First assume $l_{f\Delta}(ac)_a \succeq l_{f\Delta}(ab)_a$. Set

$$v_f = \alpha_f[\,\frac{1}{2}[l_{f\Delta}(ac)_a, m_{ab}], m_{ac}].$$

See figure 8. Triangulate the a-polygon of f with v_f, i.e. connect v_f to m_{ac}, m_{ab}, and any vertex on the chain from $l_{f\Delta}(ab)_a$ to $l_{f\Delta}(ac)_a$. If $l_{f\Delta}(ab)_a \succ l_{f\Delta}(ac)_a$, the construction is similar, with m_{ab} and m_{ac} interchanged and $l_{f\Delta}(ab)_a$ substituting for $l_{f\Delta}(ac)_a$. The other two polygons are triangulated in a similar fashion.

Lemma 6.2 *Let Δ be a triangle of T and $f, f' \in F_\Delta$.*

1. *$d(l_f(\Delta), f) < \kappa$.*
2. *If $f \preceq f'$, then $l_f(\Delta) \preceq l_{f'}(\Delta)$.*
3. *Every vertex coordinate of $l_f(\Delta)$ is an integral multiple of $1/2^{\lceil \log_2 n \rceil + 2}$.*
4. *$l_f(\Delta)$ has $O(n)$ cells.*

Proof:

1. Let Δ have vertices a, b, c. Every vertex of $l_f(\Delta)$ is within the convex hull of $\{l_{f\Delta}(ab),\ l_{f\Delta}(ac),\ l_{f\Delta}(bc)\}$. The claim follows using lemma 5.3(1) and proposition 3.1.

2. We can assume that $l_{f\Delta}(ac)_a \succeq l_{f\Delta}(ab)_a$. Using lemma 5.3(2), we must have

$$l_{f'\Delta}(ac)_a \succeq l_{f\Delta}(ac)_a \text{ and } l_{f'\Delta}(ab)_a \succeq l_{f\Delta}(ab)_a.$$

If $l_{f'\Delta}(ab)_a \succ l_{f\Delta}(ac)_a$, then the result is immediate, since the convex hull of $\{l_{f\Delta}(ab), l_{f\Delta}(ac)\}$ and the convex hull of $\{l_{f'\Delta}(ab), l_{f'\Delta}(ac)\}$ have disjoint interiors. Hence we can assume that

$$l_{f'\Delta}(ac)_a \succeq l_{f\Delta}(ac)_a \succeq l_{f'\Delta}(ab)_a \succeq l_{f\Delta}(ab)_a.$$

Let i_f and $i_{f'}$ be the a-indices of f and f' respectively. If $i_f = i_{f'}$, then the a-polygons for f and f' are identical. Otherwise, $i_f > i_{f'}$ since $f \preceq f'$. Let s be the edge connecting $v_{f'}$ to the midpoint of $l_{f'\Delta}(ac)$; clearly we have $\pi_{xy}(v_f) \in \pi_{xy}(s)$ since $\alpha_f > \alpha_{f'}$. Furthermore we have $v_f \prec s$, since $l_{f\Delta}(ab) \preceq l_{f'\Delta}(ab)$ and $l_{f\Delta}(ac) \preceq l_{f'\Delta}(ac)$ with inequality holding in at least one case. $l_f(\Delta) \preceq l_{f'}(\Delta)$ follows easily.

3., 4. Immediate. □

6.2 Vertical ordering

It is tempting to define $\sigma(f) = \bigcup_{\Delta \in T_f} l_f(\Delta)$. This definition would preserve or collapse vertical order (in the sense of theorem 2.1) over triangles of T, but not necessarily over edges and vertices. Hence we develop an alternate definition of $\sigma(f)$.

Let $E_f(v)$ be the set of all endpoints over v of all edges $l_{f\Delta}(e)$, where edge e of T is incident to v and triangle Δ of T is incident to e. Define

$$\begin{aligned} a_f(v) &= \min\{\sigma(v^*) : v^* \in P_v^* \text{ and } v^* \succeq f\} \cup E_f(v) \\ b_f(v) &= \max\{\sigma(v^*) : v^* \in P_v^* \text{ and } v^* \preceq f\} \cup E_f(v). \end{aligned}$$

The *lifting of vertex v for facet f*, $l_f(v)$, is the subchain of $VC(v)$ connecting $a_f(v)$ and $b_f(v)$.

For e an edge of T_f, let $E_f(e)$ be all edges $l_{f\Delta}(e)$, where Δ varies over triangles in T_f incident to e. Define

$$\begin{aligned} a_f(e) &= \min\{\sigma(e^*) : e^* \in P_e^* \text{ and } e^* \succeq f\} \cup E_f(e) \\ b_f(e) &= \max\{\sigma(e^*) : e^* \in P_e^* \text{ and } e^* \preceq f\} \cup E_f(e). \end{aligned}$$

The *lifting of edge e for facet f*, $l_f(e)$, is all edges and vertices w of $VC(e)$ satisfying $b_f(e) \succeq w$ and $w \succeq a_f(e)$.

Lemma 6.3 *Suppose w is a vertex or edge of T, $w^* \in P_w^*$, and f is a facet of P. Then $w^* \preceq f$ implies $\sigma(w^*) \preceq l_f(w)$ and $w^* \succeq f$ implies $\sigma(w^*) \succeq l_f(w)$.*

Proof: By construction. □

Lemma 6.4 *Let f be a facet of P and w an edge of T_f. Then $d(l_f(w), f) \le \kappa$.*

Proof: Similar to the proof of lemma 5.1. □

For each facet f of P, define

$$\sigma(f) = \bigcup_{w \in T_f} l_f(w),$$

where w varies over vertices, edges, and triangles. It is easy to check that $\sigma(f)$ is a subdivision.

Lemma 6.5 *If f, f' are cells of P and $f \prec f'$, then $\sigma(f) \preceq \sigma(f')$.*

Proof: The lemma follows from lemmas 6.1 and 6.3 if one of f and f' is a vertex or edge. So suppose both are facets. For each triangle Δ in both T_f and $T_{f'}$, $l_f(\Delta) \preceq l_{f'}(\Delta)$ by lemma 6.2. Suppose e is an edge in both T_f and $T_{f'}$. If there is a triangle Δ in both T_f and $T_{f'}$ incident to e, lemma 6.2 again implies $l_f(e) \preceq l_{f'}(e)$. Otherwise, up to symmetry, there is an edge $e^* \in P_e^*$ bounding f with $e^* \preceq f'$, so by lemma 6.3, $\sigma(e^*) \preceq l_{f'}(e)$. Since $e^* \preceq f$, $\sigma(e^*) \subseteq l_f(e)$, and $l_f(e) \preceq l_{f'}(e)$. A similar argument shows that if v is a vertex in both T_f and $T_{f'}$, then $l_f(v) \preceq l_{f'}(v)$. Hence $\sigma(f) \preceq \sigma(f')$. □

6.3 The subdivision Q

Let $Q = \bigcup_f \sigma(f)$, where f varies over all facets of P. It is easy to check that Q is a subdivision and that σ is an embedding of P into Q.

Lemma 6.6 *Q has $O(n^4)$ cells and can be computed in time $O(n^4)$.*

Proof: For each facet f of P, T_f has $O(n^2)$ triangles Δ. By lemma 6.2, $l_f(\Delta)$ has $O(n)$ cells. Hence $\sigma(f)$ has $O(n^3)$ cells, for a total of $O(n^4)$ over all facets of f. Q can easily be computed in the same time. □

6.4 Hausdorff distance

It follows immediately from lemmas 6.2 and 6.4 that $d(\sigma(f), f) \leq \kappa$. The proof that $d_H(\sigma(f), f) \leq \kappa$ requires a proof that $d(f, \sigma(f)) \leq \kappa$. This proof has a topological flavor, using $d(e, \sigma(e)) \leq \kappa$ for each edge e bounding f. The proof is omitted due to lack of space.

7 Discussion

It may be possible to improve the worst-case bounds given in theorem 2.1. For example, the $O(n^4)$ bound on the size of Q could be an artifact of vertical projection; perhaps an $O(n^3)$ bound could be obtained by using different projection directions in different places, each tuned to the local configuration. Obtaining a worst-case bound below $O(n^3)$ seems very challenging. It would also be desirable to remove the extra $\lceil \log_2 n \rceil + 2$ bits needed for vertex coordinates; again, this may be an artifact of vertical projection.

The algorithm of theorem 2.1 adds many vertices, far more than are necessary unless the input subdivision has been chosen by an adversary. Another challenge is to devise a straightforward algorithm that adds vertices only to nearby features, just enough to avoid self-intersections and to maintain combinatorial ordering. Presumably, most subdivisions would need far fewer new vertices than the bounds given in theorem 2.1.

A programmer would probably prefer a simple rounding algorithm, even at the expense of degraded worst-case bounds, as long as the typical-case bounds are reasonable. One reason that the rounding algorithm is complicated is the need to avoid edge crossings. Milenkovic [17] suggests rounding existing vertices to integer coordinates. If two rounded edges cross, then a vertex of intersection is added, with coordinates computed exactly. This would require a constant-factor increase in the bit-length of some vertex coordinates, and hence of some predicate evaluations. However, the maximum required bit-length is still bounded, and perhaps the increased-length calculations are relatively infrequent. It may be that this approach can lead to a practical rounding algorithm.

References

[1] H. Brönnimann, M. Yvinec, Efficient exact evaluation of signs of determinants, *Proc. Thirteenth Ann. Symp. Comp. Geom*, pp. 166-173, 1997.

[2] H. Brönnimann, I. Emiris, V. Pan, S. Pion, Computing exact geometric predicates using modular arithmetic with single precision, *Proc. Thirteenth Ann. Symp. Comp. Geom*, pp. 174–182, 1997.

[3] K.L. Clarkson, Safe and effective determinant evaluation, *33th Symp. on Found. Comp. Sci. 387–395, 1992.*

[4] S. Fortune, Robustness issues in geometric algorithms, *Applied computational geometry: towards geometric engineering*, pp. 9–14, Lecture notes in computer science 1148, M. Lin, D. Manocha, eds., Springer-Verlag, 1996.

[5] S. Fortune, Polyhedral modelling with multiprecision integer arithmetic, *Computer-Aided Design*, **20**:123–133, 1997.

[6] S. Fortune, C. Van Wyk, Static analysis yields efficient exact integer arithmetic for computational geometry, *ACM Trans. Graphics*, 15(3), pp. 223–248, July 1996.

[7] M. Goodrich, L. Guibas, J. Hershberger, P. Tanenbaum, Snap-rounding line segments efficiently in two and three dimensions, *Proc. Thirteenth Ann. Symp. Comp. Geom*, pp. 284–293, 1997.

[8] D. Greene, F. Yao, Finite-resolution computational geometry, *Proc. 27th IEEE Symp. Found. Comp. Sci.*, pp. 143–152, 1986.

[9] D. Greene, Integer line segment intersection, unpublished manuscript.

[10] L. Guibas, D. Marimount, Rounding arrangements dynamically, *Proc. Eleventh Ann. Symp. Comp. Geom*, pp. 190–199, 1995.

[11] J. Hobby, Practical segment intersection with finite precision output, *Computational geometry: theory and applications*, to appear.

[12] C. Hoffmann, The problems of accuracy and robustness in geometric computation, *Computer* **22**:31–42, 1989.

[13] P. Jaillon, Proposition d'une arithmétique rationnelle paresseuse et d'un outil d'aide à la saise d'objets en synth'èse, Ecole Nationale Superieure des Mine de Saint-Etienee, 1993.

[14] M. Karasick, D. Lieber, L. Nackman, Efficient Delaunay triangulation using rational arithmetic, *ACM Trans. Graphics* **10**(1):71–91, 1990.

[15] V. Milenkovic, Verifiable implementations of geometric algorithms using finite precision arithmetic, *Artificial intelligence*, **37**:377–401, 1988.

[16] V. Milenkovic, Rounding face lattices in d dimensions, Abstract for *Second Canadian Computational Geometry Conference*, 1990.

[17] V. Milenkovic, private communication, 1996.

[18] V. Milenkovic, Shortest path geometric rounding, submitted.

[19] V. Milenkovic, L. Nackman, Finding compact coordinate representations for polygons and polyhedra, *IBM J. Res. Dev.* **34**(5):752–768, 1990. A version also appeared in *Proc. Sixth Ann. Symp. Comp. Geom.* 244–252, 1990.

[20] J. Shewchuk, Robust adaptive floating-point geometric predicates, *Proc. Twelfth Ann. Symp. Comp. Geom*, pp. 141–150, 1997.

Exact Algorithms for Circles on the Sphere

Marcus Vinícius A. Andrade
University of Viçosa
marcus@dpi.ufv.br

Jorge Stolfi
University of Campinas
stolfi@dcc.unicamp.br

Abstract

We develop exact algorithms for geometric operations on general circles and circular arcs on the sphere, using integer homogeneous coordinates. The algorithms include testing a point against a circle, computing the intersection of two circles, and ordering three arcs out of the same point. These operations allow robust manipulation of maps on the sphere, providing a reliable framework for GIS, robotics, and other geometric applications.

1 Introduction

A *spherical map* is a partition of the sphere's surface in three kinds of *elements*: *vertices* (points), *edges* (circles and arcs of circles, not necessary geodesic ones), and *faces* (open regions). Most maps handled in geographical information systems (GIS) are of this type: examples include geodesic polygons, latitude-longitude grids, polygonal maps under stereographic projection, sattelite images, etc. [6].

Our ultimate goal is to develop an exact representation of spherical maps, and robust algorithms to operate on them. Here we describe a basic ingredient of that goal: an exact representation of points, circles and circular arcs on the sphere, and exact formulas for various geometric operations on such objects. In particular, we give exact formulas for computing the points of intersection of two circles, the location of a point with respect to a circle, the ordering of circular arcs around a point and the circular ordering of three points on a circle. We have implemented and tested all these algorithms in Modula-3 [7], using the *GNU Multiple Precision* (GMP) arithmetic library [4].

Exactness may seem a pointless luxury in GIS applications, since GIS data is by its nature approximate, and approximate results are sufficient for all practical purposes. However, most geometric algorithms used in GIS, such as point location and map overlay, become much more complex and prone to failure if their basic operations are subject to rounding errors, no matter how small [5, 11, 10]. Consider for example a distributed application that cuts a map into smaller submaps, handles each piece to a separate processor, and combines the partial results into a single map. If the cutting step is exact, the final step needs only to identify common boundary edges between the partial results, and remove them. The task becomes much harder if the cutting step is affected by rounding errors: the partial results may overlap, or may be separated by gaps. The pasting operation is then almost impossible to specify, let alone to implement [2, 1].

The representation that we define here allow us to represent any spherical map with arbitrary accuracy, without increasing the number of elements of the original map; and also to compute exactly the overlay [9] of two such maps. Moreover, the resulting maps can be overlaid again and again without intermediate rounding or simplification steps. We note that most geometric operations used in GIS (region intersection, point location, feature extraction, clipping etc) can be reduced to map overlay.

To be sure, there are many geometric operations that cannot always be carried out exactly in our model: for instance, constructing a circle through three vertices, or centered at one vertex. But such objects can still be approximated in our model with arbitrary accuracy, and then exactly manipulated by our algorithms.

2 Spherical geometry

2.1 Circles on the sphere

Let $\mathbb{S}^2$ be the *unit sphere* of $\mathbb{R}^3$, that is, the sphere with unit radius centered at the origin O of $\mathbb{R}^3$. In this paper, we are concerned with oriented circles on the sphere $\mathbb{S}^2$, which we call *S-circles*.

Note that every S-circle corresponds to the intersection of $\mathbb{S}^2$ with an oriented plane. More precisely, given an S-circle c there is a unique oriented plane α such that $c = \alpha \cap \mathbb{S}^2$ and vice-versa. We say that α is the *supporting plane* of c, and write $\alpha = \operatorname{spln}(c)$. Conversely, we also say that c is the S-circle defined by α, and write $c = \text{scrc}(\alpha)$. By definition, the *positive sense of travel* along $\text{scrc}(\alpha)$ is the one that agrees with the orientation of the plane α. See figure 1.

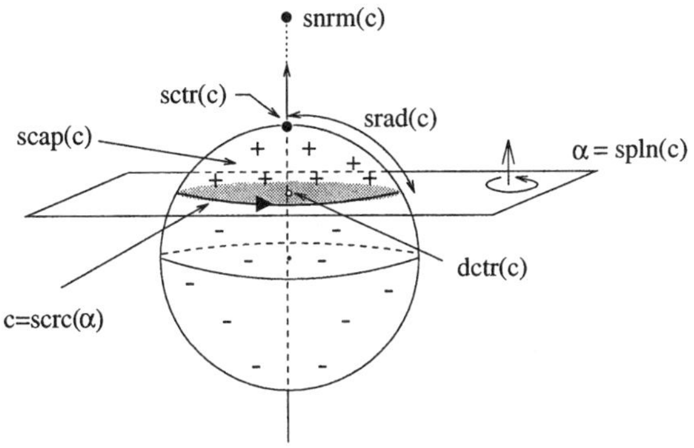

Figure 1: Elements of an S-circle.

Thus, we can unambiguously represent an S-circle by the coefficients $\langle\alpha_0, \alpha_1, \alpha_2, \alpha_3\rangle$ of the oriented plane α: four real numbers such that a point p with homogeneous coordinates $[w, x, y, z]$ is on the plane α if and only if $\alpha_0 w + \alpha_1 x + \alpha_2 y + \alpha_3 z = 0$; and, moreover, p is on the *positive* (resp. *negative*) *side* of α if and only if $\alpha_0 w + \alpha_1 x + \alpha_2 y + \alpha_3 z$ is positive (resp. negative)[1].

The plane α defines an S-circle (i.e., intersects the sphere $\mathbb{S}^2$) if and only if $\alpha_0^2 \leq \alpha_1^2 + \alpha_2^2 + \alpha_3^2$. In this case, the S-circle defined by α will be denoted by $(\!(\alpha_0, \alpha_1, \alpha_2, \alpha_3)\!)$. If $\alpha_0^2 = \alpha_1^2 + \alpha_2^2 + \alpha_3^2$, then α is tangent to $\mathbb{S}^2$, and $\text{scrc}(\alpha)$ reduces to a point.

The direction orthogonal to $\text{spln}(c)$ and pointing into its positive side is the *normal* of c, denoted $\text{snrm}(c)$. It is convenient to think of $\text{snrm}(c)$ as the point at infinity $[0, \alpha_1, \alpha_2, \alpha_3]$.

An S-circle c divides $\mathbb{S}^2$ into two open regions, one of which may be empty. By definition, the *positive cap* of c is the part of $\mathbb{S}^2$ that lies on the positive side of $\text{spln}(\alpha)$. Also, an S-circle has two centers: the *S-center* on its positive cap, and the *D-center* on its supporting plane. Their homogeneous coordinates are:

$$\begin{aligned} \text{sctr}(c) &= \left[\sqrt{\alpha_1^2 + \alpha_2^2 + \alpha_3^2}, \alpha_1, \alpha_2, \alpha_3\right] \\ \text{dctr}(c) &= [\alpha_1^2 + \alpha_2^2 + \alpha_3^2, -\alpha_0\alpha_1, -\alpha_0\alpha_2, -\alpha_0\alpha_3] \end{aligned}$$

In particular, if c is a single point, its positive cap is either empty or the whole $\mathbb{S}^2$, and $\text{dctr}(c) = \pm\,\text{sctr}(c)$.

The length of the spherical arc that joins $\text{sctr}(c)$ to any point on c is the *S-radius* of c, given by $\text{srad}(c) = \arccos(-\alpha_0/(\sqrt{\alpha_1^2 + \alpha_2^2 + \alpha_3^2}\,)) \in [0 \;..\; \pi]$. Conversely, the S-circle c whose S-center is $p = [p_0, p_1, p_2, p_3]$ and whose S-radius is $\theta \in [0 \;..\; \pi]$ is $c = (\!(-p_0\cos\theta, p_1, p_2, p_3)\!)$.

[1]We use the notation of Stolfi [8] where the homogeneous coordinates of a point (x, y, z) of $\mathbb{R}^3$ are $[1, x, y, z]$ with positive scalar multiples identified; that is, for $\lambda > 0$, $[w, x, y, z]$ and $[\lambda w, \lambda x, \lambda y, \lambda z]$ are the same point. In particular, the point $[0, x, y, z]$ is assumed to be at infinity in the direction of the Cartesian vector (x, y, z).

2.2 Arcs of S-circles

Two distinct points p and q on an S-circle $c = \text{scrc}(\alpha)$ divide it into two connected parts called S-arcs. See figure 2. By definition, *the S-arc of c from p to q*, denoted by $\text{sarc}(p, q, c)$, is the set of points encountered as we move from p to q on c, along its positive sense of travel. Given $A = \text{sarc}(p, q, c)$, we write $c = \text{scrc}(A)$, $p = \text{org}(A)$ and $q = \text{dst}(A)$.

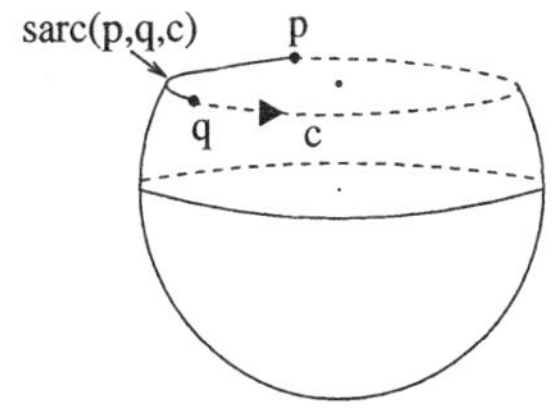

Figure 2: The S-arc of c determined by p and q.

Geometrically, $\text{sarc}(p, q, c)$ corresponds to the part of c that lies on the positive side of the plane $\beta = p \vee q \vee \text{snrm}(c)$, assuming $\vee$ as defined by Stolfi [8].

3 Exact spherical geometry

3.1 Rational points

A point is said to be *rational* if all its Cartesian coordinates are rational numbers. Note that the set of rational points is a dense subset of $\mathbb{R}^3$.

Since a point does not change when its homogeneous coordinates are scaled by a positive factor, any rational point can be represented by four integers.

3.2 Exactly representable points of $\mathbb{S}^2$

There are two sets of points of $\mathbb{S}^2$ that have obvious exact representations. One is the set $\mathcal{A}$ consisting of the rational points of $\mathbb{S}^2$, that is

$$\mathcal{A} = \left\{[a_0, .. a_3] \mid a_1^2 + a_2^2 + a_3^2 = a_0^2 \text{ and } (a_0, .. a_3) \in \mathbb{Z}^4\right\}$$

Another is the set $\mathcal{B}$ consisting of the points $p = [w, x, y, z]$ whose homogeneous coordinates x, y, and z are rational. Since p lies on $\mathbb{S}^2$, the weight coordinate w can be determined by the formula $w = \sqrt{x^2 + y^2 + z^2}$, and can be left implicit.

$$\mathcal{B} = \left\{[b_0, .. b_3] \mid b_1^2 + b_2^2 + b_3^2 = b_0^2 \text{ and } (b_1, b_2, b_3) \in \mathbb{Z}^3\right\}$$

Note that $\mathcal{B}$ is the radial projection onto $\mathbb{S}^2$ of all rational points of $\mathbb{R}^3$. Obviously, $\mathcal{A} \subset \mathcal{B}$.

3.3 Rational Planes and S-circles

We say that a plane α is *rational* if its coefficients $\langle\alpha_0, \alpha_1, \alpha_2, \alpha_3\rangle$ are rational—or, equivalently, integers. An S-circle is *rational* if its supporting plane is rational.

We propose to consider only rational S-circles and S-arcs thereof. This is not a significant restriction for practical purposes, since, in the Hausdorff metric, the set of rational S-circles is dense in the set of all S-circles. In fact, let $a = ((a_0, a_1, a_2, a_3))$ be any S-circle, and let ϵ be a real number between 0 and 1. One can verify that

$$b = \left(\left(\left\lceil \frac{a_0}{\delta} \right\rceil, \left\lceil \frac{a_1}{\delta} \right\rceil, \left\lceil \frac{a_2}{\delta} \right\rceil, \left\lceil \frac{a_3}{\delta} \right\rceil\right)\right)$$

where $\delta = \epsilon^2 \sqrt{a_1^2 + a_2^2 + a_3^2}/36$, is a rational S-circle such that $||a - b|| \leq \epsilon$ in the Hausdorff metric.

Now, let $c = ((c_0, .. c_3))$ be a rational S-circle. By formulas of section 2.1, we have $\text{sctr}(c) \in \mathcal{B}$, $\text{dctr}(c) \in \mathcal{A}$ and $\text{snrm}(c) \in \mathcal{A}$, so these elements can be computed exactly. Also, given any $p \in \mathcal{B}$, and any $q \in \mathcal{A}$, there is a rational S-circle with S-center p passing through q.

Finally, given three $\mathcal{A}$-points p, q and r, the S-circle passing through these points is $\text{scrc}(\alpha)$ where $\alpha = p \vee q \vee r$. The coefficients of α are 3×3 determinants on the coordinates of p, q and r [8] and therefore are rational numbers. Note that the circle passing through three $\mathcal{B}$-points may not be rational; fortunately, this operation is not necessary for map overlay.

3.4 Exact intersection of rational S-circles

An essential step in the map overlay operation is to compute the intersection of two S-circles—which, in general, is either empty or consists of two points. To make this operation unambiguous, we define *the (canonical) meeting point* of two S-circles a and b as being the point $a \wedge b$ where a crosses b from its positive side into its negative side. See figure 3(a).

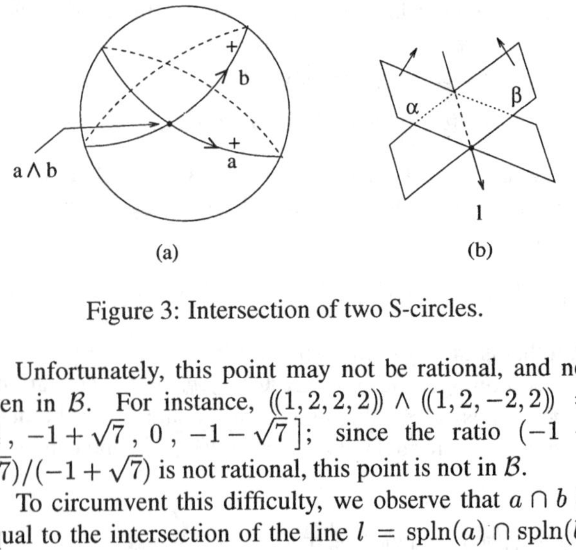

Figure 3: Intersection of two S-circles.

Unfortunately, this point may not be rational, and not even in $\mathcal{B}$. For instance, $((1, 2, 2, 2)) \wedge ((1, 2, -2, 2)) = [4, -1 + \sqrt{7}, 0, -1 - \sqrt{7}]$; since the ratio $(-1 - \sqrt{7})/(-1 + \sqrt{7})$ is not rational, this point is not in $\mathcal{B}$.

To circumvent this difficulty, we observe that $a \cap b$ is equal to the intersection of the line $l = \text{spln}(a) \cap \text{spln}(b)$ and $\mathbb{S}^2$. See figure 3(b). Thus, the problem of representing $a \cap b$ reduces to that of representing the line l. Moreover, if we consider l to be oriented, we can use its orientation to unambiguously select $a \wedge b$.

3.5 Plücker coefficients of a line

By definition, given two (oriented) planes α and β, their oriented intersection $\alpha \wedge \beta$ is the line $\alpha \cap \beta$, oriented in the direction $normal(\alpha) \times normal(\beta)$—see figure 3(b). That line can be unambiguously represented by the six coefficients $\langle l_{01}, l_{02}, l_{12}, l_{03}, l_{13}, l_{23} \rangle$ where $l_{ij} = \alpha_i\beta_j - \alpha_j\beta_i$. These six numbers are the *Plücker coefficients* of l [8, 3]. We write $l = \langle l_0, l_1, l_2, l_3, l_4, l_5 \rangle$, with the determinants implicitly ordered as above.

These numbers are not independent: a non-zero sextuple $\langle l_0, .. l_5 \rangle$ represents a line of $\mathbb{R}^3$ if and only if $l_0 l_5 - l_1 l_4 + l_2 l_3 = 0$. On the other hand, the Plücker coefficients of a line are homogeneous, that is, $\langle l_0, .. l_5 \rangle$ and $\langle \lambda l_0, .. \lambda l_5 \rangle$ are the same line, for any $\lambda > 0$.

The *direction* of a finite line l is the point at infinity $\text{dir}(l) = [0, l_5, -l_4, l_2]$. Thus the lines $l = \langle l_0, .. l_5 \rangle$ and $m = \langle -l_0, .. -l_5 \rangle$, which are incident to the same points, can be distinguished by their orientation. We say that m is the *opposite* of l, denoted by $m = \neg l$.

Let $p = [p_0, .. p_3]$ and $q = [q_0, .. q_3]$ be two points, $\alpha = \langle \alpha_0, \alpha_1, \alpha_2, \alpha_3 \rangle$ and $\beta = \langle \beta_0, \beta_1, \beta_2, \beta_3 \rangle$ be two planes and $l = \langle l_0, .. l_5 \rangle$ be a line. The basic geometric operations $\vee$ (join) and $\wedge$ (meet), as defined by Stolfi [8], involving lines are given by these formulas:

$$\begin{aligned}
\alpha \wedge \beta &= \langle \alpha_0\beta_1 - \alpha_1\beta_0, \alpha_0\beta_2 - \alpha_2\beta_0, \alpha_1\beta_2 - \alpha_2\beta_1, \\
&\qquad \alpha_0\beta_3 - \alpha_3\beta_0, \alpha_1\beta_3 - \alpha_3\beta_1, \alpha_2\beta_3 - \alpha_3\beta_2 \rangle \\
l \wedge \alpha &= [-l_2\alpha_3 + l_4\alpha_2 - l_5\alpha_1, l_1\alpha_3 - l_3\alpha_2 + l_5\alpha_0, \\
&\qquad -l_0\alpha_3 + l_3\alpha_1 - l_4\alpha_0, l_0\alpha_2 - l_1\alpha_1 + l_2\alpha_0] \\
p \vee l &= \langle \ l_0 p_1 + l_1 p_2 + l_3 p_3, -l_0 p_0 + l_2 p_2 + l_4 p_3, \\
&\qquad -l_1 p_0 - l_2 p_1 + l_5 p_3, -l_3 p_0 - l_4 p_1 - l_5 p_2 \rangle \\
p \vee q &= \langle p_2 q_3 - p_3 q_2,\ p_3 q_1 - p_1 q_3,\ p_0 q_3 - p_3 q_0, \\
&\qquad p_1 q_2 - p_2 q_1,\ p_2 q_0 - p_0 q_2,\ p_0 q_1 - p_1 q_0 \rangle
\end{aligned} \tag{1}$$

Note that these formulas return all-zero tuples when the operation is not defined; for instance, $l \wedge \alpha = [0, 0, 0, 0]$ when l lies on the plane α.

The line l passes through the origin O if and only if $l_0 = l_1 = l_3 = 0$. More generally, the Euclidean distance between l and O is

$$\text{dist}(l, O) = \sqrt{\frac{l_0^2 + l_1^2 + l_3^2}{l_2^2 + l_4^2 + l_5^2}} \tag{2}$$

The *bissector* of l is the plane $\text{bisect}(l)$ perpendicular to l, passing through O and oriented such that $\text{dir}(l)$ is on its positive side; its coefficients are $\langle 0, l_5, -l_4, l_2 \rangle$.

A line is said to be *rational* if its Plücker coefficients are rational or, equivalently, integers. Formulas (1) imply that the intersection of two rational planes is a rational line, and so is the line passing through two rational points.

3.6 Enter and exit points of a line

With the above conventions, the meeting point $a \wedge b$ of two S-circles is the point $\text{ext}(l)$ where the oriented line $l = \text{spln}(a) \wedge \text{spln}(b)$ exits the sphere $\mathbb{S}^2$. See figure 4.

The point where l enters the sphere will be denoted $\text{ent}(l)$. The midpoint of $\text{ent}(l)$ and $\text{ext}(l)$ is $\text{mid}(l) = l \wedge \text{bisect}(l)$, which is the point of l closest to the origin

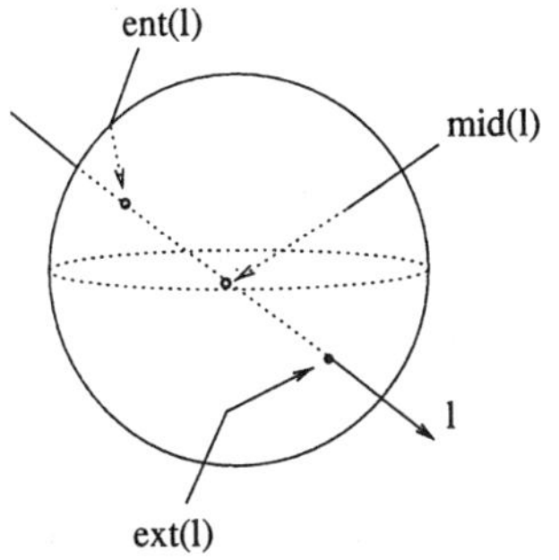

Figure 4: The points $\mathrm{ent}(l)$, $\mathrm{mid}(l)$ and $\mathrm{ext}(l)$.

O. If $l = \langle l_0, .. l_5 \rangle$,

$$
\begin{aligned}
\mathrm{ext}(l) &= [\, \mu, \quad -l_1 l_2 - l_3 l_4 + l_5\sqrt{\delta}, \\
&\qquad\quad l_0 l_2 - l_3 l_5 - l_4\sqrt{\delta}, \\
&\qquad\quad l_0 l_4 + l_1 l_5 + l_2\sqrt{\delta}\ ] \\
\mathrm{mid}(l) &= [\, \mu, \quad -l_1 l_2 - l_3 l_4, \\
&\qquad\quad l_0 l_2 - l_3 l_5, \\
&\qquad\quad l_0 l_4 + l_1 l_5\ ] \\
\mathrm{ent}(l) &= [\, \mu, \quad -l_1 l_2 - l_3 l_4 - l_5\sqrt{\delta}, \\
&\qquad\quad l_0 l_2 - l_3 l_5 + l_4\sqrt{\delta}, \\
&\qquad\quad l_0 l_4 + l_1 l_5 - l_2\sqrt{\delta}\ ]
\end{aligned}
\tag{3}
$$

where $\mu = l_2^2 + l_4^2 + l_5^2$ and $\delta = \mu - (l_0^2 + l_1^2 + l_3^2)$.

Observe that $\mathrm{ent}(l) = \mathrm{ext}(l)$ if and only if l is tangent to $\mathbb{S}^2$. By formula (2), this happens if and only if $l_2^2 + l_4^2 + l_5^2 = l_0^2 + l_1^2 + l_3^2$.

3.7 Sub-rational points of $\mathbb{S}^2$

From the previous sections, we can conclude that the meeting points of two rational S-circles belong to the set of *sub-rational* points of $\mathbb{S}^2$

$$\mathcal{C} = \{\, \mathrm{ext}(l) \mid l \text{ is a rational line} \,\}$$

Note that $\mathcal{A} \subset \mathcal{B} \subset \mathcal{C}$. Every point p of $\mathcal{C}$ can be represented exactly by the six integer coefficients of any rational line l such that $p = \mathrm{ext}(l)$. The relationship between $\mathcal{A}$ and $\mathcal{C}$ is clarified by the following theorems (whose proofs are given in appendix A):

Theorem 1 *Let $l = \langle l_0, l_1, l_2, l_3, l_4, l_5 \rangle$ be a rational line. Then, $\mathrm{ext}(l)$ is in $\mathcal{A}$ if and only if $(l_2^2+l_4^2+l_5^2) - (l_0^2+l_1^2+l_3^2)$ is a perfect square.*

Theorem 2 *A point p lies in $\mathcal{A}$ if and only if there are two distinct rational lines l and m such that $\mathrm{ext}(l) = p = \mathrm{ext}(m)$.*

Theorem 3 *A point p of $\mathbb{R}^3$ is in $\mathcal{C}$ if and only if $p = [a_0, a_1 + b_1\sqrt{c}, a_2 + b_2\sqrt{c}, a_3 + b_3\sqrt{c}\,]$ where a_0, a_1, a_2, a_3, b_1, b_2, b_3, and c are integers satisfying*

(i) $a_0 \neq 0$ *and* $b_1^2 + b_2^2 + b_3^2 \neq 0$

(ii) $a_1 b_1 + a_2 b_2 + a_3 b_3 = 0$

(iii) $a_0(a_0 - c) = a_1^2 + a_2^2 + a_3^2$

In that case, $p = \mathrm{ext}(l)$ *where* $l = \langle a_2 b_3 - a_3 b_2,\ a_3 b_1 - a_1 b_3,\ a_0 b_3,\ a_1 b_2 - a_2 b_1,\ -a_0 b_2,\ a_0 b_1 \rangle$.

3.8 Canonical representation of points

Theorem 2 says that a point p belongs to $(\mathcal{C} \setminus \mathcal{A})$ if and only there is a unique rational line l such that $p = \mathrm{ext}(l)$. In that case, we denote l by $\mathrm{stab}(p)$. On the other hand, if $p \in \mathcal{A}$, there are infinitely many lines such that $p = \mathrm{ext}(l)$. In particular, for any rational point q inside $\mathbb{S}^2$, the line $q \vee p$ will have this property. In this case, we define $\mathrm{stab}(p)$ as being the line $O \vee p$.

The geometric algorithms become much simpler if we consistently represent every point $p \in \mathcal{C}$ by the integer Plücker coefficients of the canonical line $\mathrm{stab}(p)$, and if these coefficients are *reduced* by eliminating all common factors. Then, for example, testing the equality of two points reduces to comparing their 6-tuples. We implement this idea by two procedures:

---

procedure *Apt*(p : $\mathcal{C}$-*point*) : $\mathcal{A}$-*point*;
  {If $p \in \mathcal{A}$ returns the homogeneous coordinates of p; otherwise returns $[0,0,0,0]$}
  $\delta \leftarrow p_2^2 + p_4^2 + p_5^2 - (p_0^2 + p_1^2 + p_3^2)$;
  if δ is a perfect square **then**
    return $\big[\mu, -p_1 p_2 - p_3 p_4 + p_5\sqrt{\delta},$
      $p_0 p_2 - p_3 p_5 - p_4\sqrt{\delta},$
      $p_0 p_4 + p_1 p_5 + p_2\sqrt{\delta}\,\big]$
  else return $[0,0,0,0]$

---

procedure *Stab*(p : $\mathcal{C}$-*point*) : $\mathcal{C}$-*point*;
  {Returns the canonical representation of the $\mathcal{C}$-point p, i.e. the line $\mathrm{stab}(p)$}
  ap $\leftarrow$ *Apt*(p);
  if *ap* $= [0,0,0,0]$ **then return** *Reduce*(p)
  else return *Reduce*($O \vee$ *ap*)

---

4 Robust algorithms

4.1 Relative position of a point and a plane

Another important predicate is the position of a point $p \in \mathcal{C}$ relative to a rational plane α, denoted by *SideOf*(p, α).

To implement this test, we consider three cases. If p is rational, then *SideOf*(p, α) can be written as *Apt*$(p) \diamond \alpha$, where $\diamond$ is the point-plane test as defined by Stolfi [8]. Otherwise, let $l = \mathrm{stab}(p)$ and $q = l \wedge \alpha$. If q is inside $\mathbb{S}^2$, then $\mathrm{ext}(l)$ and the rational point $\mathrm{dir}(l)$ are on the same side of α. On the other hand, if q is outside $\mathbb{S}^2$, then $\text{ext}(l)$ and $\text{ent}(l)$ are on the same side of α, and so is $\text{mid}(l)$ (which is rational).

---

procedure *SideOf*$(p : \mathcal{C}\text{-point};\ \alpha : \text{Plane}) : \text{Sign}$;

{Returns the relative position of a $\mathcal{C}$-point p and a plane α}

$ap \leftarrow Apt(p)$;

if $ap \neq [0,0,0,0]$ **then return** $ap \diamond \alpha$

else

$l \leftarrow \text{stab}(p);\quad q \leftarrow l \wedge \alpha$;

$\sigma \leftarrow \text{sign}(q_1^2 + q_2^2 + q_3^2 - q_0^2)$;

$\mu \leftarrow \text{mid}(l) \diamond \alpha;\quad \delta \leftarrow \text{dir}(l) \diamond \alpha$;

if $\sigma = -1$ **then return** δ

elsif $\sigma = 1$ **then return** μ

else if $\mu = \delta$ **then return** μ **else return** 0

Note that this operation also allow us to test the position of p relative to a rational S-circle c, by evaluating *SideOf*$(p, \text{spln}(c))$.

5 Circular order

5.1 Ahead and behind

Many algorithms for circles on the sphere rely on the *circular ordering* of points along a circle. This concept can be defined in terms of a basic predicate $\otimes_c(p,q)$ that takes two points p and q on an S-circle c, and returns $+1$ if q is *ahead of* p or -1 if q is *behind* p. See figures 5(a,b).

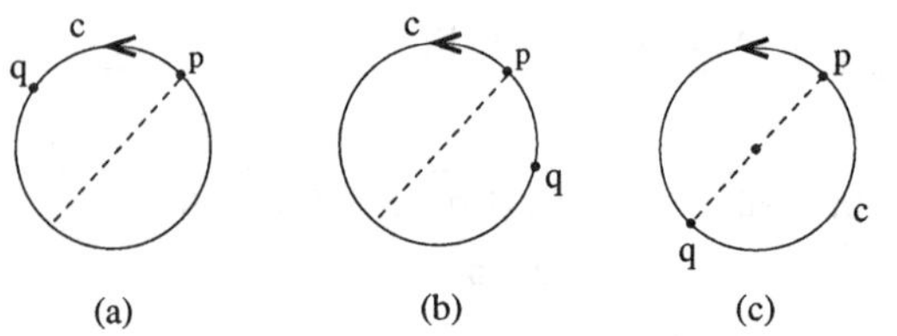

Figure 5: (a) $\otimes_c(p,q) = +1$, (b) $\otimes_c(p,q) = -1$, (c) $q = \oslash_c(p)$.

This predicate describes the direction of the "natural" or "shortest" route from p to q. For each point p on the circle c we must define the "diametrally opposite" point $\oslash_c(p)$. See figure 5(c). Then the "shortest" path from p to q is the one that does not go through $\oslash_c(p)$. Note that the direction of the shortest path is ambiguous when $q = p$, or $q = \oslash_c(p)$; in those cases, we define $\otimes_c(p,q)$ as 0 (indeterminate). As one can see, $\otimes_c(q,p) = \otimes_{\neg c}(p,q) = -\otimes_c(p,q)$.

Observe that the predicate $\otimes_c(p,q)$ corresponds to checking whether the line $p \vee q$ is oriented positively or negatively with respect to the axis of the S-circle c, which is defined as the line $O \vee \text{snrm}(c)$. Thus, $\otimes_c(p,q)$ is the the orientation of the tetrahedron $O, \text{snrm}(c), p, q$. Therefore,

$$\otimes_c(p,q) = \text{sign}\begin{vmatrix} 1 & 0 & 0 & 0 \\ c_0 & c_1 & c_2 & c_3 \\ p_0 & p_1 & p_2 & p_3 \\ q_0 & q_1 & q_2 & q_3 \end{vmatrix} = \text{sign}\begin{vmatrix} c_1 & c_2 & c_3 \\ p_1 & p_2 & p_3 \\ q_1 & q_2 & q_3 \end{vmatrix}$$

Since the points p or q may not be rational, we cannot compute the determinant exactly. Instead, let l be $\text{stab}(p)$, let α be the $\text{spln}(c)$, and let $\beta = \langle \beta_0, \beta_1, \beta_2, \beta_3 \rangle$ be the plane $l \vee \text{snrm}(c)$.

If β passes through O, then $\otimes_c(p,q)$ reduces to *SideOf*(q, β). See figure 6(a). On the other hand, if O is not on β, let γ be the reflection of the plane β across the origin O, that is, $\gamma = \langle \beta_0, -\beta_1, -\beta_2, -\beta_3 \rangle$. Then, $\oslash_c(p) = \text{ext}(\alpha \wedge \gamma)$. Firstly, suppose that O is on the positive side of β. Then $\otimes_c(p,q)$ is $+1$ if and only if

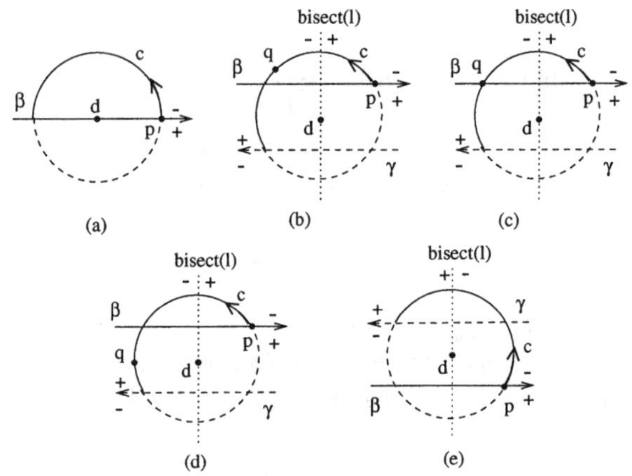

Figure 6: Ahead/Behind of two points on an S-circle; $d = \text{dctr}(c) = (O \vee \text{snrm}(c)) \wedge \text{spln}(c)$

- q is on the negative side of β (figure 6(b)); or
- q is on β and $p \neq q$ (figure 6(c)); or
- q is on the positive side of β and on the positive side of γ and also, on the negative side of the plane $\text{bisect}(l)$ (figure 6(d)).

Secondly, if O is on the negative side of β, as in figure 6(e), we can reduce this case to the previous one, using the identity $\otimes_c(p,q) = \otimes_{\neg c}(q,p)$. Note that inverting c corresponds to inverting β and γ. The complete algorithm is

---

procedure *Ahead*(c : Plane; p, q : $\mathcal{C}$-point) : Sign;
  {Computes $\otimes_c(p,q)$}
  $\beta \leftarrow Stab(p) \vee \mathrm{snrm}(c)$; $d_c \leftarrow SideOf(O, \beta)$;
  if $d_c = 0$ **then return** $-SideOf(q, \beta)$;
  case $d_c \cdot SideOf(q, \beta)$ **of**
  $-1 : d_m \leftarrow +1$;
   0 : **if** $p = q$ **then** $d_m \leftarrow 0$ **else** $d_m \leftarrow +1$
  $+1$: **if** $SideOf(q, \mathrm{bisect}(l)) \geq 0$
    then $d_m \leftarrow -1$
    else
     $\gamma \leftarrow \langle \beta_0, -\beta_1, -\beta_2, -\beta_3 \rangle$;
     $d_m \leftarrow dc \cdot SideOf(q, \gamma)$
  return $d_c \cdot d_m$

5.2 Circular order of three points on an S-circle

The concept of circular order can be formalized by a predicate $\otimes_c(p,q,r)$ that returns $+1$ if the three points p, q, r occur in that order along the oriented circle c; -1 if they occur in the opposite order; and 0 if any two of the points coincide. In other words, $\otimes_c(p,q,r)$ is the orientation $\Delta(p,q,r,\mathrm{snrm}(c))$ of the triangle pqr relative to $\alpha = \mathrm{spln}(c)$, that is

$$\otimes_c(p,q,r) = \mathrm{sign}\begin{vmatrix} p_0 & p_1 & p_2 & p_3 \\ q_0 & q_1 & q_2 & q_3 \\ r_0 & r_1 & r_2 & r_3 \\ c_0 & c_1 & c_2 & c_3 \end{vmatrix}$$

Again, since p, q, or r may not be rational, we cannot compute the determinant exactly. Thus, we will determine $\otimes_c(p,q,r)$ reducing it to three calls of the ahead/behind predicate $\otimes_c$.

Let u be a point on the plane α. The *signature* of u relative to the triangle *pqr* and the plane α is the sign sequence $\sigma_0\sigma_1\sigma_2$, where $\sigma_0 = \Delta(u,q,r,c)$, $\sigma_1 = \Delta(p,u,r,c)$, $\sigma_2 = \Delta(p,q,u,c)$ [8]. Given the orientation of the triangle *pqr* relative to α, each σ_i tells whether the point u and the ith vertex of the triangle are both on the same side of the line determined by the other two vertices. See figure 7. In particular, if u is inside the triangle *pqr*, its signature is '$+++$' or '$---$', depending on the triangle's orientation relative to α.

Moreover, let c be a circle passing through the points p, q, and r. It can be seen that the signature of any point u inside that circle is dominated by the orientation of the triangle *pqr* relative to α, that is $\Delta(p,q,r,\alpha) = \mathrm{sign}(\sigma_0 + \sigma_1 + \sigma_2)$. In particular, if we use $u = \mathrm{dctr}(c)$ we have $\sigma_0 = \Delta(\mathrm{dctr}(c), q, r, \mathrm{snrm}(c))$. Since $\mathrm{dctr}(c) \vee \mathrm{snrm}(c) = O \vee \mathrm{snrm}(c)$ then we can replace $\mathrm{dctr}(c)$ with O in this formula. Therefore, $\sigma_0 = \Delta(O,q,r,\mathrm{snrm}(c)) = \otimes_c(q,r)$.

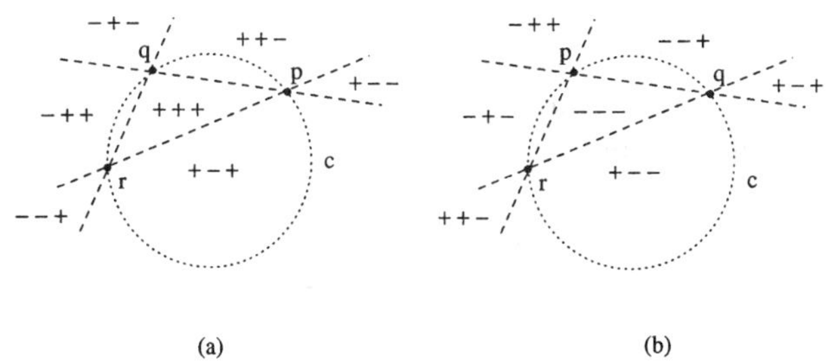

Figure 7: Signatures of a point relative to the triangle p, q, and r on the plane α.

By a similar reasoning, we get $\sigma_1 = \otimes_c(r,p)$ and $\sigma_2 = \otimes_c(p,q)$. The test then becomes

---

procedure *COrder*(c : S-circle; p, q, r : $\mathcal{C}$-point) : Sign;
  {Computes $\otimes_c(p,q,r)$}
  return $\mathrm{sign}(Ahead(c,p,q) + Ahead(c,q,r) +$
      $Ahead(c,r,p))$

5.3 Circular ordering of three arcs around a common origin

In the construction of spherical maps, a frequent subproblem is to compute the order in which three S-arcs leave a common origin vertex. That is, given three S-arcs $A = \mathrm{sarc}(v,p,a)$, $B = \mathrm{sarc}(v,q,b)$ and $C = \mathrm{sarc}(v,r,c)$, we wish to compute $\otimes_v(A,B,C)$, the circular ordering in which the three arcs cross a sufficiently small S-circle centered at v. See figure 8. Informally the result $\otimes_v(A,B,C)$ should be $+1$ if the order is counterclockwise, as seen from $\mathrm{dir}(O \vee v)$; -1 if it is clockwise; and 0 if any two of the arcs are parts of the same oriented circle.

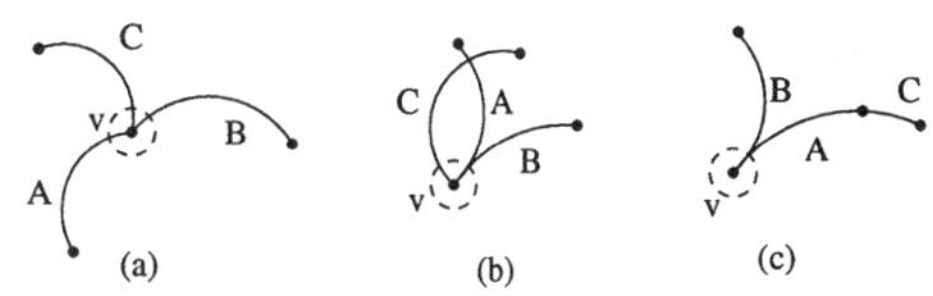

Figure 8: Circular order of three S-arcs.

An essential step of this algorithm is to compute the circular order $\otimes_l(\alpha,\beta,\gamma)$ of three planes α, β and γ around a common line l. We solve this problem by the same principle used to compute $\otimes_c(p,q,r)$, namely we reduce $\otimes_c$ to three ahead/behind tests, each involving two of the planes and the line l. In this case, the ahead/behind predicate $\otimes_l(\alpha,\beta)$ reduces to testing whether $\alpha \wedge \beta$ is the line l or its opposite; the "diametral opposite" of α is then the plane $\neg\alpha$.

Now, to compute $\otimes_v(A,B,C)$, we only need to order the directions in which the three arcs leave the point v. However, we must overcome two difficulties: the direction vectors may not be rational, and two distinct arcs may leave the point v in the same direction (as in figure 8(b)).

If v is not a rational point, then, by theorem 2, v is the exit point of a unique rational line l, which is not tangent to the sphere. In that case, the supporting planes of the three arcs must contain l; and the order of the arcs around v is the same as the order of the three planes around l, i.e. $\otimes_l(\text{spln}(a), \text{spln}(b), \text{spln}(c))$.

If v is a rational point, the three planes may not have a common line. In that case, let σ be the (rational) plane tangent to $\mathbb{S}^2$ at v oriented so that $\text{dir}(O \vee v)$ is on its positive side. So, the direction of A at v is $d_A = \text{dir}(\text{spln}(a) \wedge \sigma)$; and similarly for B and C. If d_A, d_B and d_C are distinct then it suffices to determine the circular order of these directions around $O \vee v$.

To complete the algorithm, we need to consider the case of two or more arcs leaving v in the same direction. In that case, we must break the tie by comparing the curvatures of the arcs—which is equivalent to ordering their planes around the common tangent line. In particular, if all three directions are equal, the result is simply the circular order of the three planes around the common tangent line. If only two directions coincide, it suffices to check the ahead/behind order of the two corresponding planes around their common tangent line.

---

procedure *ArcOrder*$(A, B, C$: S-arc$)$: Sign;
  {Computes $\otimes_v(A, B, C)$,
   where $v = \text{org}(A) = \text{org}(B) = \text{org}(C)$.}
  $\alpha \leftarrow \text{spln}(\text{scrc}(A))$;
  $\beta \leftarrow \text{spln}(\text{scrc}(B))$;
  $\gamma \leftarrow \text{spln}(\text{scrc}(C))$;
  $v \leftarrow \text{org}(A)$; $p \leftarrow Apt(v)$;
  if $p = [0, 0, 0, 0]$ **then return** $\otimes_{\text{stab}(v)}(\alpha, \beta, \gamma)$
  else
    {σ is the plane tangent to $\mathbb{S}^2$ at p }
    $\sigma \leftarrow \langle -p_0, p_1, p_2, p_3 \rangle$;
    $d_A \leftarrow \text{dir}(\alpha \wedge \sigma)$;
    $d_B \leftarrow \text{dir}(\beta \wedge \sigma)$;
    $d_C \leftarrow \text{dir}(\gamma \wedge \sigma)$
    if $d_A \neq d_B$ **and** $d_B \neq d_C$ **and** $d_C \neq d_A$
    then return $\otimes_{\sigma \wedge \Omega_2}(d_A, d_B, d_C)$
    elsif $d_A = d_B = d_C$
    then return $\otimes_{d_A}(\alpha, \beta, \gamma)$
    else {only two tangents are equal.}
      while $d_A \neq d_B$ **do**
        $(\alpha, \beta, \gamma) \leftarrow (\beta, \gamma, \alpha)$;
        $(d_A, d_B, d_C) \leftarrow (d_B, d_C, d_A)$
      return $\otimes_{d_A}(\alpha, \beta)$

---

6 Conclusions

In this paper, we developed a basic collection of exact algorithms for manipulating circles and circular arcs on the sphere. These algorithms are exact, in the sense that they only use integer arithmetic. We have implemented all these algorithms in Modula-3 [7], using the *GNU Multiple Precision* (GMP) arithmetic library [4] providing a reliable framework for GIS, robotics, and other geometric applications.

7 Acknowlegdements

We would like to thank Prof. Leo Guibas who hosted the first author as a visiting student at Computer Science Department of Stanford University, where part of this work was performed. This work was partially supported by PICDT/CAPES and CNPq (grant 301016/92-5).

References

[1] A. Aho, D. S. Jonhson, R. M. Karp, S. R. Kosaraju, C. C. McGeoch, C. H. Papadimitriou, and P. Pevzner. Theory of computing: Goals and directions. Manuscript, 1996.

[2] CG Impact Task Force. Applications challenges to computational geometry. Technical Report TR-521-96, Princeton University, 1996.

[3] J. Little D. Cox and D. O'Shea. *Ideals, Varieties, and Algorithms*. Springer-Verlag, 1992.

[4] T. Granlund. The GNU multiple precision arithmetic library. Technical report, Free software foundation, 1996.

[5] C. M. Hoffmann. The problems of accuracy and robustness in geometric computation. *IEEE Computer*, (22):31–42, 1989.

[6] D. J. Maguire, M. F. Goodchild, and D. Rhind. *Geographical Information Systems - principles and applications*. John Wiley & Sons, 1991.

[7] Greg Nelson, editor. *Systyms programming with Modula-3*. Prentice Hall, 1991.

[8] J. Stolfi. *Oriented Projetive Geometry - A framework for geometric computations*. Academic Press, 1991.

[9] Peter Y. F. Wu and Wm. Randolph Franklin. A logic programming approach to cartographic map overlay. *Canadian Computational Intelligence Journal*, 6(2):61–70, 1990.

[10] C. K. Yap. Towards exact geometric computation. In *Proc. 5th Canad. Conf. Comput. Geom.*, pages 405–419, 1993.

[11] C. K. Yap and T. Dubé. The exact computation paradigm. In D.-Z. Du and F. K. Hwang, editors, *Computing in Euclidean Geometry*, volume 1 of *Lecture Notes Series on Computing*, pages 452–492. World Scientific Press, Singapore, 2nd edition, 1995.

A Appendix: Proofs of all theorems

Theorem 1 *Let $l = \langle l_0, l_1, l_2, l_3, l_4, l_5\rangle$ be a rational line. Then,* $\mathrm{ext}(l)$ *is in $\mathcal{A}$ if and only if $(l_2^2+l_4^2+l_5^2)-(l_0^2+l_1^2+l_3^2)$ is a perfect square.*

PROOF: By formula (3),

$$\begin{aligned}\mathrm{ext}(l) \;=\; [\,\mu,\;& -l_1l_2 - l_3l_4 + l_5\sqrt{\delta},\\ & l_0l_2 - l_3l_5 - l_4\sqrt{\delta},\\ & l_0l_4 + l_1l_5 + l_2\sqrt{\delta}\;]\end{aligned}$$

where $\mu = l_2^2 + l_4^2 + l_5^2$ and $\delta = \mu - (l_0^2 + l_1^2 + l_3^2)$.

Obviously, if δ is a perfect square then $\mathrm{ext}(l) \in \mathcal{A}$.

Conversely, suppose that δ is not a perfect square. Since l intersects $\mathbb{S}^2$, we know by formula (2) that $(l_2, l_4, l_5) \neq (0,0,0)$. Then, one of the homogeneous coordinates of $\mathrm{ext}(l)$ have the form $a_i + l_i\sqrt{\delta}$, with a_i rational and $l_i \neq 0$. Since μ is rational the ratio $\mu/(a_i + l_i\sqrt{\delta})$ is irrational which implies that $p \notin \mathcal{A}$. □

Lemma 4 *Two lines $l = \langle l_0, .. l_5\rangle$ and $m = \langle m_0, .. m_5\rangle$ intersect if and only if*

$$l_0m_5 - l_1m_4 + l_2m_3 + l_3m_2 - l_4m_1 + l_5m_0 = 0$$

In that case, if the lines are not parallel,

$$\begin{aligned} l \cap m = [\,& \kappa_{24}^2 + \kappa_{25}^2 + \kappa_{45}^2,\\ & (l_3m_2 - l_1m_4 + l_5m_0)\kappa_{24} - \kappa_{15}\kappa_{25} - \kappa_{35}\kappa_{45},\\ & (l_0m_5 + l_3m_2 - l_4m_1)\kappa_{25} + \kappa_{04}\kappa_{24} + \kappa_{34}\kappa_{45},\\ & (l_0m_5 - l_1m_4 + l_2m_3)\kappa_{45} - \kappa_{02}\kappa_{24} - \kappa_{12}\kappa_{25}\,]\end{aligned} \tag{4}$$

where

$$\kappa_{ij} = \begin{vmatrix} l_i & l_j \\ m_i & m_j \end{vmatrix}$$

Theorem 2 *A point p lies in $\mathcal{A}$ if and only if there are two distinct rational lines l and m such that* $\mathrm{ext}(l) = p = \mathrm{ext}(m)$.

PROOF: First, suppose that $p \in \mathcal{A}$. Let q_1 and q_2 be two distinct points in $\mathcal{A}$ such that $q_1 \neq p \neq q_2$. Then, the two lines $l = q_1 \vee p$ and $m = q_2 \vee p$ are rational, are distinct and $\mathrm{ext}(l) = p = \mathrm{ext}(m)$.

Conversely, let l and m be two distinct rational lines such that $\mathrm{ext}(l) = p = \mathrm{ext}(m)$. If $l \neq \neg m$ then formula (4) shows that $p = l \cap m$ is rational. On the other hand, if $l = \neg m$ then $\mathrm{ext}(l) = \mathrm{ext}(m)$ which implies that l and m are tangent to $\mathbb{S}^2$. In that case $p = \mathrm{mid}(l)$ which is rational.

□

Theorem 3 *A point p of $\mathbb{R}^3$ is in $\mathcal{C}$ if and only if $p = [a_0, a_1 + b_1\sqrt{c}, a_2 + b_2\sqrt{c}, a_3 + b_3\sqrt{c}\,]$ where a_0, a_1, a_2, a_3, b_1, b_2, b_3, and c are integers satisfying*

(i) $a_0 \neq 0$ and $b_1^2 + b_2^2 + b_3^2 \neq 0$

(ii) $a_1b_1 + a_2b_2 + a_3b_3 = 0$

(iii) $a_0(a_0 - c) = a_1^2 + a_2^2 + a_3^2$

In that case, $p = \mathrm{ext}(l)$ where $l = \langle\, a_2b_3 - a_3b_2,\; a_3b_1 - a_1b_3,\; a_0b_3,\; a_1b_2 - a_2b_1,\; -a_0b_2,\; a_0b_1\,\rangle$.

PROOF: Let p be a point in $\mathcal{C}$. Thus, there is a rational line $l = \langle l_0, .. l_5\rangle$ such that $\mathrm{ext}(l) = p$. Choosing

$$\begin{aligned} a_0 &= l_2^2 + l_4^2 + l_5^2 & b_1 &= l_5\\ a_1 &= -l_1l_2 - l_3l_4 & b_2 &= -l_4\\ a_2 &= l_0l_2 - l_3l_5 & b_3 &= l_2\\ a_3 &= l_0l_4 + l_1l_5 & c &= a_0 - (l_0^2 + l_1^2 + l_3^2)\end{aligned}$$

we have

(i) $a_0 = b_1^2 + b_2^2 + b_3^2 \neq 0$, since $l_2 = l_4 = l_5 = 0$ only for lines at infinity—formula (2).

(ii) $$\begin{aligned} a_1b_1 + a_2b_2 + a_3b_3 &= l_5(-l_1l_2 - l_3l_4) -\\ & \quad l_4(l_0l_2 - l_3l_5) +\\ & \quad l_2(l_0l_4 + l_1l_5) \qquad = 0\end{aligned}$$

(iii) $$\begin{aligned} a_1^2 &= (-l_1l_2 - l_3l_4)^2 = l_1^2l_2^2 + l_3^2l_4^2 + l_1l_2l_3l_4 + l_1l_2l_3l_4\\ a_2^2 &= (l_0l_2 - l_3l_5)^2 \;= l_0^2l_2^2 + l_3^2l_5^2 - l_0l_2l_3l_5 - l_0l_2l_3l_5\\ a_3^2 &= (l_0l_4 + l_1l_5)^2 \;= l_0^2l_4^2 + l_1^2l_5^2 + l_0l_1l_4l_5 + l_0l_1l_4l_5\end{aligned}$$

so

$$\begin{aligned} a_1^2 + a_2^2 + a_3^2 &= l_0^2(l_2^2 + l_4^2) + l_1^2(l_2^2 + l_5^2) +\\ & \quad l_3^2(l_4^2 + l_5^2) + l_2l_3(l_1l_4 - l_0l_5) +\\ & \quad l_0l_5(l_1l_4 - l_2l_3) + l_1l_4(l_2l_3 + l_0l_5)\end{aligned}$$

and, applying the Plücker condition to the last three terms, we have

$$\begin{aligned} a_1^2 + a_2^2 + a_3^2 &= l_0^2(l_2^2 + l_4^2) + l_1^2(l_2^2 + l_5^2) +\\ & \quad l_3^2(l_4^2 + l_5^2) + l_2l_3(l_2l_3) +\\ & \quad l_0l_5(l_0l_5) + l_1l_4(l_1l_4)\\ &= l_0^2(l_2^2 + l_4^2 + l_5^2) + l_1^2(l_2^2 + l_4^2 + l_5^2) +\\ & \quad l_3^2(l_2^2 + l_4^2 + l_5^2)\\ &= (l_0^2 + l_1^2 + l_3^2)(l_2^2 + l_4^2 + l_5^2)\\ &= a_0(a_0 - c)\end{aligned}$$

Moreover, by formula (3), $p = \mathrm{ext}(l) = [a_0, a_1 + b_1\sqrt{c}, a_2 + b_2\sqrt{c}, a_3 + b_3\sqrt{c}\,]$.

Conversely, let a_0, a_1, a_2, a_3, b_1, b_2, b_3, and c be integers such that (i), (ii) and (iii) are satisfied. Let $l = \langle\, a_2b_3 - a_3b_2,\; a_3b_1 - a_1b_3,\; a_0b_3,\; a_1b_2 - a_2b_1,\; -a_0b_2,\; a_0b_1\,\rangle$. By condition (i), $a_0 \neq 0$ and there exists $i \in \{1,2,3\}$ such that $b_i \neq 0$. Therefore, $a_0b_i \neq 0$ which implies that at least one component of l is not zero. Moreover,

$$\begin{aligned} l_0l_5 - l_1l_4 + l_2l_3 &= a_0b_1(a_2b_3 - a_3b_2) -\\ & \quad (-a_0b_2)(a_3b_1 - a_1b_3) +\\ & \quad a_0b_3(a_1b_2 - a_2b_1) \qquad = 0\end{aligned}$$

which implies that the Plücker condition is satisfied. Thus, l is a valid rational line. By formula (3),

$$\begin{aligned}\mathrm{ext}(l) \;=\; [\,\mu,\;& -l_1l_2 - l_3l_4 + l_5\sqrt{\delta},\\ & l_0l_2 - l_3l_5 - l_4\sqrt{\delta},\\ & l_0l_4 + l_1l_5 + l_2\sqrt{\delta}\;]\end{aligned}$$

where

$$\mu = l_2^2 + l_4^2 + l_5^2 = a_0^2 b_3^2 + a_0^2 b_2^2 + a_0^2 b_1^2 = a_0^3$$

and $\delta = \mu - (l_0^2 + l_1^2 + l_3^2)$. Now

$$\begin{aligned} l_0^2 + l_1^2 + l_3^2 &= (a_2b_3 - a_3b_2)^2 + (a_3b_1 - a_1b_3)^2 + \\ &\quad (a_1b_2 - a_2b_1)^2 \\ &= a_2^2b_3^2 + a_3^2b_2^2 - 2a_2a_3b_2b_3 + a_1^2b_3^2 \\ &\quad a_3^2b_1^2 - 2a_1a_3b_1b_3 + a_1^2b_2^2 + a_2^2b_1^2 - \\ &\quad 2a_1a_2b_1b_2 \\ &= a_1^2(b_2^2 + b_3^2) + a_2^2(b_1^2 + b_3^2) + \\ &\quad a_3^2(b_1^2 + b_2^2) - 2a_3b_3(a_2b_2 + a_1b_1) \\ &\quad -2a_1a_2b_1b_2 \end{aligned}$$

Using condition (ii), we get

$$\begin{aligned} l_0^2 + l_1^2 + l_3^2 &= a_1^2(b_2^2 + b_3^2) + a_2^2(b_1^2 + b_3^2) + \\ &\quad a_3^2(b_1^2 + b_2^2) + 2a_3^2b_3^2 - 2a_1a_2b_1b_2 \\ &= a_1^2(b_1^2 + b_2^2 + b_3^2) + a_2^2(b_1^2 + b_2^2 + b_3^2) + \\ &\quad a_3^2(b_1^2 + b_2^2 + b_3^2) \\ &= (a_1^2 + a_2^2 + a_3^2)(b_1^2 + b_2^2 + b_3^2) \\ &= a_0(a_1^2 + a_2^2 + a_3^2) \end{aligned}$$

Therefore $\delta = a_0(a_0^2 - a_1^2 - a_2^2 - a_3^2)$.
Since

$$\begin{aligned} -l_1l_2 - l_3l_4 &= -(a_3b_1 - a_1b_3)a_0b_3 - \\ &\quad (a_1b_2 - a_2b_1)(-a_0b_2) \\ &= -a_0a_3b_1b_3 + a_0a_1b_3^2 + a_0a_1b_2^2 - \\ &\quad a_0a_2b_1b_2 \\ &= -a_0b_1(a_3b_3 + a_2b_2) + a_0a_1(b_1^2 + b_3^2) \\ &= a_0a_1b_1^2 + a_0a_1(b_1^2 + b_3^2) \\ &= a_0a_1(b_1^2 + b_2^2 + b_3^2) = a_0^2a_1 \end{aligned}$$

then $-l_1l_2 - l_3l_4 - l_5\sqrt{\delta}$ is given by

$$\begin{aligned} & a_0^2a_1 + a_0b_1\sqrt{a_0(a_0^2 - a_1^2 - a_2^2 - a_3^2)} \\ &= a_0^2\left(a_1 + b_1\sqrt{\frac{a_0(a_0^2 - a_1^2 - a_2^2 - a_3^2)}{a_0^2}}\right) \\ &= a_0^2(a_1 + b_1\sqrt{c}) \end{aligned}$$

Analogously, the third and fourth homogeneous coordinates of $\mathrm{ext}(l)$ are respectively,

$$\begin{aligned} l_0l_2 - l_3l_5 - l_4\sqrt{\delta} &= a_0^2(a_2 + b_2\sqrt{c}) \\ l_0l_4 + l_1l_5 + l_2\sqrt{\delta} &= a_0^2(a_3 + b_3\sqrt{c}) \end{aligned}$$

So,

$$\begin{aligned} \mathrm{ext}(l) = [\, a_0^3 \quad & a_0^2(a_1 + b_1\sqrt{c}) \\ & a_0^2(a_2 + b_2\sqrt{c}) \\ & a_0^2(a_3 + b_3\sqrt{c}) \,] \end{aligned}$$

By condition (i), we know that $a_0 \neq 0$. Then we have

$$\mathrm{ext}(l) = \left[a_0,\ a_1 + b_1\sqrt{c},\ a_2 + b_2\sqrt{c},\ a_3 + b_3\sqrt{c}\right]$$

□

Cutting Cycles of Rods in Space *

Alexandra Solan†

Abstract

We study several variants of the problem of cutting line segments in space in order to eliminate all cycles in their depth order. Our results include: (a) an output-sensitive algorithm for the general case (its performance depends on the minimum number of cuts needed to eliminate all cycles); (b) an $O(n^{9/5} \log n)$ algorithm for the case of 'bipartite weavings'; (c) an $O(n^{3/2+\varepsilon})$ algorithm for the c-oriented case and for some other special cases.

1 Introduction

Hidden surface removal is an important problem arising in computer graphics. In this problem we are given a set of nonintersecting (convex) objects in space and a view point, and want to compute which parts of the objects are visible from the view point.

One common technique for solving this problem is the *painter's algorithm*; see, e.g., [3]. In this algorithm we sort the objects by depth, that is, if P and Q are two objects and P obstructs Q (or parts of Q) as seen from the view point, then P precedes Q in order and we denote this relation by $P < Q$ (that is, there exists a ray that emerges from the view point and intersects both P and Q such that it intersects P before it intersects Q). After this sorting the painter's algorithm proceeds by painting the objects from back to front.

This algorithm will work only if such an order (called a *depth order*) on the objects exists, which is not always true. We may have a *cycle* in this relation, i.e., a sequence of objects $P_1, P_2, ..., P_k$, such that P_{i+1} obstructs parts of P_i, for $i = 1, 2, ..., k-1$, and P_1 obstructs parts of P_k; in our notations: $P_1 < P_2 < ... < P_k < P_1$. See Figure 1.

Figure 1: A cycle of length 3

In this case we may cut some objects into subparts in order to eliminate all such cycles, and then apply the painter's algorithm to the subparts. There are two main problems that this approach raises. One is to obtain combinatorial bounds on the minimum number of cuts that are needed to eliminate all cycles, and the other is to devise efficient algorithms for eliminating cycles by cutting. These algorithms should be efficient in terms of both the number of cuts that they perform and their running time. In spite of the significance of these problems, only relatively few results have been obtained so far, and the problems are still largely unsolved.

We consider here the simple case where the given objects are n nonintersecting line segments in space; we will refer to them as *rods*. Also, without loss of

*This paper is part of the author's M.Sc. Thesis, prepared at Tel Aviv University under the supervision of Professor Micha Sharir.

†Department of Computer Science, Tel Aviv University, Tel Aviv 69978, Israel

SCG 98 Minneapolis Minnesota USA
Copyright ACM 1998 0-89791-973-4/98/ 6...$5.00

generality, we will assume that the view point lies at $z = -\infty$. Clearly, if we project the given rods onto the xy-plane, and cut the rods at each point that projects to a crossing between the projected rods, we eliminate all cycles in the depth order. However, the number of cuts that we make is $\Theta(n^2)$ in the worst case. In contrast, the best lower bound on the number of cuts that might be required in the worst case is $\Omega(n^{3/2})$, which follows from a construction given in [8] (see also Section 4).

No subquadratic upper bound on the number of cuts is known in general. However, in the special case of a complete bipartite *weaving* of rods (informally, this case arises when the xy-projections of the rods form a grid-like pattern; a formal definition is given in Section 3), it was shown by Chazelle et al. [1] that $O(n^{9/5})$ cuts are sufficient to eliminate all cycles. (As remarked above, this bound is not known to be tight; in fact, we conjecture that it is not tight.) However, they supply no algorithm to actually perform those cuts (or any subquadratic number of cuts).

Even when the given set of objects has no cycles, the problem of computing a depth order on them is a nontrivial task, because usually not every pair of objects can be compared (they can be compared if and only if their xy-projections intersect). de Berg et al. present in [2] an algorithm for constructing a depth order on a set of n rods in 3-space, if the set contains no cycles, or detect a cycle when one exists. The running time of the algorithm is $O(n^{4/3+\varepsilon})$, for any $\varepsilon > 0$. Of course, we can now apply this algorithm iteratively: when the algorithm detects a cycle we eliminate it by cutting one of the rods participating in the cycle, and then run the algorithm again on the new set of rods, and keep doing so until no more cycles are detected. However, this approach has no control over the number of cuts that it performs, as compared with the minimum number of cuts that are actually needed.

In this paper we make several contributions to these problems. In Section 2 we consider the general problem and show how to eliminate cycles in a set of n rods in space in time $O(n^{4/3+\varepsilon}\mu^{1/3})$, for any $\varepsilon > 0$, where $\mu > 0$ is the minimal number of cuts needed to eliminate all cycles (the number of cuts that the algorithm performs is $O(n^{1+\varepsilon}\mu^{1/2})$). In Section 3 we give an algorithm for cutting rods in space which form a bipartite weaving in order to eliminate all the cycles. The algorithm makes at most $O(n^{9/5}\log n)$ cuts; See Theorem 3.6 for more precise performance bounds. In Section 4 we give the a bound of $O(n^{3/2+\varepsilon})$ on the number of cuts needed to eliminate cycles in the case where the rods have a total of c orientations, for $c \geq 3$ a constant. This number of cuts is nearly optimal in the worst case. We also present an algorithm that runs in $O(n^{3/2+\varepsilon})$ time and performs these many cuts. Some additional special cases are also considered.

2 The General Case of Rods in Space

In this section we consider the following problem. Given a set S of n rods in 3-space containing cycles, we want to eliminate all the cycles by cutting some rods into two (or more) parts, so that the resulting set of new rods admits an (acyclic) depth order. Ideally we want to minimize the number of cuts. Alternatively, we want the number of cuts to be a good approximation of the minimum number of cuts, that is required to eliminate all the cycles.

A *cutting set* for S is a finite set of points on the rods of S such that if we cut the rods of S at these points, we eliminate all the cycles. The *cutting number* of S is the size of the minimal cutting set for S. The cutting number is denoted by μ.

We use as a subroutine the algorithm of [1] which can determine, in $O(n^{4/3+\varepsilon})$ time, for any $\varepsilon > 0$, whether there exists a cycle in a given set of n rods, and report a cycle when one exists. Our algorithm runs as follows:

1. If the number of rods is smaller than some fixed constant, apply a brute-force technique to eliminate all cycles, e.g., cut each rod at each point whose xy-projection overlaps the xy-projection of another rod.
2. Otherwise run the cycle detection algorithm of [1].
3. If no cycles are detected then stop.
4. If a cycle is detected, proceed as follows:
    - Denote by S^* the set of vertical projections of the rods into the xy-plane.
    - Construct a $(1/r)$-*cutting*, for a sufficiently large constant parameter r, on the set S^*, using, e.g., the algorithm of [6]. The output of the algorithm is a partition of the xy plane into $O(r^2)$ trapezoidal cells, where each cell is cut by at most n/r rods of S^*.
    - Cut each rod of S at the points where its xy-projection crosses the boundaries of the cells of the cutting.
5. Continue recursively at each cell with the collection of rod portions that project onto the cell.

The correctness of the algorithm is obvious – after cutting each rod at the boundaries of the cells of the cutting, the only cycles that are still possible are cycles that involve only rod portions that project into a single cell, and thus the problem reduces to eliminating the cycles over each cell separately, which is taken care of recursively. Let $T(n, \mu)$ denote the maximum running time of the algorithm on sets of n rods with cutting number μ. Since the cycle detection algorithm of [1] takes time $O(n^{4/3+\varepsilon})$ for any $\varepsilon > 0$, and since a $(1/r)$-cutting for a constant r can be computed in $O(n)$ time [6], we obtain the following recurrence for $T(n, \mu)$:

$$T(n,\mu) = O(n^{4/3+\varepsilon}) + \sum_{i=1}^{ar^2} T(\frac{n}{r}, \mu_i), \qquad (1)$$

for an appropriate constant a (independent of r), where μ_i is the cutting number of the set of rod portions that project into the ith cell of the cutting.

We next claim that

$$\sum_{i=1}^{ar^2} \mu_i \le \mu.$$

Indeed, take a cutting set P for S, such that $|P| = \mu$. Let P_i be the subset of P that projects onto the ith cell. We claim that P_i is a cutting set for the rod portions of S that are projected to the ith cell. To see this, cut these rods at the points of P_i and suppose to the contrary that there is a cycle C of rod portions that project onto the ith cell. Since P is a cutting set for S, C must contain a point of $p \in P$, but p must be projected onto the ith cell because it is contained in a cycle that projects into this cell. Therefore $p \in P_i$, hence C cannot remain a cycle after cutting the rods at the points of P_i. In other words, we have

$$\sum_{i=1}^{ar^2} \mu_i \le \sum_{i=1}^{ar^2} |P_i| = \mu. \qquad (2)$$

We claim that the solution to the recurrence (1) with the condition (2) is $T(n, \mu) \le An^{4/3+\varepsilon}\mu^{1/3}$, for any $\varepsilon > 0$ and for an appropriate constant A that depends on ε.

We prove it by induction on n. We choose r to be a sufficiently large constant (as a function of ε; see below). For $n \le r$ the bound holds for A sufficiently large. For $n > r$, we have, by the induction hypothesis,

$$T(n,\mu) \le A\Big(\frac{n}{r}\Big)^{4/3+\varepsilon} \sum_{i=1}^{ar^2} \mu_i^{1/3} + Bn^{4/3+\varepsilon}. \qquad (3)$$

By Hölder's inequality :

$$\begin{aligned}\sum_{i=1}^{ar^2} \mu_i^{1/3} &= \sum_{i=1}^{ar^2} (1^{2/3} \cdot \mu_i^{1/3}) \le (ar^2)^{2/3} (\sum_{i=1}^{ar^2} \mu_i)^{1/3} \\ &\le a^{2/3} r^{4/3} \mu^{1/3}. \qquad (4)\end{aligned}$$

Thus

$$\begin{aligned}T(n,\mu) &\le a^{2/3}(A/r^{\varepsilon}) n^{4/3+\varepsilon}\mu^{1/3} + Bn^{4/3+\varepsilon} \\ &\le An^{4/3+\varepsilon}\mu^{1/3}, \qquad (5)\end{aligned}$$

provided that r is chosen sufficiently large (as a function of ε).

If $\mu = \Theta(n^2)$ then $T(n, \mu) = \Theta(n^2)$ but otherwise we have a subquadratic bound on the running time of the algorithm.

Similarly, we can bound the maximum number $f(n, \mu)$ of cuts that we actually perform. This number satisfies the recurrence:

$$f(n,\mu) \le \sum_{i=1}^{ar^2} f(\frac{n}{r}, \mu_i) + O(nr). \qquad (6)$$

This recurrence formula solves to

$$f(n,\mu) = O(n^{1+\varepsilon}\mu^{1/2}), \qquad (7)$$

which can be shown by an analysis similar to the one just given. We thus obtain the following theorem:

THEOREM 2.1 *Let S be a set of n rods in 3-space, with cutting number μ. Then, given $\varepsilon > 0$, one can compute a cutting set for S of size $O(n^{1+\varepsilon}\mu^{1/2})$ in time $O(n^{4/3+\varepsilon}\mu^{1/3})$.*

Remarks: (i) The preceding technique has several additional applications. First, we may apply it to construct cutting sets for sets of more general curves in space. For this we have to use a more expensive cycle detection algorithm, for example, by finding explicitly all relations $P < Q$ between pairs of objects, and then applying to the resulting graph an algorithm for cycle detection in directed graphs that runs in $O(n^2)$ time, where n is the number of curves. That yields an algorithm whose running time is $O(n^{2+\varepsilon})$, for any $\varepsilon > 0$, but the bound for the number of cuts performed is the same as in Theorem 2.1.
(ii) Another application of the technique described above is to the following problem: Given a set of n curves in the plane, each pair intersecting in at most c points, for a constant c, cut the curves such that the resulting subcurves will form a collection of *pseudo-segments*, i.e., each pair of them will intersect at most once. We do it exactly as in our algorithm, but instead of using a cycle detection algorithm we use an algorithm that determines whether a set of curves is a set of pseudo segments (which can be easily done in time $O(n^2)$). The number of cuts that we perform is $O(n^{1+\varepsilon}\mu^{1/2})$, where μ is the optimal number of points needed to cut the curves into pseudo segments. For a set of *pseudo parabolas*, for example, we have by a result of [10], $\mu = O(n^{5/3})$, thus we have an algorithm that performs $O(n^{11/6})$ cuts, which is a subquadratic bound.

3 Cutting Cycles in the Bipartite Case

In this section we consider the special case of *bipartite weavings*. See [1] for more details. A *bipartite weaving* is a pair of sets of rods in 3-space, $H = \{h_1, h_2, ..., h_n\}$ and $V = \{v_1, v_2, ..., v_m\}$, whose xy-projections, $H' =$

$\{h'_1, h'_2, ..., h'_n\}$ and $V' = \{v'_1, v'_2, ..., v'_m\}$, are such that each line segment in H' meets all the line segments in V' in the order $v'_1, v'_2, ..., v'_m$ and each line segment in V' meets the line segments in H' in the order $h'_1, h'_2, ..., h'_n$. Moreover, the line segments in H' are pairwise disjoint as are the line segments in V'. see Figure 2.

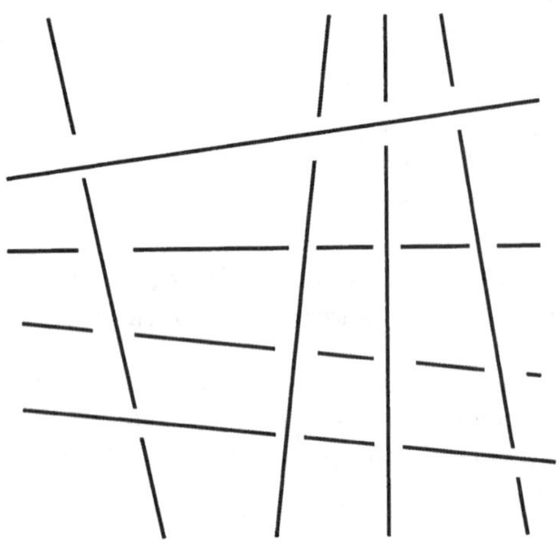

Figure 2: A bipartite weaving

We want to cut the rods of $H \cup V$ so as to eliminate all the cycles using a small number of cuts. We will use here the term *H-cutting set* to denote a set of points P on the rods of H, such that cutting the rods of H at the points of P eliminates all the cycles.

In [1] the following theorem is proved, for $|H| = n$, $|V| = m$:

THEOREM 3.1 *Let $1 \leq r \leq n$. For each $h_i \in H$ let $p_{ik} \in h_i$ be a point whose xy-projection lies between the crossings of h'_i with v'_{kr} and with v'_{kr+1}, for $k = 1, 2, ..., \lfloor n/r \rfloor$. Similarly for each $v_j \in V$ let q_{jk} be a point whose xy-projection lies between the crossings of v'_j with h'_{kr} and with h'_{kr+1} for $k = 1, 2, ..., \lfloor m/r \rfloor$. If we cut each rod $v_j \in V$ at all points q_{jk}, and cut each rod $h_i \in H$ at all points p_{ik}, thus creating nm/r^2 "squares" (and making nm/r cuts), then $O(r^{3/2}m^{1/2}n + r^3m)$ additional cuts are sufficient to eliminate all the cycles, and all the additional cuts may be made only at rods of H (or only at rods of V). Of course, the symmetric bound of $O(r^{3/2}n^{1/2}m + r^3n)$ on the number of required additional cuts also holds.*

The theorem implies that if we take $r = n^{1/5}$ (or $m^{1/5}$ if $m > n$), then $O(n^{4/5}m)$ or $O(m^{4/5}n)$ cuts are sufficient to eliminate all cycles. However no algorithm is given in [1] to actually perform the cuts. The algorithm we describe here computes an H-cutting set whose size is $O(\mu_H \log n)$ when μ_H is the size of the minimal H-cutting set. Thus, if we cut the rods as prescribed in the first step of Theorem 3.1, for $r = n^{1/5}$, and then apply our algorithm on each $r \times r$ "square" separately we perform at most $O(n^{4/5}m \log n)$ (or $O(m^{4/5}n \log m)$) cuts.

We now describe an algorithm whose input are the sets H and V whose sizes are n, m respectively, as above. Assume without loss of generality, that $m \leq n$. The algorithm proceeds as follows:

1. Solve the problem recursively for H and $\{v_1, v_2, ..., v_{\lfloor m/2 \rfloor}\}$.
2. Solve the problem recursively for H and $\{v_{\lfloor m/2 \rfloor+1}, v_{\lfloor m/2 \rfloor+2}, ..., v_m\}$.
3. Merge the two solutions.

We prove now some lemmas that are helpful for describing the merge step. The merge step itself is described in the next subsection.

After the first two steps the only cycles left are cycles whose xy-projections cross $v'_{\lfloor m/2 \rfloor}$. We prove the following lemma:

LEMMA 3.2 *After the cuts in the two recursive calls have been made, there exists an H-cutting set of size μ_H so that the xy-projections of all its points are on $v'_{\lfloor m/2 \rfloor}$.*

Proof: First, it is obvious that after steps 1, 2 of the algorithm there is still an H-cutting set of size μ_H – by cutting some of the rods we can not increase μ_H. We use the following lemma, proved in [1].

LEMMA 3.3 *For every cycle C in the weaving there exists a cycle T of length 4 such that the portions of the line segments of H that participate in T are portions of the line segments of H that participate in C.*

Lemma 3.3 implies that if we cut only rods that belong to H we keep the invariant that each remaining cycle "contains" a cycle of length 4 (in the sense of Lemma 3.3). Thus it suffices to cut all cycles of length 4 whose xy–projection crosses $v'_{\lfloor m/2 \rfloor}$.

Let P be an H-cutting set of the rods. Replace each point $p \in P$ with a point p' that belongs to the same rod but is shifted so that its xy-projection lies on $v'_{\lfloor m/2 \rfloor}$. Let P' be the set of all the shifted points p'; note that $|P'| \leq |P|$. We claim that P' is also an H-cutting set. Indeed, if a cycle is left then it has a sub-cycle of length 4 which must contain a point of P on one of its rods, h, that belongs to H. Hence the cycle contains a point of P' on h, because the xy-projection of the portion of h that participates in the cycle must cross $v'_{\lfloor m/2 \rfloor}$, and cutting at this point eliminates the cycle. ■

3.1 The merge step

Denote by R_H the set of portions of rods of H (after the cuts in the recursive calls have been made) whose xy-projections cut $v'_{\lfloor m/2 \rfloor}$. In the merge step we process only the sets V and R_H. We can do it because we need to eliminate only cycles of length 4 whose xy-projections cross $v'_{\lfloor m/2 \rfloor}$, and a portion of a rod of H not belonging to R_H cannot participate in such a cycle. This guarantees that the number of rods we deal with in the merge step is bounded by $|H| + |V|$ (recall that the rods in V have not been cut at all). The merge step then proceeds as follows.

1. Run the cycle detection algorithm of [2] on $V \cup R_H$.
2. If no cycle is detected then stop.
3. If a cycle is detected find a sub-cycle of length 4. See subsection 3.2 for details.
4. Cut the two rods of H of the sub-cycle at the places where their xy-projections cross $v'_{\lfloor m/2 \rfloor}$.
5. Repeat steps $1-4$ until termination.

Let P be an H-cutting set whose xy-projection is contained in $v'_{\lfloor m/2 \rfloor}$. In every round of the algorithm we cut at least one of the points of P, because any cycle of length 4 must contain a point of P at one of the two points where the xy-projections of the rods of H that form the cycle cross $v'_{\lfloor m/2 \rfloor}$. However, since we cut both points, it follows that at least one point of P is cut. Therefore the overall number of cuts is at most $2\mu_H$.

Denote the overall number of cuts that we perform by $f(n, m, \mu)$, where $\mu = \mu_H$. This number satisfies the following recurrence:

$$f(n, m, \mu) = f(n, \lfloor m/2 \rfloor, \mu_1) + f(n, \lceil m/2 \rceil, \mu_2) + O(\mu),$$

where μ_1 and μ_2 are the *H-cutting numbers* (sizes of minimal H-cutting sets) of the two sub-problems of step 1 and 2 respectively. Arguing as in section 2, we have $\mu_1 + \mu_2 \leq \mu$.

The recursion produces $O(\mu)$ cuts in each level of the recursion tree and since there are $\log m$ such levels we perform $O(\mu \log m) = O(\mu_H \log m)$ cuts.

Let us now bound the maximum running time $T(n, m, \mu)$ of the algorithm for input sets H, V whose sizes are n, m respectively. The running time of the cycle detection routine of [2] is $O((n+m)^{4/3+\varepsilon})$. As we will show in the next subsection, this is also the cost of extracting a sub-cycle of length 4. The recurrence for $T(n, m, \mu)$ is thus:

$$\begin{aligned} T(n, m, \mu) &= T(n, \lfloor m/2 \rfloor, \mu_1) + T(n, \lceil m/2 \rceil, \mu_2) \\ &\quad + O(\mu (n+m)^{4/3+\varepsilon}), \end{aligned}$$

where $\mu_1 + \mu_2 \leq \mu$. In each level of the recursion tree we are doing at most $O(\mu(n+m)^{4/3+\varepsilon})$ operations, for a total of

$$T(n, m, \mu) = O(\mu (n+m)^{4/3+\varepsilon} \log m) = O(\mu (n+m)^{4/3+\varepsilon'})$$

operations, for a slightly larger but still arbitrarily small $\varepsilon' > 0$.

The only thing left to do is to describe the algorithm for extracting a sub-cycle of length 4 out of a longer cycle, which is needed for step 3 of the algorithm. We describe it in the next subsection.

3.2 Extracting a sub-cycle of length 4

In this subsection we solve the following problem. Suppose we have a cycle

$$a_1 < a_2 < \ldots < a_n < a_1,$$

where the a_i's are portions of rods that belong to a weaving, and we wish to find a sub-cycle of length 4, whose existence is guaranteed by Lemma 3.3. This is done as follows.

Step 1

Find a minimal index t such that there exists an index $p < t$ with $a_t < a_p$. Since $a_n < a_1$, such a t exists. The way to find t is to construct a dynamic data structure $D_<$, similar to that used in [2], that maintains a set of rods S, and, given a rod a, returns a rod $b \in S$ such that $a < b$ or returns *nil* if no such rod exists. We first add only a_1 to $D_<$. In the tth step we query $D_<$ with a_t and if the answer is *nil* we add a_t to $D_<$ and continue. Otherwise we stop and return as an output t (and p, where a_p the answer of $D_<$ to the query).

Step 2

We next find a maximal index s, $p \leq s < t$, such that $a_t < a_s$. We do it by simply comparing a_t with each of $a_{t-1}, a_{t-2}, \ldots, a_p$. Clearly s exists and is $\geq p$.

LEMMA 3.4 *There exists a sub-cycle of length 4 consisting of rods from the cycle $a_s < a_{s+1} < \ldots < a_t < a_s$ that contains the rods a_s and a_t, i.e., a sub-cycle of the form $a_s < a_q < a_{q'} < a_t < a_s$.*

Proof: By our assumptions and by Lemma 3.3 there exists a sub-cycle of length 4 whose rods belong to $a_s, a_{s+1}, \ldots, a_t$. Suppose that this sub-cycle does not contain both a_s and a_t. Then there exist two indices g, h such that $g < h$ and $a_h < a_g$, and either $s < g$ or $h < t$. The latter inequality is impossible because it contradicts the minimality of t. Hence $h = t$, but then $g > s$ is impossible because it contradicts the maximality of s. ■

LEMMA 3.5 *Any pair of indices q, q' such that (i) $s < q < q' < t$, (ii) the xy-projections of $a_q, a_{q'}$ intersect, (iii) the xy-projections of a_s and a_q intersect, and (iv) the xy-projections of $a_{q'}$ and a_t intersect, satisfy the inequalities $a_s < a_q < a_{q'} < a_t$ (and thus form a cycle of length 4). Moreover, there exists at least one pair of indices with these properties.*

Proof: We have to prove the three inequalities

- $a_s < a_q$
- $a_q < a_{q'}$
- $a_{q'} < a_t$.

We will prove the first inequality; the proofs of the others are similar. We are given that a_s and a_q are comparable. It is impossible that $a_s > a_q$ because this is a contradiction to the minimality of t. The second part of the lemma follows from the existence of a cycle as in Lemma 3.4. ∎

Step 3

We now find a pair of indices $q < q'$ as in Lemma 3.5, thereby obtaining a sub-cycle of length 4. This is done as follows. We construct the set A (resp. B) of all the rods in $\{a_{s+1}, a_{s+2}, \ldots, a_{t-1}\}$ whose xy-projection crosses that of a_s (resp. of a_t). Our goal is to find a pair of rods $a_q \in A$, $a_{q'} \in B$ with intersecting xy-projections, such that $q < q'$. We employ the following divide-and-conquer approach. We split A and B into the subsets

$$\begin{aligned} A^- &= \{a_q \in A \mid q \leq \lfloor (s+t)/2 \rfloor\} \\ A^+ &= \{a_q \in A \mid q > \lfloor (s+t)/2 \rfloor\} \\ B^- &= \{a_{q'} \in B \mid q' \leq \lfloor (s+t)/2 \rfloor\} \\ B^+ &= \{a_{q'} \in B \mid q' > \lfloor (s+t)/2 \rfloor\}. \end{aligned}$$

We then take the xy-projections of the rods in $A^- \cup B^+$, and run on them a standard segment intersection detection algorithm (see [9]). If an intersection has been detected, it must be between the projections of a rod of A^- and a rod of B^+. We have thus found a pair of indices that we seek, and we stop. Otherwise, we recurse on $A^- \cup B^-$ and on $A^+ \cup B^+$. It is clear that this procedure outputs a desired pair of indices upon termination. The overall running time of this procedure is $O(n \log^2 n)$.

This completes the description of our algorithm. Using the technique of [2], step 1 can be implemented in time $O(n^{4/3+\varepsilon})$, for any $\varepsilon > 0$. Since this dominates the running time of the other two steps, we conclude that the procedure for finding a sub-cycle of length 4 runs in time $O(n^{4/3+\varepsilon})$, for any $\varepsilon > 0$.

In summary, we have shown:

THEOREM 3.6 *Let $H \cup V$ be a bipartite weaving of a total of n rods in space. We can eliminate all cycles in the weaving by performing at most $O(\mu_H \log n)$ cuts, where μ_H is the H-cutting number of $H \cup V$. In the worst case the number of cuts is $O(n^{9/5} \log n)$. The running time of the algorithm is $O(\mu_H n^{4/3+\varepsilon})$, for any $\varepsilon > 0$. In the worst case this bound is $O(n^{47/15+\varepsilon})$.*

Remark: An interesting open problem is to improve the upper bound $O(n^{9/5})$ of [1], for the number of cuts needed to eliminate all cycles in the bipartite weaving. We strongly believe that this bound is far from being tight.

4 Rods with c Orientations

In this section we deal with the following problem. We are given n rods that have only c different orientations in 3-space, for some constant $c \geq 3$. We design an algorithm for eliminating all cycles, which performs $O(n^{3/2+\varepsilon})$ cuts and takes $O(n^{3/2+\varepsilon})$ time, for any $\varepsilon > 0$. The cases $c = 1, 2$ are trivial since we need at least three distinct orientations to form a cycle.

The algorithm proceeds as follows:

We perform c iterations. In the ith iteration we find a set of points P_i such that after cutting the rods at the points of P_i, there is no cycle that contains a rod of the ith orientation. The cutting set P is then defined by $P = \bigcup_{i=1}^{c} P_i$ (actually, $P = \bigcup_{i=1}^{c-2} P_i$ would suffice).

We describe now an algorithm to eliminate cycles that contain a rod of the ith orientation.

The idea is to construct a linear transformation $T : R^3 \to R^2$, with the following two properties :

1. Each rod of the ith orientation is mapped to a single point.
2. Any two points (x, y, z) and (x, y, z') with the same x and y-coordinates are mapped to two respective points (u, v) and (u, v') with the same u-coordinate, and $z > z'$ iff $v > v'$. This implies that for any pair of rods $s_1 > s_2$ there exist two points p_1, p_2 on their respective images s'_1, s'_2 such that p_1 is above p_2.

Let $\vec{N}$ be the ith orientation. To satisfy (1) we require that $T(\vec{N}) = 0$. Consider the transformation that moves each point in the direction of $\vec{N}$ until it hits the yz-plane (we are ignoring the case where $\vec{N}$ is vertical), and take (u, v) to be the (y, z) coordinates of this hitting point. It is equivalent to use the following matrix as the transformation:

$$\begin{pmatrix} -n_2/n_1 & 1 & 0 \\ -n_3/n_1 & 0 & 1 \end{pmatrix},$$

where

$$N = \begin{pmatrix} n_1 \\ n_2 \\ n_3 \end{pmatrix}.$$

It is easy to see that this transformation satisfies the requirements above.

The second step is to construct a $(1/r)$-cutting recursively in the image plane of T on the set of T-images of the rods. We stop recursing on a cell if either it contains no points (images of rods of the ith orientation) or it does not intersect any line segment (images of the other rods). In each step we cut each rod s (not of the ith orientation) at the points whose T-images lie on the boundaries of the cells of the cutting. We then recurse on (preimages of) the segment portions within each cell of the cutting. This completes the description of our algorithm.

Let us bound the maximum number $f(n, m)$ of cuts that the algorithm performs in the ith step, where m is the number of rods of the ith orientation and n is the number of all the rods. We perform $O(nr)$ cuts in each step and then recurse over at most ar^2 cells, for an appropriate absolute constant a. The recurrence formula is thus:

$$f(n,m) = \sum_{i=1}^{ar^2} f(n/r, m_i) + O(nr),$$

where m_i is the number of image points in the ith cell, thus,

$$\sum_{i=1}^{ar^2} m_i \le m,$$

and $f(n, 0) = f(0, m) = 0$. We saw already in Section 2 that this recurrence formula solves to $O(n^{1+\varepsilon}m^{1/2})$, for any $\varepsilon > 0$. The running time $T(n, m)$ of the algorithm obeys the recurrence formula

$$T(n,m) = \sum_{i=1}^{ar^2} T(n/r, m_i) + O(nr + m \log r),$$

where $\sum_{i=1}^{ar^2} m_i \le m$, and $T(n, 0) = T(0, m) = O(1)$. The solution of this recurrence is, as in Section 2, $O(n^{1+\varepsilon}m^{1/2} + m^{1+\varepsilon})$, for any $\varepsilon > 0$. Summing it over all c orientations makes both bounds (running time and number of cuts) be $O(n^{3/2+\varepsilon})$. The correctness of the algorithm is a consequence of the following lemma.

LEMMA 4.1 *After the ith iteration of the algorithm no cycle containing a rod of the ith orientation is left.*

Proof: Suppose to the contrary that such a cycle exists. Its T-image is a circular chain of portions of the original projected line segments and points, $s'_1, s'_2, ..., s'_k$, such that segment s'_i contains a point above s'_{i-1}, for $i = 2, ..., k$, and s'_1 contains a point above s'_k, and such that at least one of these elements, say s'_1, is a point. We claim that all these projected segment portions and points are contained in a single cell, which contradicts the termination condition of the recursion.

Let us consider the sequence of cells containing the segments $s'_1, s'_2, ..., s'_k$, such that if two consecutive segments are contained in the same cell we take only one copy of it. Denote the cells by $C_1, C_2, ..., C_l$. We define an order on these cells so that $C \prec C'$ if the cell C' contains a point above a point in the cell C. As is well known, this relation is acyclic, for pairwise disjoint planar convex regions, as follows from the results of [5, 7, 4]. This leads to an immediate contradiction and thus implies the lemma. ■

We have thus shown:

THEOREM 4.2 *Given a set of n rods in space with a total of c orientations, for a fixed constant $c \ge 3$, we can eliminate all cycles in this set in time $O(n^{3/2+\varepsilon})$, by performing $O(n^{3/2+\varepsilon})$ cuts, for any $\varepsilon > 0$. The performance of the algorithm is nearly optimal in the worst case.*

Proof: The construction given in [8] involves a total of n rods with three different orientations, which requires $\Theta(n^{3/2})$ cuts to eliminate all cycles. This establishes the near–optimality of the algorithm. ■

Remark: We can apply the same method, with similar performance bounds, to a set of rods $H \cup V$, such that all the rods of H have the same orientation and there is no cycle composed only of rods of V.

Acknowledgments: I deeply thank Professor Micha Sharir for his help in preparing this work. I also thank Sariel Har-Peled for his useful comments. In particular, the remarks following Theorem 2.1 are a consequence of his comments.

References

[1] B. Chazelle, H. Edelsbrunner, L. J. Guibas, R. Pollack, R. Seidel, M. Sharir, and J. Snoyink, Counting and cutting cycles of lines and rods in space, *Comput. Geom. Theory Appls.* 1 (1992), 305–323.

[2] M. de Berg, M. Overmars, and O. Schwarzkopf, Computing and verifying depth orders, *SIAM J. Comput.* 23 (1994), 437–446.

[3] J. D. Foley and A. van Dam, *Fundamentals of Interactive Computer Graphics*, Addison-Wesley, Reading, MA, 1982.

[4] L. Guibas, M. Overmars and M. Sharir, Ray Shooting, Implicit Point Location, and Related Queries in Arrangements of Segments, *Tech. Report 433, Computer Science Department, New York University*, March 1989.

[5] L. J. Guibas and F. F. Yao, On translating a set of rectangles, *Proc. 12th Annu. ACM Sympos. Theory Comput.*, 1980, 154–160.

[6] J. Matoušek, Approximation and geometric divide and conquer, *J. Comput. Syst. Sci.* 50 (1995), 203–208.

[7] T. Ottman and P. Widmayer, On translating a set of line segments, *Comp. Vision, Graphics & Image Processing*, 24 (1983), 154–160.

[8] M. S. Paterson and F. F. Yao, Optimal binary space partitions for orthogonal objects, *J. Algorithms* 13 (1992), 99–113.

[9] F. P. Preparata and M. I. Shamos, *Computational Geometry, An Introduction*, Springer-Verlag, 1987 p. 278.

[10] H. Tamaki and T. Tokuyama, How to cut pseudo-parabolas into segments, *Proc. 11th ACM Symp. On Computational Geometry,* 1995, 230–237.

Fly Cheaply: On the Minimum Fuel-Consumption Problem

Alon Efrat* Sariel Har-Peled†

1 Introduction

Assume we wish to plan a path of a flight, starting at an airport s (i.e. Tel-Aviv), and our destination is the airport t (i.e. Minneapolis). Our plane is able to carry enough fuel, so a direct flight from s to t is possible. However, we might find it beneficial to stop at some airport(s) on our way, even though this increases somehow the length of our path. One reason to do so is that the longer we fly without a stop, the more fuel we need to take, the weight of the plane increases, so the cost of a kilometer of flight increases. It is thus possible that we can reduce fuel consumption by flying along two cheaper segments instead of flying directly from s to t. We thus face the problem of planning the cheapest flight path, given s, t, and a set of intermediate airports at which we can (but do not have to) make a stop.

We can abstract and formalize this problems as follows: Let P be a set of n points in the plane, containing two given special points s and t. We are also given a positive cost-function $l : P \times \mathbb{R}^2 \to \mathbb{R}^+$, such that $l(p,q)$ (for $p, q \in \mathbb{R}^2$) is the cost of a direct flight from p to q. We call a sequence $\pi = (v_0, v_1, \ldots v_k)$, where each $v_i \in P$, a *path* from v_0 to v_k. The *cost* of π, is the sum $l(\pi) = \sum_{i=1}^{k} l(v_{i-1}, v_i)$. The *minimum fuel-consumption problem* (MFC problem in short) is the following: given P, s, t and $l(\cdot,\cdot)$ as above, find a path π from s to t, whose cost is minimal, among the costs of all paths from s to t.

Throughout the paper, we make the following assumptions[1].

- The function $l(\cdot,\cdot)$ has *constant description complexity*, in the sense that it is algebraic and defined in terms of a constant number of polynomial equalities and inequalities of constant maximum degree.
- Given $p \in P, q \in \mathbb{R}^2$ we can compute $l(p,q)$ in time $O(1)$.
- $l(p,q) < l(p,q')$ if and only if $|p-q| < |p-q'|$, where $|p-q|$ denote the Euclidean distance from p to q.

Note that this setup allows us to consider also applications where landing at an airport u incurs an additional penalty $c(u)$ and several other variants.

Clearly, we can solve the MFC problem in time $O(n^2)$ by computing all $\Theta(n^2)$ values $l(p,q)$ between all pairs $p, q \in P$, and apply Dijkstra's algorithm for the complete graph whose vertices are the points of P, and the cost of an edge pq is $l(p,q)$.

In this paper, we show that it is possible to do better; In Section 2, we present an $O(n^{1.5+\varepsilon})$-time algorithm[2] for any $\varepsilon > 0$, in the case of arbitrary positive cost function $l(\cdot,\cdot)$. In 3 we show to solve MFC problem in time $O(n \log n)$, in the special case that that $l(p,q) = |p-q|^2 r(|p-q|)$ for all $p, q \in P$, where $r(\cdot)$ is a positive monotonically increasing function.

For $u, v \in P$, let $\delta(u,v)$ denote the cost of the cheapest path from u to v, and let $\delta(v) = \delta(s,v)$. For two points $p, q \in \mathbb{R}^2$, we denote by $e = pq$ the straight line segment whose endpoints are p and q. The length of this segment is denoted $|pq|$. We also define $l(e) = l(p,q)$.

2 The case of a General Cost Function

We show in this section how to compute $\delta(t)$, the cost of the cheapest path from s to t under a general cost function satisfying the properties assumed in the introduction. We omit the description of computing the path itself, which can be obtained from the algorithm in a standard manner (see e.g. [3]). The algorithm we present is a variant of Dijkstra's algorithm. Recall that Dijkstra's algorithm maintains a set $S \subseteq P$, its complement $\bar{S} = P \setminus S$, and a value $d[v]$ for each point $v \in P$. This value is the cost of the cheapest path found so far from s to v. Initially, $S = \emptyset$, $d[s] = 0$ and $d[v] = \infty$, for all other points of P. The algorithm repeatedly chooses a vertex $u \in \bar{S}$ that minimizes $d[u]$ over $\bar{S}$, adds u to S, and set

$$d[v] = \min\{d[v], d[u] + l(u,v)\}, \quad (1)$$

for each other vertex $v \in \bar{S}$. It can be shown that at that instance $d[u] = \delta(u)$. The algorithm stops once $u = t$.

*School of Mathematical Sciences, Tel-Aviv University, Tel-Aviv 69982, Israel. alone@cs.tau.ac.il

†School of Mathematical Sciences, Tel-Aviv University, Tel-Aviv 69982, Israel. sariel@cs.tau.ac.il

[1] There is an additional technical assumption which is described in Section 2.

[2] Throughout the paper, ε stands for an arbitrarily small positive constant parameter; the constant of proportionality in bounds that include ε depends on ε, and usually tends to ∞ as $\varepsilon \to 0$

SCG 98 Minneapolis Minnesota USA

In our variant of the algorithm, we follow the same operation, except that the we update $d[v]$ only for the vertices v of a (relatively small) subset of $\bar{S}$, which we describe shortly. From time to time we stop the process, and update $d[v]$ for all vertices of $\bar{S}$.

One might regard our algorithm, as a lazy implementation of Dijkstra's algorithm: During regular execution, it lags behind Dijkstra's algorithm, performing only part of the update operations (1). Once in a while it stops, and catches up with the standard Dijkstra's algorithm, by performing (some of the) update operations it skipped. Fortunately, because of the underlying geometric structure of the function $l(\cdot,\cdot)$, this bulk update stage can be performed efficiently. Moreover, it can be shown that the updates (1) that were skipped do not effect the correctness of the algorithm.

We next describe the algorithm in details. The data structures and their initialization is the same as in Dijkstra's algorithm. The algorithm consists of $\sqrt{n}$ *stages*. Each stage consists of two *phases*. The first phase, called the *growing phase*, consists of adding $\sqrt{n}$ points to S. The second phase, called the *re-evaluation phase*, consists of updating $d[v]$ for each point $v \in \bar{S}$. We next describe the pseudocode of the main operations of the algorithm. Details will be supplied later.

In order to find $N(u)$, we maintain all points of $\bar{S}$ in the data structure of Agarwal and Matoušek [1]. This data structure enables us to find the nearest neighbor to a query point q in time $O(\log n)$, and to delete such a point in time $O(n^{\epsilon})$. Thus to find $N(u)$, we repeatedly find the nearest neighbor of u, and delete it from the data structure, until $\sqrt{n}$ such points have been found. After $N(u)$ have been computed, we restore the data structure as it was before the deletions took place. Note that $N(u)$ is also the set of $\sqrt{n}$ nearest neighbors of u under the cost function $l(u,\cdot)$. The time needed to compute $N(u)$ is clearly $O(n^{0.5+\epsilon})$, which sums, over all $u \in P$, to $O(n^{1.5+\epsilon})$.

The re-evaluation phase: Let $S_{new} \subseteq S$ denote the $\sqrt{n}$ points added to S in the preceding growing phase. For each $w \in S_{new}$, let $f_w(x) = d[w] + l(w,x)$ (for $x \in \mathbb{R}^2$). We assume non-degeneracy of the family of the set of functions $\mathcal{F} = \{f_w(x) \mid w \in S_{new}\}$, such that each triple of functions of $\mathcal{F}$ intersects only at a constant number of points, etc. Let $f_{min}(x) = \min\{f_w(x) \mid w \in S_{new}\}$ be the lower envelope of $\mathcal{F}$. It was shown by Halperin and Sharir [5], that the complexity of $f_{min}(\cdot)$ is $O(|\mathcal{F}|^{2+\epsilon}) = O(n^{1+\epsilon})$. We compute $f_{min}(\cdot)$ in time $O(n^{1+\epsilon})$ using the algorithm of [2], thus obtaining a subdivision M of the xy-plane into $O(n^{1+\epsilon})$ faces, such that in each face the function of $\mathcal{F}$ that determines $f_{min}(\cdot)$ is the same for all points in the face. We prepare a point-location data structure for M [6, 8], which enables us to find in time $O(\log n)$ the cell of M containing a query point q, and thus to compute $f_{min}(q)$. The preprocessing and storage required for the data structure is $O(n^{1+\epsilon})$. We next query each $v \in \bar{S}$ to find $f_{min}(v)$, (which equals $\min_{w \in S_{new}} d[w] + l(w,v)$) and set $d[v]$ to be $\min\{d[v], f_{min}(v)\}$. The time needed for this phase is thus $O(n^{1+\epsilon})$, and there are $O(\sqrt{n})$ such phases. Thus, the running time of the algorithm is $O(n^{1.5+\epsilon})$.

For each $v \in \bar{S}$ (for the value of $\bar{S}$ at the re-evaluation phase), define $d'_S(v) = \min\{d[u] + l(u,v) \mid u \in S\}$. The correctness of the algorithm is a consequence of the following lemma:

Lemma 2.1 *Assume we are at step (*) of the algorithm. Then, for the point u chosen at this step, and for the values of S and $\bar{S}$ just before this step, we have*

(i) $d[u] = \min\{d'_S(u') \mid u' \in \bar{S}\}$,

(ii) *For each* $w \in S \cup \{u\}$, $\delta(w) = d[w]$.

Proof:. We prove the lemma by induction on the sequence of insertions into S. The claims are clearly true when $S = \emptyset$. We assume by induction that $d[w] = \delta[w]$ for each $w \in S$, and show that the claims also hold for the next point u to be inserted at step (*). To show (i), assume that $u' \neq u$ is a point of current $\bar{S}$ minimizing $d'_S(u)$, and that $d[u] \neq d'_S(u')$. Observe that for each $u'' \in \bar{S}$ we have that $d[u''] \geq d'_S(u'')$ (this follows from the induction hypothesis and from the fact that $d'_S(u'')$ is the value of $d[u'']$ computed by the standard Dijkstra's algorithm, where as our $d[u'']$ is computed by fewer minimum operations), so $d[u] \geq d'_S(u) \geq d'_S(u')$, and thus $d[u'] \geq d[u] > d'_S(u')$.

Let $w \in S$ be a point for which $d'_S(u') = d[w] + l(w,u')$. Since $d[u'] > d'_S(u')$ it follows that w joined S after the last re-evaluation phase. Otherwise, during the last re-evaluation phase we must have set $d[u']$ to be equal to the current value of $d'_{S^0}(u')$, where S^0 equals the set S as it was in the last re-evaluation stage. By the same argument, $u' \notin N(w)$. Thus, for each point $x \in N(w)$, we have that $|w - x| < |w - u'|$, which implies by the constraints on $l(\cdot,\cdot)$ that

$$l(w,x) < l(w,u') .$$

Thus

$$d[x] \leq d[w] + l(w,x) < d[w] + l(w,u') = d'_S(u') < d[u] .$$

Since the number of points added to S since the last re-evaluation phase is strictly smaller than $\sqrt{n}$, there must be a point $x \in N(w) \cap \bar{S}$, and by the above discussion, $d[x] < d[u]$, a contradiction to the assumption that $d[u]$ is minimal. Thus (i) is proven. By the induction hypothesis for (ii), $d[w] = \delta[w]$ for every $w \in S$, thus $d[u] = \min\{\delta(w) + l(w,v) \mid w \in S, v \in \bar{S}\}$.

As easily verified, u would be the next point to be added to S, and by the arguments used to show the validity of Dijkstra's algorithm, $d[u] = \delta(u)$. This proves (ii). ■

Thus we have shown:

Theorem 2.2 *Given P, s, t and $l(\cdot,\cdot)$ as above, we can compute $\delta(t)$ in time $O(n^{1.5+\epsilon})$ for any $\epsilon > 0$.*

3 Quadratic and Superquadratic Cost Functions

In this section, we present simple and efficient algorithms for the case where our cost function is of the form $l(x,y) = |xy|^2 r(|xy|)$, where $r(a)$ is a monotone increasing function.

Let $\pi = (s = p_1, \ldots, p_k = t)$ be the cheapest path between s and t, under the given cost function, and let $e_i = p_i p_{i+1}$, for $i = 1, \ldots, k-1$. Let D_i be the disk with e_i as a diameter.

We claim that, for $i = 1, \ldots, k-1$, the disk D_i does not contain points of P in its interior. Assume for the sake of contradiction, that D_m (for $1 \leq m \leq k-1$) does contain the point $p_j \in P$ in its interior. Then,

$$|p_m p_j|^2 + |p_j p_{m+1}|^2 < |p_m p_{m+1}|^2 .$$

Hence,

$$l(p_m, p_j) + l(p_j, p_{m+1})$$

```
Initialization:   S = ∅; d[s] = 0; d[v] = ∞, ∀v ∈ P \ {s}.
While t ∉ S Do
Begin
  The growing phase:
    While less than √n points were added to S since the last re-evaluation phase
    Begin
  (*) Choose a point u ∈ S̄ that minimizes d[u] and add u to S.
      Find N(u), the set of the √n Euclidean nearest neighbors of u in S̄.
      For each point v ∈ N(u) Do
        d[v] = min{d[v], d[u] + l(u, v)}
    End
  The re-evaluation phase:
    /* Let S_new ⊆ S be the set of points added to S
       during the preceding growing phase. */
    For each point v ∈ S̄ Do
      d[v] = min {d[v], min{d[u] + l(u, v) | u ∈ S_new}}
End
```

$$
\begin{aligned}
&= |p_m p_j|^2 r(|p_m p_j|) + |p_j p_{m+1}|^2 r(|p_j p_{m+1}|) \\
&\leq (|p_m p_j|^2 + |p_j p_{m+1}|^2) r(|p_m p_{m+1}|) \\
&< |p_m p_{m+1}|^2 r(|p_m p_{m+1}|) \\
&= l(p_m, p_{m+1}),
\end{aligned}
$$

since $|p_m p_j|, |p_j p_{m+1}| \leq |p_m p_{m+1}|$. Hence,

$$
\begin{aligned}
\delta(s, p_{m+1}) &= \delta(s, p_m) + l(p_m, p_{m+1}) \\
&> \delta(s, p_m) + l(p_m, p_j) + l(p_j, p_{m+1}) \\
&\geq \delta(s, p_{m+1})
\end{aligned}
$$

which is a contradiction.

It follows that all the edges of the cheapest path are edges of the so-called *Gabriel graph* $\mathcal{G}(P)$ consisting of all edges (p, q), for which the disk with pq as a diameter does not contain any point of P in its interior (see [4]). As is well known, $\mathcal{G}(P)$ is a subgraph of the Delaunay triangulation of P.

Hence, to solve the cheapest path problem in our setting, we compute the Delaunay triangulation $\mathcal{D}(P)$ of P, in time $O(n \log n)$ [7], and solve the shortest path problem on the graph $\mathcal{D}(P)$ under the cost function $l(\cdot, \cdot)$, using the standard Dijkstra's algorithm. Since the size of $\mathcal{D}$ is $O(n)$ this implies that the overall running time of the algorithm is $O(n \log n)$. To summarize,

Theorem 3.1 *Given P, s, t as above, and a cost function $l(x, y) = |xy|^2 r(|xy|)$, where $r(\cdot)$ is a monotonically increasing function, we can compute $\delta(t)$ in time $O(n \log n)$.*

References

[1] P. K. Agarwal and J. Matoušek, *Dynamic half-space range reporting and its applications*, Algorithmica, 13 (1995), pp. 325–345.

[2] P. K. Agarwal, O. Schwarzkopf, and M. Sharir, *The overlay of lower envelopes and its applications*, Discrete Comput. Geom., 15 (1996), pp. 1–13.

[3] T. H. Cormen, C. E. Leiserson, and R. L. Rivest, *Introduction to Algorithms*, The MIT Press, Cambridge, Mass., 1990.

[4] J. E. Goodman and J. O'Rourke, eds., *Handbook of Discrete and Computational Geometry*, CRC Press LLC, 1997.

[5] D. Halperin and M. Sharir, *New bounds for lower envelopes in three dimensions, with applications to visibility in terrains*, Discrete Comput. Geom., 12 (1994), pp. 313–326.

[6] K. Mulmuley, *A fast planar partition algorithm, II*, J. ACM, 38 (1991), pp. 74–103.

[7] J. O'Rourke, *Computational Geometry in C*, Cambridge University Press, Cambridge, 1994.

[8] R. Seidel, *A simple and fast incremental randomized algorithm for computing trapezoidal decompositions and for triangulating polygons*, Comput. Geom. Theory Appl., 1 (1991), pp. 51–64.

Designing a Data Structure for Polyhedral Surfaces

Lutz Kettner*, ETH Zürich, Switzerland.

Abstract

Design solutions for a program library are presented for combinatorial data structures in computational geometry, such as planar maps and polyhedral surfaces. Design issues considered are genericity, flexibility, time and space efficiency, and ease-of-use. We focus on topological aspects of polyhedral surfaces. Edge-based representations for polyhedrons are evaluated with respect to the design goals. A design for polyhedral surfaces in a halfedge data structure is developed following the generic programming paradigm known from the Standard Template Library STL for C++. Connections are shown to planar maps and face-based structures managing holes in facets.

1 Introduction

Combinatorial structures, such as planar maps, are fundamental in computational geometry. In order to use computational geometry in practice, a solid library must provide generic and flexible solutions as one of its fundamental cornerstones. Other design criteria are time and space efficiency. Ease-of-use is necessary to make the power of a design accessible and to attract users. We report a solution proposed for the Computational Geometry Algorithms Library CGAL[1], which is a joint effort of seven academic institutes in Europe [7, 6, 27].

We focus on edge-based representations of three-dimensional polyhedral surfaces and illustrate connections to planar maps and face-based structures, which may have holes in their facets. We concentrate on the topological aspects and derive solutions applicable to other data structures as well. In particular, we want to vary the internal storage organization and the kind of incidences that are actually stored. Additional user data can be integrated easily. A top-level interface ensures ease-of-use and combinatorial integrity. On the other hand, a protected access to the internal representation is granted.

*Part of this research was carried out while the author was at the Freie Universität Berlin with a scholarship from the Graduiertenkolleg "Algorithmische Diskrete Mathematik", DFG We 1265/2-1. Support from the ESPRIT IV LTR Project No. 21957 (CGAL) is also acknowledged.

[1] `http://www.cs.ruu.nl/CGAL/`

SCG 98 Minneapolis Minnesota USA

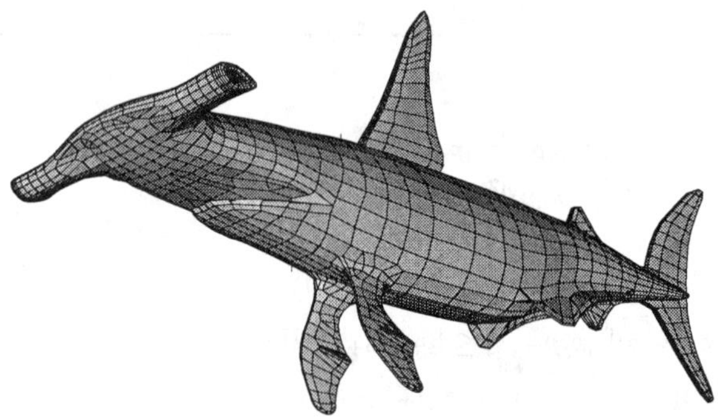

Figure 1: Hammerhead, an orientable 2-manifold of 2560 vertices. This one is homeomorphic to a sphere.

In the first part of the paper we define polyhedral surfaces and review known edge-based boundary representations. In the second part we start with a short introduction to the modern design principles available in C++ and known as the generic programming paradigm from the Standard Template Library, STL [4, 25, 29]. We derive design goals and evaluate previous work. We continue with an overview of our design, present several aspects in more detail and conclude with its evaluation. The two main advantages of our design are: The flexibility is completely handled at compile time, i.e. there is no runtime overhead due to the flexibility, and memory is only allocated for the features actually used. For example, a polyhedron with no information in facets does not allocate facet nodes and facet pointers at all.

2 Polyhedral Surfaces

A boundary representation of a polyhedral surface consists of a set of vertices V, a set of edges E, a set of facets F and an incidence relation on them. Introductions can be found in [13, 21]. For a living example see Figure 1.

The two types of boundary representations are 2-manifold and non-manifold surfaces. A 2-manifold surface is a surface where for each point on the surface there exists a neighborhood that is homeomorphic to the open disc. Non-manifold examples are two tetrahedrons glued together at a single vertex or a common edge. The next distinction is between orientable and non-orientable 2-manifold surfaces. Without going into details, a surface is orientable if a consistent orientation can be assigned to each facet such that for each edge the two incident facets have opposite orientations at this edge. An example of a non-orientable 2-manifold is

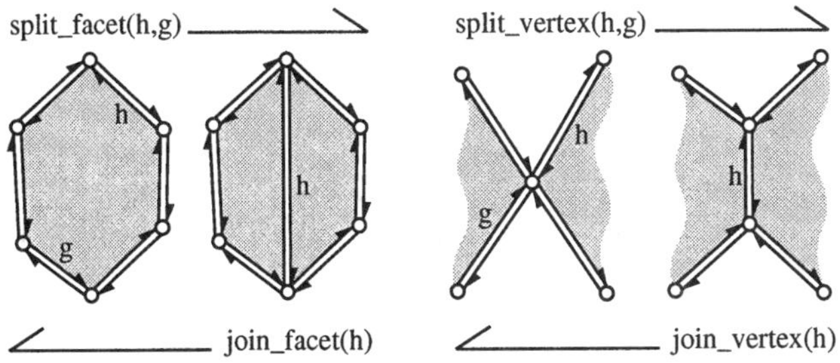

Figure 2: Euler operator examples for polyhedral surfaces.

the Klein bottle. We consider only orientable 2-manifolds.

The natural operations under which 2-manifolds are closed are Euler operations; Four of them are shown in Figure 2. The principal characteristic of an Euler operation is the invariance of the Euler-Poincaré formula. A sufficiency proof for a specific set of Euler operations can be found in [21]. Note that 2-manifolds are not closed under (regularized) boolean operations.

The class of representable surfaces is further restricted by the kind of geometry associated with vertices, edges and faces. Vertices map to points in $\mathbb{R}^3$. For polyhedra the edges are typically the straight line segments between their two endpoints and the facets are simple, planar polygons. Other classes might allow curved surfaces as facets.

We now present a definition for polyhedral surfaces following Steinitz [30]. It is the basis for the combinatorial integrity definition of the polyhedral surface data structure and will lead to a stricter class of representable surfaces, which provides more insight in the combinatorial structure of the representation.

Definition 2.1. A *structural complex* is a union $C = V \cup E \cup F$ of three disjoint sets together with an incidence relation. We call V the vertices, E the edges and F the facets of the structural complex. The incidence relation on C must be symmetric. No two elements from the same set V, E or F are incident. If $v \in V$ is incident to $e \in E$ and e is incident to $f \in F$ then v is incident to f.

Definition 2.2. A *polyhedral complex* is a structural complex with four additional conditions.

(1) Every edge is incident to two vertices.

(2) Every edge is incident to two facets.

(3) For every incident pair v, f, there are exactly two edges incident to both.

(4) Every vertex and every facet is incident to at least one other element.

The *neighborhood* of a vertex is the set of edges and facets incident to the vertex. If we restrict the incidence relation to this neighborhood then each facet is incident to exactly two edges and each edge is incident to exactly two facets. The neighborhood decomposes into disjoint cycles. As for the dual, the neighborhood of a facet is the set of incident edges and vertices and decomposes into cycles too. Assuming that the neighborhood of each facet is a single cycle (geometrically speaking: no holes in the facet), we can define a polyhedral complex as *oriented* if each cycle around a facet is oriented and if, for each edge, the two cycles of its two incident facets are oriented in opposite directions. A polyhedral complex is *orientable* if there exists such an orientation.

Definition 2.3. The *boundary representation of a polyhedron* is a polyhedral complex with a mapping $V \rightarrow \mathbb{R}^3$. This extends to the edges by mapping them to the open, straight line segments between their two incident endpoints. The following additional conditions must hold.

(5) The neighborhood of each vertex and each facet is a single cycle.

(6) The polyhedral complex is orientable.

(7) The mapping of the cycle of the neighborhood of each facet is the boundary of a simple, planar polygon. The mapping extends for F to the open region of these polygons.

(8) The images of V, E and F are pairwise disjoint.

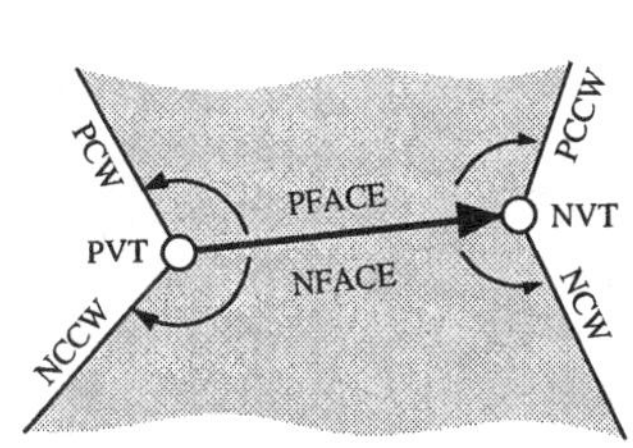

Figure 3: A winged-edge.

The surface defined by such a boundary representation is an orientable 2-manifold where the neighborhoods of two vertices have at most one edge and two facets in common, the edge and vertex graphs are connected within each connected component of the surface and where each facet has at least three edges on its boundary. The smallest possible configuration is a tetrahedron.

The closed surfaces considered so far can be extended to surfaces with boundaries by two changes in the definition: Condition (2) can be relaxed to allow edges that are incident to one facet; they are called *border edges*. This induces a modification of (5): The neighborhood of a vertex decomposes into either a cycle or a collection of open paths going from border edge to border edge. Although the surface is no longer closed, the orientation still defines a "solid" side of the surface. The minimal configuration for surfaces with boundaries is a triangle. The data structures we will describe can be used for polyhedrons as well as for surfaces with boundaries with a simple extension denoting "empty" facets.

A suitable data structure based on the Definition 2.3 for polyhedral surfaces has been used successfully for three years in a project on contour-edge-based polyhedron visualization where we take advantage of the strict properties imposed by the definitions: For example the definition for contour-edges is based on the orientable 2-manifold property, and the lack of holes in facets simplifies certain algorithms[2] [18]. An initial implementation of the data structure made it easy to compute the silhouette for a polyhedral surface [14]. The extension of this data structure design and their advantages are presented in the following sections.

3 Data Structures for Boundary Representations

The following survey of edge-based data structures addresses their sufficiency for modeling topology and the efficiency of their primi-

[2] And holes are not represented in the file-formats that occur usually in visualization, for example VRML [12], Open Inventor [34] or the Object File Format OFF [28]. These consist of a list of vertices followed by a list of facets. Each facet is a list of indices denoting a subset of the points. Edges are not explicitly stored but can be derived from the vertices shared by facets. These formats are not strict enough for our purpose since they can represent non-manifold configurations where three or more facets are incident to a single edge, non-orientable 2-manifolds, and also violate condition (3) for polyhedral complexes. But they cannot represent holes in facets.

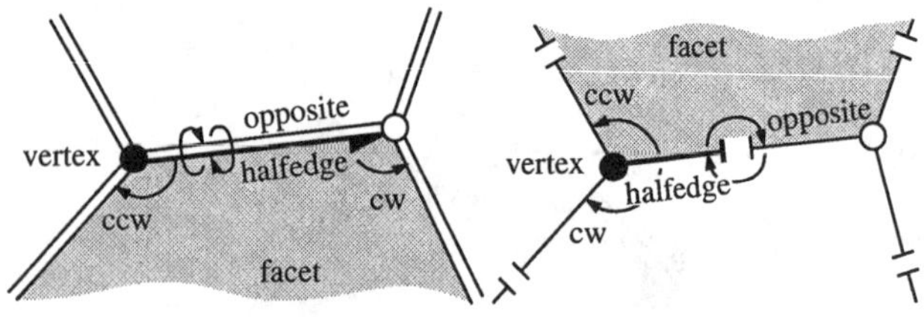

Figure 4: FE-structure (left) and VE-structure (right).

tive operations and storage costs. The representative example chosen is the traversal around a vertex to the next counterclockwise edge.

Winged-Edge Data Structure. The winged-edge data structure [1, 10] stores, for each oriented edge, eight references: two vertices (PVT, NVT), two faces (PFACE, NFACE) and four incident edges that share the same faces and vertices (PCW, PCCW, NCW and NCCW), the so-called *wings*, see Figure 3. An edge is oriented from the source vertex PVT to the target vertex NVT. The face PFACE is to the left of the oriented edge when the surface is seen from the outside.

This data structure is able to model orientable 2-manifolds. It is even sufficient for curved-surface environments where loops and multi-edges are allowed [33]. The basic operations include traversal around a vertex and around a facet. High-level operations maintaining integrity are Euler operators. The next edge counterclockwise around a vertex `v` for an edge `e` is equal to `e->PCW` if `e->PVT == v` and `e->NCW` otherwise.

Variants are possible where vertex and facet pointers can even be omitted without loosing the traversal capabilities knowing the edge visited previously. However, all four edge pointers must remain if loops or multi-edges are allowed since otherwise the traversal around a vertex or facet is no longer uniquely defined [33]. The winged-edge data structure where the wings PCCW and NCCW are omitted has been called **Doubly Connected Edge List** (DCEL) by [24] though this name is now more commonly used for the halfedge data structure [5]. [3]

The two symmetric parts in the winged-edge correspond to the two possible orientations of the edge. The inefficient case distinction in the traversal computation results from the fact that a pointer to an edge does not encode the orientation it is currently used with. One extension of the winged-edge maintains an additional bit with each edge-pointer to code the orientation, but this leads to cumbersome storage layouts and function interfaces.

Halfedge Data Structure. The orientation problem can be solved for the winged-edge data structure by splitting the edge into the two symmetric records, called *halfedges*, and adding mutual links to each other [33]. There are two ways of splitting the edge, which are actually dual to each other. In both situations the halfedge contains a pointer to an incident vertex, an incident facet and the opposite halfedge. It is a matter of convention whether the source or target vertex is the one chosen to be stored in a halfedge or whether the facet to the left or the right is stored. In [33] the source vertex and the facet to the right were chosen. The FE-structure in Figure 4 additionally stores a pointer to the next clockwise halfedge and optionally a pointer to the previous counterclockwise halfedge around the facet. It is therefore biased towards traversals around the incident facet. The dual VE-structure is depicted in Figure 4 on the right. Its next and optional previous pointer refer to halfedges counterclockwise and clockwise around the incident vertex. The traversal operation that is not directly accessible with a single pointer access is available through the opposite halfedge. For example the next halfedge around the incident source vertex for the FE-structure is `opposite()->next()`. The different conventions are not independent. If the convention defines the halfedge order around a facet to be clockwise, the halfedge order around the vertex will be counterclockwise, and vice versa.

[3] In order to avoid confusion we will not use the name DCEL since it turned out to be ambiguous. In fact, the name is misleading when denoting halfedges and the possible variants of single linking.

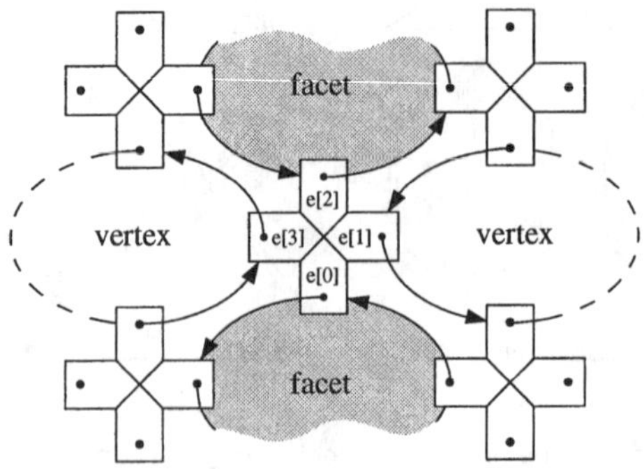

Figure 5: Quad-edge data structure.

The halfedge data structure is able to model orientable 2-manifolds. It is sufficient for modeling topology even in the presence of loops and multi-edges, which can occur in curved-surface environments [33]. High-level operations maintaining integrity are again Euler operators. The solid modeling system GWB [21] is based on a halfedge-data structure, though it uses an additional edge record between two opposite halfedges, which makes this access less efficient. The Minimal Rendering Tool MRT [2] uses a halfedge data structure for polygonal surfaces.

Quad-Edge Data Structure. If we perform both halving steps for the halfedge data structure, we end up with the quad-edge data structure [11]. It provides a fully symmetric view on the primal and the dual graph, as can be seen in Figure 5. Instead of using opposite pointers, a two bit counter r is used to address a slot in an edge record of four quad-edges. With an additional bit f per edge for the flipped status the quad-edge data structure is able to model non-orientable 2-manifolds.

A quad-edge data structure is defined as an edge algebra with three operations: `Onext()`, `Rot()` and `Flip()`. An edge is represented as a triple (e, r, f) with $r \in \{0, 1, 2, 3\}$ and $f \in \{0, 1\}$. e is the base pointer to the quad-edge record with the four incident edges $e[0]$ to $e[3]$. The operations are implemented as follows with a calculus modulus 4 for r and modulus 2 for f:

$$\begin{aligned} \texttt{Rot}(e, r, f) &= (e, r + 1 + 2f, f), \\ \texttt{Flip}(e, r, f) &= (e, r, f + 1), \\ \texttt{Onext}(e, r, f) &= \texttt{Flip}^f(\texttt{Rot}^f(e[r + f])) . \end{aligned}$$

Four different orientations of an edge are considered: two orientations from vertex to vertex and two orientations for the dual edge from facet to facet. The `Rot` operator rotates the edge by 90 degrees, oscillating between the primal and the dual view of the structure. For non-orientable 2-manifolds an edge can additionally be seen from above or below the surface, which is encoded in the f bit. The `Flip` operation changes the view from above to below or vice versa. The `Onext` operation gives the next quad-edge in counterclockwise order around the source vertex (origin), or the next quad-edge in clockwise order if f is equal to one. The values for `Onext` are simply stored in the record for each edge (i.e. four

	Winged-Edge	Half-Edge	Quad-Edge
Modeling space	orientable 2-manifold		2-manifold
Operations	Euler operator		`Splice()`
Duality	at compile time (with adaptor)		at runtime (`Rot` operator)
Holes in facets	yes	yes	no
Basic traversal	case distinction	direct access	modulus operation
Min size per edge	4 ptr	4 ptr	2 ptr + 2 bits
Max size per edge	8 ptr	10 ptr	8 ptr + 12 bits

Table 1: Comparison of the edge-based data structures.

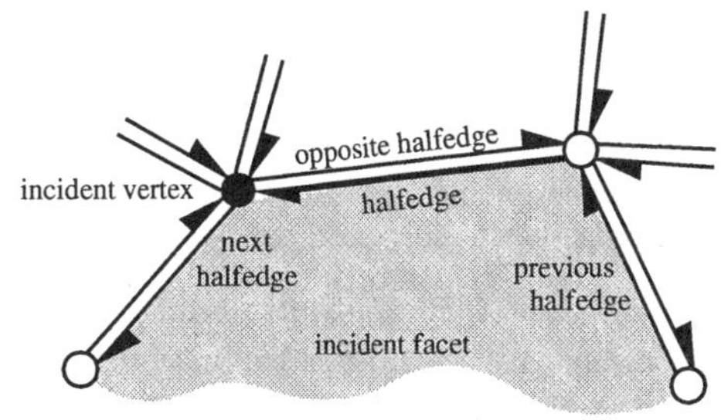

Figure 6: Halfedge data structure.

pointers and four times three bits for r and f). The operations simplify considerably for orientable 2-manifolds. They can be further simplified if the dual graph is not necessary. This reduces to the winged-edge data structure enriched with a bit to encode orientation.

The single high-level operation that modifies a quad-edge data structure is the `Splice()` operation. It is its own dual. The usual Euler operators can be implemented in terms of `Splice()`. The quad-edge data structure provides a unified view for the primal and dual graph. This implies that vertices and facets cannot be distinguished with strong type checking at compile time. The definition used for duality implies, furthermore, that the facets must have a single connected boundary. Holes in facets are not allowed. If strong type checking is desired, the `Splice()` operation is needed twice, once for the primal view and once for the dual view. `Splice()` can also be provided for the halfedge data structure.

Comparison of Edge-Based Representations. The main differences of these edge representations are captured in Table 1. The differences in the basic traversal capabilities are not negligible, especially when considering modern microprocessor architectures where conditional branching can be an order of magnitude slower than computing. The storage size requirements are quite similar. Our design will focus on the flexibility of trading runtime against storage costs. We are interested in the minimal and maximal configurations for the halfedge data structure and the space efficiency of the quad-edge data structure. Another issue is the preference for strong type checking at compile time. Polyhedral surfaces have different information stored in the vertices and facets, namely points and plane equations. These can be treated as duals of each other, but in a strongly-typed geometry kernel (like the one CGAL provides) they are different types and might even be represented differently. Additional information, like color, will finally destroy the typeless symmetry of the duality assumed by the quad-edges. We consider non-orientability as not so important since three-dimensional surfaces of solid objects are always orientable.

The choice for our design is a halfedge data structure like the FE-structure. The conventions used are depicted in Figure 6. We have `next()`, `opposite()` and `prev()` pointers for the halfedges. The incident vertex is the target vertex of the oriented halfedge. The incident facet is to the left of the halfedge which implies a counterclockwise ordering of the halfedges around the facet and a clockwise ordering around the vertex when seen from the outside. This complies with the right-hand rule for out-facing normals of plane equations for facets.

4 Generic and Object-Oriented Programming

The major design issues considered for polyhedral surfaces are genericity, flexibility, time efficiency, space efficiency and ease-of-use. Two techniques are available in C++ for realizing generic and flexible designs: *Object-oriented programming*, using inheritance from base classes with virtual member functions, and *generic programming*, using class templates and function templates.

The flexibility in the *object-oriented programming paradigm* is achieved with a virtual base class, which defines an interface, and as many derived classes as different actual implementations of the interface are present in a system. The technique of so-called virtual member functions and runtime type information allows a user to select any of the derived classes wherever the base class is required – even at runtime. Generic functionality can be programmed in terms of the base class without knowing all possible derived implementations beforehand.

The advantages are the clear definition of the interface and the flexibility at runtime. There are four main disadvantages: This paradigm cannot provide strong type checking at compile time, enforces tight coupling through the inheritance relationship [19], it adds additional memory to each object derived from the base class (the so-called *virtual function table pointer*) and it adds an indirection through the virtual function table for each call to a virtual member function [20]. The latter one is of particular interest when considering runtime performance since virtual member functions can usually not be made *inline* and are therefore not subject to code optimization within the calling function. Modern microprocessor architectures[4] can optimize at runtime, but, besides that runtime predictions are difficult, these mechanisms are more likely to fail for virtual member functions. These effects are negligible for larger functions, but small functions will suffer a loss in runtime of one or two orders of magnitude. Significant examples are coordinate access and arithmetic for low-dimensional geometric objects and traversals of combinatorial structures. Vertices, edges and facets for polyhedrons are anticipated to be small objects with simple member functions. The space and runtime overhead introduced through virtual member functions would not be negligible.

The *generic programming paradigm* features what is known in C++ as *class templates* and *function templates*. Templates are program recipes where certain types are only given symbolically, the so called *template arguments*. The compiler replaces these arguments with actual types where the program recipe is actually used, at the place of the *template instantiation*. The recipe transforms to a normal part of a program. For function templates this can even be done automatically by the compiler, since the types of the function parameters are known to the compiler. Examples are a generic list class for arbitrary item types or a swap function exchanging variable values for all possible types. The following definitions would enable us to use `list<int>` as a list of integers or to swap two integer variables `x` and `y` with `swap(x,y)`.

```
template <class T>  class list {
  // ... , uses T as item type.
};
```

[4] Pipelining, branch prediction, speculative execution and reordering, global optimizers using runtime statistics and the interplay with the cache architecture.

```
template <class T>  void swap( T& a, T& b) {
    T tmp = a; a = b; b = tmp;
}
```

The example of the swap function illustrates that a template usually assumes some properties to hold for the template arguments, here that variables of those type can be assigned to each other. These *requirements* are not expressed within C++, only in the accompanying documentation. An actual type used in the template instantiation must fulfill the requirements of the template argument in order of the template to work properly. Requirements can be classified into syntactical ones, there must be an assignment operator, and semantical ones, the implementation of the operator must really do what it is supposed to do. Syntactical requirements will be checked by the compiler at instantiation time of the template. Semantical requirements cannot be checked. In certain situations it might be wishful to stress semantical requirements with additional syntactical, i.e. checkable, requirements, e.g. symbolic tags.

For class templates exist the specialty that different member functions might impose different requirements on the template arguments, but a certain instantiation of the class template uses only a subset of the member functions. Here, the arguments must only fulfill the requirements imposed by the member functions actually used. In particular, the compiler is only allowed to instantiate those member functions of an *implicit instantiation* of a class template that are actually used [4]. This enables us to design class templates with optional functionality that impose additional requirements on the template arguments if and only if this functionality is used.

A good example for the generic programming paradigm is the Standard Template Library [4, 25, 29]. The main source of its generality and flexibility stems from the separation of *concepts* and *models* [29]. For example, an iterator is an abstract *concept* defined in terms of requirements. A certain class is said to be a *model* of the concept if it fulfills the requirements. The iterator concept is a generalization of a pointer and the usual C-pointer is a model of an iterator. Iterators serve two purposes: They refer to an item and they traverse over the sequence of items in a container class. Container classes manage collections of items. Different categories are defined for iterators: input, output, forward, bidirectional and random-access iterators. They differ mainly in their traversal capabilities. The usual C-pointer is a random-access iterator. *Generic algorithms* in the STL are not written for a particular *container class* but for a pair of iterators instead. The so called *range* `[first,beyond)` of two iterators denotes the sequence of all iterators obtained by starting with `first` and advancing `first` until `beyond` is reached, but does not include `beyond`. A container is supposed to provide a type, which is a model of an iterator, and two member functions: `begin()` returns the start iterator of the sequence and `end()` returns the iterator referring to the 'past-the-end'-position of the sequence. A generic `contains` function could be written as follows and will work for any model of an input iterator.

```
template <class InputIterator, class T>
bool contains( InputIterator first,
               InputIterator beyond,
               const T& value)
{
    while ((first != beyond) && (*first != value))
        ++first;
    return (first != beyond);
}
```

The advantages of the generic programming paradigm are strong type checking at compile time during the template instantiation, no need for extra storage or additional indirections during function calls, and full support of inline member functions and code optimization at compile time [32]. One disadvantage is the lack of a formal scheme in the language for expressing the requirements of template arguments, the equivalent to the virtual base class in the object-oriented programming paradigm. This is left to the program documentation. Another disadvantage is that the flexibility is only available at compile time. Polymorphic lists at runtime cannot be implemented in this way.

In many places we follow in CGAL the generic programming paradigm to gain flexibility and efficiency. Important is the compliance with the STL to promote the re-use of existing generic algorithms and container classes, but – more important – to unify the look-and-feel of the design of CGAL with the C++ Standard. It is therefore easy to learn and easy to use for users familiar with the STL. In a few places we make use of the object-oriented programming paradigm, for example the protected access to the internal representation, see Section 9.

5 Design Goals for Polyhedral Surfaces

We define the concept *polyhedron* similar to STL container classes to be responsible of managing the items of a polyhedral surface and their combinatorial structure. We have identified the following design issues:

1. The edge-based data structures discussed in the previous section have a natural notion of the edges around a vertex or around a facet. It would be costly to provide iterators for these kind of circular sequences since the notion of ranges and the 'past-the-end' value do not extend naturally. We propose a concept similar to iterators – what we call *circulator* – for this kind of structure.

2. STL containers base their interface on iterators. For polyhedral surfaces the order of the stored items is not well-defined in certain situations, e.g. after Euler operations. Here we fall back on the concept of *handles*, which is the item-denoting part of iterators and ignore the traversal capabilities. In particular, any model of an iterator or circulator is a model of a handle.

3. The actual storage organization of the vertices, edges and facets influences the space and runtime efficiency. A doubly-connected list representation allows random insertion and removal while providing bidirectional iterators that enumerate all items. A more space efficient storage uses an STL vector which allows only the efficient removal of items at the end of the vector but provides random-access iterators. Other variants, like managing chunks of memory or simple allocation on the heap without any iterators over all items could be anticipated too.

4. The necessary incidence information might depend on the application. The minimum needed for traversals are `next()` and `opposite()` pointers. The `prev()` pointer can be simulated with a search around the vertex or facet. For triangulations this is still a simple expression, i.e. `prev()` $\equiv$ `next()->next()`, and for constant degree a constant time operation. If no information needs to be attached to vertices or facets, no storage should be allocated for them, including the referencing pointer in the edges. In its extreme the data structure reduces to an undirected graph.

5. It should be easy to add additional information to the different items, e.g. color to facets. Geometry will be attached using the same technique. Modifying one item should not hinder the re-use of the other items, for example, adding color to facets should not imply that a new vertex type must be declared.

6. The data structure should provide an easy-to-use high-level interface. This interface should protect the internal combinatorial integrity of the data structure as given in Definition 2.3. Advanced algorithms concerned with efficiency, e.g. a file format scanner, should be allowed to access the internal structure in a controlled fashion.

7. The management of connected components and containment relations, e.g. holes in facets or shells, is seen as an independent functionality with its own layer. Different solutions can be envisioned.

We concentrate on the combinatorial aspects of the polyhedral surface. Additional issues will appear when considering geometry, for example flexibility with the point type and the geometric predicates. One technique explored for this in CGAL is an extension of the traits classes [26] known from the C++ standard library, see [7, 9, 3].

6 Previous Work

The Library of Efficient Datatypes and Algorithms (LEDA) [22, 23] contains no data structure tailored for three-dimensional polyhedrons, but it provides a general data structure for graphs and one for planar maps derived from graphs. Additional information can be attached either by parameterized graphs or by node arrays and edge arrays. These are associative arrays (hash tables) which allows the easy addition of information even for temporary purposes. The disadvantage of the parameterized graph is that one must always specify both parameters. The disadvantages of the node and edge arrays are the additional costs for the lookup operations and additional storage requirements. A more subtle disadvantage is that a reference to a node is not sufficient to retrieve its associated attributes. The array must be known too. The current size of graph nodes is equivalent to 13 pointers and for halfedges it is 11 pointers. There is no flexibility for obtaining smaller graph structures. LEDA has its own notion of iterators and a rich body of algorithms for working on graphs. At the moment it is not compliant to the STL. LEDA is very homogeneous and easy-to-use. Its own framework is generic and flexible but (currently) monolithic when combined with other libraries like the STL. It does not reach the flexibility in runtime and space efficiency tuning achieved with the approach presented in this paper.

The Minimal Rendering Tool MRT uses a halfedge data structure to represent polyhedral surfaces [2]. It is implemented as a C++ class hierarchy and provides Euler operations to maintain combinatorial integrity. The internal representation is accessible at construction time and protected thereafter. No other access is granted. It separates geometry and topology except for vertices where a point is incorporated just at the combinatorial level for efficiency reasons. Flexibility and genericity are achieved with virtual member functions for geometric properties. No flexibility is available at the topological level. Facets are responsible of storing the ring of halfedges of their boundary. Summarizing, this approach leads to larger nodes for vertices, halfedges and facets and slower functions for geometric properties than the solution we developed.

7 Circulators

Our new concept of *circulators* reflects the fact that combinatorial structures often lead to circular sequences, in contrast to the linear sequences supported with iterators and container classes in the STL. For example polyhedral surfaces and planar maps give rise to the circular sequence of edges around a vertex or a facet. Implementing iterators for circular sequences is possible, but not straightforward, since no natural past-the-end situation is available. An arbitrary sentinel in the cyclic order would break the natural symmetry in the configuration, which is in itself a bad idea, and will lead to cumbersome implementations. Another solution stores, within the iterator, a starting edge, a current edge, and a kind of winding-number that is zero for the `begin()`-iterator and one for the past-the-end iterator[5]. No solution is known to us that would provide a light-weight iterator as it is supposed to be (in terms of space and efficiency). Therefore we introduced in CGAL the similar concept of *circulators*, which does allow light-weight implementations. The CGAL support library provides adaptor classes that convert between iterators and circulators, thus integrating this new concept into the framework of the STL.

Circulators share most requirements with iterators. Three circulator categories are defined: forward, bidirectional and random-access circulators. Given a circulator `c` the operation `*c` denotes the item the circulator refers to. The operation `++c` advances the circulator by one item and `--c` steps a bidirectional circulator one item backwards. For random-access circulators `c+n` advances the circulator by `n` where `n` is a natural number. Two circulators can be compared for equality.

Circulators develop different notions of reachability and ranges than iterators. A circulator `d` is called *reachable* from `c` if `c` can be made equal to `d` with finitely many applications of the operator `++c`. Due to the circularity of the data structure this is always true if both circulators refer to items of the same data structure. In particular, `c` is always reachable from `c`. Given two circulators `c` and `d`, the range `[c,d)` denotes all circulators obtained by starting with `c` and advancing `c` until `d` is reached, but does not include `d` if `d` $\neq$ `c`. So far it is the same range definition as for iterators. The difference lies in the use of `[c,c)` for denoting all items in the circular data structure, whereas for an iterator `i` the range `[i,i)` denotes the empty range. As long as `c != d` the range `[c,d)` behaves like an iterator range and could be used in STL algorithms. It is possible to write just as simple algorithms that work with iterators as well as with circulators, including the full range definition [16]. An additional test `c == NULL` is now required that is true if and only if the data structure is empty. In this case the circulator `c` is said to have a *singular value*. For the complete description of the requirements for circulators we refer to [16].

We repeat the example for the generic `contains` function from Section 4 for a range of circulators. The main difference is the use of a `do-while` loop instead of a `while` loop.

```
template <class InputCirculator, class T>
bool contains( InputCirculator c,
               InputCirculator d,
               const T& value)
{
    if ( c != NULL) {
        do {
            if ( *c == value)
                return true;
        } while (++c != d);
    }
    return false;
}
```

8 Design Overview

The global picture of the design is given in Figure 7. The design strictly separates topology and geometry. Vertices, halfedges and facets carry both kinds of information. The `Halfedge_data_structure` is the container managing these three items and their topological relations. The `Topological_planar_map` is a face-based representation. It maintains holes in facets and is able to enu-

[5]This is currently implemented in CGAL as an adaptor class which provides a pair of iterators for a given circulator.

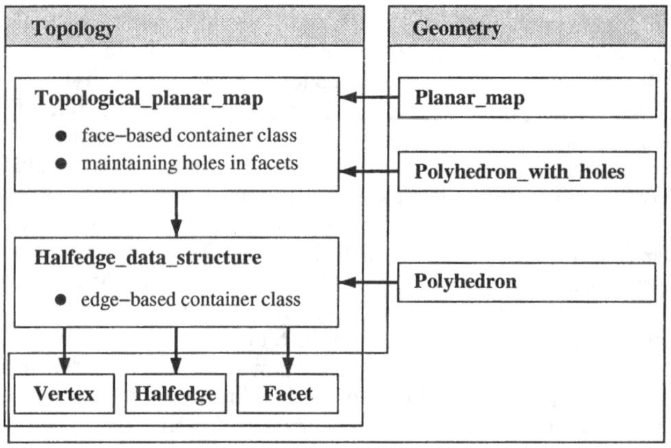

Figure 7: Design overview.

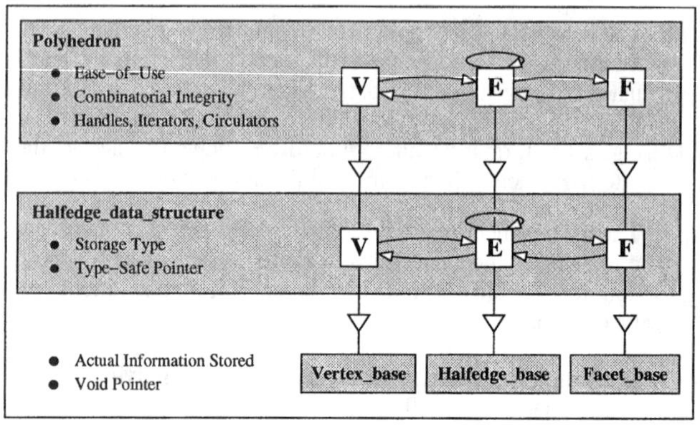

Figure 8: Responsibilities of the different layers in the design.

merate outer and inner boundaries of a facet. It uses the edge-based data structure. The `Polyhedron` uses the `Halfedge_data_structure` and adds geometric operations. It imposes further restrictions on the data structure as defined for the polyhedral complex above, for example, that an edge always has two distinct endpoints. The `Planar_map` and a possible `Polyhedron_with_holes` are based on the topological planar map since they will maintain holes in facets.

All entities in this picture are sets of requirements. Each can have multiple models. There are many different possibilities for vertices, edges and facets. Currently two different models are provided for the `Halfedge_data_structure`. Many combinations are possible and result in different polyhedron data structures. We make use of the implicit instantiation of template classes in C++ as described above. The requirement sets consist of a mandatory part that every model must comply with and certain optional parts a model must only comply with if the corresponding functionality is actually used. For example, a vertex is allowed to be empty. If we want to use it for the polyhedral surface then the normal vector computation imposes the additional requirements that the vertex must contain a three-dimensional point and must give access to it with the member function `point()`.

9 Design of the Polyhedron

A more refined picture of the design is shown in Figure 8. At the bottom we start with base classes for vertices, halfedges and facets. Their responsibilities are the actual storage of the incidences in terms of `void`-pointers, the geometry and other attributes. Especially the storage of incidences with `void`-pointers allows, for example, the facet class to be exchanged without changing the halfedge class. The advantage of strong type checking will be reestablished for the `void`-pointers in the next layer. Implementations for vertices, edges and facets are provided that fulfill the minimal set of requirements. They can be used as base classes for own extensions. Richer implementations are provided as defaults; for polyhedrons they provide a three-dimensional point in the vertices and a plane equation in the facets.

The `Halfedge_data_structure` is responsible of the storage organization of the vertices, halfedges and facets. Currently implementations are provided that use a bidirectional list or an STL vector internally. The `Halfedge_data_structure` derives new classes for vertices, halfedges and facets. They replace the `void`-pointer incidence information with type-safe pointers at the interface. Additional information besides the incidence information simply stays unaffected and will be inherited from the base classes.

For the `Halfedge_data_structure` different models are possible (two are already available). Thus the set of requirements for the `Halfedge_data_structure` is kept small. To support the implementation of high-level operations, a helper class `Halfedge_data_structure_decorator` is provided, which is not shown in Figure 8 but would be placed at the side of the `Halfedge_data_structure` since it broadens that interface but does not hide it. It adds Euler operations and adaptive functionality. For example, if the `prev()` function is not provided for halfedges, a `find_prev()` function searches the previous halfedge along the facet. If the `prev()` function is now implemented, the `find_prev()` function simply calls it. This distinction can be resolved at compile time with a technique called *compile-time tags*, similar to iterator tags in [31].

The `Polyhedron` layer adds ease-of-use in terms of high-level functions, high-level concepts for accessing the items, i.e. handles, iterators and circulators (pointers are no longer visible at this interface), and the protection of the combinatorial integrity. It derives new vertices, halfedges and facets to provide the handles and to hide the pointers.

Algorithms with invalid intermediate states need access to the internal representation. A protected access is granted for classes derived from `Modifier_base` following the strategy pattern [8]. The example in Figure 9 depicts the class design for a file format scanner for polyhedrons. The back-door provided here is a kind of callback-function embedded in a class object. The `Polyhedron` accepts a modifier class with the `delegate()` member function and calls its virtual `operator()` member function with the internal halfedge data structure. The `Scanner` class derives from the `Modifier_base` and implements the `operator()` function where it can access the internal representation. The achievement is here that the `delegate()` function of the `Polyhedron` can verify the validity of its own internal representation after the `operator()` function has done its work. The `Scanner` class is in charge of returning from execution only with a valid representation, even in the case of a failure. This approach is also known from database systems as transactions. The special task the `Scanner` accomplishes (only creation of new items) enables us to implement the transaction scheme efficiently with a simple rollback function that deletes all items created so far in the case of a failure. In general the rollback would be more costly.

10 Evaluation of the New Design

We will illustrate in the following that our design not only meets the design goals formulated in Section 5 but that it is also still easy to use. A certain familiarity with the look-and-feel of C or C++ will help in this section.

```
#include <CGAL/Cartesian.h>
#include <CGAL/Halfedge_data_structure_polyhedron_default_3.h>
#include <CGAL/Polyhedron_default_traits_3.h>
#include <CGAL/Polyhedron_3.h>

typedef CGAL_Cartesian<double>                                    R;
typedef CGAL_Halfedge_data_structure_polyhedron_default_3<R>  HDS;
typedef CGAL_Polyhedron_default_traits_3<R>                       Traits;
typedef CGAL_Polyhedron_3<Traits,HDS>                             Polyhedron;

int main() {
    Polyhedron P;
    P.make_tetrahedron();
    return 0;
}
```

Figure 10: Example program illustrating a default polyhedron instantiation in CGAL.

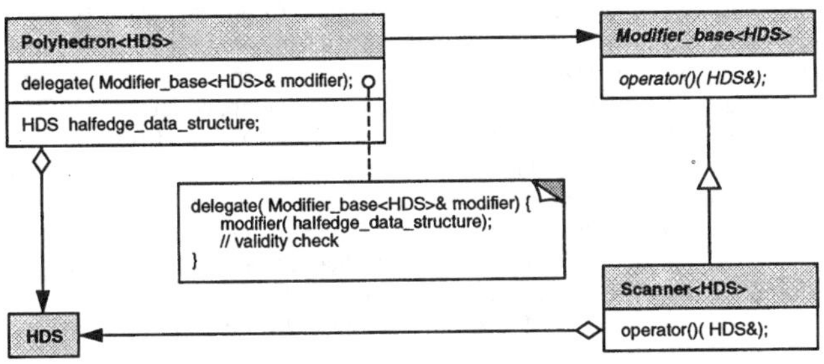

Figure 9: Class diagram of the polyhedron design illustrating the safe access to the internal representation using the strategy pattern.

We start with a complete program in Figure 10 using a default polyhedron instantiation in CGAL[6]. The `#include` directives provide the types used in the example. It is convenient to use `typedef`'s to create the nested type declarations one by one. CGAL supports different representation types; `CGAL_Cartesian` is one of them, parameterized with the coordinate type `double`. The default halfedge data structure for polyhedrons uses three-dimensional points for the vertices and a plane equation for the facets as determined by `R`. The `CGAL_Polyhedron_3` is parameterized with a traits class for the geometric parameterization and the halfedge data structure. The `main()` function declares a variable `P` for the polyhedron and creates a combinatorial tetrahedron in `P` where space for points and plane equations is reserved in the vertices and facets (even though the space is not used here). More information about representation classes and traits classes in CGAL can be found in [7].

The class `Polyhedron` provides the claimed ease-of-use with handles, iterators, circulators and high-level operations, thus addressing design goals (1) and (2). This default representation is actually equivalent to:

```
typedef  CGAL_Halfedge_data_structure_using_list<
    CGAL_Vertex_max_base< CGAL_Point_3<R> >,
    CGAL_Halfedge_max_base,
    CGAL_Polyhedron_facet_max_base<R> >     HDS;
```

The internal storage organization can be easily changed with the class `CGAL_Halfedge_data_structure_using_vector` to a vector-based one. This satisfies design goal (3). The requirements that a self-written class must fulfill to be a model of a halfedge data structure are documented in [17].

If we exchange the default base classes for the minimal base classes `CGAL_Vertex_min_base`, `CGAL_Halfedge_min_base` and `CGAL_Facet_min_base`, we get a data structure for undirected graphs; vertices and facets are empty (besides a few compile-time tags) and the minimal halfedge base class stores only a `next()` and an `opposite()` pointer. This makes four pointers per edge. See [17] for the actual short implementations of these base classes. In analogy to the quad-edge data structure we can replace the `opposite()` pointer internally by a single bit knowing that opposite halfedges are always stored in consecutive places by our halfedge data structures. Knowing C, this bit can be put into the least-significant bit of the `next()` pointer, which is always zero on todays systems. This yields an implementation with two pointers per edge! The easy realization of this idea in our design is shown in the manual [17].

We can add a color variable to the default facet:

```
template <class R>
struct Facet
: public CGAL_Polyhedron_facet_max_base<R> {
    CGAL_Color color;
};
```

This facet can be used instead of the default and we have a polyhedron with colored facets. This satisfies design goal (5). Similarly easy is the realization of data structure with incidences that lie between the minimal and maximal supported incidences. We add a previous pointer to the minimal halfedge in the following example. The type `Supports_halfedge_prev` indicates that the class now supports a previous pointer.

```
class Halfedge : public CGAL_Halfedge_min_base {
    void* prv;
public:
    typedef CGAL_Tag_true Supports_halfedge_prev;
    void*       prev()          { return prv;}
    const void* prev() const    { return prv;}
    void        set_prev( void* h) { prv = h;}
};
```

The whole spectrum of incidences is easily available in the design presented, fulfilling design goal (4). Design goals (6) and (7) are also met. The example for the easy use of the modifier mechanism and the transaction scheme for the scanner can be found in the manual [15].

Summing up, we have met all design goals presented in Section 5 without imposing runtime and storage overhead. Predefined implementations are easy to use and different solutions within the possible flexibility can be achieved with little effort.

[6] CGAL uses the prefix `CGAL_` for all global names, which will be replaced by a namespace, and the suffix `_3` for three-dimensional entities.

11 Conclusion

We have presented a design framework for combinatorial data structures, such as planar-maps and polyhedral surfaces. It extends to curved-surface environments and can also be applied to other combinatorial data structures, such as triangle-based structures for triangulations. We have identified important fundamentals for such a design: A proper definition of the modeling space, strong type checking, time and space efficiency.

The adaption of the generic programming paradigm used in the STL has led to an easy-to-use and flexible high-level interface for polyhedral surfaces featuring handles, iterators, the new concept circulators and Euler operators. The internal representation can be chosen from a wide range of different halfedge data structures exploiting many tradeoffs between time and storage efficiency, iterator categories and modifiability. Additional attributes are easy to add. Other solutions, such as dynamic type checking at runtime, generic attribute pointers or templates, can still be added within this design. The integrity of the internal representation is protected and a mechanism is available that grants safe access to it. We expect a continuation of this approach in CGAL.

Acknowledgments. The author wishes to thank Emo Welzl for the freedom and encouraging guidance throughout these studies, Nora Sleumer and Martin Will for reading and commenting on previous versions, and Andreas Fabri, Geert-Jan Giezeman, Stefan Schirra and Sven Schönherr for the intensive discussions on C++ design and software engineering since we started with the CGAL kernel almost three years ago. The CGAL project has grown and the author want to thank all members of the project for the inspiring environment, especially the people attending the CGAL meeting at INRIA, Sophia Antipolis, where the connection between the planar map and the polyhedral surface design has been discussed. Thanks also to Jean-Daniel Boissonnat for encouragement.

References

[1] Bruce G. Baumgart. A Polyhedron Representation for Computer Vision. In *National Computer Conference*, pages 589–596, Anaheim, CA, 1975. AFIPS.

[2] Heinzgerd Bendels, Dieter W. Fellner, and Sven Havemann. Modellierung der Grundlagen: Erweiterbare Datenstrukturen zur Modellierung und Visualisierung polygonaler Welten. In D. W. Fellner, editor, *Modeling – Virtual Worlds – Distributed Graphics*, pages 149–157, Bad Honnef/Bonn, 27.–28. November 1995. `http://www.graphics.uni-bonn.de/CompGraph.ResearchProjects.MRT`.

[3] Hervé Brönnimann, Andreas Fabri, Stefan Schirra, and Remco Veltkamp, editors. *CGAL Reference Manual. Part 2: Basic Library*. 1998. CGAL R1.0. `http://www.cs.ruu.nl/CGAL`.

[4] Working Paper for Draft Proposed International Standard for Information Systems – Programming Language C++. ANSI X3, Information Processing Systems, December 1996. `http://www.maths.warwick.ac.uk/c++/pub/`.

[5] Mark de Berg, Marc van Krefeld, Mark Overmars, and Otfried Schwarzkopf. *Computational Geometry: Algorithms and Applications*. Springer Verlag, 1997.

[6] Andreas Fabri, Geert-Jan Giezeman, Lutz Kettner, Stefan Schirra, and Sven Schönherr. The CGAL Kernel: A Basis for Geometric Computation. In M. C. Lin and D. Manocha, editors, *ACM Workshop on Applied Computational Geometry*, pages 191–202, Philadelphia, Pennsylvenia, May, 27–28 1996. Lecture Notes in Computer Science 1148.

[7] Andreas Fabri, Geert-Jan Giezeman, Lutz Kettner, Stefan Schirra, and Sven Schönherr. On the Design of CGAL, the Computational Geometry Algorithms Library. In Dorothea Wagner, editor, *Algorithm Engineering*, Trends in Software. John Wiley & Sons, New York, 1997. to appear.

[8] E. Gamma, R. Helm, R. Johnson, and J. Vlissidis. *Design Patterns – Elements of Reusable Object-Oriented Software*. Addison-Wesley, 1995.

[9] Geert-Jan Giezeman, Remco Veltkamp, and Wieger Wesselink. *Getting Started with CGAL*, 1998. CGAL R1.0.

[10] Andrew S. Glassner. Maintaining Winged-Edge Models. In James Arvo, editor, *Graphics Gems II*, pages 191–201. Academic Press, 1991.

[11] Leonidas Guibas and Jorge Stolfi. Primitives for the Manipulation of General Subdivisions and the Computation of Voronoi Diagrams. *ACM Transaction on Graphics*, 4(3):74–123, July 1985.

[12] Jed Hartman and Josie Wernecke. *The VRML 2.0 Handbook: Building Moving Worlds on the Web*. Addison-Wesley, 1996.

[13] Christoph M. Hoffmann. *Geometric and Solid Modeling – An Introduction*. Morgan Kaufmann, 1989.

[14] Michael Hoffmann. Line-Sweep auf einem Gitter, 1996. Diplomarbeit. Freie Univ. Berlin, Germany.

[15] Lutz Kettner. 3D-Polyhedral Surfaces. In Hervé Brönnimann, Andreas Fabri, Stefan Schirra, and Remco Veltkamp, editors, *CGAL Reference Manual. Part 2: Basic Library*. 1998. CGAL R1.0.

[16] Lutz Kettner. Circulators. In Hervé Brönnimann, Andreas Fabri, Stefan Schirra, and Remco Veltkamp, editors, *CGAL Reference Manual. Part 3: Support Library*. 1998. CGAL R1.0.

[17] Lutz Kettner. Halfedge Data Structure. In Hervé Brönnimann, Andreas Fabri, Stefan Schirra, and Remco Veltkamp, editors, *CGAL Reference Manual. Part 2: Basic Library*. 1998. CGAL R1.0.

[18] Lutz Kettner and Emo Welzl. Contour Edge Analysis for Polyhedron Projections. In Wolfgang Straßer, Reinhard Klein, and René Rau, editors, *Geometric Modeling: Theory and Practice*. Springer Verlag, 1997. (Proc. Int. Conf. Theory and Practice of Geometric Modeling in Blaubeuren, Germany, Oct. 1996).

[19] John Lakos. *Large Scale C++ Software Design*. Addison-Wesley, 1996.

[20] Stanley B. Lippman. *Inside the C++ Object Model*. Addison-Wesley, 1996.

[21] Martti Mäntylä. *An Introduction to Solid Modeling*. Computer Science Press, Rockville, Maryland, 1988.

[22] Kurt Mehlhorn and Stefan Näher. LEDA: A Platform for Combinatorial and Geometric Computing. *Communications of the ACM*, 38(1):96–102, January 1995.

[23] Kurt Mehlhorn, Stefan Näher, and Christian Uhrig. *The LEDA User Manual, Version 3.5*. LEDA Software GmbH, 66123 Saarbrücken, Germany, 1997.

[24] D. E. Muller and F. P. Preparata. Finding the Intersection of two Convex Polyhedra. *Theoretical Computer Science*, 7:217–236, 1978.

[25] David R. Musser and Atul Saini. *STL Tutorial and Reference Guide: C++ Programming with the Standard Template Library*. Addison-Wesley, 1996.

[26] Nathan C. Myers. Traits: a New and Useful Template Technique. *C++ Report*, June 1995.

[27] Mark H. Overmars. Designing the Computational Geometry Algorithms Library CGAL. In M. C. Lin and D. Manocha, editors, *ACM Workshop on Applied Computational Geometry*, Philadelphia, Pennsylvenia, May, 27–28 1996. Lecture Notes in Computer Science 1148.

[28] Mark Phillips. *Geomview Manual: Geomview Version 1.5*. The Geometry Center, University of Minnesota, October 1994.

[29] Silicon Graphics Computer Systems, Inc. Standard Template Library Programmer's Guide. `http://www.sgi.com/Technology/STL/`, 1997.

[30] E. Steinitz and H. Rademacher. *Vorlesung über die Theorie der Polyeder (unter Einschluß der Elemente der Topologie)*. Springer, Berlin, 1934.

[31] Alexander Stepanov and Meng Lee. The Standard Template Library. `http://www.cs.rpi.edu/~musser/doc.ps`, October 1995.

[32] Bjarne Stroustrup. *The C++ Programming Language*. Addison-Wesley, 3rd edition, 1997.

[33] Kevin Weiler. Edge-Based Data Structures for Solid Modeling in Curved-Surface Environments. *IEEE Computer Graphics and Application*, 5(1):21–40, January 1985.

[34] Josie Wernicke. *The Inventor Mentor: Programming Object-Oriented 3D Graphics with Open Inventor, Release 2*. Addison-Wesley, 1994.

Improved Algorithms for Robust Point Pattern Matching and Applications to Image Registration

David M. Mount* Nathan S. Netanyahu† Jacqueline Le Moigne‡

Abstract

Given two images of roughly the same scene, image registration is the process of determining the transformation that most nearly maps one image to another. This problem is of particular interest in remote sensing applications, where it is known that two images correspond to roughly the same geographic region, but the exact alignment between the images is not known. There are many approaches to image registration. We will consider an approach based on extracting a set of point features from each of the two images, and thus reducing the problem to a point pattern matching problem.

Because of measurement errors and the presence of outlying data points in either of the images, it is important that the distance measure between two point sets be robust to these effects. We will measure distances using the partial Hausdorff distance. An important element of image registration applications is that the search begins with *a priori* information on the bounds of transformation, and a good algorithm should be able to take advantage of this information.

Point matching can be a computationally intensive task, and there have been a number of algorithms and approaches proposed for solving this problem, both from theoretical and applied standpoints. One common approach is based on a geometric branch-and-bound search of transformation space and another is based on using point alignments to derive the matching transformation. The former has the advantage that it can provide guarantees on the accuracy of the final match, and that it naturally uses any *a priori* information to bound the search. The latter has the advantage of simplicity and speed.

We introduce a new practical approximation algorithm, which we call *bounded alignment*, for robust point pattern matching. This algorithm is a novel combination of these two approaches. Our algorithm operates within the framework of the branch-and-bound, but employs point-to-point alignments to accelerate the search. We show that this combination retains many of the strengths of branch-and-bound search, but provides significantly faster search times by exploiting alignments. We have implemented the algorithm and have demonstrated its performance on both synthetically generated data points and actual satellite images.

1 Introduction

In the image registration problem, we are given two images of roughly the same scene, and are asked to determine the transformation that most nearly maps one image to the other. This problem is of particular interest in remote sensing and medical imaging. Our interest in the problem stemmed from remote sensing applications, where it is known that two satellite or aerial images correspond to roughly the same geographic region, but the exact alignment between the images is not known. This is a heavily studied problem in the field of computer vision. See, for example, Brown's survey [Bro92] of image registration algorithms.

One common method for solving the image registration problem is through the extraction of a set of *feature points* from each of the images, and then finding the transformation that most nearly matches these point sets. Features could be control points, corners, line segments, etc., which have been extracted by of a number of standard procedures. We assume that the set of transformations is a subspace of the set of affine transformations, which includes translation, rotation, scaling along each of the axes, and shearing. This problem is called the *point pattern matching problem.* See the papers by Alt and Guibas [AG96] and Ton and Jain [TJ89], for a discussion of the complexity of feature-based methods.

Unlike general point pattern matching, one important element of image registration applications is that the search typically begins with *a priori* information on the bounds of transformation. This information is provided from bounds

*Department of Computer Science and Institute for Advanced Computer Studies, University of Maryland, College Park, Maryland. Email: mount@cs.umd.edu. The support of the National Science Foundation under grant CCR–9712379 is gratefully acknowledged.

†Center for Automation Research, University of Maryland, College Park, and Center of Excellence in Space Data and Information Sciences (CESDIS), Code 930.5, Space Data and Computing Division, NASA Goddard Space Flight Center, Greenbelt, Maryland. Email: nathan@cfar.umd.edu. The support of the Applied Information Sciences Branch (AISB), Code 935, NASA/GSFC, under contract NAS 5555-37 is greatfully acknowledged.

‡Universities Space Research Association/CESDIS, Code 930.5, Space Data and Computing Division, NASA/GSFC, Greenbelt, Maryland. Email: lemoigne@cesdis.gsfc.nasa.gov. The support of the AISB, Code 935, NASA/GSFC is greatfully acknowledged.

on the uncertainty in the exact position or direction of the imaging system. Thus, unlike model-based pattern matching applications, rather than searching large spaces for a small number of likely matches, in image registration the focus is on producing a single match with high reliability and accuracy.

Our focus is on producing practical problem formulations and algorithms of practical efficiency. Researchers at NASA/GSFC have been constructing a toolbox of registration programs and procedures [XMT+97], and one of our ultimate goals is to provide an efficient registration algorithm for this toolbox, based on the principals developed in the field of computational geometry.

1.1 Problem Formulation and the Nature of Errors

The point pattern matching problem can be defined abstractly as follows. We are given two points sets A and B lying in (conceptually at least) two different spaces. We are given a space $\mathcal{T}$ of transformations mapping one space to the other. We are also given some measure of the distance between any two point sets. The problem is to find the transformation $t \in \mathcal{T}$ that minimizes the distance between $t(A)$ and B. There are many issues that need to be resolved in order to reduce the image registration problem to the point pattern matching problem. As described by Brown [Bro92], these include the following elements:

Feature extraction: How are feature points to be extracted from the images, and what (if any) additional attributes are to be assigned to feature points?

Search space: What is the class of transformations to be considered in making the match?

Similarity measure: What is the measure of similarity or distance between two sets of points?

Search strategy: What search strategy is to be employed in finding the optimum transformation?

In this paper we will not consider the feature extraction issue in detail. Although this is a very important issue, it is beyond the scope of this study. However, there are two very realistic assumptions about feature extraction, which which will play a critical role in the rest of our development. The first is that the extracted feature points are subject to a *perturbation error*, which results from a combination of the image digitization process, expansion or shrinkage of objects due to variations in lighting conditions, and the vagaries of the feature extraction algorithm. Depending on the feature extraction algorithms used, the magnitude of this error is typically predictable (e.g., proportional to the size of a pixel in the image).

The second source of error is the presence of *outliers*, that is, feature points from either image that are simply not present in the other image. Outliers can result from many sources: the fact that the two images cover different regions, the presence of clutter or occluding objects in either image, the sensitivity of the feature extraction algorithm to variations in lighting, point of view, or other aspects of the imaging process. Unlike perturbation errors, the errors resulting from outliers can be unpredictably large.

1.2 Partial Hausdorff Distance

In this paper we will consider a well-known robust measure of similarity, called the *partial Hausdorf distance* [HR92, HR93, HKR93]. Consider the set of distances resulting by considering each point in one set, and then finding the nearest point in the other set. Rather than taking the sum or the maximum of these distances, which may be affected by outliers, we consider the median or generally the kth smallest distance. More formally, given two points sets A and B, and a parameter k, $1 \leq k \leq |A|$, define the *directed partial Hausdorff distance* from A to B to be

$$H_k(A,B) = K^{th}_{a \in A} \min_{b \in B} \mathrm{dist}(a,b),$$

where K^{th} returns the kth smallest element of the set, and where $\mathrm{dist}(a,b)$ is the Euclidean distance from point a to b. (Readers familiar with the work of Huttenlocher et al. should note that we reverse the roles of A and B.) The parameter k is typically based upon *a priori* bounds of the number of points of A that are expected to find a close match in B under the optimum transformation. These are the inliers.

It is often more natural to express k in terms of a quantile. For $0 < q \leq 1$, define $H_q(A,B)$ to be $H_k(A,B)$ where $k = \lceil q|A| \rceil$. If $q = 0.5$, then partial Hausdorff distance is a 50% breakdown-point estimator, since the quality of the final match cannot be affected by any subset of outliers of size less than $|A|/2$.

1.3 Prior Work

There have been a large number of papers written on the point pattern matching problem, arising from the fields of computer vision and pattern recognition, as well as from computational geometry. Most of the formulations of the point matching problem from computational geometry are not suitable for use in noisy, cluttered images either because they require exact matches [dRL95], they require 1–1 matches [AMWW88, HS94], or because they assume that every point from one set find a close match in the other set by assuming, say, the (standard) Hausdorff distance [AAR94, CGH+97, GMO94, HKS93]. Even under these relatively restrictive assumptions, the computational complexity can be quite high. For example, the best known algorithm for determining the translation and rotation that minimizes the Hausdorff distance between sets of sizes m and n runs in $O(m^3n^2\log^2 mn)$ time [AAR94, CGH+97]. This complexity may be unacceptably high for applications involving hundreds to thousands of feature points.

Numerous point matching algorithms have been also proposed and applied by researchers in the fields of computer vision and pattern recognition. These range from relaxation-based methods (e.g., [RR80, TJ89]) to cluster detection in transformation space (by computing point-to-point correspondences) [SKB82, GS85, GSP86], to hierarchical decomposition of transformation space coupled with the application of a robust similarity measure [HR93, HKR93, OH97, HV97]. Most of the techniques presented in these papers are computationally intensive (in a worst case theoretical sense), or could take a long time to run, in practice.

There are two robust distance measures that have been studied from an algorithmic perspective. The first is the method of partial Hausdorff distance matching, introduced by Huttenlocher, Klanderman and Rucklidge [HR92, HR93, HKR93]. The other is the recently introduced *absolute difference* by Hagedoorn and Veltkamp [HV97]. Both algorithms are based on a branch-and-bound search of transformation space, which we will discuss in greater detail below.

In an attempt to circumvent the high complexities of point pattern matching, some researchers have considered

Monte Carlo algorithms, which combine the use of sampling and transformations based on point-to-point alignments. The basic idea is that the images of three non-collinear points determine a unique affine transformation in the plane. Thus, by sampling a number of triples of points (or fewer, depending on the number of degrees of freedom in the transformation space), and enumerating all possible matching triples in the other point set, one is likely to encounter a good match. This is the basis of many matching algorithms, including those of Goshtasby and others, [SKB82, GS85, GSP86], Goodrich et al. [GMO94], and is the basis for methods based on geometric hashing [IR96, LW88].

1.4 Contributions

In this paper, we propose a new practical algorithm for robust point pattern matching, called *bounded alignment*, which we feel combines the stronger points of the branch-and-bound and alignment methods. Like alignment it is a Monte-Carlo algorithm, but the user can make the probability of failure arbitrarily small. The algorithm is based on the branch-and-bound search framework, but it uses point alignments to speed up the search. In particular, when the search reaches a point where a significant number of point-to-point correspondences can be inferred, it starts applying alignments to these pairs. Unlike the brute force alignment method described above, the correspondences are known, and so only a small number of alignments need be checked. (The number is dependent on the fractions of outliers, the dimension of transformation space, and the desired probability of failure, but not on the total number of feature points). We provide empirical evidence that with high probability, at least one these alignments will allow the search algorithm to "zoom-in" rapidly to the neighborhood of the optimum transformation. This acceleration can lead to significantly faster overall search times, especially when the final Hausdorff distance is small relative to the distances between the feature points.

2 The Branch-and-Bound Algorithm

Our registration algorithm is based on a geometric branch-and-bound framework. This framework was first developed by Huttenlocher and Rucklidge for performing approximate point pattern matching using the partial Hausdorff distance [HR93, Ruc95], and was also used by Hagedoorn and Veltkamp in their algorithm for performing approximate matching using the absolute difference similarity measure [HV97]. Our algorithm was developed independently of that Hagedoorn and Veltkamp, but the overall structure is quite similar. Because the bounded alignment algorithm is based on the branch-and-bound algorithm, we will describe the branch-and-bound algorithm here.

Recall that we are given two points sets A and B and a space of transformations $\mathcal{T}$. We are also given a distance quantile q, $0 < q \leq 1$. The problem is to find $t \in \mathcal{T}$ that minimizes the generalized Hausdorff distance $H_q(t(A), B)$. Since A and B will be fixed for the remainder of the discussion, let us call this $sim_q(t)$, for the similarity measure of t. (Note that the better the match, the smaller the similarity measure.) Let t_{opt} denote this optimum transformation, and let sim_{opt} be the optimum similarity.

2.1 Transformation Space

There are a number of different ways to describe the space $\mathcal{T}$ of transformations. In our implementation we have assumed that $\mathcal{T}$ is a bounded subset of affine transformations. Assuming that the point sets A and B are both in 2-dimensional space, any affine transformation t applied to a point $a \in A$ can be expressed as a linear transformation M followed by a translation c:

$$t(a) : Ma + c,$$

where a and c are 2-element column vectors, and M is a 2×2 matrix. Since there are six parameters needed to define the transformation, this naturally leads to a 6-dimensional *transformation space*. The algorithm can be applied to any reasonable parameterization of this space.

2.2 Approximation Error

There are three natural ways to define the approximation error. It may be suboptimal by some relative error factor, it may be suboptimal by some absolute additive error, and it may be less robust than the optimum, by using a somewhat smaller quantile in the Hausdorff distance. Rather than define three separate approximation problems, we provide three nonnegative parameters which allow the user to control of these tradeoffs:

- ϵ_r, the relative error bound,
- ϵ_a, the absolute error bound,
- ϵ_q, the quantile error bound.

Define $q' = (1 - \epsilon_q)q$, to be the *weak quantile*. Note that since $q' \leq q$, we have $sim_{q'}(t) \leq sim_q(t)$, for any $t \in \mathcal{T}$. We say that a transformation t is *approximately optimal* relative to these parameters if either of the following *approximate correctness conditions*:

$$\begin{aligned} sim_{q'}(t) &\leq (1+\epsilon_r)sim_{opt} \qquad \text{or} \\ sim_{q'}(t) &\leq sim_{opt} + \epsilon_a. \end{aligned}$$

Thus, the approximate solution is slightly less robust, in that it considers only the weak quantile rather than the true quantile, and may exceed the optimum similarity by a relative error of ϵ_r or an absolute error of ϵ_a. Later we show that as long as $\epsilon_a > 0$ or $\epsilon_r sim_{opt} > 0$, then termination is guaranteed. The branch-and-bound search of [HR93] and [HV97] implicitly provide for ϵ_a but not the other two parameters.

2.3 Overview

As mentioned earlier, the algorithm is based on a geometric branch-and-bound search of transformation space. We construct a search tree, where each node of this tree is identified with the set of transformations contained in some axis-aligned hyperrectangle in the 6-dimensional transformation space. These hyperrectangles are called *cells*. Each cell T is represented by a pair of transformations, (t_{lo}, t_{hi}), whose coordinates are the upper and lower bounds on the transformations of the cell. Any transformation whose coordinates lie between the corresponding coordinates of t_{lo} and t_{hi} lies in this cell.

The algorithm begins with an initial cell, which is assumed to contain the optimum transformation. This is supplied by the user, based on *a priori* knowledge of the nature of the transformation (e.g., bounds on the possible translation and rotation). A cell is either *active*, meaning that it is a candidate to provide the optimum transformation, or *killed*, meaning that it cannot provide the optimum solution. At any time the union of the active and killed cells forms a cover of the initial cell into subcells with disjoint interiors. The algorithm proceeds by selecting one of the active cells to be *processed*. After processing, a cell is either killed, or it is split into two disjoint subcells, each of which is made active. We will discuss termination conditions later.

Let us consider the elements of the algorithm in somewhat greater detail. For each cell T that we process, we are interested in the transformation of this cell that has the smallest similarity measure. We compute an upper bound $sim_{hi}(T)$ and a lower bound $sim_{lo}(T)$ on this smallest similarity (discussed below). For each upper bound, there will be a transformation that serves as a witness to this upper bound. The algorithm maintains the best similarity sim_{best} of that it has encountered thus far in the search, and the associated transformation t_{best}. To process a cell, we first compute its lower bound. If the lower bound exceeds sim_{best} then we kill this cell. Otherwise, we compute the upper bound, and if it is less than sim_{best} then we update the the current best.

To compute the upper bound, we may sample any transformation from within the cell. In our implementation this is done by simply taking the midpoint t of the cell, and then computing $sim_q(t)$. In particular, this involves computing the image of each point of A under t, and then computing the nearest neighbor in B for each image point. Since the cell's optimum similarity is smaller than any sampled point, this is an upper bound on the minimum similarity of the cell. Nearest neighbors are computed by storing the points of B in a kd-tree data structure, and applying known efficient search techniques [FBF77, AM93, AMN⁺94].

To compute the lower bound we use a similar technique as described in [HR93] and [HV97]. Given any cell $T \subset \mathcal{T}$, and given any point $a \in A$, consider the image of a under every $t \in T$. It is an easy matter to compute a bounding rectangle for this set. For example, if the coordinates of a are nonnegative, then this is the rectangle defined by the corner points $t_{lo}(a)$ and $t_{hi}(a)$. (The general formula involves a case analysis on the signs of the coordinates of a.) We call this bounding rectangle the *uncertainty region* of a relative to cell T. (This is essentially the same concept as the traced volumes introduced by Hagedoorn and Veltkamp.) In this way, each cell is associated with a collection of uncertainty regions, one for each point of A. (See Fig. 1 for an illustration of uncertainty regions.)

Define the distance between an uncertainty region and a point $b \in B$ to be the minimum distance between b and any part of the uncertainty region. If b lies inside the uncertainty region, then the distance is zero. To derive our lower bound for T, we compute the distances from each uncertainty region to its closest point in B, and take the qth smallest such distance. Call this $sim_{lo}(T)$. This can be done by a straightforward generalization of the kd-tree-based nearest neighbor method described above [AM95].

Observe that the image of a point $a \in A$ under any transformation $t \in T$ lies within the corresponding uncertainty region, and hence this distance is a lower bound on the distance from $t(a)$ to its nearest neighbor in B. This implies that $sim_{lo}(T) \leq sim_q(t)$, for any $t \in T$. In other words, this is indeed a lower bound on the optimum similarity in this cell.

Figure 1: Uncertainty regions. The points of A are shown in white, each transformed by the midpoint of the cell, the points of B are shown in black, and the uncertainty regions are shown as rectangles.

2.4 Cell Processing

Next we consider how cells are processed. We split each cell into two subcells (rather than $2^6 = 64$ subcells as in [HR93]), and choose the cell to be processed and the splitting dimension so as to reduce the size of the uncertainty region as much as possible. In particular, define the *size* of an uncertainty region to be its longest side. Define the *size* of a cell to be largest size among the associated uncertainty regions for each point in A. (Note that cell size is *not* defined in terms of the size of the hyperrectangle in transformation space.) Our implementation computes a fast upper bound on this size, by first computing the bounding box for the points of A, and then computing the maximum size of the uncertainty region generated by any point in this bounding box.) The active cells are stored in a priority queue, ordered by cell size. The largest cell, that is, the cell with the largest uncertainty region, is chosen for processing at each stage.

If a cell survives processing, then it is split along one of its sides into two smaller cells. Consider a cell T defined by a lower and upper bound range (t_{lo}, t_{hi}). To determine which side to split, we determine which of the six components of the transformation range contributes most to the size of the uncertainty region, and then we split the cell through its midpoint along this dimension. This is explained in the full version of the paper.

Lemma 1

(i) *After six consecutive splits, the size of a cell (that is, the maximum side length of its largest uncertainty region) decreases by at least half.*

(ii) *If there are currently m active cells, then after at most $64m$ instances of cell processing, the size of the largest cell will decrease by at least half.*

2.5 Approximation Algorithm

This completes the description of the exact algorithm. We modify this to an approximation algorithm, as follows. First, we compute the upper bound $sim_{hi}(T)$ on the similarity of a cell using the weak quantile q', rather than the true quantile q. Thus we have $sim_{hi}(T) \geq sim_{q'}(t)$, for any $t \in T$. Second, the condition for killing a cell is not that the cell's lower bound be greater than the current best, but that it not be significantly lower than the current best. Given a cell T with lower bound $sim_{lo}(T)$, it is killed if either:

$$sim_{lo}(T) > \frac{sim_{best}}{1+\epsilon_r} \qquad \text{or} \qquad sim_{lo}(T) > sim_{best} - \epsilon_a.$$

The algorithm terminates when all cells have been killed or when $sim_{best} < \epsilon_a$.

The correctness of the algorithm is established by the following theorem.

Theorem 1 *Assuming that $\epsilon_a > 0$ or $\epsilon_r sim_{opt} > 0$, then the this approximation will eventually terminate, and upon termination t_{best} satisfies the approximate correctness conditions given above.*

3 Bounded Alignment

There are many nice features of the branch-and-bound algorithm, but its main drawback is its relatively high running time. This occurs especially when high accuracy is required and the optimum similarity is very good. The reason for this is that the algorithm must decompose a (typically 2–6 dimensional) search space into very small cells in order for the uncertainty regions to be small enough that the lower bounds are good enough to kill unpromising cells.

For this reason, we introduce an additional process called *bounded alignment* to help accelerate the search. Throughout this section we assume that the transformation space is the 6-dimensional space of affine transformations. Generalizations to other transformation spaces are straightforward.

3.1 Intuition

Before discussing what bounded alignment is, let us consider why the approximation algorithm may run slowly, and how it could be sped up. It is argued in the proof of Theorem 1 that the algorithm terminates when the cell sizes decrease to around $\max(\epsilon_a, \epsilon_r sim_{opt})$. If this quantity is small compared to the initial uncertainty bounds, then the algorithm may require a great deal of splitting before achieving this size. Of course, not all cells need to be split to this small size, but at the very least, observe that if the uncertainty region of a cell contains even a single point of B, then this region contributes a distance of zero to the set of distances used in computing the lower bound for this cell. Thus to even get a nonzero lower bound, a significant fraction (at least q) of uncertainty regions must contain no points of B. As cell sizes shrink, before we arrive at a stage in which many cells have nonempty uncertainty regions, we will first reach a stage where many cells have uncertainty regions with at most a single point in their uncertainty regions. When this happens, we can determine that there is a unique corresponding matching point for each of these uncertainty regions. It is at this point that alignment can be applied.

To illustrate the idea, consider the example shown in Fig. 2 below. To simplify things, consider the case of translation alone, let $k = 4$ ($q = 0.5$), and suppose that there is a match with $H_4 = 0$. There are 8 points in A (shown as white circles), and there is translation in which at least 4 of them match (actually 5) match exactly.

On the left we show an example where the cell contains the optimum transformation. Suppose that the search has progressed to a stage where most of the uncertainty regions associated with the points of A contain at most one point of B. Consider the points of A that have exactly one point of B in their uncertainty region. We sample one such point at

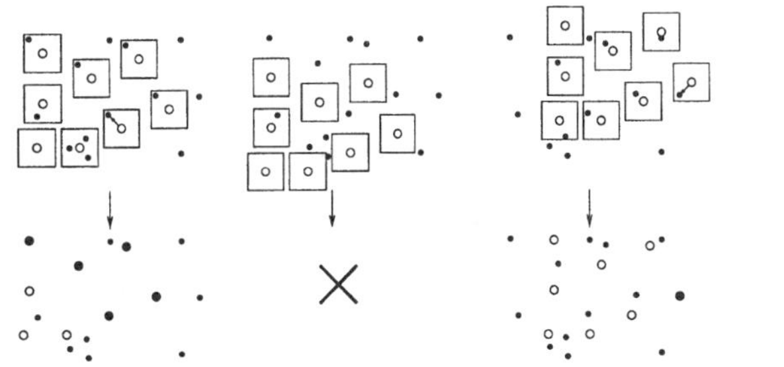

Figure 2: Alignment applied to three cells in transformation space. Points of A are indicated in white and points of B in black.

random and compute the unique transformation that maps this point to the corresponding point of B (indicated with an arrow in the figure). There is a good probability that the point of A that we picked is an inlier (at least 1/2), implying that the point B in its uncertainty region will be its matching point. The resulting transformation is the desired optimum (or in the case of errors should be close to the optimum).

On the other hand, if we consider a cell that does not contain the optimum (or any transformation close to the optimum) then consider two cases. In the middle of the figure we show a case in which most of the uncertainty regions contain no matching point of B. In this easy case, even the branch-and-bound algorithm will report that no match exists and kill the cell. However, consider the case shown on the right. Here a majority of uncertainty regions contain a point of B. As a result the lower bound associated with this cell will be zero, and it is not possible for the branch-and-bound algorithm to distinguish this cell from the optimum cell on the left at this stage. However, if we apply the same sampling that we did to the cell on the left (one is shown with an arrow in the figure), then we would expect to find that every sample returns a rather poor similarity measure. After making a number of samples and seeing repeatedly poor similarities, we may take this as evidence that this cell does not contain the optimum and kill it. This is the intuitive basis of bounded alignment—using the bounds provided from the branch-and-bound framework to guide the alignment process (rather than simply enumerating all possible alignments as in [GMO94]), and then using the results of these alignments to guide the branch-and-bound search.

3.2 Alignment

Let us think of the points of A as being partitioned into two classes, *inliers*, the points that are mapped by the optimum transformation to lie within distance sim_{opt} of their nearest neighbor in B, and *outliers*, all of the remaining points of A. By the definition of sim_{opt} as a q-quantile Hausdorff distance, at least a fraction q of points of A are inliers. (Normally the concept of an inlier is based on the underlying point distribution, but this functional definition is sufficient for our purposes.) To simplify the explanation, suppose for the moment that the inliers are completely free of any error, that is, the q-quantile Hausdorff distance is zero for the optimum transformation, t_{opt}. If we were able to find three inliers of A and match them against the corresponding elements of B, and further — if these three points of A were affinely independent (i.e., not collinear) — then these three point correspondences would uniquely determine the affine transformation that achieves a Hausdorff distance of zero. The process whereby triples from A are matched against prospective corresponding triples from B in order to determine a candidate transformation is called *alignment.*

Of course, in noisy environments it is unreasonable to assume that points could be matched free from any error or that the numerical extraction of the affine transformation would be error free. To generalize this to noisy environments, suppose that for each inlier $a \in A$ there is a point of B that lies within some small distance σ of its optimum image $t_{opt}(a)$. We assume that an upper bound on σ is provided to the search algorithm. In practice for registration applications, such an estimate would be based on knowledge of the accuracy of the feature extraction process. More generally, it could be determined through binary search. The final Hausdorff distance will be no larger than σ. If σ is small relatively to the size of the image, and if the three points are "geometrically well distributed", then we would expect that the match that results from alignment should be fairly close to the optimum. Intuitively, a geometrically well distributed triple should define a large, "fat" triangle among the points of A. To make this more formal, we present the following definition.

Definition: Given a set of points A in the plane and $\alpha \geq 1$, a triple $\{a_1, a_2, a_3\} \subseteq A$ is *well-distributed* relative to α if every point $a \in A$ can be expressed as an affine combination $a = \alpha_1 a_1 + \alpha_2 a_2 + \alpha_3 a_3$, such that, $|\alpha_i| \leq \alpha$.

Given this definition, the following lemma follows easily from the triangle inequality. It analogous to bounds proven in [GMO94].

Lemma 2 *Given a planar point set A, a well-distributed triple $\{a_1, a_2, a_3\}$ relative to some $\alpha > 1$, and an affine transformation t that maps each of the points a_i to a point that lies within distance σ of a_i, then for any point $a \in A$, the distance from $t(a)$ and a is at most $6\alpha\sigma$.*

This lemma implies that if alignment is applied to a well-distributed triple of points and if the images of the aligned points are perturbed by a small amount σ, then the perturbation induced by the resulting transformation will be proportional to σ.

In general, the number of subsets that may have to be tested may be quite large. However, branch-and-bound search provides a particularly simple way of identifying the most promising correspondences to try. In particular, as the search proceeds and cells are split, the search arrives eventually at a stage at which a significant fraction of uncertainty regions will contain at most a single point of B. An uncertainty region is said to be *"alignable"* if there is at most one point of B in the region, or if the region is empty and there is at least one point of B within distance σ of the region. If the current cell has a significant fraction of alignable uncertainty regions, then we say that this cell is *eligible for aligning.* (See Fig. 3.)

Consider the cell containing the optimum transformation. If it is eligible for aligning, then for each of its alignable uncertainty regions, there is a single obvious candidate for the corresponding point of A to be mapped to. That is, if the uncertainty region is nonempty, then there is a point of B lying within the uncertainty region, and otherwise it is the nearest neighbor of the uncertainty region. Further, if $a \in A$ is an inlier, then the nearest neighbor in B should lie within distance σ of a, and hence within σ of the uncertainty region.

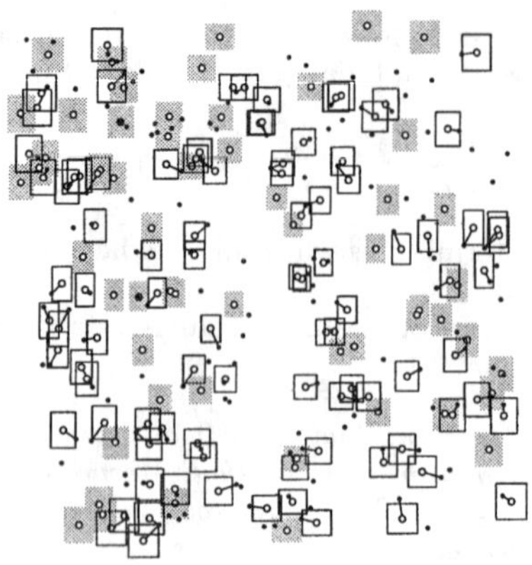

Figure 3: Alignable uncertainty regions. The transformed points of A are shown in white, points of B are shown in black. Alignable uncertainty regions are shown with solid lines, and unalignable uncertainty regions are shaded. For each alignable uncertainty region, a line segment joins the associated point of A with its nearest neighbor in B.

To apply aligning we need to make the assumption that the uncertainty regions of outliers are not more likely to be alignable than inliers. (Or at least, if this assumption is violated, then we should have a bound on the difference in the probabilities.) Under this assumption, if we randomly sample three alignable uncertainty regions of A (without replacement), then with probability roughly q^3, all three points will be inliers. Thus, the three respective nearest neighbors of B will be within distance σ from each of these points. Further, it follows from Lemma 2 that if these points of A are well-distributed, then the aligning transformation that results from considering this correspondence will be a reasonable approximation to the optimum, in the sense that the Hausdorff distance should not exceed the optimum Hausdorff distance by an amount greater than a constant c times σ (as determined by Lemma 2).

Thus, under the above assumptions, and depending on q, if we take a sufficiently large number N of random samples, then the probability of generating a aligning transformation whose distance from optimum is more than $c\sigma$ is at most $(1-q^3)^N$. (To see this, observe that $(1-q^3)$ is the probability that all three sampled points of A are outliers, and for failure this event must be repeated N times.) To reduce the chances of failing to find such a good triple to some small failure probability, p_f, the number of well-distributed samples should satisfy

$$N(q, p_f) \geq \frac{\ln p_f}{\ln(1-q^3)}.$$

For example, to reduce the probability failure below 0.01, and for $q = 0.5$, roughly 35 sampled triples are sufficient. Note that this constant, while large, is independent of the number of feature points in either of the images.

If on the other hand, after $N(q, p_f)$ well-distributed sampled triples we find that all exceed the current best similarity by at least $c\sigma$, then we conclude that this cell does not contain the optimum transformation, and kill it. This is the basis of aligning. Observe that this is a Monte Carlo process, in the sense that with probability p_f we may erroneously kill the cell containing the optimum transformation.

3.3 Processing Alignments

Now let us describe the aligning process in greater detail. First, we need make the following assumption upon which aligning relies.

Alignment Assumption: For a cell containing the optimum transformation, the probability that an uncertainty region associated with this cell is alignable, and the probability that a alignable triple is well-distributed is not greater for outliers than for inliers.

As we mentioned above, it is not necessarily realistic to require that this assumption holds. But if it is possible to estimate the difference in the conditional probabilities for inliers and outliers, and these probabilities are not significantly different, then the number of samples derived below can be modified accordingly.

We can incorporate alignment into the existing branch-and-bound approach as follows. We supply the algorithm with three additional pieces of information, i.e., the expected perturbation σ between inliers and their corresponding points in the final match, and two quantities called the sampling quantile q_s and a minimum sample size N_s. First, we augment the procedure that computes the distance from each uncertainty region to its nearest neighbor to compute also the number of points of B that lie within the uncertainty region. This can be done by modifying the kd-tree nearest neighbor algorithms [AM95]. If this number is at most one, and the nearest neighbor is within σ of the uncertainty region, then we flag this region as being alignable.

After processing all the uncertainty regions, if the fraction of alignable uncertainty regions exceeds q_s, then we declare that the cell itself is eligible for alignment. If so, we sample N_s triples of points from A, such that each triple is well-distributed (relative to some given constant). We generate the aligning transformation mapping each sampled triple to the nearest neighbor of the corresponding uncertainty region, and compute the similarity of this transformation. If the similarity of all the N_s transformations exceeds the current best sim_{best} by an additive amount of $c\sigma$, then we kill this cell. Otherwise, the cell remains active, and the normal processing resumes.

Since we test the similarity of each sampled transformation, if we find that any sampled transformation is better than the current best (and the transformation satisfies the additional allowability conditions), then we take this transformation to be the current best. Thus alignment not only helps in improving lower bounds, but also helps in generating better upper bounds.

Our implementation differs slightly from this description. We do not check that the points are well-distributed. Instead we apply a somewhat more precise test. We generate the aligning transformation and then test whether this transformation lies within or is sufficiently close to the current cell. The definition of "sufficiently close" is that the translational component should be within some fixed constant factor of σ, and the entries of the matrix M should be within a constant factor of the ratio of σ to the diameter of the point set A. Our reasoning is that points do not need to be well-distributed, as long as they generate a transformation that is reasonably "true" to the cell. Because of its Monte Carlo nature, the danger in alignment is that the cell containing the optimum may be killed accidentally. If the current cell contains the optimum transformation, then we expect that the transformation generated by a triple of inliers should either lie inside or close to the cell. If not, then we conclude that the triple is unacceptable. Because it is easy to test this condition (once the transformation has been generated), we generate triples until this condition is satisfied. If we fail after a repeated number of tries (10 in the current implementation), then we treat this sample as a failure.

3.4 Analysis

Under what circumstances does alignment offer an advantage over pure branch-and-bound? At this point we do not have a theoretical analysis or even an exhaustive empirical analysis explaining when this happens. However, based on our empirical experiences with the algorithm on uniformly distributed point sets (see Section 4), in many instances alignment can offer significant improvements in running time over the pure branch-and-bound search. In particular, we have observed that alignment seems to offer the greatest advantage when the deviation in the location of feature points σ is relatively small with respect to the expected distance δ from a random point and its nearest neighbor in B.

We provide some informal justification for this empirical observation. Assume that the points of B are uniformly distributed, and for simplicity suppose that the center of each uncertainty region is sufficiently random that the expected distance to the nearest point of B is δ. When the sizes of the uncertainty regions decrease to roughly δ, then alignment becomes possible, because many uncertainty regions of the cell containing the optimum transformation should contain at most one point of B. Also note that at this point, the current best similarity is expected to be proportional to δ. This is because we cannot expect to locate the optimum much more precisely than the cell size (unless we just happen to be very lucky in sampling a transformation). However, by the reasoning made earlier in this section, once we reach this stage, and alignment can be applied to the optimum cell, we expect to discover a transformation whose similarity is not greater than $sim_{opt} + c\sigma$. Recall that $sim_{opt} \leq \sigma$, and so sim_{best} is now proportional to the optimum. Thus, if the ratio δ/σ is large, then the alignment algorithm gains a big advantage as the tighter upper bound makes it possible to kill more unpromising cells.

Also consider the conditions needed to kill a nonoptimal cell. Assuming, as before, that sim_{best} is proportional to δ at this stage of the search, then (considering just the relative error) a cell T cannot be killed until

$$sim_{lo}(T) > \frac{sim_{best}}{1+\epsilon_r} \geq \frac{\delta}{1+\epsilon_r}.$$

The distance of an uncertainty region to its nearest neighbor in B is expected to be at most δ minus the radius of the uncertainty region. It follows from simple algebra that if the cell's size is larger than $2\epsilon_r\delta$, then this condition is not likely to be satisfied by most of the cell's uncertainty regions. But alignment already applies when the cell's size is roughly δ. Thus the pure branch-and-bound approach must continue to split the cell until its size is reduced by an amount proportional to $2\epsilon_r$ before it can be killed. By Lemma 1, this may require $6\log_2(1/2\epsilon_r)$ additional splits, and could result in the generation of $(1/2\epsilon_r)^6$ additional cells. For even moderately small values of ϵ_r, this value may be significantly larger than the N_s samples generated by alignment, and hence alignment may offer a significant improvement over pure branch-and-bound.

4 Empirical Study

We have implemented search algorithms for both standard branch-and-bound search, as well as branch-and-bound with bounded alignment search in C++. We also implemented a driver program to run a series of tests, and to gather statistics about the progress and performance of the two algorithms. Nearest neighbor and range queries were performed using kd-trees as generated by the **ANN** approximate nearest neighbor library [MA97]. The driver program can either read input sets from a file or generate inputs randomly from a number of distributions.

We ran experiments on both real satellite images, to test the general performance of our algorithms, and on synthetically generated data sets to test the relationship between branch-and-bound both with and without bounded alignment.

4.1 Synthetic Experiments

For the synthetic experiments, the point sets were generated as follows. First the user specifies the number of points in the sets A and B, bounding rectangles for A and B, the desired fraction of inliers of A, and the standard deviation σ of a random perturbation to be applied to each image point. (Here σ is not the same as the upper bound σ of the previous section, although the two are related.) Then the program generates an affine transformation t at random from a collection of bounds on the x and y translation, rotation, x and y scale, and shearing. The program then generates the desired numbers of uniformly distributed inliers (with Guassian perturbations) and outliers. The given transformation t is called the *target transformation.* It will generally not be the optimum transformation (unless $\sigma = 0$), but should be close enough to the optimum that we may use it for validating the results of the experiments.

The user supplies the desired quantile q for the Hausdorff distance, the error factors ϵ_r, ϵ_a, and ϵ_q, and the bounds on the initial cell. If alignment is being applied we also supply two additional parameters, q_s, the fraction of alignable uncertainty regions needed invoke aligning for a cell, and N_s, the number of well-distributed triples to sample from the alignable regions. The program can be run using either the standard branch-and-bound algorithm or using bounded alignment. It is also possible to select from two different parameterizations of transformation space. The first is full 6-dimensional space (although lower dimensional searches are possible by allowing some of the parameter ranges degenerate to width 0). The other allows searches of 3-dimensional space of rigid transformations by specifying a range of angles and two ranges of translations.

The program measures a number of statistics on its own performance, including CPU seconds, the number of cells expanded, and the number of times the Hausdorff-based similarity was computed. Statistics are also gathered on a level-by-level basis for the search tree. In addition, the program computes a quantity called the *average error.* Normally, this should be done by comparing the algorithms performance against the optimum, but since we do not know the optimum transformation, we measured the error against the target transformation, which was used to generate the data sets. The algorithm computes the similarity of the target transformation sim_{targ} and compares it against the similarity returned by the program sim_{best}. Both are computed relatively to the weak quantile $q(1 - \epsilon_q)$. The average error is defined to be the minimum of the relative error of sim_{best}

$$e_r = \frac{sim_{best} - sim_{targ}}{sim_{targ}},$$

and the absolute error

$$e_a = sim_{best} - sim_{targ}.$$

Note that because sim_{targ} is not the optimum, it is actually possible for the algorithm to return a (typically small) negative average error. For the branch-and-bound algorithm, the average error cannot exceed $\max(\epsilon_r, \epsilon_a)$. Since the bounded alignment algorithm is a Monte Carlo algorithm this might happen in the unlikely event that the optimum cell was killed.

We conjecture that bounded alignment should perform best when the Hausdorff distance is small relative to the nearest neighbor distance, since in this case the best alignments are generated. To test this we generated points with varying degrees of perturbation error, keeping the number of points and the number outliers fixed. The perturbation errors tested were $\sigma \in \{0.1, 0.2, 0.5, 1, 2, 5\}$. We generated 300 points for the set A uniformly distributed in a square $[-400, 400]^2$ of which 180 were inliers. We applied to these points a random rotation in the interval $[40, 45]$ degrees, and a random translation in $[-10, 10]^2$. We mapped 180 inliers of A under this transformation and added a Gaussian noise of standard deviation σ to each x- and y-coordinate. We generated an additional 280 outlying points for the set B at random in a square $[-500, 500]^2$. (Different bounding squares were used to minimize boundary effects.) We fixed $\epsilon_q = \epsilon_r = 0.2$, and computed the median Hausdorff distance. (Out of the roughly 70 experiments run, the largest observed error relative to the initial transformation was only 10%, and most of the errors were less than 2%.) The absolute error ϵ_a was set equal to σ, since this is a lower bound on the expected Hausdorff error. The initial search interval was a 3-dimensional rigid transformation with 10 degrees of variation in rotation and 40 units of translation is each of x and y. Both branch-and-bound search and bounded alignment were invoked 10 times for value of σ, and averaged. Experiments were compiled using `g++ -O3`, and run on a Sun SPARC 5, running SunOS 5.5.

From the graphs of Fig. 5 observe that bounded alignment offered a significant advantage over standard branch-and-bound both in terms of CPU time and the number of cells processed when the perturbation error is small, implying that the final Hausdorff distance is small as well (typically $\sqrt{2}\sigma$). When the Hausdorff distance was large (implying that there is no obvious strong match) both algorithm perform roughly equally, but both are quite efficient in this case. Thus, bounded alignment performed well in near exact match situations where alignment would do well, and also in poor match situations where branch-and-bound would do well. (Note that the plots are on log-log scale.)

To get a closer insight into this behavior, we computed the number of cells processed and the average best similarity for each level of the search tree for each value of σ. We show the case for a relatively good match at $\sigma = 0.2$. Observe that, as expected, bounded alignment finds transformations with smaller Hausdorff distances earlier in the search than does standard branch-and-bound. As a result of this and the ability to do better pruning on cells, a smaller number of cells is processed per level of the tree. (Note that the y-axis is on a log scale.)

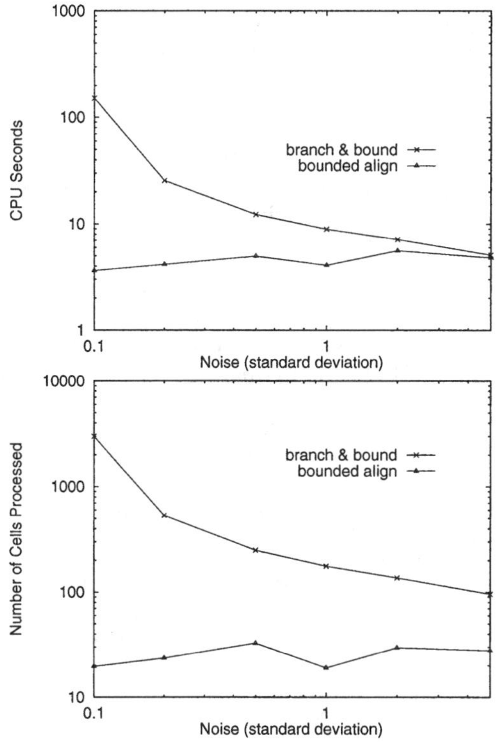

Figure 4: Execution time and number of cells processed for pure branch-and-bound and bounded alignment as a function of error perturbation σ.

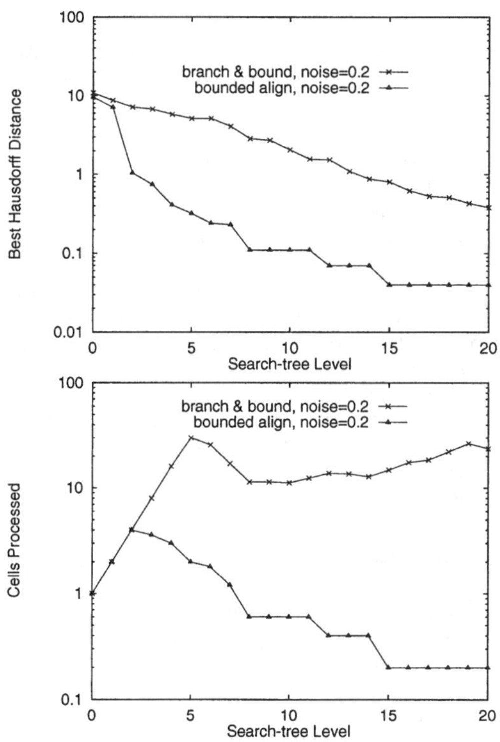

Figure 5: Best similarity and number of cells processed per level of the search tree for $\sigma = 0.2$.

4.2 Experiments on Satellite Imagery

For our second experiment we took two Landsat/TM images (one from a region in the Pacific northwest (Pac) and the other from the area around Washington DC (DC)), a pair of GOES images over Baja, and a pair of AVHRR images over South Africa (SA). Due to space limitations these images are not shown. They will appear in the full version of this paper [MNM98].

The first TM image had $|A| = 1756$ and $|B| = 1845$ over a 128×128 gray scale image. Its counterpart was generated by applying a -18 degree rotation to the first image and then applying a feature extractor based on a wavelet decomposition. We ran the algorithms with an initial search interval with angular range of 2 degrees and a translational range of 5 pixels on an ULTRA-SPARC. The results are summarized in the table below. Note that bounded alignment (B. A.) outperformed branch-and-bound (B&B) by a significant factor, while producing roughly the same similarity. The target similarity at exactly 18 degrees is 0.81, worse than either of the similarities derived from the algorithm.

	Alg	Cells	Sec	Angle	t_x	t_y	Sim
Pac	B&B	423	67	18.7	0.38	-0.38	0.42
Pac	B. A.	145	41	18.7	0.37	-0.69	0.43
DC	B&B	195	14	-0.1	32.1	32.0	0.25
DC	B. A.	51	11	-0.1	32.1	32.0	0.25
Baja	B&B	484	16	-0.1	9.9	-9.9	0.17
Baja	B. A.	6	0.2	0.0	10.0	-10.0	0.00
SA	B&B	2037	153	0.03	4.9	-5.8	0.18
SA	B. A.	110	19	0.0	5.0	-6.0	0.00

The TM image of DC had $|A| = 763$ and $|B| = 766$. One image was generated by applying a translation of $(32.5, 32.5)$ to the other. The initial search cell allowed 10 degrees of rotation and 5 pixels of translation. The target similarity was 0.71. Note that although bounded alignment visits fewer cells, the running times are quite similar. We believe that this is due to the nature of the feature detector, which produced more feature "clumps". Bounded alignment requires isolated feature points for good performance.

The GOES subimages of Baja had $|A| = 326$ and $|B| = 503$. These subimages were cropped from the original data, so as to avoid undesired boundary effects. The target transformation $(0.0, 10.0, -10.0)$ results in a target similarity of 0.0. Likewise, the (cropped) AVHRR subimages of SA had $|A| = 872$ and $|B| = 927$, and the target similarity was 1.0. In both cases, bounded alignment outperformed B&B significantly, converging on a perfect match with similarity of 0.0.

It can be seen that in all instances bounded alignment produced comparable or better results than branch-and-bound, and the running times both in terms of CPU seconds and the number of cells visited were smaller.

Acknowledgements

The second and third authors would like to thank William J. Campbell and Robert F. Cromp of Code 935, NASA/GSFC, for their ongoing support and for sustaining the activities of the image registration group at NASA/GSFC.

References

[AAR94] H. Alt, O. Aichholzer, and G. Rote. Matching shapes with a reference point. In *Proc. 10th*

Annu. ACM Sympos. Comput. Geom., pages 85–92, 1994.

[AG96] H. Alt and L. J. Guibas. Discrete geometric shapes: Matching, interpolation, and approximation. Technical Report 96–11, Institut für Informatik, Freie Unversität Berlin, Berlin, Germany, 1996.

[AM93] S. Arya and D. M. Mount. Algorithms for fast vector quantization. In J. A. Storer and M. Cohn, editors, *Proc. of DCC '93: Data Compression Conference*, pages 381–390. IEEE Press, 1993.

[AM95] S. Arya and D. M. Mount. Approximate range searching. In *Proc. 11th Annu. ACM Sympos. Comput. Geom.*, pages 172–181, 1995.

[AMN+94] S. Arya, D. M. Mount, N. S. Netanyahu, R. Silverman, and A. Wu. An optimal algorithm for approximate nearest neighbor searching. In *Proc. 5th ACM-SIAM Sympos. Discrete Algorithms*, pages 573–582, 1994.

[AMWW88] H. Alt, K. Mehlhorn, H. Wagener, and E. Welzl. Congruence, similarity and symmetries of geometric objects. *Discrete Comput. Geom.*, 3:237–256, 1988.

[Bro92] L. G. Brown. A survey of image registration techniques. *ACM Computing Surveys*, 24:325–376, 1992.

[CGH+97] L. P. Chew, M. T. Goodrich, D. P. Huttenlocher, K. Kedem, J. M. Kleinberg, and D. Kravets. Geometric pattern matching under Euclidean motion. *Comput. Geom. Theory Appl.*, 7:113–124, 1997.

[dRL95] P. J. de Rezende and D. T. Lee. Point set pattern matching in d-dimensions. *Algorithmica*, 13:387–404, 1995.

[FBF77] J. H. Friedman, J. L. Bentley, and R. A. Finkel. An algorithm for finding best matches in logarithmic expected time. *ACM Transactions on Mathematical Software*, 3:209–226, 1977.

[GMO94] M. T. Goodrich, J. S. Mitchell, and M. W. Orletsky. Practical methods for approximate geometric pattern matching under rigid motion. In *Proc. 10th Annu. ACM Sympos. Comput. Geom.*, pages 103–112, 1994.

[GS85] A. Goshtasby and G. C. Stockman. Point pattern matching using convex hull edges. *IEEE Transactions on Systems, Man, and Cybernetics*, 15:631–637, 1985.

[GSP86] A. Goshtasby, G. C. Stockman, and C. V. Page. A region-based approach to digital image registration with subpixel accuracy. *IEEE Transactions on Geoscience and Remote Sensing*, 24:390–399, 1986.

[HKR93] D. P. Huttenlocher, G. A. Klanderman, and W. J. Rucklidge. Comparing images using the Hausdorff distance. *IEEE Transactions on Pattern Analysis and Machine Intelligence*, 15:850–863, 1993.

[HKS93] D. P. Huttenlocher, K. Kedem, and M. Sharir. The upper envelope of Voronoi surfaces and its applications. *Discrete Comput. Geom.*, 9:267–291, 1993.

[HR92] D. P. Huttenlocher and W. J. Rucklidge. A multi-resolution technique for comparing images using the Hausdorff distance. Technical Report 1321, Dept. of Computer Science, Cornell University, Ithaca, NY, 1992.

[HR93] D. P. Huttenlocher and W. J. Rucklidge. A multi-resolution technique for comparing images using the Hausdorff distance. In *Proc. IEEE Conf. Comput. Vision Pattern. Recogn.*, pages 705–706. IEEE, 1993.

[HS94] P. J. Heffernan and S. Schirra. Approximate decision algorithms for point set congruence. *Comput. Geom. Theory Appl.*, 4:137–156, 1994.

[HV97] M. Hagedoorn and R. C. Veltkamp. Reliable and efficient pattern matching using an affine invariant metric. Technical Report RUU-CS-97-33, Dept. of Computing Science, Utrecht University, The Netherlands, 1997.

[IR96] S. Irani and P. Raghavan. Combinatorial and experimental results for randomized point matching algorithms. In *Proc. 12th Annu. ACM Sympos. Comput. Geom.*, pages 68–77, 1996.

[LW88] Y. Lamdan and H. J. Wolfson. Geometric hashing: A general and efficient model-based recognition scheme. In *2nd Inter. Conf. on Comput. Vision*, pages 238–249, 1988.

[MA97] D. M. Mount and S. Arya. Ann: A library for approximate nearest neighbor searching. CGC 2nd Annual Fall Workship on Computational Geometry, URL: `http://www.cs.umd.edu/~mount/ANN.`, 1997.

[MNM98] D. M. Mount, N. S. Netanyahu, and J. Le Moigne. Improved algorithms for robust point pattern matching and applications to image registration. Full paper version, in preparation, 1998.

[OH97] C. F. Olson and D. P. Huttenlocher. Automatic target recognition by matching oriented edge pixels. *IEEE Transactions on Image Processing*, 6:103–113, 1997.

[RR80] S. Ranade and A. Rosenfeld. Point pattern matching by relaxation. *Pattern Recognition*, 12:269–275, 1980.

[Ruc95] W. J. Rucklidge. Locating objects using the Hausdorff distance. In *Proc. 5th Int'l Conf. on Computer Vision*, pages 457–464. IEEE, 1995.

[SKB82] G. C. Stockman, S. Kopstein, and S. Benett. Matching images to models for registration and object detection via clustering. *IEEE Transactions on pattern Analysis and Machine Intelligence*, 4:229–241, 1982.

[TJ89] J. Ton and A. K. Jain. Registering landsat images by point matching. *IEEE Transactions on Geoscience and Remote Sensing*, 27:642–651, 1989.

[XMT+97] W. Xia, J. Le Moigne, J. C. Tilton, B-T. Lerner, E. Kaymaz, J. Pierce, S. Raghavan, S. Chettri, T. El-Ghazawi, M. Manohar, N. S. Netanyahu, W. J. Campbell, and R. F. Cromp. A registration toolbox for multi-source remote sensing applications. In *Proc. Int'l Conf. on Earth Observation and Environmental Information*, 1997.

Interval Arithmetic Yields Efficient Dynamic Filters for Computational Geometry*

Hervé Brönnimann[†] Christoph Burnikel[‡] Sylvain Pion[†]

Abstract

We discuss interval techniques for speeding up the exact evaluation of geometric predicates and describe an efficient implementation of interval arithmetic that is strongly influenced by the rounding modes of the widely used IEEE 754 standard. Using this approach we engineer an efficient floating point filter for the computation of the sign of a determinant that works for arbitrary dimensions. Furthermore we show how to use our interval techniques for exact linear optimization problems of low dimension as they arise in geometric computing. We validate our approach experimentally, comparing it with other static, dynamic and semi-static filters.

1 Introduction

Numerical inaccuracy in the evaluation of arithmetic predicates is one of the main obstacles in implementing geometric algorithms robustly. There are numerous approaches to get the problem under control, of which the immediate solution of exact computation stands out because of its generality. For faster yet exact computation, arithmetic filters were proposed in [9, 15] and showed to be efficient both in practice [10, 18] and in theory [7].

The work done so far mainly concerns static and semi–static filters where the error bounds, or at least parts of it, are determined at compile time. Static filters are restricted to integral expressions of small bounded depth and, which is even worse, require that good upper bounds on the input variables are known in advance. Semi-static filters remedy most of the mentioned problems, but divisions and square roots can only be handled at the price of a significantly reduced quality of the resulting error bounds. Moreover, to use semi-static filters the complete structure of a computation has to be preprocessed in advance. Dynamic filtering on the other hand is often considered too inefficient for use in computational geometry. First advances have been made to incorporate dynamic filters into the LEDA reals [5].

In this paper, we propose to use *interval analysis* [16, 17, 13] for more efficient dynamic filters. The technique is based on carefully engineered interval arithmetic. It is very simple to use and yields the most flexible dynamic floating-point filters we know: divisions can be handled as well as square roots and hence the technique is not limited to rational expressions. With the IEEE 754 standard for floating point computations [14], the computed intervals are locally optimal in the sense that every single operation results in the smallest possible interval. Consequently, the produced filters have the maximal achievable probability of success. On the other hand, interval arithmetic is still relatively fast, being roughly 3-8 times slower than floating-point evaluation. Our implementation of interval arithmetic is heavily influenced by that of the rounding modes of the IEEE 754 standard.

Many geometric predicates boil down to computing the sign of a determinant. Much effort has already been made towards the exact evaluation of signs of determinants, using various specific solutions such as Clarkson's or the lattice method [6, 1, 3], or using general solutions such as exact integer arithmetic [9] and modular arithmetic [2]. For $d \times d$ determinants, the complexities range from $O(d^3 \log d)$ to $O(d^4 \log d)$ with a potentially large constant in the asymptotic bounds. For all these methods we observe that they are, practically, several orders of magnitude slower than the straightforward, inexact floating point evaluation. In this paper we engineer a fast floating-point filter for computing the sign of a determinant in time $O(d^3)$ with a small constant. Our

*This research was partially supported by the ESPRIT IV LTR Project No. 21957 (CGAL).

[†]{Herve.Bronnimann,Sylvain.Pion}@sophia.inria.fr . INRIA Sophia-Antipolis, BP 93, 06902 Sophia-Antipolis cedex, France.

[‡]burnikel@mpi-sb.mpg.de. Max-Planck-Institut für Informatik, Im Stadtwald, 66123 Saarbrücken, Germany.

SCG 98 Minneapolis Minnesota USA
Copyright ACM 1998 0-89791-973-4/98/ 6...$5.00

filter fails only for matrices which are singular or nearly singular. There is also a simplified filter for small dimensions, which has a weaker probability of success but is very fast. As this version of the filter crucially uses divisions, semi-static or static error computation cannot be used here.

Another class of problems that we consider are geometric optimization problems solvable by linear programming. Here the central predicate is computing the sign of a dot product over vectors whose coordinates are determined by matrix operations, including matrix inversion. In [11], Gärtner describes an exact implementation of the simplex algorithm using a semi-static floating-point filter for better efficiency. However, Gärtner needs to compute the inverses of the base matrices with exact arithmetic. In [13] it is shown how to compute a verified inclusion for the solution of a linear system $Ax = b$ and the method is applied to a partially correct implementation of the simplex algorithm. We present another non-iterative filter method having the advantage that one can state an explicit condition under which the method is applicable. Here we can avoid to compute an interval inclusion for the inverse A^{-1}.

There are two intrinsic limitations to the use of interval arithmetic. First, interval arithmetic may fail to detect degenerate instances because of the roundoff errors. When the operations are performed exactly, the interval is reduced to a point and this certifies that there is indeed a degeneracy. However, this only happens when the bit length of the input data is small with respect to the machine precision. Second, although the computed intervals are optimal for every single operation, there can be a huge overestimation of the error for a cascaded sequence of operations. For the specific problem of evaluating the sign of a determinant, the latter limitation is none: We show how to combine interval arithmetic with a posteriori error computation to a filter whose effectiveness decreases in practice only very slowly with the dimension. This means that our filter can come into effect for matrices of almost arbitrarily large dimensions. The same is true for the predicates used in the simplex algorithm.

Our paper is organized as follows. In section 2 we lay out a classification of filters into static, semi-static, and dynamic filters and we discuss their usage in precompiled, hand-coded and fully packaged cascaded computation. In section 3 we introduce the basic notions of interval analysis and give an overview of the efficient methods to obtain verified inclusions. Here we introduce a heuristic measure of the effectiveness of interval analysis for a given expression $\mathcal{E}$. In section 4 we present a new filter for computing the sign of a determinant using a posteriori error analysis and interval arithmetic. In section 5 we sketch how the simplex algorithm can be implemented using techniques similar to those used in section 4. In section 6 the approach is validated experimentally with a new implementation of interval arithmetic that relies on the rounding modes of the IEEE 754 standard. Beside the mentioned applications we also consider other geometric predicates, such as those encountered in optimization or Delaunay sweep algorithms.

2 Arithmetic filters

Filters are used when determining the sign of a fixed expression $\mathcal{E} = \mathcal{E}(x_1, \dots, x_n)$. They allow to evaluate the sign in a robust manner, while being far quicker than the complete exact evaluation, in most cases. A filter never returns a wrong answer, but may fail to return a meaningful answer at all (in this case returning NO_IDEA).

We will focus on single precision[1] floating point filters, because they have a speed comparable to the simple floating point evaluation. We distinguish mainly three kinds of such filters, described below.

Fully static: An upper bound $|x_i| \leq X_i$ is known for each i, and $\mathcal{E}$ contains only $+, -, \cdot, \sqrt{\cdot}$. Then E is computed such that the error on computing $\mathcal{E}$ is bounded by E for all inputs. For a particular input, the filter fails if $|\mathcal{E}| \leq E$, otherwise the sign of $\mathcal{E}$ is known safely.

Semi-static: Sometimes it is impossible to specify a good bound on the entries, but there is a simple formula $E = E(x_1, \dots, x_n)$ having a structure similar to $\mathcal{E}$ that gives a valid error bound for a particular input, even when E is evaluated with single precision. E is computed dynamically, and the filter fails if $|\mathcal{E}| \leq E$; otherwise the sign of $\mathcal{E}$ is known safely.

Dynamic: the computation of E is carried along with the computation of $\mathcal{E}$. Typically, for each operation of $\mathcal{E}$, a simple rule determines the error bound for the result of that operation based on the operands and error bound on them.

Exact computation can be considered the last of filters, which never fails. The cost of the total computation can then be expressed as a combination of the cost of the different filters multiplied by their conditional probability of success.

Static filters are implemented for instance in LN [10], semi-static in [1, 4], and dynamic filters in EXPR [19] and LEDA [5]. Also Schewchuk approximates $\mathcal{E}$ up to first order error terms, then up to second order errors etc., until the sign can be safely determined. This procedure combines a dynamic filter according to our

[1] Single precision here means a precision of 53 bits, used by IEEE 754 doubles.

description with exact computation, since it can reduce the error to zero if needed. A static filter does the floating point evaluation of $\mathcal{E}$ plus one extra comparison, whose running time is usually neglectable. The cost of a semi-static filter exceeds that of a static filter by the cost of computing the error bound E, which is typically about as much as for the computation of $\mathcal{E}$. Finally, the cost of a dynamic filter is a constant factor times that of the floating point evaluation.

We now describe in detail a particular filter, which is dynamic, based on interval arithmetic.

3 Interval arithmetic

The major tool used within our filter is *interval arithmetic.* The use of interval arithmetic in the context of matrix operations was originally proposed by Moore [16] and further promoted through a research group directed by Kulisch; see [13] for a recent survey of the available computational methods. In [13], interval arithmetic is successfully applied to many basic computation tasks in numerical linear and nonlinear algebra. However, the problem considered in the present paper seems to have been overlooked so far.

Interval arithmetic deals with *intervals* $[x] = [\underline{x}, \overline{x}]$ of real numbers. These intervals may arise from uncertainty in the input (e.g., when the input is subject to imprecise measurement) as well as from approximate intermediate calculations. In fact, our interval methods can be applied when the input matrices have interval entries but in our applications the coefficients of A are given exactly. This is because we want to use interval arithmetic as a preliminary stage for *exact* computation. Note that for interval–type input matrices e.g. the determinant does not necessarily have a unique sign; this fact cannot be altered if we later use exact arithmetic.

The basic interval operations are defined essentially as in [13]. Namely, if both operands $[x] = [\underline{x}, \overline{x}]$, $[y] = [\underline{y}, \overline{y}]$ are finite intervals we set

$$\begin{aligned}
[x]+[y] &= [\underline{x}+\underline{y}, \overline{x}+\overline{y}] \\
[x]-[y] &= [\underline{x}-\overline{y}, \overline{x}-\underline{y}] \\
[x]\cdot[y] &= [\min\{\underline{x}\underline{y}, \underline{x}\overline{y}, \overline{x}\underline{y}, \overline{x}\overline{y}\}, \max\{\underline{x}\underline{y}, \underline{x}\overline{y}, \overline{x}\underline{y}, \overline{x}\overline{y}\}] \\
[x]/[y] &= \begin{cases} [x]\cdot[1/\overline{y}, 1/\underline{y}] & , \ 0 \notin [y] \\ \mathbb{R} & , \ \text{otherwise} \end{cases} \\
[x]^{1/2} &= \begin{cases} [\underline{x}^{1/2}, \overline{x}^{1/2}] & , \ 0 \notin [x] \\ \mathbb{R} & , \ \text{otherwise} \end{cases}
\end{aligned}$$

If one of the intervals $[x]$, $[y]$ is infinite, we set the resulting interval to $\mathbb{R} = [-\infty, \infty]$. Since the computed intervals $[x]$ in general have bounds $\underline{x}$, $\overline{x}$ that are not contained in the given finite set $\mathbb{F}$ of floating point numbers, we compute in each arithmetic step the smallest interval $\diamond[x] = [\nabla\underline{x}, \Delta\overline{x}]$ that encloses $[x]$ such that $\nabla\underline{x}$ and $\Delta\overline{x}$ are contained in $\mathbb{F}$. This means that $\nabla\underline{x}$ and $\Delta\overline{x}$ are the numbers in $\mathbb{F}$ next to $\underline{x}$ and $\overline{x}$ when rounding downwards (resp. upwards). Then the approximate floating-point interval operations are given by

$$\begin{aligned}
[x]\oplus[y] &= \diamond([x]+[y]) \\
[x]\ominus[y] &= \diamond([x]-[y]) \\
[x]\odot[y] &= \diamond([x]\cdot[y]) \\
[x]\oslash[y] &= \diamond([x]/[y]).
\end{aligned}$$

This notation is adapted to interval functions like $f([x]) = (e^{[x]}-1)[x]$ by writing $f_\diamond([x])$ for the smallest floating-point interval containing $f([x])$. Let $\mathcal{E}$ be an arithmetic expression over the operations $\{+,-,\cdot,/,\sqrt{\ }\}$. If the expression is applied to intervals $[x_1],\ldots,[x_n]$ instead of real numbers we denote the resulting interval expression by $[\mathcal{E}]$ for clarity. For the rest of the paper we assume that interval expressions are evaluated using the approximate interval operations $\oplus, \ominus, \odot, \oslash$ and $\text{sqrt}_\diamond([x]) = \diamond([x]^{1/2})$.

It turns out that the interval evaluation of expressions is not always effective, depending on the particular structure of the expression. This notion of effectiveness is related to the size of the resulting intervals, not to the time necessary to evaluate the interval expression. This is because the interval evaluation incurs only a constant overhead over the usual floating–point approximation[2]. Important types of expressions that are well suited for interval evaluation are *dot products* or *inner products* of vectors and the derived operations of matrix-matrix product and matrix-vector product. The following *interval degree Ideg($\mathcal{E}$)* $\in \mathbb{Z}$ is a heuristic, asymptotic measure for the average number of uncertain bits of $[\mathcal{E}]$ and hence for the quality of the interval evaluation of $\mathcal{E}$. For an expression $\mathcal{E}$ consisting of a single input number z we set $\text{Ideg}(\mathcal{E}) = 0$ and inductively, if $\mathcal{E}$ is computed from expressions $\mathcal{X}$, $\mathcal{Y}$ by the operations $\{+,-,*,/,\sqrt{\ }\}$ we set

$$\begin{aligned}
\text{Ideg}(\mathcal{X}+\mathcal{Y}) &= \max\{\text{Ideg}(\mathcal{X}), \text{Ideg}(\mathcal{Y})\} \\
\text{Ideg}(\mathcal{X}-\mathcal{Y}) &= \max\{\text{Ideg}(\mathcal{X}), \text{Ideg}(\mathcal{Y})\} \\
\text{Ideg}(\mathcal{X}\cdot\mathcal{Y}) &= 1+\max\{\text{Ideg}(\mathcal{X}), \text{Ideg}(\mathcal{Y})\} \\
\text{Ideg}(\mathcal{X}/\mathcal{Y}) &= 1+\max\{\text{Ideg}(\mathcal{X}), \text{Ideg}(\mathcal{Y})\} \\
\text{Ideg}(\mathcal{X}^{1/2}) &= \text{Ideg}(\mathcal{X}).
\end{aligned}$$

In this notation, the mentioned products all have degree 1. On the other hand, many of the basic operations in linear algebra have larger degree. For example, the degree of computing $\det(A)$, the degree of computing decompositions $P \cdot A = L \cdot U$, and the degree of solving triangular systems for a d dimensional matrix are all $\Theta(d)$. Typically, the computed intervals are useless if $\text{Ideg}(\mathcal{E})$ has the same order of magnitude than the used mantissa length p of the floating–point numbers and are very useful if $\text{Ideg}(\mathcal{E})$ is a small constant. As an example

[2]This overhead depends on the particular platform given by hardware architecture, programming language and compiler.

for the calculation of the degree we prove the following lemma.

Lemma 1 *The interval solution of a triangular system $T \cdot [x] = b$ with exact input data has degree d.*

Proof: Without loss of generality, we consider forward substitution with a lower triangular matrix $T = (t_{i,j})_{i,j}$. In the basic step $d = 0$ we have $x_0 = b_0/t_{0,0}$ which has degree 1. Let for $d > 0$ $x_0, \ldots, x_{d-1}$ have degree $\leq d$ and $\text{Ideg}(x_{d-1}) = d$. From the formula

$$x_d = b_d/l_{d,d} - \sum_{j=0}^{d-1} x_j(l_{d,j}/l_{d,d})$$

we see that since $\text{Ideg}(l_{d,j}/l_{d,d}) = 1$ each product $x_j(l_{d,j}/l_{d,d})$ has degree $\leq d+1$. Because the latter term has degree $d+1$ for $j = d$, the whole sum has degree $d+1$. Our statement follows by induction. ∎

Experiments with a randomly chosen matrix A show that the interval solution of a system $A[x] = b$ using a LU decomposition of A in fact incurs an uncertainty in the last $\Theta(d)$ places, even if A is tridiagonal. Another simple lemma is that Gauss elimination is of degree $2d$.

Let us stress again that our notion of degree gives a *purely heuristic and asymptotic* estimate of the usefulness of interval arithmetic. Our point is that interval arithmetic can be useful for expressions with small bounded degree or else if the problem has small dimension[3]. Nevertheless, our degree is logarithmically related to the index of [4] (if we do not count additions), and this index gives a bound on the relative error of a computation.

As we just said, many of the most important computations in linear algebra have non-constant degree. How can we save the interval method? There are several powerful complementary approaches of which we cite the following three.

1. *A posteriori error analysis:* Compute first a candidate for the solution by conventional floating-point arithmetic and then use this candidate to compute an interval inclusion for the solution by interval arithmetic.

2. *Preconditioning* First bring the problem in a form that is better suited for interval computation, by applying an appropriate preconditioning transformation.

3. *Fixpoint theorems:* Reformulate the problem such that the solution is expressed as the limit of an iteration process $x \leftarrow f(x)$ where $f(x)$ has small degree in the unknown x. Try to find an interval I such that $f(I) \subset I$ can be shown (again by interval arithmetic). Now use an appropriate fixpoint theorem like the classical one by Banach or the more sophisticated by Brouwer [17] and try to prove that the iteration converges to a solution contained in I.

[3]For problems of large degree in large dimensions, interval arithmetic can still be useful if the input data has an appropriate (sparse) structure.

These methods work well for many of the basic problems in numerical mathematics, including solutions of linear and nonlinear systems, global optimization and automatic differentiation [13]. However, applying the methods to the computation of a determinant is not straightforward. Why? First, the fix-point method does not seem to work: If A is a symmetric matrix, then we could compute the determinant as the product of the eigenvalues of A. Given an approximate eigenvector x' and an approximate eigenvalue λ' of A, we could further use a Newton iteration that converges to the desired solution of $(A - \lambda I)x = 0$ as in [13]. However, in the case of a non-symmetric matrix A, only the absolute value of $\det(A)$ is expressible by the product of the singular values of A. Hence we cannot handle the general case in this way. Second, the standard way of preconditioning in the case of a linear problem is to first multiply the matrix A with an approximate inverse A_{inv} of A and then apply e.g. Gaussian elimination to the resulting interval matrix. This is not helpful in our case because it is no easier to compute the determinant sign for the approximate inverse of A than for A itself. However we will discuss another preconditioning transformation in the next section. The best solution that we propose uses a posteriori error analysis.

4 Computing the sign of a determinant

The problem that we consider in this section is the following. Let $\mathbb{F}$ be a set of *fixed precision* floating–point numbers. Given a matrix $A \in \mathbb{F}^{d,d}$ over $\mathbb{F}$, compute the exact sign of $\det(A)$. This is an important problem in computational geometry since many geometric predicates are expressible by determinants. The following methods are available to compute the exact sign of *any* determinant $A \in \mathbb{F}^{d,d}$:

- Exact integer or floating-point arithmetic
- Exact modular arithmetic [2]
- Clarkson's re–orthogonalization method [6, 3]
- The lattice method [1, 3]

In order to apply some of these methods, it is necessary to make the matrix entries integral by multiplying the matrix with a large enough power of 2. Note that this scaling does not change the determinant's sign, but may generate integers whose bit length is more than single precision. This imposes severe restrictions especially on Clarkson's method and on the lattice method, because they are designed with the intention to *avoid* multiple precision. The first method has a running time of $d^3M(d)$, $M(d)$ being the time required to compute the product of two numbers of word-length d (words in $\mathbb{F}$ have length 1). The modular variant runs in time $O(d^4 \log(d))$, Clarkson's algorithms runs in time $O(d^3 \log(d))$ [3]. The last algorithm has exponential worst-case running time but is reported in [3] to take only time $O(d^3)$ in practice, like Clarkson's algorithm. It should be noted that in the general case, the four methods above all require multi-precision, hence there is a potentially large constant hidden in these asymptotic bounds. If the entries are substantially smaller than the maximum integer representable in $\mathbb{F}$, however, Clarkson's algorithm incurs little overhead on top of direct floating point computation. If the matrix is reasonably orthogonal, the running time of the last two methods is close to $O(d^3)$, these algorithms are said to be adaptive; they behave like a filter. The first two methods, however, handle all cases similarly and are not adaptive.

In this paper we design fast *floating-point filters*, with running time $O(d^3)$ comparable to floating point computation, without imposing any restriction on the input. The filters are effective also if the input is not representable in $\mathbb{F}^{d,d}$. The filters fail only for matrices that are singular or nearly singular.

4.1 A posteriori method

One of the standard methods in numerical analysis to compute a determinant $\det(A)$ uses the LU decomposition $P \cdot A = L \cdot U$ where P is a permutation matrix, L is lower triangular and U is upper triangular. Of course, using fixed precision arithmetic we will get only an approximate decomposition $P \cdot A \approx L \cdot U$. The determinant of P is ± 1 and can be computed without rounding error, and $\det(L)$ is 1 since the diagonal elements of L are all equal to 1. So the product of the diagonal elements $u_{i,i}$ of U gives an approximation of $\pm\det(A)$. For well-conditioned matrices this approximation is usually quite reliable. In the rest of the section we shall investigate under what circumstances we can conclude that $\det(P \cdot A)$ has the same sign as $\prod_i \operatorname{sign}(u_{i,i})$. Since $P \cdot A \approx L \cdot U$ we expect that $A^{-1} \approx U^{-1}L^{-1}P$. Because we cannot evaluate U^{-1} and L^{-1} exactly, we invert U and L numerically to matrices $U_{inv} \approx U^{-1}$ and $L_{inv} \approx L^{-1}$ which gives the approximate inverse

$$B := U_{inv} L_{inv} P$$

of A. Note that by the triangular structure of L and U, L_{inv} still has all diagonal elements equal to 1 and the diagonal elements of U_{inv} are $1 \oslash u_{i,i}$. This shows that $\det(B)$ has the same sign as $\det(U)\det(P)$. Since $\det(A) = \det(B^{-1})\det(B \cdot A)$ it follows that $\det(A)$ has the same sign as $\det(B)$ if we can show that $\det(B \cdot A) > 0$. Here we use the following lemma, valid for any given matrix norm.

Lemma 2 *Let $\|F\| < 1$ for some fixed matrix norm. Then $\|(I+F)^{-1}\| \le \frac{1}{1-\|F\|}$ and $\det(I+F) > 0$.*

Proof: $I + F$ is a regular matrix since for a vector $x \neq 0$ we have

$$\begin{aligned} \|(I+F)x\| &= \|x + Fx\| \ge \|x\| - \|Fx\| \\ &\ge \|x\| - \|F\|\|x\| \\ &= (1 - \|F\|)\|x\| > 0. \end{aligned}$$

This also shows that $\|(I+F)^{-1}\| \le \frac{1}{1-\|F\|}$. Consider the mapping $D : t \mapsto \det(I + t \cdot F), t \in [0,1]$. D is continuous and $D(0) = 1$. Since all matrices $I + t \cdot F, t \in [0,1]$ are regular by the argumentation above, $D(t)$ is always nonzero. By continuity of D, this means that $D(1) = \det(I+F) > 0$. ■

All we have to do now is to check whether the defect matrix $E = I - B \cdot A$ has norm < 1 for some matrix norm of our choice. To this end, evaluate $[I - B \cdot A] = ([e_{i,j}])_{i,j}$ by interval arithmetic. Using the norm $\|\cdot\|_\infty$ we have to test whether the sum of the maximal possible absolute values of $e_{i,j}$ is smaller than 1 in every row. We denote the maximal row sum in this calculation by $\|[I - B \cdot A]\|_\infty$. The description of our algorithm for the computation of $\operatorname{sign}(\det(A))$ is as follows. Note that the algorithm either returns $+1$, -1, or the string NO_IDEA (if the filter fails).

Algorithm 1 *(returns the sign of $\det(A)$, if successful):*

1. *Compute a numerical decomposition $PA \approx LU$. If this is not possible for numerical reasons, return NO_IDEA.*

2. *Compute numerical inverses U_{inv} of U and L_{inv} of L. In case of exponent overflow in the floating-point computation, return NO_IDEA.*

3. *Compute $\|[I - BA]\|_\infty$ by interval arithmetic where $B = U_{inv} L_{inv} P$*

4. *If $\|[I - BA]\|_\infty < 1$, return $\det(P) \prod_i \operatorname{sign}(u_{i,i})$. Otherwise return NO_IDEA.*

Remarks:

- The computation of the numerical LU decomposition in Step 1 can fail if all elements of a pivot column are zero. In practice this occurs if A is singular or very nearly singular.

- If Step 1 was correctly completed, Step 2 can only fail because of exponent overflow.

- There seems to be no easy way to improve the approximate inverse B if the quality does not turn out to be sufficient in Step 3. The reason is that B does not only have to be a good approximate inverse of A, but also sign(det(B)) has to be computable exactly.

Lemma 3 *Algorithm 1 takes at most $d^3+O(d^2)$ floating-point operations and $d^3+O(d^2)$ interval operations (1 operation = 1 addition + 1 multiplication).*

Proof: The computation of the LU decomposition takes $d^3/3+O(d^2)$ operations, and so does each inversion of a triangular matrix. Computing the product of a full matrix with a triangular matrix takes $d^3/2+O(d^2)$ operations. All other computations require only $O(d^2)$ operations. ■

Let us now give a rough estimate for the running time of algorithm 1 with respect to the unfiltered evaluation of $\text{sign}(\det(A)) \approx \text{sign}(\det(P))\prod_i \text{sign}(u_{i,i})$ ("naive algorithm"). Hammer et al. [13] report that interval computations roughly take double time than ordinary floating-point calculations. In our own C++ implementation, this (clearly optimal) overhead was for matrix computations only nearly achieved; an overhead factor of about $3-4$ is realistic, though. Our measurements showed that algorithm 1 takes in fact $9-12$ times longer than the naive algorithm.

4.2 Naive interval arithmetic

Algorithm 1 is applicable even for very large dimensions; it was successfully tested up to $d=800$. On the other hand, many applications, e.g. in computational geometry, are small–dimensional ($d \leq 25$). For small dimensions, an overhead factor of about 10 is not always better than the overhead that we get using one of the exact algorithms. However, remember that for those cases the straight evaluation of the determinant in interval arithmetic is possible. Hence we propose the following alternative algorithm.

Algorithm 2 *(returns the sign of det(A), if possible; for small dimensions)*

1. *Compute $PA = [L]\cdot[U]$ in interval arithmetic.*
2. *If one of the intervals $[r_{i,i}]$ on the diagonal of $[U]$ contains zero, return NO_IDEA. Otherwise, return $sign(det(P))\cdot\prod_i sign([r_{i,i}])$ where $sign([r_{i,i}])$ is the sign of any number in $[r_{i,i}]$.*

Remarks:

- Step 1 can always be completed, since the division by a "pivot" interval containing 0 is a regular operation that results in $[-\infty,\infty]$.
- The lower triangular part $[L]$ does not have to be computed explicitly, since it is not needed in the sign calculation.

If algorithm 2 returns a result, it always outperforms algorithm 1. We measured that algorithm 2 is only at most 2.6 times slower than with naive floating point calculation, for random entries and dimensions $5-40$. On the other hand, algorithm 1 has a higher probability of success since the computed intervals are smaller. See section 6 for detailed measurements.

Lemma 4 *Algorithm 2 takes at most $d^3/3+O(d^2)$ interval operations.*

Algorithm 2 is best applied to matrices of small dimension. However, for large matrices with a special structure, like certain sparse matrices, diagonally dominant matrices and block matrices of small block size, it might still be useful. Note that algorithm 2 cannot be enhanced by the standard trick of multiplying A with an approximate inverse A_{inv} to make it diagonally dominant. As already mentioned, it is as hard to compute the sign of $\det(A_{inv})$ as that of $\det(A)$. Another possible preconditioning matrix is $L_{inv}P$ where L_{inv} is an approximate inverse of L whose determinant is known to be 1. In fact, the resulting interval inclusion $[\tilde{U}]$ of $L_{inv}PA$ is nearly upper triangular. Unfortunately, the following Gaussian elimination that makes $[\tilde{U}]$ truly upper triangular incurs intervals that are even larger than those obtained by doing Gaussian elimination directly on A. We do not know how to find a preconditioning matrix whose determinant is both easy to compute and leads to reduced interval sizes.

For the sake of completeness, we sketch how to compute not only the sign of the determinant but also a verified enclosure for the *value* of the determinant. We need a slightly stronger version of Lemma 2 that uses the Euclidean matrix norm.

Lemma 5 *For $\|F\|_2 < 1$ we have $|det(I+F)-1| < (1+\|F\|_2)^d - 1$.*

Proof: Let $\sigma_i, i=1,\dots,d$ be the (nonnegative) singular values of the matrix $I+F$. As we already know by Lemma 2, $\det(I+F)>0$ and hence $\det(I+F)=\prod_i \sigma_i$. Now recall that the singular values of I (which are all 1) and the σ_i can differ by at most $\|F\|_2$, see Corollary 8.6.2 of [12]. This implies that

$$|\det(I+F)-1| = |\prod_i \sigma_i - 1| \leq (1+\|F\|_2)^d - 1. \qquad ■$$

We can apply Lemma 5 in the context of Algorithm 1 with $F = B\cdot A - I$. If the coefficients of F are all bounded by φ, we can bound the 2-norm of F by $\|F\|_2 < d\varphi$ and this bound is hopefully $\ll 1$. Hence we get $|1-\det(I+F)| = |1-\det(B\cdot A)| \lessdot d^2\varphi$. From

$\det(A) = \det(B \cdot A)/\det(B)$ we get an enclosure for $\det(A)$, evaluating $\det(B) = \prod_i b_{i,i}$ with interval arithmetic.

4.3 Discussion

Our new algorithm 1 is an effective mechanism to filter out "easy" cases of the sign determination. The algorithm is attractive because of its numerical quality (it fails only for nearly singular matrices) and because it is not restricted in the dimension (other filters work only for small dimensions). If the input matrix A has arbitrary floating–point entries, it is often much more efficient than the exact methods.

There are also certain restrictions to the interval method. First of all, interval arithmetic almost always fails if the matrix is singular or nearly singular. In such cases the user should apply one of the exact algorithms. Second, the interval method of algorithm 1 can in general not benefit (much) from sparsity in the matrix A. This is because it computes an approximate inverse of A, which is usually a dense matrix. On the other hand, algorithm 2 can often profit from sparsity, but it is essentially restricted to small dimensions. It is an open problem how to design a more efficient filter for computing the determinant sign of general large sparse matrices.

5 Filters for the exact simplex algorithm

In a paper to appear in SODA'98 [11], Gärtner discusses an exact implementation of the revised simplex algorithm. The optimization problem he considers is in the standard form

$$\text{(LP)} \quad \begin{array}{ll} \text{maximize} & z = c^T \cdot x \\ \text{subject to} & Ax = b \\ \text{and} & x \geq 0, \end{array}$$

where A is an $m \times n$ matrix, c, x are n-vectors, b is an m-vector, and we assume $m \ll n$. The latter assumption is realistic for geometric optimization problems such as the smallest enclosing annulus of n points in $\mathbb{R}^d$ ($m = d + 2$ variables, n constraints), or the largest disk in the kernel of a simple n-vertex polygon ($m = 3$).

One iteration of the revised simplex algorithm consists of three parts, the *pricing step* where an index i is chosen that enters the basis, the *ratio test* where an index j is chosen that leaves the basis to be replaced by i, and the *update step* where the auxiliary data structures are adapted to the new basis.

Gärtner achieves the correctness of his implementation by using exact computation for the critical steps of the simplex algorithm. These steps are (1) checking the consistency of the chosen candidate for the index i in the pricing step, and (2) the ratio test. To achieve efficiency, Gärtner computes a candidate for the index i by straightforward floating point computation. To make this approach even more appealing, Gärtner uses the pricing strategy *partial reduced cost pricing* going back to Dantzig and Clarkson. With this strategy, the pricing step dominates the running time of the simplex algorithm and hence the fraction of exact computation is minimized, taking advantage of $m \ll n$. Moreover, this pricing strategy allows to use a semi-static floating point filter to reduce the time penalty of exact computation.

However, Gärtner always computes the inverse of the base matrix by exact arithmetic. This is because in the computation of the inverse, as Gärtner puts it, "the error bounds obtained from interval arithmetic are typically a gross overestimate, making them too large to be useful even for small values of m." In fact, it does not make much sense to compute an interval inclusion of the base inverse using e.g. direct Gauss-Jordan elimination. Again, the trick is to reduce the arithmetic degree, i.e. the interval degree of the computation. Here one can choose among the three basic interval methods mentioned in the section 3, which are (1) a posteriori error analysis, (2) preconditioning, and (3) the use of appropriate fixed point theorems. The last option has already been discussed extensively in the literature on interval computation [13] and has shown to be effective. In this paper we sketch the application of the first method that relies on the following lemma.

Lemma 6 *Let $A, B \in \mathbb{R}^{d,d}$ where B is an approximate inverse of A such that $\|I - BA\| < 1$. Then for any vectors $x, b \in \mathbb{R}^d$ the approximate solution $y = Bx$ of the system $Ax = b$ satisfies the estimate*

$$|y - A^{-1}b| \leq \frac{1}{1 - \|I - BA\|} |B(Ay - b)|.$$

Here $\|\cdot\|$ is the matrix norm corresponding to the vector norm $|\cdot|$.

Proof: From

$$|y - A^{-1}b| = |(BA)^{-1}B(Ay - b)| \leq \|(BA)^{-1}\| |B(Ay - b)|$$

the statement follows by applying Lemma 2 to the matrix $F = I - BA$. ∎

The previous lemma allows the following strategy for finding an inclusion of the solution of $Ax = b$. First, compute an approximate floating-point inverse B of A and an approximate solution y of $Ax = b$. Then, compute an interval inclusion for the defect $[I - B \cdot A]$ and the quantity $[z] = [B(Ay - b)]$. If $\|[I - B \cdot A]\| < 1$, we get the a posteriori error bound

$$|y - A^{-1}b| \leq \frac{1}{1 - \|[I - D]\|} [z].$$

Although the revised simplex method can be implemented using the method sketched above, it has to be refined for better efficiency. The expensive step here is the computation of the defect $[I - B \cdot A]$ which takes $d^3 + O(d^2)$ operations.[4] Since the update of a basis matrix M is done by $M' = M \cdot \Lambda$ with a matrix Λ of the form

$$\Lambda = \begin{pmatrix} 1 & & & \lambda_1 & & & \\ & \ddots & & \vdots & & & \\ & & 1 & \lambda_{i-1} & & & \\ & & & \lambda_i & & & \\ & & & \lambda_{i+1} & 1 & & \\ & & & \vdots & & \ddots & \\ & & & \lambda_m & & & 1 \end{pmatrix},$$

we can update the matrix $[D]$, $D = B \cdot A$ by $[D'] = [\Lambda^{-1}][D][\Lambda]$. It is easy to see that the update needs only $O(d^2)$ interval operations and the interval degree grows by at most 2. After some number of iterations (say, after 20 iterations) it is necessary to recompute the defect matrix $[I - D]$ from scratch.

We remark that in principle we could use the matrix $[D]$ to replace the system $A \cdot x = b$ by its "preconditioned" interval version $[D] \cdot x = [B \cdot b]$ and then solve the latter system. This is not a great idea, though, because then we would have to use Gaussian elimination at least once for *each* iteration of the simplex algorithm. As Lemma 6 shows, this effort would be wasted.

Another possible interval approach to linear programming is to maintain an inclusion $[M^{-1}]$ for the inverse of the base matrix M. Using plain Gaussian elimination we would get an interval degree $2d$, so this is not an option. It turns out that we can use the same argumentation as in Lemma 6 to get the desired inclusion. The following lemma is a variant of [17], Theorem 4.1.11.

Lemma 7 *Let $A, B \in \mathbb{R}^{d,d}$ where B is an approximate inverse of A such that $\beta = ||I - BA|| < 1$ for any matrix norm $||\cdot||$. Then*

$$||A^{-1} - B|| \leq \frac{\beta}{1-\beta}||B||.$$

Proof:

$$\begin{aligned} ||A^{-1} - B|| &= ||(A^{-1} - B)B^{-1}B|| \\ &= ||(BA)^{-1}(I - BA)B|| \\ &\leq ||(BA)^{-1}|| \cdot ||I - BA|| \cdot ||B|| \\ &\leq \frac{1}{1-\beta} \cdot \beta \cdot ||B|| \end{aligned}$$

where the last estimate again follows by applying Lemma 2 to the matrix $F = I - BA$. ■

[4] In the computation of the defect interval we do not strictly need our packaged interval arithmetic since it is enough to evaluate BA once rounding towards $+\infty$ and once rounding towards $-\infty$.

Here the advantage is that is that the inclusion of the inverse can be updated by a single interval matrix multiplication $[\Lambda^{-1}][M^{-1}]$. A disadvantage is that, if we only have the inclusion $[M^{-1}]$ at hand and not the defect matrix $[I - D]$, we cannot use semi-static filtering in the pricing step and in the ratio test. This is because a semi-static filter needs the input quantities exactly or rounded to the next floating point value. In contrast, by the first method above using $[I - D]$ we can start with a semi-static filter similar to that used by Gärtner; only when this filter fails, we use dynamically computed intervals.

It is not clear which of the two mentioned strategies of a posteriori error analysis is superior for interval linear programming: we believe that it is necessary to carefully implement and compare both to decide that question. Note that the intervals obtained with each of the two methods are much smaller than if we used interval arithmetic for direct Gaussian elimination. In practice, we do not expect to resort to exact computation except for degenerate cases. We therefore hope to keep the cost almost the same as that of the floating point implementation.

6 Experimental Validation

We experiment our filter in a variety of settings encountered by geometric algorithms. In this paper, we have not investigated the practical efficiency of our filters for solving linear problems.

Implementation of basic interval operations. We implemented interval operations by C++ operators, using IEEE standard 754 rounding. Since we maintain here the upper interval bound and the *negative* of the lower interval bound, we can always round upwards. For example, if $[x] = [-\underline{x}, \overline{x}]$ and $[y] = [-\underline{y}, \overline{y}]$ we compute the smallest interval $[z] = [-\underline{z}, \overline{z}]$ such that $[x]+[y] \subset [z]$ by setting $\underline{z} = \triangle(\underline{x} + \underline{y})$ and $\overline{z} = \triangle(\overline{x} + \overline{y})$. Subtraction is implemented very similarly. Multiplications and divisions are slightly more complex because they require some case distinctions for the various possible signs of $\underline{x}, \overline{x}, \underline{y}, \overline{y}$. The advantage of taking the negative of the lower interval bounds is that within a sequence of operations the rounding mode never has to be adjusted, if initially set to $+\infty$. For a geometric predicate this means that that the rounding mode is set manually *outside* the interval operators in the beginning of the computation and reset afterwards to avoid side effects. We call this implementation "OutFilter". We also provided an implementation called "InFilter" that is more safe against misuse since here the rounding mode is automatically set and reset *within* each operator call.

Isolated geometric predicates. This section presents some benchmarks on isolated well known low-dimensional predicates, to show the overhead caused by the use of

method, n	5	6	7	8	9	10	11	12	13	14	15	20	24	28	32	44	52	68
naive	47	46	45	44	43	42	41	40	39	37	37	32	28	24	21	11	2	–
a posteriori	46	45	45	44	43	43	43	42	42	42	40	40	39	39	39	38	37	36

Table 1: The number of perturbed bits for which the interval filters fail for at least 50% of the cases.

various filters, compared to the pure floating point computation.

We evaluate the semi-static filters given in [4], the semi-static error bound being evaluated once only for the entire predicate in "Semi-static," and once for each operation in "FpFilter."

Beside our new filter types "OutFilter" and "InFilter" we also tested two other, completely different dynamic filters called "AbsFilter" and "RelFilter" that use standard error analysis to dynamically compute absolute and relative errors, respectively.

The considered predicates are the standard orientation and in-sphere predicates, and a specific in-circle predicate for the Voronoi diagram of line segments that asks whether a point in the circle circumscribed to two lines and a point site. The latter predicate involves three square root operations but no divisions. The benchmarks of table 2 were made on a Ultra Sparc 200 MHz, with the GNU compiler using the flag -O2.

	Insphere 3D	Orient 3D	Orient 2D	Incircle 2D (llp_p)
Semi-static	2.28	1.63	1.45	1.58
FpFilter	3.36	2.55	1.77	2.17
OutFilter	5.15	2.98	1.92	3.38
InFilter	10.58	5.85	3.13	4.89
AbsFilter	20.00	10.68	5.38	7.42
RelFilter	94.00	102.5	69.70	9.23

Table 2: Running time overhead caused by the use of various filters, compared to the the static filter (pure floating-point computation). The time taken by other computations in case of filter failure is not taken into account.

As our experiments show, the variant "OutFilter" is always the fastest. Hence in all other experiments, we use only this filter variant. Table 2 documents only differences in the running times, without accounting for the other computations in case of failure. Moreover, the filters have different probabilities of success. For a fair comparison, one needs to evaluate them within some geometric algorithm. This is the topic of both sections below.

Computing determinants. We investigate the efficacy of our interval filter for computing signs of nearly degenerate determinants. Our test determinants have all floating point entries near to 1 but with a random perturbation of size $\leq 2^{-p}$, i.e., after the pth bit. A meaningful parameter to investigate the quality of the filter is the average value k of the parameter p such that failure rate gets above 50%. In table 1 we display this number k, first for the naive method and then for the a posteriori method. A dashed entry means "fails always."

And in a sweep algorithm for Voronoi. We have incorporated a floating point filter into our implementation of the sweep algorithm for building Voronoi diagrams [8]. The predicates involve orientation tests, comparing the ordinates of a point and of an intersection of two parabolae, and comparing between elements of a set of abscissae of points or maximum abscissae of circumscribed circles. The latter is the more demanding predicate as it uses square roots and has Ideg 4, but its exact computation with integers would require 20-fold precision.

We ran the program on a variety of configurations. The first configuration is random; the second is a perturbed grid and the third is a perturbed circle; finally, we ran the program on a degenerate circle. All numbers were generated on 53 bits between $\frac{1}{2}$ and 1, and perturbed by $\varepsilon \in [10^{-7}, 10^{-2}]$. We compare our dynamic filter with the semi-static filter of [4].

There was no great difference in the overall running time between using the dynamic filter and using the semi-static filter. The time spent specifically in the arithmetic operations was only twice/three times more for the interval filter than for the semi-static filter. Neither filter made a mistake on the random cases, nor did the standard floating point implementation. More interestingly, on perturbed grid or circle, we clearly demonstrate the efficiency of the interval filter, which rarely fails, whereas the static filter shows the weakness of its bounds. For perturbed circular configurations, the interval filter needs a perturbation smaller than 10^{-10} to fail at least once consistently. We give in table 3 the number of failures of both filters on these two cases. Since we didn't use exact arithmetic, the dashes indicate when the floating point computation failed. Interestingly, we saw that the median value of the interval may be a more stable approximation than the floating point computation.

In the case of degenerate points, neither filter can of course detect the degeneracies, but they have similar running times.

	circular $+\varepsilon$ 1000 points		grid $+\varepsilon$ 70×70 points	
	static	dynamic	static	dynamic
10^{-2}	11	0	20	0
10^{-3}	39	0	22	0
10^{-4}	43	0	41	0
10^{-5}	76	0	53	0
10^{-6}	106	0	–	8
10^{-7}	–	1	–	62

Table 3: Number of failures for the filters on different almost degenerate distributions.

7 Conclusions

The best prior art proposes static filters to low-dimensional primitives dealing only with integer entries [1, 3, 6, 10], or to rational 2d and 3d primitives when dealing with floating-point entries [18]. Our solution is powerful as it avoids the overestimation of errors induced by static bounds and is efficient, i.e., always less than an order of magnitude slower than the straightforward floating-point implementation. Unlike previous filters, it can handle square roots and divisions. Square roots are needed for instance in computing the Voronoi diagram of a set of points or line segments, and divisions are needed in the efficient computation of the sign of a large determinant. For the latter problem, our filter is the first that works for arbitrary dimensions.

For low-dimensional geometry, we have packaged this filter for the CGAL library. Experiments show that it is only little slower than the semi-static filter when used in a predicate and with a precompiler such as [4], but that it rarely fails on non-degenerate instances that make the semi-static filter fail. Hence, we recommend interval arithmetic as the ultimate level of filter before resorting to efficient exact arithmetic. In most cases, we expect that resorting to exact arithmetic will not be needed.

References

[1] F. Avnaim, J.-D. Boissonnat, O. Devillers, F. Preparata, and M. Yvinec. Evaluating signs of determinants using single-precision arithmetic. *Algorithmica*, 17:111–132, 1997.

[2] H. Bronnimann, I. Emiris, V. Pan, and S. Pion. Computing exact geometric predicates using modular arithmetic with single precision. In *Proc. 13th Annu. ACM Sympos. Comput. Geom.*, pages 174–182, 1997.

[3] H. Bronnimann and M. Yvinec. Efficient exact evaluation of signs of determinants. In *Proc. 13th Annu. ACM Sympos. Comput. Geom.*, pages 166–173, 1997.

[4] C. Burnikel, S. Funke, and M. Seel. Exact arithmetic using cascaded computations. In *Submitted to this conference.* ACM, 1998.

[5] Christoph Burnikel, Jochen Könnemann, Kurt Mehlhorn, Stefan Näher, Stefan Schirra, and Christian Uhrig. Exact geometric computation in LEDA. In *Proc. 11th Annu. ACM Sympos. Comput. Geom.*, pages C18–C19, 1995.

[6] K. L. Clarkson. Safe and effective determinant evaluation. In *Proc. 33rd Annu. IEEE Sympos. Found. Comput. Sci.*, pages 387–395, 1992.

[7] O. Devillers and F. P. Preparata. A probabilistic analysis of the power of arithmetic filters. Technical Report CS-96-27, Center for Geometric Computing, Dept. Computer Science, Brown Univ., 1996.

[8] S. Fortune. Voronoi diagrams and Delaunay triangulations. In D.-Z. Du and F. K. Hwang, editors, *Computing in Euclidean Geometry*, volume 1 of *Lecture Notes Series on Computing*, pages 225–265. World Scientific, Singapore, 2 edition, 1992.

[9] S. Fortune and C. J. Van Wyk. Efficient exact arithmetic for computational geometry. In *Proc. 9th Annu. ACM Sympos. Comput. Geom.*, pages 163–172, 1993.

[10] S. Fortune and C. J. van Wyk. Static analysis yields efficient exact integer arithmetic for computational geometr y. *ACM Trans. Graph.*, 15(3):223–248, July 1996.

[11] B. Gärtner. Exact arithmetic at low cost - a case study in linear programming, proc. 9th annual acm-siam symp. on discrete algorithms. In *Proc. 9th Annual ACM-SIAM Symp. on Discrete Algorithms (SODA), pages 157–166, 1998.*

[12] G.H. Golub and C.F. van Loan. *Matrix Computations.* Johns Hopkins University Press, third edition, 1996.

[13] R. Hammer, M. Hocks, U. Kulisch, and D. Ratz. *C++ Toolbox for Verified Computing.* Springer, 1995.

[14] IEEE. IEEE standard 754-1985 for binary floating-point arithmetic. Reprinted in SIGPLAN *Notices* 22(2):9–25, 1987.

[15] K. Mehlhorn and St. Näher. The implementation of geometric algorithms. In *13th World Computer Congress IFIP94*, volume 1, pages 223–231. Elsevier Science B.V. North-Holland, Amsterdam, 1994.

[16] R.E. Moore. *Interval Analysis.* Prentice-Hall, 1966.

[17] A. Neumaier. *Interval Methods for Systems of Equations.* Cambridge University Press, 1990.

[18] Jonathan R. Shewchuk. Robust adaptive floating-point geometric predicates. In *Proc. 12th Annu. ACM Sympos. Comput. Geom.*, pages 141–150, 1996.

[19] C. K. Yap and T. Dubé. The exact computation paradigm. In *Computing in Euclidean Geometry*, D.-Z. Du and F. K. Hwang, eds., Lecture Notes Series on Computing, volume 1, pages 452–492. World Scientific, Singapore, 2nd edition, 1995.

Exact Geometric Predicates using Cascaded Computation*

Christoph Burnikel Stefan Funke Michael Seel
Max-Planck-Institut für Informatik
Im Stadtwald
66123 Saarbrücken
Germany
{burnikel,funke,seel}@mpi-sb.mpg.de

Abstract

In this paper we talk about a new efficient numerical approach to deal with inaccuracy when implementing geometric algorithms. Using various floating-point filters together with arbitrary precision packages, we develop an easy-to-use expression compiler called EXPCOMP. EXPCOMP supports all common operations $+, -, \cdot, /, \sqrt{}$. Applying a new semi-static filter, EXPCOMP combines the speed of static filters with the power of dynamic filters. The filter stages deal with all kinds of floating-point exceptions, including underflow. The resulting programs show a very good runtime behaviour.

1 Introduction

When computer scientists design geometric algorithms, they usually assume the availability of exact arithmetic on real numbers. Since no computer directly provides exact arithmetic on real numbers, programmers implementing these algorithms must find some substitution. Quite commonly, they resort to floating-point arithmetic due to its support by hard- and software as well as its convenient use. The resulting programs may not behave as expected, though. There are several ways a geometric algorithm may behave when exact arithmetic is replaced by floating-point arithmetic.

Some algorithms produce usable results in spite of some incorrect decisions – these algorithms are called stable (e.g. [14])– but most do not, they either produce completely inconsistent results, crash or loop. Hence for most problems it is crucial to ensure correctness of every predicate evaluation.

There are several exact arithmetic schemes designed specifically for computational geometry; most of them are methods for exactly evaluating the sign of a determinant using IEEE double precision floating-point arithmetic, and hence can be used to perform e.g. the orientation and incircle tests or even the insphere test [1]. Difficulties arise, if the tests to be performed involve previously computed geometric objects which require extended precision to be exactly represented.

A more general approach, which is not specific to determinants or even predicates, are multiprecision packages [2, 9, 13]. They allow arbitrary precision arithmetic on integers, fixed-point numbers or floating-point numbers. Of course, exact arithmetic with these packages has its cost, which is considerably higher than floating-point arithmetic. Depending on the input bit-length the arbitrary precision primitives are at least about 10-100 times slower than their floating-point counterparts.

If a predicate, expressed as the sign of an arithmetic expression, is to be computed, an obvious technique to reduce this overhead is trying to decide the predicate using floating-point arithmetic first. Only if no guarantee for the correctness of the outcome can be given, one resorts to arbitrary precision arithmetic. This technique is called a *floating-point filter* (also see [5]).

The focus of this paper is the conception of such a floating-point filter to be applied to both, integer and algebraic arithmetic; important issues are efficacy and ease of use.

In section 2 we focus on the problem of deriving error bounds for floating-point arithmetic at a reasonable cost. The approach presented is not restricted to integer values and handles all possible situations such as underflow exceptions and errors in the computation of the error bounds themselves. In section 3 we discuss packaging the results of the first part in a way such that it can be easily used within C++ programs. We design and implement an expression compiler called EXPCOMP which gives transparent access to the floating-point filter techniques. Finally, section 5 presents some results of applying the expression compiler – both with synthetic benchmarks and existing implementations of geometric algorithms.

Prior results in the field of floating-point filter techniques are mostly restricted to integer computation without divisions and square roots or neglect the problems of underflow. More importantly, most of these floating-point filter techniques are still unavailable for everyday programming

*This research was partially supported by the ESPRIT IV LTR Project No. 20244 (Alcom-IT).

work as they have to be encoded manually. There are similar approaches of packaging the filter techniques such as the expression compiler LN by Fortune/vanWyk ([6]), which is restricted to integer computation with fixed bit-lengths only, though. LN not only generates the code implementing the filter techniques but also the exact evaluation code, whereas EXPCOMP relies on an external datatype for the exact evaluation — we used the LEDA datatypes for exact computation.

2 A floating-point filter

If the floating-point arithmetic on a machine complies to the IEEE standard, we can guarantee certain bounds for the error occurring in *one particular* operation, see [8]. But as we evaluate complex expressions involving more than one operator, errors are propagated in some way from the "earlier stages" of computation to the final result. Assuming we have evaluated a complex expression e with floating-point arithmetic to $\widetilde{e}$, what we then want is an upper bound *err* for the error of this value:

$$|e - \widetilde{e}| \leq err$$

There exist several approaches for computing *err*. There are dynamic error estimation schemes, where the error bound is computed completely at runtime, hence it can make use of the actual input values to get the best error estimation possible. This may lead to a considerable overhead, though. See [9] and [3] for example.

On the other hand, there are static error estimation schemes, see [6] for example, which compute an upper bound for the error before runtime. This requires some knowledge about the actual input values at runtime, in particular the bit-length. The problem with this scheme is that if the presumed bit-length is chosen too long, the resulting error bounds are pessimistic and the filter is likely to fail. On the other hand, if the bit-length is chosen too short, using the program for larger input values requires recompiling.

For our purposes, we will divide the computation of the error bound in a static part and a dynamic part. The static part can be precomputed before runtime without any knowledge of the actual values of the expressions, whereas the dynamic part is computed during runtime.

For every expression e we not only compute the floating-point approximation $\widetilde{e}$ but also an upper bound $\widetilde{e_{sup}}$ for $|\widetilde{e}|$, called the supremum of $\widetilde{e}$, and an integer ind_e – the index of e – , such that the following bound for the absolute error of the floating-point approximation is true at all times (assuming that the index ind_e never exceeds $2^{\frac{p-2}{2}}$, which is not a critical constraint in practice, though):

$$|\widetilde{e} - e| \leq \widetilde{e_{sup}} \cdot ind_e \cdot 2^{-p} \tag{1}$$

where p is the mantissa length which is 52 for the standard C/C++ datatype **double**. An input value x exactly representable by a **double** has the floating-point approximation $\widetilde{x} = x$, the supremum $\widetilde{x_{sup}} = |\widetilde{x}|$ and the index 0. An input value not exactly representable by a **double** has the floating-point approximation $\widetilde{x} = round(x)$, the supremum $\widetilde{x_{sup}} = |\widetilde{x}| = |round(x)|$ and the index 1.

The index ind_e may be computed statically whereas $\widetilde{e}$ and $\widetilde{e_{sup}}$ must be computed at runtime using the inductively given rules in table 1. In the following $+, -, \cdot, /, \cdot^{\frac{1}{2}}$ denote exact addition, subtraction, multiplication, division and square root, whereas $\oplus, \ominus, \odot, \oslash, \surd$ denote their floating-point counterparts. Detailed proofs can be found in [7]. A similar error bound is also shown in [10].

Note that the computation of the supremum is quite similar to the computation of the floating-point approximation itself and therefore the implied overhead is reasonably small. In case of $+, -, \cdot$, only twice as much operations are needed, three times as much in case of $\surd$. The division has a considerably higher overhead but its use often can be avoided in practice.

With this error bound, it is now easy to decide if a floating-point evaluation $\widetilde{e}$ of an expression e has the correct sign. This is obviously the case if

$$|\widetilde{e}| > \widetilde{e_{sup}} \cdot ind_e \cdot 2^{-p}$$

Another assumption even allows a statement about the numerical exactness of $\widetilde{e}$. If the input values are of integer type and $+, -, \cdot$ the only operations used,

$$\widetilde{e_{sup}} \cdot ind_e \cdot 2^{-p} < 1$$

implies $\widetilde{e} = e$.

Nothing has been said yet about what happens if an underflow occurs during the computation of $\widetilde{e}$ or $\widetilde{e_{sup}}$. Normally this is a pathological point of many error estimation schemes. For our scheme, we have proven that in case of addition, subtraction and square root underflow can be neglected, and in case of the division and multiplication it can be compensated by adding a small constant ($MinDbl$) to $\widetilde{e_{sup}}$. If the "original" input values are of integer type and no divisions occur – which is often the case in practice – underflow can totally be neglected. Apart from that, our scheme also deals with second order errors, i.e. errors which occur during the computation of the error bounds themselves.

Looking at the rules for computing the suprema it turns out that if only $+, -, \cdot$ occur, and upper bounds for $\lceil |x| \rceil$ of all input values x exist, an upper bound for $\widetilde{e_{sup}}$ can be statically computed, which then leads us to the following evaluation strategy:

1. evaluate the expression using floating-point arithmetic to get the approximation $\widetilde{e}$

2. if only $+, -, \cdot$ occur and upper bounds for the input values are available, check $\widetilde{e}$ with the statically determined error bound $e_{sup_{static}} \cdot ind_e \cdot 2^{-p} \Rightarrow$ *fully-static filter*

expr. e	approx. $\widetilde{e}$	supremum $\widetilde{e_{sup}}$	index ind_e
$x+y$	$\widetilde{x} \oplus \widetilde{y}$	$\widetilde{x_{sup}} \oplus \widetilde{y_{sup}}$	$1+MAX(ind_x, ind_y)$
$x-y$	$\widetilde{x} \ominus \widetilde{y}$	$\widetilde{x_{sup}} \oplus \widetilde{y_{sup}}$	$1+MAX(ind_x, ind_y)$
$x \cdot y$	$\widetilde{x} \odot \widetilde{y}$	$\widetilde{x_{sup}} \odot \widetilde{y_{sup}}$	$1+ind_x+ind_y$
x/y	$\widetilde{x} \oslash \widetilde{y}$	$\frac{(\lvert\widetilde{x}\rvert \oslash \lvert\widetilde{y}\rvert) \oplus (\widetilde{x_{sup}} \oslash \widetilde{y_{sup}})}{(\lvert\widetilde{y}\rvert \oslash \widetilde{y_{sup}}) \ominus (ind_y+1) \cdot 2^{-p}}$	$1+MAX(ind_x, ind_y+1)$
$x^{\frac{1}{2}}$	$\sqrt{\widetilde{x}}$	$\begin{cases} (\widetilde{x_{sup}} \oslash \widetilde{x}) \odot \sqrt{\widetilde{x}} & if\ \widetilde{x}>0 \\ \sqrt{\widetilde{x_{sup}}} \odot 2^{\frac{p}{2}} & if\ \widetilde{x}=0 \end{cases}$	$1+ind_x$

Table 1: Rules for computing approximations, suprema and indices.

3. if 2. fails, compute dynamically the supremum $\widetilde{e_{sup}}$ and check $\widetilde{e}$ with the error bound $\widetilde{e_{sup}} \cdot ind_e \cdot 2^{-p} \Rightarrow$ *semi-static filter*.
4. if 2. and 3. fail, evaluate the expression using arbitrary precision arithmetic

3 Packaging the floating-point filter

Manually encoding the filter mechanisms shown in the last chapter is much too cumbersome. Not only has a programmer to implement a given predicate evaluation several times – at least once for the floating-point filter stage and once for the exact evaluation– he also has to compute the indices manually. Even small changes in a predicate then require a considerable effort. There are basically two ways of packaging the floating-point filter:

1. a C++ class wrapper
2. an expression compiler

In the following we will discuss both approaches and show their advantages and drawbacks.

3.1 Class Wrapper

As an example of a class wrapper we implement a class `FpFilt` replicating the functionality of the C/C++ number types. Internally though, an error bound is computed for each expression using the methods derived in section 2. If the sign of such a value is asked for using the `::sign()` member function, it may return -1, 0, +1 or NO_IDEA if no guarantee for the exactness of the sign can be given.

We will illustrate the use of this new class with a simple example occurring in many geometric algorithms – the 2d orientation test.

Given three points $p, q, r \in \mathbf{Z}^2$ with $p = (p_x, p_y)$, $q = (q_x, q_y)$, $r = (r_x, r_y)$; to which side of the oriented line $\overrightarrow{pq}$ is r located ?

The following pseudo-code evaluates this predicate using the LEDA datatype for arbitrary precision integers:

```
integer DX ← p_x − q_x
integer DY ← p_y − q_y
integer TL ← q_x · DY − q_y · DX
integer RES ← DX · r_y − DY · r_x + TL
ressign ← sign(RES)
```

The resulting sign in *ressign* indicates, where with respect to $\overrightarrow{pq}$ the point r is located: left of (*ressign*=-1), right of (*ressign*=1) or on the line (*ressign*=0). The same predicate, when enhanced with a floating-point filter using the new C++ class `FpFilt`:

```
FpFilt DX ← FpFilt(p_x) − FpFilt(q_x)
FpFilt DY ← FpFilt(p_y) − FpFilt(q_y)
FpFilt TL ← FpFilt(q_x) · DY − FpFilt(q_y) · DX
FpFilt RES ← DX · FpFilt(r_y)
           −DY · FpFilt(r_x) + TL
ressign ← sign(RES)
if (ressign ≡ NOIDEA)
{integer DX ← p_x − q_x
 integer DY ← p_y − q_y
 integer TL ← q_x · DY − q_y · DX
 integer RES ← DX · r_y − DY · r_x + TL
 ressign ← sign(RES)
}
```

We see that the "filter" code looks just like a reimplementation of the predicate in the filter datatype – something which could be automated. Ideally the floating-point filter should be invisible for the programmer. Nevertheless, with this approach, the code evaluating the predicate can be interleaved or nested arbitrarily within other language constructs such as if-statements or loops. This allows a maximum of flexibility in the use of the floating-point filter. Apart from that, no preprocessing of the source file is necessary.

But there are disadvantages as well. Wrapping functionality in a C++ class always implies some overhead as we will see in our experimental results. This agrees with the observations made in [6]. Apart from that, this approach obviously does not allow static precomputations before runtime, i.e. in our case both the computation of the suprema and the indices have to take place at runtime.

3.2 Expression compiler

An expression compiler remedies most of the drawbacks mentioned above. It parses marked code segments within a source file which represent the evaluation of a predicate and replaces them with the equivalent two-level filter–arbitrary precision scheme.

The source code for the orientation test example from above may then look like that:

```
BEGIN_PREDICATE
{
integer DX ← p_x − q_x
integer DY ← p_y − q_y
integer TL ← q_x · DY − q_y · DX
integer RES ← DX · r_y − DY · r_x + TL
ressign ← sign(RES)
}
END_PREDICATE
```

Note that the use of a floating-point filter is almost invisible to the programmer. The only difference are the BEGIN_PREDICATE ... END_PREDICATE directives around the original code, which are required by the expression compiler to recognize the predicate evaluation. The output may look like that:

```
<floating-point filter evaluation>
<with result in ressign>
if (ressign ≡ NOIDEA)
{integer DX ← p_x − q_x
 integer DY ← p_y − q_y
 integer TL ← q_x · DY − q_y · DX
 integer RES ← DX · r_y − DY · r_x + TL
 ressign ← sign(RES)
}
```

Before generating code, the expression compiler may also perform some static precomputations, such that less work is done at runtime. In particular, if we use the semi-static error analysis, the expression compiler could precompute the indices of all (sub)expressions.

There are disadvantages as well, though. The programmer has to restrict to a reasonably small subset of all C/C++ language constructs, to keep parsing and code generation manageable. Loops for example would complicate static analysis considerably.

In an actual implementation of the expression compiler, the programmer must provide some information about the types of the identifiers (such as p_x, p_y, q_x ...) referenced within the marked code sequence.

The advantages and drawbacks of both approaches suggest the use of an expression compiler which we will describe in the following.

4 EXPCOMP – an expression compiler

In this section we sketch our implementation of an expression compiler as suggested in the last section. We designed it in such a way that it can be easily used with the datatypes for arbitrary precision provided by the LEDA software library (Library of Efficient Datatypes and Algorithms) (see [9]). In LEDA the programmer has access to three C++ classes for integer, rational and algebraic computation (**integer**, **rational** and **real**). They can be used very similarly to the standard C/C++ number datatypes due to operator overloading, but also have additional functions for conversion, sign determination etc. Adapting EXPCOMP to other arbitrary precision datatypes is not difficult, though.

The expression compiler EXPCOMP is basically a preprocessing tool which is invoked with a source and a target filename. The source file contains C++ code mixed with special statements which allow the expression compiler to identify the code sequences to be replaced using the floating-point filter techniques introduced in section 2.

There are two ways in which the expression compiler can be used:

1. Predicate evaluation
2. Value computation

For simple programs where no extensive numeric precomputations takes place, i.e. the numeric part of computation mainly consists of evaluating predicates which do not reference precomputed values, it is sufficient to apply the expression compiler to the predicate evaluations. In programs though, where large parts of the numeric computation consist of computing values for the later use in predicates, the expression compiler can also be applied. In both cases only small modifications of the original code are necessary.

4.1 Predicate evaluation

When used to speed up predicate evaluation, the following modifications have to be made:

- marking the code sequences which represent the predicate evaluations using BEGIN_PREDICATE { ... } END_PREDICATE
- declaring the types of all identifiers/functions referenced within the code sequence
- optionally giving the expression compiler more information about the input values, such as bit-length or sign

In the previous section we have come across the orientation test occurring in many geometric algorithms. The following code fragment evaluates this predicate using LEDA **integer** arithmetic, provided the points P, Q, R have their x- and y-coordinate are stored as $P.X$, $P.Y$, ...:

```
integer DX=P.X-Q.X;
integer DY=P.Y-Q.Y;
integer TL=Q.X*DY-Q.Y*DX;
integer RES=DX*R.Y-DY*R.X+TL;
ressign=sign(RES);
```

To make use of the floating-point filter, this code sequence is mixed with special expression compiler statements. In our example we tell EXPCOMP that all input values of the predicate are nonnegative LEDA **integer**s which have at most bit-length 16:

```
BEGIN_PREDICATE
{
DECLARE_ATTRIBUTES integer_type pos_val FOR
    P.X P.Y Q.X Q.Y R.X R.Y;
DECLARE_BITLENGTH 16 FOR
    P.X P.Y Q.X Q.Y R.X R.Y;
integer DX=P.X-Q.X;
integer DY=P.Y-Q.Y;
integer TL=Q.X*DY-Q.Y*DX;
integer RES=DX*R.Y-DY*R.X+TL;
ressign=sign(RES);
}
END_PREDICATE
```

This code is then replaced by the expression compiler by the appropriate 2-level code. Additionally, if a LEDA **real** predicate involves square roots and divisions, and all input values are of integer type, EXPCOMP can generate code in which an intermediate filter stage is introduced. Then the evaluation scheme is first to try making a decision based on floating-point arithmetic; if this fails, the predicate is tested using LEDA **integer** arithmetic. Note that this requires checking after every square root and division if the result is really an integer value. Only if this second stage also fails, LEDA **real** arithmetic is used. In practice this has proven to be very efficient for detecting degenerate cases (see also [4]).

4.2 Value precomputation

So far, we have assumed that the input variables for a predicate are available as exact values. In more complex applications though, one often decides to store the results of intermediate calculations for later use. The predicates are then expressed using these precomputed values, which possibly cannot be exactly computed using **double** precision. As an example we will have a look at the incircle test occurring in the construction of the Voronoi diagram of line segments.

> Let v be a Voronoi vertex defined by three sites s_1, s_2, s_3 (points or lines), of which at least s_3 is a point p. Given a point q, does q touch, lie inside or outside the clearance circle of v ?

Usually the programmer would have created a C++ class **vnode** which on instantiation by its three defining sites computes the coordinates of this Voronoi node. If these coordinates are used within a predicate evaluation later on, they are accessed using the appropriate member functions of the **vnode** class.

In the first place, all of the above is implemented using an arbitrary precision datatype such as LEDA **real**. If we want to introduce a floating-point filter not only to the predicate evaluation but also to the computation of the Voronoi coordinates, we have to use a lazy-evaluation scheme.

First, the **vnode** class is extended such that it has member functions not only for returning the coordinates as LEDA **real** (e.g. **real** ::X()), but also for returning the **double** approximation (::XD()) and the supremum (::XDS()) respectively. Most importantly none of the coordinates are computed on instantiation of a **vnode** object, but only at the time they are requested first, usually by a predicate evaluation. On instantiation, only the three defining sites are stored such that the actual coordinates can be computed at a later time using the required datatype.

With this scheme, a **vnode** instantiation v which is never asked for its **real** coordinates, because all predicates involving v can be decided by **double** arithmetic, never computes the coordinates by arbitrary precision arithmetic.

One can imagine that the code sequences which compute the coordinates of a Voronoi node – one using **double** arithmetic, the other using LEDA **real** arithmetic – are very similar, as it was the case for predicate evaluation. So again, EXPCOMP can be used to generate these two "versions" of the code. The programmer only has to provide a "template" which is basically the original computation with LEDA **real**s enriched with some special expression compiler statements. The usage is very similar to the predicate evaluation, for more details see [7]. The expression compiler also propagates the indices of precomputed values to the predicates where they are referenced. As above this scheme easily extends to a 3-level evaluation scheme with an intermediate **integer** stage, which tries to decide the predicate if the filter stage fails.

Of course, this scheme is not restricted to our Voronoi node example. Many implementations of geometric algorithms which compute intermediate results for later use, can be modified in the same manner. As the expression compiler statically computes the indices, the only restriction is that the stream of computation must be completely known at compile-time. In particular, loops and recursive calls during the calculation are not allowed.

5 Experimental Results

5.1 Synthetic benchmarks

In this section we will determine experimentally to which extent the use of the expression compiler affects the performance of synthetic benchmarks as well as actual implementations of geometric algorithms. As test environment we used a 200 MHz UltraSparc running Solaris 2.5. All programs are compiled using g++ 2.7.2 with the -O option set.

Bit-length	int.	real	FpFilt	int. filter	real filter	stat. filter
k=8	42.8	24.1	2.9	1.9	2.2	2.6
k=12	46.6	24.6	2.9	1.8	2.3	2.7
k=16	46.2	56.1	2.8	1.9	2.3	2.7
k=24	61.5	62.6	2.8	1.7	2.3	2.6
k=32	65.7	64.9	2.9	1.9	2.2	2.5
k=40	123.3	64.6	2.9	1.8	2.3	1.5
k=48	125.1	60.3	2.9	1.8	2.3	1.5

Table 2: Det3x3, random data

Bit-length	int.	real	FpFilt	int. filter	real filter	stat. filter
k=8	37.9	24.0	2.4	1.4	26.3	2.0
k=12	44.5	24.1	2.7	1.4	26.3	2.3
k=16	45.3	1378	2.4	1.4	1378	2.0
k=24	54.3	1719	56.3	56.5	1721	54.7
k=32*	56.3	2633	56.5	58.4	2636	59.3
k=40*	117.4	4353	119.4	117.5	2644	117.3
k=48*	135.2	5151	136.5	135.1	5157	136.9

Table 3: Det3x3, degenerate data

Signs of 3x3 determinants

In this experiment we measure the runtimes of several number types and the code generated by the expression compiler when computing 100000 times the sign of 3x3 determinants with integer entries.

The following implementations are tested and compared to **double** arithmetic evaluation (i.e. the time overhead $\frac{runtime\ impl}{runtime\ double}$ is measured):

- LEDA **integer** [int.]
- LEDA **real** [real]
- LEDA **integer** with FpFilt class (error bounds as in section 2) [FpFilt]
- LEDA **integer** tuned with EXPCOMP [int. filter]
- LEDA **real** tuned with EXPCOMP (implies underflow "countermeasures") [real filter]
- LEDA **integer** tuned with EXPCOMP 2-level full- and semi-static filter for a presumed bit-length of 48 [stat. filter]

All implementations ensure that the delivered sign result is exact, which is not the case for the **double** evaluation — so the comparison is somewhat unfair.

We will test both, matrices with random entries and non-regular matrices where the determinant is zero. The determinants are evaluated according to Sarrus' rule. The test results are listed in tables 2 and 3.

As we can see, all tests with random data can be decided by the filter-stages. The fully static filter only comes into effect when the actual bit-length is near the presumed one, otherwise the second stage – the semi-static filter – succeeds. The differences between the integer filter and the static filter for bit-lengths < 40, can be explained by the fact that in case of the integer filter, the instructions for computing $\widetilde{e}$ and $\widetilde{e_{sup}}$ are interleaved, otherwise they aren't. Due to the compiler's optimizer and the architecture of the FPU this may lead to differences in runtime.

The degenerate cases could only be decided by integer filter for short bit-lengths, so most of the time the arbitrary precision calculation dominates the runtime. It is remarkable how badly LEDA **real**s perform in degenerate cases.

Bit-length	real	3-level filter	2-level filter
k=8	20.2	2.4	2.5
k=12	20.3	2.4	2.5
k=16	21.1	2.4	2.5
k=24	21.7	2.5	2.6
k=32	29.2	2.5	2.6
k=40	28.8	2.5	2.6
k=48	28.3	2.4	2.5

Table 4: Incircle, random data

Incircle test

In this experiment we use a predicate occurring in the construction of the Voronoi diagram of line segments and measure the different implementations. The code is taken from [4], page 59.

> Let $v = (v_x, v_y, v_z)$, $v_z > 0$ be a Voronoi vertex defined by three sites p, l_1, l_2, where $p = (0,0,0)$ is the origin and l_1, l_2 are lines. Given a line l_3, does l_3 touch, intersect or miss the clearance circle of v ?

Assuming that the input variables are all of integer type, we will test three implementations (which all guarantee the exact sign result) of this incircle test and compare them with the (possibly faulty) **double** arithmetic implementation:

- LEDA **real**
- EXPCOMP 3-level (**filter–integer–real**)
- EXPCOMP 2-level (**filter–real**)

We will measure both, incircle tests with randomly chosen as well as degenerate input. For the degenerate cases we construct three lines that touch a common circle passing through $p = (0,0)$, as described in [4] page 59. The test results are listed in tables 4 and 5.

As for the tests with signs of determinants, it turns out that LEDA **real**s perform badly in degenerate cases. In the non-degenerate cases, the overhead of LEDA **real** compared to **double** is by far not as large as for the determinants' tests.

Bit-length	real	3-level filter	2-level filter
k=8	14.7	31.7	16.7
k=12	37.5	53.5	37.4
k=16	115.5	43.7	115.8
k=24	367.5	70.3	366.3
k=32	673.3	76.9	674.3
k=40	888.3	106.4	889.1
k=48	1169	127.9	1169

Table 5: Incircle, degenerate data

This is due to the expensive square root operation which dominates the runtime cost.

5.2 Real-world benchmarks

When measuring the runtime of real-world applications, care has to be taken, since not only time spent in the numeric part is measured but also the time spent in the symbolic part. Therefore the total speedup when using floating-point filter techniques is usually much smaller than the results of the previous section may suggest. Of course, the user is only interested in the total speed-up of the program, but for us, a closer examination may be interesting.

The runtime T of a program consists of a symbolic part S and a numeric part N:

$$T = T_{(1)} = S + N$$

If we modify the program such that each numeric computation is performed $c > 1$ times, e.g. $c = 2$, the runtime of this modified program should be

$$T_{(c)} = S + c \cdot N$$

With these two measurements it is easy to get approximate values for S and N.

Triangulation of simple polygons

We implemented an algorithm for the incremental randomized construction of triangulations of simple polygons according to Seidel ([12]). The algorithm first computes the trapezoidal decomposition induced by a set S of n non-crossing line segments in the plane. If S is a simple polygonal chain the expected runtime of the algorithm is O(n log* n). Given the trapezoidation of a simple polygon it is easy to get its triangulation in O(n) time. Hence the triangulation of a simple polygon can be computed in O(n log*n) time with this algorithm.

We are interested in the geometric predicates used in the algorithm. The algorithm is quite simple and hence there is only one non-trivial predicate — the sidedness test which we have already come across in the previous sections.

We assume that the input data (the coordinates) is of integer type and has **double** precision, i.e. it can be exactly represented by **doubles**.

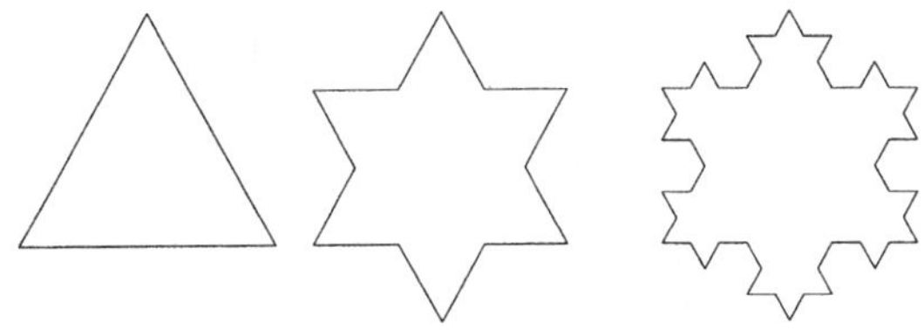

Figure 1: Koch-isles of recursion depth 0, 1 and 2

bit-length	double	integer	filter
16	4.25	20.91	5.09
24	4.36	20.86	5.19
32	4.31	21.38	5.71
40	4.25	24.93	8.20
48	-	26.69	8.81

Table 6: PolyTriang: time (in seconds)

Runtime experiments were made with Koch-isles (see figure 1) of several recursion depths where the corner coordinates are rounded to the next integers.

The following tables show the average runtime for different implementations and different input data bit-lengths. We measured three implementations of the predicate:

- **double** arithmetic
- LEDA **integer** arithmetic
- LEDA **integer** arithmetic with EXPCOMP

and several input bit-lengths between 16 and 48 bit for a "snowflake" of recursion depth 7, i.e. a polygon with 49152 nodes. The results are shown in table 6.

Note though that the implementations do not only differ in the way the predicate is evaluated. Each implementation is optimized with respect to its predicate evaluation. In particular, each implementation stores the coordinates of the input data in the most suitable type, which is **double** for the filter implementation.

Up to an input data bit-length of 24, all predicates can be decided by the floating-point filter. With increasing bit-length more and more decisions require exact (**integer**) arithmetic, so the overhead increases. For bit-length 48 the **double** implementation crashed due to predicates which are evaluated incorrectly.

For a more precise analysis we examine the number of "difficult" tests that require **integer** arithmetic in the filter implementation. Table 7 shows the results.

In our example, even with bit-length 48 less than six percent of all tests require resorting to **integer** arithmetic. Most of these were degenerate cases, which are indeed more difficult to decide.

Finally we measure how much time is actually spent in the symbolic and in the numeric part of the computation using the method explained at the beginning of this section. We

bit-length	# of tests	diff.	diff. and "0"
16	373610	0	0
24	374558	0	0
32	385855	14213	13954
40	384232	18339	12443
48	387932	21425	13386

Table 7: PolyTriang: #of tests

	double	integer	filter
total time	4.43	22.94	5.90
symbolic	4.18	12.45	4.84
numeric	0.25	10.49	1.06

Table 8: PolyTriang, bit-length 32: time (in seconds)

take the example with 49152 nodes and input bit-length 32. Table 8 shows the results. One does not expect differences in the symbolic runtimes; but as each implementation is optimized with respect to its predicate evaluations, i.e. for example that the integer implementation uses LEDA **integer**s to store the original data, whereas the other implementations use **double**s, their symbolic runtimes may vary.

Voronoi diagram of points and line segments

In this last experiment, the expression compiler was used to tune an implementation of an randomized incremental algorithm for computing the Voronoi diagram of points and line segments (see[11]).

As to be expected there are much more predicates than in the previous algorithm. What is also different from the previous algorithm is the fact that the input values for some of the predicates are values that have been precomputed. Moreover, these values may be non-rational, for the coordinates of Voronoi nodes are usually non-rational if at least one defining site is a segment.

As a very first step, we only concentrated on the predicates mentioned above, such as the incircle test. By applying the expression compiler, we replaced the exact evaluations by two-level evaluations, where we first try to evaluate the predicates using **double** arithmetic and only if this fails, exact arithmetic is used. The result was quite disappointing, performance improved by less than 15 % only.

This can be explained by the fact, that the "input values" of these predicates – the coordinates of the Voronoi nodes – are still computed using exact (LEDA **real**) arithmetic.

Hence we incorporated the computation of the Voronoi node coordinates into the floating-point filter mechanism using the expression compiler as sketched in the last section. Moreover we introduced a third intermediate filter stage where we try to decide a predicate using **integer** arithmetic if the floating-point filter fails. We have measured five different implementations with different sets of input data:

Bit-length	orig.	r	dr	dir	d
12	189.2	218.8	85.5	82.3	-
16	204.1	242.8	91.5	86.7	-
24	240.0	279.8	101.4	93.8	80.4
32	336.4	387.4	183.9	136.5	-
40	459.0	511.7	282.7	200.3	-
48	494.2	548.6	288.0	187.7	-

Table 9: VoroDiag: time (in seconds)

	r	dir	dr	d
total time	278.9	96.8	99.3	80.9
symbolic	76.8	82.1	81.9	71.0
numeric	202.1	14.7	17.4	9.9

Table 10: VoroDiag, bit-length 24: time(in seconds)

orig	original impl. using LEDA **real**
r	lazy-eval, no filter stages
dr	lazy-eval, with fp-filter
dir	lazy-eval, with 3-level filter
d	lazy eval, only fp filter (not always exact)

For each set of input data we not only measure the runtimes of the different implementations but also analyze, how much time is spent in the symbolic/numeric part of computation as we did for the triangulation program. See tables 9 and 10 for the results. Unfortunately, due to the internal representation of LEDA **real**, a second computation of a predicate does not take twice as much time. So in these cases, the runtime of the numeric part of computation is *underestimated.*

Furthermore we will analyze all occurring instances of the incircle test — how many can be decided at which stage and whether they are degenerate. See table 11.

As for the triangulation algorithm we use the "Koch isle" for benchmarking the Voronoi program. Due to the considerably more complex computations of the RIC-AVD algorithm, we restrict to a recursion depth of 4 (768 nodes).

5.3 Summary

The synthetic (low-level) benchmarks showed that the floating-point filter itself only introduces an overhead of

	double	integer	real	$\sum$
non deg.	497166 (99.22%)	3042 (0.61%)	66 (0.01%)	500274 (99.84%)
deg.	0 (0%)	718 (0.14%)	62 (0.01%)	780 (0.16%)
$\sum$	497166 (99.22%)	3760 (0.75%)	128 (0.03%)	501054 (100%)

Table 11: VoroDiag, bit-length 24: #of tests

about factor 2-3 compared to **double** arithmetic, whereas the LEDA datatypes for arbitrary precision introduce an overhead of at least factor 20. Of course, when applying the expression compiler to actual applications, the speed-up is not as high as the low-level benchmarks may suggest, since the algorithms also spend some time in symbolic computations, which is not or far less affected by the use of a floating-point filter. Nevertheless, the two examples where the expression compiler was used, showed a considerable speed-up. After a closer examination, the speed-up of the numerical part of computation showed to be near the one suggested by the low-level benchmarks. It also showed that LEDA **real**s, which perform quite badly in degenerate cases, can be efficiently tuned using an intermediate **integer** stage.

6 Conclusion

The expression compiler we implemented has proven to be very useful for tuning implementations of geometric algorithms. In contrast to existing solutions it supports all operations ($+, -, \cdot, /, \sqrt{}$), storing intermediate results for later use in predicates and various filter stages. Existing code can be reused and converted in a very transparent manner. In contrast to Fortune/vanWyk's LN, no auxiliary language is introduced. The performance of the resulting programs is very good. Nevertheless, correctness of the optimized code is always ensured, e.g. possible underflow conditions are not neglected.

Our implementation can easily be adapted to other non-LEDA datatypes for arbitrary precision numbers, in the future it may also be possible to let the expression compiler generate directly the arbitrary precision code (which for example Fortune/vanWyk's LN does) and not use an external datatype. Introducing more filter-stages, such as the fully-dynamic approach shown in [3] is also possible with little effort.

The original work ([7]) and the source code can be downloaded from *http://www.mpi-sb.mpg.de/~funke*.

References

[1] F. Avnaim et al.: *Evaluating signs of determinants using single-precision arithmetic*, (INRIA, 1995)

[2] D. H. Bailey: *A Portable High Performance Multiprecision Package*, (Technical Report RNR-90-022, NASA Ames Research Center, May 1993)

[3] H. Brönnimann, C. Burnikel, S. Pion: *Interval Analysis Yields Efficient Arithmetic Filters for Computational Geometry*, (Proceedings of the 14th Symposium on CG 1998, ACM Press)

[4] C. Burnikel: *Exact Computation of Voronoi Diagrams and Line Segment Intersections*, (PhD thesis, Max-Planck-Institut für Informatik, 1996)

[5] O. Devillers, P. Preparata: *A Probabilistic Analysis of the Power of Arithmetic Filters*, (Technical Report CS 96-27, Center for Geometric Computing, Dept. Computer Science, Brown Univ)

[6] Steven Fortune, Christopher J. Van Wyk: *Static analysis yields efficient exact integer arithmetic for computational geometry,*, (ACM Trans. Graphics, 15(3), pp. 223–248, July 1996)

[7] S. Funke: *Exact Arithmetic using Cascaded Computation*, (Master-Thesis, Max-Planck-Institut für Informatik, 1997)

[8] *IEEE standard 754-1985 for binary floating-point arithmetic*, reprinted in SIGPLAN 22, 2:9-25, 1987

[9] K. Mehlhorn, S. Näher, C. Uhrig: *The LEDA User Manual (Version R 3.4)* (Max-Planck-Institut für Informatik, 1996)

[10] K. Mehlhorn, S. Näher: *The Implementation of Geometric Algorithms*, Proc. 13th World Computer Congress IFIP94, vol.1, 223-231

[11] M. Seel: *Eine Implementierung abstrakter Voronoidiagramme*, (Master Thesis, Universität des Saarlandes 1994)

[12] R. Seidel: *A simple and fast incremental randomized algorithm for computing trapezoidal decompositions and for triangulating polygons*, (Computational Geometry: Theory and Applications 1 (1991) 51-64)

[13] B. Serpette, J. Vuillemin, J.C. Herve: *BigNum: a portable and efficient package for arbitrary-precision arithmetic*, (INRIA)

[14] K. Sugihara, M. Iri: *A robust Topology-Oriented Incremental Algorithm for Voronoi diagrams*, (Internat. J. Comput. Geom. Appl. 4 (1994) 179-228

Geometric graphs with few disjoint edges

*Géza Tóth** *Pavel Valtr†*

Abstract

A *geometric graph* is a graph drawn in the plane so that the vertices are represented by points in general position, the edges are represented by straight line segments connecting the corresponding points.

Improving a result of Pach and Törőcsik, we show that a geometric graph on n vertices with no $k+1$ pairwise disjoint edges has at most $k^3(n+1)$ edges. On the other hand, we construct geometric graphs with n vertices and approximately $\frac{3}{2}(k-1)n$ edges, containing no $k+1$ pairwise disjoint edges.

We also give an $O(m^2)$ algorithm for finding at least $(m/(n+1))^{1/3}$ pairwise disjoint edges in a geometric graph with n vertices and m edges ($m > n$).

1 Introduction

A *geometric graph* G is a graph drawn in the plane by (possibly crossing) straight line segments, i.e., it is defined as a pair $G = (V, E)$, where V is a set of points in general position in the plane and E is a set of closed segments whose endpoints belong to V.

The following question was raised by Avital and Hanani [AH], Kupitz [K1], Erdős and Perles. Determine the smallest number $e_k(n)$ such that any geometric graph with n vertices and $m > e_k(n)$ edges contains $k+1$ pairwise disjoint edges. By a result of Hopf and Pannwitz [HP] and Erdős [E], $e_1(n) = n$. Alon and Erdős [AE] showed that $e_2(n) \leq 6n$ which was improved by Goddard, Katchalski and Kleitman [GKK] to $e_2(n) \leq 3n$. The best known lower bound, $e_2(n) \geq 2.5n - 4$, is due to Perles (see [PA]). In [GKK] it was also shown that $3.5n - 6 \leq e_3(n) \leq 10n$ and that $e_k(n) \leq c_k n(\log n)^{k-3}$ ($k \geq 4$). For any fixed k, Pach and Törőcsik [PT] were first to prove that $e_k(n)$ is linear in n; their bound was $e_k(n) \leq k^4 n$. On the other hand, $e_k(n) \geq kn$ for $n \geq 2k$ follows from a result of Kupitz [K1]. In this paper we further improve both the upper and lower bounds for general k.

*Courant Institute, NYU and DIMACS Center, Rutgers University. E-mail: geza@dimacs.rutgers.edu. Supported by OTKA-T-020914, OTKA-F-22234 and NSF grant CCR-94-24398.

†Charles University, Prague and DIMACS Center, Rutgers University. E-mail: valtr@kam.mff.cuni.cz. Supported by the Czech Republic grant GAČR 0194 and Charles University grants GAUK 193, 194.

SCG 98 Minneapolis Minnesota USA

Theorem 1 *For $n \geq 2k$,*

$$\frac{3}{2}(k-1)n - 2k^2 \leq e_k(n) \leq k^3(n+1).$$

In any geometric graph G, one can find a set of disjoint edges, at least as large as guaranteed by the upper bound in Theorem 1, in $O(m^2)$ time, where m is the number of edges of G.

Theorem 2 *There is an $O(m^2)$ time algorithm for finding at least $\left(\frac{m}{n+1}\right)^{1/3}$ pairwise disjoint edges in a geometric graph with n vertices and m edges ($m > n$).*

For $k = 3$, the above mentioned bounds can be slightly improved.

Theorem 3 *For any $n \geq 6$,*

$$4n - 9 \leq e_3(n) \leq 8.5n.$$

Theorems 1, 2 and 3 are proved in Sections 2, 3 and 4, respectively. Throughout the paper, we do not make any notational distinction between an edge and the segment representing it.

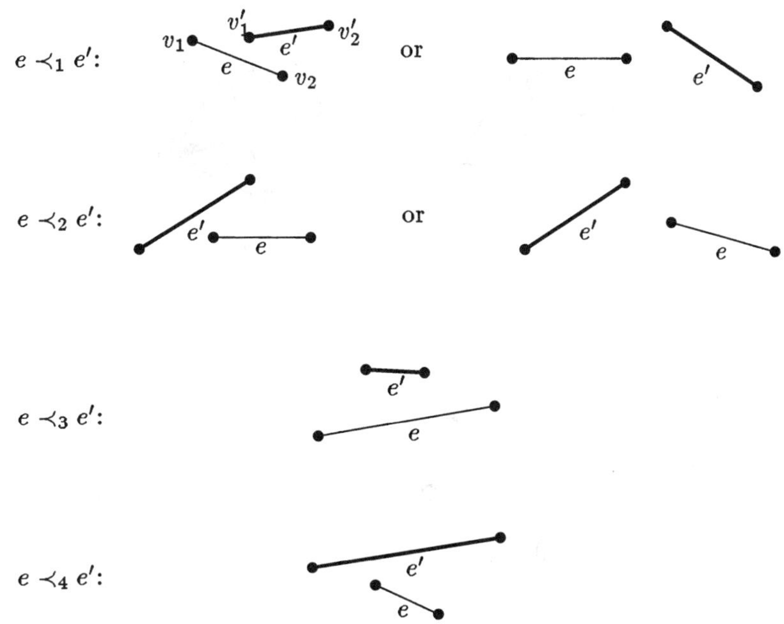

Figure 1: The relations $\prec_i$.

2 The general case

2.1 The upper bound

Our proof, as the proof of Pach and Töröcsik [PT], is based on Dilworth's theorem [D].

Dilworth's Theorem. *Let P be a partially ordered set containing no chain (totally ordered subset) of size $k+1$. Then P can be covered by k antichains (subsets of pairwise incomparable elements).*

Let $G=(V,E)$ be a geometric graph on n vertices, containing no $k+1$ pairwise disjoint edges. For a vertex v, let $x(v)$ and $y(v)$ denote its x- and y-coordinate, respectively. We can assume without loss of generality that no two vertices have the same x-coordinate.

An edge e is said to *lie below* an edge e', if no vertical line crossing both e and e' crosses e strictly above e'. Finally, let $\pi(e)$ denote the orthogonal projection of e to the x-axis.

Define four binary relations $\prec_i$ $(i=1,\ldots,4)$ on the edge set E as follows (see also [PT, PA]). Let $e=v_1v_2, e'=v_1'v_2'$ be two disjoint edges of G, where $x(v_1)<x(v_2)$ and $x(v_1')<x(v_2')$. Then (see Fig. 1)

$e \prec_1 e'$, if

$x(v_1)<x(v_1'),\ x(v_2)<x(v_2')$, and e lies below e',

$e \prec_2 e'$, if

$x(v_1)>x(v_1'),\ x(v_2)>x(v_2')$, and e lies below e',

$e \prec_3 e'$, if

$x(v_1)<x(v_1'),\ x(v_2)>x(v_2')$, and e lies below e',

$e \prec_4 e'$, if

$x(v_1)>x(v_1'),\ x(v_2)<x(v_2')$, and e lies below e'.

Obviously, each of the relations $\prec_i$ is a partial ordering, and any pair of disjoint edges in G is comparable by at least one of them.

Since G does not contain $k+1$ disjoint edges, $(E,\prec_1)$ does not contain a chain of length $k+1$. Therefore, by Dilworth's theorem, E can be covered by k antichains with respect to $\prec_1$. Let E_1 be the largest of these antichains, thus $|E_1|\geq|E|/k$. Applying Dilworth's theorem on $(E_1,\prec_2)$, we similarly get an antichain (with respect to $\prec_2$) $E_2\subseteq E_1$ of size $|E_2|\geq|E_1|/k\geq|E|/k^2$. In the rest of the proof we estimate the size of E_2 from above.

Since E_2 is an antichain with respect to $\prec_1$ and $\prec_2$, $\pi(e)\cap\pi(e')\neq\emptyset$ for any $e,e'\in E_2$. Therefore, $\bigcap_{e\in E_2}\pi(e)\neq\emptyset$, so there is a vertical line ℓ which intersects all edges in E_2.

Let $\overrightarrow{G_2}=(V,\overrightarrow{E_2})$ be a directed geometric graph obtained from (V,E_2) by replacing each edge $e=v_1v_2$ in

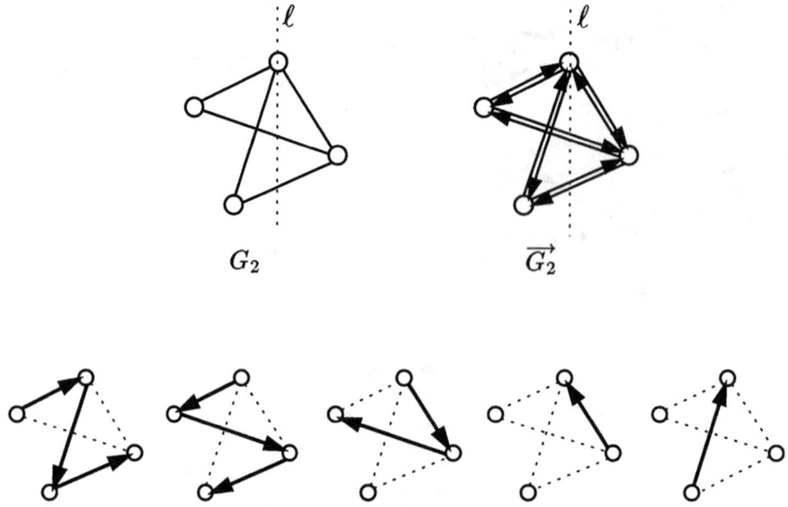

all 5 maximal zigzag paths in $\overrightarrow{G_2}$

Figure 2: Maximal zigzag paths.

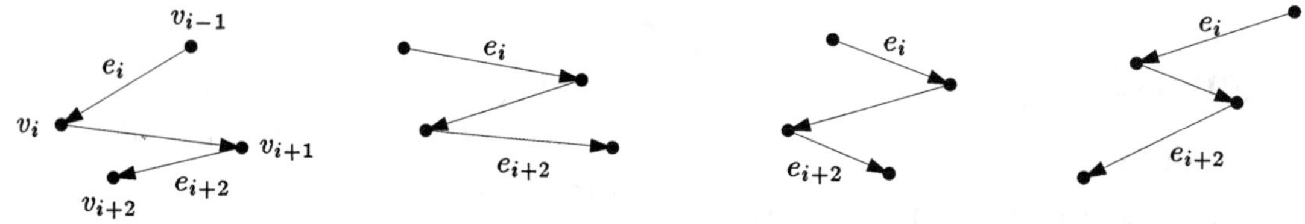

Figure 3: $e_{i+2} \prec_1 e_i$ or $e_{i+2} \prec_2 e_i$.

E_2 by the two oriented edges $\overrightarrow{v_1v_2}$ and $\overrightarrow{v_2v_1}$. For two edges $e_1 = \overrightarrow{v_0v_1}$, $e_2 = \overrightarrow{v_1v_2}$ forming a path in $\overrightarrow{G_2}$, we say that e_2 is a *zag of* e_1, if the following two conditions hold:

(i) $\pi(e_1) \cap \pi(e_2)$ has positive length,

(ii) for any $z \in (\pi(e_1) \cap \pi(e_2)) \setminus \{\pi(v_1)\}$, the vertical line through z intersects e_2 below e_1, and it intersects no other edge going from v_1 between e_1 and e_2.

Observe that each edge in $\overrightarrow{G_2}$ has at most one zag. We call an oriented path $e_1e_2 \ldots e_r$ in $\overrightarrow{G_2}$ a *zigzag path*, if e_{i+1} is the zag of e_i, for each $i = 1, \ldots, r-1$ (see Fig. 2).

Lemma 4 *Every zigzag path in* $\overrightarrow{G_2}$ *has at most* $2k$ *edges.*

Lemma 5 $\overrightarrow{G_2}$ *has at most* $n+1$ *maximal zigzag paths.*

Lemmas 4 and 5 immediately give Theorem 1: every edge in $\overrightarrow{G_2}$ is contained in a maximal zigzag path; therefore, by Lemmas 4 and 5,

$$|\overrightarrow{E_2}| \le 2k(n+1),$$

and, consequently,

$$|E| \le k^2|E_2| = k^2|\overrightarrow{E_2}|/2 \le k^3(n+1).$$

It remains to prove Lemmas 4 and 5.

Proof of Lemma 4. Let $P = e_1e_2 \ldots e_r$ be a zigzag path in $\overrightarrow{G_2}$, and let $e_i = \overrightarrow{v_{i-1}v_i}$ for $i = 1, \ldots, r$. We need the following claim.

Claim One of the two sequences

$$S_0 = x(v_0), x(v_2), \ldots \qquad S_1 = x(v_1), x(v_3), \ldots$$

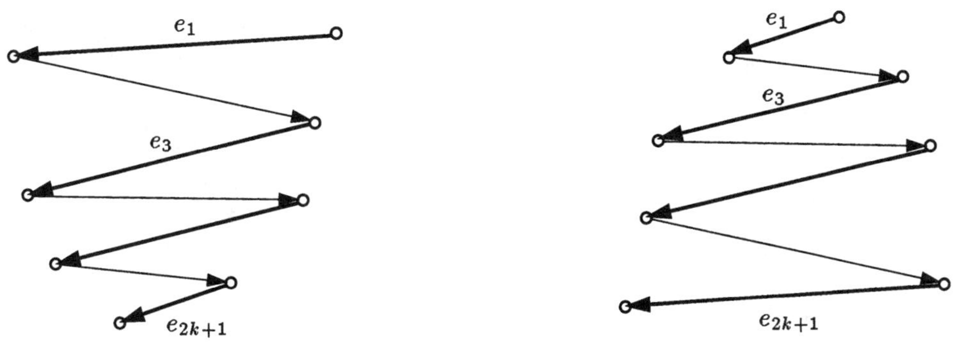

Figure 4: $e_1, e_3, \ldots, e_{2k+1}$ are disjoint.

is decreasing, and the other one is increasing.

Proof of Claim. Suppose the claim is false. Then there is an index i such that either $x(v_{i-1}) < x(v_{i+1})$ and $x(v_i) < x(v_{i+2})$, or $x(v_{i-1}) > x(v_{i+1})$ and $x(v_i) > x(v_{i+2})$. Consequently, $e_{i+2} \prec_1 e_i$ or $e_{i+2} \prec_2 e_i$ (see Fig. 3 for all four possible cases), which is a contradiction with the definition of E_2. The claim is proved. □

The length of P is at most $2k$, otherwise $e_1, e_3, \ldots, e_{2k+1}$ would be $k+1$ disjoint edges according to Claim (see Fig. 4). □

Proof of Lemma 5. Let $v \in V$ be a vertex lying to the right of ℓ, and let P_1, P_2 be two different zigzag paths in $\overrightarrow{G_2}$ ending in v. If the slope of the last edge in P_1 is, say, smaller than the slope of the last edge in P_2, then the last edge of P_1 has a zag and, consequently, P_1 can be extended to a longer zigzag path. It follows that at most one maximal zigzag path ends in v, and this is similarly true for any vertex not lying on ℓ. Similarly, at most two zigzag paths end in any vertex v on ℓ (one coming to v from the left, the other one from the right). The lemma now follows from the assumption that no pair of vertices lies on a vertical line. □

2.2 The lower bound

For simplicity, suppose that k is even and n is odd. Set $z = (n-k+1)/2$. Let P be a set of z points $p_1, \ldots, p_z$ placed equidistantly in this order from left to right on a horizontal line p. Let $Q = \{q_1, \ldots, q_z\}$ be a translation of P such that q_i always corresponds to p_i and $p_1 p_z q_z q_1$ is a square. Let q be the line containing Q, and let r be the line parallel to p and q, halfway between them. Let E_1 be the set of all segments (edges) $p_i q_j$ with $-k/2 \le i+j-(z+1) \le k/2$ (see Fig. 5).

The segments of E_1 intersect the line r in $k+1$ points $r_{-k/2}, r_{-k/2+1}, \ldots, r_{k/2}$ such that r_t always lies on the segments $p_i q_j$ with $i+j-(z+1) = t$. Let R be the set of the centers of the segments $r_t r_{t+1}$ $(t = -k/2, -k/2+1, \ldots, k/2-2)$, and let E_2 be the set of all edges (segments) joining vertices of R with vertices of $P \cup Q$ (see Fig. 5).

We now show that the geometric graph $G = (P \cup Q \cup R, E_1 \cup E_2)$ gives the lower bound in Theorem 1.

First, observe that G has $z + z + (k-1) = n$ vertices and

$$|E_1| + |E_2|$$

$$= \frac{(n-k+1)(k+1) - 4(1+2+\ldots+k/2)}{2}$$

$$+(n-k+1)(k-1)$$

$$= (n-k+1)(\frac{3k}{2} - \frac{1}{2}) - \frac{k(k-2)}{4} > \frac{3}{2}(k-1)n - 2k^2$$

edges. It remains to show that G contains no $k+1$ pairwise disjoint edges. Suppose that D is a set of disjoint edges in G. If $|D \cap E_1| \le 1$, then $|D| \le |R| + 1 = k$. Otherwise, let the leftmost edge of $D \cap E_1$, e_1, intersects r in a point r_s, and the rightmost one, e_2, in a point r_t. Since there are $s + k/2$ vertices of R to the left of e_1, D contains at most $s + k/2$ edges to the left of e_1. Similarly, D contains at most $k/2 - t$ edges to the right of e_2. Since there are $t - s - 2$ vertices of $P \cup Q$ between e_1 and e_2, D contains at most $t - s - 2$ edges between e_1 and e_2. Altogether, D contains at most $(s + k/2) + (k/2 - t) + (t - s - 2) + 2 = k$ edges. □

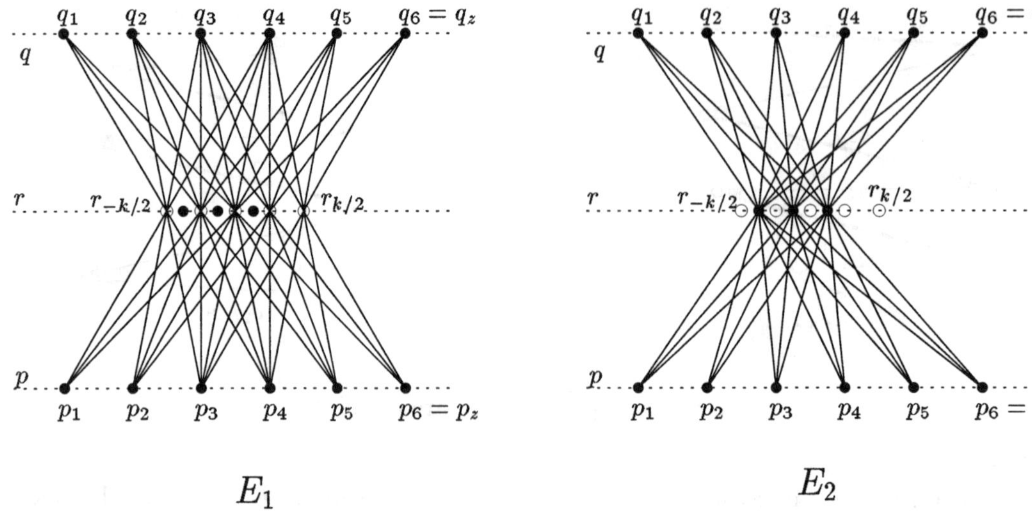

Figure 5: The edges of E_1 and E_2 (for $k = 4$, $n = 15$, $z = 6$).

3 Algorithm

Set $k := \left(\frac{m}{n+1}\right)^{1/3}$. Let $G = (V, E)$ be a geometric graph with n vertices and m edges. Consider the set E partially ordered by $\prec_1$. For each edge $e \in E$, let $\text{Rank}_1(e)$ be the length of the longest chain in $(E, \prec_1)$ with the largest element e. We can find the numbers $\text{Rank}_1(e), e \in E$, in $O(m^2)$ time as follows. We denote the edges of E by $e_1, e_2, \ldots, e_m$ according to increasing x-coordinate of the leftmost point (vertex). Obviously, $\text{Rank}_1(e_1) = 1$ and, for $i > 1$,

$$\text{Rank}_1(e_i) = 1 + \max\{\text{Rank}_1(e_j) \mid j < i \text{ and } e_j \prec_1 e_i\},$$

unless $j \not\prec_1 i$ for all $j < i$ in which case we have $\text{Rank}_1(e_i) = 1$. Using this, we can determine the number $\text{Rank}_1(e)$ one by one for the edges $e_1, e_2, \ldots, e_m$ in this order. Obviously, this can be done in $O(m^2)$ time. If one of the numbers $\text{Rank}_1(e_i)$ is at least k, then we have a chain with respect to $\prec_1$ consisting of at least k, and its members are at least k disjoint edges. Otherwise, in $O(m)$ time we find an i, $1 \leq i < k$, for which $|\{e \in E : \text{Rank}_1(e) = i\}|$ is maximal. For this i, let $E_1 = \{e \in E : \text{Rank}_1(e) = i\}$. The set E_1 has size at least m/k and forms an antichain with respect to $\prec_1$.

Doing the same procedure for E_1 and $\prec_2$, we get either at least k edges which form a chain with respect to $\prec_2$ and thus pairwise disjoint, or we get a subset $E_2 \subset E_1$ of size at least m/k^2 which forms an antichain with respect to $\prec_2$ (and also in $\prec_1$). In the first case we are done, in the latter case we proceed as follows below. This step can be done again in $O(m^2) + O(m) = O(m^2)$ time.

Following the proof of Theorem 1, we build the oriented geometric graph $\overrightarrow{G_2} = (V, \overrightarrow{E_2})$ from E_2 in $O(m)$ time. Further, we order the edges from each vertex in clockwise order (say) and consequently determine the zag of each edge in $\overrightarrow{G_2}$. This can be done in $O(m \log m)$ time. Then, we partition $\overrightarrow{E_2}$ into (at most $n+1$) maximal zigzag paths, and take every other edge in the largest maximal zigzag paths ($O(m)$ time). This gives us at least $(m/k^2)/(n+1) = k$ pairwise disjoint edges.

4 The case $k = 3$

4.1 The upper bound

We use the following result.

Lemma 6 ([GKK]) *If a geometric graph G of n vertices does not contain 4 pairwise disjoint edges and there is a line which intersects every edge of G and contains no vertex of G, then G has at most $7n$ edges.*

Lemma 6 is not stated in [GKK] explicitly. However, its proof (relatively long case-analysis) is readily contained in the proof of Theorem 2 in [GKK].

Let $G = (V, E)$ be a geometric graph without 4 pairwise disjoint edges. Denote the vertices by $v_1, \ldots, v_n$

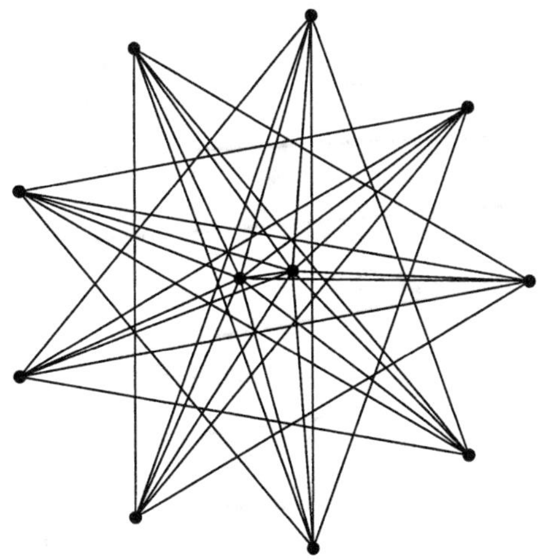

Figure 6: A geometric graph with no 4 pairwise disjoint edges.

from left to right, and assume that no pair of them lies on a vertical line. For any $1 \leq i < n$ let $\overline{G}_i$ be the subgraph of G which contains only those edges $v_\alpha v_\beta$ of G where $\alpha \leq i < \beta$.

It follows from Lemma 6 and the assumption that for any $1 \leq i < n$ $|E(\overline{G}_i)| \leq 7n$. For any $1 \leq i < n$, let G_i^- and G_i^+ be the subgraph of G induced by the vertices $v_1, \ldots, v_i$ and by $v_i, \ldots, v_n$, respectively. Let

$$j = \max\left\{i \mid G_i^- \text{ does not contain 2 disjoint edges}\right\}.$$

Since G_{j+1}^- contains two disjoint edges, G_{j+2}^+ does not. Suppose without loss of generality that $j < n/2$. Then

$$E(G) = E(G_j^-) \cup E(G_{j+2}^+) \cup E(\overline{G}_{j+1}) \cup \{v_i v_{j+1} \in E(G) \mid i \leq j\}.$$

Therefore,

$$|E(G)| \leq e_1(j) + e_1(n-j-1) + 7n + j < 8.5n.$$

□

4.2 The lower bound

First, let n be odd. Take the vertices of a regular $(n-2)$-gon, and join each of them with the furthermost 4 vertices. Add two vertices near the center of the $(n-2)$-gon, and join each of them by an edge to all the other vertices. The resulting geometric graph on n vertices has $4(n-2)/2 + 2(n-1) - 1 = 4n - 7$ edges. It is easy to see that no 4 edges are pairwise disjoint, see Fig. 6.

For n even, we take the above geometric graph on $n+1$ vertices, and remove a vertex and the six edges incident to it. The resulting graph on n vertices has $4n - 9$ edges. □

5 Remarks

By a little modification, it is possible to slightly improve the upper bound of Theorem 1.

Theorem 7 *For any $k \leq n/2$,*

$$e_k(n) \leq \frac{4}{21}k^3 n + O(k^2 n).$$

Proof. (Sketch) We use the same partial orderings, $\prec_1, \prec_2, \prec_3, \prec_4$, as in the proof of Theorem 1 (see Fig. 1). For any edge e of a geometric graph $G = (V, E)$, let $\text{Rank}_3^+(e)$ be the maximum number $a \geq 0$ such that there exist $e_1, e_2, \ldots e_a, \in E$, $e \prec_3 e_1 \prec_3 e_2 \prec_3 e_3 \cdots \prec_3 e_a$, and let $\text{Rank}_4^-(e)$ be the maximum number $b \geq 0$ such that there exist $e_1, e_2, \ldots, e_b \in E$, $e_1 \prec_4 e_2 \prec_4 e_3 \cdots \prec_4 e_b \prec_4 e$.

Lemma 8 *Let $G = (V, E)$ be a geometric graph with n vertices and with no $k+1$ pairwise disjoint edges. Suppose that there is a line which avoids all vertices and crosses all edges of G. Then*

$$|E| \leq \frac{1}{6}k^3 n + O(k^2 n).$$

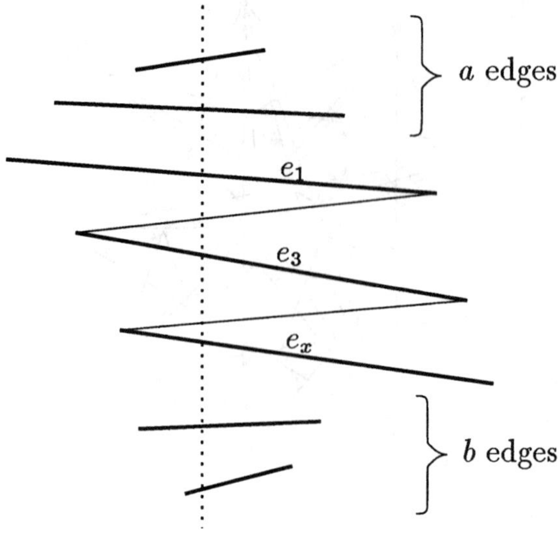

Figure 7: $k+1$ disjoint edges.

Proof of Lemma 8. For any $0 \le a, b \le k-1$ let

$$E_{a,b} = \{e \in E \mid \text{Rank}_3^+(e) = a, \text{Rank}_4^-(e) = b\}.$$

Clearly, $E_{a,b}$ is an antichain with respect to $\prec_3$ and $\prec_4$. We can define $\overrightarrow{E_{a,b}}$ and its maximal zigzag paths just as in the proof Theorem 1. It is easy to see that there are at most n maximal zigzag paths and every edge of $\overrightarrow{E_{a,b}}$ is contained in at least one of them (in fact, in exactly one of them, but here we do not need that). Suppose that there is a zigzag path $\overrightarrow{e_1}, \overrightarrow{e_2}, \ldots, \overrightarrow{e_x}$ of $x = 2(k-a-b)+1$ edges. Then $\overrightarrow{e_1}, \overrightarrow{e_3}, \overrightarrow{e_5}, \ldots, \overrightarrow{e_x}$ are $k-a-b+1$ pairwise disjoint edges. Moreover, we can add a edges above $\overrightarrow{e_1}$ and b edges below $\overrightarrow{e_x}$ to get $k+1$ pairwise disjoint edges, contradicting our assumption. (See Fig. 7.)

Therefore, every maximal zigzag path of $\overrightarrow{E}_{a,b}$ has at most $2(k-a-b)$ edges, so

$$|E_{a,b}| \le n(k-a-b),$$

$$|E| = \sum_{\substack{0 \le a,b \\ a+b<k}} |E_{a,b}| \le n \sum_{\substack{0 \le a,b \\ a+b<k}} (k-a-b) = \binom{k+2}{3} n.$$

□

Return to the proof of Theorem 7. It is not hard to see that there is a line ℓ which avoids all vertices of G and on one side of ℓ there are $\lceil n/2 \rceil$ vertices and at most $\lceil k/2 \rceil$ pairwise disjoint edges, while on the other side of ℓ there are $\lfloor n/2 \rfloor$ vertices and at most $\lfloor k/2 \rfloor$ pairwise disjoint edges of G. We get the recursion

$$e_k(n) \le \frac{1}{6}k^3 n + O(k^2)n + e_{\lceil k/2 \rceil}(\lceil n/2 \rceil) + e_{\lfloor k/2 \rfloor}(\lfloor n/2 \rfloor)$$

and Theorem 7 follows. □

If the number of edges in a geometric graph is at least $\Omega(n^2)$, then Theorem 1 guarantees $\Omega(n^{1/3})$ pairwise disjoint edges. This is improved by the following result of Pach [P].

Theorem 9 ([P]) *For any $c > 0$ there is a $c' > 0$ such that every geometric graph of n vertices and at least cn^2 edges has at least $c'\sqrt{n}$ pairwise disjoint edges.*

References

[AE] N. Alon and P. Erdős, Disjoint edges in geometric graphs, *Discrete and Computational Geometry* **4** (1989), 287-290.

[AH] S. Avital and H. Hanani, Graphs, *Gilyonot Lematematika* **3** (1966), 2-8 (in Hebrew).

[D] R. P. Dilworth, A decomposition theorem for partially ordered sets, *Annals of Mathematics* **51** (1950), 161-166.

[GKK] W. Goddard, M. Katchalski, and D. J. Kleitman, Forcing disjoint segments in the plane, *European Journal of Combinatorics* **17** (1996), 391-395.

[HP] H. Hopf and E. Pannwitz, Aufgabe No. 167, *Jahresbericht der Deutschen Mathematiker-Vereinigung* **43** (1934), 114.

[K1] Y. Kupitz, Extremal problems in combinatorial geometry, *Aarhus University Lecture Notes Series,* **53**, (1979), Aarhus University, Denmark.

[LMPT] D. Larman, J. Matoušek, J. Pach, and J. Törőcsik, A Ramsey-type result for planar convex sets, *The Bulletin of the London Mathematical Society* **26** (1994), 132-136.

[P] J. Pach, personal communication.

[PA] J. Pach and P.K. Agarwal, *Combinatorial Geometry,* John Wiley, New York, 1995.

[PT] J. Pach and J. Törőcsik, Some geometric applications of Dilworth's theorem, *Discrete and Computational Geometry* **12** (1994), 1-7.

Results on k-Sets and j-Facets via Continuous Motion*

Artur Andrzejak
ETH Zürich

Boris Aronov
Poly Univ New York

Sariel Har-Peled
Tel Aviv Univ

Raimund Seidel
Univ des Saarlandes

Emo Welzl
ETH Zürich

Abstract

Let P be a set of n points in $\mathbb{R}^d$ in general position, i.e., no $i+1$ points on a common $(i-1)$-flat, $1 \leq i \leq d$. A *k-set of P* is a set S of k points in P that can be separated from $P \setminus S$ by a hyperplane. A *j-facet of P* is an oriented $(d-1)$-simplex spanned by d points in P which has exactly j points from P on the positive side of its affine hull.

If P is a planar point set and n is even, a *halving edge* is an undirected edge between two points, such that the connecting line has the same number of points on either side. The number of $(n/2)$-sets is twice the number of halving edges. Inspired by Dey's recent proof of a new bound on the number of k-sets we show that

$$C + \sum_{p \in P} \binom{(\deg p + 1)/2}{2} = \binom{n/2}{2}$$

where $\deg p$ is the number of halving edges incident to point p and C is the number of crossing pairs of halving edges. The identity allows us, among other things, to determine the maximum number of halving edges in a set of 12 points. An analogous identity holds for j-facets.

For P in $\mathbb{R}^3$ we show that for $j \leq n/4 - 2$ the number of $(\leq j)$-facets (i.e., i-facets with $0 \leq i \leq j$) is maximized for sets in convex position, where this number is known to be

$$(j+1)(j+2)n - 2(j+1)(j+2)(j+3)/3\,.$$

For $k \leq n/4 - 1$, $k^2 n - k(k-1)(2k+5)/3$ is the tight upper bound for the number of $(\leq k)$-sets (i.e., i-sets with $1 \leq i \leq k$).

Finally we discuss the relation between the vector of numbers of k-sets, $k = 1, \ldots, n-1$ and the vector of numbers of j-facets, $j = 0, \ldots, n-d$ for a given point set. In the plane the number of k-sets equals the number of $(k-1)$-facets. In $\mathbb{R}^3$ the k-set vector determines the j-facet vector (and vice versa) by a linear relation. There is no such relation in $\mathbb{R}^d$ for d exceeding 3.

These results can be obtained by arguments via continuous motion of one point set to another while observing certain quantities related to k-sets and j-facets. For the relation between k-sets and j-facets in $\mathbb{R}^3$, we give a more direct argument via so-called k-set polytopes.

1 Introduction and basics

Let P be a set of n points in $\mathbb{R}^d$ in general position, i.e., no $i+1$ points on a common $(i-1)$-flat, $1 \leq i \leq d$.

k-Sets. A *k-set of P* is a set S of k points in P that can be strictly separated from $P \setminus S$ by a hyperplane. We denote by $e_k(P)$ the number of k-sets of P, and by $E_k(P)$ the number of $(\leq k)$-sets, i.e., $E_k(P) := \sum_{i=1}^{k} e_i(P)$. If P is understood, we write e_k and E_k short for $e_k(P)$ and $E_k(P)$, respectively.

In the plane, an upper bound of $O(n\sqrt{k})$ on e_k was given in [ELSS] (see also [Lo]), where also sets with $e_k = \Omega(n \log(k+1))$ were described (see [EW] for an alternative construction). After the improvement of the upper bound to $O(n\sqrt{k}/\log^*(k+1))$ in [PSS] in 1989, Dey recently provided a further substantial improvement of the bound to $O(n\sqrt[3]{k})$ (building upon considerations in [AAS]). In $\mathbb{R}^3$, the best upper bound is $O(nk^{5/3})$ from [AACS], for general k, and from [DE], for $k = \lfloor n/2 \rfloor$ (improving on bounds in [BFL, ACEGSW, Ep, AAS]). In $\mathbb{R}^d$ bounds of the form $O(n^{\lfloor d/2 \rfloor} k^{\lceil d/2 \rceil - \varepsilon_d})$ for some small $\varepsilon_d > 0$ have been obtained recently in [AACS] (see [ZV, AAS] for previous steps).

The situation is much better understood for $(\leq k)$-sets, where [CS] provide an asymptotically tight bound of $O(n^{\lfloor d/2 \rfloor} k^{\lceil d/2 \rceil})$ (this bound is attained by points on the moment curve), and in the plane there is even a tight upper bound of kn for $k < n/2$ [AG, Pe].

---

*Part of this work was performed while R.S. and E.W. were visiting the DIMACS center in November 1989, while R.S. visited FU Berlin in 1992, while E.W. visited Tel Aviv University in February 1994, while A.A. and E.W. were still at FU Berlin, and while B.A. was visiting ETH in April 1997. B.A. has been partially supported by a Sloan Research Fellowship. E.W. has been partially supported by a Max-Planck Research Prize.

SCG 98 Minneapolis Minnesota USA

For a more complete account of the history of the problem, for related notions and applications in computational geometry, see the survey [AW]. Remarkably, almost all algorithmic applications and also almost all proofs of bounds proceed actually via a different notion (and its dual) which is described next.

j-Facets. A *j-facet of P* is an oriented $(d-1)$-simplex spanned by d points in P which has exactly j points from P on the positive side of its affine hull.

We denote by $g_j(P)$ the number of j-facets of P, and by $G_j(P)$ the number of $(\leq j)$-facets, i.e., $G_j(P) := \sum_{i=0}^{j} g_i(P)$. If P is understood, we write g_j and G_j short for $g_j(P)$ and $G_j(P)$, respectively. The term "facet" is justified by the fact that the 0-facets of P are exactly the facets of the polytope conv$P$. There is a relationship between the maximum number of k-sets and the maximum number of $(k \pm d)$-facets of sets of n points, although this relation has never really been worked out carefully. In addition, the reader should be aware of ambiguities concerning the notion of j-facets in the literature.

Results. In Section 2, we give an identity for the planar case concerning the number of crossings between j-facets (which we call *j-edges* in the plane) and the sequence of numbers of j-edges incident to points $p \in P$. In this extended abstract we restrict the proof to halving edges—these are the undirected versions of $(n/2-1)$-facets under the assumption that n is even; see Theorem 1. The discovery of the relations was inspired by Dey's recent proof of the new $O(n\sqrt[3]{k})$ bound for planar k-sets. In fact, this bound follows also directly from the identities via a known lemma on the number of edges of graphs which can be embedded in the plane with few crossings ([ACNS, Lei]).

Section 3 provides exact upper bounds on the number of $(\leq j)$-facets and the number of $(\leq k)$-sets in $\mathbb{R}^3$, provided j and k is not too large (roughly $n/4$); see Theorem 2.

Finally, in Section 4 we show that the vectors $\vec{e} = (e_k)_{k=1}^{n-1}$ and $\vec{g} = (g_j)_{j=0}^{n-d}$ determine each other in $\mathbb{R}^3$ by a linear relation (for $d=2$ this is simple); see Theorem 3. In fact, this allows us to infer the tight bound on $(\leq k)$-sets directly from the bound on $(\leq j)$-facets, so we have to provide a proof for the latter only.

Back to j-facets. A sequence $(p_1, \ldots p_d)$ of d linearly independent points in $\mathbb{R}^d$ partitions the space into two open halfspaces and a hyperplane (the affine hull of $\{p_1, \ldots p_d\}$): Points p_{d+1} for which the sign of the determinant of the matrix with rows $(p_i\ 1)$, $i = 1, \ldots d+1$, is positive, points for which the sign is negative, and points for which the sign is 0. We denote this sign by $\operatorname{sgn}(p_1, \ldots p_d, p_{d+1})$. So a j-facet can be specified by a sequence $(p_1, \ldots p_d)$ of d points in P such that $\operatorname{sgn}(p_1, \ldots p_d, p) > 0$ for exactly j points p in P. In fact, we will use the notation $[p_1, \ldots p_d]$ for j-facets, meaning all permutations of $(p_1, \ldots p_d)$ which can be obtained from that sequence via an even number of transpositions of adjacent elements (since this will not change the sign of the determinant with a $d+1$-th point p). An odd number of transpositions in this sequence will change a j-facet into an $(n-j-d)$-facet.

For our proofs we want to make explicit that the structure of j-facets of $P = \{p_1, \ldots p_n\}$ is completely determined by the $\binom{n}{d+1}$ signs $\operatorname{sgn}(p_{i_1}, \ldots p_{i_{d+1}})$, $1 \leq i_1 < \ldots < i_{d+1} \leq n$. If we move the points in P, then no j-facet will change its index (i.e., the value j), unless one of these signs changes. If the sign changes for exactly one such tuple τ, then only j-facets composed of d of the points in τ change their index, either to $(j-1)$ or to $(j+1)$.

Moving around. In our proofs we consider the changes in $\vec{g}$ while the point set moves continuously. We will assume that during the motion the set stays in general position, except for a finite number of discrete instances, where the sign of exactly one $(d+1)$-tuple changes.

Of course, this paradigm is not new for the analysis of configurations in combinatorial geometry. Tverberg's original proof of his famous generalization of Radon's theorem is a prominent example [Tv]. Recently, continuous motions were used also in the context of k-sets by Gullikson and Hole [GH].

2 Planar identities

Let us briefly recapitulate the set-up for this section. We are given a set P of n points in the plane, n even, such that no three points lie on a common line. A *halving edge* is an undirected edge between two points, such that the connecting line has the same number of points on either side; sometimes, when we refer to such an edge we mean the straight line segment connecting its endpoints. Two halving edges *cross*, if their segments intersect in a single point in their relative interiors.

Theorem 1

$$C + \sum_{p \in P} \binom{(\deg p + 1)/2}{2} = \binom{n/2}{2} \tag{1}$$

where $\deg p$ is the number of halving edges incident to p (this number is always odd), and C is the number of pairwise crossings of halving edges.

Implications. Before we proceed to the proof, let us exhibit three implications of this identity, most prominently Tamal Dey's recent bound on the number of halving edges (which relates to k-sets by the fact that the number of $(n/2)$-sets is twice the number of halving edges).

Corollary 1 ([De]) *The number of halving edges of P is bounded by $O(n^{4/3})$.*

This bound is a direct consequence of the inequality $C < n^2/8$ which follows from Identity (1) and the fact that a graph with n vertices embedded in the plane with c edge crossings cannot have more than $O(\max(n, \sqrt[3]{cn^2}))$ edges ([ACNS, Lei], see also [PT, PA]).

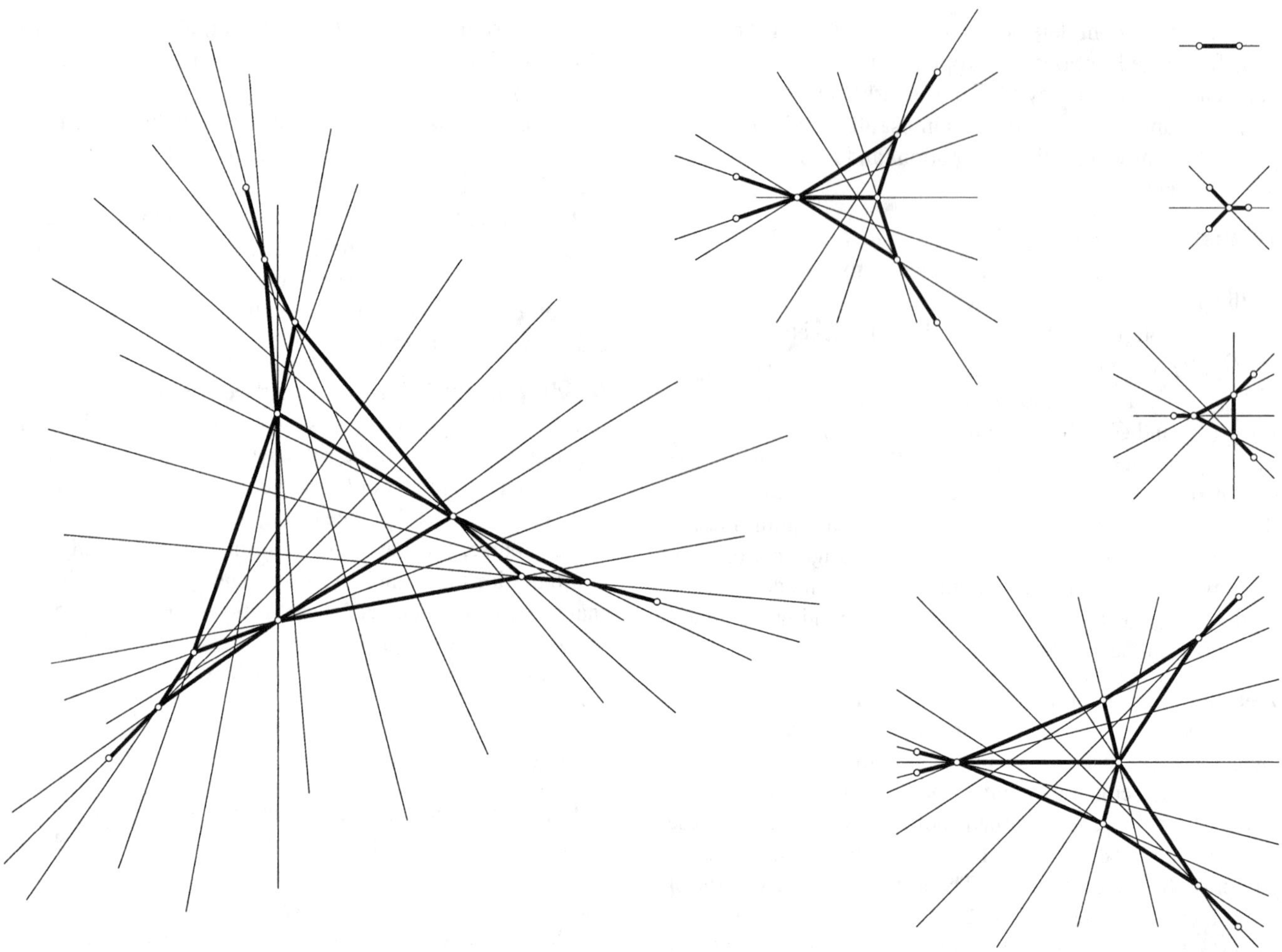

Figure 1: Point configurations which maximize the number of halving edges.

Corollary 2 *If h_n, n even, denotes the maximum possible number of halving edges of n points in the plane, then $h_2 = 1$, $h_4 = 3$, $h_6 = 6$, $h_8 = 9$, $h_{10} = 13$, and $h_{12} = 18$.*

The numbers for $n \leq 10$ were previously known (see e.g., [Fe], where h_{10} was determined by a computer-aided enumeration of all possible combinatorial configurations of 10 points in the plane). Figure 1 displays configurations which realize the quantities claimed in the corollary. All upper bounds can be readily derived from the inequality $\sum_{p \in P} \binom{(\deg p+1)/2}{2} \leq \binom{n/2}{2}$ implied by Identity (1), in conjunction with the facts—implied by Lemma 1 below—that every point is incident to an odd number of halving edges, and that there are at least three points incident to exactly one halving edge (extreme points must satisfy this condition). For example for 12, this implies that observing these constraints the sum $\sum_{p \in P} \deg p$ is maximized for the "degree sequence" $(1,1,1,3,3,3,3,3,3,5,5,5)$.

Corollary 3 ([PS]) *P allows a perfect* cross-matching *(a partition into edges such that any pair of such edges crosses), iff P has exactly $n/2$ halving edges.*

First observe that in a perfect cross-matching, every edge is halving. Identity (1) tells us, that if there are already $\binom{n/2}{2}$ crossings among halving edges, then $\deg p = 1$ for all $p \in P$, and thus there cannot be more halving edges beyond those in the perfect cross-matching. On the other hand, if there are $n/2$ halving edges, then $\deg p = 1$ for all p (since $\deg p$ has to be at least 1 in any case), and thus there must be $\binom{n/2}{2}$ crossings. That is, the $n/2$ halving edges pairwise cross and form a perfect cross-matching.

Next we prepare two ingredients for the proof of Theorem 1.

Lovász' Lemma. Let ℓ be a line through point $p \in P$ which misses all other points in P. Then there is a unique side of ℓ which contains the majority of points from $P \setminus \{p\}$. Call this side the *large side* of ℓ, and the other open halfplane determined by ℓ the *small side* of ℓ.

Lemma 1 ([Lo]) *If a line ℓ contains a unique point p in P, and there are x halving edges incident to p emanating into the small side of ℓ, then there are $x + 1$ halving edges emanating into the large side of ℓ.*

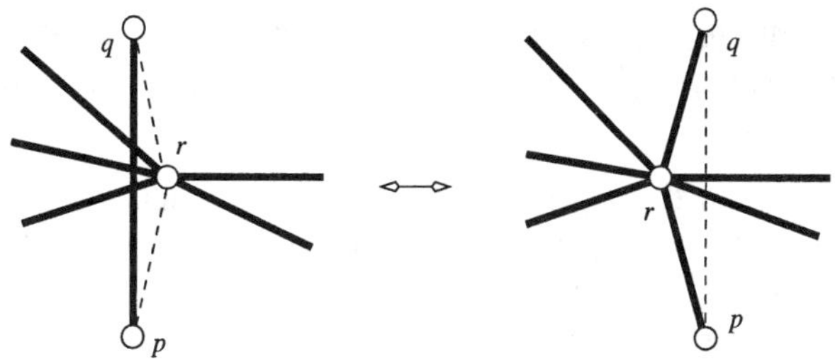

Figure 2: Local changes during mutation.

The lemma can be proven by rotating a line λ about point p starting in position ℓ until it coincides with ℓ again. The halving edges incident to p are encountered in alternation on the large and small side of ℓ, starting and ending on the large side.

In fact, the lemma completely characterizes the graph of halving edges of a point set. Simple implications of the lemma which we have mentioned before are that the number of halving edges incident to a point in P is always odd, and that there is exactly one halving edge incident to each extreme point of P.

Mutations while moving. Recall from the introduction, that if we start moving the points in P, the graph of halving edges will not change unless a triple (p, q, r) of points becomes collinear and changes its orientation. Even then, the graph of halving edges will not change except for edges on $\{p, q, r\}$. Following the terminology of oriented matroids (cf. [BLSWZ]), we call such a change of orientation a *mutation*.

Let us investigate such a mutation on three points $\{p, q, r\}$. We assume that this is the only mutation that occurred (i.e., there is no other simultaneous mutation), and that the points stayed disjoint when they passed though collinearity. First we consider only the case when $\{p, q\}$ is a halving edge before mutation, and that r lies on the segment connecting p and q at the moment of collinearity (see Figure 2). Hence, $\{q, r\}$ and $\{r, p\}$ are not halving edges before mutation. After mutation, $\{p, q\}$ is not halving, but $\{q, r\}$ and $\{r, p\}$ are. That is, the number of halving edges increased by one, and no degree in the graph of halving edges changed except for point r, whose degree increased by 2.

What happened to crossings of halving edges? If we ignore edges incident to r, then nothing changes. Crossings with the edge $\{p, q\}$ are replaced by crossings with $\{q, r\}$ or $\{r, p\}$ after mutation. As for the edges incident to r, let ℓ be a line through r parallel to the segment connecting p and q. The halfplane of ℓ containing $\{p, q\}$ is the large side of ℓ, before and after mutation. If x is the number of halving edges incident to r emanating into the large side of ℓ before mutation, then these edges were responsible for x crossings with $\{p, q\}$. These crossings disappear after mutation. No new crossings appear.

Let $\deg r$ and $\deg' r$ denote the number of halving edges incident to r before and after mutation, respectively. Note that $\deg' r = \deg r + 2$ and $x = (\deg r + 1)/2$. Let C and C' denote the number of crossings of halving edges before and after mutation, respectively. We have $C' = C - x$, and so

$$C + \binom{(\deg r + 1)/2}{2} = C + \binom{x}{2} =$$

$$C - x + \binom{x+1}{2} = C' + \binom{(\deg' r + 1)/2}{2}$$

which proves that the validity of Identity (1) is not affected by the mutation, since no degree other than $\deg r$ changes during mutation.

Now recall that we assumed that $\{p, q\}$ was a halving edge before mutation. However, the mutation described, and its inverse, are the only types of mutations affecting the graph of halving edges and its crossings.

Proof of Theorem 1. First observe that for all even n there is a set P of n points which satisfies Identity (1). The vertices of a regular n-gon, or the vertices of a regular $(n-1)$-gon together with its center are easy examples. Now it remains to use the fact that any two sets of n points in general position can be continuously transformed into each other in such a way that the points remain pairwise distinct, they never have more than one triple of points collinear, and such a collinearity occurs only finitely often. □

Other identities. A simple algebraic manipulation allows us to rewrite Identity (1) as

$$8C + \sum_{p \in P} (\deg p)^2 = n(n-1)\,.$$

Let $\deg_j p$ denote the number of j-edges[1] emanating from point p (which equals the number of incoming j-edges). Let C_j denote the number of crossings of j-edges, and for $i \neq j$, let $C_{i,j}$ denote the number of crossings between i-edges and j-edges. Then (reading G_{-1} as 0) we have

$$C_j + \sum_{p \in P} \binom{\deg_j p}{2} = G_{j-1}\,,$$

for $0 \le j \le n/2 - 1$, and

$$C_{i,j} + \sum_{p \in P} \deg_i p(\deg_j p - 1) = 2G_{i-1}\,,$$

for $0 \le i < j \le n/2 - 1$. (The latter identity allows improvements of previous bounds in [We] on $\sum_{k \in K} e_k$, for $K \subset \{1, 2, \ldots n\}$.) Proofs follow from an analysis of mutations as in the proof given here, and will be given in the full version of the paper.

[1] Recall, that we use "j-edges" for j-facets in the plane.

3 ($\leq j$)-facets and ($\leq k$)-sets in $\mathbb{R}^3$

Theorem 2 *Let P be a set of n points in general position in $\mathbb{R}^3$. Then*

$$G_j \leq (j+1)(j+2)n - 2(j+1)(j+2)(j+3)/3$$

for $j \leq n/4 - 2$, and

$$E_k \leq k^2 n - k(k-1)(2k+5)/3$$

for $k \leq n/4 - 1$. Both bounds are attained for sets in convex position.

Recall that in the plane a tight bound of $E_k \leq kn$ and $G_j \leq (j+1)n$ is known for $k < n/2$ and $j < n/2 - 1$ ([AG, Pe]). In $\mathbb{R}^3$, the number of j-facets is $2(j+1)(n-j-2)$ for every set of n points in convex position ([Lee, CS, Sh]), and by Theorem 3 below this implies that the number of k-sets is $n + 2(n-k-1)(k-1)$ (see also [GH]). The theorem above quotes the resulting numbers of ($\leq j$)-facets and ($\leq k$)-sets for point sets in convex position. In our proof we will show that we can always move a point set in $\mathbb{R}^3$ into convex position while the numbers G_j, $j \leq n/4 - 2$, never decrease. This gives the result claimed for ($\leq j$)-sets. Theorem 3 below yields $E_k = (G_{k-1} + G_{k-2})/2 + 2k$, so the bound for ($\leq k$)-sets can be easily obtained, too.

So we consider a set P of $n \geq 3$ points in $\mathbb{R}^3$, and analyze the effect of mutations of P on the vectors $\vec{g}$ and $\vec{G}$.

A mutation is the situation that four points become coplanar and change their sign as discussed in the introduction. More formally, a *mutation* is a triple (P^-, P^0, P^+), where $P^- = (p_1^-, \ldots, p_n^-)$, $P^0 = (p_1^0, \ldots, p_n^0)$, and $P^+ = (p_1^+, \ldots, p_n^+)$ are ordered point sets, P^- and P^+ in general position, such that: (i) $\text{sgn}(p_1^-, \ldots, p_4^-) < 0$, $\text{sgn}(p_1^0, \ldots, p_4^0) = 0$, and $\text{sgn}(p_1^+, \ldots, p_4^+) > 0$. (ii) For $0 \leq a < b < c < d \leq n$, $(a,b,c,d) \neq (1,2,3,4)$, we have

$$\text{sgn}(p_a^-, p_b^-, p_c^-, p_d^-) = \text{sgn}(p_a^0, p_b^0, p_c^0, p_d^0) = \text{sgn}(p_a^+, p_b^+, p_c^+, p_d^+)\,.$$

(iii) There is a continuous motion from P^- to P^0 and from P^0 to P^+ with all intermediate stages in general position.

If $[p_1^-, p_2^-, p_3^-]$ is a j-facet in P^-, then we call j the *index* of the mutation.

Convex mutation. We call a mutation a *convex mutation*, if the sequence $(p_1^0, \ldots, p_4^0)$ forms a convex quadrilateral in its plane (referring to the notation as set above); see Figure 3. This scenario is characterized by the fact that for all points $p \in \mathbb{R}^3$, p is on same side of all oriented facets $[p_1^0, p_2^0, p_3^0]$, $[p_1^0, p_2^0, p_4^0]$, $[p_1^0, p_3^0, p_4^0]$, and $[p_2^0, p_3^0, p_4^0]$. In fact, because of (ii) and (iii) above, this is true for all points in $P^+ \setminus \{p_1^+, \ldots, p_4^+\}$, even if we replace the "0" by "+" in the superscripts.

It follows that $[p_1^-, p_2^-, p_3^-]$ is a j-facet that turns into a $(j+1)$-facet $[p_1^+, p_2^+, p_3^+]$, since it "gained" p_4^+ on its positive side, $[p_1, p_3, p_4]$ turns from a j- to a $(j+1)$-facet, and $[p_1, p_2, p_4]$ and $[p_2, p_3, p_4]$ turn from $(j+1)$- to j-facet. Hence, vector $\vec{g}$ (and thus $\vec{G}$) does not change during a convex mutation.

Mutation through triangle. We call a mutation a *mutation through triangle*, if p_4^0 is in the convex hull of $\{p_1^0, p_2^0, p_3^0\}$; see Figure 4. This is characterized by the fact that for all points $p \in \mathbb{R}^3$, p is on the same side of all oriented facets $[p_1^0, p_2^0, p_3^0]$, $[p_1^0, p_2^0, p_4^0]$, $[p_1^0, p_4^0, p_3^0]$, and $[p_2^0, p_3^0, p_4^0]$.

Hence, if $[p_1^-, p_2^-, p_3^-]$ is a j-facet, then $[p_1, p_2, p_3]$ turns from a j-facet to a $(j+1)$-facet, while $[p_1, p_2, p_4]$, $[p_1, p_4, p_3]$, and $[p_2, p_3, p_4]$ turn from $(j+1)$-facets to j-facets. That is, the number of j-facets increases by two, while the number of $(j+1)$-facets decreases by the same amount. There is also a mirrored change in the number of $(n-j-3)$– and $(n-j-4)$-facets, which may lead to interferences if j is close to $n/2$. This is taken care of in the following lemma.

Lemma 2 *For a mutation through triangle with index j, with the four numbers $j, j+1, n-j-3$ and $n-j-4$ distinct, we have*

$$\begin{aligned} g_{n-j-3}(P^+) &= g_j(P^+) \\ &= g_j(P^-) + 2 = g_{n-j-3}(P^-) + 2\,, \quad \textit{and} \\ g_{n-j-4}(P^+) &= g_{j+1}(P^+) \\ &= g_{j+1}(P^-) - 2 = g_{n-j-4}(P^-) - 2\,. \end{aligned}$$

If $j+1 = n-j-4$, then

$$\begin{aligned} g_{n-j-3}(P^+) &= g_j(P^+) \\ &= g_j(P^-) + 2 = g_{n-j-3}(P^-) + 2\,, \quad \textit{and} \\ g_{j+1}(P^+) &= g_{j+1}(P^-) - 4\,. \end{aligned}$$

If $j = n-j-3$, then

$$\begin{aligned} g_j(P^+) &= g_j(P^-) + 4\,, \quad \textit{and} \\ g_{n-j-4}(P^+) &= g_{j+1}(P^+) \\ &= g_{j+1}(P^-) - 2 = g_{n-j-4}(P^-) - 2\,. \end{aligned}$$

If $j \neq n-j-4$, then

$$\begin{aligned} G_j(P^+) &= G_j(P^-) + 2\,, \quad \textit{and} \\ G_{n-j-4}(P^+) &= G_{n-j-4}(P^-) - 2\,. \end{aligned}$$

No other changes than those indicated above occur in $\vec{g}$ and $\vec{G}$. In particular, no changes occur if $j = n-j-4$. □

REMARK. The lemma can be readily generalized to a description of changes of the vectors $\vec{e}$ and $\vec{E}$, and to dimensions $d > 3$. Note that there is one type of mutations in the plane and there are two types ("convex" and "through triangle") in $\mathbb{R}^3$. In $\mathbb{R}^d$ such mutations can be classified by a parameter m, $1 \leq m \leq \lceil d/2 \rceil$. Details follow in the full version.

We extract the implication of Lemma 2 relevant for our purposes.

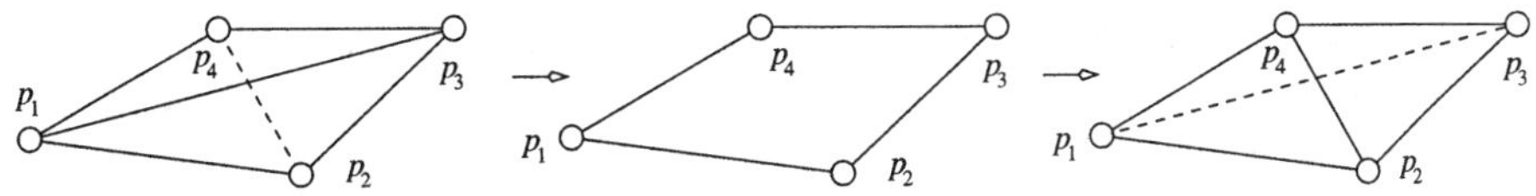

Figure 3: Convex mutation (point of view on the positive side of $[p_1^-, p_2^-, p_3^-]$).

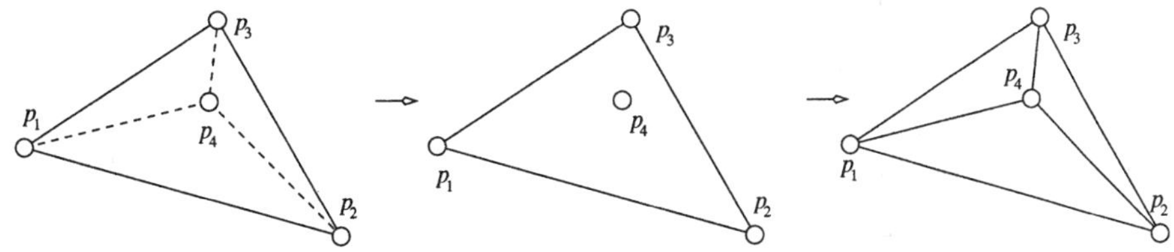

Figure 4: Through-triangle-mutation (seen from the positive side of $[p_1^-, p_2^-, p_3^-]$).

Corollary 4 *In a mutation through triangle with index j no entry in the vector $(G_0, G_1, \ldots, G_{n-j-5})$ decreases.* □

Proof of Theorem 2. We show now that we can continuously transform every point set into a set in convex position such that G_J decreases for no $J \leq n/4 - 2$. We know that it suffices to ensure that the index j of each mutation through triangle satisfies $n - j - 5 > n/4 - 2$, or equivalently, $j + 3 < 3n/4$. For that purpose let us recall that every point set P has a *centerpoint* c (not necessarily in P) with the property that any open halfspace that misses c contains at most $3n/4$ points from P (a consequence of Helly's Theorem, cf. [Ed]). First we observe that if c is a point in P, then for $j \leq n/4 - 2$, neither c is on the positive side of a j-facet nor c participates in a j-facet (in both cases we can find open halfspaces disjoint from c which contain more than $3n/4$ points). That is, for $J \leq n/4 - 2$, $G_J(P \setminus \{c\}) = G_J(P)$ and we can apply induction to prove the theorem (starting with a set of 4 points).

Hence, we restrict ourselves to the case that c is not in P, and without loss of generality we assume that the origin o is a centerpoint of P. For a real number $\lambda > 0$, we define $P(\lambda) := \{\mu p | p \in P, \mu = \min(1, \lambda/||p||)\}$. Note that for $\lambda_0 = \max_{p \in P} ||p||$, $P(\lambda_0) = P$, and for $\lambda_1 = \min_{p \in P} ||p||$, $P(\lambda_1)$ lies on a sphere of radius λ_1, and thus it is in convex position. The motion of $P(\lambda)$ for λ from λ_0 to λ_1 can be visualized by a sphere initially containing P, and then contracting the sphere while dragging points towards to the origin as soon as they appear on the sphere. Throughout the whole process, the origin stays a centerpoint of the moving point set.

We can always perturb the set P (without changing j-facets) in such a way that no two points lie on a common line with c, and during the whole motion the set is either in general position or there is a unique 4-tuple of points which is coplanar (and in general position in its plane). As we have learned, the mutations where the coplanar points are in convex position do not affect $\vec{G}$. Now consider the case when a point p moves through a triangle determined by three points q, r, s. Point p is still in the interior of the contracting sphere, otherwise it could not be be in the convex hull of three other points. In fact, it is easily seen that p is on the same side of the triangle spanned by $\{q, r, s\}$ as the origin before mutation, and on the opposite side after mutation. Now we have to recall the definition of a mutation through triangle and its index j. This index was determined by the number of points on the side of the triangle opposite to the point which is about to move through the triangle. Hence, the index of our mutation is the number j of points on the side of qrs opposite to p before mutation, i.e., opposite to the origin. Consequently, when p becomes coplanar with q, r, s, there is an open halfspace that misses the origin which contains these j points and p, q, r, s. Because of the centerpoint property of the origin, $j + 4 \leq 3n/4$, or $j + 3 < 3n/4$ as it was necessary to prove for the monotonicity of G_J, $J \leq n/4 - 2$. This concludes the proof of the theorem. □

4 Relation between k-sets and j-facets

Theorem 3 *For a set P of n points in $\mathbb{R}^3$ we have*

$$\begin{aligned} e_1 &= g_0/2 + 2, \quad e_{n-1} = g_{n-3}/2 + 2, \quad \textit{and} \\ e_k &= (g_{k-1} + g_{k-2})/2 + 2 \ \textit{for} \ 2 \leq k \leq n-2\,. \end{aligned}$$

Of course, we can also deduce how $\vec{g}$ determines $\vec{e}$. The corresponding relation of $e_k = g_{k-1}$ in the plane can be easily proved. In $\mathbb{R}^d$, $d > 3$, $\vec{g}$ does not determine $\vec{e}$ (see full version). For the proof of the theorem, we could simply observe the changes in $\vec{g}$ and $\vec{e}$ under continuous motion (see remark after Lemma 2). Such a proof provides some insight why such a relation does not generalize to $\mathbb{R}^d$, $d > 3$. We give here an alternative proof.

k-Set polytope. We employ the notion of a k-set polytope from [EVW]. Given a set P of n points and $1 \leq k \leq n-1$, the *k-set polytope of P* is

$$Q_k = \operatorname{conv}\left\{\sum_{p \in T} p \mid T \in \binom{P}{k}\right\},$$

where $\binom{P}{k}$ denotes the set of all subsets of P of cardinality k. The vertices of $\mathcal{Q}_k$ are in one-to-one correspondence to the k-sets of P. We briefly recapitulate the argument. A k-set S can be separated from $P \setminus S$ by a hyperplane. That is, there is a vector c and a real number b such that $\langle c, p \rangle > b$ for $p \in S$ and $\langle c, p \rangle < b$ for $p \in P \setminus S$. Clearly, this implies that $\langle c, \sum_{p \in S} p \rangle$ exceeds $\langle c, \sum_{p \in T} p \rangle$ for all T with $S \neq T \in \binom{P}{k}$ and so $\sum_{p \in S} p$ is a vertex of $\mathcal{Q}_k$.

On the other hand, if $\sum_{p \in S} p$ is a vertex of $\mathcal{Q}_k$, then there is a vector c such that $\langle c, \sum_{p \in S} p \rangle$ exceeds $\langle c, \sum_{p \in T} p \rangle$ for all T with $S \neq T \in \binom{P}{k}$. Set $b' = \min_{p \in S} \langle c, p \rangle$. If $\langle c, q \rangle \geq b'$ for $q \in P \setminus S$, then $\langle c, \sum_{p \in S} p \rangle < \langle c, \sum_{p \in S \setminus \{p'\} \cup \{q\}} p \rangle$, for p' a point in S with $\langle c, p' \rangle = b'$; a contradiction. Hence $\langle c, q \rangle < b'$ for all $q \in P \setminus S$ and S can be separated from $P \setminus S$ by a hyperplane with normal vector c.

Proof of Theorem 3. So far we have not referred to the dimension of P. Now let us assume that P is a set in general position in $\mathbb{R}^3$. The 1-sets of P are the vertices of $\operatorname{conv} P$ and the 0-facets of P are the facets of $\operatorname{conv} P$. Hence, $g_0 = 2e_1 - 4$ follows directly from Euler's relation, and $e_1 = g_0/2 + 2$ and $e_{n-1} = g_{n-3}/2 + 2$ hold.

For $2 \leq k \leq n-2$ we want to show that the set of $(k-1)$- and $(k-2)$-facets of P are in one-to-one correspondence to the facets of $\mathcal{Q}_k$ which will entail the remaining relations.

Let $[q_1, q_2, q_3]$ be a $(k-1)$-facet of P with U the set of $k-1$ points on its positive side. Let c be a vector and b be a real number such that $\langle c, q_1 \rangle = \langle c, q_2 \rangle = \langle c, q_3 \rangle = b$, $\langle c, p \rangle > b$ for $p \in U$, and $\langle c, p \rangle < b$ for $p \in P \setminus (U \cup \{q_1, q_2, q_3\})$. It follows that $S_i = U \cup \{q_i\}$, $i = 1, 2, 3$ are the only sets T in $\binom{P}{k}$ which maximize $\langle c, \sum_{p \in T} p \rangle$. Consequently, the points $\sum_{p \in S_i} p$ are vertices of a triangular facet of $\mathcal{Q}_k$. Similarly, if $[q_1, q_2, q_3]$ is a $(k-2)$-facet of P with U the set of $k-2$ points on its positive side, then the sets $S_i = (U \cup \{q_1, q_2, q_3\}) \setminus \{q_i\}$ give rise to a facet of $\mathcal{Q}_k$. A reverse argument shows that this mapping from $(k-1)$- and $(k-2)$-facets of P to facets of $\mathcal{Q}_k$ is already a bijection. In particular, all facets of $\mathcal{Q}_k$ are triangular, and $g_{k-1} + g_{k-2} = 2e_k - 4$ is implied by Euler's relation. □

References

[AACS] Pankaj K. Agarwal, Boris Aronov, Timothy M. Chan, and Micha Sharir, On levels in arrangements of lines, segments, planes, and triangles, *Discrete Comput Geom* (1998), to appear

[AAS] Pankaj K. Agarwal, Boris Aronov, and Micha Sharir, On levels in arrangements, segments, planes, and triangles, *in* "Proc 13th Ann ACM Symp on Comput Geom" (1997) 30–38

[ACNS] Miklós Ajtai, Vasek Chvátal, M. M. Newborn, and Endre Szemerédi, Crossing-free subgraphs, *Ann Discrete Math* **12** (1982) 9–12

[AG] Noga Alon and Ervin Győri, The number of small semispaces of a finite set of points in the plane, *J Combin Theory Ser A* **41** (1986) 154–157

[AW] Artur Andrzejak and Emo Welzl, k-Sets and j-facets – A tour of discrete geometry, in preparation (1997)

[ACEGSW] Boris Aronov, Bernard Chazelle, Herbert Edelsbrunner, Leonidas Guibas, Micha Sharir, and Rephael Wenger, Points and triangles in the plane and halving planes in space, *Discrete Comput Geom* **6** (1991) 435–442

[BFL] Imre Bárány, Zoltan Füredi, and László Lovász, On the number of halving planes, *Combinatorica* **10** (1990) 175–183

[BLSWZ] Anders Björner, Michel Las Vergnas, Bernd Sturmfels, Neil White, and Günter Ziegler, *Oriented Matroids*, Cambridge University Press, Cambridge UK (1993)

[CS] Kenneth L. Clarkson and Peter W. Shor, Applications of random sampling in computational geometry, II, *Discrete Comput Geom* **4** (1989) 387–421

[De] Tamal K. Dey, Improved bounds for planar k-sets and related problems, *in* "Proc 38th Ann IEEE Symp on Foundations of Comput Sci" (1997) 156–161

[DE] Tamal K. Dey and Herbert Edelsbrunner, Counting triangle crossings and halving planes, *Discrete Comput Geom* **12** (1994) 281–289

[Ed] Herbert Edelsbrunner, *Algorithms in Combinatorial Geometry*, Springer Verlag, Berlin, Heidelberg, New York (1987)

[EVW] Herbert Edelsbrunner, Pavel Valtr, and Emo Welzl, Cutting dense point sets in half, *Discrete Comput Geom* **17** (1997) 243–255

[EW] Herbert Edelsbrunner and Emo Welzl, On the number of line separations of a finite set in the plane, *J Combin Theory Ser A* **38** (1985) 15–29

[ELSS] Paul Erdős, László Lovász, A. Simmons, and E. G. Straus, Dissection graphs of planar point sets, *in* A Survey of Combinatorial Theory (Eds. Jagdish N. Srivastava et al.), North Holland Publishing Company (1973) 139–149

[Ep] David Eppstein, Improved bounds for intersecting triangles and halving planes, *J Combin Theory Ser A* **62** (1993) 176–182

[Fe] Stefan Felsner, On the number of arrangements of pseudolines, *Discrete Comput Geom* **18** (1997) 257–267

[GH] Tor Gullikson and Arne Hole, On the number of linearly separable subsets of finite sets in R^n, *Discrete Comput Geom* **18** (1997) 463–472

[Lee] D. T. Lee, On k-nearest neighbor Voronoi diagrams in the plane, *IEEE Trans Comput* **C-31** (1982) 478–487

[Lei] Tom Leighton, *Complexity Issues in VLSI*, Foundation of Computing Series, MIT Press, Cambridge (1983)

[Lo] László Lovász, On the number of halving lines, *Ann Universitatis Scientarium Budapest, Eötvös, Sectio Mathematica* **14** (1971) 107–108

[PA] János Pach and Pankaj K. Agarwal, *Combinato-*

rial Geometry, Wiles-Interscience Series in Discrete Mathematics and Optimization, John Wiley & Sons, Inc. (1995)

[PSS] János Pach, William Steiger, and Endre Szemerédi, An upper bound on the number of planar k-sets, *Discrete Comput Geom* **7** (1992) 109–123

[PS] János Pach and József Solymosi, Halving lines and perfect cross-matchings, unpublished manuscript (1997)

[PT] János Pach and Géza Toth, Graphs drawn with few crossings per edge, *Lecture Notes in Comput Sci* **1190** (1997) 345–354

[Pe] G. W. Peck, On k-sets in the plane, *Discrete Math* **56** (1985) 73–74

[Sh] Micha Sharir, k-Sets and random hulls, *Combinatorica* **13** (1993) 483–495

[Tv] Helge Tverberg, A generalization of Radon's Theorem, *J London Math Soc* **41** (1966) 123–128

[We] Emo Welzl, More on k-sets of finite sets in the plane, *Discrete Comput Geom* **1** (1986) 95-100

[ZV] Rade T. Živaljević and Sinisa T. Vrećica, The colored Tverberg's problem and complexes of injective functions, *J Combin Theory Ser A* **61** (1992) 309–318

Point-sets with few k-sets

Helmut Alt
Institut für Informatik
Freie Universität Berlin
Takustr. 9
14195 Berlin, Germany
alt@inf.fu-berlin.de

Stefan Felsner
Institut für Informatik
Freie Universität Berlin
Takustr. 9
14195 Berlin, Germany
felsner@inf.fu-berlin.de

Ferran Hurtado*
Departament de Matemàtica Aplicada II
Universitat Politècnica de Catalunya
Pau Gargallo 5
08028-Barcelona, España
hurtado@ma2.upc.es

Marc Noy*
Departament de Matemàtica Aplicada II
Universitat Politècnica de Catalunya
Pau Gargallo 5
08028-Barcelona, España
noy@ma2.upc.es

Abstract

A k-set of a finite set S of points in the plane is a subset of cardinality k that can be separated from the rest by a straight line. The question of how many k-sets a set of n points can contain is a longstanding open problem where a lower bound of $\Omega(n \log k)$ and an upper bound of $O(nk^{1/3})$ are known today.

Under certain restrictions on the set S, for example, if all points lie on a convex curve, a *linear* upper bound can be shown. Here, we will generalize this observation by showing that if the points of S lie on a constant number of convex curves, the number of k-sets remains linear in n.

Keywords: convex curve, k-set, Lovász' procedure.

1 Introduction and definitions

Let S be a set of n points in the plane. A *k-set* ($1 \leq k \leq n-1$) is a subset of S of cardinality k that can be separated from the rest by a straight line. The simple and natural question of how many k-sets a set of n points can contain has been considered for more than 25 years and has inspired considerable research. Nevertheless, it is still not solved completely and remains one of the most prominent open problems in combinatorial geometry.

The first upper bound is due to Lovász [4] he showed that the number of k-sets can be at most $O(n\sqrt{k})$ which is $O(n^{3/2})$. Erdős et al [3] constructed a family of sets having $\Omega(n \log k)$ k-sets which is $\Omega(n \log n)$ for suitable values of k. For a long time the gap between lower and upper bound could not be narrowed until Pach, Steiger and Szemerédi [5] showed an upper bound of $O(n\sqrt{k}/\log^* k)$. More significant progress was made only recently by Dey [2] who proved an upper bound of $O(n\sqrt[3]{k})$ which is $O(n^{4/3})$.

Various generalizations of the k-set problem or its dual formulation, namely determining the number of cells at the *k-level* of the arrangement of a set of n straight lines, have been considered. Among them are k-sets in higher dimensions or k-levels of curves in two dimensions and surfaces in three dimensions [1, 6].

Under certain restrictions on the set S it is possible to prove better upper bounds on the number of k-sets. For example, if all points of S lie on a straight line, there are only 2 k-sets. If all points of S lie on a convex curve, S contains at most n k-sets. Here, we will generalize the latter observation by showing that if the points of S lie on a constant number of convex curves, the number of k-sets remains *linear* in n.

In the following let S be a fixed set of n points in the plane and let $k \leq n-2$. An oriented straight line l is called a *k-line* of S exactly if the open halfplane right of l contains k points of S. As easily can be seen, to any k-set M of S there exists either a corresponding k-line incident to two points $p, q \in S \setminus M$ or a $(k-1)$-line incident to points $p \in M$ and $q \in S \setminus M$ having $M \setminus \{p\}$ as its corresponding $(k-1)$-set. Let us call any line segment $\overline{pq}$ with the former property a *k-segment* of S (a segment with the latter property, then, is a $(k-1)$-segment). The idea is to obtain an upper bound on the number of k-segments and $(k-1)$-segments and, thus, on the number of k-lines. We will make use of a powerful lemma, due to Lovász, that was shown in the classical

*Partially supported by Proyecto MEC-DGES-SEUID PB96-0005-C02-02.

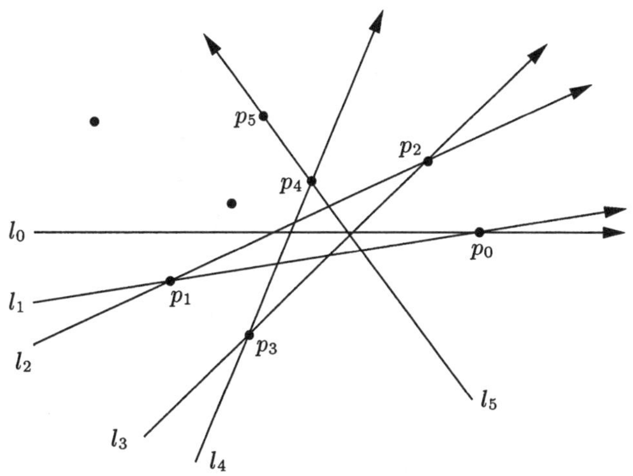

Figure 1: Initial stages of Lovász' procedure for $k = 2$.

articles [3, 4].

Lovász' Lemma: *Let S be a set of n points and l a straight line containing no points of S and dividing S into two subsets of m and $n - m$ points, respectively. Then for any $k \leq n - 2$ the number of k-segments intersecting l is at most $2\min\{m, n - m, k + 1\}$.*

A simple consequence of Lovász Lemma is that the number of k-segments intersecting a straight line l is $O(n)$ (in fact, it is at most n). The lemma is proven by giving a procedure that enumerates all $(k-1)$- and k-segments. For convenience we briefly review *Lovász procedure*:

Let l be an oriented line and let S_l be the set of points of S on the (open) right side of l. Assume without loss of generality that no two points in S have the same y-coordinate. Then there is a unique horizontal line l_0 containing a point p_0 of S such that S_{l_0} is of cardinality exactly k, i.e., there are k points of S below l_0. Let $l = l_0$ and repeat the following step: rotate l counterclockwise around p_i until it hits a point p_{i+1}, let l_{i+1} be the line through p_i and p_{i+1} and increment i. Figure 1 illustrates the initial stages of this procedure on a set of 6 points with $k = 2$. This is done until the line returns into the initial horizontal position, i.e., $l_i = l_0$. Let $l_1, l_2, \ldots, l_z$ be the different lines generated by this process. Each line l_i contains a segment $\overline{p_{i-1}p_i}$, we call these segments the *L-segments* of the process. Lovász observed the following fact.

Fact 1. The L-segments are the union of the $(k-1)$- and the k-segments.

So, by the observation above, any upper bound on the number of L-segments will be one on the number of k-sets.

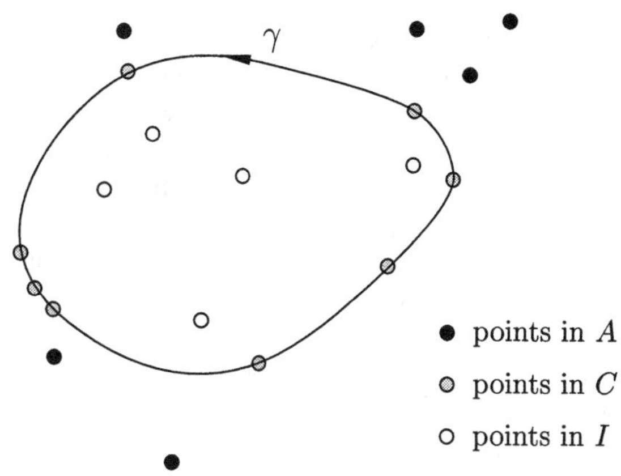

Figure 2: Curve γ and sets A, C and I

2 The main result

In the following we will consider point sets lying on a fixed collection of curves in the plane. Each curve can be closed or not, bounded or unbounded. We call a curve *convex* exactly if it lies completely on the boundary of its convex hull. Our main result is stated in the following theorem.

Theorem *Consider a fixed, finite collection of p pairwise disjoint convex curves in the plane. Then there is a constant $c_p > 0$ such that any set S of n points, each lying on one of the curves has at most $c_p n$ k-sets.*

We first observe that without loss of generality we may assume that the points in S are in *general position* in the sense that no three of them lie on a straight line. In fact, let $K \subset S$ be a k-set. Then there exists a line l not containing any point of S which separates K and $S \backslash K$. Consequently, under slight perturbations of the points in S any k-set K still remains a k-set. So the points of S can be brought into general position without decreasing the number of k-sets.

The main step towards the proof of the theorem is the following proposition which considers one fixed convex curve and gives an upper bound on the number of k-segments with at least one endpoint on the curve.

Let γ be a convex curve, i.e., the boundary curve of some convex set Γ in the plane. Partition the n points of S into the subset C of points on γ, the subset I of points in the open interior of Γ and the subset A of points outside γ, i.e., $A = S \backslash (C \cup I)$. Figure 2 illustrates the situation.

Proposition 1 *Let r be the number of k-segments of S with at least one endpoint in C and let r_A be the number of k-segments with one endpoint in A and one in C. Then*

$$r \leq 4|C| + 8r_A.$$

Proof. Let us apply Lovász' procedure to the set S and let $p_0, \ldots, p_z$ be the sequence of points and $l_1, \ldots, l_z$

the sequence of lines traversed. We will enumerate all L-segments containing a point of C. For any straight line l intersecting γ we call the first intersection point of l with γ when following its orientation its *entry-point*. We will prove an upper bound on the number ρ of L-segments where the entry-point is in C, i.e., on the number of L-segments $\overline{p_i q_i}$ such that the first point of intersection between γ and the oriented line l supporting $\overline{p_i q_i}$ is either p_i or q_i and hence in C. For reasons of symmetry the same bound holds for the L-segments with the dually defined exit-point in C.

Let $l_1^*, l_2^* \dots l_\rho^*$ be the subsequence of $l_1, \dots, l_z$ of those lines l_i^* with entry-point $q_i \in C$. Let γ be oriented such that the convex region Γ is to the left of γ. The orientation of γ induces a cyclic ordering of the points in C. If during Lovász' procedure every line l between l_i^* and l_{i+1}^* intersects γ we call the pair l_i^*, l_{i+1}^* a *standard step*.

Claim 1. If l_i^*, l_{i+1}^* is a standard step then either q_{i+1} is the point immediately preceding q_i in the cyclic order on C or q_{i+1} is the point immediately following q_i or $q_{i+1} = q_i$.

Proof. During Lovász' procedure the entry-point describes a continuous movement on curve γ. △

We classify types of standard steps: A standard step l_i^*, l_{i+1}^* is a *trivial step* if it corresponds to a rotation around the entry-point $q_i = q_{i+1}$. Note that every second step is a trivial step. A non-trivial standard step is called a *forward step* if q_{i+1} is the point following q_i in the cyclic order on C else, i.e., if $q_{i+1} = q_i$ or if q_{i+1} is the immediate predecessor of q_i in the cyclic order on C, the step is called a *backward step*. A step which is not standard is called a *special step*. If s is the number of special steps, a the number of forward steps and b the number of backward steps, then we have for the total number ρ of steps

$$\rho/2 = a + b + s. \tag{1}$$

Let l_i^*, l_{i+1}^* be a backward step and observe the entry-point when leaving q_i. When the point moves backward on γ the rotation of line l is a rotation around some point $p \in A$. In this case we "charge" the step l_i^*, l_{i+1}^* to the L-segment $\overline{p q_i}$. When the entry-point moves forward on γ after leaving from q_i then it must be $q_{i+1} = q_i$ and the entry point must move backward and return eventually. Again this backward motion is due to rotation around some point p in A and we charge the step to the L-segment $\overline{p q_{i+1}}$. This shows

$$b \le r_A. \tag{2}$$

We now turn to a closer discussion of special steps. To simplify the exposition we assume that γ has non-vanishing curvature everywhere (this can be assumed w.l.o.g. since we presume that S is in general position).

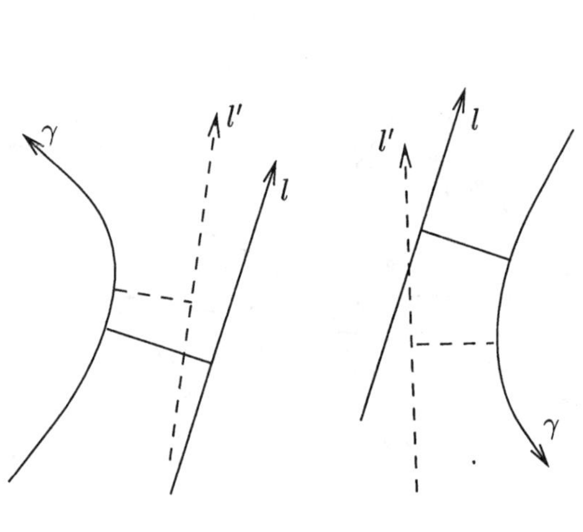

(a) γ left of l (b) γ right of l

Figure 3: Rotating from l to l' the nigh-point moves forward on γ

Suppose some line l' between l_i^* and l_{i+1}^* has no entry-point on γ. Then there is a moment when the rotating line l leaves γ, i.e., l is tangent to γ. For a line with less than two points of intersection with γ we define the ***nigh-point of** l* as the (unique!) closest point to l on γ. There are two important observations about the nigh-point:

Fact 2. For every line l the entry-point or the nigh-point are defined. If they are both defined then they coincide with the tangent point. Observing whichever is defined during Lovász' procedure we, hence, obtain a continuously moving *en-point* for l on γ.

Fact 3. While l rotates outside of γ the nigh-point always moves forward in the orientation of γ. This is easily seen to be true in both possible cases: γ left of l and γ right of l (see Figure 3).

As a consequence of Fact 3 we obtain that the en-point can move backwards only when it is the entry-point. Therefore, if l_i^*, l_{i+1}^* is a special step then either q_{i+1} is the point immediately preceding q_i in the cyclic order on C or $q_{i+1} = q_i$ or the nigh-point sweeps over a piece of curve γ containing all points of C that are between q_i and q_{i+1} in the cyclic order.

We now refine the classification and add the attribute *backward* to special steps with either q_{i+1} preceding q_i in the cyclic order on C or $q_{i+1} = q_i$. The attribute *forward* is given to all other special steps. Let s_a be the number of forward special steps and s_b be the number of backward special steps, then

$$s = s_a + s_b \tag{3}$$

As in the argument preceding Inequality (2) we find an L-segment involving a point p of A and a point of C for every backward special step. This improves Inequality 2 to

$$b + s_b \le r_A. \tag{4}$$

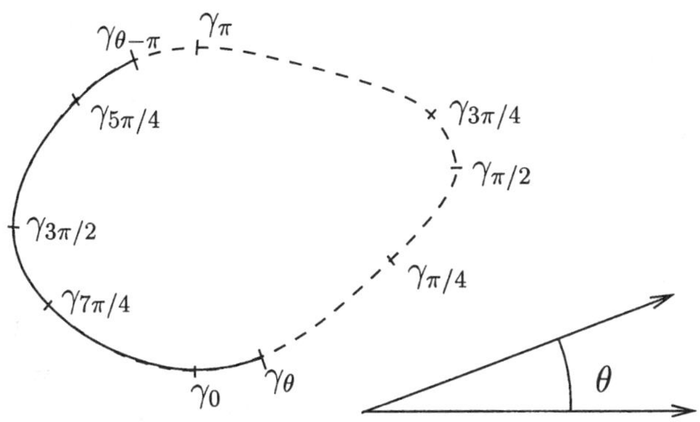

Figure 4: The potential locations for the en-point of a line at angle θ cover the part of γ parametrized by $[\theta - \pi, \theta]$.

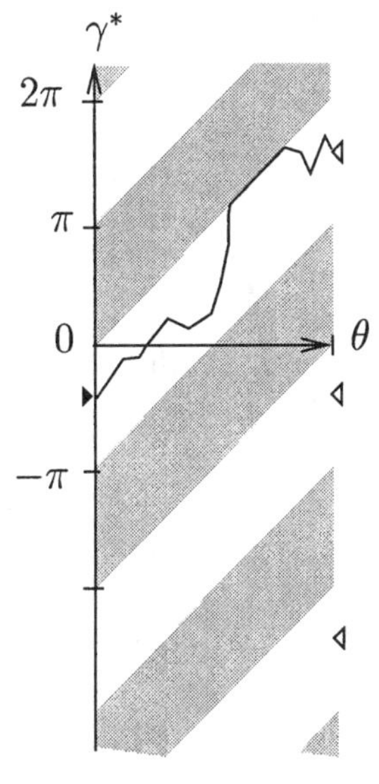

Figure 5: Possible motion of the en-point

The next lemma will show that the winding number of the en-point is +1.

Lemma 1 *During Lovász' procedure the effect of the movement of the en-point is one full rotation around* Γ.

Proof. For a fixed angle θ consider the en-points of all θ-oriented lines. These points cover one half of γ if we parametrize the curve by tangential-angle (see Figure 4).

Consider the lifting-space γ^* of curve γ, a doubly infinite spiral where each point $\gamma(t)$ on γ has a copy $\gamma^*(t + k2\pi)$ for every integer k. We identify γ^* with the reals by $t \leftrightarrow \gamma^*(t)$. During Lovász' procedure the rotating line sweeps through all angles θ from 0 to 2π. For every θ the tangential-angle of the en-point of $l(\theta)$ is in the interval $[\theta - \pi, \theta]$. Lines $l(0)$ and $l(2\pi)$ and hence their en-points equal each other. Therefore in the lifting-space γ^* the en-points of $l(0)$ and $l(2\pi)$ differ by multiples of 2π. Together this shows that as a function from the angle into the lifting-space the en-point starts at some value $x \in [-\pi, 0]$, stays in the white area in Figure 4 and ends in one of the copies of x. A possible value for x is indicated by the black triangle in Figure 4; the copies of x on the right side are the white triangles. Since the function stays in the white strip containing x we know at which of the white triangles the function ends. It follows that in the lifting-space of γ the en-point of l moves up exactly one level during Lovász' procedure, this is another way of saying that the en-point makes one full rotation around Γ. △

This lemma will be used next to bound the difference between forward and backward steps during the procedure.

Claim 2. If $|C| = t$ then

$$(a + s_a) - (b + s_b) \leq t. \quad (5)$$

Proof. To see this recall three already proven facts: (1) Every forward step moves forward at least one element in the cyclic order on C. (2) Every backward step moves backward at most one element in the cyclic order on C. (3) As a consequence of Lemma 1 the overall surplus of forward steps cannot exceed $|C|$. △

We are ready now to come to an end with the proof of Proposition 1.

$$\begin{aligned} \rho/2 &= (a + s_a) + (b + s_b) && \text{by (1) and (3)} \\ &\leq t + 2(b + s_b) && \text{by (5)} \\ &\leq t + 2r_A && \text{by (4)} \end{aligned}$$

For the number ρ' of L-segments with exit-point in C the dual argument shows that $\rho' \leq 2t + 4r_A$. With $r = \rho + \rho'$ we obtain the bound claimed in the proposition. □

Next we extend the result to sets of nested curves.

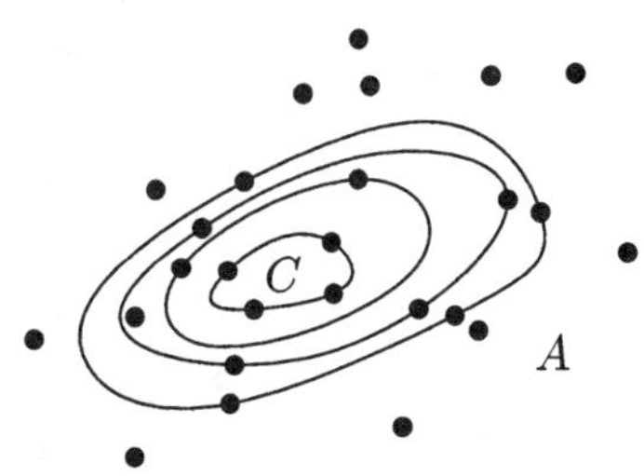

Figure 6: Nested convex curves.

Lemma 2 *Let* $\gamma_1 \ldots, \gamma_p$ *be closed convex curves where* γ_i *is contained inside* γ_{i-1}, $i = 2, \ldots, p$. *The points of* S *may lie on the curves (subset* C*) or outside of* γ_1 *(subset* A*) see Figure 6. There is a constant* c_p *such that* $r_C \leq c_p(|C| + r_A)$ *when* r_C *denotes the number of* k*-segments incident to at least one point in* C *and* r_A *the number of* k*-segments with one endpoint in* A *and one in* C.

Proof. Let r_i be the number of k-segments from γ_i to some $\gamma_j, j \geq i, i = 1, \ldots, p-1$ and let $r_0 = r_A$ be the number of k-segments from A to some γ_i, $j = 1, \ldots, p$.

Then by Proposition 1

$$r_i \leq c\,(n + s_i)$$

for some constant c, where s_i is the number of segments with one endpoint in $A \cup \gamma_1 \cup \ldots \cup \gamma_{i-1}$ (the outside of γ_i) and one on γ_i.
So $s_i \leq r_0 + \ldots + r_{i-1}$ and we have the recursive inequality

$$\begin{aligned} r_i &\leq c\,(n + \sum_{j=0}^{i-1} r_j) \;\; i = 1, \ldots, p \text{ with} \\ r_0 &= r_A. \end{aligned}$$

Certainly $r_i \leq t_i$, $i = 0, 1, \ldots$, where the sequence (t_i) is defined by the same recursive system, except that the $\leq$-inequalities are replaced by equalities.

We have $t_i - t_{i-1} = c\,t_{i-1}$, i.e., $t_i = (c+1)\,t_{i-1}$ and therefore $t_i = (c+1)^{i-1} t_1 = (c+1)^{i-1} c(n + r_A)$ is an upper bound on r_i.

We want an upper bound on the number of segments with at least one endpoint on $\gamma_1 \cup \ldots \cup \gamma_p$, i. e. on

$$\begin{aligned} \sum_{i=0}^{p} r_j &\leq r_A + c(n + r_A) \sum_{i=1}^{p} (c+1)^{i-1} \\ &= r_A + [(c+1)^p - 1](n + r_A) \end{aligned}$$

which proves Lemma 2. □

Next assume that the points of the set S lie on a finite number of closed convex curves which form p nested "clusters" each of the form described in Lemma 2 (see Figure 7). We claim

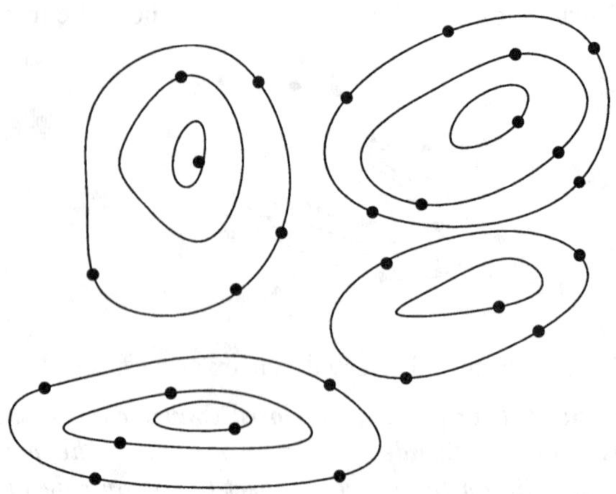

Figure 7: Four clusters of nested convex curves.

Lemma 3 *If the points of S lie on constantly many fixed nested clusters of closed convex curves, then there are $O(n)$ k-sets.*

Proof. Consider one cluster C of the p clusters and let r_C be the number of segments with at least one endpoint on C. C can be separated from any other cluster by a straight line, so by Lovász' Lemma there can be only $O(n)$ segments between the two clusters, so the number r_A of segments with one endpoint in C and one outside of C is $O(n)$. With Lemma 2 we obtain $r_C = O(n + r_A)$ which is $O(n)$. The lemma follows since there are only constantly many clusters. □

Proof **[Theorem]**. Let us consider p disjoint convex curves, which may contain the points of the set S. Since S is finite we can assume that all curves are bounded. To the curves we add all vertical lines through their endpoints and all vertical tangents (see Figure 8) The

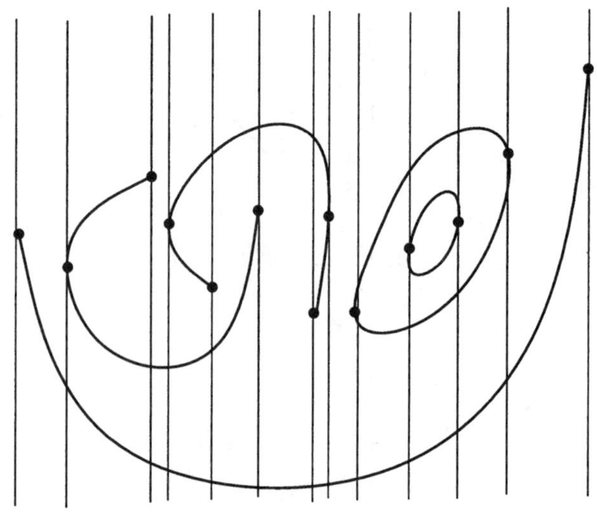

Figure 8: Vertical lines added to a set of convex curves.

vertical lines decompose the plane into constantly many slabs where each slab contains constantly many upward or downward convex segments of curves (see Figure 9). We now add line segments to the curve segments in a slab as shown in Figure 9: If a downward convex segment lies directly above an upward convex segment we connect the endpoints of each of them with a line segment. If the topmost (bottommost) segment is upward (downward) convex we also add the line segment between its endpoints.

Furthermore, from each downward convex segment we construct a closed convex curve by connecting its endpoints by a vertical line segment to the nearest endpoint below (above) of a line segment constructed in the previous phase. Thus, the curve segments obtained by introducing slabs are transformed into a set of clusters of convex curves . The number of clusters and the number of curves within each cluster are bounded by constants. Observe that, without changing the point set S, the curves can be slightly modified so that they do not touch each other any more. So we can apply Lemma 3 and conclude that S has only $O(n)$ k-segments which

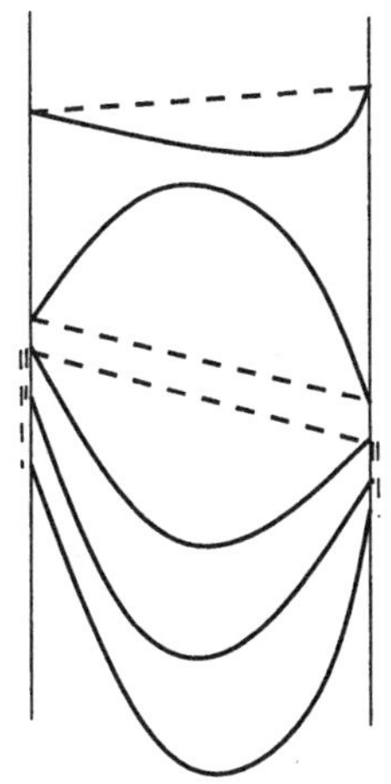

Figure 9: Curve segments and added line segments within a slab.

proves the theorem. □

Acknowledgment. We would like to thank Michael Godau and Jordi Saludes for helpful discussions concerning this research.

References

[1] P. K. AGARWAL, B. ARONOV, AND M. SHARIR, *On levels in arrangements of lines, segments, planes, and triangles*, in Proc. 13th Annu. ACM Sympos. Comput. Geom., 1997.

[2] T. K. DEY, *Improved bounds on planar k-sets and k-levels*, in Proc. 38rd Annu. IEEE Sympos. Found. Comput. Sci., 1997.

[3] P. ERDÖS, L. LOVÁSZ, A. SIMMONS, AND E. G. STRAUS, *Dissection graphs of planar point sets*, in A Survey of Combinatorial Theory, North-Holland, 1973, pp. 139–149.

[4] L. LOVÁSZ, *On the number of halving lines*, Ann. Univ. Sci. Budapest, Eötvös, Sect. Math., 14 (1971), pp. 107–108.

[5] J. PACH, W. STEIGER, AND E. SZEMERÉDI, *An upper bound on the number of planar k-sets*, Discr. Comput. Geom., 7 (1992), pp. 109–123.

[6] M. SHARIR, *On k-sets in arrangements of curves and surfaces*, Discr. Comput. Geom., 6 (1991), pp. 593–613.

On the Union of κ-Curved Objects*

Alon Efrat † Matthew J. Katz ‡

Abstract

A (not necessarily convex) object C in the plane is κ-*curved* for some constant κ, $\kappa < 1$, if it has constant description complexity, and for each point p on the boundary of C, one can place a disk B whose boundary passes through p, its radius is $\kappa \cdot diam(C)$ and it is contained in C. We prove that the combinatorial complexity of the boundary of the union of a set $\mathcal{C}$ of n κ-curved objects (e.g., fat ellipses or rounded heart-shaped objects) is $O(\lambda_s(n) \log n)$, for some constant s.

1 Introduction

Let C be a not necessarily convex object in the plane, and let κ, $\kappa < 1$, be a constant. We say that C is κ-*curved* if

(i) C has constant description complexity s_0. (This implies, for example, that there are at most s_0 local minima or maxima on its boundary, and that the number of intersection points between the boundary of C and the boundary of another object C' with constant description complexity s_0 is at most s_0 (assuming general position).)

(ii) For each point p on ∂C, we can place a disk B whose boundary passes through p, its radius is $\kappa \cdot diam(C)$, and it is contained in C (see Figure 1). We say that the radius $\kappa \cdot diam(C)$ is the *critical radius* of C, and that the disk B is a *critical disk* of C at p.

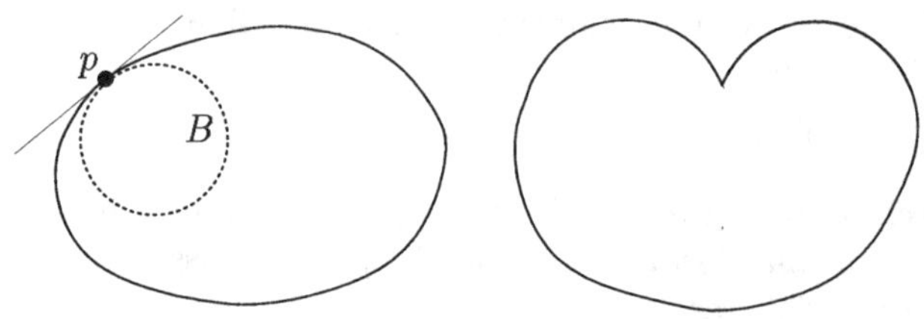

Figure 1: Two κ-curved objects

The second condition is similar to bounding the curvature of the boundary of C, but is more general (see Figure 1). It can be illustrated as follows: Imagine a car moving along the boundary of C such that the interior of C is to its left. Then the car is allowed to make very sharp right turns, but when turning left, the radius of the turn is bounded from below by some fraction of the diameter of C.

Regarding the second condition, if C has a tangent at a point p on its boundary, then there exists only one critical disk B of C at p, and its tangent at p coincides with the tangent of C at p. However, C may have a constant number of points on its boundary at which the tangent is not defined. At these points, though, C does have a left and a right tangent.

In this paper we prove that for a set $\mathcal{C} = \{C_1, \ldots, C_n\}$ of n κ-curved objects, the number of vertices on the boundary of the union U of the objects in $\mathcal{C}$ is only $O(\lambda_s(n) \log n)$, for some constant s.

We say that an object C is α-fat, for a constant $\alpha > 1$, if $r_1/r_2 \leq \alpha$, where r_1 is the radius of a smallest disk containing C, and r_2 is the radius of a largest disk contained in C. Obviously, a κ-curved object is α-fat for an appropriate constant α, but the opposite statement is false. Fat objects received much attention

*Work by the second author has been supported by the Israel Science Foundation founded by the Israel Academy of Sciences and Humanities.

†Department of Computer Science, Tel-Aviv University, Tel-Aviv 69978, Israel. *alone@math.tau.ac.il*, *http://www.math.tau.ac.il/~alone*

‡Department of Mathematics and Computer Science, Ben-Gurion University of the Negev, Beer-Sheva 84105, Israel. *matya@cs.bgu.ac.il*, *http://www.cs.bgu.ac.il/~matya*

SCG 98 Minneapolis Minnesota USA

in recent years. One of the first papers on fat objects is by Matoušek et al. [10] who showed that a set of n triangles determines only a linear number of "holes," and that the number of vertices on the boundary of its union is only $O(n \log\log n)$. Since then many authors considered various definitions of fatness (which are all more or less equivalent—at least for convex objects), and obtained either interesting combinatorial results or efficient geometric algorithms for various classes of fat objects (see e.g. [2, 3, 5, 7, 8, 9, 11, 13, 14]). However, the question which properties suffice so that the number of vertices on the boundary of the union of a set of planar objects having these properties is always subquadratic remained open for quite a few years. Recently, Efrat and Sharir [6] showed that if the objects are convex, fat, and the boundaries of each pair intersect at most a constant number of times, then the boundary of their union consists of only $O(n^{1+\varepsilon})$ vertices, for any constant $\varepsilon > 0$. [1] In a preliminary version of their paper, it was shown that if, in addition, the objects have bounded curvature and are more or less of the same size, then the number of vertices on the union's boundary is only $O(\lambda_s(n))$, for some constant s.

Our result improves upon the result of [6] for convex κ-curved objects such as fat ellipses, and complements it for objects that are non-convex (but κ-curved). We prove the following theorem.

Theorem 1.1 *The combinatorial complexity of the boundary of the union of n κ-curved objects is $O(\lambda_s(n)\log n)$, for some constant s.*

In the proof we use a well known data structure, namely, a segment tree, and its properties. We project the input κ-curved objects on the y-axis, and construct a segment tree $\mathcal{T}$ for these projections. We then insert the objects into $\mathcal{T}$ according to their projection on the y-axis. As usual, we associate with a node μ of $\mathcal{T}$ its canonical y-interval, which we think of as an horizontal slab. Now, roughly speaking, the vertices on the boundary of the union of the input objects are distributed among the nodes of $\mathcal{T}$, so that, if a vertex w ends up at a node μ, then w lies in the canonical slab of μ and is formed by a pair of objects that are stored at μ. (The objects that are stored at μ consist of the objects in the canonical subset of μ and the objects in the canonical subsets of all the descendants of μ.) By proving a connection between the number of vertices that end up at μ and the number of objects that are stored at μ, and by summing over all nodes in $\mathcal{T}$, we obtain the claimed bound.

The paper is organized as follows. In Section 2 we establish an auxiliary result, which is later used (Claim 3.2) in the proof of Theorem 1.1. The proof of Theorem 1.1 is given in Section 3, except for the proof of a "key lemma," which is stated in this section and proven in Section 4. Section 5 concludes the paper.

[1] Throughout the paper, ε stands for a positive constant which can be chosen arbitrarily small with an appropriate choice of other constants of the big-O notation.

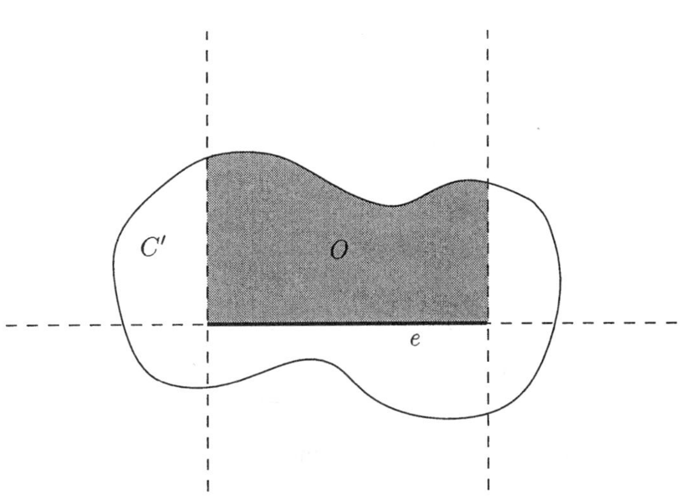

Figure 2: C' is function-defined with respect to e (from above)

2 Partitioning the Boundary of an Object

Let ρ be a horizontal strip of width δ that is divided into squares $\Sigma = \{\sigma_1, \sigma_2, \ldots\}$ of edge-length δ. Let ℓ_{top} and ℓ_{bottom} denote the top and bottom horizontal lines defining ρ. Let C be a κ-curved object whose diameter is at least $3\delta/\kappa$. (The radius of a critical disk of C is therefore at least 3δ). In this section we show that it is possible to obtain from C a constant number of smaller objects (called *parts*), such that (i) each part is contained in C and has some desirable properties, and (ii) for each point p on $\rho \cap \partial C$, there exists a part such that p lies on its boundary. We use this result in the proof of Theorem 1.1.

We need the following definition.

Definition: Consider an object C' in the plane, and a horizontal segment e. We cut off the part of C' that lies below the line containing e, and the parts of C' that lie outside the vertical strip defined by e. Let O denote the remaining part of C'. We say that C' is *function-defined* with respect to e (from above), if for every point p on e, the intersection of O with the vertical ray emanating from e and directed upwards is a closed segment whose bottom endpoint is p (see Figure 2). In other words, the top boundary of O is the graph of some function defined on e, and O is the area enclosed between this graph and the segment e. We define in an analogous manner the

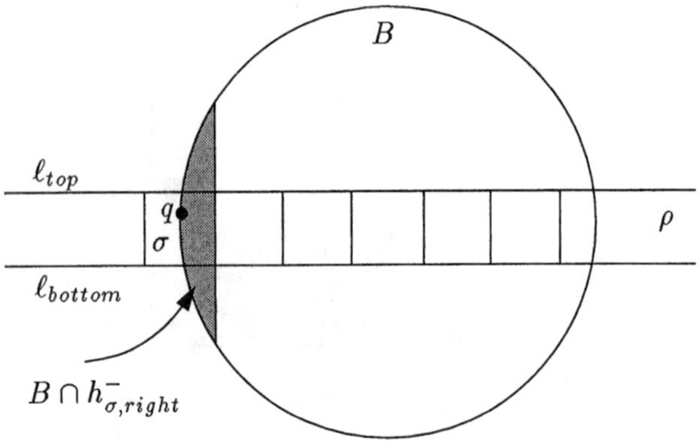

Figure 3: $B \cap h^-_{\sigma,right}$ is function defined with respect to $\ell_{\sigma,right}$

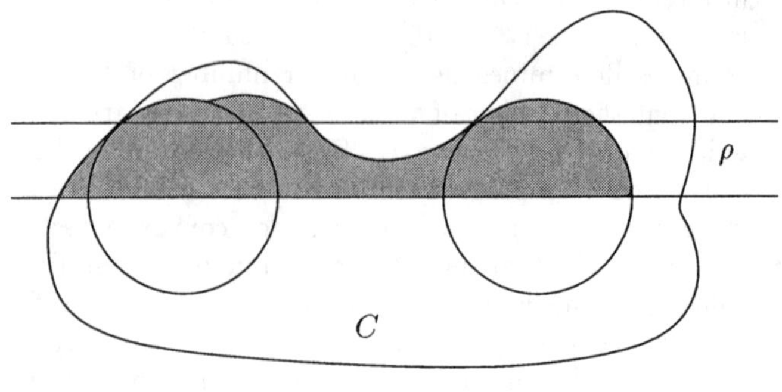

Figure 4: The part C_{bottom} (in gray) and two of its defining disks

statements: C' is *function-defined* with respect to e from below, or, for a vertical segment e, C' is *function-defined* with respect to e from the left (alternatively, from the right).

For each point p on ∂C, let $\mathcal{B}(p)$ denote the collection of disks of radius $\kappa \cdot diam(C)$ that are contained in C and whose boundaries pass through p. (As mentioned, if ∂C has a tangent at p, then $\mathcal{B}(p)$ consists of a single disk.) For a disk $B \in \mathcal{B}(p)$, let B' denote the disk of radius 3δ obtained from B by moving its center towards p while maintaining the contact with p. Let h^+_{bottom} (resp. h^-_{top}) denote the halfplane bounded by ℓ_{bottom} (resp. ℓ_{top}) and lying above ℓ_{bottom} (resp. below ℓ_{top}). Similarly, for a square $\sigma \in \Sigma$, let $\ell_{\sigma,left}$ and $\ell_{\sigma,right}$ denote the lines containing the left and right edges of σ, respectively, and let $h^+_{\sigma,left}$ (resp. $h^-_{\sigma,right}$) denote the halfplane bounded by $\ell_{\sigma,left}$ (resp. $\ell_{\sigma,right}$) and containing σ. We use the following simple but important observation (see Figure 3).

Claim 2.1 *Let B be a disk of radius at least 3δ that intersects ρ, and let q be a point on $\rho \cap \partial B$. Let σ be the cell of Σ containing q. Then either $B \cap h^+_{bottom}$ is function defined with respect to ℓ_{bottom}, or $B \cap h^-_{top}$ is function defined with respect to ℓ_{top}, or $B \cap h^+_{\sigma,left}$ is function defined with respect to $\ell_{\sigma,left}$, or $B \cap h^-_{\sigma,right}$ is function defined with respect to $\ell_{\sigma,right}$.*

We next define the set $C_{bottom} \subseteq C$.

$$C_{bottom} = \bigcup\{B' \cap h^+_{bottom} \mid B \text{ is a disk of } \mathcal{B}(p) \text{ for some } p \in \rho \cap \partial C, \text{ and } B' \cap h^+_{bottom} \text{ is function-defined for } \ell_{bottom}\}\ .$$

The set C_{top} is defined analogously. It is easy to see that the sets C_{bottom} and C_{top} are function-defined with respect to ℓ_{bottom} and ℓ_{top}, respectively, and that they have constant description complexity. For each $\sigma \in \Sigma$ we define

$$C_{\sigma,left} = \bigcup\{B' \cap h^+_{\sigma,left} \mid B \text{ is a disk of } \mathcal{B}(p) \text{ for some } p \in \sigma \cap \partial C, \text{ and } B' \cap h^+_{bottom} \text{ is not function-defined for } \ell_{bottom}, \text{ and } B' \cap h^+_{top} \text{ is not function-defined for } \ell_{top}, \text{ and } B' \cap h^+_{\sigma,left} \text{ is function-defined for } \ell_{\sigma,left}\}\ .$$

The set $C_{\sigma,right}$ is defined analogously. It is easy to see that the set $C_{\sigma,left}$ (resp. $C_{\sigma,right}$) is function-defined with respect to $\ell_{\sigma,left}$ (resp. $\ell_{\sigma,right}$), that it has constant description complexity, and that its x-span is contained in the x-span of the union of σ and the six cells immediately to its right (resp. left).

Define the function f_{bottom} on ℓ_{bottom} as follows: For $x \in \ell_{bottom}$, if there is no point in C_{bottom} that lies vertically above x, then $f_{bottom}(x)$ is not defined; otherwise, $f_{bottom}(x) = y$, where y is the y-coordinate of the highest among the points in C_{bottom} that lie vertically above x. The graph of the function f_{bottom} is actually the upper envelope of C_{bottom}. The functions f_{top}, $f_{\sigma,left}$, $f_{\sigma,right}$ are defined analogously. From Claim 2.1 it follows that $\rho \cap \partial C$ is contained in the union of the graph of f_{bottom}, the graph of f_{top}, and the graphs of $f_{\sigma,left}$ and of $f_{\sigma,right}$ taken over all $\sigma \in \Sigma$.

Next we claim that $C_{\sigma,left}$ and $C_{\sigma,right}$ are not empty only for a constant number of cells σ. Indeed, let p be a point on $\sigma \cap \partial C$ for which there exists a disk $B \in \mathcal{B}(p)$ such that B' is function defined say for the left edge of σ, but not for ℓ_{bottom} nor ℓ_{top}. Analyzing the relative positions of p and of the center of B', we deduce that B' fully contains the left edge of σ, or the left edge of the cell σ^l immediately to the left of σ. Therefore, either ∂C has a (locally) leftmost point in σ or in σ^l (i.e., ∂C turns rightwards), or ∂C intersects either ℓ_{bottom} or ℓ_{top} within σ or σ^l. Thus in both cases some event occurs either in σ or in σ^l. However, our assumptions concerning the boundary of C imply that both these types of events may occur only a constant number of times, and therefore the number of non-empty sets of the form $C_{\sigma,left}$ or $C_{\sigma,right}$ is bounded by some constant.

3 Proof of Theorem 1.1

Let $\mathcal{C}$ be a set of n κ-curved objects, and let U denote the union of the objects in $\mathcal{C}$. We prove that the combinatorial complexity of ∂U is $O(\lambda_s(n)\log n)$, for some constant s.

Project the objects in $\mathcal{C}$ on the y-axis, and construct a segment tree $\mathcal{T}$ for these projections. Insert the objects of $\mathcal{C}$ into $\mathcal{T}$ according to their projection on the y-axis. For a node μ of $\mathcal{T}$, let y_μ denote the canonical y-interval that is associated with μ, and let $\mathcal{C}_\mu$ be the canonical subset that is stored at μ. We think of y_μ as the horizontal slab whose top (resp. bottom) defining line passes through the top (resp. bottom) endpoint of the y-interval denoted by y_μ. We also store at μ a second subset $\mathcal{D}_\mu$ which is the union of all canonical subsets stored at (the proper) descendants of μ. It is well known (see e.g. [4]) that

$$\sum_\mu (|\mathcal{C}_\mu| + |\mathcal{D}_\mu|) = O(n \log n) .$$

Let U_μ denote the union of the objects in $\mathcal{C}_\mu \cup \mathcal{D}_\mu$ restricted to the slab y_μ. We first prove the following easy (and known) claim.

Claim 3.1 *Let w be a vertex on ∂U that is an intersection point of the boundaries of two objects $C_1, C_2 \in \mathcal{C}$, then there exists a node μ of $\mathcal{T}$ such that w lies in the slab y_μ, and either*

1. *both C_1 and C_2 are in $\mathcal{C}_\mu$, or*

2. *one of them is in $\mathcal{C}_\mu$ and the other is in $\mathcal{D}_\mu$.*

Moreover, w is also a vertex on ∂U_μ.

Proof: The first part (i.e., there exists such a node μ) follows from basic properties of segment trees, since the projection of w on the y-axis lies in both the projection of C_1 and the projection of C_2. The second part is also obvious, since w is a vertex on the boundary of the union of any subset of $\mathcal{C}$ that includes both C_1 and C_2. □

We thus distinguish between two types of vertices on ∂U_μ. A vertex of type I is an intersection point between the boundaries of two objects in $\mathcal{C}_\mu$, and a vertex of type II is an intersection point between the boundaries of an object in $\mathcal{C}_\mu$ and an object in $\mathcal{D}_\mu$. Let u_μ be the number of vertices on ∂U_μ of type I and type II. We

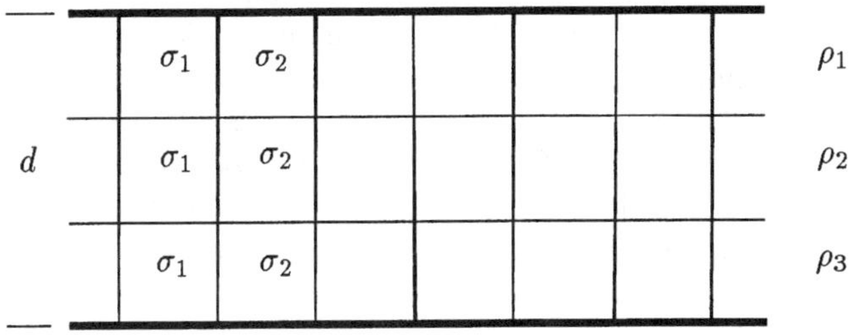

Figure 5: The slab y_μ partitioned into 3 strips ρ_1, ρ_2, ρ_3

prove that $u_\mu = O(\lambda_s(|\mathcal{C}_\mu| + |\mathcal{D}_\mu|))$, for some constant s, and therefore

$$\sum_\mu u_\mu = \sum_\mu O(\lambda_s(|\mathcal{C}_\mu| + |\mathcal{D}_\mu|)) = O(\lambda_s(n)\log n) .$$

This equation together with Claim 3.1 yields the desired result, i.e., the number of vertices on ∂U is $O(\lambda_s(n)\log n)$.

Consider a node μ of $\mathcal{T}$, and let d be the width of the slab y_μ. We partition the slab y_μ into $3/\kappa$ horizontal strips each of width $\frac{\kappa}{3}d$. We partition each of these strips into disjoint squares $\sigma_1, \sigma_2, \ldots$ of edge-length $\frac{\kappa}{3}d$, by adding vertical walls (see Figure 5). Consider any one of the strips ρ, and let l_1 (resp. l_2) denote its lower (resp. upper) defining line. We show that the number of vertices on ∂U_μ of type I and type II that lie in ρ is $O(\lambda_s(|\mathcal{C}_\mu| + |\mathcal{D}_\mu|))$, and thus obtain that $u_\mu = O(\lambda_s(|\mathcal{C}_\mu| + |\mathcal{D}_\mu|))$ (since y_μ was partitioned into a constant number of strips).

Clearly any object in $\mathcal{C}_\mu$ has diameter at least d. Let C be an object in $\mathcal{C}_\mu \cup \mathcal{D}_\mu$ whose diameter is at least d. In Section 2, we proved the following claim.

Claim 3.2 *It is possible to obtain from C a constant number of (not necessarily connected) parts, such that (i) each of the parts is function-defined with respect to either l_1, l_2, or a line containing a vertical wall in ρ, (ii) each of the parts has constant description complexity, (iii) those parts that are function-defined with respect to a line containing a vertical wall e, are contained in a vertical slab defined by a section of ρ that begins at e and is six squares wide, and (iv) if p is a point on $\rho \cap \partial C$, then p lies on the (appropriate) envelope of one of the parts (i.e., either on the upper envelope of C_{bottom}, or on the lower envelope of C_{top}, etc.).*

Let $\mathcal{E}_\mu$ be the set of all objects in $\mathcal{C}_\mu \cup \mathcal{D}_\mu$ with diameter at least d, and let $\mathcal{F}_\mu$ be the set $(\mathcal{C}_\mu \cup \mathcal{D}_\mu) \setminus \mathcal{E}_\mu$. We partition all objects in $\mathcal{E}_\mu$ as in Claim 3.2. Let γ_1 (resp. γ_2) denote the upper envelope (resp. lower envelope) of all parts that are function-defined with respect to l_1 (resp. l_2). The combinatorial complexity of γ_i is $O(\lambda_{s_0}(m_i))$, where $m_i = O(|\mathcal{E}_\mu|) = O(|\mathcal{C}_\mu| + |\mathcal{D}_\mu|)$ is the number of parts that are function-defined with respect to l_i, $i = 1, 2$; [12]. For each square σ, let σ_l (resp. σ_r) denote the right envelope (resp. left envelope) of all parts that are function-defined with respect to the left (resp. right) edge of σ. The combinatorial complexity of σ_l (resp. σ_r) is $O(\lambda_{s_0}(m_l))$ (resp. $O(\lambda_{s_0}(m_r)))$, where m_l (resp. m_r) is the number of parts that are function-defined with respect to the left (resp. right) edge of σ. From Claim 3.2 we know that

$$\sum_{\sigma}(m_l + m_r) = O(|\mathcal{E}_\mu|) = O(|\mathcal{C}_\mu| + |\mathcal{D}_\mu|) .$$

Consider now the objects in $\mathcal{F}_\mu$, i.e., the objects in $\mathcal{C}_\mu \cup \mathcal{D}_\mu$ with diameter less than d. Each such object intersects only a constant number of squares of ρ. For each square σ, let $\mathcal{F}_\sigma \subseteq \mathcal{F}_\mu$ be the subset of objects that intersect σ; we have $\sum_\sigma |\mathcal{F}_\sigma| = O(|\mathcal{F}_\mu|)$. Recall that our goal is to bound the number of vertices on ∂U_μ of type I and II that lie in ρ. We bound the number of vertices that appear when considering various pairs of envelopes, and various pairs consisting of an envelope and a subset of $\mathcal{F}_\mu$. That is, when considering a pair of envelopes we count the number of intersection points between the envelopes, or, in other words, if $\mathcal{X}$ and $\mathcal{Y}$ are the two underlying sets of parts, then we count the number of *bichromatic* vertices on the boundary of the union of the objects in $\mathcal{X} \cup \mathcal{Y}$, where a vertex is bichromatic if it lies on the boundary of an object of $\mathcal{X}$ and on the boundary of an object of $\mathcal{Y}$. And when considering a pair consisting of an envelope and a subset $\mathcal{F}'$ of $\mathcal{F}_\mu$, we count the number of *bichromatic* vertices on the boundary of the union of the objects in $\mathcal{X} \cup \mathcal{F}'$, where $\mathcal{X}$ is the set of parts underlying the envelope. More precisely, we bound the number of vertices that appear when considering the following pairs.

(a) (γ_1, γ_2)

(b) For each square σ,
(γ_1, σ_l), (γ_1, σ_r), (γ_2, σ_l), (γ_2, σ_r).

(c) For each square σ,
(σ_l, σ_r).

(d) $(\gamma_1, \mathcal{F}_\mu)$, $(\gamma_2, \mathcal{F}_\mu)$

(e) For each square σ,
$(\sigma_l, \mathcal{F}_\sigma)$, $(\sigma_l, \mathcal{F}_\sigma)$.

We now claim that all 'interesting' vertices appear.

Claim 3.3 *If w is a vertex on ∂U_μ of type I or II that lies in ρ, then either (i) w is a vertex of one of the envelopes considered, or (ii) w appears when one of the above pairs is considered.*

Proof: If w is of type I, that is, w is an intersection point of the boundaries of two objects in $\mathcal{C}_\mu$. Then clearly w is either a vertex of one of the envelopes γ_1, γ_2, or σ_l, σ_r, for a square $\sigma \in \Sigma$, or a vertex that appears when considering one of the pairs listed in **(a)**, **(b)**, and **(c)** above.

If w is of type II, that is, w is an intersection point of the boundaries of an object in $\mathcal{C}_\mu$ and an object in $\mathcal{D}_\mu$, then we distinguish between two cases. If the object from $\mathcal{D}_\mu$ is large, i.e., it is in $\mathcal{E}_\mu$, then, as before, w is either a vertex of an envelope or appears when considering a pair of envelopes. Otherwise, the object from $\mathcal{D}_\mu$ is small (i.e., it is in $\mathcal{F}_\mu$), and w is a vertex that appears when considering one of the pairs listed in **(d)** and **(e)**. □

Notice that we also count many 'uninteresting' vertices such as vertices that are formed by two objects in $\mathcal{D}_\mu$, or vertices that 'do not make it' to the boundary of the full union.

In the next section we prove a key lemma stating that if γ is an envelope as those defined above, and $\mathcal{A}$ is a set of κ-curved objects, then the number of 'visible' bichromatic vertices on γ for which the larger object (of the two objects forming the vertex) comes from the set $\mathcal{X}$ underlying γ is $O(\lambda_{s_0}(|\mathcal{X}|) + \lambda_s(|\mathcal{A}|))$, where a visible vertex is a vertex on the boundary of the union of $\mathcal{X} \cup \mathcal{A}$, and the size of a part in $\mathcal{X}$ is the size of the object to which it belongs.

We employ this lemma to bound the number of vertices that appear when considering the pairs listed above. We can immediately apply the lemma to the two pairs of **(d)** above, since each object in the set underlying γ_i is larger than all objects in $\mathcal{F}_\mu$. Thus we obtain an $O(\lambda_s(|\mathcal{C}_\mu| + |\mathcal{D}_\mu|))$ bound for these two pairs. Similarly, we can apply the lemma to the pairs of **(e)**. Recalling that the total complexity of the envelopes corresponding to vertical walls in ρ is $O(\lambda_{s_0}(|\mathcal{C}_\mu| + |\mathcal{D}_\mu|))$, and that $\sum_\sigma |F_\sigma| = O(|\mathcal{D}_\mu|)$, we obtain a bound of $O(\lambda_s(|\mathcal{C}_\mu| + |\mathcal{D}_\mu|))$ for all the pairs of **(e)** together.

In order to apply the lemma to the pairs of **(b)**, we first observe that when a pair (γ_i, σ_z), $i \in \{1, 2\}$, $z \in \{l, r\}$, is considered, we may restrict γ_i to the square σ. However, there is still a problem, since it is not true anymore that the larger object (of the two objects forming a countable vertex) always comes from the same underlying set. We thus consider a pair (γ_i, σ_z) twice. First we bound the number of vertices on γ_i for which the smaller object (of the two objects forming it) comes from the set underlying σ_z, by applying the lemma, and then we bound the number of vertices on σ_z for which the smaller object comes from the set underlying γ_i, again by applying the lemma. In this way we bound all vertices that appear when considering a pair (γ_i, σ_z), and obtain a bound of $O(\lambda_s(|\mathcal{C}_\mu| + |\mathcal{D}_\mu|))$ for all the pairs of **(b)** together.

We can immediately bound the number of vertices that appear when considering the pair of **(a)** or the pairs of **(c)**, and obtain in both cases a bound of $O(\lambda_{s_0}(|\mathcal{C}_\mu|+|\mathcal{D}_\mu|))$.

We thus conclude that the number of vertices of ∂U_μ of type I or II that lie in ρ is $O(\lambda_s(|\mathcal{C}_\mu|+|\mathcal{D}_\mu|))$, leading as detailed above to the main theorem.

Theorem 3.4 *The combinatorial complexity of ∂U is $O(\lambda_s(n)\log n)$, for some constant s.*

4 Proof of the Key Lemma

Let ρ be a strip of width δ and denote by l its bottom boundary. Let $\mathcal{S} = \{S_1, \ldots, S_m\} \subseteq \mathcal{C}$ be a set of m large input objects, that is, the diameter of S_i is at least $3\delta/\kappa$, $i = 1, \ldots, m$. Apply the process described in Section 2 to the objects $S_1, \ldots, S_m$, and let $O_1, \ldots, O_m$ be the m bottom parts that are obtained. Consider γ, the upper envelope of $O_1, \ldots, O_m$, and denote by R the region between γ and l. Let $\mathcal{A} \subseteq \mathcal{C}$ be a set of k input objects. We wish to bound the number of bichromatic vertices on the boundary of $V = R \cup (\cup\mathcal{A})$ that lie on γ and for which the larger of the two appropriate objects comes from $\mathcal{S}$.

We divide each $A \in \mathcal{A}$ into a constant number of primitive objects $\alpha_1, \alpha_2, \ldots$ by vertically decomposing A. That is, for each of the locally x-extreme points p on ∂A, if we remain in A when moving slightly upwards or downwards from p, then we draw a vertical segment beginning at p and directed upwards (downwards), until it hits ∂A. Denote by $\mathcal{A}'$ the set of primitive objects that is obtained; $|\mathcal{A}'| = O(k)$. (Notice that a primitive object is trapezoid-like, it is defined by (at most) two vertical walls and by two x-monotone curves, a top curve and a bottom curve.)

When walking along γ from left to right, let L_{top} (resp. L_{bot}) be the sequence of names of primitive objects in $\mathcal{A}'$ corresponding to the bichromatic vertices of V that lie on top (resp. bottom) boundaries of primitive objects in $\mathcal{A}'$. In the remainder of this section we prove the following lemma.

Lemma 4.1 (Key Lemma) $|L_{top}| = |L_{bot}| = O(\lambda_{s_0}(m)+\lambda_s(k))$, *for some constant s.*

Notice that whenever there are more than s_0 consecutive occurrences of the same name, there must be a vertex of γ somewhere in between. Thus, if we replace in the sequence L_{top} (resp. L_{bot}) all consecutive occurrences of a name by a single representative occurrence of that name, we remain with a sequence L'_{top} (resp. L'_{bot}), and $|L_{top}| = |L'_{top}| + O(\lambda_{s_0}(m))$ (resp. $|L_{bot}| = |L'_{bot}| + O(\lambda_{s_0}(m))$). In the claim below we prove that $|L'_{top}| = |L'_{bot}| = O(\lambda_s(k))$, for some constant s, and therefore $|L_{top}| = |L_{bot}| = O(\lambda_{s_0}(m) + \lambda_s(k))$.

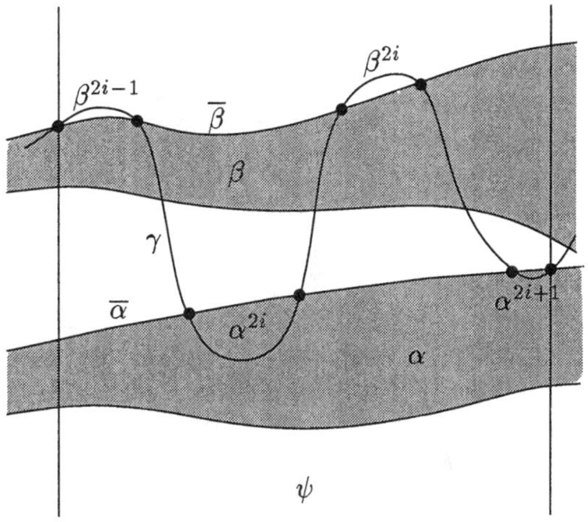

Figure 6: β is above α in ψ

Proposition 4.2 *The sequence L'_{top} (alternatively L'_{bot}) is a Davenport-Schinzel sequence [12] of order $2s_0 + 1/\kappa + c$, where c is a small constant.*

Proof: Consider first the sequence L'_{top}. Assume that there are two primitive objects $\alpha, \beta \in \mathcal{A}'$ with top boundaries $\overline{\alpha}$ and $\overline{\beta}$, respectively, for which there exists a long subsequence of L'_{top} of the form $\alpha^1\beta^1\alpha^2\beta^2\ldots\alpha^t\beta^t$ (or $\alpha^1\beta^1\alpha^2\beta^2\ldots\alpha^t\beta^t\alpha^{t+1}$). We focus on the x-interval whose endpoints are β^1 and α^t (or more precisely the first occurrence in the sequence of occurrences represented by β^1 and the last occurrence in the sequence represented by α^t). Both top boundaries $\overline{\alpha}$ and $\overline{\beta}$ are defined over the entire interval.

Consider four consecutive representatives β^{2i-1}, α^{2i}, β^{2i}, α^{2i+1}. (If t is even then we disregard the last two representatives β^{t-1} and α^t.) We restrict our attention to the vertical slab ψ whose left bounding line passes through the first occurrence in the sequence of occurrences represented by β^{2i-1}, and whose right bounding line passes through the last occurrence in the sequence represented by α^{2i+1}. Let A and B be the objects of $\mathcal{A}$ from which α and β were obtained.

If α and β intersect within ψ, then we ignore this quadruple, since this implies that the boundaries of A and B intersect within ψ, and therefore there are at most s_0 such quadruples. Thus we assume that either α is above β in ψ, or vice versa. We show that the width of ψ (under this assumption) is at least $2\kappa\cdot diam(C)$, where C is the smaller object among A and B, and therefore there can be at most $\frac{1}{2\kappa}$ such quadruples.

Assume first that β is above α (see Figure 6). We restrict our attention further to the triple $\alpha^{2i}, \beta^{2i}, \alpha^{2i+1}$. Consider p the vertex corresponding to the first occurrence in the sequence represented by β^{2i}, and let $S \in \mathcal{S}$ be the object to which the arc of γ that passes through

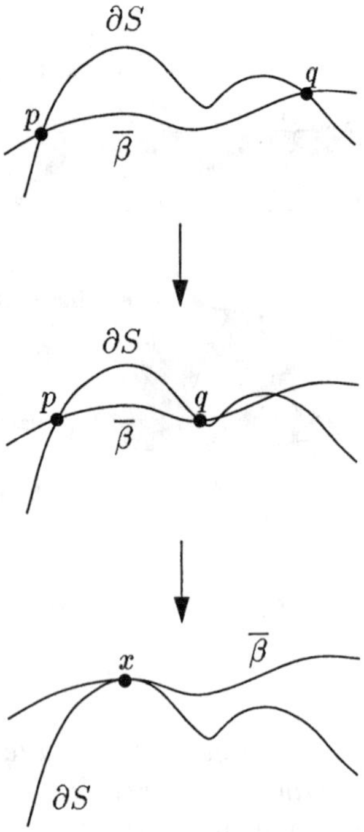

Figure 7: Translating S downwards

p belongs. Assume that ∂S exits β at p, and let q be the first point to the right of p on $\overline{\beta}$ where ∂S enters β (see Figure 7, top). (If ∂S enters β at p, then we define q to be the first point to the left of p on $\overline{\beta}$ where ∂S exists β, and proceed similarly.) We now think of α^{2i} as the rightmost intersection point corresponding to it, and of α^{2i+1} as the leftmost intersection point corresponding to it.

We move γ rigidly downwards, varying the points α^{2i}, p, q and α^{2i+1} accordingly, until p and q coincide at a point x on $\overline{\beta}$ (see Figure 7). In other words, during this process, p is the (constantly moving rightwards) exit point of ∂S and q is the (constantly moving leftwards) entrance point of ∂S, α^{2i} is the rightmost intersection point of γ and $\overline{\alpha}$ to the left of p, and α^{2i+1} is the leftmost intersection point of γ and $\overline{\alpha}$ to the right of q. Notice that the path traced by p (alternatively, q) on $\overline{\beta}$ is not necessarily connected (see Figure 7). At the end of this process, ∂S passes through x and lies below $\overline{\beta}$ in a small neighborhood of x. Clearly the final location of α^{2i} is more to the right than the initial location of α^{2i}, and the final location of α^{2i+1} is more to the left than its initial location.

If $\overline{\beta}$ does not have a tangent at x, then we may ignore this quadruple, since there are at most s_0 such points on ∂B. Therefore, we assume that $\overline{\beta}$ does have a tangent at x, and let D be the critical disk for the point x on ∂B, i.e., D is a disk of radius $\kappa \cdot \mathit{diam}(B)$, D is contained in B and its boundary passes through x. We now claim

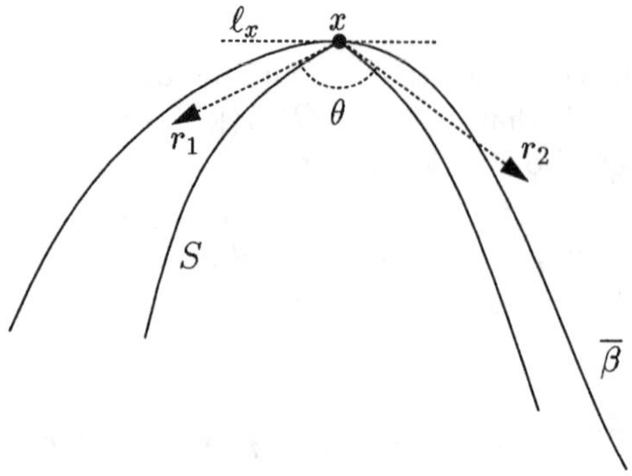

Figure 8: Proof of $D \subseteq S$

that the disk D is also contained in S, and therefore it is contained in the region lying below γ.

Observe that if (as we assume) ∂B has a tangent at x, then so does ∂S. Assume this is false, and let r_1 and r_2 be the two rays tangent from the left and from the right, respectively, to ∂S at x (see Figure 8). S lies locally below both of them. Let θ be the inward angle between r_1 and r_2, and let ℓ_x be the tangent to ∂B at x. If $\theta < \pi$, then it is impossible to draw a disk that is contained in S and whose boundary passes through x. If, on the other hand, $\theta > \pi$, then either r_1 or r_2, say r_2, is above ℓ_x, but then all points of ∂B to the right of x and close enough to x, are below ∂S, which contradicts the way in which S was translated. Thus we conclude that $\theta = \pi$, and ∂S has a tangent at x. Moreover, this tangent is necessarily ℓ_x. Since S is larger than B, D is contained in the (unique) critical disk D' of S for the point x.

The last claim implies that D is contained in $\beta \cap \psi$, since if it is not, then the boundary of D must intersect one of the bounding lines of ψ at two points lying between the bottom and top boundaries of β. But if so γ cannot intersect the top boundary of α within the slab ψ on both sides of x (since $D \subseteq S$).

We now claim that (the current) α^{2i} lies completely to the left of D, and (the current) α^{2i+1} lies completely to the right of D (and therefore this is surely true for the initial α^{2i} and α^{2i+1}). Therefore the horizontal distance between the initial α^{2i} and α^{2i+1} is at least $2\kappa \cdot \mathit{diam}(B)$. The claim is correct since α lies below β in ψ, D is contained in $\beta \cap \psi$ and D is contained in the region below γ, and $\overline{\alpha}, \overline{\beta}$, and γ are x-monotone. If α is above β, then we consider the triple $\beta^{2i-1}, \alpha^{2i}, \beta^{2i+1}$ and treat this case analogously.

Consider now the sequence L'_{bot}. If β is above α we consider the triple $\alpha^{2i}, \beta^{2i}, \alpha^{2i+1}$, and if α is above β we consider the triple $\beta^{2i-1}, \alpha^{2i}, \beta^{2i}$. In both cases, we translate γ until S just touches (locally) the top boundary of the lower object, and essentially continue as for the sequence L'_{top}. We describe in detail the case where β is above α, so we consider the triple $\alpha^{2i}, \beta^{2i}, \alpha^{2i+1}$. Consider p the vertex corresponding to the first occurrence in the sequence represented by β^{2i}, and let $S \in \mathcal{S}$ be the object to which the arc of γ that passes through p belongs. Assume that ∂S enters β at p, and let q be the first point to the right of p on the bottom curve of β where ∂S exits β. (If ∂S exits β at p, then we define q to be the first point to the left of p on the bottom curve of β where ∂S enters β, and proceed similarly.) We now think of α^{2i} as the rightmost intersection point corresponding to it, and of α^{2i+1} as the leftmost intersection point corresponding to it. We translate γ rigidly downwards, until ∂S touches $\overline{\alpha}$ at a point x, to the right of p and to the left of q, and ∂S lies below $\overline{\alpha}$ at a neighborhood of x. As above, if A has a tangent at x, then so does S and the two tangents coincide. We now distinguish between two cases. If $diam(B) < diam(A)$, then at x we may draw a disk of radius $\kappa \cdot diam(B)$ which is surely contained in A and in S. Again we claim that the points α^{2i} and α^{2i+1} are now closer to each other and that they are to the left and to the right of the disk we drew. This means that the horizontal distance between the initial α^{2i} and α^{2i+1} is at least $2\kappa \cdot diam(B)$. If however $diam(B) > diam(A)$, then at x we draw a disk of radius $\kappa \cdot diam(A)$, which is also contained in S since $diam(S) > diam(B)$, and the horizontal distance between the initial α^{2i} and α^{2i+1} is at least $2\kappa \cdot diam(A)$. □

5 Conclusion

We have proven that the combinatorial complexity of the boundary of the union of a set of n κ-curved objects is $O(\lambda_s(n) \log n)$, for some constant s. This bound improves the recent bound of Efrat and Sharir [6] for the case of convex κ-curved objects (e.g., fat ellipses). (They obtained a bound of $O(n^{1+\varepsilon})$ for convex fat objects.) This bound is also the first non-trivial bound for the case of non-convex κ-curved objects (e.g., rounded heart-shaped objects).

In the full version of this paper, an improved (dynamic) data structure for point location queries for a set of n κ-curved objects is presented. In a point location query, one must determine whether a query point q lies in the union of the input objects, and, if so, report a witness object containing q, or, alternatively, report all k objects containing q. The data structure is similar to those described in [5]; its size is $O(n^{1+\varepsilon})$, the cost of an insertion or deletion of an object is $O(n^{\varepsilon})$, and the cost of a query is $O(\log^3 n)$ (alternatively, $O(\log^3 n + k)$).

References

[1] P.K. Agarwal, M.J. Katz and M. Sharir, Computing depth orders and related problems, *Comp. Geom. Theory and Appls* 5 (1995), 187–206.

[2] H. Alt, R. Fleischer, M. Kaufmann, K. Mehlhorn, S. Näher, S. Schirra and C. Uhrig, Approximate motion planning and the complexity of the boundary of the union of simple geometric figures, *Algorithmica* 8 (1992), 391–406.

[3] M. de Berg, Linear size binary space partitions for fat objects, *Proc. 3rd Annu. European Symp. Algorithms*, 1995, 252–263.

[4] B. Chazelle, H. Edelsbrunner, L. Guibas and M. Sharir, Algorithms for bichromatic line segment problems and polyhedral terrains, *Algorithmica* 11 (1994), 116–132.

[5] A. Efrat, M.J. Katz, F. Nielsen and M. Sharir, Dynamic Data Structures for Fat Objects and their Applications, *Proc. 5th Workshop Algorithms and Data Struct.*, 1997, 297–306.

[6] A. Efrat and M. Sharir, On the complexity of the union of fat objects in the plane, *Proc. 13th ACM Symp. on Computational Geometry*, 1997, 104–112.

[7] M.J. Katz, 3-D vertical ray shooting and 2-D point enclosure, range searching, and arc shooting, amidst convex fat objects, *Comp. Geom. Theory and Appls* 8 (1997), 299–316.

[8] M.J. Katz, M.H. Overmars and M. Sharir, Efficient hidden surface removal for objects with small union size, *Comp. Geom. Theory and Appls* 2 (1992), 223–234.

[9] M. van Kreveld, On fat partitioning, fat covering, and the union size of polygons, *Proc. 3rd Workshop Algorithms Data Struct.*, 1993, 452–463.

[10] J. Matoušek, J. Pach, M. Sharir, S. Sifrony and E. Welzl, Fat triangles determine linearly many holes, *SIAM J. Comput.* 23 (1994), 154–169.

[11] M.H. Overmars and A.F. van der Stappen, Range searching and point location among fat objects, *J. Algorithms* 21 (1996), 629–656.

[12] M. Sharir and P.K. Agarwal, *Davenport-Schinzel Sequences and Their Geometric Applications*, Cambridge University Press, New York, 1995.

[13] F. van der Stappen, *Motion planning amidst fat obstacles*, Ph.D. Dissertation, Utrecht University, 1994.

[14] J. Vleugels, *On fatness and fitness — realistic input models for geometric algorithms*, Ph.D. Dissertation, Utrecht University, 1997.

Features of Deformation Grids: An Approach via Singularity Theory

Fred L. Bookstein
Institute of Gerontology, University of Michigan
Ann Arbor, Michigan 48109–2007 USA
fred@brainmap.med.umich.edu

Abstract. Biological shape differences often are represented as diffeomorphisms of a Cartesian coordinate grid. Typically such maps are encoded in a spatially distributed parameterization, such as a list of paired point-locations. The problem addressed here is the extraction of spatially discrete, localized features of such transformation grids. This paper suggests that the features ultimately interesting to a biomedical scientist, for instance the details that suggest underlying developmental or pathological processes, can often be identified with variants of the singularity $(x, y) \to (x, x^2y + y^3)$ that are visually evident in extrapolations. This is a nongeneric singularity at which a pair of cusps appears as a function of a parameter for "magnification." I introduce canonical coordinates for such singularities and show how they may be used to make sense out of thin-plate splines depicting effects of two different types of brain disease, schizophrenia and Fetal Alcohol Syndrome, on brain form in the midsagittal plane (plane of symmetry).

The primal scene

Consider the deformation function below in Figure 1. (Note that the scales of the axes here are unequal.) This figure is a transformation of originally square Cartesian axes according to the composition of three functions. Two are uninteresting: the map $(x, y) \to (x - 2y^2, y)$, which bends lines $x = c$ into parabolas of vertical vertex tangent, and the map $(x, y) \to (x+8y, y)$, which imposes a shear along the x-axis. The core of the transform, however, is quite interesting: the map $(x, y) \to (x, x^2y+y^3)$, in the upper panel of the figure, with that curious singularity at $(0, 0)$.

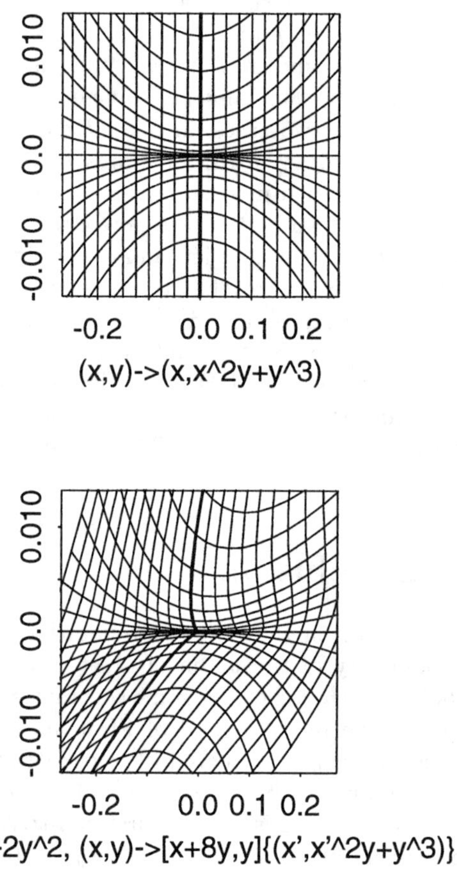

Figure 1. The basic singularity and a generalization.

The generic singularities of maps $R^2 \to R^2$ are folds and cusps (Whitney's theorem; see Bruce & Giblin, 1992), having canonical forms $(x, y) \to (x, y^2)$ and $(x, y) \to (x, xy+y^3)$, respectively. Clearly the singularity here is neither of these. Lines $y = c$ are transformed here into a nested family of parabolas $cx^2 + c^3$ whose spacing shrinks faster than curvatures as they approach the real axis. In both panels of the figure, the heavy line is the image of the original locus $x = 0$. Its singularity is concealed in the upper image—it falls to speed zero as it traverses the axis—but, unfolded below, we can see how when symmetry is broken it can change direction there.

Nongeneric singularities can be usefully approached as limiting cases of generic ones. In the present instance, the map $(x, y) \to (x, x^2y + y^3)$ can be viewed as the limit as $a \to 0$ of maps $(x, y) \to (x, (x^2 - a)y + y^3)$, Figure 2. For $a > 0$, each of these has two ordinary cusps at $\pm\sqrt{a}$. The Jacobian of the map is negative in the lip-shaped region between the cusps. As a passes through 0, the two cusps momentarily fuse in a higher-order singularity, then vanish.

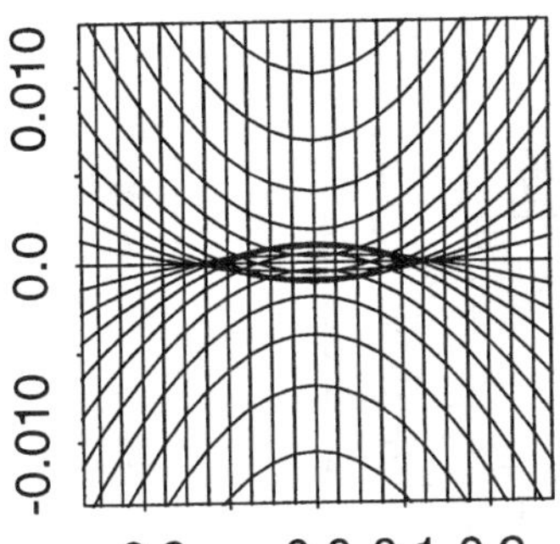

(x,y)->((x^2-a)y+y^3), a=.02>0, two cusps

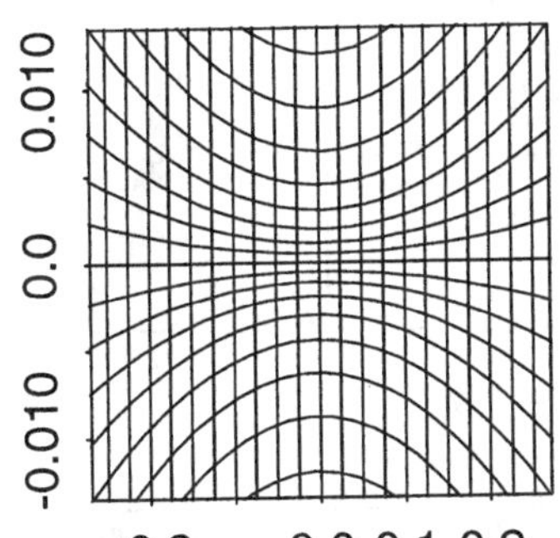

a=-.02<0, no singularity

Figure 2. Limit of paired cusps: $(x, y) \to ((x^2 - a)y + y^3, y)$ for a near 0.

How these arise in data

The manner in which singularities like these arise as descriptors of biological shape phenomena is, needless to say, not as the limit of the parametrized mappings in Figure 2. Realistic maps of one biological form onto another do not have cusps or folds. Instead, our singularity is encountered when realistic deformations are extrapolated to arbitrarily high multiples.

The maps we are exploring arise as expressions of statistical contrasts corresponding to a variety of applied biometric questions, such as dimensions of greatest natural variations in shape or aspects of shape that optimally covary with exogenous causes or effects of shape. Both examples here illustrate the simplest such context, the visualization of the difference between two group average shapes. In this version of morphometrics, shape data are formalized via ordered n-tuples of *landmark points* in two- or three-dimensional Cartesian space. Landmark points, such as "bridge of the nose" or "tip of the chin," have operational definitions permitting their location on specimens one by one but also are presumed to correspond on biological grounds across all the specimens of a sample. The shape of a set of landmarks is the equivalence class of their configuration under the ordinary Euclidean similarity group of translations, rotations, and changes of scale. This space is a Riemannian manifold under the celebrated *Procrustes metric,* submersion of the original Euclidean sum-of-squares (Kendall, 1984). For the purpose of this paper, we are not interested in variation of specimens within samples, but restrict our attention to representations of group average shapes, where the "average" is the form of least summed squared Procrustes distance to all the forms of the subsample. The computational geometry of the Procrustes methods has been reviewed elsewhere (cf. Bookstein, 1997a); its details are not important for the analysis of singularities to follow.

Rather more important is the algebra of the thin-plate splines that we use for smooth interpolation between different shapes, such as averages of different subgroups of one data set. The material to follow has appeared in many previous publications (Bookstein, 1991, 1997a), but is nevertheless worth reprinting here. Let U be the function $U(\vec{r}) = r^2 \log r$, and consider a reference shape (in practice, some Procrustes average shape) with landmarks $P_i = (x_i, y_i)$, $i = 1, \ldots, k$. Writing $U_{ij} = U(P_i - P_j)$, build up matrices

$$K = \begin{pmatrix} 0 & U_{12} & \ldots & U_{1k} \\ U_{21} & 0 & \ldots & U_{2k} \\ \vdots & \vdots & \ddots & \vdots \\ U_{k1} & U_{k2} & \ldots & 0 \end{pmatrix}, \quad Q = \begin{pmatrix} 1 & x_1 & y_1 \\ 1 & x_2 & y_2 \\ \vdots & \vdots & \vdots \\ 1 & x_k & y_k \end{pmatrix},$$

$$L = \begin{pmatrix} K & Q \\ Q^t & O \end{pmatrix}, \quad (k+3) \times (k+3),$$

where O is a 3×3 matrix of zeros. The thin-plate spline $f(P)$ having heights (values) h_i at points $P_i = (x_i, y_i)$, $i = 1, \ldots, k$, is the function $f(P) = \sum_{i=1}^{k} w_i U(P - P_i) + a_0 + a_x x + a_y y$ where $W = (w_1, \ldots, w_k, a_0, a_x, a_y)^t = L^{-1}H$ with $H = (h_1, h_2, \ldots, h_k, 0, 0, 0)^t$. Then we have $f(P_i) = h_i$, all i: f interpolates the heights h_i at the landmarks P_i. Moreover, the function f has minimum **bending energy** of all functions that interpolate the heights h_i in that way: the minimum of

$$\iint_{\mathbf{R}^2} \left(\left(\frac{\partial^2 f}{\partial x^2}\right)^2 + 2\left(\frac{\partial^2 f}{\partial x \partial y}\right)^2 + \left(\frac{\partial^2 f}{\partial y^2}\right)^2 \right)$$

where the integral is taken over the entire picture plane. The value of this bending energy is $\frac{1}{8\pi}W^tH = \frac{1}{8\pi}H_k^tL_k^{-1}H_k$, where L_k^{-1}, the *bending energy matrix*, is the $k \times k$ upper left submatrix of L^{-1}, and H_k is the corresponding k-vector of "heights" $(h_1, h_2, \ldots, h_k)$.

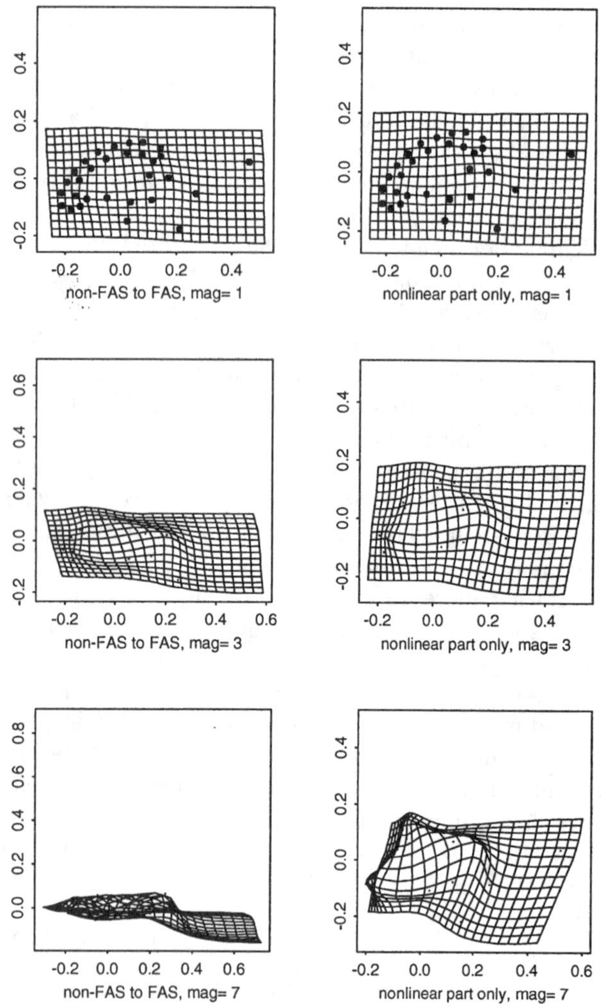

Figure 3. A two-group contrast, with ordinary and affine-free extrapolates.

A plane-to-plane interpolation is couched as a Cartesian pair (f_x, f_y) of these functions based in the same matrix L and for which f_x uses a vector H_x of x-coordinates of a target form and f_y uses a vector H_y of y-coordinates. The bending energy being minimized is now the quadratic form $H_x^tL_k^{-1}H_x + H_y^tL_k^{-1}H_y$.

The spline helps visualize statistical summaries based in these Procrustes coordinates in a remarkably effective manner. Any vector computed in the course of a multivariate analysis—a mean difference, for example—can be visualized as a deformation in this way. That the double integral of second derivatives is minimized means that the spline fits shape changes with the smallest possible variation of affine derivative (shapes of the little grid cells in these figures). If a given change of affine derivative can be managed over a larger interval, its contribution to the integral of squares is lower; hence the spline tries to represent deformations "as globally as possible." Global features of such changes are plainly visible in the grids without further parameterization. Furthermore, because the spline's coefficients $L^{-1}H$ are linear in H, the deformed configuration, we can unambiguously construct an α-fold extrapolation (magnification) of the map $S_1 \to S_2$ as the map $S_1 \to S_1 + \alpha(S_2 - S_1)$. We will see shortly how these extrapolations lead the eye to a remarkably reproducible assessment of local features of change.

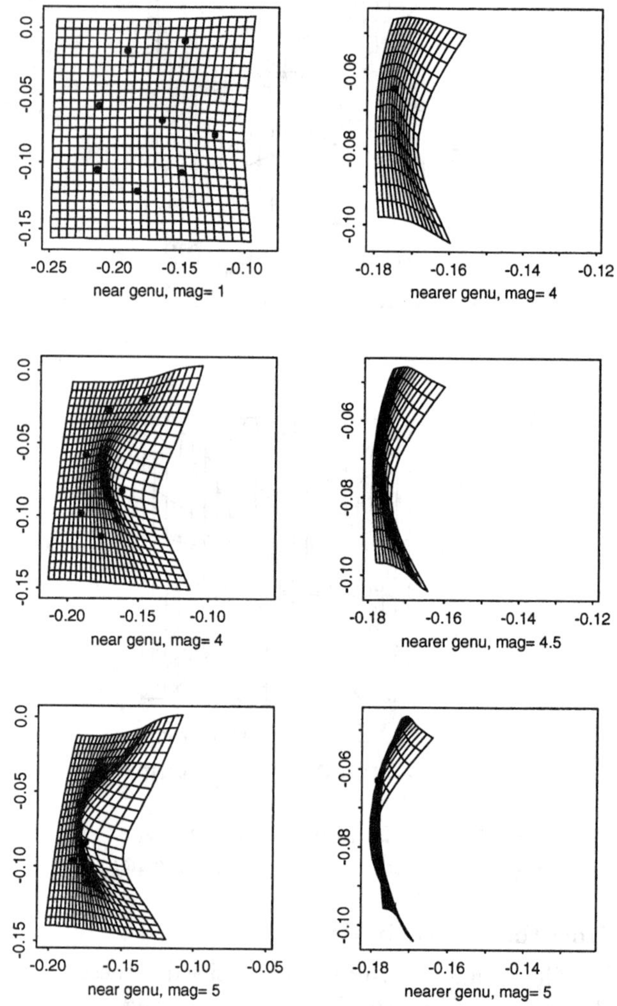

Figure 4. Zooming in on the singularity in genu, at two different scales.

In the three-dimensional version, the kernel $r^2 \log r$ is replaced by $|r|$, so that the spline is no longer smooth at the landmarks. The matrix Q gains a fourth column of z-coordinates, the vectors H and W gain $(k+4)$-th terms 0 and a_z, respectively, and the integral is now a triple integral having six terms instead of three.

The bending energy minimized by the spline is now $-(H_x^t L_k^{-1} H_x + H_y^t L_k^{-1} H_y + H_z^t L_k^{-1} H_z)$.

An instructive example

Figure 3 shows, at upper left, a deformation grid that typifies an averaged landmark configuration from the midsagittal plane of the brain in 4 adults with Fetal Alcohol Syndrome (FAS) as a deformation of an average over 8 persons not so affected. (Fetal Alcohol Syndrome is an irreversible process of prenatal brain damage discovered in 1973 and known to be caused by exposure to very high levels of alcohol early in fetal life.) Subjects are facing left. The curving arch from left to top of the picture is the *corpus callosum*, the principal conduit of information between the cerebral hemispheres; cerebellum is at the lower right. The left part of the callosum is called "genu"; the bulb at the right, "splenium"; the narrowing just to its left, "isthmus."

The grid at upper left in Figure 3 is obviously not affine, but it is difficult to put into words exactly what it is instead. Down the left column of the figure are extrapolations of this transformation by factors of 3 and then 7. The bending is becoming more patent, but the collapsing of the shape at large extrapolations is interfering with our understanding. In the right column we apply an elementary stratagem from the contemporary morphometric toolkit, removing the affine term by linear projection in the Procrustes geometry (Bookstein, 1997a, 1997b). Now it is clear that there is an extended extremum of hypoplasia (compression) across the entire length of the corpus callosum. Somewhere between the threefold extrapolation and the sevenfold, the deformation goes singular in two different regions. We will analyze them separately.

Figure 4 enlarges regions around genu near the magnification at which the singularity appears. In the left column are shown the actual deformation of the region in question and also multiples of the original transformation by four and by five; the singularity appears somewhere in between. In the right column, a further enlargement localizes the singularity quite near the multiple of 4.5. The higher magnification at lower right shows the expected paired-cusp structure quite nicely.

Something here is not yet as symmetrical as it might be. In the left center panel of Figure 4, count down about eight grid lines on the left and the right margins of the little grid cell. You will find that their images are not aligned—none of the grid lines drawn appear to be transversals of the singularity. We can make this a bit clearer, Figure 5, by rotating the starting grid through a range of orientations. At some orientations, lines slew upward along the singularity; at others, they slew downward. There is thus some orientation along which they do not slew at all, the orientation exploited at lower left in the figure. Here grid lines approximate a proper corner where they cross the tangent to the singularity just upstream or just downstream.

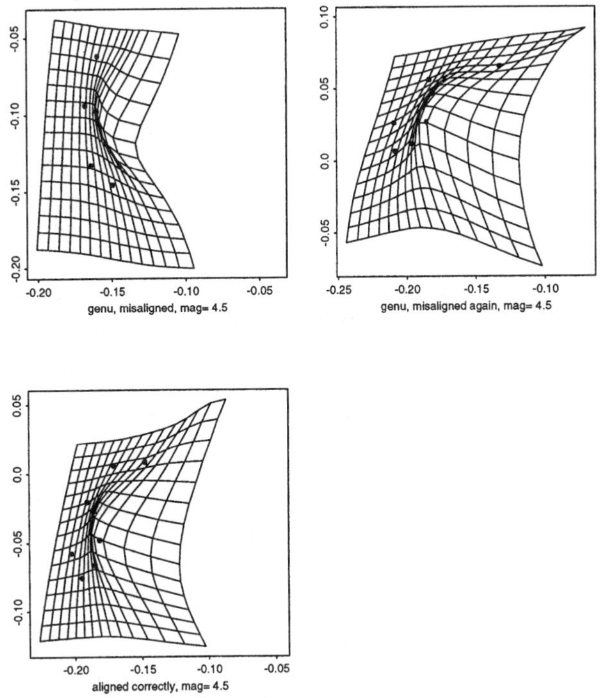

Figure 5. Rotating the grid through the singularity.

What is the meaning of such a finding? I suggest that it constitutes a *finite feature*, a salient organizing focus for description of the map in the large. Figure 3, lower right, shows this interpretation quite clearly. We have located the center of horizontal compression of a vertically organized dysmorphogenic field that, extended, generates the cusp visible there right up to the boundaries of the next such structure (to be encountered in Figure 7). That the singularity appears at magnification 4.5 means that the largest compression in any direction is by 1/4.5=22% of the original length, and the field of compression extends for a considerable distance upward from this feature. (Notice, for instance, the continuing fold above the cusp at this magnification of 7: the tissue remains compressed by 15% or so all the way up the front half of the callosum.)

These grids look remarkably like the prototype I led you to expect (recall Figure 1). We can come to understand their structure better by intentionally misaligning the axes in the prototype. In Figure 6, the map is a composition of the same $(x, y) \to (x - 2y^2, y)$ and $(x, y) \to (x, x^2y + y^3)$, now without the final shear, but the starting grid is rotated by 45° from how it lay in Figure 1. What appear to be cusps in the singularity structure, right center, are actually proper transversals instead, which appear at upper right as highly elongated but nonsingular regions of the grid. The effect is clearer if we turn from Cartesian coordinates to polar coordinates, lower row. The curves here are the image of a pencil of lines evenly spaced in angle around $(0,0)$. (Note the twentyfold increase of vertical scale; we are very near indeed to the singular locus.) In the map $(x,y) \to (x, x^2y + y^3)$, without the quadratic bending, the heavy vertical line is the only true transversal of the singular locus (the only one having a limiting tangent direction not horizontal). At lower right, we see the effect of premultiplying by the map $(x,y) \to (x - 2y^2, y)$ that sends lines $x = c$ to parabolas. The vertical compression in the vicinity of the singularity foreshortens each hemiparabola into an ordinary vector; the transversal has different limiting tangents at 0^+ and 0^-.

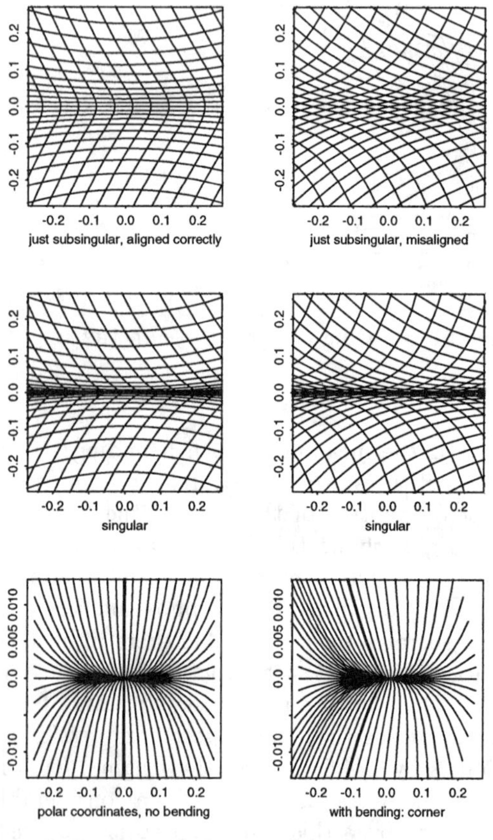

Figure 6. In general, traversal is at a proper corner.

The parameters of one of these features total eight: two for location, one for direction of the axis of the singularity, two for the two angles between the tangent to the singular locus and the principal transversal at its "corner," one for the first derivative of the map in the tangent direction, one for the second derivative in the transversal direction, and one for the extrapolation at which all this goes singular. In principle these parameters are subject to permutation-based methods of statistical inference analogous to those now solidly in place for global features (Bookstein, 1996).

This decomposition contrasts considerably with the existing methods for feature extraction from deformations. The method of biorthogonal grids, for instance (Bookstein, 1978), computes principal strains of the affine derivative at every point, but its singularities are umbilics. It offers no formalism for finding the local extrema of strain rate analogous to the analysis here. And studies that compare areas or that search for sources or sinks have hitherto been scalar-based, lacking the very important directional features of this parameterization.

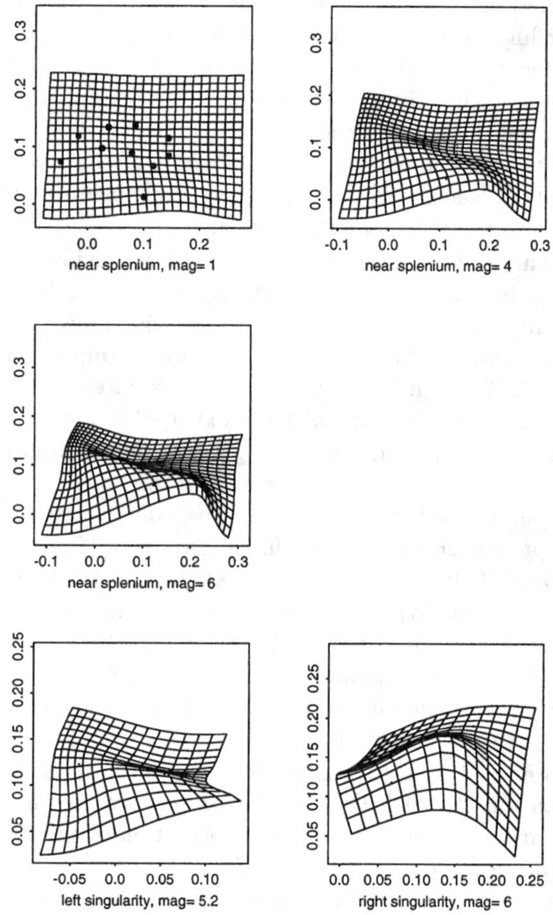

Figure 7. Near splenium, there are two more singularities, nearly collinear.

Returning to the full transformation, Figure 3 (lower right), let us track down the structure of the other salient compression, at its upper center. As Figure 7 shows, this is a composite of two singularities at about the same maximal compression ($\frac{1}{5}$ to $\frac{1}{6}$) but different principal axes. In the rightmost of the two singularities, the corner of the transversal makes rather different angles with the tangent to the singularity on its two sides, as Figure 1 already prototyped (by shear along the tangent line). All along this arc of tissue (a histological locus, not a mathematical one), the average FAS case falls short of normal by about $1/5.2$=19% in callosal thickness—that's a lot of neural tissue missing.

The pair of findings here can be interpreted as owing to the same alcohol-induced embryological disorganization that leads to the characteristic facial features by which FAS is diagnosed (Bookstein and Streissguth, 1998).

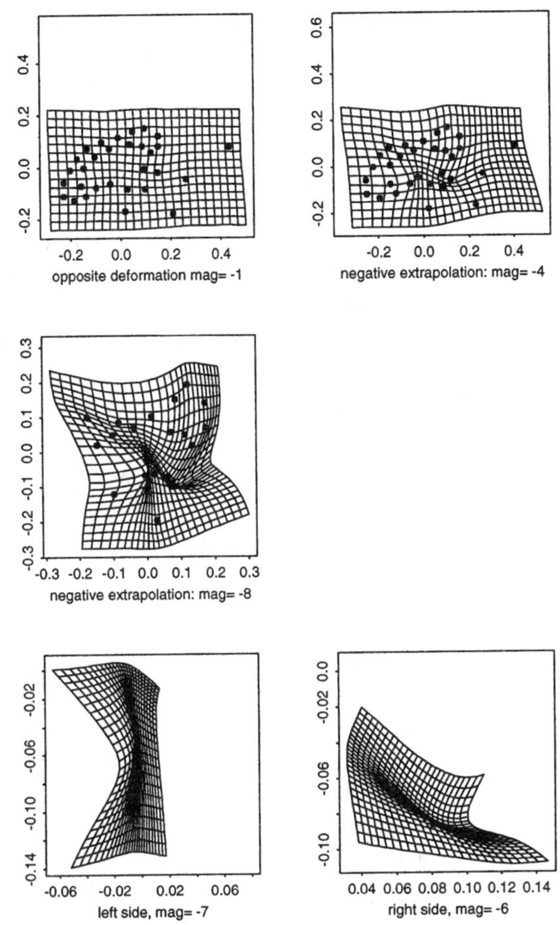

Figure 8. Negative extrapolation of the affine-free FAS grid shows two foci of expansion.

To complete the analysis of this transformation, we need to look at the singularities of the opposite transformation (negative extrapolations of the same map). Figure 8 shows that this transformation has two singularities of its own, rather close together, both of the same familiar structure. Because these are folds of the opposite map, they are local extrema of expansion, not compression, of the original comparison; together they account for the great white area at the lower right in Figure 3. These loci of greatest expansion are both within the fluid compartment of this brain section, an expansion that does the victim of the syndrome no good at all.

A second example

Figure 9 introduces another example that has already been intensively analyzed in Bookstein, 1997c: the contrast of the form of the corpus callosum between 12 doctors and 13 of their schizophrenic patients. The difference here is statistically significant at about the 1% level by the standard significance tests that apply to data sets like these. The moderate (eightfold) extrapolation of the transformation shows obvious displacement of the narrowing (the isthmus) posteriorly and upward and a compression of the genu oriented somewhat vertically. Further magnification at genu, left center panel, shows a singularity of the form we have now encountered many times. But that at isthmus, right center panel, seems a new sort of beast, a continuous bell-shaped line of folding with the delimiting cusps quite a distance away at the very onset.

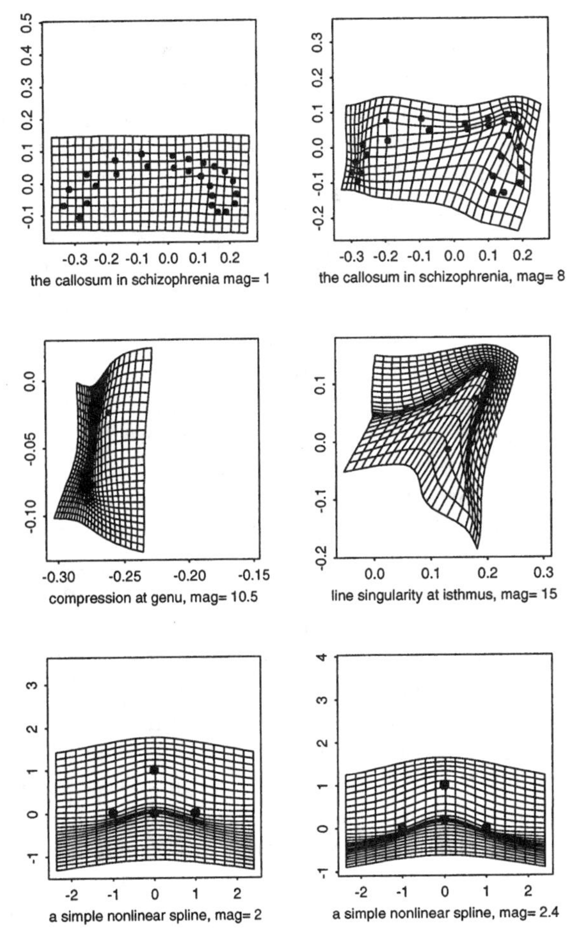

Figure 9. Another data set seems to include a higher-order singularity. (See text.)

This is, in fact, a hint about a different singularity, a "more singular" one. In simplest form it arises even in configurations of a mere four landmarks, the simplest nonlinear thin-plate spline. This spline is shown at lower left in the figure for a realistic case (square-to-kite, Bookstein, 1991) and, at lower right, just as it starts to go singular. When it does, it folds almost simultaneously all along a line. The displacement "piles up" along that symmetric pair of neighboring landmarks to produce three roughly collinear cusps at very nearly the same time. In the real data, it is this displacement of the isthmus region, not its folding, that accounts for the statistical significance of the comparison as a whole.

How is this computational geometry?

The detection and parametrization of these singularities arises from practical problems apparently unfamiliar to your field, and the issues of representation and computation it generates are of a complexity that I cannot estimate at present. The fundamental data structure here, the discretely sampled displacement field, is simply the vectorization of a scalar field. Scalar fields are the usual domain of "empirical Morse theory," the search for peaks, pits, passes, ridges, courses, and watersheds of images of real landscapes. The theorems that guarantee the taxonomy of those beasts are not of much use in the algorithmics of their discrete approximations—in the search for nongeneric points or nongeneric configurations there is at present no substitute for an intelligent viewing eye (Koenderink, 1990). Similarly, the theorems of Whitney and others about singularities of maps of surfaces to surfaces specify the variety of beasts we are likely to find, but do not much help in the localization of the nongeneric species.

Displacement fields have certain simplicities that should commend them to the computational geometer's attention. While less commonly encountered than polygonal outlines or arbitrary point-sets, displacement fields remain somewhat easier to notate, and the crucial construct for the intuition of the computations here is the same second derivative that is familiar to every student who navigates in a Newtonian world of uniformly accelerated motion under gravity. The detection of the singularities here is the spatial equivalent of sensing the bottom of a roller-coaster trajectory or spying the fold in a piece of imperfectly draped clothing. Still, like any other problem of discrete feature extraction, the detection of these singularities is not a lossless geometric transformation such as a Voronoi operation. In this it strongly resembles certain other problematic maneuvers at the root of applied computational geometry, such as the search for branch points of a medial axis, that likewise scan for local extrema of the discrete counterpart of a continuous differential descriptor, in the hope that they will convey a disproportionately large share of the scientifically relevant information in a disproportionately small fraction of the available bandwidth. The problem of knot selection for representation of arbitrary curves, too, shares some of this spirit whenever the simplest report is desired that permits reliable decisions to be made. That the thin-plate spline interpolant here is driven by a very sparse set of discrete landmarks is not essential to the conditioning of this problem. Any alternative source of adequately smooth displacement data will do.

Future work

The visualizations and the computational geometry reported here are all at very early stages of development. The representations here were produced slowly and interactively by hand; we need very fast algorithms for locating these structures in real data so that the permutation-based statistical tests for reliability of these features can go forward in practical time. The extended singularity in Figure 9, although it arises quite simply in the square-to-kite prototype, needs to be understood in a manner more independent of the algebra of this particular interpolation engine. The list of generic singularities for data in three dimensions is longer than the list in two (see Whitney, 1992), and, while we already have three-dimensional landmark data sets that can be analyzed by extrapolation to singularity (cf. Buckley et al., 1998), we thus far lack visualization tools that will let us explore the organization of the singularities at which they start to fold, focus by focus.

The thin-plate spline itself, because it minimizes the integral of summed squared second derivatives, is "aware" of the extrema that we are tracking here (Bookstein, 1997d). Thus the spline may be of considerable service in affording analyses of the ways that foci of compression interact with foci of expansion and other aspects of the global structure of these very interesting representations.

The import for computational geometry thus emerges as a specific challenge: given that the displacement field is a perfectly acceptable data structure, of considerable importance in applications from medical imaging to, say, earthquake prediction or dam design, to experiment with a new class of algorithms that will produce the hierarchy of these local maxima as a rigorously defined geometric object. I will be most interested in any responses to this challenge by the members of this community during the course of this conference or at any later date.

Acknowledgements. The work reported here was supported in part by NIH grants DA–09009 and GM–37251 to Fred L. Bookstein. The former grant is supported jointly by NIDA, NIMH, and NIA as part of the Human Brain Project. I am grateful to Hemant Tagare (Yale) for a crucial suggestion early in the gestation of this work and to my longtime collaborator Bill Green, now of Western Washington University, for the exploratory software incorporated in his `ewsh` package that permits free play with magnification and grid placement. The FAS data set was gathered as pilot data for a study sponsored by NIH grant AA–10836 to Ann P. Streissguth of the University of Washington.

Literature Cited

Bookstein, F. L. *The Measurement of Biological Shape and Shape Change. Lecture Notes in Biomathematics,* vol. 24. Springer-Verlag, 1978.

Bookstein, F. L. *Morphometric Tools for Landmark Data.* New York: Cambridge University Press, 1991.

Bookstein, F. L. Biometrics, biomathematics, and the Morphometric Synthesis. *Bulletin of Mathematical Biology* 58:313–365, 1996.

Bookstein, F. L. Shape and the information in medical images: a decade of the morphometric synthesis. *Computer Vision and Image Understanding* 66:97–118, 1997a.

Bookstein, F. L. Biometrics and brain maps: the promise of the Morphometric Synthesis. Pp. 203–254 in S. Koslow and M. Huerta, eds., *Neuroinformatics: An Overview of the Human Brain Project. Progress in Neuroinformatics,* vol. 1. Hillsdale, NJ: Lawrence Erlbaum, 1997b.

Bookstein, F. L. Landmark methods for forms without landmarks: Localizing group differences in outline shape. *Medical Image Analysis* 1:225–243, 1997c.

Bookstein, F. L. Quadratic variation of deformations. Pp. 15–28 in J. S. Duncan and G. Gindi, eds., *Information Processing in Medical Imaging, 15th International Conference. Lecture Notes in Computer Science,* v. 1230. Berlin: Springer-Verlag, 1997d.

Bookstein, F. L., and A. P. Streissguth. Midline anomalies of brain shape in Fetal Alcohol Syndrome: a suggestive preliminary finding. To be submitted, 1998.

Bruce, J. W., and P. J. Giblin. *Curves and Singularities,* second edition. Cambridge University Press, 1992.

Buckley, P. F., D. Dean, F. Bookstein, L. Friedman, D. Kwon, J. Lewin, J. Kamath, and C. Lys. Three-dimensional MR-based morphometrics and ventricular dysmorphology in schizophrenia. *Biological Psychiatry,* to appear, 1998.

Kendall, D. G. Shape-manifolds, procrustean metrics, and complex projective spaces. *Bulletin of the London Mathematical Society* 16:81–121, 1984.

Koenderink, J. *Solid Shape.* Cambridge: the M.I.T. Press, 1990.

Whitney, H. Singularities of mappings of Euclidean spaces. In his *Collected Papers,* Birkhäuser, 1992, vol. 1, pp. 436–452.

Effective Nearest Neighbors Searching on the Hyper-Cube, with Applications to Molecular Clustering

F.Cazals[1]

[1]Algorithms project, INRIA Rocquencourt, F-78153 Le Chesnay
and
Prisme project, INRIA Sophia, F-06902 Sophia-Antipolis
email: Frederic.Cazals@inria.fr

Abstract

Let M be a set of n words of length d on the binary alphabet $\{0, 1\}$, or equivalently a set of n points on the hyper-cube $H_d = \{0, 1\}^d$. Define the similarity $sim(m_i, m_j)$ between two words m_i and m_j as the number of common 1 bits. Also define the size s_i of a word m_i as its number of 1 bits. Given two positive real numbers $\alpha \in [0, 1]$ and $\beta \in [1, \infty)$, we address the following question: report for each $m_i \in M$ all or part of the set of m_j such that $sim(m_i, m_j) \geq \alpha\ min(s_i, s_j)$ and $max(s_i, s_j)/min(s_i, s_j) \leq \beta$. This problem arises in computational chemistry where the planning of experiments in drug design requires finding molecules that are structurally close among large data bases —for this reason we shall also call the m_i molecules. Since practically $d \geq 1000$, the usual geometric constructions such as box trees or (approximate) Voronoi diagrams seem to be useless. This problem therefore falls in the scope of the CG Task Force: theoretically challenging and important for practitioners!

The solution we propose relies on two new randomized data structures, the Directional Skip List and the Recursive Directional Skip List —DSL and RDSL. A DSL is a skip-list like data structure on the directions of H_d. In other words it is a sequence of nested approximations of the hyper-cube. A RDSL is a recursive DSL aiming at splitting M into small enough subsets by assigning the molecules to the hyper-cube approximations. In addition to the description of the DSL and RDSL data structures, we show that exactly analyzing the RDSL is related to Sperner theory and thus difficult. We also present experiments on real-world datasets illustrating the efficiency of the construction.

1 Introduction

When drug designers are asked to devise a medicine for a given pathology, they usually do not start a program from scratch. Instead, if its symptoms are "somewhat" related to those of already well known diseases, they first investigate the molecules used to cure those diseases. But this may even be too much work if there are too many such candidates. A common strategy is therefore to randomize as follows: try a few molecules at random; then, investigate more systematically the molecules which are "structurally close" from the most promising ones in the sample. Reporting the molecules "structurally close" to a given molecule is therefore a prerequisite to this experimental planning, and this is the problem we address in this paper. But before doing so, we introduce some terminology and formalize it.

SCG 98 Minneapolis Minnesota USA

1.1 Molecules and fragments: definitions

The input of the problem is a set of n molecules $M = \{m_1, \ldots, m_n\}$ where each molecule is defined as a multi-set of a set of molecular fragments $\{f_1, \ldots, f_d\}$. In other words, each molecule contains from 0 to any number of every fragment, and for the sake of clarity we identify the fragments to their indices $1, \ldots, d$. A molecule m_i is therefore a sequence of d boolean values $m_i = m_{i,1}m_{i,2} \ldots m_{i,d}$ with $m_{i,j} = 1$ if m_i contains the jth fragment and $m_{i,j} = 0$ otherwise. The number of terms equal to 1 in this description is called the size of the molecule and denoted by $s_i = size(m_i)$; we also define $s_{min} = \min_{i=1..n} s_i$ and $s_{max} = \max_{i=1..n} s_i$. At last, we let N_f stand for the total number of fragments used to describe the molecules in M that is $N_f = \sum_{i=1}^{n} s_i$.

This formulation fails in reporting multiples occurrences of a given fragment in a molecule, but it has the nice property that a molecule is represented by a point on the hyper-cube $H_d = \{0, 1\}^d$ or equivalently by a word of length d on the binary alphabet $\{0, 1\}$. Given two such words m_i and m_j, by $m_i \subset m_j$ we mean that every 1 digit in m_i is also 1 is m_j; and $m_i \subsetneq m_j$ means in addition that m_j contains at least one more 1 digit than m_i. Also, we call their ***similarity*** the number of common fragments that is the quantity $sim(m_i, m_j) = \sum_{k=1}^{d} m_{i,k}.m_{j,k}$. The reason for using this quantity rather than the usual Hamming distance [1] is just a choice of chemists —see however table 3 in section 3.

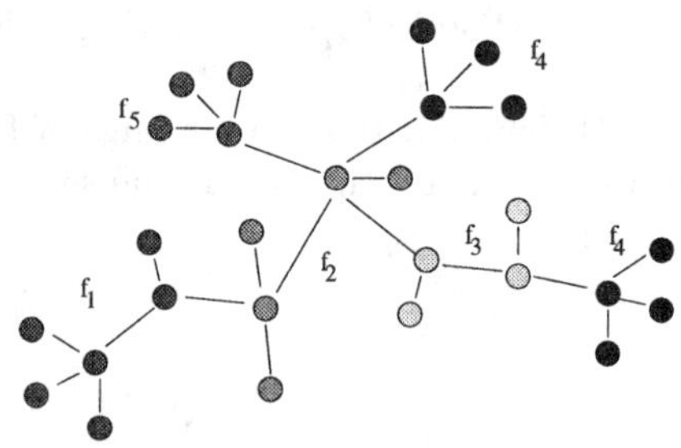

Figure 1: Molecule decomposed into fragments

1.2 Molecular databases

Practically, we shall deal in this paper with a molecular database, TLS1296, characterized by $n = 11046$, $d = 1616$ and $N_f = 252325$. This particular example shows that we actually have a very sparse distribution on H_{1616} since this hyper-cube has

[1]the Hamming distance of two binary strings counts the number of digits they disagree on.

about 10^{486} vertices. The reason why it is so is illustrated on fig. 2 where the fragments of 50 molecules are depicted —one line corresponds to a molecule and one column to a fragment. On this example, almost every molecule introduces at least one new fragment, so that the actual number of molecules if far from 2^d.

	1	2	...	d
m_1	0	1		1
m_2	1	0		1
$\vdots$				
m_n	1	1		0

Table 1: Molecular database as a 01 matrix

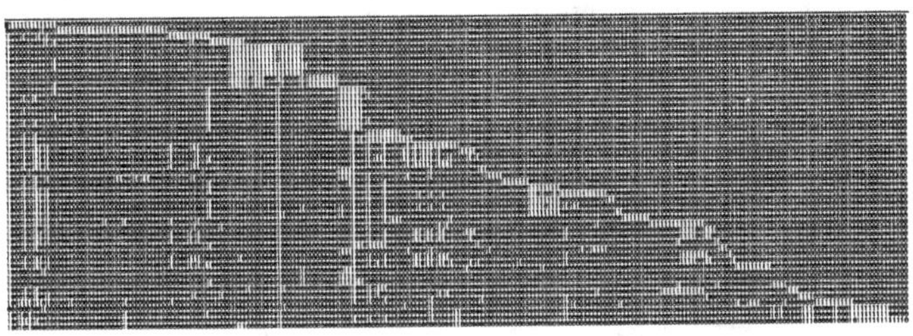

Figure 2: Fifty molecules in the database TLS1296

1.3 Problem description and algorithm overview

As already mentioned, we are interested in reporting pairs of molecules that are close enough. More precisely:

Definition 1 *Let α and β be two real numbers in $[0,1]$ and $[1,\infty)$. Two molecules m_i and m_j are called (α,β)-similar if $sim(m_i,m_j) \geq \alpha\ min(s_i,s_j)$ and $max(s_i,s_j)/min(s_i,s_j) \leq \beta$.*

The α part of definition 1 aims at taking into accounts the molecules' sizes in the similarity: setting a fixed lower bound for two molecules to be similar would condemn the pairs in which one molecule is smaller than this bound. In the same vein, β aims at not reporting the pairs satisfying the α condition but with a size mismatch. This is important since one cannot expect a big molecule and a small one to have the same biological effects.

In order to outrun the naive algorithm that tests every pair for similarity and whose complexity is $\Theta(n^2)$, a natural way to proceed consists in splitting the input set M into small enough subsets within which all the pairs can be checked. But an immediate drawback of this divide-and-conquer strategy is to reduce the number of parties a molecule is tested for similarity with. To formalize these ideas we introduce the following notations:

Definition 2 *Let M be a dataset of n molecules and denote $\pi(M) = \cup_{i=1}^{c} M_i$ a partition of M. Given a molecule m_i and its class M_k in π, the set of (α,β)-neighbors of m_i in M_k is defined by:*

$$N_{\alpha,\beta,\pi}(m_i) = \{m_j \in M_k \mid m_i \text{ and } m_j \text{ are } (\alpha,\beta)-similar\}.$$

The number of (α,β)-similar pairs captured by π is $PC(\pi) = \frac{1}{2}\sum_{i=1}^{n} \mid N_{\alpha,\beta,\pi}(m_i) \mid$. And the number of pairwise comparisons associated to π is defined by $s_2(\pi(M)) = \sum_{i=1}^{c} \mid M_i \mid^2 /n$.

Definition 3 *Let $\pi_1,\dots,\pi_T$ be a sequence of partitions built independently, with T a natural integer. For any $t \in 1,\dots,T$, the number of pairs captured up to the time t, denoted $PCt(t)$, is the number of* different *pairs captured by the π_i for $i = 1,\dots,t$.*

The number of pairs captured $PC(\pi)$ should be compared to the total number of pairs $PC(M)$ which is obtained when the partition has a single class, namely the whole set itself. But as already observed we want to avoid performing $\Theta(n^2)$ comparisons. The s_2 statistic which is sought to be $o(n)$ is therefore a way to detect those partitions with "too crowded" classes. This is also the reason why it may be convenient to look for the (α,β)-similar pairs over a sequence of partitions, each having a "small" s_2 statistic.

1.4 Paper overview

In section 2 we briefly review some geometric constructions in high dimension with a particular emphasis on the problem of the constants. In section 3 we present precise statistics on TLS1296. In section 4 we show that coming up with a combinatorial or probabilistic model for a molecular database is a difficult question. In section 5 we introduce the DSL and RDSL data structures. Section 6 is devoted to experimental results on the database of section 3. And in section 7 we conclude and list future work.

2 Previous work for d-dimensional geometric problems

A large body a work has been devoted to reporting (approximate) nearest neighbors over a set M of size n in a d-dimensional space. We briefly review the main trends [2] with a particular attention to the problems of the constants.

Geometric decompositions

In [AM93], a data structure called the Randomized Neighborhood Graph is used. The key step consists for each point in M in covering the whole space by narrow cones —Yao's cones [Yao82]. Unfortunately, one needs an exponential number of them. A different approach based on hierarchies of bounding boxes is proposed in [A+94]. But reporting a nearest neighbor still requires visiting an exponential number of cells. At last in [Cla94], the problem is reduced to polytope approximation for certain Voronoi regions. The approximation algorithm uses randomization and skip list like techniques, but the number of facets introduces constants exponential in d.

The conclusion may be that high-dimensional geometric objects have too many faces which intrinsically yields exponential constants. In this context the breakthrough of [Kle97] consists in retrieving the proximity information from the projections of M onto a large enough number of random lines going trough the origin. In particular this gives an approximation algorithm whose time-space complexity is polynomial in d. But the constants are still huge due to the use of Vapnik-Chervonenkis sample sizes.

Tree-like approaches

Given these difficulties, algorithms using trees have been explored. By tree we mean a data structure using an input point to split and organize the input dataset. The most famous are probably the the $k-d$ trees introduced in [FBF77] and refined up to [Spr91]. But range searching with this data structure results in approximating an hyper-sphere by an hyper-rectangle which yields constants exponential in d due to the volume ratio between these geometries. Moreover, a $k-d$ tree for points on H_d makes less sense than for continuous data since it is not clear how to prune the search from the locality between the query point and the splitting hyper-planes.

Another example which is based on the triangular inequality verified by any metric is the vantage tree of [Yia93]. Here the input dataset is recursively split apart one of its elements, the vantage point, with reference to the distance to that point. But as for the $k-d$ trees, using one-dimensional projections eventually yields constants exponential in d.

Of a different flavor is [EW82]. Let P_k a partition of the d directions into k subsets. The method builds a binary tree of

[2]See also the forthcoming paper [IM98].

height k by assigning an element of P_k to each level. Then, a point is recursively assigned to the left subtree if its keys are zero along the directions attached to the current node, and to the right subtree otherwise. Unfortunately, the way P_k is built is not specified. This method can however be considered as an ancestor of the Recursive Directional Skip List.

Bucket like approaches

As opposed to a tree, a bucket-like data structure splits the space containing the input dataset rather than this set itself – see [Dev86, G+93, CP97]. But splitting means at least into two, so that these strategies usually produce an exponential number of buckets.

Hashing via error correcting codes

Given a set M of n binary words, [Riv74] analyses a hashing algorithm aiming at reporting the string with minimum Hamming distance to a query string. It is shown under certain hypothesis that using the Golay perfect code $G_{23,12}$ as hashing function is optimal. It is not clear how one may choose the minimum distance $2\lambda + 1$ between code words and how our problem may accommodate the number of buckets, but this technique deserves more attention. Recent work of the same flavor is [Mil98].

3 A few observations on the database TLS1296

3.1 Molecules' size

The histogram and the cumulated distribution function of the molecules size for TLS1296 are depicted on fig. 3. The conventions are as follows: x axis for the number of fragments; leftmost y scale for the number of molecules of a given size; rightmost y scale for the cumulated percentage of molecules smaller than a given size. The corresponding average and standard deviation are $\mu = 22.84$ and $\sigma = 8.73$. For example, about 80% of the molecules are made of 30 fragments or less.

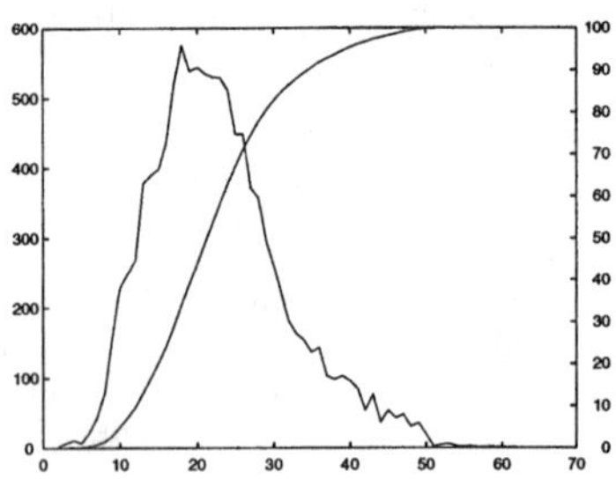

Figure 3: Molecules sizes: #molecules=f(molecular size)

3.2 Axis' importance

Let us define the following quantities:

FHN, NFHN for a given fragment, FHN which stands for Fragment Hit Number is the number of molecules using that fragment. The number of fragments with a given hit number is denoted NFHN;

%Frags, %CFrags for a given FHN, we set %Frags=NFHN/d and %CFrags is the associated cumulated value from the smallest FHN;

%Hits, %CHits for a given FHN, %Hits=FHN*NFHN/N_f, and %CHits is the associated cumulated value also from the smallest FHN.

Figure 4 depicts for each fragment its hit number together with the cumulated function. The conventions are: x axis for the indices of the 1616 fragments; leftmost y scale for the number of hits of a given fragment; rightmost y scale for the cumulated percentage of the number of hits. The average and standard deviation of FHN are $\mu = 156.14$ and $\sigma = 530.14$. It turns out that the higher the index of the fragment, the lower the hit number. One gets a somewhat different picture by grouping the fragments with a given hit number as shown on table 2:

1. The first two rows show that a great number of fragments are not used more than a few times in the description of the database. In particular, 291 fragments which is about 18% of the total number of fragments are just used once. Of course these can be discarded since they cannot contribute to the similarity of any two molecules.
2. The second set of lines initiates those fragments whose hit number is greater than 10, and this is 100-51.92=48.08% of the directions. This shows that the dimensionality of the problem cannot be reduced significantly without damaging the information related to small groups of very specific molecules.
3. The third sequence of values refers to the fragments whose contribution to the number of hits is about 90%, and corresponds to 17.02% of fragments. This sequence starts with 3 fragments used in 139 molecules.

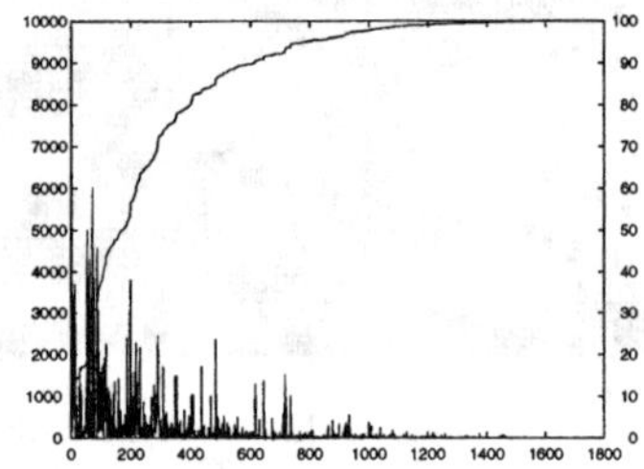

Figure 4: Number of hits per fragment

FHN	NFHN	%Frags	%CFrags	%Hits	%CHits
1	291	18.01	18.01	0.1153	0.1153
2	169	10.46	28.47	0.134	0.2493
⋮					
10	27	1.671	51.92	0.107	1.046
11	25	1.547	53.47	0.109	1.155
⋮					
139	3	0.1856	82.86	0.1653	10.03
141	2	0.1238	82.98	0.1118	10.14
⋮					
9878	1	0.06188	100	3.915	100

Table 2: Grouping the fragments by increasing hit number

3.3 Similarities and (α, β)-similar neighbors

The similarity of any two molecules is smaller than the largest molecule size s_{max}. The average and standard deviation computed over the $\binom{n}{2}$ pairs are respectively $\mu = 4.04$ and $\sigma = 3.69$. The corresponding histogram and cumulated distribution function are depicted on figure 5. These values justify the definition 1. Indeed, since 80% of the pairs of molecules have between 0 and 6 fragments in common, a similarity threshold of say 12-14 would have condemned the pairs with a similarity less than 12.

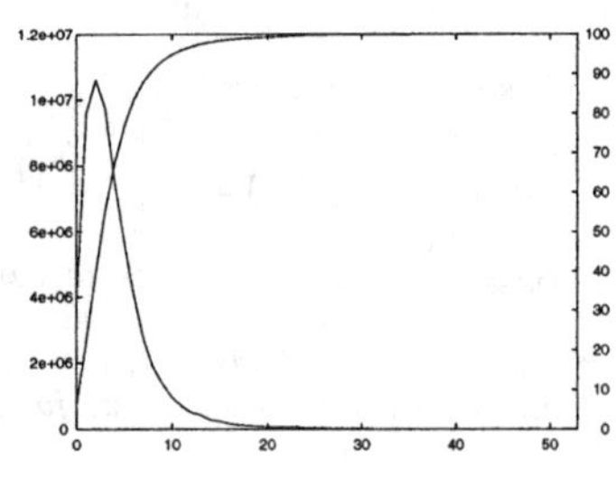

Figure 5: Molecules similarity: #pairs=f(similarity)

Table 3 also reports the average and std. dev. of the common number of 0 bits between two molecules, as well as the Hamming distance.

The number of molecules having a given number of (α, β)-similar parties is depicted on figs. 6(a)(b) for $(\alpha, \beta) = (0.9, 1.1)$ and $(\alpha, \beta) = (0.9, \infty)$. For these two situations, the average molecular size over all the molecules having a given number of parties is depicted on figs. 7(a)(b). For example 50% of the molecules have less than 250 $(0.9, \infty)$-similar neighbors and the average size of the molecules having 250 parties is about 24 fragments. The observation that legitimates the β part of definition 1 is the following: for $\beta = \infty$, about 10% of molecules have over 600 parties and four of them "agree" with about half the database. All these molecules have less than 15 fragments. Smallest values of β correct this problem and the total number of similar pairs together with the percentage of molecules having at least one party are given on table 4.

	sim	sim_0	H_d
μ	4.04	1574.25	37.69
σ	3.69	11.14	10.98

Table 3: Average and std. dev. of the common number of 1, common number of 0, and Hamming distance

β	$PC(M)$	%mols
1.1	115479	0.80
1.2	192530	0.88
1.3	249272	0.91
∞	568170	0.98

Table 4: Total number of $(0.9, \beta)$-similar pairs

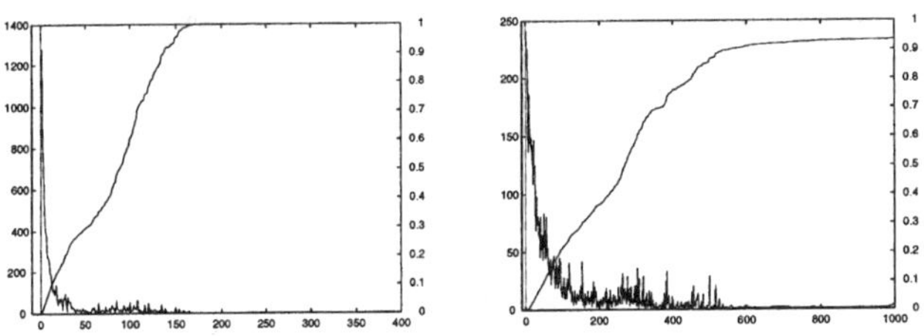

Figure 6: (a)#molecules=f(# of (0.9,1.1)-similar pairs) (b)#molecules=f(# of (0.9,∞)-similar pairs)

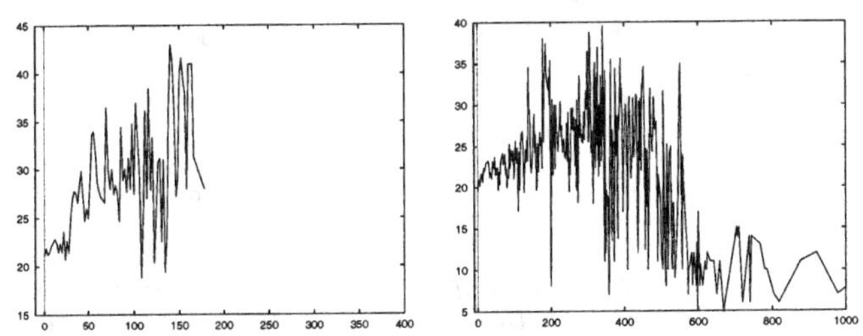

Figure 7: (a)molecular size=f(# of (0.9,1.1)-similar pairs) (b)molecular size=f(# of (0.9,∞)-similar pairs)

4 Some combinatorial properties

In this section we investigate some combinatorial properties of a database containing n molecules of equal size s and such that any pair has a similarity equal to $p \leq s$. In particular, we are interested in the relation between the values of n, s, d and p. The reason for doing so is twofold. First, one could expect some insight on how to store and manipulate molecules with similarity constraints from an algorithm building them. Second, we need such a model to analyze any algorithm manipulating molecules. Unfortunately, these are difficult questions which have an extremal set theory —also known as Sperner theory— flavor [Eng97, vLW92], and the properties proved thereafter are not constructive. The difficulty to enumerate exactly sets of molecules with similarity constraints leads us to consider two simpler randomized models.

4.1 A few counting results

Proposition 1 *Under the previous hypothesis, the maximum and minimum number of fragments used are* $d_{min} \geq \frac{ns^2}{s+(n-1)p}$ *and* $d_{max} = p + n(s-p)$.

Proof: For the upper bound, observe that for a given molecule and with reference to any set of molecules, one cannot have strictly more than $s - p$ new fragments. Indeed, since the length is s this would result in a similarity strictly less than p. On the other hand, having the same p fragments common to all the molecules and adding $s - p$ new fragments for each molecule yields d_{max} fragments.

For the lower bound, denote χ_i the characteristic function of the ith molecule, i.e. the function such that $\chi_i(f_j) = 1$ if m_i contains the fragment f_j and 0 otherwise. Also, let c_j the number of times the jth fragment is used to define the n molecules that is $c_j = \sum_{i=1}^{n} \chi_i(f_j)$. At last, let the $\int$ sign be used to count the number of fragments used by a given function, e.g. $\int \chi_i = \sum_{j=1}^{d} \chi_i(f_j)$. We have $\int \chi_i = s$ and $\int \chi_i \chi_j = p$. Also

$$\sum_{j=1}^{d} c_j^2 = \int (\sum_{i=1}^{n} \chi_i)^2 = \sum_i \chi_i + 2 \sum_{j \neq i} \chi_i \chi_j = ns + n(n-1)p.$$

But from the Cauchy-Schwartz inequality, we get

$$ns = \sum_{j=1}^{d} c_j.1 \leq \sqrt{d}\sqrt{n(s+(n-1)p)}$$

■

Another way to look at the lower bound is as follows: given a certain number of fragments d, how many molecules of length s such that any pair has similarity p can one build? We can prove:

Proposition 2 *Let* $M = \{m_1, \ldots, m_n\}$ *a collection of* n *distinct* s*-subsets of* $\{1, 2, \ldots, d\}$ *such that* $| m_i \cap m_j | = p$ *for* $i \neq j$ *and* $s \leq \lfloor \frac{d+p}{2} \rfloor$. *Then*

$$\binom{n}{2} \leq p \binom{d-1}{p-1} \frac{(d-p)!}{((s-p)!)^2 (d+p-2s)!}$$

Proof: Place the integers $1..d$ on a circle and consider the family $\mathcal{F} = \{F_1, \ldots, F_d\}$ of all consecutive s-tuples on the circle, i.e. $F_i = \{i, i+1, \ldots, i+s-1\}$, where the integers are taken mod d. We have $| M \cap \mathcal{F} | \leq 2$. Indeed, if $F_i = m_u$ for some i and u, since any two subsets share exactly p elements and $s \leq \lfloor \frac{d+p}{2} \rfloor$, it turns out that at most another element m_v from M can be matched to F_{i+p-s} or $F_{i+s+p-1}$. In other words, the fragments shared by m_u and m_v are located on the slots $\{i, \ldots, i+p-1\}$ or $i+s+p-1, \ldots, i+s-1$. This counting argument still holds if we replace $\{1, 2, \ldots, d\}$ by any permutation of the symmetric group S_d, so that if $\mathcal{F}^\pi$ denotes the family $\mathcal{F}$ for the permutation π

$$\Sigma = \sum_{\pi \in S_d} | M \cap \mathcal{F}^\pi | \leq 2d!$$

To count Σ in another way, let us fix two subsets m_u, m_v and some F_i. The number of permutations such that the matches of m_u and m_v are located for example on $\{i, \ldots, i+p-1\}$ is $p!(s-p)!(s-p)!(d-p-2(s-p))!$. So that

$$\Sigma = \sum_{\pi \in S_d} | M \cap \mathcal{F}^\pi | = d\binom{n}{2} 2p!((s-p)!)^2(d+p-2s)!$$

■

An interesting question would be to come up with a constructive lower bound $n \geq f(d, s, p)$, for example using a greedy strategy. Meanwhile we define two models where the cardinality of the subsets' intersections is a consequence and not a constraint of the construction.

4.2 Two simplified models

As observed in sec. 3, the average and std. dev. of the similarity between two molecules are $\mu = 4.04$ and $\sigma = 3.69$. To shed some light on this fact, we define two simplified models for which these values can be theoretically computed:

Model 1 *Define a molecule as a subset of a set of size d obtained by flipping independently d Bernoulli random variables of parameter p.*

Proposition 3 *Under model 1, the expectation and variance of the similarity between two random molecules are given by* $\mathbf{E}[sim] = d\ p^2$ *and* $\mathbf{Var}[sim] = d\ p^2(1-p^2)$.

Proof: The number of common 1 bits is $\sum_{i=1}^{d} X_i^2$ where the X_i are d iid Bernoulli random variables of parameter p. Whence the result. ■

Model 2 *Define a molecule as a random subset of size s of a set of size d, the $\binom{d}{s}$ subsets being equally likely.*

Proposition 4 *Under model 2, the expectation and variance of the similarity between two random molecules are given by*

$$\mathbf{E}[sim] = \frac{s^2}{d} \quad \text{and} \quad \mathbf{Var}[sim] = \frac{s^2(s-1)^2}{d(d-1)} + \frac{s^2}{d} - \frac{s^4}{d^2}$$

Proof: This is best proved using trivariate generating functions as follows. Let U and V be two subsets of the set of size d. For a given item, let us use the marker variable u if it is put in U, v if it put in V, y if it is put in U and V, and 1 otherwise. Enumerating all the configurations for U and V results in deciding for a given item where we put it. And the independence between the d choices results in the generating function

$$G(u, v, y) = (1 + u + v + yuv)^d.$$

Its coefficient [3] $[u^s v^s]G$ is a polynomial of degree s in y counting the configurations where U and V have size s. Therefore, the sum of the intersection sizes between all the pairs of subsets of size s is given by

$$[u^s v^s]\frac{\partial G}{\partial y}_{|y=1} = [u^s v^s]d(1+u+v+yuv)^{d-1}uv$$
$$= [u^{s-1}v^{s-1}]d(1+u)^{d-1}(1+v)^{d-1} = d\binom{d-1}{s-1}^2$$

From which $\mathbf{E}[sim] = d\binom{d-1}{s-1}^2 / \binom{d}{s}^2$. The variance is computed in the same way from $\frac{\partial^2 G}{\partial y^2}_{|y=1}$ and the formula $V(X) = E(X(X-1)) + E(X) - E(X)^2$. ■

	$\mathbf{E}[s]$	$\mathbf{E}[sim]$	$\mathbf{Var}[sim]$
$p = 0.013$	21.00	0.27	0.52
$p = 0.050$	80.80	4.04	2.00

	s	$\mathbf{E}[sim]$	$\mathbf{Var}[sim]$
$s = 20$	20	0.24	0.24
$s = 80$	80	3.96	3.58

Table 5: Values for models 1 and 2

[3] As usual, if G is the polynomial $G = \sum_{i=1}^{n} a_i x^i$, we have $[x^k]G = a_k$.

As shown on table 5, it turns out that these two models are equivalent which is not surprising since in both cases all the directions are independent and equally likely. If one sets the main parameter —p or s— to match the average molecule size, then the corresponding similarity is too small. On the other hand, tuning p or s to get the right similarity corresponds to molecules of size about 80. This corresponds to the non independence of the fragments and shows the interest of enumerations a la Sperner. In the lack of a better solution, we shall however use these models for analyzing our data structures.

5 Directional and Recursive Directional Skip Lists

5.1 Definitions

As already observed in sections 1 and 3, we are interested in sets of molecules of the "same" size and using the "same" fragments. But as depicted on fig. 8, if we pick at random a fragment in the pool of fragments used by a set of such molecules, chances are that this will be a common fragment since by definition there are more of them!

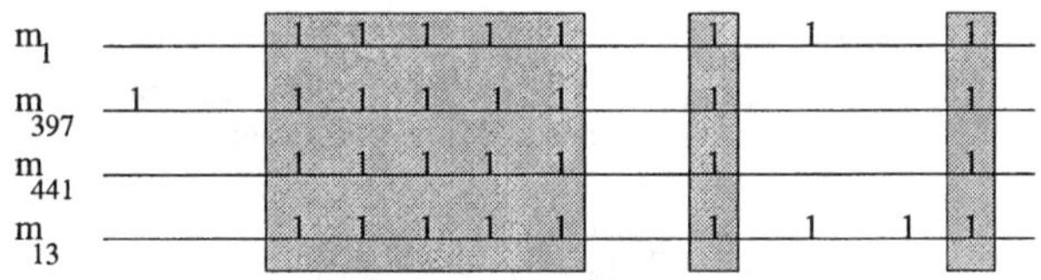

Figure 8: Discarding directions

This provides the intuitive basis for the Directional Skip List(DSL) data structure which is a skip-list-like [4] data structure:

Definition 4 *Let M be a set of molecules with $dim(M) = d$. A* Directional Skip List *for the set M is a nested hierarchy of hyper-cubes*

$$H_d = H_{i_0} \supsetneq H_{i_1} \supsetneq \cdots \supsetneq H_{i_r}$$

with

$$d > i_1 > \cdots > i_r.$$

A molecule $m \in M$ is said to be hooked at level $k \in 0..r$ if H_{i_k} is the smallest hyper-cube such that $m \subset H_{i_k}$. The set of molecules hooked at level k is denoted M_k. If $\mid M_k \mid = 0$ the level is said to be empty.

Each hyper-cube is encoded by its direction so that by analogy with usual skip lists, one gets a nested hierarchy by iteratively discarding directions. The motivation behind this definition is the computation of a partition of the input set $M = \cup_{i=0}^{r} M_i$ such that if the molecule m is hooked at level i, so are its nearest neighbors. Of course, the natural way to assign a molecule to a level is to choose the one with the minimum number of directions. But as observed in sec. 1.3, since the goal pursued is to split the input set into "small-enough" subsets within which all the pairwise comparisons can be computed, we refine the previous definition as follows:

Definition 5 *Given a set of molecules M and a positive integer $MAXM$, a Recursive Directional Skip List (RDSL) is a "tree-like" data structure built as follows:*

1. *store the set M in a DSL,*
2. *recurse for all the DSL levels containing more than $MAXM$ molecules.*

[4] Skip lists [Pug90] are an alternative to balanced binary search trees for searching in dimension one since they enable dynamic queries in $\tilde{O}(\log n)$ time taking $\tilde{O}(n)$ space. By $f(n) = \tilde{O}(g(n))$ we mean $f(n) = O(g(n))$ with a high probability. That is, for some constant $c > 0$, $f(n) < cg(n)$ with probability $1/p(n)$ where $p(n)$ is a polynomial whose degree depends on c, and this degree can be made arbitrarily high by choosing c large enough.

By recursivity level rL *we call the depth of a given DSL in the tree whose nodes are DSLs, the root being at depth 0. A leaf is a non-empty level of a DSL that did not undergo recursion.*

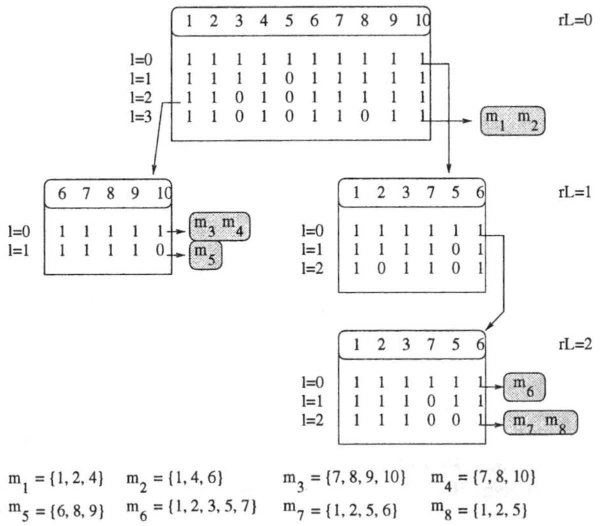

Figure 9: Recursive Directional Skip List: example

An example RDSL is depicted on fig. 9 for a set of 8 molecules using 10 fragments and $MAXM = 2$. This RDSL contains 1,2 and 2 leaves at the recursivity levels 0,1 and 2. For each DSL we have represented in the arcbox the indices of the fragments of interest.

Given the definition of a DSL, it is implicitly assumed that no DSL in a RDSL has a direction with a null hit number. On our example, this accounts for the fact that after to have hooked m_6, m_7, m_8 at the 0th level of the root node —since these molecules contain the 5th fragment which is no present at level 1, the 4th direction is dropped when recursing at level 1. This operation called *dimension reduction* is described in sec. 5.3.3.

5.2 Elementary properties

Before specifying more precisely this construction, let us discuss the relation of the DSL to usual skip lists. The first difference is that we do not encode the objects we are interested in, but an approximation of the space containing them. The second one is that we do not require the last vector of the sequence to be the null vector. In any case this happens with high probability for $r = O(\log d)$. But in our case, a necessary condition for the rth level not to be empty is $i_r \geq s_{min}$. If a single 1 digit is dropped at each stage, the number of terms in the sequence is therefore upper-bounded by $d - s_{min} + 1$. The following heuristic gives some intuition on the minimum number of direction $i_r = p$ ensuring that H_{i_r} will hook at least one molecule.

Suppose the molecules are drawn according to model 2 and call s the size of a molecule m. The number of favorable configurations for $m \subset H_{i_r}$ is $\binom{d-s}{p-s}$, so that the probability for m not to be hooked by H_{i_r} is

$$Pr[m \not\subset H_{i_r}] = 1 - \binom{d-s}{p-s} / \binom{d}{p}$$

If all the molecules are supposed to be independent, we get

$$Pr[\mid M_r \mid > 0 \mid p] = 1 - \prod_{i=1}^{n} \left(1 - \frac{\binom{d-s_i}{p-s_i}}{\binom{d}{p}}\right).$$

And since $\binom{d-s}{p-s}$ is a decreasing function of s:

$$Pr[s_{max}] \leq Pr[\mid M_r \mid > 0 \mid p] \leq Pr[s_{min}]$$

with

$$Pr[s] = 1 - \left(1 - \frac{\binom{d-s}{p-s}}{\binom{d}{p}}\right)^n.$$

On one hand, this computation gives pretty good results as depicted on fig. 10: choosing $p = \lfloor 3d/4 \rfloor$ results in H_{i_r} being non empty with a high probability, which turns out to be coherent with experiments. On the other hand, it is a coarse approximation since the counting argument does not take into account the fact that a single DSL is randomly drawn for all the molecules, and that these molecules certainly share fragments.

Observe however that small values of r are not such a big deal since too crowded M_r are taken care of by the recursivity induced by $MAXM$.

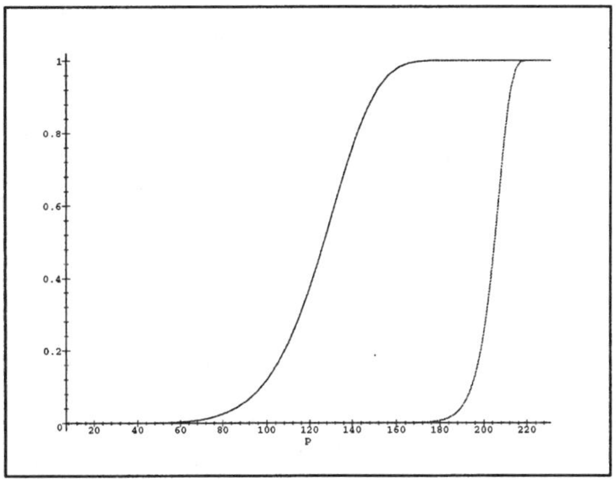

Figure 10: Upper and lower bound for $Pr[\mid M_r \mid > 0 | p]$ on the sample database of fig. 2: $n = 50, d = 231, s_{min} = 7$, $s_{max} = 33$, and $\lfloor 3d/4 \rfloor = 173$.

5.3 Constructing a RDSL

5.3.1 Building a DSL

We now get to the construction of a DSL. For each level, we need to drop some directions from the previous ones. As for usual skip lists, this can be done by flipping a coin with the rule that if an output turns out to be smaller than some probability flp, the corresponding digit is switched to 0. By tuning the probability flp one can get more or less "smooth" transitions between the levels, the smoothest one corresponding to a single digit dropped at each stage. This particular way to build a DSL does not require drawing a random real number since we just have to choose a direction at random, all the directions being equally likely. When running experiments with different values of flp, we shall refer to this particular scheme by $flp = -1$.

5.3.2 Hooking a molecule

To hook a molecule into $H_{i_l}, \ldots, H_{i_r}$, it suffices to scan the DSL right to left to find the smallest hyper-cube that contains it. But one can do better by the following dichotomic-like algorithm: first examine the median hyper-cube H_{i_j} with $j = \lfloor (l+r)/2 \rfloor$; if it does not contain the molecule, iterate on $H_{i_l}, \ldots, H_{i_{j-1}}$; if it does and if does so $H_{i_{j+1}}$, iterate on $H_{i_{j+1}}, \ldots, H_{i_r}$; otherwise done. Hooking a molecule therefore has a worst-case complexity logarithmic in the number of levels.

5.3.3 Dimension reduction

Suppose we are given a subset N of M, with $dim(M) = d$. Some directions in $1, \ldots, d$ may have a null hit number for the molecules in N. We therefore reduce the length of the molecules in N by packing to the right all the zeroes. Suppose for example we have three molecules defined as follows over five fragments: $m_1 = 10010, m_2 = 11010, m_3 = 11000$. Since the directions 3 and 4 have null hit numbers, the molecules can be written as words of length 3, which is easily done by swapping the digits of indices 3 and 5. More generally, if p is the number of directions with 0 hit numbers, we just have to switch the bits located in the range $1, \ldots, d-p$ with a zero hit number with those located in the range $d-p+1, \ldots, d$ with a non-zero hit number. This operation is actually compulsory in order for the 1-to-0 bit flipping operation to work correctly. Indeed, for subsets of molecules with a dimension significantly less than d, by randomly switching bits to 0 one would reach a majority of bits with null hit numbers, thus prolongating the separation process.

6 Experimental results

We eventually present a few experimental results obtained with various RDSLs on TLS1296. For more details, see [Caz98]. When naming an experiment, **flp** refers to the flip probability and **MAXM** to the recursion termination condition with the convention $MAXM_\infty$ for a DSL. For example, TLS1296.MAXM50.flp-1 refers to results obtained on TLS1296 with a RDSL having at most 50 molecules per leaf and whose DSLs' 01 vectors are derived from one another by switching a single digit to 0. The depth of the DSLs was chosen according to the discussion of section 5.2.

6.1 The s_2 statistic

The averages and std. dev. of the s_2 statistic computed over a sequence of 100 runs on TLS1296 for various RDLSs are displayed on table 6. Also, for the sake of illustration, the distribution of s_2 for $RDSL_{100}$ has been depicted on fig. 11. In any case, this distribution seems to be unimodal.

We first observe that without any recursion —$MAXM = \infty$, $RDSLs$ are not very interesting: s_2 is elevated and the corresponding standard deviation is huge. Otherwise, one can expect reasonable savings. For example a $RDSL.MAXM_{500}$ is expected to run $11046/168.3 \sim 65$ faster than the naive algorithm which checks the $\binom{n}{2}$ pairs. And more generally, the number of pairwise comparisons checked is roughly proportional to the leaf size $MAXM$: for example when $MAXM$ goes from 50 to 200, the s_2 statistic is multiplied by about 4. This is understandable since reducing $MAXM$ yields a smaller number of comparisons per leaf but also more leaves.

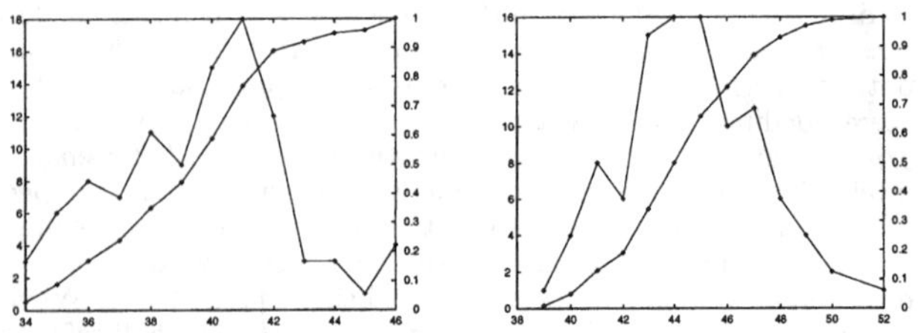

Figure 11: s_2 statistic over 100 runs for $RDSL.MAXM_{100}$: (a)$flp = -1$ (b)$flp = 0.01$

$RDSL$	flp	μ_{s_2}	σ_{s_2}
$MAXM_\infty$	-1	1069.03	674.33
	0.001	1259.99	998.86
	0.01	2249.98	1394.35
$MAXM_{500}$	-1	168.3	23.35
	0.001	180.19	24.32
	0.01	202.57	23.41
$MAXM_{200}$	-1	71.44	5.79
	0.001	72.49	5.91
	0.01	84.37	4.78
$MAXM_{100}$	-1	39.60	2.82
	0.001	40.32	2.50
	0.01	44.61	2.54
$MAXM_{50}$	-1	21.87	0.95
	0.001	23.02	1.79
	0.01	24.68	1.54

Table 6: s_2 statistic for various RDSLs

6.2 Evolution of the number of (α, β)-similar pairs captured

The evolution —wrt the total number of pairs computed off-line— of the number PCt of $(0.9, \beta)_{,\beta=1.1,\infty}$ similar pairs captured by 5 independent RDSLs with parameters $flp = -1$ and $MAXM = 100, 500$ is depicted on fig. 12 to 15. As an example, let us examine carefully fig. 12:

1. the left chart depicts the percentages of the total number of (0.9,1.1) similar pairs captured after constructing 1,2,3,4 and 5 RDSLs. Thus after 5 runs we have reported about 82% of the total number of pairs.
2. the right one reports the percentage of parties that were reported after 5 runs for all the molecules having a given number of parties. For example, among all the molecules having 50 (0.9,1,1)-similar parties, about 85% of them were found by 5 RDSLs.

A first observation is that for $\beta = \infty$, small values of $MAXM$ —e.g. $MAXM = 100$— make reporting an elevated number of pairs difficult. The molecules which are most missing parties are those with a large number of neighbors, i.e. the small ones, and this is so because the $MAXM$ constraint prevents them from being tested with a large enough number of adversaries. But as already mentioned, large values of β are not interesting from a biological standpoint.

More interesting is the ability of the different RDSLs to capture the similar pairs for small values of β. To see why, let us compare $MAXM = 100$ against $MAXM = 500$ for the (0.9,1.1) situation:

1. as we have just seen, 4 and 5 runs with $MAXM = 100$ respectively result in 76% and 82% of the total number of pairs reported. This corresponds to a unit construction cost of $s_2 = 39.60$ —see table 6.
2. as opposed to that, a single RDSL with $MAXM = 500$ returns about 70% of pairs, and the corresponding s_2 is 168.3.

In other words, one may prefer 4 or 5 RDSLs with $MAXM = 100$ than a single one with $MAXM = 500$. Coming up with a theoretical way to determine the right value of $MAXM$ as a function of α, β is of prime interest: too small values of $MAXM$ require a large number of constructions to report a significant percentage of the total number of pairs; too big ones correspond to a huge s_2 statistic and do not provide the running time gain expected. It should be noticed that this question has some similarity with the optimal file size below which recursive quick-sort should be switched to insertion sort.

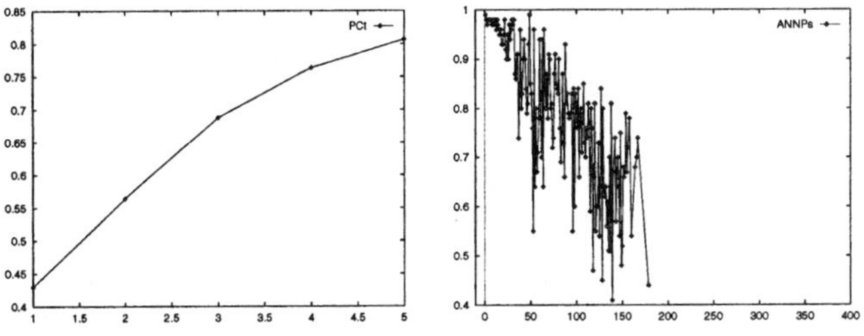

Figure 12: (0.9,1.1)-similar pairs for TLS1296.MAXM100.nbR5

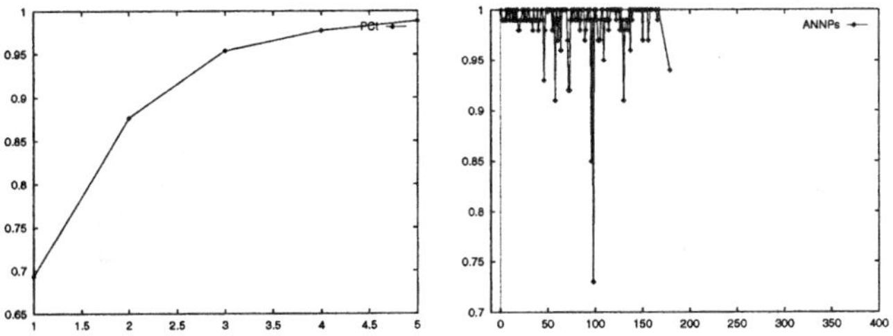

Figure 13: (0.9,1.1)-similar pairs for TLS1296.MAXM500.nbR5

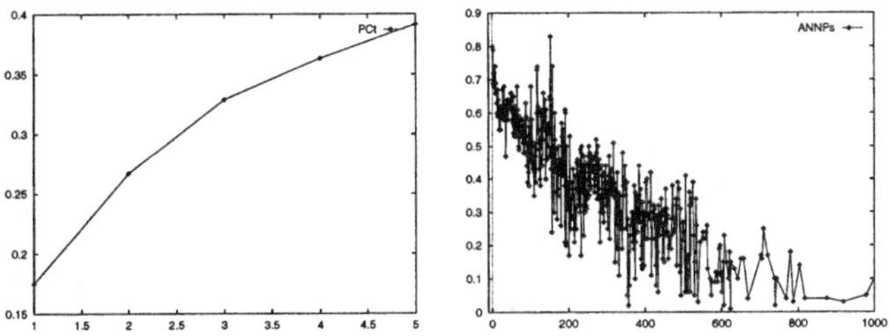

Figure 14: (0.9,∞)-similar pairs for TLS1296.MAXM100.nbR5

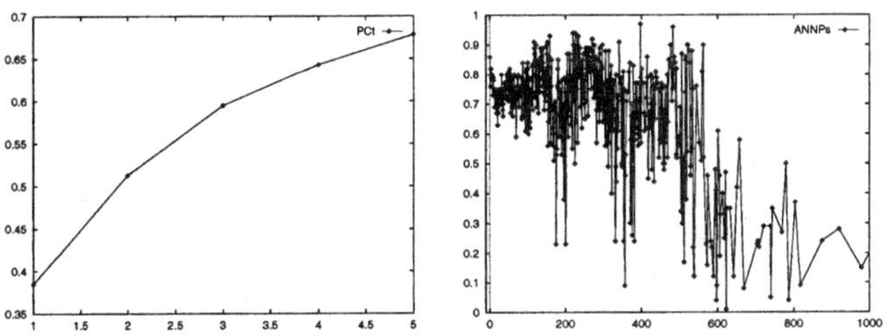

Figure 15: (0.9,∞)-similar pairs for TLS1296.MAXM500.nbR5

6.3 Similarities and dimensions

To get more insight on the RDSL data structure, we now present statistics describing the partitioning of M induced. More precisely, we define for **each recursivity level** the following quantities:

μ_d the average dimension of the DSLs,

μ_{nb1} the average number of 1 digits in the 01 strings defining the non-empty leaves,

μ_{nbM} the average number of molecules hooked in leaves containing **at least** 2 molecules,

μ_{sim} the average of the average similarity between two molecules hooked in non-empty leaves,

μ_s the average of the average size of the molecules hooked in non-empty leaves,

μ_{ss} the ratio μ_{sim}/μ_s

Results obtained for typical runs are depicted on the tables 7, 8 and 9. Let us review in detail an entry of these, for example the third line of the first one. It turns out that 2770 molecules were hooked at a recursivity depth of two. The average dimension of the DSLs at this recursivity level is of 261.1, and the average number of 1 in the 01 vectors defining the non-empty levels of this recursivity level is 244.1. The average number of molecules per leaf of this recursivity level is 9.89. The average of the averages sizes of the molecules is 21.9 while the average of the averages similarities is 16.11. The entries with the $\emptyset$ symbol correspond to recursivity levels where all the leaves hook a single molecule and for which the average statistics are irrelevant. General comments based on observations gathered for several values of flp and $MAXM$ are the following:

- in any case very few molecules are isolated in leaves containing a single molecule;
- the bigger $MAXM$, the smaller on average the ratio μ_{sim}/μ_s;
- values of flp greater than 0.05 result in bad partitions: the not-so-smooth transitions between the 01 vectors makes the splitting process longer. For example the combination $MAXM = 50, flp = 0.1$ result in hierarchies of depth several hundreds;
- with the right combination of parameters —see table 7, one gets a μ_{sim}/μ_s statistic bigger than 0.73 for 90% of the molecules, while it is of 0.17 for the whole database;

r_L	#*leaves*	#mols	μ_d	μ_{nb1}	μ_{nbM}	μ_{sim}	μ_s	μ_{ss}
0	98	579	1616.0	1503.0	8.9	10.9	20.7	0.53
1	556	3463	388.4	358.8	9.1	14.7	21.4	0.69
2	394	2770	261.1	244.1	9.9	16.1	21.9	0.74
3	175	1218	125.8	119.8	9.1	19.0	23.0	0.83
4	97	686	109.6	105.2	10.1	22.8	25.8	0.88
5	60	551	115.9	111.8	12.2	24.8	28.9	0.86
6	66	550	83.3	79.5	11.1	22.4	24.6	0.91
7	43	246	81.6	77.9	8.5	22.8	24.9	0.91
8	11	100	57.5	56.4	12.1	22.1	24.7	0.90
9	15	197	62.7	61.5	17.6	25.3	27.6	0.92
10	10	145	47.8	46.2	14.5	23.4	24.5	0.95
11	4	45	43.0	41.8	14.7	15.6	16.6	0.94
12	5	62	56.0	55.4	15.2	25.6	27.1	0.94
13	13	174	46.1	44.9	14.4	23.9	25.2	0.95
15	2	2	64.0	63.5	$\emptyset$	$\emptyset$	$\emptyset$	$\emptyset$
16	1	2	49.0	49.0	2.0	33.0	33.0	1.00
17	2	52	56.0	55.5	26.0	46.7	47.3	0.99
18	2	73	59.0	58.5	36.5	35.9	37.4	0.96
21	1	3	45.0	42.0	3.0	28.0	29.3	0.95
24	4	52	42.0	40.5	25.0	29.6	30.0	0.99
25	1	2	55.0	53.0	2.0	40.0	41.0	0.98
28	1	10	55.0	54.0	10.0	38.2	39.7	0.96
36	1	1	55.0	55.0	$\emptyset$	$\emptyset$	$\emptyset$	$\emptyset$
37	2	63	50.0	49.5	31.5	44.1	44.8	0.99
	1564	11046						

Table 7: TLS1296.MAXM50.flp-1

r_L	#*leaves*	#mols	μ_d	μ_{nb1}	μ_{nbM}	μ_{sim}	μ_s	μ_{ss}
0	172	2211	1616.0	1464.0	17.1	10.9	18.9	0.58
1	536	4463	461.6	424.3	11.9	14.6	21.4	0.68
2	205	2251	249.3	235.8	16.0	17.6	23.6	0.75
3	106	1106	172.7	161.4	14.2	19.8	23.9	0.83
4	24	275	92.4	89.2	18.9	21.8	25.0	0.87
5	19	347	78.5	75.2	20.3	29.1	31.2	0.93
6	10	24	121.8	118.7	3.3	26.9	28.5	0.94
7	6	121	92.3	91.3	20.2	30.1	32.1	0.94
8	4	121	83.0	80.8	40.0	28.7	31.4	0.92
9	1	3	56.0	56.0	3.0	24.7	26.3	0.94
10	1	11	55.0	55.0	11.0	25.6	27.4	0.94
11	4	113	53.0	51.5	28.2	25.4	27.6	0.92
	1088	11046						

Table 8: TLS1296.MAXM100.flp0.05

r_L	#*leaves*	#mols	μ_d	μ_{nb1}	μ_{nbM}	μ_{sim}	μ_s	μ_{ss}
0	127	2059	1616.0	1504.0	22.5	11.2	20.7	0.54
1	468	6212	561.8	520.4	18.7	14.1	21.4	0.66
2	143	1826	279.1	263.3	18.0	17.3	23.5	0.74
3	37	736	242.8	231.4	25.1	13.8	20.2	0.68
4	9	213	55.0	50.8	26.5	10.2	11.5	0.88
	784	11046						

Table 9: TLS1296.MAXM200.flp-1

6.4 More on the influence of flp and $MAXM$

Most of the experimental results presented so far correspond to $flp = -1$. The tables below seem to indicate that given a molecular database such as TLS1296, this is indeed the best strategy to build nested hierarchies of hyper-cubes.

$flp/MAXM$	50	100	200	500
-1	0.61	0.83	0.96	0.99
0.01	0.60	0.81	0.91	0.99
0.05	0.62	0.83	0.94	0.98
0.10	0.63	0.79	0.94	0.97

Table 10: % of $(0.9, 1.1)$-similar pairs reported after 5 runs

$flp/MAXM$	50	100	200	500
-1	0.52	0.73	0.92	0.99
0.01	0.52	0.76	0.91	0.98
0.05	0.52	0.72	0.90	0.98
0.10	0.53	0.74	0.85	0.96

Table 11: % of $(0.9, 1.2)$-similar pairs reported after 5 runs

$flp/MAXM$	50	100	200	500
-1	0.46	0.69	0.86	0.98
0.01	0.47	0.66	0.84	0.97
0.05	0.46	0.66	0.83	0.95
0.10	0.45	0.63	0.81	0.92

Table 12: % of $(0.9, 1.3)$-similar pairs reported after 5 runs

$flp/MAXM$	50	100	200	500
-1	0.26	0.40	0.55	0.70
0.01	0.25	0.39	0.53	0.66
0.05	0.24	0.37	0.47	0.60
0.10	0.24	0.35	0.45	0.55

Table 13: Percentage of $(0.9, \infty)$-similar pairs after 5 runs

7 Conclusion and future work

The main contribution of this paper is to introduce two new data structures, the Directional Skip List and Recursive Directional Skip List, for effectively solving problems on the hypercube $H_d = \{0,1\}^d$ when the order of magnitude of d is several hundreds. Unlike the usual geometric algorithms which at some point yield constants exponential in d, our construction eludes this difficulty by a combinatorial argument. More precisely, the idea consists in first randomly building a sequence of nested approximations of H_d, and second assigning each point processed to the smallest hyper-cube containing it. This way, neighbor points are expected to be hooked together with high probability.
The RDSL efficiency is validated on a computational chemistry problem —finding molecules that are structurally close— which reduces to finding similar strings of length d on the alphabet $\{0,1\}$. For a database with $n = 11046$ molecules in dimension $d = 1616$, the running time is improved within a factor from 30 to 60 with reference to the naive $\Theta(n^2)$ algorithm and even more with respect to neural networks techniques.

Several important issues still remain. First, a probabilistic analysis of the RDSL must be performed, if possible in conjunction with Sperner theory. This should tell if the data structure is worth being derandomized and also if it works in some output-sensitive mode. Second, it should be determined whether or not the RDSL also makes sense for points inside H_d, probably under some convenient sparsity condition –that is relatively few non null coordinates per point.
At last, RDSLs may also find other application areas such as for example information retrieval on the web or [BYF92, Kle98]. An interesting test-suite may also be the next DIMACS Implementation Challenge on near neighbor searches [5].

Acknowledgments: The author wishes to thank Jean-Marie Bernassau from Elf-Sanofi Research for submitting this problem and following the subsequent developments. Also special thanks to the members of the Algorithms project for helpful discussions and in particular: Philippe Dumas for proving the lower bound of prop. 1; Philippe Flajolet for the symbolic method used to prove prop. 4; and Bruno Salvy for his advice on solving algebraic equations of elevated degree.

This work was supported by the Long Term Research Project Alcom-IT #20244 of the European Union.

References

[A+94] S. Arya et al. An optimal algorithm for approximate nearest neighbour searching. In *ACM SODA*, 1994.

[AM93] S. Arya and D. Mount. Approximate nearest neighbour queries in fixed dimensions. In *ACM SODA*, 1993.

[BYF92] R. Baeza-Yates and B. Frakes. *Information Retrieval: Data Structures and Algorithms.* Prentice-Hall, 1992.

[Caz98] F. Cazals. Effective nearest neighbors searching on the hyper-cube, with applications to molecular clustering. Technical Report To appear, INRIA, 1998.

[Cla94] K. Clarkson. An algorithm for approximate closest-point queries. In *ACM SoCG*, 1994.

[CP97] F. Cazals and C. Puech. Bucket-like space partitioning data structures with applications to ray-tracing. In *ACM SoCG*, Nice, France, 1997.

[Dev86] L. Devroye. *Lectures notes on bucket algorithms.* Birkhauser, 1986.

[Eng97] K. Engel. *Sperner Theory.* Cambridge University Press, 1997. Encyclopedia of Mathematics and its Applications.

[EW82] C. Eastman and S. Weiss. Tree structures for high dimensionality nearest neighbour searching. *Information Systems*, 7(2), 1982.

[FBF77] J. Friedman, JL. Bentley, and R. Ari Finkel. An algorithm for finding best matches in logarithmic expected time. *ACM Trans. on Math. Software*, 3(3), 1977.

[G+93] M. Golin et al. Randomised data structures for the dynamic closest-pair problem. In *ACM SODA*, Austin, TX, USA, 1993.

[IM98] P. Indyk and R. Motwani. Approximate nearest neighbors: towards removing the curse of dimensionality. In *ACM STOC*, Dallas, TX, USA, (To appear) 1998.

[Kle97] J. Kleinberg. Two algorithms for nearest-neighbour search in high dimension. In *ACM STOC*, El Paso, Texas, USA, 1997.

[Kle98] J. Kleinberg. Authoritative sources in a hyperlink environment. In *ACM SODA*, San Francisco, CA, USA, 1998.

[Mil98] P.B. Miltersen. Error correcting codes, perfect hashing circuits, and deterministic dynamic dictionaries. In *ACM SODA*, San Francisco, CA, USA, 1998.

[Pug90] W. Pugh. Skip lists: a probabilistic alternative to balanced trees. *Com. ACM*, 33(6), 1990.

[Riv74] R. Rivest. On the optimality of elias's algorithm for performing best matches searches. *Information Processing*, North-Holland Pub. Co.:678–681, 1974.

[Spr91] R. Sproull. Refinements to nearest-neighbour searching in k-dimensional trees. *Algorithmica*, 6, 1991.

[vLW92] J.H. van Lint and R.M. Wilson. *A course in Combinatorics.* Cambridge University Press, 1992.

[Yao82] A. Yao. On constructing minimum spanning trees in k-dimensional spaces and related problems. *SIAM J. on Computing*, 11(4), 1982.

[Yia93] P. Yianilos. Data structures and algorithms for nearest neighbour search in general metric spaces. In *ACM SODA*, Austin, TX, USA, 1993.

[5] see http://www.dimacs.rutgers.edu/Challenges/Sixth/

Matching 2D Patterns of Protein Spots *

Frank Hoffmann† Klaus Kriegel† Carola Wenk†

Department of Computer Science
Freie Universität Berlin

Abstract

A new algorithmic approach to comparing 2D patterns of protein spots obtained by the 2D gel electrophoresis technique is presented. Both the matching of a local pattern vs. a full 2D gel image and the global matching between full images are discussed. Preset slope and length tolerances of pattern edges serve as matching criteria. The local matching algorithm relies on a data structure derived from the incremental Delaunay triangulation of a point set and a 2–step hashing technique. The approach for the global matching uses local matching for landmark settings, which in most previous algorithmic solutions has been done interactively by the user.

1 Introduction

1.1 Point Pattern Matching

The matching problem for geometric point patterns has been subject of intensive research in the last decade. Given a point pattern P and another target point set T one wants to compute all occurrences of P in T. Usually, an admissible space $\mathcal{A}$ of transformations (e.g. translations, rigid motions and/or scalings) is given which can be used to map the pattern into or as close as possible to the point set T. Most common is to consider the Hausdorff H distance, see [10]. Additionally we have a distance measure d between patterns. In general, we want to find such an $f \in \mathcal{A}$ and a pattern Q in T for which $d(f(P), Q) \leq \epsilon$, where ϵ is a prescribed error tolerance. We distinguish between exact matchings (ϵ is zero) and approximate matching solutions, otherwise. The latter are important in most practical applications. Like in our concrete application it is sometimes only possible to find partial matchings, i.e., we will be looking for as large as possible sub-patterns of P which have an approximate matching pattern in T.

A survey on several variants of the general geometric matching problem, different geometric approaches, and various algorithmic techniques can be found in [2]. In particular, for most settings there are algorithms which solve the problem well from a theoretical point of view but which are hardly applicable in practice because of their time complexity.

Here we want to mention two approaches which have proved to be useful for our application, too: the alignment method and geometric hashing.
The alignment method is based on the observation that any similarity transformation is determined up to reflection by the mapping of a single line segment. Thus, we will choose two points a, b in the pattern P and map the edge $\overline{ab}$ to all edges $\overline{uv}$ in the target set T. For each mapping we check whether it induces a (large partial approximate) matching of P. Note, in the special situation of partial matchings it is not sufficient to consider only one edge $\overline{ab}$. So in a worst case scenario one has to map all pattern edges to all edges in T. The situation is much easier to handle if scalings are not allowed (or strongly restricted). Then for a given edge $\overline{ab}$ it is sufficient to search for all target edges of approximately the same length. Analogously, if rotations are forbidden it suffices to search for all target edges with approximately the same slope as $\overline{ab}$.
An essential speed up of the alignment method can be obtained if the points in both the pattern and the target are labeled with positive values (intensities) such that for any valid matching the intensity ordering of the pattern is consistent with the intensity ordering in the target. Then the quadratic size search space of all target edges can under certain circumstances be reduced to the set of all edges in the history of the incremental Delaunay triangulation of T which has expected linear size. The details of this idea will be discussed in the next section.
The main drawback of the alignment method consists in the waste of time caused by the fact that one tries to construct

---

* Part of a joint research project with Deutsches Herzzentrum Berlin, supported by Deutsche Forschungsgemeinschaft, grant FL 165/4–1.

† Institut für Informatik, Freie Universität Berlin, Takustr. 9, D-14195 Berlin, email:*name@inf.fu-berlin.de*

a matching for each legal pair of edges $\overline{ab}$ in P and $\overline{uv}$ in T, but finally only few of these attempts will be successful. One can avoid this by making use of geometric hashing. This method requires some preprocessing for a hash table. After that the number of pattern points which are matched by a transformation (induced by a legal edge pair) can be counted very efficiently. Finally, one has to compute the matchings only for the best transformations.

Point pattern matching has not only been studied in computational geometry. It is such a fundamental and natural task that it comes up in various fields. Of special interest for our application is the rich pattern recognition and image processing literature on the topic, compare with [17], [7], [8]. Moreover, it is no surprise that basic ideas and principles, like the alignment paradigm, have been rediscovered several times. The same applies to the use of Delaunay triangulation graphs for the matching task, see for example [13], [11]. However, the novelty of the algorithmic solution presented below is, that, firstly, the way to construct the Delaunay triangulation graph rather than the final graph itself is used for the matching process and, secondly, that the approach works in the case of noise.

1.2 2D Gel Electrophoresis

Surveying the protein components of cells is a fundamental task in molecular biology, see [19]. For this purpose the 2–dimensional gel electrophoresis technique for protein separation was introduced by O'Farrell in 1975. With a resolution separation of several thousand spots in real samples it is almost two orders of magnitude better than competing techniques available for proteins. A 2D gel is the product of two separations performed sequentially in acrylamide gel media: isoelectric focusing as the first dimension and a separation by molecular size as the second dimension. A 2D pattern of spots each representing a protein is the result of that process. Eventually, spots are detected by staining or radiographic methods.

In Figure 1 two typical examples of 2–dimensional spot images are shown. The left image shows a gel image of a human heart ventricle tissue sample. It contains about 1500 protein spots, while the right one is an image produced in another laboratory with the same technology but with a total of about 3000 spots. Their original size is about 23cm by 29cm.

Comparing visually such an image with a master gel image (which is available in different databases on the Internet) is one way for putative protein spot identification without using expensive sequencing techniques. Another interesting possible application is related to the fact that with some diseases there are associated typical deviations of certain protein spots compared to standard spot size/intensity, an example is described in [14]. To detect such deviations is of great importance for example in view of a possible drug design.

However, for the visual comparison substantial difficulties arise from the fact that images are – due to inaccuracies in the complicated electrophoresis process itself – distorted and noisy. They have usually (especially in the inter–laboratory comparison) different separation resolution, different spot expression etc. To a much greater extent this applies to the computer assisted comparison, where one first applies an algorithm to detect spots and to extract their features like spot size, spot intensity, or spot shape, see [4] for definitions. This so called spot detection stage is a necessary preprocessing step but, at the same time, a severe error source for the subsequent spot matching problem.

Starting with the early eighties there is an extensive literature on solutions for a computer assisted gel image comparison, see [19]. However, many of the proposed algorithms make use of so–called *landmarks* and a general alignment of the images by warping techniques, [12]. Landmarks are spot pairs interactively marked in both images by the user and selected as putative matching pairs. Using various heuristics one then algorithmically extends this partial relation to a full matching.

One can do the spot matching either relying primarily on the pixel level information (like the Melanie software system [4] using spot areas) or on derived geometric information like in our solution.

For the purpose of illustration in Figure 2 more details are shown of the small rectangular window regions marked in Figure 1.

2 Our Approach to the Matching Problem; Basic Algorithmic Ideas

We address the following algorithmic *Local Matching Problem*:

Given a local spot pattern P selected from a 2D gel source image S find all local spot patterns in a target image T that resemble at least partially both the geometric shape and the spot intensities of P.

The algorithm we are going to describe starts from two assumptions. Firstly, the geometry of spot patterns is given by point patterns; and a single real value represents both spot size and spot intensity. Secondly, the algorithm should not use the relative position of the selected pattern within the source image or its location with respect to possibly given landmarks. Such information, if available, can be used to speed up our solution considerably, for example by restricting the search range in the target image.

To illustrate an instance of the local matching problem consider the example of Figure 2: Given the indicated pattern of eight spots drawn from the small rectangle in the left image find all its matching counterparts in the right image. In Figure 3 the spot point sets (without intensities) and, again, one partial matching corresponding to Figure 2 is shown. Remark that the algorithm computed a partial matching on six

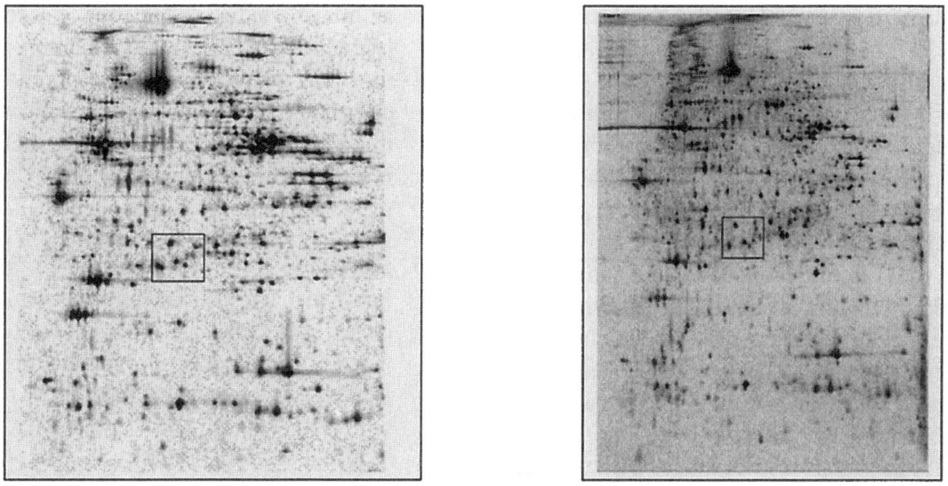

Figure 1: Two gel images of comparable samples

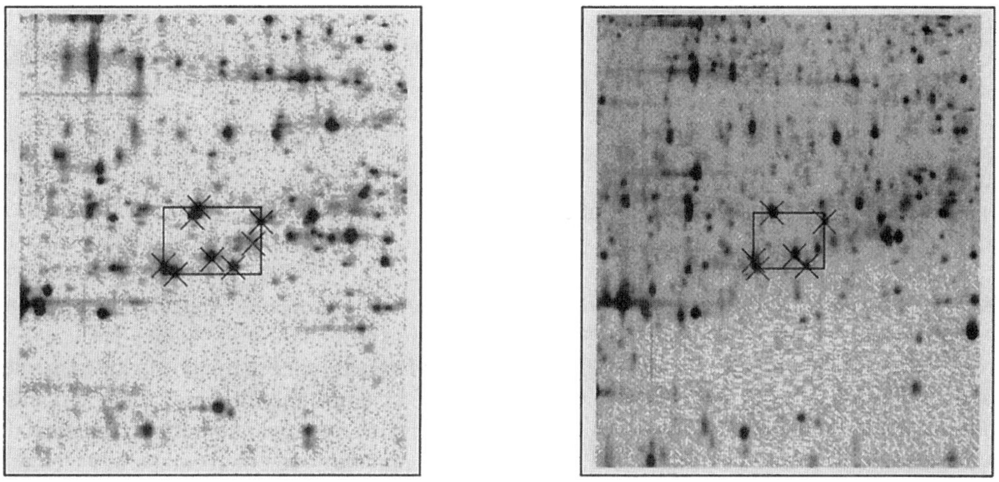

Figure 2: Detailed local images of Fig.1, a selected pattern on the left side and a partial matching

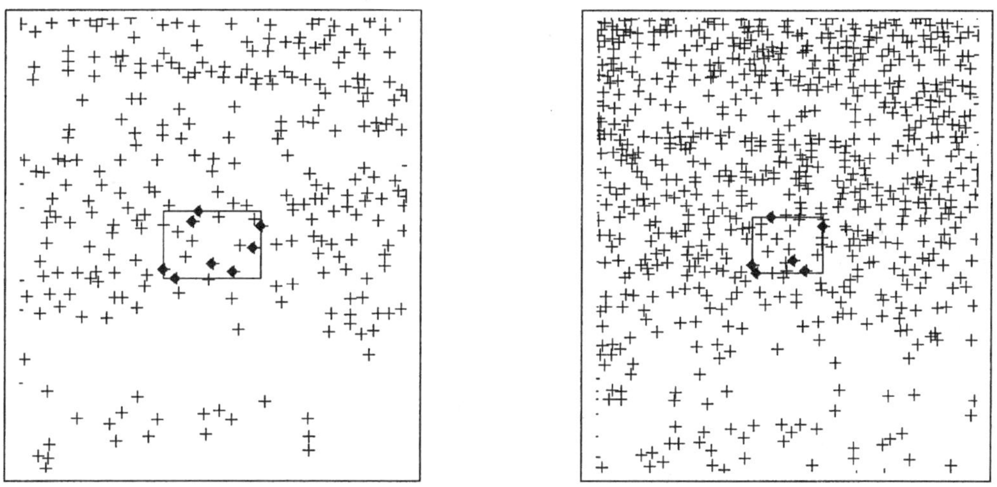

Figure 3: Spot point sets detected with indicated partial matching

spots only although there are candidate spots for the remaining two pattern spots. The reason is that their intensities differ too much from those of the pattern spots.
The problem is formalized as follows. During a first feature extracting preprocessing step a spot detection algorithm is applied to both source and target image. As a result we obtain for each gel image lists of spots. Now a spot is simply a vector $(x(s), y(s), i(s))$ consisting of its nonnegative point coordinates $(x(s), y(s))$ in the Euclidean plane and a positive number $i(s)$ describing its intensity. Observe that with this special representation we necessarily lose a lot of visual information the original gel image carries like geometric spot shapes etc. Moreover, source and target image are assumed to have the same bounding box, otherwise they are linearly scaled accordingly. The spot intensities induce a linear order in the spot list.

Next we fix the admissible transformation space and a distance measure to evaluate matchings based on the following observations.

1. Assume we want to choose a pattern P from a small rectangular window in the source image S. Source and target image can have significantly different spot numbers but since intensive spots tend to appear first it makes only sense to choose and restrict oneself to such patterns P that consist of the locally most intensive spots.

2. On the other side, a matching pattern P' in the target T should also consist of locally intensive spots. Moreover, we should also accept solutions in which P' resembles only a large portion of P. This way we can also try to correct certain errors made by the spot detection algorithm, which tends to have difficulties to interpret spots that are very close to each other correctly.

3. To model the pure geometric resemblance between P and P' we use the following simple rule. We call two line segments $\overline{st}$ and $\overline{s't'}$ (λ, α)*–similar* if their absolute slope difference is smaller than α and for their lengths we have:

$$1 - \lambda \leq |\overline{st}|/|\overline{s't'}| \leq 1 + \lambda$$

Two point patterns P and P' are (λ, α)–similar if there is a bijection f between the point sets such that $\overline{st}$ and $\overline{f(s)f(t)}$ are (λ, α)–similar for all $s, t \in P$.
In sum, from the application point of view we want to find (λ, α)–matchings between as large as possible subpatterns $P'' \subset P$ and target patterns P', compare also [13].

4. To model the intensity resemblance between spots we do not use directly the absolute intensity values. Instead, we apply the following very robust heuristic ranking rule that assigns to each spot a discrete intensity integer between 1 and 10. The 500 most intensive spots in an image are distributed equally according to cardinality between 10 and 6; the remaining spots are assigned to values ≤ 5 such that the total intensity sum in each class is the same. For the matching we use the criterion that a pattern spot s can only be matched to a spot in P' if their discrete intensities differ by at most 2. (This is the default value in the implementation.)

5. Since the edge similarity constants λ and α are small we know each (λ, α)–matching between $P'' \subset P$ and P' is close to a translation t, more exactly the Hausdorff distance $\tilde{H}(t(P''), P')$ between the translated P'' and P' is in worst case bounded from above by $max_{s,t \in P''} \epsilon |\overline{st}|$ with $\epsilon = \sqrt{2(1 + \lambda)(1 - \cos \alpha) + \lambda^2}$ in worst case.

Besides the size of P'' another criterion for evaluating the matching could be the Euclidean distance of the center of P' from the expected center position of the transformed P in the target image, provided its position in the source is known.

Given this general setting our local matching algorithm is based on the following key idea that was first time used in [18].
Let's call a triple of spots in a gel image *intensive* if its circumcircle does not contain a spot that is more intensive. An edge connecting spots s, t is *intensive* if there exists a third spot forming together with s and t an intensive triple.
This concept of intensive edges is very strongly related to the Delaunay triangulation construction of a point set. A triangulation of a point set S in the plane is called *Delaunay triangulation* if for each triangle in the triangulation its circumcircle contains only the three triangle points. One can construct such a triangulation in an incremental way by adding one point after the other, compare [9]. Now the main observation in [18] reads in our terminology:

Proposition 1: *Assume that the Delaunay triangulation of a gel image is computed incrementally by inserting spots in order of decreasing intensity. Then the set of all Delaunay triangles and edges occurring during the history of that process is exactly the set of intensive triangles and intensive edges.*

Let Hist(T) be a data structure representing all intensive edges. Its usage for the matching problem stems from the following observation.
If a pattern P of locally intensive spots occurs in T, then we can expect that, despite the possible noise, at least a few of the edges connecting spots in P will be (λ, α)–similar to edges in Hist(T).
This is the point where our approach and that one from [18] branch. While in [18] according to the alignment technique one tries to extend each occurrence of a pair of similar edges to a matching of the complete pattern we have to opt for a different strategy. The main reason for this is the small but nevertheless considerable length and slope tolerance (in the

implementation the default values are $\lambda = 0.2$ and $\alpha = 0.2$) that imply a search range that is too large.

The alternative to the alignment technique in [18] that we choose is a 2–step variant of geometric hashing , see [2].
In a first step we want to compute all locations within the target image where a good matching with the pattern is likely to occur. The actual local matchings are computed subsequently in a second step.
As pointed out above, if there is a good partial matching between pattern P and a pattern P' in T then a portion of edges connecting spots in P will have (λ, α)–similar matching edges in Hist(T).
Even more, if we associate with each occurrence of a similar edge the translation vector of the edge midpoints then a matching pattern will correspond to a cluster of vectors in the translation space. It turns out that the clusters can be computed and evaluated efficiently. For the best candidate clusters we then recompute locally an actual matching.
Another problem to be addressed is how to proceed in the case that there is a more severe distortion between the pattern P and its counterpart P' in the target image.

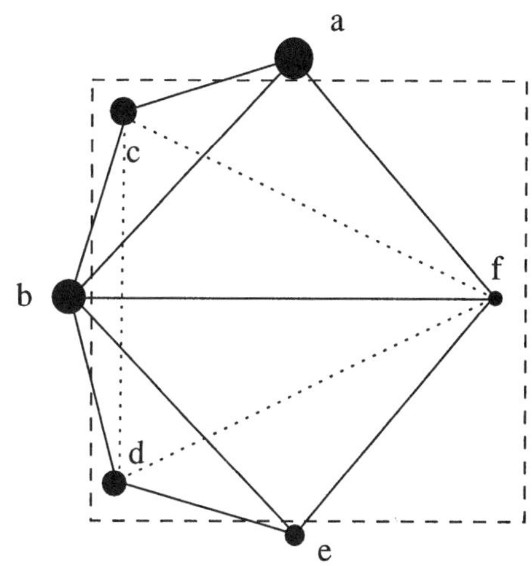

Figure 4: Pattern edges that are not Delaunay edges

3 Details of the Local Matching Algorithm

3.1 Preprocessing the Target Image

As indicated above we triangulate the underlying point set of T using the incremental Delaunay triangulation algorithm of [9] inserting spots according to decreasing absolute spot intensities. When a new spot is inserted a few edges are deleted from the current triangulation, a few new ones added. We call these edges Delaunay edges. Additionally we consider all edges connecting the new spot with opposite spots in neighboring triangles. Let us call these edges *flipped diagonals*. We store all Delaunay edges and flipped diagonals occurring during that process in a data structure Hist$^*(T)$, that describes the *extended history* of the incremental Delaunay triangulation of T. With each object in Hist$^*(T)$ we store also its length and its slope. For this purpose range trees are the appropriate -at least theoretically- data structure, see for example [5].

Using Seidel's backward analysis technique, see [16], we have:

Proposition 2: *The expected number of edges in the extended history of a randomized incremental Delaunay triangulation of a point set in the plane is bounded by* $12n$*, where* n *is the number of points.*

The main question is to which extent do the history, respectively extended history allow to recognize a local pattern ? Assume we are given an identical copy P' of the pattern P in the target. The matching between P and P' is done via Delaunay edges connecting spots of P'. However, how many edges actually belong to the history depends not only on P' but also on its context in the target, since the 'local' history of the Delaunay triangulation can be strongly influenced by intensive spots in the neighborhood of P', as demonstrated by the example in Figure 4. In Figure 4 let c, d and f be spots in the target triangle pattern P' corresponding to pattern spots in P. The pattern was chosen from a window corresponding to the box drawn with dashed lines. Now consider the incremental Delaunay triangulation which inserts the spots in alphabetical order. We observe that none of the edges forming the pattern triangle occurs in the history of the triangulation, but all of them are flipped diagonals in Hist$^*(T)$. Therefore, using Hist$^*(T)$ as search space we still have the advantage of its expected linear size and at the same time we increase the probability to include edges from P'.

This also applies to the case that there is some noise in the target. To illustrate these facts we have run computer experiments that mimic the situation and quantities given in our approach to the local gel image matching. Assume we randomly draw from a unit square B a pattern P' of k ($k = 8, k = 12$) spots. Moreover, we take a 7×7 square T that contains B and generate $48k$ random points in $T \setminus B$. Next we simulate overall noise in T by adding l new random points. Finally, by picking at random a permutation of all $49k + l$ points in T we fix a linear intensity order. For each value of l we compute 1000 random instances and count in each corresponding incremental Delaunay triangulation the fraction of those edges in the simple, respectively extended history that connect pattern spots. In Figure 5 we summarize the data which clearly indicate that Hist$^*(T)$ is well suited for identifying a local pattern in images with up to one third of noisy points.

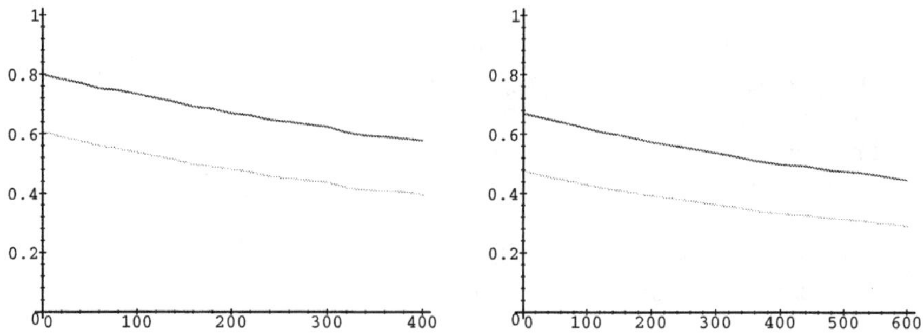

Figure 5: The fraction of pattern edges that are elements of Hist(T) (lower curves) and Hist$^*(T)$ in the presence of l random distortion spots (abscissa) and random intensity order for $k = 8$ (left) and $k = 12$

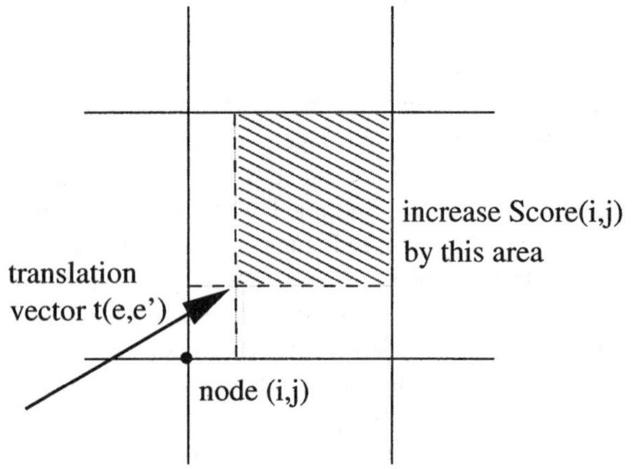

Figure 6: Updating the score for a pair e, e' of similar edges

3.2 Approximating the Matching Locations

After the preprocessing we are able to answer queries of the type: Given a pattern edge e find all target edges e' in Hist$^*(T)$ that meet the tolerance bounds with respect to length, slope, and discrete spot intensities.

However, we do not store the results of such a query as an edge list. We first compute the vector $t(e, e')$ that translates the midpoint of edge e to the midpoint of e'. For all these we maintain a scoring list that indirectly stores translation vectors and yields at the same time clusters of such vectors. Observe that such clusters correspond to possible matching locations. This is done as follows.

The bounding box of T is interpreted as the possible space for translation vectors $t(e, e')$. Next we overlay a regularly spaced grid on the translation space and maintain a data structure for integer scores, initially all zero, which are defined for each grid node.

Each translation vector $t(e, e')$ increases the score of the four grid nodes defining the grid cell the vector falls into. $t(e, e')$ subdivides this cell into four rectangles as depicted in Figure 6. Each of the four grid nodes adds to its current score an amount proportional to the area of the opposite rectangle given the total area by 100. Let Score(i, j) be the total score accumulated in grid node (i, j) after probing all $\binom{|P|}{2}$ pattern edges. All local maxima that are greater than a threshold value depending on $|P|$ are considered to correspond to potential matching locations.

Eventually, we can approximate the actual center (i_c, j_c) of the vector cluster stemming from a local maximum at node (i, j) by computing a weighted average of the scores at (i, j) and all scores at neighboring grid nodes:

$$(i_c, j_c) = \frac{\sum_{k=i-1}^{i+1} \sum_{l=j-1}^{j+1} \text{Score}(k, l) \cdot (k, l)}{\sum_{k=i-1}^{i+1} \sum_{l=j-1}^{j+1} \text{Score}(k, l)}$$

Figure 7 illustrates the result of the scoring procedure in the neighborhood of the actual matching position in the target in Figure 2.

3.3 Verifying and Evaluating a Local Matching

After the scoring procedure we are given a list of putative locations of matching pattern centers c. Next we recompute the actual patterns that define the matchings. To this end we consider the bounding box of the pattern P, scale it by length tolerance factor $1 + \lambda$ and for each center c we compute the rectangular sub-image T_c centered at c of the target image with this size.

Next we have a *voting* procedure to compute a partial (λ, α)–matching between P and some pattern P' in T_c. This voting is very similar to the scoring procedure above. We compute the extended history Hist$^*(T_c)$. For each pattern edges $\overline{st}$ in P we search for all (λ, α)–similar edges $\overline{s't'} \in$ Hist$^*(T_c)$. But this time we insert each found spot s' in a *candidate list* for s and t' in a corresponding list for t. We say that s gets a vote from s' and t from t', respectively.

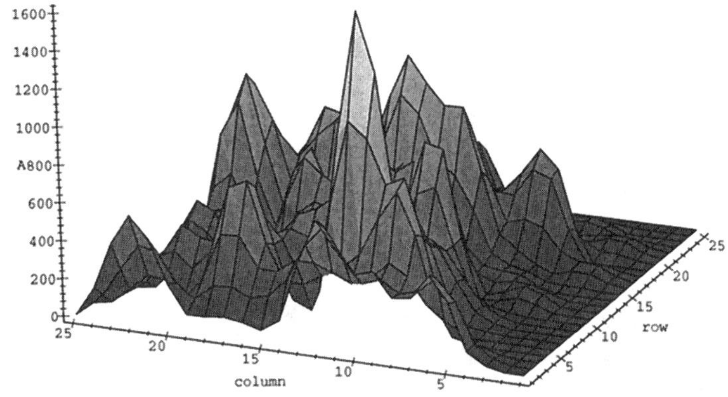

Figure 7: Local result of the scoring procedure

Eventually, we form a tentative partial matching between all pattern spots s and their matching candidates that accumulated a maximum number of votes and exceed a certain threshold depending on the pattern size.

This tentative matching is neither necessarily a 1-1–matching nor a (λ, α)–matching. Such a situation especially occurs if two pattern spots s, t are very close to each other but in the target there is simply only one sufficiently intensive spot in that corresponding place, or there is a spot pair s', t' that violates the slope tolerance. Therefore there is a final *clearing* step that chooses a maximum size sub-matching of the tentative matching that is both 1–1 and meets the tolerance bounds.

How should we evaluate a found local matching? This is clearly application depending. If source and target gel image represent non–comparable samples then the matching cardinality is the only criterion. Otherwise, a ranking that combines both the cardinality and the distance from the expected location is possible. In the latter case we have also the possibility to test the consistency of a local matching result by the following simple iterative method. We accept a local matching for P only if it is confirmed by local matchings for neighboring patterns Q that have nontrivial intersections with P. This idea serves as the basis for a *global matching* algorithm, see below.

3.4 Strongly Distorted Patterns

The local matching algorithm described before is robust to the respect that it indeed computes for each pattern a list of partial matchings within the prescribed tolerance bounds. Another problem is how to proceed in the case that there is a more severe geometric distortion between the pattern P and its counterpart P' in the target image. Obviously, one solution would be to increase the values for slope and length tolerances yielding increased time bounds for the geometric hashing process. What we are doing instead is a standard heuristic trick, compare [17]; we distort the pattern P iteratively in a typical (application depending) way and then we search for the distorted pattern while keeping similarity tolerances small. In our case these distortions are combinations of independent $x - y$–scalings and shifts that transform rectangular regions into parallelograms. Let us denote by Distort(P) the list of all patterns derived from P that way.

The other more elegant way is the following. We consider the 4–dimensional transformation space of translations and x– resp. y–scalings. Given upper bounds on the single coordinates we can implement a scoring procedure that accumulates scores in nodes of a sufficiently refined 4–dimensional grid for each pair of similar edges.

3.5 Estimating the Time Complexity

First of all, we have to remark that the results of the following worst case analysis do not completely reflect the performance of the algorithms for real world input data. However, this analysis is useful in order to compare our algorithm with previous approaches and to make clear where the progress comes from.

Let k and n denote the sizes of the pattern P and of the target point set T.

The running time of the alignment method is the product of the number $sep(P,T)$ of (λ, α)–similar edge pairs and the costs $m_t(P,T)$ to compute a matching under a translation t induced by such an edge pair. In the general setting we have a trivial but tight worst case upper bound of $sep(P,T) = \mathcal{O}(k^2n^2)$, compare also page 153ff. in [1], and $m_t(P,T) = \mathcal{O}(k \log n)$. Switching to the extended Delaunay history Hist* we get an expected value of $sep(P,T) = \mathcal{O}(k^2n)$.

This implies a total expected upper bound of $\mathcal{O}(k^3 n \log n)$ which was obtained in [18]. Our geometric hashing variant (i. e., the voting procedure) allows to remove the $k \log n$ factor and to replace it by some additive terms. The general upper bound for geometric hashing of $\mathcal{O}(n^3)$ is also very bad, even without taking into account the preprocessing and final computation of the matching.
In contrast, our voting procedure requires $\mathcal{O}(n \log n) +$ $+\mathcal{O}(k^2 n) + \mathcal{O}(|G|)$ time, where the first term represents the costs of the incremental Delaunay triangulation, while the second one bounds the number of similar edge pairs. In the third term $|G|$ denotes the constant size of the grid used to store the votes. It is of smaller order and can be ignored. Finally, computing the C best matchings costs $C \cdot m_t(P, T)$. Thus, altogether for computing the best matchings for one given pattern we achieve an expected $\mathcal{O}(n(\log n + k^2))$ upper bound which has to be multiplied by the number of patterns in Distort(P).

3.6 Global Matching via Local Matching

Previous algorithms for the global matchings of gel images are based mainly on landmarks set by the user. In this context a landmark is a pair of points, one in the source image S, the other one in the target T. Thus, the user fixes a partial matching on a sufficiently large set of so called support points $S_{supp} \subset S$. Triangulating S_{supp} and constructing the corresponding triangulation in the target image one gets a piecewise affine transformation f defined on the triangles. Finally, for any source point $p \in S$ one has to search for the nearest neighbor of $f(p)$ in T where the distance is a combination of the euclidian distance and the intensity difference.
Our aim is to avoid interactive landmark setting by making use of the local matching solution. The problem is that this algorithm computes several matchings of a chosen pattern and in general it is not clear which is the right one. There are two approaches to improve the confidence in a found matching. The first idea is to insert a spot $p \in S$ into different patterns $P_1, \ldots, P_m$ and to compute their best local matchings independently. We considered four patterns extending a bounding box from p into the four quadrants. Then a point $q \in T$ will be accepted as an image of p if for all patterns there is at least one (of the best) matching mapping p to q. It turns out that answers found this way are sufficiently satisfying. However, due to the strong restrictions within this approach it may happen that no answer will be replied.
The second approach consists in covering the source image by patterns in a grid like fashion. Computing the best local matchings for all patterns one has to look for a consistent choice of matchings. Let P_1 and P_2 be neighboring patterns in S. A matching of P_1 to P_1' in T will be called consistent with a matching of P_2 to P_2' if for their centers $c(P)$ holds that the edges $\overline{c(P_1), c(P_2)}$ and $\overline{c(P_1'), c(P_2')}$ are (λ', α')-similar. Here, we have to enlarge the tolerance bounds to reflect the fact that for each single pattern P_i the matching can stem from the list Distort(P_i).

3.7 Implementation and User Interface

The matching algorithms have been implemented and are part of the Carol software system [3]. It has essentially two parts:
The first part, the combinatorial and geometrical kernel of the matching algorithms, has been implemented in C++. It makes essential use of the Standard Template Library (STL) and of the Computational Geometry Algorithms Library (CGAL), [6]. The latter library provides several geometric data structures and functions and especially an implementation of the incremental Delaunay triangulation. The second part of the Carol system is the graphical user interface which has been implemented in Java. It can be run as an applet started out of an internet browser or as an application. The communication with the algorithmical program part is established via internet sockets, whereby the C++-program works as a server which waits for matching requests from the Java-client, performs the computation and sends eventually back the results to the client. The program will be eligible to match gel images from databases all over the internet. This feature is strongly supported by the possibility to run the user interface as an applet and furthermore by the client-server architecture of the program.

The user has the possibility to set parameters like tolerance bounds, pattern size etc., for more details of the user interface see *http://gelmatching.inf.fu–berlin.de*.
An unavoidable and critical preprocessing step is the spot detection stage. It is planned to include into the Carol system a spot detection algorithm that has been recently developed at Deutsches Herzzentrum Berlin, see [15].

The local matching algorithm run on a Sun Sparc Ultra 1 computes the best 9 matchings for a pattern of 8 spots in about 3 seconds including the preprocessing of a 3000 spot target image. Each further pattern in the list Distort(P) increases this time on the average by about 0.3 seconds.

4 Conclusions and Directions for Further Work

We have presented the underlying ideas for an algorithmic solution of the local matching problem of 2D patterns of protein spots in electrophoresis images. Its main features are:

1. Local matchings for a source pattern are found in the target image without knowledge of its context.

2. The local matching algorithm works for locally intensive patterns. There are standard techniques like point location combined with affine approximation and nearest neighbor search that extend the solution to other spots.

3. The local matching algorithm can be used as a basic step for the global matching problem for gel images. In fact, local matching is used then like landmark setting.

4. The central idea for the algorithm stems from the use of the extended history of the incremental Delaunay triangulation, which proved to be a suitable structure for the local matching problem because of its expected linear size and its robustness in the presence of noise.

There are several issues for theoretical investigations raised by our approach. One topic for further work is certainly the analysis of the 'local' history of a random incremental Delaunay triangulation and its dependency on noise.
Last, but not least, there is the question for other applications of our local matching algorithm which uses only a minimum of specific application knowledge.

Acknowledgement: We would like to thank Klaus–Peter Pleißner from Deutsches Herzzentrum Berlin, Helmut Alt, Christian Knauer and Sven Schönherr from Freie Universität Berlin for their generous help.

References

[1] P. Agarwal, J. Pach, Combinatorial Geometry, John Wiley 1995

[2] H. Alt, L. Guibas, Discrete Geometric Shapes: Matching, Interpolation, and Approximation, A Survey, TR B 96-11, Inst. f. Informatik, Freie Universität Berlin, to appear in J. Urrutia, J.–R. Sack, eds., Handbook for Computational Geometry (North Holland)

[3] H. Alt, F. Hoffmann, K. Kriegel, C. Wenk, K.-P. Pleißner, CAROL - New Algorithmic Tools for Comparing Two–Dimensional Electrophoretic Gel Images, Proceedings Electrophorese Forum '97, Strasbourg, P21

[4] R. Appel, J. Vargas, P. Palagi, D. Walther, D. Hochstrasser, Melanie II - a third generation software package for analysis of two–dimensional electrophoresis images: II. Algorithms, Electrophoresis 18 (1997), 2735–2748

[5] M. de Berg, M. van Kreveld, M. Overmars, O. Schwarzkopf, Computational Geometry: Algorithms and Applications, Springer Verlag, 1997

[6] CGAL, The Computational Geometry Algorithms Library, http://www.cs.ruu.nl/CGAL

[7] S.–H. Chang, F.–H. Cheng, W.–H. Hsu, G.–Z. Wu, Fast Algorithm for Point Pattern Matching: Invariant to Translation, Rotations and Scale Changes, Pattern Recognition 30(2) (1997), 311–320

[8] F.–H. Cheng, Point Pattern Matching Algorithm Invariant to Geometrical Transformation and Distortion, Pattern Recognition Letters 17 (1997), 1429–1435

[9] L. Guibas, D. Knuth and M. Sharir, Randomized Incremental Construction of Delaunay and Voronoi Diagrams, Algorithmica 7(4) (1992) 381–413

[10] D. P. Huttenlocher, G. Klanderman and W Rucklidge, Comparing images using the Hausdorff distance, IEEE Trans. Pattern Analysis and Machine Intelligence 15 (1993) 850–863

[11] A. Finch, R. Wilson, E. Hancock, Matching Delaunay Graphs, Pattern Recognition 30(1) (1997), 123–140

[12] P. F. Lembkin, Comparing two–dimensional electrophoretic gel images across the Internet, Electrophoresis 18(3–4) (1997), 461–470

[13] H. Ogawa, Labeled Point Pattern Matching by Delaunay Triangulation and Maximal Cliques, Pattern Recognition 19(1) (1986), 35–40

[14] K.–P. Pleissner, V. Regitz–Zagrosek, C. Weise, M. Neuss, B. Kruedewagen, P. Soeding, K. Buchner, F. Hucho, A. Hildebrandt, E. Fleck, Chamber–specific expression of human myocardial proteins detected by two–dimensional gel electrophoresis, Electrophoresis 16 (1995) 841–850

[15] K.–P. Pleissner, A. Sahlström, S. Wegner, H. Oswald, E. Fleck, Protein spot detection in WWW–accessible 2–D gel images using the Watershed Transformation, Proceedings Electrophorese Forum '97, Strasbourg, P20

[16] R. Seidel, Backwards Analysis of Randomized Geometric Constructions, in J. Pach, ed., New Trends in Discrete and Computational Geometry (Springer Verlag), 1993

[17] M. Sonka, V. Hlavac, R. Boyle, Image Processing, Analysis and Machine Vision, Chapman and Hall, 1994

[18] G. Weber, L. Knipping and H. Alt, Point Pattern Matching in Astronautics, J.Symbolic Computation 17 (1994) 321–340

[19] M. R. Wilkins, K. L. Williams, R. D. Appel, D. F. Hochstrasser (eds.) Proteome Research: New Frontiers in Functional Genomics, Springer-Verlag Berlin Heidelberg New York, 1997

Multiresolution Banded Refinement to Accelerate Surface Reconstruction from Polygons

James D. Fix* Richard E. Ladner†

Abstract

We propose a method for constructing a tiling between a pair of planar polygons. Our technique uses multiresolution: tilings of lower resolution polygons are used to construct a tiling for the full resolution polygons. The tilings are constructed using *banded refinement*, by restricted dynamic programming, in roughly linear time and space. By contrast, the optimal dynamic programming method requires quadratic time and space. In our empirical study of surface reconstruction of brain contours our algorithm exhibited significant speedup over the optimal dynamic program, yet nearly always found an optimal reconstruction. Our approach appears to be generalizable to other geometric problems solvable by dynamic programming, and flexible enough to be tuned for varying data set characteristics.

1 Introduction

A typical subproblem of many surface reconstruction algorithms is the fitting of a surface to a set of parallel contours. This problem arises, for example, in medical visualization applications, where organ outlines are culled from cross-sectional images, with the goal of constructing a three dimensional model of the organ from those contours. Typically, the outlines from a cross-sectional slice are modeled as polygons, and the reconstruction is modeled as a polyhedral surface that interpolates the polygonal slices (see, for example, [4, 5, 2, 16]). The surface is chosen to be piecewise linear and composed of triangular tiles. Figures 9 and 10 give an example of such a reconstruction from contours of a portion of a human brain.

In the simplest formulation of the problem, there is only one contour per slice, and the reconstruction method independently computes a tiling between adjacent contour pairs [12, 9, 14]. The reconstructed surface is topologically equivalent to a cylinder. In this case, surface reconstruction from contours is reduced to the *polygon tiling problem*: find a surface composed of triangular tiles between a pair of planar polygons subject to some optimization criteria. Finding a polygon tiling is the main focus of this paper.

Fuchs, Kedem, and Uselton [9] devised a dynamic programming algorithm that computes an optimal tiling in $O(n^2 \log n)$ time and $O(n^2)$ space using minimum surface area as their criterion. Their optimal method can be too costly when the polygons consist of many vertices, especially when used in interactive applications. As a result, faster methods were proposed that use greedy heuristics to find approximate solutions [7, 10]. While extremely efficient, the surfaces found by the heuristic methods are often unsatisfactory, even on simple inputs [15]. Meyers [14] proposed a multiresolution method that uses an optimal tiling of low resolution approximations of the polygons to guide a search for an optimal tiling. Based on local search, it finds near-optimal tilings quickly but often fails to find optimal ones (see figure 6).

We propose an approximation methodology for computing a polygon tiling using multiresolution techniques that is based on the dynamic programming formulation of Fuchs et al. [9]. The method proceeds as follows: we model the inputs as piecewise linear curves and perform a wavelet decomposition. This yields a sequence of approximations of decreasing resolution, hence fewer sample points. We compute an optimal tiling of the lowest resolution curves using dynamic programming. This low resolution tiling is used to create a tiling for the next higher resolution polygon. The tiling is further refined using a restricted dynamic program. We call the restricted dynamic program *banded refinement*:

*Department of Computer Science and Engineering, University of Washington, Seattle, WA 98195, fix@cs.washington.edu.

†Department of Computer Science and Engineering, University of Washington, Seattle, WA 98195, ladner@cs.washington.edu.

SCG 98 Minneapolis Minnesota USA

Figure 1: Multiresolution tiling. From left to right: the original polygons at full resolution; a low resolution tiling; an intermediate resolution tiling; and a full resolution tiling.

a bound b (called the *band width*) is placed on how far the dynamic program searches from the low resolution solution. Repeating this process yields a tiling for the original, full-resolution polygons. The tilings in figure 1 are snapshots of this process.

The main advantage of our method is its time and space complexity. For example, when b is fixed for each resolution level, our method uses time proportional to $O(bn \log b + \tilde{n}^2 \log \tilde{n})$ where $\tilde{n} \leq n$ is the size of the low resolution approximations. In a typical application, $\tilde{n} = O(\sqrt{n/\log n})$ and b is a constant, yielding a linear time approximation algorithm. The tradeoff is that the method is not guaranteed to give the optimal tiling. We provide empirical results that show that our method almost always finds the optimal tiling using a small band width. For example, with $b = 2$ in our empirical study nearly all tilings found were within one percent of the optimal surface area and at least eighty percent were optimal.

1.1 Related Work

Multiresolution analysis has had a strong impact on computer algorithms, most notably in the areas of signal processing and compression, numerical methods, and computer graphics [8, 20]. Used for compression and function approximation, multiresolution analyses and wavelet decompositions yield compact representations that mostly characterize the original functions. When applied to numerical optimization problems, multiresolution methods find solutions quickly by providing a coarse solution which can then be refined with higher levels of detail [13].

Our method is a synthesis of the optimal tiling algorithm of Fuchs et al. [9] and the multiresolution tiling method of Meyers [14]. Meyers' tiling algorithm was one of the first compelling applications of multiresolution and wavelet analysis to computational geometry and computer graphics. Like our method, his method computed tilings from lower resolution tilings, but employed a local search heuristic to refine the tilings. Section 6 compares this method to our tiling algorithm: our method yields better quality tilings with only a slight performance penalty.

Much work has been done on solving the general problem of surface reconstruction from contours. In the general case, a surface my have holes or branches, and thus the cross sections might contain multiple contours. While our tiling method does not solve the general problem, many methods have been developed that use polygon tiling as a subcomponent. [16, 7, 10, 19]. These methods first find a correspondence between the polygons of adjacent layers. Structures are added that bridge merging or branching contours so that they form one polygon and can be tiled as a single pair of polygons. Our polygon tiling algorithm can be invoked to perform this tiling. Meyers et al. [16] provide a good survey of these methods.

When a pair of polygons to be tiled is dissimilar, the surface computed by polygon tiling is often of poor quality, and many of the above techniques fail to generate good solutions [2]. Other general techniques exist that do not rely on solution of the polygon tiling problem. Boissonnat [5], and Boissonnat and Geiger [6] compute a tiling by Delauney triangulation. Barequet and Sharir [4] and Barequet et al. [3] use curve similarity techniques to tile portions of contours, then decompose the remaining untiled portions into three-dimensional polygons and tile them using polygon triangulation. This methodology is followed by Bajaj et al. [2], providing a more rigorous formulation and techniques for handling dissimilar contours. While we have not developed this idea, multiresolution techniques similar to ours might be applied to accelerate the matching and polygon triangulation portions of these algorithms.

There are many problems that are closely related to the polygon tiling problem as we formulate it here. Most similar are those that arise in curve comparison applications: finding figure correspondences for polygon morphing [18], matching curve features [23, 11], dynamic time-warping for handwriting and speech recog-

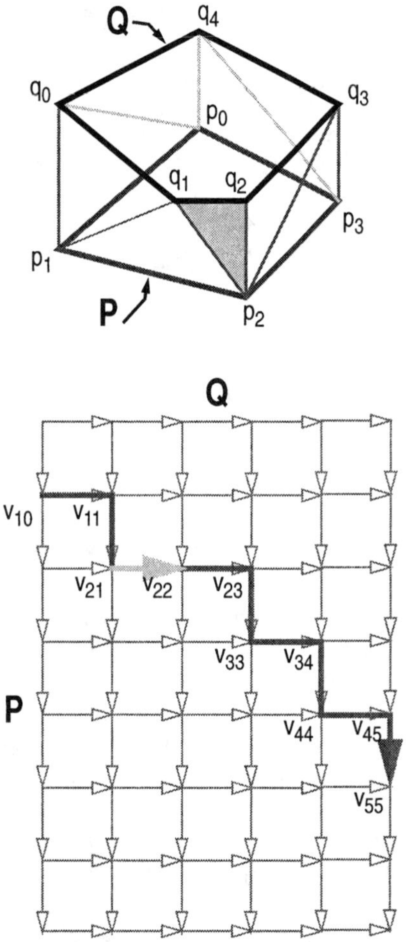

Figure 2: A tiling of two polygons and its associated tiling path in the tiling graph. The shaded triangle corresponds to the highlighted edge in the graph.

nition [17]. All of these can be formulated and solved using dynamic programming— there seems to be great potential in generalizing our method to accelerate their solutions as well. The formulation of Alt and Godau for determining whether two polygonal curves resemble each other shares some features with these problems, and might also benefit from a multiresolution approach [1].

1.2 Paper Overview

We begin by defining the parallel polygon tiling problem, and describe its solution using dynamic programming. In section 3 we describe multiresolution analysis of polygons, and motivate its application to the polygon tiling problem by giving a high-level overview of our algorithm. Section 4 presents our main contribution: refinement of a tiling by banded dynamic programming. In section 6 we present empirical evidence of the efficacy of our method, by applying it to a complex medical imaging data set.

2 Optimal Polygon Tiling

A typical subproblem of many surface reconstruction algorithms is the fitting of a surface to two parallel contours represented as polygons. Our discussion of the polygon tiling problem and its optimal solution follows that of Fuchs et al. [9].

Let $P = \{p_0, \ldots, p_{m-1}\}$ and $Q = \{q_0, \ldots, q_{n-1}\}$ be polygons in R^3 lying in parallel planes. A *triangular tile* of P and Q is a triple of the form (p_i, q_j, p_{i+1}) or of the form (p_i, q_j, q_{j+1}). A *tiling* of P and Q is a sequence of triangular tiles $T = \{t_0, t_1, \ldots, t_{n+m-1}\}$ with the following property: for each tile t_k, if $t_k = (p_i, q_j, p_{i+1})$ then $t_{k+1} = (p_{i+1}, q_j, r)$ for some r and if $t_k = (p_i, q_j, q_{j+1})$ then $t_{k+1} = (p_i, q_{j+1}, r)$ for some r.[1] A polygonal tiling can be viewed as a correspondence between vertices of polygons P and Q with sequences of edges of P and Q that respects the ordering of their vertices and edges. Figure 2 shows a legal tiling of two polygons.

For any triangle $t \in T$ define its weight $w(t) \in \mathcal{R}$, which represents the cost of including that triangle in T. The cost of a tiling is then

$$w(T) = \sum_{i=0}^{m+n-1} w(t_i).$$

Given polygons P and Q, the polygonal tiling problem is to find a tiling T^* of P and Q where

$$w(T^*) = \min\{w(T) \mid T \text{ is a tiling of } P \text{ and } Q\}.$$

For example, a natural cost measure for $w(t)$ is the surface area of triangle t, and so T^* would be the tiling with minimum surface area.

A minimum tiling of P and Q can be computed by considering an equivalent graph problem. Given the two polygons given above, consider the *tiling graph* $G(P, Q) = (V, E)$ where

$$\begin{aligned} V &= \{v_{ij} \mid 0 \le i < 2m \text{ and } 0 \le j \le n\} \\ E &= \{(v_{ij}, v_{i+1,j}) \mid 0 \le i < 2m - 1\} \\ &\quad \cup \{(v_{ij}, v_{i,j+1}) \mid 0 \le j < n\}. \end{aligned}$$

Each graph vertex v_{ij} corresponds to a possible spanning edge from point p_i to q_j in a tiling of P and Q. The graph edges correspond to possible tiles. For example the graph edge $(v_{ij}, v_{i+1,j})$ corresponds to the tile (p_i, q_j, p_{i+1}) and $(v_{ij}, v_{i,j+1})$ corresponds to the tile (p_i, q_j, q_{j+1}).[2] Because of this correspondence, a directed path from a vertex v_{i0} to $v_{i+m,n}$ corresponds to a

[1] For this and the remainder of the paper, the indices of points in P and Q and the indices of tiles in a tiling T are assumed to be taken modulo m,n, and $m + n$, respectively.

[2] Note that the indices of vertices v_{ij} should not be taken modulo m and n.

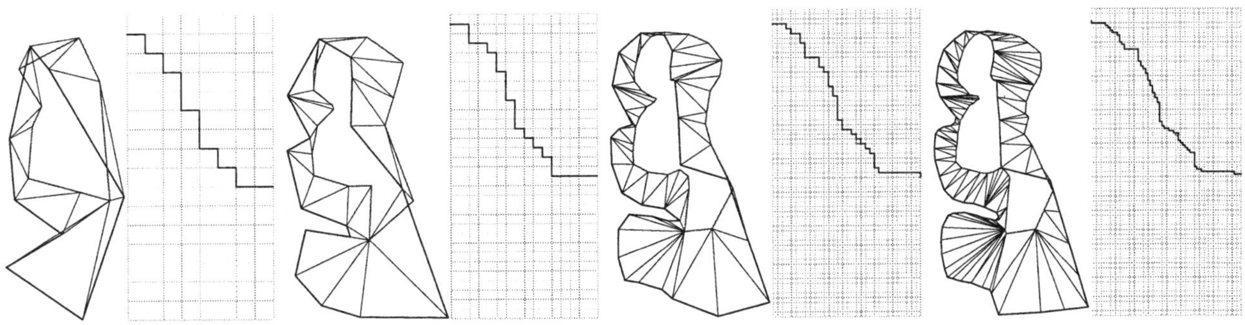

Figure 3: A sequence of tilings and their tiling paths at different resolutions.

tiling of P and Q. The lower half of figure 2 shows the equivalent tiling graph for the polygons in the figure. The highlighted edge of the tiling graph corresponds to the shaded tile of the tiling, and the darkened path corresponds to the tiling. There is some redundancy in the graph representation, for example edges $(v_{ij}, v_{i,j+1})$ and $(v_{i+m,j}, v_{i+m,j+1})$ represent the same tile. The graph is an unfolding of a toroidal graph that more directly represents the problem. The unfolded representation is easier to manipulate in the implementation of the algorithms and will be useful later when we discuss banded refinement in section 4.

If we associate weights to edges equal to the weights of the corresponding tiles, a minimum tiling can be found by finding the shortest path between vertices v_{i0} and $v_{i+m,n}$ over all $0 \leq i < m$. Using a straightforward dynamic program, this minimal path can be found in $O(m^2 n)$ time and $O(mn)$ space. By some clever observations about optimal paths in $G(P, Q)$, Fuchs, Kedem, and Uselton discovered a dynamic programming strategy that used only $O(mn \log m)$ time [9].

3 Multiresolution Tiling

Our tiling method hinges on the ability to construct approximations of polygons, and construct a tiling based on the tilings of those approximations.

3.1 Multiresolution Analysis of Polygons

Multiresolution analysis provides a fast and effective way for computing approximations of a polygon, where the points of a lower resolution polygon are computed by taking linear combinations of the points of the next higher resolution polygon. As an example, given polygon P, consider the polygon whose vertices are the midpoints of every other edge of P. Continuing this yields a trivial multiresolution analysis of P.

This scheme can be generalized: let a be a finite sequence $\{a_{-m}, \ldots, a_{-1}, a_0, a_1, \ldots, a_m\} \subset \mathcal{R}$, let k be a positive integer and assume polygon P has n points where n is divisible 2^k. Our goal is to compute a sequence of approximations $P_k, P_{k-1}, \ldots, P_0$ where each polygon P_l at *resolution* l has $n_l = \frac{n}{2^{k-l}}$ vertices. At the highest resolution k, let $P_k = P$. Let the vertices of each polygon be $\{p_0^l, p_1^l, \ldots, p_{n_l}^l\}$. Then, given P_{l+1}, the next lower resolution polygon P_l is given by:

$$p_i^l = \sum_{j=-m}^{m} a_j p_{2i+j}^{l+1}$$

where $0 \leq i < n_l$. The sequence $P_0, P_1, \ldots, P_k$ is a *multiresolution analysis* of P, and the sequence a is called the *analysis filter*. For example, in the "midpoint" multiresolution scheme above, a would be defined by: $\{a_{-1} = 0, a_0 = 1/2, a_1 = 1/2\}$. Sweldens and Schröder [21] describe techniques for designing analysis filters a for various applications. For polygons, we use the analysis filter of Meyers: $\{a_0 = 19/28, a_{\pm 1} = 9/28, a_{\pm 2} = -1/8, a_{\pm 3} = -1/14, a_{\pm 4} = 1/28\}$ [14]. This was chosen so that the polygon P_l is nearly a least squares approximation of P_{l+1}. Example approximations given by this scheme can be seen in the polygons of figure 1.

Performing a multiresolution analysis of P is very efficient. Since each polygon in the sequence has half the number of points than the next higher resolution, the total time to compute the entire sequence takes only $O(n)$ time. Some applications need also maintain the error of the approximations in the form of *wavelet* or *detail* coefficients. In this case, a second filter is used and the process is called a *wavelet decomposition*. Our current strategy does not use the error.

3.2 Algorithm Overview

Given this multiresolution analysis framework, consider the following question: what is the relationship between an optimal tiling of a pair of polygons and an optimal tiling of a lower resolution approximation of those polygons? Figure 3 gives a sequence of four optimal tilings, starting with one for a low resolution approximation on the left and ending with the full resolution one on the right. The corresponding paths of the tilings are shown beside the tilings. The similarity of the paths suggests that, for reasonably well-behaved data, optimal tilings

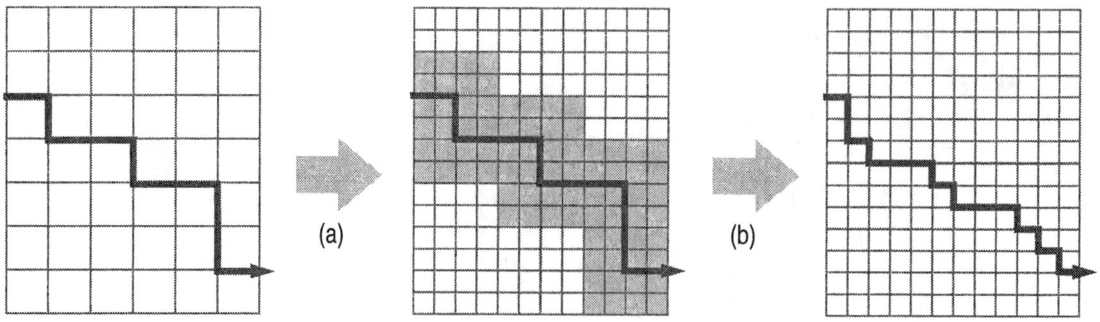

Figure 4: (a) Raising the resolution of a tiling by splitting tile path edges. (b) Refining a tiling by using banded refinement with band width $b = 2$.

at lower resolutions could be useful in finding tilings of higher resolution polygons. In our multiresolution tiling algorithm, we use the path of a lower resolution tiling to guide dynamic programming to construct a higher resolution tiling.

Meyers [14] multiresolution tiling algorithm provides a framework for our technique. The algorithm works as follows: let P and Q be polygons with m and n points, and let k be a non-negative integer. Assume that m and n are divisible by 2^k. First, perform a multiresolution analysis of P and Q yielding approximations $P_0, P_1, \ldots, P_{k-1}$ and $Q_0, Q_1, \ldots, Q_{k-1}$, respectively. Using the optimal tiling algorithm, compute a minimum tiling T_0 for P_0 and Q_0. Generally for $l > 0$, given a tiling T_{l-1} of P_{l-1} and Q_{l-1}, first construct a tiling U_l by simple process of raising the resolution of the tiling T_{l-1}. Then starting with the tiling U_l and the polygons P_l and Q_l refine U_l to a tiling T_l. The final tiling is T_k. This algorithm is summarized below:

```
MULTITILE(P,Q)
    compute multiresolution analysis P_0, P_1, ... P_k of
        polygon P = P_k
    compute multiresolution analysis Q_0, Q_1, ... Q_k of
        polygon Q = Q_k
    T_0 ← OPTIMALTILE(P_0, Q_0)
    for l ← 1 to k do
        U_l ← RAISETILING(T_{l-1})
        T_l ← REFINETILING(U_l,P_l,Q_l)
    return T_k;
```

Here, the procedure OPTIMALTILE computes the optimal tile using the optimal dynamic programming algorithm, the procedure RAISETILING computes a tiling of a polygons P_l, Q_l based on the tiling of P_{l-1}, Q_{l-1}, and procedure REFINETILING attempts to improve its given tiling. The latter procedure is described in detail in the next section.

The procedure RAISETILING is a very simple process. Suppose we have the tiling T_{l-1} of polygons P_{l-1} and Q_{l-1}. By the multiresolution analysis described above, the l-th resolution polygons P_l and Q_l will each have twice the number of vertices. We can view this resolution increase as a splitting of the edges of each polygon with single vertices, and adjustment of the positions of all the vertices due to the contribution of the wavelets at that resolution. Meyers gave a straightforward way of tiling P_l and Q_l by splitting each tile of T_{l-1} [14]. For example, let superscripts of p_i and q_j denote the resolution levels of polygon vertices as before. If $(p_i^{l-1}, q_j^{l-1}, p_{i+1}^{l-1})$ is a tile in T_{l-1}, then $(p_{2i}^l, q_{2j}^l, p_{2i+1}^l)$ and $(p_{2i+1}^l, q_{2j}^l, p_{2i+2}^l)$ are tiles in the tiling produced by RAISETILING(T_{l-1}). In the tiling graph, this corresponds to splitting the edges of the corresponding solution path in the tiling graph. Thus, if graph edge $(v_{ij}, v_{i+1,j})$ is in the tiling path for T_{l-1} the raised tiling's associated path would have graph edges $(v_{2i,2j}, v_{2i+1,2j})$ and $(v_{2i+1,2j}, v_{2i+2,2j})$. This is illustrated in figure 4a.

4 Banded Refinement

The key to the multiresolution approach is the ability to construct good solutions based on a solution to a lower resolution version of the problem. For the polygon tiling method presented so far, the RAISETILING step is not satisfactory, as it is oblivious to the higher resolution polygons it is tiling. The REFINETILING step remedies this. One approach to refinement would be to make local improvements to a tiling— for example, one could apply "edge flips" (see figure 5a) that improve the tiling, until no improvement can be made. This was the method applied by Meyers [14]. Figure 5b illustrates the weakness of this technique: it is possible for a tiling to be close to an optimal tiling, yet have an edge flip operation lead the refinement to a sub-optimal tiling that is locally optimal with respect to the edge flip operation.

Our method of refining a tiling is based on the observation that the optimal tiling in the preceding example could have been found by searching among all tiling paths near the current tiling path. Let U be a tiling of polygons P and Q. Let V be the set of vertices of tiling graph $G(P, Q)$, and $S \subset V$ be the vertices on the tiling

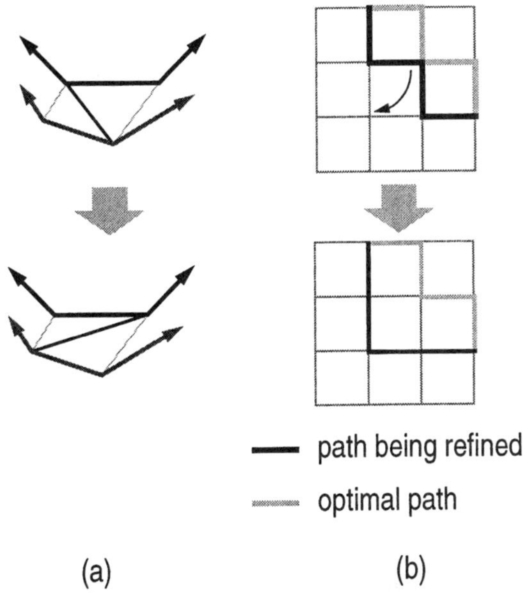

Figure 5: (a) Refining a tiling by an edge flip. (b) Local search using edge flips leading to a sub-optimal path.

path corresponding to tiling U. Let b be a positive integer. Define $R_b(S)$ to be the set

$$\{v_{ij} \in V \mid |i - i'| \leq b, |j - j'| \leq b \text{ for some } v_{i'j'} \in S\}.$$

$R_b(S)$ defines a banded region around S in the tiling graph. This is illustrated for a band width $b = 2$ for the tiling path in figure 4b: the vertices in $R_2(S)$ for S on that path are those on or within the shaded region. Our procedure REFINETILING given U, computes the shortest path S' among all paths starting from a vertex of the form $v_{i0} \in R_b(S)$ and ending at a vertex $v_{i+m,n} \in R_b(S)$ in the tiling graph $G(P, Q)$ restricted to $R_b(S)$, and returns tiling T corresponding to S'. We call b the *band width* of the refinement. The refined tiling T is not guaranteed to be an optimal tiling of P and Q. However, if U is constructed from near-optimal tilings of lower resolution approximations of P and Q, the hope is that T will also be close to optimal.

5 Odd polygon sizes

As we've described them so far, the multiresolution analysis and RAISETILING portions of our algorithm rely strongly on the number of vertices in each polygon being divisible by 2^k, where k is the number of resolution levels. We have a simple procedure for handling a polygon with an odd number of vertices: we find the longest polygon edge from an *even-indexed* vertex and add its midpoint. With an even number of a vertices, we can reduce resolution of the polygon by one level. At most one vertex is introduced at each resolution level, and the additional points are only required for lower resolutions. They are "removed" during the raise-refine steps. When RAISETILING is applied to tiling T_{l-1}, and at least one of P_l or Q_l contains an odd number of vertices, then a tile of T_{l-1} will not have to be split to generate U_l. For example, if $t = (p_i^{l-1}, q_j^{l-1}, p_{i+1}^{l-1})$ is a tile of T_{l-1} and edge (p_{2i}^l, p_{2i+1}^l) was split to compute polygon P_{l-1} from P_l, then t would induce only a single tile $(p_{2i}^l, q_{2j}^l, p_{2i+1}^l)$ in U_l.

This method for handling odd numbers of vertices is efficient, only requiring an additional linear time overhead during the polygon analysis phase, and is effective for finding tilings. It is not standard multiresolution analysis, and might not be appropriate in other applications.

6 Discussion and results

There are two parameters governing our multiresolution tiling algorithm— the band width b and the number of resolution levels k— and these parameters allow us to tune the algorithm in the quality of the tilings produced and the performance cost in space and running time. We demonstrate these tradeoffs in the experiments below, but first we consider the running time of the algorithm REFINETILING. Our analysis assumes that b is fixed at the refinement at each resolution level. This is only a simplification, as numerous schemes could be devised that vary the bandwidth, for example, as a function of the resolution level, or the approximation error.

Let n be the size of the polygons given as input to MULTITILE. The optimal tiling of the low resolution polygons can be computed in $O(n^2\, 2^{-2k} \log(n2^{-k}))$ time. At each resolution l, RAISETILING uses $O(n2^{l-k})$ time and REFINETILING uses $O(bn\, 2^{l-k} \log b)$ time, so in total the raise-refine steps use $O(bn \log b)$ time. If we choose k so that $n^2 2^{-2k} \log(n2^{-k}) = O(n)$, then the total running time is $O(bn \log b)$. With this choice of k, the algorithm uses only $O(bn)$ space.

6.1 Empirical results

We applied multiresolution tiling methods to medical imaging data from the University of Washington Digital Anatomist project. This data set consists of 121 parallel cross-sections of the frontal lobe portion of a human brain cortex, ranging in complexity from 58 to 316 points. The contours form a single tube, thus there is only one polygon contour per slice. These contours are a significant challenge to the tiling algorithms because of the intricacy of the folds on the cortex surface. We compared the three tiling methods described in this paper:

Banded: our multiresolution tiling algorithm that uses banded refinement. The band width was fixed at some value b at all resolution levels, and the number of resolution levels k was $logn/2$.

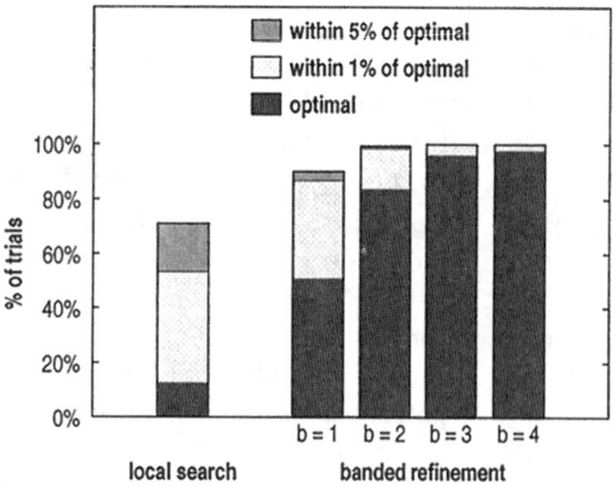

Figure 6: Comparison of the quality of tilings found by multiresolution tiling using banded and local search against the optimal.

Optimal: our implementation of the optimal dynamic programming tiling algorithm [9].

Local search: our implementation of the multiresolution algorithm that refines using local search by edge flip operations [14].

Figures 9 and 10 show the tilings that resulted from applying the banded method with $b = 8$ to these data sets.

Figure 6 shows the quality of the tilings produced by the tiling algorithms: for local search (the first bar) and for banded refinement using various choices of the band width parameter b (the remaining bars). The height of each bar is the percentage of tilings produced by the algorithm that were within 5% of optimal. Each bar is then subdivided into the percentage of tilings that were optimal, within 1% of optimal, and within 5% of optimal. The results clearly indicate that the band width refinement methods are superior to local search. The local search method found an optimal tiling in less than 13% of the trials. By using a band width of only $b = 2$, the banded refinement algorithm finds an optimal solution for 83% of the trials, all tilings were within 5% of the optimal tiling, and nearly all were within 1% of the optimal tiling. Using a band width $b = 4$, all but 2.5% were optimal. These are corrected by using a band width $b = 8$.

Figure 7 shows the speedups achieved by the multiresolution algorithms over the optimal algorithm as a function of problem size (the total number of vertices in both polygons). There is significant variability in the local search results since the time for a local search is highly data dependent. The running time for the optimal and banded methods are mostly insensitive to polygon shape. The multiresolution algorithms are significantly faster than the optimal algorithm, achieving linear speedups. In addition, the low band width variants

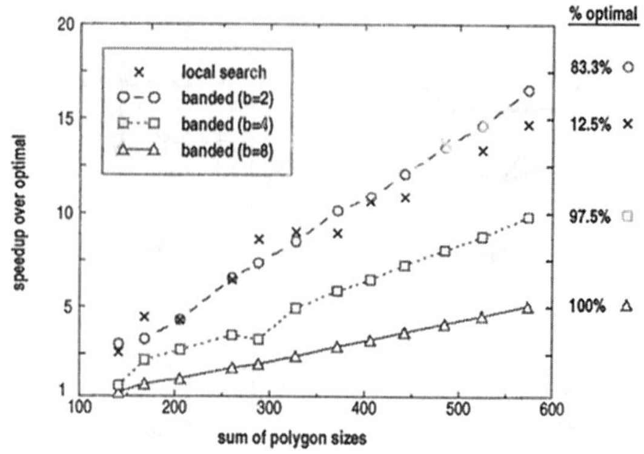

Figure 7: Running time speedup of the multiresolution methods over the optimal meth

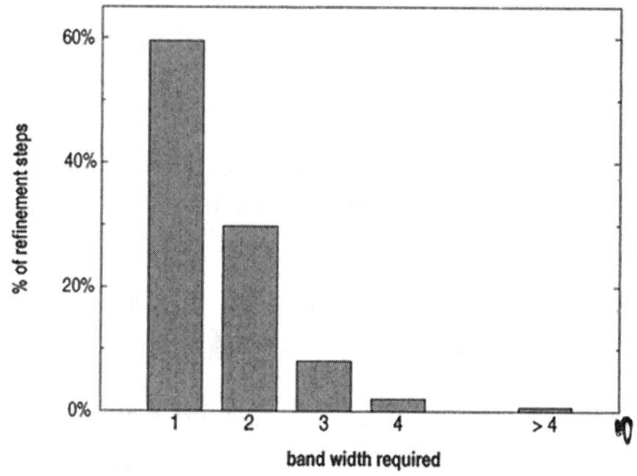

Figure 8: Distribution of band widths required to maintain optimality in a banded refinement.

of the multiresolution algorithm are competitive with the local search variant. To the right of the graph is a summary of figure 6, showing the percentage of the data set for which optimal tilings were found by each method. For polygon pairs where the number of vertices in each was around 300 points, both local search and banded refinement with $b = 2$ were about fifteen times faster than the optimal method, however the banded method found the optimal tiling nearly seven times more frequently than local search.

An outstanding issue is how one chooses the band width parameter b when applying the multiresolution banded method to a new data set. Ideally, one would like to have the algorithm automatically determine the appropriate band width from the input. As a preliminary experiment towards this goal, we performed the following: given two polygons' approximations at successive resolutions, compute the optimal tiling at each resolution and determine the minimum band width necessary for a raise-refine step to find the higher resolution tiling from the lower. Figure 8 summarizes these results, giving the distribution of band widths over every raise-refine step when applied to the brain data set. These results complement figure 6, suggesting that band widths of four or less are adequate at least 99% of the time. It would be interesting to measure this band width profile for data sets with different characteristics.

In summary our results have demonstrated that for solving the surface reconstruction problem from polygonal contours the multiresolution banded refinement algorithm is very fast and yields very high quality solutions.

7 Acknowledgments

We thank John Sundsten, Jim Brinkley, Kevin Hinshaw of the University of Washington Structural Informatics Group for providing the data set used in our study. We also thank David Meyers and Adam Finkelstein for several inspiring conversations, and the referees for their helpful comments.

References

[1] H. Alt and M. Godau. Measuring the resemblance of polygonal curves. In *Proceedings of the 8th ACM Symposium on Computational Geometry*, pages 102–109, 1992.

[2] C.L. Bajaj, E.J. Coyle, and K-N. Lin. Arbitrary topology shape reconstruction from planar cross sections. *Graphical Models and Image Processing*, 58(6):524–543, November 1996.

[3] G. Barequet, D. Shapiro, and A. Tal. History consideration in reconstructing polyhedral surfaces from parallel slices. In *Proceedings of Visualization*, pages 149–156, October 1996.

[4] G. Barequet and M. Sharir. Piecewise-linear interpolation between polygonal slices. In *Proceedings of the 10th ACM Symposium on Computational Geometry*, pages 93–102, 1994.

[5] J.D. Boissonnat. Shape reconstruction from planar cross sections. *Computer Vision, Graphics and Image Processing*, 44:1–29, 1988.

[6] J.D. Boisssonnat and B. Geiger. Three dimensional reconstruction of complex shapes based on the Delauney triangulation. In *Procedings of Biomedical Image Processing and Biomedical Visualization*, volume 1905, pages 964–975. SPIE Press, 1993.

[7] H.N. Christiansen and T.W. Sederberg. Conversion of complex contour line definitions into polygonal element mosaics. *Computer Graphics*, 12(2):187–192, August 1978.

[8] M. Cohen, T. DeRose, D. Salesin, P. Schröder, and W. Sweldens. Wavelets in computer graphics, 1996. ACM SIGGRAPH Course Notes.

[9] H. Fuchs, Z.M. Kedem, and S.P. Uselton. Optimal surface reconstruction from planar contours. *Communications of the ACM*, 20(10):693–702, October 1977.

[10] S. Ganapathy and T.G. Dennehy. A new general triangulation method for planar contours. *Computer Graphics*, 16(3):69–75, July 1982.

[11] A. Kalvin, E. Schonberg, J.T. Schwartz, and M. Sharir. Two-dimensional model-based boundary matching using footprints. *International Journal of Robotics Research*, 5:38–55, 1986.

[12] E. Keppel. Approximating complex surfaces by triangulation of contour lines. *IBM Journal of Research and Development*, 19:2–15, 1975.

[13] Z. Liu, S. Gortler, and M. Cohen. Heirarchical spacetime control. In *Computer Graphics Proceedings*. ACM SIGGRAPH, 1994.

[14] D. Meyers. Multiresolution tiling. *Computer Graphics Forum*, 13(5):325–340, December 1994.

[15] D. Meyers. *Reconstruction of Surfaces from Planar Contours*. PhD thesis, University of Washington, July 1994.

[16] D. Meyers, S. Skinner, and K. Sloan. Surfaces from contours. *ACM Transactions on Graphics*, 11(3):228–258, July 1992.

[17] D. Sankoff and J. Kruskal, editors. *Time Warps, String Edits, and Macromolecules: the Theory and Practice of Sequence Comparison*. Addison-Wesley, Reading, Mass., 1983.

[18] T.W. Sederberg and E. Greenwood. A physically based approach to 2D shape blending. In *Computer Graphics Proceedings*. ACM SIGGRAPH, 1992.

[19] M. Shantz. Surface definition for branching contour defined objects. *Computer Graphics*, 15:242–270, July 1981.

[20] E. Stollnitz, T. DeRose, and D. Salesin. *Wavelets for computer graphics: Theory and Applications*. Morgan Kaufmann, San Francisco, 1996.

[21] W. Sweldens and P. Schröder. Building your own wavelets at home. Technical Report 1995:5, Department of Mathematics, University of South Carolina, 1995.

[22] E. Welzl and B. Wolfers. Surface reconstruction between simple polygons via angle criteria. *Journal of Symbolic Computation*, 17:351–369, July 1994.

[23] H.J. Wolfson. On curve matching. *IEEE Transactions on Pattern Matching and Machine Intelligence*, 12(5):483–489, May 1990.

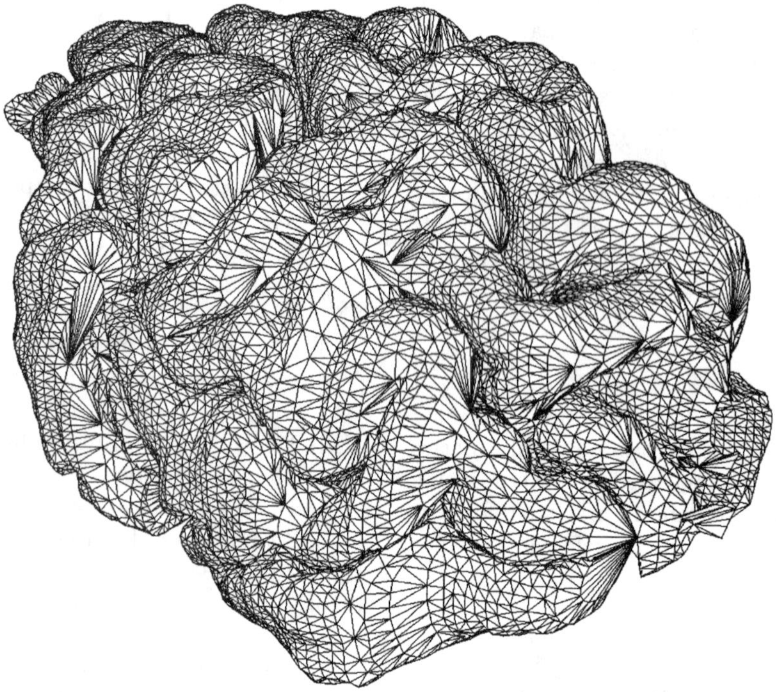

Figure 9: Reconstruction of the data set used for our study rendered as tiles.

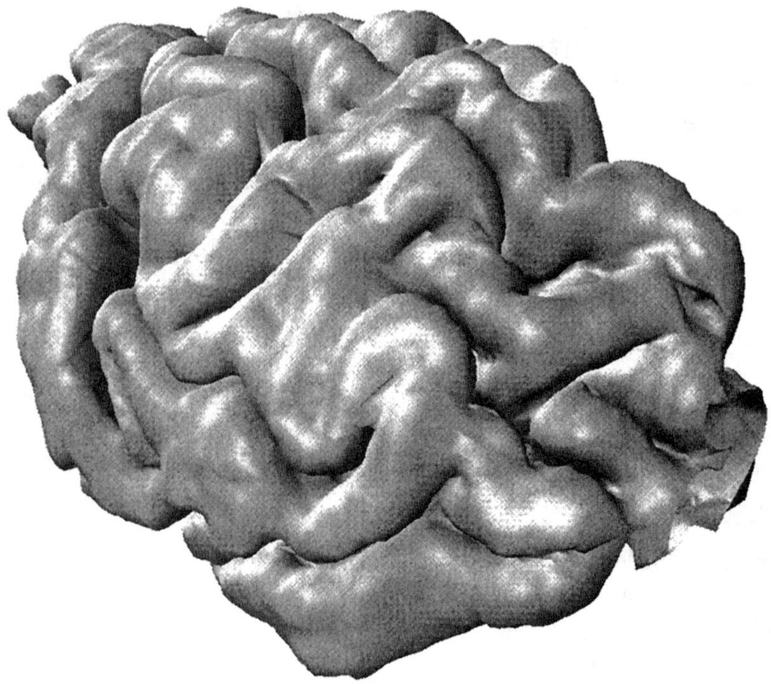

Figure 10: Same reconstruction with smooth shading.

Degenerate convex hulls on-line in any fixed dimension

Hervé Brönnimann*

Abstract. The main goal of this paper is to show how a simple modification of the formulation of an algorithm by Clarkson and Shor can handle the case of degenerate sets of points for computing convex hulls on-line in $\mathbb{R}^d$, $d \geq 2$. At the same time, a simple analysis allows to maintain the expected time complexity within $O(n^{\lfloor d/2 \rfloor})$. This analysis uses connections between regions, pivots, flags, and barycentric triangulations.

1 Introduction

Given n points in $\mathbb{R}^d$, to enumerate the facets of their convex hull is a classical problem of optimization and computational geometry [1, 2, 8, 13]. The scenario in which we work here is that d is constant and small (say, $d \leq 10$), but the number of points is large, they are inserted on-line, and they may not be in general position.

Several algorithms have been given in various cases. In the off-line case, the points can be sorted and their affine span computed. Then the algorithms of [2, 8, 13] can all be applied. The on-line algorithms that use a conflict graph or an influence graph [5, 11] are all limited to non-degenerate cases. This is due to the definition of the problem and the definition commonly adopted makes these algorithms, which are optimal for non-degenerate inputs, perform badly (up to quadratic slowdown) on degenerate instances.

The problems of degeneracies is well known and several approaches are possible. Perturbation methods (e.g. [13]) remove degeneracies but the problem solved is not the original problem. This can be an issue when counting facets and not simply computing a volume. If we insist on solving the original problem, the off-line algorithms can usually be adapted easily [11, 13, 16]. To solve the one-line case, the only algorithm available in any dimension is given by Burnikel, Mehlhorn, and Schirra [7]. Although it is a fairly small modification of an algorithm that works on non-degenerate inputs, their analysis is fairly involved and they do not provide the same upper bounds for degenerate and non-degenerate cases. In particular, it is not known if their algorithms runs in expected time $O(n^{\lfloor d/2 \rfloor})$ on all instances. As they put it, in degenerate cases, the "running time depends also on the structural properties of lower dimensional faces," which then becomes hard to analyze.

We propose an extension of the algorithm of [11] which is on-line, can handle arbitrarily degenerate instances, and is fairly easy to analyze thanks to its connections to flag vectors [3]. Its expected running time is $O(n^{\lfloor d/2 \rfloor})$ on all instances. Since the problem is usually the space needed by the data structure, we propose a space-efficient version which can be abbreviated further by the techniques of [19]. Interestingly, the bounds we give can be used to analyze many pivoting algorithms in degenerate cases, such as that of [16].

We organize the paper around the various triangulation schemes in section 2. Section 3 gives our formulation of the problem, and in section 4, we devise a persistent data-structure for storing the incidence lattice of a polytope which implicitly contains the structure of section 3. We conclude on the limitations and questions raised by our approach.

2 Triangulations of d-polytopes

We assume that the reader is familiar with the notions of polytope, incidence of faces, simplicial complex, triangulation [5, 13, 14]. Here we denote by $\mathcal{F}(\mathcal{P})$ the set of faces of a polytope $\mathcal{P}$, and by $\mathcal{F}_k(\mathcal{P})$ the set of its k-faces. In particular, $\mathcal{F}_0(\mathcal{P})$ is the set of vertices of $\mathcal{P}$. For a fixed dimension d, the *upper bound theorem* states that a d-polytope with n vertices or n facets has $O(n^{\lfloor d/2 \rfloor})$ faces and pairs of incident faces.

*INRIA Sophia-Antipolis, 2004 Route des Lucioles, B.P. 93, 06902 Sophia-Antipolis Cedex, FRANCE. Herve.Bronnimannsophia.inria.fr

SCG 98 Minneapolis Minnesota USA

2.1 Canonical triangulations

A *canonical map* for the polytope $\mathcal{P}$ is a map π from the set $\mathcal{F}(\mathcal{P})$ of faces of $\mathcal{P}$ onto the set $\mathcal{F}_0(\mathcal{P})$ of its vertices that maps each face to one of its vertices. The *canonical* subdivision $\mathcal{T}_{c(\pi)}(\mathcal{P})$ of $\mathcal{P}$ is a set of d-simplices obtained by recursively building the canonical triangulation of all the facets f of $\mathcal{P}$ that do not contain $\pi(\mathcal{P})$, forming a set $\cup_f \mathcal{T}_{c(\pi)}(f)$ of simplices, and lifting all these simplices towards $\pi(\mathcal{P})$. It is clear that each simplex in $\mathcal{T}_{c(\pi)}(\mathcal{P})$ is d-dimensional. The number $t^{(d)}_{c(\pi)}(\mathcal{P})$ of d-simplices in the canonical subdivision of $\mathcal{P}$ verifies $t^{(0)}_{c(\pi)}(\mathcal{P}) = 1$ if $\mathcal{P}$ is a 0-polytope, and if $\mathcal{P}$ is a d-polytope

$$t^{(d)}_{c(\pi)}(\mathcal{P}) = \sum_{\substack{f\in\mathcal{F}_{d-1}(\mathcal{P})\\ f\not\ni\pi(\mathcal{P})}} t^{(d-1)}_{c(\pi)}(f), \qquad (1)$$

where f ranges over all the facets of $\mathcal{P}$ which do not contain the vertex $\pi(\mathcal{P})$.

One may think that the canonical subdivision is a triangulation, but this is not so. It is a triangulation if and only if the canonical map verifies a *compatibility condition*: for any faces f and g, if g contains f, and $\pi(g)$ is in f, then $\pi(g) = \pi(f)$. The usual bottom-vertex triangulation satisfies this condition. In fact, given any ordering of the vertices, the canonical map which sends any face to the minimum vertex it contains verifies this condition. It also is not hard to see that any map that verifies the compatibility condition defines such an ordering for the vertices.

That any canonical subdivision of a d-polytope with n facets has at most $O(n^{\lfloor d/2\rfloor})$ d-simplices was proved in [9] for the more general case of line-free polyhedra, by pushing the facets, a technique dual to pulling the vertices [7]. The dual statement seems to be part of the folklore, but it applies only to the compatible maps, and hences to canonical triangulations. The triangulation is then constructed by pulling the vertices in the corresponding order. Anticipating slightly, we can also prove the following extension, whose proof requires the next section, but which is not used there.

Theorem 1 *Any canonical subdivision of a polytope with n vertices has at most $O(n^{\lfloor d/2\rfloor})$ simplices.*

Proof outline. By Theorem 2, we know that $t_{c(\pi)}(\mathcal{P})$ is bounded by $t_b(\mathcal{P})$. Let $\bar{\mathcal{P}}$ denote the dual of $\mathcal{P}$, so that it has n facets. By duality and Theorem 2,

$$t_b(\mathcal{P}) = t_b(\bar{\mathcal{P}}) \le (d+1)!t_{c(\pi)}(\bar{\mathcal{P}}) = O(n^{d/2}),$$

where the last bound follows from Clarkson's result [9]. □

2.2 Barycentric triangulation

Assume we have already computed the center of gravity of $\mathcal{P}$, called the *barycenter*, and of all its faces. The *barycentric* triangulation is obtained by recursively triangulating all the facets of $\mathcal{P}$ (with respect to their barycenters) and lifting all these simplices towards the barycenter of the polytope. The number $t^{(d)}_b(\mathcal{P})$ of d-simplices in a barycentric triangulation of a d-polytope $\mathcal{P}$ satisfies $t^{(0)}_b(\mathcal{P}) = 1$ if $\mathcal{P}$ is a 0-polytope, and

$$t^{(d)}_b(\mathcal{P}) = \sum_{f\in\mathcal{F}_{d-1}(\mathcal{P})} t^{(d-1)}_b(f), \qquad (2)$$

where F ranges over all the facets of $\mathcal{P}$.

Barycentric triangulations have been advocated by Brisson [6] as a uniform representation for structured manifolds. In the case of polytopes, it is easy by a perturbation argument to prove that $t^{(d)}_b(\mathcal{P}) = O(n^{\lfloor d/2\rfloor})$ for any d-polytope $\mathcal{P}$. Below, we discuss a bijection between barycentric triangulations and chains in the incidence lattice [3], that implies that the number of chains in the lattice of a d-polytope is also $O(n^{\lfloor d/2\rfloor})$. The following theorem relates the sizes of barycentric triangulations and canonical subdivisions, and gives a finer bound on the size of a barycentric triangulation for a given d-polytope.

Theorem 2 *For any d-polytope $\mathcal{P}$ and any canonical map π from $\mathcal{F}(\mathcal{P})$ onto $\mathcal{F}_0(\mathcal{P})$, we have*

$$t^{(d)}_{c(\pi)}(\mathcal{P}) \le t^{(d)}_b(\mathcal{P}).$$

Conversely, there exists a canonical map π such that

$$t^{(d)}_b(\mathcal{P}) \le (d+1)!\, t^{(d)}_{c(\pi)}(\mathcal{P}).$$

The factor $(d+1)!$ cannot be improved, as it is tight for a d-simplex.

Proof outline. The first inequality, namely $t^{(d)}_{c(\pi)}(\mathcal{P}) \le t^{(d)}_b(\mathcal{P})$, follows from the fact that $\mathcal{T}_{c(\pi)}(\mathcal{P})$ can be obtained from $\mathcal{T}_b(\mathcal{P})$ by dragging the barycenters of the faces towards their canonical vertex. Clearly, this process may collapse some simplices, but never creates other simplices, and hence the number of d-simplices in any canonical triangulation is less than the corresponding number in the barycentric triangulation. We call such a process a *dragging* process.

We now prove the second part of the theorem. More precisely, we prove that it is possible to choose $(d+1)!$ canonical maps π_j such that no d-simplex is collapsed during all the corresponding $(d+1)!$ dragging processes. Therefore, we have

$$t^{(d)}_b(\mathcal{P}) \;\le\; \sum_{j=0}^{(d+1)!} t^{(d)}_{c(\pi_j)}(\mathcal{P}) \le (d+1)!\, t^{(d)}_{c(\pi)}(\mathcal{P})$$

where π is the π_j that maximizes $t^{(d)}_{c(\pi_j)}(\mathcal{P})$.

We now show how to construct the canonical maps π_j. For each $k = 0, \ldots, d$, and each k-face f of $\mathcal{P}$, pick $k+1$ affinely independent vertices of f and number them $p_0(f), \ldots, p_k(f)$. Call these vertices the *marked* vertices.

Each π_j is entirely determined by a sequence $(j_0, j_1, \ldots, j_d)$, where j_k is comprised between 0 and k, by setting $\pi_j(f) = p_{j_k}(f)$ for any face f whose dimension is k. There are exactly $(d+1)!$ choices for π_j. The proof that no d-simplex can collapse during all the corresponding dragging processes follows by induction on k of the fact that no k-simplex can be collapsed during all the corresponding dragging processes. Indeed, this is true for $k = 0$ and $k = 1$. We note that a k-simplex in a k-face f is the convex hull of the barycenter of f and of a $(k-1)$-simplex Δ in a facet of f. By induction, this simplex is not collapsed for some choice of $j_0, j_1, \ldots, j_{k-1}$. The facet of f containing Δ has dimension $k-1$, so it cannot contain all the $k+1$ points $p_{j_k}(f)$; for at least one choice of j_k, the simplex $\operatorname{conv}(p_{j_k}(f) \cup \Delta)$ does not collapse. The proof is complete by induction. □

There is a bijection between the d-simplices in the barycentric triangulation of a polytope $\mathcal{P}$ and the maximal chains in the incidence lattice of $\mathcal{P}$, which are sometimes called *flags* [3]. Indeed, each d-simplex is, by construction, the convex hull of the centers of gravity of the d-face, of a facet, of a $(d-2)$-face, etc., until it all comes down to a vertex which is its own barycenter. This succession of facets gives a chain in the incidence graph, that uniquely corresponds to the d-simplex. This bijection extends between all the simplices in the barycentric triangulation and the (possibly not maximal) chains in the incidence graph. The bijection is depicted in figure 1.

This bijection shows that there are as many d-simplices in the barycentric triangulation of a polytope as in that of its dual. With the previous theorem, we have thus proven that the number of monotone chains in the incidence graph of any polytope is $O(n^{\lfloor d/2 \rfloor})$. This could also be obtained by a perturbation arguments (pulling the barycenters).

Corollary 3 *The barycentric triangulation of a d-polytope $\mathcal{P}$ has $O(n^{\lfloor d/2 \rfloor})$ d-simplices, where n is either the number of vertices or the number of facets of $\mathcal{P}$. There is a bijection between the simplices of this triangulation and the ascending chains in the incidence graph.*

2.3 Output-sensitivity

The complexities of the triangulations given above all follow from the upper bound theorem [13, 14]. They do not, however, take into account the actual complexity of the incidence graph of the polytope. In view of the results above, I venture the following conjecture, already stated in [19] in slightly different form:

Conjecture 4 *For a fixed dimension d, the number of simplices in any canonical or barycentric triangulation of a polytope, hence the number of maximal chains in the incidence graph of a polytope, is at most proportional to the complexity of the incidence graph.*

As observed by Raimund Seidel (private communication), the constant of proportionality must depend superexponentially on d. For instance, a d-dimensional cube has $2d$ facets, 2^d vertices, and 3^d faces of all dimension; the number of incidences between these faces is $O(d3^d)$, since the cube is a simple polytope. Yet, any triangulation of its boundary requires at least $(d/2)^{d/2}$ $(d-1)$-simplices, which is superexponentially larger.

We now give some support to the conjecture. As mentioned in [19] (and reference [8] by Billera and Lee therein), the conjecture is verified for simple and simplicial polytopes in any dimension d. In these cases, the number of chains is even more precisely $\Theta(d^{\lfloor d/2 \rfloor}(n_0 + n_{d-1}))$, where n_k is the number of k-faces for any k. The number of chains in a quasi-simplicial d-polytope is $(d-1)!$ times the number of chains that link a $(d-2)$-face to the d-face of the polytope; this number is at exactly twice the number n_{d-2} of ridges, and so there are $2(d-1)!n_{d-2}$ chains in a quasi-simplicial d-polytope.

It is easy to verify that the conjecture is verified for d-polytopes, with any $d \leq 3$. As argued by Raimund Seidel (private communication), it is still valid for 4-polytopes. Indeed, the 2-faces are contained in exactly two 3-faces of the 4-polytope, and in the unique 4-face. The 1-faces contain exactly two 0-faces, contained in the unique empty face. Thus, if $n_{1,2}$ is the number of incidences between the 1-faces and the 2-faces, there are exactly $4n_{1,2}$ chains in the incidence graph, hence 4-simplices in the barycentric triangulation.

For 5-polytopes, the number of chains is also easily shown to be exactly equal to four times the number of chains from any 1-face to any 3-face. Denoting by $\operatorname{sub}(f)$ and $\sup(f)$ the numbers of sub- and super-faces of a face f of a polytope $\mathcal{P}$, this latter number equals the "second moment" sum

$$K_2(\mathcal{P}) = \sum_{f \in \mathcal{F}_2(\mathcal{P})} \operatorname{sub}(f)\sup(f).$$

It is known that this parameter $K_2(\mathcal{P})$ is maximized for cyclic polytopes [1], for which it equals $12n_3(\mathcal{P})$. Nevertheless, it is not known if this parameter K_2 is linear in the complexity of the incidence graph of any 5-polytope $\mathcal{P}$.

The conjecture is still open (as far as I know) for general d-polytopes, with a fixed $d \geq 5$.

3 Convex hulls

Some familiarity with incremental randomized algorithms [10, 5] is assumed in this section. In this framework,

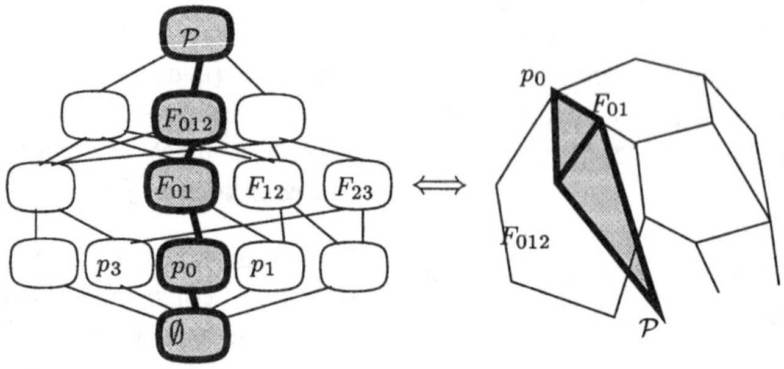

Figure 1: Bijection between chains in the incidence graph and simplices in the barycentric triangulation.

objects are inserted in a set of regions, and the set of regions which do not conflict with the objects is maintained. For convex hulls, the objects are the input points, the regions defined by the points are simplices whose vertices are the points, and the conflict relationship is such that the regions defined and without conflicts over the input points represent exactly the facets of the convex hull. The incremental algorithm maintains a data structure, a *conflict graph* [10] an *influence graph* [4, 5] under any insertion order. The theorem of the influence graph [4, 5] shows that, if a certain update condition is satisfied, the expected time complexity is $O(n \log n + n^{\lfloor d/2 \rfloor})$ and its average space complexity $O(n^{\lfloor d/2 \rfloor})$, over all insertion orders (uniformly). This framework is only valid when the points are in general position.

In this section, we present a new definition of objects, regions and conflicts which, when applied to the formalism of influence graphs, yields an on-line algorithm for computing the d-dimensional convex hull of n points in $\mathbb{R}^k$. The points can be in arbitrary degenerate position. The expected time complexity of this algorithm is $O(n \log n + n^{\lfloor d/2 \rfloor})$ and its average space complexity $O(n^{\lfloor d/2 \rfloor})$. The time needed to insert a point, averaged over all insertion orders of the previous n points, is $O(\log n + n^{\lfloor d/2-1 \rfloor})$. Note that d denotes the dimension of the convex hull and not the dimension k of the ambient space. We explicit the algorithm in the three-dimensional case.

3.1 Objects, regions, and conflicts

As is done in [10, 5], we define the *objects* to be the points in $\mathbb{R}^k$. A *region* is defined as any ordered set of points in affinely independant position. To refine the terminology, we define a *d-region* as a $(d+1)$-tuple $(p_0, \dots, p_d)$ of points in affinely independent position. In $\mathbb{R}^k$, at most $k+1$ points may be in affinely independent position, thus a region is a k-region for some $d = 0, \dots, k$. Therefore, a region is determined by at most $k+1$ points, and a subset of d objects determines at most $(d+1)!$ regions, for any $0 \leq d \leq k$.

The definition of conflict is recursive, based on the dimension. We say that a point q conflicts with a d-region determined by $(p_0, \dots, p_d)$ if one of the following conditions is fulfilled:

1. q does not belong to the d-dimensional space $\text{aff}(p_0, \dots, p_d)$ or

2. q and p_d are on opposite sides of the hyperplane[1] passing through the points $p_0, \dots, p_{d-1}$, or

3. q and p_0 are on opposite sides of the hyperplane passing through the points $p_1, \dots, p_d$, or

4. if q is on the hyperplane passing through the points $p_0, \dots, p_{d-1}$, then q conflicts with the $(d-1)$-region determined by $(p_0, \dots, p_{d-1})$ in this hyperplane, or

5. if q is on the hyperplane passing through the points $p_1, \dots, p_d$, then q conflicts with the $(d-1)$-region determined by $(p_1, \dots, p_d)$ in this hyperplane.

The reader can check that this definition does not change compared to [10] if the set $\{q, p_0, \dots, p_d\}$ is in general position. It introduces a refined conflict for points in degenerate position. Since the definition is recursive, we must define the notion of conflict with a 0-region. Following the convention that a region does not conflict with the points that determine it, we agree that a point q that belongs to the 0-dimensional space that is the affine hull of a point p_0 does not conflict with the 0-region determined by p_0, because q and p_0 coincide.

This definition implies that two distinct points always conflict. A point conflicts with a 1-region (an oriented segment) if and only if it does not lie on the line that supports the edge or else it is not contained in the edge or one of its vertices. The domains of influence of a point, of an edge (p_0, p_1), of a triangle (p_0, p_1, p_2) and of a tetrahedra (p_0, p_1, p_2, p_3) are illustrated in figure 2. For commodity, the shaded region is the complement of the domain of influence; in the 3D case, the portion that conflicts on the boundary of the wedge is nevertheless lightly shaded as if the shading of the inside of the wedge showed through. The reader is invited to picture the domain of influence of a d-region for a higher

[1] The notion of a hyperplane is relative to the d-dimensional flat spanned by $(p_0, \dots, p_d)$, and the two (open) sides are half-spaces of dimension d in this d-dimensional space.

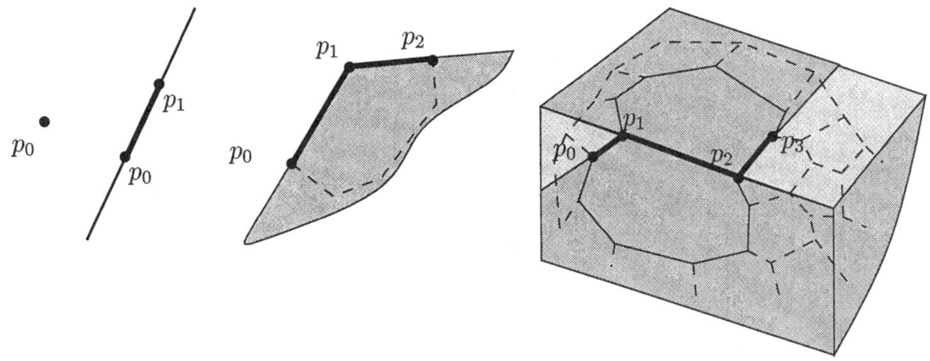

Figure 2: The complement of the domain of influence is shown shaded in 0, 1, 2 and 3 dimensions.

dimension d.

3.2 How to efficiently compute a conflict

To test a conflict using the straightforward inductive procedure suggested by the definition requires computing the orientation of $d+1$ points in $\mathbb{R}^d$ at most 2^{d+1} times. Using dynamic programming and the sub-lattice structure, the maximal number of orientation tests reduces to $d(d-1)/2$. We now show that a conflict can be tested by computing $d+1$ orientation tests.

The on-line algorithms demand only to compute the orientation of $d+1$ points in $\mathbb{R}^d$, an operation that can be implemented using determinant sign computation. Given $(d+1)$ points $p_0, \ldots, p_d$, if p_i has homogeneous coordinates $(p_{i,j})_{j=0,\ldots d}$ with $p_{i,0} > 0$, we let the function $\text{sign}(p_0, \ldots, p_d)$ compute the sign of the $(d+1)$-determinant of the matrix $(p_{i,j})_{0 \le i,j \le d}$. In the following, even though only points p_0 to p_k are given by the region, we assume that some points p_{k+1} to p_d are known such that $p_0, \ldots, p_d$ are affinely independent.

The proof of the following lemma is omitted for brevity in this version. (See the appendix.) Note that all the s_j's cannot be null simultaneously since $p_0, \ldots, p_k$ are affinely independent.

Lemma 5 *Let $s_j = \text{sign}(p_0, \ldots, p_k)$ where p is substituted for p_j. Then there is a value j such that $s_0, s_1, \ldots, s_{j-1}$ are all null and $s_j \neq 0$; similarly, there is a value j' such that $s_k, s_{k-1}, \ldots, s_{j'+1} = 0$ and $s_{j'} \neq 0$. If $s_j = s_{j'} = \text{sign}(p_0, \ldots, p_d)$, then p does not conflict with the k-region $(p_0, \ldots, p_k)$, otherwise it does.*

Therefore, computing a conflict requires computing the sign of 2 $(d+1)$-determinants in the non-degenerate case, and at most $d+1$ such determinants in any case.

3.3 Regions defined and without conflict

Now we come to the characterization of the regions defined and without conflict over a set of points. A region is *defined and without conflict* over a set $\mathcal{S}$ of points if it is determined by points in $\mathcal{S}$ and if it does not conflict with any point in $\mathcal{S}$. Recall that a region does not conflict with the points that determine it. For conciseness, when the set $\mathcal{S}$ is understood, we speak of a *free* region.

The lemma below offers a characterization of regions defined and without conflict over a set of points. Its proof is given for completeness in the appendix. Recall that $\text{aff}(\mathcal{S})$ denotes the flat spanned by a set $\mathcal{S}$ of points.

Lemma 6 *A region $(p_0, \ldots, p_d)$ is defined and without conflict over a set $\mathcal{S}$ of points in $\mathbb{R}^k$ if and only if $\text{aff}(\mathcal{S})$ has dimension d and if, for any pair (i,j), $0 \le i < j \le d$, the $(j-i)$ dimensional space $\text{aff}(p_i, \ldots, p_j)$ contains a unique $(j-i)$-face $F_{i,j}$ which contains $p_i, \ldots, p_j$.*

In particular, the faces $(F_{0,j})_{j=0,\ldots,d}$ form an ascending sequence of incident faces, i.e. a chain in the incidence graph. Moreover, $F_{i+1,j}$ can be obtained by pivoting from $F_{i,j-1}$ around $F_{i+1,j-1}$ inside $F_{i,j}$, and thus all the $F_{i,j}$'s can be retrieved from knowledge of the chain of $F_{0,j}$'s. This shows a bijection between the free regions, the maximal chains in the incidence graph, and the maximal-dimensional simplices in the barycentric triangulation, which is depicted in figure 3.

Theorem 7 *Let $\mathcal{S}$ be a set of n points in $\mathbb{R}^k$, not necessarily in general position. If $\text{conv}(\mathcal{S})$ is a d-polytope, $d \le k$, then the number of regions defined and without conflict over $\mathcal{S}$ is $O(n^{\lfloor d/2 \rfloor})$.*

This theorem is an immediate consequence of those bijections. If our conjecture is true, then the upper bound can be sharpened to the complexity of the incidence graph of the polytope $\text{conv}(\mathcal{S})$, a truly output-sensitive result.

3.4 An on-line algorithm in 3D

With this definition of objects, regions and conflict, an on-line algorithm can be given by using an influence graph for any dimension. For lack of space, and because we propose a variant for any dimension in the next section, we focus here on the important three-dimensional

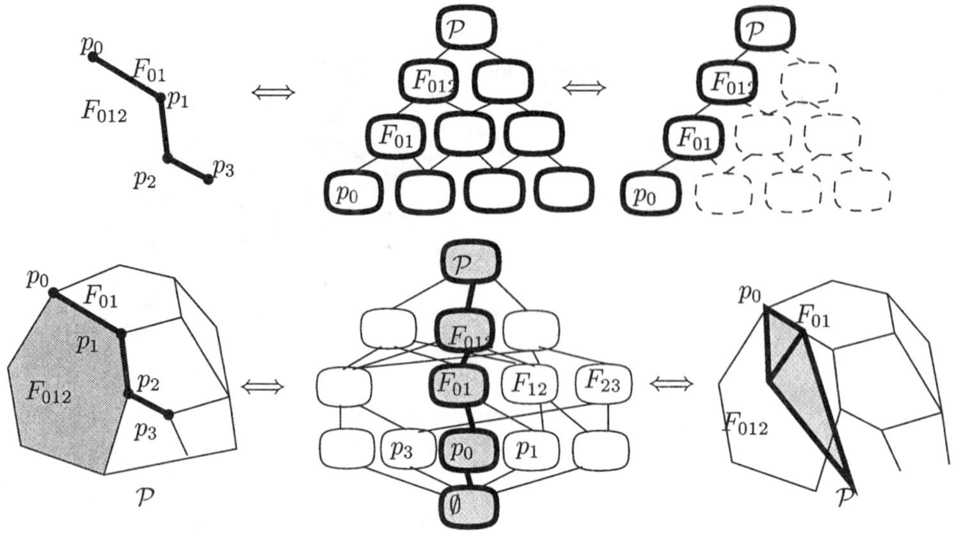

Figure 3: Bijection between free regions, maximal chains in the incidence graph, and maximal-dimensional simplices in the barycentric triangulation. On the top, the bijection between the free region, the sub-lattice of the $F_{i,j}$'s, and the chain of the $F_{0,j}$'s.

case, which admits a representation of its own: in space, a polyhedron is often represented by a DCEL or half-edge data structure [15]. We will work with the latter, which is also equivalent to that given in [5, 13].

Each edge e is oriented and has four pointers: $org(e)$ to the vertex at the origin of the edge, $sym(e)$ to the edge with opposite orientation, $pred(e)$ and $succ(e)$ to the edges before and after e on the face to the left of e, oriented clockwise. For convenience, one may assume also $dest(e) = org(sym(e))$ pointing to the other vertex (not the origin).

Each oriented edge $e = p_1p_2$ corresponds to two free regions (p_0, p_1, p_2, p_3) where $p_0 = org(prec(e))$ and $p_3 = org(prec(sym(e)))$, or $p_0 = dest(succ(sym(e)))$ and $p_3 = dest(succ(e))$. Moreover, the oriented edges $p1p2$ and p_2p_1 lead to the same region in either case. We may thus store only two region per pair of opposite oriented edges. This lets us speak for instance of a conflict with an edge. According to the methodology, all the edges created during the succesive insertions are stored as nodes in the influence graph. Initially there is one node corresponding to the empty region. Out of each edge go at most two arcs, and the nodes have the *inclusion property*: the domain of influence of a node is contained in the union of domains of influence of its parents. The leaves of the graph are the free regions, hence the edges of the current $conv(S)$.

The algorithm processes the n points of $\mathcal{S}$ one by one, adding them in the order $p_1, \dots, p_n$. At step r, all the points $p_1, \dots, p_{r-1}$ have been inserted. The algorithm maintains the convex hull of $p_1, \dots, p_{r-1}$ upon adding p_r. Note that lower dimensional cases (all points belonging to a single plane) are only easier to process and can be incorporated in the algorithm without modification.

Upon inserting a new point, the algorithm performs the following steps. For the purpose of the discussion, when adding a new point q to the current set $\mathcal{S}$, a facet is colored red if its supporting hyperplane separates $conv(S)$ from q, yellow if it contains q, and blue otherwise (see the beneath-beyond method [13] for an explanation of this coloring scheme). A face is colored according to the colors of the facets that contain it: blue, red, yellow, purple (blue+red), green (blue+yellow), orange (red+yellow), brown (blue+red+yellow).

Locating. The algorithm finds all the leaves that conflict with the new point by traversing the influence graph from its root, recursively visiting the children of a node which conflict with the new point. These leaves are said to be killed by q.

Updating. If no edge is found to conflict with the new point q, then q lies inside the current convex hull. Otherwise, the purple, yellow and green edges are found. Then each correspond to an edge e, or equivalently a region (p_0, p_1, p_2, p_3), which is killed by q, giving way to a new region (p_0, p_1, p_2, q). This new region is added as a leaf in the influence graph, and hooked to a single parent (p_0, p_1, p_2, p_3): one verifies easily that this preserves the inclusion property. This also creates a new edge p_2q (more exactly two half-edges p_2q and qp_2). See figure 4 for details.

Pointers are created for the new edges: all are straightforward, only the difficult part is to identify $succ(p_2q)$ and $prec(qp_2)$. For this, note that the face containing (p_1, p_2, p_3) is red or yellow, and that one may traverse

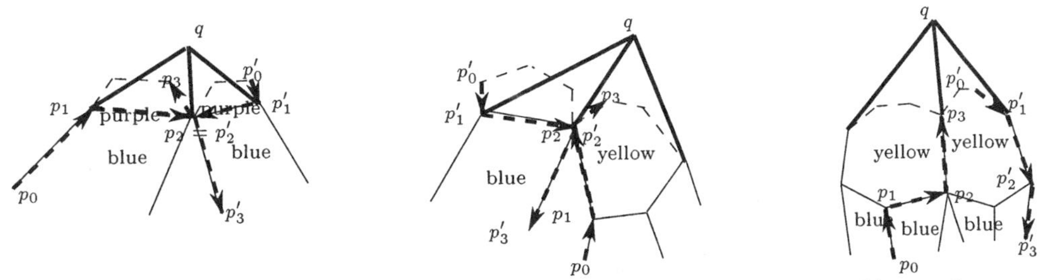

(a) e is purple (b) e is green (c) e is yellow

Figure 4: The three cases when inserting a point q in $\mathbb{R}^3$. In all three cases, the successor of p_2q is qp'_1; the new region (p_1, p_2, q, p'_1) is hooked to the two killed regions (p_0, p_1, p_2, p_3) and to (p'_0, p'_1, p'_2, p'_3), and the inclusion property is verified. The update are similar for the other successor, predecessors of p_2q and qp_2, and the other new region created by qp_2.

the red or orange edges along its boundary, starting at p_2p_3 (see the figure below). When a purple or green edge $p'_0p'_1$ is reached, we know that qp'_1 is the succcessor of p_2q. Similarly for $prec(qp_2)$. This is depicted in figure 4.

A new region is then created for each new edge. Since the edge pointers have been updated, this region and its at most two parents are found easily (see the caption of the figure for the parents in the different cases). These regions are then hooked in the influence graph. This concludes the updating step.

Analysis. The analysis is a consequence of the theorem of the influence graph of [10] once it is noted that the work performed at step r is proportional to the number of edges killed or new at step r: in particular, it is important to traverse the red or orange edges, and not the green or purple edges, to find the successor of p_2q and the predecessor of qp_2. Once an edge is killed, it will never receive other children any more. Moreover, each node has at most two children, which verify the inclusion property. Hence, the algorithm verifies the update condition for algorithms that use an influence graph [10]. Hence the theorem of the influence graph shows that:

Theorem 8 *The algorithm maintains the convex hull of a set of n points (not necessarily in general position) in $\mathbb{R}^3$ on-line, in average time $O(n \log n)$ and average space $O(n)$. The time needed to insert a new point, averaged over all the insertion orders of the previous n points, is $O(\log n)$.*

This also applies to the representation of the Delaunay triangulation or a Voronoi diagram as a half-edge encoded planar map, by the standard duality with respect to the paraboloid.

4 Randomized beneath-beyond

In this section, we present an algorithm that only maintains the incidence graph of the polytope under the insertion of a new point, using the beneath-beyond method [13]. In the beneath-beyond method, the points need to be inserted in lexicographic order, so as to find the red facets efficiently. The influence graph allows to perform the same function while allowing a random order of insertion. Nevertheless, we show that we do not need an influence graph; instead, we maintain the history of the insertions in a persistent data structure. This allows to randomize the order of insertion of the points in the beneath-beyond method. We then show that the analysis of the algorithm given in the previous section can be used to show the optimality of the randomized beneath-beyond algorithm. We assume that the reader is familiar with the beneath-beyond method as presented in [13].

The algorithm. We introduce a *persistent incidence graph* as a data structure to maintain efficiently the incidence graph of a polytope under insertion of new vertices (see also [12, 17] for other examples of persistent data structures). As observed in [13], when adding a point p to a polytope $\mathcal{P}$, two cases can happen: the point may or may not belong to the affine span of the polytope. The case of a pyramidal update (p does not belong to the affine span of $\mathcal{P}$) is easily taken care of by a copy-and-link procedure.

When p belongs to the affine span of $\mathcal{P}$, the incidence graph can be updated once a red facet is known (coloring the facets is done as in the 3D case above). We explain how to find this red facet by using the persistent graph below. First, we outline the changes in the incidence graph. Inserting p introduces new faces, and new incidence arcs between new faces or between an old and a new face. Indeed, an old face f may induce a new face $conv(f \cup \{p\})$ which contain it, and which is incident to it. As explained in [13], the insertion can be carried out by finding all the non-bl e facets and ridges, then recursively finding all the re aining non-blue faces in decreasing order of their dir ension, then updating the arcs incident to the faces in ' ıeasing ·r-der of their dimension. The only modificat. ı is that we are careful not to remove the old faces that are not in the new convex hull; we say that those faces and their incident arcs are killed. Instead, we update the *time interval* of the incidence arc: each arc has a time interval that remembers the step that the arc was created and the step that it was killed. Killed arcs are not removed, they remain in the persistent graph, but they are not followed in subsequent insertions. Thus when looking for the superfaces of a face, one is careful to follow only the arcs that have not yet been killed.

Let the sequence of polytopes $\mathcal{P}_1, \dots, \mathcal{P}_n$ be obtained incrementally and stored in the persistent incidence graph, and consider the addition of a new vertex q to $\mathcal{P}_n$. We assume that at the update phase yet another set of pointers called the *killer pointers* have been maintained for the ridges of the successive $\mathcal{P}_i$'s: when adding a point p_i, a new ridge is always the convex hull of p_i and of a purple, green or brown $(d-3)$-face of $\mathcal{P}_{i-1}$. Since the set of purple or yellow faces is isomorphic to a $(d-1)$-polytope [13], this $(d-3)$-face is incident to exactly two purple or green ridges of $\mathcal{P}_{i-1}$. A killer pointer is added to link these two ridges to the new ridge of $\mathcal{P}_i$. It is easily checked that a node has at most $d-1$ killer pointers leaving it during an incremental step.

There could be arbitrarily incidence arcs and killer pointers leaving a node after several subsequent incremental steps. To deal with this, these pointers also have a *time stamp*, and are maintained by the node-copying technique of [12, 17]. Let us focus for instance on killer pointers. The killer pointers are stored into a ridge in the incidence graph, into an array that can store $t+2$ pointer, where t is a fixed integer, and each killer pointer possesses its stamped version which is the number of the current incremental step. Whenever an new ridge is killed by a node, a new pointer from the array is taken to store this killer link, and stamped appropraitely. When all $t+2$ pointers are taken, a copy of the node is created, with identical pointers to the subfaces and other information, except for the $t+2$ pointers to the killer nodes, which are all free for use. The stamped version of the copy is the number of the current incremental step. In order to avoid copying all the pointers to the subfaces (which could be huge), the copy node has a special pointer to the original node, and only has the pointers to subfaces that are created in subsequent incremental steps. Finally, all the pointers to the copied face in the superfaces of this face are updated to point towards the copy.

We can now finish the description of the process that retrieves a red facet of the current hull. The red, green or purple ridges of $\mathcal{P}_i$ are obtained by following the killer pointers stamped by step i of the red, green or purple ridges of $\mathcal{P}_{i-1}$. When $i = n$, a purple ridge of $\mathcal{P}_n$ is found, and is incident to a red facet of $\mathcal{P}_n$. Then the other steps of the beneath beyond method are performed, not forgetting to update the killer pointers for the new convex hull.

Analysis. Analyzing this algorithm is a difficult task to perform directly. The node-copying can be shown to require constant amortized time by an analysis similar to that given in [12, 17]. Fortunately, the persistent incidence graph contains *implicitely* all the chains, therefore all the free regions as defined in the previous section. The time to find a red facet is proportional to the number of red and yellow facets of all the polytopes $\mathcal{P}_1, \dots, \mathcal{P}_n$. These facets are contained in the regions that would be traversed in the influence graph to find the free regions killed by the new point. The time to update the incidence graph is bounded by the number of red, yellow or new facets of $\mathcal{P}_n$. This number is upper-bounded by the number of killed or new regions. Therefore the number of operations performed by the algorithm is upper-bounded by the complexity of a (fictitious) algorithm that uses an influence graph and verifies the update condition. The theorem of the influence graph of [10] shows that:

Theorem 9 *The algorithm maintains on-line the convex hull of any set of n points in $\mathbb{R}^k$ in average time $O(n \log n + n^{\lfloor d/2 \rfloor})$ and average space $O(n^{\lfloor d/2 \rfloor})$. The time needed to insert any point, averaged over all insertion orders of the previous n points, is $O(\log n + n^{\lfloor d/2-1 \rfloor})$.*

5 Conclusion and bibliographic notes

We have given an on-line algorithm that maintains the convex hull or the Voronoi diagram of a set of points in any dimension, which works and is optimal even if the points are in degenerate position. An important extension is to make the algorithms fully dynamic (deletions as well as insertions). Hopefully, this should not be very hard by using the paradigm presented in [10, 20].

The limiting factor in computing enumerating facets of polytopes, is the storage [1]. Our data structure, like its analogues, allows to rapidly locate new conflicts at the expense of storing a huge graph. Thus the method is only limited to a domain where d is small, or n is small as a function of d. Nevertheless, computing four-dimensional convex hulls is equivalent to computing 3D Voronoi diagrams and Delaunay tesselations so our techniques find applications in computation geometry. For diminishing the storage, Swart [19] has shown how to abbreviate the incidence graph to take into account the non-simplicial faces only.

It is also shown in [19, 16] how the running times of several algorithms depend on the parameter K_2 introduced above, and on the number of monotone chains in the abbreviated incidence graph. The bounds we supply here prove partially the conjecture stated in [19], in that $\gamma_d(n, f) = O_d(n^{\lfloor d/2 \rfloor})$ for fixed d (in the terminology used there). To our surprise, this provided a bound to the complexity of several pivoting algorithms [19, 16]. Unfortunately, we cannot prove the full conjecture in dimensions higher than 4 and we ask again: is $\gamma_d(n, f) = O_d(n + f)$ for $d \geq 5$? Answering this conjecture is important to assess the output-sensitivity of all these algorithms and would provide some insight on the complexity of linear programming and facet enumeration.

References

[1] D. Avis, D. Bremner, and R. Seidel. How good are convex hull algorithms? *Comput. Geom. Theory Appl.*, 7:265–302, 1997.

[2] D. Avis and K. Fukuda. A pivoting algorithm for convex hulls and vertex enumeration of arrangements and polyhedra. *Discrete Comput. Geom.*, 8:295–313, 1992.

[3] L. Billera and A. Björner. Face numbers of polytopes and complexes. In *The Handbook Of Discrete And Computational Geometry*, J. E. Goodman and J. O'Rourke, eds., CRC Press LLC, Boca Raton, FL, 1997, pages 251–310.

[4] J.-D. Boissonnat, O. Devillers, R. Schott, M. Teillaud, and M. Yvinec. Applications of random sampling to on-line algorithms in computational geometry. *Discrete Comput. Geom.*, 8:51–71, 1992.

[5] J.-D. Boissonnat and M. Yvinec. *Algorithmic geometry.* Cambridge University Press, 1998. Also in French: *Géométrie algorithmique*, Ediscience, 1995.

[6] E. Brisson. Representing geometric structures in d dimensions: Topology and order. In *Proc. 5th Annu. ACM Sympos. Comput. Geom.*, pages 218–227, 1989.

[7] C. Burnikel, K. Mehlhorn, and S. Schirra. On degeneracy in geometric computations. In *Proc. 5th ACM-SIAM Sympos. Discrete Algorithms*, pages 16–23, 1994.

[8] D. R. Chand and S. S. Kapur. An algorithm for convex polytopes. *J. ACM*, 17:78–86, 1970.

[9] K. L. Clarkson. A randomized algorithm for closest-point queries. *SIAM J. Comput.*, 17:830–847, 1988.

[10] K. L. Clarkson and P. W. Shor. Applications of random sampling in computational geometry, II. *Discrete Comput. Geom.*, 4:387–421, 1989.

[11] M. de Berg, M. van Kreveld, M. Overmars, and O. Schwarzkopf. *Computational Geometry: Algorithms and Applications.* Springer-Verlag, Berlin, 1997.

[12] J. R. Driscoll, N. Sarnak, D. D. Sleator, and R. E. Tarjan. Making data structures persistent. *J. Comput. Syst. Sci.*, 38:86–124, 1989.

[13] H. Edelsbrunner. *Algorithms in Combinatorial Geometry*, volume 10 of *EATCS Monographs on Theoretical Computer Science.* Springer-Verlag, Heidelberg, West Germany, 1987.

[14] B. Grünbaum. *Convex Polytopes.* Wiley, New York, NY, 1967.

[15] L. Kettner. *Polyhedra in half-edge data-structures.* These proceedings.

[16] G. Rote. Degenerate convex hulls in high dimensions without extra storage. In *Proc. 8th Annu. ACM Sympos. Comput. Geom.*, pages 26–32, 1992.

[17] N. Sarnak and R. E. Tarjan. Planar point location using persistent search trees. *Commun. ACM*, 29:669–679, 1986.

[18] R. Seidel. Output-size sensitive algorithms for constructive problems in computational geometry Ph.D. Thesis. Technical Report TR 86-784, Dept. Comput. Sci., Cornell Univ., Ithaca, NY, 1986.

[19] G. F. Swart. Finding the convex hull facet by facet. *J. Algorithms*, 6:17–48, 1985.

[20] M. Teillaud. *Towards dynamic randomized algorithms in computational geometry*, volume 758 of *Lecture Notes Comput. Sci.* Springer-Verlag, 1993.

A Convex hulls

Lemma 10 *A region $(p_0, \ldots, p_d)$ is defined and without conflict over a set $\mathcal{S}$ of points in $\mathbb{R}^k$ if and only if aff($\mathcal{S}$) has dimension d and if, for any pair (i, j), $0 \leq i < j \leq d$, the $(j - i)$ dimensional space aff$(p_i, \ldots, p_j)$ contains a unique $(j-i)$-face $F_{i,j}$ which contains $p_i, \ldots, p_j$.*

Proof. First, assume that a region $(p_0, \ldots, p_d)$ is defined and without conflict over $\mathcal{S}$. We prove by induction on the dimension $j - i$, from d downto 0, that there exists a $(j - i)$-face $F_{i,j}$ of conv($\mathcal{S}$) contained in aff$(p_i, \ldots, p_j)$, and that $p_i, \ldots, p_j$ are vertices of $F_{i,j}$. First observe that $F_{0,d} =$ conv($\mathcal{S}$), and thus aff($\mathcal{S}$) has dimension d. By assumption, no point belongs to one side of the hyperplane aff$(p_0, \ldots, p_{d-1})$, so that aff$(p_0, \ldots, p_{d-1})$ supports a face $F_{0,d-1}$. In aff$(p_0, \ldots, p_{d-1})$, the region $(p_0, \ldots, p_{d-1})$ is defined and without conflict over $\mathcal{S} \cap$ aff$(p_0, \ldots, p_{d-1})$ by the definition of conflicts, therefore by induction any aff$(p_i, \ldots, p_j)$ supports a face $F_{i,j}$ of $F_{0,d-1}$, and $F_{i,j}$ is also a face of conv($\mathcal{S}$), for $0 \leq i < j \leq d - 1$; also, p_k is a vertex of $F_{i,j}$ for any $0 \leq i < j \leq d - 1$, and $i \leq k \leq j$ by induction. Similarly, aff$(p_1, \ldots, p_d)$ supports a face $F_{1,d}$, and by induction any aff$(p_i, \ldots, p_j)$ supports a face $F'_{i,j}$ of $F_{0,d-1}$, with $p_i, \ldots, p_j$ being vertices of $F'_{i,j}$. In fact, when $0 < i < j < d$, $F_{i,j}$ and $F'_{i,j}$ coincide, since $F_{i,j}$ is uniquely defined as the $(j - i)$-face which contains $p_i, \ldots, p_j$. Finally, it follows that p_i is a vertex of conv($\mathcal{S}$) for any $0 \leq i \leq d$ because p_i is a vertex of $F_{i,i}$ and $F_{i,i}$ is a subface of either $F_{0,d-1}$ or $F_{1,d}$.

Let us now assume that, for any pair (i, j), $0 \leq i < j \leq d$, the $(j - i)$ dimensional space aff$(p_i, \ldots, p_j)$ supports a $(j - i)$-face $F_{i,j}$ of conv($\mathcal{S}$), and $p_i, \ldots, p_j$ are vertices of $F_{i,j}$. Then again $F_{i,j}$ is uniquely defined as the $(j - i)$-face which contains $p_i, \ldots, p_j$. We prove that $(p_i, \ldots, p_j)$ is defined and without conflict over $\mathcal{S} \cap$ aff$(p_i, \ldots, p_j)$ by induction on $j - i$, but this time from 0 to d. If $j = i$, this is trivial as the 0-region (p_i) is defined and without conflict over $\mathcal{S}$ by the very definition of conflicts. A point q cannot conflict with the $(j - i)$-region $(p_i, \ldots, p_j)$, if we assume that both $(j-i-1)$-regions $(p_i, \ldots, p_{j-1})$ and $(p_{i+1}, \ldots, p_j)$ are defined and without conflict over $\mathcal{S}$. Indeed, if q does not belong to aff$(p_i, \ldots, p_j)$, it does not conflict with $(p_i, \ldots, p_j)$. Otherwise, q and p_j cannot be on different sides of aff$(p_i, \ldots, p_{j-1})$ in aff$(p_i, \ldots, p_j)$ because aff$(p_i, \ldots, p_{j-1})$ contains a face $F_{i,j-1}$ of conv($\mathcal{S}$), and similarly q and p_i cannot be on different sides of $\text{aff}(p_{i+1}, \dots, p_j)$ in $\text{aff}(p_i, \dots, p_j)$. If q belongs to $\text{aff}(p_i, \dots, p_{j-1})$, then the induction hypothesis shows that q does not conflict with $(p_i, \dots, p_{j-1})$. Similarly, If q belongs to $\text{aff}(p_{i+1}, \dots, p_j)$, then the induction hypothesis shows that q does not conflict with $(p_{i+1}, \dots, p_j)$. Finally, q does not fulfill either of the four conditions that define a conflict with $(p_i, \dots, p_j)$. □

Randomized External-Memory Algorithms for Some Geometric Problems

A. Crauser P. Ferragina K. Mehlhorn U. Meyer E. Ramos

Max-Planck-Institut für Informatik

Im Stadtwald, 66123 Saarbrücken, Germany

{crauser,paolo,mehlhorn,umeyer,ramos}@mpi-sb.mpg.de

Abstract

We show that the well-known random incremental construction of Clarkson and Shor [14] can be adapted via *gradations* to provide efficient external-memory algorithms for some geometric problems. In particular, as the main result, we obtain an optimal randomized algorithm for the problem of computing the trapezoidal decomposition determined by a set of N line segments in the plane with K pairwise intersections, that requires $\Theta(\frac{N}{B}\log_{M/B}\frac{N}{B}+\frac{K}{B})$ expected disk accesses (I/Os), where M is the size of the available internal memory and B is the size of the block transfer. The approach is sufficiently general to obtain algorithms for the problems of 2-d and 3-d convex hulls, 2-d abstract Voronoi diagrams and batched point location in a planar subdivision, which require an optimal expected number of I/Os and are simpler than the ones previously known. The results extend to a external-memory model with multiple disks.

1 Introduction

There is a growing interest in algorithms working on sets of data that are too large to be fit in *internal memory*, and that consequently need to perform input/output accesses to *external storage devices*, like disks and CD-ROMs (see e.g. [4, 11, 19, 21, 29, 38]). These devices are roughly 10^6 times slower than internal memory in terms of access time. In many applications, this disparity has given rise to an input/output (or *I/O*) bottleneck, in which the time spent on moving data between internal and external memory dominates the overall execution time [20]. Such an I/O bottleneck is increasing in significance since the gap between the speed of (mechanical) disks versus (electronic) internal memories is growing, especially with more and more use of parallel computers [30]. Therefore, it is more than ever urgent to minimize the I/O communication in large-scale applications.

Computer graphics [33] as well as Geographical Information Systems [16, 23] are nowadays rich sources of large-scale computational problems. Therefore, they need the design of appropriate external-memory techniques and data structures to efficiently cope with the enormous amount of spatial data which have to be searched, stored and manipulated. In those applications, most of the subproblems require the processing of geometric primitives, and so research on geometric algorithms in the external-memory model is important.

SCG 98 Minneapolis Minnesota USA

Problems. A first goal of our paper is the design of an I/O-efficient algorithm for the *segment intersections* problem which consists of computing the *arrangement* of a set of N line segments in the plane with K pairwise intersections in an *output sensitive* manner. Computing the intersections I/O-optimally has been posed as an open problem in [5]. A second goal is to investigate the applicability of random sampling to the design of efficient geometric algorithms in the external-memory setting, and to develop a general randomized approach suitable to solve I/O-efficiently not only the segment intersections problem, but also several others geometric problems (e.g. convex hulls, Voronoi diagrams, point location).

We study these problems in the *external-memory model*, introduced in [38]. Here a computer consists of a processing unit, an internal memory of size M and an (unbounded) external memory which is partitioned into blocks of size B, $B \leq M$. Each access to the external memory transfers from/to the internal memory one block of B items (e.g., integers, pointers, characters). The goal is to design algorithms which take advantage of the block transfer and thus exhibit *locality of reference*. The complexity of an algorithm is evaluated in this model by providing asymptotical bounds for the total number of disk accesses (I/Os) performed by the various operations, and for the number of internal operations executed (CPU time). For the sake of presentation we adopt the notation $n = N/B$, $m = M/B$, $k = K/B$.

Previous Work. Several (internal memory) algorithms for the segment intersection problem are known that execute an optimal number of (internal) operations $O(N\log N+K)$, both sequentially (deterministic [9] and randomized [14, 26]) and in parallel (deterministic [2] and randomized [13]). In the external-memory model, however, no I/O-optimal solution is known and the best algorithm has been devised by Arge et al. [5]. It is deterministic, involved, requires $O((n+k)\log_m n)$ I/Os, which is non-optimal, and computes the arrangement *only* for the case $k = 0$ (i.e., no crossings). Goodrich et al. [21] and Arge et al [5] presented some external-memory techniques that solve I/O-optimally other geometric problems. In particular, 2-d and 3-d convex hulls (using $O(n\log_m n)$ I/O operations), and the batched point location (using $O((n+k)\log_m n)$ I/O operations where now K is the number of queries). However, the resulting algorithms are complicated and require the use of several non-trivial I/O-optimal subroutines to solve some specific subproblems like 3-d maxima and 2-d convex hull. Furthermore, the approaches are not easily extensible to other geometric problems, like the segment intersection problem in which we are interested.

Recent results [11, 35, 17] have shown a close relationship between parallel algorithms and external-memory algorithms, so that it is natural to look at the work on parallel geometric solutions to see if some of those results can be adapted to work efficiently in the external-memory model. The parallel algorithms cited above for the segment intersections problem are based on *random sampling* (the one in [2] is obtained via derandomization), and follow a divide-and-conquer approach (originated in [14]) in which a random sample is used to divide the problem into subproblems that are then solved independently. The main difficulty which arises with this approach is the need to bound the total size of the subproblems; as a result, some problem-dependent tricks are usually needed to achieve optimal results—like *filtering* [32] (also called *pruning*) and *vertex-accounting* [8]. The necessity of bounding the subproblems size was overcome in [13] by limiting the divide-and-conquer approach to only one level of recursion and using suboptimal algorithms for the induced subproblems. Among these parallel approaches, only the algorithm in [32] has been adapted so far to work in external memory by Goodrich et al. [21] thus achieving I/O-optimal solutions for the 3-d halfspace intersection problem (hence for 3-d convex hull and 2-d Euclidean Voronoi diagrams, see comments above).

Another approach based on random sampling is the random incremental construction, RIC (also originated in [14]). This approach adds the *objects* (e.g. segments) one at time, $S_i = S_{i-1} \cup \{s\}$, with the choice made at random, and updates the arrangement $\mathcal{T}(S_i)$ at each step accordingly. Although the RIC approach does not have the problem of controlling the total size of the subproblems, it is inherently sequential in its original formulation. However, there is a natural way to parallelize it using a *gradation*, that is, a geometrically growing random sequence of subsets of the input objects, $S_0 \subseteq \cdots \subseteq S_l = S$. In this case, $\mathcal{T}(S_i)$ is constructed from $\mathcal{T}(S_{i-1})$ by adding the objects in $S_i - S_{i-1}$. Gradations have been mostly used to build data structures, as in the work of Mulmuley [27] and Mulmuley and Sen [28] (in fact, each *level* of the gradation is constructed from scratch using a regular RIC algorithm). Chazelle [8] and Brönnimann et al. [7] used a gradation approach as an intermediate step to derandomize the incremental construction of convex hulls, and this was followed by Amato et al. [2] to devise a parallel algorithm. The RIC approach with gradations has not received much attention otherwise, perhaps because sequentially it provides no advantage over the standard RIC algorithms, and in parallel it gives only expected bounds while there the emphasis is most often on high probability bounds.

Our Results. We show that the RIC approach can be adapted via *gradations* to provide a randomized incremental technique for external memory that results in expected I/O optimal algorithms for several geometric problems. The algorithms are also optimal in the expected number of internal operations performed (CPU time). By a suitable choice of the parameters of the gradation, the incremental step from $\mathcal{T}(S_{i-1})$ to $\mathcal{T}(S_i)$ can be performed by using efficient internal memory algorithms together with simple I/O efficient procedures—like straightforward movement of data in blocks or, in the most sophisticated case, external-memory sorting routines [29].

As the main result, we obtain an I/O-optimal randomized algorithm for the segment intersections problem that requires $O(n \log_m n + k)$ expected I/O operations. Since the algorithm necessarily computes the pairwise intersections among the input segments, this settles a question posed in [5]. The algorithm can also be adapted to handle degeneracies (particularly, to achieve output sensitivity with respect to the number of actual intersection points, not of pairwise intersections).

Our technique allows also to design algorithms with optimal expected I/O-bounds for the problems of 2-d and 3-d convex hulls, 2-d abstract Voronoi diagrams and batched point location in a planar subdivision. Although I/O-optimal algorithms for these problems are already available (but for 2-d Voronoi diagrams, only the Euclidean case), our RIC approach via gradations represents a uniform and simpler solution to those problems.[1]

Our algorithms apply to all values of the model parameters B and M and problem size N. Some steps of the algorithms are only needed to cope with extreme values of M, N and B, e.g., $B \leq M/\log^2 M$ and $\log n \geq (MB)^{1/4}$. For practical values of N, M, and B, an implementation would require only modules that are, for example, available in LEDA ([25]), TPIE([36, 37]), and LEDA-SM ([15]), namely optimal internal memory algorithms for trapezoidal decompositions, point location in trapezoidal decompositions, and topological sorting of graphs, and optimal external-memory algorithms for building, scanning, and sorting lists.

Finally, our results also extend to the external-memory model with D disks, in which an I/O-operation can transfer D disk blocks simultaneously, one for each disk. It seems likely that the approach is also efficient for a model with multiple CPU's and multiple disks.

Basic Preliminaries. The RIC approach is valid with great generality within the framework of *configuration spaces* [14, 27]. However, for the sake of simplicity, the presentation in Sections 2, 3 and 4 is restricted to the problem of computing the *trapezoidal decomposition* of an arrangement of N segments with $K(S)$ pairwise intersections. This decomposition is obtained by extending each segment endpoint and each intersection point between segments vertically upward and downward until it hits a segment; the trapezoids are the resulting connected regions of the plane. Given a subset $R \subseteq S$ of segments (*objects*), the set of resulting trapezoids (*cells*) is denoted by $\mathcal{T}(R)$. For $\sigma \in \mathcal{T}(R)$, S_σ denotes those

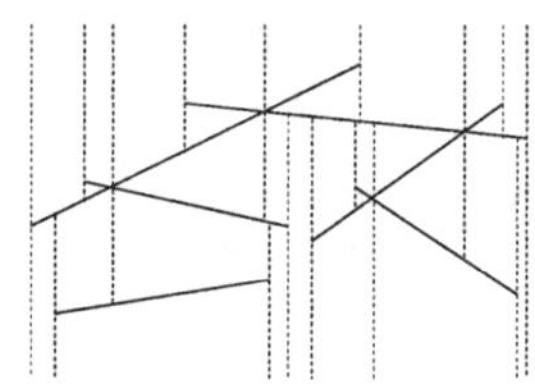

Figure 1: Trapezoidal decomposition

segments in S that intersect the interior of σ, and is called the *conflict list*. Let $N_\sigma = |S_\sigma|$. A *p-sample* R from S is obtained by choosing every $s \in S$ into R independently with

[1]The external-memory RIC approach also extends to other higher dimensional problems like convex hulls, hyperplane arrangements, etc. However, in those cases, the need for a *sorting step* leads to algorithms with I/O bounds which are suboptimal by a factor $O(\log_m n)$. For simplicity of exposition and importance/applicability of achieved results, we prefer to limit our presentation to the lower dimensional problems.

probability p.[2] Note that $f(p,S) = p|S| + p^2 K(S)$ is the expected size of $\mathcal{T}(R)$ for a p-sample S (for simplicity of presentation, throughout we ignore multiplicative constants). There are two main properties of this sampling process that are relevant to the analysis of the algorithms [14, 12, 27]. First, *the average conflict list size is at most* $1/p$. More precisely, for a constant $C > 0$:[3]

$$\mathbf{E}\left[\sum_{\sigma\in\mathcal{T}(R)} N_\sigma\right] \le C\frac{f(p,S)}{p}. \tag{1}$$

Second, *the deviation of the conflict list size is* $O(\log s)$ *with high probability*. More precisely, for $s \ge pN$, given $c > 0$ there is $C > 0$, such that with probability at least $1 - 1/s^c$:

$$\max_{\sigma\in\mathcal{T}(R)} N_\sigma \le C\frac{\log s}{p}. \tag{2}$$

The size function $f(p,S) = p|S| + p^2K(S)$ is also valid for the other problems we will investigate, by setting $K(S) = 0$. Though somewhat greater generality is possible, the final analysis that we present in Appendix B is restricted to this function. For the optimal algorithms under general conditions, the concept of a $(1/r)$*-cutting* for S is needed [10, 24]:[4] *A decomposition of the underlying space into a set* T *of disjoint cells such that* $\max_{\sigma\in T} N_\sigma \le N/r$. In particular, the following fact is used (see Appendix A.2 for more details):

Fact 1.1 *There is a randomized algorithm that, given a set* S *of* N *objects, constructs a* $(1/r)$*-cutting for* S *of size* $O(r^c)$*, requiring* $O(r^cN)$ *expected operations, where* c *is a small constant depending on the problem.*

Contents. In Sections 2 and 3 we present the RIC approach via gradations and its implementation in the external memory model, using the segment intersection problem as a concrete example. In Section 4 we describe simpler algorithms under restricted but practical conditions. In Section 5, we describe the changes needed to include other applications and state some of them. In the last section, we indicate the extension to multiple disks. In the Appendix, state some additional facts on sampling in configuration spaces, provide the analysis for the RIC approach via gradations, and for the segment intersection problem give some algorithmic details omitted in the main part.

2 RIC via Gradations

We present the RIC approach via gradations using the segment intersection problem as a concrete example, for the sake of simplicity. The extension to other problems will be clear, except that some problem dependent details will have to be provided in each case (e.g. clean-up in Step 3). To facilitate the presentation we use the following additional notation: Given a set of objects X and a set of cells T, we write $T[X]$ to denote the set of conflict lists X_σ for $\sigma \in T$, and write $|T[X]|$ to denote $\sum_{\sigma\in T}|X_\sigma|$.

2.1 Gradation

The algorithm is a variant of the RIC approach [14] and follows [8, 7]. Given parameters μ and μ', it choses a sequence

$$\emptyset = S_0 \subseteq S_1 \subseteq \ldots \subseteq S_{l-1} \subseteq S_l = S$$

of subsets of S, called a *gradation*, where S_{i-1} is a $(1/\mu)$-sample from S_i for $i < l$ and S_{l-1} is a $(1/\mu')$-sample from $S_l = S$.[5] Then, it iteratively constructs the decomposition $\mathcal{T}(S_i)$, for all i, $1 \le i \le l$. At the beginning of the i-th *round*, the decomposition $T_{i-1} = \mathcal{T}(S_{i-1})$ and its conflict lists $T_{i-1}[S]$ are available (initially, $S_0 = \emptyset$ and thus T_0 consists of a single "unbounded" cell with conflict list S), then the algorithm constructs the new decomposition $T_i = \mathcal{T}(S_i)$ by using T_{i-1} and $T_{i-1}[S]$. We refer to the l-th round as the *last* round, and to all the others as *early* rounds. Let R_i be $S_i - S_{i-1}$, the set of objects added in the i-th round, and for each cell $\sigma \in T_{i-1}$, let $R_{i,\sigma}$ be the subset of the objects in R_i which are conflicting with σ. Because of the random sampling, the sets $R_{i,\sigma}$ will be well balanced on the average. More precisely, for early rounds, the average is at most μ, and for the last round is at most μ' (recall Eqn. (1)).

2.2 A Round

The i-th round computes T_i and $T_i[S]$ in three steps:

1. *Intermediate decomposition:* For each $\sigma \in T_{i-1}$, identify $R_{i,\sigma}$ by scanning S_σ and taking from it the segments which belong to R_i. Then compute[6] the restriction of $\mathcal{T}(R_{i,\sigma})$ to σ,[7] denoted T_σ. This results in an intermediate decomposition $T_i^I = \bigcup_{\sigma\in T_{i-1}} T_\sigma$.
2. *Intermediate conflict lists:* For each $\sigma \in T_{i-1}$, compute the conflict lists $T_\sigma[S]$ by taking each $s \in S_\sigma$ at a time and after an *initial search* in T_σ to locate an endpoint of s, determine the conflicts of s with the cells in T_σ by appropriately walking through it. These are the intermediate conflict lists $T_i^I[S]$.
3. *Clean-up:*[8] Obtain T_i and its conflict lists $T_i[S]$ from T_i^I and its conflict lists $T_i^I[S]$. Observe that $\tau \in T_i$ can be *chopped* into pieces $\tau\cap\sigma$, for $\sigma \in T_{i-1}$. So we need to stitch together τ from its pieces $\tau\cap\sigma$ and also build its conflict list from the conflict lists of its pieces.[9]

[2] The sampling result in Appendix A.1 also hold in the sampling model where a subset R of size $r = pN$ is chosen at random among all subsets of that size. Therefore, the complexity of our algorithms also hold in that model.

[3] A more general version is actually needed in the analysis and reviewed in the Appendix A.1.

[4] As shown in Sections 4.1 and 4.2, making use of the high probability bound in Eqn. (2), the use of cuttings in our algorithms can be avoided for a large, and practically reasonable, range of values of the parameters N, B, M. In these cases, we can obtain a considerably simpler (practical) algorithm that achieves optimal I/O bounds. Thus, cuttings are a useful tool only for our theoretical result.

[5] The need for two parameters will become clear when discussing the external-memory implementation.

[6] If $|R_{i,\sigma}|$ is $O(1)$ then a nonoptimal polynomial algorithm would suffice here. In our external-memory implementation, $|R_{i,\sigma}|$ is relatively large and we will require that an optimal internal memory algorithm be used.

[7] Although it holds here, for other applications the cells determined by X when restricted to σ are not necessarily the cells determined by X, without the restriction, intersected with σ.

[8] We need to recover $\mathcal{T}(S_i)$ so that we can apply the sampling results for S_i with respect to S. If the clean-up is not performed, then we could only use the sampling results locally in each cell for S_i with respect to S_{i-1}, and we cannot prove optimality (although the algorithm might still be optimal).

[9] This is not the case for other applications; that is, in general $\tau \in T_i$ cannot be recovered from pieces $\tau\cap\sigma$. This step is very problem dependent.

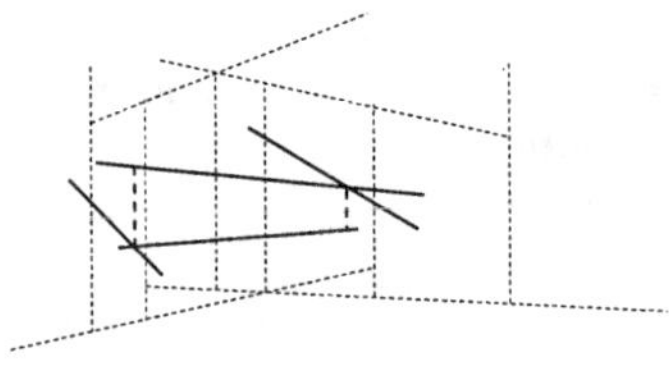

Figure 2: A chopped trapezoid

Analysis of internal running time. For simplicity, here as well as in later analysis, we ignore multiplicative constants in the bounds. The steps are implemented so that: (i) in Step 1, an optimal algorithm is used, requiring a number of operations at most $t_0(X) = |X| \log |X| + K(X)$; (ii) in Step 2, the initial search requires $O(\log |R_{i,\sigma}|)$ operations, and the walk to determine the conflicts requires a number of operations proportional to the number of conflicts found; and (iii) Step 3 requires a number of operations proportional to the total size of the intermediate conflict lists $T_i^I[S]$, and to the total size of the resulting conflict lists $T_i[S]$. It is well-known how to achieve (ii) and (iii) [14, 26]. Then the total number of operations required in the i-th round is

$$\sum_{\sigma \in T_{i-1}} \left(K(R_{i,\sigma}) + N_\sigma \log |R_{i,\sigma}| + \sum_{\tau \in T_\sigma} N_\tau \right). \quad (3)$$

The last round is simpler in that Step 2 is not needed at all, and in Step 3 no conflict lists are computed. In the expression above for the cost, only the first two terms are needed (note that $R_{l,\sigma} = S_\sigma$). In Appendix B, the expectation of the total number of operations performed by the algorithm is computed for $f(p, S) = p|S| + p^2 K(S)$, thus showing the following.

Theorem 2.1 *The RIC approach via gradations solves the trapezoidal decomposition problem using an optimal expected number of operations $O(N \log N + K)$.*

3 Algorithm in External Memory

Recall the notation $m = M/B$, $n = N/B$ and $k = K/B$. We present an I/O-efficient implementation of the algorithm in the previous section. First, as a technical point, we assume that the gradation is constructed before the algorithm starts, so that for each object $s \in S$ there is an associated *tag* that indicates the round in which it is *inserted.* The tag is carried by each copy of s (in each conflict list) so that, in the i-th round, the sets $R_{i,\sigma}$ can be easily determined by a scanning of the conflict lists S_σ. This is important in an efficient external memory implementation where we cannot assume random access to data.

The choice of parameters μ and μ' is done as follows: $\mu = m^{1/2}$ and $\mu' = \max\{B, M^{1/2}\}$. Therefore, the expected number of levels in the gradation is $l = O(\log_\mu(N/\mu')) = O(\log_m n)$.[10] The main observations that lead to an I/O efficient implementation are:

(i) The choice of μ' implies that $f(1/\mu', S) = O(f(S)/B)$. As a result, in the i-th round, the algorithm does not need to handle T_{i-1} in an I/O-efficient manner: even incurring *one* I/O per operation of the internal memory algorithm is acceptable. Only the last round must handle $T_i = T_l$ in an I/O-efficient manner, but then this is aided by the knowledge of T_{l-1}.
(ii) The choice of μ implies that in each early round the average size of T_σ is at most $\mu^2 = M/B$ (since the average size of $R_{i,\sigma}$ is at most μ). Therefore, ignoring deviations, T_σ can be computed in internal memory and one block of memory allocated for each cell of T_σ. The latter fact allows the conflict lists to be written in an I/O-efficient manner. Similarly, in the last round, the choice of μ' implies that, also ignoring deviations, the average size of $R_{l,\sigma} = S_\sigma$ is at most μ'. If $B^2 \leq M$ then $\mu' = M^{1/2}$ and an optimal internal memory algorithm can be used to construct T_σ. If $B^2 > M$ then $\mu' = B$ and an I/O-optimal external-memory algorithm that can handle small inputs of size less than B is required.

To simplify the presentation, we describe the algorithm in two stages. First, we ignore the deviations from the average values of $|R_{i,\sigma}|$ and describe an algorithm that performs an optimal number of I/Os. Specifically, we assume $|R_{i,\sigma}| \leq \mu$ for $i < l$ and $|R_{l,\sigma}| \leq \mu'$. Then, using a refinement approach of Chazelle and Friedman [10], we obtain an algorithm that always works, at the cost of complicating the algorithm. In the next section, we will show that the original simpler but incomplete algorithm actually leads to an algorithm with optimal I/O bounds under certain reasonable conditions on N, M and B.

3.1 Ignoring Deviations

We discuss the I/O operations needed in each of the three steps of the basic algorithm when implemented in external memory. We will refer later to this as Algorithm **I**. The number of internal memory operations remains as analyzed in the previous section. As before, to simplify the equations, we ignore multiplicative constants.

Step 1. Each cell $\sigma \in T_{i-1}$ in turn and its conflict list S_σ are loaded into *internal memory,* so that $R_{i,\sigma}$ is determined by checking the tags, and $T_\sigma = \mathcal{T}(R_{i,\sigma})$ is computed. Since the size of T_σ is at most $\mu^2 \leq M/B$, then it can be computed by an optimal internal memory algorithm without performing any I/Os. Thus, the number of I/Os required by this step is:

$$\sum_{\sigma \in T_{i-1}} \left(\left\lceil \frac{N_\sigma}{B} \right\rceil + \left\lceil \frac{|R_{i,\sigma}|}{B} \right\rceil + \left\lceil \frac{|T_\sigma|}{B} \right\rceil \right) \leq |T_{i-1}| + \frac{|T_{i-1}[S]|}{B} + \frac{|T_i^I|}{B}.$$

Step 2. Since $|T_\sigma| \leq m^2 = M/B$, by assumption, we can reserve in internal memory a buffer of size B for each cell $\tau \in T_\sigma$. The conflict lists $T_\sigma[S]$ for $\sigma \in T_{i-1}$ are computed by scanning S_σ and walking through T_σ in internal memory. As a conflict of s with some τ is determined, it is written into the buffer corresponding to τ. As a buffer becomes full, it is written to external memory. In this way, the number of I/Os is proportional to the size of the scanned and returned conflict lists, divided by B. Therefore, the number of I/Os required by this step is:

$$\sum_{\sigma \in T_{i-1}} \left\lceil \frac{N_\sigma}{B} \right\rceil + \sum_{\sigma \in T_{i-1}} \sum_{\tau \in T_\sigma} \left\lceil \frac{N_\tau}{B} \right\rceil \leq$$

$$\frac{|T_{i-1}[S]|}{B} + \frac{|T_i^I[S]|}{B} + |T_i^I| + |T_{i-1}| \leq \frac{|T_i^I[S]|}{B} + |T_i^I|.$$

Step 3. We have the intermediate decomposition T_i^I and need to determine the decomposition T_i. As noted earlier,

[10] This also holds with high probability $1 - 1/n^c$, since $n(1/\mu)^{(c+1)\log_\mu n} = 1/n^c$.

a trapezoid $\tau \in T_i$ can be chopped into pieces $\tau \cap \sigma$, for $\sigma \in T_{i-1}$. To achieve optimal bounds, the stitching of pieces $\tau \cap \sigma$ has to be performed using a number of I/Os proportional to the total number of blocks of size B required to hold the conflict lists $T_i^I[S]$, and the resulting conflict lists $T_i[S]$. Achieving this goal presents some difficulties because we cannot afford to use sorting;[11] still we can do it in an I/O efficient manner. The idea is to traverse the decomposition T_{i-1} following a linear ordering consistent with the partial ordering between trapezoids induced by their vertical adjacencies. The linear ordering is obtained by *topologically sorting* the partial order. A difficulty is that we do not know how to perform this operation with a "linear" number of I/Os, that is, $|T_{i-1}|/B$. Fortunately, $|T_{i-1}|$ is "small", and one can afford to use an optimal internal memory algorithm that has a bad I/O performance (it could even perform one I/O per internal memory operation). As T_{i-1} is traversed, the chopped trapezoids of T_i are put together by maintaining a list of those that cross the right vertical boundary of the trapezoid in T_{i-1} currently considered, and matching this list to the list of chopped trapezoids in T_i that cross the left vertical boundary of the adjacent trapezoid in T_{i-1} when it comes under consideration. Further details are given in Ap-

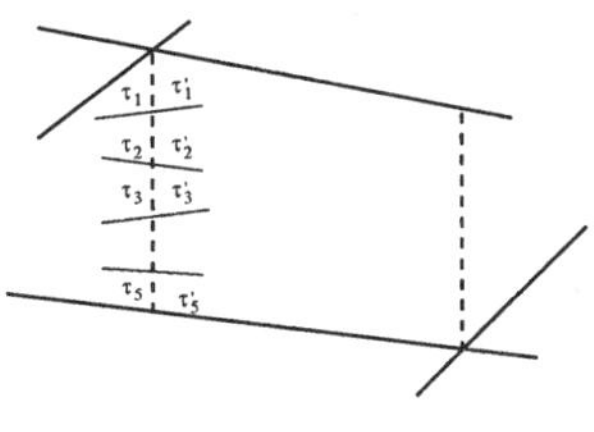

Figure 3: Trapezoids crossing a vertical edge

pendix C. The conclusion is that Step 3 can be performed using a number of I/Os proportional to the total number of blocks of size B required to hold the conflict lists $T_i^I[S]$ and the resulting conflict lists $T_i[S]$:

$$\sum_{\sigma \in T_{i-1}} \sum_{\tau \in T_\sigma} \left\lceil \frac{N_\tau}{B} \right\rceil + \sum_{\sigma \in T_i} \left\lceil \frac{N_\sigma}{B} \right\rceil \leq$$

$$\frac{|T_i^I[S]|}{B} + \frac{|T_i[S]|}{B} + |T_i^I| + |T_i| \leq \frac{|T_i^I[S]|}{B} + |T_i^I|.$$

Essentially, Steps 1–3 require an *I/O-efficient* handling of the conflict lists $T_{i-1}[S]$, $T_i^I[S]$; that is, we need to devise external-memory algorithms that manage them by executing a number of I/Os which is proportional to their sizes divided by the block size B. On the other side, the implementation of Steps 1–3 allow us to handle in an *I/O-inefficient* way the decompositions T_{i-1}, T_i^I and T_i; that is, we can design *inefficient* external-memory algorithms that manage those decompositions by executing a number of I/Os which is proportional to their whole size.

The Last Round. The goal is now to compute the final decomposition $T_l = \mathcal{T}(S)$ from T_{l-1}. We recall that S_{l-1} is a $(1/\mu')$-sample from S, and we assumed that $|R_{l,\sigma}| \leq \mu'$. We need an optimal algorithm that handles the small case, that is, it computes $\mathcal{T}(S)$ for $|S| \leq \mu' \leq M$ requiring $t_1(S) = (|S|/B)\log_m(|S|/B) + K(S)/B$ I/Os. This is trivial if $K(S) = 0$ because we can use an internal-memory algorithm that require linear space. But when $K(S) \neq 0$, $\mathcal{T}(S)$ might not fit in internal memory at once and a proper *optimal external-memory* algorithm is needed. Such an algorithm can be obtained again using sampling: Take a sample of size $\sqrt{M}$, compute its decomposition using an internal memory algorithm, compute its conflict lists, compute each of the resulting subproblems again in internal memory, and finally put together the result. Further details are given in Appendix C.

1.1 For each trapezoid $\sigma \in T_{l-1}$, compute its decomposition T_σ according to the segments $R_{l,\sigma} = S_\sigma$ using the small size case algorithm, since $|R_{l,\sigma}| = N_\sigma \leq \mu' \leq M$.

1.2 Obtain $\mathcal{T}(S)$ from the collection T_σ, for all $\sigma \in T_{l-1}$.

Thus, Step 1.1 takes $O((N_\sigma/B)\log_m(N_\sigma/B) + K_\sigma/B)$ I/Os on each $\sigma \in T_{l-1}$. Summing over all the trapezoids $\sigma \in T_{l-1}$, and taking expectation, it is shown that the expected number of I/Os is $O(n \log_m n + k)$.[12] Step 1.2 is indeed exactly the same as Step 3 of the i-th early round (above), with the further simplification that now there are no conflict lists to merge. We therefore just need to stitch appropriately the trapezoids in the collections T_σ's, for $\sigma \in T_{l-1}$. Hence, the I/O-cost of Step 1.2 is $|T_{l-1}^I[S]|/B + |T_{l-1}^I|$.

Finally, using the analysis for the basic algorithm (Appendix B), with $f(p,S) = p|S| + p^2K(S)$, the following theorem is obtained. Note that by ignoring the deviations and then proceeding with the analysis, we are actually assuming that the internal memory can accomodate $R_{i,\sigma}$, its decomposition and corresponding buffers independently of its deviations.

Theorem 3.1 *Let us assume that the internal memory can always accomodate $R_{i,\sigma}$, its decomposition T_σ and the corresponding $|R_{i,\sigma}|$ buffers of size B. Then Algorithm **I** solves the trapezoidal decomposition problem using $O(n\log_m n + k)$ expected I/Os, which is optimal.*

The algorithm can be modified so that it handles degeneracies while still achieving an optimal I/O bound: $O(n \log_m n + i)$ where $i = I/B$ with I the number of intersection points (which can be much smaller than the number of pairwise intersections). See Appendix C for details.

3.2 Handling Deviations

Now we show how to remove the assumptions made on $|R_{i,\sigma}|$, and cope with its *deviations*. Specifically, this is achieved by using a refinement approach of Chazelle and Friedman [10, 24] (see Fact A.2 in Appendix A.2 for details). For $i < l$, the idea is to construct using Fact 1.1, for each $\sigma \in T_{i-1}$, a $(1/t_\sigma)$-cutting $\hat{T}_\sigma$ for $R_{i,\sigma}$ restricted to σ, where t_σ is the excess of σ (i.e., $t_\sigma = |R_{i,\sigma}|/\mu$). Since each cell $\tau \in \hat{T}_\sigma$ satisfies the desired constraint on the size of its conflict list, we can apply to it the Steps 1 and 2 of the i-th round in Algorithm **I**. Then, to take care of the independent refinement performed by the cutting process on each cell $\sigma \in T_{i-1}$, we perform *two levels of clean-up*; both levels are similar to Step 3 in Algorithm **I**. The first clean-up obtains T_σ and $T_\sigma[S]$ from the T_τ's and $T_\tau[S]$'s with $\tau \in \hat{T}_\sigma$, and the second one obtains the final decomposition $T_i = \mathcal{T}(S_i)$ and its conflict lists $T_i[S]$ from T_i^I and $T_i^I[S]$.[13] Similarly, for the last round.

[11] See Section 5 for further discussion.

[12] Using Equation (4) in Appendix A.1 with $g(x) = (x/B)\log_m(x/B)$.

[13] In some other applications, e.g. 3-d halfspace intersection, the clean-up can be easily performed in one level.

Complete i-th round algorithm. The modified algorithm for the i-th early round (i.e., $i < l$) is described below. For simplicity of exposition, we also specify between square brackets the small changes which are needed to get the modified last round. We will refer to this algorithm as Algorithm **H**. Note that Step 2.1 uses an algorithm for the special case $N \leq M$ which is described in Appendix C.

1. For each $\sigma \in T_{i-1}$, use Fact 1.1 to construct a $(1/t_\sigma)$-cutting $\hat{T}_\sigma$ for $R_{i,\sigma}$ restricted to σ, where $t_\sigma = |R_{i,\sigma}|/\mu$ is the excess of σ. Thus each $\tau \in \hat{T}_\sigma$ satisfies the desired relation $N_\tau \leq \mu$. Additionally, compute the conflict lists S_τ in $\hat{T}_\sigma[S]$ by checking each $s \in S_\sigma$ versus each $\tau \in \hat{T}_\sigma$. This computation allows also to determine the list $R_{i,\tau}$. [In the last round, we have $t_\sigma = |R_{i,\sigma}|/B$ so that $N_\tau \leq \mu' = B$.]
2. For each $\sigma \in T_{i-1}$ do the following:
   - 2.1. For each $\tau \in \hat{T}_\sigma$, load $R_{i,\tau}$ in internal memory and compute $T_\tau = \mathcal{T}(R_{i,\tau})$. [In the last round, apply the algorithm for $N \leq M$ (see Appendix C).]
   - 2.2. For each $\tau \in \hat{T}_\sigma$, compute the conflict lists $T_\tau[S]$, by scanning S_τ and walking in T_τ. [In the last round, this step is not executed.]
   - 2.3. Obtain $T_\sigma = \mathcal{T}(R_{i,\sigma})$ and its conflict lists $T_\sigma[S]$ from the collection of T_τ's and their conflict lists $T_\tau[S]$, where $\tau \in \hat{T}_\sigma$. [In the last round, we just obtain T_σ.]
3. Obtain T_i and its conflict lists $T_i[S]$ from the intermediate decomposition T_i^I and its conflict lists $T_i^I[S]$. [In the last round, we just obtain $T_l = \mathcal{T}(S)$.]

Steps 2.1–2.3 are just like the corresponding Steps 1–3 of Algorithm **I**, and Step 3 here is like the corresponding Step 3 there. From Fact 1.1, the size of a $(1/t_\sigma)$-cutting for S_σ is $O(t_\sigma^c)$ and the expected number of required operations to construct it is $O(t_\sigma^c N_\sigma)$. Consequently, it turns out to be essential for the overall I/O-efficiency of our complete algorithm that on the average t_σ^c behaves *as a constant* when multiplied by N_σ^c, by $|R_{i,\sigma}|$ and by $\sum_{\tau \in T_\sigma} N_\tau$. More precisely, the following relations can be proved using Eqn. (4). The important point is that the upper bounds on the right are the same as for the quantities on the left without the t_σ factors. Here $\mathbf{E}_i$ and $\mathbf{E}_{i-1,i}$ denote expectations over the random choices of S_i, and of both S_{i-1}, S_i respectively (see Appendix B).

Fact 3.2 *The following relations hold for* $f(S) = |S| + K(S)$*:*

(i) $\mathbf{E}_{i-1}\left[\sum_{\sigma \in T_{i-1}} t_\sigma^c N_\sigma^e\right] \leq \frac{1}{q_{i-1}^e} f(q_{i-1}, S)$.

(ii) $\mathbf{E}_{i-1,i}\left[\sum_{\sigma \in T_{i-1}} t_\sigma^c |R_{i,\sigma}|\right] \leq \mu f(q_{i-1}, S)$.

(iii) $\mathbf{E}_{i-1,i}\left[\sum_{\sigma \in T_{i-1}} t_\sigma^c \sum_{\tau \in T_\sigma} N_\tau\right] \leq \frac{1}{q_i} f(q_i, S)$

These relations allow us to prove that, within a constant factor, the I/O cost of Algorithm **H** is the same as for Algorithm **I**. We verify this for the i-th round below.

Step 1. By Fact 1.1, the decomposition $\hat{T}_\sigma$ (and its conflict lists with $R_{i,\sigma}$) has size $O(t_\sigma^c)$ and is computed in expected $O(t_\sigma^c |R_{i,\sigma}|)$ operations, hence within the same I/O-bound (at most one I/O per operation). Fact 3.2(ii) shows that this I/O-bound can be afforded. From Fact 3.2(i), with $d = c$ and $e = 1$, we infer that one can also afford to compute the conflict lists by brute force.

Step 2. From Fact 3.2(i-iii), we conclude that the I/O bounds obtained for Steps 1-3 of the Algorithm **I** also apply to Steps 2.1–2.3 of Algorithm **H**. In fact, by Fact 3.2(iii), the overall conflict lists S_τ are not larger than the S_σ's within a constant factor.

Step 3. This is the same as Step 3 of Algorithm **I**, so its I/O-bound also applies here.

Theorem 3.3 *Algorithm* **H** *solves the trapezoidal decomposition problem using* $O(n \log_m n + k)$ *expected I/Os, which is optimal.*

4 Simplified Algorithms Under Practical Conditions

4.1 First Version

We want to obtain an algorithm that achieves the optimal I/O bounds but does not make use of cuttings. We can achive this goal under some conditions on the parameters N, M and B that hold for current computers and applications. Recall that Algorithm **I** performs an optimal expected number of internal operations and that, under certain assumptions, it also performs an optimal expected number of I/Os. We modify Algorithm **I** as follows when executed on an internal memory of size M: (i) the gradation parameters are $\mu = m^{1/4}/C$ and $\mu' = M/(C^2 \log M \log n)$, where C is determined by Eqn. (2) with $c = 2$; (ii) if at a given round $|R_{i,\sigma}|^2 > m$, then it performs I/Os as necessary to access data that does not fit in internal memory (up to one I/O per internal memory operation); (iii) the small size case is $N \leq M/C \log M$ and the algorithm used is the one described in Appendix C.2 (without the use of cuttings as the resulting subproblems have size at most $\sqrt{M}$). Let us call this modified version Algorithm **I'**. The following theorem establishes conditions for N, M, B under which Algorithm **I'** performs, within a constant factor, the same expected number of I/Os as Algorithm **I**.

Lemma 4.1 *Algorithm* **I'** *solves the segment intersection problem using an optimal expected number of I/Os, under the conditions* $\max\{B, 2\log m\} \leq n$*,* $B \leq M/\log^2 M$ *and* $\log n \leq m^{1/4}$.

Proof. First, note that $\mu' \geq 1$ because $\log n \leq M/C \log M$ is weaker than the assumption $\log n \leq m^{1/4}$. For the i-th round, let us denote by I_i (resp. I_i') the expected number of I/Os performed by Algorithm **I** (resp. Algorithm **I'**), and let W_i be the expected number of internal operations performed. If ρ_i is the probability that the assumptions do not hold during the i-th round (i.e., the probability that the algorithm starts executing an I/O per internal memory operation), then $I_i' \leq I_i + \rho_i W_i$. Then, adding over all the rounds, $I' \leq I + \sum_i \rho_i W_i \leq I + \alpha B I \max_i \rho_i$, where $\alpha = \log N / \log_m n$ (using $I = n \log_m n + k$ and $W = N \log N + K$). So we only need to establish the conditions under which $\max_i \rho_i \leq 1/\alpha B$ holds, and then $I' = O(I)$ will follow.

Notice that $1 - \rho_i$ (resp. $1 - \rho_l$) is the probability that $|R_{i,\sigma}| \leq m^{1/4}$ (resp. $|R_{l,\sigma}| \leq M/C \log M$). Since S_{i-1} is a $(1/\mu)$-sample (resp. $(1/\mu')$) from S_i, using Eqn. (2) with $c = 2$, we obtain that $|R_{i,\sigma}| \leq C\mu \log s$ (resp. $|R_{l,\sigma}| \leq C\mu' \log s$) with probability at least $1 - 1/s^2$, provided that $s \geq |S_i|/\mu$. Using the assumption $B \leq n$ and setting $s = \max\{n, B\} = n$, it suffices to enforce $C\mu \log s \leq m^{1/4}$,

$C\mu' \log s \leq M/C \log M$ and $1/n \leq 1/\alpha$ to guarantee that the constraint on $|R_{i,\sigma}|$ fails with probability at most $1/\alpha B$. These conditions hold under the additional assumptions $\log n \leq m^{1/4}$ and $2 \log m \leq n$ (since $\log N \leq 2 \log n$ because $B \leq n$).

Still, we need to check that the choice of μ' does not affect the optimality of the algorithm because the size of T_{l-1} becomes too large (since we allow a number of I/Os linear in the size of T_l). So we need $N/\mu' + K/\mu'^2 \leq C'(n \log_m n + k)$ or, splitting into two inequalities, $B \leq C\mu' \log_m n$ and $B \leq C^2 \mu'^2$. This holds with the additional condition $B \leq M/\log^2 M$ (together with $\log n \leq M^{1/2}$ which is weaker than $\log n \leq m^{1/4}$).

Finally, the small size case $N \leq M/C \log M$ does not need to use cuttings because for a sample of size $\sqrt{M}$, each of the trapezoids in its decomposition have conflict list size at most $C(M/C \log M) \log M/\sqrt{M} = \sqrt{M}$ with probability at least $1/2$, so we can repeat the sampling until a good one is obtained. ■

The conditions $2 \log m \leq n$ and $B \leq M/\log^2 M$ are weaker than the others for practical values.

4.2 Improved Conditions

We take advantage of the averaging results on the excess t_σ, to modify Algorithm $\mathbf{I}'$ so that it performs well under less restrictive conditions. The new Algorithm $\mathbf{I}''$ proceeds as Algorithm $\mathbf{I}'$ with the exception that in Step 2 of the early rounds it uses buffers of size $b_\sigma = \max(1, B/t_\sigma^2)$, instead of B, to produce the conflict lists $T_\sigma[S]$ (see Section 2.2). Since $|R_{i,\sigma}|^2 B/t_\sigma^2 \leq M$, now the buffers can be allocated in the internal memory as long as $|R_{i,\sigma}|^2 \leq M$. As long as this is the case, since t_σ^2 behaves as a constant on the average, the expected total number of I/Os required by Step 2 is equal to the one required by Algorithm $\mathbf{I}'$. But now the algorithm performs a high number of I/Os when $\mathcal{T}(R_{i,\sigma})$ does not fit in the memory of size M, because then it could need even one I/O per internal memory operation.

Theorem 4.2 *Algorithm* $\mathbf{I}''$ *solves the segment intersection problem using an optimal expected number of I/Os, under the conditions* $\max\{B, 2\log m\} \leq n$, $B \leq M/\log^2 M$ *and* $\log n \leq (MB)^{1/4}$.

Proof. Fact 3.2 is used to conclude that reducing the size of the buffers to b_σ does not affect the expected number of I/Os performed by Step 2. Therefore, as long as the internal memory can hold $R_{i,\sigma}$ and its decomposition, that is as long as $|R_{i,\sigma}| \leq M^{1/2}$, the expected total number of I/Os is equal to that for Algorithm $\mathbf{I}$ within a constant factor.

The effect of $|R_{i,\sigma}| \leq M^{1/2}$ not holding is analized following an argument similar to that for Lemma 4.1. We want $C\mu \log s \leq M^{1/2}$ (in the early rounds) and $C\mu' \log s \leq M/C \log M$ (in the last round), where $s = \max(B, n) = n$. From the first requirement, we find the condition $\log n \leq M^{1/2}/m^{1/4} = (MB)^{1/4}$. The other conditions remain as in Lemma 4.1. ■

The last condition is a considerable improvement over the one in Lemma 4.1 for Algorithm $\mathbf{I}'$. The conditions hold for the values of N, M, B in current computers and applications; e.g. if $B = 8$Kb and $M = 64$Mb, they hold for N greater than 64Mb and as large as 2^{500}.

5 General Algorithm and Other Applications

5.1 General Algorithm

The RIC approach via gradations extends to many other problems in the framework of configuration spaces [14, 27]. The outline of the basic algorithm is the same as that in Section 2. In most cases the resulting algorithm is optimal in the expected number of (internal) operations performed. The external memory implementation is also just as outlined in Section 3 (Algorithms **I** and **H**). Only the Clean-up Step requires some additional comments.

This is a very problem dependent step. In general, it requires some data movement that must be performed efficiently in the number of I/Os required. However, operations like integer sorting, connectivity, and graph traversal that can be performed using $O(X)$ operations in internal memory cannot be performed with $O(X/B)$ I/Os in external memory [11]. Unfortunately, in some applications these expensive operations cannot be avoided and they are therefore implemented in external memory by means of a sorting step which takes $\mathrm{sort}(X) = (X/B) \log_m (X/B)$ I/Os to sort X items on the disk [29]. Therefore, in those cases, we have that in addition to the linear cost needed in the case of trapezoidal decomposition, Step 3 needs also a sorting step for T_i^I. Thus, the following term must be added to the I/O-complexity of that step

$$\mathrm{sort}\left(\sum_{\sigma \in T_{i-1}} |T_\sigma|\right) = \mathrm{sort}(|T_i^I|).$$

Note that this cost involves the decompositions, not the conflict lists which are not involved in the sorting.

This sorting cost in the Clean-up step is acceptable in the applications where $f(S) = |S|$ and the corresponding desired I/O bound is $n \log_m n$. Although the RIC approach via gradations extends to other problems like higher dimensional convex hulls and hyperplane arrangements, the bound that can be achieved is a factor $\log_m n$ away from optimality because a sorting step seems necessary. In the next section we describe some problems for which the resulting algorithm is optimal.

5.2 Other Applications

We state the problems and results. Details are omitted in this extended abstract.

Theorem 5.1 *The 3-d halfspace intersection problem is solved optimally by Algorithm* **H** *executing* $O(n \log_m n)$ *expected I/Os, and taking* $O(N \log N)$ *expected internal operations. Under the conditions* $B \leq n$ *and* $\log n \leq (MB)^{1/2}$, *the optimality is also achieved by the simpler Algorithm* $\mathbf{I}''$.

By standard geometric transformations [18], this also solves the 3-d convex hull and 2-d Euclidean Voronoi diagram problems. Clearly, a somewhat simpler algorithm also solves the 2-d convex hull problem.

In the batched planar point location one is given a set S of N interior disjoint edges that define a decomposition of the plane into *regions*, and for each point in a set P of K query points one must determine the region that contains it. This result has already been solved optimally by Arge et al. [5].

Theorem 5.2 *Algorithm* **H** *solves the batched planar point location problem using* $O((n+k)\log_m n)$ *expected I/Os, and taking* $O(N \log N + K)$ *expected internal operations. Under the conditions* $B \leq n$ *and* $\log n \leq (MB)^{1/2}$*, the optimality is also achieved by the simpler Algorithm* $\mathbf{I}''$.

Abstract Voronoi diagrams were introduced in [22]. We can deal with a subset of them considered in [3] that still includes a large number of concrete examples, e.g. Voronoi diagrams of line segments (the important point is that there is a cell decomposition).

Theorem 5.3 *Abstract Voronoi diagrams (as defined in [3]) are computed optimally by Algorithm* **H** *executing* $O(n \log_m n)$ *expected I/Os, and taking* $O(N \log N)$ *expected internal operations. Under the conditions* $B \leq n$ *and* $\log n \leq (MB)^{1/2}$*, the optimality is also achieved by the simpler Algorithm* $\mathbf{I}''$.

6 Multiple Disks

We discuss the modifications required to extend our results to systems with D disks. In the D disk model it is assumed that an I/O-operation can transfer D disk blocks, one from each disk. This implies that the main memory has size $M = \Omega(DB)$. In the D-disk model a set of L items can be scanned with $\lceil L/(DB) \rceil$ I/O-operations provided the L/B disk blocks containing the items are evenly spread over the disks. Under the same assumption the set can be sorted with $L/(DB)\log_m(L/(DB))$ I/O-operations, see [29].

We modify our algorithms such that the conflict list of every trapezoid spreads nearly evenly over the disks. This requires the following changes to steps two and three. We maintain a global buffer area of size DB in main memory. In step two we reserve as before a buffer area of size B for each trapezoid. When the buffer of a trapezoid is full, we write the block to the global buffer area, and when the global buffer area is full, we choose a random permutation π of the integers 1 to D and write the i-th block in the global buffer area to disk $\pi(i)$. This guarantees that the conflict list of every trapezoid spreads nearly evenly over the disks (assuming the conflict list size is $\Omega(DB \log D)$) [6]. The conflict lists in step three are handled in a similar way. Step three requires to topologically sort T_{l-1}. This may require an I/O-operations for every internal operation. We therefore set $\mu' = \max(DB \log D, M^{1/2})$ (the $\log D$ factor is so that conflict list sizes are $\Omega(DB \log D)$) and obtain an I/O-bound of $O(N/(DB)\log_m(N/(DB)) + K/(DB))$.

It seems likely that our approach is also efficient for a model with multiple disks and multiple CPUs; such as the models in [17, 38]. Steps one and two parallelize trivially by working on p trapezoids concurrently, where p is the number of CPUs. Step 3 is more difficult to parallelize. If T_{l-1} is sufficiently small, topological sorting may be performed with a single processor. Otherwise, since T_{l-1} is a planar graph, it follows from the planar separator theorem that T_{l-1} can be divided into about p pieces of size $|T_{l-1}|/p$ each by removing no more than $\sqrt{|T_{l-1}| \cdot p}$ adjacencies between trapezoids. We assign each piece to a single processor and let the processor do the clean-up across adjacencies within the piece. We then do the clean-up across all removed adjacencies on a single processor. As long as $|T_{l-1}|/p \geq \sqrt{|T_{l-1}| \cdot p}$ or $p \leq |T_{l-1}|^{1/3}$ this should result in perfect speed-up.

References

[1] A. Aggarwal and J. S. Vitter. The Input/Output complexity of sorting and related problems. *Communications of the ACM*, 31(9):1116-1127, 1988.

[2] N. M. Amato, M. T. Goodrich, and E. A. Ramos. Computing faces in segment and simplex arrangements. In *Proc. 27th Annu. ACM Sympos. Theory Comput.*, 672–682, 1995.

[3] N. M. Amato and E. A. Ramos. On computing Voronoi diagrams by divide-prune-and-conquer. In *Proc. 12th Annual ACM Sympos. Comput. Geom.*, 1996. 672–682, 1995.

[4] L. Arge, P. Ferragina, R. Grossi, and J. S. Vitter. On sorting strings in external memory. *Proc. ACM Symposium on the Theory of Computing (STOC '97)*, 540–548, 1997.

[5] L. Arge, D. E. Vengroff, and J. S. Vitter. External-memory algorithms for processing line segments in geographic information systems. In *Proc. Annual European Symposium on Algorithms, LNCS 979*, 295–310, 1995.

[6] R.D. Barve, E.F. Grove and J.S. Vitter. Simple randomized mergesort on parallel disks. In *Proc. IEEE Symp. on Foundations of Comp. Sci.*, 1996.

[7] H. Brönnimann, B. Chazelle, and J. Matoušek. Product range spaces, sensitive sampling, and derandomization. In *Proc. 34th Annu. IEEE Sympos. Found. Comput. Sci.*, 1993, 400–409.

[8] B. Chazelle. An optimal convex hull algorithm in any fixed dimension. *Discrete Comput. Geom.* **10** (1993) 377–409.

[9] B. Chazelle and H. Edelsbrunner. An optimal algorithm for intersecting line segments in the plane. *J. ACM* **39** (1992) 1–54.

[10] B. Chazelle and J. Friedman. A deterministic view of random sampling and its use in geometry. *Combinatorica* **10** (1990) 229–249.

[11] Y.-J. Chiang, M. T. Goodrich, E. F. Grove, R. Tamassia, D. E. Vengroff, and J. S. Vitter. External-memory graph algorithms. In *Proc. ACM-SIAM Symp. on Discrete Algorithms*, 139–149, 1995.

[12] K.L. Clarkson. Randomized geometric algorithms. In D.-Z. Du and F.K.Hwang, eds., *Computing in Euclidean Geometry*, Vol.1 of *Lecture Notes Series on Computing*, pages 117–162. World Scientific, Singapore, 1992.

[13] K. L. Clarkson, R. Cole, and R. Tarjan. Randomized parallel algorithms for trapezoidal diagrams. *International Journal of Computational Geometry & Applications*, pp. 117–1 33, 1992.

[14] K. L. Clarkson and P. W. Shor. Applications of random sampling in computational geometry, II. *Discrete Comput. Geom.*, **4** (1989) 387–421.

[15] A. Crauser and K. Mehlhorn. LEDA-SM, a platform for secondary memory computation, 1998, see WEB-pages of the authors.

[16] R. F. Cromp. An intelligent information fusion system for handling the archiving and querying of terabyte-sized spatial databases. In *S. R. Tate ed., Report on the Workshop on Data and Image Compression Needs and Uses in Scientific Community, CESDIS TR-93-99*, 75–84, 1993.

[17] F. Dehne, W. Dittrich, and D. Hutchinson. Efficient external memory algorithms by simulating coarse-grained parallel algori thms. *ACM Symposium on Parallel Algorithms and Architectures*, pp. 106–115, 1997.

[18] H. Edelsbrunner. *Algorithms in Combinatorial Geometry*. Springer Verlag, Heidelberg, 1987.

[19] P. Ferragina and R. Grossi. A fully-dynamic data structure for external substring search. In *Proc. 27th Annu. ACM Sympos. Theory Comput.*, 693–702, 1995.

[20] G. A. Gibson, J. S. Vitter, and J. Wilkes. Report of the working group on storage I/O issues in large-scale computing. *ACM Computing Surveys*, 28(4), December 1996.

[21] M. T. Goodrich, J.-J. Tsay, D. E. Vengroff, and J. S. Vitter. External-memory computational geometry. In *Proc. IEEE Symp. on Foundations of Comp. Sci.*, 714–723, 1993.

[22] R. Klein. *Concrete and Abstract Voronoi diagrams*. LCNS 400, Springer-Verlag, 1988.

[23] R. Laurini and A. D. Thompson. *Fundamentals of Spatial Information Systems.* A.P.I.C. Series, Academic Press, NY 1992.

[24] J. Matoušek. Cutting hyperplane arrangements. *Discrete Comput. Geom.* **6** (1991) 385–406.

[25] K. Mehlhorn and S. Näher. LEDA: A platform for combinatorial and geometric computing. *Communications of the ACM* **38** (1995) 96–102.

[26] K. Mulmuley. A fast planar partition algorithm, I. In *Proc. 29th Annu. IEEE Sympos. Found. Comput. Sci.*, 1988, 580–589.

[27] K. Mulmuley. *Computational Geometry: An Introduction Through Randomized Algorithms.* Prentice Hall, Englewood Cliffs, NJ, 1993.

[28] K. Mulmuley and S. Sen. Dynamic point location in arrangements of hyperplanes. *Discrete Comput. Geom.*, **8** (1992) 335–360.

[29] M. H. Nodine and J. S. Vitter. Greed sort: An optimal sorting algorithm for multiple disks. *Journal of the ACM* (1995) 919–933.

[30] Yale N. Patt. The I/O subsystem—a candidate for improvement. *Guest Editor's Introduction in IEEE Computer*, 27(3):15–16, 1994.

[31] F. P. Preparata and M. I. Shamos. *Computational Geometry: An Introduction.* Springer-Verlag, New York, 1985.

[32] J.H. Reif and S. Sen. Optimal parallel randomized algorithms for three-dimensional convex hulls and related problems. *SIAM J. Comput.* **21** (1992) 466–485.

[33] H. Samet. *Applications of Spatial Data Structures: Computer Graphics, Image Processing, and GIS.* Addison Wesley, MA 1989.

[34] R. Seidel. Backward analysis of randomized geometric algorithms. In *New Trends in Discrete and Computational Geometry* (J. Pach, ed.), *Algorithms and Combinatorics*, Vol. 10, Springer-Verlag, 1993, pp. 37–67.

[35] J. Sibeyn and M. Kaufmann. BSP-like external-memory computation. *Proc. of Italian Conference on Algorithms and Complexity*, 1997.

[36] D.E. Vengroff. *TPIE User Manual and Reference.* Duke University, Technical Report 1995.

[37] D.E. Vengroff and J. Vitter. I/O-efficient scientific computation using TPIE. IEEE Symp. on Parallel and Distributed Computing, 1995.

[38] J. S. Vitter and E. A. M. Shriver. Algorithms for parallel memory, I: Two-level memories. *Algorithmica*, 12(2–3):110–147, 1994.

A Sampling in Configuration Spaces

A.1 Sampling

We consider geometric problems formulated in terms of *objects* and *cells*. A set S of objects determines a decomposition (of the underlying space) into *cells*, $\mathcal{T}(S)$. These objects and cells are assumed to form a *configuration space* satisfying the *bounded degree* and *locality* properties [26, 27]. Recall that $f(p, S)$ is an upper bound on the expected size of $\mathcal{T}(R)$ where R is a p-sample from S. We assume that $f(p/2, S) = O(f(p, S))$. Let $N_\sigma = |S_\sigma|$. The following generalization of Eqn. (1) is needed in the analysis of the algorithms [14, 26, 27]:

Fact A.1 *There is a constant $c > 0$ such that if the function g satisfies $\sum_{t=1}^{\infty} g(tx)e^{-ct} = O(g(x))$, then there is a constant C such that for a finite set S of objects, if R is a p-sample from S, then:*

$$\mathbf{E}\left[\sum_{\sigma\in\mathcal{T}(R)} g(N_\sigma)\right] \leq C\, g(1/p)\, f(p, S). \qquad (4)$$

The left hand side of Eqn. (4) is a functional average of the conflict list sizes, and thus this equation indicates in a strong sense that the *average* conflict list size is at most $1/p$. Using $g(x) = x$ in Eqn. (4), we obtain Eqn. (1).

A.2 Cuttings

Fact 1.1 is well known but we sketch the proof as the procedure is essential in our optimal algorithms.

Proof. Let D be so that x objects determine $O(x^D)$ cells (bounded degree property). Obtain a p-sample R from S, with $p = (Cr)^2/N$ and compute $\mathcal{T}(R)$ and its conflict lists. This takes $O(r^{2D}N)$ time using brute force (there are at most $O(r^{2D})$ cells to check). Repeat until no conflict list is greater than $|S|/r$. From Eqn. (2), with probability at least $1/2$, $\max_{\sigma\in\mathcal{T}(R)} N_\sigma \leq |S|/r$. Thus, the expected number of trials is $O(1)$ and, hence, the expected number of operations required for the construction is $O(r^{2D}N)$. ■

Optimal size cuttings. Neither the bound on the average of Eqn. (1) nor the bound on the deviation of Eqn. (2) are sufficient in our applications. Fortunately, there is the following result due to Chazelle and Friedman [10, 24]. Again, we sketch the proof.

Fact A.2 *There is a randomized algorithm that, given a set S of N objects, constructs a $(1/p)$-cutting for S of expected size $O(f(p, S))$, requiring $O((1/p)f(p, S))$ expected operations.*

Proof. Let R be a p-sample from S. For each $\sigma \in \mathcal{T}(R)$, construct a $(1/t_\sigma)$-cutting T_σ for S_σ restricted to σ, where $t_\sigma = pN_\sigma$ (t_σ is called the *excess* of σ), using the algorithm in Fact 1.1. Then the cells in $T = \bigcup_\sigma T_\sigma$ have conflict list size at most $1/p$ because for each $\tau \in T_\sigma$, $N_\tau \leq N_\sigma/t_\sigma = 1/p$, and by using Eqn. (4), one can verify the claims on size and expected number of operations. ■

B Analysis of RIC via gradations

In this section we analyze the number of operations executed by the RIC via gradations. For the purpose of analysis, we look at the computation either from a *backward* or from a *forward* point of view. In the former case, S_{i-1} is seen as a $(1/\mu)$-sample from S_i for $i < l$ and as a $(1/\mu')$-sample for $i = l$, and this implies that S_i is a q_i-sample from S where $q_i = 1/\mu'\mu^{l-1-i}$. In the latter case, S_i is obtained from S_{i-1} by adding a p_i-sample R_i taken from $S - S_{i-1}$, where p_i satisfies $q_i = q_{i-1} + (1 - q_{i-1})p_i$ (i.e. $p_i = (q_i - q_{i-1})/(1 - q_{i-1})$). Note that $p_i \approx q_i$. In the analysis, we use $\mathbf{E}_i$ to denote the expectation for sampling S_i from S, and $\mathbf{E}_{i-1,i}$ to denote the expectation for sampling S_i and S_{i-1} from S. The expectations $\mathbf{E}_i$ can be evaluated using Eqn. (4). To compute the expectations $\mathbf{E}_{i-1,i}$ one can take advantage of either the forward or backward views, respectively:

$$\mathbf{E}_{i-1,i}[X] = \mathbf{E}_{i-1}[\mathbf{E}_i[X|S_{i-1}]] = \mathbf{E}_i[\mathbf{E}_{i-1}[X|S_i]].$$

The computation of the double expectation $\mathbf{E}_{i-1,i}$ presents some difficulty if we keep using an arbitrary function $f()$. So, we concentrate on $f(p, S) = p|S| + p^2K(S)$, $K(S) = |\mathcal{K}(S)|$, where $\mathcal{K}(S)$ is additive over cells and $|\mathcal{K}(S)| = O(|S|^2)$ ($\mathcal{K}(S)$ is the set of pairwise intersection points in the segment intersection problem).

Lemma B.1 *Let $f(p, S) = p|S| + p^2K(S)$. We have:*

(i) $\mathbf{E}_i[|T_i|] \leq f(q_i, S)$ *and* $\mathbf{E}_i[|T_i[S]|] \leq \dfrac{f(q_i, S)}{q_i}$

(ii) $\mathbf{E}_{i-1,i}[|\mathcal{K}(R_{i,\sigma}) \cap \sigma|] \leq f(q_i, S)$

(iii) $\mathbf{E}_{i-1,i}[N_\sigma \log |R_{i,\sigma}|] \leq \log\left(\dfrac{q_i}{q_{i-1}}\right) \dfrac{f(q_{i-1}, S)}{q_{i-1}}$

(iv) $\mathbf{E}_{i-1,i}[|T_i^I|] \leq f(q_i, S)$ *and* $\mathbf{E}_{i-1,i}[|T_i^I[S]|] \leq \dfrac{f(q_i, S)}{q_i}.$

Bound on Internal Operations. Using (i-iii) in the previous lemma, in Eqn. (3) and adding over all rounds, we obtain the following bound for the expected number of operations performed (ignoring constant factors)

$$(\mu' \log \mu') f(q_{l-1}, S) + f(S) + \log \mu \sum_{i=1}^{l} \frac{f(q_{i-1}, S)}{q_{i-1}}.$$

Finally, substituting $f(p, S) = p|S| + p^2K(S)$, we obtain the bound $|S| \log |S| + K(S)$.

Bound on I/O Operations. In the case $K(S) \neq 0$, a sorting operation is not used in the Clean-up; then the total number of I/Os in the i-th round is bounded by $|T_i^I| + \frac{|T_i^I[S]|}{B}$. In the case $K(S) = \emptyset$, a sorting operation is used in the Clean-up; then the total number of I/Os in the i-th round is bounded by $\text{sort}(|T_i^I|) + \frac{|T_i^I[S]|}{B}$. In both cases, using the results above and adding up over all rounds, the bound $n \log_m n + k$ results.

C Arrangement of Segments in the Plane

C.1 Step 3

Let us assume that the graph G_{i-1} describing the vertical adjacencies between the trapezoids of T_{i-1} is available (a *node* in G_{i-1} corresponds to a trapezoid in T_{i-1} and an *arc* in G_{i-1} corresponds to an adjacency through a vertical edge between two trapezoids in T_{i-1}). Let us consider an arc in G_{i-1} directed according to the left to right ordering of the corresponding trapezoids. This is an acyclic ordering, so we can choose a linear extension (i.e., *topological sort*) and traverse G_{i-1} according to that order. Usually, performing such an ordering on a generic graph is not I/O-efficient, but here we exploit the fact that in these earlier rounds ($i < l$), the underlying graph G_{i-1} has *small size* and thus we can afford to pay even *one* I/O per visited node. Hence, we can compute this ordering by adopting a standard internal-memory algorithm which pays *one I/O* per step, and thus requires a number of I/Os linear in the size of the graph G_{i-1}, that is $|G_{i-1}| = |T_{i-1}|$ I/Os (this term is hidden by the next I/O-bounds). We can then proceed to visit the trapezoids

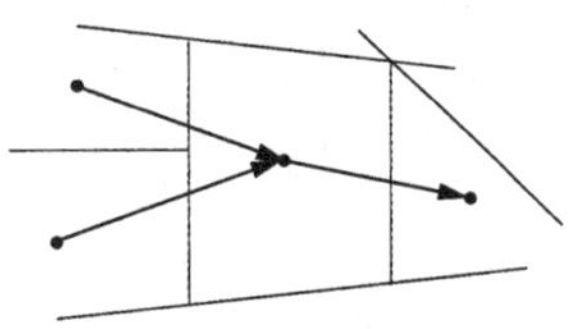

Figure 4: Adjacency graph

according to the induced order. At a generic step of the traversal, we say that an arc of G_{i-1} is *waiting* if its starting node has been visited but its ending node is yet to be visited; we also say that the corresponding vertical edge is waiting and that the trapezoid currently under consideration is *active*. We assume that the trapezoids τ in T_σ are stored with a particular order: First appear those trapezoids that touch the left vertical edge in order from top to bottom, then those that do not touch the vertical edges, and then those that touch the right vertical edge from top to bottom. During the visit of G_{i-1}, as the next active trapezoid $\sigma \in T_{i-1}$ is considered, we take the (ordered) list of trapezoids which are still *not completed* and thus are associated with the one or the two waiting vertical edges e_1 and e_2 (the neighbors of σ on the left). We merge these lists with the (ordered) list of trapezoids of T_σ, by taking $O((|T_\sigma|/B) + 1)$ I/Os, since the first two lists are surely smaller than the latter. As a result, some new trapezoids in T_i are completed, some are completely inside σ, some other are started, and still others continue (i.e., stab σ). The not completed trapezoids are stored on the disk, associated with the (at most two) vertical waiting edges to the right of σ (thus preserving the invariant). Therefore, the overall I/O-cost of merging trapezoids in the T_σ's to get T_i is $\sum_{\sigma \in T_{i-1}} \lceil \frac{|T_\sigma|}{B} \rceil$.

Within the same I/O-bound, it is possible to compute the next graph G_i, thus preserving the invariant. We are left with the problem of computing the conflict lists for $\mathcal{T}(S_i)$, which are large, and thus must be managed properly. As before, we assume that the trapezoids τ in T_σ are stored in a particular order and now have associated their conflict lists. The algorithm visits G_{i-1} according to the induced order. When a trapezoid $\sigma \in T_{i-1}$ becomes active, the algorithm proceeds by matching the trapezoids in the one or the two corresponding waiting vertical edges e_1 and e_2 (the neighbors of σ on the left), and updating their conflict lists. Namely, if τ was incomplete in e_i and matches with $\tau' \in T_\sigma$, then we *merge* τ and τ' by augmenting the conflict list of τ with those conflicts of τ' that do not intersect e_i. These conflicts are guaranteed to refer to *new* segments which don't intersect τ and thus have been not yet inserted in its conflict list. This approach allows to merge the conflict lists without *rescanning* their segments over and over. Clearly, we must pay at least *one* disk access per trapezoid in T_σ but this is not much of a problem due to the small size of G_{i-1} (see considerations above). In conclusion, the number of I/Os necessary to do this merging process is bounded by the number of trapezoids in T_{i-1}, in T_i^I and in T_i, and by the size of all the corresponding conflicts divided by B. Thus, Step 3 can be performed using a number of I/Os proportional to the total number of blocks of size B required to hold the conflict lists $T_i^I[S]$, and the resulting conflict lists $T_i[S]$.

C.2 An Algorithm for $N \leq M$

We describe an algorithm for a set of segments S of size $N \leq M$. This computation is non-trivial since K can be as large as M^2 and, hence, we cannot afford to even construct $\mathcal{T}(S)$ completely in internal memory. For simplicity, in describing the algorithm we will ignore (in Step 1 below) the deviations in the conflict list sizes resulting from sampling. We can nevertheless cope with them by adopting the *cutting* technique as we have done for the general algorithm in Section 3.2.

1. Take a p-sample R from S where $p = 1/\sqrt{M}$ and compute $T_0 = \mathcal{T}(R)$ in internal memory. Here, we assume that $\max_{\sigma \in T_0} N_\sigma \leq 1/p = \sqrt{M}$ and this assumption is removed in Section 3.1. Also compute the adjacency graph G_0 of T_0 and a topological sort L_0 (in internal memory).

2. For each $s \in S$ compute the conflicts with T_0 by walking on it in internal memory and by storing the conflicts on the disk as they are determined. Then sort all these conflicts to bring together the ones corresponding to S_σ, for each $\sigma \in T_0$, according to ordering L_0.

3. For each $\sigma \in T_0$ compute $T_\sigma = \mathcal{T}(S_\sigma)$ in internal memory (see assumption in Step 1).

4. Obtain $\mathcal{T}(S)$ from the collection of T_σ's, where $\sigma \in T_0$. First, traverse G_0 according to the ordering L_0, and assign addresses to the trapezoids $\tau \in \mathcal{T}(S)$ by loading in turn the T_σ's (if a trapezoid $\tau \in \mathcal{T}(S)$ is divided into many pieces in T_0, then only the piece appearing first is given an address). Then, the different pieces of each $\tau \in \mathcal{T}(S)$ are brought together using sorting. At this point, all the pieces of $\tau \in \mathcal{T}(S)$ can find out the final address of τ by a simple scan. Subsequently, we undo the sort so that each trapezoid can determine the address of its neighbors. This information is propagated to all trapezoid pieces by again sorting, scanning and undoing the sorting. Finally, a traversal of G_0 according to the ordering L_0, is used to write the resulting adjacency graph of $\mathcal{T}(S)$.

We show that the total number of I/Os is $O(n \log_m n + k)$. The critical part is to verify that the sort in Steps 2 and 4 can be performed within this bound. Indeed, the number of sorted items is equal to the number of conflicts and its expected value is $N + (1/\sqrt{M})K$. Hence, using the sorting algorithm in [29], the expected number of I/Os is (ignoring constants):

$$n + k/\sqrt{M}) \log_m(n + k/\sqrt{M}) = \\ n \log_m(n + k/\sqrt{M}) + k/\sqrt{M} \log_m(n + k/\sqrt{M}).$$

First note that the second term is always $O(k)$ because $\log_m(n + k/\sqrt{M})/\sqrt{M} = O(1)$, and that this term is dominant if $k/\sqrt{M} > n$. On the other hand, if $k/\sqrt{M} \leq n$, then the first term is $O(n \log_m n)$. Therefore, the expected number of I/Os is indeed $O(n \log_m n + k)$.

Geometric Applications of a Randomized Optimization Technique

Timothy M. Chan*

Abstract

We describe general randomized reductions of certain geometric optimization problems to their corresponding decision problems. These reductions increase the expected time complexity by only a constant factor and eliminate extra logarithmic factors in previous approaches. Improved algorithms are thus obtained for a variety of problems in computational geometry: finding minimal k-point subsets, matching point sets under translation, and solving problems related to p-centers and linear programming with violations.

1 Introduction

Consider the classic randomized algorithm for finding the minimum of r numbers $\min\{A[1], \ldots, A[r]\}$:

Algorithm RAND-MIN

1. randomly pick a permutation $\langle i_1, \ldots, i_r \rangle$ of $\langle 1, \ldots, r \rangle$
2. $t \leftarrow \infty$
3. for $k = 1, \ldots, r$ do
4.   if $A[i_k] < t$ then
5. $t \leftarrow A[i_k]$
6. return t

By a well-known fact [21, 36], the expected number of times that step 5 is executed is bounded by the harmonic number $1 + 1/2 + \cdots + 1/r \leq \ln r + 1$. If a decision $A[i] < t$ can be made in D time but an evaluation of $A[i]$ takes E time, then algorithm RAND-MIN runs in $O(Dr + E \log r)$ expected time, which is an improvement over the obvious $O(Er)$ worst-case bound when $D \ll E$.

This simple observation suggests a general strategy to solve a given optimization problem: express the solution as a minimum of the solutions of several subproblems and apply the above algorithm to find the minimum. If the decision versions of the subproblems are easier to solve than the subproblems themselves, then a faster algorithm for the optimization problem may be obtained with randomization.

In the next section, we formulate the idea more precisely and propose a recursive version of the algorithm in which we use a constant number r of subproblems at each node of the recursion. This recursive algorithm can be regarded as a generalization of prune-and-search. Assuming that the size of the subproblems is reduced by a constant factor at each level, we show that the optimization problem can be solved within the same asymptotic expected time bound as the decision problem. Previous reduction of the optimization problem to the decision problem is usually obtained by some kind of binary or parametric search, which increases the running time by a polylogarithmic factor.

New Results. This simple technique is applicable to a surprisingly diverse range of geometric optimization problems. Besides re-deriving some old results (Section 3), we are able to prove a number of new ones (Sections 4–6), including the following:

- The minimum L_∞-diameter k-point subset of a planar n-point set [7, 22, 28] can be found in $O(n \log n)$ expected time for any $1 \leq k \leq n$.
- The minimum L_∞-Hausdorff distance between two planar n-point sets under translations [16] can be found in $O(n^2 \log n)$ expected time.
- The two-dimensional rectilinear 5-center problem [50, 53] can be solved in $O(n \log n)$ expected time.
- The two-dimensional linear programming problem with k violations [41, 49] can be solved in $O(n \log n)$ expected time in the feasible case for any $0 \leq k \leq n$. (For a wide range of k, the time bound is actually linear.)
- Given an infeasible two-dimensional linear program with n constraints, the smallest number k of violations that makes it feasible [29] can be found in $O(nk)$ expected time.

*Department of Mathematics and Computer Science, University of Miami, Coral Gables, FL 33124–4250, USA, E-mail: tchan@cs.miami.edu

SCG 98 Minneapolis Minnesota USA

- The three-dimensional Euclidean discrete 1-center problem [18] can be solved in $O(n \log n)$ expected time.

In the first five cases, the time bounds improve previous ones by a logarithmic factor. This list of problems is not meant to be exhaustive. Rather, it is used to illustrate the various ways in which the technique can be applied. We expect that more applications will follow.

Previous Approaches. One of the most general approaches to reduce geometric optimization problems to their decision problems is *parametric search*, invented by Megiddo [44]. The reduction requires an efficient parallel version of the decision algorithm and involves simulating this parallel algorithm sequentially. Running time typically increases by logarithmic factors, even when an improvement by Cole [20] is applicable. Furthermore, as many researchers have commented, the resulting algorithms tend to be complicated and impractical; see the survey by Agarwal and Sharir [3]. In contrast, our randomized reductions use the decision algorithms as "black boxes" only, avoid the extra logarithmic factors in the running time, and are easier to implement.

A number of alternatives to parametric search have been proposed in the geometry literature [3]. First, if the search space has linear size, then an ordinary binary search is sufficient. For many rectilinear problems, the search space forms a matrix with sorted rows/columns, and one can use Frederickson and Johnson's selection algorithm [30] to carry out the binary search [32]. In other instances, one can employ constructions of expander graphs by a technique of Katz and Sharir [35]. Without additional ideas, all of these techniques increase the running time by logarithmic factors.

Randomized techniques have also been suggested as an alternative in several isolated cases [2, 10, 23, 27, 37, 40]. Unfortunately, since these techniques are not unified and as straightforward to apply as parametric search, potential applications are sometimes missed. (One important exception though is the class of *LP-type* optimization problems [52], where general randomized solutions have been developed.) We hope that our randomized technique will partially rectify this shortcoming.

One other alternative to parametric searching that has been noted [8] involves the application of geometric cutting tools. Some of these tools (in a very elementary form) will be of use in several of our randomized reductions.

In Section 7, we discuss limitation of our technique and some issues regarding possible derandomization.

2 The Technique

We describe our technique in a general setting. Let Π represent the *problem space*. Given a problem $P \in \Pi$, let $w(P) \in \mathbb{R}$ be its *solution*. Denote the *size* of P by $|P|$ (a positive integer). We assume that the solution of a problem of constant size can be computed in constant time. Furthermore, we assume that there is an algorithm to decide whether $w(P) < t$ for any $P \in \Pi$ and $t \in \mathbb{R}$ in $D(|P|)$ time.

Lemma 2.1 *Let $\alpha < 1$, $\varepsilon > 0$, and r be constants. Given any problem $P \in \Pi$, suppose we can construct r subproblems $P_1, \ldots, P_r \in \Pi$, each of size at most $\lceil \alpha|P| \rceil$, within $D(|P|)$ time, so that*

$$w(P) = \min\{w(P_1), \ldots, w(P_r)\}.$$

Then for any $P \in \Pi$, we can compute the solution $w(P)$ in $O(D(|P|))$ time, assuming that $D(n)/n^\varepsilon$ is a nondecreasing function of n.

Proof: We compute $w(P)$ by applying Algorithm RAND-MIN to the numbers $w(P_1), \ldots, w(P_r)$. Deciding $w(P_i) < t$ takes $D(|P_i|)$ time. Evaluating $w(P_i)$ is done recursively, unless $|P_i|$ drops below a certain constant. Note this procedure not only computes $w(P)$ but identifies a constant-size subproblem attaining the minimum.

Let $T(P)$ be the random variable corresponding to the time needed to compute $w(P)$ by this procedure. Let $N(P_i)$ be a 0-1 random variable, having value 1 if and only if $w(P_i)$ is evaluated. We have

$$T(P) \le \sum_{1 \le i \le r} N(P_i) T(P_i) + O(rD(|P|)).$$

As noted earlier, the expected number of evaluations by Algorithm RAND-MIN is $E[\sum_{1 \le i \le r} N(P_i)] \le \ln r + 1$.

Define $T(n) = \max_{|P| \le n} E[T(P)]$. Since $N(P_i)$ and $T(P_i)$ are independent, we have

$$\begin{aligned} E[T(P)] &\le \sum_{1 \le i \le r} E[N(P_i)]\, E[T(P_i)] + O(rD(|P|)) \\ &\le (\ln r + 1)\, T(\lceil \alpha|P| \rceil) + O(rD(|P|)), \end{aligned}$$

implying that $T(n) \le (\ln r + 1) T(\lceil \alpha n \rceil) + O(D(n))$. If we assume

$$(\ln r + 1)\, \alpha^\varepsilon < 1, \tag{1}$$

then by induction this recurrence solves to $T(n) \le C \cdot D(n)$, for an appropriate constant C (depending on α, r, and ε).

To enforce condition (1), we compress ℓ levels of the recursion into one before applying Algorithm RAND-MIN, where ℓ is a sufficiently large constant. Then r increases to r^ℓ and α decreases to α^ℓ. To finish the proof, just note that $\lim_{\ell \to \infty} (\ln r^\ell + 1)\, \alpha^{\ell\varepsilon} = 0$. □

For $r = 1$, randomization is not required: our recursive algorithm reduces to the standard prune-and-search algorithm. For a larger constant r, the worst-case running time is at least $\Omega(n^{\log r/\log(1/\alpha)})$. So the efficiency of a deterministic algorithm depends crucially on the values of the constants α and r. In contrast, by Lemma 2.1, these constants are unimportant in bounding the randomized complexity; we can thus afford a crude scheme to divide a problem into subproblems. Such a scheme is easily obtainable for certain classes of problems, as we point out in the next section; however, it may be less apparent for others.

3 Closest Pairs and Ray Shooting

We now illustrate our technique in its simplest form on some abstract closest-pair and ray-shooting problems. Although no specific new results are obtained, the applications are instructive.

Closest pairs. Let U be a collection of objects. Given a *distance function* $d : U \times U \to \mathbb{R}$, the *closest-pair problem* is to compute $w(P) = \min_{p,q \in P} d(p,q)$ for a given set $P \subset U$ of size n. The *closest-pair decision problem* is to determine whether $w(P) < t$ for a given P and $t \in \mathbb{R}$.

Theorem 3.1 *If the closest-pair decision problem can be solved in $D(n)$ time, then the closest-pair problem can be solved in $O(D(n))$ expected time, assuming that $D(n)/n$ is nondecreasing.*

Proof: Arbitrarily partition P into three subsets P_1, P_2, P_3 of roughly equal size. Then

$$w(P) \;=\; \min\{w(P_1 \cup P_2),\, w(P_1 \cup P_3),\, w(P_2 \cup P_3)\}.$$

This divides the problem into three subproblems, each of size roughly $2n/3$. Now apply Lemma 2.1. □

Rabin [47] was the first to apply randomized techniques to the standard closest-pair problem, where U is a fixed-dimensional Euclidean space and $d(\cdot,\cdot)$ is the Euclidean metric. Clarkson and Shor [19] studied the *Euclidean diameter problem* in $U = \mathbb{R}^3$, where $d(\cdot,\cdot)$ is the negation of the Euclidean distance. Agarwal and Sharir [2] considered a certain closest-pair problem that arises in finding the width of a point set in $\mathbb{R}^3$ and in other problems; the elements of U are lines in $\mathbb{R}^3$, each colored red or blue (with $d(p,q) = \infty$ when p and q have the same color). In all three papers, randomized algorithms are obtained by modifying certain implicit decision algorithms. Our technique gives alternative randomized algorithms.

Ray shooting. Let U be a collection of objects and V be a collection of rays. Let $\tau : U \times V \to \mathbb{R}$ be an *ordering function*, where $\tau(p_1, q) < \tau(p_2, q)$ means that ray q hits object p_1 before p_2. The *ray-shooting problem* is to preprocess a given set $P \subset U$ of size n into a data structure that answers queries of the following type: given $q \in V$, compute $w(P,q) = \min_{p \in P} \tau(p,q)$. (The object p that attains this minimum represents the first object hit by the query ray q.) In the *ray-shooting decision problem*, a query has the type: given any $q \in V$ and $t \in \mathbb{R}$, determine whether $w(P,q) < t$.

Theorem 3.2 *If the ray-shooting decision problem can be solved with $P(n)$ preprocessing and $D(n)$ query time, then the ray-shooting problem can be solved with $O(P(n))$ preprocessing and $O(D(n))$ expected query time, assuming that $P(n)/n^{1+\varepsilon}$ and $D(n)/n^{\varepsilon}$ are nondecreasing for some $\varepsilon > 0$.*

Proof: In the preprocessing, partition P into two subsets P_1, P_2 of roughly equal size, build the decision data structures for P_1 and P_2, and recursively preprocess P_1 and P_2. The new preprocessing time $P'(n)$ satisfies the recurrence

$$P'(n) = 2P'(n/2) + O(P(n)),$$

which solves to $P'(n) = O(P(n))$ if $P(n)/n^{1+\varepsilon}$ is nondecreasing. To compute $w(P,q)$ for a given $q \in V$, we can divide the problem into two subproblems, each of size roughly $n/2$: $w(P,q) = \min\{w(P_1,q),\, w(P_2,q)\}$. Now we can apply Lemma 2.1. □

Previously, Agarwal and Matoušek [1] described a general deterministic reduction of the above ray shooting problem to the decision problem (which they called *segment emptiness*). The reduction uses parametric search and is not likely to be practical (besides increasing the query time by a polylogarithmic factor).

In an earlier paper [10], the author gave a randomized reduction of linear programming queries to half-space range reporting. This reduction, when specialized to ray shooting, yields a different approach in which the preprocessing algorithm employs random sampling. (Arya and Mount [6] noted a similar reduction in the context of nearest neighbor searching.) In contrast, in the proof of Theorem 3.2, randomization is used in the query algorithm but not the preprocessing. The random sampling approach however achieves high-probability bounds in many instances and is susceptible to derandomization.

4 Rectilinear Applications

For many geometric problems, the process of dividing a problem into subproblems is usually more involved than the simple applications from the previous section. We take a number of specific problems in the rectilinear plane and demonstrate how this division can be accomplished in each case. In what follows, all squares and rectangles are implicitly assumed to be axis-parallel. Given two rectangles R_1, R_2, we let $R_1 \vee R_2$ denote the smallest rectangle enclosing their union.

Minimal k-point subsets. Motivated by applications in clustering and statistical analysis, a number of researchers [5, 7, 22, 26, 28, 40] have recently looked at problems of the type: given an n-point set P, select a "minimal" k-point subset. We illustrate our technique on one specific case, where the point set S is planar, and the measure of minimality is the L_∞-diameter (another case will be mentioned in Section 7). This particular problem is identical to:

> Given an n-point set $P \in \mathbb{R}^2$ and $2 \le k \le n$, find the smallest square enclosing at least k points of P.

Eppstein and Erickson [28] observed that the decision version of the problem can be solved in $O(n \log n)$ time by a standard plane-sweep algorithm. (The decision problem reduces to computing the *depth* of an arrangement of squares.) Using the search technique of Frederickson and Johnson [30], we can then solve the optimization problem in $O(n \log^2 n)$ time.

Datta, *et al.* [22] described a general process of dividing the problem into $O(n/k)$ subproblems each of size $O(k)$, so that the solution is the minimum of the solutions of these subproblems; this division takes $O(n \log n)$ time. Immediately, the $O(n \log^2 n)$ time bound reduces slightly to $O(n \log n + (k \log^2 k)(n/k))$. If we apply Algorithm RAND-MIN to find the minimum, we get a randomized time bound $O(n \log n + (k \log k)(n/k) + (k \log^2 k) \log(n/k))$; this bound is recently observed by Bhattacharya and ElGindy [7], using a more cumbersome derivation. The randomized bound matches the $O(n \log n)$ decision time bound when $k = O(n/(\log n \log\log n))$. We give a different algorithm that runs in $O(n \log n)$ expected time for *all* values of k.

Theorem 4.1 *The smallest (axis-aligned) square containing k of n given planar points can be computed in $O(n \log n)$ expected time.*

Proof: Before applying Lemma 2.1, we find it necessary to extend the problem (and its decision problem) slightly: given an n-point set $P \subset \mathbb{R}^2$ and a rectangle R, we will compute $w(P, R, k)$, the side length of the smallest square S^* that contains at least k points of P and, in addition, contains R. The decision algorithm by Eppstein and Erickson [28] can be modified for this extended problem. Alternatively, we can directly reduce the extended problem to the original problem, since $w(P, R, k) = w(P', \emptyset, k + 4n)$, if we let P' to be the union of P with n copies of the four corner points of R.

Our division process is as follows. First, draw vertical lines at the $\lceil n/5 \rceil$-th smallest and $\lceil n/5 \rceil$-th largest x-coordinates of the points in P. Similarly, draw horizontal lines at the $\lceil n/5 \rceil$-th smallest and $\lceil n/5 \rceil$-th largest y-coordinates. Let R_0 be the rectangle bounded by these four lines. Let $H_1, \ldots, H_4$ be the four bounding half-planes of R_0. The optimal square S^* must belong to one of two cases:

Case 1. S^* contains R_0. Then $w(P, R, k) = w(P, R \vee R_0, k) = w(P \setminus R_0, R \vee R_0, k - |P \cap R_0|)$.

Case 2. $S^* \subset H_i$ for some $i \in \{1, \ldots, 4\}$. Then $w(P, R, k) = w(P \cap H_i, R, k)$.

In any case, one can check the identity:

$$w(P, R, k) = \min\{ w(P \setminus R_0, R \vee R_0, k - |P \cap R_0|), w(P \cap H_1, R, k), \ldots, w(P \cap H_4, R, k) \}.$$

Each of the five subproblems $P \setminus R_0$, $P \cap H_1, \ldots, P \cap H_4$ has size at most $4\lceil n/5 \rceil$. The theorem follows. □

For k very close to n, the time bound can be further reduced to $O(n + (n - k) \log n)$; we leave the proof as an easy exercise. The technique can be extended to higher dimensions [28] or to the problem of finding the smallest homothet of a fixed convex polygon enclosing k points [26].

Matching point sets. Numerous papers (e.g., [14, 15, 16, 34]) have dealt with the problem of matching point sets under a class of transformations by minimizing the Hausdorff distance. We investigate the case where the point sets are two-dimensional, the allowable transformations are translations, and the metric is L_∞: given two n-point sets $A, B \subset \mathbb{R}^2$, find $v \in \mathbb{R}^2$ minimizing the *directed Hausdorff distance*

$$H(A + v, B) = \max_{a \in A} \min_{b \in B} \| a + v - b \|_\infty .$$

(Our algorithm can be easily modified if instead the *undirected Hausdorff distance* $\min\{H(A + v, B), H(B, A + v)\}$ is minimized.)

Chew and Kedem [16] showed that the decision version of the above problem can be solved in $O(n^2 \log n)$ time. Frederickson and Johnson's search technique then yields an $O(n^2 \log^2 n)$ time bound for the optimization problem. We show how to remove the extra $\log n$ factor.

Theorem 4.2 *The translation that minimizes the directed L_∞-Hausdorff distance between two planar n-point sets can be found in $O(n^2 \log n)$ expected time.*

Proof: Consider the set $P = B - A$ of $N = O(n^2)$ points. Assign colors to each point of P so that two points have the same color if and only if they belong to $B - a$ for a common $a \in A$. Let $c(p)$ denote the color of a point p and $c(P) = \{c(p) : p \in P\}$. Our problem reduces to the following with $k = n$:

> Given a set P of N colored points in $\mathbb{R}^2$, find the smallest square S^* that contains k different colors, i.e., $|c(P \cap S^*)| \geq k$.

Chew and Kedem's algorithm solves the decision problem (finding the *color-depth* of an arrangement of squares) using a plane sweep in $O(N \log N)$ time. We show how to solve the optimization problem in $O(N \log N)$ expected time by Lemma 2.1.

Again, we need to extend the problem slightly: we will compute $w(P, R, k)$, the side length of the smallest square S^* such that $|c(P \cap S^*)| \geq k$ and, in addition, contains R. We can easily get rid of the extra constraint about R: $w(P, R, k) = w(P', \emptyset, k+4)$, if we let P' be the union of P with the four corner points of R and assign the four corner points four new colors.

Construct the rectangle R_0 and its four bounding halfplanes $H_1, \ldots, H_4$ as in the proof of Theorem 4.1. As before, the optimal square S^* belongs to one of two cases:

Case 1. S^* contains R_0. Then $w(P, R, k) = w(P, R \vee R_0, k) = w(P_0, R \vee R_0, k_0)$, where we let

$$\begin{aligned} P_0 &= \{p \in P \setminus R_0 : c(p) \notin c(P \cap R_0)\}, \text{ and} \\ k_0 &= k - |c(P \cap R_0)|. \end{aligned}$$

Case 2. $S^* \subset H_i$ for some $i \in \{1, \ldots, 4\}$. Then $w(P, R, k) = w(P \cap H_i, R, k)$.

In any case, one can prove the identity:

$$\begin{aligned} w(P, R, k) &= \min\{\, w(P_0, R \vee R_0, k_0), \\ &\quad w(P \cap H_1, R, k), \ldots, w(P \cap H_4, R, k)\,\}. \end{aligned}$$

Each of these five subproblems has size at most $4\lceil N/5 \rceil$. □

Chew, *et al.* [14] discusses the extension of the problem to higher dimensions. Our technique again improves their running time by a logarithmic factor.

Rectilinear p-centers. A class of facility location problems known as *p-center problems* has received much attention in the computational geometry literature (e.g., see [4, 24, 27, 51, 53]). We consider the case where p is a constant, the dimension is two, and the metric is L_∞: given an n-point set $P \subseteq \mathbb{R}^2$, find p congruent squares of the smallest size covering P.

The decision problem reduces to a *rectangular p-piercing problem*: given a set of n rectangles in the plane, determine whether there exists a set of p points that intersects every rectangle. Sharir and Welzl [53] described efficient algorithms for this piercing problem for $p \leq 5$. As they noted, reduction from the p-center problem to the p-piercing problem can be accomplished by Frederickson and Johnson's technique [30], which increases the running time by a factor of $\log n$. We show:

Theorem 4.3 *If the rectangular p-piercing problem can be solved in $D_p(n)$ time, then the rectilinear p-center problem can be solved in $O(D_p(n))$ expected time, where the hidden constant depends on p.*

Proof: We again need to consider a generalized problem (which includes the rectilinear p-center problem with "additive weights"):

> Given a set $\mathcal{R}$ of n rectangles in $\mathbb{R}^2$, find p congruent squares $S_1^*, \ldots, S_p^*$ of the smallest size such that for each $R \in \mathcal{R}$, there exists an S_i^* that contains R.

Let $w(\mathcal{R})$ denote the side length of the optimal squares. Observe that the decision $w(\mathcal{R}) < t$ reduces to the above p-piercing problem.

The division process is more involved than in the previous two proofs. First list the $2n$ x-coordinates of $\mathcal{R}$ and draw a vertical line at the $i\lfloor \frac{n}{5p} \rfloor$ smallest x-coordinate for $i = 1, \ldots, 10p$. Similarly, list the $2n$ y-coordinates and draw a horizontal line the $j\lfloor \frac{n}{5p} \rfloor$ smallest y-coordinate for $j = 1, \ldots, 10p$. Call the above $O(p)$ vertical and horizontal lines *grid lines* and call a rectangle bounded by four grid lines a *grid rectangle*. Clearly, there are $O(p^4)$ possible grid rectangles.

A rectangle is said to be *δ-dense* if it contains at least δn rectangles of $\mathcal{R}$. We first prove:

> *Claim.* One of the optimal squares S_i^* contains a $(\frac{1}{5p})$-dense grid rectangle R_0.

The claim can be seen from the following argument. Since every rectangle of $\mathcal{R}$ is contained in S_i^* for some $i \in \{1, \ldots, p\}$, there exists an S_i^* that is $(1/p)$-dense. Now, take the largest grid rectangle R_0 inside S_i^* (see Figure 1). A rectangle of $\mathcal{R}$ that is contained in S_i^* but not in R_0 must have a vertex inside $S_i^* \setminus R_0$; by our construction of the grid, there are at most $\frac{4n}{5p}$ such rectangles of $\mathcal{R}$. Thus, R_0 is $(\frac{1}{5p})$-dense.

From the claim, we can easily show the identity

$$w(\mathcal{R}) = \min_{R_0} w(\{R \in \mathcal{R} : R \not\subseteq R_0\} \cup \{R_0\}),$$

where the minimum is taken over all $O(p^4)$ of the $(\frac{1}{5p})$-dense grid rectangles R_0. The size of each subproblem is bounded roughly by $(1 - \frac{1}{5p})n$, so we can apply Lemma 2.1. □

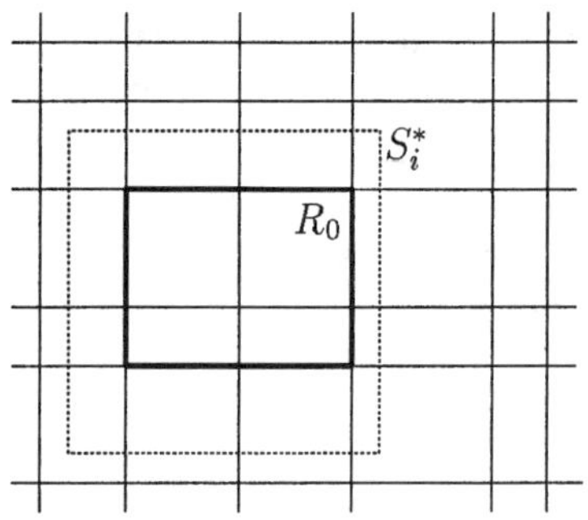

Figure 1: Grid lines and the grid rectangle R_0.

We get one particular new result for the case $p = 5$: Sharir and Welzl [53] showed that $D_5(n) = O(n \log^4 n)$. Their $O(n \log^5 n)$ bound for the rectilinear 5-center problem can therefore be reduced to $O(n \log^4 n)$ with randomization. Recently, Segal [50] improved the bound for rectangular 5-piercing to $D_5(n) = O(n \log n)$. This implies an $O(n \log n)$ randomized algorithm for rectilinear 5-centers.

5 Linear Programming with Violations

As we have seen in the previous section, the division of a problem into subproblems can usually be accomplished by elementary means for rectilinear applications. For non-rectilinear problems, we often need to resort to tools for geometric divide-and-conquer known as *cuttings*.

Given a collection H of n hyperplanes in $\mathbb{R}^d$, a *δ-cutting of size s* is a partition of space into s (possibly unbounded) simplices $\{\Delta_1, \ldots, \Delta_s\}$ such that the interior of each simplex Δ_i intersects at most δn of the hyperplanes of H. We need a (rather weak) lemma on cuttings:

Lemma 5.1 *Given n hyperplanes in $\mathbb{R}^d$ for a sufficiently large n, a δ-cutting of constant size can be constructed in $O(n)$ time for some constant $\delta < 1$.*

The earliest proof of the lemma for $d = 2$ can be traced back to papers by Dyer [25] and Megiddo [45]; they design prune-and-search algorithms for three-dimensional linear programming using a construction of a $(7/8)$-cutting of size 4. The construction has been extended to higher dimensions (with worse constants) [46]. A simpler randomized method was suggested by Clarkson [17]: take a random sample $R \subset H$ of size $O((1/\delta)\log(1/\delta))$ and canonically triangulate the arrangement of R. Derandomization of this method and refinements on the constants are discussed in several papers [12, 33, 38, 42]. For a variant known as *shallow cuttings* [39], better bounds are still possible. Since the size of the cutting is not important when applying Lemma 2.1 (as long as it is a constant), the original construction of Dyer and Megiddo is sufficient for our purposes.

Feasible case. As a first example, we consider the *linear programming problem with $\leq k$ violations* [41] in the two-dimensional feasible case: given $0 \leq k \leq n$ and a collection H of n closed halfplanes where the origin is in the interior of $\bigcap H$, find the lowest point in $L_k(H)$, the *$(\leq k)$-level of H*:

$$L_k(H) = \{p \ : \ p \text{ violates at most } k \text{ halfplanes of } H\}.$$

(A point p *violates* a halfplane h if $p \notin h$.) Set $L_k(H) = \emptyset$ if $k < 0$. Note that for $k = 0$, this reduces to the standard linear programming problem.

As Matoušek [41] and Roos and Widmayer [49] independently observed, an $O(n \log^2 n)$-time solution to the problem can be obtained by parametric search or slope-selection methods. (Roos and Widmayer also claimed an improved bound of $O(n \log n + k \log^2 k)$, but the validity of their argument is questionable.) A different approach of Matoušek [41] solves the problem in $O(n \log n)$ time when k is roughly smaller than $\sqrt{n}$. This $O(n \log n)$ bound can be improved to $O(n \log k)$ by a technique of the author [11].

We derive here the expected time bound $O(n \log n)$ for *all* values of k. Surprisingly, if $k = O(n/\log^\beta n)$ for a constant $\beta > 1$, our bound strengthens to $O(n)$, matching the complexity of the standard linear programming problem and improving all previous results.

Theorem 5.2 *The lowest point in $L_k(H)$ can be computed in $O(n + k(n/k)^\varepsilon \log n)$ expected time in the feasible case for any constant $\varepsilon > 0$.*

Proof: To avoid some (minor) technical complications, we assume that the given halfplanes are in general position. Consider the slightly generalized problem of computing $w(H, \Delta, k)$, the smallest y-coordinate inside $L_k(H) \cap \Delta$ for a given triangle Δ. We first show how to decide whether $w(H, \Delta, k) \leq t$ for a given $t \leq 0$. The decision $w(H, \Delta, k) < t$ can be made by a modification of the procedure below.

Let Δ' be the intersection of Δ with the halfplane $y \leq t$. We have $w(H, \Delta, k) \leq t$ if and only if $L_k(H)$ intersects Δ'. Now, observe that (i) $L_k(H)$ is a connected region (since $\bigcap H$ is nonempty), and (ii) $L_k(H)$ is not completely contained in Δ' (since $L_k(H)$ contains a point above the x-axis). Therefore, $w(H, \Delta, k) \leq t$ if and only if $L_k(H)$ intersects one of the (at most four) edges of Δ'.

Deciding whether $L_k(H)$ intersect a line segment is equivalent to the following one-dimensional problem:

> Let I be a collection of n half-infinite intervals. Given a, b, decide whether there exists a point in $[a, b]$ that is contained in all but at most k of the intervals of I.

This problem can be solved as follows. Write $I = \{[a_i, \infty)\} \cup \{(-\infty, b_j]\}$, where $a_1 \geq a_2 \geq \cdots$ and $b_1 \leq b_2 \leq \cdots$. One can verify that the answer to the problem is yes if and only if there exists some $i \in \{0, \ldots, k\}$ such that

$$\max\{a, a_{i+1}\} \leq \min\{b, b_{k-i+1}\}.$$

The condition can be checked in $O(k)$ time, once we have computed $a_1, \ldots, a_k$ and $b_1, \ldots, b_k$. This computation requires sorting the k smallest/largest elements in a list and takes $O(n + k \log n)$ time (for instance, by a modified heapsort).

Having solved the decision problem in $O(n+k \log n)$ time, we now proceed to solve the optimization problem. First compute a δ-cutting of the n bounding lines of H by Lemma 5.1. Intersect the cutting triangles with Δ and triangulate to form a new collection of triangles $\{\Delta_i\}$. For each Δ_i, let H_i be the set of halfplanes of H whose bounding lines intersect Δ_i and let k_i be the number of halfplanes of H not intersecting Δ_i. Then

$$w(H, \Delta, k) = \min_i w(H, \Delta_i, k) = \min_i w(H_i, \Delta_i, k-k_i).$$

This divides the problem into a constant number of subproblems each of size roughly δn. We can now apply Lemma 2.1 with the upper bound $D(n) = O(n + k^{1-\varepsilon} n^{\varepsilon} \log n)$. □

We leave as an open question whether the optimization problem can be solved in $O(n + k \log n)$ expected time. Extending the technique to higher dimensions is also possible, but it is unclear at the moment how much is gained by examining the decision problem.

Infeasible case. In the case that $\bigcap H = \emptyset$, our technique can be applied to solve a different problem: given a set H of n halfplanes in $\mathbb{R}^2$, find the smallest k such that there exists a point in $L_k(H)$. The problem is related to finding a maximal consistent subset of halfplanes. The dual is related to a certain line separation/transversal problem [29].

An $O(n^2)$-time algorithm is immediate after constructing the arrangement of the n bounding lines. Everett, *et al.* [29] gave a "quality-sensitive" algorithm for this problem that runs in $O(n \log n + nk \log k)$ time. The algorithm first solves the decision problem (given k, is $L_k(H)$ empty?) in $O(n \log n + nk)$ time, and then uses a binary search to find the optimal k. With our technique (and cuttings), we can solve the original problem in $O(n \log n + nk)$ expected time, answering one of Everett, *et al.*'s open problems. Due to space limitation, details are omitted in this version. We note that when k is small, a different algorithm of Matoušek [41], with a refinement by the author [11], yields a better running time of $O(n \log k + k^3 \log^2 n)$. It is open what the optimal time bound should be for the entire range of k.

6 The Discrete 1-Center Problem in $\mathbb{R}^3$

As a final example that uses cuttings, we examine a variant of the standard Euclidean 1-center problem known as the *discrete 1-center problem*: given an n-point set P, find the smallest ball enclosing P whose center belongs to P.

Recently, the discrete 2-center problem in the plane was studied by Agarwal, *et al.* [4]. As they noted, the discrete 1-center problem in the plane is easily solvable in $O(n \log n)$ time by constructing the farthest-point Voronoi diagram. However, for points in three dimensions, the size of the Voronoi diagram can be $\Omega(n^2)$. A better approach in $\mathbb{R}^3$ is to first solve the decision problem, which involves the construction of an intersection of congruent balls. This intersection has $O(n)$ size and can be computed in $O(n \log n)$ time by the randomized method of Clarkson and Shor [19] or the deterministic method of Brönnimann, *et al.* [9]. Like the Euclidean diameter problem, the discrete 1-center problem in $\mathbb{R}^3$ can then be solved by parametric search in $O(n \operatorname{polylog} n)$ time [9, 13, 43, 48].

We now show how parametric search can be replaced by a randomized search, solving the three-dimensional discrete 1-center problem in $O(n \log n)$ expected time. The randomized reduction to the decision problem is more involved than in the diameter problem [19] (Section 3). In particular, we need to borrow a certain prune-and-search idea that was used previously by Megiddo [45] for the standard 1-center problem.

We have learned just recently of a randomized algorithm of Clarkson [18] that solves the same problem without using such a prune-and-search technique; its expected running time is however slightly slower ($O(n \log n \log\log n)$).

Theorem 6.1 *The discrete 1-center problem in* $\mathbb{R}^3$ *can be solved in* $O(n \log n)$ *expected time.*

Proof: Given an m-point set P and an n-point set Q in $\mathbb{R}^3$, we solve a "min-max" problem: compute

$$w(P, Q) = \min_{p \in P} \max_{q \in Q} d(p, q),$$

where $d(\cdot, \cdot)$ is the Euclidean metric. The radius of the optimal ball for the discrete 1-center problem is $w(P, P)$. (Note that the "max-max" problem corresponds to the diameter problem.)

To decide whether $w(P, Q) < t$, we compute the intersection of the balls centered at the points of Q with radius t. If there exists a point of P in the interior of this intersection, then we return yes. By the algorithm of Clarkson and Shor [19] or Brönnimann, *et al.* [9], the decision can be made in $O((n + m) \log n)$ time.

Now, we apply Lemma 2.1 to solve the optimization problem. First, it is easy to reduce the size of P

by a constant factor: arbitrarily partition P into two subsets P_1, P_2 of roughly size $m/2$, and observe that

$$w(P,Q) = \min\{w(P_1,Q), w(P_2,Q)\}. \quad (2)$$

To reduce the size of Q by a constant factor, we arbitrarily pair points of Q and construct the bisector (a plane) for each of the $\lfloor n/2 \rfloor$ pairs. By Lemma 5.1, construct a δ-cutting $\{\Delta_i\}$ of these $\lfloor n/2 \rfloor$ bisectors.

For each tetrahedron Δ_i, we build a point set Q_i. Initially, Q_i is set to Q. Examine all pairs (q_1, q_2) whose bisectors avoid the interior of Δ_i. Say q_1 lies on the same side of the bisector as Δ_i. Then any point in Δ_i is closer to q_1 than q_2. Remove q_1 from Q_i. As approximately $(1-\delta)n/2$ points are pruned, each Q_i has size bounded by roughly $(1+\delta)n/2$. Furthermore, $\max_{q\in Q} d(p,q) = \max_{q\in Q_i} d(p,q)$ for any $p \in \Delta_i$. Hence,

$$w(P,Q) = \min_i w(P\cap\Delta_i, Q) = \min_i w(P\cap\Delta_i, Q_i). \quad (3)$$

By applying equations (2) and (3) alternately, we can reduce both m and n by a constant factor using a constant number of subproblems. The result follows. □

Clearly, the same reduction extends to the discrete 1-center problem in higher dimensions. The reduction also works for "max-min" problems, such as the computation of the Hausdorff distance of two point sets; unfortunately, in this case, the decision problem does not seem to be any easier to solve.

7 Discussions

We have obtained improved expected time bounds (optimal in some cases) for several geometric optimization problems using a common randomized technique. The technique reduces these problems to their corresponding decision problems with only a constant-factor increase in the running time. The constants are quite large though (especially for the rectilinear p-center problem), and it is of practical interest to lower them.

Take for example the closest-pair application (Section 3). The reduction in the proof of Theorem 3.1 partitions the given set into three subsets and as a result reduces the problem into $r = 3$ subproblems of a fraction $\alpha = 2/3$ of the original size. When applying Lemma 2.1, we need to compress $\ell = 5$ levels of the recursion, in effect, raising the number of subproblems to 3^5. A better approach is to modify the reduction directly by partitioning the given set into q subsets for a constant $q > 3$. Then we can create $\binom{q}{2}$ subproblems of $2/q$ times the size. By choosing $q = 10$, we avoid the compression and have only 45 subproblems. A similar approach can be taken to optimize the constants in some of our other applications.

We see two more directions for further research. The first is to find more geometric applications (or perhaps non-geometric ones). The second is to find deterministic algorithms matching the performance of our randomized algorithms.

Limitation. Regarding the first direction, we note two requirements in the application of our technique, as shown by many of our examples: (i) an appropriate extension of the optimization and the decision problems, and (ii) a way to divide a problem so as to reduce the problem size by a constant factor.

Requirement (ii) needs more technical consideration but actually can be dealt with in a fairly general way using existing tools on geometric cuttings. For an illustration, take the well-studied *planar Euclidean two-center problem*, where there is an $O(n \log n)$ solution for the decision problem and an $O(n \log^2 n)$ randomized solution for the optimization problem from the recent work of Sharir [51] and Eppstein [27]. We can transform the problem into one involving an arrangement of n surfaces in five dimensions (using four variables for the coordinates of the two centers). Constructing a cutting for this collection of piecewise-algebraic surfaces establishes (ii). However, we need to solve an extension of the decision problem where the solution is restricted to lie inside a given Tarski cell in the transformed five-dimensional space. Unfortunately, the $O(n \log n)$ decision algorithm does not work for this extended problem, so requirement (i) is not satisfied.

Derandomization? Regarding the second direction, we point out that derandomization of our generic recursive algorithm (Section 2) is probably hard, if at all possible. We cannot even achieve good high-probability bounds. The problem lies in the fact that algorithm RAND-MIN is applied to a constant number of elements.

Consider one special case where we want to find the minimum $t^* = \min\{A[1], \ldots, A[r]\}$, knowing that t^* belongs to a given search space of size U, say $\{1, \ldots, U\}$. Suppose that a decision $A[i] < t$ takes D time. Then evaluating an $A[i]$ takes $O(D \log U)$ time by binary search, and algorithm RAND-MIN runs in $O(Dr + D \log U \log r)$ expected time.

Here is a simple deterministic algorithm that beats the obvious $O(Dr \log U)$ time bound for small values of U:

Algorithm DET-MIN

1. $i \leftarrow 1$
2. while $i \leq r$ do
3. if $A[i] \geq U$ then $i \leftarrow i+1$ else $U \leftarrow U-1$
4. return U

The correctness of the algorithm is easily seen from the invariant $\min\{A[1],\ldots,A[i-1]\} \geq U$. The running time of the algorithm is clearly $O(D(r+U))$.

For larger values of $U > r$, the process can be sped up as follows. Apply Algorithm DET-MIN to find $t_0 = \min_i \lfloor A[i]\cdot(r/U)\rfloor$. As t_0 lies in $\{1,\ldots,r\}$, it can be found in $O(Dr)$ time. Now, the minimum t^* lies between t_0U/r and $(t_0+1)U/r$, so we just need to solve the problem recursively on a search space of size U/r. The depth of the recursion is $O(\lceil\log_r U\rceil)$, so the minimum can be found in $O(Dr\lceil\log_r U\rceil)$.

The deterministic complexity of this minimization problem is therefore the same as the randomized complexity—namely, $O(Dr)$—if the size of the search space U is polynomial in r. For still larger U, one can apply a more complicated procedure due to Gao, *et al.* [31].

We are only able to find one application of this deterministic approach:

> Given a set of n points in the plane, find a k-point subset of the minimum Euclidean diameter.

For k close to a fraction of n, the best algorithm for the above problem is due to Eppstein and Erickson [28] and runs in $O(n^3\log^2 n)$ time. This algorithm decomposes the problem into n subproblems. Each subproblem is solved in $O(n^2\log^2 n)$ time by applying binary search to an $O(n^2\log n)$-time decision algorithm. The search space is the set of all interpoint distances, enumerable in $O(n^2\log n)$ time easily. Thus, we have here $r = n$, $D = O(n^2\log n)$, and $U = O(n^2)$, so we can solve the problem in $O(n^3\log n)$ worst-case time, a $\log n$ factor improvement over Eppstein and Erickson's bound.

For smaller values of k, one can combine this result with known techniques [22, 28] to derive the improved deterministic time bound of $O(n\log n + nk^2\log k)$. The corresponding randomized bound was recently discovered by Bhattacharya and ElGindy [7].

References

[1] P. K. Agarwal and J. Matoušek. Ray shooting and parametric search. *SIAM J. Comput.*, 22:764–806, 1993.

[2] P. K. Agarwal and M. Sharir. Efficient randomized algorithms for some geometric optimization problems. *Discrete Comput. Geom.*, 16:317–337, 1996.

[3] P. K. Agarwal and M. Sharir. Efficient algorithms for geometric optimization. Technical Report CS-1996-19, Department of Computer Science, Duke University, Durham, NC, 1996.

[4] P. K. Agarwal, M. Sharir, and E. Welzl. The discrete 2-center problem. In *Proc. 13th ACM Sympos. Comput. Geom.*, pages 147–155, 1997.

[5] A. Aggarwal, H. Imai, N. Katoh, and S. Suri. Finding k points with minimum diameter and related problems. *J. Algorithms*, 12:38–56, 1991.

[6] S. Arya and D. M. Mount. Approximate nearest neighbor queries in fixed dimensions. In *Proc. 4th ACM-SIAM Sympos. Discrete Algorithms*, pages 271–280, 1993.

[7] B. K. Bhattacharya and H. ElGindy. Biased search and k-point clustering. In *Proc. 9th Canad. Conf. Comput. Geom.*, 1997.

[8] H. Brönnimann and B. Chazelle. Optimal slope selection via cuttings. In *Proc. 6th Canad. Conf. Comput. Geom.*, pages 99–103, 1994.

[9] H. Brönnimann, B. Chazelle, and J. Matoušek. Product range spaces, sensitive sampling, and derandomization. In *Proc. 34th IEEE Sympos. Found. Comput. Sci.*, pages 400–409, 1993.

[10] T. M. Chan. Fixed-dimensional linear programming queries made easy. In *Proc. 12th ACM Sympos. Comput. Geom.*, pages 284–290, 1996.

[11] T. M. Chan. Output-sensitive results on convex hulls, extreme points, and related problems. *Discrete Comput. Geom.*, 16:369–387, 1996.

[12] B. Chazelle. Cutting hyperplanes for divide-and-conquer. *Discrete Comput. Geom.*, 9:145–158, 1993.

[13] B. Chazelle, H. Edelsbrunner, L. Guibas, and M. Sharir. Diameter, width, closest line pair and parametric searching. *Discrete Comput. Geom.*, 10:183–196, 1993.

[14] L. P. Chew, D. Dor, A. Efrat, and K. Kedem. Geometric pattern matching in d-dimensional space. In *Proc. 2nd European Sympos. Algorithms*, Lect. Notes in Comput. Sci., vol. 979, Springer-Verlag, pages 264–279, 1995.

[15] L. P. Chew, M. T. Goodrich, D. P. Huttenlocher, K. Kedem, J. M. Kleinberg, and D. Kravets. Geometric pattern matching under Euclidean motion. *Comput. Geom. Theory Appl.*, 7:113–124, 1997.

[16] L. P. Chew and K. Kedem. Improvements on geometric pattern matching problems. In *Proc. 3rd Scand. Workshop Algorithm Theory*, Lect. Notes in Comput. Sci., vol. 621, Springer-Verlag, pages 318–325, 1992.

[17] K. L. Clarkson. New applications of random sampling in computational geometry. *Discrete Comput. Geom.*, 2:195–222, 1987.

[18] K. L. Clarkson. Algorithms for the minimum diameter of moving points and for the discrete 1-center problem. Manuscript, 1997.

[19] K. L. Clarkson and P. W. Shor. Applications of random sampling in computational geometry, II. *Discrete Comput. Geom.*, 4:387–421, 1989.

[20] R. Cole. Slowing down sorting networks to obtain faster sorting algorithms. *J. ACM*, 34:200–208, 1987.

[21] T. H. Cormen, C. E. Leiserson, and R. L. Rivest. *Introduction to Algorithms.* MIT Press, Cambridge, Mass., 1990.

[22] A. Datta, H.-P. Lenhof, C. Schwarz, and M. Smid. Static and dynamic algorithms for k-point clustering problems. *J. Algorithms*, 19:474–503, 1995.

[23] M. Dillencourt, D. Mount, and N. Netanyahu. A randomized algorithm for slope selection. *Int. J. Comput. Geom. Appl.*, 2:1–27, 1992.

[24] Z. Drezner. On the rectangular p-center problem. *Naval Res. Logist. Quart.*, 34:229–234, 1987.

[25] M. E. Dyer. Linear time algorithms for two- and three-variable linear programs. *SIAM J. Comput.*, 13:31–45, 1984.

[26] A. Efrat, M. Sharir, and A. Ziv. Computing the smallest k-enclosing circle and related problems. *Comput. Geom. Theory Appl.*, 4:119–136, 1994.

[27] D. Eppstein. Faster construction of planar two-centers. In *Proc. 8th ACM-SIAM Sympos. Discrete Algorithms*, pages 131–138, 1997.

[28] D. Eppstein and J. Erickson. Iterated nearest neighbors and finding minimal polytopes. *Discrete Comput. Geom.*, 11:321–350, 1994.

[29] H. Everett, J.-M. Robert, and M. van Kreveld. An optimal algorithm for the $(\leq k)$-levels, with applications to separation and transversal problems. *Int. J. Comput. Geom. Appl.*, 6:247–261, 1996.

[30] G. N. Frederickson and D. B. Johnson. The complexity of selection and ranking in $X + Y$ and matrices with sorted rows and columns. *J. Comput. Sys. Sci.*, 24:197–208, 1982.

[31] F. Gao, L. J. Guibas, D. G. Kirkpatrick, W. T. Laaser, and J. Saxe. Finding extrema with unary predicates. *Algorithmica*, 9:591–600, 1993.

[32] Alex Glozman, Klara Kedem, and Gregory Shpitalnik. On some geometric selection and optimization problems via sorted matrices. In *Proc. 4th Workshop Algorithms Data Struct.*, Lect. Notes in Comput. Sci., vol. 955, Springer-Verlag, pages 26–37, 1995.

[33] S. Har-Peled. Constructing cuttings in theory and practice. In these proceedings.

[34] D. P. Huttenlocher, K. Kedem, and M. Sharir. The upper envelope of Voronoi surfaces and its applications. *Discrete Comput. Geom.*, 9:267–291, 1993.

[35] M. Katz and M. Sharir. An expander-based approach to geometric optimization. *SIAM J. Comput.*, 26:1384–1408, 1997.

[36] D. E. Knuth. *Fundamental Algorithms*, volume 1 of *The Art of Computer Programming.* Addison-Wesley, 1968.

[37] J. Matoušek. Randomized optimal algorithm for slope selection. *Inform. Process. Lett.*, 39:183–187, 1991.

[38] J. Matoušek. Efficient partition trees. *Discrete Comput. Geom.*, 8:315–334, 1992.

[39] J. Matoušek. Reporting points in halfspaces. *Comput. Geom. Theory Appl.*, 2:169–186, 1992.

[40] J. Matoušek. On enclosing k points by a circle. *Inform. Process. Lett.*, 53:217–221, 1995.

[41] J. Matoušek. On geometric optimization with few violated constraints. *Discrete Comput. Geom.*, 14:365–384, 1995.

[42] J. Matoušek. On constants for cuttings in the plane. *Discrete Comput. Geom.*, to appear.

[43] J. Matoušek and O. Schwarzkopf. A deterministic algorithm for the three-dimensional diameter problem. *Comput. Geom. Theory Appl.*, 6:253–262, 1996.

[44] N. Megiddo. Applying parallel computation algorithms in the design of serial algorithms. *J. ACM*, 30:852–865, 1983.

[45] N. Megiddo. Linear time algorithms for linear programming in R^3 and related problems. *SIAM J. Comput.*, 12:759–776, 1983.

[46] N. Megiddo. Linear programming in linear time when the dimension is fixed. *J. ACM*, 31:114–127, 1984.

[47] M. O. Rabin. Probabilistic algorithms. In *Algorithms and Complexity* (J. F. Traub, ed.), pages 21–30, Academic Press, New York, NY, 1976.

[48] E. Ramos. Construction of 1-d lower envelopes and applications. In *Proc. 13th ACM Sympos. Comput. Geom.*, pages 57–66, 1997.

[49] T. Roos and P. Widmayer. k-Violation linear programming. *Inform. Process. Lett.*, 52:109–114, 1994.

[50] M. Segal. On piercing of axis-parallel rectangles and rings. In *Proc. 5th European Sympos. Algorithms*, Lect. Notes in Comput. Sci., vol. 1284, Springer-Verlag, 1997.

[51] M. Sharir. A near-linear algorithm for the planar 2-center problem. *Discrete Comput. Geom.*, 18:125–134, 1997.

[52] M. Sharir and E. Welzl. A combinatorial bound for linear programming and related problems. In *Proc. 9th Sympos. Theoret. Aspects Comput. Sci.*, Lect. Notes in Comput. Sci., vol. 577, Springer-Verlag, pages 569–579, 1992.

[53] M. Sharir and E. Welzl. Rectilinear and polygonal p-piercing and p-center problems. In *Proc. 12th ACM Sympos. Comput. Geom.*, pages 122–132, 1996.

On Enumerating and Selecting Distances

Timothy M. Chan*

Abstract

Given an n-point set, the problems of enumerating the k closest pairs and selecting the k-th smallest distance are revisited. For the enumeration problem, we give simpler randomized and deterministic algorithms with $O(n \log n + k)$ running time in any fixed-dimensional Euclidean space. For the selection problem, we describe an approach to obtaining k-sensitive time bounds. We also point out output-sensitive results for halfspace range counting that are of use in more general distance selection problems. None of our algorithms requires parametric search.

1 Introduction

Finding the *closest pair* of an n-point set has a long history in computational geometry. In the plane, the problem can be solved in $O(n \log n)$ time using the Delaunay triangulation. In an arbitrary fixed dimension d, the first $O(n \log n)$ algorithm, based on divide-and-conquer, was described by Bentley and Shamos [5]. Another $O(n \log n)$ algorithm of Vaidya [33] can actually find the nearest neighbor to each of the given points. Back in 1976, Rabin [31] suggested a random sampling method that requires only $O(n)$ expected time assuming a RAM model of computation that supports the floor function and constant-time hashing (see also [12, 17, 22]). Under the more usual algebraic decision tree model, linear randomized complexity is still attainable when the points have been pre-sorted along each of the d coordinates.

The more general problem of *enumerating the k closest pairs* (or enumerating the first k smallest distances) has also received much attention. One of the earliest reported algorithms is by Dickerson, *et al.* [13], who used the Delaunay triangulation to enumerate the k closest pairs in $O((n + k) \log n)$ time in two dimensions. (Their algorithm actually enumerates the distances in sorted order.) Salowe [32] was the first to give an $O(n \log n + k)$ algorithm for any fixed dimension. His algorithm employs techniques such as parametric searching [28] or Vaidya's all-nearest-neighbor method. Lenhof and Smid [23] subsequently pointed out a much simplified algorithm; the only geometric structure needed is a grid. Alternatives that apply more advanced geometric structures have been proposed by Dickerson and Eppstein [14] (higher-dimensional Delaunay triangulations) and Arya and Smid [4] (spanners).

In this paper, we add two more distance enumeration algorithms to the list:

- A further simplification of Lenhof and Smid's algorithm, running in $O(n \log n + k)$ time. In their algorithm, parametric search is avoided by using instead an implicit binary search on a matrix with sorted rows and columns. This binary search is carried out in the style of Frederickson and Johnson [19] and requires repeated calls to a linear-time subroutine for weighted medians. We replace these median computations with a simpler "approximate-and-refine" scheme. A complete description of the new algorithm fits in less than two pages (see Lemma 3.2 and Theorem 3.5).

- An even simpler algorithm based on random sampling. This can be regarded as an extension of Rabin's closest-pair algorithm. The expected running time is $O(n \log n + k)$ and can be brought down to $O(n + k)$ for pre-sorted point sets. The approach follows from a general result from Section 2.1, which also has an application to enumerating k farthest pairs (see Section 4).

Our algorithms work under the algebraic decision tree model and are thus optimal in this model. Distances are not enumerated in sorted order.

A related problem is how to *select the k-th smallest distance*. In the plane (under the Euclidean metric), Agarwal, *et al.* [1] solved a decision version of the problem in $O(n^{4/3} \log^{2/3} n)$ expected time. Applying parametric search, they then showed that the k-th smallest distance can be selected in $O(n^{4/3} \log^{8/3} n)$ expected time in dimension 2. Deterministic variants were con-

*Department of Mathematics and Computer Science, University of Miami, Coral Gables, FL 33124–4250, USA, E-mail: tchan@cs.miami.edu

SCG 98 Minneapolis Minnesota USA
Copyright ACM 1998 0-89791-973-4/98/ 6...$5.00

sidered by Goodrich [18] and Katz and Sharir [21]. Improvements in the logarithmic factors were noted briefly by Matoušek [24]; more importantly, he showed how parametric search can be avoided to obtain simpler algorithms that are randomized.

Our contributions to the distance selection problem can be summarized as follows:

- A randomized reduction to the decision problem without parametric search. Matoušek only described in detail his randomized technique for the slope selection problem; we state and prove the result under a general setting in Section 2.2.

- A way to obtain running time sensitive to k. Agarwal, *et al.*'s bound is worse than the $O(n \log n + k)$ bound for the enumeration problem when $k \ll n^{4/3}$. We show that a hybrid approach in the planar case yields an expected running time of $O(n \log n + n^{2/3}k^{1/3} \log^{5/3} n)$, improving both results for most values of k between 1 and $\Theta(n^2)$. Any improvement to the k-insensitive bound automatically implies an improvement to the k-sensitive bound (see Theorem 3.6).

In passing, we note that de Berg and Schwarzkopf [6] and Pellegrini [30] have obtained similar bounds—near $O(n + n^{2/3}k^{1/3})$—for the problem of counting the k intersecting pairs of n given line segments. It is plausible that their techniques could lead to the same result for distance selection. However, our approach exploits special properties of distances and thus avoids "intersection-sensitive cuttings;" furthermore, our approach extends to higher dimensions.

Finally, Section 4 mentions connection of more general distance selection problems to *halfspace range counting*: how to preprocess a given n-point set so that one can quickly count the number k of points inside a given query halfspace. The problem has been studied by a number of researchers; see [2, 27] for surveys. In the planar case, $O(n^{1/2} \operatorname{polylog} n)$ query time is attainable after $O(n \log n)$ preprocessing time [25]. On the other hand, results on halfspace range reporting imply an output-sensitive query time of $O(\log n + k)$ with the same preprocessing time [7]. We point out that combining previous approaches properly yields a query time bound of $O(k^{1/2}n^{\varepsilon})$ for any constant $\varepsilon > 0$ in this planar case. The precise output-sensitive bounds in higher dimensions, including preprocessing/query-time tradeoffs, are given by Theorem 4.1. Although the proof of these output-sensitive bounds requires no new ideas (the main ingredients being Matoušek's "shallow cutting lemma" and "partition theorem for shallow hyperplanes" [26]), they are worth noting due to the fundamental nature of range searching.

2 General Randomized Reductions

Let U represent the *object space* and $d : U \times U \to \mathbb{R}$ represent the *distance function*, where $d(p, q) = d(q, p)$ for any $p, q \in U$.

Notation. Given sets $P, Q \subset U$, define the multiset $d(P, Q) = \{d(p, q) : (p, q) \in P \times Q\}$. For any integer $1 \le k \le |P| \cdot |Q|$, let $R(P, Q, k)$ be the k-th smallest element in $d(P, Q)$. For any $r \in \mathbb{R}$, let $K(P, Q, r) = |d(P, Q) \cap (-\infty, r]|$. Observe that $R(P, Q, k) \le r$ iff $K(P, Q, r) \ge k$. For notational convenience, set $d(P) = d(P, P)$, $R(P, k) = R(P, P, k)$, and $K(P, r) = K(P, P, r)$.

Given as input a set $P \subset U$ of size n, an integer k, and a number $r \in \mathbb{R}$, we study the following problems:

Enumeration. Return the elements $R(P, 1)$, $R(P, 2)$, $\ldots$, $R(P, k)$ in any order.

Reporting. Return the elements in $d(P) \cap (-\infty, r]$ in any order.

Selection. Return the element $R(P, k)$.

Counting. Return the number $K(P, r)$.

The reporting problem is also known as "fixed-radius nearest neighbor search;" the counting problem is the same as determining the "rank" of a given distance.

The above formulation of the problems differs slightly from usual convention. For example, the standard selection problem asks for the k-th smallest element $d(p, q)$ over all unordered pair $\{p, q\} \subset P$. The two versions are basically equivalent. Indeed, in most applications, all distances are nonnegative and $d(p, p) = 0$. So, the k-th smallest distance of all unordered pairs is identical to the $(n + 2k)$-th smallest distance of all ordered pairs; in the time bounds, we just replace k with $\Theta(n + k)$.

2.1 From Enumeration to Reporting

It is a straightforward exercise to exhibit an efficient reduction from the reporting problem to the enumeration problem. In this subsection, we show that a general reduction in the reverse direction is possible via random sampling.

Lemma 2.1 *In $O(n)$ time, we can select an r such that $R(P, k) \le r \le R(P, k')$ with a positive-constant probability, where $k' = 6(n + k)$.*

Proof: Choose n elements $r_1, \ldots, r_n$ randomly from $d(P)$ and select the m-th smallest element r in the multiset $\{r_1, \ldots, r_n\}$ with $m = 3\lceil k/n \rceil$. It suffices to show that sum of the two probabilities $\Pr\{r < R(P, k)\}$ and $\Pr\{r > R(P, k')\}$ is bounded by a constant strictly less than 1.

If $r < R(P,k)$, then $N = |\{i : r_i < R(P,k)\}|$ is at least m. The random variable N satisfies a binomial distribution with n trials, where each trial has success probability less than k/n^2. The mean $\mu = E[N]$ is less than k/n. Therefore, by Markov's inequality,

$$\Pr\{r < R(P,k)\} \leq \Pr\{N \geq m\} \leq \Pr\{N > 3\mu\} \leq 1/3.$$

If $r > R(P,k')$, then $N' = |\{i : r_i \leq R(P,k')\}|$ is less than m. The random variable N' satisfies a binomial distribution with n trials, where each trial has success probability at least k'/n^2. The mean $\mu' = E[N']$ is at least $k'/n = 6(k/n+1)$. Therefore,

$$\begin{aligned}\Pr\{r > R(P,k')\} &\leq \Pr\{N' < m\} \leq \Pr\{N' < \mu'/2\} \\ &< e^{-\mu'/8} < e^{-3/4} < 1/2,\end{aligned}$$

by a standard Chernoff bound (e.g., [29, p. 70]). □

The above idea of taking a random sample of size n to approximate a set of size $O(n^2)$ is hardly new. A similar approach was taken by Matoušek [24] in the context of slope selection and was named "randomized interpolating search." One of the earliest uses of the idea was actually by Rabin when discussing the closest-pair problem in his seminal paper on randomized algorithms [31].

Suppose that an algorithm for the reporting problem is available and runs in $T_R(n, K(P,r))$ time. Assume that $T_R(an, b(n+k)) = O(T_R(n,k))$ for any constants a, b. We now apply the above lemma to get a simple randomized algorithm for the enumeration problem.

The possibility of degeneracy (i.e., duplicate distances) creates some technical complications, which we will overcome by introducing an *infinitesimal* δ: its value is assumed to be arbitrarily small but positive. For example, applying the reporting algorithm on $r - \delta$ should produce all distances in the open interval $(-\infty, r)$. (Symbolic simulation can usually be avoided though.)

Theorem 2.2 *The enumeration problem can be solved in $O(T_R(n,k))$ expected time.*

Proof: Construct r by Lemma 2.1. Verify that $R(P,k) \leq r \leq R(P,k')$. If not, repeat for a different r; the expected number of trials is bounded by a constant. To verify that $r \geq R(P,k)$, simulate the reporting algorithm on r for $T_R(n,k)$ time steps; the computation should either fail or return a list of k or more distances. To verify that $r \leq R(P,k')$, simulate the reporting algorithm on $r-\delta$ for $T_R(n,k')$ time steps; the computation should succeed with a list of fewer than k' elements.

Now, report all elements in $\Delta = d(P) \cap (-\infty, r)$ and select the first k smallest elements. In case that r is precisely equal to $R(P,k)$, add $\max\{k - |\Delta|, 0\}$ copies of r to the output. As $|\Delta| < k'$, the running time is $O(T_R(n,k'))$. □

2.2 From Selection to Counting

An ordinary binary search reduces the counting problem to the selection problem. In this subsection, we explore the reverse direction. As Agarwal, *et al.* [1] noted, a fairly general reduction of selection to counting is possible by parametric search. We point out a simpler approach by randomization.

Our approach is similar to a scheme that Matoušek [24] called "randomized halving" (see also [16]). The idea is a binary search that repeatedly cuts the search interval using a random element in the interval. We describe a recursive procedure to produce such a random element efficiently in general. (Matoušek's procedure is specifically designed for the slope selection problem.)

Suppose that an algorithm is available for the counting problem and runs in time $T_C(n, K(P,r))$. Assume that $T_C(n,k)/n^\varepsilon$ is an increasing function in both n and k for some constant $\varepsilon > 0$. Assume also that $T_C(an, b(n+k)) = O(T_C(n,k))$ for constants a, b. As before, degeneracy is handled by using an infinitesimal δ. Note that we can compute $K(P,Q,r)$ using the counting algorithm by the identity

$$\begin{aligned}K(P,Q,r) &= (K(P \triangle Q, r) + K(P,r) + K(Q,r))/2 \\ &\quad - K(P \setminus Q, r) - K(Q \setminus P, r), \qquad (1)\end{aligned}$$

where $P \triangle Q$ denotes the symmetric difference of P and Q.

Lemma 2.3 *Given $A < B$, one can select an element uniformly at random from $d(P,Q) \cap (A,B)$ in $O(T_C(n,k))$ time, where $n = |P| = |Q|$ and $k = K(P \cup Q, B - \delta)$.*

Proof: Define $N(P,Q) = |d(P,Q) \cap (A,B)| = K(P,Q,B-\delta) - K(P,Q,A)$, which can be computed in $O(T_C(n,k))$ time by (1). Partition $P = P_1 \cup P_2$ and $Q = Q_1 \cup Q_2$, with $|P_1|, |P_2|, |Q_1|, |Q_2| \leq \lceil n/2 \rceil$. Compute $N(P_i, Q_j)$ for each i, j. Generate a random pair $(i,j) \in \{1,2\} \times \{1,2\}$ such that (i,j) is chosen with probability $N(P_i,Q_j)/N(P,Q)$. Now, recursively select an element uniformly at random from $d(P_i,Q_j) \cap (A,B)$. The running time of this process is $O(T_C(n,k) + T_C(n/2,k) + T_C(n/4,k) + \cdots) = O(T_C(n,k))$. □

Theorem 2.4 *The selection problem can be solved in $O(T_C(n,k)\log n)$ expected time.*

Proof: We will maintain an interval $(A,B]$ containing $R(P,k)$. Initially, we can set $A = 0$ and B with the value r from Lemma 2.1. As in the proof of Theorem 2.2, we can verify that $R(P,k) \leq r \leq R(P,k')$ (this time, by simulating the counting algorithm); if this is not true, repeat for an expected constant number of trials.

Now pick a random element $r \in d(P) \cap (A, B)$ by Lemma 2.3. If no such element exists, then return B. Compute $K(P, r)$ in $T_C(n, k')$ time. If $K(P, r) < k$, then replace A with r; otherwise, replace B with r.

It suffices to show that the expected number of iterations is logarithmic. Let $N = |d(P) \cap (A, B)|$. Since r is equally likely to be any element in $d(P) \cap (A, B)$, the probability that $|d(P) \cap (A, r)| \geq 2N/3$ and the probability that $|d(P) \cap (r, B)| \geq 2N/3$ are both at most $1/3$. The expected number of iterations to reduce N by a factor of $2/3$ is thus bounded by 3, and consequently, the expected number of iterations to reduce N to 0 is $O(\log n)$. □

Removing the extra $\log n$ factor seems to require stronger oracles than just the counting algorithm.

3 Algorithms for Natural Distances

In this section, we give specific enumeration and selection algorithms for the following kind of distance functions, where objects are points from a fixed-dimensional space $U = \mathbb{R}^d$:

Definition. A distance function $d : \mathbb{R}^d \times \mathbb{R}^d \to \mathbb{R}$ is *natural* if $\| p - q \|_\infty \leq d(p, q) \leq c \| p - q \|_\infty$ for some constant c. Here, $\| \cdot \|_\infty$ denotes the L_∞ norm.

Obviously, all L_p distance functions—in particular, the Euclidean metric—are natural under this definition. The following combinatorial property of Salowe will be important. (The original proof was for the L_∞ metric only, but readjusting constants immediately imply the statement for any natural distance function.)

Proposition 3.1 (Salowe [32]) *If the distance function is natural, then $K(P, ar) = O(|P| + K(P, r))$ for any constant a.*

The enumeration algorithms of Salowe [32] and Lenhof and Smid [23] are based on a lemma similar to the below; a proof is included for completeness. We will use this in our solutions to both the enumeration and the selection problem. (*Pre-sorting* here refers to the computation of the sorted order of the points of P in each of the d coordinates; it can be done in $O(n \log n)$ time.)

Lemma 3.2 *Suppose the distance function is natural. Let $r > 0$ and $P \subset \mathbb{R}^d$ be a given pre-sorted n-point set. In linear time, one can construct subsets $\{P_i\}$ and $\{Q_i\}$ of total size $O(n)$, such that*

(i) *the P_i's are disjoint;*

(ii) *any pair $(p, q) \in P \times P$ with $d(p, q) \leq r$ belongs to some $P_i \times Q_i$;*

(iii) *any pair $(p, q) \in P_i \times Q_i$ has $d(p, q) < 2cr$ for a constant c.*

Proof: We follow the approach of Lenhof and Smid [23] to build a "degraded grid." For each $j \in \{1, \ldots, d\}$, let X_j denote the set of j-th coordinates of the n points. Cover X_j by $\leq n$ disjoint open intervals of length r which we call *j-intervals*. This can be accomplished easily in linear time by scanning the coordinates in sorted order. Define the *neighborhood* of an interval $I = (A, B)$ to be $\widehat{I} = (A - r, B + r)$. Observe that a coordinate can belong to the neighborhoods of at most three j-intervals.

Define a *grid cell* to be a box of the form $h = I_1 \times \cdots \times I_d$ where each I_j is a j-interval. Its *neighborhood* is $\widehat{h} = \widehat{I_1} \times \cdots \times \widehat{I_d}$. A point can belong to the neighborhoods of at most 3^d grid cells. For each nonempty grid cell h, construct corresponding subsets $P_i = P \cap h$ and $Q_i = P \cap \widehat{h}$. One can verify properties (i)–(iii) easily: if a pair (p, q) has L_∞-distance $\leq r$, then it is in some $P_i \times Q_i$; conversely, any pair in $P_i \times Q_i$ has L_∞-distance $< 2r$.

Regarding the construction time, we can label the j-intervals by integers from $\{1, \ldots, n\}$ and grid cells by d-tuples from $\{1, \ldots, n\}^d$. For each j-th coordinate x, find the label of the j-th interval containing x as well as the labels of the j-th intervals whose neighborhoods contain x. This can be done by a linear scan. For each point p, we thus have the label of the grid cell containing p and the labels of the grid cells whose neighborhoods contain p. A radix sort then generates the collection of points inside each nonempty grid cell and the collection of points inside its neighborhood. □

3.1 Enumeration

One immediate corollary from Lemma 3.2 is an algorithm for the reporting problem: simply list all the elements in $\Delta = \bigcup_i d(P_i, Q_i)$ and return $\Delta \cap (-\infty, r]$. The running time after pre-sorting is $O(n + |\Delta|) = O(n + K(P, 2cr)) = O(n + K(P, r))$ by Proposition 3.1.

By Theorem 2.2, we have immediately a randomized enumeration algorithm:

Theorem 3.3 *For a pre-sorted n-point set with a natural distance function, the enumeration problem can be solved in $O(n + k)$ expected time.*

Including the time to pre-sort, the expected running time is therefore $O(n \log n + k)$. We now propose a deterministic alternative.

Because of the reporting algorithm, all we need to solve the enumeration problem is a good approximation r to $R(P, k)$. We will obtain such an r by an approximate binary search.

Lemma 3.4 *Under the same setup as in Lemma 3.2, one can reach one of the two conclusions in $O(n)$ time: (a) $R(P,k) > r$, or (b) $R(P,k) < 2cr$.*

Proof: Compute $S = \Sigma_i |P_i| \cdot |Q_i|$. If $S < k$, then (a) is true; if $S \geq k$, then (b) is true. □

Lenhof and Smid [23] observed that an approximation of $R(P,k)$ can be found by searching among all $O(n^2)$ L_∞-distances. To avoid generating the entire search space of quadratic size, one performs an implicit binary search, using repeated weighted-median computations like in the techniques of Frederickson and Johnson [19] and Cole [11].

We suggest a simpler approach by observing that an approximation can be found in a reduced search space of linear size. The approximation factor is quite high ($O(n)$), but can be refined easily afterwards. As a result, we bypass the use of a linear-time median-finding subroutine (except at the very end).

Theorem 3.5 *The enumeration problem for a natural distance function can be solved in $O(n \log n + k)$ time.*

Proof: Fix $j \in \{1, \ldots, d\}$. Let $x_1 \leq \cdots \leq x_n$ be the j-th coordinates of the points in sorted order. Define

$$\Xi_j = \{x_2 - x_1,\ x_3 - x_2,\ \ldots,\ x_n - x_{n-1}\}.$$

Construct the set $\Xi = \Xi_1 \cup \cdots \cup \Xi_d$. Let $\xi_1 \leq \cdots \leq \xi_N$ be its elements in sorted order ($N = O(n)$).

We keep indices ℓ and h while maintaining the following invariant:

$$\xi_\ell / 2c \ <\ R(P,k)\ <\ \xi_h.$$

Initially, we can set $\ell = 0$ and $h = N+1$ with $\xi_0 = 0$ and $\xi_{N+1} = \infty$. By applying Lemma 3.4 with $r = \xi_{\lfloor (\ell+h)/2 \rfloor}/2c$, we can cut $h - \ell$ by half in $O(n)$ time. Consequently, in $O(n \log n)$ time, we obtain:

$$\xi_\ell / 2c \ <\ R(P,k)\ <\ \xi_{\ell+1}.$$

We know that $\sigma \leq R(P,k) \leq c\sigma$ for some L_∞-distance σ. Since σ is a sum of at most n elements of Ξ, there must exist some ξ_i with $\xi_i \leq \sigma \leq n\xi_i$. It follows that $\xi_i \leq R(P,k) < \xi_{\ell+1}$ and thus $i \leq \ell$. It also follows that $R(P,k) \leq cn\xi_i \leq cn\xi_\ell$.

Next we keep numbers A and B while maintaining the following invariant:

$$A/2c \ <\ R(P,k)\ \leq\ B.$$

Initially, we can set $A = \xi_\ell$ and $B = cn\xi_\ell$. By applying Lemma 3.4 with $r \approx \sqrt{AB}/2c$, we can reduce B/A to approximately $\sqrt{B/A}$ in $O(n)$ time. Consequently, in $O(n \log\log n)$ time, we obtain:

$$A/2c \ <\ R(P,k)\ \leq 2A.$$

Now, we can solve the enumeration problem by reporting all elements in $d(P) \cap [0, 2A]$ and selecting the first k smallest elements, in time $O(n + K(P, 2A)) = O(n + K(P, A/2c)) = O(n+k)$ by Proposition 3.1. □

The above idea also simplifies the parallel version of Lenhof and Smid's enumeration algorithm [23].

We leave as an open problem whether deterministically we can achieve $O(n+k)$ running time for the enumeration problem when the point set is pre-sorted. We do not know, for that matter, a linear deterministic algorithm to find the closest pair of a pre-sorted point set. In two dimensions, an $O(n \log\log n)$ method exists for the closest pair: first construct the L_∞-Voronoi diagram in $O(n \log\log n)$ time by an algorithm of Chew and Fortune [10] (using van Emde Boas trees); find the smallest L_∞-distance r in linear time by examining pairs that form bisectors in the diagram; then report all pairs of distances $\leq cr$ in $O(n + K(P, cr)) = O(n + K(P, r)) = O(n)$ time by Proposition 3.1.

3.2 k-Sensitive Selection

We now describe a general strategy to derive k-sensitive bounds for the selection problem when the distance function is natural. We first solve the counting problem and then apply Theorem 2.4.

Theorem 3.6 *Suppose we have an $O(nf(n))$-time algorithm for the counting problem, where f is a concave nondecreasing function. For a natural distance function, we can solve the selection problem in $O(n \log n + nf(k/n) \log n)$ expected time.*

Proof: First note that $K(P,Q,r)$ can be evaluated in $O(nf(n))$ time by (1) if $|P|, |Q| \leq n$. For uneven sizes $|P| = n$ and $|Q| = m$, we can partition Q into subsets $Q_1, \ldots, Q_{\lceil m/n \rceil}$ each of size $\leq n$; then $K(P,Q,r) = \sum_i K(P, Q_i, r)$ can be evaluated in $O(\lceil m/n \rceil \cdot nf(n)) = O((m+n)f(n))$ time.

A k-sensitive algorithm to compute $k = K(P,r)$ can be obtained from Lemma 3.2:

$$K(P,r) \ =\ \sum_i K(P_i, Q_i, r),$$

where the sizes $n_i = |P_i|$ and $m_i = |Q_i|$ satisfy the bounds

$$\sum_i (m_i + n_i) \ =\ O(n), \quad \text{and}$$

$$\sum_i m_i n_i \ \leq\ K(P, 2cr) \ =\ O(n+k),$$

by Proposition 3.1. Without loss of generality, assume $m_i \geq n_i$. Then we can calculate all $K(P_i, Q_i, r)$ in time of the order of

$$\sum_i m_i f(n_i) \leq \left(\sum_i m_i\right) \cdot f\left(\frac{\sum_i m_i n_i}{\sum_i m_i}\right) = O(nf((n+k)/n)),$$

by Jensen's inequality (since f is concave).

Thus, the counting problem can be solved in $O(n(1 + f(K(P, r)/n)))$ time after pre-sorting. The selection problem can then be solved by Theorem 2.4 in $O(n(1 + f(k/n)) \log n)$ expected time. (Note that the subproblems that occur in the proof of Lemma 2.3 are already pre-sorted.) □

Agarwal, *et al.* [1] showed that the counting problem for Euclidean distances in the plane can be solved in $O(n^{4/3} \log^{2/3} n)$ expected time. Applying the above theorem, we have:

Corollary 3.7 *The selection problem for the Euclidean distance function in the plane can be solved in $O(n \log n + n^{2/3} k^{1/3} \log^{5/3} n)$ expected time.*

4 Halfspace Range Counting

Enumerating the k farthest pairs (or k largest distances) fits into the framework of Section 2 if we simply negate all the distances. Unfortunately, the resulting distance function is no longer natural and the algorithms of Section 3 cannot be applied. The same can be said for variations of the enumeration/selection problem in the bichromatic case (where we take the distance of two points of the same color to be infinite), and in the weighted case (where the actual distance is multiplied by or added to the weight of a point). Therefore, these problems have to be solved by other means.

Take for an example the k farthest pairs problem in the plane under the Euclidean metric. A point (ξ, η) has distance r away from (a, b) if and only if the halfspace $\{(x, y, z) : \xi^2 + \eta^2 - 2\xi x - 2\eta y + z \geq r^2\}$ contains the lifted point $(a, b, a^2 + b^2)$. Thus, the distance reporting problem for a given $r > 0$ can be solved using data structures for halfspace range reporting queries in $\mathbb{R}^3$. By known range-searching results [8], we can report the k distances $\geq r$ in $O(n \operatorname{polylog} n + k)$ time, and by Theorem 2.2, we can enumerate the k farthest pairs in $O(n \operatorname{polylog} n + k)$ expected time. (For this enumeration problem, Katoh and Iwano [20] described a generally less efficient algorithm, running in $O(n \log n + k^{4/3})$ time; for uniformly distributed points, Dickerson and Shugart [15] gave a faster method.)

The above approach can also enumerate the k bichromatic closest pairs in the plane. More generally, distance enumeration reduces to halfspace range reporting if it is possible to "linearize" the region $\{q \in U : d(p, q) \leq r\}$ using a fixed number of variables for a given $p \in U$ and $r \in \mathbb{R}$. Similarly, distance selection reduces to halfspace range counting under the same condition using Theorem 2.4.

We now point out an output-sensitive result for the halfspace range counting problem. Details are deferred to an appendix.

Notation. $f = O^*(g)$ if $f = O(g \cdot n^\varepsilon)$ for an arbitrarily small constant $\varepsilon > 0$.

Theorem 4.1 *Given a parameter $m \geq n$, we can preprocess an n-point set $P \subset \mathbb{R}^d$ in $O^*(m)$ time and space such that the number k of points inside a query halfspace can be determined in time*

$$O^*\left(1 + \frac{n}{m^{1/\lfloor d/2 \rfloor}} + \left(\frac{n^{\lfloor d/2 \rfloor} k^{\lceil d/2 \rceil}}{m}\right)^{1/d}\right).$$

The previous query bounds are $O^*(1 + n/m^{1/d})$ [9, 25] and $O^*(1 + n/m^{1/\lfloor d/2 \rfloor} + k)$ [26].

Choosing the tradeoff parameter m appropriately, guessing the output size k, and applying Hölder's inequality, we get:

Corollary 4.2 *An online sequence of n halfspace range counting queries on a given n-point set in $\mathbb{R}^d$ require time*

$$O^*(n^{2-2/(\lfloor d/2 \rfloor + 1)} + (n^{2\lfloor d/2 \rfloor} k^{\lceil d/2 \rceil})^{1/(d+1)}),$$

where k is the sum of the counts returned by the queries ($k \leq n^2$).

If we want to select the k-th largest distance among n points in the Euclidean plane, then the above result implies an $O^*(n + \sqrt{nk})$ expected time algorithm, since the lifted points have dimension $d = 3$. Improvements could be possible as the actual dimension of the problem is smaller than this "linearization dimension" [3]. We will not go further in this direction here: the main point is that k-sensitive results are obtainable if we just re-examine previous range-searching techniques more carefully.

References

[1] P. K. Agarwal, B. Aronov, M. Sharir, and S. Suri. Selecting distances in the plane. *Algorithmica*, 9:495–514, 1993.

[2] P. K. Agarwal and J. Erickson. Geometric range searching and its relatives. To appear in *Discrete and Computational Geometry: Ten Years Later* (B. Chazelle, J. E. Goodman, and R. Pollack, ed.), AMS Press.

[3] P. K. Agarwal and J. Matoušek. On range searching with semialgebraic sets. *Discrete Comput. Geom.*, 11:393–418, 1994.

[4] S. Arya and M. Smid. Efficient construction of a bounded-degree spanner with low weight. *Algorithmica*, 17:33–54, 1997.

[5] J. L. Bentley and M.I. Shamos. Divide-and-conquer in multidimensional space. In *Proc. 8th ACM Sympos. Theory of Comput.*, pages 220–230, 1976.

[6] M. de Berg and O. Schwarzkopf. Cuttings and applications. *Int. J. Comput. Geom. Appl.*, 5:343–355, 1995.

[7] B. Chazelle, L. Guibas, and D. T. Lee. The power of geometric duality. *BIT*, 25:76–90, 1985.

[8] B. Chazelle and F. P. Preparata. Halfspace range searching: An algorithmic application of k-sets. *Discrete Comput. Geom.*, 1:3–93, 1986.

[9] B. Chazelle, M. Sharir, and E. Welzl. Quasi-optimal upper bounds for simplex range searching and new zone theorems. *Algorithmica*, 8:407–429, 1992.

[10] L. P. Chew and S. Fortune. Sorting helps for Voronoi diagrams. *Algorithmica*, 18:217–228, 1997.

[11] R. Cole. Slowing down sorting networks to obtain faster sorting algorithms. *J. ACM*, 34:200–208, 1987.

[12] M. Dietzfelbinger, T. Hagerup, J. Katajainen, and M. Penttonen. A reliable randomized algorithm for the closest-pair problem. *J. Algorithms*, 25:19–51, 1997.

[13] M. T. Dickerson, R. L. S. Drysdale, and J.-R. Sack. Simple algorithms for enumerating interpoint distances and finding k nearest neighbors. *Int. J. Comput. Geom. Appl.*, 2:221–239, 1992.

[14] M. T. Dickerson and D. Eppstein. Algorithms for proximity problems in higher dimensions. *Comput. Geom. Theory Appl.*, 5:277–291, 1996.

[15] M. T. Dickerson and J. Shugart. A simple algorithm for enumerating longest distances in the plane. *Inform. Process. Lett.*, 45:269–274, 1993.

[16] M. Dillencourt, D. Mount, and N. Netanyahu. A randomized algorithm for slope selection. *Int. J. Comput. Geom. Appl.*, 2:1–27, 1992.

[17] M. Golin, R. Raman, C. Schwarz, and M. Smid. Simple randomized algorithms for closest pair problems. *Nordic J. Comput.*, 2:3–27, 1995.

[18] M. T. Goodrich. Geometric partitioning made easier, even in parallel. In *Proc. 9th ACM Sympos. Comput. Geom.*, pages 73–82, 1993.

[19] G. N. Frederickson and D. B. Johnson. The complexity of selection and ranking in $X + Y$ and matrices with sorted rows and columns. *J. Comput. Sys. Sci.*, 24:197–208, 1982.

[20] N. Katoh and K. Iwano. Finding k farthest pairs and k closest/farthest bichromatic pairs for points in the plane. *Int. J. Comput. Geom. Appl.*, 5:37–51, 1995.

[21] M. Katz and M. Sharir. An expander-based approach to geometric optimization. In *Proc. 9th ACM Sympos. Comput. Geom.*, pages 198–207, 1993.

[22] S. Khuller and Y. Matias. A simple randomized sieve algorithm for the closest-pair problem. *Inform. Comput.*, 118:34–37, 1995.

[23] H.-P. Lenhof and M. Smid. Sequential and parallel algorithms for the k closest pairs problem. *Int. J. Comput. Geom. Appl.*, 5:273–288, 1995.

[24] J. Matoušek. Randomized optimal algorithm for slope selection. *Inform. Process. Lett.*, 39:183–187, 1991.

[25] J. Matoušek. Efficient partition trees. *Discrete Comput. Geom.*, 8:315–334, 1992.

[26] J. Matoušek. Reporting points in halfspaces. *Comput. Geom. Theory Appl.*, 2:169–186, 1992.

[27] J. Matoušek. Geometric range searching. *ACM Comput. Surveys*, 26:421–461, 1994.

[28] N. Megiddo. Applying parallel computation algorithms in the design of serial algorithms. *J. ACM*, 30:852–865, 1983.

[29] R. Motwani and P. Raghavan. *Randomized Algorithms.* Cambridge University Press, New York, 1995.

[30] M. Pellegrini. On counting pairs of intersecting segments and off-line triangle range searching. *Algorithmica*, 17:380–398, 1997.

[31] M. O. Rabin. Probabilistic algorithms. In *Algorithms and Complexity* (J. F. Traub, ed.), pages 21–30, Academic Press, New York, 1976.

[32] J. S. Salowe. Enumerating interdistances in space. *Int. J. Comput. Geom. Appl.*, 2:49–59, 1992.

[33] P. M. Vaidya. An $O(n \log n)$ algorithm for the all-nearest-neighbors problem. *Discrete Comput. Geom.*, 4:101–115, 1989.

A Appendix: Proof of Theorem 4.1

We describe how halfspace range counting can be solved within the bounds stated in Theorem 4.1. The details are technical and are only sketched here. Basically, the underlying data structures are the same as those of Matoušek [26] for halfspace range reporting; to make query bounds output-sensitive, these are combined with known simplex range counting results. We first examine the near-linear space case and then consider tradeoffs between space and query time.

Near-Linear Space. Given an n-point set $P \subset \mathbb{R}^d$, a *simplicial partition of size* s is a collection $\{(P_1, \Delta_1), \ldots, (P_s, \Delta_s)\}$ such that (i) the P_i's are disjoint subsets whose union is P, (ii) the Δ_i's are relatively open simplices, and (iii) $P_i \subset \Delta_i$ for each i. The *class size* refers to the maximum of the sizes of $\{P_i\}$. The *crossing number* of a halfspace γ refers to the number of simplices from $\{\Delta_i\}$ that the bounding hyperplane $\partial\gamma$ crosses. (A hyperplane h *crosses* Δ if $h \cap \Delta \neq \emptyset$ and $\Delta \not\subseteq h$.) We say that γ is ℓ-*shallow* if $|P \cap \gamma| \leq \ell$.

Lemma A.1 (Matoušek's Partition Theorem for Shallow Hyperplanes [26]) *Given any* $1 \leq r \leq n$, *one can find a simplicial partition of size* $O(r)$ *such that the class size is* $O(n/r)$ *and the crossing number of any* (n/r)-*shallow halfspace is* $O(r^{1-1/\lfloor d/2 \rfloor} + \log r)$. *The running time is* $O(n)$ *if* r *is a constant.*

A near-linear-space data structure for P can be obtained as follows. Store the cardinality of P and build a linear-space data structure for simplex range counting [9, 25]. Then construct a simplicial partition $\{(P_i, \Delta_i)\}$ using an arbitrarily large constant for r and recursively build data structures for the subsets $\{P_i\}$. The preprocessing time and space of the data structure satisfy the recurrence

$$T(n) \;=\; O(r)\,T(n/r) \,+\, O^*(n),$$

which solves to $T(n) = O^*(n)$.

To count $k = |P \cap \gamma|$ for a query halfspace γ, we determine the simplices $\{\Delta_i\}$ that are crossed by $\partial\gamma$. If the number of such simplices exceeds $C(r^{1-1/\lfloor d/2 \rfloor} + \log r)$ for an appropriate constant C (independent of r), then γ is not (n/r)-shallow, and therefore, $k > n/r$. In this case, we will answer the query directly by simplex range counting [9, 25] in time $O^*(n^{1-1/d}) = O^*((rk)^{1-1/d})$. Otherwise, we recursively count $|P_i \cap \gamma|$ for each subset P_i corresponding to a crossing simplex Δ_i. We can trivially count $|P_i \cap \gamma|$ for each subset P_i with $\Delta_i \subseteq \gamma$. The total count is k.

To analyze the query time, we examine its recursion tree. Let k_ν be the count returned at leaf ν of the tree. We can bound the sum $\sum_\nu k_\nu^{1-1/\lfloor d/2 \rfloor}$ by $O^*(n^{1-1/\lfloor d/2 \rfloor})$, since the recurrence

$$t(n) \;=\; O(r^{1-1/\lfloor d/2 \rfloor} + \log r)\,t(n/r) \,+\, n^{1-1/\lfloor d/2 \rfloor}$$

solves to $t(n) = O^*(n^{1-1/\lfloor d/2 \rfloor})$. Now, the actual cost of a query is asymptotically bounded by the number of nodes in the recursion tree plus the sum $\sum_\nu k_\nu^{1-1/d}$. The first term is $O^*(n^{1-1/\lfloor d/2 \rfloor})$, whereas the second term is by Hölder's inequality:

$$\sum_\nu k_\nu^{1-1/d} \;\leq\; \left(\sum_\nu k_\nu^{1-1/\lfloor d/2 \rfloor}\right)^{\lfloor d/2 \rfloor / d} \left(\sum_\nu k_\nu\right)^{\lceil d/2 \rceil / d}$$
$$=\; O^*((n^{1-1/\lfloor d/2 \rfloor})^{\lfloor d/2 \rfloor / d}\, k^{\lceil d/2 \rceil / d}).$$

We conclude that the halfspace range counting problem can be solved using $O^*(n)$ preprocessing time and space and $O^*(n^{1-1/\lfloor d/2 \rfloor} + (n^{\lfloor d/2 \rfloor - 1} k^{\lceil d/2 \rceil})^{1/d})$ query time.

Tradeoffs. Let H be a set of n hyperplanes in $\mathbb{R}^d$. The *level* of a point refers to the number of hyperplanes below it. The *conflict list* $H(\Delta)$ of a simplex Δ consists of all hyperplanes that intersect the interior of Δ. We say that a collection of simplices $\{\Delta_1, \ldots, \Delta_s\}$ is a $(1/r)$-*cutting for the* $(\leq \ell)$-*level of size* s if (i) the Δ_i's cover all points of level $\leq \ell$, and (ii) $|H(\Delta_i)| \leq n/r$ for each Δ_i.

Lemma A.2 (Matoušek's Shallow Cutting Lemma [26]) *Given any* $1 \leq r \leq n$, *one can find a* $(1/r)$-*cutting for the* $(\leq \ell)$-*level of size* $O(r^{\lfloor d/2 \rfloor} q^{\lceil d/2 \rceil})$, *where* $q = \ell(r/n) + 1$. *The running time is* $O(n)$ *if* r *is a constant.*

Dualization reduces the halfspace range counting problem to the following: preprocess a set H of n hyperplanes so that given a query point, one can quickly compute its level k. A data structure for processing such queries on H can be obtained as follows. First fix a parameter $p > 0$. If $n \leq p$, build the above primal near-linear-space structure. Otherwise, construct a $(1/r)$-cutting $\{\Delta_i\}$ for the $(\leq n/r)$-level of size $O(r^{\lfloor d/2 \rfloor})$ for an arbitrarily large constant r. Store the number of hyperplanes strictly below each Δ_i, and build a data structure for simplex range counting [9, 25] with $m(n) = (n/p)^{\lfloor d/2 \rfloor} p$ space. Then recursively build data structures for each $H(\Delta_i)$. The preprocessing time and space satisfy the recurrence

$$T(n) = \begin{cases} O^*(n) & \text{if } n \leq p \\ O(r^{\lfloor d/2 \rfloor})T(n/r) + O^*((n/p)^{\lfloor d/2 \rfloor} p) & \text{if } n > p \end{cases}$$

which solves to $T(n) = O^*(n + (n/p)^{\lfloor d/2 \rfloor} p)$.

We determine the level k of a query point q as follows. If $n \leq p$, then use our primal data structure to answer the query in time $O^*(1 + p^{1-1/\lfloor d/2 \rfloor} + (p^{\lfloor d/2 \rfloor - 1} k^{\lceil d/2 \rceil})^{1/d})$. Otherwise, find a simplex Δ_i that contains q. If none exists, then $k > n/r$; in this case, we answer the query directly by simplex range counting [9, 25] in time $O^*(1 + n/m(n)^{1/d}) = O^*(1 + (n^{\lceil d/2 \rceil} p^{\lfloor d/2 \rfloor - 1})^{1/d}) = O^*(1 + ((rk)^{\lceil d/2 \rceil} p^{\lfloor d/2 \rfloor - 1})^{1/d})$. If we succeed in finding Δ_i, then we recursively compute the level of q with respect to $H(\Delta_i)$. Adding to this the number of hyperplanes strictly below Δ_i, we get k. Since the depth of the recursion is logarithmic, we conclude that the halfspace range counting problem can be solved using $O^*(n + (n/p)^{\lfloor d/2 \rfloor} p)$ preprocessing time and space and

$$O^*(1 + p^{1-1/\lfloor d/2 \rfloor} + (p^{\lfloor d/2 \rfloor - 1} k^{\lceil d/2 \rceil})^{1/d})$$

query time. Finally setting $p^{\lfloor d/2 \rfloor - 1} = n^{\lfloor d/2 \rfloor}/m$ yields the bounds stated in Theorem 4.1.

Drawing Planar Partitions I: LL-Drawings and LH-Drawings

Therese C. Biedl *

Abstract

In this paper, we study how to draw a planar partition, i.e., a planar graph with a given partition of the vertices. The goal is to obtain a drawing without crossings such that the partition is clearly visible. Previously, only the special case of bipartite graphs has been studied.

For two models of displaying the partition, we show necessary and sufficient conditions for the existence of a drawing, and how to test them in linear time. We also present linear-time algorithms to create such drawings, if possible.

1 Introduction

The problem of *graph drawing* involves finding a geometric representation of a graph (typically using points and segments) that facilitates understanding the graph. It has been studied intensely in the last few years, see for example the annotated bibliography by Di Battista et al. [7], and the proceedings of the annual conference on Graph Drawing, e.g. [18]. Graph Drawing has a wide range of applications, in particular in the area of visualization of information. Specific examples include CASE tools, Data Base Design, telecommunications and network management.

In most applications, the original input is more than a graph, it is a structure from which a graph can be extracted. Thus, it may be that a graph drawing produced without consideration of the underlying structure may not yield the desired results, since the user subconsciously expected the algorithm to understand the structure, and certain constraints that are placed on the graph drawing due to this structure.

A very interesting problem therefore is *graph drawing under constraints*. Systems have been designed that deal with geometric constraints, such as the minimum distance between vertices. Unfortunately, to solve such problems one either has to apply brute-force methods, which may be costly in time, or one has to restrict the attention to special graph classes, such as trees and series-parallel graphs. We refer to [6] and the references therein for further information.

In this paper, we deal with a very different type of constraint: We assume that the graph has a partition structure, i.e., the vertices V are partitioned into two sets A and B. We search for drawings of the graph that display this partition clearly. The precise definition of "clearly" has to be specified through different models.

Our study of this question was motivated by Graph B of the Graph Drawing Competition 1996 (see [18]), which is a graph of telephone calls and had a natural partition structure. Other applications can be found in transportation problems, which are represented by bipartite graphs, or in cuts, as they arise in many algorithms for combinatorial optimization, for example maximum flows and Dijkstra's algorithm.

Two typical drawings of partitions are shown in Fig. 1 and 2. The subject of this paper is to study whether we can create such without crossing, which is possible only if the graph is planar.

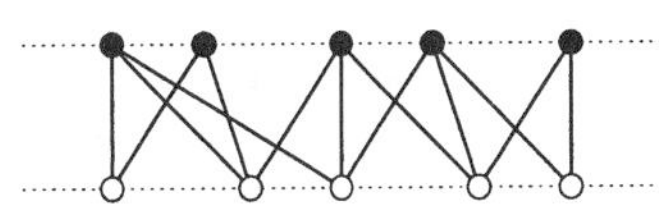

Figure 1: Drawing a partition using two lines.

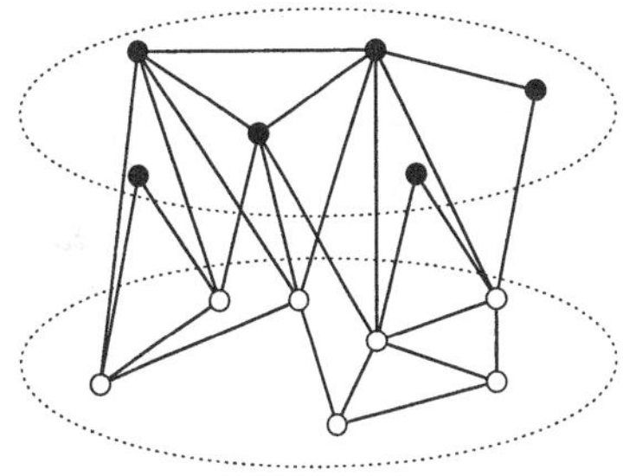

Figure 2: Drawing a partition using two disjoint convex sets.

*School of Computer Science, McGill University, 3480 University St., Montreal, Quebec H3A2A7, Canada. The author would like to thank Endre Boros, Michael Kaufmann, and Petra Mutzel for inspiring discussions.

SCG 98 Minneapolis Minnesota USA
Copyright ACM 1998 0-89791-973-4/98/ 6...$5.00

1.1 Existing results

Our topic combines two different objectives: drawing a planar graph without a crossing, and displaying a partition. We consider *straight-line* and *poly-line drawings*, i.e., vertices are drawn as points, and edges are drawn either as straight lines or as sequences of straight-line segments.[1]

There are many algorithms to create planar straight-line drawings of planar graphs. The first one to achieve area $\mathcal{O}(n^2)$ was presented by de Fraysseix, Pach and Pollack [11]. Subsequent improvements can for example be found in [4, 15]. However, these algorithms permit little control on the vertex placement, and a given partition may not be easy to see.

Drawings that clearly display the partitions are easy to create: place the vertices on two parallel lines (Fig. 1), or in two disjoint sets (Fig. 2). However, these drawings may have unnecessarily many crossings. For example, the graphs shown in Fig. 1 and 2 are planar, but are drawn with crossings in order to display the partition.

Drawings that combine both aspects have been studied only for bipartite graphs. A bipartite graph has a drawing as in Fig. 1 if and only if it is a caterpillar [12, 8]. Algorithms to minimize the number of crossings in such a drawing are known as well [14].

Two other models of displaying the partition, and some surprising applications in boolean functions, were studied by Knuth and by Kratochvil and Křivanek [16, 17]. Independently, Fößmeier and Kaufmann introduced these two and some more models, and derived tests and drawing algorithms [10]. However, for one of these models, the testing algorithms takes $\mathcal{O}(n^2)$ time, and the authors did not provide any area-bounds for the resulting drawings.

1.2 Our work

To our knowledge, all previous papers studied only bipartite graphs, and not arbitrary partitions, as they may arise for example in a flow network. This topic is the object of study in this paper. Our contribution is three-fold:

First, we review some models, list the exact extend of existing results, and introduce the model of Fig. 2. We assign new names to the various models, as to achieve a unifying naming scheme.

Secondly, we study two of the models in detail. We provide algorithms that determine in linear time whether a given planar partition has a drawing in this model. The algorithms work by identifying necessary and sufficient conditions for the existence of such drawings. The sufficiency proof yields an algorithm to construct a drawing in linear time in a grid of at most $\mathcal{O}(n^2)$ area. Thus, we improve the running time of the existing algorithm, provide good area-bounds on the resulting drawings, and extend the results to arbitrary partitions.

[1] Throughout this paper, vertices must be placed on points with integer coordinates, i.e., on a grid.

Thirdly, testing the necessary and sufficient conditions reduces to the so-called *face-boundary constrained embedding problem*, which we think interesting in its own right: Given a planar graph and a set of cycles $\mathcal{C}$, does there exist a planar embedding such that the cycles are boundaries of faces? We show how to answer this question in linear time.

2 Definitions

We assume familiarity with graphs and algorithms, and refer for example to [9] for further reference. So let $G = (V, E)$ be a graph, which we assume to be finite and simple (no loops and multiple edges). Denote $n = |V|$ and $m = |E|$.

The graph G is called *planar* if it can be drawn in the plane without edge crossings. A planar drawing of G splits the plane into connected pieces called *faces*, the unbounded face is called the *outer-face*. A specific planar drawing of a planar graph determines the *planar embedding*, that is, the cyclic order of edges around each vertex in the incidence list.

A planar graph has at most $3n - 6$ edges, a planar bipartite graph has at most $2n - 4$ edges. In a planar graph with $3n - 6$ edges, every face is a triangle, and the graph is called *triangulated*. Such a graph is triconnected and has a unique planar embedding. Given a planar embedding, a triangle that is not the boundary of a face is called a *separating triangle*.

A *planar partition* is a planar graph $G = (V, E)$ together with a partition of the vertices $V = A \cup B$, and will be denoted $G = (A \cup B, E)$. We assume that both A and B are not empty. A planar partition is called *bipartite* if the input graph is bipartite, and the vertex classes of the partition correspond to the vertex classes of the bipartite graph. In other words, in a bipartite partition all edges have the form (a, b) with $a \in A$ and $b \in B$. Partitions that may or may not be bipartite will be called *non-bipartite*.[2]

3 Models for drawing partitions

In this section, we review the existing models of drawing a planar partition $G = (A \cup B, E)$, and introduce one new model. We name the models according to whether a line (L) or a half-space (H) is reserved for each vertex class.

3.1 LL-drawings

The classical style of drawing a bipartite graph will be called *LL-drawing*: place the vertices of A on a line, and the vertices of B on a parallel line. After rotation and translation, we may assume that the vertices in A have y-coordinate 0, while the vertices in B have y-coordinate 1. See Fig. 1.

Harary and Schwenk [12], and independently Eades, McKay, and Wormald [8] showed that a connected bipartite planar partition has a planar LL-drawing if and only if it is a *caterpillar*, i.e., a graph which is a path after removing all vertices of degree 1.

[2] The term *non-bipartite* is something of a misnomer, since a bipartite partition is also a non-bipartite partition. The reader is invited to use the mnemonic *not necessarily bipartite*.

In Section 4, we study planar LL-drawings of non-bipartite planar partitions, and provide necessary and sufficient conditions for the existence of a planar LL-drawing which can be tested in $\mathcal{O}(n)$ time. We also give a linear-time algorithm to create, if possible, a planar LL-drawing of area $\mathcal{O}(n)$.

3.2 LLL-drawings

In [10], Fößmeier and Kaufmann introduced *LLL-drawings*, which they called the *two-sides-constrained-model*: place the vertices on three horizontal lines, i.e., the vertices of A have y-coordinate 0, while the vertices of B have y-coordinate 1 or -1.

The authors studied only bipartite planar partitions, and provided necessary and sufficient conditions for the existence of planar LLL-drawings, which can be tested in linear time [10]. It remains an open problem to generalize this result to non-bipartite planar partitions.

3.3 LH-drawings

An *LH-drawing* is a drawing where the vertices of A have y-coordinate 0, while the vertices of B have y-coordinate > 0. This model was introduced, under a different name, by Knuth [16] and Kratochvil and Křivanek [17], who studied it in the context of boolean functions as follows:

If $f(x_1,\ldots,x_n) = \bigwedge_{j=1}^{m} C_j(x_1,\ldots,x_n)$ is a boolean function, where each clause C_j has the form $C_j(x_1,\ldots,x_n) = \bigvee_{x\in P_j} x_j \vee \bigvee_{x\in N_j}(1-x_j)$, then a variable x_i and a clause C_j are called *incident* if $i \in P_j \cup N_j$. These incidences define a bipartite partition G_f with vertex classes $X = \{x_1,\ldots,x_n\}$ and $C = \{C_1,\ldots,C_m\}$. A boolean function f is called *nested* if G_f has a planar LH-drawing with respect to (X, C), and *co-nested* if it has a planar LH-drawing with respect to (C, X). For both nested and co-nested boolean functions, the satisfiability problem can be solved in linear time [16, 17].

LH-drawings were studied independently by Fößmeier and Kaufmann [10], who called it the *one-side-free model*.

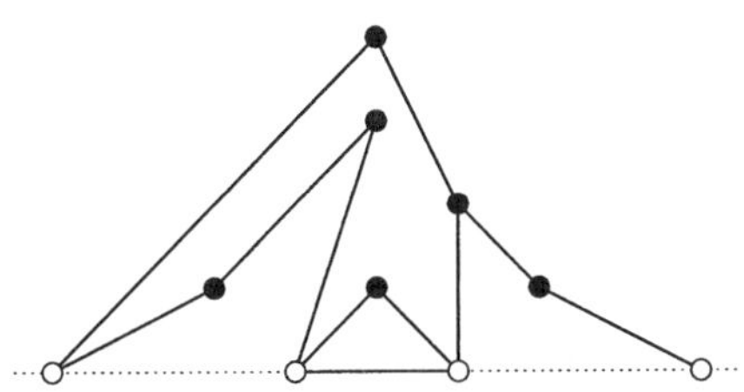

Figure 3: An LH-drawing.

All previous papers considered only bipartite partitions. Kratochvil and Křivanek gave necessary conditions for the existence of a planar LH-drawing which can be tested in $\mathcal{O}(n)$ time. They claimed that these conditions are sufficient as well, but provide no details [17]. Fößmeier and Kaufmann presented necessary and sufficient conditions for the existence of a planar LH-drawing which can be tested in $\mathcal{O}(n^2)$ time. They provide a quadratic-time algorithm to construct such drawings, if possible, but give no bounds on the area of the resulting drawing [10].

In Section 5, we study planar LH-drawings of non-bipartite planar partitions. We provide necessary and sufficient conditions for the existence of a planar LH-drawing which can be tested in $\mathcal{O}(n)$ time; for bipartite partitions they are the same as the ones by Kratochvil and Křivanek. We also give a linear-time algorithm to create, if possible, a planar LH-drawing of area $\mathcal{O}(n^2)$. In comparison to the results by Fößmeier and Kaufmann therefore, we improve the time complexity, establish area bounds, and generalize the results to non-bipartite partitions.

3.4 LHH-drawings

An *LHH-drawing* is a drawing where the vertices of A have y-coordinate 0, while the vertices of B have y-coordinate $\neq 0$. Kratochvil and Křivanek [17], and independently Fößmeier and Kaufmann [10] studied LHH-drawings, terming them *doubly-nested boolean formulas* and *two-sides-free model*, respectively. Both sets of authors proved that testing the existence of a planar LHH-drawing is $\mathcal{NP}$-complete, by reducing it to the Hamiltonian cycle problem in the dual graph. See [17] and [10] for details.

3.5 HH-drawings

An *HH-drawing* is a drawing where the vertices of A have y-coordinate < 0, while the vertices of B have y-coordinate > 0. See Fig. 2. This seems to us a very intuitive way of drawing a partition, but to our knowledge it has not been considered yet.

We study planar HH-drawings in the second paper of this series [3], and provide necessary and sufficient conditions for the existence of a planar HH-drawing, which can be tested in $\mathcal{O}(n)$ time. We also give a linear-time algorithm to create, if possible, a planar HH-drawing of area $\mathcal{O}(n^2)$. As a special case, we obtain the surprising result that any bipartite planar partition has a planar HH-drawing.

3.6 Monotone vs. non-monotone drawings

All previous papers on drawing partitions considered only straight-line drawings. However, we did not succeed for HH-drawings to find an algorithm for straight-line drawings, and in fact, proved that this is impossible unless we allow an exponential area [3]. Hence we also permit poly-line drawings. As it turns out, it then makes a considerable difference whether we require edges to be drawn monotone or not. The precise definitions are as follows: An edge is drawn *y-monotone* if, with a suitable choice of one endpoint, the y-coordinate does not decrease when walking from this endpoint to the other endpoint. A drawing is *y-monotone* if all edges are drawn y-monotone. A drawing that is not necessarily y-monotone will be called *non-monotone*.[3]

[3] A similar remark as in footnote 2 applies here.

This concept is closely related to upward drawings for directed graphs: A drawing of a directed graph is called *upward* if for each edge the y-coordinate increases as we walk from the tail to the head. However, as opposed to upward drawings, horizontal lines are permitted in y-monotone drawings, hence any straight-line drawing is y-monotone.

The existing results on straight-line drawings transfer to monotone, but not to non-monotone poly-line drawings. In particular, for each model of drawing a planar partition, there exist graphs that do not have a planar y-monotone drawing in this model. Opposed to that, any planar partition has a non-monotone planar LL-drawing with only a constant number of bends per edge (see [3]). Since LL-drawings are the strongest of all models, any planar partition has a non-monotone planar drawing in any of the presented models.

We believe monotonicity facilitates following the drawing of an edge, hence y-monotone drawings are preferable, From now on, unless specifically mentioned otherwise, all drawings are required to be y-monotone.

4 LL-drawings

In this section, we study planar LL-drawings. As mentioned, not every planar partition has a planar LL-drawing. One counter-example is $K_{2,2}$, since a planar LL-drawing of $K_{2,2}$ could be completed to a planar drawing of $K_{3,3}$. So we are interested in algorithms to test whether a given planar partition has a planar LL-drawing. To that end, we develop necessary and sufficient conditions for the existence of a planar LL-drawing.

4.1 Necessary condition

Assume that we have a planar LL-drawing of a planar partition $G = (A \cup B, E)$, thus the vertices of A have y-coordinate 0, while the vertices in B have y-coordinate 1. After possible translation all x-coordinates are bigger than 0. Add a vertex v_A at $(0, -1)$ and a vertex v_B at $(0, 2)$. Connect v_A with the vertices in A, v_B with the vertices in B, and add the edge (v_A, v_B). The resulting graph G^+_{LL} then is planar, since all edges are drawn y-monotone. See also Fig. 4.

Number the vertices of A as $a_1, \ldots, a_k$, sorted by increasing x-coordinate. Consider an arbitrary triangle containing v_A, by construction it has the form $\{v_A, a_i, a_j\}$, for some $1 \leq i < j \leq k$. But edge (a_i, a_j) can exist only if $j = i + 1$, for otherwise it either overlaps vertex a_{i+1}, or it is drawn non-monotone. Furthermore, the triangle $\{a_i, a_{i+1}, v_A\}$ is a face. We conclude that in this planar embedding, any triangle containing v_A is a face, and similarly any triangle containing v_B is a face.

Lemma 4.1 *A planar y-monotone LL-drawing of a planar partition $G = (A \cup B, E)$ can exist only if*

- *the graph G^+_{LL} defined above is planar, and*
- *there exists a planar embedding of G^+_{LL} such that any triangle containing v_A or v_B is a face.*

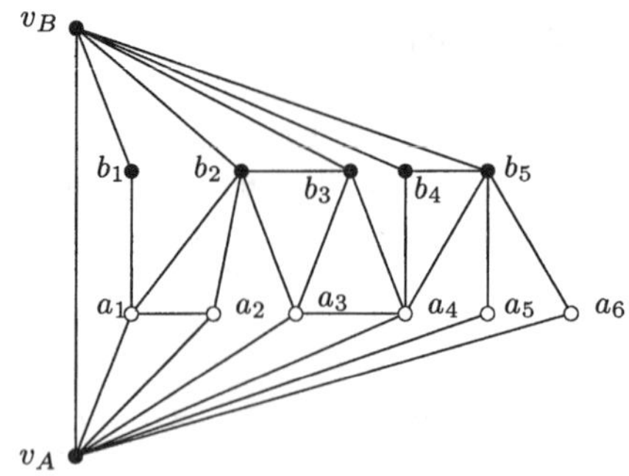

Figure 4: G^+_{LL} is planar and each triangle containing v_A or v_B is a face.

Planarity testing can be done in linear time [13]. Finding all triangles containing v_A or v_B can be done by marking the neighbors of v_A and v_B and then scanning all edges, this takes $\mathcal{O}(m)$ time. Testing whether a planar graph can be embedded without a separating triangle at all can be tested in linear time [2]. However, in this context we need to test whether there exists an embedding such that a specific subset of triangles are faces. This is a new problem, which we call the *face-boundary constrained embedding problem.* We show how to test the existence of a face-boundary constrained embedding in linear time in Section 6.[4]

4.2 Sufficiency of the necessary conditions

We show now that the conditions of Lemma 4.1 are sufficient. Assume that we have a planar embedding of G^+_{LL} such that any triangle containing v_A or v_B is a face. Enumerate the vertices of A as $\{a_1, \ldots, a_k\}$ in counter-clockwise order around v_A, according to this embedding of G^+_{LL}, starting after the edge (v_A, v_B). Enumerate the vertices of B as $\{b_1, \ldots, b_l\}$ in clockwise order around v_B, according to this embedding of G^+_{LL}, starting after the edge (v_A, v_B).

Place a_i at $(0, i)$, $1 \leq i \leq k$, and b_j at $(1, j)$, $1 \leq j \leq l$. We claim that if we draw all edges as straight lines, we obtain a planar LL-drawing.

To prove this claim, we add edges to the graph G^+_{LL} while maintaining planarity (note that this is necessary for the proof of correctness only, not for the actual construction of the drawing). If our drawing had either a crossing or overlap between an edge and a vertex, this contradicts the planarity of the obtained graph.

Add first the edges $(a_1, a_2), \ldots, (a_{k-1}, a_k)$. These can be routed very close to the path $\{a_i, v_A, a_{i+1}\}$, $1 \leq i \leq k$, therefore the graph stays planar. Similarly, the graph stays planar after adding the edges $(b_1, b_2), \ldots, (b_{l-1}, b_l)$. Finally, add the edges (a_1, b_1) and (a_k, b_l). These can be routed very close to the paths $\{a_1, v_A, v_B, b_1\}$ and $\{a_k, v_A, v_B, b_l\}$, respectively, so the graph stays planar. See also Fig. 5.

[4] One can develop a linear-time testing algorithm for the existence of LL-drawings without using the face-boundary constrained embedding problem (Mutzel, private communication). However, since the latter is needed for LH-drawings (see Section 5.1), we prefer this shorter description.

We add these edges only if they did not exist already, so that the resulting graph G_{LL}^{++} is planar and simple. Observe that if we added triangles containing v_A or v_B, then these are faces by construction.

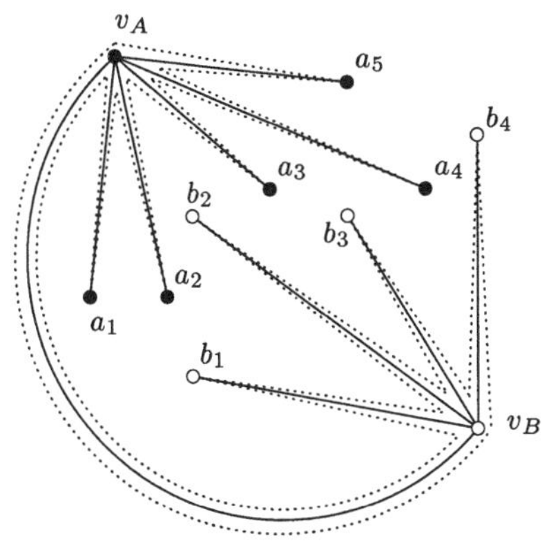

Figure 5: Adding edges to obtain G_{LL}^{++}.

Lemma 4.2 *G contains no edges (a_{i_1}, b_{j_1}) and (a_{i_2}, b_{j_2}) with $i_1 < i_2$ and $j_1 > j_2$.*

Proof: Assume that we had edges (a_{i_1}, b_{j_1}) and (a_{i_2}, b_{j_2}) with $i_1 < i_2$ and $j_1 > j_2$. Then in G_{LL}^{++}, a_{i_1} is connected to b_{j_2} along the path $a_{i_1}, a_{i_1-1}, \ldots, a_1, b_1, \ldots, b_{j_2-1}, b_{j_2}$; and a_{i_2} is connected to b_{j_1} along the path $a_{i_2}, a_{i_2+1}, \ldots, a_k, b_l, \ldots, b_{j_1+1}, b_{j_1}$. These two paths are disjoint by the conditions, so there is a subdivision of $K_{3,3}$ in the planar graph G_{LL}^{++} using the vertices $\{v_B, a_{i_1}, a_{i_2}\}$ and $\{v_A, b_{j_1}, b_{j_2}\}$, a contradiction. See also Fig. 6. □

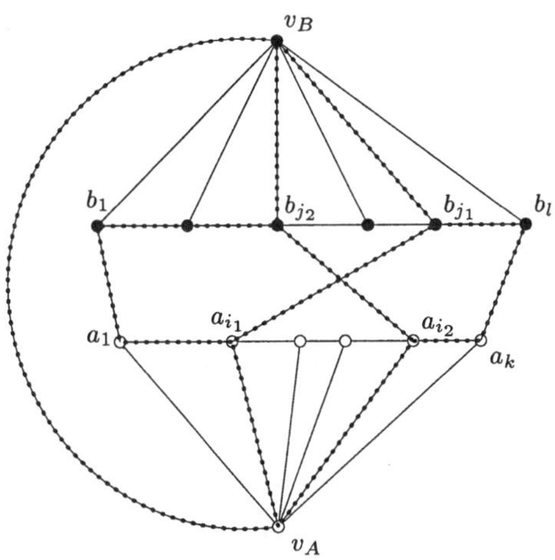

Figure 6: For Lemma 4.2.

Lemma 4.3 *G contains no edge (a_i, a_j) with $j > i+1$.*

Proof: Assume that we had such an edge (a_i, a_j), $j > i+1$. Then $\{v_A, a_i, a_j\}$ is a triangle in G_{LL}^{++}, and by the conditions on triangles containing v_A, this triangle is a face of G_{LL}^{++}. Without violating planarity, we can add a new vertex a^* inside this face and connect it to v_A, a_i and a_j. In G_{LL}^{++}, a_{i+1} is connected to a_i and a_j along the edge (a_{i+1}, a_i) and the path $a_{i+1}, a_{i+2}, \ldots, a_j$, respectively. Vertex b_1 is connected to a_i and a_j along the paths $b_1, a_1, a_2, \ldots, a_i$ and $b_1, b_2, \ldots, b_l, a_k, a_{k-1}, \ldots, a_j$, respectively. These paths are disjoint by $j > i$. Finally, vertex b_1 is connected to v_A along the path b_1, v_B, v_A. Thus, there is a subdivision of $K_{3,3}$ in the planar graph $G_{LL}^{++} \cup \{a^*\}$, using the vertices $\{v_A, a_i, a_j\}$ and $\{b_1, a_{i+1}, a^*\}$, a contradiction. See also Fig. 7. □

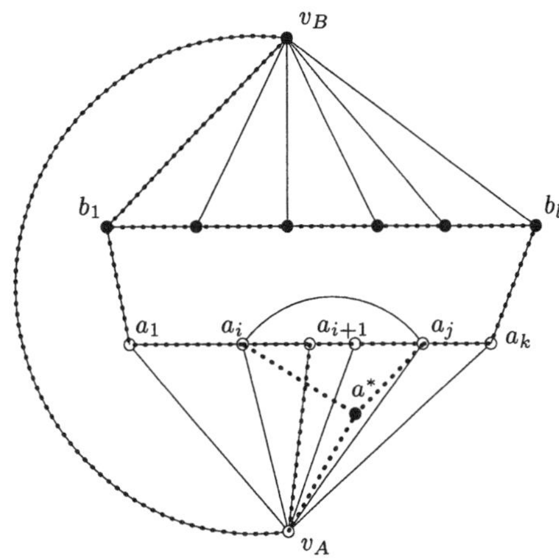

Figure 7: For Lemma 4.3.

Thus, if we place the vertices as described above, i.e., a_i at $(0, i)$ and b_j at $(1, j)$, and draw the edges as straight lines, then we have no crossing by Lemma 4.2 and no overlap between an edge and a vertex by Lemma 4.3, and get the desired planar LL-drawing. The algorithm to compute this drawing only enumerates vertices according to the planar embedding, so it clearly works in linear time.

Theorem 1 *Any planar partition $G = (A \cup B, E)$ satisfying the necessary condition of Lemma 4.1 has a planar straight-line LL-drawing in a $\max\{|A|, |B|\} \times 2$-grid. It can be computed in linear time.*

5 LH-drawings

In this section, we study planar LH-drawings. As mentioned, not every planar partition has a planar LH-drawing. One counter-example is $K_{3,2}$, for if it had a planar LH-drawing, we could complete it to a planar drawing of $K_{3,3}$. So we are interested in algorithms to test whether a given planar partition has a planar LH-drawing. To that end, we develop necessary and sufficient conditions for the existence of a planar LH-drawing.

5.1 Necessary condition

Assume that we have a planar LH-drawing of a planar partition $G = (A \cup B, E)$, thus the vertices of A have y-coordinate 0, while the vertices in B have a larger y-coordinate.

Add a vertex v_A at $(0, -1)$ and connect it with the vertices in A. The resulting graph G^+_{LH} then is planar, since all edges are drawn y-monotone. See also Fig. 8.

Number the vertices of A as $a_1, \ldots, a_k$, sorted by increasing x-coordinate. Consider an arbitrary triangle containing v_A, by construction it has the form $\{v_A, a_i, a_j\}$, for some $1 \leq i < j \leq k$. But edge (a_i, a_j) can exist only if $j = i + 1$, for otherwise it either overlaps vertex a_{i+1}, or it is drawn non-monotone. Furthermore, the triangle $\{a_i, a_{i+1}, v_A\}$ is a face. We conclude that in this planar embedding, any triangle containing v_A is a face.

Finally, since any edge (a_i, a_j) has the form $j = i + 1$, the vertices $\{a_1, \ldots, a_k\}$ induce a collection of paths.

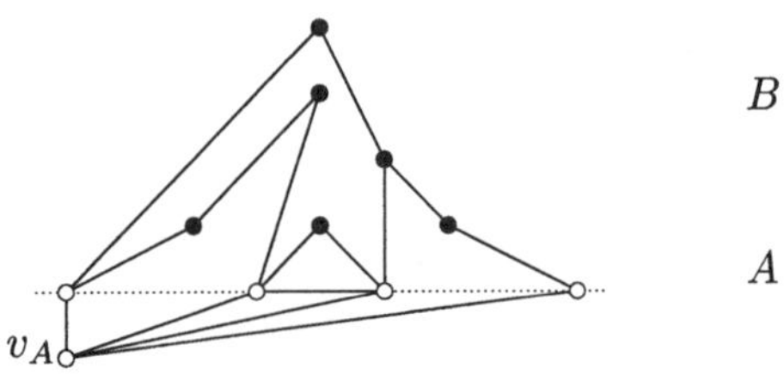

Figure 8: The graph G^+_{LH} is planar and each triangle containing v_A is a face.

Lemma 5.1 *A planar y-monotone LH-drawing of a planar partition $G = (A \cup B, E)$ can exist only if*

- *the vertices of A induce a collection of paths,*
- *the graph G^+_{LH} defined above is planar, and*
- *there exists a planar embedding of G^+_{LH} such that any triangle containing v_A is a face.*

Testing whether A induces a collection of paths and whether G^+_{LH} is planar can be done in linear time. Finding all triangles containing v_A can be done in $\mathcal{O}(m)$ time. The third condition amounts to the face-boundary constrained embedding problem, which can be tested in linear time (see Section 6). For a bipartite partition, the conditions of Lemma 5.1 reduce to the planarity of G^+_{LH}, which is the necessary condition given by Kratochvil and Křivanek [17].

5.2 Sufficiency of the necessary conditions

We show now that the conditions of Lemma 5.1 are sufficient, by presenting an algorithm in three steps. First, we add edges to the input graph while maintaining planarity (these edges are necessary for the algorithm). Then we compute a special vertex order of the resulting graph G_{LH}. Using it, we produce an LH-drawing of $G_{LH} - e$, where e is one of the added edges. This gives a planar LH-drawing of G after deleting the other added edges.

Assume that the conditions of Lemma 5.1 are satisfied with some specific planar embedding of the graph G^+_{LH}. If $|A| \leq 2$, then we add $3 - |A|$ singleton vertices to the graph. It is easily verified that the resulting graph also satisfies the conditions of Lemma 5.1, hence we can create an LH-drawing as shown below, and delete the added vertices. So we may assume in the following that $|A| \geq 3$. Note that we also assumed $|B| \geq 1$, therefore $|A| < n$.

Enumerate $A = \{a_1, \ldots, a_k\}$, $3 \leq k < n$, in clockwise order around v_A with respect to the planar embedding of G^+_{LH}. Here, choose a_1 such that there is no edge (a_1, a_k); this is possible since A induces a collection of paths.

5.2.1 Adding edges

To extend the input graph G to the graph G_{LH} which will actually be drawn, start with G^+_{LH}, and add the edges (a_1, a_2), (a_2, a_3), $\ldots$, (a_{k-1}, a_k) and (a_k, a_1) if they did not exist before. The resulting graph G^{++}_{LH} is planar, since for $1 \leq i < k$ edge (a_i, a_{i+1}) can be routed very close to the path $\{a_i, v_A, a_{i+1}\}$, and edge (a_k, a_1) can be routed very close to the path $\{a_k, v_A, a_1\}$. Every newly created triangle containing v_A is a face.

We triangulate G^{++}_{LH} to get G^{+++}_{LH} such that afterwards the triangles containing v_A are faces, or in other words, no separating triangle contains v_A. To do so, we use the algorithm by Biedl, Kant and Kaufmann [2]. They showed that it is almost always possible to triangulate a graph without adding a separating triangle. The one exception adds a separating triangle containing v_A only if v_A is connected to all other vertices. This implies $A = V$, which we excluded, so we can indeed triangulate G^{++}_{LH} without adding a separating triangle containing v_A, and obtain G^{+++}_{LH}.

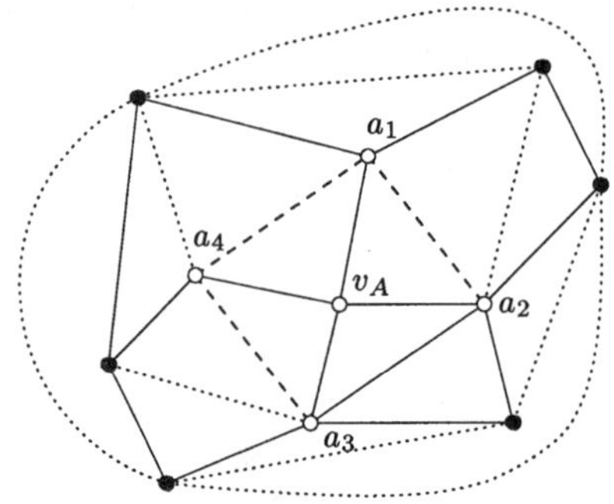

Figure 9: The transformation from G^+_{LH} to G^{++}_{LH} (add dashed edges) and G^{+++}_{LH} (add dotted edges).

Let G_{LH} be $G^{+++}_{LH} - \{v_A\}$. In G_{LH}, the edges (a_1, a_2), $\ldots$, (a_{k-1}, a_k), (a_k, a_1) form a cycle, since $|A| \geq 3$, and this cycle is a face since otherwise there would have been a separating triangle containing v_A in G^{+++}_{LH}. Call this face F_A. The vertices of F_A are $a_1, \ldots, a_k$, in this order. Face F_A does not have a chord, i.e., an edge between two non-consecutive vertices on F_A, since a chord corresponds to a separating triangle containing v_A in G^{+++}_{LH}. All faces $\neq F_A$ are triangles. So G_{LH} is a simple graph, which is triangulated except for one face, and this one non-triangle face has no chord. Such a graph is triconnected.

5.2.2 An ordering of the vertices

To create a planar LH-drawing of $G_{LH}^{+++} - (a_1, a_k)$, we modify the algorithm by Chrobak and Kant [4], which is based on the canonical ordering by Kant [15].

Definition 1 *[15] Let G be a planar graph with a fixed planar embedding. A vertex order $V = \{v_1, \dots, v_n\}$ is called a* canonical ordering *if the following holds:*

- *(v_1, v_2) is an edge and belongs to the outer-face.*
- *v_n is on the outer-face and has at least three neighbors.*
- *For $3 \le j \le n-1$, v_j is in the outer-face of $G(j-1)$, the graph induced by $v_1, \dots, v_{j-1}$, and one of the following holds:*
  - *Either, v_j has at least two neighbors in $G(j-1)$, at least one neighbor in $G - G(j)$, and $G(j)$ is biconnected. We call v_j a* single vertex.
  - *Or, there exists i, k, $i \le j \le k$, such that for $i < l < k$, v_l is adjacent to v_{l-1} and v_{l+1}, has no other neighbor in $G(k)$, and at least one neighbor in $G - G(k)$. Furthermore, v_i and v_k have each exactly one neighbor in $G(i-1)$ and at least one neighbor in $G - G(k)$, and $G(k)$ is biconnected. We call the set $\{v_i, \dots, v_k\}$ a* chain.

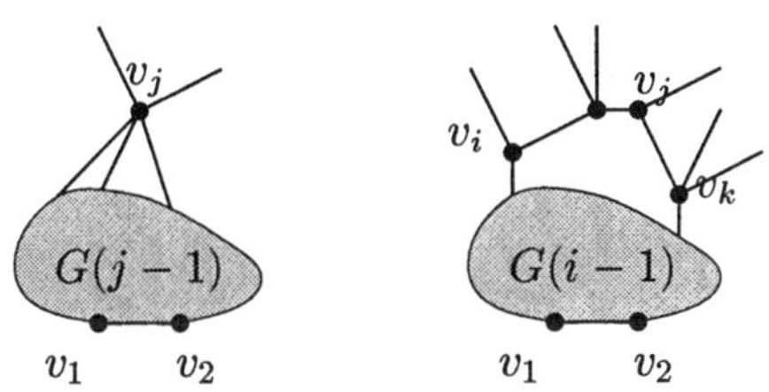

Figure 10: The two different possibilities for v_j.

Kant showed the existence of such an ordering for any triconnected planar graph, given any choice of the outer-face and any choice of an edge (v_1, v_2) and a vertex $v_n \neq v_1, v_2$ on the outer-face. We choose these three vertices in a particular fashion to obtain a planar LH-drawing later.

Recall that F_A is the face containing the vertices of A, and in particular the edge (a_1, a_k). Let F_o be the other face incident to the edge (a_1, a_k). Change the planar embedding of G_{LH} such that F_o is the outer-face. Let v_n be the vertex $\neq a_1, a_k$ on the triangular face F_o. Compute Kant's canonical ordering with a_1, a_k as the first two and v_n as last vertex.

A *vertex group* is either a single vertex or a chain in the canonical ordering. Each addition of a vertex group adds at least one face to the graph. The face created by the second vertex group of the canonical ordering contains the edge $(v_1, v_2) = (a_1, a_k)$, and must be F_A, since F_o is the outer-face. Therefore, the second vertex group consists of the vertices $\{a_2, \dots, a_{k-1}\}$.

5.2.3 An algorithm for straight-line-drawings

Our algorithm to compute a planar straight-line LH-drawing is a modification of the algorithm by Chrobak and Kant [4], from now on called CK-algorithm. They build straight-line drawings of triconnected planar graphs by adding the vertex groups according to a canonical ordering. Vertices v_1 and v_2 are placed at $(0,0)$ and $(1,0)$. Throughout the additions of vertex groups, the following invariants of the drawing are maintained: (1) the vertices on the outer-face of the current graph have strictly increasing x-coordinates when walking from v_1 to v_2, and (2) the edges of the outer-face are drawn with slope in the set $\{-1, 0\} \cup [1, \infty)$.

To add a new vertex v_{i+1}, Chrobak and Kant consider its leftmost and rightmost predecessor c_α and c_β. They shift c_β and the vertices to the right of it by one unit to the right. They draw a line of slope -1 from c_β, and embed v_{i+1} at the crossing of this line with the vertical line through $x(c_\alpha) + 1$. See also Fig. 11. A new chain $v_i, \dots, v_l$ is added similarly, shifting some vertices by $l - i + 1$ units. See [4] for details.

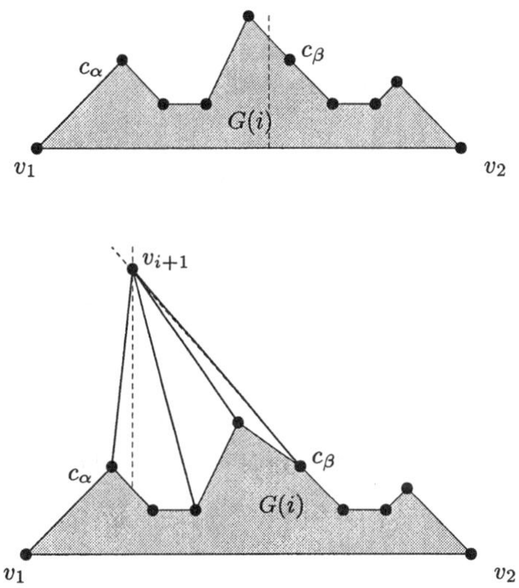

Figure 11: A sketch of the CK-algorithm. We shift the vertices to the right of and including c_β; then v_{i+1} can be placed without crossings and the invariants are maintained.

Our algorithm to create a planar LH-drawing varies only in the placement of the first two vertex groups, which were shown to consist of $a_1, \dots, a_k$. For $1 \le i \le k$, place a_i at $(i-1, 0)$, and draw the edges $(a_1, a_2), \dots, (a_{k-1}, a_k)$ as horizontal lines. We do not draw the edge (a_1, a_k) (this edge was not in the original graph G by choice of a_1). Since $a_1, \dots, a_k$ form the first two vertex groups of the canonical ordering, and since all edges have slope 0, the invariants of the CK-algorithm are satisfied. Subsequent vertex groups are placed exactly as in the CK-algorithm.

Theorem 2 *Any planar partition $G = (A \cup B, E)$ satisfying the necessary condition of Lemma 5.1 has a planar straight-line LH-drawing in an $(n-2) \times (n-2)$-grid. It can be computed in linear time.*

Proof: Add edges to G, compute a canonical ordering and apply the CK-algorithm with the described modifications. We need to show the correctness, the time-complexity, and the area bounds.

The first two vertex groups, i.e., the vertices of A, have y-coordinate 0. In subsequent steps, the vertices of B are added with y-coordinate > 0. Vertices are shifted in x-direction only, so the y-coordinates stay unchanged, and the resulting drawing is a planar LH-drawing.

Adding the edges can be done in linear time [2]. The canonical ordering can be computed in linear time [15]. The CK-algorithm works in linear time [4].

After drawing $a_1, \ldots, a_k$, the width is $k-1$. The width increases by one unit for each additional vertex, so the final width is $n-1$. Vertex v_n is adjacent to a_1 and a_k, hence it is placed at the intersection of the line of slope -1 emanating from $(n-1, 0)$ and the vertical line with x-coordinate 1, hence at point $(1, n-2)$. Since the triangle $\{a_1, a_k, v_n\}$ is the outer-face, the height of the final drawing is $n-2$. The width can be improved to $n-2$ by not shifting for the last vertex, see [4] for details. □

6 Face-constrained embedding problems

We showed that the conditions for LL-drawings (Lemma 4.1) and for LH-drawings (Lemma 5.1) can be tested in linear time, except for questions concerning whether there exists a planar embedding such that a given set of triangles are faces. This motivates the following problem:

Problem 1 *Given a graph G, and a set $\mathcal{C}$ of (simple) cycles in G. Does there exist a* face-boundary constrained *embedding of G, i.e., a planar embedding of G such that for each cycle C in $\mathcal{C}$, the edges of C form the boundary of a face?*

For LL-drawings and LH-drawings, we could restrict the attention to the case that all cycles are triangles. However, we think the face-boundary constrained embedding problem interesting in its own right. and treat the most general version possible.

A closely related problem could be called the *face-points constrained embedding problem*: Given a collection $\mathcal{L}$ of set of vertices, find an embedding such that to each $L \in \mathcal{L}$ there is a uniquely assigned face containing the vertices in L. This problem can be solved easily: Add a vertex v_L for every $L \in \mathcal{L}$, and connected v_L to all vertices in L. If the resulting graph is planar, then its planar embedding, after deleting the added vertices, is a face-points constrained embedding.

Solving the face-boundary constrained embedding problem is done similarly, but we need to add more edges, and argue more carefully why we obtain a face-boundary constrained embedding. In particular, we will encounter problems since we require that each cycle $C \in \mathcal{C}$ corresponds *exactly* to the boundary of the face, i.e., no additional edges are allowed in the face. Obtaining such an embedding is easy only if the graph is biconnected, hence we study graphs with cut-vertices separately later.

6.1 Biconnected graphs

Assume that G is biconnected, and that a set $\mathcal{C}$ of simple cycles in G is given. Two conditions must be satisfied, otherwise there cannot be a face-boundary constrained embedding. First, $|\mathcal{C}|$ must be no more than the number of faces $f = m - n + 2 \leq 2n - 4$. Secondly, each edge of G belongs to at most two cycles of $\mathcal{C}$. Assume that these conditions are satisfied. We transform G into a larger graph G_{aug}, and claim that G can be embedded such that each cycle in $\mathcal{C}$ is a face if and only if G_{aug} is planar.

To transform G, subdivide all edges, and add for each cycle $C = \{v_1, \ldots, v_k\} \in \mathcal{C}$ a new vertex v_C connected to $v_1, \ldots, v_k$ and to the subdivision vertices of the edges in C. See also Fig. 12. Call the resulting graph G_{aug} the *cycle-augmentation with respect to $\mathcal{C}$.*

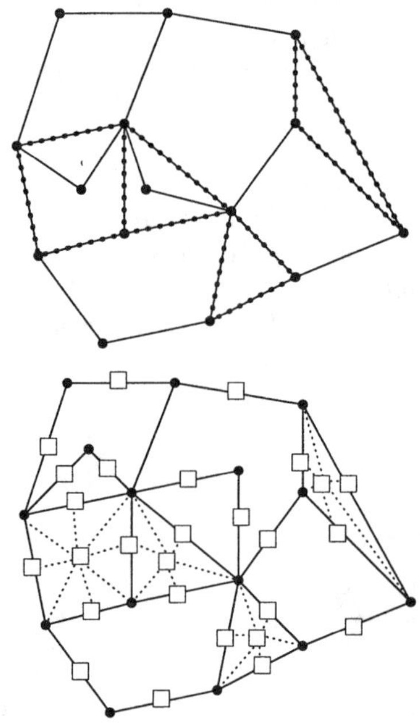

Figure 12: In this example, the cycle-augmentation G_{aug} is planar after changing the embedding of G.

Lemma 6.1 *Given a biconnected graph G and a set of cycles $\mathcal{C}$, there exists a face-boundary constrained embedding if and only if the cycle-augmentation G_{aug} is planar.*

Proof: Assume that G has a face-boundary constrained planar embedding. Subdividing edges maintains planarity. Every cycle $C \in \mathcal{C}$ is a face, so we can add the vertex v_C and its incident edges inside this face without introducing a crossing, therefore G_{aug} is planar.

Conversely, if we are given a planar embedding of G_{aug}, then, as we will show, removing the added vertices yields a face-boundary constrained embedding of G. Roughly speaking, for every cycle $C \in \mathcal{C}$ the incident edges of v_C prevent any vertex from being inside C by biconnectivity. The precise proof is as follows:

For a cycle $C \in \mathcal{C}$, let G_C be the subgraph consisting of the vertices and subdivided edges of C and the vertex v_C connected to them. The graph G_C is triangulated, except for the outer-face which has no chord, so G_C is triconnected. See also the left picture of Fig. 13.

We claim that every face of G_C is "empty" in any embedding of the cycle-augmentation G_{aug}. The precise statement is as follows:

Claim: For every triangular face F of G_C, the triangle T_F that is the boundary of F in G_C is a face in any planar embedding of G_{aug}.

To prove this claim, note that by construction, one vertex of T_F was v_C, the vertex added for C, one vertex was v_e, the subdivision vertex of an edge $e \in C$, and one vertex was v_r, an endpoint of e. Observe that all neighbors of v_C are in G_C; and that the neighbors of v_e are in G_C or in some other graph $G_{C'}$ added for a cycle $C' \in \mathcal{C}$ that also contained e.

If there are two cycles $C, C' \in \mathcal{C}$ containing e, then $G_C \cup G_{C'}$ is triconnected, since G_C and $G_{C'}$ are triconnected and they share three vertices (the two endpoints and the subdivision vertex of e). Define $G(e) = G_C \cup G_{C'}$ if e is in two cycles $C, C' \in \mathcal{C}$, and $G(e) = G_C$ otherwise. Then $G(e)$ is triconnected and planar, and all neighbors of v_c and v_e are in $G(e)$. See also Fig. 13.

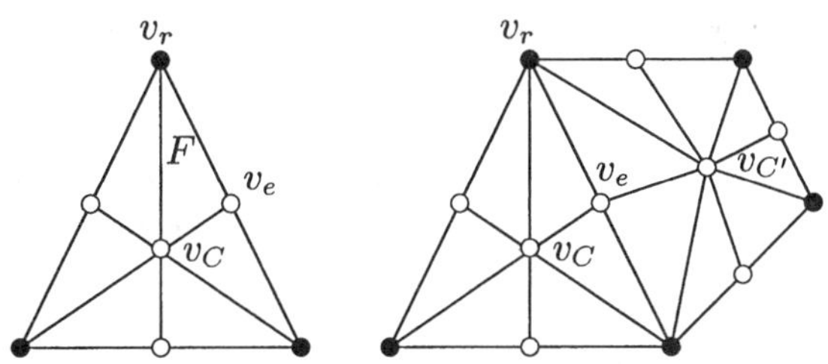

Figure 13: The graph $G(e)$ if e is in one and two cycles of $\mathcal{C}$, respectively.

Assume that there exists a planar embedding of the cycle-augmentation G_{aug} in which T_F is not a face. In the induced embedding of $G(e)$, T_F is a face, since $G(e)$ is triconnected and has only the one planar embedding shown in Fig. 13. Therefore the vertices of $G(e)$ are on one side of T_F, say, outside T_F, and v_r is the only vertex of T_F that may have neighbors inside T_F. There must be vertices inside T_F, since T_F is a triangle and not a face, so v_r is a cut-vertex, a contradiction to biconnectivity. This proves the claim.

Remove the added vertices v_C, $C \in \mathcal{C}$, and undo the subdivision of the edges. For each cycle $C \in \mathcal{C}$ the triangular faces of G_C were faces in G_{aug}, so C is now the boundary of a face. Thus, we found a face-boundary constrained embedding of G, and Lemma 6.1 is proved. □

Since $|\mathcal{C}| \leq 2n-4$, we add $m+|\mathcal{C}| = \mathcal{O}(n)$ new vertices. Planarity can be tested in linear time [13], so we can test the existence of a face-boundary constrained embedding for biconnected graphs in linear time.

6.2 Connected graphs

To test the existence of a face-boundary constrained embedding for a connected graph G, we proceed by induction on the number of cut-vertices of G. If there are none, then G is biconnected, and we apply the algorithm of the previous section. So assume that G has a cut-vertex v, and let $G_1, \ldots, G_s$ be the *cutting components* with respect to v, i.e., the maximal subgraphs for which v is not a cutvertex. Every cycle in $\mathcal{C}$ must be entirely inside one cutting component G_i, $i = 1, \ldots, s$, since otherwise it would use v twice, and would be a circuit, not a cycle. For $i = 1, \ldots, s$, let $\mathcal{C}_i$ be the cycles in $\mathcal{C}$ that are contained in G_i.

Lemma 6.2 *If G has a face-boundary constrained embedding, then for $i = 1, \ldots, s$, if d_i is the degree of the cut-vertex v in the cutting component G_i, then there are at most $d_i - 1$ cycles in $\mathcal{C}_i$ that contain v.*

Proof: Fix a face-boundary constrained embedding of G. For $i = 1, \ldots, s$, there exists a face F_i that contains edges of G_i and of some other cutting component. The boundary of F_i contains v twice, and is not a cycle, so in particular $F_i \notin \mathcal{C}$. The induced embedding of G_i is face-boundary constrained with respect to $\mathcal{C}_i$, and the boundary of the face that was part of F_i is not in $\mathcal{C}_i$. Since there are d_i faces in G_i that contain v, at most $d_i - 1$ of them can be in $\mathcal{C}_i$. □

We test the existence of a face-boundary constrained embedding of G as follows. For $i = 1, \ldots, s$, G_i has fewer cut-vertices than G, so by induction we test the existence of a face-boundary constrained embedding of G_i with respect to $\mathcal{C}_i$. If G_i has no face-boundary constrained embedding, then G cannot have a face-boundary constrained embedding either, and we are done. Otherwise, we test whether in the face-boundary constrained embedding of G_i there exists a face F_i incident to v that is not in $\mathcal{C}_i$. If not, then by Lemma 6.2 there is no face-boundary constrained embedding of G, and we are done. Otherwise, we make F_i the outer-face of the planar drawing of G_i.

If this procedure succeeds for all graphs G_i, $i = 1, \ldots, s$, then we construct a face-boundary constrained embedding of G by identifying the drawings of v such that the faces $F_1, \ldots, F_s$ become one face. See also Fig. 14. The only cycles that are not faces any longer are the boundaries of F_i, $i = 1, \ldots, s$. By choice of these faces, they are not in $\mathcal{C}$, so we obtain a face-boundary constrained embedding of G.

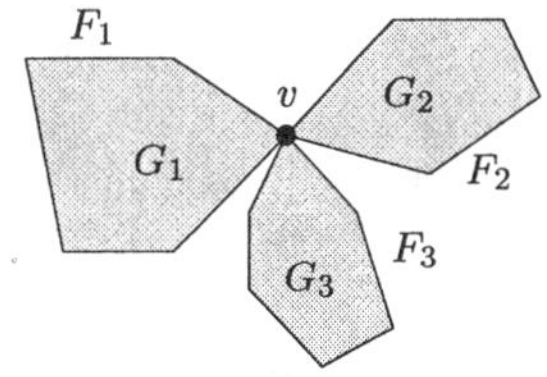

Figure 14: We merge $G_1, \ldots, G_s$ such that $F_1, \ldots, F_s$ become one face.

This algorithm can be implemented in linear time, by computing all cut-vertices and biconnected components beforehand, and processing them in an appropriate order. For each biconnected component B, we need $\mathcal{O}(|B|)$ time, and to merge all biconnected components containing the cut-vertex v, we need $\mathcal{O}(\deg(v))$ time. Hence the total time is $\mathcal{O}(m) = \mathcal{O}(n)$.

Theorem 3 *Given a graph G and a set of simple cycles $\mathcal{C}$ in G, there exists a linear-time algorithm to test the existence of a planar embedding of G such that the cycles in $\mathcal{C}$ are boundaries of faces.*

Two corollaries are worth mentioning. First, it follows that we can test in linear time whether a graph G can be embedded without separating triangle, since we can enumerate all triangles linear time [5], and then apply the above algorithm. This result was shown before [2], but the here presented approach is simpler, since it avoids splitting the graph into its triconnected components. Knowledge of such an embedding enables us to make a planar graph 4-vertex-connected by adding edges while maintaining planarity.

Secondly, for any given $k < n$, we can test whether G can be embedded with at most k non-empty triangles in $\mathcal{O}(n(2n+k)^k)$ time, which is polynomial if k is a constant. We do this by enumerating all possible subsets of at most k triangles, and for each of them apply the above algorithm with $\mathcal{C}$ as all remaining triangles. This result contrasts to the fact that embedding a graph with the minimum number of non-empty triangles is $\mathcal{NP}$-complete [2].

7 Conclusion

We presented the first in a series of papers on how to draw a planar partition. We gave an overview of models of displaying a planar partition and existing results. For two of the models, we provided necessary and sufficient conditions for the existence of a crossing-free drawing. The sufficiency-proof yields a linear-time algorithm to create such a straight-line drawing in a grid of small area. No such results were known before for arbitrary planar partitions. The necessary and sufficient conditions can be tested in linear time.

In the other papers of this series, we study a third model, and again provide necessary and sufficient conditions, which can be tested in linear time, an algorithm to create a poly-line drawing of small area, and a lower-bound for the area of straight-line drawings [3, 1].

We leave some open problems. First, what are other models of drawing partitions, and what can be said about them? Secondly, how should a k-partition, $k \geq 3$ be drawn? It is easy to define suitable models, but how quickly can we test the existence of such drawings, and construct them?

References

[1] T. Biedl. Drawing planar partitions III: Two constrained embeddings. Technical Report RRR 13-98, RUTCOR, Rutgers University, 1998.

[2] T. Biedl, G. Kant, and M. Kaufmann. On triangulating planar graphs under the four-connectivity constraint. *Algorithmica*, 19(4):427–446, 1997.

[3] T. Biedl, M. Kaufmann, and P. Mutzel. Drawing planar partitions II: HH-drawings. Technical Report RRR 11-98, RUTCOR, Rutgers University, 1998.

[4] M. Chrobak and G. Kant. Convex grid drawings of 3-connected planar graphs. *Internat. J. Comput. Geom. Appl.*, 7(3):211–223, 1997.

[5] N. Chiba and T. Nishizeki. Arboricity and subgraph listing algorithms. *SIAM J. Computing*, 14(1):210–223, 1985.

[6] I. Cruz and A. Garg. Drawing graphs by example efficiently: Trees and planar acyclic digraphs. In *Graph Drawing '94*, volume 894 of *Lecture Notes in Computer Science*, pages 404–415. Springer-Verlag, 1995.

[7] G. Di Battista, P. Eades, R. Tamassia, and I. Tollis. Algorithms for drawing graphs: an annotated bibliography. *Comp. Geometry: Theory and Applications*, 4(5):235–282, 1994.

[8] P. Eades, B. D. McKay, and N. Wormald. On an edge crossing problem. In *ACSC 9, 9th Australian Computer Science Conference*, pages 327–334, 1986.

[9] S. Even. *Graph Algorithms*. Computer Science Press, 1979.

[10] U. Fößmeier and M. Kaufmann. Nice drawings for planar bipartite graphs. In *CIAC '97*, volume 1203 of *Lecture Notes in Computer Science*, pages 122–134. Springer-Verlag, 1997.

[11] H. de Fraysseix, J. Pach, and R. Pollack. How to draw a planar graph on a grid. *Combinatorica*, 10:41–51, 1990.

[12] J. Harary and A. Schwenk. A New Crossing Number for Bipartite Graphs. *Utilitas Mathematica*, 1:203–209, 1972.

[13] J. E. Hopcroft and R. E. Tarjan. Efficient planarity testing. *Journal of the Association for Computing Machinery*, 21(4):549–568, October 1974.

[14] M. Jünger and P. Mutzel. 2-layer straightline crossing minimization: Performance of exact and heuristic algorithms. *J. Graph Alg. Appl.*, 1(1):1–25, 1997.

[15] G. Kant. Drawing planar graphs using the canonical ordering. *Algorithmica*, 16:4–32, 1996.

[16] D. E. Knuth. Nested satisfiability. *Acta Informatica*, 28:1–6, 1990.

[17] J. Kratochvil and M. Křivanek. Satisfiability of co-nested formulas. *Acta Informatica*, 30(4):397–403, 1993.

[18] S. North, editor. *Symposium on Graph Drawing 96*, volume 1190 of *Lecture Notes in Computer Science*. Springer Verlag, 1997. For the contest, see *http://www.-research.att.com/conf/gd96/contest.html*.

[19] R. Tamassia. On embedding a graph in the grid with the minimum number of bends. *SIAM J. Computing*, 16(3):421–444, 1987.

Approximation Algorithms for Multiple-Tool Milling

Sunil Arya* Siu-Wing Cheng† David M. Mount‡
Department of Computer Science
The Hong Kong University of Science & Technology
Clear Water Bay
Hong Kong

Department of Computer Science &
Institute for Advanced Computer Studies
University of Maryland
College Park, Maryland

Abstract

Milling is the mechanical process of removing material from a piece of stock through the use of a rapidly spinning circular milling tool in order to form some desired geometric shape. An important problem in computer-aided design and manufacturing is the automated generation of efficient milling plans for computerized numerically controlled (CNC) milling machines. Among the most common milling problems is simple 2-dimensional pocket milling: cut a given 2-dimensional region down to some constant depth using a given set of milling tools. Most of the research in this area has focused on generating such milling plans assuming that the machine has a tool of a single size. Since modern CNC milling machines typically have access to a number of milling tools of various sizes and the ability to change tools automatically, this raises the important optimization problem of generating efficient milling plans that take advantage of this capability to reduce the total milling time.

We consider the following *multiple-tool milling problem*: Given a region in the plane and a set of tools of different sizes, determine how to mill the desired region with minimum cost. The problem is known to be NP-hard even when restricted to the case of a single tool. In this paper, we present an efficient approximation algorithm for the multiple-tool milling problem. The running time and approximation ratio of our algorithm depend on the simple cover complexity of the milling region.

*Research supported in part by RGC CERG HKUST 736/96E.
†Research supported in part by RGC CERG HKUST160/95E.
‡Research supported in part by the National Science Foundation under grant CCR–9712379.

SCG 98 Minneapolis Minnesota USA
Copyright ACM 1998 0-89791-973-4/98/ 6...$5.00

1 Introduction.

Milling is one of the most important methods used in the manufacturing of mechanical parts in computer-aided manufacturing (CAM). It is applied to a *workpiece* of (typically metal) stock sometimes called the *part* or *billet*, which is clamped to a moving platform that is then translated under a rapidly spinning circular-shaped *milling tool*. It is somewhat more natural to think of the stock as remaining stationary and the tool translating above it. In this way material is removed, or *milled*, from the part. The overall problem is how to construct milling plans in order to achieve a given final geometric shape within the shortest amount of time.

There are several kinds of milling depending on the numbers of degrees of freedom possessed by the tool relative to the workpiece. In this paper we focus on the simplest case, but one that is common in practice, where continuous tool movement is possible in one plane and the direction normal to it is used only for retracting the tool. This situation is commonly referred to as 2D milling or *pocket machining*. (Pocket refers to the region being milled).

There has been a lot of research on the subject of automatic generation of tool paths for computerized numerically controlled (CNC) pocket machining. However, much of this study, both theoretical and practical, has focused on machining pockets using a single tool; the question of how to machine pockets efficiently using more than one tool has been largely ignored, and seems to be considerably deeper and richer than the single-tool problem. Modern milling machines have the capability of automatically loading different milling tools of a wide range of radii. Using a larger tool when possible offers a significant advantage in terms of milling time. In this paper we propose a cost model for describing multiple-tool milling problem, and present an approximation algorithm.

Previous results. Within the computer-aided design and manufacturing community there has been a considerable amount of study of various heuristics for the automatic generation of tool paths for pocket machining. The most common general strategies are *contour-parallel* (also known as *window-pane milling*) [Bru82, HA92, Per78, Pre89, PK83, PK85, SL90], in which the tool spirals inwards from (or outwards to) the boundary of the region, and *axis-parallel milling* (also known as *zig-zag* or *staircase milling*) [Bru82, DBW93, Kra92, PGW90, W^+87], in which the milling tool moves back and forth cutting parallel strips. Multiple tool milling has been considered, see for example [BC91], but there is no theoretical analysis of the performance of the heuristics proposed.

Held [Hel89, Hel91a, Hel91b] made a comprehensive study of milling heuristics from a computational geometry perspective. For the single tool case, he presented efficient algorithms to find a feasible tool path, given the shape of the pocket to be milled and the size of the tool. On the theoretical side, Arkin, et al. [AFM93, AFM95] and Iwano, et al. [IRT94] have given constant-factor approximation algorithms for finding shortest paths for the single-tool milling problem and for the closely related problem of lawnmowing. Arkin, et al. [AHS96] have also given approximation algorithms for minimizing the number of retractions for the zig-zag milling problem, subject to the constraint that one is not allowed to mill the same region again. The problem is known to be NP-hard even when restricted to the case of a single tool [AFM93]. We know of no theoretical work considering the use of multiple tools in milling.

Domain and tools. We model the pocket machining problem as follows. Tools are changed at a designated location called the *tool-change center.* Thus the path for each tool is assumed to start and end from this center. The input to our problem provides a planar domain P to be milled, a set of tools of different sizes, and the location of the tool-change center. We make the realistic assumptions that the ratio between consecutive tool sizes is bounded (for simplicity, we assume this ratio to be bounded by 2), and that the smallest tool can mill P without the need to be lifted. (In fact, our analysis can be generalized when the assumptions do not hold, increasing the approximation ratio by at most a logarithmic factor). The tools are disks with different radii and the domain P is a connected region (possibly with holes) bounded by straight line segments and circular arcs. We call these segments and arcs *domain edges.* The tools can be moved arbitrarily. The input size n refers to the number of vertices in the domain and the parameter k refers to the number of different tools given.

Cost model. A milling plan is a sequence of tours, each for a particular tool size. In a tour, a tool alternates between being engaged with the material (milling path) or being retracted (transport path in air). Due to stress on the tool, the speed with which the tool can be moved, called the *feed rate* is typically much smaller for milling than for transport. Consider a milling plan Ψ. Define $mill(\Psi)$ to be the total milling path length for Ψ, $transport(\Psi)$ to be the total transport path length for Ψ, and $ntools(\Psi)$ to be the number of tools used by Ψ. The total milling cost in this model is $cost(\Psi) = \alpha \cdot mill(\Psi) + \beta \cdot transport(\Psi) + \gamma \cdot ntools(\Psi)$, where α, β, and γ are arbitrary nonnegative values supplied by the user as part of the input. Each milling path with tool of size t includes a cost of $2t$ in addition to the length of the path. This additional component is included to account for the time to place the cutting tool within the material. This might be done either by milling in from the side or by drilling a hole and milling down into the material. This assumption is added to prevent ridiculous solutions based on using the tool like a cookie-cutter to stamp out disks without paying any milling cost at all. From a practical standpoint, plunging the tool into material induces considerable stresses on the milling tool, and is not used in practice, or only after the time has been spent to drill a hole where the center of the milling tool is to be placed.

Output representation. As observed in [AFM95], the milling path of a tool may require a combinatorially very large description even if the size of the region milled is combinatorially very small, e.g. a small tool milling a large circle. Following the approach in [AFM95], we use a succinct representation of milling paths instead. In our output, we represent the points milled by each tool as a collection of simple regions of regular structure. If desired, the actual milling paths can be extracted by contour-parallel milling or zig-zagging within each output regions. We also output the cost of our approximate milling plan.

Outline of paper. Section 2 gives an overview of our approximation algorithms and states the main result. In Section 3 and 4, we show how to discretize the problem. Section 5 shows that our discretization method indeed approximates the milling problem to within a constant factor. Section 6 shows how to reduce the milling problem to a weighted set cover problem and describes the approximation algorithm.

2 Overview and summary of results

Milling the domain is equivalent to covering the points in the domain with copies of the tools available. Each copy of a tool used will incur some cost. This suggests that milling is related to the discrete optimization problem of weighted set-cover. Indeed, we will transform our problem to a weighted set-cover problem and then solve the weighted set-cover problem using a greedy strategy to obtain a logarithmic ratio. To construct the transformation, we need to define the elements in the base set and the weighted subsets. The transformation is not straightforward for two reasons. First, it is infeasible to use points as set elements directly as there is an infinite number of them. We overcome this by discretizing the domain into simple regions and use these simple regions as set elements instead. A nice property is that we can assume that each simple region is milled with only one tool, while increasing the approximation ratio only by a constant. Second, to produce all the weighted subsets explicitly, we need to enumerate an exponential number of combinations of these simple regions and the tools used to mill them. This is too time-consuming. We overcome this by using an approximate greedy strategy that does not require the weighted subsets to be explicitly provided. Such an approximate greedy strategy will incur another constant factor in the approximation ratio and it is based on the Euclidean k-TSP problem [Aro97]. We will be bounding the running time, space, and approximation factor in terms of k and the size of the discretization of the domain.

We first compute an $O(n)$ set of line segments for subdividing P into monotonous regions [Hel91b]. The endpoints of these line segments lie on the boundaries of P and we introduce them as extra vertices. This will split some domain edges of P into several pieces. Let P^* denote the modified domain. The size of our discretization is equal to the simple cover complexity of P^*. The *simple cover complexity*, or $scc(P^*)$, is an intrinsic measure of the geometric complexity of P^* [MMS94]. It is defined as follows. A disk is *simple* if it intersects at most 2 edges of P^*. Given any $\epsilon > 0$, we say that a ball of radius r is *ϵ-strongly simple* if the ball with the same center and radius $(1+\epsilon)r$ is simple. Given ϵ, a *strongly simple cover* of a region P^* is a collection of ϵ-strongly simple balls whose union contains P^*. Given any fixed ϵ (for example $\epsilon = 1/2$), the simple cover complexity of P^* is defined to be the cardinality of the smallest strongly simple cover of P^*. Our main result is:

Theorem 2.1 *Given a domain P of n vertices, and k circular tools, let $N = scc(P^*)$ denote the simple cover complexity of P^*. Let α, β, and γ, be nonnegative cost parameters (where $\beta < \alpha$). Then an $O(\log(kN))$-factor approximation to the optimum cost milling plan for P can be computed in time that is polynomial in n, N and k. (Constant factors hidden by the "big-Oh" do not depend on any of the input parameters.)*

One can also define a rectilinear version of the problem in which tools are squares and the domain is rectilinear. For this version, we obtain an approximation algorithm by applying similar ideas to a modification of the quadtree-based subdivision described by Mitchell, Mount, and Suri [MMS94]. The running time is polynomial in n and k. Details are omitted in this version.

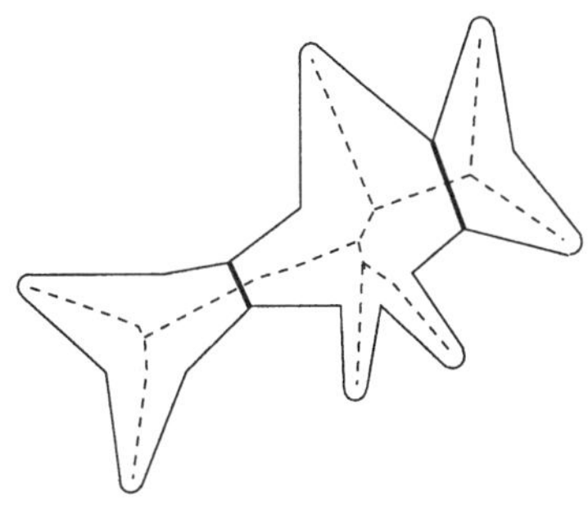

Figure 1: The dashed curves show the Voronoi diagram for the domain. The bold lines are the two cuts.

3 Subdividing the domain

Consider the Voronoi diagram for P. We define a distance function $\alpha(p)$ to ∂P from every point p on the Voronoi diagram. For each point p on the Voronoi diagram such that $\alpha(p)$ is locally minimal, we call the two nearest points on ∂P to p *bottleneck points.* We call the line segment joining them the *bottleneck segment.* We introduce all the bottleneck points to P. Each bottleneck point splits a domain edge into two smaller domain edges. We denote by P^* the resulting domain. Figure 1 shows the bottleneck segments for a simple domain.

We perform a quadtree decomposition of P^*. For any box x, we denote its side length by *width*(x). We denote by cx a box with the same center as x whose side length is $c \cdot$ *width*(x).

The goal of the decomposition is to cover P^* with a set X of boxes such that the scene within each box is extremely simple. X is generated as follows. We first enclose P^* in its bounding box. Then we apply the following splitting rule. For any box x, we call $3x$ the *buffer zone* of x and denote it by *buf*(x). Take any box x, if *buf*(x) intersects more than two domain edges, then split x through its center into four identical boxes each of half the size. When the splitting process stops (which will occur eventually since each vertex is adjacent to at most two domain edges), we obtain a set of boxes covering P^*.

A *cell* is a connected component of $P^* \cap x$ for some box $x \in X$. Given a cell C, we denote by *box*(C) the box $x \in X$ that contains C. We define the *size* of C, *size*(C), to be *width*(*box*(C)). The size of our subdivision of P^* is bounded by the *simple cover complexity* [MMS94].

Lemma 3.1 *There are $O(scc(P^*))$ boxes covering P^*. Each box x has a constant number of neighboring boxes and each neighbor has width $\Omega(width(x))$.*

4 Basic milling actions

Let $d(p, r)$ denote a disk of radius r centered at a point p. Given a tool t, we also use t to denote its radius. Whenever we put the center of t at a point p, we call

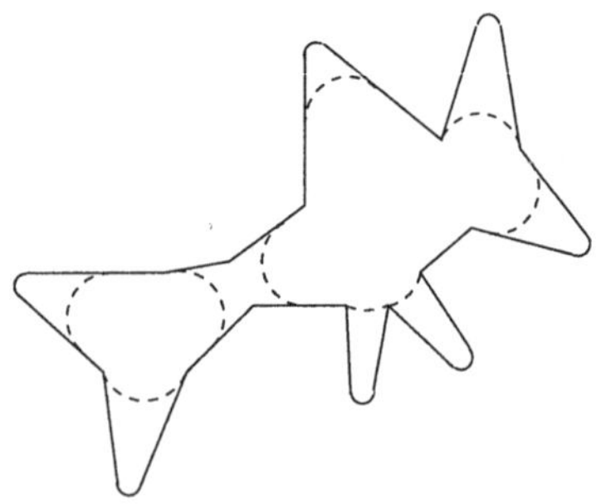

Figure 2: The accessibility arcs for a particular tool size are shown in dashed arcs.

$d(p,t)$ a *placement of* t. A placement of t is *free* if it stays within P^*. The *free space* of t is the locus of centers of all free placements of t. We denote it by $\mathcal{F}(t)$ and formally $\mathcal{F}(t) = P^* \ominus t$, where $\ominus$ is the Minkowski difference operator. Thus, $\mathcal{F}(t) \oplus t$ is the set of points in P^* that can be covered by a free placement of t, where $\oplus$ is the Minkowski sum operator.

At each vertex p of $\mathcal{F}(t)$, $d(p,t)$ touches ∂P^* at several points. If the arc α between two consecutive contact points on the boundary of $d(p,t)$ is less than a semicircle, then α belongs to the boundary of $\mathcal{F}(t) \oplus t$ and we call it an *accessibility arc*. The collection of accessibility arcs for a tool t is denoted by $\mathcal{A}(t)$. See Figure 2.

A subset K of $\mathcal{F}(t)$ and a subset $M \subseteq K \oplus t$ define a *milling action*. Specifically, t moves with center in K to remove all points in M. If K is not connected, then we have to pick t up and this will incur a charge of $O(t)$ per connected component in K plus the transport cost to visit all components in K. We say that a point in M is *milled* by this action. Note that, physically speaking, a larger set than M may be removed by t's movement. However, we only consider points in M as milled and we will have to deploy other milling actions to mill points not in M. Imagine that an arbitrary sequence of milling actions (possibly of different tools) has been applied to the collection of cells defined in the previous section. Given a cell C, we call the set of points in C not milled the *unmilled region*, and we call a connected component of the unmilled region an *unmilled component*.

Given a tool t, we restrict ourselves to two kinds of milling actions of t on a cell C. We say that t is *large* for a cell C if $t > size(C)/4$ and *small* otherwise. Conversely, we say that a cell C is small for t if $t > size(C)/4$ and large otherwise.

4.1 Large tool basic milling action

Overlay a square grid $\mathcal{G}_t$ of side length δt on P^*, where δ is any positive constant. Define $\mathcal{B}_t$ to be the set of grid squares b such that $b \oplus 4t$ overlaps some quadtree box of width less than $4t$. Any grid square $b \in \mathcal{B}_t$ induces a (possibly empty) set of milling actions as follows. If t is the smallest tool, let K be a connected component of $(b \oplus 11t) \cap \mathcal{F}(t)$; otherwise, let K be a connected component of $(b \oplus 11t) \cap \mathcal{F}(t)$ that contains some component of $b \cap \mathcal{F}(2t)$. Define $\mathcal{C}_b$ to be the collection of cells C such that t is large for C and C overlaps $b \oplus 4t$. Define M to be $K \oplus t \cap \bigcup_{C \in \mathcal{C}_b} C$. Then K and M define a *large tool basic milling action* denoted by *large_tool*(t, K, M), i.e., move t along the surface with center in K to mill points in M. We shall abbreviate basic milling action as BMA.

Lemma 4.1 *Let C be a cell small for tool t. For any large_tool(t, K, M), either $M \cap C$ is empty or the boundary of $M \cap C$ consists of points in ∂C and $\mathcal{A}(t)$.*

The above result shows that after a large tool BMA by t, new boundaries of any resulting unmilled component are accessibility arcs of t. Now, we prove that the number of large tool BMAs by all tools is polynomial in N and k. Also, the cost of a large tool BMA for tool t is $O(t)$.

Lemma 4.2 *There are $O(1)$ large tool BMAs by t induced by any grid square $b \in \mathcal{B}_t$. The cost of each such BMA is $O(t)$.*

Proof (Sketch) If t is the smallest tool, let $\mathcal{K}$ be the set of connected components of $b \oplus 11t \cap \mathcal{F}(t)$; otherwise let $\mathcal{K}$ be the set of connected components of $b \oplus 11t \cap \mathcal{F}(t)$ that contain some component of $b \cap \mathcal{F}(2t)$. Since one large tool BMA for tool t is defined for each component in $\mathcal{K}$, we need to show that the number of components in $\mathcal{K}$ is $O(1)$.

Consider the case when t is not the smallest tool. Let K be any component in $\mathcal{K}$. Since K contains a component of $b \cap \mathcal{F}(2t)$, there is a free placement $d(p, 2t)$ where $p \in K \cap b$. Clearly $d(p,t) \subseteq \mathcal{F}(t)$ since a disk of radius t with center in $d(p,t)$ lies completely within $d(p, 2t)$. Further, since $p \in b$, $d(p,t) \subseteq b \oplus 11t$. It follows that $d(p,t) \subseteq K$. By a packing argument, the number of components in $\mathcal{K}$ is $O(1)$.

If t is the smallest tool, then by our assumption $\mathcal{F}(t)$ is one connected component. It is straightforward to show that this implies that the number of components in $\mathcal{K}$ is $O(1)$. Details are omitted in this version.

Placing t with center on ∂K costs $O(t)$. Then we move t along the boundary of K and then zig-zag inside K along vertical segments that are separated by a distance at least $2t$ apart. By the result in [AFM95], the cost of this is proportional to the sum of the boundary length of K and the total length of the vertical segments. The total length of the vertical segments is $O(t)$. The boundary length of K can be shown to be $O(t)$. Thus, the cost of each *large_tool*(t, K, M) is $O(t)$. □

Corollary 4.1 *There are $O(kN)$ large tool BMAs for all tools.*

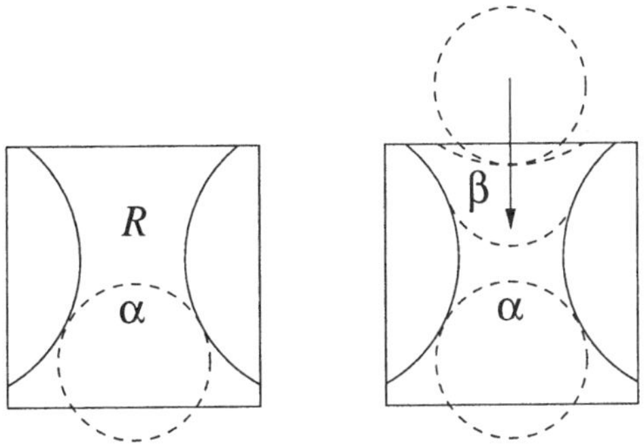

Figure 3:

Proof For each tool t, we define large tool BMAs for each $b \in \mathcal{B}_t$. Since a quadtree box of width less than $4t$ can overlap $b \oplus 4t$ for at most a constant number of grid squares b in $\mathcal{G}_t$, the number of squares in $\mathcal{B}_t$ is $O(N)$. By Lemma 4.2, the number of large tool BMAs induced by each $b \in \mathcal{B}_t$ is $O(1)$. It follows that there are $O(N)$ large tool BMAs defined for a given tool t. Summing over all tools, we have the desired result. □

4.2 Small tool basic milling action

Given a set S, we use $\text{cl}(S)$ to denote its closure. Let U be an unmilled component inside C and t be small for C. Let K be a connected component of $\text{cl}(U) \oplus t \cap \mathcal{F}(t)$. Define M to be $K \oplus t \cap U$. K and M define a small tool BMA by t, denoted *small_tool*(t, K, M). This action will move t on the surface with center in K to mill points in M. We show that after a small tool BMA by t, each resulting unmilled component is bounded by *exactly one* accessibility arc for t.

Lemma 4.3 *For any small_tool(t, K, M) acting on an unmilled component U, any resulting unmilled component is bounded by exactly one accessibility arc and some points in ∂C. Moreover, the accessibility arc belongs to $\mathcal{A}(t)$.*

Proof (Sketch) Let C be the cell containing U. Let *small_tool*(t, K, M) be a small tool BMA applied to an unmilled component U in C. Let W be a resulting unmilled component and so W is bounded by an accessibility arc α of radius t. Since t is small for C, α must have its endpoints on ∂P^* inside *buf*(*box*(C)). If both endpoints of α lie on one domain edge, the lemma is clearly true. Otherwise, α has endpoints on two domain edges e_1 and e_2 as shown on the left of Figure 3 and W lies in the region R shown in the figure. Assume to the contrary that W is bounded by one more accessibility arc β of some radius t'. Suppose that $t' \geq t$. Then it implies that we can advance another disk of radius t from β towards α. This disk will eventually be stopped by e_1 and e_2 as shown on the right of Figure 3. This implies that e_1 and e_2 contain a bottleneck vertex, contradiction. The case where $t' < t$ can be handled symmetrically. □

4.3 Complexity of unmilled region

In this section, we establish that after any sequence of BMAs the combinatorial complexity of an unmilled region inside a cell is always bounded by a constant. In the sequence, tools can change and large tool and small tool BMAs can interleave. The consequence is that within each cell, the number of unmilled components is always bounded by a constant and each unmilled component has constant complexity.

Lemma 4.4 *Let U be an unmilled component in a cell C after a sequence of BMAs on C. Let t be a tool small for C. After applying small_tool(t, K, M) on U, for each resulting unmilled subset W of U,*

(i) $\partial\text{cl}(W)$ overlaps with some domain edge intersecting C or contains a vertex of box(C).

(ii) W has constant combinatorial complexity.

Proof If W is not bounded by ∂P^*, then for any maximal connected component s of a side of *box*(C) bounding W, s cannot be incident to two accessibility arcs as there is only one by Lemma 4.3. Thus, s is incident to a vertex of *box*(C). The complexity of W is bounded by the complexity of the arrangement of sides of *box*(C), domain edges of P^*, and accessibility arcs bounding W. There is a constant number of elements in each category and hence W has a constant complexity. □

Lemma 4.5 *After any sequence of BMAs on a cell C, the unmilled region in C has constant combinatorial complexity.*

A consequence of Lemma 4.5 is that there is a polynomial number of small tool BMAs.

Lemma 4.6 *There are $O((kN)^{O(1)})$ small tool BMAs by all tools.*

5 Approximating optimal milling using BMAs

We assume that with each BMA *large_tool*(t, K, M) or *small_tool*(t, K, M) there is an associated *starting point*, which may be any point in K. Conceptually, milling begins at the starting point and at the end the tool center is returned to this point.

The main result of this section is to show that any milling plan can be converted into a milling plan consisting entirely of BMAs while sacrificing at most a constant factor in cost.

Define $mill(t, K, M)$ to be the milling cost of the milling operation defined by the BMA $large\_tool(t, K, M)$ or $small\_tool(t, K, M)$. Given a set of BMAs S, define S_t to be the subset of S which uses tool t. Define $TSP(S)$ to be the length of a minimum Euclidean traveling salesman tour on the starting points of S and the tool change center. The cost of S is composed of two elements: the time required to perform each of its milling operations and the time to move from the starting point of one to the starting point of another. Define: $mill(S) = \sum_{(x,t,t')\in S} mill(x, t, t')$ and $transport(S) = \sum_t TSP(S_t)$.

We immediately have the following.

Lemma 5.1 *Let S be a set of BMAs which mills P^*. Then there exists a milling path Ψ, using no tools other than those used in S, such that*

$$mill(\Psi) = mill(S), \qquad transport(\Psi) = transport(S).$$

Conversely we assert that any milling plan Ψ can be transformed into a set of BMAs which mills P^*, whose milling and moving times are comparable to those of Ψ.

Theorem 5.1 *Let Ψ be a milling plan. Then there exists a set S of BMAs which mills P^* using at most a constant factor more tools as those used in Ψ, such that $mill(S) = O(mill(\Psi))$ and $transport(S) = O(mill(\Psi) + transport(\Psi))$.*

Proof (Sketch)

We show how to find a set S of BMAs corresponding to Ψ, and show how to charge each BMA to some portion of Ψ.

Large tool BMAs:

For each tool t, let Ψ_t denote the portion of the milling path only involving tool t. We think of Ψ_t as the path along which the center of the tool moves. For each continuous curve η_t in Ψ_t, we find a set of large tool BMAs using t', where t' is the tool just smaller than the smallest tool in the range $[t/2, t]$. If there is no such tool, we let t' be the smallest tool. Since successive tools differ in size by a factor of at most 2, either t' is not the smallest tool and lies between $t/4$ and $t/2$ or it is the smallest tool and lies between $t/4$ and t.

Let $\mathcal{X}_{t'} \subseteq \mathcal{B}_{t'}$ be the set of squares through which η_t passes. For each $b \in \mathcal{X}_{t'}$, we add to S all the large tool BMAs $large\_tool(t', K, M)$ induced by b such that K contains a point on η_t inside b (i.e. $K \cap \eta_t \cap b$ is not empty).

We claim the following:

(1) The total milling cost of all the large tool BMAs is no more than $O(mill(\Psi))$.

(2) Let q be a point in some cell C which is milled by Ψ using a tool of size greater than $size(C)$. Then q is milled by some large tool BMA in S.

Let L be the length of η_t. It is straightforward to show that η_t passes through at most $O(L/t'+1)$ squares in $\mathcal{G}_{t'}$. Thus $|\mathcal{X}_{t'}|$ is $O(L/t' + 1)$. By Lemma 4.2, the number of BMAs $large\_tool(t', K, M)$ induced by any grid square $b \in \mathcal{X}_{t'}$ is bounded by a constant, and thus the total number of BMAs $large\_tool(t', K, M)$ generated by η_t is $O(L/t' + 1)$. From Lemma 4.2, we know that the milling cost of a $large\_tool(t', K, M)$ is $O(t')$, which implies that the total milling cost of all these BMAs is $O(L + t')$. This is within a constant factor of the total of the length of η_t and the placement cost of $2t$. This implies (1).

To prove (2), let $q \in C$ be milled in Ψ by t centered at point p on η_t. Let $b \in \mathcal{G}_{t'}$ be the grid square that contains p. We will show that $b \in \mathcal{B}_{t'}$, and there is a BMA $large\_tool(t', K, M)$ induced by b that is added to S and which mills q.

Since $p \in b$ and $d(p, t)$ overlaps C and $t < 4t'$, this implies that $b \oplus 4t'$ overlaps C. Also, since $size(C) < t$ and $t < 4t'$, $size(C) < 4t'$. Thus, by definition, $b \in \mathcal{B}_{t'}$, and b is used to define large tool BMAs for t'.

Let K be the connected component of $b \oplus 11t \cap \mathcal{F}(t')$ that contains point p. We claim that $large\_tool(t', K, M)$ is a valid BMA, is added to S, and it mills point q.

If t is the smallest tool, then $large\_tool(t', K, M)$ is a valid BMA. Otherwise, if t is not the smallest tool then, since $p \in \mathcal{F}(t)$ and $t > 2t'$, $p \in \mathcal{F}(2t')$. Thus K contains a component of $b \cap \mathcal{F}(2t')$ and, therefore, $large\_tool(t', K, M)$ is a valid BMA.

Since $K \cap b$ contains point p, $large\_tool(t', K, M)$ must be added to set S. To see that $large\_tool(t', K, M)$ mills point q, observe first that there is a placement $d(p', t')$ which touches point q and is contained within $d(p, t)$, and second, $p' \in K$ since we can move a disk of radius t' along the line segment from p to p', while it remains entirely within $d(p, t)$.

Call this Step 1. This puts all the large tool BMAs needed to S.

Small tool BMAs:

Next we show how to find a sequence of BMAs corresponding to Ψ for the small tools that mill some cell. This part of the proof involves a complex charging argument, and due to space limitations we only sketch the argument. Let C be a cell. Let us assume that all the large tool BMAs have already been added by the argument given above.

At this time we assert that for any point $q \in C$, if q is milled by Ψ using a tool t whose size is greater than $size(C)$, then q is milled by some large tool BMA by tool t' (where t' lies between $t/4$ and t) already added by the construction given above. Let U be any unmilled component within cell C. Any point in U can only be milled by Ψ using tool t of size less than $size(C)$, otherwise it would have been handled already by a large tool BMA. From Lemma 4.5 it follows that the unmilled region within C consists of a constant number of components, and each component is bounded by edges of the following two types:

(a) accessibility arcs generated by large tool BMAs (and later by small tool BMAs)

(b) unmilled portions of the cell boundary, including domain edges.

To account for the cost for the small tool BMAs, it will be helpful to endow each of the types of boundary elements along the unmilled component with a *charge*, proportional to its length. We show that the sum of these charges over all cells is proportional to the cost of Ψ, and then we show that the cost of the small tool BMAs for cell C is proportional to the sum of these charges for cell C. We can justify these charges as follows.

(a) The length of the accessibility arcs generated by a large tool BMA can be shown to be at most proportional to the radius of the tool. Thus we can account for these just as we accounted for the milling cost of the large tool BMA.

(b) In order to mill a portion of the boundary of C requires a milling path whose length is at least proportional to the length of the milled boundary. We charge this portion of the boundary to this milling action in Ψ. Since this is being milled with a tool that is small for C, since each cell contains a constant number of unmilled components, and since the subdivision is smooth, it follows that any segment of Ψ can be charged at most a constant number of times.

Now we proceed with the accounting for the small tool BMAs. Let U be an unmilled component inside cell C after all the large tool BMAs have acted upon it. Let T denote the set of tools that Ψ uses to mill any point in U. We add the small tool BMAs from T in decreasing order of tool size, updating the unmilled region U with each addition. Let t denote the largest of the remaining tools in T. We add the small-tool BMAs defined by U and t to set S. That is, for each component K of $U \oplus t \cap \mathcal{F}(t)$, we add a BMA *small_tool*(t, K, M). It is easy to show that there are only a constant number of such BMAs.

To account for the placement costs for these BMAs, observe that Ψ either places a copy of tool t in the vicinity, or has brought t from somewhere distant into this cell. We charge the BMA's placement against Ψ's placement or Ψ's milling cost needed to approach this cell, respectively. Again because of smoothness and the fact that the tool is small, it follows that these elements are charged a constant number of times.

To account for the milling cost of these BMAs, we make the following simple observation. Since large tools mill more efficiently than small tools, we lose nothing by making the greatest use of the largest tool that Ψ brings into an unmilled region. To make this more formal, consider the introduction of the BMA for a tool of size t, after all larger tool BMAs have already been added. Let η denote the intersection of all the milling paths in Ψ (necessarily of this and smaller tools) with K, where K is a component of $U \oplus t \cap \mathcal{F}(t)$. We charge the BMA for tool t against the length of η. We claim that the disconnected pieces of η can be easily "stitched together" using pieces from the boundary of K to get an Eulerian tour and combining this with a tour of the boundary of K, we obtain a new tour η' for t. Any point in U that can be milled by the BMA *small_tool*(t, K, M), can be milled by moving tool t along η'. The length of the BMA *small_tool*(t, K, M) is no more than the length of tour η'.

Thus, we only need to account for the length of tour η'. Observe that η' is composed of pieces from η and from the boundary of K. The pieces of η are charged against Ψ. To account for the length of the boundary of K, we need to distinguish between two types of segments on the boundary of K: (1) the boundary of $\mathcal{F}(t)$ and (2) the boundary of $U \oplus t$. Segments of type (1) are proportional in length to portions of the domain edges that get milled, and hence can be accounted for by using up the charge on this portion of the domain edge. Segments of type (2) are proportional in length an accessibility arc on the boundary of U and the cell boundary, and hence can be accounted for by using up the charge on the accessibility arc and cell boundary.

Next, we argue that accessibility arcs continue to carry charge even if after small tool BMAs have acted upon cell C. This is easy because a small tool BMA generates just one accessibility arc whose charge (which will be used in the future) can be accounted for by Ψ's placement of tool t in the vicinity of C.

Finally, we show that $\textit{transport}(S) = O(\textit{mill}(\Psi) + \textit{transport}(\Psi))$. Let S_1 and S_2 be the set of large-tool and small-tool BMAs corresponding to the milling paths of tool t in Ψ. By our construction, the large-tool BMAs in S_1 use tool t', where t' lies between $t/4$ and t, and the small-tool BMAs in S_2 use tool t. Again, by our construction, for each such BMA, Ψ places a tool copy of size t that overlaps the region covered. Therefore, the transport for tool t in Ψ is as long as a TSP tour visiting the covered regions. S_1 and S_2 bring tool t' and tool t, respectively, on a tour to visit a subset of the same covered regions, but possibly at different points. Since two points in a covered region can be connected by a local milling path, we see that $\textit{TSP}(S_1) = O(\textit{mill}(S_1) + \textit{transport}(\Psi_t))$ and $\textit{TSP}(S_2) = O(\textit{mill}(S_1) + \textit{transport}(\Psi_t))$. Summing over all tools, we conclude that $transport(S) = O(mill(S) + transport(\Psi)) = O(mill(\Psi) + transport(\Psi))$ as $mill(S) = O(mill(\Psi))$. This completes the proof.

□

6 Greedy Approximation

We reduce the problem of finding an optimal milling plan to a *weighted set cover problem*: given a set S and a family $\mathcal{Z}$ of some subsets of S where each $X \in \mathcal{Z}$ is associated with weight $w(X)$, the objective is to find $\mathcal{Y} \subseteq \mathcal{Z}$ such that $S = \bigcup_{X \in \mathcal{Y}} X$ and $\sum_{X \in \mathcal{Y}} w(X)$ is minimized. Denote such an instance by $(S, \mathcal{Z})$. We approximate the weighted set cover problem with a greedy algorithm and thus achieving the claimed logarithmic factor approximation. The difficulty is that, as we will see, the instance of our weighted set cover problem is too large to be described explicitly. Therefore, we run an approximate greedy algorithm on a succinct problem description. We will show that this only adds an extra constant factor.

6.1 Reduction

We refine the subdivision of P^* by overlaying the accessibility arcs of $\mathcal{F}(t)$ for all t on it. The set of faces in the refined subdivision is the set S in our weighted set cover instance. It remains to define $\mathcal{Z}$.

For each *large_tool*(t, K, M) or *small_tool*(t, K, M), M is a subset of faces in S. We call K a *t-locus* and M a *t-region*. A *t-subset* is union of some t-regions and so a t-subset is a set of faces in S. For each t-subset X, we denote by X_M the set of t-regions forming X and we denote by X_K the collection of corresponding t-loci. The weight $w(X)$ of X is the sum of three terms. The first term is the milling costs of t-regions in X_M. The second term is the length of the MST connecting the tool change center and the t-loci in X_K. The third term is the sum of placement costs $O(|X_K| \cdot t)$. These terms give the total cost of milling faces in X using the BMAs induced by X_K.

We first include in $\mathcal{Z}$ the collection of all possible t-subsets for each tool t. Then we prune $\mathcal{Z}$ such that for each tool t, $X \in \mathcal{Z}$, and any cell C, if there are *small_tool*(t, K, M) and *small_tool*(t, K', M') where $K, K' \in X_K$, $M, M' \in X_M$, and $M, M' \subseteq C$, then *small_tool*(t, K, M) and *small_tool*(t, K', M') act on unmilled components of the same unmilled region of C. By our proof of Theorem 5.1, solving $(S, \mathcal{Z})$ will return at least a constant factor approximation of the milling problem.

6.2 Approximate greedy algorithm

If $(S, \mathcal{Z})$ is described explicitly, then we can use the following greedy heuristic to obtain an $O(\log kN)$ approximation factor. Initialize the cover $\mathcal{Y}$ to be empty. Compute the t-subset $X \in \mathcal{Z}$ that minimizes the *average weight* defined to be the ratio $w(X)$ divided by the number of faces in X that are currently in S. Then include X in $\mathcal{Y}$ and remove all the faces in S contained in X. Repeat until S becomes empty. This achieves an approximation factor of $\ln |S| = O(\log kN)$. Unfortunately, it is impossible to describe $(S, \mathcal{Z})$ explicitly as $|\mathcal{Z}|$ is exponential in the number of faces in Z. Instead we only store the set of t-regions for all tools and compute a t-subset of approximately minimum average weight (within a constant factor) by solving a series of instances of a variant of j-TSP. Hence, we still solve $(S, \mathcal{Z})$ within a logarithmic approximation factor.

The m-WTSP. We briefly review the j-TSP problem and introduce a variant that will be solved repeatedly as a subproblem. Given l points in the plane, the j-TSP is to find a tour of minimum length that visits any j of the l given points. The j-TSP problem is NP-hard but it can be approximated to within any constant factor in polynomial time [Aro96, Mit96]. All of the algorithms for j-TSP are based on showing that the approximate tour has a particular structure. Then the optimum tour of the particular form can be computed using dynamic programming. We generalize the j-TSP to find a tour that collects coins at the visited points. Each point p is given a table *table*(p) and each entry of *table*(p) is a coins-cost pair which tells the cost of collecting the associated number of coins at p. The m-WTSP is to find a tour that collects m coins so that the tour length plus the sum of the cost of collecting coins at visited points is minimized. This is defined to be the weight of the tour. The dynamic programming paradigm for approximating the j-TSP is powerful enough to approximate the m-WTSP within a constant factor in polynomial time.

Strategy and difficulty. Our strategy is to compute, for each tool t, a t-subset of approximately minimum average weight, and then return the one of the least average weight. Consider computing this t-subset for tool t. Given a t-region, fix a point p in the corresponding t-locus. Let f_p be the number of faces covered by M. Let c_p be the sum of milling cost and placement cost for the BMA by t that mills M. Create the table *table*(p) which contains only one entry, namely, (f_p, c_p). Repeat this for all other t-regions. This yields a collection of points and their associated tables. Let F_t be the number of faces currently in S that are covered by some t-regions. For $1 \leq m \leq F_t$, find the approximate m-WTSP and compute its average weight. Afterwards, select the tour with minimum average weight and this corresponds to the t-subset with approximately minimum average weight. The flaw in the above strategy is that the number of faces covered when visiting a set of points is not simply the sum of numbers of faces covered when visiting each point. This is due to possible overlap among t-regions. We describe below two transformations to get around this problem and obtain the desired series of instances of m-WTSP.

Collapsing small tool BMAs. Let *large_cell*(t) denote the set of cells for which t is small. Given a cell $C \in$ *large_cell*(t), all the t-regions inside C are generated by small tool BMAs by t. If $2box(C)$ does not contain the tool change center, then we put a representative point $pt(C)$ at the center of $box(C)$ as the common point for all t-regions in C. If $2box(C)$ contains the tool change center, then we put a representative point $pt(C)$ at the tool change center. (If $box(C)$ contains two cells, then we can put these two points slightly apart at the center of $box(C)$.) Let $\mathcal{M}$ be any milling plan. We modify $\mathcal{M}$ as follows. First, if t is transported to C and M is the first t-region in C visited by t, then we first transport t to $pt(C)$ and then to M to start milling. Afterwards, if t is transported to $pt(C)$ several times, then we transport t to $pt(C)$ exactly once and then mill all the t-regions in C that t should mill before going to another cell. Let $\mathcal{M}'$ denote the modified milling plan.

Lemma 6.1 *The tour length of t in $\mathcal{M}'$ is within a constant factor of the tour length of t in $\mathcal{M}$.*

Proof (Sketch) If we can show that detouring via the common point $pt(C)$ of a cell C increases the tour length of t by a constant factor, then it is clear that we can make t being transported to $pt(C)$ exactly once. Thus, we focus on proving that the detour is not expensive. Let $\mathcal{M}^*$ denote the modified $\mathcal{M}$ with the detour. Let T and T^* be a tour of t in $\mathcal{M}$ and its modified version in $\mathcal{M}^*$ respectively. We first modify T^* as follows. In T^*, if there is an edge e from a t-locus K_1 for a t-region in cell C_1 to $pt(C_2)$ and then to the t-locus K_2 for a t-region in cell C_2, then we replace e by a path of two edges: from K_1 to K_2 and then to $pt(C_2)$. Let $\tilde{T}$ denote the modified T^* and the length of $\tilde{T}$ is no less than the length of T^* by triangle inequality. We call an edge in $\tilde{T}$ between a t-locus for a t-region in a cell C and $pt(C)$ a detour edge. By construction, $\tilde{T}$ contains all the edges in T and some detour edges. Let e be a detour edge from a point p to $pt(C)$ for some cell C. If $2box(C)$ contains the tool change center, then $pt(C)$ is the tool change center. Then we charge e to the entire T. Suppose that $2box(C)$ does not contain the tool change center. In T, after leaving the point p, the tour has to leave $2box(C)$ eventually. Let ρ be the path in T starting at p and ending at the boundary of $2box(C)$. The length of ρ is $\Omega(size(C))$. We charge the length of e in $\tilde{T}$ to line segments on ρ proportional to their lengths. By smoothness, it is not hard to see that only a constant number of boxes can have detour edges charged to ρ. The result then follows since there is a constant number of detour edges for each box. □

For each cell $C \in$ *large_cell*(t), we associate a table *table*($pt(C)$) with $pt(C)$ to reflect all the possible effects of small tool BMAs by t in C and the corresponding cost. We enumerate all possible unmilled regions in C and for each unmilled region, we enumerate all possible combinations of small tool BMAs on unmilled components. Note that each unmilled component induces at most a constant number of small tool BMAs. For each such combination of small tool BMAs by t, we find out the number of faces covered and the cost (sum of costs of the small tool BMAs, MST length connecting $pt(C)$ and the t-loci, and the placement cost). The table *table*($pt(C)$) has $O((kN)^{O(1)}2^{O(1)})$ entries.

Separating large tool BMAs. After collapsing small tool BMAs by a tool t, we have one representative point for each cell $C \in$ *large_cell*(t). We now address the overlapping among t-regions of large tool BMAs. We divide such t-regions into groups so that within a group, no two t-regions overlap. The division is done by coloring an induced graph as follows. Consider the grid squares that induce the large tool BMAs by t. There is a constant number of large tool BMAs induced by each grid square. Since each grid square has width $\Omega(t)$, the t-region of a large tool BMA cannot overlap more than a constant number of grid squares. Thus, the t-region of a large tool BMA overlaps at most a constant number of t-regions of other large tool BMAs. This induces a graph of maximum degree bounded by a constant Δ. Such a graph is colorable using at most $\Delta + 1$ colors which yields at most $\Delta + 1$ groups of t-regions. We fix a point p in each t-region M and associate with it a table *table*(p) of a single entry, namely, the faces covered by M and cost (cost of the corresponding large tool BMA plus placement cost). Thus, we obtain at most $\Delta + 1$ groups of points each associated with a table. We denote these groups by $G_i(t)$ for $1 \le i \le \Delta + 1$. Finally, we add one last group $G_{\Delta+2}(t)$ which contains $pt(C)$ along with *table*($pt(C)$) for each cell $C \in$ *large_cell*(t).

Lemma 6.2 *There is a milling plan $\mathcal{A}''$ in which each tour of a tool t beginning and ending at the tool change center visits only points in some $G_i(t)$. Moreover, $\mathcal{A}''$ approximates the optimal milling plan within a constant factor.*

The complete algorithm. The following is the complete algorithm. Recall that F_t is the number of faces currently in S covered by some t-regions.

```
while S ≠ ∅ do
  for each tool t do
    for 1 ≤ m ≤ F_t do
      for each G_i(t) do
        solve the m-WTSP on G_i(t)
      end for
      T_m(t) = the minimum weight tour;
```

end for
$T(t)$ = minimum average weight $T_m(t)$;
end for
T = minimum average weight $T(t)$;
remove faces covered in T from S;
update F_t for all t;
update the numbers of faces in all tables;
end while

Theorem 6.1 *An approximate milling plan can be computed in $O((kN)^{O(1)})$ time with cost within $O(\log kN)$ of the optimal.*

Acknowledgments

We thank Ajay Joneja of the Industrial Engineering Department at the Hong Kong University of Science and Technology for suggesting the multiple-tool milling problem and for valuable discussions on the practical aspects of milling.

References

[AFM93] E. Arkin, S. Fekete, and J. Mitchell. The lawnmower problem. In *Proc. 5th Canad. Conf. Comput. Geom.*, pages 461–466, 1993.

[AFM95] E. Arkin, S. Fekete, and J. Mitchell. Approximation algorithms for lawnmowing and milling. Report, SUNY Stony Brook, 1995.

[AHS96] Esther M. Arkin, Martin Held, and Christopher L. Smith. Optimization problems related to zigzag pocket machining. In *Proc. 7th ACM-SIAM Sympos. Discrete Algorithms*, pages 419–428, 1996.

[Aro96] S. Arora. Polynomial time approximation schemes for Euclidean TSP and other geometric problems. In *Proc. 37th Annu. IEEE Sympos. Found. Comput. Sci.*, pages 2–11, 1996.

[Aro97] S. Arora. Nearly linear time approximation schemes for Euclidean TSP and other geometric problems. In *Proc. 38th Annu. IEEE Sympos. Found. Comput. Sci.*, 1997.

[BC91] M. Bala and T.-C. Chang. Automatic cutter selection and optimal cutter path generation for prismatic parts. *Int. J. Prod. Res*, 29:2163–2176, 1991.

[Bru82] L. K. Bruckner. Geometric algorithms for 2.5d roughing process of sculptured surfaces. In *Proc. Joint Anglo-Hungarian Seminar on Computer-Aided Geometric Design*, 1982.

[DBW93] A. V. Deshmukh, M. M. Barash, and H. P. Wang. On selection of tool path orientations for generating prismatic features. Report, IE Dept, Purdue U, 1993.

[HA92] A. Hansen and F. Arbab. An algorithm for generating NC tool paths for arbitrarily shaped pockets with islands. *ACM Trans. Graph.*, 11(2):152–182, April 1992.

[Hel89] M. Held. Geopocket – A sophisticated computational geometry solution of geometrical and technological problems arising from pocket machining. In F. Kimura and A. Rolstadås, editors, *Proc. Computer Applications in Production and Engineering*, pages 283–293. North-Holland, October 1989.

[Hel91a] M. Held. A geometry-based investigation of the tool path generation for zigzag pocket machining. *Visual Comput.*, 7(5–6):296–308, September 1991.

[Hel91b] M. Held. *On the Computational Geometry of Pocket Machining*, volume 500 of *Lecture Notes Comput. Sci.* Springer-Verlag, June 1991.

[IRT94] K. Iwano, P. Raghavan, and H. Tamaki. The traveling cameraman problem, with applications to automatic optical inspection. In *Proc. 5th Annu. Internat. Sympos. Algorithms Comput.*, Lecture Notes Comput. Sci., Beijing, 1994. Springer-Verlag.

[Kra92] T. R. Kramer. Pocket milling with tool engagement detection. *J. of Maufacturing Systems*, 11(2):114–123, 1992.

[Mit96] Joseph S. B. Mitchell. Guillotine subdivisions approximate polygonal subdivisions: A simple new method for the geometric k-MST problem. In *Proc. 7th ACM-SIAM Sympos. Discrete Algorithms*, pages 402–408, 1996.

[MMS94] J. S. B. Mitchell, D. M. Mount, and S. Suri. Query-sensitive ray shooting. In *Proc. 10th Annu. ACM Sympos. Comput. Geom.*, pages 359–368, 1994.

[Per78] H. Persson. NC machining of arbitrarily shaped pockets. *Comput. Aided Design*, 10(3):169–174, May 1978.

[PGW90] P. V. Prabhu, A. K. Gramopadhye, and H. P. Wang. General mathematical model to optimizing NC tool path for face milling of flat convex polygonal surfaces. *Int. J. Production Research*, 28(1):101–130, 1990.

[PK83] K. Preiss and E. Kaplansky. Automatic mill routing from solid geometry information. In *Proc. Computer Applications in Production and Engineering.* North-Holland, 1983.

[PK85] K. Preiss and E. Kaplansky. Automated CNC milling by artificial intelligence methods. *J. of Maufacturing Systems*, 4(1):51–63, 1985.

[Pre89] K. Preiss. Automated mill pocketing computations. In *Advanced Geometric Modeling for Engineering Applications.* North-Holland, 1989.

[SL90] Y. S. Suh and K. Lee. NC milling tool path generation for arbitrary pockets defined by sculptured surfaces. *Comput. Aided Design*, 22(5):273–284, June 1990.

[W+87] H.-P. Wang et al. On the efficiency of NC tool path planning for face milling operations. *Trans. of the ASME, J. of Engineering for Industry*, 109(4):370–376, November 1987.

Resource-Constrained Geometric Network Optimization

Esther M. Arkin* Joseph S. B. Mitchell† Giri Narasimhan‡

Abstract

We study a variety of geometric network optimization problems on a set of points, in which we are given a resource bound, B, on the total length of the network, and our objective is to maximize the number of points visited (or the total "value" of points visited). In particular, we resolve the well-publicized open problem on the approximability of the *rooted* "orienteering problem" for the case in which the sites are given as points in the plane and the network required is a cycle. We obtain a 2-approximation for this problem. We also obtain approximation algorithms for variants of this problem in which the network required is a tree (3-approximation) or a path (2-approximation). No prior approximation bounds were known for any of these problems.

We also obtain improved approximation algorithms for geometric instances of the unrooted orienteering problem, where we obtain a 2-approximation for both the cycle and tree versions of the problem on points in the plane, as well as a 5-approximation for the tree version in edge-weighted graphs. Further, we study generalizations of the basic orienteering problem, to the case of multiple roots, sites that are polygonal regions, etc., where we again give the first known approximation results.

Our methods are based on some new tools which may be of interest in their own right: (1) some new results on m-guillotine subdivisions in the plane, strengthening the original approximation bound, and (2) some new combinatorial results on partitioning trees.

Our study helps to highlight where geometry "helps" in being able to approximate these hard combinatorial optimization problems.

*Department of Applied Mathematics and Statistics, State University of New York, Stony Brook, NY 11794-3600; estie@ams.sunysb.edu. Partially supported by NSF grant CCR-9504192.

†Department of Applied Mathematics and Statistics, State University of New York, Stony Brook, NY 11794-3600; jsbm@ams.sunysb.edu. This work was partially conducted while J. Mitchell was supported as a Fulbright Scholar at Tel Aviv University. J. Mitchell is also partially supported by NSF grant CCR-9504192, and by grants from Boeing Computer Services, Bridgeport Machines, Hughes Aircraft, and Sun Microsystems.

‡Department of Mathematical Sciences, University of Memphis, Memphis, TN 38152; giri@msci.memphis.edu. Partially funded by NSF grant CCR-940-9752 and by a grant from Cadence Design Systems. Much of this work was conducted while this author was visiting the University at Stony Brook.

SCG 98 Minneapolis Minnesota USA

1 Introduction

Consider a traveling salesperson who has a given amount of gasoline (allowing travel up to distance B) and wants to maximize the number of sites visited before running out of gasoline. This problem is one of a class of *orienteering* problems that require us to design a network that visits a maximum number of nodes (sites), subject to an upper bound on the total length of the network.

In contrast with the classical traveling salesperson problem (TSP), which asks for a minimum-length cycle visiting *all* sites, in the orienteering problem there is a resource constraint on the length of the cycle and the problem is to do the "best possible" in the sense of maximizing the number of sites visited. In a sense, the orienteering problem is dual to the problem of minimizing the length of a cycle, subject to meeting a quota on the number of sites visited (in the "k-TSP" version) or the sum of the values of the sites visited (the "quota-driven salesperson" problem).

The orienteering problem has also been called the "bank robber" problem (a thief wishes to maximize the "haul", with a single tank of gas in the get-away car) and the "generalized traveling salesperson problem" [16, 30]. The name "orienteering" comes from the outdoor sport by the same name, in which each player is given a compass and a map of a given outdoor terrain, and has the goal to accumulate the largest score within a fixed amount of time. Players set out from a given root (the "home base"), and obtain a certain score (value) by visiting a subset of the n sites ("control points") before returning to the root.

In this paper, we distinguish between different versions of the orienteering problem according to the type of network that is to be constructed: In the *tree-orienteering* (*path-orienteering*, *cycle-orienteering*) problem, we must construct a tree (path, cycle, resp.) visiting a subset of a given set of n *sites*, subject to an upper bound B on the total length of the network. Additionally, we may have one or more *roots*, which are "essential" sites that we require the network to visit.

The *rooted* orienteering problem has, for several years, been a particularly challenging problem from the point of view of approximation algorithms — in fact, prior to our work, *no* approximation algorithms have been known for the problem, either in graphs or in geometry. Here, we provide *constant*-factor approximation algorithms for the geometric instances of the problem, in which the sites are given as points in the plane.

Summary of Main Results:

1. We give a 2-approximation algorithm for the *rooted* cycle- and path-orienteering problems on a set of points in the plane. This result is the first of its kind, as no approximation algorithms were previously known.
2. We give a 3-approximation algorithm for the rooted tree-orienteering problem, for points in the plane. Again, no prior approximation algorithms were known.
3. We give improved approximation algorithms for the unrooted instances of the orienteering problem. These results include a 2-approximation for the geometric path- and cycle-orienteering, a 5-approximation for tree-orienteering in edge-weighted graphs.
4. We study the *multiply-rooted* versions of the orienteering problems in the plane, providing both an approximation algorithm when the number of roots is a constant, and a proof of hardness of approximability when the number of roots is part of the input.
5. We study geometric instances of the orienteering problems in which the sites are given as a set of *regions* (polygons) in the plane. We give $O(\log n)$-approximation algorithms for these versions.

Key to our methods are some new tools which may be of independent interest:

(1) We obtain new results on approximating an arbitrary connected planar subdivision with an m-guillotine subdivision, strengthening the original analysis given in Mitchell [23]; these new theorems seem to be crucial in addressing the types of approximation problems that arise in the rooted cases, and they may be more widely applicable.

(2) We obtain a variety of new combinatorial results on partitioning trees; these lemmas are critical in obtaining some of our improved bounds on the unrooted versions of the orienteering problems, as well as in analyzing the generalizations to non-point sites.

Related Work

There has been a wealth of work on the related traveling salesperson problem (TSP), both in networks and in geometric settings [9, 18, 29, 17]. TSP is NP-hard, even for points in the Euclidean plane. Until recently, the best approximation algorithm known for the Euclidean TSP was the Christofides heuristic (see, e.g., [18]), which yields a 1.5-approximation for the more general case of undirected networks whose edge lengths obey the triangle inequality. Polynomial-time approximation schemes (PTAS) for geometric (e.g., Euclidean) instances were discovered recently by Arora [4, 5] and by Mitchell [21, 23, 24]; most recently, Rao and Smith [27] give a deterministic PTAS with running time $O(n \log n)$ in any fixed dimension. The methods of [21, 23] employ "m-guillotine subdivisions" in the plane; this tool is also essential for some of our results, which are based on a strengthening of some of the analysis in [21, 23].

A related problem is the k-MST problem, in which we are given a graph on n vertices with nonnegative distances on the edges, and an integer $k \leq n$, and our goal is to find a tree of least total weight that spans some subset of k vertices. The problem is known to be NP-hard, both in general graphs and in the Euclidean plane [12, 31, 28]. The current best approximation algorithm for general edge-weighted graphs is a 3-approximation by Garg [14], which applies also to the "rooted" case (the tree is required to include a given node); this has been improved to a 2.5-approximation, by Arya and Ramesh [6], if the tree is not "rooted." The Euclidean k-MST now has a PTAS [4, 5, 21, 23], as does the k-TSP, which asks for a shortest cycle visiting some subset of k points. For the graph version of the k-TSP, with edge weights obeying triangle inequality, Garg's method yields a 3-approximation for both the rooted and unrooted version; the result of Arya and Ramesh [6] does not apply.

Another related problem is the *quota-driven salesperson* problem, in which each site has an associated integral value, w_i, and a salesperson has a given integer quota, R. The objective is to find a shortest possible cycle such that the sum of the values for the sites visited is at least R. By replicating each site (w_i times), a k-MST approximation algorithm also gives an approximation for this problem. In the *prize-collecting salesman* problem, as studied by Balas [8] (see also [10]), the setup is the same as in the quota-driven salesman problem, except that, in addition to "values" w_i, there are non-negative penalties associated with each site, and now the objective is to minimize the sum of the distances traveled *plus* the sum of the penalties on the points *not* visited, subject to satisfying the quota R. As discussed in [7], an approximation algorithm follows from concatenating a cycle obtained for the quota-driven salesman, with the 2-approximation cycle given by the algorithm of Goemans and Williamson [15] (which considers the effect of penalties, but does not use the quota constraint).

The class of problems we address in this paper is based on the *orienteering problem* (or *bank robber* problem). The problem is essentially the dual of the quota-driven salesman problem: Given an upper bound on distance, B, maximize the total value of all points visited, for a salesperson that departs from a given root site. Heuristics were provided (without approximation guarantees) by [16, 30]. Awerbuch et al. [7] give an approximation algorithm for the *unrooted* cycle case, based on approximations to the k-TSP. These previous bounds are discussed in more detail later, as we compare them to our own results. For the original (rooted) version of the problem, it has been an intriguing open problem to obtain *any* nontrivial approximation algorithm.

Our work is also related to problems in bicriteria path optimization, in which we are to minimize one measure of "length", subject to an upper bound on another notion of "length"; see [2, 3, 11, 25], as well as the surveys by Mitchell [20, 22].

Preliminaries

We assume that the input to our problems is given by a set of n points (*sites*) in the Euclidean plane. For the path and cycle cases, we also consider a discrete metric space (an edge-weighted graph whose weights obey the triangle inequality), while for the tree case, we consider graphs with general (nonnegative) edge weights. We use the terms "length" and "weight" interchangeably. We let $d(p, q)$ denote the distance between p and q.

We let $n(T)$ and $w(T)$ denote the number of sites visited by network T and the total weight of T, respectively.

We let $B > 0$ be the given resource bound; B is an upper

bound on the total length of the allowed network (tree, path, or cycle).

A *root* is a site that is required to be covered by the network.

Sites may have associated weights ("values"). Assuming these weights are integral, we can replicate point sites, and thereby reduce the problem to the unweighted case, in which our objective is to maximize the *number* of sites visited by the network. (This approach results in time bounds that are pseudopolynomial, in n and the sum of the weights.) Thus, we focus on the unweighted case in the remainder of this paper.

In this paper, all algorithms are *polynomial-time* approximation algorithms, whose worst-case complexity is polynomial in n. We omit the term "polynomial-time," as well as the actual time bounds, in our statements of results. A c-approximation algorithm for a maximization (resp., minimization) problem is one which guarantees that the solution obtained has an objective value that is at least OPT/c (resp., at most $c \cdot OPT$), where OPT is the optimal value of the objective function.

We will have use of the definition of an "m-guillotine subdivision" from [23]; roughly speaking, an "m-guillotine subdivision" is a polygonal subdivision with the property that there exists a line ("cut"), whose intersection with the subdivision edges consists of a small number ($O(m)$) of connected components, and the subdivisions on either side of the line are also m-guillotine. The upper bound on the number of connected components allows one to apply dynamic programming to optimize over m-guillotine subdivisions, as there is a succinct specification of how subproblems interact across a cut.

More precisely, we consider a polygonal subdivision ("planar straight-line graph") S that has n edges (and hence $O(n)$ vertices and facets). Let E denote the union of the edge segments of S, and let V denote the vertices of S. We can assume (without loss of generality) that S (E) is restricted to the unit square.

In the following definitions, we fix attention on a given *window*, W, which is simply a closed, axis-aligned rectangle, with W in the unit square. We let $\overline{W} \subseteq W$ denote the minimal bounding box (axis-aligned rectangle) containing the vertices $V \cap W$ within W. A horizontal or vertical line ℓ is a *cut* for E (with respect to W) if $\ell \cap int(W) \neq \emptyset$. The intersection, $\ell \cap (E \cap int(W))$, of a cut ℓ with $E \cap int(W)$ (the restriction of E to the window W) consists of a discrete (possibly empty) set of subsegments of ℓ. (Some of these "segments" may be single points, where ℓ crosses an edge.) The ξ endpoints of these subsegments are called the *endpoints along* ℓ (with respect to W) and are denoted by $p_1, \ldots, p_\xi$, in order along ℓ.

For a positive integer m, we define the *m-span*, $\sigma_m(\ell)$, of ℓ (with respect to W) as follows. If $\xi \leq 2(m-1)$, then $\sigma_m(\ell) = \emptyset$; otherwise, $\sigma_m(\ell)$ is defined to be the (possibly zero-length) line segment, $p_m p_{\xi-m+1}$, joining the mth endpoint, p_m, with the mth-from-the-last endpoints, $p_{\xi-m+1}$.

A (horizontal or vertical) cut ℓ is an *m-perfect cut with respect to* W if $\sigma_m(\ell) \subseteq E$. In particular, if $\xi \leq 2(m-1)$, then ℓ is trivially an m-perfect cut (since $\sigma_m(\ell) = \emptyset$). Similarly, if $\xi = 2m-1$, then ℓ is m-perfect (since $\sigma_m(\ell)$ is a single point). Otherwise, if ℓ is m-perfect, and $\xi \geq 2m$, then $\xi = 2m$.

Finally, we say that S is an *m-guillotine subdivision with respect to window* W if either (1) $V \cap int(W) = \emptyset$; or (2) there exists an m-perfect cut, ℓ, with respect to the minimal window, $\overline{W} \subseteq W$, such that S is m-guillotine with respect to windows $W \cap H^+$ *and* $W \cap H^-$, where H^+, H^- are the closed halfplanes induced by ℓ. We say that S is an *m-guillotine subdivision* if S is m-guillotine with respect to the unit square.

2 Rooted Orienteering Problems for Points in the Plane

We begin by considering the *rooted* case of the orienteering problems, in which one particular site, r, is given, and the tree/path/cycle is constrained to contain r.

The effect of having a root is considerable, as far as approximation algorithms are concerned. The difficulty stems from the fact that an optimal network may visit a huge number of sites that lie in a very small cluster at a distance from r that is nearly B, making it difficult to visit at least a constant fraction of these sites, unless the network is very efficient in its use of length.

In fact, no previous approximation algorithms, even with a logarithmic approximation factor, were known for rooted orienteering problems. Here, by exploiting geometric structure, and tightening some analysis of m-guillotine subdivision approximations, we are able to obtain constant-factor approximations.

2.1 Rooted Tree-Orienteering Problem

We obtain a 3-approximation for points in the plane as follows.

(1) For each value of k, $k = 1, \ldots, n$, we find a shortest possible connected m-guillotine subdivision, rooted at r, spanning k or more points. We tabulate the lengths of the resulting graphs, for each k.

(2) For each value of k, $k = 1, \ldots, n$, and for each vertex v, we find a shortest possible connected m-guillotine subdivision, rooted at v, spanning k or more points; if the resulting graph does not contain r, then we add the vertex r and the edge rv to the graph. We tabulate the lengths of the resulting graphs, for each k.

(3) Among the graphs tabulated, the algorithm selects a graph with the maximum number of points and whose total length is less than the specified bound B. The shortest tree spanning the sites in this subgraph is then output.

In steps (1) and (2), the algorithm tries every possible value of k, since the number of vertices spanned by an optimal solution to the tree-orienteering problem is not known. Also, since any tree spanning k vertices can be transformed into an m-guillotine subdivision spanning k vertices with only a small increase in total length ([23]), in step (1), the algorithm searches for a shortest connected m-guillotine subdivision spanning k vertices including the root r. Step (2) is not an obvious step; the enumeration in this step is needed, as shown in the proof below. The goal of this step is to search for additional "good" trees that span k (or $k+1$) vertices, including the root. This is achieved by searching for trees rooted at an arbitrary vertex v (for all possible vertices v) and spanning k vertices and then adding the edge rv to it. Since such trees can also be transformed into m-guillotine subdivisions spanning the same set of vertices, the algorithm looks for a shortest subdivision spanning k vertices including vertex v and then adds edge rv, if neccessary. Step (3) is a straightforward step that invloves finding a shortest connected subgraph (i.e., tree) spanning the sites in the best m-guillotine subdivision found in one of the earlier steps.

The general approach used above is also used in other algorithms that follow in this paper. The approach consists of three steps: (a). showing that any optimal geometric network satisfying a property $\mathcal{P}$ can be transformed into a m-guillotine subdivision that satisfies property $\mathcal{Q}$; (b). applying dynamic programming to compute in polynomial time a shortest m-guillotine subdivision satisfying property $\mathcal{Q}$; and then (c). showing that any m-guillotine subdivision satisfying property $\mathcal{Q}$ can be mapped back into a network satisfying property $\mathcal{P}$. In the above rooted tree-orienteering problem, property $\mathcal{P}$ is that the network is a tree spanning k vertices and having total length at most B, while property $\mathcal{Q}$ is that the network is a connected m-guillotine subdivision spanning k vertices.

Finally, we note that, in the above algorithm, it will suffice to set $m = 2$. Let T^* denote an optimal rooted tree, spanning the maximum number, $n(T^*)$, of points, while having weight at most B. Now, we claim that:

Lemma 1 *Among the graphs tabulated of types (1) and (2), there must be one whose weight is less than B, while the cardinality, k, of the point set spanned is at least $n(T^*)/3$.*

Proof. By standard results on separators in trees (e.g., see [26], Theorem 9.1, page 150), we know that there exists a partitioning of the optimal tree T^* into two subtrees, T_1 and T_2, sharing a common vertex v of T^*, so that $(2/3)n(T^*) \geq n(T_i) \geq (1/3)n(T^*)$. Without loss of generality, we assume that $r \in T_1$. We now consider two cases:

(i) Case: $w(T_1) - d(r,v) > w(T_2)$

Let T' be the union of $T_{2,G}$ and the segment rv, where $T_{2,G}$ is an m-guillotine subdivision containing T_2. By Theorem 3.3 of Mitchell [23], we know that there exists an m-guillotine subdivision, $T_{2,G}$ so that

$$w(T_{2,G}) \leq (1+\epsilon)w(T_2),$$

where $\epsilon = \frac{\sqrt{2}}{m}$, which implies that

$$\begin{aligned} w(T') &\leq (1+\epsilon)w(T_2) + d(r,v) \\ &< \epsilon w(T_2) + w(T_1) \\ &< w(T_2) + w(T_1) \\ &= w(T^*) \leq B \end{aligned}$$

where we assume that $m \geq 2$, so that $\epsilon < 1$. Thus, T' will be among the subdivisions considered in case (2) of the algorithm: The algorithm will enumerate a tree having the same number of vertices as T', while containing the root r and having weight at most $w(T')$, which, as shown above, is at most B.

(ii) Case: $w(T_1) - d(r,v) \leq w(T_2)$

Let $T'' = T_{1,G}$ be an m-guillotine subdivision containing T_1, and let $\epsilon = \frac{\sqrt{2}}{m}$. Again, by Theorem 3.3 of Mitchell [23], we know that there exists an m-guillotine subdivision, $T_{1,G}$ so that

$$w(T_{1,G}) \leq (1+\epsilon)w(T_1).$$

However, it turns out that the above inequality is not strong enough for our purposes here. Theorem 4 (see Section 2.4 below) is a strengthening of Theorem 3.3 of Mitchell [23]. Corollary 1 of Theorem 4 proves the following inequality:

$$w(T_{1,G}) \leq d(r,v) + (1+\epsilon)(w(T_1) - d(r,v)).$$

With this result, we then obtain

$$\begin{aligned} w(T'') = w(T_{1,G}) &\leq d(r,v) + (1+\epsilon)(w(T_1) - d(r,v)) \\ &\leq d(r,v) + 2(w(T_1) - d(r,v)) \\ &= w(T_1) + w(T_1) - d(r,v) \\ &\leq w(T_1) + w(T_2) = w(T^*) \leq B, \end{aligned}$$

where we have again assumed that $m \geq 2$, so that $\epsilon < 1$. Thus, T'' will be among the trees of type (1) considered by the algorithm.

In conclusion, both T' and T'' are among the trees considered in cases (1) and (2) of the algorithm; thus, the algorithm enumerates trees having at least $n(T^*)/3$, while having weight at most B. □

Theorem 1 *There is a 3-approximation polynomial-time algorithm for the rooted tree-orienteering problem on a set of points in the plane.*

2.2 Rooted Path-Orienteering Problem

The algorithm for the path-orienteering problem is very similar to the one for the tree-orienteering problem. The main difference is that instead of looking for connected m-guillotine subdivisions in the first two steps, the algorithm looks for connected m-guillotine subdivisions from which a path induced on the same set of vertices can be extracted. The proof of the approximation guarantee is also similar, except for one essential difference: there exists a partitioning of the optimal path T^* into two subpaths, sharing a common vertex, each having at least $n(T^*)/2$ vertices. This enables us to prove a performance guarantee of 2 for rooted path-orienteering. We omit details here.

Theorem 2 *There is a 2-approximation polynomial-time algorithm for the rooted path-orienteering problem on a set of points in the plane.*

2.3 Rooted Cycle-Orienteering Problem

We obtain a 2-approximation for points in the plane as follows. The algorithm computes for each value of k, $k = 1, \ldots, n$, a shortest possible connected "bridge-doubled" m-guillotine subdivision that includes the root vertex r and at least $k-1$ other points, and where every vertex that is not on a bridge has degree two. It then extracts a tour from each such subdivision, using the property that bridge-doubling allows one to obtain a subset of the edges which is Eulerian (as outlined in Section 4.2 of Mitchell [23]). The algorithm tabulates the lengths of the resulting tours for each k, and among the tours with length bounded by B, the algorithm selects one having the maximum number of vertices.

In the algorithm, it will suffice to set $m \geq 3$ (so that $\epsilon = 2\sqrt{2}/m < 1$). Let T^* denote an optimal tour visiting the maximum number, $n(T^*)$, points, while having weight at most B. Now, we claim that:

Lemma 2 *Among the tours tabulated by the algorithm described above, there must be one whose weight is less than B, while the cardinality, k, of the point set visited is at least $n(T^*)/2$.*

Proof. The length, $w(T^*)$, of an optimal tour, T^*, is bounded by B. Let v be a point on the tour (not necessarily a vertex) such that the points r and v partition the tour into two paths of weight $w(T^*)/2$. Let the two paths be called

T_1 and T_2. Without loss of generality, assume that the path T_1 has at least $n(T^*)/2$ vertices on it.

Let T' be the union of T_1 and the segment rv. Let T'_G be a (bridge-doubled) m-guillotine subdivision containing T'. Once again, we need a stronger version of Theorem 3.3 from Mitchell [23]. Here, since T' contains a tour on its induced vertices, we apply Corollary 1 from Section 2.4 below, which provides the appropriate strengthening. Thus, we have:

$$\begin{aligned} w(T'_G) &\leq w(T') + \epsilon(w(T') - 2d(r,v)) \\ &= w(T_1) + d(r,v) + \epsilon(w(T_1) - d(r,v)) \\ &= (1+\epsilon)w(T_1) + (1-\epsilon)d(r,v) \\ &\leq (1+\epsilon)w(T_1) + (1-\epsilon)w(T_1) = 2w(T_1) = B \end{aligned}$$

Although v may not be a vertex, it is clear that there exists a m-guillotine subdivision T''_G that contains the same set of vertices as T'_G and with length no more than that of T'_G. Thus, T''_G will be among the tours considered by the algorithm; we will obtain a tour visiting as many vertices as T''_G, while having weight at most B. □

Theorem 3 *There is a 2-approximation polynomial-time algorithm for the rooted cycle-orienteering problem on a set of points in the plane.*

2.4 A Strengthened m-Guillotine Subdivision Theorem

We now supply a theorem on m-guillotine subdivisions, which strengthens the original analysis in [23]. The crucial new aspect to the theorem is that, when converting an arbitrary subdivision into an m-guillotine subdivision, there is some portion of the length of the original subdivision that does not need to be "charged." This allows us, in the corollary below, to bound the increase in length (between an original subdivision and its guillotine version) by a small fraction of the *difference* between the length of the original subdivision and the length of an appropriate subgraph that is known to exist in the subdivision.

First, we need some notation. We consider a line segment pq to have two *sides* — a top side, and a bottom side. For a set E of line segments, we say that the top side of a segment pq on an edge of E is *i-exposed above* if there are exactly i connected components in the intersection of E and the upward-pointing ray from any point on pq. (The ray includes its endpoint, so that the intersection with pq is counted among the i components.) See Figure 2.4 for an example of a portion of E that is 2-exposed above. If there are exactly i connected components in the intersection of E and the rightward-pointing ray from any point on pq, then the top (resp., bottom) side of pq is said to be *i-exposed right*, if pq has negative (resp., positive) slope. We similarly define i-exposed *below* and i-exposed *left*. Let $\xi^{(m)}(E)$ be the sum of the lengths of all the sides of subsegments of E that are i-exposed in some direction (above, below, left, or right), for some $i \leq m$. Since we distinguish between the sides of a segment, it is possible that the length of pq is counted twice in the total $\xi^{(m)}(E)$ (e.g., if the top side of pq is i-exposed above *and* the bottom side is i'-exposed below, for $i, i' \leq m$).

Theorem 4 *Let S be a polygonal subdivision, with edge set E, of length L. Then, for any positive integer m, there exists an m-guillotine subdivision, S_G, of length at most $L + \frac{\sqrt{2}}{m}(L - \frac{\xi^{(m)}(E)}{2})$ whose edge set, E_G, contains E.*

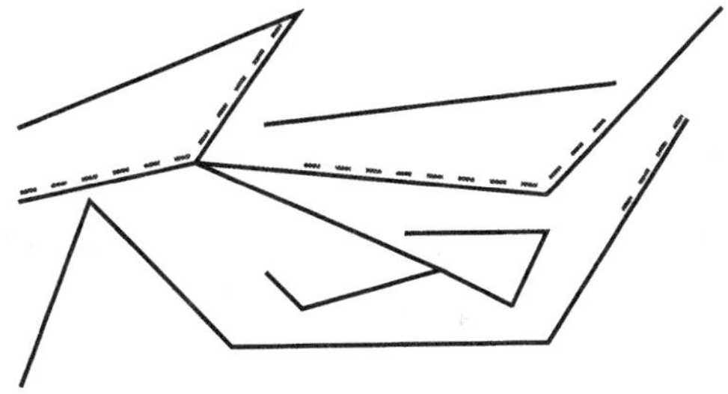

Figure 1: The portion of the tops of the edge set E that is 2-exposed above is highlighted.

Proof. (Sketch) We follow much of the proof of Mitchell [23], while making use of the fact that the portion of E that is within m levels of the top (resp., bottom, left, right) of the boundary box of E is never "charged" from above (resp., from the bottom, left, right). Thus, of the total side length $2L$ for E, there is side length $\xi^{(m)}(E)$ that never gets charged. Hence, the total length of all bridges constructed in the proof of [23] is at most $\frac{\sqrt{2}}{2m}(2L - \xi^{(m)}(E))$. □

The following corollary is required in proofs from Section 2.1, Section 2.2, and Section 2.3 above.

Corollary 1 *Let S be a polygonal subdivision, with edge set E, of length L. If S connects two vertices r and v in the subdivision, then, for any positive integer m, there exists an m-guillotine subdivision, S_G, whose edge set contains E, and is of length at most $L + \frac{\sqrt{2}}{m}(L - d(r,v))$. Furthermore, if S contains a cycle through the vertices r and v, then, for any positive integer $m \geq 2$, there exists an m-guillotine subdivision, S_G, whose edge set contains E, and is of length at most $L + \frac{\sqrt{2}}{m}(L - 2d(r,v))$ (or, $L + \frac{2\sqrt{2}}{m}(L - 2d(r,v))$, in the case of bridge-doubled m-guillotine subdivisions)*

Proof. (Sketch) The two statements in the corollary follow from Theorem 4, after showing the following: if S connects vertices r and v, then $\xi^{(m)}(E) \geq \xi^{(1)}(E) \geq 2d(r,v)$; if S contains a cycle involving the vertices r and v, and $m \geq 2$, then $\xi^{(m)}(E) \geq \xi^{(2)}(E) \geq 4d(r,v)$. In particular, it is not hard to show that for a simple polygon P, $\xi^{(1)}(P)$ is at least as large as the perimeter of the convex hull of P. □

3 Unrooted Orienteering Problems

The following observation was made in [7], for the cycle-orienteering problem. A similar proof holds also for the path and tree versions:

Theorem 5 *Given a c-approximation for the k-TSP we can obtain a $2c$-approximation for the unrooted orienteering problem (tree-, path-, or cycle-orienteering).*

Proof. We first prove the cycle-orienteering bound: For all $k = 1, 2, \ldots, n$ find T_k an approximate k-TSP. Let k be the number of points such that $T_k \leq cB$ and $T_{k+1} > cB$. Take the cycle T_k and break it into $2c$ pieces of length at most $B/2$ each. Take the piece that contains the most points, double it to create a cycle of length at most B containing at least $k/(2c)$ points. Note that since $T_{k+1} > cB$ we must have that the optimal cycle of $k+1$ points $T^*_{k+1} > B$, therefore the optimal solution to the orienteering problem with bound B is k.

The proof for the path- and tree-orienteering problems is similar: For all $k = 1, 2, \ldots, n$ find T_k an approximate k-TSP. Let k be the number of points such that $T_k \leq 2cB$ and $T_{k+1} > 2cB$. Take the cycle T_k and break it into $2c$ pieces of length at most B each. Take the piece that contains the most points, it is a path, therefore also a tree of length at most B containing at least $k/(2c)$ points. Note that since $T_{k+1} > 2cB$ we must have that the optimal cycle of $k+1$ points $T^*_{k+1} > 2B$, and hence the best path and the best tree on $k+1$ points are of length $> B$, and therefore the optimal solutions to the path- and tree-orienteering problems with bound B is k. □

Remark. We note that the method in the proof of the above theorem does not apply to the *rooted* cases.

Garg's [14] $c = 3$ bound for metric spaces for the k-TSP thus yields an approximation factor 6 for the unrooted tree-, path-, and cycle-orienteering problems in metric spaces. The PTAS of Mitchell [23] or Arora [4] imply $c = 1 + \epsilon$, and therefore an approximation factor $2 + \epsilon$ for the unrooted tree-, path-, and cycle-orienteering problems for points in the plane. Next, we improve these results.

3.1 Unrooted Tree-Orienteering

We begin with a result on the geometric version of the problem:

Theorem 6 *There is a 2-approximation polynomial-time algorithm for the unrooted tree-orienteering problem on a set of points in the plane.*

Proof. Our algorithm begins by computing a $(1+\epsilon)$-approximate k-MST, for each of the n possible values of k ($k = 1, 2, \ldots, n$); this can be done in polynomial time, using the algorithms of Mitchell [23] or Arora [4]. Then, among these n trees, we simply output the maximum cardinality (maximum k) tree whose weight is at most B.

We now argue that this algorithm yields a tree having at least $\lceil k^*/2 \rceil$ vertices, where k^* is the optimal value of k (the maximum number of points that can be spanned with a tree of weight at most B).

For each value of $k = 1, 2, \ldots, n$, let B_k be the length of an *optimal* k-MST, and let A_k be the length of the approximately optimal tree produced by the algorithm. We know that $A_k \leq B_k(1+\epsilon)$, for each value of k. By Theorem 9 (below), we know that $B_{\lceil k/2 \rceil} \leq (2/3)B_k$. If we choose ϵ such that $\epsilon < 1/2$, then it is clear that $A_{\lceil k/2 \rceil} \leq B_{\lceil k/2 \rceil}(1+\epsilon) \leq (2/3)B_k(1+\epsilon) < B_k$, for all k. Thus, since k^* is the optimal value of k, meaning that $B_{k^*} \leq B < B_{k^*+1}$, we know that $A_{\lceil k^*/2 \rceil} < B_{k^*} \leq B$. This implies that our algorithm will select a tree having a value of k that is *at least* as large as $\lceil k^*/2 \rceil$. □

Next we provide a 5-approximation algorithm for graphs, improving the previous factor of 6, while removing the requirement that the edge lengths satisfy the triangle inequality.

Theorem 7 *There exists a 5-approximation algorithm for the (unrooted) tree-orienteering problem in an edge-weighted graph (even without edge lengths satisfying the triangle inequality).*

Proof. The algorithm remains the same as in the proof of the previous theorem. However, no PTAS is known for this problem. There exists a 2.5-approximate algorithm for the k-MST problem [6].

For each value of $k = 1, 2, \ldots, n$, let B_k be the length of an *optimal* k-MST, and let A_k be the length of the approximately optimal tree produced by the algorithm. We know that $A_k \leq (5/2)B_k$, for each value of k. By Theorem 10 below, with $K = 5/2$, we know that $B_{\lceil k/5 \rceil} \leq (2/5)B_k$. Hence, $A_{\lceil k/5 \rceil} \leq (5/2)B_{\lceil k/5 \rceil} \leq (2/5)(5/2)B_k = B_k$, for all k. Thus, since k^* is the optimal value of k, meaning that $B_{k^*} \leq B < B_{k^*+1}$, we know that $A_{\lceil k^*/5 \rceil} \leq B_{k^*} \leq B$. This implies that our algorithm will select a tree having a value of k that is *at least* as large as $\lceil k^*/5 \rceil$. □

3.2 Unrooted Path- and Cycle-Orienteering Problems

Our improvement to the previously known approximation bounds is made in the geometric case, where, using the results on the *rooted* version, we achieve a factor 2, improving the previous bound of $2 + \epsilon$. We simply apply the rooted result for each choice of root. (We remark that our combinatorial tools (of the next section) do not lead to improved bounds.)

Theorem 8 *There is a 2-approximation polynomial-time algorithm for the unrooted path- and cycle-orienteering problems on a set of points in the plane.*

3.3 Combinatorial Tools

(Some proofs are deferred to the Appendix.)

Lemma 3 (Number-Partition Lemma) *If S is a set of n real numbers $x_1, x_2, \ldots, x_n$ such that their sum is B and $x_i < B/3, \forall i = 1, 2, \ldots, n$, then there exists a partition of S into two subsets, S_1 and S_2, such that $B/3 \leq \sum_{x_j \in S_i} x_j \leq 2B/3, i = 1, 2$.*

Lemma 4 (Tree-Partition Lemma) *Let T be an edge-weighted tree. The following statements are true:*

1. *There exists a partition of the edges of T into an edge e and two subtrees T_1 and T_2 such that e is incident on both the subtrees, and $w(T_i) < w(T)/3$, $i = 1, 2$ (so $w(e) > w(T)/3$); OR*
2. *There exists a partition of the edges of T into two subtrees T_1 and T_2 such that $w(T)/3 \leq w(T_i) \leq 2w(T)/3$, $i = 1, 2$.*

Proof. See Appendix. □

Theorem 9 *Given an edge-weighted tree T spanning n vertices, the following statements are true:*

1. *There exists a subtree T' of T with $w(T') \leq 2w(T)/3$ and spanning at least $n/2$ vertices, AND*
2. *There exists a subtree T' of T with $w(T') \leq w(T)/2$ and spanning at least $n/3$ vertices.*

Proof. See Appendix. □

We generalize Theorem 9 as follows.

Theorem 10 *Given an edge-weighted tree T spanning n vertices, and any $K > 1$, there exists a subtree T' of T with $w(T') \leq w(T)/K$ and spanning at least $n/(2K)$ vertices.*

Proof. Start with the tree T spanning n vertices. Now apply the partition lemma on T. If the first case holds, then among the two subtrees T_1 and T_2 "pick" the subtree with more vertices. If the second case holds, then "pick" the subtree with the larger ratio of number of vertices to the total weight. If the subtree "picked" has weight at most $w(T)/K$, then return it. If the subtree "picked" has weight larger than $2w(T)/K$, then iterate the process with the "picked" subtree. Let the subtree at the end of the j-th iteration be denoted by $T^{(j)}$. Let the above process terminate after $i-1$ iterations and let $T^{(i-1)}$ be the resulting tree with weight at most $2w(T)/K$.

The idea here is that whenever the partition lemma is applied, it produces a subtree whose ratio of number of vertices to the total weight is no less than that at the start of the iteration. Let $T_1^{(j)}$, $T_2^{(j)}$ be the two subtrees of $T^{(j)}$, and assume without loss of generality that $n(T_1^{(j)})/w(T_1^{(j)}) \geq n(T_2^{(j)})/w(T_2^{(j)})$, then by cross multiplying we get

$$\frac{n(T_1^{(j)})}{w(T_1^{(j)})} \geq \frac{n(T_1^{(j)}) + n(T_2^{(j)})}{w(T_1^{(j)}) + w(T_2^{(j)})} = \frac{n(T^{(j)}) + 1}{w(T^{(j)})}.$$

Therefore, since the ratio of the the number of vertices to the weight never decreases, $n(T^{(i-1)}) \geq n \cdot w(T^{(i-1)})/w(T) \geq n/K$.

The last step of the algorithm has two cases. First, if $w(T^{(i-1)}) \leq 3w(T)/(2K)$ then by the first part of Theorem 9, there exists a subtree T^i such that

$$w(T^i) \leq \frac{2w(T^{(i-1)})}{3} \leq \frac{w(T)}{K},$$

with $n(T^i) \geq n(T^{(i-1)})/2 \geq n/(2K)$. On the other hand, if $3w(T)/(2K) < w(T^{(i-1)}) \leq 2w(T)/K$, then by the second part of Theorem 9, there exists a subtree T^i such that

$$w(T^i) \leq \frac{w(T^{(i-1)})}{2} \leq \frac{w(T)}{K},$$

with $n(T^i) \geq n(T^{(i-1)})/3 \geq n/(2K)$. □

In the Appendix (Lemmas 6 and 7), we show that the bounds above are tight for the metric case (for $K > 2$, in Theorem 10).

4 Generalizations and Extensions

4.1 Multiply-Rooted Orienteering

In the multiply-rooted orienteering problems, we are given, in addition to the length bound B, a set R of $\rho \geq 1$ required *root* sites (they must be visited by the network) in addition to the n original sites, S. The goal is to design a network of length at most B that visits all ρ of the root sites and that maximizes the number of non-root sites visited by the network.

Since our results for rooted versions of the orienteering problem are strongly based on geometric methods, we again restrict ourselves to the Euclidean plane.

The multiply-rooted tree-orienteering problem is closely related to the (planar) Euclidean Steiner tree problem, which is known to be NP-hard [13]. In the Steiner tree problem, one is to determine a tree of minimum total length whose vertices are a *superset* of a given set of points. The hardness of the Steiner tree problem implies that it is NP-hard to decide if there exists a tree spanning all of the given set, R, of roots, whose length is at most B, since this amounts to asking if the length, L^*, of a minimum Steiner tree satisfies $L^* \leq B$.

One might ask whether the problem becomes "easier" if we are told that $B > L^*$. We show that even for such instances, if we can approximate the multiply-rooted tree-orienteering problem within a constant factor, then we can solve the Euclidean Steiner tree problem exactly, within the same time bound. Hardness also applies to approximating the path and cycle versions of the problem.

Theorem 11 *Unless P=NP, there is no polynomial-time algorithm achieving a constant-factor approximation of the number of additional (non-root) sites visited by a multiply-rooted tree (or cycle or path) of length at most B, where B is known to exceed the length of a Steiner tree on the set of roots.*

Proof. We give here the proof for the tree version of the problem; proofs for the cycle and path versions are similar, from Euclidean TSP. Given an instance of the Euclidean Steiner tree problem on a set P of points, we create an instance of the tree-orienteering problem in which the set of roots is $R = P$. For the Steiner tree problem, we must decide if $L^* \leq L$, where L^* is the length of the minimum Steiner tree on R. We add one additional non-root site v, at distance $B - L$ from a point $p \in R$ on the convex hull of R (so that the distance of v to R is minimized at p). Clearly, a tree of length at most B that visits v, as well as all roots, exists if and only if a Steiner tree of length at most L exists for the points R. Thus, if $L^* \leq L$, then a c-approximation algorithm for tree-orienteering must return a tree that includes v; otherwise, it may return a tree with zero additional (non-root) points. □

Remark. We note that the hardness proof applies also to the problem of approximating the maximum *total* number of sites visited (including the roots), by replacing v with many copies of itself.

We obtain a 3-approximation for the multiply-rooted tree-orienteering problem, provided the number, ρ, of roots is *constant.* (The running time is $O(n^{O(\rho)})$.) Additionally, we obtain 2-approximations for the path and cycle versions of the problem.

Theorem 12 *Assuming that the number (ρ) of roots is constant, there is a polynomial-time 2-approximation (resp., 3-approximation) algorithm for the multiply-rooted path- and cycle-orienteering problems (resp., tree-orienteering problem) on a set of points in the plane.*

The proof of this theorem is based on some of the same ideas as in Theorems 1, 2, and 3, but with the application of Corollary 2 below.

We concentrate here only on the tree-orienteering case. The algorithm does the following: For each choice of

(a). integer k, $k = 1, \ldots, n$,

(b). site v,

(c). partition of R into two sets, R_1 and R_2,

we find a shortest possible connected m-guillotine subdivision (with $m \geq \rho - 1$), visiting site v, spanning k or more non-root sites, and spanning all of the roots R_1. We then append this tree to the minimum (Steiner) tree spanning $\{v\} \cup R_2$, and tabulate the lengths of all such trees. The tree whose length is at most B that maximizes the number of sites visited is a tree that we claim approximates the multiply-rooted tree-orienteering problem.

Lemma 5 *Among the graphs tabulated by the algorithm, there must be one whose weight is less than B, while the cardinality of the point set spanned is at least $n(T^*)/3$.*

Proof. As in the proof of Lemma 1, we know that there exists a partitioning of an optimal tree T^* into two subtrees, T_1 and T_2, sharing a common vertex v of T^*, so that $(2/3)n(T^*) \geq n(T_i) \geq (1/3)n(T^*)$. Let S_i denote the Steiner tree spanning the roots that are contained in subtree T_i.

Without loss of generality, we may assume that $w(T_1) - w(S_1) \leq w(T_2) - w(S_2)$. Now, consider the tree $T' = T_{1,G} \cup S_2$, where $T_{1,G}$ is an m-guillotine subdivision, with $m \geq \rho - 1$, containing the edge set of T_1. By Corollary 2, $w(T_{1,G}) \leq w(T_1) + \epsilon(w(T_1) - w(S_1))$. Then, T' spans all of R, and it is among the class of trees considered by the algorithm. Thus, we have only to show that its weight is bounded by B:

$$\begin{aligned} w(T') &= w(T_{1,G}) + w(S_2) \\ &\leq w(T_1) + \epsilon(w(T_1) - w(S_1)) + w(S_2) \\ &\leq w(T_1) + \epsilon(w(T_2) - w(S_2)) + w(S_2) \\ &\leq w(T_1) + (w(T_2) - w(S_2)) + w(S_2) \\ &= w(T_1) + w(T_2) \leq B \end{aligned}$$

where we assume that $m \geq 2$, so that $\epsilon < 1$, and we have used Corollary 2, below, to obtain the first inequality. □

The following corollary to Theorem 4 is based again on a more careful analysis of the charging scheme in [23], together with the following generalization of the observation used in Corollary 1: If E spans the set R of ρ roots, then E must contain a (Steiner) spanning tree of R; thus, the total $\leq (\rho - 1)$-exposed side length, $\xi^{(\rho-1)}(E)$, of E must be at least twice the length of the Steiner minimum spanning tree of R.

Corollary 2 *Let S be a polygonal subdivision, with edge set E, of length L, If E spans the set, R, of ρ roots, then, for any positive integer $m \geq \rho - 1$, there exists an m-guillotine subdivision, S_G, whose edge set contains E, and is of length at most $L + \frac{\sqrt{2}}{m}(L - L^*(R))$, where $L^*(R)$ is the length of a Steiner minimum spanning tree of R.*

4.2 Polygonal Sites

In the case that the n sites are given to us as a collection of *regions* rather than points, we get a problem closely related to the "TSP with Neighborhoods" (TSPN) problem. We assume that the regions are given as a collection of n simple polygons, having a total of N vertices.

For the TSPN, Arkin and Hassin [1] have obtained $O(1)$-approximation algorithms for "well behaved" (possibly overlapping) regions (e.g., regions are disks or have roughly equal-length and parallel diameter segments), while Mata and Mitchell [19] have obtained an $O(\log n)$-approximation algorithm for n general (possibly overlapping) regions, based on "guillotine rectangular subdivisions".

We prove the following theorem, giving the first approximation results for the orienteering version of the problem:

Theorem 13 *There is an $O(\log n)$-approximation algorithm for the tree-orienteering (and path- and cycle-orienteering) problems on a set of polygonal sites in the plane. The running time is polynomial in n and N.*

Proof. (Sketch) We give here the proof only for the tree version of the problem. Let T^* denote an optimal rooted tree, spanning the maximum number, $n(T^*)$, points, while having weight at most B.

Our approximation algorithm is to tabulate, for each value of k, $k = 1, \ldots, n$, a shortest possible connected *guillotine rectangular subdivision* spanning k or more regions. (This can be done in time polynomial in n and N, using dynamic programming, as in [19].) We then select, from among these, the subdivision that visits the maximum number k of regions, subject to its weight being less than the bound B.

We now have to prove that, among the subdivisions tabulated, there must be one, T, whose weight is less than B, while the cardinality, k, of the set of regions spanned is at least $n(T^*)/O(\log n)$. Our proof is based on our earlier combinatorial theorem, Theorem 10, which guarantees that there exists a subtree, T', of T^*, with $w(T') \leq w(T^*)/C \log n$ and spanning at least $n(T^*)/2C \log n$ regions, for any constant C. Then, using the fact ([19]) that any rectilinear edge set is contained within the edge set of a guillotine rectangular subdivision, having length within factor $O(\log n)$, we get that there must exist a guillotine rectangular subdivision, G, spanning the same set of regions as does T' (and thus, spanning at least $n(T^*)/2C \log n$ regions), such that the length of G is at most $w(T') \cdot C \log n \leq w(T^*) \leq B$, for some constant C. This proves that our algorithm will indeed discover a tree spanning at least $n(T^*)/O(\log n)$ regions. □

5 Conclusion

An outstanding open problem that remains is to obtain an approximation algorithm (or prove that none exists) for the rooted orienteering problems in graphs.

Also, it is interesting to note that, in the geometric case, our current best bound (2) for the rooted cycle- and path-orienteering problems is better than the bound (3) for the rooted tree-orienteering problem. Can the tree version be improved? (In contrast, in the unrooted graph version, our methods give a factor of 5 for trees and 6 for cycles and paths.) Is the factor 2 best possible for the cycle and path versions?

Acknowledgements

We thank Yi-Jen Chiang and Steve Skiena for useful discussions on the subject of this paper. We also thank John Hershberger, who provided valuable suggestions that improved the presentation.

References

[1] E. M. Arkin and R. Hassin. Approximation algorithms for the geometric covering salesman problem. *Discrete Appl. Math.*, 55:197–218, 1994.

[2] E. M. Arkin, S. Khuller, and J. S. B. Mitchell. Geometric knapsack problems. *Algorithmica*, 10:399–427, 1993.

[3] E. M. Arkin, J. S. B. Mitchell, and C. D. Piatko. Bicriteria shortest path problems in the plane. In *Proc. 3rd Canad. Conf. Comput. Geom.*, pages 153–156, 1991.

[4] S. Arora. Polynomial time approximation schemes for Euclidean TSP and other geometric problems. In *Proc. 37th Annu. IEEE Sympos. Found. Comput. Sci. (FOCS 96)*, pages 2–12, 1996.

[5] S. Arora. Nearly linear time approximation schemes for Euclidean TSP and other geometric problems. In *Proc.*

38th Annu. IEEE Sympos. Found. Comput. Sci. (FOCS 97), 1997.

[6] S. Arya and H. Ramesh. A 2.5 factor approximation algorithm for the k-MST problem. *Inform. Process. Lett.*, 65:117–118, 1998.

[7] B. Awerbuch, Y. Azar, A. Blum, and S. Vempala. Improved approximation guarantees for minimum-weight k-trees and prize-collecting salesmen. In *Proc. 27th Annu. ACM Sympos. Theory Comput. (STOC 95)*, pages 277–283, 1995.

[8] E. Balas. The prize collecting traveling salesman problem. *Networks*, 19:621–636, 1989.

[9] J. L. Bentley. Fast algorithms for geometric traveling salesman problems. *ORSA J. Comput.*, 4(4):387–411, 1992.

[10] D. Bienstock, M. X. Goemans, D. Simchi-Levi, and D. Williamson. A note on the prize collecting traveling salesman problem. *Math. Prog.*, 59:413–420, 1993.

[11] D. Eppstein, M. Overmars, G. Rote, and G. Woeginger. Finding minimum area k-gons. *Discrete Comput. Geom.*, 7:45–58, 1992.

[12] M. Fischetti, H. W. Hamacher, K. Jørnsten, and F. Maffioli. Weighted k-cardinality trees: Complexity and polyhedral structure. *Networks*, 24:11–21, 1994.

[13] M. R. Garey, R. L. Graham, and D. S. Johnson. The complexity of computing Steiner minimal trees. *SIAM J. Appl. Math.*, 32:835–859, 1977.

[14] N. Garg. A 3-approximation for the minimum tree spanning k vertices. In *37th Annual Symposium on Foundations of Computer Science*, pages 302–309, Burlington, Vermont, October 14-16 1996.

[15] M. Goemans and D. Williamson. General approximation technique for constrained forest problems. In *Proc. 3rd ACM-SIAM Sympos. Discrete Algorithms (SODA '92)*, pages 307–315, 1992.

[16] B. L. Golden, L. Levy, and R. Vohra. The orienteering problem. *Naval Res. Logistics*, 34:307–318, 1991.

[17] M. Jünger, G. Reinelt, and G. Rinaldi. The traveling salesman problem. In M. O. Ball, T. L. Magnanti, C. L. Monma, and G. L. Nemhauser, editors, *Network Models*, Handbook of Operations Research/Management Science, pages 225–330. Elsevier Science, Amsterdam, 1995.

[18] E. L. Lawler, J. K. Lenstra, A. H. G. Rinnooy Kan, and D. B. Shmoys, editors. *The Traveling Salesman Problem*. Wiley, New York, NY, 1985.

[19] C. Mata and J. S. B. Mitchell. Approximation algorithms for geometric tour and network design problems. In *Proc. 11th Annu. ACM Sympos. Comput. Geom.*, pages 360–369, 1995.

[20] J. S. B. Mitchell. Shortest paths and networks. In J. E. Goodman and J. O'Rourke, editors, *Handbook of Discrete and Computational Geometry*, chapter 24, pages 445–466. CRC Press LLC, 1997.

[21] J. S. B. Mitchell. Guillotine subdivisions approximate polygonal subdivisions: A simple new method for the geometric k-MST problem. In *Proc. 7th ACM-SIAM Sympos. Discrete Algorithms*, pages 402–408, 1996.

[22] J. S. B. Mitchell. Geometric shortest paths and network optimization. In J.-R. Sack and J. Urrutia, editors, *Handbook of Computational Geometry*. Elsevier Science, Amsterdam, 1998, to appear.

[23] J. S. B. Mitchell. Guillotine subdivisions approximate polygonal subdivisions: A simple polynomial-time approximation scheme for geometric TSP, k-MST, and related problems. Manuscript, 1996. *SIAM J. Comput.*, to appear.

[24] J. S. B. Mitchell. Guillotine subdivisions approximate polygonal subdivisions: Part III – Faster polynomial-time approximation schemes for geometric network optimization. Manuscript, 1997.

[25] J. S. B. Mitchell, C. Piatko, and E. M. Arkin. Computing a shortest k-link path in a polygon. In *Proc. 33rd Annu. IEEE Sympos. Found. Comput. Sci.*, pages 573–582, 1992.

[26] T. Nishizeki and N. Chiba. Planar graphs: Theory and algorithms. *Ann. Discrete Math.*, 32, 1988.

[27] S. B. Rao and W. D. Smith. Improved approximation schemes for traveling salesman tours. *Proc. 30th Annu. ACM Sympos. Theory Comput.*, to appear, 1998.

[28] R. Ravi, R. Sundaram, M. V. Marathe, D. J. Rosenkrantz, and S. S. Ravi. Spanning trees short and small. In *Proc. 5th ACM-SIAM Sympos. Discrete Algorithms*, pages 546–555, 1994.

[29] G. Reinelt. Fast heuristics for large geometric traveling salesman problems. *ORSA J. Comput.*, 4:206–217, 1992.

[30] T. Tsiligirides. Heuristic methods applied to orienteering. *J. of the Operational Research Society*, 35(9):797–809, 1984.

[31] A. Zelikovsky and D. Lozevanu. Minimal and bounded trees. In *Tezele Cong. XVIII Acad. Romano-Americane, Kishinev*, pages 25–26, 1993.

Appendix

Proof of the Tree-Partition Lemma 4:

Assume that the first case does not happen, i.e., there does not exist a large weight edge e such that the edges of T can be partitioned into edge e and two subtrees T_1 and T_2 with $w(T_i) < w(T)/3, i = 1, 2$. We will then show that there must exist a partition of the edges of T into two subtrees T_1 and T_2 such that $w(T)/3 \leq w(T_i) \leq 2w(T)/3, i = 1, 2$. We display an algorithm that will, in fact, find such a partition.

The algorithm is as follows. Start at any vertex v with incident edges $e_1, e_2, \ldots, e_t$. Consider a partition of T into subtrees $T_i, i = 1, 2, \ldots, t$, such that subtree T_i includes edge e_i and all the subtrees share only one vertex in common, namely vertex v.

Case 1: If there exists a subtree T_i such that $w(T)/3 \leq w(T_i) \leq 2w(T)/3$, then the lemma is proved by the following argument. Let T_1 be the subtree T_i. Let T_2 be the union of the other subtrees after identifying vertex v. The subtrees T_1 and T_2 satisfy the inequality $w(T)/3 \leq w(T_i) \leq 2w(T)/3, i = 1, 2$.

Case 2: If there exists no subtree T_i such that $w(T)/3 \leq w(T_i)$, then by Lemma 3, there exists a partition of the subtrees into two sets such that the sum of the weights of the subtrees in each of the sets lies between $w(T)/3$ and $2w(T)/3$.

Case 3: If there exists a subtree T_i such that $w(T_i) > 2w(T)/3$, and if the edge e_i is incident on vertices v and v', then we restart the algorithm from vertex v'.

It is clear that if Cases 1 or 2 occur at any time, the algorithm terminates after correctly finding the subtrees T_1 and T_2 needed for the proof of the lemma. We are left to show that Case 3 can occur only a finite number of times. As long as the algorithm does not return to a vertex v, it is bound to hit a leaf and terminate in at most n steps. If the algorithm ever returns to vertex v from vertex v', it will oscillate forever between vertices v and v'. We will now show that this cannot happen. For the sake of contradiction, assume that the algorithm oscillates between adjacent vertices v and v'. Let e be the edge connecting them. Let the union of all subtrees incident on v (v') that do not include edge e be called T_{i1} and T_{i2}. Define the subtrees $T_i = T_{i1} \cup \{e\}$ and $T_i' = T_{i2} \cup \{e\}$. Since the algorithm moves from vertex v to v', we know that $w(T_i) > 2w(T)/3$. Similarly, since the algorithm moves from vertex v' to v, we know that $w(T_i') > 2w(T)/3$. Since $w(T_{i1}) + w(T_i') = w(T) = w(T_{i2}) + w(T_i')$, we get that $w(T_{i1}) < w(T)/3$ and $w(T_{i2}) < w(T)/3$, contradicting our assumption that Case 1 does not occur. □

Proof of Theorem 9:

By the tree-partition lemma (Lemma 4), the tree T has one of two possible decompositions. If the first case occurs, then T can be partitioned into two subtrees (T_1 and T_2) and an edge with both subtrees of weight at most $w(T)/3$. Since T_1 and T_2 also partition the vertex set of T, one of them must contain at least $n/2$ vertices. Without loss of generality we assume that T_1 contains at least $n/2$ vertices. Then T_1 satisfies both the statements of the theorem.

If, on the other hand, the first case does not occur, then T can be partitioned into two subtrees (T_1 and T_2) with both subtrees of weight between $w(T)/3$ and $2w(T)/3$. Since one of them must contain at least $n/2$ vertices, this subtree must satisfy the first statement of the theorem. We are left only with proving the second statement of the theorem for this case. Without loss of generality assume that $w(T_1) \geq w(T_2)$. Thus $w(T_2) \leq w(T)/2$. If $n(T_2) \geq n/3$ vertices, then T_2 satisfies the requirements of the theorem. However, if $n(T_2) < n/3$, then $n(T_1) > 2n/3$. Now we apply the partition lemma on tree T_1.

If the first case occurs, then T_1 can be partitioned into two subtrees (T_{11} and T_{12}) and an edge, with both subtrees of weight at most $w(T_1)/3$. Let $n(T_{11}) \geq n(T_{12})$. Thus, $n(T_{11}) \geq n(T_1)/2 \geq (1/2)(2n/3) = n/3$. Also, $w(T_{11}) < w(T_1)/3 < w(T)/3$. Thus, T_{11} satisfies the requirements of the theorem.

Finally, if the first case does not hold for T_1, then T_1 can be partitioned into two subtrees, T_{11} and T_{12}, each of weight at most $2w(T_1)/3$. Without loss of generality, assume that $n(T_{11}) \geq n(T_{12})$. Thus $n(T_{11}) \geq n(T_1)/2 \geq (1/2)(2n/3) = n/3$. Also, $w(T_{11}) \leq 2w(T_1)/3 \leq (2/3)(2w(T)/3) < w(T)/2$. Thus, T_{11} satisfies the requirements of the theorem. □

Lemma 6 *For $K > 2$, there exists a metric space on a set S of $Kt + 1$ vertices such that any tree spanning at least $t+2$ of the vertices has total weight at least $2L/K$, where L is the weight of a minimum spanning tree of S. For $K = 2$, there exists a metric space on a set S of $2t$ vertices such that any tree spanning at least $t + 1$ of the vertices has total weight L, where L is the weight of a minimum spanning tree connecting S.*

Proof. For $K = 2$, consider a configuration of n points that consists of two groups of vertices (each with t vertices) where any two vertices in the same group are at distance 0 from each other and the distance between a pair of vertices in two different groups is L, which is also the length of a minimum spanning tree on the given vertices. Clearly, any tree that spans $t + 1$ vertices or more, must have length at least L.

For $K > 2$ we modify the construction slightly. Let s be a special vertex. Consider K groups of vertices (each with t vertices) where any two vertices in the same group are at distance 0 from each other and each of the vertices is at distance L/K from vertex s. The distance between a pair of vertices in two different groups is defined by the metric space and is thus $2L/K$. Any tree that spans $t+2$ vertices or more, must span vertices from at least two different groups and consequently must have length at least $2L/K$. □

Lemma 7 *There exists a set S of $3t+1$ points in the plane such that any tree spanning at least $t + 2$ of the vertices has total weight at least $L/\sqrt{3}$, where L is the weight of a minimum spanning tree connecting S.*

Proof. Let s be a special vertex. Consider 3 groups of vertices (each with t vertices) with each group located at the 3 corners of an equilateral triangle with s at its center. The distance from a corner of the triangle to its center is $L/3$. Any two points in the same group is at distance 0 from each other. The distance between a pair of vertices in two different groups is $L/\sqrt{3}$. □

The above lemma also shows that there exist configurations for which one needs a tree of weight at least $L/\sqrt{3}$ in order to span at least $n/2$ points.

Efficiently Approximating Polygonal Paths in Three and Higher Dimensions*

Gill Barequet[†] Danny Z. Chen[‡] Ovidiu Daescu[‡]
Michael T. Goodrich[†] Jack Snoeyink[§]

Abstract

We present efficient algorithms for solving polygonal-path approximation problems in three and higher dimensions. Given an n-vertex polygonal curve P in $\mathbb{R}^d$, $d \geq 3$, we approximate P by another polygonal curve P' of $m \leq n$ vertices in $\mathbb{R}^d$ such that the vertex sequence of P' is an ordered subsequence of the vertices of P. The goal is to either minimize the size m of P' for a given error tolerance ε (called the *min-# problem*), or to minimize the deviation error ε between P and P' for a given size m of P' (called the *min-ε problem*). Our techniques enable us to develop efficient near-quadratic-time algorithms in 3-D and sub-cubic-time algorithms in 4-D for solving the min-# and min-ε problems. We discuss extensions of our solutions to d-dimensional space, where $d > 4$.

*Work on this paper by the first and the fourth authors has been supported in part by the U.S. Army Research Office under Grant DAAH04-96-1-0013. Work by the second and third authors has been supported in part by the National Science Foundation under Grant CCR-9623585. Work by the fourth author has been supported also by NSF grant CCR-96-25289. Work by the fifth author has been supported by NSERC, CIES, and by the Geometry Center at Johns Hopkins University.

[†]Center for Geometric Computing, Dept. of Computer Science, Johns Hopkins University, Baltimore, MD 21218. E-mail: [barequet|goodrich]@cs.jhu.edu

[‡]Dept. of Computer Science and Engineering, University of Notre Dame, Notre Dame, IN 46556. E-mail: [chen|odaescu]@cse.nd.edu

[§]Dept. of Computer Science, The University of British Columbia, 201-2366 Main Mall, Vancouver, B.C., Canada V6T 1Z4. E-mail: snoeyink@cs.ubc.ca

SCG 98 Minneapolis Minnesota USA

1 Introduction

In this paper we consider the problem of approximating an arbitrary n-vertex polygonal path P in 3-dimensional space by another polygonal path P' whose vertices form an ordered subsequence of the vertices in the original path P. We further discuss this problem in the d-dimensional space, where $d \geq 4$.

This problem arises in many applications, such as robotics, image processing, computer graphics, cartography, and data compression. In these applications it is often desirable to approximate a complex graphical or geometric object, which is specified by a polygonal path, by a simpler object that captures the essence of the original object yet achieves a certain degree of data compression [3,23, 24]. For example, we have been asked to compress 3D path data for optic nerves acquired in medical studies (Kelly Booth, personal communication, 1995) and of river networks in geographic information systems analysis of spawning habitat [20]. Path compression is also important for devices with low bandwidth, from pen plotters to the WWW.

In general, the goal of the path-approximation problem is to replace a (very) high-resolution path P with an approximate path P' that achieves significant time and space improvements while suffering only a small approximation error. Two factors come to play in such an approximation: the number of vertices in the approximation and the error between the approximation and the original path. These two factors give rise to two different versions of the path-approximation problem—the min-ε version and the min-# version. In the min-ε version one is given a parameter $m \leq n$ and asked to find an approximating path P' with m vertices whose *error*, ε, from the original path is minimized.

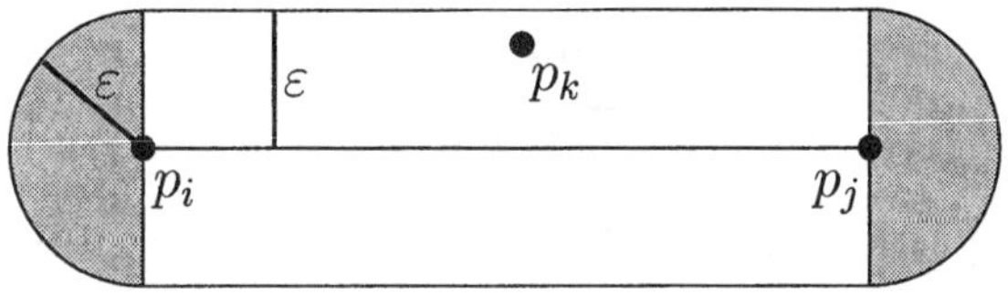

Figure 1: The error tolerance zone of a single segment $\overline{p_i p_j}$ in 2-D.

This version of the problem is motivated by a desire to obtain the best approximation possible that achieves a certain degree of data compression. In the min-# version of the path-approximation problem one is instead given a parameter $\varepsilon > 0$ and asked to find the smallest-vertex path P' that is within *error* ε of the original path. This version of the problem is motivated by a desire to obtain the smallest-complexity path that maintains a certain level of accuracy. We study both versions of the d-dimensional path-approximation problem, for $d \geq 3$, in this paper.

One issue that immediately arises in the path approximation problem is that one must formalize the "error" between the original path P and the approximation P'. We believe that one of the most natural definitions is the *tolerance zone* error measure [3, 17, 18, 22]. Define the ε*-tolerance zone* (for $\varepsilon > 0$) of a line segment $\overline{pq}$ to be the region of space (or "*zone*") that is the union of all radius-ε balls centered at points along the segment $\overline{pq}$ (see Figure 1). This definition is, of course, parameterized by the metric used to define radius-ε balls. We consider the usual Euclidean metric, L_2, as well as the L_∞ and L_1 distance metrics. We may then formalize the path-approximation problem as follows:

Polygonal Path Approximation:
Given a polygonal path $P = (p_1, p_2, \ldots, p_n)$, find a path $P' = (p_{i_1}, p_{i_2}, \ldots, p_{i_m})$, such that, for each $j \in \{1, 2, \ldots, m-1\}$:

- $i_j \in \{1, 2, \ldots, n\}$, for $j = 1, 2, \ldots, m$;
- $i_j < i_{j+1}$; and
- the subpath $(p_{i_j}, p_{i_j+1}, \ldots, p_{i_{j+1}})$ is completely contained in the ε-tolerance zone for the segment $\overline{p_{i_j} p_{i_{j+1}}}$.

If the subpath $P_{i_j, i_{j+1}} = (p_{i_j}, p_{i_j+1}, \ldots, p_{i_{j+1}})$ is completely contained in the ε-tolerance zone for the segment $\overline{p_{i_j} p_{i_{j+1}}}$, we call $\overline{p_{i_j} p_{i_{j+1}}}$ an *approximating line segment* for $P_{i_j, i_{j+1}}$.

In the min-ε version, we are given $m < n$ and wish to minimize ε; in the min-# version, we are given $\varepsilon > 0$ and wish to minimize m.

We restrict our attention to the version of this problem that requires that $p_{i_1} = p_1$ and $p_{i_m} = p_n$. In the general version we only require that $i_1 \leq i_m$, and that the subpaths $P_{1, i_1 - 1}$ and $P_{i_m + 1, n}$ are completely contained in the ε-balls centered at P_{i_1} and P_{i_m}, respectively. We explain later in the introduction how to reduce the general problem to the restricted version of it.

1.1 Related Previous Work

Imai and Iri [17,18] and Melkman and O'Rourke [22] study the 2-D min-# and min-ε polygonal-path approximation problems. Using the tolerance zone measure of error, as defined above, they achieve algorithms whose running times are $O(n^2 \log n)$ and $O(n^2 \log^2 n)$, respectively, and using $O(n^2)$ space, for the two problems. Chan and Chin [4] reduce the time complexity of both results by a logarithmic factor. Chen and Daescu [7] further show that the algorithms of [4] can use only $O(n)$ space without increasing their running times.

Varadarajan [27] studies the min-# and min-ε problems for 2-D polygonal paths that are monotone, i.e., any line parallel to the y axis intersects a path in a point. Given a monotone approximation of a monotone path, Varadarajan measures the error as the maximum distance between the intersection points with some line parallel to the y axis. He gives $O(n^{4/3+\delta})$ time and space algorithms for both problems, where $\delta > 0$ is an arbitrarily small constant. However, those algorithms cannot be extended to the 3-D version of the problem or to the tolerance zone error criterion in 2-D. The recursive simplification heuristic of Douglas and Peucker [8], which is popular in GIS, does extend to curves in 3 dimensions in $O(n^2)$ worst-case time, but solves neither the min-# nor the min-ε problems.

There has been other less-related, but still significant, work done on other metrics for comparing 2-dimensional polygonal paths [18] that do not necessarily use the same set of vertices. Arkin et al. [2], Alt and Godau [1], and Rote [25] describe several metrics for comparing polygonal curves. Guibas et al. [15] study different error measures for the min-# problem, with an eye toward those that could be implemented in $O(n \log n)$ time by greedy algorithms. Fleischer et al. [11] approximate polygonal shapes by inner and outer polygons and show their applications to the problems of polygon containment and of planar motion planning.

The only work on the general problem, that we are aware of, achieves super-cubic algorithms for the 3-D path-approximation problem. Ihm and Naylor [16] give an $O(n^3 \log m)$ algorithm for the min-ε problem. Eu and Toussaint [10] restrict the polygonal chains to be monotonic with respect to one of the three coordinate axes, and define the approximation in terms of an *infinite beam* measure of error. In this measure of error one requires that each subpath $(p_{i_j}, p_{i_j+1}, \ldots, p_{i_{j+1}})$ be completely contained in the ε-tolerance zone of the *infinite line* containing the segment $\overline{p_{i_j} p_{i_{j+1}}}$, called the infinite beam. When infinite beams are defined in terms of the L_1 or L_∞ metrics, Eu and Toussaint show how to achieve running times of $O(n^2)$ and $O(n^3)$ for the min-# and min-ε problems, respectively, for monotone 3-D polygonal paths. Under the Euclidean (L_2) metric, they consider the 3-dimensional version of the problem in [27] and give algorithms that run in $O(n^3)$ and $O(n^3 \log n)$ time, respectively, for the min-# and min-ε problems on monotone 3-D polygonal paths.

1.2 Our Results

In this paper we give the first efficient algorithms for the d-dimensional path approximation problem, where $d \geq 3$, for general (non-monotonic) paths, using the tolerance-zone measure of error for segments. We believe that this notion of error better captures the intuitive concept of the shape of a d-dimensional polygonal path. We illustrate in Figure 2 the difference between approximating a 2-dimensional path with tolerance zones and with infinite beams.

We summarize the running times of our methods in Table 1. Note that all of our bounds are sub-cubic (if not nearly quadratic). Fairly straightforward bounds of $O(n^3)$ for min-# and $O(n^3 \log n)$ for min-ε follow by adapting previous 2-dimensional results [4, 17, 18, 22]. Even so, our approach still follows the general framework of these earlier algorithms. To solve the min-# problem, we first build a graph G on the vertices of the input path

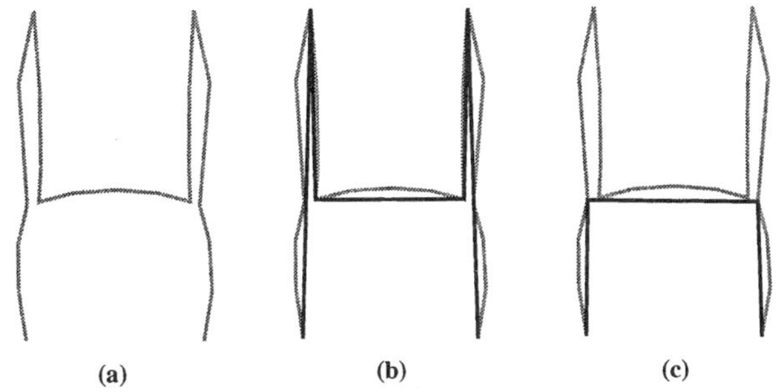

Figure 2: An illustration of the difference between the tolerance zone measure of error and the infinite beam measure of error, for a polygonal path in 2-D, with $\varepsilon = 3$mm. The path is shown in (a), its approximation using the tolerance zone in (b), and its approximation using infinite beams in (c).

P, where each edge represents a valid "shortcut," and we then perform a shortest-path computation in G. To solve the min-ε problem we perform a "binary-search-like" computation using the min-# algorithm as a "probe."

Our algorithms differ from previous approaches in some important ways, however. One of the most fundamental differences is that we develop a d-dimensional generalization of the 2-dimensional approach of Chan and Chin [4] for constructing the "shortcut" graph G. Rather than constructing this graph directly, we instead construct it as the intersection of two other graphs, $\overrightarrow{G}$ and $\overleftarrow{G}$, which respectively define tolerance zones in terms of "semi-infinite" beams (pointing in opposite directions). The difficult part of this computation is that, fixing an index $1 \leq i < n$, we must determine, for each index j, with $i < j \leq n$, if the ray $\overrightarrow{p_i p_j}$ originating at p_i and passing through p_j intersects each radius-ε ball centered at a vertex p_k for $i < k < j$. We reduce this problem to the following data structuring problem:

> **Off-Line Ball Inclusion Testing:** Maintain a collection of different-radius balls subject to a given sequence of the following operations:
>
> - `insert`(B): add the ball B to the collection.
> - `inside`(p): determine if the point p is inside the common intersection of the balls.

Metric	3 Dimensions Min-#	3 Dimensions Min-ε	$d \geq 4$ Dimensions Min-#	$d \geq 4$ Dimensions Min-ε
L_2	$O(n^2 \log n)$	$O(n^2 \log^3 n)$	$O(n^{3-2/(\lfloor \frac{d}{2} \rfloor+1)} \text{polylog}\, n)$	$O(n^{7/3} \text{polylog}\, n)^{\clubsuit}$
L_1 & L_∞	$O(n^2)$	$O(n^2 \log n)$	$O(n^2)$	$O(n^2 \log n)$

$^{\clubsuit}$For $d = 4$ only.

Table 1: Summary of our min-# and min-ε results.

For $d = 3$, we refer to the problem above as the off-line disk-inclusion testing problem (the balls are disks on a 3-D sphere) and provide a data structure for solving this problem in amortized $O(\log n)$ time for inserts and queries, even though each disk insertion can cause $\Theta(n)$ structural changes to the common intersection. We use this data structure to derive an $O(n^2 \log n)$ time algorithm for the min-# path approximation problem. To turn this into an efficient algorithm for the min-ε problem we give an application of the parametric search paradigm that is based upon a non-trivial parallel version of our off-line inclusion testing data structure.

When the ε-tolerance zones are defined using the L_1 or L_∞ metrics, we can do better than this. We present an $O(n^2)$-time algorithm for the min-# problem and an $O(n^2 \log n)$-time algorithm for the min-ε problem. For the latter algorithm we avoid using parametric search.

For $d > 3$, we show how to achieve a subcubic time for the min-# problem under the L_2 metric. We also give $O(n^2)$ and $O(n^2 \log n)$-time algorithms (for a fixed $d > 3$) for the min-# and min-ε problems, respectively, under the L_1 and L_∞ metrics. The constants hidden in the big-Oh notation depend on d and on the used metric: they are $O(d^2)$ for L_∞ and $O(d2^d)$ for L_1.

The general version of the problem can be reduced to the version solved in this paper (where $i_1 = 1$ and $i_m = n$) by adding to G two vertices p_0 and p_{n+1}, and edges p_0p_i (resp., p_ip_{n+1}), for $1 \leq i \leq n$, if all the vertices p_j, for $1 \leq j < i$ (resp., $i < j \leq n$) are contained in an ε-ball centered at p_i. This additional processing takes $O(n^2)$ time, and we then solve the restricted problem for p_0 and p_{n+1}.

We outline the main ideas behind our results in the remaining sections of this extended abstract.

2 Solving the Min-# Problem in 3-D

In this section we present the main ideas behind our $O(n^2 \log n)$-time algorithm for solving the 3-D min-# problem. The important part of this computation involves a reduction to the off-line inclusion testing problem, which we solve in Section 3.

Suppose then that we are given a 3-D polygonal curve $P = (p_1, p_2, \ldots, p_n)$, and an error value $\varepsilon > 0$, and we wish to find the smallest $m \leq n$ such that there is a subpath $P' = (p_{i_1}, p_{i_2}, \ldots, p_{i_m})$ that satisfies the ε-tolerance zone criterion (with $i_1 = 1$ and $i_m = n$). As outlined in the introduction, we solve this problem by the following general approach:

1. Construct a graph $G = (V, E)$, where V is the vertex set of P and E consists of directed edges (p_i, p_j), $1 \leq i < j \leq n$, such that $\overline{p_ip_j}$ is an approximating line segment of the subchain $(p_i, p_{i+1}, \ldots, p_j)$ of P;
2. Find a shortest path from p_1 to p_n in G (i.e., a path with the fewest number of edges). This path gives us the sought approximating curve P'.

Since the second step is easily solved by a breadth-first search computation, let us concentrate on an efficient technique for constructing G from P.

As mentioned above, we construct G as the intersection of two graphs $\overrightarrow{G}$ and $\overleftarrow{G}$. Say that edge (p_i, p_j) is in the graph $\overrightarrow{G}$ if the ray $\overrightarrow{p_ip_j}$ originating at p_i and passing through p_j intersects each radius-ε ball $B(p_k, \varepsilon)$, centered at a vertex p_k, for $i < k < j$. Likewise, say that edge (p_i, p_j) is in the graph $\overleftarrow{G}$ if the ray $\overleftarrow{p_ip_j}$ originating at p_j and passing through p_i intersects each $B(p_k, \varepsilon)$, for $i < k < j$. It is an easy observation that an edge (p_i, p_j) is in G if and only if it is in both $\overrightarrow{G}$ and $\overleftarrow{G}$. Since the constructions of these two graphs are

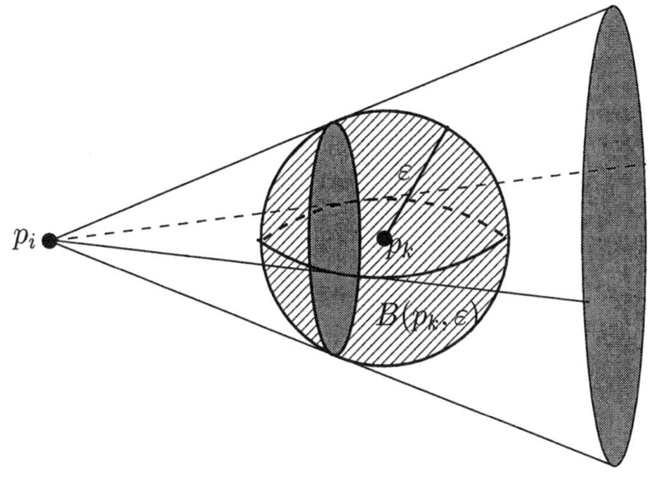

Figure 3: The projection of the radius-ε ball $B(p_k, \varepsilon)$ centered at p_k onto the sphere S_i.

symmetric computations, let us concentrate further upon the construction of the graph $\overrightarrow{G}$.

In this section we show how to reduce the problem of determining the edges in $\overrightarrow{G}$ to the off-line disk-inclusion problem. Let us fix an index i, with $1 \leq i < n$, and focus on computing all the edges of the form (p_i, p_j) in $\overrightarrow{G}$, for $i < j \leq n$. Say that the ray $\overrightarrow{p_ip_j}$ *properly approximates* the polygonal chain $(p_i, \ldots, p_j)$ if it intersects every ball $B(p_k, \varepsilon)$ for $i < k < j$. Now imagine a "large" sphere S_i centered at p_i, and project from p_i all the balls $B(p_k, \varepsilon)$ $(i < k < j)$ onto this sphere. Each ball $B(p_k, \varepsilon)$ is then projected to a disk D_k on the sphere S_i (see Figure 3).

Lemma 1 *The ray $\overrightarrow{p_ip_j}$ is a proper approximation if and only if:*

1. *The intersection of the set of disks $\mathcal{D}(i,j) = \{D_{i+1}, \ldots, D_{j-1}\}$ on S_i is nonempty.*
2. *The projection of p_j from p_i onto S_i is found within the intersection of the disks in $\mathcal{D}(i,j)$.*

Proof: Omitted in this extended abstract. □

This lemma gives us the ability to reduce the construction of all edges of the form (p_i, p_j) in $\overrightarrow{G}$, for our fixed p_i, to the off-line disk-inclusion problem. Specifically, we define the operations:

- `insert`(D_j): Inserts the disk D_j to a data structure, where D_j is the projection of $B(p_j, \varepsilon)$ onto the sphere S_i using p_i as the center of projection.
- `inside`(p'_j): Returns "true" or "false," depending on whether the point p'_j, the projection of p_j onto S_i using p_i as the center of projection, is contained by all the disks currently in the data structure.

We then construct a sequence Σ_i of `insert` and `inside` operations, as follows:

Initialize Σ_i to insert a disk containing the entire sphere S_i.
for $j = i + 1$ to n **do**
  Append the operation `inside`(p'_j) to Σ_i.
  Append the operation `insert`(D_j) to Σ_i.
end for

Lemma 2 *The edge (p_i, p_j) is in $\overrightarrow{G}$ if and only if the response to the operation* `inside`(p'_j) *in Σ_i is "true."*

Proof: Omitted in this extended abstract. □

Thus, we have reduced the problem of constructing $\overrightarrow{G}$ to the problem of solving $n - 1$ independent instances of the off-line disk-inclusion problem. We outline in the next section how we can efficiently solve this problem.

3 The Off-Line Disk Inclusion Problem

Suppose we are given a sequence Σ of n `insert`(D_i) and `inside`(p_i) operations defining an instance of the off-line disk-inclusion problem. In this section we show how to determine the answers to all the `inside`(p_i) queries in $O(n \log n)$ total time. Before we provide our solution we remark that Sharir [26] gives a data structure for maintaining the on-line intersection of a collection of equal-radius disks. This data structure cannot be applied to our disk-inclusion problem, however, as the disks in our problem do not in general have equal radii.

Let us sketch our solution to the off-line disk-inclusion problem. We begin by building a complete binary tree T such that each leaf of T is associated with a different disk given in the sequence Σ. We order the leaves of T from left to right in the order they are given in the sequence Σ. For each internal node v in T, let $I(v)$ denote a representation of the common intersection of the disks stored at leaf-descendents of v. In the full version of this paper we show how $I(v)$ can be constructed, for all vertices v in T in $O(n \log n)$ total time. The important observation to make at this point is that each leaf of T is associated with a different index in Σ. Thus, if an `inside` query appears in position i in Σ, we can form a search path π_i from the root of T to the leaf corresponding to the first insertion after position i. Say that a node w of T is on the *left fringe* of π_i if w is not on π_i but is a left child of a node on π_i. Noting that the `inside` query is a decomposable search problem, we can answer an `inside`(p) query for position i by determining if p is inside $I(w)$ for each w that is a left fringe node of π_i (answering "true" if and only if it is inside them all). In the full version of this paper we use this observation, together with the fractional cascading technique of Chazelle and Guibas [5, 6], to answer each `inside` query in Σ in $O(\log n)$ time. This gives us the following:

Theorem 3 *Given a sequence Σ of n disk* `insert` *updates and point* `inside` *queries, one can answer all the* `inside` *queries in Σ in $O(n \log n)$ time.*

Combining this with the discussion from the previous section immediately gives us the following:

Theorem 4 *Given an n-vertex polygonal path P in 3-dimensional space, and a parameter $\varepsilon \geq 0$, one can find the fewest-vertex approximating path P' to P in $O(n^2 \log n)$ time under the ε-tolerance zone-error criterion.*

4 The Min-ε Problem in 3-D

Let us now turn to the min-ε problem. Suppose then that we are given a 3-D polygonal curve $P = (p_1, p_2, \ldots, p_n)$, and an integer parameter $m < n$, and we wish to find the smallest $\varepsilon \geq 0$ such that there is a subpath $P' = (p_{i_1}, p_{i_2}, \ldots, p_{i_m})$ that satisfies the ε-tolerance zone criterion. For solving the min-ε problem, we parallelize the min-# algorithm and invoke it as a subroutine in a parametric search algorithm, where the sought solution ε^* is the minimum ε for which there exists an approximating curve of no more than m vertices.

In fact, we don't really have to parallelize the entire min-# algorithm described above. We need only parallelize the portions of the algorithm that depend upon the parameter ε. In the case of our min-# algorithm, the only place where ε plays a role is in the sizes of the disks that are used in the off-line disk-inclusion algorithm. Thus, the only part of the min-# algorithm that we need to parallelize is the off-line disk-inclusion algorithm itself.

The following problem is a key to our algorithm: In the plane, given n different-radius discs $d_1, d_2, \ldots, d_n$ and n points $a_1, a_2, \ldots, a_n$, determine for every $j = 1, 2, \ldots, n$, whether $a_j \in \Pi_j = \cap_{i=1}^{j} d_i$. We now give an $O(\log n)$ time, $O(n)$ processor algorithm for solving this problem in the parallel comparison model.

Our algorithm is based on Goodrich's paradigm [12] for computing the upper envelope of n surfaces in the k-intersecting class. The algorithm is sketched below. We first build an n-leaf complete binary tree T. The ith leaf of T stores d_i and a_i. Every internal node v of T is associated with (1) the intersection $I(v)$ of the discs stored at the leaves of the subtree T_v rooted at v (of course, $I(v)$ needs to be computed); and (2) the point set $A(v)$ consisting of all the points a_i stored at the leaves of T_v. Observe that, for every j, the question of whether $a_j \in \Pi_j = \cap_{i=1}^{j} d_i$ can be answered by checking whether $a_j \in I(v)$ for $O(\log n)$ nodes v of the tree T.

One could use a naive parallel algorithm for computing the $I(v)$'s and checking whether $a_j \in I(v)$: Go up the tree T level by level, starting from the leaves; for every node v at each level, first compute $I(v)$ by merging the sorted sequences of the boundary vertices of $I(x)$ and $I(y)$, where x and y are the left and right children of v, and then check whether the points of $A(y)$ belong to $I(x)$. This checking can also be done by merging a sorted sequence of $A(y)$ with the boundary vertices of $I(x)$. At the root of T, we have the answers for all questions of whether $a_j \in \Pi_j$. This naive algorithm can be easily implemented in $O(\log n \log \log n)$ time and $O(n/ \log \log n)$ processors in the CREW PRAM model.

To obtain a faster algorithm in the parallel comparison model, we pipe-line the merge operations in computing $I(v)$ from $I(x)$ and $I(y)$ and in checking whether $a_j \in I(v)$, as in Goodrich's paradigm. A logarithmic number of stages are used for this pipe-lined procedure. Every $I(v)$ is maintained as a linked list. At the end of each stage t, a list $I_t(v)$ is stored at v. We say v is *full* when $I_t(v) = I(v)$. The merge at the children of v is pipe-lined to v by maintaining a list which is an *approximately-uniform* subsequence of $I_t(v)$, and $I_{t+1}(v)$ is defined as $I_t(x) \cup I_t(y)$. The information for the points in $A(v)$ can be maintained and pipe-lined in a similar fashion as for $I(v)$. These structures at every node v can be maintained in $O(1)$ time per stage. By using the techniques of [12] we can show that the problem of checking whether $a_j \in \Pi_j$ for every j is solvable in $O(\log n)$ time using n processors in the parallel comparison model. More details our found in the full version of the paper.

We then use this parallel algorithm to drive a parametric-search algorithm [21]. We simulate each step of this parallel algorithm sequentially, noting that some of the computations in this step depend upon whether the optimal value ε^* falls in specific intervals $[\varepsilon_1, \varepsilon_2]$. We find the median such ε_i in linear time and use the min-# algorithm as a subroutine, running in $O(n^2 \log n)$ time, to determine if ε^* is above or below this median. The answer to this question resolves half of the comparisons for this parallel step. We can therefore repeat this process recursively and in $O(\log n)$ calls to the min-# algorithm we will have resolved all the comparisons for this parallel step. We then repeat this process for the next parallel step. Since our parallel algorithm takes time $O(\log n)$, this implies that the total running time for our min-ε algorithm is $O(n^2 \log^3 n)$. Thus, we have the following:

Theorem 5 *Given an n-vertex polygonal path P in 3-dimensional space, and an integer parameter $m < n$, one can find the best m-vertex approximating path P' to P in $O(n^2 \log^3 n)$ time under the ε-tolerance zone-error criterion.*

The above discussion assumed that the underlying metric for defining radius-ε balls was the L_2 metric. In a later section we explore improvements and simplifications that can be achieved if we use the L_1 or L_∞ metrics for this purpose.

5 Min-# and Min-ε Algorithm for the L_2 Metric in d-D ($d \geq 4$)

As in three dimensions, the key to solving the the min-# problem in d dimensions is to reduce it to the off-line ball-inclusion problem in a lower dimensional space and apply decomposability to the ball-inclusion problem. We sketch below the main ideas of this technique.

In order to approximate a d-D curve, we solve n off-line ball inclusion problems, each performing $O(n)$ inclusion tests of points in $(d-1)$-balls. By a standard lifting map, testing if a point is in a set of $(d-1)$-balls is equivalent to asking whether a point in d-D is below the lower envelope of a collection of d-D hyperplanes: The query point is lifted to a paraboloid, and the balls are mapped to hyperplanes; the intersections of the d-D hyperplanes with the paraboloid, projected into $d-1$ dimensions, are the original balls.

Thus, to solve one instance of off-line ball inclusion, we build a static data structure for answering ray-shooting queries (see [9, 19]), using a parameter s to control a tradeoff between query time and space/preprocessing time. In particular, we build a ray-shooting data structure in $O(s \operatorname{polylog} n)$ space and preprocessing time, that can answer each ray-shooting query in $O(n \log n / s^{1/\lfloor \frac{d}{2} \rfloor})$ time. In the full paper we give a more complete discussion of space/time tradeoffs for the off-line ball inclusion problem.

Now, to solve a min-# problem, we have to build n data structures and ask $O(n^2)$ queries. If we balance the time required for the preprocessing and for performing the queries, we find out that $s = O(n^{2-2/(\lfloor \frac{d}{2} \rfloor + 1)} \text{polylog} n)$ and consequently obtain an $O(n^{3-2/(\lfloor \frac{d}{2} \rfloor + 1)} \text{polylog}\, n)$-time algorithm. For $d = 4$ or 5, for example, this gives us an algorithm for the min-# problem (under the L_2 metric) that runs in $O(n^{7/3} \text{polylog}\, n)$ time. Only for $d > 19$ does the running time of the algorithm exceed $O(n^{2.8} \text{polylog}\, n)$.

Since each ray-shooting data structure has subquadratic space, and we only need one structure at a time, the amount of space required by this algorithm is still dominated by the maximum size of the graph G, which is $\Theta(n^2)$. Thus, for d-dimensions we obtain:

Theorem 6 *Given an n-vertex polygonal path P in d-dimensional space, and an integer $m < n$, one can find the best m-vertex approximating path P' to P in $O(n^{3-2/(\lfloor \frac{d}{2} \rfloor+1)}\text{polylog}\, n)$ time by using $O(n^2)$ space.*

For the min-ε problem in 4-D, we need only observe (as for the 3-D version of the problem) that we have to parallelize only the steps in the sequential min-# algorithm that depend on ε. For this purpose we use the ray-shooting data structure of [13, 14]. With some preprocessing, the construction of this data structure does not depend on the value of ε. In this data structure the points are sorted according to their respective coordinates in parallel. As in the 3-D case, we perform the ray-shooting queries in parallel, which costs us an $O(\log^2 n)$ factor in the running time. Over all, we obtain an $O(n^{7/3}\text{polylog}\, n)$-time algorithm for the min-ε 4-D problem under the L_2 metric. The details are found in the full version of the paper. Note that this method does not extend to dimensions higher than 4.

6 Min-# and Min-ε Algorithms for L_1 and L_∞ Metrics ($d \geq 3$)

In this section we present our algorithms for the min-# and min-ε problems under the L_∞ and L_1 metrics. Decomposability of on-line ball inclusion, and the fact that the unit balls in these metrics are polytopes of constant complexity for constant d allow us to give efficient general solutions in d dimensions. We describe our algorithm for the ε-tolerance zone measure of error.

Recall that to compute the graph $\overrightarrow{G}$ we had to determine, for a fixed index $1 \leq i < n$ and each $i < j \leq n$, whether the ray $\overrightarrow{p_i p_j}$ intersects each radius-ε ball centered at a vertex p_k for $i < k < j$. As Figure 3 illustrates, the rays from p_i that intersect the radius-ε ball around p_k form a *cone*. If the ball at p_k contains p_i then the cone is the entire space.

Under L_∞, L_1, or any metric whose ball is a polytope whose complexity depends only on d, we can reduce the problem of constructing the graph $\overrightarrow{G}$ to a constant number of instances of the 1-D version of Ball Inclusion, which is the 1-D version of Off-Line Interval Inclusion Testing.

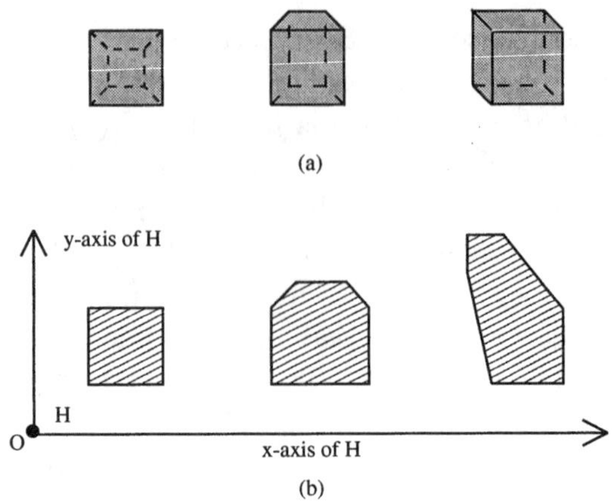

Figure 4: (a) Three different views of a cube from a point. (b) Three corresponding cross-sections for a cone.

Lemma 7 *In d dimensions, the question of whether a ray $\overrightarrow{p_i p_j}$ intersects each ball $B(p_k, \varepsilon)$, for $i < k < j$, reduces to $O(d^2)$ instances of Off-line Interval Inclusion Testing for the L_∞ metric or $O(d2^d)$ instances for the L_1 metric.*

Proof: Construct the cones of the rays from p_i through radius-ε balls centered at each p_k. The ray $\overrightarrow{p_i p_j}$ intersects all balls if and only if it is in the intersection of these cones. The key is to observe that halfspaces that define the boundary of these cones can be separated into a constant number of families, where each family projects to a halfplane in 2-D.

Since the metric is defined by a convex polytope, each cone is convex and polyhedral—it is the intersection of hyperplanes through p_i and the $(d-2)$-dimensional faces, called *ridges*, that are on the silhouette of the ball at p_k. For example, ridges in 3-D are one-dimensional edges of the 3-cube. Different views of a 3-cube give cones of different cross-sections, as shown in Figure 4, but all boundary facets of the corresponding cone are planes through p_i and a ridge of the 3-cube.

To determine whether a point p_j is in the intersection of cones defined by balls at p_k for $i < k < j$, we form an instance of Off-Line Interval Inclusion Testing for each ridge: A translated version of the ridge r generates a halfspace whenever it appears on the silhouette when viewed from p_i; since all halfspaces generated by translates of r are bounded by hyperplanes parallel to r that contain p_i, we can project each translate onto the circle at infinity that is contained in the 2-flat through p_i and orthogonal to ridge r. (Knowing the projection, the ridge, and p_i we can reconstruct the hyperplane that forms one of the cone's boundary facets.)

By decomposability, point p_j is in the intersection of cones if and only if it is in the intersection of their halfspaces, which happens if and only if the projection of p_j is included in each instance of Interval Inclusion Testing. Thus, $\vec{G}$ can be constructed by solving an instance of Interval Inclusion Testing for each ridge and intersecting the solutions.

We complete the proof by counting ridges in L_∞ and L_1 polytopes. In the d-cube, every pair of non-parallel planes generates a ridge, giving $2d(d-1)$ ridges. In the d-crosspolytope (a d-D L_1 ball), a ridge is obtained by choosing positive or negative vertices for $d-1$ of the d coordinate directions, giving $d2^{d-1}$ ridges. Not every ridge is on the silhouette—for the cube, at most $2d$ will be—ridges not on the silhouette and ridges whose cubes contain p_i do not contribute constraints. □

It is now straightforward to solve the min-# problem in $O(n^2)$ time for any constant dimension d with a constant-sized polytope for the metric. Note that the reduction to a number of 2-D problems allows us to use only $O(n)$ space, as in [7].

Theorem 8 *Given an n-vertex polygonal path P in d-D, and a parameter $\varepsilon \geq 0$, one can find the fewest-vertex approximating path P' to P in $O(n^2)$ time and $O(n)$ space under the ε-tolerance zone-error criterion using the L_∞ or L_1 error metric.*

We solve the d-D min-ε problem in $O(n^2 \log n)$ time without using parametric search. Instead, we can explicitly construct the set, *ERR*(P), of all candidate optimal ε values and then perform a simple binary search in this set using the min-# algorithm as the "probe." We use the fact that the optimal ε value must be determined by some edge (p_i, p_j) in G such that the farthest point p_k, with $i < k < j$, from the segment $\overline{p_i p_j}$ is at distance *exactly* ε from $\overline{p_i p_j}$. In the full version of the paper we show how, fixing the starting point p_i, we can determine the farthest such point p_k for each edge $\overline{p_i p_j}$ in $O(n \log n)$ time. Thus, we can compute the set *ERR*(P) in $O(n^2 \log n)$ time, and then perform a binary search in this quadratic-size set in $O(n^2 \log n)$ time using our quadratic-time min-# algorithm as the "probe."

We can actually achieve an $O(n)$ space bound, but to do so we cannot afford to maintain the error set *ERR*(P) explicitly (since $|ERR(P)| = O(n^2)$). Instead, we make use of the sampling technique of [7], which allows us to store only $O(n)$ errors of *ERR*(P), while still being able to perform binary search on the $O(n^2)$ errors. Therefore, we can achieve the following:

Theorem 9 *Given an n-vertex polygonal path P in d-D, and an integer parameter $m \leq n$, one can find the best m-vertex approximating path P' to P in $O(n^2 \log n)$ time under the ε-tolerance zone-error criterion using the L_∞ or L_1 metric.*

References

[1] H. Alt and M. Godau, Measuring the resemblance of polygonal curves, *Proc. 8th Ann. ACM Symp. on Computational Geometry*, Berlin, Germany, 102–109, 1992.

[2] E.M. Arkin, L.P. Chew, D.P. Huttenlocher, K. Kedem, and J.S.B. Mitchell, An efficiently computable metric for comparing polygonal shapes, *IEEE Trans. on Pattern Analysis and Machine Intelligence*, 13(3):209–216, 1991.

[3] B. Buttenfield, Treatment of the cartographic line, *Cartographica*, 22:1–26, 1985.

[4] W.S. Chan and F. Chin, Approximation of polygonal curves with minimum number of line segments or minimum error, *Int. J. of Computational Geometry and Applications*, 6(1):59–77, 1996.

[5] B. Chazelle and L.J. Guibas, Fractional cascading: I. A data structuring technique, *Algorithmica*, 1:133–162, 1986.

[6] B. Chazelle and L.J. Guibas, Fractional cascading: II. Applications, *Algorithmica*, 1:163–191, 1986.

[7] D.Z. Chen and O. Daescu, Space-efficient algorithms for approximating polygonal curves in two-dimensional space. Manuscript, 1997.

[8] D.H. Douglas and T.K. Peucker, Algorithms for the reduction of the number of points required to represent a digitized line or its caricature, *Canadian Cartographer*, 10(2):112–122, 1973.

[9] J. Erickson, Space-time tradeoffs for emptiness queries, *Proc. 13th Ann. ACM Symp. on Computational Geometry*, Nice, France, 304–313, 1997.

[10] D. Eu and G.T. Toussaint, On approximation polygonal curves in two and three dimensions, *CVGIP: Graphical Models and Image Processing*, 56(3):231–246, 1994.

[11] R. Fleischer, K. Mehlhorn, G. Rote, E. Welzl, and C.-K. Yap, Simultaneous inner and outer approximation of shapes, *Algorithmica*, 8:365–389, 1992.

[12] M.T. Goodrich, Using approximation algorithms to design parallel algorithms that may ignore processor allocation. *Proc. 32nd Ann. IEEE Symp. on Foundations of Computer Science*, San Juan, Puerto Rico, 711–722, 1991.

[13] M.T. Goodrich and R.Tamassia, Dynamic ray shooting and shortest paths via balanced geodesic triangulations, *Proc. 9th Ann. ACM Symp. on Computational Geometry*, San Francisco, CA, 318–327, 1993.

[14] M.T. Goodrich and R. Tamassia, Dynamic trees and dynamic point location, *SIAM J. Computing*, to appear.

[15] L.J. Guibas, J.E. Hershberger, J.S.B. Mitchell, and J.S. Snoeyink, Approximating polygons and subdivisions with minimum link paths, *Int. J. of Computational Geometry and Applications*, 3(4):383–415, 1993.

[16] I. Ihm and B. Naylor, Piecewise linear approximations of digitized space curves with applications, in N. M. Patrikalakis (ed.), *Scientific Visualization of Physical Phenomina*, Springer-Verlag, Tokyo, 545–569, 1991.

[17] H. Imai and M. Iri, Computational-geometric methods for polygonal approximations of a curve, *Computer Vision, Graphics, and Image Processing*, 36:31–41, 1986.

[18] H. Imai and M. Iri, Polygonal approximations of a curve-formulations and algorithms, *Computational Morphology*, 71–86, North–Holland, Amsterdam, 1988.

[19] J. Matoušek and O. Schwarzkopf, On ray shooting in convex polytopes, *Discrete & Computational Geometry*, 10:215–232, 1993.

[20] M. McAllister and J. Snoeyink, Medial axis generalisation of hydrology networks, *AutoCarto 13: ACSM/ASPRS Ann. Convention Technical Papers*, Seattle, WA, 164–173, 1997.

[21] N. Megiddo, Applying parallel computation algorithms in the design of serial algorithms, *J. of the ACM*, 30:852–865, 1983.

[22] A. Melkman and J. O'Rourke, On polygonal chain approximation. *Computational Morphology*, 87–95, North–Holland, Amsterdam, 1988.

[23] B.K. Natarajan, On comparing and compressing piecewise linear curves, Technical Report, Hewlett Packard, 1991.

[24] B.K. Natarajan and J. Ruppert, On sparse approximations of curves and functions, *Proc. 4th Canadian Conf. on Computational Geometry*, 250–256, 1992.

[25] G. Rote, A new metric between polygons, and how to compute it, *Proc. 19th Int. Colloq. Automata Lang. Program., Lecture Notes in Computer Science*, 623:404–415, Springer-Verlag, 1992.

[26] M. Sharir, A near-linear algorithm for the planar 2-center problem, *Proc. 12th Ann. ACM Symp. on Computational Geometry*, Philadelphia, PA, 106–112, 1996.

[27] K.R. Varadarajan, Approximating monotone polygonal curves using the uniform metric, *Proc. 12th Ann. ACM Symp. on Computational Geometry*, Philadelphia, PA, 106–112, 1996.

Constructing Cuttings in Theory and Practice*

Sariel Har-Peled†

Abstract

We present a new randomized incremental algorithm for computing a cutting in an arrangement of lines in the plane. The algorithm produce cuttings whose expected size is $O(r^2)$, and the expected running time of the algorithm is $O(nr)$. Both bounds are asymptotically optimal for nondegenerate arrangements. The algorithm is also simple to implement, and we present empirical results showing that the algorithm and some of its variants perform well in practice. We also present another efficient algorithm (with slightly worse time bound) that generates small cuttings whose size is guaranteed to be close to the known upper bound of [9].

1 Introduction

A natural approach for solving various problems in computational geometry is the divide-and-conquer paradigm. A typical application of this paradigm to problems involving a set L of n lines in the plane, is to fix a parameter $r > 0$, and to partition the plane into regions $R_1, \ldots, R_m$ (those regions are usually vertical trapezoids, or triangles), such that the number of lines of L that intersect the interior of R_i is at most n/r, for any $i = 1, \ldots, m$. This allows us to split the problem at hand into subproblems, each involving the subset of lines intersecting a region R_i. Such a partition is known as a $(1/r)$-*cutting* of the plane. See [1] for a survey of algorithms that use cuttings. For further work related to cuttings, see [4].

The first (though not optimal) construction of cuttings, is due to Clarkson [6]. Chazelle and Friedman [5] showed the existence of $(1/r)$-cuttings with $m = O(r^2)$ (a bound that is worst-case tight). They also showed that such cuttings, consisting of vertical trapezoids, can be computed in $O(nr)$ time. Although this construction is asymptotically optimal, it does not seem to produce a practically small number of regions. Coming up with the smallest possible number of regions (i.e., reducing the constant of proportionality) is important for the efficiency of (recursive) data structures that use cuttings. Currently, the best lower bound on the number of vertical trapezoids in a $(1/r)$-cutting in an arrangement of lines, is $2.54(1 - o(1))r^2$, and the optimal cutting has at most $8r^2 + 6r + 4$ trapezoids, see [9]. Improving the upper and lower bounds on the size of cuttings is still open, indicating that our understanding of cuttings is still far from being satisfactory. In Section 3, we outline Matoušek construction for achieving the upper bound and show a slightly improved construction (see below for details).

In spite of the theoretical importance of cuttings (in the plane and in higher dimensions), we are not aware of any implementation of efficient algorithms for constructing cuttings. In this paper we propose a new and simple randomized incremental algorithm for constructing cuttings, and prove the expected worst-case tight performance bounds of the new algorithm, as stated in the abstract. We also present empirical results on several algorithms/heuristics for computing cuttings that we have implemented. They are mostly variants of our new algorithm, and they all perform well in practice. An $O(r^2)$ bound on the expected size of the cuttings for some of those variants can be proved, while for the others no formal proof of performance is currently available. We leave this as an open question for further research.

Matoušek [9] gave an alternative construction for cuttings, showing that there exists a $(1/r)$-cutting with at most (roughly) $8r^2$ vertical trapezoids. Unfortunately, this construction relies on computing the whole arrangement, and its computation thus takes $O(n^2)$ time. We present a new randomized algorithm that is based on Matoušek's construction; it generates a $(1/r)$-cutting of size $\leq (1+\varepsilon)8r^2$, in $O\left(\frac{nr}{\varepsilon}\log^2 n\right)$ time, where $0 < \varepsilon \leq 1$ is a prescribed constant.

In Section 2, we present the new algorithm, and analyze its expected running time and the expected number of trapezoids that it produces. Specifically, the expected running time is $O(nr)$ and the expected size of the output cutting is $O(r^2)$. In Section 3 we present our variant of Matoušek's construction. In Section 4 we present our empirical results, comparing the new algorithm with several other algorithms/heuristics for constructing cuttings. These algorithms are mostly variants of our main algorithm, but they also include a variant of the older algorithm of Chazelle and Friedman. The cuttings generated by the new algorithm and its variants are of size, roughly, $15r^2$ in the worst case. (The algorithm generates smaller cutting when r is small.

*This work has been supported by a grant from the U.S.–Israeli Binational Science Foundation. This work is part of the author's Ph.D. thesis, prepared at Tel-Aviv University under the supervision of Prof. Micha Sharir.

†School of Mathematical Sciences, Tel Aviv University, Tel Aviv 69978, Israel; sariel@math.tau.ac.il

SCG 98 Minneapolis Minnesota USA

For example, for $r = 2$ the constant is about 9.) In contrast, the Chazelle-Friedman algorithm generates cuttings of size roughly $70r^2$. We conclude in Section 5 by mentioning a few open problems.

2 Incremental Randomized Construction of Cuttings

Given a set $\hat{S}$ of n lines in the plane, let $\mathcal{A}(\hat{S})$ denote the arrangement of $\hat{S}$; namely, the partition of the plane into faces, edges, and vertices as induced by the lines of $\hat{S}$. Let $\mathcal{A}^{||}_{\text{Mrg}}(\hat{S})$ denote the partition of the plane into vertical trapezoid, obtained by erecting two vertical segments up and down from each vertex of $\mathcal{A}(\hat{S})$, and extending each of them until it either reaches a line of $\hat{S}$, or all the way to infinity.

Computing the decomposed arrangement $\mathcal{A}^{||}_{\text{Mrg}}(\hat{S})$ can be done as follows. Pick a random permutation $S =< s_1, \ldots, s_n >$ of $\hat{S}$. Put $S_i =< s_1, \ldots, s_i >$, for $i = 1, \ldots, n$. We compute incrementally the decomposed arrangements $\mathcal{A}^{||}_{\text{Mrg}}(S_i)$, for $i = 1, \ldots, n$, by inserting the i-th line s_i of S into $\mathcal{A}^{||}_{\text{Mrg}}(S_{i-1})$. To do so, we compute the zone Z_i of s_i in $\mathcal{A}^{||}_{\text{Mrg}}(S_{i-1})$, which is the set of all trapezoids in $\mathcal{A}^{||}_{\text{Mrg}}(S_{i-1})$ that intersect s_i. We split each trapezoid of Z_i into at most 4 trapezoids, such that no trapezoid intersects s_i in its interior, as in [14]. Finally, we perform a pass over all the newly created trapezoids, merging vertical trapezoids that are adjacent, and have identical top and bottom lines. The merging step guarantees that the resulting decomposition is $\mathcal{A}^{||}_{\text{Mrg}}(S_i)$, independently of the insertion order of elements in S_i; see [7].

However, if we decide to skip the merging step, the resulting structure, denoted as $\mathcal{A}^{||}(S_i)$, depends on the order in which the lines are inserted into the arrangement. In fact, $\mathcal{A}^{||}(S_i)$ is $\mathcal{A}^{||}_{\text{Mrg}}(S_i)$ with additional superfluous vertical walls. Each such vertical wall is a fragment of a vertical wall that was created at an earlier stage and got split during a later insertion step.

Definition 2.1 Let $\hat{S}$ be a set of n lines in the plane, and let $c > 0$ be a constant. A *c-cutting* of $\hat{S}$ is a partition of the plane into regions $R_1, \ldots, R_m$, such that, for each $i = 1, \ldots, m$, the number of lines of $\hat{S}$ that intersect the interior of R_i is at most cn.

A region C in the plane is *c-active*, if the number of lines of $\hat{S}$ that intersect the interior of C is larger than cn.

A $(1/r)$*-cutting* is thus a partition of the plane into m regions such that none of them is $(1/r)$-active. Chazelle and Friedman [5] showed that one can compute, in $O(nr)$ time, a $(1/r)$-cutting that consists of $O(r^2)$ vertical trapezoids.

We propose a new algorithm for computing a cutting that works by incrementally computing the arrangements $\mathcal{A}^{||}(S_i)$, using a random insertion order of the lines. The new idea in the algorithm, is that any "light" trapezoid (i.e., a trapezoid that is not $(1/r)$-active) constructed by the algorithm is immediately added to the final cutting, and the algorithm does not maintain the arrangement inside such a trapezoid from this point on. In this sense, one can think of the algorithm as being greedy; that is, it adds a trapezoid to the cutting as soon as one is constructed, until the whole plane is covered. The algorithm, called `CutRandomInc`, is depicted in Figure 1.

If `CutRandomInc` outputs C_k, for some $k < n$, then C_k has no $(1/r)$-active trapezoids, and it is thus a $(1/r)$-cutting. Otherwise, if C_n is output, then again it has no $(1/r)$-active trapezoids, because any such trapezoid must have been processed and split earlier (when one of the lines crossing the trapezoid is inserted). Thus C_n is a $(1/r)$-cutting. This implies the correctness of `CutRandomInc`.

The covering C_i of the plane maintained by `CutRandomInc` depends heavily on the order in which the lines are inserted into the arrangement. To bound the expected running time of `CutRandomInc`, and the expected size of the cutting that it computes, we adapt the analysis of Agarwal et al. [3] to our case.

The following elegant argument, due to Agarwal et al. [2], shows that the expected complexity of $\mathcal{A}^{||}(S_i)$ remains quadratic:

Lemma 2.2 *Let $\hat{S}$ be a set of n lines, and let S be a random permutation of $\hat{S}$. Then the expected complexity of $\mathcal{A}^{||}(S)$ is $O(n^2)$, where $\mathcal{A}^{||}(S)$ is the decomposed arrangement computed by the incremental algorithm outlined at the beginning of the section,* without *performing merging.*

Proof: Let V be the set of all intersection points of pairs of lines of $\hat{S}$ (the vertices of $\mathcal{A}(\hat{S})$). For $v \in V$, let $D(v,i)$ (resp. $U(v,i)$) be an indicator variable, such that $D(v,i)$ (resp. $U(v,i)$) is 1 if the vertical wall emanating from v still exists in $\mathcal{A}^{||}(S)$ as we go downward (resp. upward) from v after crossing i lines. Clearly, the complexity of $\mathcal{A}^{||}(S)$ is proportional to

$$\sum_{v \in V} \sum_{i=0}^{n-2} (D(v,i) + U(v,i)).$$

However, if $D(v,i) = 1$ then the two lines defining v must appear in S before the first i lines that intersect the downward ray emanating from v. Thus, $Pr\left[D(v,i) = 1\right] \le \frac{2}{(i+1)(i+2)}$, and the same inequality holds for $U(v,i)$. Therefore, the expected complexity of $\mathcal{A}^{||}(S)$ is

$$O\left(\sum_{v \in V} \sum_{i=0}^{n-2} \frac{2}{(i+1)(i+2)}\right) = O\left(\sum_{v \in V} 1\right) = O(n^2).$$

■

The analysis is applied in the following abstract framework. Let $\hat{S}$ be a set of n objects (in our case the objects are lines in $\mathbb{R}^2$). A *selection* of $\hat{S}$ is an ordered sequence of distinct elements of $\hat{S}$. Let $\sigma(\hat{S})$ denote, as above, the set of all selections of $\hat{S}$. For a permutation S of $\hat{S}$, let S_i denote the subsequence consisting of the first i elements of S, for $i = 1, \ldots, n$. For each $R \in \sigma(\hat{S})$, we define a collection $CT(R)$ of 'regions', each defined by a small subset of R (in our case, this is the collection of $(1/r)$-active vertical trapezoids, after the lines of R are inserted by `CutRandomInc`). Let $\mathcal{T}(S) = \cup_{R \in \sigma(\hat{S})} CT(R)$ denote the set of all possible regions.

We associate two subsets $D(\Delta), K(\Delta) \subseteq S$ with each region $\Delta \in T$, where $D(\Delta)$ is the *defining set* of Δ, in the sense that $\Delta \in CT(R)$ only if $D(\Delta) \subseteq R$. (In our case, $D(\Delta)$ defines the four sides of Δ, as illustrated in Figure 2.) The size of $D(\Delta)$ is assumed to be bounded by a fixed constant (6 in our case, see Figure 2). The set $K(\Delta)$ is the *killing set* of Δ; namely, if $K(\Delta) \cap R \neq \emptyset$, then $\Delta \notin CT(R)$. (In our settings, $K(\Delta)$ is the set of lines of $\hat{S}$ intersecting

ALGORITHM CutRandomInc($\hat{S}$, r)

Input: A set $\hat{S}$ of n lines, a positive integer r

Output: A $(1/r)$-cutting of $\hat{S}$ by vertical trapezoids

begin

Choose a random permutation $S =< s_1, s_2, \ldots, s_n >$ of $\hat{S}$.

$\mathcal{C}_0 \leftarrow \{\mathbb{R}^2\}$.

$i \leftarrow 0$.

while there are $(1/r)$-active trapezoids in $\mathcal{C}_i$ **do**

$i \leftarrow i + 1$

$Zone_i \leftarrow$ The set of $(1/r)$-active trapezoids in $\mathcal{C}_{i-1}$ that intersect s_i.

$Zone_i' \leftarrow \cup_{T \in Zone_i} split(T, s_i)$,

where $split(T, s)$ is the operation of splitting a vertical trapezoid T crossed by a line s into at most four vertical trapezoids, as in [7], such that the new trapezoids cover T, and they do not intersect s in their interior.

$\mathcal{C}_i \leftarrow (\mathcal{C}_{i-1} \setminus Zone_i) \cup Zone_i'$.

end while

return $\mathcal{C}_i$

end CutRandomInc

Figure 1: Algorithm for constructing a $(1/r)$-cutting of an arrangement of lines

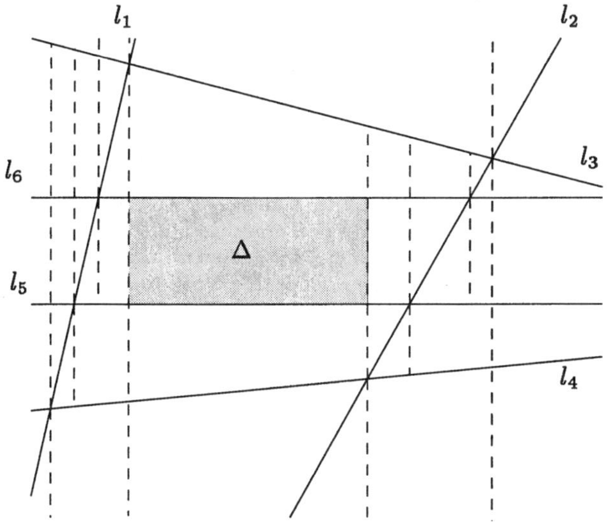

Figure 2: Each vertical trapezoid is defined by at most 6 lines, two containing its top and bottom edges, and two pairs defining its vertical sides. For example, the vertical walls of the trapezoid Δ are defined by the intersection points $l_1 \cap l_3$ and $l_2 \cap l_4$, and its bottom and top lines are l_5, l_6, respectively.

the interior of Δ.) Let $w(\Delta) = |K(\Delta)|$ denote the *weight* of Δ.

Let $\hat{S}$ be a set of objects such that, for any sequence $R \in \sigma(\hat{S})$, the following axioms hold:

(A) For any $\Delta \in CT(R)$, we have $D(\Delta) \subseteq R$, and $K(\Delta) \cap R = \emptyset$.

(B) If $\Delta \in CT(R)$, then for any subsequence R' of R, such that $D(\Delta) \subseteq R'$, we have $\Delta \in CT(R')$.

The above is a natural extension of the settings of Agarwal et al. [3], where the insertion order of objects in the sample is important. For any natural number k, define

$$\mathcal{T}_k(S) = \left\{ \Delta \in \mathcal{T}(S) \;\middle|\; w(\Delta) \geq k \right\}.$$

Intuitively, as the execution of CutRandomInc progresses, the number of trapezoids with heavy weight becomes smaller. The following key lemma asserts that the expected number of heavy trapezoids decreases exponentially as a function of their weight.

Lemma 2.3 *Given a set $\hat{S}$ of n objects, let R be a random sequence of $r \leq n$ distinct elements of $\hat{S}$, where each such sequence (of size r) is chosen with equal probability, and let t be a parameter, $1 \leq t \leq r/d$, where $d = \max_\Delta |D(\Delta)|$. Assuming that $\hat{S}$ satisfies Axioms (A) and (B), we have*

$$E\left[|CT(R) \cap \mathcal{T}_k(S)| \right] = O(2^{-t}) \cdot E\left[|CT(R')| \right],$$

where R' is a random subsequence of $\hat{S}$, as above, of size $r' = \lfloor r/t \rfloor$, and $k = t \lceil n/r \rceil$.

Proof: The proof is a straightforward adaptation of the proof of Lemma 2.2 of [3] to our ordered sampling. ■

Note that Axioms (A) and (B) hold for the active trapezoids in $\mathcal{C}_i$, the set of trapezoids computed in the i-th iteration of CutRandomInc.

Theorem 2.4 *The expected size of the $(1/r)$-cutting generated by* CutRandomInc *is $O(r^2)$, and the expected running time is $O(nr)$, for any integer $1 \le r \le n$.*

Proof: Omitted. Will appear in the full-version. ■

Remark 2.5 A natural open problem that arises is the following: What if we perform merging in CutRandomInc? While in practice the resulting algorithm yields smaller cuttings (see Section 4), the above analysis is no longer valid, because Axiom (B) may fail. See Figure 3 for an example that shows that axiom (B) may be violated if we use merging. Can one prove that the expected size of the cutting and the expected running time are still $O(r^2)$ and $O(nr)$ in this case?

The algorithm CutRandomInc works also for planar arrangements of segments and x-monotone curves (such that the number of intersection of each pair of curves is bounded by a constant). This follows immediately by observing that Lemma 2.2 can be extended for those cases, and that Axioms (A) and (B) hold for the vertical decomposition of such arrangements.

Lemma 2.6 *Let Γ be a set of x-monotone curves such that each pair intersects in at most a constant number of points. Then the expected size of the $(1/r)$-cutting generated by* CutRandomInc *for Γ is $O(r^2)$, and the expected running time is $O(nr)$, for any integer $1 \le r \le n$.*

However, the arrangement of a set of n segments or curves might have subquadratic complexity (since the number of intersection points might be subquadratic). This raises the question whether CutRandomInc generates smaller cuttings for such sparse arrangements.

Definition 2.7 Let Γ be a set of curves in the plane. We denote by $\kappa(\Gamma)$ the number of intersection points between pairs of curves of Γ.

Lemma 2.8 *Let $\hat{\Gamma}$ be a set of n curves, so that each pair of curves from Γ have at most $O(1)$ intersection points, and let Γ be a random permutation of $\hat{\Gamma}$. Then the expected complexity of $\mathcal{A}^{||}(\Gamma)$ is $O(n \log n + \kappa(\hat{\Gamma}))$, where $\mathcal{A}^{||}(\Gamma)$ is the decomposed arrangement computed by the incremental insertion algorithm,* without *performing merging.*

Proof: Note that any intersection point of a pair of curves of $\hat{\Gamma}$, induces an upward and downward vertical "walls", and the expected number of vertical walls in $\mathcal{A}^{||}(\Gamma)$ is $O(\kappa(\hat{\Gamma}))$, arguing as in the proof of Lemma 2.2.

Additionally, there are vertical walls defined by the endpoints of the curves of Γ. Let p be an endpoint of a curve $\gamma \in \hat{\Gamma}$. The probability that the vertical upward ray v_p emanating from p will introduce i vertical walls in $\mathcal{A}^{||}(\Gamma)$, is the probability that γ will be chosen before the first i curves of Γ that this vertical ray intersects. Thus, the expected number of superfluous vertical walls introduced by v_p is

$$\sum_{i=0}^{n} \frac{1}{i+1} = O(\log n).$$

Thus, the total number of vertical walls in $\mathcal{A}^{||}(\Gamma)$ introduced by the endpoints of arcs in $\hat{\Gamma}$ is $O(n \log n)$, and the Lemma readily follows. ■

Corollary 2.9 *Let $\hat{\Gamma}$ be a set of n curves, so that each pair of curves from $\hat{\Gamma}$ have at most $O(1)$ intersection points, and let Γ_r be a random selection of r elements of $\hat{\Gamma}$. Then the expected complexity of $\mathcal{A}^{||}(\Gamma_r)$ is $O(r \log r + \frac{r^2}{n^2}\kappa(\hat{\Gamma}))$.*

Proof: We note that the probability of an intersection point of $\mathcal{A}(\hat{\Gamma})$ to appear in $A = \mathcal{A}(\Gamma_r)$ is $\frac{r(r-1)}{n(n-1)}$. Hence, the expected number of intersection points of arcs of $\hat{\Gamma}$ in A is

$$O\left(\frac{r^2}{n^2}\kappa(\hat{\Gamma})\right).$$

The lemma now readily follows by applying Lemma 2.8 to $\mathcal{A}^{||}(\Gamma_r)$. ■

Theorem 2.10 *Let $\hat{\Gamma}$ be a set of n curves, such that each pair of curves of $\hat{\Gamma}$ intersect in at most a constant number of points. Then the expected size of the $(1/r)$-cutting generated by* CutRandomInc*, when applied to Γ, is*

$$O\left(r \log r + \frac{r^2}{n^2}\kappa\right),$$

and the expected running time is $O(n \log^2 r + r\kappa/n)$, for any integer $1 \le r \le n$, where $\kappa = \kappa(\hat{\Gamma})$.

Proof: The proof is a tedious extension of the proof of Theorem 2.4. We derive similar recurrences to the ones used in the proof of Theorem 2.4. In deriving and solving those recurrences, we repeatedly apply the bounds stated in Lemma 2.6. We omit the details. ■

Remark 2.11 An interesting question is whether CutRandomInc can be extended to higher dimensions. If we execute CutRandomInc in higher dimensions, we need to use a more complicated technique in decomposing each of our "vertical trapezoids" whenever it intersects a newly inserted hyperplane. Chazelle and Friedman's algorithm uses bottom vertex triangulation for this decomposition. However, in our case, it is easy to verify that CutRandomInc might generate simplices that are no longer defined by a constant number of hyperplanes. This implies that Lemma 2.3 can no longer be applied to CutRandomInc in higher dimension. We leave the problem of extending CutRandomInc to higher dimensions as an open problem for further research.

3 Generating Small Cuttings

In this section, we present an efficient algorithm that generates cuttings of guaranteed small size. The algorithm is based on Matoušek's construction of small cuttings [9]. We first review his construction, and then show how to modify it for building small cuttings efficiently.

Definition 3.1 **([9])** Let L be a set of n lines in the plane in general position, i.e., every pair of lines intersect in exactly one point, no three have a common point, no line is vertical or horizontal, and the x-coordinate of all intersections are pairwise distinct. The *level* of a point in the plane is the number of lines of L lying strictly below it. Consider the set E_k of all edges of the arrangement of L having level k (where $0 \le k < n$). These edges form an x-monotone polygonal line, which is called the *level k* of the arrangement of L.

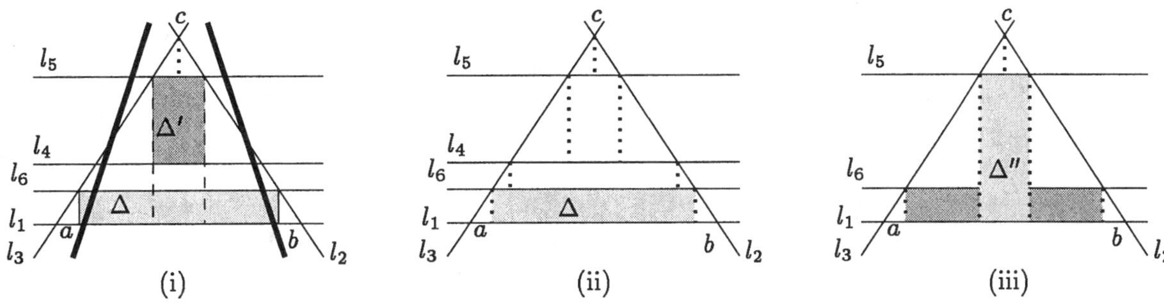

Figure 3: Axiom (B) fails if merging is used: The thick lines represent two sets of 100 parallel lines, and we want to compute a (1/10)-cutting. We execute CutRandomInc with the first 6 lines $l_1, \ldots, l_6$ in this order. Note that any trapezoid that intersects a thick line is active. The first trapezoid Δ' inside $\triangle abc$ that becomes inactive is created when the line l_5 is being inserted; see parts (i) and (ii). However, if we skip the insertion of the line l_4 (as in part (iii)), the corresponding inactive trapezoid Δ'' will extend downwards and intersect Δ. However, the trapezoid Δ'' is inactive, implying that the decomposition of the plane inside Δ'' is no longer maintained. In particular, this implies that the trapezoid Δ will not be created, since it is being blocked by Δ'', and no merges involving areas inside Δ'' will take place. This is a contradiction to axiom (B), since l_4 does not belong to the killing or defining sets of Δ.

Definition 3.2 ([9]) Let E_k be the level k in the arrangement $\mathcal{A}(L)$ with edges $e_0, e_1, \ldots, e_t$ (from left to right), and let p_i be a point in the interior of the edge e_i, for $i = 0, \ldots, t$. The *q-simplification* of the level k, for an integer parameter $1 \le q \le t$, is defined as the x-monotone polygonal line containing the part of e_0 to the left of the point p_0, the segments $p_0p_q, p_qp_{2q}, \ldots, p_{\lfloor (t-1)/q \rfloor q}p_t$ and the part of e_t to the right of p_t. Let $\mathrm{simp}_q(E_k)$ denote this polygonal line.

Let L be a set of n lines, and let $\mathcal{E}_{i,q}$ denote the union of the levels $E_i, E_{i+q}, \ldots, E_{i+\lfloor (n-1-i)/q \rfloor q}$, for $i = 0, \ldots, q-1$. Let $\mathrm{simp}_q(\mathcal{E}_{i,q})$ denote the set of edges of the q-simplifications of the levels of $\mathcal{E}_{i,q}$.

Matoušek showed that the vertical decomposition of the plane induced by $\mathrm{simp}_q(\mathcal{E}_{i,q})$, where $q = n/(2r)$ (we assume that n is divisible by $2r$), is a $(1/r)$-cutting of the plane, for any $i = 0, \ldots, q-1$. Moreover, the following holds:

Theorem 3.3 ([9]) *Let L be a set of n lines in general position, and let r be a positive integer, and let $q = n/(2r)$. Then the subdivision of the plane defined by the vertical decomposition of $\mathrm{simp}_q(\mathcal{E}_{m,q})$ is a $(1/r)$-cutting of $\mathcal{A}(L)$, where $0 \le m < q$ is the index $i \in \{0, \ldots, q-1\}$ for which $|\mathcal{E}_{i,q}|$ is minimized. Moreover, the cutting generated has at most $8r^2 + 6r + 4$ trapezoids.*

Remark 3.4 (i) Matoušek's construction can be slightly improved, by noting that the leftmost and rightmost points in a q-simplification of a level can be placed at "infinity"; that is, we replace the first and second edges in the q-simplification by a ray emanating from p_q which is parallel to e_0. We can do the same thing to the two last edges of the simplified level. We denote this improved simplification by simp'_q. It is easy to prove that using this improved simplification results in a $(1/r)$-cutting of $\mathcal{A}(L)$ with at most $8r^2 + 6r + 4 - 4(2r-1) = 8r^2 - 2r + 8$ vertical trapezoids.

(ii) Inspecting Matoušek's construction, we see that if we can only find an i such that $|\mathcal{E}_{i,q}| \le cn^2/q$, where $c > 1$ is a prescribed constant, then the vertical decomposition induced by $\mathrm{simp}_q(\mathcal{E}_{i,q})$ is a $(1/r)$-cutting having $\le c(8r^2 - 2r + 8)$ trapezoids.

Let $n_i = |\mathcal{E}_{i,q}|$, for $i = 0, \ldots, q-1$. Matoušek's construction is carried out by computing the numbers $n_0, \ldots, n_q$, and picking the minimal number n_i, which is guaranteed to be no larger than the average n^2/q. Unfortunately, implementing this scheme directly, requires computing the whole arrangement $\mathcal{A}(L)$, so the resulting running time is $O(n^2)$. Let assume for the moment that one can compute the number n_i quickly, then, as the following lemma testifies, one can compute a number n_i which is $\le (1+\varepsilon)n^2/q$, without computing all the n_is.

Lemma 3.5 *Let $n_0, \ldots, n_{q-1}$ be q positive integers, such that their sum $m = \sum_{i=0}^{q-1} n_i$ is known in advance, and let $\varepsilon > 0$ be a prescribed constant. One can compute an index $0 \le k < q$, such that $n_k \le \lceil (1+\varepsilon)m/q \rceil$, by repeatedly picking uniformly and independently a random index $0 \le i < q$, and by checking whether $n_i \le \lceil (1+\varepsilon)m/q \rceil$. The expected number of iterations required is $\le 1 + 1/\varepsilon$.*

Proof: Let Y_i be the random variable which is the value of n_i picked in the i-th iteration. Using Markov inequality[1], one obtain:

$$Pr\left[Y_i \ge (1+\varepsilon)\frac{m}{q}\right] \le \frac{E\left[Y_i\right]}{(1+\varepsilon)\frac{m}{q}}.$$

Since, $E[Y_i] = m/q$, we have that the probability for failure in the i-th iteration is

$$Pr\left[Y_i \ge (1+\varepsilon)\frac{m}{q}\right] \le \frac{1}{1+\varepsilon}.$$

Let X denote the number of iterations required by the algorithm. Then $E[X]$ is bounded by the expected number of trials to the first success in a binomial distribution with probability $p \ge 1 - \frac{1}{1+\varepsilon}$. Thus, the expected number of iterations is bounded by

$$E[X] \le \frac{1}{p} \le \frac{1}{1 - \frac{1}{1+\varepsilon}} = 1 + \frac{1}{\varepsilon}.$$

■

To apply Lemma 3.5 in our setting, we need to supply an efficient algorithm for computing the level of an arrangement of lines in the plane.

[1] The inequality asserts that $Pr\left[Y \ge t\right] \le \frac{E[Y]}{t}$, for a random variable Y that assumes only nonnegative values.

Lemma 3.6 ([8]) *Let L be a set of n lines in the plane. Then one can compute, in $O((n+h)\log^2 n)$ time, the level k of $\mathcal{A}(L)$, where $h = |E_k|$ is the complexity of the level.*

Combining Lemma 3.5 and Lemma 3.6, we have:

Theorem 3.7 *Let L be a set of n lines in the plane, and let $0 < \varepsilon \leq 1$ be a prescribed constant. Then one can compute a $(1/r)$-cutting of $\mathcal{A}(L)$, having at most $(1+\varepsilon)(8r^2 - 2r + 8)$ trapezoids. The expected running time of the algorithm is $O\left(\left(1+\frac{1}{\varepsilon}\right) nr \log^2 n\right)$.*

Proof: By the above discussion, it is enough to find an index $0 \leq i \leq q-1$, such that $|\mathcal{E}_{i,q}| \leq M = (1+\varepsilon)\frac{n^2}{q} \leq 2(1+\varepsilon)nr$, where $q = \lceil n/(2r) \rceil$. By Remark 3.4 (ii), the vertical decomposition of $\mathrm{simp}'_q(\mathcal{E}_{i,q})$, is a $(1/r)$-cutting of the required size.

Picking i randomly, we have to check whether $|\mathcal{E}_{i,q}| \leq M$. We can compute $\mathcal{E}_{i,q}$, by computing the levels $E_i, E_{i+q}, \ldots, E_{i+\lfloor (n-i-1)/q \rfloor q}$ in an output sensitive manner, using Lemma 3.6. Note that if $|\mathcal{E}_{i,q}| > M$, we can abort as soon as the number of edges we computed exceeds M. Thus, checking if $|\mathcal{E}_{i,q}| \leq M$ takes $O((1+\varepsilon)nr\log^2 n)$ time. By Lemma 3.6, the expected number of iterations the algorithm performs until the inequality $|\mathcal{E}_{i,q}| \leq M$ will be satisfied is $\leq 1 + 1/\varepsilon$. Thus, the expected running time of the algorithm is

$$O\left(\left(1+\frac{1}{\varepsilon}\right) nr \log^2 n\right),$$

since the vertical decomposition of $\mathrm{simp}'_q(\mathcal{E}_{i,q})$ (which is the resulting cutting) can be computed in $O((1+\varepsilon)nr)$ time. In fact, one can also compute, in $O((1+\varepsilon)nr)$ time, for each trapezoid in the cutting, the lines of L that intersect it. ■

4 Empirical Results

4.1 The Implemented Algorithms

We have implemented the algorithm presented in Section 2, as well as several other algorithms for constructing cuttings. We have also experimented with the algorithm of Section 3. In this section, we report on the experimental results that we obtained.

Most of the algorithms we have implemented are variants of `CutRandomInc`. The algorithms implemented are the following:

Parallel Incremental: Let C_i be the covering generated in the i-th iteration of the algorithm. For each active trapezoid $\Delta \in C_i$, pick a random line from $K(\Delta)$, and insert it into Δ (i.e., splitting Δ). Continue until there are no active trapezoids. Note that unlike `CutRandomInc` the insertion operations are performed locally inside each trapezoid, and the line chosen for insertion in each trapezoid is independent of the lines chosen for other trapezoids. It is easy to verify that the proof of Theorem 2.4 also works for this algorithm, implying that the expected size of the cutting it computes is $O(r^2)$, and the expected running time is $O(nr)$.

Classical: This is a variant of the algorithm of Chazelle and Friedman [5] for constructing a cutting. We pick a sample $R \subseteq \hat{S}$ of r lines, and compute its arrangement $A = \mathcal{A}^{||}_{\mathrm{Mrg}}(R)$. For each active trapezoid $\Delta \in A$, we pick a random sample $R_\Delta \subseteq K(\Delta)$ of size $6k\log k$, where $k = \lceil r|K(\Delta)|/n \rceil$, and compute the arrangement of $\mathcal{A}^{||}_{\mathrm{Mrg}}(R_\Delta)$ inside Δ. If $\mathcal{A}^{||}_{\mathrm{Mrg}}(R_\Delta)$ is not a $(1/r)$-cutting, then the classical algorithm performs resampling inside Δ until it reaches a cutting. Our implementation is more naive, and it simply continues recursively into the active subtrapezoids of $\mathcal{A}^{||}_{\mathrm{Mrg}}(R_\Delta)$.

Cut Randomized Incremental: This is `CutRandomInc` without merging, as described in Figure 1.

The following four heuristics, for which we currently do not have a proof of any concrete bound on the expected size of the cutting that they generate, perform well in practice.

Randomized Incremental: This is `CutRandomInc` with merging.

Greedy Trapezoid: This is a variant of `CutRandomInc` where we try to be "smarter" about the line inserted into the partition in each iteration. Let V_i be the set of trapezoids of C_i with maximal weight. We pick randomly a trapezoid Δ out of the trapezoids of V_i, and pick randomly a line s from $K(\Delta)$. We then insert s into C_i.

Greedy Line: Similar to **Greedy Trapezoid**, but here we compute the set $\mathcal{U}$ of lines of $\hat{S}$, for which $w'(s)$ is maximal, where $w'(s)$ is the number of active trapezoids in C_i that intersect the line s. We pick randomly a line from $\mathcal{U}$ and insert it into the current partition of the plane.

Greedy Weighted Line: Similar to **Greedy Line**, but our weight function is:

$$w'(s) = \sum_{\Delta \in C_i, s\cap\Delta \neq \emptyset, w(\Delta) > n/r} \left\lfloor \frac{w(\Delta)}{\lfloor \frac{n}{3r} \rfloor} \right\rfloor;$$

namely, we give a higher priority to lines that intersect heavier $(1/r)$-active trapezoids.

4.2 Implementation Details

As an underlying data-structure for our testing, we implemented the history-graph data-structure [12]. Our random arrangements were constructed by choosing n points uniformly and independently on the left side of the unit square, and similarly on the right side of the unit square. We sorted the points, and connected them by lines in a transposed manner. This yields a random arrangement with all the $\binom{n}{2}$ intersections inside the unit square.

We had implemented our algorithm in C++. We had encountered problems with floating point robustness at an early stage of the implementation, and decided to use exact arithmetic instead, using LEDA rational numbers [10]. While this solved the robustness problems, we had to deal with a few other issues:

- Speed: Using exact arithmetic instead of floating point arithmetic resulted in a slow down by a factor 20–40.[2] The time to perform an operation in exact arithmetic is proportional to the bit-sizes of the numbers involved. To minimize the size of the numbers used in the computations, we normalized the line equations so that the coefficients are integer numbers (in reduced form).

[2] This is the reason why we do not present running times in our tables.

- Memory consumption: A LEDA rational is represented by a block of memory dynamically allocated on the heap. In order to save, both in the memory consumed and the time used by the dynamic memory allocator, we observe that in a representation of vertical decomposition the same number appears in several places (i.e., an x-coordinate of an intersection point appears in 6 different vertical trapezoids). We reduce memory consumption, by storing such a number only once. To do so, we use a repository of rational numbers generated so far by the algorithm. Whenever we compute a new x-coordinate, we search it in the repository, and if it does not exist, then we insert it. In particular, each vertical trapezoid is represented by two pointers to its x_{left} and x_{right} coordinates, and pointers to its top and bottom lines.

  The repository is implemented using Treaps [13].

4.3 Results

The empirical results we got for the algorithms/heuristics of Subsection 4.1, are depicted in Tables 1 - 5.

For each value of r, and each value of n, we computed a random arrangement of lines inside the unit square, as described above. For each such arrangement, we performed 10 tests for each algorithm/heuristic. The tables present the size of the minimal cutting computed in those tests. Each entry is the size of the output cutting divided by r^2. In addition, each table caption presents a range containing the size of the cutting that can be computed by Matoušek's algorithm [9].

As noted in Remark 2.5, it is an interesting question whether or not using merging results in smaller cuttings generated by CutRandomInc. We tested this empirically, and the results are presented in Table 6. As can be seen in Table 6, using merging does generate smaller cuttings, but the improvement in the cutting size is not dramatic. The difference in the size of the cuttings generated seems to be less than $2r^2$.

4.4 Implementing Matoušek's Construction

In Table 7, we present the empirical results for Matoušek's construction, comparing it with the slight improvement described in Remark 3.4. For small values of r the improved version yields considerably smaller cuttings than Matoušek's construction, making it the best method we are aware of for constructing small cuttings.

We had implemented Matoušek's algorithm naively, using quadratic space and time. Currently, this implementation can not be used for larger inputs because it runs out of memory. Implementing the more efficient algorithm described in Theorem 3.7 is non-trivial since it requires the implementation of the data-structure of Overmars and van Leeuwen [11]. However, it is critical to reduce the size of a cutting for large inputs, the algorithm of Theorem 3.7 seems to be the best available option.

5 Conclusions

In this paper, we presented a new approach, different from that of [5], for constructing cuttings. The new algorithm is rather simple and easy to implement. We proved the correctness and bounded the expected running time of the new algorithm, while demonstrating that the new algorithm performs much better in practice than the algorithm of [5].

Moreover, the empirical results show that the size of the cutting constructed by the new algorithm is not considerably larger than the cuttings that can be computed by the currently best theoretical algorithm (too slow to be useful in practice due to its $O(n^2)$ running time) of Matoušek [9]. The empirical constants that we obtain are generally between 10 and 13. Moreover, the various variants of CutRandomInc seem to produce constants that are rather close to each other. There is no obvious winner among those methods, and the Parallel Incremental version appears to be slightly worse. As noted above, the method described in Remark 3.4 generates the smallest cuttings (but requires quadratic time). Given this, the method that we recommend for use is CutRandomInc (without merging, as this is the only "non-trivial" part in the implementation of the algorithm), because of the relative simplicity of its implementation, and its speed in practice.

While we did not collect statistics about the running time of the programs (mainly because of the need to use exact arithmetic), it seems that the running time of all the heuristics and algorithms, for the same value of n, is proportional to the size of cuttings they compute. However, it seems that the fastest algorithm is the classical algorithm of Chazelle and Friedman. The second fastest algorithm seems to be CutRandomInc, with all the rest of the algorithms seem to be considerably slower.

There are several interesting open problems for further research:

- Can one bound the expected size of the cutting generated by CutRandomInc, when merging is being used? While this has minor importance from practical point of view, as indicated by our implementation (see Table 6), such a proof might provide a new insight into the problem of constructing cuttings.
- Can one prove the existence of a cutting smaller than the one guaranteed by the algorithm in Remark 3.4 for specific values of r? For example, Table 1 suggests a smaller cutting should exist for $r = 2$. In particular, the test results hint that a cutting made out of 32 vertical trapezoid should exist, while the cutting size guaranteed by Matoušek's algorithm [9] is 48.
- Can one generate smaller cuttings by modifying CutRandomInc to be smarter in its decision when to merge trapezoids?

Acknowledgments

The author wishes to thank Pankaj Agarwal, Boris Aronov, Jiri Matoušek and Micha Sharir for helpful discussions concerning the problems studied in this paper and related problems.

References

[1] Agarwal, P. Geometric partitioning and its applications. In *Computational Geometry: Papers from the DIMACS special year*, J. E. Goodman, R. Pollack, and W. Steiger, Eds. American Mathematical Society, 1991.

[2] Agarwal, P., Erickson, J., and Guibas, L. Kinetic binary space partitions for triangles. In *9th ACM-SIAM Sympos. Discrete Algorithms* (1998), pp. 107–116.

[3] Agarwal, P., Matoušek, J., and Schwarzkopf, O. Computing many faces in arrangements of lines and segments. In *Proc. 10th Annu. ACM Sympos. Comput. Geom.* (1994), pp. 76–84.

[4] Agarwal, P. K., and Matoušek, J. Dynamic half-space range reporting and its applications. *Algorithmica 13* (1995), 325–345.

[5] Chazelle, B., and Friedman, J. A deterministic view of random sampling and its use in geometry. *Combinatorica 10*, 3 (1990), 229–249.

[6] Clarkson, K. L. New applications of random sampling in computational geometry. *Discrete Comput. Geom. 2* (1987), 195–222.

[7] de Berg, M., van Kreveld, M., Overmars, M., and Schwarzkopf, O. *Computational Geometry: Algorithms and Applications.* Springer-Verlag, 1997.

[8] Edelsbrunner, H., and Welzl, E. Constructing belts in two-dimensional arrangements with applications. *SIAM J. Comput. 15* (1986), 271–284.

[9] Matoušek, J. On constants for cuttings in the plane. *Discrete Comput. Geom.* (1998).

[10] Mehlhorn, K., and Näher, S. LEDA: a platform for combinatorial and geometric computing. *Commun. ACM 38* (1995), 96–102.

[11] Overmars, M. H., and van Leeuwen, J. Maintenance of configurations in the plane. *J. Comput. Syst. Sci. 23* (1981), 166–204.

[12] Seidel, R. A simple and fast incremental randomized algorithm for computing trapezoidal decompositions and for triangulating polygons. *Comput. Geom. Theory Appl. 1* (1991), 51–64.

[13] Seidel, R., and Aragon, C. R. Randomized search trees. *Algorithmica 16* (1996), 464–497.

[14] Sharir, M., and Agarwal, P. *Davenport-Schinzel Sequences and Their Geometric Applications.* Cambridge University Press, New York, 1995.

Lines	Parallel Inc	Classical	Randomized Inc	Greedy Trapezoid	Greedy Line	Greedy Weighted Line
4	1.50	1.50	1.50	1.50	1.50	1.50
8	2.00	12.50	2.75	2.25	2.25	2.50
16	4.75	25.75	4.25	4.75	4.00	4.50
32	4.75	24.75	7.00	5.50	6.00	6.25
64	6.75	28.25	6.50	7.50	7.50	7.25
128	8.75	26.75	7.75	7.25	8.25	8.50
256	8.75	30.75	6.50	8.00	6.75	7.25
512	6.00	36.50	8.25	9.75	8.50	8.00
1024	9.25	26.00	10.00	7.75	7.75	9.00
2048	7.50	28.50	9.00	7.25	8.75	9.75
4096	7.50	35.25	8.50	8.25	7.75	7.50
8192	8.25	36.75	7.00	7.75	6.25	7.50
16384	9.00	30.25	8.75	8.00	8.50	8.75
32768	10.25	33.00	7.75	9.00	6.25	7.75
65536	10.00	31.25	6.50	6.75	6.75	8.50

Table 1: Results for 1/2-cuttings. Each entry is the size of the minimal cutting computed, divided by $2^2 = 4$. The corresponding value in Matoušek's construction [9] is smaller than 12.00

Lines	Parallel Inc	Classical	Randomized Inc	Greedy Trapezoid	Greedy Line	Greedy Weighted Line
8	1.88	2.69	1.56	2.00	1.62	1.62
16	3.69	20.31	3.44	3.62	3.50	3.62
32	6.12	31.12	6.56	5.50	5.75	5.50
64	8.25	35.94	8.06	7.69	7.12	8.06
128	9.19	37.12	9.75	9.94	8.62	7.88
256	11.00	44.06	9.62	9.00	10.25	8.19
512	11.75	48.75	9.81	10.31	9.75	10.12
1024	12.75	46.88	12.12	11.25	10.25	10.62
2048	11.19	37.00	11.50	11.00	10.81	10.50
4096	11.81	44.38	10.94	11.19	10.62	10.50
8192	12.19	52.00	10.19	11.06	10.25	10.00
16384	12.31	43.88	10.94	11.25	10.31	10.88
32768	11.50	42.69	10.69	11.81	11.31	10.25

Table 2: Results for 1/4-cuttings. Each entry is the size of the minimal cutting computed, divided by $4^2 = 16$. The corresponding value in Matoušek's construction [9] is smaller than 9.75

Lines	Parallel Inc	Classical	Randomized Inc	Greedy Trapezoid	Greedy Line	Greedy Weighted Line
16	2.23	4.08	2.06	1.89	1.86	1.83
32	4.80	23.17	4.19	3.92	3.91	3.53
64	7.67	37.25	6.64	6.27	6.31	6.12
128	10.44	45.50	8.84	8.83	8.80	8.31
256	11.56	44.12	9.91	10.53	9.91	9.94
512	12.77	50.14	11.36	11.11	10.86	11.20
1024	13.31	47.58	11.61	11.33	10.95	11.31
2048	13.42	54.84	12.36	11.17	12.38	11.17
4096	15.00	53.17	12.08	12.22	12.08	11.98
8192	13.98	51.75	11.73	12.19	12.67	12.27

Table 3: Results for 1/8-cuttings. Each entry is the size of the minimal cutting computed, divided by $8^2 = 64$. The corresponding value in Matoušek's construction [9] is smaller than 8.81

Lines	Parallel Inc	Classical	Randomized Inc	Greedy Trapezoid	Greedy Line	Greedy Weighted Line
32	2.70	5.54	2.18	2.09	2.04	2.11
64	5.38	24.27	4.52	4.21	4.36	4.20
128	8.16	44.00	7.16	7.00	6.66	6.55
256	10.88	56.30	9.34	9.25	9.16	8.48
512	12.61	66.60	11.26	10.85	10.37	10.18
1024	14.02	66.64	12.30	11.40	11.23	11.10
2048	14.24	67.25	12.51	11.84	11.98	???

Table 4: Results for 1/16-cuttings. Each entry is the size of the minimal cutting computed, divided by $16^2 = 256$. The corresponding value in Matoušek's construction [9] is smaller than 8.39

Lines	Parallel Inc	Classical	Randomized Inc	Greedy Trapezoid	Greedy Line	Greedy Weighted Line
64	2.84	5.45	2.26	2.19	2.14	2.14
128	5.48	24.73	4.54	4.44	4.33	4.22
256	8.89	52.62	7.52	7.05	6.68	6.61
512	11.26	67.72	9.48	9.54	8.98	8.87
1024	13.25	74.47	11.63	10.86	10.23	10.34

Table 5: Results for 1/32-cuttings. Each entry is the size of the minimal cutting computed, divided by $32^2 = 1024$. The corresponding value in Matoušek's construction [9] is smaller than 8.19

	Value of r									
Number of Lines	2		4		8		16		32	
		Merge		Merge		Merge		Merge		Merge
4	1.50	1.50	—	—	—	—	—	—	—	—
8	2.25	2.75	1.88	1.50	—	—	—	—	—	—
16	4.75	4.00	3.44	3.56	2.27	2.03	—	—	—	—
32	5.00	5.75	6.44	6.25	4.86	4.31	2.56	2.14	—	—
64	7.25	7.75	8.25	9.00	7.36	6.45	5.30	4.54	2.78	2.29
128	8.00	7.50	8.94	9.38	9.56	9.00	8.48	7.19	5.44	4.52
256	8.00	8.00	11.94	9.88	11.86	10.81	10.84	9.73	8.26	7.29
512	5.75	9.00	10.25	10.62	13.05	11.48	12.36	11.29	11.17	9.55
1024	8.25	6.75	12.62	10.12	12.69	11.83	13.80	11.96	13.07	11.24
2048	7.50	8.50	11.31	10.31	13.06	12.66	14.11	13.44	14.04	12.38
4096	8.75	9.25	12.31	11.12	13.81	12.39	13.95	13.13	14.61	12.96
8192	7.50	8.00	12.00	12.12	13.34	12.97	14.67	12.67	14.72	13.18
16384	9.25	8.50	11.38	10.56	12.69	12.33	14.53	13.61	15.02	13.30
32768	7.25	7.50	12.12	10.44	13.00	12.59	13.96	13.00	15.17	13.46

Table 6: Comparing the size of cuttings computed by CutRandomInc, with or without using merging. Each entry is the size of the minimal cutting computed, divided by r^2.

	Value of r									
Number of Lines	2		4		8		16		32	
		Impr'		Impr'		Impr'		Impr'		Impr'
4	8.75	5.25	—	—	—	—	—	—	—	—
8	10.75	6.25	8.25	6.56	—	—	—	—	—	—
16	10.25	7.25	9.19	7.44	8.14	7.27	—	—	—	—
32	10.00	7.00	9.38	7.56	8.64	7.64	8.06	7.63	—	—
64	10.00	7.25	9.81	7.25	8.66	7.73	8.32	7.82	8.04	7.81
128	10.50	6.50	9.31	7.06	8.72	7.70	8.38	7.90	8.15	7.91
256	11.00	7.50	9.81	7.44	8.81	7.73	8.37	7.90	8.18	7.92
512	11.00	6.75	10.06	7.69	8.81	7.84	8.41	7.88	8.17	7.94

Table 7: Comparing the size of cuttings computed by Matoušek's method, to the slightly improved method described in Section 3.

Point Set Labeling with Sliding Labels*

Marc van Kreveld
Dept. of Computer Science
Utrecht University
marc@cs.ruu.nl

Tycho Strijk
Dept. of Computer Science
Utrecht University
tycho@cs.ruu.nl

Alexander Wolff
Dept. of Computer Science
Free University Berlin
awolff@inf.fu-berlin.de

Abstract

This paper discusses algorithms for labeling sets of points in the plane, where labels are not restricted to some finite number of positions. We show that continuously sliding labels allows more points to be labeled both in theory and in practice. We define six different models of labeling, and analyze how much better—more points get a label—one model can be than another. Maximizing the number of labeled points is NP-hard, but we show that all models have a polynomial-time approximation scheme, and all models have a simple and efficient factor-$\frac{1}{2}$ approximation algorithm. Finally, we give experimental results based on the factor-$\frac{1}{2}$ approximation algorithm to compare the models in practice.

Figure 1: Rectangular labels of cities of the U.S.A.

1 Introduction

Annotating sets of points is a common task to be performed in Geographic Information Systems. Cities on small-scale maps are shown as points with the city's name attached (Figure 1 shows names as rectangles), points of altitude usually are small "+"-signs with a value, and in point pattern analysis [2], points in a plot are labeled with a sequence number. In (spatial) statistics [15], point sets are also common in data postings of field measurements, scatterplots of principal component analysis, and variograms, for instance.

Generally it is assumed that a point label can be seen as an axis-parallel rectangle, the bounding box of the text (see Figure 1). Several algorithms for point feature labeling have been described in the automated cartography literature and in computational geometry (far too many to list; for bibliographies see [8, 22]). The more general problem of map labeling includes line feature labeling (roads, rivers) and area labeling (countries) as well.

A good label placement for the points of a given set has two basic requirements. A label should be placed with the point to which it belongs, and two labels should not overlap each other. For high quality cartographic label placement, more requirements can be formulated such as unambiguity [14, 23]. Given these basic requirements, an algorithm can try to either label as many points as possible, or find the largest possible font such that all points can be labeled. Labeling as many points as possible already give rise to a computationally intractable problem under the two basic conditions [9, 10, 16].

Nearly all of the existing algorithms for point annotation limit the placement of a label with respect to its point to a finite number of possible positions. Algorithms described before usually allow four label positions (the point is at one of the four corners) [9, 21, 20], eight (many papers in the automated cartography literature), or any constant number [1]. We call restrictions of the allowed label positions the *model* that is used by the algorithm. Models that allow a finite number of positions per label are *fixed-position models.*

In this paper we drop the restriction that a label can only be placed at a finite number of positions. Instead, we allow any position on the edges of the rectangle to coincide with the point, see Figure 1. Such a model is called a *slider*

*This research is supported by the Dutch Organization for Scientific Research (N.W.O.), the ESPRIT IV LTR Project No. 21957 (CGAL), and by the Deutsche Forschungsgemeinschaft (DFG) under grant Wa 1066/3-1.

SCG 98 Minneapolis Minnesota USA

model. We will study how many more labels can be placed with slider models than with fixed-position models, and to what extent slider models require more difficult algorithms. We generally assume that labels have equal height but not necessarily the same width. This is a natural assumption if labels contain text or numbers of a fixed font size. We consider the rectangle that represents a label to be closed, which implies that labels are not allowed to touch.

Slider models have been used in two previous papers. In Hirsch's paper [11], repelling forces are defined for overlapping labels and computes translation vectors for them. After translation, this process is repeated and hopefully, a labeling with few overlaps appears after a number of iterations. This is completely different from our approach, which is combinatorial. The paper by Doddi et al. [7] contains a couple of labeling problems and algorithms, each using a different labeling model. One of the problems is solved in a slider model, where each label is allowed to rotate about the point to be labeled. The labels must be equal-size squares (or other regular polygons). In that paper, the objective is to maximize the label *size*.

This paper is structured as follows. Section 2 introduces the six models—three fixed-position and three slider—that are compared in this paper. We analyze how many more labels can be placed in one model than another, in theory. In Section 3, we show that the slider models allow a simple factor-$\frac{1}{2}$ approximation algorithm that uses $O(n)$ space and $O(n \log n)$ time. This was already known for the fixed-position models. In Section 4, a polynomial time approximation scheme is given, showing that for any constant $\epsilon > 0$, there is a polynomial time algorithm that labels a fraction of at least $1 - \epsilon$ of the optimal number of labels that can be placed. Again, this result was already known for the fixed-position models but not for the slider models. We remark that our algorithms can be adapted for labels of varying height, but the approximation factors don't hold any more.

Section 5 contains a comparison of the six different models in practice. For these models, the factor-$\frac{1}{2}$ approximation algorithms have been implemented and tested on real world data. We used three different data sets. One contains 1000 cities of the U.S.A., another contains a data posting with 236 measurements, and the third contains 75 points in a scatterplot near a regression line. Here the labels are the sequence numbers of the points. We give tables showing how many points are labeled in each model. It appears that the slider models produce about 10–15% more labels than the fixed-position models. This improvement is significant, because more labels are placed in the difficult areas.

2 Comparing label models

This paper considers three fixed-position models and three slider models for point feature label placement. These are shown in Figure 2. It is obvious that a maximum labeling of a set P of points in the 4-slider model must have as least as many labels as in any of the other models. Similarly, the

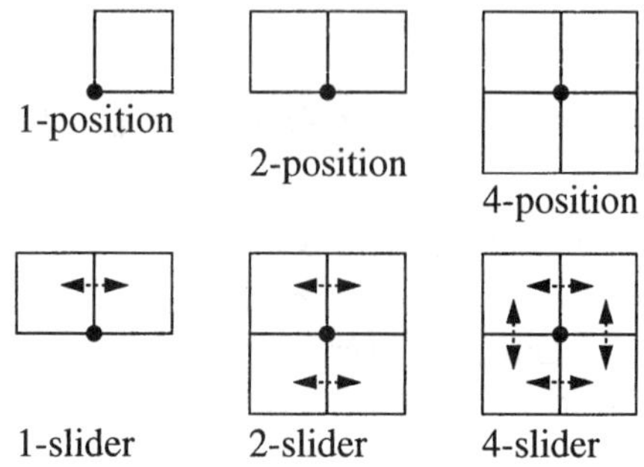

Figure 2: Fixed-position and slider models.

1-slider model is at least as good as the 2-position model. The partial order representing the generality of the models, shown in Figure 3, presents itself immediately. It is clear that

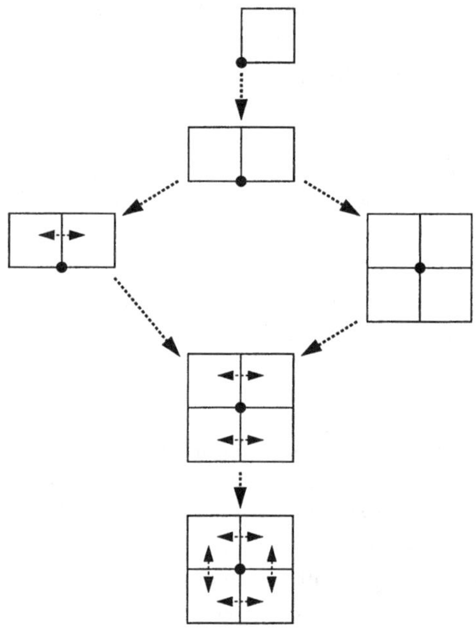

Figure 3: Partial order on the models.

there are point sets where all points can be labeled in any of the models. It is more difficult to answer the question "How many more points can be labeled in one model than another, for *some* point set?". This section deals with this question. For simplicity, all labels in this section are squares of unit size.

Let P be a set of n points in the plane. Let M_1 and M_2 be two models for labeling P, and let $\text{opt}_{M_1}(P)$ and $\text{opt}_{M_2}(P)$ be the maximum number of points of P that receive a label in the models M_1 and M_2, respectively. Then the (M_1, M_2)-*ratio* is the supremum of the ratio $\text{opt}_{M_1}(P)/\text{opt}_{M_2}(P)$ for $n \to \infty$ and maximized over all point sets P with n points. Figure 4 shows, for instance, that the 2-position model can be twice as good as the 1-position model, which is not really surprising. We can also show that the ratio of 2 is tight. Con-

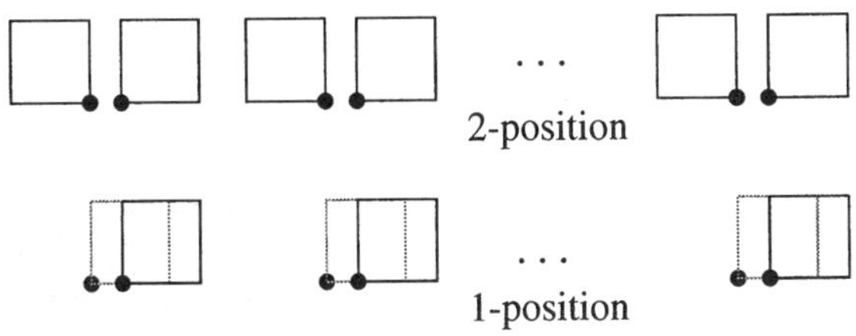

Figure 4: Optimal labeling of P in two models.

sider an maximum labeling of k points from a point set P with n points, using the 2-position model. Move the labels of these k points into the position allowed in the 1-position model. Either $\lceil k/2 \rceil$ of the labels were already in this position, and they form a nonintersecting set, or $\lceil k/2 \rceil$ of the labels were in the other position and have been *flipped*. Since these squares were nonintersecting before flipping, they must also be nonintersecting after flipping, which proves that the ratio of 2 is tight. The same simple idea gives a ratio of 2 between the 1-slider and 2-slider models, and the 2-position and 4-position models. To analyze the ratios between other models, more complicated variations of this idea are needed.

To analyze the ratios between other models M_1 and M_2, we use the following strategy. We want to bound the ratio ρ by which more labels can be placed in the model with more degrees of freedom, say M_1. We assume an optimal label placement in M_1. Then we canonically re-label the labeled points by moving every label into a position which is valid in the more restrictive model M_2. This may cause some labels to intersect. We determine the maximum number δ_{left} of M_2-labels that intersect the leftmost M_2-label l. Then we put l into a set S, remove l and all its conflict partners from the instance and repeat until no labels remain. At the end of the process, S contains at least $\text{opt}_{M_1}(P)/(\delta_{\text{left}} + 1)$ nonintersecting M_2-labels, where $\text{opt}_{M_1}(P)$ is the size of the assumed optimal M_1-placement. The size of S is a lower bound for the size of an optimal M_2-placement, thus $\delta_{\text{left}} + 1$ is an upper bound for the (M_1, M_2)-ratio.

To prove a lower bound for the ratio of two models, we need to give an example of arbitrary size for which any M_2-placement is worse by this ratio.

Lemma 1 *Given two labeling models M_1 and M_2 such that the labels in M_1 can be slid into some position of M_2 using only horizontal sliding or only vertical sliding (not both). Let ρ be the (M_1, M_2)-ratio. Then $\rho \leq 3$.*

Proof: Again we consider an optimal M_1-labeling of an arbitrary instance. Assume that we can slide M_1- into M_2-label positions vertically (for instance, M_1 is the 4-slider model and M_2 is the 2-slider model). We canonically slide all M_1-labels that are not yet in an M_2-label position upwards, until they arrive in an M_2-label position. We show that the leftmost M_2-label l_2 can then intersect at most two other M_2-labels. This yields the upper bound of 3 for ρ.

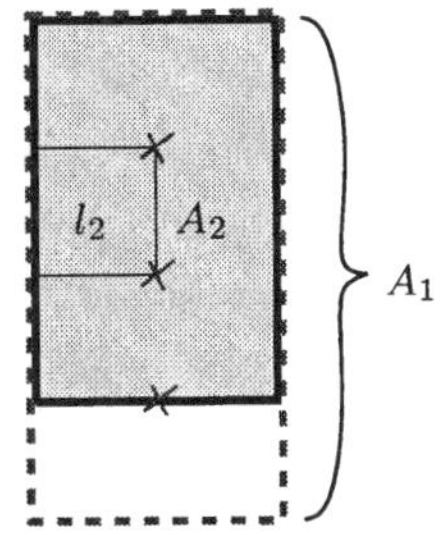

Figure 5: If M_1 can be slid into M_2 then the leftmost M_2-label l_2 cannot intersect more than two M_2-labels.

M_2-labels intersecting l_2 can only lie within area A_2 in Figure 5 since l_2 is leftmost. The corresponding M_1-labels are restricted to area A_1. Every label in A_1 must contain one of the three gridpoints in the interior of A_1 (marked by crosses in Figure 5). Thus A_1 can contain only three nonintersecting M_1-labels including the M_1-counterpart of l_2. It follows that l_2 cannot intersect more than two M_2-labels, and that $\rho \leq 3$. ∎

A simple example gives a lower bound of 2 for the ratio of, for instance, the 2-slider model and the 4-slider model. Consider the set of n points $\{(0, i) \mid 0 \leq i \leq n-1\}$. Then the 2-slider model can label at best every other point, giving $\lceil n/2 \rceil$ labels. The 4-slider model can label the point at the origin right and below, and every next point with a label shifted $1 + 1/(n-1)$ upwards with respect to the label below. This example makes explicit use of the assumption that squares be closed, but this is not necessary: use two columns of points at x-distance $\frac{1}{2}$, and use $1-\epsilon$ as y-distance between the points.

In the full paper [19], we show the upper and lower bounds on the ratios between two models given in Figure 6. The lower bound $2\frac{1}{4}$ stems from an example, where the 1-position model gives a 9-cycle of intersecting squares, but the 1-slider model allows a non-intersecting placement.

Finally, we wish to note that slider models can be $3/2$ times as good as *any finite approximation* of that slider model. For example, consider the 1-slider model and a model that allows a fixed set of one hundred different positions where the bottom edge of the square coincides with the point to be labeled. Then we take three points $(0,0)$, $(x,0)$, $(1+\epsilon,0)$, choose $\epsilon > 0$ to be very small, and choose x such that none of the one hundred allowed positions fits between the outer two points. This is always possible. The 1-slider model can label all three points regardless of the choice of ϵ and x. By copying these three points $n/3$ times we get the ratio $3/2$.

3 Greedy approximation algorithms

In the following sections we consider algorithms for point feature labeling in the slider models. Labeling the maximum number of points is intractable in the fixed-position

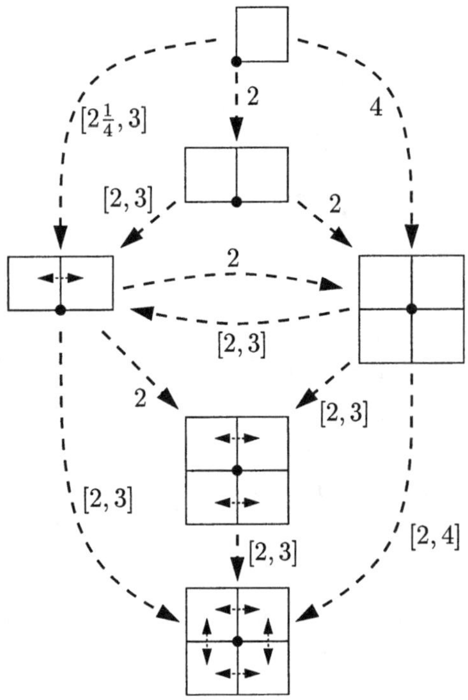

Figure 6: Upper and lower bounds for the ratios of two models.

models [9, 10, 16]. In the full version [19] we show this result for the slider models, so we have to be content with approximation algorithms. In this section we describe an $O(n \log n)$ time algorithm for the slider models which approximates an optimal solution in the following sense. If the maximum number of labels that can be placed is K, then our algorithm places at least $K/2$ labels: a factor-$\frac{1}{2}$ approximation algorithm. For most data sets, however, we expect to come much closer to the optimum.

For the fixed position models, simple $O(n \log n)$ time, factor-$\frac{1}{2}$ approximation algorithms were described recently by Agarwal et al. [1]. We obtain the same result for the slider models. We'll only describe the most general 4-slider algorithm; it is an extension of the 1-slider and 2-slider algorithms. It is based on a greedy strategy. For convenience we'll do as if labels were allowed to touch, unlike in the previous section. In the full paper we show that simple adaptations can be made to obtain non-touching labels.

Given a set of points with labels that have already been placed, and a set of points that don't have a label yet, define the *leftmost label* to be the label whose right edge is leftmost among all possible label positions of unlabeled points. So by definition, the leftmost label doesn't intersect any label that has been placed. The strategy is the same as for fixed position labels [1] but the algorithm is quite different. The proof of the following lemma is given in the full paper.

Lemma 2 *Given labels of fixed height and any of the slider models, the greedy strategy of repeatedly choosing the leftmost label finds a labeling of at least half the number of points labeled in an optimal solution.*

A brute-force algorithm for this simple strategy would need $O(n^3)$ steps. In order to achieve an $O(n \log n)$ time bound, we must use some common geometric data structures.

Let $\{p_1, \ldots, p_n\}$ be the set of points that has to be labeled. The label of p_i is denoted by l_i, and the reference point of a label is its lower left vertex. The possible positions of the reference point of a point p_i are represented by four line segments. Two are horizontal, h_{2i-1} and h_{2i}, and two are vertical, v_{2i-1} and v_{2i}. Their position is exactly the position of the edges of the label l_i if it were placed left and below p_i. The width of l_i is denoted w_i, and the height is always 1 (we can normalize to this situation).

If a label l_i has been placed, then no reference point position inside l_i is possible. The same holds for reference points inside the rectangle l'_i precisely one unit below l_i (since any label extends one unit above its reference point). Furthermore, since labels are placed from left to right, no reference point positions to the left of l_i and l'_i will still be accepted by the algorithm. Suppose a subset of the points has already received labels by the algorithm. The *right envelope* of all labels l and their copies l' outlines all reference point positions that are impossible, or cannot occur any more, see Figure 7. We call this right envelope the *frontier* and denote it by F.

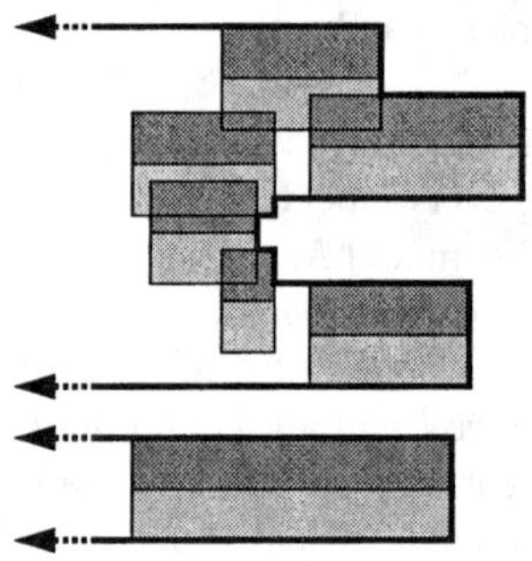

Figure 7: Frontier of the placed labels (dark grey) and their lowered copies (light grey).

To determine the next leftmost label, we only have to consider the frontier F and the segments $h_{2i-1}, h_{2i}, v_{2i-1}$, and v_{2i} of the points p_i to the right of F that don't have a label yet. Given a horizontal segment h and the frontier F, there are three possibilities: (i) h lies completely left of F. Then h can be discarded; a point on it cannot be a reference point for a label that doesn't overlap another label. (ii) h lies completely right of F. Then the leftmost point on h is a candidate for the next leftmost label. (iii) h intersects F. Then a point just right of the intersection point is the candidate. For a vertical segment v, a similar situation occurs. If v lies left of F, it can be discarded; if v lies right of F, any point on v

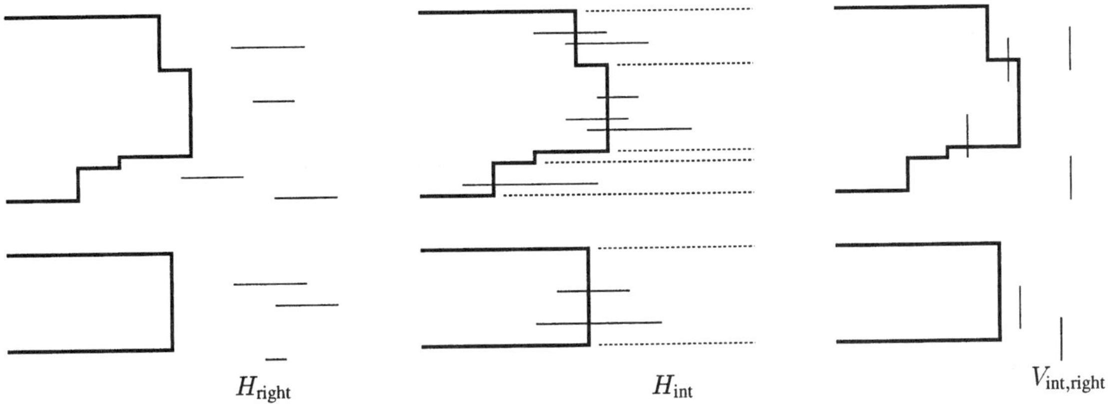

Figure 8: The sets H_{right}, H_{int}, and $V_{\text{int,right}}$. The dashed lines in the middle picture separate the segments of H_{int} that are in different red-black trees $\mathcal{T}_i$.

can be chosen; and if v and F intersect, then any point on v right of F can be chosen as a candidate.

Let H be the set of all horizontal segments that represent reference points of the labels. Similarly, let V be the set of the corresponding vertical segments. Let $H_{\text{right}} \subseteq H$ be the subset of all horizontal segments that lie completely right of F. See Figure 8. Let $H_{\text{int}} \subseteq H$ be the subset of all horizontal segments that intersect F. Let $H_{\text{left}} \subseteq H$ be the subset of all horizontal segments that lie completely left of F (these cannot give a valid label any more). Let $V_{\text{int,right}} \subseteq V$ be the subset of all vertical segments that contain at least some point right of F.

To maintain the frontier and the candidates for the best reference point we'll use a few data structures. Some of the data structures are used to find the next leftmost label; other data structures are only used to update the former ones efficiently. The data structures are red-black trees $\mathcal{T}$, heaps $\mathcal{H}$, and priority search trees $\mathcal{P}$ [17]. These are described in standard textbooks on algorithms [5] and computational geometry [6]. We use three data structures to find the leftmost label position among the ones represented by H_{int}, H_{right}, and $V_{\text{int,right}}$.

Leftmost label query structures.

1. For each segment in H_{right} we store the x-coordinate of its right endpoint. This corresponds to the right edge of a label whose reference point is the left endpoint of the segment. These values are stored in a heap, where the root stores the minimum.

2. The subset H_{int} is stored as follows. For each vertical segment f_i of F, we maintain a red-black tree $\mathcal{T}_i$ with the segments in H_{int} that intersect f_i (see the middle picture of Figure 8). These are stored in the leaves sorted on y-coordinate. With each leaf we also store the width of the corresponding label. We augment each red-black tree by storing at each internal node the minimum width label in the subtree of that node [5]. We use a heap $\mathcal{H}_{\text{int}}$ to have fast access to the segment in H_{int} that allows the leftmost label placement. $\mathcal{H}_{\text{int}}$ stores for each $\mathcal{T}_i$ the sum of the x-coordinate of f_i and the minimum width of the segments in $\mathcal{T}_i$. Thus the root of $\mathcal{H}_{\text{int}}$ corresponds to the leftmost label among the labels represented by H_{int}.

3. For the vertical segments in V, we don't maintain the set $V_{\text{int,right}}$ but some set V' for which $V_{\text{int,right}} \subseteq V' \subseteq V$. The x-coordinate of each segment of V' is stored in a heap. The heap may return as the minimum some segment that lies completely left of F, so it may also contain labels that cannot be placed. After extracting the minimum from the heap, we test if it is in $V_{\text{int,right}}$. If not, we discard it and extract the next minimum from the heap, until we find one in $V_{\text{int,right}}$.

We query the three heaps described above. Among their answers, one corresponds to the leftmost label. This is the label we place. Then we must update the frontier F and several of the data structures described above. This is not so easy. We'll use some more data structures that help to do the updating after the frontier has changed. Let f_{new} be the union of the right edges of the newly placed label l and its copy l'. f_{new} is a vertical line segment of length 2. The new frontier F is the right envelope of the old frontier and f_{new}, see Figure 9.

Update assistance structures.

1a. To determine which segments move from H_{right} to H_{int} or H_{left} when the frontier changes, we use a priority search tree $\mathcal{P}_{\text{left}}$ on the left endpoints of segments in H_{right}. After placing a label, we query $\mathcal{P}_{\text{left}}$ with the region left of f_{new} (grey in Figure 9) to locate the left endpoints of all segments that are no longer in H_{right}. We delete these endpoints from $\mathcal{P}_{\text{left}}$, and we delete the corresponding segments from the heap for H_{right}. For each deleted segment we test whether its right endpoint

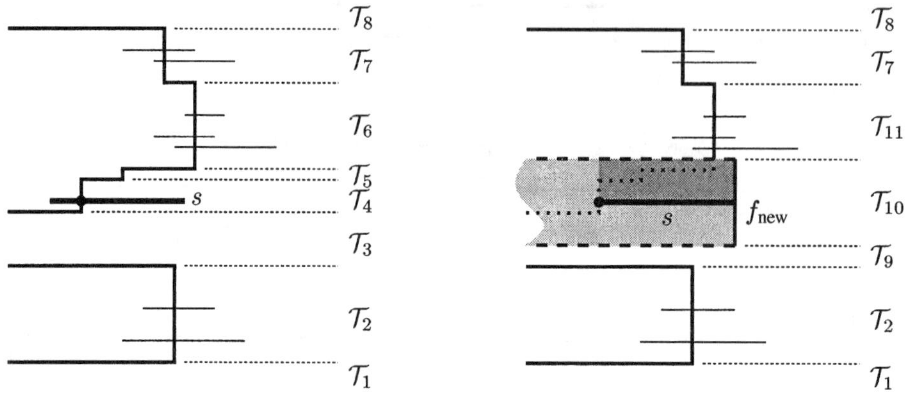

Figure 9: When the fat horizontal segment s from H_{int} is chosen, the frontier becomes the right envelope of f_{new} and the old frontier. The new label is dark grey. The grey range (light and dark) is the one with which queries in the priority search trees are done.

is right of the frontier. If so, that segment is in H_{int}, and we insert it in the data structures for H_{int}. If not, the segment is in H_{left} and can be discarded.

2a. To determine which segments move from H_{int} to H_{left} when the frontier changes, we use a priority search tree $\mathcal{P}_{right}$ on the right endpoints of segments in H_{int}. After placing a label, we query $\mathcal{P}_{right}$ with the region left of f_{new} (grey in Figure 9) to locate all right endpoints of segments that have moved from H_{int} to H_{left}. Then we delete the entries corresponding to these segments from the trees $\mathcal{T}_i$, from the heap $\mathcal{H}_{int}$ and from $\mathcal{P}_{right}$ itself.

When the frontier changes, we must also reorganize the red-black trees and $\mathcal{H}_{int}$ as a whole. Recall that we use a red-black tree $\mathcal{T}_i$ for each vertical segment of F. At most three new vertical segments can arise when the frontier changes, but many more vertical segments may cease to exist. We use the trees of the destroyed vertical segments of F to construct the new red-black trees. This is done by the operations SPLIT and CONCATENATE, which are standard for red-black trees. In Figure 9 the trees $\mathcal{T}_3$, $\mathcal{T}_4$, $\mathcal{T}_5$, and $\mathcal{T}_6$ are reorganized to the new trees $\mathcal{T}_9$, $\mathcal{T}_{10}$, and $\mathcal{T}_{11}$. The heap $\mathcal{H}_{int}$ is updated by removing the value of each destroyed tree, and inserting the value of each new tree.

3a. We don't need any additional data structures to update the heap on the vertical segments. However, we need to decide whether an extracted minimum from the heap really is in $V_{int,right}$. We use an augmented red-black tree for this test. The leaves of this tree store the vertical segments of the frontier sorted from bottom to top. Each leaf also stores the x-coordinate of its segment. Each internal node is augmented with a value that represents the minimum x-coordinate in its subtree. For each y-interval, the augmented red-black tree reports the minimum x-coordinate of the frontier in this y-interval.

The algorithm is given below. Due to lack of space, details have been omitted.

Algorithm.

While there are still segments in any of the heaps for V', H_{right}, or H_{int}, do the following steps:

1. Let v be the vertical segment that corresponds to the minimum of the heap for V'. Search in the augmented red-black tree on F with v to see if v has some point right of F. If not, remove v from the heap and repeat this step.

2. Determine the smallest among the minima of the three heaps for V', H_{right}, and H_{int}. Remove this minimum from its heap. Let l_i be the label position of point p_i corresponding to this minimum. Choose l_i as the next label to be placed.

3. Determine f_{new}, the right edge of l_i extended one unit downwards. Update the frontier F with f_{new}. Update the augmented red-black tree from **3a** with f_{new}. Search with the region horizontally left of f_{new} (grey in Figure 9) in the priority search trees of **1a** and **2a** and update the structures of **1**, **1a**, **2**, and **2a** accordingly.

Analysis.

The basic structures used by the algorithm are heaps, red-black trees, augmented red-black trees, and priority search trees. All of these structures require $O(n)$ space for a set of size n. Also, these structures can be updated in $O(\log n)$ time per insertion or deletion, or extract-min for heaps. Red-black trees allow SPLIT and CONCATENATE in $O(\log n)$ time. The queries on the red-black trees take $O(\log n)$ time, and the queries on the priority search trees take $O(k + \log n)$ time, where k is the number of points found in the query range.

The efficiency of the algorithm is established by the following observations. Any vertical segment f_{new} creates one vertical edge in the frontier F, and changes at most two of them. It follows that throughout the whole algorithm, at most $3n - 2$ different vertical edges appear in F. Therefore at most $3n - 2$ vertical edges have to be destroyed. This bounds the total number of red-black trees from **2** that can appear, the total number of SPLIT operations, and the total number of CONCATENATE operations to $O(n)$. Since SPLIT and CONCATENATE operations take $O(\log n)$ time each, at most $O(n \log n)$ time is spent on splitting and concatenating. The augmented red-black tree of **3a** can be maintained in $O(n \log n)$ time for the same reasons.

For each new label placed, one query is done on each of the two priority search trees. Such a query takes $O(k+\log n)$ time, where k is the number of points in the range. These points are always deleted from the priority search tree, so later on, the algorithm doesn't spend time on reporting them again. The priority search trees are initialized with one point for each horizontal segment, and we never add more points to them. So in total, at most $O(n \log n)$ time is spent for querying and updating the priority search trees.

We conclude:

Theorem 1 *Given n points in the plane, and for each point a rectangular label with fixed height and some given width. Then for each of the fixed-position and slider models, there is an $O(n \log n)$ time and linear space algorithm which places at least half the maximum number of labels.*

Remark 1. For the 1-slider and 2-slider models, we can omit the data structures listed under **3** and **3a**.

Remark 2. For fixed position models, the algorithm can be implemented using only one priority search tree and one heap. We initialize the priority search tree with the reference points of all labels that may still be placed. In the heap, we store the sum of x-coordinate and label width for each reference point. When the label corresponding to the heap's minimum is chosen, we query in the priority search tree with the appropriate range to find the reference points that are no longer valid. We remove these from heap and priority search tree, and repeat by selecting the minimum from the heap.

4 Polynomial time approximation scheme

A polynomial time approximation scheme for a labeling problem means that for any constant $\epsilon > 0$, there is an algorithm that runs in $O(n^c)$ time and places at least a fraction of $1 - \epsilon$ of the number of labels in an optimal placement. Here c is a constant that may depend on ϵ. Polynomial time approximation schemes (PTAS) have been developed for problems that are NP-hard, so all known algorithms producing an optimal solution require exponential time. A good survey of approximation algorithms for NP-hard geometric problems is by Bern and Eppstein [3].

For a set of unit squares, Hunt et al. [13] gave a PTAS to find the largest size subset of squares that don't intersect. This implies that there is a PTAS for labeling as many points as possible in any fixed-position model. This result was extended by Agarwal et al. [1] to labels with unit height but arbitrary width. In this section we show that the same result holds for the slider models. The algorithm is mainly of theoretical interest, so we sketch the ideas only briefly. First we outline the approach for the 1-position model and unit height labels [1]. Then we extend to the 1-slider model and other models.

Let P be a set of n points and assume the 1-position model. We begin by stabbing the set of n possible labels by horizontal lines at distance greater than 1, and such that each label is stabbed by some horizontal line. Let these lines be $l_1, \ldots, l_m$ from top to bottom, and let $P_1, \ldots, P_m$ be the subsets such that P_i contains the points with labels intersecting l_i. The idea is to discard every t-th subset and solve the problem for $P_1, \ldots, P_{t-1}$ optimally, then solve the problem for $P_{t+1}, \ldots, P_{2t-1}$ optimally, and so on. These optimal solutions to subproblems can simply be joined to an optimal solution for $P_1, \ldots, P_{t-1}, P_{t+1}, \ldots, P_{2t-1}, \ldots$, since no label of a point in $P_1, \ldots, P_{t-1}$ can intersect a label of a point in $P_{t+1}, \ldots, P_{2t-1}$. For the whole set P, these joined optimal solutions to subsets form a suboptimal solution. The labeling problem for $t - 1$ consecutive subsets can be solved optimally by dynamic programming in time polynomial in n but exponential in t.

Then we apply the Shifting Lemma idea [12] and discard the subsets $P_{t-1}, P_{2t-1}, P_{3t-1}, \ldots$, and compute an optimal solution for the remaining subset. Next we discard $P_{t-2}, P_{2t-2}, P_{3t-2}, \ldots$, and so on. One of the t solutions that we find is a $\frac{t-1}{t}$-approximation of the optimal solution for the whole set by the pigeon hole principle. For any constant t we have a polynomial time algorithm, so we obtain a polynomial time approximation scheme.

Now assume the 1-slider model. The idea of stabbing the labels by horizontal lines still works. So does the idea of combining the solutions of $P_1, \ldots, P_{t-1}$, of $P_{t+1}, \ldots, P_{2t-1}$ and so on. But we don't know how to solve the problem for $P_1, \ldots, P_{t-1}$ optimally in time polynomial in n. So the idea is to *approximate* the optimal solution for $P_1, \ldots, P_{t-1}$. Roughly, we set $k = t^2$ to be another constant, and determine the *leftmost* labeling of any k points of $P_1, \ldots, P_{t-1}$ by brute force as follows. Since k is constant, we can test every subset of k points and label them leftmost. If we cannot label the k points of some subset without intersection, we discard the subset. We choose the subset that gives the labeling that is leftmost among all subsets without intersections.

Let l be the leftmost vertical line that has the interiors of the k chosen labels to its left. We discard all label positions that intersect l or lie left of it. Then we repeat to find the next k leftmost labels. The observation is that the optimal solution cannot have a line left of our leftmost line with k labels to its left. So from the optimal solution, we discard at most one label from each of the $t-1$ subsets in our approximation. For the subsets $P_1, \ldots, P_{t-1}$, we approximate the optimal solution with a factor of $\frac{k}{k+t-1}$. The Shifting Lemma idea of discarding every t-th subset approximates the optimal solution with a factor of $\frac{t-1}{t}$. So the approximation factor for the whole algorithm is the product of the two: $\frac{k}{k+t-1} \cdot \frac{t-1}{t} = \frac{t^2}{t^2+t-1} \cdot \frac{t-1}{t} = \frac{t^2-t}{t^2+t-1}$. By increasing the value of t we get an approximation factor that is arbitrarily close to 1.

Assume that $P_1, \ldots, P_{t-1}$ together contain n' points. To find the leftmost k labels we try every $\binom{n'}{k}$ subsets of k points. We label these from left to right, and leftmost. This is easy in $O(k^2)$ time.

In the 4-slider model, we take the same approach, but it takes more time to find the leftmost labeling of k points. For each such subset, we test all possibilities of taking the top, bottom, left, or right slider for each label. This gives 4^k possibilities for one subset. Top and bottom sliders are put leftmost; left and right sliders are put bottommost. We must also try each of the $k!$ insertion orders of the k labels. So testing one subset can be done in $O(4^k \cdot k! \cdot k^2)$ time.

Theorem 2 *For each of the slider models and for any constant $\epsilon > 0$, there is a polynomial time algorithm which labels at least $(1-\epsilon)$ times the maximum number of input points that can be labeled.*

5 Implementation and test results

In this section we compare experimentally how many labels are placed by the greedy algorithms of Section 3 in each of the six models. We implemented the algorithms in C++ and used some data structures of LEDA, the Library of Efficient Data types and Algorithms [18]. Since LEDA doesn't have priority search trees, we used orthogonal range trees instead. Our implementation is simpler than the one described here in two respects. Firstly, the red-black trees $\mathcal{T}_i$ of **2** can be expected to contain only a few horizontal segments of H_{int} at any moment. So we replaced them by lists. Secondly, the augmented red-black tree doesn't profit much from the augmentation in practice. When searching for the minimum x-coordinate of F in a y-interval, we simply scan all leaves of the red-black tree in that interval. One can expect to visit only a few leaves, since the y-interval is only twice the unit height.

The first of the three data sets contains 1000 cities of the U.S.A. that must be labeled with their name. We used several different font sizes, and labeled the cities with the bounding boxes of their names. The results are shown in Table 1. The codes 1P, 2P, and 4P are shorthand for the 1-, 2-, and 4-position models. The codes 1S, 2S, and 4S are shorthand for the slider models. The values in the second table show the percentages with respect to the 4-position labeling.

The second data set contains the 236 points of a data posting. The labels are measurement values and come from a book on geostatistics [15]. Figure 10 shows the labeled data set and the number of labels placed in each model.

No. of labels placed

font	1P	2P	4P	1S	2S	4S
5	851	950	971	990	993	999
6	777	910	952	967	982	986
7	705	852	901	932	964	972
8	686	845	896	918	952	958
9	607	758	817	836	890	902
10	554	704	769	787	853	872
11	520	657	721	735	805	831
12	500	637	709	719	796	813
13	448	570	638	649	716	734
14	433	557	624	637	695	712
15	382	494	550	556	627	645

Percentage w.r.t. 4-position model

font	1P	2P	4P	1S	2S	4S
5	87	97	100	101	102	102
6	81	95	100	101	103	103
7	78	94	100	103	106	107
8	76	94	100	102	106	106
9	74	92	100	102	108	110
10	72	91	100	102	110	113
11	72	91	100	101	111	115
12	70	89	100	101	112	114
13	70	89	100	101	112	115
14	69	89	100	102	111	114
15	69	89	100	101	114	117

Table 1: 1000 cities on a large map.

The third data set contains the 75 points of a regression analysis. Here the points are clustered near a regression line, and the labels are simply identification numbers. Figure 11 shows the labeling.

The bottom tables of Figures 10 and 11 show that the 4-slider model sometimes places 10–15% more labels than the 4-position model. This improvement is significant, since it is always caused by a better labeling of the areas that are difficult to label. We also created artificial, pseudo-random data sets where all areas are difficult to label. Here we indeed found higher improvements: up to 92%.

Efficiency was not the main motivation for these experiments. Still it appeared that the label placement was computed in a few seconds for all data sets we tried, up to 2500 points. A plot shown on a computer screen seldom contains more than 1000 labeled points.

6 Conclusions and extensions

We have compared six different models for labeling a set of points with rectangular labels of fixed height. New in our paper (except for the work of Hirsch [11] and Doddi et al. [7]) is that we don't restrict the placement of labels to a finite number of positions, but only require that labels touch the point they belong to. All previous papers either limited the number of possible positions for a label to a constant, didn't give any time bounds, or had worse (if any) approximation factors. Some had the objective to maximize the label size rather than the number of labeled points, or used a model that allows an arbitrary orientation of the labels.

We proved that slider models for label placement are con-

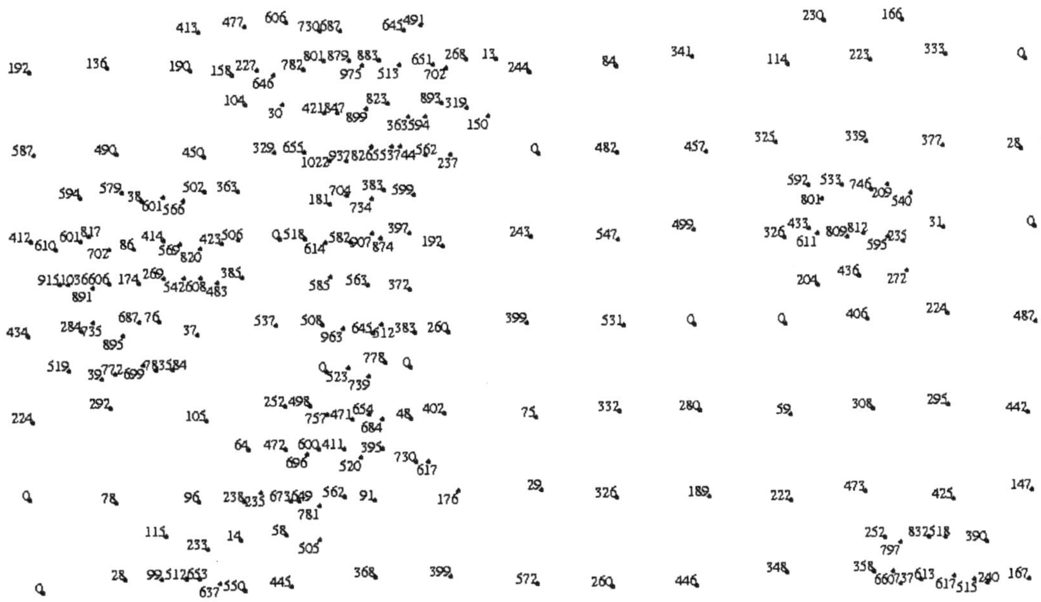

font	1P	2P	4P	1S	2S	4S
	No. of labels placed					
5	229	236	236	236	236	236
6	216	235	235	236	236	236
7	197	219	230	236	236	236
8	197	219	230	236	236	236
9	185	205	218	235	236	236
10	175	193	207	223	231	230
11	174	189	200	213	221	224
12	174	189	200	213	221	224
13	169	180	188	203	212	212
14	169	180	188	203	212	212
15	157	170	176	192	200	203

font	1P	2P	4P	1S	2S	4S
	Percentage w.r.t. 4-position model					
5	97	100	100	100	100	100
6	91	100	100	100	100	100
7	85	95	100	102	102	102
8	85	95	100	102	102	102
9	84	94	100	107	108	108
10	84	93	100	107	111	111
11	87	94	100	106	110	112
12	87	94	100	106	110	112
13	89	95	100	107	112	112
14	89	95	100	107	112	112
15	89	96	100	109	113	115

Figure 10: Labeling of the data posting in 9pt font using the 4-slider model (scaled to fit), and tables with the performance.

siderably better than fixed-position models in theory. Secondly, we showed that for each of the slider models, there is a simple factor-$\frac{1}{2}$ approximation algorithm that requires $O(n \log n)$ time and linear space. Thirdly, we sketched a polynomial time approximation scheme for each of the slider models. Finally, we compared the six different models on various data sets experimentally, using the factor-$\frac{1}{2}$ approximation algorithms. We observed, for instance, that on real-world data, the algorithm for the 4-slider model can place 10–15% more labels than the corresponding algorithm for the 4-position model. Improvements are higher for pseudo-random data.

The simple approximation algorithms we gave only apply to labels of the same height. We also have variations of our algorithms that work for labels of varying heights; these are described in the full paper [19].

We haven't compared our results with the extensive tests of Christensen et al. [4], since our main objective was to mutually compare the different models. Still it would be interesting to see how the 4-slider model performs compared to the various algorithms tested by Christensen et al. [4]. Since we don't have an exact algorithm to compute an optimal labeling, we don't know how much better the greedy algorithm performs than its approximation guarantee of $\frac{1}{2}$ suggests.

Another extension of our work would be to produce label placements that have high quality according to other criteria as well, like avoiding ambiguity, and giving preference to certain label positions over others.

References

[1] P. Agarwal, M. van Kreveld, and S. Suri. Label placement by maximum independent set in rectangles. In *Proceedings of the 9th Canadian Conference on Computational Geometry*, pages 233–238, 1997.

[2] T. Bailey and A. Gatrell. *Interactive Spatial Data Analysis*. Longman, Harlow, 1995.

[3] M. Bern and D. Eppstein. Approximation algorithms for geometric problems. In D. S. Hochbaum, editor, *Approximation Algorithms for NP-Hard Problems*, pages 296–345. PWS Publishing Company, Boston, MA, 1997.

[4] J. Christensen, J. Marks, and S. Shieber. An empirical study of algorithms for point-feature label placement. *ACM Transactions on Graphics*, 14(3):203–232, 1995.

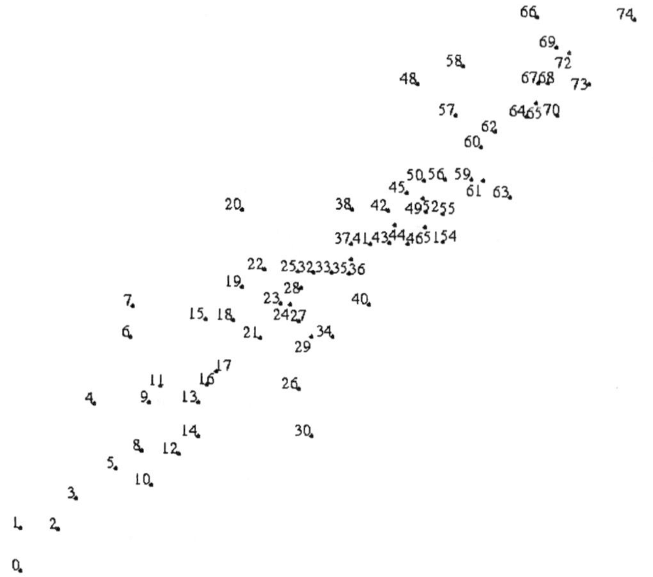

No. of labels placed

font	1P	2P	4P	1S	2S	4S
5	75	75	75	75	75	75
6	75	75	75	75	75	75
7	70	74	74	75	75	75
8	70	74	74	75	75	75
9	60	69	70	73	74	74
10	58	65	68	72	72	72
11	55	61	66	66	70	70
12	55	61	66	66	70	70
13	51	58	64	63	68	71
14	51	58	64	63	68	71
15	50	56	61	62	67	68

Percentage w.r.t. 4-position model

font	1P	2P	4P	1S	2S	4S
5	100	100	100	100	100	100
6	100	100	100	100	100	100
7	94	100	100	101	101	101
8	94	100	100	101	101	101
9	85	98	100	104	105	105
10	85	95	100	105	105	105
11	83	92	100	100	106	106
12	83	92	100	100	106	106
13	79	90	100	98	106	110
14	79	90	100	98	106	110
15	81	91	100	101	109	111

Figure 11: Labeling of the scatterplot in 11pt font using the 4-slider model (scaled to fit), and tables with the performance.

[5] T. H. Cormen, C. E. Leiserson, and R. L. Rivest. *Introduction to Algorithms.* MIT Press, Cambridge, MA, 1990.

[6] M. de Berg, M. van Kreveld, M. Overmars, and O. Schwarzkopf. *Computational Geometry - Algorithms and Applications.* Springer-Verlag, 1997.

[7] S. Doddi, M. V. Marathe, A. Mirzaian, B. M. Moret, and B. Zhu. Map labeling and its generalizations. In *Proc. 8th ACM-SIAM Symposium on Discrete Algorithms*, pages 148–157, 1997.

[8] J. Doerschler and H. Freeman. A rule-based system for dense-map name placement. *Communications of the ACM*, 35:68–79, 1992.

[9] M. Formann and F. Wagner. A packing problem with applications to lettering of maps. In *Proc. 7th Annu. ACM Sympos. Comput. Geom.*, pages 281–288, 1991.

[10] R. J. Fowler, M. S. Paterson, and S. L. Tanimoto. Optimal packing and covering in the plane are NP-complete. *Inform. Process. Lett.*, 12(3):133–137, 1981.

[11] S. A. Hirsch. An algorithm for automatic name placement around point data. *The American Cartographer*, 9(1):5–17, 1982.

[12] D. Hochbaum and W. Maass. Approximation schemes for covering and packing problems in image processing and VLSI. *J. ACM*, 32:130–136, 1985.

[13] H. Hunt III, M. Marathe, V. Radhakrishnan, S. Ravi, D. Rosenkrantz, and R. Stearns. A unified approach to approximation schemes for NP- and PSPACE-hard problems for geometric graphs. In *Proc. 2nd Europ. Symp. on Algorithms*, volume 855 of *Lect. Notes in Comp. Science*, pages 424–435, 1994.

[14] E. Imhof. Positioning names on maps. *The American Cartographer*, 2(2):128–144, 1975.

[15] E. H. Isaaks and R. M. Srivastava. *An Introduction to Applied Geostatistics.* Oxford University Press, New York, 1989.

[16] J. Marks and S. Shieber. The computational complexity of cartographic label placement. Technical report, Harvard CS, 1991.

[17] E. M. McCreight. Priority search trees. *SIAM J. Comput.*, 14:257–276, 1985.

[18] K. Mehlhorn and S. Näher. LEDA: a platform for combinatorial and geometric computing. *Commun. ACM*, 38:96–102, 1995.

[19] M. van Kreveld, T. Strijk, and A. Wolff. Point set labeling with sliding labels. Technical report, Utrecht University, to appear, 1998.

[20] F. Wagner and A. Wolff. Map labeling heuristics: Provably good and practically useful. In *Proc. 11th Annu. ACM Sympos. Comput. Geom.*, pages 109–118, 1995.

[21] F. Wagner and A. Wolff. A practical map labeling algorithm. *Computational Geometry: Theory and Applications*, 7:387–404, 1997.

[22] A. Wolff. A map labeling bibliography, 1997. `http://www.inf.fu-berlin.de/map-labeling/bibliography.html`.

[23] P. Yoeli. The logic of automated map lettering. *The Cartographic Journal*, 9:99–108, 1972.

A Unified Approach to Labeling Graphical Features *

Konstantinos G. Kakoulis and Ioannis G. Tollis
Department of Computer Science
The University of Texas at Dallas
Richardson, TX 75083-0688
email: kostant@utdallas.edu, tollis@utdallas.edu

Abstract

The automatic placement of text or symbol labels corresponding to graphical objects is critical in several application areas such as Cartography, Geographical Information Systems, and Graph Drawing. In this paper we present a general framework for solving the problem of assigning text or symbol labels to a set of graphical features in two dimensional drawings or maps. Our approach does not favor the labeling of one type of graphical feature (such as a node, edge, or area) over another. Additionally, the labels are allowed to have arbitrary size and orientation. We have applied our framework to drawings of graphs. We have implemented our techniques and have performed extensive experimentation on hierarchical and orthogonal drawings of graphs. The resulting label assignments are very practical and indicate the effectiveness of our approach.

1 Introduction

An important aspect of information visualization is the automatic placement of text or symbol labels corresponding to graphical objects of two dimensional maps and drawings. The problem of automatic label placement is critical in several application areas such as Cartography [3], Geographical Information Systems [12], and Graph Drawing [4]. It is worth noting that the placement of labels is primarily performed manually, especially in map production [1, 24]. Also, according to the ACM Computational Geometry Task Force [2], label placement is an important area of research in Computational Geometry. Most of the previous research concentrated on labeling points in geographical and technical maps [3]. In this paper we present a unified framework for the general labeling problem, where objects to be labeled may be points, lines, or areas, in two dimensional maps and drawings.

In map production, labeling is typically partitioned into three tasks: (a) labeling points (e.g., cities), (b) labeling lines (e.g., roads or rivers), and (c) labeling areas (e.g., lakes or oceans). Progress has been made in solving the problem of assigning labels to a set of points or nodes, the *Node Label Placement* (NLP) problem [3]. The problem of assigning labels to a set of lines or edges, also known as the *Edge Label Placement* (ELP) problem, has not been well resolved [26, 29]. A successful technique for solving the ELP problem for hierarchical and force directed drawings of graphs is presented in [19]. It is worth noting that both the NLP [11, 20, 22] and ELP [18] problems are NP-Hard.

Christensen, Marks, and Shieber present a comprehensive survey of algorithms for the labeling problem in [3]. A variety of types of algorithms have been used in order to solve the labeling problem: Greedy algorithms [3, 16], exhaustive search algorithms [7, 8, 13], algorithms that simulate physical models (i.e., Simulated Annealing [3]), algorithms that reduce the labeling problem to a variant of 0-1 integer programming [29], and algorithms that restrict the labeling problem to a variant of 2-SAT problem [11, 27]. Many of these algorithms address only the NLP problem [11, 16, 29]. Others, even though they address the general labeling problem, are based on exhaustive search algorithms [7, 8, 13]. While these algorithms perform well for very small problems, they are impractical due to their exponential nature.

Approximation algorithms for restricted versions of the NLP problem are presented in [6, 11, 27]. Specifically, the approach of [11, 27] assigns labels of *equal* size to all points while attempting to maximize the size of the assigned labels. A similar approach has been taken in [6]. This works

*Research supported in part by NIST, Advanced Technology Program grant number 70NANB5H1162. A patent on these and related results is pending.

very well when one has to label points with labels of equal size. In effect, finding the maximum label size is equivalent to finding the smallest factor by which the map has to be zoomed out such that each point has a label assigned to it. Unfortunately these techniques cannot be used in most applications, where either the label size depends heavily on each object, or the objects are not necessarily points.

Almost all of the above techniques start with a rather small initial set of potential label positions from which they derive a final label assignment. This precondition works well when solving the NLP problem. For example, each point is given at most four potential label positions. However, such a limited label space will not work well when solving either the ELP problem, or the general labeling problem. This is because the initial set of label positions plays a critical role in a final labeling solution. Also, none of the above techniques addresses successfully the ELP problem for general drawings and maps.

In this paper we present a unified approach for solving the general labeling problem. We will refer to this as the *Graphical Feature Label Placement* (GFLP) problem, where a graphical feature can be a node, edge, or area. Our approach does not favor the labeling of one type of graphical feature (e.g., node, edge, or area) over another. Additionally, the labels may have arbitrary size and orientation. Our approach consists of three major steps: (a) find a large set of potential labels for each graphical feature, (b) refine the set of label positions by deleting a number of heavily overlapping labels, and (c) assign a label to each graphical feature so that there are no overlaps. Each step can be implemented using various algorithms. Hence, from this point of view, our paper presents a framework for solving the GFLP problem. We also present heuristics that solve the GFLP problem following this general framework. In addition, we have implemented these heuristics for labeling edges and nodes of graph drawings, and we have done extensive experiments, with very encouraging results.

Our paper is organized as follows: In Section 2 we define the GFLP problem formally. In Section 3 we introduce the main ideas of our approach, and present some sufficient conditions that guarantee an optimal label assignment with no overlaps. In Section 4 we present more details on possible algorithms for solving each step. In Section 5 we present methods for improving the final label assingment. Finally, in Section 6 we present experimental results of an implementation of our approach.

2 The GFLP problem.

Let Γ be a drawing of a graph $G(V, E)$, and F be the set of graphical features of Γ to be labeled. A solution to the GFLP problem assigns text labels to the graphical features of drawing Γ such that the relevant information is communicated in the best possible way. Hence it is very important that a visual inspection of a label assignment be sufficient to explain the semantics of any label. This can be achieved by positioning the labels in the most appropriate place.

Cartographers, through the years have devised some general rules that a good label placement must follow.

Basic rules for labeling quality [17, 28]:

- No overlaps of a label with other labels or other graphical features of the drawing are allowed.
- Each label can be easily identified with exactly one graphical feature of the drawing.
- Each label must be placed in the best possible position (among all acceptable positions).

The first 2 rules evaluate the clarity of the association of the labels to their corresponding graphical features. The third rule defines a ranking among all label positions that typically captures the aesthetic preference for specific labels. This is an essential criterion for the quality of labeling of geographical and to some extent technical maps. It also allows the introduction of problem specific constraints (i.e., if the graphical feature is an edge of a graph drawing, the label of an edge must be closer to the source or destination node).

For a given drawing Γ of a graph G we define the following: (i) Λ_f is the set of all label positions for graphical feature f of Γ; (ii) Λ is the set of all label positions for all graphical features of Γ to be labeled; (iii) $\lambda : F \rightarrow \Lambda$ is a function that assigns to each graphical feature f in F a label position $\lambda(f) = \lambda_f \in \Lambda_f$.

When we consider the input drawing to be fixed, there are instances where even an optimal assignment produces labels that do not strictly follow the above rules. Therefore, each label position λ_f that is part of a final label assignment is associated with a cost. $COST : \Lambda \rightarrow \mathcal{N}$ is a function that gives the cost of assigning label λ_f to f in the final label assignment. $COST(\lambda_f)$ is typically a linear combination of two (or more) components: (i) A cost which reflects the severity of the violation of the first two basic rules for label λ_f; (ii) a cost associated with the ranking of label λ_f.

In $Fig.$ 1(a), the graphical feature to be labeled is a node. A corner of a label position touches the node. Label positions are not allowed to overlap the node that are associated with, or any other graphical feature. In $Fig.$ 1(b), the graphical feature to be labeled is an edge. Labels like A, B and D are preferable. On the other hand, labels like C, that overlap their associated edge, should be avoided. However,

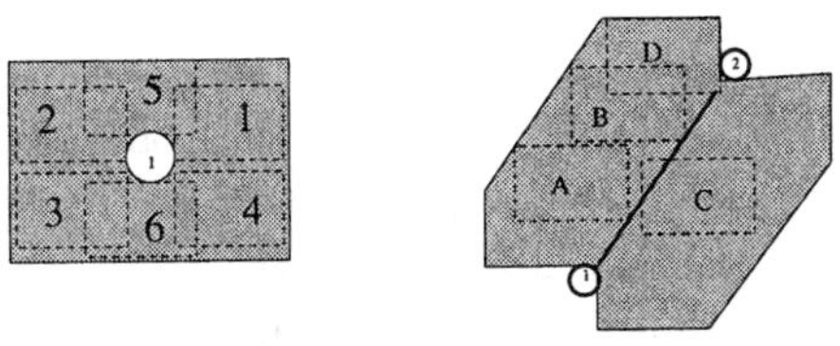

(a) (b)

Figure 1: (a) Labeling space of a node; (b) Labeling space of an edge.

if the drawing is very dense, they can be acceptable with some appropriate cost assigned to them.

The GFLP problem can be viewed as an optimization problem where the objective is to find a label assignment of minimum total cost.

The GFLP Problem : Let F be a set of graphical features of a drawing Γ.

Question : Find a label assignment that minimizes the following function:

$$\sum_{i \in F} \sum_{j \in \Lambda_i} COST(\lambda(i)) P(i,j)$$

Where:

$$P(i,j) = \begin{cases} 1, & if\ \lambda(i) = j, \\ 0, & otherwise \end{cases}$$

and

$$\sum_{i \in F} \sum_{j \in \Lambda_i} P(i,j) = |F|$$

Where:

$$\sum_{j \in \Lambda_i} P(i,j) = 1, \qquad i \in F.$$

□

3 Sufficient conditions for solving the GFLP problem

In this section we present the general framework of our technique along with some sufficient conditions for solving the GFLP problem optimally.

We transform the GFLP problem into a matching problem where each graphical feature f to be labeled is matched to one of its label positions in Λ_f. This transformation is a two step process. First, we select a set of label positions Λ. Then we reduce and partition Λ into subsets or clusters, such that each remaining label position is a member of exactly one cluster of Λ. Finally, graphical features are matched to these clusters such that, in the final label assignment, at most one label position from each cluster is assigned to a graphical feature. In our technique we cluster together label positions that overlap each other. Each label that belongs to a cluster can potentially overlap at least one other label within the cluster. We formalize this idea by introducing an equivalence relation $\mathcal{R}$. We denote with ${}_x\mathcal{R}_y$ that x and y are members of the same equivalence class in $\mathcal{R}$.

Definition 1 *For any label positions x, y in Λ, if ${}_x\mathcal{R}_y$ then $\forall z$, where z is a label position in Λ, z overlaps x* if and only if *z overlaps y as well.*

Relation $\mathcal{R}$ restricts the number of qualified label positions that can be included in a class of $\mathcal{R}$. By the definition of $\mathcal{R}$ we observe the following: (i) A label position is a member of a class r in $\mathcal{R}$ if and only if it overlaps every other label position which is a member of r; (ii) a label position forms a single class in $\mathcal{R}$ if and only if it does not overlap any other label position in Λ; (iii) in a label assignment with no overlaps there can be at most one label position from each class in $\mathcal{R}$.

To further clarify the main idea of our technique we introduce the *overlap* and *matching* graphs:

Definition 2 *Given a drawing Γ, a set F of graphical features to be labeled, and a set Λ of label positions for F, the overlap graph $G_o(V_f, V_l, E_o)$ is defined as follows:*

- *Each node in V_f corresponds to a graphical feature in F.*
- *Each node in V_l corresponds to a label position in Λ.*
- *Each edge (f, l) in E_o connects a node f in V_f, to a node l in V_l, if and only if f in F has a label position l in Λ_f. Also, if labels l_1 and l_2 overlap then there is an edge (l_1, l_2) in E_o.*

Definition 3 *Given a drawing Γ, a set F of graphical features to be labeled, and a set Λ of label positions for F, the matching graph $G_m(V_f, V_c, E_m)$ is defined as follows:*

- *Each node f in V_f corresponds to a graphical feature f in F.*
- *Each node r in V_c corresponds to a class in $\mathcal{R}$.*
- *Each edge (f, r) in E_m connects a node in V_f, to a node r in V_c, if and only if f in F has a label position that is a member of class r of $\mathcal{R}$.*

In order to form an overlap graph representing an instance of the GFLP problem the set Λ of label positions must be known. By representing the GFLP problem as a graph, the inherent hardness of the problem is revealed. According to [21], the labeling problem is closely related to the *independent set* problem. Indeed, if there is only one potential label position for each graphical feature in the drawing, the problem of assigning labels to the maximum number of graphical features is equivalent to finding a maximum size independent set.

Given a drawing Γ and a set F of graphical features to be labeled, we construct an overlap graph by selecting a finite set of label positions Λ. Next we remove some label positions until each label position remaining in Λ is a member of some class in $\mathcal{R}$. Then we construct the matching graph G_m from the set of classes in $\mathcal{R}$. Notice that G_m is a bipartite graph and the cost of assigning label l to graphical feature f is the weight of edge (f, l) in G_m. Therefore, a maximum cardinality minimum weight matching for graph G_m will give us an optimal (maximum number of labels with minimum cost) label assignment with no overlaps with respect to the *reduced* label set Λ. Hence, we have the following theorem:

Theorem 1 *Let Λ be a set of label positions for all graphical features to be labeled. If for every label position λ, $\lambda \in \Lambda$, there exists a class r in $\mathcal{R}$, such that $\lambda \in r$, then a maximum cardinality minimum weight matching of the corresponding matching graph G_m produces an optimal solution to the GFLP problem with respect to Λ, with no overlaps.*

The main difficulties of optimally solving the GFLP problem are: (a) the construction of a sufficient set Λ of label positions, and (b) the reduction of Λ such that each label position is a member of some class in $\mathcal{R}$. In fact, this is exactly where the hardness of the problem lies. In the following sections we discuss some heuristics that attempt to solve these two steps.

4 Solving the GFLP problem

We solve the GFLP problem by constructing the sufficient conditions for Theorem 1. This is a technique that consists of three major steps:

Algorithm_GFLP
INPUT: A drawing Γ and a set F of graphical features to be labeled.
OUTPUT: A label assignment.
1. Find label positions for each graphical feature in F.
2. Reduce the set of label positions found in Step 1 such that each one of the remaining label positions is a member of exactly one class in $\mathcal{R}$.
3. Make the final label assignment by solving the matching problem.

□

Next, we expand on the three basic steps of the labeling algorithm:

STEP 1
To find a set of label positions for each graphical feature we can use a number of heuristics (e.g., for points see [3], for areas see [13, 26]). Next, we present a simple heuristic when the graphical features to be labeled are edges. As *Fig.* 11 illustrates, we define a number of equally spaced points on each edge. Each assigned label position λ_i is associated with exactly one of these points i, such that one of the corners of the label λ_i coincides with point i. In addition, label λ_i does not overlap its corresponding graphical feature or any other graphical feature (except other label positions). A global approach for finding an initial set of label positions for non-horizontal edges can be found in [19]. The size of the initial set of label positions must be kept reasonably small since it affects the performance of any labeling algorithm.

A more general approach is the following: First we find the solution space for each object to be labeled (i.e., the shaded regions in *Fig.* 1). Then we split the solution space of each object into sets of subregions: (a) those that intersect subregions of other objects, and (b) those that are free of intersections. For the first case we assign label positions according to the simple heuristic described above. For the latter case we do the following: If a label is free of overlaps and is included inside a subregion, then we assign that label to its corresponding object and remove it from the list of objects to be labeled.

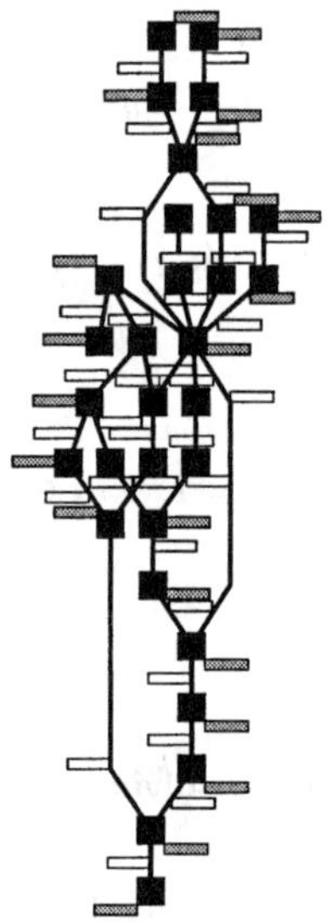

Figure 2: Sample hierarchical drawing; labels are parallel to the horizontal axis; the grey boxes are node labels and the white boxes are edge labels.

STEP 2
One important aspect of any labeling algorithm is the effi-

cient detection of overlaps. If label positions are parallel to the axis then we can detect overlaps very efficiently using the techniques in [9, 10]. Otherwise, the techniques of [15] that detect intersections between convex polygons are used in order to find overlaps of labels with arbitrary orientation.

In order to reduce the set of label positions we do the following: First we group together label positions that overlap into clusters. Let $G_R(V_l, E_c)$ be the graph obtained from the overlap graph by removing all nodes corresponding to graphical features along with their incident edges (e.g., the graph in *Fig.* 13 is the G_R graph of the overlap graph in *Fig.* 12). Then, each connected component in G_R is a cluster of overlapping labels. From the definition of the equivalence relation $\mathcal{R}$ it is clear that a cluster c is a class in $\mathcal{R}$ if and only if each label position in c overlaps all other label positions in c (i.e., c is a complete graph). We call such clusters *complete*. Next we give an algorithm that finds a set of classes in $\mathcal{R}$:

Algorithm_Find_Classes_In_R

INPUT: A set of label positions corresponding to graphical features to be labeled.

OUTPUT: A set of classes in $\mathcal{R}$.

1. Create an overlap graph corresponding to the input set of label positions.
2. Obtain graph $G_R(V_l, E_c)$ from the overlap graph.
3. Group incident nodes of G_R into clusters.
4. *For* each cluster c that is not complete *do*:
   Remove a number of nodes from c until either c becomes complete, or c is split into two or more clusters.
5. Repeat Step 4 until all remaining clusters are complete.

□

The optimal solution would be one with the maximum number of minimum size complete clusters, with the additional constraint that each graphical feature has a large number of label positions as part of some clusters. In reality we want to find an independent set of complete clusters of label positions. We can use heuristics based on techniques that solve the independent set problem. Next we present a simple, and very successful (according to our experiments) technique. If a cluster c must be split, then we remove from c the node with the highest degree, unless that node corresponds to a label position of some graphical feature with very few label positions. In that case we remove the next highest degree node from c. Repeat this process until either c is split into at least two disjoint clusters, or c is complete. One can make this process more efficient by heuristically removing label positions that add to the complexity of the problem without potentially improving the solution. For example, if a label position l of a graphical feature f is free of overlaps, then we can safely remove all label positions for f with lower ranking than l.

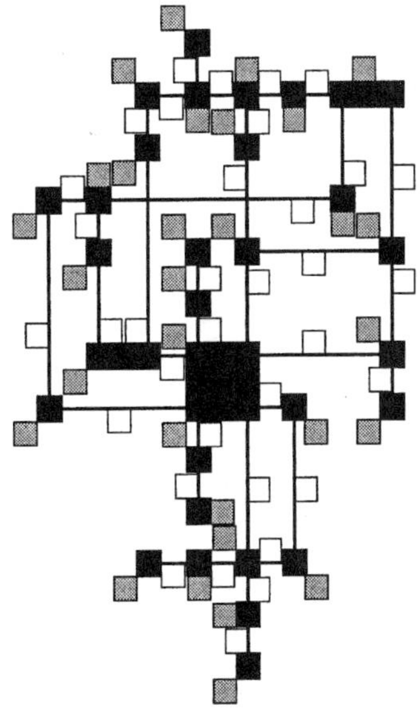

Figure 3: Sample orthogonal drawing; labels are parallel to the horizontal axis; the grey boxes are node labels and the white boxes are edge labels.

STEP 3

Once we have the set of classes in $\mathcal{R}$, the construction of the matching graph is trivial. By solving the maximum cardinality minimum weight matching problem (see [14, 25] for efficient algorithms) for graph G_m we find a final label assignment. The size of the matching graph depends not only on the size of the input drawing, but also on the size of the set Λ of labels and the density of overlaps.

If the matching graph has a particular structure then we might be able to solve the matching problem more efficiently. For example, if each label position overlaps at most one other label position then we have devised an algorithm that finds a maximum matching with low total weight in linear time with respect to the size of G_m [19].

A simple example that illustrates the major steps of the labeling framework presented in this section can be found in the appendix.

5 Extensions and Improvements

The framework presented in this paper does not always produce an optimal solution because it considers a subset of the solution space, since label positions must be discrete and must not overlap other graphical objects, except other label positions. In many instances further improvements of the label assignment are possible by performing a postprocessing step. In the postprocessing step we employ techniques

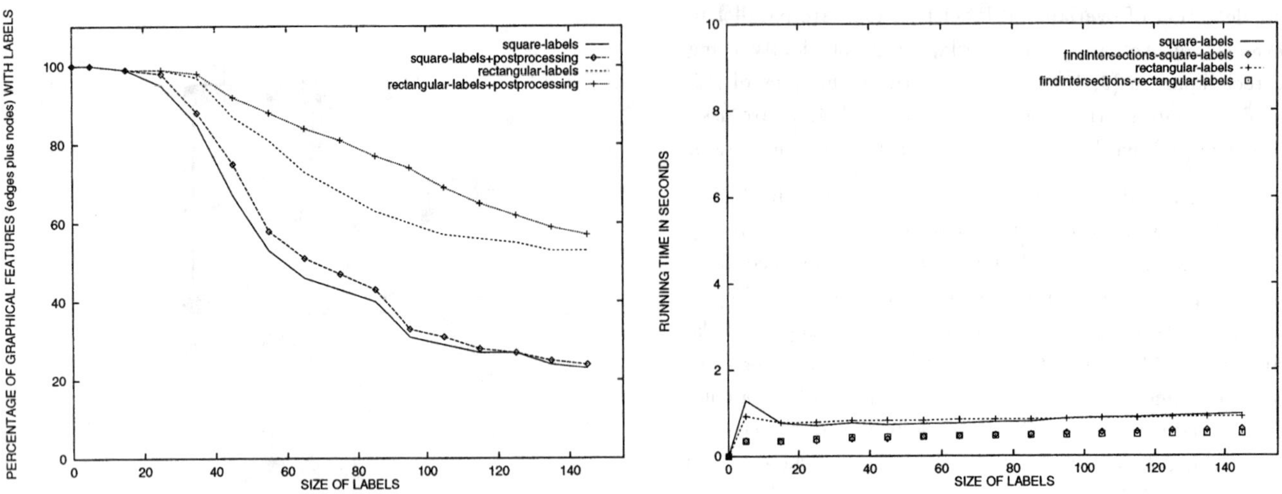

Figure 4: Experimental results for hierarchical drawings of 30 nodes and on the average 36 edges.

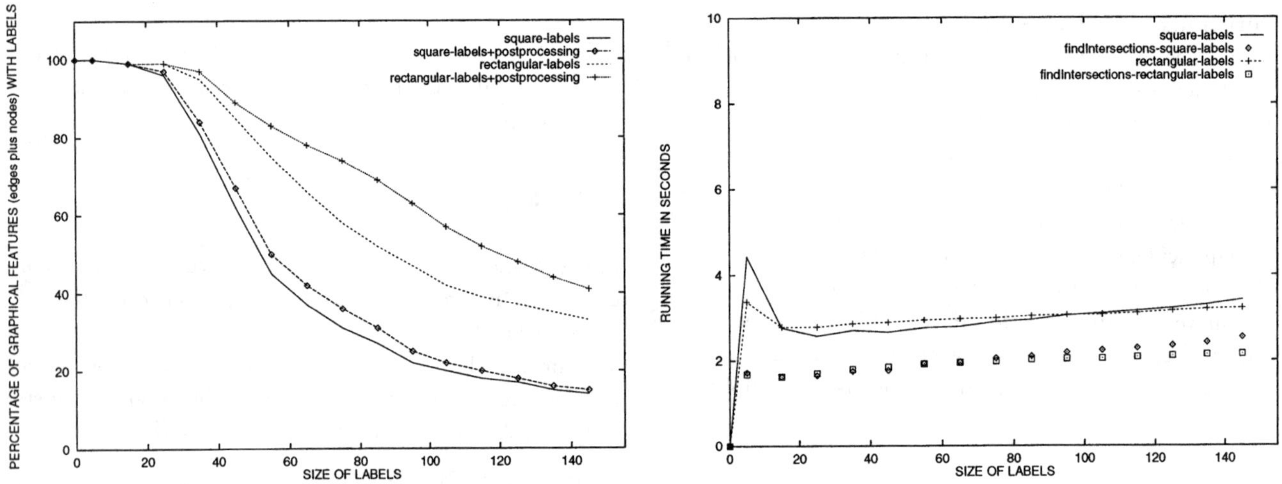

Figure 5: Experimental results for hierarchical drawings of 60 nodes and on the average 79 edges.

based on local and exhaustive search. We explore the solution space in three ways: (*i*) Locally shift assigned labels to create space for a new label to be assigned; (*ii*) search the solution space to find if there is enough space to assign a label after repositioning already assigned labels; (*iii*) search the solution space after relaxing the restrictions on the quality of the label assignment by allowing labels to overlap their associated graphical features and/or other labels.

In [3], a survey of search algorithms for the labeling problem is presented. Because of the exponential nature of these approaches, only a constant number of steps are allowed for backtracking. We can speedup the postprocessing step considerably by bounding unnecessary backtracking by using the information obtained in the main algorithm, which reveals the conflict areas (with respect to labeling) of the drawing.

6 Experimental results.

We have implemented a version of the labeling framework presented in this paper on top of the Graph Layout Toolkit (GLT[1]) in C++. We ran our experiments on a SPARC 5 unix based machine. The test data are graphs taken from the collection used in [5].

In our implementation the graphical features to be labeled are edges and nodes of graph drawings produced us-

[1]The Graph Layout Toolkit is a product of Tom Sawyer Software, Inc. (http://www.tomsawyer.com).

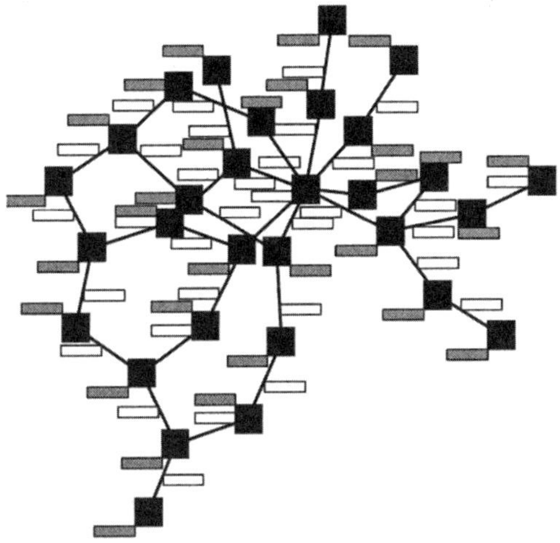

Figure 6: Sample force-directed drawing; labels are parallel to the horizontal axis; the grey boxes are node labels and the white boxes are edge labels.

ing hierarchical[2] and orthogonal[3] drawing algorithms implemented in GLT. For the experiments, labels are rectangles of equal size, parallel to the horizontal axis, and all label positions of the same object have the same ranking. Of course our implementation allows variable size and ranking; it also allows edge labels to follow the direction of their corresponding edges (see *Fig.* 9). Furthermore, it successfully produces label assignments for graph drawings of any style. Examples include hierarchical (Figures 2 and 9), orthogonal (Fig. 3), force-directed (Fig. 6), and circular (Fig. 10).

The initial set of label positions has been produced using the simple heuristic presented in Step 1 of the main algorithm. Each graphical feature f has been assigned the maximum number of potential labels with the restriction that no label position overlaps any other label positions of f. We reduce the overlap graph to the matching graph using the method described in the Step 2 of the algorithm. We used the bipartite matching algorithm from the LEDA library [23] to find the maximum cardinality matching of the matching graph. For a postprocessing step we used a simple greedy heuristic that searches the solution space to find if there is enough space to assign a label.

In the final label assignment produced by the GFLP algorithm, each assigned label does not overlap any other label, edge, or node (even though it is allowed to overlap its associated edge but not node). If we allow labels in the final assignment to overlap other graphical features in the drawing, then the success rate of the GFLP algorithm is much greater. However, there will be a significant decrease in the quality of the label assignment.

We ran 2 sets of experiments; (*i*) Labels are square, and in successive runs of the algorithm we increase the size of the square; (*ii*) Labels are rectangles, and in successive runs of the algorithm we increase only the width of the rectangle.

We are interested in how the success rate (percentage of graphical features with assigned labels) decreases as we increase the size of the labels. All drawings are grid drawings, that is nodes and bends have integer coordinates. The label size reported is related to the grid-size. For example, a label with width 20 is a label with width equal to 20 grid units.

In Figures 4, 5, 7, and 8 we present the results of our experiments. The plain lines represent the results after running only the main algorithm. The marked lines represent the results after running the algorithm with the postprocessing step. In each figure the left plot shows the success rate of the algorithm (the size of the labels is represented in the x-direction, and the percentage of graphical features with assigned labels is represented in the y-direction). The right plot shows the running time of the set of experiments that are represented in the left plot. From these figures it is clear that our algorithm performs very well, since it successfully assigns labels to all objects, for labels up to size 40. Furthermore, the success rate of assigning rectangular labels for orthogonal drawings is much better than hierarchical. This is significant since previous techniques for labeling graph drawings (see [19]) could not efficiently assign labels to the horizontal edges of orthogonal drawings. As expected, the performance of our technique on both types of drawings is much better for rectangular than for square labels. In addition, once the size of the labels increases beyond a certain threshold the labeling quality deteriorates rapidly.

[2] A *hierarchical* drawing maps each edge into a polygonal chain; the nodes and bends are constrained to lie on a set of equally spaced horizontal lines, called *layers*.

[3] An *orthogonal* drawing maps each edge into polygonal chains consisting of horizontal and vertical segments.

References

[1] D. Beus and D. Crockett. Automated production of 1:24,000 scale quadrangle maps. In *Proc. 1994 ASPRS/ACSM Ann. Convention and Exposition*, volume 1, pages 94–99, 1994.

[2] B. Chazelle et al. Application challenges to computational geometry: CG impact task force report. Technical Report TR-521-96, Princeton Univ., April 1996.

[3] J. Christensen, J. Marks, and S. Shieber. An empirical study of algorithms for Point Feature Label Placement. *ACM Trans. on Graphics*, 14(3):203–232, July 1995.

[4] G. Di Battista, P. Eades, R. Tamassia, and I. G. Tollis. Algorithms for drawing graphs: an annotated bibliography. *Comput. Geom. Theory Appl.*, 4:235–282, 1994.

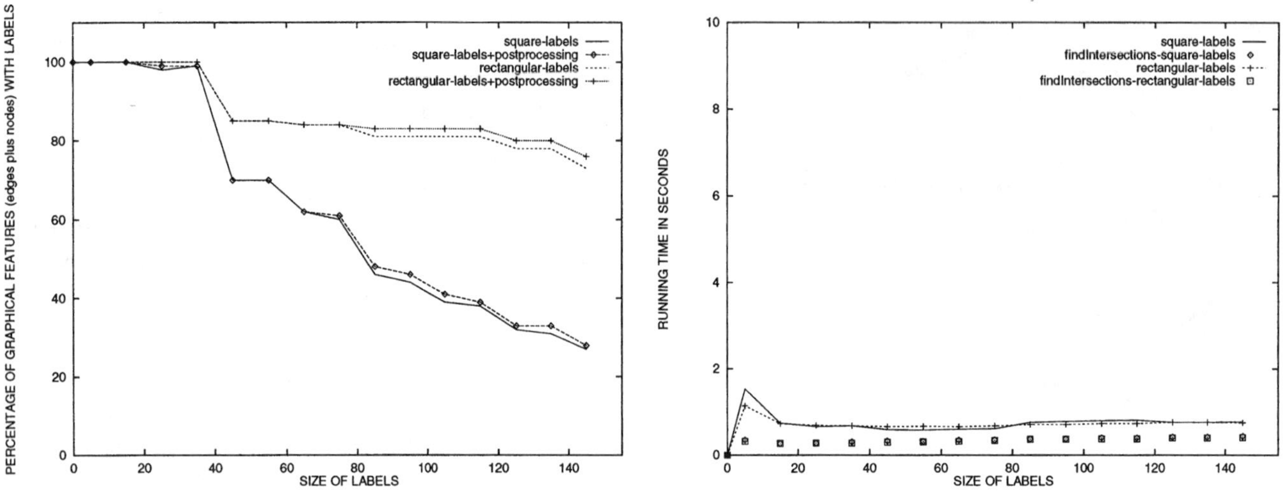

Figure 7: Experimental results for orthogonal drawings of 30 nodes and on the average 36 edges.

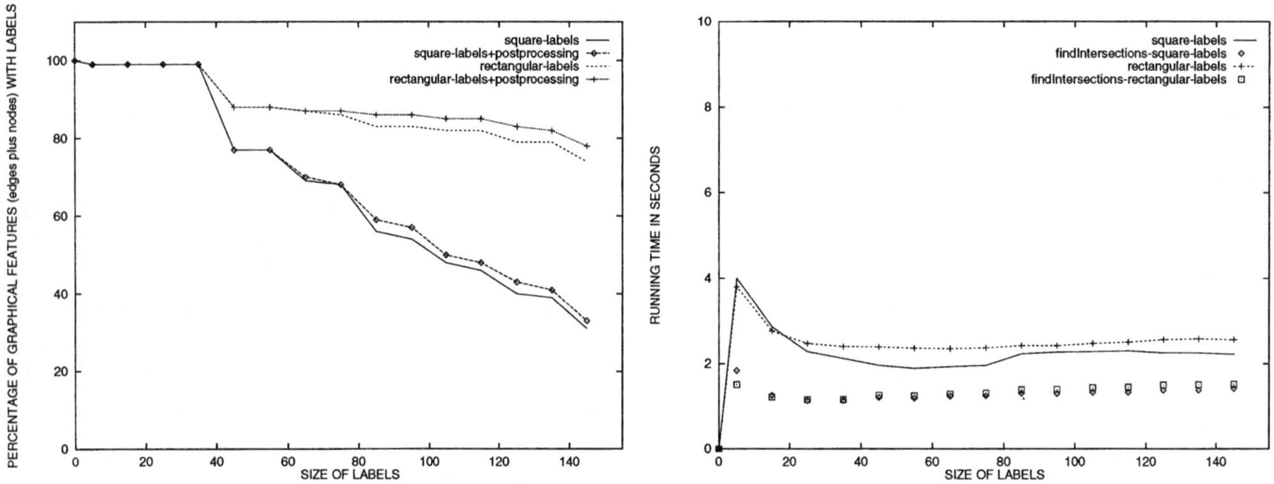

Figure 8: Experimental results for orthogonal drawings of 60 nodes and on the average 79 edges.

[5] G. Di Battista, A. Garg, G. Liotta, R. Tamassia, E. Tassinari, and F. Vargiu. An experimental comparison of four graph drawing algorithms. *Comput. Geom. Theory Appl.*, 7(5-6):303–325, 1997.

[6] S. Doddi, M. V. Marathe, A. Mirzaian, B. M. Moret, and B. Zhu. Map Labeling and Its Generalizations. In *Proc. 8th ACM-SIAM Sympos. Discrete Algorithms*, pages 148–157, 1997.

[7] J. S. Doerschler and H. Freeman. A rule based system for dense map name placement. *Communications of ACM*, 35(1):68–79, 1992.

[8] L. R. Ebinger and A. M. Goulete. Noninteractive automated names placement for the 1990 decennial census. *Cartography and Geographic Informaton Systems*, 17(1):69–78, January 1990.

[9] H. Edelsbrunner. A new approach to rectangle intersections, Part I. *Internat. J. Comput. Math.*, 13:209–219, 1983.

[10] H. Edelsbrunner. A new approach to rectangle intersections, Part II. *Internat. J. Comput. Math.*, 13:221–229, 1983.

[11] M. Formann and F. Wagner. A packing problem with applications to lettering of maps. In *Proc. 7th Annu. ACM Sympos. Comput. Geom.*, pages 281–288, 1991.

[12] H. Freeman. Computer name placement. In D. J. Maguire, M. F. Goodchild, and D. W. Rhind, editors, *Geographical Information Systems: Principles and Applications*, pages 445–456. Longman, London, 1991.

[13] H. Freeman and J. Ahn. On the problem of placing names in a geographical map. *Int. J. of Pattern Rec. and Artificial Intelligence*, 1(1):121–140, 1987.

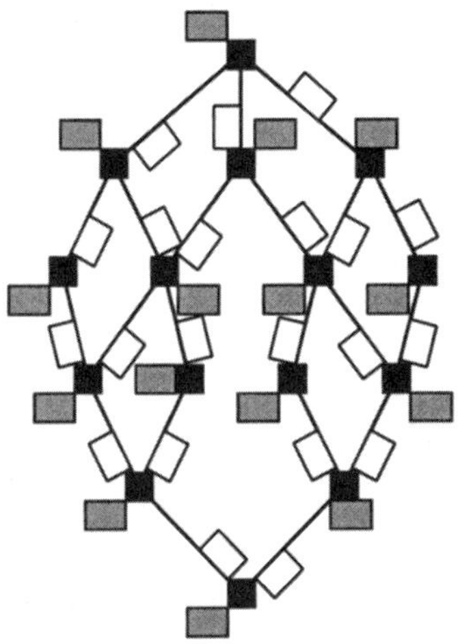

Figure 9: Sample hierarchical drawing; labels of nodes are parallel to the horizontal axis; labels of edges follow the direction of the edges; the grey boxes are node labels and the white boxes are edge labels.

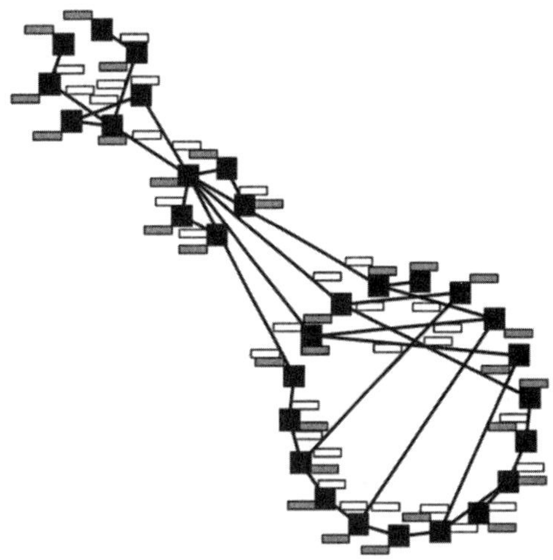

Figure 10: Sample circular drawing; labels are parallel to the horizontal axis; the grey boxes are node labels and the white boxes are edge labels.

[14] A. V. Goldberg and R. Kennedy. An Efficient Cost Scaling Algorithm for the Assignment Problem. *Mathematical Programming*, 71:153–178, December 1995.

[15] P. Gupta, R. Janardan, and M. Smid. Efficient algorithms for counting and reporting pairwise intersections between convex polygons. In *Proc. 8th Canadian Conference on Computational Geometry*, pages 8–13. Carleton University Press, 1996.

[16] S. A. Hirsch. An algorithm for automatic name placement around point data. *The American Cartographer*, 9(1):5–17, 1982.

[17] E. Imhof. Positioning names on maps. *The American Cartographer*, 2(2):128–144, 1975.

[18] K. G. Kakoulis and I. G. Tollis. On the Edge Label Placement Problem. In S. North, editor, *Graph Drawing (Proc. GD '96)*, volume 1190 of *Lecture Notes in Computer Science*, pages 241–256. Springer-Verlag, 1997.

[19] K. G. Kakoulis and I. G. Tollis. An Algorithm for Labeling Edges of Hierarchical Drawings. In G. Di Battista, editor, *Graph Drawing (Proc. GD '97)*, volume 1353 of *Lecture Notes in Computer Science*, pages 169–180. Springer-Verlag, 1998.

[20] T. Kato and H. Imai. The NP-completeness of the character placement problem of 2 or 3 degrees of freedom. In *Record of Joint Conference of Electrical and Electronic Engineers in Kyushu*, pages 11–18, 1988. In Japanese.

[21] D. Knuth and A. Raghunathan. The problem of compatible representatives. *SIAM J. Disc. Math.*, 5:36–47, 1992.

[22] J. Marks and S. Shieber. The computational complexity of cartographic label placement. Technical Report 05-91, Harvard University, 1991.

[23] K. Mehlhorn and S. Näher. LEDA: a platform for combinatorial and geometric computing. *Commun. ACM*, 38:96–102, 1995.

[24] A. H. Robinson, J. L. Morrison, P. C. Muehrcke, A. J. Kimerling, and S. C. Guptill. *Elements of Cartography*. John Willey & Sons, Inc., (pages 420-421), 6th edition, 1995.

[25] R. E. Tarjan. *Data Structures and Network Algorithms*, volume 44 of *CBMS-NSF Regional Conference Series in Applied Mathematics*. Society for Industrial Applied Mathematics, 1983.

[26] J. W. van Russel. An algorithm for locating candidate labeling boxes within a polygon. *The American Cartographer*, 16(3):201–209, 1989.

[27] F. Wagner and A. Wolff. Map labeling heuristics: Provably good and practically useful. In *Proc. 11th Annu. ACM Sympos. Comput. Geom.*, pages 109–118, 1995.

[28] P. Yoeli. The logic of automated map lettering. *The Cartographic Journal*, 9(2):99–108, 12 1972.

[29] S. Zoraster. The solution of large 0-1 integer programming problems encountered in automated cartography. *Operation Research*, 38(5):752–759, September-October 1990.

APPENDIX

A A simple example of the labeling algorithm through pictures.

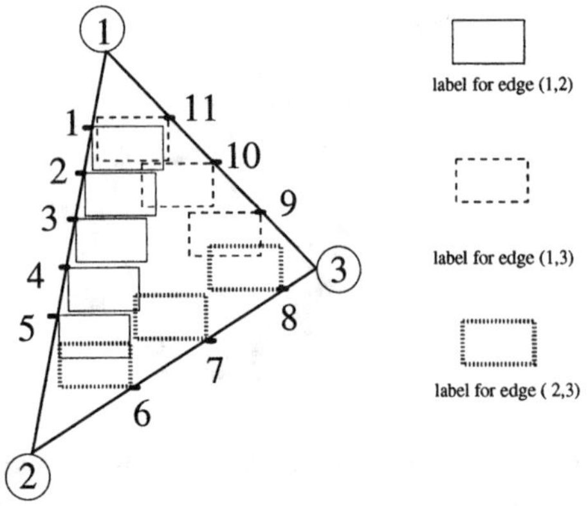

Figure 11: The input drawing with label positions assigned to each edge of the drawing.

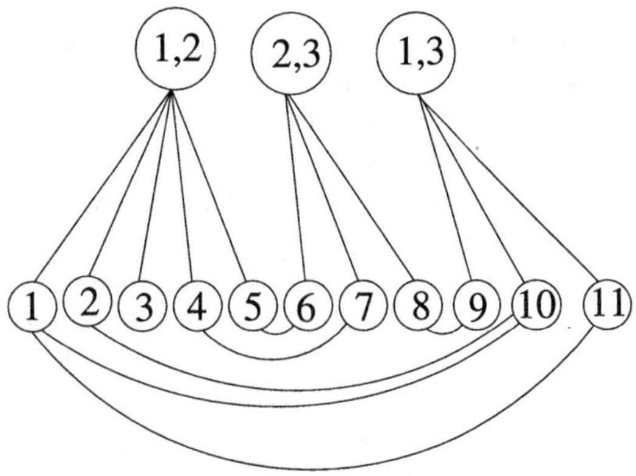

Figure 12: The corresponding overlap graph.

Figure 13: The G_R graph obtained from the overlap graph.

Figure 14: The corresponding clusters.

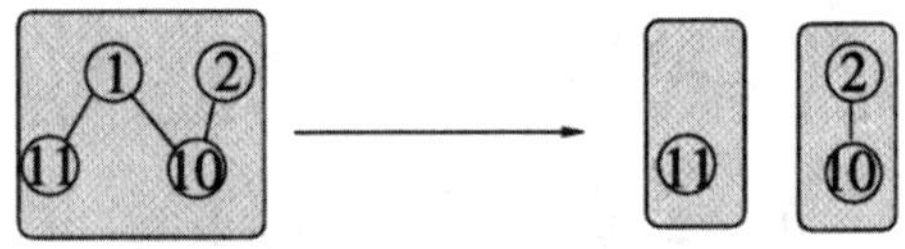

Figure 15: Splitting not complete clusters.

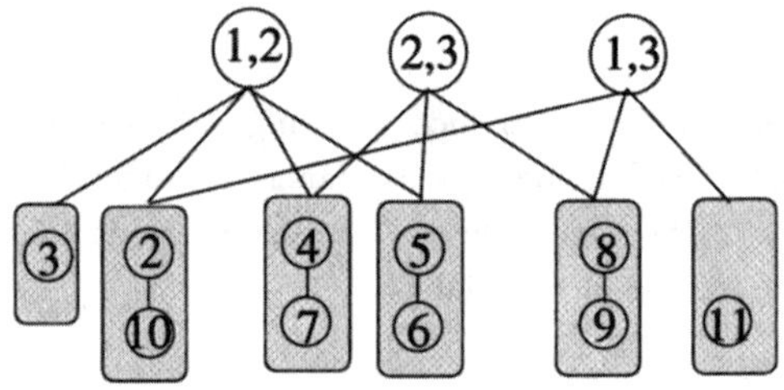

Figure 16: The corresponding matching graph.

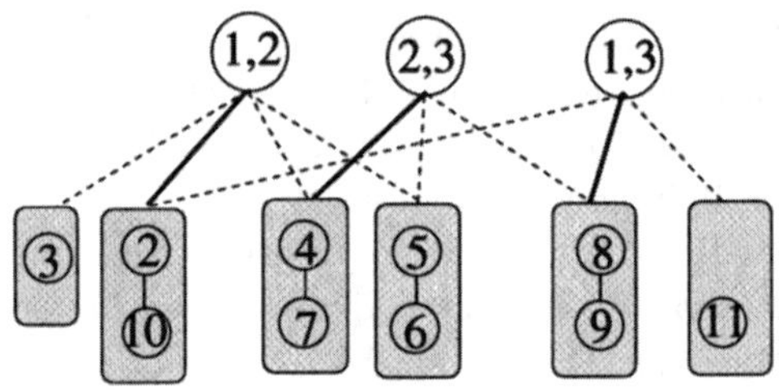

Figure 17: A maximum cardinality matching.

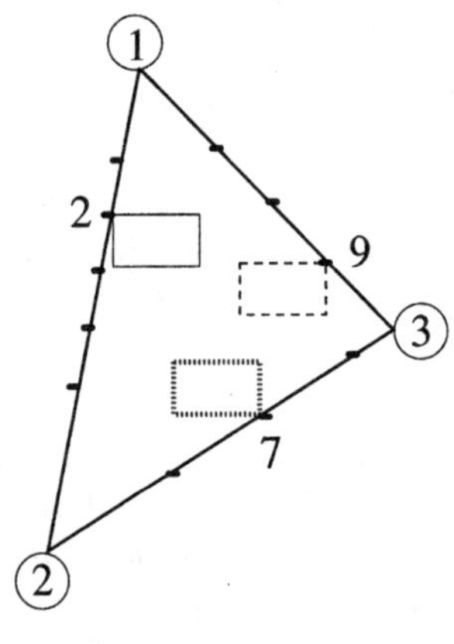

Figure 18: The corresponding label assignment.

An Output Sensitive Algorithm for Discrete Convex Hulls*

Sariel Har-Peled†

Abstract

Given a convex body C in the plane, its discrete hull is $C^0 = \text{ConvexHull}(C \cap \mathcal{L})$, where $\mathcal{L} = \mathbb{Z} \times \mathbb{Z}$ is the integer lattice. We present an $O(|C^0| \log \delta(C))$-time algorithm for calculating the discrete hull of C, where $|C^0|$ denotes the number of vertices of C^0, and $\delta(C)$ is the diameter of C. Actually, using known combinatorial bounds, the running time of the algorithm is $O(\delta(C)^{2/3} \log \delta(C))$. In particular, this bound applies when C is a disk.

1 Introduction

Let C be a planar convex body which we assume to be *sufficiently round*, in the sense that the following condition holds: Let $\mathcal{L} = \mathbb{Z} \times \mathbb{Z}$ denote the planar integer lattice. We require that $C \cap \mathcal{L}$ is *lattice-connected*; that is, the union of all the horizontal and vertical unit line segments that connect between pairs of points in $C \cap \mathcal{L}$ is connected; see Figure 1. The ***discrete hull*** C^0 of C is defined as the convex hull of $C \cap \mathcal{L}$; see figure 2 for an illustration.

The discrete hull arises in several applications. For example, in computer graphics, C^0 is a natural polygonal representation of the convex shape C. See [BV92, EG94, KV86, Thi91] for related work.

The discrete hull has the useful property that the number of its vertices is relatively small; more precisely, this number is $O(\delta(C)^{2/3})$, where $\delta(C)$ is the diameter of C. This is a special case of a more general bound, established in [And63], for convex lattice polytopes in arbitrary dimension. For the sake of completeness, we provide a simple proof for the planar case (see Lemma 3.1 below). A similar proof has also been given in [KV86]. This property makes the discrete hull an attractive and useful tool for approximating convex shapes (see [HPSV96] for an application of discrete hulls in three dimensions for computing approximate shortest paths on convex polytopes).

Solving a problem raised by Pankaj Agarwal, we present in this paper an output-sensitive algorithm for calculating the discrete hull C^0 of a sufficiently round convex body C. The algorithm runs in time $O(|C^0| \log \delta(C))$, where $|C^0|$ is the number of vertices of C^0. By the above results, this bound is also $O(\delta(C)^{2/3} \log \delta(C))$.

Suppose we associate a value with each point of G, where G is all the points of the lattice inside a square of size $N \times N$. Katz and Volper [KV86] present a data-structure for an intermixed sequence of updates and retrievals of the sum of values of the points of G within a disk, such that a retrieval or update operation takes $O(N^{2/3} \log^3 N)$ time. (An update operation modifies the value associated with a vertex of G) The first stage in answering a query, in the algorithm of [KV86], is the computation of the discrete hull of the query disk. Katz and Volper do not present an efficient algorithm for computing the discrete hull of a disk, and their bound holds when counting only the semi-group operations (see [KV86]). By our main result, their algorithm can be extended such that their bound holds in the regular RAM model as well.

Our algorithm employs a variant of the continued-fraction technique used to approximate a real number by rationals with small denominators. We first review the machinery needed for this algorithm in Section 2, and give a geometric interpretation of it that we will later exploit. The connection between continued fractions and the discrete hull of a half-plane was already observed by Klein in 1895 (see [Dav83]). Furthermore, [GY86, LC92, KS96] use continued fractions to calculate convex polygonal chains

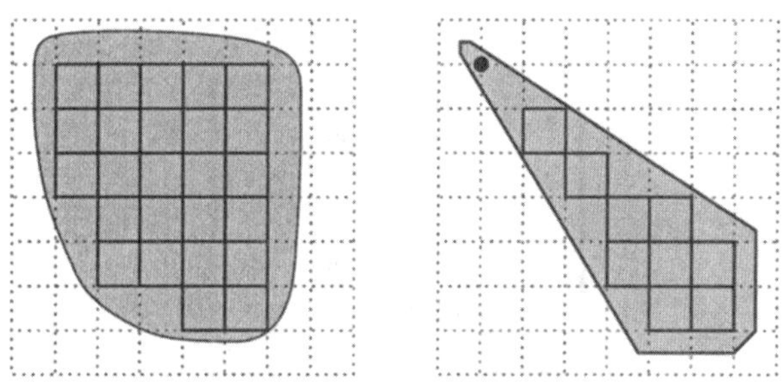

Figure 1: On the left, a convex body C such that $C \cap \mathcal{L}$ is lattice-connected; on the right, a convex body C such that $C \cap \mathcal{L}$ is not lattice-connected

*This work has been supported by a grant from the U.S.–Israeli Binational Science Foundation. This work is part of the author's Ph.D. thesis, prepared at Tel-Aviv University under the supervision of Prof. Micha Sharir.

†School of Mathematical Sciences, Tel Aviv University, Tel Aviv 69978, Israel; sariel@math.tau.ac.il

SCG 98 Minneapolis Minnesota USA
Copyright ACM 1998 0-89791-973-4/98/ 6...$5.00

that are discrete hulls of some specific shapes. We extend their techniques to handle the more general case of an arbitrary convex body.

Our discrete hull algorithm is presented and analyzed in Section 3. First, in Section 3.1, we derive several combinatorial bounds on the complexity of the discrete hull. Next, we describe, in Section 3.2, our algorithm in detail, and present several variants of the algorithm.

We have implemented this algorithm for the special case of a disk. The experimental results are reported in Section 4.

2 The Geometric Interpretation of Continued Fractions

In this section we review the machinery developed for the computation of continued fractions and discuss its geometric interpretation. For a review of continued fractions see [Dav83].

Given a positive real number r, it has a unique representation as a simple continued fractions:

$$r = q_0 + \cfrac{1}{q_1 + \cfrac{1}{q_2 + \cfrac{1}{q_3 + \cdots}}} = q_0 + \frac{1}{q_1 +}\frac{1}{q_2 +}\frac{1}{q_3 +}\cdots,$$

where $q_i > 0$, for $i > 0$, $q_0 \geq 0$, and $q_0, q_1, \ldots$ are integers. If $r = a/b$ is a rational (in reduced form), the continued fraction representation of r is finite, namely $r = q_0 + \frac{1}{q_1 +} \cdots \frac{1}{q_n}$. Consider in what follows such an r. Moreover, the following recurrence holds

$$q_i = \lfloor x_i \rfloor, x_{i+1} = \frac{1}{x_i - q_i}, \text{for } i = 0, \ldots, n,$$

where $x_0 = r$. In particular, the terms are byproduct of the calculation of the $gcd(a,b)$, and thus can be calculated in $O(\log \min(a,b))$ time (See [CLR90]).

Given a reduced fraction $r' = a/b$, we define $G(r') = (b, a)$, a point of the integer lattice corresponding to r'.

The intermediate continued fractions, $r_i = q_0 + \frac{1}{q_1 +} \cdots \frac{1}{q_i}$, for $i = 0, \ldots, n$, are called the *convergents* of r. Let $\frac{a_i}{b_i}$ be the reduced form of r_i, for $i = 0, \ldots, n$. The convergent r_i is an optimal approximation to r, in the sense that if $\frac{c}{d}$ is between r and r_i and $\left|\frac{c}{d} - r\right| < \left|\frac{a_i}{b_i} - r\right|$ then $d > b_i$, for $i = 1, \ldots, n$. See [Dav83].

Let $p = G(r) = (b, a)$ be the point corresponding to r, and let $p_i = G(r_i) = (b_i, a_i)$ be the planar grid point

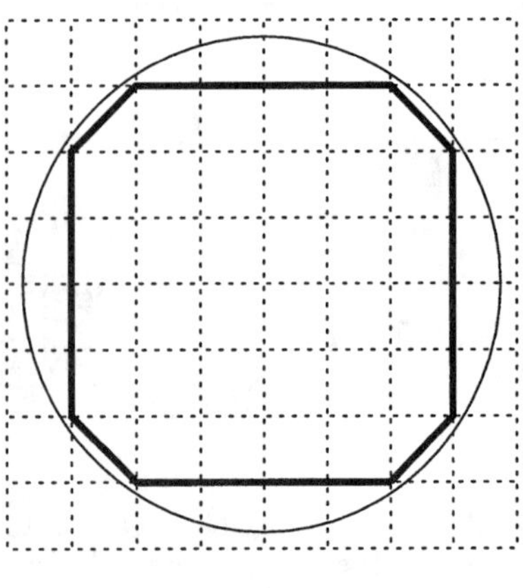

Figure 2: A disk and its corresponding discrete convex hull

Algorithm: GEOMETRIC-GCD(r)
Input: A rational number r.
Output: The convergents of r: $p_0, \ldots, p_n$.

(I) Set $p_{-2} \leftarrow (1,0), p_{-1} \leftarrow (0,1)$, $i \leftarrow 0$.
Let l be the line $y = rx$.

(II) Let $dray_i = dray(p_{i-2}, p_{i-1})$. If $dray_i \cap l \neq \emptyset$ then let q_i be the index of the point lying on l, namely $p_i = q_i p_{i-1} + p_{i-2} \in l$. Clearly, p_i is the last convergent of r and we are done.
Otherwise, let q_i be the largest integer such that $p_i = q_i p_{i-1} + p_{i-2}$ and p_{i-1} lie on different sides of l.

(III) $i \leftarrow i + 1$. Goto step (II).

Figure 3: The algorithm for computing the discrete-hull simulates the above algorithm, that calculates the convergents of a rational number in a geometric manner.

corresponding to r_i, for $i = 0, \ldots, n$. We then have $p_i = q_i p_{i-1} + p_{i-2}$, for $i = 0, \ldots, n$, where $p_{-2} = (1,0)$ and $p_{-1} = (0,1)$. Furthermore, the only points of $\mathcal{L}$ in $\Delta o p_i p_{i+1}$ are its vertices, for $i = 0, \ldots, n-1$ (see [Dav83]). We refer to $p_0, \ldots, p_n$ as the (geometric) convergents of p. All this is illustrated in Table 1.

Given two vectors $u, u' \in \mathcal{L}$, where the coordinates of u' are relatively prime, let $r = ray(u, u')$ be the ray emanating from u in the direction of u'. Let $p(r, i) = u + iu'$ be the i-th lattice point of the ray, for $i \geq 0$. We denote by $r[i]$ the closed segment on r between $p(r,i)$ and $p(r, i+1)$, for $i \geq 0$. We denote by $dray(u, u')$ the *discrete ray* emanating from u in the direction u':

$$dray(u, u') = \left\{ u + iu' \;\middle|\; i \in \mathbb{Z}, i > 0 \right\}.$$

The execution of the *gcd* algorithm can be interpreted as a geometric algorithm in the following manner: The intermediate numbers a_i, b_i in the i-th stage of the *gcd* algorithm are the coefficients of $p = G(r)$ in its representation in the base $\{p_{i-1}, p_{i-2}\}$. Each iteration refines the base by replacing its shorter vector by a longer one (the new convergent), such that p remains 'in the middle' (i.e., p remains a positive combination of the two base vectors). This process continues until the last convergent is equal to p.

Thus, calculating the convergents of r can be done in a geometric setting. The algorithm GEOMETRIC-GCD that does so is presented in Figure 3, and an example of its performance is shown in Figure 4. Note that in the algorithm, the division operation of the *gcd* is replaced by a discrete ray-shooting.

If $r > 0$ is irrational, the odd convergents $p_{-1}, p_1, \ldots$ are the vertices of the discrete hull of the region that lies above the line l in the positive quadrant of the plane, where l is the line $y = rx$. Similarly, the even convergents $p_{-2}, p_0, \ldots$ are the vertices of the discrete hull of the region that lies below l in the x positive half-plane. Those two polygonal chains lie in the half-planes defined by l, and they intersect l only if r is rational (see [Dav83]).

3 The Discrete Hull

In this section, we introduce the notion of discrete hull and provide an algorithm for its computation. In Section 3.1,

a	b	$\lfloor a/b \rfloor$	Current representation of $(5,8)$	The convergent r_i
8	5	1	$8(0,1)+5(1,0)$	1
5	3	1	$5(1,1)+3(0,1)$	$1+\frac{1}{1}$
3	2	1	$3(1,2)+2(1,1)$	$1+\frac{1}{1+\frac{1}{1}}$
2	1	2	$2(2,3)+1(1,2)$	$1+\frac{1}{1+\frac{1}{1+\frac{1}{2}}}$

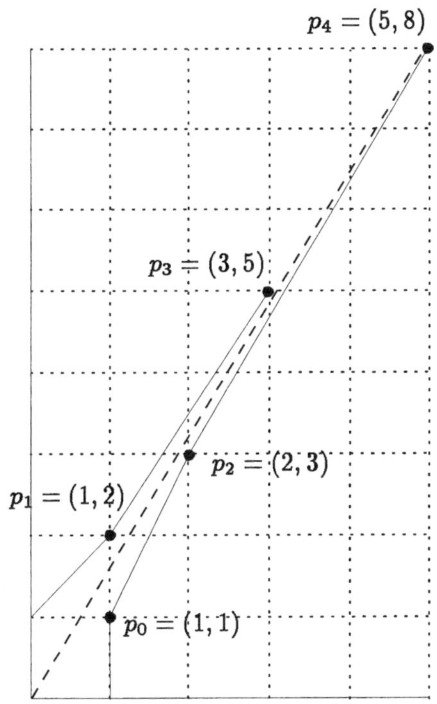

Figure 4: The convergents generated by GEOMETRIC-GCD(8/5)

we prove several bounds on the complexity of the discrete hull. In Section 3.2, we describe algorithm for computing the discrete hull.

3.1 Combinatorial Bounds on the Complexity of the Discrete Hull

Given a convex body C in the plane, we define the *discrete hull* C^0 of C as the convex hull of $C \cap \mathcal{L}$, where $\mathcal{L} = \mathbb{Z} \times \mathbb{Z}$ is the integer lattice. The following lemma is well known [And63, BB91, KV86, Thi91]. For the sake of completeness we provide here a simple proof. A similar proof has been given in [KV86].

Lemma 3.1 *Let C be a convex polygon in the plane, all of whose vertices belong to $\mathcal{L}$, and let D be its diameter. Then C has $O(D^{2/3})$ vertices.*

Proof: The polygon C is contained in an axis-parallel square of size D, thus the perimeter $P(C)$ of C is at most $4D$.

Let $p_1, \ldots, p_m$ be the vertices of C in their counterclockwise order around C.

Let $V = \{v_1, \ldots, v_m\}$ be the set of vectors connecting consecutive vertices on ∂C; that is, $v_i = p_{i+1} - p_i$, for $i = 1, \ldots, m-1$, and $v_m = p_1 - p_m$.

Let l be a parameter to be specified shortly. By definition, $P(C) = \sum_{v \in V} |v|$. Thus, there are at most $4D/l$ vectors in V of length larger than or equal to l.

Clearly, V cannot contain three collinear vectors, for otherwise one of the points p_i would have to lie in the relative interior of an edge of ∂C, so it cannot be a vertex of C. This argument also implies that any pair of collinear vectors in V must have opposite orientations. Thus the number of vectors in V of length smaller than l is at most the number of lattice points at distance smaller than l from the origin, which in turn is at most $4l^2$.

Thus the number of vertices of C is upper bounded by $\frac{4D}{l} + 4l^2$. Set $l = (D/2)^{1/3}$, the bound becomes $6 \cdot \sqrt[3]{2} D^{2/3}$ and the lemma follows. ■

For the a case of disk, we have the following extension of Lemma 3.1.

Lemma 3.2 *Let D be a disk of radius $r > 1$ in the plane. Then $|\mathcal{L} \cap \partial \mathrm{DiscreteHull}(D)| = O(r^{2/3})$.*

Proof: Let $D^0 = \mathrm{DiscreteHull}(D)$. For an edge $e = \vec{pq}$ of D^0, we define its *weight* to be $w(e) = |\mathcal{L} \cap e| - 1$. We assume that e is oriented so that int D^0 lies to the left of e. Let $v = v(e)$ be the direction vector of e; namely, the shortest lattice vector which is parallel to e and points in the same direction. Let $c = ||v||$. Clearly, $w(e) = ||e||/c$.

Let l be the line passing through e, and let H^- denote the half-plane bounded by l that does not contain the center of D. Let D' denote the disk of radius r whose boundary passes through p and q, and whose center lies in the same side of l as the center of D.

Clearly, $H^- \cap D' \subseteq H^- \cap D$. Let s denote the point on $H^- \cap \partial D'$ with a tangent to D' that is parallel to e. Let h be the distance between s and e. See Figure 5.

We claim that if $w = w(e) > 1$ then $h \leq 2/c$. Indeed, let l' be the translation of l by distance $1/c$ into H^-. The line l' contains points of the lattice $\mathcal{L}$, and the distance between two consecutive lattice points on l' is c (See [HW65, Chap. 3]). Thus, if $h \geq 2/c$ then $||l' \cap D|| \geq ||l' \cap D'|| \geq c$. This

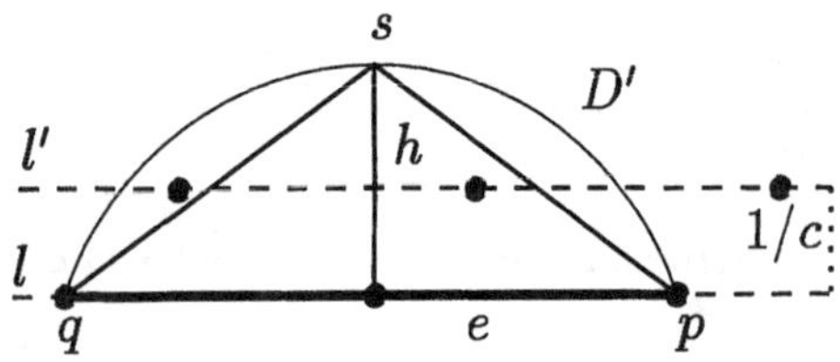

Figure 5: Illustrating the proof that $h \leq 2/c$ when $w(e) > 1$.

however imply that there must be a point of $\mathcal{L}$ inside $H^- \cap D'$, contradicting the fact that e is an edge of D^0.

An easy calculation shows that $h = r - \sqrt{r^2 - w^2c^2/4}$. Thus, if $w > 1$ then $r - \sqrt{r^2 - w^2c^2/4} \leq 2/c$, implying that

$$\frac{w^2c^2}{2r} \leq \frac{w^2c^2}{r + \sqrt{r^2 - \frac{w^2c^2}{4}}} \leq \frac{8}{c}.$$

Thus $w(e) = w \leq \frac{4\sqrt{r}}{c^{3/2}}$. This implies that, for $w \geq 2$, the maximum possible value of c is $c \leq \lceil \sqrt[3]{4r} \rceil$.

The number of lattice vectors of length between c and $c+2$ is $O(c)$, and, by Lemma 3.1, the number of edges of D^0 is $O(r^{2/3})$. Hence, the number of lattice points lying on the edges of D^0 is bounded by the number of edges of weight one, plus the number of lattice points lying on edges of weight larger than 1; that is, this number is

$$O\left(r^{2/3}\right) + \sum_{c=1}^{\lceil \sqrt[3]{4r} \rceil} O(c) \cdot \frac{4\sqrt{r}}{c^{3/2}}$$

$$= O\left(r^{2/3}\right) + O\left(\sqrt{r} \cdot \sum_{c=1}^{\lceil \sqrt[3]{4r} \rceil} \frac{1}{\sqrt{c}}\right) = O\left(r^{2/3}\right).$$

■

Remark 3.3 Lemma 3.2 was proved independently by Bárány [Bár96] (but was not published). Furthermore, Bárány and Larman [BL96] showed that the number of k-dimensional faces of the discrete hull of the ball of radius r centered at the origin (in dimension d) is $\Theta(r^{\frac{d(d-1)}{d+1}})$, for $k = 0, \ldots, d-1$. All these results were obtained recently and independently after the original preparation of this paper.

Remark 3.4 For a compact convex and smooth shape C in the plane, such that the curvature of C is bounded from below by $c > 0$, the number points $|\mathcal{L} \cap \partial \mathrm{DiscreteHull}(C)|$ is $O((1/c)^{2/3})$. This follows by a straightforward extension of the proof of Lemma 3.2.

Lemma 3.5 *Given a triangle T, such that two of its vertices belong to $\mathcal{L}$, the number of vertices of the discrete hull of T is at most $2\lceil \log_\phi \delta(T) \rceil + 7$, where $\phi = \frac{\sqrt{5}+1}{2}$ is the Golden Ratio.*

Proof: Kahan and Snoeyink showed in [KS96] how to compute the discrete hull inside such a triangle in $O(\log \delta(T))$ time. A careful inspection of their algorithm reveals that it generates at most $2\lceil \log_\phi \delta(T) \rceil + 7$ vertices. ■

Lemma 3.6 *Given a polygon P with n edges in the plane, the complexity of the discrete hull of P is at most $(2\lceil \log_\phi \delta(T) \rceil + 7)\, n$, where $\delta(P)$ is the diameter of P.*

Proof: Let P^0 be the discrete hull of P, and p a vertex of P. Let r^+, r^- denote the two rays that emanate from p and are tangent to P^0, and let v_p^+, v_p^- be the two vertices of P^0, furthest from p, that lie on r^+, r^-, respectively. Clearly, the vertices of P^0 inside $T = \Delta p v^- v^+$ belong to the discrete hull of T. By Lemma 3.5, the number of those vertices (including v^- and v^+), which, in particular, include all vertices of P^0 that are visible from p, is at most $2\lceil \log_\phi \delta(T) \rceil + 7$.

Applying the above argument to all the vertices of P, we have that the number of vertices of P^0 is bounded by $(2\lceil \log_\phi \delta(T) \rceil + 7)\, n$, since each vertex of P^0 is visible from at least one of the vertices of P. ■

3.2 The Discrete Hull Algorithm

In this section, we present an output sensitive algorithm for computing the discrete hull. The main procedure that we present here receives as input a sufficiently round convex body C, and a vertex p of C^0. It computes the next vertex p' of C^0 counterclockwise from p. The full algorithm first computes an initial (e.g., the lowest) vertex p_0 of C^0, and then applies repeatedly the above procedure until all vertices of C^0 are obtained.

We assume a model of computation in which C is given implicitly so that (a) we are given a lattice point inside C, and (b) we can obtain, in constant time, the intersection points of any query line with ∂C, when these points exist. Moreover, we also include the floor function in in our model of computation. This allows us also to perform, in constant time, discrete ray shootings, where, for a query discrete ray $dray(u, v)$, we want to find its first point inside or outside C.

Let $T = \Delta ovw$ be a right-angle triangle, such that o is the origin, $v = (x(v), y(v))$ lies in the positive quadrant of the plane, and $w = (x(v), 0)$ lies on the x-axis. The point u of the (upper) edge ou of the discrete hull of T is a multiple of u', where u' is the largest (even) convergent of v that still lies inside T. (See Figure 4 for an illustration: Let v be a point slightly below p_4; then op_2 is the upper edge of the discrete hull.) This observation (or a variant of it) has been used in calculating parts of the discrete hull in some special cases, where the body is polygonal or given in an explicit form. See [GY86, LC92, KS96]. In fact, u is calculated in [KS96] by performing ray-shootings inside and outside T, an idea that we extend to handle our more general settings.

Unfortunately, these methods fail when the body is either given in an implicit form or is not polygonal. We present an extension to the above techniques, where we compute the convergents of the required edge direction by performing a sequence of ray-shootings on the given body.

Lemma 3.7 *Let $\rho = (a, b)$ be a point of the lattice $\mathcal{L}$, such that $a, b > 0$ and $gcd(a, b) = 1$. Let $\rho_0, \ldots, \rho_n = \rho$ be the sequence of convergents of ρ.*

Then the segment $s = \rho_i \rho_i'$ must intersect the segment op, where o denote the origin and $\rho_i' = \rho_i + \rho_{i-1}$, for $i = 0, \ldots, n$.

Proof: Let l be the line passing through o and ρ. By definition, $\rho_i = q_i \rho_{i-1} + \rho_{i-2}$ and q_i is maximal such that ρ_i lies on the same side of l as ρ_{i-2}. Thus, $\rho_i \rho_i'$ must intersect l, or q_i would not be maximal.

Moreover, $\rho_i\rho_i'$ must intersect $o\rho$, or else the vector ρ_{i+1} would be longer than ρ, which is impossible. ∎

Let e' be the (currently unknown) vector connecting p with p'. Let e be the reduced vector (i.e. $gcd(x(e), y(e)) = 1$) having the same direction as e'. Given e, we can easily obtain p' by calculating the last intersection between C and the discrete ray $dray(p, e)$. Without loss of generality, we assume that both coordinates of e are positive. This issue is discussed later in more detail.

In the following, we assume, for simplicity, that p is the origin. The algorithm, FIND-EDGE-DIRECTION, calculates the convergents of e, and thus e itself, by simulating the execution of GEOMETRIC-GCD on e. The idea is to replace the half-plane ray-shooting of GEOMETRIC-GCD by ray shootings on ∂C. FIND-EDGE-DIRECTION generates several candidates for the vector e, because, unlike GEOMETRIC-GCD, the vector e is not given explicitly, and the algorithm does not know when to stop. The required vector e is simply the clockwise-most vector out of the group of generated candidates.

Let l be the line passing through e. Denote by H^+ the open half-plane lying to the left of e, and by H^- the open half-plane lying to the right of e.

The odd convergents of e are the approximations to e from 'inside' C (namely, these convergents lie inside H^+ and may lie inside C), and the even convergents are the approximations to e from the 'outside' of C (namely, these convergents do not lie inside C, and lie inside H^-, except maybe for the last convergent that may lie on l).

We now describe FIND-EDGE-DIRECTION in detail. The algorithm starts with $p_{-2} = (1,0)$ and $p_{-1} = (0,1)$. In the i-th stage the algorithm calculates the intersection between $ray_i = ray(p_{i-2}, p_{i-1})$ and C. By the relation $p_i = q_ip_{i-1} + p_{i-2}$ it follows that the next convergent of e lies on the discrete ray $dray_i = dray(p_{i-2}, p_{i-1})$. If ray_i does not intersect C then we stop the procedure, since by Lemma 3.7 we know that all the convergents of e were computed. Otherwise, we apply Lemma 3.7 to decide what is the next convergent. There are two cases:

- **i is odd:** Let k be maximal such that $ray_i[k] \cap C \neq \emptyset$. If $k = 0$ then we are done, in the sense that all convergents of e have already been computed (otherwise, we would have $p_i = p_{i-2}$, which is impossible, for $i > 0$, because all the convergents are distinct). Set $p_i \leftarrow p(ray_i, k)$. If $p_i \in C$ then p_i might be e, and thus it is a candidate for the vector e.
- **i is even:** Let k be minimal such that $ray_i[k] \cap C \neq \emptyset$. If $p(ray_i, k+1) \in C$, then it is a candidate for the vector e, otherwise $p(ray_i, k)$ is a convergent of e. If $k = 0$ and $i > 0$ then we are done. Otherwise, set $p_i \leftarrow p(ray_i, k)$.

The algorithm returns the clockwise-most vector out of the candidates generated. See Figure 6.

Let $p_0, \ldots, p_n$ be the convergents of e. The correctness of FIND-EDGE-DIRECTION follows from the fact that it performs exactly the same ray-shootings as GEOMETRIC-GCD(e) performs, up to the $(n-1)$-th stage, and selects the correct points on these rays to be the convergents of e. It is easy to verify, that if one of the results of these $n-1$ ray-shootings of FIND-EDGE-DIRECTION differs from the corresponding $(n-1)$ ray-shootings of GEOMETRIC-GCD(e), then e is not the direction of the next edge of the discrete hull, because there exists a vector of $\mathcal{L}$ that lies inside C and is

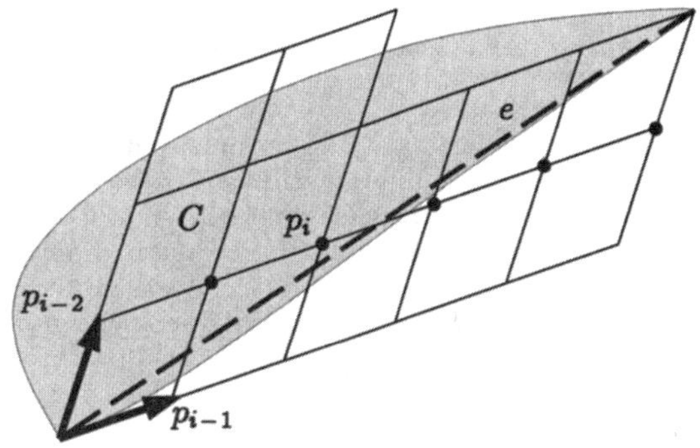

Figure 6: Illustrating an iteration of FIND-EDGE-DIRECTION

clockwise to e. This implies that e is not an edge of the discrete hull, which is a contradiction.

As for the running time of the algorithm, we notice that the coordinates of the sequence of convergents that FIND-EDGE-DIRECTION generates, are a "super-Fibonacci" sequence since $x(p_i) \geq x(p_{i-1}) + x(p_{i-2})$ and $y(p_i) \geq y(p_{i-1}) + y(p_{i-2})$. Thus $||p_i|| \geq |F_i| = \lfloor \phi^k/\sqrt{5} \rfloor$, where F_i is the i-th Fibonacci number and $\phi = \frac{\sqrt{5}+1}{2}$ is the Golden Ratio (See [CLR90]).

Thus, the $(2 + \lceil \log_\phi \delta(C) \rceil)$-th convergent generated by FIND-EDGE-DIRECTION must lie outside C, and the ray created in the next step will not intersect C, implying that the algorithm will have terminated by this iteration. Each iteration of the algorithm takes $O(1)$ time, and, as just argued, there are at most $O(\log \delta(C))$ iterations. Thus, the total running time of FIND-EDGE-DIRECTION is $O(\log(\delta(C)))$. However, it might be that no candidate was generated by FIND-EDGE-DIRECTION. This implies that our assumption that $x(e), y(e) \geq 0$ is false, and e does not lie in the positive quadrant of the plane.

Thus, during the calculation of the discrete hull, the algorithm can either maintain the current quadrant of the plane containing e, or alternatively, perform a search for e in each quadrant of the plane. Picking e out of the (at most) four candidates generated can be done in $O(1)$ time, by computing their convex hull C_H, where e is the first edge of C_H encountered when tracing ∂C_H in a counterclockwise direction starting from o.

We thus obtain:

Theorem 3.8 *Given (i) a compact convex shape C in the plane such that the intersection between C and a line can be calculated in $O(1)$ time, and (ii) a vertex v_0 of the discrete hull C^0 of C, then C^0 can be calculated in $O(|C^0| \log \delta(C))$ time and $O(1)$ space, where $|C^0|$ denotes the number of vertices of C^0.*

Although the assumption in Theorem 3.8 on the availability of a vertex of the discrete hull seems to be self-defeating, it can be replaced by a more natural condition, as follows.

Definition Let $\hat{G} = (\mathcal{L}, E)$ be the *lattice graph*, whose edges connect every pair of lattice points at distance 1. A subset S of $\mathcal{L}$ is *lattice-connected* if the induced subgraph $\hat{G}_S$ is connected.

Lemma 3.9 *Given a compact convex body C in the plane, such that $C \cap \mathcal{L}$ is lattice-connected, and a point $v \in C \cap \mathcal{L}$,* *then a point on the discrete hull of C can be calculated in $O(\log \delta(C))$ time.*

Proof: The idea is to perform an unbounded search for the lowest vertex of the discrete hull of C. We intersect C with the horizontal lines l_i passing through $v + (0, -2^i)$, for $i = 1, 2, \ldots$, and stop as soon as we encounter a horizontal line, l_k, that does not intersect $C \cap \mathcal{L}$. We then perform a binary search for the bottom horizontal line of the grid that intersects C, between the horizontal line passing through v and l_k. The lattice-connectivity of $C \cap \mathcal{L}$ guarantees the correctness of this procedure.

Since $k \leq 1 \lceil \log \delta(C) \rceil$, it follows that the binary search is performed on an interval of size $O(\delta(C))$.

Given this line, we can calculate a vertex of the discrete hull by performing a single ray-shooting. ∎

We thus obtain the main result of this paper:

Theorem 3.10 *Given a compact convex shape C in the plane such that the intersection between C and any line can be calculated in $O(1)$ time, such that $C \cap \mathcal{L}$ is lattice-connected and such that a point in $C \cap \mathcal{L}$ is available, then the discrete hull C^0 of C can be calculated in $O(|C^0| \log \delta(C))$ time and $O(1)$ space, where $|C^0|$ denotes the number of vertices of C^0.*

In particular, we have the following corollary:

Corollary 3.11 *The discrete hull of a disk D of radius r in the plane can be calculated in $O(r^{2/3} \log r)$ time and $O(1)$ space.*

Proof: In $O(1)$ time we can find a point $v^0 \in D \cap \mathcal{L}$, by finding the closest point in the lattice to the center of D. Unless $r < 1/\sqrt{2}$, $D \cap \mathcal{L}$ is not empty and is always lattice-connected, so, by Lemma 3.9, a vertex of the discrete hull of D can be calculated in $O(\log r)$ time. The corollary is now an immediate consequence of Theorem 3.10 and Lemma 3.1. ∎

Given a convex polygon P with n edges in the plane, the algorithm of Lee and Chang [LC92] computes, in $O(n + \log \delta(P))$ time, a vertex of the discrete hull. All the algorithms described so far, assume that ray-shooting on C can be computed in $O(1)$ time. In general, if $T(n)$ is the time for answering ray-shooting queries on C, then the running time increases by a factor of $T(n)$. For example, we have the following:

Corollary 3.12 *Let P be a convex polygon with n edges in the plane. The discrete hull P^0 of P can be computed in $O(n + |P^0| \log \delta(P) \log n)$ time, where $|P^0|$ denotes the number of vertices of P^0.*

Proof: Using the algorithm of [LC92], we compute in $O(n + \log \delta(P))$ time a vertex v_0 of the discrete hull. We preprocess P in $O(n)$ time to answer ray-shooting queries, in $O(\log n)$ time, by computing the Dobkin-Kirkpatrick hierarchical decomposition of P [DK85].

We then compute the discrete hull using the algorithm of Theorem 3.10. Since each ray-shooting takes $O(\log n)$ time, the resulting algorithm runs in $O(|P^0| \log \delta(P) \log n)$ time. ∎

Corollary 3.13 *The discrete hull of a convex n-gon P can be computed in time $O(n \log n \log^2 \delta(P))$.*

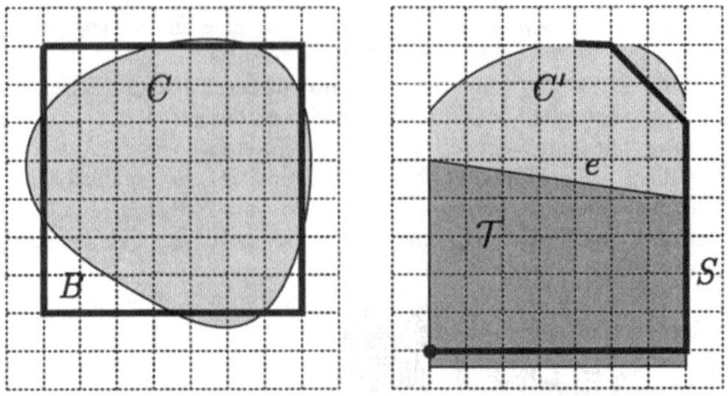

Figure 7: Computing a vertex of the discrete hull of C.

Proof: Readily follows from Corollary 3.12 and Lemma 3.6. ∎

Using Lemma 3.6, it is possible to replace the conditions in Theorem 3.10 by a more natural condition.

Theorem 3.14 *Given a compact convex shape C in the plane and a point $p \in C$, such that the intersection between C and any line can be calculated in $O(1)$ time, the discrete hull C^0 of C can be calculated in $O(\log^2 \delta(C) + |C^0| \log \delta(C))$ time and $O(1)$ space, where $|C^0|$ denotes the number of vertices of C^0.*

Proof: Let $B(C)$ denote the axis-parallel bounding box of C, and let B denote the discrete hull of $B(C)$ (which is an axis parallel rectangle). Using unbounded search (starting from a lattice point near p), similar to the one in the proof of Lemma 3.9, one can compute B, in $O(\log \delta(C))$ time, by performing a sequence of vertical and horizontal ray-shootings on C.

Let $C' = C \cap B$. Clearly DiscreteHull(C) = DiscreteHull(C'), and one can compute the intersection of C' with a line, in $O(1)$ time, by computing the corresponding line-intersections with C and B.

Let p_0, p_1 be two x-extreme points of C', such that p_0 and p_1 lie on different sides of B, and let $e = p_0p_1$. If $|e| < 100$ then compute the discrete hull of C, in $O(1)$ time, by performing at most $2 \cdot 100$ vertical ray-shootings.

Otherwise, let $\mathcal{T}$ be the vertical trapezoid such that e is its upper edge, the length of its shorter vertical edges is δ, and its bottom edge is horizontal, where δ is the diameter of B. See Figure 7.

Using the algorithm of Theorem 3.8, one can compute, in $O(\log^2 \delta(C))$ time, a chain S of m adjacent vertices of the discrete hull of $C' \cup \mathcal{T}$, starting from the lowest lattice point lying on the left edge of $\mathcal{T}$, where $m = 1 + 4\left(2\left\lceil \log_\phi(2\delta(C)) \right\rceil + 7\right)$. If $C' \setminus \text{int}\, \mathcal{T}$ contains a vertex of C^0, then S must contain such a vertex. Otherwise, by Lemma 3.6, we have $S = \text{DiscreteHull}(\mathcal{T}) = \text{DiscreteHull}(\mathcal{T} \cup C')$.

If no such vertex was found, apply the above procedure to the corresponding symmetric vertical trapezoid having e as its lower edge, in order to compute a vertex of C^0 below e (if such a vertex exists). Thus, we can compute a vertex of C^0, in $O(\log^2 \delta(C))$ time, if such a vertex exists.

Starting from this vertex of the discrete hull, one can compute C^0, in $O\left(|C^0| \log \delta(C)\right)$ time, by the algorithm of Lemma 3.8. ∎

4 Experimental Results

We have implemented the discrete hull algorithm of Section 3, for the special case where C is a disk. The algorithm was implemented in C++ on a Sun Sparc-10. The implementation is straightforward, and consists of about 150 lines of source code.

As a competing, naive algorithm, we implemented a variant of the Graham convex-hull algorithm (See [O'R94, pp. 80–96]), whose input consists of $O(r)$ lattice points, namely, the two extreme points on every vertical lattice line that intersects C. This algorithm generates the points "on demand", instead of calculating them in advance (thus using only $O(1)$ storage). This algorithm takes $O(r)$ time to calculate the discrete hull of a disk of radius r.

See Tables 2, 3, 4 for results from experimenting with the algorithm, where IG and DCH denote the naive and the new algorithm, respectively. As Table 2 testifies, the new algorithm is efficient in practice, and is faster than the 'naive' one, for $r \geq 100$. Despite the uncertainty in execution timing on different platforms, algorithm implementation, the result indicate a factor of > 6 in the running time of the new and naive algorithm, indicating that the new algorithm is considerably faster for large values of r.

As the results of Table 3 indicate, the average number of vertices of the discrete hull of a disk of radius r is about $3.45r^{2/3}$. The constant 3.45 is within the range 0.33...5.54 proven in [BB91]. Moreover, Balog also performed experimental testing for the case of disks centered at the origin, and got similar results [Bár96]. Thus, our results suggest that $\lim_{r\to\infty} |DH(D_r)|/r^{2/3}$ exists and lies in the range 3.44...3.46, where D_r is the disk of radius r centered at the origin. The existence of this limit was posed as an open question in [BB91].

We are not aware of any work that gives a rigorous analysis of this constant, and we leave it as an open problem for further research. Nevertheless, Balog and Deshoullier proved the existence of $\lim_{r\to\infty} |N(r)|/r^{2/3}$, where $N(r)$ is the average of $|DH(D_r)|$ over a small interval $[r, r+H]$ for some $H < 1$ [Bár96].

As the results of Table 4 indicate, the ratio between the average number of lattice points on the boundary of the discrete hull of a disk of radius r and $r^{2/3}$, appears to be monotone increasing. On the other hand, by Lemma 3.2 this ratio is $O(1)$. Thus, we conjecture that this ratio converges to a constant, as r tends to infinity.

Acknowledgments

I wish to thank my thesis advisor, Micha Sharir, for his help in preparing this manuscript. I also wish to thank Pankaj Agarwal, Imre Bárány, Alon Efrat, and Günter Rote for helpful discussions concerning this and related problems.

The author also wish to thank Gill Barequet and the referees for their comments and suggestions.

References

[And63] G. E. Andrews. A lower bound for the volume of strictly convex bodies with many boundary lattice points. *Trans. Amer. Math. Soc.*, 106:270–279, 1963.

[Bár96] I. Bárány. personal communication. 1996.

[BB91] A. Balog and I. Bárány. On the convex hull of the integer points in a disc. In *Proc. 7th Annu. ACM Sympos. Comput. Geom.*, pages 162–165, 1991.

[BL96] I. Bárány and D. G. Larman. The convex hull of the integer points in a large ball. manuscript, 1996.

[BV92] I. Bárány and A.M. Vershik. On the number of convex lattice polytopes. *Geometry and Functional Analysis*, 2:381–393, 1992.

[CLR90] T. H. Cormen, C. E. Leiserson, and R. L. Rivest. *Introduction to Algorithms.* The MIT Press, Cambridge, Mass., 1990.

[Dav83] H. Davenport. *The Higher Arithmetic: Introduction to the Theory of Numbers.* Oxford University Press, Oxford, 1983.

[DK85] D. P. Dobkin and D. G. Kirkpatrick. A linear algorithm for determining the separation of convex polyhedra. *J. Algorithms*, 6:381–392, 1985.

[EG94] A. Efrat and C. Gotsman. Subpixel image registration using circular fiducials. *Int. J. Comput. Geom. Appls.*, 4(4):403–422, 1994.

[GY86] D. H. Greene and F. F. Yao. Finite-resolution computational geometry. In *Proc. 27th Annu. IEEE Sympos. Found. Comput. Sci.*, pages 143–152, 1986.

[HPSV96] S. Har-Peled, M. Sharir, and K. R. Varadarajan. Approximate shortest paths on a convex polytope in three dimensions. In *Proc. 12th Annu. ACM Sympos. Comput. Geom.*, pages 329–338, 1996.

[HW65] G. Hardy and E. Wright. *The Theory of Numbers.* Oxford University Press, London, England, 4th edition, 1965.

[KS96] S. Kahan and J. Snoeyink. On the bit complexity of minimum link paths: Superquadratic algorithms for problems solvable in linear time. In *Proc. 12th Annu. ACM Sympos. Comput. Geom.*, pages 151–158, 1996.

[KV86] M. D. Katz and D. J. Volper. Data structures for retrieval on square grids. *SIAM J. Comput.*, 15(4):919–931, 1986.

[LC92] H. S. Lee and R. C. Chang. Approximating vertices of a convex polygon with grid points in the polygon. In *Proc. 3rd Annu. Internat. Sympos. Algorithms Comput. (ISAAC '92)*, volume 650 of *Lecture Notes in Computer Science*, pages 269–278. Springer-Verlag, 1992.

[O'R94] J. O'Rourke. *Computational Geometry in C.* Cambridge University Press, Cambridge, 1994.

[Thi91] T. Thiele. Diplomarbeit. Free University Berlin, August 1991.

a	b	$\lfloor a/b \rfloor$	Representation of $(14,31)$	The convergent r_i
31	14	2	$31(0,1)+14(1,0)$	2
14	3	4	$14(1,2)+3(0,1)$	$2+\frac{1}{4}$
3	2	1	$3(4,9)+2(1,2)$	$2+\frac{1}{4+\frac{1}{1}}$
2	1	2	$2(5,11)+1(4,9)$	$2+\frac{1}{4+\frac{1}{1+\frac{1}{2}}}$

Table 1: The convergents of the number $\frac{31}{14}$ can be obtained by calculating $gcd(31,14)$ by Euclid algorithm.

r	Avg. Number of Points on Discrete Hull	Max. Number of Iterations Per Edge	Time in Seconds DCH	IG
10	16	5	0.0015	0.0010
100	74	6	0.0097	0.0097
1,000	345	7	0.0550	0.0890
10,000	1,603	9	0.2935	0.8991
100,000	7,442	11	1.5259	9.0403
1,000,000	34,535	13	7.9662	90.4785
10,000,000	160,317	14	43.0855	906.9861

Table 2: Experimental results for the discrete hull of a disk D of radius r. The center of the disk was randomly chosen in the unit square, one hundred times for each indicated radius.

	Points on DH			$\lvert DH\rvert/r^{2/3}$		
r	Min	Max	Avg.	Min	Max	Avg.
10	15	20	16	3.750	5.000	4.143
100	68	85	74	3.238	4.048	3.555
1,000	333	360	345	3.330	3.600	3.450
10,000	1,568	1,636	1,603	3.379	3.526	3.456
100,000	7,355	7,521	7,442	3.415	3.492	3.455
1,000,000	34,358	34,720	34,535	3.436	3.472	3.454
10,000,000	159,879	160,674	160,317	3.445	3.462	3.454

Table 3: Experimental results on the size of the discrete-hull of a disk of radius r.

	Points on DHS			$\lvert DHS\rvert/r^{2/3}$		
r	Min	Max	Avg.	Min	Max	Avg.
10	32	46	40	8.000	11.500	10.123
100	216	251	235	10.286	11.952	11.216
1,000	1,075	1,278	1,217	10.750	12.780	12.173
10,000	5,697	6,260	5,986	12.278	13.491	12.902
100,000	28,081	29,748	28,807	13.037	13.811	13.374
1,000,000	133,964	139,767	137,535	13.396	13.977	13.754
10,000,000	637,281	657,949	647,339	13.730	14.175	13.947

Table 4: Experimental results on the number of lattice points on the boundary of the discrete hull of a disk of radius r.

Asymmetric Rendezvous on the Plane

Edward J. Anderson
Australian Graduate School of Management
University of New South Wales
Sydney 2052
AUSTRALIA
eddiea@agsm.unsw.edu.au

Sándor P. Fekete
Center for Parallel Computing
Universität zu Köln
D–50923 Köln
GERMANY
fekete@zpr.uni-koeln.de

Abstract

We consider rendezvous problems in which two players move on the plane and wish to cooperate in order to minimise their first meeting time. We begin by considering the case when they know that they are a distance d apart, but they do not know the direction in which they should travel. We also consider a situation in which player 1 knows the initial position of player 2, while player 2 is only given information on the initial distance of player 1. Finally we give some results for the case where one of the players is placed at an initial position chosen equiprobably from a finite set of points.

1 Introduction

Rendezvous problems are a type of search game in which two players cooperate in order to minimise their meeting time. Probably the first discussion of a rendezvous problem appears in Schelling [29] who discusses a situation in which two parachutists land in a field and wish to find each other as quickly as possible.

It is possible to formulate rendezvous problems in a wide variety of different settings. For example, rendezvous problems can involve two people meeting each other on discrete locations (Anderson and Weber [9], Alpern and Howard [6]), on an infinite line (Alpern and Beck [4], Anderson and Essegaier [8], Alpern and Gal [5], Alpern and Lim [24], Baston [14], Baston and Gal [15], Gal [19], Howard [23]), or on a circle [23]. Alpern [1] gives a comprehensive introduction to these problems. Other aspects of the problem that have been considered include rendezvous with more than two players (Lim, Alpern, and Beck [26], Lim and Alpern [25]), rendezvous with bounded resources (Alpern and Beck [2, 3]), and rendezvous evasion games (Alpern and Lim [7]). On a less serious note, see [22].

Despite the recent increase in interest in these problems, we know of no work that addresses the rendezvous problem in the plane, even though this was the original setting of Schelling.

In this paper we will consider a number of different scenarios in the plane. In each case two players are placed at different positions and each is given some information about the other's position. The aim is to choose a trajectory for each player in such a way as to minimise the expected meeting time or minimise the worst-case time, given maximum speeds v_1 and v_2 at which the players can move. Other aspects that we discuss are concerned with bounded resources, where player i can travel a total distance of f_i. We concentrate on *asymmetric* rendezvous search, where both players have agreed in advance on their role in a combined strategy. For example, they may agree on one player keeping her position, while the other player does all the searching. This is different to *symmetric* rendezvous search, where the players did not have a chance to agree on their roles in a combined strategy. Furthermore we can make different assumptions on the sense of direction: if each player has a compass that enables her to tell the direction of "north", we speak of an *oriented* problem, otherwise the problem is *unoriented.* In any case we assume that both players can tell a clockwise orientation from a counterclockwise one.

In a two-dimensional setting there are several different scenarios: The players may be required to meet at the same point at the same time ("zero visibility"), or they may be able to see each other when sufficiently close to each other (within a "visibility radius r"). They may also have some information on the other's initial position ("distance d"). We also consider situations with *asymmetric information* (as opposed to asymmetric strategies): For example player 1 may know the initial position of player 2, while player 2 does not know more than the initial distance of player 1.

The final part of the paper is concerned with the problem in which the players are constrained to move on a grid or network. This enables an optimal strategy to be derived for some problems when the version in which unconstrained

SCG 98 Minneapolis Minnesota USA

movement is possible turns out to be frustratingly difficult. The problem on a grid also has interest because optimal policies for rendezvous on a grid may provide an approximation procedure for certain problems with a visibility radius.

The rest of the paper is organised as follows. In Section 2 we give an overview of some basics. Section 3 considers rendezvous search when the initial distance is known; we describe optimal strategies for scenarios at zero visibility, unbounded and bounded resources, as well as problems arising from limited visibility. In Section 4 we describe an asymptotically optimal strategy for a scenario in which the initial distance is unknown. In Sections 5 and 6 we discuss a variety of problems in which each player knows that the other player's initial location is taken from a finite set of candidate points.

2 Problem Setting and Fundamentals

The reader is referred to Alpern [1] for a discussion of the ways in which rendezvous problems can be formalised. We will depart from the formulation given by Alpern who assumes that both players are given initial positions which are independently chosen and uniform over the whole of a metric space X.

Since the rendezvous search takes place on the plane, it is only the relative position of the two players at the start which is important. Hence we may suppose, without loss of generality, that the initial position of player 1 is at the origin. The initial position of player 2 is one of a given set of points Y, which may be infinite. We shall assume that each point in Y is equally likely to be chosen as the initial position of player 2.

Care is needed to specify the information that is given to the two players. Following Alpern we can describe the information available to the players by specifying a subgroup of the isometry group on the plane. In fact we shall consider a group G of isometries with the property that $G(Y) = Y$ and also assume that the group G acts transitively on Y (i. e., any point in Y is taken into any other point in Y by some element of the group).

A solution to the problem will take the form of the specification of two policies, one for each player. Since we are considering the asymmetric rendezvous problem we can assume that each policy consists of a single trajectory. Because the players do not know their initial positions, the same trajectory is followed from each of the starting positions in Y. We write $s_i(t)$ for the trajectory for player i, $i = 1, 2$. We assume that $s_1(0)$ is the origin and $s_2(0)$ is in Y. Further we assume that player 1 can move with (maximal) speed v_1, while player 2 can move with speed v_2.

Every time the rendezvous search takes place one element $g \in G$ is chosen with an equal chance for each of the elements of G. Then the meeting takes place at the first time t for which $s_1(t) = g(s_2(t))$. The expected meeting time $T(s_1, s_2)$ is the average of this over G. Since G acts transitively on Y the averaging over G includes an average over different starting points in Y.

A key part of the problem formulation is the question of whether the two players share a common sense of orientation in the plane. We may assume that player 1 has a compass, so that she may travel North if she chooses. We will consider two different types of problem, which we call oriented and unoriented. In an oriented problem player 2 is also given a compass, in an unoriented problem she is not. In an unoriented problem the choice of a trajectory for player 2 leads to the player setting off with an equal chance of going in each of the possible directions from the starting point.

In terms of the subgroup G, an oriented problem has nothing other than translation moves included in G, while an unoriented problem also includes all rotation moves.

Alpern has pointed out that we also need to distinguish between an unoriented arrangement, in which the players share a common notion of "clockwise" on the surface, and one in which they do not. We will not consider this latter case, which corresponds to the inclusion of reflections in the group G.

The lemma below shows that, for zero visibility, we need only consider problems in which Y has zero measure in the plane. Y captures the uncertainty about the initial positions of the two players. If this uncertainty is too great, as happens when Y has non-zero measure, the players do not have sufficient information to guarantee a meeting.

Lemma 2.1 *If Y has non-zero measure then for any strategy the probability of the players meeting in finite time is zero.*

Proof: Observe that the probability of meeting before time t given a strategy pair (s_1, s_2) in an unoriented problem is the same as that which can be achieved in the oriented problem if player 2 randomises over all possible directions at the start. Thus it is enough to consider an oriented problem.

Consider the movement of player 1 relative to player 2. This is then equivalent to the problem in which player 2 has an initial uniform distribution over Y and player 1 searches for player 2 with a maximum speed of $v_1 + v_2$. The probability of a meeting before time t is the expected value (over player 1 trajectories) of the measure of the area traced out within Y by the player 1 trajectory up to time t, normalised by the area of Y. Since Y has non-zero measure and the measure of the trajectory is zero we are done. □

3 Rendezvous with a Known Distance

In this section we consider the case where both players know their original distance d, but do not know the direction of the other player. We assume that all directions are equally likely, meaning that from the point of view of either player, the location of the other player is chosen uniformly at random from the circle of radius d. The players meet when they occupy the same point at the same time. In addition to this basic scenario, we consider other scenarios involving resource constraints, an objective function of expected rendezvous time as well as worst-case rendezvous time, and a visibility radius r.

3.1 The Basic Scenario

Suppose the players start with the knowledge of their original distance d, and the visibility is zero. This situation has been considered by Alpern and Gal [5] for the one-dimensional case, with both players moving at unit speed; they showed that, if there is no common sense of direction for both players, then there is an optimal strategy that results in an expected rendezvous time of $\frac{13d}{8}$. In addition when there is a common sense of direction (i. e., a notion of "left" and "right"), it is easy to show that an optimal strategy yields an expected rendezvous time of d: player 1 moves a distance of $\frac{d}{2}$ to the right, while player 2 moves the same distance to the left. If the players do not meet (which happens with probability $\frac{1}{2}$), then both players reverse their directions and meet after moving a further distance of d.

In the following we assume, without loss of generality, that the maximal speed for player 1, v_1, is no smaller than the maximal speed for player 2, v_2. Both players have a common 2-dimensional sense of direction, i.e., a compass and a notion of "clockwise". We will denote by $t_d = \frac{d}{v_1+v_2}$ the minimum time for a meeting of the two players.

There is an easy asymmetric strategy that yields finite bounds: player 2 stays put, while player 1 does all the searching. (This strategy has been dubbed "waiting for momma" by Gal.) Clearly, this yields a worst case time of $\frac{d}{v_1}(1+2\pi)$ and an expected time of $\frac{d}{v_1}(1+\pi)$.

In the following lemma, we describe a better, "kissing circles" strategy:

Lemma 3.1 *There is a strategy that yields a worst case time of $t_d(1+2\pi)$ and an expected time of $t_d(1+\pi)$.*

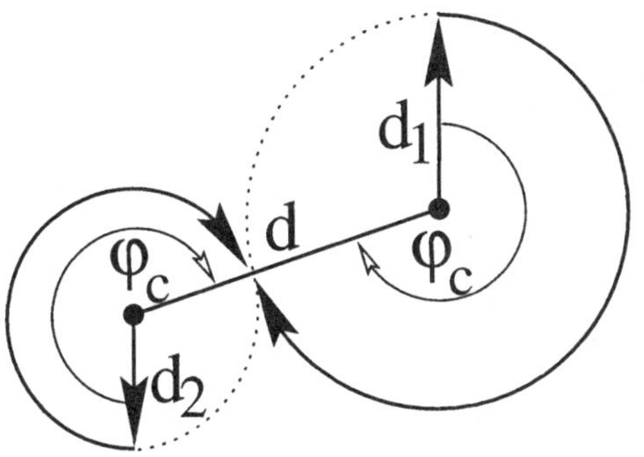

Figure 1: The "kissing circles" strategy

Proof: See Figure 1. Player 1 moves a distance of $d_1 = \frac{dv_1}{v_1+v_2}$ north, then at time $t_d = \frac{d_1}{v_1}$ she follows a clockwise circle around her original location. Similarly, player 2 moves a distance of $d_2 = \frac{dv_2}{v_1+v_2}$ south, then (at the same time $t_d = \frac{d_2}{v_2}$) she starts to follow a clockwise circle around her original location. At any given time $t \geq t_d$, player i has traveled an angle of $\varphi_i(t) = \frac{v_i}{d_i}(t - t_d)$. Let φ_c be the (clockwise) angle between the north direction and the directed line from player 1 to player 2 at time $t = 0$. Clearly, φ_c is also the clockwise angle between the south direction and the directed line from player 2 to player 1 at time $t = 0$. Since $\frac{v_i}{d_i} = \frac{v_1+v_2}{d}$, both players will have traveled the angle φ_c at the same time, and therefore meet (at $t = \frac{(1+\varphi_c)d}{v_1+v_2}$) at the point where their respective circles touch. This yields the values claimed for worst case and expected times. □

We can show that the kissing circles strategy is optimal:

Theorem 3.2 *The "kissing circles" strategy yields optimal worst case time and optimal expected time for a rendezvous.*

Proof: We have to show that no strategy can beat a worst case rendezvous time of $t_d(1+2\pi)$ or an expected rendezvous time of $t_d(1+\pi)$. Consider the position of player 2 in coordinates that are fixed at player 1, i.e., the vector $p(t) = s_2(t) - s_1(t)$. $p(0)$ is chosen uniformly at random from the circle C_p with radius d around the origin. By definition, the velocity $\dot{p}(t)$ has norm no greater than $v_1 + v_2$. When following $p(t)$, we have to search the circle C_p. Clearly, we have to travel a distance of at least d before encountering any points of C_p; since C_p has measure $2\pi d$ and the search speed is bounded by v_1+v_2, the worst case rendezvous time is bounded from below by $\frac{d(1+2\pi)}{v_1+v_2} = t_d(1+2\pi)$, and the expected rendezvous time is bounded from below by $\frac{d(1+\pi)}{v_1+v_2} = t_d(1+\pi)$. □

After the preceding discussion, it is straightforward to give a more general result for oriented rendezvous search: by using the orientation of the problem, we can reduce the rendezvous search problem with speeds v_1 and v_2 to a search problem with speed v_1+v_2. This is achieved in the following way:

Theorem 3.3 *Let $S(t)$ be an optimal trajectory for player 1, starting at the origin and searching with maximum speed $v_1 + v_2$ for a stationary player 2, positioned with a given probability distribution on a set Y. Then $S(t)$ induces an optimal pair of trajectories for the rendezvous search with mutual distribution Y.*

Proof: $S(t)$ can be converted into a pair of trajectories for the rendezvous search problem by taking

$$s_1(t) = \frac{v_1}{v_1+v_2}S(t)+s_1(0) \text{ and } s_2(t) = \frac{-v_2}{v_1+v_2}S(t)+s_2(0).$$

This guarantees that in a coordinate system centered at player 2, player 1 will follow the search path $S(t)$. □

3.2 Bounded Resources

Next consider the situation where both players have bounded resources: player i can travel at most a distance f_i before running out of fuel. In this scenario, the prime objective is to maximise the probability of a meeting. Problems of this type have been considered by Alpern and Beck [3] for the one-dimensional case. In the following, we describe a strategy that is probability-optimal; furthermore, among all strategies that are probability-optimal, it is also time-optimal: if there is a rendezvous, it will happen with smallest expected time. (See Alpern and Beck [2] for the one-dimensional case.)

Lemma 3.4 *There is a strategy that yields a probability of $\min\left\{\frac{f_1+f_2-d}{2\pi d}, 1\right\}$ for a rendezvous, provided that $f_1 + f_2 \geq d$. Conditional on there being a rendezvous, the worst case time is $\max\left\{\frac{f_1}{v_1}, \frac{f_2}{v_2}\right\}$; with $\frac{f_j}{v_j} \geq \frac{f_i}{v_i}$, the expected time is $\frac{\frac{f_i^2}{v_i}+\frac{f_j^2}{v_j}-dt_d}{2(f_1+f_2-d)}$ if $\frac{f_i}{v_i} \geq t_d$, and $\frac{f_j-f_i+d}{2v_j}$ otherwise.*

Proof: See Figure 2. Let $t_j = \frac{f_j}{v_j} \geq \frac{f_i}{v_i} = t_i$. Both players follow the kissing circles strategy until player i runs out of fuel, at time t_i, and has to stay at position $s_i(t_i)$. At time t_i, player j modifies her strategy: instead of a circle of radius d_j around her original position $s_j(0)$, she follows a circle of radius d around position $s^* = s_j(0) + s_i(t_i) - s_i(0)$. If $f_i \geq d_i$, i.e., $t_i \geq t_d$ and both players have already started along their kissing circles when player i runs out of fuel, then

$$\begin{aligned} & \|s_j(t_i) - s^*\| \\ = & \|(s_j(t_i) - s_j(0)) - (s_i(t_i) - s_i(0))\| \\ = & \|s_j(t_i) - s_j(0)\| + \|s_i(t_i) - s_i(0)\| \\ = & d_j + d_i \\ = & d \end{aligned}$$

by construction of the kissing circles. Furthermore, the new circle for player j will have a common tangent with her kissing circle, at position $s_j(t_i)$.

If $f_i \leq d_i$, then player j continues along her straight path until both players have moved a combined total of d, before following a circle of radius d around s^*. Since player j has traveled $f_i \frac{v_j}{v_i}$ at time t_i, while player i has traveled f_i, this additional distance is $d - f_i(1 + \frac{v_j}{v_i})$.

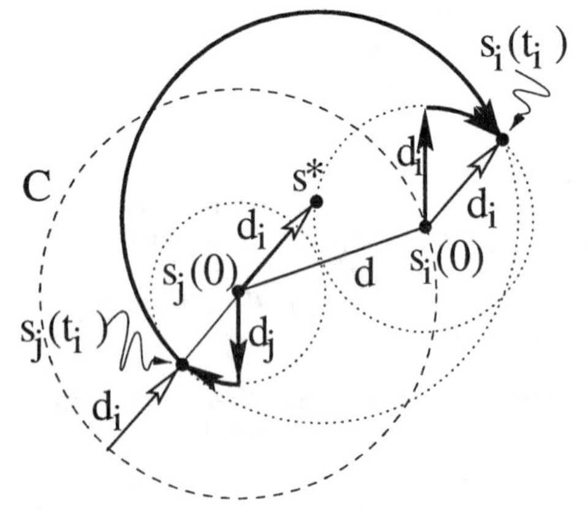

Figure 2: The strategy for bounded resources

Let C be the circle of radius d around $s_j(0)$. If the players have not met at time t_i, then player i is positioned on the circle $C + s_i(t_i) - s_i(0)$, i. e., on the circle of radius d around s^* that is searched by player j. If $2\pi d \geq f_1 + f_2 \geq d$, the described rendezvous strategy guarantees a rendezvous for a fraction of $\frac{f_1+f_2-d}{2\pi d}$ of the perimeter of C. For a uniform distribution along C, we get the claimed probability.

Now consider the set of positions where a rendezvous happens. In the worst case, this is at time $t_j = \frac{f_j}{v_j}$, when player j is just running out of fuel. For the expected time, first consider the case where $t_i \geq t_d$: for a time period of $t_i - t_d = \frac{f_i}{v_i} - t_d$, an arc of C is searched with speed $v_1 + v_2$, for a length of $\ell_1 = f_i \frac{v_1+v_2}{v_i} - d$; the rest of the search is done with speed v_j, covering an arc of length $\ell_2 = f_j - f_i \frac{v_j}{v_i}$. This yields an expected time of

$$\frac{\frac{t_d+t_i}{2}\ell_1 + \frac{t_i+t_j}{2}\ell_2}{\ell_1 + \ell_2} =$$

$$= \frac{\left(t_d + \frac{f_i}{v_i}\right)\left(f_i \frac{v_1+v_2}{v_i} - d\right) + \left(\frac{f_i}{v_i} + \frac{f_j}{v_j}\right)\left(f_j - f_i \frac{v_j}{v_i}\right)}{2(f_1 + f_2 - d)}$$

$$= \frac{\frac{f_i^2}{v_i} + \frac{f_j^2}{v_j} - dt_d}{2(f_1 + f_2 - d)}.$$

Now consider the case $t_d > t_i$. Player j reaches the search circle at time $t_i + \frac{d - t_i(v_1+v_2)}{v_j}$ and searches with speed v_j until time t_j, for an expected time of

$$\frac{t_i + \frac{d}{v_j} - f_i \frac{v_1+v_2}{v_i v_j} + t_j}{2} = \frac{\frac{f_i}{v_i} + \frac{d}{v_j} - \frac{f_i}{v_j} - \frac{f_i}{v_i} + \frac{f_j}{v_j}}{2} = \frac{f_j - f_i + d}{2v_j}.$$

□

Again, we can argue along the lines of Theorem 3.2 and Theorem 3.3 that this strategy is optimal:

Theorem 3.5 *The modified kissing circles strategy yields the maximum probability for a rendezvous. Among all strategies with maximum probability, it has smallest worst case time and optimal expected time for a rendezvous, assuming the players meet.*

Proof: It is impossible to search more than a portion of $f_1 + f_2 - d$ of the circle C, so the modified kissing circles strategy is probability-optimal. It is clear that the worst case time cannot be improved. To see that the expected time in case of a meeting is optimal, note that for any given time $t \geq t_d$, no search strategy can have searched a larger portion of C. □

3.3 Limited Visibility

Now let us consider the case of limited visibility, where the players meet when they come within distance r of each other. Theorem 3.3 allows us to focus on the search with a search radius r. Problems of this type have been dubbed "lawn mower problems" by Arkin, Fekete, and Mitchell [10]. If we consider the set Y to be a (possibly disconnected) piece of lawn, the problem of finding a shortest (closed) tour for a (circular) lawn mower of radius r is equivalent to finding a trajectory that minimises the worst case time for a rendezvous. Being a generalisation of the well-known Travelling Salesman Problem, the lawn mower problem is NP-hard, so one cannot hope to find an efficient algorithm that computes an optimal tour for any input Y. In [11], Arkin et al. give a number of efficient approximation algorithms for several versions of the problem.

However, the scenario of rendezvous search with a known distance d and search radius r is less general, so we do not have to deal with all the difficulties of the lawn mower problem. In fact, Gluss [20] has given a solution to this question when considering the min-max problem of "Searching for a circle in the plane". See Figure 3 for the case $d = 2r = 2$; the figure shows a trajectory which is optimal for the worst case. Note that a straightforward generalisation of the kissing circles strategy for zero visibility (yielding $(1+2\pi)(d-r)$) is not optimal, due to boundary effects.

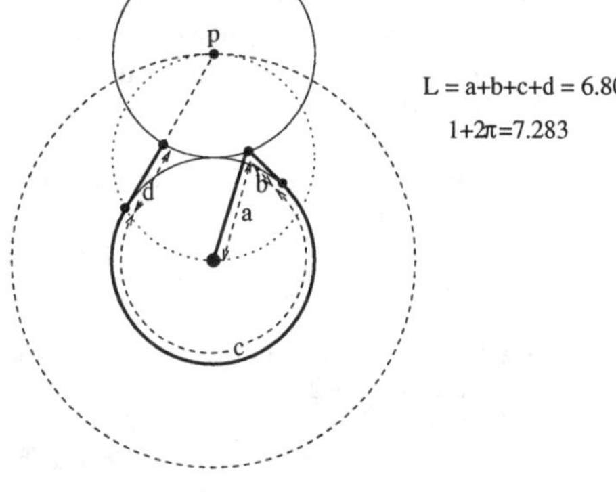

Figure 3: An optimal search trajectory with limited visibility

Things are much trickier if we want to minimise the *expected time* for a rendezvous. We think that an optimal solution will involve a spiral instead of the straight line labelled d in Figure 3, but an exact optimal solution is hard to determine.

4 Rendezvous with an Unknown Distance

Before we pursue the question of finding trajectories that minimise the expected time for a rendezvous, we will take a brief look at the situation where both players do *not* know each other's initial distance d. As before, it is possible to reduce the rendezvous search problem to a pure search problem. This search problem has been solved for the one-dimensional case by Beck and Newman [16]; a good overview can be found in [18]; see also [13].

In the two-dimensional scenario with an unknown distance, we cannot hope for a rendezvous if the visibility is zero – see Lemma 2.1. However, we can consider the two-dimensional case for a visibility radius of r by considering trajectories that will guarantee a rendezvous for any initial distance d. The rendezvous time (worst case or expected) can then be given as a function of d. Since we cannot impose any bounds on d, it makes sense to consider *asymptotically optimal* search trajectories: the value of a trajectory is $(1+\varepsilon_d)L_d$, where ε_d tends to zero as d grows to infinity, and L_d is a lower bound for the optimal value with distance d. This type of optimality has been discussed by Gal [18] for various types of search problems. He also uses ideas somewhat similar to the following.

In Figure 4 we show a "semi-circular spiral", a search trajectory which is asymptotically optimal for both worst case and expected time. Starting at the origin, the trajectory consists of semi-circles of radius $r, 2r, 3r, \ldots$ Each of the semi-circles "mows" an area of width $2r$. It is easy to see that (except for a set of points of measure zero at the "seams" of the semi-circles) no area is mown more than once. By travelling a distance of L, we can mow at most an area of $2rL+\pi r^2$. This means that we need at least a distance of $\frac{\pi d^2-\pi r^2}{2r}$ to mow all points within d of the origin. It is straightforward to see that the semi-circular spiral trajectory mows the circle of radius d while travelling a distance of $\frac{\pi d^2+O(d)}{2r}$. This implies that this trajectory is asymptotically optimal.

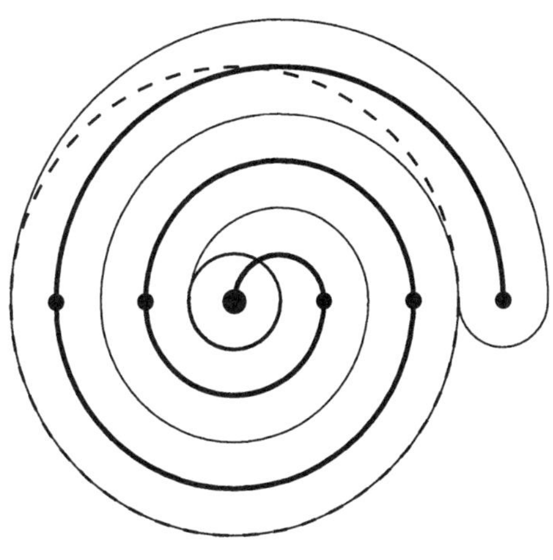

Figure 4: An asymptotically optimal trajectory when the distance is unknown

5 Rendezvous with a Fixed Point

In the following two sections we will discuss scenarios which are more complicated, owing to a combination of weaker assumptions on orientation, asymmetric information, or the more complex objective "expected time". We will show that, even for the special case of identical speeds, matters become very difficult.

We begin our discussions by considering the case where Y is a single point. Without loss of generality we take $Y=(d,0)$. In this case the oriented problem is trivial, since the two players can move towards each other meeting at time $d/(v_1+v_2)$. Thus we consider the unoriented problem in which player 2 does not know the direction of the starting point for player 1 (though she knows the distance). Notice that the problem is not the same for the two players: player 1 does know the initial position of player 2. For this problem G is simply the group of rotations about Y.

In practice rendezvous problems are usually solved by both players moving to some common focal point. This idea of coordination through some common point was the starting point for work in this area (see Schelling [29], Mosteller [28]). The problem in which Y is a single point also has the characteristic of having a unique point which both players can find: the initial position of player 2.

One easy solution is for player 2 to remain at her initial point and for player 1 to move on the trajectory $s_1(t)=(v_1t,0)$ so that the players meet at time d/v_1. As we shall see the players can do better than this.

Because of the special status of the initial point for player 2, it is better to make this the origin; so the initial position of player 1 becomes $(-d,0)$. It is easier to use polar coordinates to describe the trajectories of the two players. We suppose that player i uses a trajectory

$$s_i(t)=(R_i(t),\Theta_i(t)),\ 0<t,\ i=1,2.$$

The actual path traced out by player 2 is $g(s_2(t))$ which is

$$(R_2(t),\Theta_2(t)+\Theta_0),\ t>0,$$

where Θ_0 is chosen with a uniform probability distribution over the interval $(0,2\pi)$.

First we prove a lemma giving the form of the optimal policies for the two players.

Lemma 5.1 *If there is an optimal solution, then an optimal solution exists in which player 1 follows a trajectory of the form (in polar coordinates):*

$$\begin{aligned} s_1(t) &= (d-v_1t,\pi),\ 0\le t\le t_d,\\ s_1(t) &= (R(t),\Theta_1(t)),\ t>t_d, \end{aligned}$$

while player 2 uses the trajectory

$$\begin{aligned} s_2(t) &= (v_2t,0),\ 0\le t\le t_d,\\ s_2(t) &= (R(t),\Theta_2(t)),\ t>t_d, \end{aligned}$$

where Θ_i are monotonic functions for $t\ge t_d$, with $\Theta_1(t_d)=\pi$ and $\Theta_2(t_d)=0$. Moreover, the players move at their maximum speeds and in opposite rotational directions.

Proof: First we show that once the two players are at the same distance from the origin, they may as well remain at the same distance from the origin (since they can only meet when they have the same R value). Suppose that at times t_1 and t_2 the two players are at the same distance from the origin, but not for any $t \in (t_1, t_2)$. Then, since the players cannot meet at $t \in (t_1, t_2)$, we can replace both trajectories with straight line trajectories for this period. Suppose, without loss of generality, that $|s_1(t_1) - s_1(t_2)|/v_1 \geq |s_2(t_1) - s_2(t_2)|/v_2$. Then we generate a new solution which is no worse than the current one by setting the speed of s_1 to be v_1 over this section of trajectory: then the new trajectory reaches $s_2(t_2)$ at time $t_3 = t_1 + |s_1(t_1) - s_1(t_2)|/v_1 \leq t_2$. Then we choose the speed of s_2 over (t_1, t_3) so that s_1 and s_2 have the same R value throughout this interval. Finally we replace both trajectories from time t_3 onwards with the previous trajectories used from time t_2.

The next step is to show that both players are moving at maximum speed and with opposite rotational directions. This is achieved by observing that the meeting time is unaltered if the players use trajectories with an additional rotational component, provided that this is the same for both players. In other words if

$$s_i(t) = (R_i(t), \Theta_i(t) + \Phi(t)),\ i = 1, 2$$

for some fixed $\Phi(t)$, then all meeting times are unaltered. This transformation is only possible when the resulting trajectories do not break the speed limit.

We write $\phi(t)$ for the derivative of $\Phi(t)$ and $\theta_i(t)$ for the derivative of $\Theta_i(t)$. Note that if $\phi(t)$ is the same sign as $\theta_i(t)$ then the speed of player i is increased, and if they are of different signs then the speed is decreased. It is thus easy to see that the two players must move with opposite rotational directions, otherwise we may choose Φ to decrease both their speeds. This then makes it possible to speed up both players, keeping their paths the same, and decrease the meeting time.

Now suppose that player 1, say, is not moving at maximum speed. We may choose ϕ in such a way that the speed of player 2 is decreased and the speed of player 1 is increased, but remains less than v_1. Then we can improve the meeting times from this point on by increasing the speed of both trajectories. This contradicts the presumed optimality of the solution and hence we have established that each player moves at maximum speed.

The continuity of the trajectories implies that $\Theta_1(t_d) = \pi$ and $\Theta_2(t_d) = 0$, so it only remains to show that each Θ_i is monotonic. Observe that, with the form of optimal trajectories established, the rate at which a meeting occurs is either 0, or is governed entirely by $|\theta_1|$ and $|\theta_2|$. Consider replacing Θ_i by $\tilde{\Theta}_i$ where $d\tilde{\Theta}_1/dt = -|\theta_1|$ and $d\tilde{\Theta}_2/dt = |\theta_2|$. This ensures that the trajectory for player 1 takes a clockwise spiral around the origin rather than being reversed at any point. The rate at which meeting takes place is never lower than it was previously, until $\tilde{\Theta}_2$ reaches π when no more meetings take place. Thus the expected meeting time is no larger, and we have established a pair of optimal trajectories with the properties we claimed. □

Figure 5 shows the behaviour of the pair of optimal rendezvous trajectories in this case. Each executes a spiral towards the origin. To find the function describing that spiral turns out to be a problem in the calculus of variations.

We consider the case when $v_1 = v_2 = 1$ and $d = 2$. Then the fact that each player moves at the same speed and remains at the same radial distance, implies $\Theta_1(t) = \pi - \Theta_2(t)$ for $t > 1$. Thus we may take the trajectory for

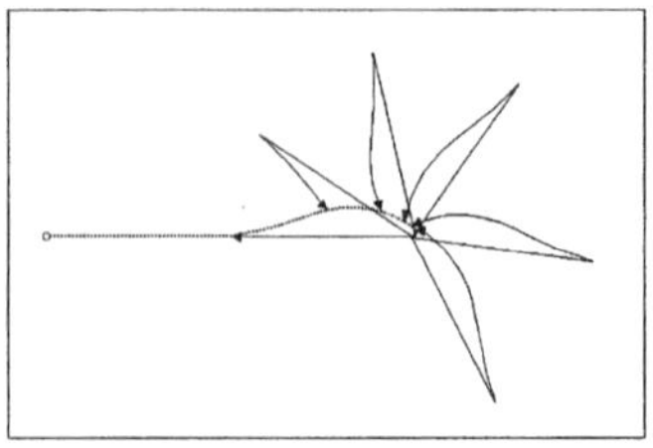

Figure 5: An optimal pair of trajectories

player 1 as given by the pair $(R(t), \pi - \Theta(t))$ for $t > 1$ with the conditions of the lemma.

It is convenient to express the trajectory in terms of the angle Θ, so we consider the function $y(\alpha), 0 \leq \alpha \leq 2\pi$ with $R(t) = y(\Theta(t))$ for $t > 1$.

Let $t(\alpha)$ be the time at which the trajectory reaches the angle α. The probability that a meeting takes place before time $1 + t(\alpha)$ is the probability that Θ_0 is in the range $(\pi - 2\alpha, \pi)$, and thus is α/π.

Hence the expected meeting time is given by the integral $\int_0^\pi (t(\alpha)/\pi) d\alpha$. Integrating by parts we see that this is

$$\int_0^\pi \left(1 - \frac{\alpha}{\pi}\right) \frac{dt}{d\alpha} d\alpha.$$

Since the trajectory moves at speed 1,

$$(dt/d\alpha)^2 = y(\alpha)^2 + (dy/d\alpha)^2.$$

So we need to minimise the integral

$$\int_0^\pi \left(1 - \frac{\alpha}{\pi}\right) \left(y(\alpha)^2 + (dy/d\alpha)^2\right)^{1/2} d\alpha$$

over choices of function y with $y(0) = 1$ and $y \geq 0$.

We can use the theory of the calculus of variations to carry out this minimisation. We wish to minimise an integral of the form $\int_0^\pi f(\alpha, y, dy/dt) d\alpha$, where

$$f(x, y, z) = \left(1 - \frac{x}{\pi}\right) (y^2 + z^2)^{1/2}$$

is convex on $[0, \pi] \times (0, \infty) \times \Re$. Hence a solution of a particular differential equation of the form $(d/dx) f_z = f_y$ with appropriate boundary conditions, will minimise the integral (see for example Troutman [30]).

The required differential equation is

$$\left(\frac{d}{d\alpha}\right) \left[\left(1 - \frac{\alpha}{\pi}\right) \left(y(\alpha)^2 + \left(\frac{dy}{d\alpha}\right)^2\right)^{-1/2} \frac{dy}{d\alpha} \right]$$

$$= \left(1 - \frac{\alpha}{\pi}\right) \left(y(\alpha)^2 + \left(\frac{dy}{d\alpha}\right)^2\right)^{-1/2} y.$$

After simplification this becomes

$$(\pi - \alpha) \left(y^3 + 2y \left(\frac{dy}{d\alpha}\right)^2 - y^2 \frac{d^2 y}{d\alpha^2}\right)$$

$$+\left(\frac{dy}{d\alpha}\right)^3 + \frac{dy}{d\alpha}y^2 = 0.$$

The only boundary condition we require in addition to setting $y(0) = 1$, is that $(dy/d\alpha)|_\pi = 0$. Any other value leads to an infinite value for $dy/d\alpha$ at π.

Figure 6 shows a numerical solution of this differential equation. Observe that $y(\pi)$ is strictly greater than zero (i. e., [illegible] will alway[illegible]

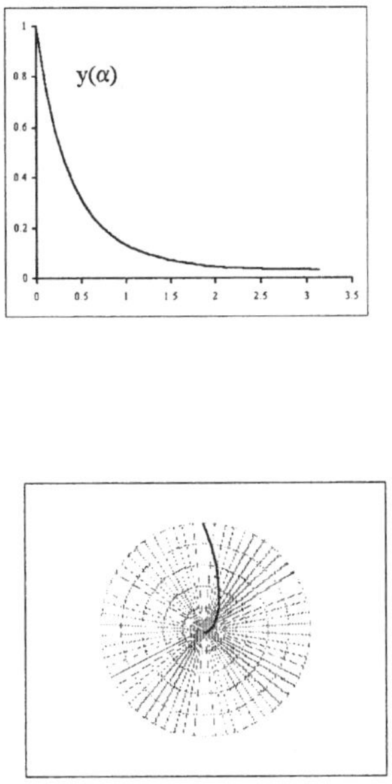

Figure 6: An optimal solution, shown in polar coordinates and Cartesian coordinates

6 Rendezvous with a Set of Points

In this section we consider the situation where Y is a finite set of points.

6.1 Oriented Rendezvous

First consider the oriented problem in which $Y = \{y_1, y_2, \ldots, y_k\}$, and we want to minimise the worst case time for a rendezvous. Suppose we already have a shortest route starting at the origin and visiting all the points in Y. Finding this is almost the same as solving the travelling salesman problem, except that we do not have to return to the origin. Since this is an NP-hard problem, we cannot hope to find such an optimal route in a short time. However, there are algorithms (see Arora [12], and Mitchell [27]) for the geometric TSP that can find a solution within $(1 + \varepsilon)$ of the optimum in polynomial time for any fixed ε.

We suppose that the points in Y have been reordered so that there is an optimal solution to the problem of a shortest path in which the points are visited in the order: first y_1, then y_2, then y_3 and so on up to y_k. Let a_j be the vector $y_j - y_{j-1}$, $j = 2, 3, \ldots k$ and let $a_1 = y_1$.

The following lemma can be proved in exactly the same way as the optimality of the kissing circles solution, when Y is a circle. We consider the position of player 1 in a coordinate system which leaves player 2 fixed. Then player 1 has a maximum velocity of $v_1 + v_2$ and by construction visits the points of Y in order.

Lemma 6.1 *The solution in which player 1 moves in direction a_1, followed by direction a_2, etc. finishing by moving in direction a_k, with the direction a_j being followed for a time $|a_j|/(v_1 + v_2)$, and in which player 2 moves at all times in the opposite direction to player 1, minimises the maximum meeting time.*

It should be noted that, when minimising the expected time instead of the worst case time, one has to consider a problem that is very similar to the so-called *minimum latency problem* (Blum et al. [17]) of finding a roundtrip that minimises the average travel time for the points. Like the travelling salesman problem, this problem is NP-hard; however, the best known approximation algorithms (see Goemans and Kleinberg [21]) have relatively high approximation factors, indicating that even the combinatorial structure of this problem is more involved than the structure of the TSP. There are additional difficulties when we consider the asymmetric case, since this brings into play the spirals that were needed in the previous section.

Now consider the problem in which the movement is restricted to a network forming a square grid, with each arc of length 1.

At this point we need to strengthen the assumptions that we make. We shall assume that $v_1 = v_2 = 1$ and that the points of Y are at the nodes of the network. Furthermore we assume that all distances between points in $Y \cup \{0\}$ are multiples of 2. Thus the points in Y are drawn from the set $\{(a, b)|\ a + b \text{ is even }\}$.

We suppose that, as before, a shortest route from the origin visiting all the points in Y exists in which the points are visited in the order $y_1, y_2, \ldots, y_k$.

Let $p(a, b)(t)$ be a particular arbitrary choice of shortest path on the square grid from position a to position b, where $t \in [0, d(a, b)]$ and $d(a, b)$ is the shortest distance from a to b on the network.

Set $x_1 = 0$. Now consider the strategy pair in which player 1 follows the path $p(x_1, y_1)$ up to time $t_1 = d(x_1, y_1)/2$ at which point player 1 is at position $x_2 = p(x_1, y_1)(t_1)$, and then the path $x_2 - y_1 + p(y_1, y_2)$ until time $t_2 = t_1 + d(y_1, y_2)/2$, and so on. Thus during stage i player 1 follows the path $x_i - y_{i-1} + p(y_{i-1}, y_i)$ until time $t_i = t_{i-1} + d(y_{i-1}, y_i)/2$ at which point she is at position $x_{i+1} = x_i - y_{i-1} + p(y_{i-1}, y_i)(t_i - t_{i-1})$. Meanwhile player 2 follows the other half of the $p(y_{i-1}, y_i)$ path during stage i. Thus player 2 moves in the direction $(d/dt)p(0, y_1)(d(0, y_1) - t)$ for $t \in (0, t_1)$, and at any time $t \in (t_j, t_{j+1})$ player 2 moves in the direction $(d/dt)p(y_{j-1}, y_j)(d(y_{j-1}, y_j) - t)$.

We need to establish that this pair of trajectories will stay on the square grid. Observe that, since all distances are even integers, all the times $t_1, t_2, \ldots t_k$ are integers. Thus if the trajectory followed up to time t_i remains on the grid, x_i will be at one of the nodes of the square grid, for $i = 1, 2, \ldots k$. But once it is established that x_i is at a node of the grid, it follows that the path $x_i - y_{i-1} + p(y_{i-1}, y_i)$ is on the grid. This shows what we require for player 1. In the same way, player 2 follows a path which at stage j is a translation of $p(y_{j-1}, y_j)(d(y_{j-1}, y_j) - t)$ and, because each t_i is even, player 2 starts each stage at a node of the grid. This is enough to show that player 2 also stays on the grid.

Lemma 6.2 *The strategy pair described above minimises the maximum time till meeting.*

Proof: Using the same technique as the lemma above, we observe that if we fix the frame of reference so that player 2 is stationary, a lower bound on the total distance travelled by player 1 is given by the TSP path which has total length $d(0, y_1) + \sum_{i=1}^{k-1} d(y_i, y_{i+1})$. In this frame of reference the maximum speed of player 1 is 2. Now observe that the total time taken under the strategy pair proposed is exactly half of the total length of the TSP path.

It only remains to be shown that this strategy pair guarantees a meeting before time t_k. If player 2 starts from position y_1 then at time t_1 she is at point x_1, by construction, and will meet player 1. Hence, if player 2 starts from position y_2 (a translation by $y_2 - y_1$), then at time t_1 she is at position $x_1 - y_1 + y_2$. Thus a meeting will take place at time t_2 at position x_2. Continuing in this way we find that if player 2 starts at y_j there will be a meeting at time t_j. □

6.2 Unoriented Rendezvous

Now we discuss an unoriented rendezvous problem on the grid. In this section we will concentrate on minimising the expected meeting time. Consider the problem where $Y = \{(1,1), (1,-1), (-1,-1), (-1,1)\}$. This is one of the simplest problems of this type which is symmetric, in the sense that the two players have the same information set at the start. Both players know that the other player is at the diagonally opposite corner of one of the 4 squares intersecting at the starting position, but have no information on which square. Notice that, even though the problem formulation is symmetric between player 1 and player 2, we are still considering asymmetric policies, in which the two players choose different strategies.

Since both players have to move along grid lines, we can show that, for an optimal policy, meetings only take palce at the nodes of the grid.

Lemma 6.3 *For the problem of minimising expected meeting time, there is an optimal strategy pair in which both players always move at speed 1, and all meetings happen at integer points and at integer times.*

Proof: We may restrict attention to strategy pairs which always move at speed 1, since other strategies can be improved by moving at speed 1 from one possible meeting point to the next, and replacing a waiting time by movement at speed 1 back and forth along one of the arcs in such a way as to reduce the time till meeting one of the trajectories for the other player. Now consider a trajectory which moves part way along an arc and then reverses direction. If this reversal happens without this also being a possible meeting time with one of the other player's trajectories, then we may replace this trajectory with another in which that section of arc is not traversed and the trajectory arives earlier at its next meeting point. But now observe that, at the first meeting that takes place part way along an arc, both players will have followed trajectories that consist of a sequence of unit length steps around the grid. This is a contradiction, since the initial distance between the players is even. Hence we may suppose that there are no reversals part way along an arc. The result follows. □

This allows us to indicate a trajectory by listing the sequence of directions in which a player travels, with each step being of unit length.

Theorem 6.4 *The expected meeting time is minimised by the strategy pair in which each player chooses an arbitrary direction to call North, and player 1 follows a trajectory NWSSEENN and player 2 follows a trajectory NSNSNSNS.*

Proof: Fix the orientation of the plane according to player 1's compass. Thus the direction that player 2 calls North may in fact be North, West, South or East. Observe that player 1 meets the 16 possible trajectories at times listed in the following Table 1.

Table 1: Meeting times along different trajectories

Player 2 initial direction	*Starting point* (1,1)	(-1,1)	(-1,-1)	(1,-1)
North	8	2	3	6
West	1	2	4	5
South	7	2	4	6
East	8	1	4	6

The meetings with these 16 trajectories occur according to the following pattern: 2 at time 1, 3 at time 2, 1 at time 3, 3 at time 4, 1 at time 5, 3 at time 6, 1 at time 7, and 2 at time 8. We represent this pattern of meetings by the sequence $(2,3,1,3,1,3,1,2)$. The proof establishes that no other feasible pattern of meetings gives a lower expected meeting time.

First observe that for any strategy pair no more than three meetings can occur at the same time. This follows from the fact that two player 2 trajectories from different starting points that move in the same initial direction remain at the same distance apart. Hence if a meeting takes place with four different trajectories, they must all have had different initial directions. Thus they must approach the meeting with player 1 from four different directions. Now we can assume that player 1 is never stationary (since we can replace a stationary period in which no meetings take place with a back and forward motion). Hence the meeting with one of the four trajectories must have taken place earlier.

Now suppose that a meeting takes place at time t^* with three trajectories for player 2. We write $x(t)$ for the player 1 trajectory, and $y_i(t)$, $y_j(t)$ and $y_k(t)$ for the three player 2 trajectories. Notice that $x(t^*)$ is equidistant from $y_i(0)$, $y_j(0)$ and $y_k(0)$. Thus if each of these is a different starting point then $x(t^*) = 0$. On the other hand if two of the three player 2 trajectories have the same starting point then they can only meet again at that starting point. But in this case all the player 2 trajectories must be at their initial positions. Hence at most one meeting can take place at the next step.

Thus the meeting with three player 2 trajectories can only occur at a starting point, either for player 1 or player 2. Now we can use the symmetry of the rendezvous game. If the two player strategies are swapped exactly the same pattern of meetings takes place. But now a situation in which three meetings take place at the origin is transformed into one in which three meetings takes place at one of the initial positions for player 2. So the observation that if three meetings takes place at time t^* at most one meeting can take place at time $t^* + 1$ will apply in all circumstances.

Now consider any feasible pattern of meetings. We show that this can be transformed into the pattern of meetings $(2,3,1,3,1,3,1,2)$ in a way which makes the expected meeting time no worse. First change any time at which there is no meeting into a time at which there is one meeting by moving forward the last meeting. Now change any time at which there is one meeting, provided it does not follow immediately after three meetings, into a time at which there are two meetings by moving forward the last meeting. Each of these changes will make the expected meeting time less. After this the pattern of meetings can be described by a sequence made up of from single elements '2' and pairs '3, 1'. The first element in the pattern will be '2' since it is impossible to meet three trajectories at time 1. Now move any '2', other than that at the start of the pattern, to the end of the pattern by repeatedly replacing the triple '2, 3, 1' by '3, 1, 2'. These changes make no difference to the expected meeting time. At the same time replace consecutive pairs '2, 2' with the pair '3, 1' except where this involves the first '2'. each of these changes will improve the expected meeting time. Clearly the end result is the pattern we require. □

Acknowledgements

The work on this joint paper originated at the CDAM workshop on rendezvous search, London School of Economics, August 1996. We would like to thank the organiser Steve Alpern, as well as the other participants Vic Baston, Anatole Beck, Carla Di Cairano, Skander Essegaier, Shmuel Gal, John Howard, and Wei Shi Lim for the productive atmosphere and useful insights. Steve and Shmuel pointed out some connections to earlier work. Thank you to Günter Rote for a motivating discussion.

References

[1] S. Alpern. The rendezvous search problem. *SIAM Journal on Control and Optimization*, **33** (1995), 673–683.

[2] S. Alpern and A. Beck. Rendezvous search on the line with bounded resources: minimizing expected meeting time. *European Journal of Operational Research*, to appear.

[3] S. Alpern and A. Beck. Rendezvous search on the line with bounded resources: maximizing the probability of meeting. LSE Mathematics Preprint series (1997).

[4] S. Alpern and A. Beck. Rendezvous search on the line is a double linear search problem. Preprint (1997).

[5] S. Alpern and S. Gal. Rendezvous search on the line with distinguishable players. *SIAM J. Control and Optimization*, **33** (1995), 1270–1276.

[6] S. Alpern and J. V. Howard. Alternating search at two locations. Preprint.

[7] S. Alpern and W. S. Lim. The symmetric rendezvous-evasion game. *SIAM Journal of Control and Optimization*, to appear.

[8] E. J. Anderson and S. Essegaier. Rendezvous search on the line with indistinguishable players. *SIAM J. Control and Optimization*, **33** (1995), 1637–1642.

[9] E. J. Anderson and R. R. Weber. The rendezvous problem on discrete locations. *J. Applied Probability*, **28** (1990), 839–851.

[10] E. M. Arkin, S. P. Fekete, and J. S. B. Mitchell. The lawnmower problem. *Proceedings of the Fifth Canadian Conference on Computational Geometry*, Waterloo 1993, 461–466.

[11] E. M. Arkin, S. P. Fekete, and J. S. B. Mitchell. Approximation algorithms for lawn mowing and milling. ZPR Report 97-255, available at `ftp://ftp.zpr.uni-koeln.de/pub/paper/zpr97-255.ps.gz`, 1997.

[12] S. Arora. Polynomial time approximation schemes for Euclidean TSP and other geometric problems. *Proceedings of the 37th Annual IEEE Symposium on the Foundations of Computer Science*, 1996, 2–12.

[13] R. A. Baeza-Yates, J. C. Culberson, and G. J. E. Rawlins. Searching in the plane. *Information and Computation*, **106** (1993), 234–252.

[14] V. J. Baston. Two rendezvous search problems on the line. Preprint. 1997.

[15] V. J. Baston and S. Gal. Minmax rendezvous on the line. Preprint. 1997.

[16] A. Beck and D. J. Newman. Yet more on the linear search problem. *Israel J. Math.*, **8** (1970), 419–429.

[17] A. Blum, P. Chalasani, D. Coppersmith, W. R. Pulleyblank, P. Raghavan, and M. Sudan, The minimum latency problem. *Proc. Ann. Symp. Theo. Comp. (STOC)*, 1994, 163–171.

[18] S. Gal. Continuous search games. Chapter 3 *in*: D. V. Chudnovsky, G. V. Chudnovsky, eds., *Search theory.* Marcel Dekker Lecture notes in pure and applied mathematics, v. 112, New York 1989.

[19] S. Gal. Rendezvous search on the line. Preprint. 1996.

[20] B. Gluss. The minmax path in the search for a circle in the plane. *Naval Res. Logist. Quart.*, **8** (1961), 357–360.

[21] M. X. Goemans and J. Kleinberg. An improved approximation ratio for the minimum latency problem. *Proc. 7th ACM-SIAM Symp. Disc. Alg. (SODA)*, 1996.

[22] T. Healey. Journeys (sometimes) end in lovers meeting. *Annals of Improbable Research*, **3** (3), 1997, 26–27.

[23] J. V. Howard. Rendezvous search on the interval and the circle. Technical Report 96-12, London School of Economics, 1996.

[24] W. S. Lim and S. Alpern. Minimax rendezvous search on the line. *SIAM J. Control and Opt.*, **34** 1996, 1650–1665.

[25] W. S. Lim and S. Alpern. Minimax triple rendezvous on the line: corrections and extensions. LSE Mathematics Preprint Series (1997).

[26] W. S. Lim., S. Alpern, and A. Beck. Rendezvous search on the line with more than two players. *Operations Research*, **45** 1997, 357–364.

[27] J. S. B. Mitchell. Guillotine subdivisions approximate polygonal subdivisions: Part II – A simple polynomial-time approximation scheme for geometric k-MST, TSP, and related problems. To appear, *SIAM Journal of Computing.* Available at `http://ams.sunysb.edu/~jsbm/jsbm.html`.

[28] F. Mosteller. *Fifty Challenging Problems in Probability with Solutions.* Addison-Wesley, New York, 1963.

[29] T. C. Schelling. *The Strategy of Conflict.* Oxford University Press, Oxford, 1960.

[30] J. L. Troutman. *Variational Calculus with Elementary Convexity.* Springer-Verlag, New York, 1983.

Motion Planning for Multiple Robots

Boris Aronov* Mark de Berg† A. Frank van der Stappen† Petr Švestka‡ Jules Vleugels†

Abstract

We study the motion-planning problem for pairs and triples of robots operating in a shared workspace containing n obstacles. A standard way to solve such problems is to view the collection of robots as one composite robot, whose number of degrees of freedom is d, the sum of the numbers of degrees of freedom of the individual robots. We show that it is sufficient to consider a constant number of robot systems whose number of degrees of freedom is at most $d-1$ for pairs of robots, and $d-2$ for triples. (For triples we need to assume that a solution with positive clearance exists.) We use this to obtain an $O(n^d)$ time algorithm to solve the motion-planning problem for a pair of robots; this is one order of magnitude faster than what the standard method would give. For a triple of robots the running time becomes $O(n^{d-1})$, which is two orders of magnitude faster than the standard method. We also apply our method to the case of a collection of bounded-reach robots moving in a low-density environment. Here the running time of our algorithm becomes $O(n \log n)$ both for pairs and triples.

1 Introduction

One of the ultimate goals of robotics research is to create robots that are capable of autonomous action, and accept and execute high-level task descriptions while requiring little or no human supervision. A fundamental task for an autonomous robot is to plan its own motion: it should be able to figure out how to get from one position (for instance, where it has picked up some object) to another position (where the object is to be delivered). In general there will be obstacles in the workspace of the robot, which it has to avoid. In many applications the situation is further complicated by the fact that the robot has to share its workspace with other robots. Examples of this are the transportation systems found at modern airports, ports, and factories. This is the setting of the motion-planning problem we study: given a collection $R_1, \ldots, R_m$ of robots with $d_1, \ldots, d_m$ degrees of freedom, respectively, and operating in a shared workspace $\mathcal{W}$ with n obstacles, find a collection of paths bringing each robot from a specified start position to a specified goal position without colliding with the obstacles or the other robots. We assume m is a small constant; we will mainly study the cases $m = 2$ and $m = 3$. The problem becomes PSPACE-complete if the number of robots is not constant [13].

Two established approaches to this problem are *decoupled planning* and *centralized planning.*

The decoupled planning approach [1, 8, 14, 21] first plans the motion of each robot individually while ignoring the existence of the other robots, then tries to combine the resulting paths by resolving possible collisions between the paths. Algorithms following this approach are usually incomplete in the sense that they are not guaranteed to find a solution if one exists.

In centralized planning [15, 17] the m robots are regarded as one multi-robot, that is, one robot with several independent body parts that are not necessarily connected to each other. Collisions between the robots R_i turn into collisions between different body parts—in other words, self-collisions—of the multi-robot. The configuration space of the multi-robot is $\mathcal{CS}_1 \times \cdots \times \mathcal{CS}_m$, where $\mathcal{CS}_i$ is the configuration space of robot R_i. The dimension of this configuration space is $d_1 + \cdots + d_m$, the sum of the dimensions of the individual configuration spaces. Using an algorithm of Basu et al. [2] for constructing roadmaps, one can thus solve the motion-planning problem in $O(n^{d_1+\cdots+d_m+1})$ time. Under certain general-position assumptions on the configuration-space obstacles one can use Canny's roadmap algorithm [4] to improve the time bound to $O(n^{d_1+\cdots+d_m} \log n)$. (In a later paper [5] Canny showed how to eliminate the general-position assumption, but unfortunately the adapted version cannot report an actual path, it can only decide on the existence of a path.) Since we wish to keep our results as general as possible, we stick to using the roadmap algorithm of Basu et al. from this point on.

Centralized approaches have the advantage that they allow for complete planners, which are guaranteed to find a solution whenever one exists. Their main drawback is that the dimension of the configuration space of the multi-robot

*Department of Computer and Information Science, Polytechnic University, Brooklyn, NY 11201-3840, USA. B.A. has been partially supported by a Sloan Research Fellowship. Part of this work was performed while B.A. was visiting Utrecht University in February 1997.

†Department of Computer Science, Utrecht University, P.O.Box 80.089, 3508 TB Utrecht, the Netherlands.

‡CWI, Kruislaan 413, 1098 SJ Amsterdam, the Netherlands.

SCG 98 Minneapolis Minnesota USA

is much higher than that of the individual robots. This increases both the combinatorial and the algebraic complexity of the problem.

Sharir and Sifrony [17] present a general centralized-planning method that is sometimes more efficient than applying the method of Basu et al. to the multi-robot. Their method requires a cell decomposition for the free portion of each individual configuration space. The cell decompositions are then combined into a representation of the free space of the multi-robot. The complexity of this representation is the product of the complexities of the individual configuration spaces. This approach can take advantage of the fact that sometimes the complexity of (the cell decomposition of) the free space of a robot R_i is significantly smaller than $\Theta(n^{d_i})$. For instance, it is well known that the complexity of the free space of a polygon translating amidst n polygonal obstacles in the plane is only $O(n)$ [12]. Thus the method of Sharir and Sifrony can plan the motion of two polygons in such a workspace in $O(n^2)$ time. Applying the method of Basu et al. to the multi-robot would lead to an $O(n^5)$ algorithm, because the number of degrees of freedom of the multi-robot is four.

We present a refinement of the centralized-planning approach for pairs and triples of robots. Our technique is quite general: it works for any type of robots (for triples of robots we need some non-degeneracy assumptions—see Section 4 for details), and it can be combined with roadmap methods and with cell-decomposition methods. The technique reduces the dimension of the configuration space one has to consider for the multi-robot. For pairs of robots it does so by stipulating that at all times either one of the robots should be at its start or goal position, or the robots should touch each other; for triples of robots the configurations of the multi-robot are constrained in a similar fashion.

The approach of reducing the dimension of the configuration space to be searched was also used by Hopcroft and Wilfong [10, 11] and Fortune et al. [9]. Hopcroft and Wilfong [10] showed that if the robots form a single connected component at their start and goal configurations and if there exists a collision-free motion of the multi-robot in which the robots need not touch, then there exists a path such that the robots form a connected component throughout the entire motion. This implies that the search in the configuration space can be limited to a lower-dimensional subspace. However, their result only holds when the configuration space is contractible to a point. In a later paper Fortune et al. [9] considered the case of two planar robot arms, each having one extendible and rotatable link. For this case, where the configuration space is not contractible to a point, they gave a proof (tailored to this special case) that it is sufficient to consider only motions in contact; in addition, they developed an algorithm to find such motions. In their second paper [11], Hopcroft and Wilfong applied the result from their first paper to translating axis-parallel polygons in $\mathbb{R}^2$. In particular, they showed that planning the motions of n rectangles in a rectangular work space is in PSPACE.

We use essentially the same approach, that is, we also show that the search of the configuration space can be limited to lower-dimensional subspaces—corresponding to configurations with specific properties. However, we do not require the configuration space to be contractible, thus generalizing the results of Hopcroft and Wilfong to arbitrary robots. For instance, our result is directly applicable to the case of two planar robot arms studied by Fortune et al. Furthermore, we not only prove the existence of a certain restricted type of multi-path, we also present a general algorithm for finding such a path efficiently. Unfortunately, we can only prove our result for two and three robots, so in this respect our results are less general than those of Hopcroft and Wilfong. More precisely, our results are as follows.

In Section 3 we prove that in order to plan the motions of two robots R_1 and R_2, one does not have to consider the entire $(d_1 + d_2)$-dimensional configuration space $\mathcal{CS}_1 \times \mathcal{CS}_2$. Instead, it is sufficient to consider a collection of five suitably linked configuration-space slices (corresponding to the constrained type of configurations mentioned above) whose dimensions are at most $d_1 + d_2 - 1$. Combining this with the method of Basu et al. we obtain a general method to solve the motion-planning problem for a pair of robots in $O(n^{d_1+d_2})$ time.

In Section 4 we extend our ideas to triples of robots. Here we show that it is sufficient to consider a constant number of configuration-space slices of maximum dimension $d_1 + d_2 + d_3 - 2$, which leads to an $O(n^{d_1+d_2+d_3-1})$ time algorithm. We need some technical assumptions to prove this result, the most important of which is that a solution with positive clearance exists. Note that one can view a collection of four or more robots as a collection of three robots, one of which is a multi-robot consisting of $m - 2$ robots. Hence, the improvement (as compared to the standard method of viewing the collection of robots as one multi-robot) of two orders of magnitude that we obtain for three robots carries over to the case of four or more robots.

Our approach becomes especially effective when the robots under consideration have bounded reach and the workspace has low density [20], as we show in Section 5. Bounded-reach robots are robots that are not too large compared to the obstacles, and a workspace has low density if any region is intersected by only a constant number of obstacles that are larger than that region. These notions were introduced to exclude unrealistic inputs—contrived workspaces and robots that induce very high free-space complexities. It is expected that in most practical situations the robots have bounded reach and the workspace they operate in has low density. Van der Stappen et al. [20] demonstrate that the complexity of the free space of a single bounded-reach robot in a low-density workspace is $O(n)$, irrespective of the number of degrees of freedom of the robot and of the dimension of its workspace. Van der Stappen et al. also give an algorithm to compute a linear-size cell decomposition of the free space, leading to an $O(n \log n)$ time algorithm to plan the motion of a single robot. Unfortunately, if we consider a pair or triple of bounded-reach robots operating in a low-density workspace, then the free-space complexity of the corresponding multi-robot can be as high as $\Theta(n^2)$ for a pair and $\Theta(n^3)$ for a triple. Nevertheless, we show how to apply our method to this case to obtain the following surprising result: for bounded-reach robots in low-density workspaces one can solve the motion-planning problem for pairs or triples of robots in the same time as for a single robot, namely in $O(n \log n)$ time. Note that the method of Sharir and Sifrony [17] can be used in this case, because a cell decomposition of the free space of the individual robots is available. This, however, leads to algorithms with considerably higher running times, namely $O(n^2)$ for a pair of robots and $O(n^3)$ for a triple.

2 Preliminaries

Although we assume some familiarity with the basic concepts in motion planning, we briefly introduce the terminology and notation used throughout the paper. A general

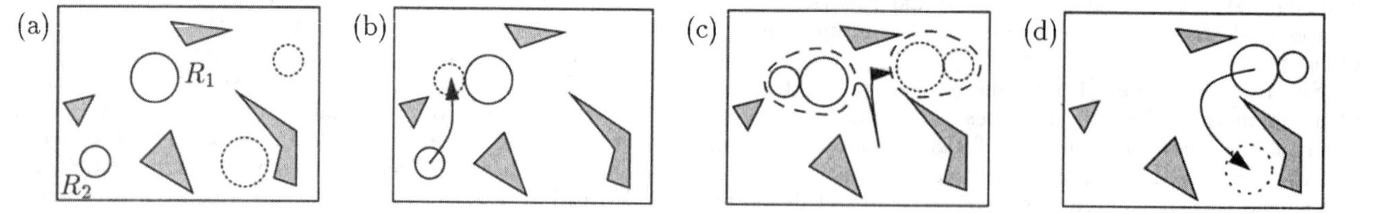

Figure 1: A feasible multi-path using permissible multi-configurations.

introduction to motion planning can be found in Latombe's book [13].

Let $\mathcal{R} = \{R_1, \ldots, R_m\}$ denote a collection of m robots. All robots operate in the same workspace $\mathcal{W}$, which contains a set $\mathcal{C} = \{C_1, \ldots, C_n\}$ of obstacles. We assume that each robot and each obstacle has constant complexity, and that they are algebraically defined. We also assume they are open sets; this means that the robot is allowed to 'slide along' an obstacle.

The *configuration space* of robot R_i is denoted by $\mathcal{CS}_i$. The dimension of $\mathcal{CS}_i$ equals d_i, the number of degrees of freedom of R_i. We assume that $d_i > 0$ for every i. Points in $\mathcal{CS}_i$—or, *configurations*—correspond to placements of R_i in the workspace; we denote the portion of the workspace occupied by R_i at configuration $p \in \mathcal{CS}_i$ by $R_i[p]$. The points in $\mathcal{CS}_i$ representing the start and goal configuration of R_i are denoted by s_i and g_i respectively. A path for R_i from its start configuration to its goal configuration corresponds to a curve in $\mathcal{CS}_i$ from s_i to g_i. We parameterize the curve by a time parameter t, with $t \in [0,1]$, so a path from start to goal configuration is a continuous mapping $\pi_i : [0,1] \rightarrow \mathcal{CS}_i$ with $\pi_i(0) = s_i$ and $\pi_i(1) = g_i$.

Each obstacle defines a subset—its *configuration-space obstacle*—in a configuration space consisting of all configurations in which the robot intersects that obstacle. The portion of the configuration space covered by the configuration-space obstacles is called the *forbidden space*, and its complement is called the *free space*. We call a path π_i for R_i *feasible* if R_i does not intersect any obstacle during its motion or, in other words, if the curve π_i lies entirely in the free space.

As stated in the introduction, we can view the collection of robots as one composite robot, or *multi-robot*, with $d := \sum_{i=1}^{m} d_i$ degrees of freedom. We refer to configurations of the multi-robot as *multi-configurations*, and we call a path for the multi-robot (which is in fact a collection of paths for the individual robots) a *multi-path*. We want to find a *feasible multi-path* for the robots $R_1, \ldots, R_m$ and their given start and goal configurations, that is, a collection of paths that brings each robot from its start configuration to its goal configuration without colliding with either the obstacles or the other robots.

3 Pairs of robots

To explain the idea of our method, we start by studying the case of a pair of robots. One way of planning the motion of a pair of robots is to view the pair as one robot with $d := d_1 + d_2$ degrees of freedom. Thus the problem can be solved in the d-dimensional configuration space $\mathcal{CS}_1 \times \mathcal{CS}_2$. Our goal is to reduce the dimension of the space we have to consider. To this end we limit the possible multi-configurations—combinations of configurations for the two robots—that we allow. Of course, we have to guarantee that a feasible multi-path continues to exist.

The multi-configurations that we allow—we call them *permissible multi-configurations*—are as follows.

- When R_1 is at its start or goal configuration, we allow any configuration of R_2.
- When R_2 is at its start or goal configuration, we allow any configuration of R_1.
- When neither R_1 nor R_2 is at its start or goal configuration, we only allow configurations where R_1 and R_2 touch each other.

We give an example. Consider the situation depicted in Fig. 1, where we have two disk-shaped robots moving amidst polygonal obstacles in the plane. The start and goal configurations of the robots are indicated in Fig. 1(a); start configurations are solid and goal configurations are dotted. A feasible multi-path for this problem that uses permissible multi-configurations is indicated in Fig. 1(b)–(d): first R_2 moves towards R_1 until it touches it, then R_2 and R_1 together move until R_2 is at its end configuration, and finally R_1 breaks off its contact with R_2 and moves to its own goal configuration.

At first sight, it seems quite severe to restrict oneself to permissible multi-configurations. Nevertheless, it turns out that solutions using only permissible multi-configurations always exist, provided a solution exists at all.

Lemma 1 *Let R_1 and R_2 be two robots operating in the same workspace. If there is a feasible multi-path for given start and goal configurations, there is also a feasible multi-path for those start and goal configurations that only uses permissible multi-configurations.*

Proof: Let $\Pi = \{\pi_1, \pi_2\}$ be a feasible multi-path. We define the *coordination diagram* for Π as follows. Let U be the unit square. We call the edges of U incident to the origin the *axes* of U. The horizontal axis, or t_1*-axis*, of U represents the configuration of R_1 along π_1; the vertical axis, or t_2*-axis*, represents the configuration of R_2 along π_2. Thus a point $(t_1, t_2) \in U$ corresponds to placing R_1 and R_2 at configurations $\pi_1(t_1)$ and $\pi_2(t_2)$ respectively. Observe that the left edge of U corresponds to multi-configurations where R_1 is at its start configuration, the top edge of U corresponds to configurations where R_2 is at its goal configuration, and so on. A point $(t_1, t_2) \in U$ is called *forbidden* if $R_1[\pi_1(t_1)]$ intersects $R_2[\pi_2(t_2)]$; otherwise it is called *free*—see Fig. 2. The coordination diagram for Π is the subdivision of U into free and forbidden regions.

Define $\mathbf{0} := (0,0)$ and $\mathbf{1} := (1,1)$ to be the lower left and top right vertex of U, respectively. We call a path in U from $\mathbf{0}$ to $\mathbf{1}$ a **0-1** *path*. Since $\{\pi_1, \pi_2\}$ is a feasible multi-path, R_1 does not intersect any obstacle along π_1 and R_1 does not

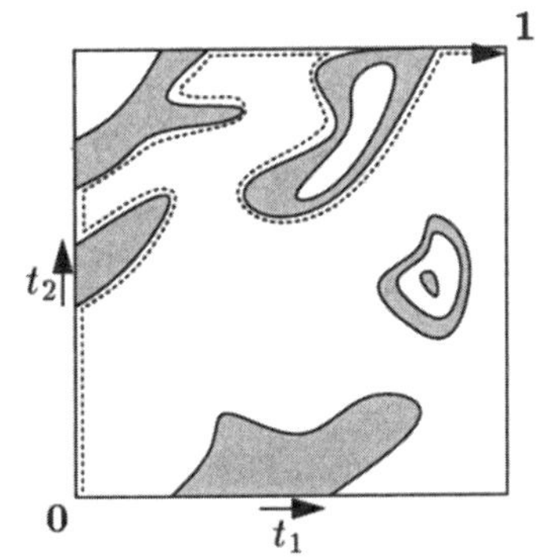

Figure 2: The coordination diagram; forbidden regions are shaded and a feasible multi-path using only permissible multi-configurations is shown dotted.

intersect any obstacle along π_2. Hence, a **0**-**1** path that lies in the free region corresponds to a feasible multi-path; we call this a *feasible* **0**-**1** *path*. Notice that the diagonal from **0** to **1** is by definition a feasible **0**-**1** path. This means that **0** and **1** lie in the same component of the free region. Since they both lie on the boundary of U, they must lie in the same component of the boundary of the free region as well. Hence, there is a feasible **0**-**1** path along the boundary of the free region, as illustrated in Fig. 2. Any point on such a **0**-**1** path corresponds to a permissible multi-configuration: the point either lies on the boundary of U, in which case one of the robots is at its start or goal configuration, or it lies on the boundary of a forbidden region, in which case the robots touch each other. □

Before we continue we make two remarks about the proof. First, the **0**-**1** path that we find is not necessarily monotone in either t_1 or t_2. This may seem like a problem, because it seems to mean that we move back in time. But what it really means it that we move back along the path π_1 or π_2, which is allowed. Second, it is important to realize that we do not have a feasible multi-path available at the start of the algorithm—otherwise we would already be done—so we cannot compute the coordination diagram used in the proof. But we can use it to prove the correctness of our approach.

We now know that we can solve the motion-planning problem by looking at only a subspace of the composite configuration space $\mathcal{CS}_1 \times \mathcal{CS}_2$. This subspace consists of five *configuration-space slices*, or *slices* for short.

- In the first slice, R_2 is free to move and R_1 is stationary at its start configuration; here we can simply consider R_1 as an additional obstacle. We denote this configuration-space slice by $\mathcal{CS}_{1,s}$; its dimension is d_2, the number of degrees of freedom of R_2.
- In $\mathcal{CS}_{1,g}$, the second slice, R_2 is again free to move and R_1 is stationary, but this time R_1 is at its goal configuration. Again, R_1 is an additional obstacle, and the dimension of the slice is d_2.
- The third slice, $\mathcal{CS}_{2,s}$, is defined analogously to $\mathcal{CS}_{1,s}$, with the roles of R_1 and R_2 reversed. Its dimension is d_1.
- The fourth slice, $\mathcal{CS}_{2,g}$, is defined analogously to $\mathcal{CS}_{1,g}$, with the roles of R_1 and R_2 reversed. Its dimension is d_1.
- The fifth slice, $\mathcal{CS}_{\text{contact}}$, is a configuration space for the *contact robot* $\langle R_1 R_2 \rangle$, which is the robot composed of R_1 and R_2 where R_1 and R_2 are required to touch each other. The requirement that R_1 and R_2 should touch implies that the number of degrees of freedom is one less than the composite robot would normally have, so the dimension of $\mathcal{CS}_{\text{contact}}$ is $d_1 + d_2 - 1$.

Linking the configuration-space slices. Of course we cannot treat the five slices completely separately; a feasible multi-path using only permissible multi-configurations will in general switch between slices a number of times. In the example of Fig. 1, for instance, the first part of the path lies in $\mathcal{CS}_{1,s}$, then (when R_2 reaches R_1) a switch is made to $\mathcal{CS}_{\text{contact}}$, and finally a switch is made to $\mathcal{CS}_{2,g}$. We have to connect the slices to make such switches possible. We do this by identifying certain *transition points* in each slice. These points correspond to configurations that are represented by a point in one of the other slices as well. For a given transition point in a slice, we call the point in another slice that corresponds to the same configuration its *twin* in the other slice. Thus if we travel along a curve in one slice and reach a transition point, we may continue in the other slice from its twin. Next we explain which transition points we use.

First we observe that no switches can occur between $\mathcal{CS}_{1,s}$ and $\mathcal{CS}_{1,g}$, because R_1 cannot go from its start to its goal configuration instantaneously. (If $s_1 = g_1$ then $\mathcal{CS}_{1,s} = \mathcal{CS}_{1,g}$, so we can discard one of these configuration-space slices.) Similarly, no switches can occur between $\mathcal{CS}_{2,s}$ and $\mathcal{CS}_{2,g}$.

There is only one point in $\mathcal{CS}_{1,s}$ where we can step to $\mathcal{CS}_{2,s}$, namely at s_2. Its twin in $\mathcal{CS}_{2,s}$ is s_1. This transition is not needed, however, because there is no reason to ever come back to the initial configuration and make a switch there. Similarly, we need not add transition points to step from $\mathcal{CS}_{1,g}$ to $\mathcal{CS}_{2,g}$, because when we arrive at that point we have solved the problem.

We do need to add transition points from $\mathcal{CS}_{1,s}$ to $\mathcal{CS}_{2,g}$, namely the point $g_2 \in \mathcal{CS}_{1,s}$ and its twin $s_1 \in \mathcal{CS}_{2,g}$. Similarly, we need to add $s_2 \in \mathcal{CS}_{1,g}$ and its twin $g_1 \in \mathcal{CS}_{2,s}$ to the collection of transition points.

The difficulty lies in defining transition points to step from $\mathcal{CS}_{\text{contact}}$ to one of the other slices. The problem is that there is an infinite number of configurations common to $\mathcal{CS}_{\text{contact}}$ and, say, $\mathcal{CS}_{1,s}$. We want to add only a limited number of transition points, while at the same time ensuring that no essential connectivity is lost. To achieve this we add the following transition points.

The free part of the slice $\mathcal{CS}_{1,s}$ consists of a number of cells (d_2-dimensional features), which are bounded by parts of constraint hypersurfaces. We call the (d_2-1)-dimensional features of a cell the *patches* of the cell. A patch is a part of the boundary of some configuration-space obstacle (recall that R_1 is now regarded as an additional obstacle); patches are by definition path-connected. Each patch corresponds to configurations where R_2, the robot which is free to move in the slice we are considering, either touches an obstacle or R_1. For each patch on the boundary of a cell in the free space that corresponds to configurations where R_1 and R_2 touch, we take an arbitrary point on the patch as a transition point (together with its twin in $\mathcal{CS}_{\text{contact}}$).

The following lemma shows that the transition points we defined are sufficient to capture the connectivity between the slices.

Lemma 2 *If there is a feasible multi-path for given start and goal configurations for R_1 and R_2, then there is a feasible multi-path whose corresponding curve in the configuration space $\mathcal{CS}_1 \times \mathcal{CS}_2$ lies entirely in the union of the five slices defined above and switches between slices at transition points.*

Proof: Let s_1, g_1, s_2, and g_2 be start and goal configurations such that there is a feasible multi-path for R_1 and R_2. Lemma 1 states that there is a feasible multi-path that uses permissible multi-configurations only, that is, lies in the union of the five slices. Consider such a multi-path $\Pi = \{\pi_1, \pi_2\}$. It is possible that Π switches between configuration-space slices at points other than transition points. We modify Π so that it only switches between slices at transition points.

Switches that are not at transition points can only occur between $\mathcal{CS}_{\text{contact}}$ and one of the other four configuration-space slices. Assume without loss of generality that Π switches to $\mathcal{CS}_{1,s}$. By definition, the switch must occur at a configuration where R_1 and R_2 touch and, moreover, R_1 is at its start configuration. Hence, this configuration is represented by a point p on a patch of some free-space cell in $\mathcal{CS}_{1,s}$. Let q be the transition point chosen for that patch. Furthermore, let p' and q' be the twins in $\mathcal{CS}_{\text{contact}}$ of p and q. Because p and q are on the same patch, there is a curve on that patch connecting them. Such a curve corresponds to a motion of R_2 that keeps it in contact with R_1. Hence, this motion is also represented by some curve connecting p' to q' in $\mathcal{CS}_{\text{contact}}$. This means that instead of stepping from $\mathcal{CS}_{\text{contact}}$ to $\mathcal{CS}_{1,s}$ at the point p', we can first follow the curve from p' to q' in $\mathcal{CS}_{\text{contact}}$, then follow the link between q' and q, and finally move back from q to p. This proves that the transition points provide all the necessary connectivity. □

The algorithm. We combine the ideas above with an algorithm of Basu et al. [2]. This algorithm computes a roadmap of a given semi-algebraic set—of the free configuration space in a motion-planning problem, for instance. A *roadmap* is a one-dimensional subspace—a graph embedded in $\mathcal{CS}$—that captures the connectivity of the free configuration space. If the number of obstacles is n and the dimension of the configuration space is d, then the algorithm constructs the roadmap in $O(n^{d+1})$ time. The algorithm by Basu et al. allows to connect a given point to the roadmap in $O(n)$ time. We will use this algorithm to include the transition points in the roadmap. In both time bounds, the constant of proportionality depends on the algebraic degree of the constraints.

Now consider the setting where we have two robots, R_1 and R_2. We use the method of Basu et al. to construct a roadmap in each of the five configuration-space slices defined above. Since the five configuration-space slices have dimensions at most $d_1 + d_2 - 1$, we can compute these roadmaps in time $O(n^{d_1+d_2})$. It remains to link the roadmaps in the five configuration-space slices. To this end, we first include in the roadmap of each configuration-space slice the set of transition points defined for that slice—taking $O(n)$ time per point. (We explain below how to find the transition points.) This gives us five graphs, each capturing the connectivity of one of the free spaces. Finally, we add links between transition points and their twins. We thus obtain one graph, which captures the connectivity of the free space in $\mathcal{CS}_1 \times \mathcal{CS}_2$.

The computation of the transition points is only interesting for transition points between $\mathcal{CS}_{\text{contact}}$ and the other configuration-space slices. Consider the surface of the configuration-space obstacle in $\mathcal{CS}_{\text{contact}}$ corresponding to R_1 at its initial configuration (that is, the configuration in which R_1 is at s_1 and in contact with R_2). This surface is an algebraic variety of dimension $d_2 - 1$. Consider the arrangement induced on it by all the configuration-space obstacles in $\mathcal{CS}_{\text{contact}}$. We place a transition point in those faces of this arrangement that correspond to free configurations of the contact robot. This is done by invoking the algorithm of Basu et al. [3], which yields in $O(n^{d_2})$ time a representative point in each face (of any dimension) of the arrangement. For each of the resulting $O(n^{d_2-1})$ representative points we determine in a brute-force manner whether the point is in the free space; the points that are in the free space are added to the set of transition points. The time needed for this is $O(n^{d_2})$. Hence, the total time to compute the transition points between $\mathcal{CS}_{\text{contact}}$ and $\mathcal{CS}_{1,s}$ is $O(n^{d_2})$.

It follows from the above discussion that the total time to compute all the transition points is $O(n^{\max(d_1,d_2)})$. The connection of all these $O(n^{\max(d_1,d_2)-1})$ transition points to their respective roadmaps takes $O(n^{\max(d_1,d_2)})$ time. As a result, the computation of the final graph, capturing the connectivity of the free space in $\mathcal{CS}_1 \times \mathcal{CS}_2$, takes $O(n^{d_1+d_2})$ time. The graph search needed to finally solve the motion-planning problem takes linear time in the size of the roadmap. As a result, the motion planning problem for a pair of robots can be solved in $O(n^{d_1+d_2})$ time. This leads to the following theorem.

Theorem 3 *Let R_1 and R_2 be two robots, with d_1 and d_2 degrees of freedom respectively, operating in a workspace with n obstacles. Then we can compute a feasible multi-path for a given pair of start and goal configurations for R_1 and R_2 in $O(n^{d_1+d_2})$ time, if it exists.*

4 Three robots

To extend the results of the previous section to the case of three robots, we have to find a suitable definition of permissible multi-configurations. We first generalize the notion of a coordination diagram, a concept from the proof of Lemma 1. We then prove the existence of a certain type of **0-1** path in the coordination diagram—such a curve represents a feasible motion—from which the definition of permissible multi-configurations follows. As in the previous section, the permissible multi-configurations induce a constant number of slices of the configuration space for the multi-robot; the motion-planning problem can then be solved in the union of these slices.

Let $\mathcal{R} = \{R_1, R_2, R_3\}$ denote a collection of three robots operating in a workspace $\mathcal{W}$, which contains a set $\mathcal{C} = \{C_1, \ldots, C_n\}$ of obstacles. Assume that a feasible multi-path exists between the given start and goal configurations of the robots. Let $\Pi = \{\pi_1, \pi_2, \pi_3\}$ denote such a feasible multi-path. The coordination diagram for Π is defined as follows. Let U denote the three-dimensional unit cube $[0,1]^3$. The edges of U incident to the origin are called the *axes* of U. Each axis represents the configuration of one of the robots; the t_1-axis represents the configuration of R_1 along π_1, the t_2-axis represents the configuration of R_2 along π_2, and the

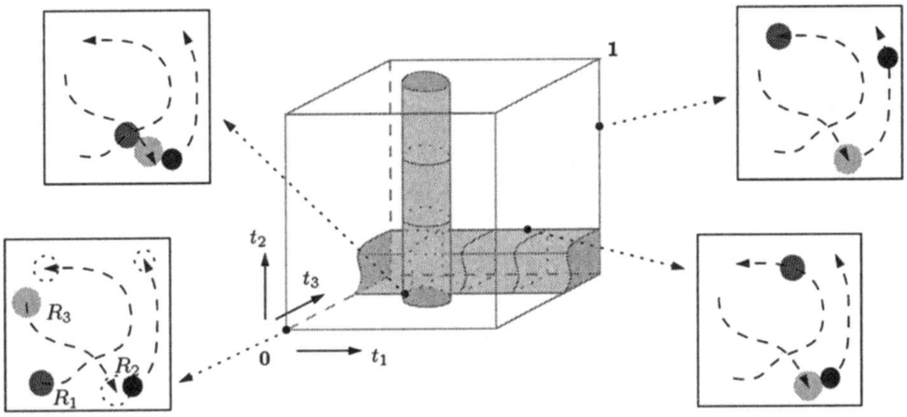

Figure 3: The coordination diagram for three robots.

t_3-axis represents the configuration of R_3 along π_3. Thus a point $(t_1, t_2, t_3) \in U$ corresponds to placing R_i at $\pi_i(t_i)$ for every $1 \leq i \leq 3$. A pair of values (α, β) with $\alpha, \beta \in [0,1]$ is called *i-j forbidden,* for some $i \neq j$, if $R_i[\pi_i(\alpha)]$ intersects $R_j[\pi_j(\beta)]$. Let the *i-j obstacle* B_{ij} be the set of all points (t_1, t_2, t_3) so that $t_i, t_j \in [0,1]$ and (t_i, t_j) is i-j forbidden. (We do not constrain the remaining coordinate to lie in the interval $[0,1]$ for technical reasons.) A point $(t_1, t_2, t_3) \in U$ is called *forbidden* if there is a pair i, j of distinct indices such that (t_i, t_j) is i-j forbidden; otherwise it is called *free.* In other words, a free point corresponds to a placement of each robot at some point along its path so that the robots do not overlap. The *coordination diagram* for Π is the subdivision of U into free and forbidden regions. Fig. 3 shows a coordination diagram for three disk-shaped robots in the plane. By definition, the forbidden region is the union of i-j obstacles truncated to within U. Each obstacle is a cylinder that is the Cartesian product of the forbidden region on some t_it_j-face of U with the remaining axis. (The *t_it_j-face* of U is the 2-dimensional face of U spanned by the t_i- and t_j-axes.) We call path-connected components of these cylinders *coordination-diagram obstacles*, or *cd-obstacles* for short. The set of all cd-obstacles is denoted by $\mathcal{B}$. Note that cd-obstacles have nothing to do with the obstacles in the workspace; they are defined using Π, the feasible multipath, and reflect possible interference between robots if they follow the paths of Π independently.

Define $\mathbf{0} := (0,0,0)$ and $\mathbf{1} := (1,1,1)$ to be the lexicographically smallest and largest vertex of U, respectively. A **0**-**1** *path* is a curve in U from **0** to **1**; it is called *feasible* if it lies in the free region $\mathcal{FR}$ defined as

$$\mathcal{FR} := U \setminus \bigcup \mathcal{B}.$$

By definition of the coordination diagram, the diagonal from **0** to **1** is a feasible **0**-**1** path. We want to prove the existence of a feasible **0**-**1** path along the boundary of the free region, because this will restrict the configurations for the collection of robots. More precisely, we want to find a **0**-**1** path along the so-called *skeleton* of $\mathcal{FR}$, which is the union of the closures of all 1-dimensional features on its boundary. We give a more formal definition below.

In the remainder of this section we will assume that all sets we manipulate are semi-algebraic and correspond to sub-complexes of a common algebraic complex, such as one produced by Collins's cylindrical algebraic decomposition procedure [6]. This "universal" complex is a purely conceptual device and need never be actually computed.

More formally, let $h_1, \ldots, h_6$ be the six planes supporting facets of U. Let H_i be the open halfspace bounded by h_i and not containing U. Put $\mathcal{H} := \{H_1, \ldots, H_6\}$.

We now make some simplifying assumptions about the interaction of the boundaries of the sets in $\mathcal{B} \cup \mathcal{H}$, which will allow us to present a fairly compact and straightforward proof of the main theorem. A less restrictive version of the argument will appear in the full paper.

Assumption A.1 Each $B \in \mathcal{B}$ is bounded by a two-dimensional surface, that is, a (not necessarily smooth) manifold.

Assumption A.2 The set of surfaces $\mathcal{S} := \{h_1, \ldots, h_6\} \cup \{\partial B : B \in \mathcal{B}\}$ has the following property: for any subset of the surfaces with a common intersection, the surfaces intersect transversally and their intersection is a manifold of appropriate dimension, that is, two surfaces intersect in a set of disjoint unbounded curves and simple cycles, three surfaces intersect in a finite set of points, and no four surfaces have a point in common.

Under these assumptions, the *1-skeleton* (or simply *skeleton*) S of $\mathcal{FR}$ has a particularly simple definition, namely:

$$S := \mathcal{FR} \cap \bigcup_{\substack{\sigma, \sigma' \in \mathcal{S} \\ \sigma \neq \sigma'}} \sigma \cap \sigma'.$$

We want to prove that there is a **0**-**1** path along the skeleton S or, in other words, that **0** and **1** are connected by a path in S. We first slightly refine the notion of a cd-obstacle. Take a pair of robots R_i, R_j. The pairs of configurations along π_i and π_j where R_i and R_j intersect define forbidden points on the t_it_j-face of U. Each maximal path-connected subset Q of the forbidden region on the t_it_j-face (such a subset Q is relatively open in the t_it_j-face) defines a cylinder, which we consider a separate cd-obstacle. The subset Q and, hence, the cylinder itself, need not be simply connected. However, any portion of the free space $\mathcal{FR}$ enclosed inside the cylinder is unreachable from **0** and hence can be disregarded completely. This means that we do not eliminate any feasible **0**-**1** paths if we 'fill up' the interior holes in Q to make every cylinder simply connected. The skeleton defined by the filled cylinders is the union of a subset of path-connected components of the skeleton of the original cylinders; it includes $\mathbf{0}$ and $\mathbf{1}$. Therefore Proposition 4 below is sufficient to prove the existence of a path along the skeleton of the coordination diagram. To be able to prove the proposition, we need to make the following additional assumption:

Assumption B (positive clearance): There is a feasible **0-1** path that stays in $\text{Int}\,\mathcal{FR}$, except at the endpoints.

This amounts to requiring that there exist a multi-path along which (except at the start and goal configurations) the robots do not touch and stay away from their starting and goal configurations. We call this a *positive-clearance multi-path.* We suspect that this assumption is not necessary, but we face some technical difficulties if we drop it. We also suspect that if there is a positive-clearance multi-path, we can enforce Assumptions A.1 and A.2 by choosing a suitable multi-path defining the coordination diagram, but we have no formal proof at this point.

Proposition 4 *Let $\mathcal{B}$ be a finite set of simply-connected cylinders as above in the three-dimensional unit cube U satisfying Assumption A and such that there is a* **0-1** *path in the free region $\mathcal{FR} := U \setminus \bigcup \mathcal{B}$ satisfying Assumption B. Then there is a* **0-1** *path along the skeleton S of $\mathcal{FR}$.*

We first sketch the line of attack. We argue by contradiction. If there is no **0-1** path along the skeleton, there must be a collection of curves (in the proof we actually use more general sets) on the surface of $\mathcal{FR}$ that separates the surface so that $\mathbf{0}$ and $\mathbf{1}$ lie in different components when the curves are removed. Then we view $\mathcal{FR}$ and its complement as two topological spaces glued together along the boundary of $\mathcal{FR}$, *with the exception of these curves.* Since the boundary of $\mathcal{FR}$ is separated by the curves, a path connecting the two points outside the cube and another one connecting them in $\mathcal{FR}$ together form a loop that cannot be contracted to a point in the topological space obtained by removal of these curves from $\mathbb{R}^3$. We continue by proving that the curves are not "tangled" in the sense that each of them can be contracted to a point in a cd-obstacle without meeting the other curves or the loop, and then removed. Thus an incremental process removes all curves, yielding $\mathbb{R}^3$. We argue that the non-contractibility of the loop we constructed above is not affected by this process—it should remain non-contractible. We then have a contradiction, since $\mathbb{R}^3$, being simply connected, admits no non-contractible loops.

We now give a more formal version of this discussion. We will need the following topological facts. Recall that we are assuming that all sets in question correspond to subcomplexes of a common algebraic complex.

Fact 5 *Let V and W be closed sets with W path-connected and $V \cup W$ contractible to a point. Then the number of path-connected components of $V \cap W$ equals the number of path-connected components of V.*

Fact 6 *Consider closed sets V, W. Let p, q be points lying in different path-connected components of $V \cap W$ and connected by a path in V and by another path in W. Then the closed path obtained by concatenating the two paths is not contractible to a point in $V \cup W$.*

Proof of Proposition 4: Put $\mathcal{FR}' = \text{Cl}(\mathbb{R}^3 \setminus \mathcal{FR})$. Note that $\mathcal{FR}'$ is necessarily path-connected, as it is the union of the exterior of the cube and a set of infinite cylinders. We apply Fact 5 to $\mathcal{FR}$ and $\mathcal{FR}'$, which is possible since both sets are closed, $\mathcal{FR}'$ is path-connected, and their union $\mathbb{R}^3$ is contractible to a point. Hence the numbers of path-connected components of $\mathcal{FR}$ and of $\partial\mathcal{FR}$ coincide. As each component of $\mathcal{FR}$ has at least one associated component of $\partial\mathcal{FR}$, this implies that the boundary of each component of $\mathcal{FR}$ is path-connected. Hence, $\mathbf{0}$ and $\mathbf{1}$ which lie in the same connected component of $\mathcal{FR}$ and on $\partial\mathcal{FR}$ must lie in the same path-connected component of $\partial\mathcal{FR}$.

Since $\mathbf{0}, \mathbf{1}$ lie on ∂U, there exists a path σ' connecting them in $\text{Cl}(\mathbb{R}^3 \setminus U) \subset \mathcal{FR}'$, and by assumption there is a path σ connecting them in $\mathcal{FR}$. Later it will be convenient to assume that σ' avoids the boundary of U except at its endpoints. Assumption B guarantees that σ may be drawn entirely in $\text{Int}\,\mathcal{FR}$ except at $\mathbf{0}$ and $\mathbf{1}$.

For a contradiction, suppose that $\mathbf{0}, \mathbf{1}$ do not lie in the same component of S. Consider $X := \mathcal{FR} \cup_S \mathcal{FR}'$, that is, the complex obtained by identifying $\mathcal{FR}$ and $\mathcal{FR}'$ along S *only*; $\partial\mathcal{FR} \setminus S$ is present in X in two copies, one as part of $\partial\mathcal{FR}$ and one as part of $\partial\mathcal{FR}'$. Apply Fact 6 to $V := \mathcal{FR}$, $W := \mathcal{FR}'$, $V \cup W = X$, $V \cap W = S$. As $\mathbf{0}$ and $\mathbf{1}$ are in different components of $V \cap W$, the paths σ, σ' that connect them in $\mathcal{FR}$ and $\mathcal{FR}'$, respectively, form a non-contractible loop η in X. Essentially, X is the same as $\mathbb{R}^3$ with (a sufficiently small open neighborhood of) $\partial\mathcal{FR} \setminus S$ removed. In other words, our assumption of no path connecting $\mathbf{0}$ to $\mathbf{1}$ in S implies that η is not contractible to a point in $\mathbb{R}^3 \setminus (\partial\mathcal{FR} \setminus S)$.

Consider a single path-connected component f of $\partial\mathcal{FR} \setminus S$. It is a two-dimensional face f of $\partial\mathcal{FR}$ contained fully in the boundary of a single cd-obstacle, say B, or in a single plane, say h_1. We start by considering the former case. Then f is a possibly non-simply-connected portion of ∂B. Suppose it has $i \geq 0$ "holes." We augment f by i topological disks, each of which has as its boundary a non-contractible simple loop 'surrounding' exactly one hole in f and is otherwise fully contained in $U \cap \text{Int}\,B$. This is possible since $U \cap B$ is simply connected and $f \subset U \cap \partial B$. We can arrange for the i disks to be pairwise disjoint. As the disks are contained in U and avoid $\mathcal{FR}$, they miss η. Let f^* be f augmented with the disks; it is contractible to a point by construction. Adding the disks to $\partial\mathcal{FR} \setminus S$ cannot make η contractible in the complement of the set, as this shrinks the complement without touching η. However, f^* avoids $(\partial\mathcal{FR} \setminus S) \setminus f$ and η, and thus can be removed, in the sense that, if η was not contractible in the complement of $f^* \cup (\partial\mathcal{FR} \setminus S)$, it must also not be contractible in the complement of $(\partial\mathcal{FR} \setminus S) \setminus f^* = (\partial\mathcal{FR} \setminus S) \setminus f$.

We now repeat this argument for the case where f is contained in the plane h_1. Here too, f need not be simply connected. However, it is fully contained in a 2-face of U. The holes in f occur in the interior of this face. As we have assumed that η avoids the boundary of U except at $\mathbf{0}$ and $\mathbf{1}$, the holes in f are not touched by η. In particular, as above, we augment f with pairwise disjoint disks, one for each hole, such that each disk interior is outside of U and does not touch η while its boundary surrounds a hole in f. Again, augmenting f with the disks does not affect contractibility of the complement, and the augmented face f^* avoids η. However, f^* is contractible to a point and disjoint from the other sets of $\partial\mathcal{FR} \setminus S$, so its removal will not affect contractibility of η.

To summarize, we have shown that a path-connected component of $\partial\mathcal{FR} \setminus S$ can be eliminated from this set without making η contractible in the complement of the set. The process can now be iterated and, as the number of components is finite, we arrive at the situation where all components have been eliminated but η is still non-contractible.

In other words, we have constructed a loop η that is not contractible to a point in $\mathbb{R}^3$. Contradiction. □

The existence of a **0-1** path along the skeleton of the coordination diagram implies the existence of a multi-path that uses certain types of multi-configurations only, as explained next. The skeleton consists of the 1-dimensional features of the free region, which are the intersections of $2-k$ boundaries of cd-obstacles with k facets of U, for some k with $0 \le k \le 2$. This means that points on the skeleton correspond to the following type of *permissible multi-configurations*:

> There are k, for some $k = 0, 1, 2$, robots placed at either their start or goal configuration, and $2-k$ pairs of robots that are in contact.

There are several different ways in which a permissible multi-configuration can be achieved. We mention a few of the possibilities. One type of permissible multi-configuration is that R_1 is at its start configuration and R_2 and R_3 form a contact robot. Another type is that R_1 is at its goal configuration, R_2 touches R_1, and R_3 is unconstrained. Although the number of types of permissible multi-configurations is fairly large, it is a constant. Each type of permissible multi-configuration gives rise to a configuration space slice, as in the previous section. Since every touching pair of robots reduces the number of degrees of freedom of the multi-robot by one, and every robot fixed at its start or goal reduces this number by at least one (namely, by the number of its degrees of freedom) the dimension of each slice is at most $d-2$.

We now proceed as in the previous section: each type of permissible multi-configuration gives a configuration-space slice in which we compute a roadmap, we connect the roadmaps at a suitable collection of transition points (when the number of configurations common to two slices is infinite, we again choose a point on each patch bounding the free parts in such slices) and search the resulting graph to finally find a feasible multi-path. We obtain the following result.

Theorem 7 *Let R_1, R_2, R_3 be three robots operating in a workspace with n obstacles. Then we can compute a feasible multi-path for a given triple of start and goal configurations for R_1, R_2, R_3 in $O(n^{d-1})$ time, where d is the sum of the degrees of freedom of the three robots, if a feasible multi-path with positive clearance exists and Assumption A is satisfied.*

5 Bounded-reach robots low-density environments

Define the *size* of an object o, denoted by size(o) to be the radius of the smallest ball enclosing it. We say that the workspace $\mathcal{W}$ with the obstacle set $\mathcal{C}$ is a *λ-low-density environment* [7, 16, 18, 20] if for any ball B, the number of obstacles $C \in \mathcal{C}$ with $\text{size(C)} \ge \text{radius(B)}$ that intersect B is at most λ. (Our definition is the one used by de Berg et al. [7], and is slightly different from the earlier definition in [18].) If λ is a (small) constant, we say that $\mathcal{W}$ has *low density*.

In this section we apply the ideas from the previous sections to obtain efficient solutions to the motion-planning problem for the so-called *bounded-reach* robots [19] moving in low-density workspaces. Bounded-reach robots are basically robots that are not too large compared to the obstacles. More precisely, they are defined as follows. Let p_R be an arbitrary reference point in a robot R. Then the *reach* of R, denoted by reach(R), is defined as the radius of the smallest ball centered at p_R that contains R, no matter in which configuration R is. For instance, if R consists of two links of length 1 that are both attached to the origin, and the reference point is the tip of one of the links, then the reach of R is 2. (If the reference point were the origin then the reach would be 1. For any two reference points in R, however, the two values reach(R) can be at most a factor of two apart.) A bounded-reach robot is now defined as a robot R with

$$\text{reach(R)} \le c \cdot \min_{C \in \mathcal{C}} \{\text{size(C)}\},$$

where c is a (small) constant.

Van der Stappen et al. [20] have shown that the free space of a bounded reach robot moving in a low-density workspace has $O(n)$ complexity, irrespective of the number of degrees of freedom of the robot or the dimension of its workspace. They also show how to compute in $O(n \log n)$ time a decomposition of the configuration space, after which a feasible path between given start and goal configurations can be found in $O(n)$ time. Hence, the total amount of time to solve the motion-planning problem is $O(n \log n)$.

If we have two or more bounded-reach robots then we cannot use the result of van der Stappen et al. directly, because the corresponding multi-robot does not have bounded reach. But when the number of robots is two or three, the multi-robots we have to consider when we restrict ourselves to permissible multi-configurations do have free configuration spaces with linear complexity.

Indeed, consider first the case of two bounded-reach robots. Clearly, when one robot is fixed at its start or goal configuration and the other robot moves, the free-space complexity of the moving robot is linear, because it has bounded reach. (The moving robot can be very large with respect to the fixed robot, which we now view as an additional obstacle, but this does not influence the asymptotic bound.) The remaining type of multi-configuration is where the two robots form a contact robot. Here the free-space complexity is also linear, because the reach of a contact robot is bounded by the reach of one of the constituent robots plus twice the reach of the other constituent robot.

For three robots a similar argument shows that the free-space complexities of all the configuration-space slices we have to consider are linear. The only case which is slightly different from the cases we have for two robots is when one robot is fixed, another robot moves in contact with the fixed robot, and the third robot is free. In this case the multi-robot formed by the second and third robot does not have bounded reach. But it follows from results of van der Stappen [18] that the robot that moves in contact with the fixed robot has only a constant number of combinatorially distinct critical configurations. Combined with the fact that the third robot had bounded reach and, hence, a linear number of critical configurations, this shows that the multi-robot consisting of the second and third robot has a free space of linear complexity.

We have argued that in all the configuration-space slices we have to consider the complexity of the free space is linear. Moreover, we can use the algorithm of van der Stappen et al. to compute decompositions of these free spaces. The nature of these decompositions is such that we can use them to compute a suitable collection of transition points. The details of this (which are not very difficult) will be given in the full paper. This leads to the following result. Notice that the time bound of our algorithm is, like in the case of a single bounded-reach robot, independent of both the numbers of degrees of freedom of the robots and of the dimension of the workspace.

Theorem 8 *(i) Let R_1 and R_2 be two bounded-reach robots operating in a low-density workspace with n obstacles. Then we can compute a feasible multi-path for a given pair of start and goal configurations for R_1 and R_2 in $O(n \log n)$ time, if it exists.*

(ii) Let R_1, R_2, and R_3 be three bounded-reach robots operating in a low-density workspace with n obstacles. Then we can compute a feasible multi-path for a given triple of start and goal configurations for R_1, R_2, and R_3 in $O(n \log n)$ time, if a feasible multi-path with positive clearance exists and Assumption A is satisfied.

6 Concluding remarks

We presented a general technique to plan the motions of pairs and triples of robots sharing the same workspace. By combining this with the roadmap algorithm of Basu et al. [2], we obtain an algorithm with $O(n^d)$ running time for a pair of robots and $O(n^{d-1})$ running time for a triple of robots, where d is the sum of the degrees of freedom of the individual robots. These bounds are one and two orders of magnitude, respectively, faster than what the standard method would give. For triples of robots we need some technical assumptions, for instance, that a solution with positive clearance exists. Our approach becomes especially effective for bounded-reach robots operating in low-density workspaces. In this case our algorithm runs in $O(n \log n)$, both for two and for three robots, irrespective of their numbers of degrees of freedom.

Any collection of $m > 3$ robots can be viewed as a triple of robots, one of which is a multi-robot consisting of $m-2$ robots. Hence, our result for triples can be applied to four or more robots. This will reduce the dimension of the configuration space one has to consider by two. Greater savings are possible if for any m there would be a **0**-**1** path along the 1-skeleton of the m-dimensional coordination diagram. If this is true (which we have been unable to prove or disprove), the resulting reduction in dimension would be $m-1$ and the motion-planning problem for m robots would be solvable in $O(n^{d-m+2})$ time.

References

[1] R. Alami, F. Robert, F. Ingrand, and S. Suzuki. Multi-robot cooperation through incremental plan merging. In *Proc. IEEE Internat. Conf. Rob. Automat.*, pages 2573–2578, Nagoya, Japan, 1995.

[2] S. Basu, R. Pollack, and M.-F. Roy. Computing roadmaps of semi-algebraic sets on a variety. In F. Cucker and M. Shub (eds.), *Foundations of Computational Mathematics*, pages 1–15, 1997. See also `http://www.math.nyu.edu/faculty/pollack/mf11-19-97.ps`.

[3] S. Basu, R. Pollack, and M.-F. Roy. On computing a set of points meeting every semi-algebraically connected component of a family of polynomials on a variety. *J. Complexity*, 13:28–37, 1997.

[4] J. Canny. *The Complexity of Robot Motion Planning.* ACM – MIT Press Doctoral Dissertation Award Series. MIT Press, Cambridge, MA, 1987.

[5] J. Canny. Computing roadmaps in general semialgebraic sets. *Comput. J.*, 36:409–418, 1994.

[6] G. E. Collins. Quantifier elimination for real closed fields by cylindrical algebraic decomposition. In *Proc. 2nd GI Conference on Automata Theory and Formal Languages*, volume 33 of *Lecture Notes Comput. Sci.*, pages 134–183. Springer-Verlag, Berlin, West Germany, 1975.

[7] M. de Berg, M. J. Katz, A. F. van der Stappen, and J. Vleugels. Realistic input models for geometric algorithms. In *Proc. 13th Annu. ACM Sympos. Comput. Geom.*, pages 294–303, 1997.

[8] M. Erdmann and T. Lozano-Pérez. On multiple moving objects. *Algorithmica*, 2:477–521, 1987.

[9] S. J. Fortune, G. Wilfon, and C. Yap. Coordinated motion of two robot arms. In *Proc. IEEE Internat. Conf. Robot. Autom.*, pages 1216–1223, 1986.

[10] J. E. Hopcroft and G. T. Wilfong. On the motion of objects in contact. Technical Report 84-602, Dept. Comput. Sci., Cornell Univ., Ithaca, NY, May 1984.

[11] J. E. Hopcroft and G. T. Wilfong. Reducing multiple object motion planning to graph searching. *SIAM J. Comput.*, 15:768–785, 1986.

[12] K. Kedem, R. Livne, J. Pach, and M. Sharir. On the union of Jordan regions and collision-free translational motion amidst polygonal obstacles. *Discrete Comput. Geom.*, 1:59–71, 1986.

[13] J.-C. Latombe. *Robot Motion Planning.* Kluwer Academic Publishers, Boston, 1991.

[14] P. A. O'Donnell and T. Lozano-Pérez. Deadlock-free and collision-free coordination of two robotic manipulators. In *Proc. IEEE Internat. Conf. Rob. Automat.*, pages 484–489, Scottsdale, AZ, 1989.

[15] J. T. Schwartz and M. Sharir. On the "piano movers" problem III: coordinating the motion of several independent bodies: the special case of circular bodies moving amidst polygonal barriers. *Internat. J. Robot. Res.*, 2(3):46–75, 1983.

[16] O. Schwarzkopf and J. Vleugels. Range searching in low-density environments. *Inform. Process. Lett.*, 60:121–127, 1996.

[17] M. Sharir and S. Sifrony. Coordinated motion planning for two independent robots. *Ann. Math. Artif. Intell.*, 3:107–130, 1991.

[18] A. F. van der Stappen. *Motion Planning amidst Fat Obstacles.* Ph.D. dissertation, Dept. Comput. Sci., Utrecht Univ., Utrecht, Netherlands, 1994.

[19] A. F. van der Stappen, D. Halperin, and M. H. Overmars. The complexity of the free space for a robot moving amidst fat obstacles. *Comput. Geom. Theory Appl.*, 3:353–373, 1993.

[20] A. F. van der Stappen. M. H. Overmars, M. de Berg, and J. Vleugels. *Motion planning in environments with low obstacle density.* Tech. Rep. UU-CS-1997-19, Dept. Comput. Sci., Utrecht Univ., 1997.

[21] C. W. Warren. Multiple robot path coordination using artificial potential fields. In *Proc. IEEE Internat. Conf. Rob. Automat.*, pages 500–505, 1990.

Constructing Approximate Shortest Path Maps in Three Dimensions*

Sariel Har-Peled†

Abstract

We present a new technique for constructing a data-structure that approximates shortest path maps in $\mathbb{R}^d$. By applying this technique, we get the following two results on approximate shortest path maps in $\mathbb{R}^3$.

(i) Given a polyhedral surface or a convex polytope $\mathcal{P}$ with n edges in $\mathbb{R}^3$, a source point s on $\mathcal{P}$, and a real parameter $0 < \varepsilon \leq 1$, we present an algorithm that computes a subdivision of $\mathcal{P}$ of size $O((n/\varepsilon)\log(1/\varepsilon))$ which can be used to answer efficiently approximate shortest path queries. Namely, given any point t on $\mathcal{P}$, one can compute, in $O(\log(n/\varepsilon))$ time, a distance $\Delta_{\mathcal{P},s}(t)$, such that $d_{\mathcal{P},s}(t) \leq \Delta_{\mathcal{P},s}(t) \leq (1+\varepsilon)d_{\mathcal{P},s}(t)$, where $d_{\mathcal{P},s}(t)$ is the length of a shortest path between s and t on $\mathcal{P}$.

The map can be computed in $O(n^2 \log n + (n/\varepsilon)\log(1/\varepsilon) \log(n/\varepsilon))$ time, for the case of a polyhedral surface, and in $O((n/\varepsilon^3)\log(1/\varepsilon) + (n/\varepsilon^{1.5})\log(1/\varepsilon)\log n)$ time if $\mathcal{P}$ is a convex polytope.

(ii) Given a set of polyhedral obstacles $\mathcal{O}$ with a total of n edges in $\mathbb{R}^3$, a source point s in $\mathbb{R}^3 \setminus int \bigcup_{O \in \mathcal{O}} O$, and a real parameter $0 < \varepsilon \leq 1$, we present an algorithm that computes a subdivision of $\mathbb{R}^3$, which can be used to answer efficiently approximate shortest path queries. That is, for any point $t \in \mathbb{R}^3$, one can compute, in $O(\log(n/\varepsilon))$ time, a distance $\Delta_{\mathcal{O},s}(t)$ that ε-approximates the length of a shortest path from s to t that avoids the interiors of the obstacles. This subdivision can be computed in roughly $O(n^4/\varepsilon^6)$ time.

1 Introduction

The *three-dimensional Euclidean shortest-path problem* is defined as follows: Given a set of pairwise-disjoint polyhedral objects in $\mathbb{R}^3$ and two points s and t, compute the shortest path between s and t which avoids the interiors of the given polyhedral 'obstacles'. This problem has received considerable attention in computational geometry. It was shown to be NP-hard by Canny and Reif [3], and the fastest available algorithms for this problem run in time that is exponential in the total number of obstacle vertices (which we denote by n) [20, 21]. The apparent intractability of the problem has motivated researchers to develop polynomial-time algorithms for computing approximate shortest paths and for computing shortest paths in special cases.

In the *approximate three-dimensional Euclidean shortest-path* problem, we are given an additional parameter $\varepsilon > 0$, and the goal is to compute a path between s and t that avoids the interiors of the obstacles and whose length is at most $(1+\varepsilon)$ times the length of the shortest obstacle-avoiding path (we call such a path an *ε-approximate path*). Approximation algorithms for the three-dimensional shortest path problem were first studied by Papadimitriou [19], who gave an $O(n^4(L + \log(n/\varepsilon))^2/\varepsilon^2)$-time algorithm for computing an ε-approximate shortest path, where L is the number of bits used in each computation. A rigorous analysis of Papadimitriou's algorithm was recently given by Choi et al. [6]. A different approach was taken by Clarkson [7], resulting in an algorithm with roughly $O(n^2/\varepsilon^4)$ running time (the precise result is stated in Theorem 4.1).

The problem of computing a shortest path between two points along the surface of a single convex polytope is an interesting special case of the three-dimensional Euclidean shortest-path problem. Sharir and Schorr [22] gave an $O(n^3 \log n)$ algorithm for this problem, exploiting the property that a shortest path on a polyhedron *unfolds* into a straight line. Mitchell et al. [16] improved the running time to $O(n^2 \log n)$; their algorithm works for non-convex polyhedra (or polyhedral surfaces) as well. Chen and Han [5] gave another algorithm with an improved running time of $O(n^2)$. It is a rather long-standing and intriguing open problem whether the shortest path on a convex polytope can be computed in subquadratic time. This has motivated the problem of finding near-linear algorithms that produce only an approximation of the shortest path. The first result in this direction is by Hershberger and Suri [12]. They present a simple algorithm that runs in $O(n)$ time, and computes a path whose length is at most $2d_P(s,t)$. Using the algorithm of [12], Agarwal et al. [1] present a relatively simple algorithm that computes an ε-approximate shortest path (i.e., a path on ∂P between two points $s, t \in \partial P$ whose length is at most $(1+\varepsilon)d_P(s,t)$), for any prescribed $0 < \varepsilon \leq 1$, where the running time of the algorithm is $O(n\log(1/\varepsilon) + 1/\varepsilon^3)$. In a companion paper [11], we present an improved algorithm,

*This work has been supported by a grant from the U.S.–Israeli Binational Science Foundation. This work is part of the author's Ph.D. thesis, prepared at Tel-Aviv University under the supervision of Prof. Micha Sharir.

†School of Mathematical Sciences, Tel Aviv University, Tel Aviv 69978, Israel; sariel@math.tau.ac.il

SCG 98 Minneapolis Minnesota USA

with $O(n)$ preprocessing time, that answers two-points ε-approximate shortest-path queries in $O((\log n)/\varepsilon^{1.5} + 1/\varepsilon^3)$ time, for *any* pair of points $s, t \in \partial P$. Recently, Varadarajan and Agarwal [24] gave a subquadratic time algorithm that computes a constant approximation to the shortest path on a polyhedral terrain. Other recent works, by Mata and Mitchell [14], and also by Lanthier at al. [13] implement various heuristics for computing approximate shortest paths on weighted terrains (i.e., each face f is being assigned a weight w_f, such that the distance between any two points $a, b \in f$ is $w_f \cdot |ab|$). Those programs give satisfactory results in practice, which are within an order of magnitude better than their worst case analysis.

In this paper, extending our work in [11], we present a new general technique for constructing a data-structure that one can use to answer ε-approximate shortest-path queries, for a source point s and approximation factor $\varepsilon > 0$ fixed in advance. Using this technique, we solve two problems involving approximate shortest path maps in $\mathbb{R}^3$.

Approximate shortest path maps. The exact algorithms of [16, 22] receive as input a convex polytope or a polyhedral surface $\mathcal{P}$, and a fixed source point s on $\mathcal{P}$, and compute a map (i.e., a subdivision of $\mathcal{P}$) of complexity $\Theta(n^2)$, that can be used to answer (exact) shortest path queries from s to any point on $\mathcal{P}$ (along $\mathcal{P}$) in $O(\log n)$ time (such a query reports the *length* of the shortest path; reporting the path itself might require more time). This shortest path map can be stored in linear space, for the case of a convex polytope, by using a persistent data-structure, see [17]. However, the time required to compute this compact representation of the shortest path map is quadratic in the worst case. This raises the problem of computing a map of near-linear size for approximate shortest-path queries from s. We show in Section 3 that this is indeed possible: Given a polyhedral surface $\mathcal{P}$ with n edges in $\mathbb{R}^3$, a source point $s \in \mathcal{P}$, and a prescribed $0 < \varepsilon \leq 1$, there exists a map (a subdivision of $\mathcal{P}$) of complexity $O((n/\varepsilon)\log(1/\varepsilon))$, such that for any $t \in \mathcal{P}$, one can compute the length of an ε-approximate shortest path between s and t on $\mathcal{P}$ in $O(\log(n/\varepsilon))$ time, by locating t in the map.

We present an algorithm that constructs such an approximation map in $O(n^2 \log n + (n/\varepsilon)\log(1/\varepsilon)\log(n/\varepsilon))$ time, for the case of a polyhedral surface, and in $O((n/\varepsilon^3)\log(1/\varepsilon) + (n/\varepsilon^{1.5})\log(1/\varepsilon)\log n)$ time, for the case of a convex polytope. Note that if $\mathcal{P}$ is a convex polytope than our previous result [11] provides an alternative structure with similar properties. However, the dependence of the query time on ε is much better in the method we present here.

Approximate spatial shortest path maps. In Section 4, we present a similar result for ε-approximate shortest paths among polyhedral obstacles in $\mathbb{R}^3$. Let $\mathcal{O}$ be a set of polyhedral obstacles in $\mathbb{R}^3$ with a total of n edges, s a source point in $\mathbb{R}^3$, and $0 < \varepsilon \leq 1$ a parameter. We show that there exists a spatial subdivision $\mathcal{M}$ of $\mathbb{R}^3$, such that for any $t \in \mathbb{R}^3$, one can compute, in $O(\log(n/\varepsilon))$ time, the length of an ε-approximate shortest path between s and t, that avoids the interiors of the obstacles, by performing a spatial point-location query with t in $\mathcal{M}$. The space needed to compute and preprocess $\mathcal{M}$ for spatial point-location is $O(n^2/\varepsilon^{4+\delta})$, for any $\delta > 0$, and the preprocessing time is

$$O\left(\frac{n^4}{\varepsilon^2}\left(\frac{\beta(n)}{\varepsilon^4}\log\frac{n}{\varepsilon} + \log(n\rho)\log(n\log\rho)\right)\log\frac{1}{\varepsilon}\right),$$

where ρ is the ratio of the length of the longest edge in $\mathcal{O}$ to the Euclidean distance between s and t, $\beta(n) = \alpha(n)^{O(\alpha(n))^{O(1)}}$, and $\alpha(n)$ is the extremely slowly growing inverse of the Ackermann function. This algorithm uses the algorithm of Clarkson [7], that computes an ε-approximate shortest path, between two given points, in

$$O\left(\frac{n^2}{\varepsilon^4}\beta(n)\log\frac{n}{\varepsilon} + n^2\log(n\rho)\log(n\log\rho)\right)$$

time.

The paper is organized as follows. Section 2 introduces the notion of a *distance function*, and show how to compute a "small-size" additive weighted Voronoi diagram that enables us to ε-approximate this function. We present two applications of this result. In Section 3, we present the algorithm mentioned above for constructing a map for approximate shortest-path queries from a fixed source on a convex polytope or a polyhedral surface in $\mathbb{R}^3$. In Section 4 we present the algorithm mentioned above for constructing a spatial subdivision for approximate shortest-path queries from a fixed source among polyhedral obstacles in $\mathbb{R}^3$. We conclude in Section 5 by mentioning a few open problems.

2 Approximating a Distance Function by a Weighted Voronoi Diagram

In this section, we introduce the notion of a distance function, and show how to compute a "small-size" additive weighted Voronoi diagram that approximates it up to a factor of $1 + \varepsilon$. We use this result in Sections 3 and Section 4 to derive our two main results.

Definition Let $\mathcal{I}$ be a subset of $\mathbb{R}^d$. A function $f : \mathcal{I} \to \mathbb{R}$ is a *distance function* on $\mathcal{I}$ if:

- $f(x) + f(y) \geq |xy|$, for any $x, y \in \mathcal{I}$.
- $f(x) + |xy| \geq f(y)$, for any straight segment $xy \subseteq \mathcal{I}$,

where $|xy|$ denotes the Euclidean distance between x and y.

Thus, a distance function has to satisfy two types of triangle inequalities. Since these inequalities are satisfied by the Euclidean distance from any fixed point, a distance function can be regarded as a certain generalization of the Euclidean distance.

Example 2.1 *Figure 1 illustrates some geometric restriction imposed on a univariate distance function.*

Example 2.2 *(i) Let $\mathcal{P}$ be a polyhedral surface in $\mathbb{R}^3$, and let s be a source point on $\mathcal{P}$. For any $t \in \mathcal{P}$, let $d_{\mathcal{P},s}(t)$ denote the length of a shortest path between s and t on $\mathcal{P}$. It is easy to verify that $d_{\mathcal{P},s}(t)$ is a distance function on $\mathcal{P}$.*

(ii) Let $\mathcal{O}$ be a collection of pairwise-disjoint polyhedral obstacles in $\mathbb{R}^3$, and let s be a point in $FP(\mathcal{O}) = \mathbb{R}^3 \setminus \bigcup_{O \in \mathcal{O}} \operatorname{int} O$. Let $FP(\mathcal{O}, s)$ denote the connected component of $FP(\mathcal{O})$ that contains s (i.e., the set of all the points in $\mathbb{R}^3$ that can be connected to s by a path that avoids the interiors of the obstacles of $\mathcal{O}$).

For any $t \in FP(\mathcal{O}, s)$, we denote by $d_{\mathcal{O},s}(t)$ the length of a shortest path between s and t, that avoids the interior of the obstacles of $\mathcal{O}$. Clearly, $d_{\mathcal{O},s}$ is a distance function on $FP(\mathcal{O}, s)$.

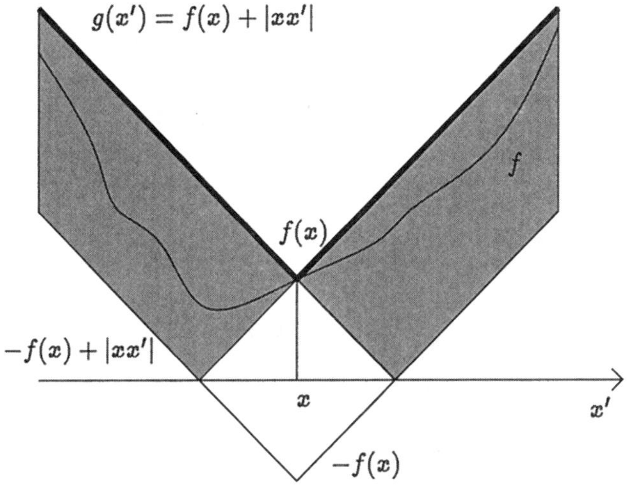

Figure 1: The graph of the distance function f must lie inside the gray area. This implies that the function $g(x') = f(x) + |xx'|$ approximates $f(x')$ "well" in a small neighborhood of x, and for sufficiently large values of x'.

Definition A pair $\mathcal{S} = (S, w)$ is a *weighted set of points* if $S = \{p_1, \ldots, p_m\}$ is a finite set of points in $\mathbb{R}^d$, and $w(\cdot)$ is a function assigning non-negative weights to the points of S. We define the distance of a point p from the point p_i to be $V_{(p_i, w(p_i))}(p) = |pp_i| + w(p_i)$. We define $V_{\mathcal{S}}(p) = \min_{i=1}^{m} V_{(p_i, w(p_i))}(p)$. The function $V_{\mathcal{S}}(p)$ induces a natural subdivision $\mathcal{V}_{\mathcal{S}}$ of $\mathbb{R}^d$ into cells, known as the (additive) *weighted Voronoi diagram* of $\mathcal{S}$, such that the i-th cell is the locus of all points closest to p_i in this distance function. As is well known, in the planar case, $\mathcal{V}_{\mathcal{S}}$ has complexity $O(m)$, and it can be computed in $O(m \log m)$ time (see [10]).

Remark 2.3 For $d \geq 3$, the complexity of an additive weighted Voronoi diagram of m points in $\mathbb{R}^d$ is $O\left(m^{\lfloor d/2 \rfloor + 1}\right)$. This follows by reducing the computation of the diagram to the computation of a convex hull in $d+2$ dimensions. Furthermore, one can compute the diagram in $O(m^{\lfloor d/2 \rfloor + 1})$ time. See [2].

We next show how to approximate a distance function $f(\cdot)$ by a weighted Voronoi diagram. First, we compute a global minimum p_0 of f. As illustrated in Figure 1, the function $V_{(p_0, f(p_0))}(p)$ approximates $f(p)$ "well" for p sufficiently close to, or sufficiently far from p_0. In other words, f is well-approximated in these regions by the weighted Voronoi diagram of the single site p_0 with weight $f(p_0)$. By adding extra sites to the diagram, we can make the distance induced by the resulting diagram an ε-approximation to f everywhere, as will be shown next.

Definition Given a point $p \in \mathbb{R}^d$, and $r \geq 0$, let $B(p, r)$ denote the closed ball of radius r centered at p, and let $\overline{B}(p, r)$ denote the set $\mathbb{R}^d \setminus B(p, r)$. For $r' > r$, let $\mathcal{A}(p, r, r')$ denote the annulus (or shell) $B(p, r') \setminus B(p, r)$.

The following sequence of technical lemmas provide the basis for approximating a given distance function by a weighted Voronoi diagram. The following lemmas are stated for arbitrary dimension d. We will apply them with $d = 1, 2$, or 3.

Lemma 2.4 *Let $\mathcal{I}$ be a convex subset of $\mathbb{R}^d$, f a distance function defined over $\mathcal{I}$, and $\mathcal{S} = (S, w)$ a weighted set, such that $S \subseteq \mathcal{I}$ and $f(x) \leq w(x)$, for all $x \in S$. Then $f(t) \leq V_{\mathcal{S}}(t)$, for all $t \in \mathcal{I}$.*

Proof: Let t be any point of $\mathcal{I}$, and let x denote the point of S realizing $V_{\mathcal{S}}(t)$. Then $V_{\mathcal{S}}(t) = w(x) + |tx| \geq f(x) + |tx| \geq f(t)$. ■

Formalizing the intuition behind Figure 1, we have the following:

Lemma 2.5 *Let $\mathcal{I}$ be a convex subset of $\mathbb{R}^d$, f a distance function defined over $\mathcal{I}$, $0 < \varepsilon \leq 1$ a parameter, p a point in $\mathcal{I}$, and w a real number such that $f(p) \leq w \leq (1+\varepsilon/8) f(p)$. Then $f(t) \leq V_{(p,w)}(t) \leq (1+\varepsilon) f(t)$, for all t in*

$$\mathcal{I} \cap \left(B\left(p, \frac{\varepsilon f(p)}{8}\right) \cup \overline{B}\left(p, \frac{6 f(p)}{\varepsilon}\right) \right).$$

Proof: The first inequality $f(t) \leq V_{(p,w)}(t)$ follows immediately from Lemma 2.4.

As for the other inequality, for $t \in \mathcal{I} \cap B\left(p, \frac{\varepsilon f(p)}{8}\right)$, we have

$$\begin{aligned} \frac{w}{(1+\varepsilon/8)} - \frac{\varepsilon w}{8} &\leq f(p) - |pt| \leq f(t) \leq V_{(p,w)}(t) \\ &= w + |pt| \leq w + \frac{\varepsilon w}{8}. \end{aligned}$$

However,

$$\begin{aligned} \frac{w + \varepsilon w/8}{\frac{w}{1+\varepsilon/8} - \varepsilon w/8} &= \frac{1+\varepsilon/8}{\frac{1}{1+\varepsilon/8} - \varepsilon/8} \leq \frac{1+\varepsilon/8}{1 - \varepsilon/8 - \varepsilon/8} \\ &= \frac{1+\varepsilon/8}{1-\varepsilon/4} \leq 1 + \varepsilon, \end{aligned}$$

since $\varepsilon \leq 1$. Thus $V_{(p,w)}(t) \leq (1+\varepsilon) f(t)$, for all $t \in \mathcal{I} \cap B(p, \varepsilon f(p)/8)$.

For $t \in \mathcal{I} \cap \overline{B}(p, 6 f(p)/\varepsilon)$, we have

$$|pt| - w \leq |pt| - f(p) \leq f(t) \leq V_{(p,w)}(t) = |pt| + w.$$

However,

$$\frac{|pt| + w}{|pt| - w} = 1 + \frac{2w}{|pt| - w} \leq 1 + \frac{2w}{3w/\varepsilon - w} \leq 1 + \frac{2w}{2w/\varepsilon} = 1 + \varepsilon,$$

since $|pt| \geq 6 f(p)/\varepsilon \geq 3w/\varepsilon$. Thus $V_{(p,w)}(t) \leq (1+\varepsilon) f(t)$, for any such t. ■

Lemma 2.6 *Let $\mathcal{I}$ be a convex subset of $\mathbb{R}^d$, f a distance function defined over $\mathcal{I}$, $0 < \varepsilon \leq 1$ a parameter, and p a point in $\mathcal{I}$. Then for any $t \in \mathcal{I} \cap B(p, \varepsilon f(p)/9)$ and any number w_t such that $f(t) \leq w_t \leq (1+\varepsilon/8) f(t)$, we have $f(p) \leq V_{(t, w_t)}(p) \leq (1+\varepsilon) f(p)$.*

Proof: Since $|pt| \leq \varepsilon f(p)/9$, it follows that $f(t) \geq f(p) - |pt| \geq f(p)(1 - \varepsilon/9)$.
Thus,

$$\frac{\varepsilon f(t)}{8} \geq \frac{\varepsilon}{8} f(p) \left(1 - \frac{\varepsilon}{9}\right) \geq \frac{\varepsilon f(p)}{9} \geq |pt|,$$

implying that $p \in B(t, \varepsilon f(t)/8)$. By Lemma 2.5, we have $f(p) \leq V_{(t, w_t)}(p) \leq (1+\varepsilon) f(p)$. ■

The preceding lemmas suggest the following strategy for constructing an approximation of a distance function f over (a convex portion of) $\mathbb{R}^d$: Pick a point p, such that $f(p)$ is close to the global minimum of f. The Voronoi diagram $\mathcal{V}_{(p,w)}$ approximates f "well" near p and outside a larger ball centered at p, where w is an approximation of $f(p)$. We approximate f in the space between those two balls by partitioning it into concentric spherical shells whose radii form an increasing geometric progression, and by covering each shell by a uniform grid (whose unit length increases with the radius of the shell). In this manner, the number of points needed is only a function of ε (the approximation factor) and d.

When approximating a distance function on a convex subset $\mathcal{I}$ of $\mathbb{R}^d$, we have to cope with the possibility that sites might be placed outside $\mathcal{I}$. We overcome this by projecting all such sites onto the boundary of $\mathcal{I}$.

Definition Let $\mathcal{I}$ be a convex subset in $\mathbb{R}^d$, and let x be a point in $\mathbb{R}^d$. Let $\nu(x, \mathcal{I})$ denote the projection of x onto $\mathcal{I}$; that is, $\nu(x, \mathcal{I})$ is the closest point (in the Euclidean distance) in $\mathcal{I}$ to x. Clearly, if $x \in \mathcal{I}$ then $\nu(x, \mathcal{I}) = x$.

When x is fixed, we call $\nu(x, \mathcal{I})$ the *hook point* of $\mathcal{I}$.

Definition Let $r \geq r' > 0$ be real numbers, let p be a point in $\mathbb{R}^d$, and let $\mathcal{I}$ be a convex set in $\mathbb{R}^d$. We denote by $S(p, \mathcal{I}, r, r')$ the set $\mathcal{I}_{r'} \cap B(p, r) \cap ((r'/\sqrt{d})\mathbb{Z}^d)$, where $\mathbb{Z}^d$ is the integer lattice, and $\mathcal{I}_{r'} = \cup_{q \in \mathcal{I}} B(q, r')$ is the set of all point in $\mathbb{R}^d$ that are at distance at most r' from some point of $\mathcal{I}$. Clearly, $|S(p, \mathcal{I}, r, r')| = O((r/r')^d)$ (with a constant of proportionality depending on d).

The following technical lemma shows how to pick the sites of the additive weighted Voronoi diagram, so that it ε-approximates a given distance function.

Lemma 2.7 *Let $\mathcal{I}$ be a convex subset of $\mathbb{R}^d$, $f : \mathcal{I} \to \mathbb{R}$ a distance function, $0 < \varepsilon \leq 1$ a parameter, c a positive constant, and p a point in $\mathcal{I}$ such that $f(t) \geq f(p)/c$, for all $t \in \mathcal{I}$. Then one can compute a set S in $\mathcal{I}$ of size $O((1/\varepsilon)^d \log(1/\varepsilon))$ (the constant of proportionality depends on c), such that $p \in S$, and for any weight function w on S satisfying $f(x) \leq w(x) \leq (1+\varepsilon/8)f(x)$, for all $x \in S$, we have $f(t) \leq V_{\mathcal{S}}(t) \leq (1+\varepsilon)f(t)$, for all $t \in \mathcal{I}$, where $\mathcal{S} = (S, w)$.*

Proof: Let w_p be any number satisfying $f(p) \leq w_p \leq (1+\varepsilon/8)f(p)$.

Let $r_i = (2^i + 1)w_p$, for $i = 1, \ldots, m$, where $m = \lceil \log_2(6/\varepsilon) \rceil$. Let $\mathcal{A}_1 = B(p, r_1)$, let $\mathcal{A}_i = \mathcal{A}(p, r_{i-1}, r_i)$, for $i = 2, \ldots, m$, and let $\mathcal{A}_{m+1} = \overline{B}(p, r_m)$. Clearly, $\mathcal{I} = \cup_{i=1}^{m+1}(\mathcal{I} \cap \mathcal{A}_i)$.

Let $r_1' = \varepsilon w_p/(18c)$, and let $r_i' = \varepsilon 2^{i-1} w_p/9$, for $i = 2, \ldots, m$. Let $S_i' = \mathcal{A}_i \cap S(p, \mathcal{I}, r_i, r_i')$, for $i = 1, \ldots, m$. Let $S = \{p\} \cup \bigcup_{i=1}^m S_i$, where $S_i = \left\{ \nu(x, \mathcal{I}) \;\middle|\; x \in S_i' \right\}$, for $i = 1, \ldots, m$. See Figure 2 for an illustration of the set S.

Let w be any weight function, such that $f(x) \leq w(x) \leq (1+\varepsilon/8)f(x)$, for any $x \in S$, and let $\mathcal{S} = (S, w)$.

We claim that $\mathcal{S}$ is the required weighted set. Indeed, let $t \in \mathcal{I}$. If $t \in \mathcal{A}_{m+1}$ then $|pt| \geq (2^m + 1)w_p \geq 6f(p)/\varepsilon$. Thus $t \in \overline{B}(p, 6f(p)/\varepsilon)$ and by Lemmas 2.4 and 2.5, we have $f(t) \leq V_{\mathcal{S}}(t) \leq V_{(p,w_p)}(t) \leq (1+\varepsilon)f(t)$.

If $t \in B(p, r_m)$, let $\mathcal{A}_i$ be the shell containing t. Let x' be the closest point to t in S_i'. Let $x = \nu(x', \mathcal{I})$. By the definition of ν and S_i' and by the convexity of $\mathcal{I}$, we have $|tx| \leq |tx'| \leq r_i'$ (see Figure 3).

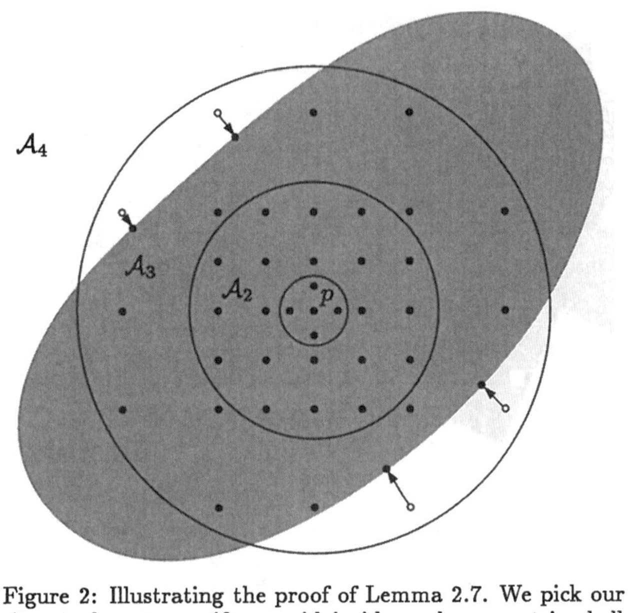

Figure 2: Illustrating the proof of Lemma 2.7. We pick our sites to be on a uniform grid inside each concentric shell around p.

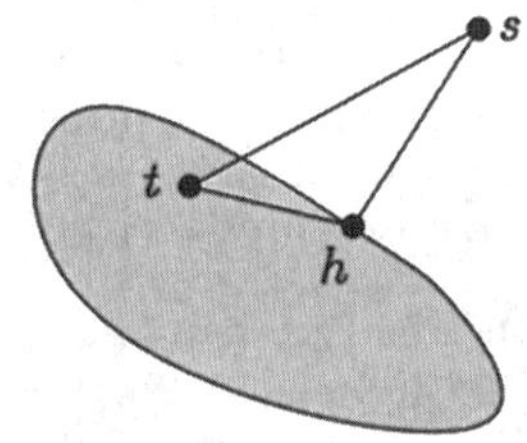

Figure 3: $|st| \geq |ht|$ for $h = \nu(s, \mathcal{I})$

If $i = 1$ then the inequality $f(t) \geq f(p)/c \geq w_p/2c$ implies that $|tx| \leq r_1' = \varepsilon w_p/(18c) \leq \varepsilon f(t)/9$. Thus $x \in B(t, \varepsilon f(t)/9)$.

If $i > 1$ then we also have $|tx| \leq r_i' = \varepsilon 2^{i-1} w_p/9 \leq \varepsilon f(t)/9$, since $f(t) \geq |pt| - f(p) \geq (2^{i-1}+1)w_p - f(p) \geq 2^{i-1} w_p$. Thus $x \in B(t, \varepsilon f(t)/9)$.

By Lemma 2.6 and Lemma 2.4, we have $f(t) \leq V_{\mathcal{S}}(t) \leq V_{(x,w(x))}(t) \leq (1+\varepsilon)f(t)$.

As for the size of S, we have

$$
\begin{aligned}
|S| &= O\left(\sum_{i=1}^{m} \left(\frac{r_i}{r_i'}\right)^d\right) \\
&= O\left(\left(\frac{3w_p}{\varepsilon w_p/(18c)}\right)^d + \sum_{i=2}^{m} \left(\frac{(2^i+1)w_p}{\varepsilon 2^{i-1} w_p/9}\right)^d\right) \\
&= O\left(\frac{1}{\varepsilon^d} \log \frac{1}{\varepsilon}\right).
\end{aligned}
$$

■

To approximate a distance function $f(\cdot)$ using the constructive proof of Lemma 2.7, we need to find a point which is, within a constant factor, a global minimum of f over the given range. The following lemma shows that this can be easily done if f has a known zero point outside the given range.

Lemma 2.8 *Let $\mathcal{I}'$ be a subset of $\mathbb{R}^d$, $f : \mathcal{I}' \to \mathbb{R}$ a distance function, $0 < \varepsilon \leq 1$ a parameter, $\mathcal{I}$ a convex subset of $\mathcal{I}'$,* *and s a point in $\mathcal{I}' \setminus \mathcal{I}$ such that $f(s) = 0$. Then $f(t) \geq f(h)/2$, for all $t \in \mathcal{I}$, where $h = \nu(s, \mathcal{I})$.*

Proof: Let t be any point in $\mathcal{I}$. Since h is the closest point in $\mathcal{I}$ to s and $\mathcal{I}$ is convex, it easily follows that $|st| \geq |ht|$, see Figure 3. Since $f(t) = f(t) + f(s) \geq |st|$, we have $f(t) \geq |ht|$. Moreover, the segment ht is a contained in $\mathcal{I}$, implying that $f(h) \leq f(t) + |ht| \leq 2f(t)$. ∎

Remark 2.9 The lemma also holds when $s \in \mathcal{I}$, but then it only yields the trivial bound $f(t) \geq 0$ for all $t \in \mathcal{I}$. Then, of course, s is the required global minimum.

Remark 2.10 Let $\mathcal{I}'$ be a set in $\mathbb{R}^d$, let $\mathcal{I}$ be a convex subset of $\mathcal{I}'$, $f : \mathcal{I}' \to \mathbb{R}$ a distance function, $0 < \varepsilon \leq 1$ a parameter, and s a point of $\mathcal{I}'$ such that $f(s) = 0$. Computing a weighted set $\mathcal{S} = (S, w)$ such that the weighted Voronoi diagram induced on $\mathcal{I}$ approximates f up to a factor of $(1 + \varepsilon)$, can be accomplished by following the proof of Lemma 2.7, in four stages:

- **(i)** Compute the point $p = \nu(s, \mathcal{I})$. By Lemma 2.8, p is "almost" a minimum of f on $\mathcal{I}$.
- **(ii)** Compute an $(\varepsilon/8)$-approximation w_p to $f(p)$ (as prescribed in the lemma).
- **(iii)** Construct the set S of points in $\mathcal{I}$, as prescribed in the proof.
- **(iv)** Approximate the distance function values of all the points of S, up to a factor of $1 + \varepsilon/8$, and use these values as the weights for the points of S.

Remark 2.11 The set of points S produced in the proof of Lemma 2.7 is made out of $O(\log(1/\varepsilon))$ subsets (i.e., $S_1, \ldots, S_m$) such that $f(x) \leq cf(t)$, for all $x, t \in S_i$, for $1 \leq i \leq m$, where c is an appropriate constant. This property enables us, in the case of shortest paths on a convex polytope, to approximate the value of the distance function to all the points of S_i simultaneously, yielding a more efficient algorithm. See Remark 3.2 for the details.

Remark 2.12 Let $\mathcal{I}'$ be a subset of $\mathbb{R}^d$, and let f, g be two distance functions defined over $\mathcal{I}'$. It is easy to verify that $h(x) = \max(f(x), g(x))$ is also a distance function. This implies that the distance function induced by any furthest neighbor Voronoi diagram of a finite set of points in $\mathbb{R}^d$, is a distance function in $\mathbb{R}^d$. Hence, by Lemma 2.7, we have:

Corollary 2.13 *Any furthest neighbor Voronoi diagram of points in $\mathbb{R}^d$, can be ε-approximated by a (nearest neighbor) weighted Voronoi diagram, having $O((1/\varepsilon^d)\log(1/\varepsilon))$ sites.*

The following is a strengthening of Lemma 2.7, by noticing that in the cases we are going to apply it to, our distance function is the length of a shortest path from a fixed source point to the given point of $\mathcal{I}$. In such cases, if the source point lies outside $\mathcal{I}$, then a shortest path connecting any point in $\mathcal{I}$ to our source point, must first pass through the boundary $\partial\mathcal{I}$. We next show that if we are able to ε-approximate the distance on the boundary of $\mathcal{I}$, then we can trivially ε-approximate the distance function to all the points of $\mathcal{I}$.

Definition Let $\mathcal{I}$ be a convex polytope in $\mathbb{R}^d$. We call a function $f : \mathcal{I} \to \mathbb{R}$ *boundary-induced* on $\mathcal{I}$, if f is a distance function, and for any $t \in \mathcal{I}$, there exists a point $x \in \partial\mathcal{I}$ such that $f(t) = f(x) + |tx|$.

Definition Given a convex polytope $\mathcal{I}$ in $\mathbb{R}^d$, we denote by $\phi(\mathcal{I})$ the set of all the facets ($(d-1)$-faces) of $\mathcal{I}$.

Lemma 2.14 *Let $\mathcal{I}$ be a convex polytope in $\mathbb{R}^d$, $f : \mathcal{I} \to \mathbb{R}$ a boundary-induced distance function, and $0 < \varepsilon \leq 1$ a parameter. For any facet F of $\mathcal{I}$, let $\mathcal{S}(F) = (S_F, w_F)$ be a weighted set of points in F, such that $f(t) \leq V_{\mathcal{S}(F)}(t) \leq (1+\varepsilon)f(t)$, for all $t \in F$. Then $f(t) \leq V_{\mathcal{S}}(t) \leq (1+\varepsilon)f(t)$, for all $t \in \mathcal{I}$, where $\mathcal{S} = (S, w) = \cup_{F \in \phi(\mathcal{I})} \mathcal{S}(F)$.*

Proof: For any $t \in \mathcal{I}$, let x be the point in $\partial\mathcal{I}$ satisfying $f(t) = f(x) + |tx|$, and let F be a facet of $\mathcal{I}$ that contains x. By Lemma 2.4, we have

$$\begin{aligned} f(x) + |tx| &= f(t) \leq V_{\mathcal{S}}(t) \leq V_{\mathcal{S}}(x) + |tx| \\ &\leq V_{\mathcal{S}(F)}(x) + |tx| \leq (1+\varepsilon)f(x) + |tx| \\ &\leq (1+\varepsilon)(f(x) + |tx|) = (1+\varepsilon)f(t). \end{aligned}$$

∎

Remark 2.15 Let $\mathcal{I}'$ be a set in $\mathbb{R}^d$, let $\mathcal{I}$ be a convex subset of $\mathcal{I}'$, $f : \mathcal{I}' \to \mathbb{R}$ a boundary-induced distance function, $0 < \varepsilon \leq 1$ a parameter, and s a point of $\mathcal{I}'$ such that $f(s) = 0$. Computing a weighted set $\mathcal{S} = (S, w)$ such that the weighted Voronoi diagram induced on $\mathcal{I}$ approximates f up to a factor of $(1+\varepsilon)$, can be done by applying the algorithm described in Remark 2.10 for each facet of $\mathcal{I}$. By Lemma 2.14, the union of all these weighted sets has the required properties.

3 Approximate Shortest-Path Map on a Polyhedral Surface in $\mathbb{R}^3$

Let $\mathcal{P}$ be a given polyhedral surface in $\mathbb{R}^3$ with n edges, let s be a source point on $\mathcal{P}$, and let $0 < \varepsilon \leq 1$ be a given parameter. In this section, we give an algorithm for constructing an approximation map on $\mathcal{P}$ of complexity $O((n/\varepsilon)\log(1/\varepsilon))$, such that given any $t \in \mathcal{P}$, one can compute in $O(\log(n/\varepsilon))$ time a distance $\Delta_{\mathcal{P}}(s,t)$ satisfying $d_{\mathcal{P}}(s,t) \leq \Delta_{\mathcal{P}}(s,t) \leq (1+\varepsilon)d_{\mathcal{P}}(s,t)$.

Although the following description is rather technical, one has to bear in mind that it is a straightforward implementation of the technique of Section 2. Namely, for each edge of our domain (polyhedral terrain, or a convex polytope) we compute a "small" set of points, we approximate the (geodesic) distance from the source point to all those points, and we construct the weighted additive Voronoi diagram that those points induce on each face of the domain.

Definition A *polyhedral surface* $\mathcal{P}$ in $\mathbb{R}^3$ is the union of a collection of planar polygonal faces, with their edges and vertices, such that each edge is incident to at most two faces and any pair of faces intersect either at a common edge, a common vertex, or not at all. A face is a simple closed polygon (i.e., it contains its boundary), and an edge is a closed segment (i.e., it contains its endpoints). Without loss of generality we assume that all the faces are triangular (since simple polygons may be triangulated in linear time [4] and the number of new edges introduced by the triangulation is linear in the number of vertices). We also assume that $\mathcal{P}$ is connected.

A *polyhedral terrain* is a polyhedral surface that intersects every vertical line in at most a single point.

Definition Given a polyhedral surface $\mathcal{P}$ in $\mathbb{R}^3$, and any two points s, t on $\mathcal{P}$, we denote by $d_{\mathcal{P},s}(t)$ the length of a shortest path between s and t on $\mathcal{P}$.

As noted in Example 2.2 (i), $d_{\mathcal{P},s}(\cdot)$ is a distance function on $\mathcal{P}$. Moreover, if F is a face of $\mathcal{P}$ and $s \notin F$ then $d_{\mathcal{P},s}$ is boundary induced on F. (If $s \in F$ then $d_{\mathcal{P},s}(t)$ is the Euclidean distance $|st|$)

The following theorem is the main result of this section.

Theorem 3.1 *Let $\mathcal{P}$ be a polyhedral surface in $\mathbb{R}^3$ with n edges, s a source point on $\mathcal{P}$, and $0 < \varepsilon \leq 1$ a real parameter. Then there exists a subdivision Π of $\mathcal{P}$ of complexity $O((n/\varepsilon)\log(1/\varepsilon))$, which facilitates ε-approximate shortest path queries from s on $\mathcal{P}$. That is, for any query point t on $\mathcal{P}$, one can compute, in $O(\log(n/\varepsilon))$ time, a distance $\Delta_{\mathcal{P}}(s,t)$, such that $d_{\mathcal{P},s}(t) \leq \Delta_{\mathcal{P}}(s,t) \leq (1+\varepsilon)d_{\mathcal{P},s}(t)$.*

The map can be computed in $O(n^2 \log n + (n/\varepsilon)\log(1/\varepsilon)\log(n/\varepsilon))$ time, if $\mathcal{P}$ is an arbitrary polyhedral surface, and in $O((n/\varepsilon^3)\log(1/\varepsilon) + (n/\varepsilon^{1.5})\log(1/\varepsilon)\log n)$ time, if $\mathcal{P}$ is a convex polytope.

The space used by the algorithm is $O((n/\varepsilon)\log(1/\varepsilon))$, if $\mathcal{P}$ is either a convex polytope or a polyhedral terrain, and $O(n^2 + (n/\varepsilon)\log(1/\varepsilon))$ otherwise.

Proof: For each face F of $\mathcal{P}$ that does not contain s, we construct a weighted Voronoi diagram that approximates $d_{\mathcal{P},s}$ on F. By Remark 2.15, this can be done by constructing a weighted Voronoi diagram on each edge of $\mathcal{P}$, as outlined in Remark 2.10.

For the general case, we compute the exact shortest-path map of s on $\mathcal{P}$, using the algorithm of [16], in $O(n^2 \log n)$ time. The exact map enables us to compute the shortest distance from s to any point of $\mathcal{P}$ in $O(\log n)$ time. Thus, computing the distances from s to the n hooks of the edges of $\mathcal{P}$ takes additional $O(n \log n)$ time. The hook point of an edge is the closest point of the edge to s, and it can be computed in $O(1)$ time.

For the convex case, we approximate the distances on $\mathcal{P}$ to all the hooks on the edges of $\mathcal{P}$, up to a factor of $(1+\varepsilon/8)$. This takes $O(n/\varepsilon^3 + (n/\varepsilon^{1.5})\log n)$ time, using the algorithm of [1, Sec. 6].

For each edge e of $\mathcal{P}$, we compute a set S_e of $O((1/\varepsilon)\log(1/\varepsilon))$ points on e, as specified in the proof of Lemma 2.7, taking the corresponding p to be the hook of e and w_p to be the approximated (exact in the non-convex case) distance along $\mathcal{P}$ from s to p. Let $S = \cup_e S_e$, taken over all edges e of $\mathcal{P}$.

We now compute (or approximate) the distances from s to all the points in S. For the non-convex case, this can be done in $O((n/\varepsilon)\log(1/\varepsilon)\log n)$ time, using the exact shortest path map.

For the convex case, we compute approximate distances from s to all points of S, up to a factor of $(1+\varepsilon/8)$. Using the observation of Remark 2.11, we partition S into $O(n\log(1/\varepsilon))$ sets, each of size $O(1/\varepsilon)$, such that the required distances to the points in each such set are within a fixed constant factor of each other (namely, for each edge e of $\mathcal{P}$, the set S_e is decomposed into $O(\log(1/\varepsilon))$ sets, as in the proof of Lemma 2.7). Using the algorithm described in Remark 3.2 below, we can compute the distances from s to all the points of S in $O((n/\varepsilon^3)\log(1/\varepsilon)+(n/\varepsilon^{1.5})\log(1/\varepsilon)\log n)$ time.

Next, we compute, for each face F of $\mathcal{P}$, the weighted Voronoi diagram induced by the weighted points of $\mathcal{S}$ that lie on ∂F. This takes $O((n/\varepsilon)\log^2(1/\varepsilon))$ overall time (see [10]), since each face contains $O((1/\varepsilon)\log(1/\varepsilon))$ points of $\mathcal{S}$. Let Π be the resulting map, consisting of the union of all those facial Voronoi diagrams.

By Lemma 2.7 and Lemma 2.14, the map Π on $\mathcal{P}$ has the required properties. Moreover, we can preprocess each face F of $\mathcal{P}$ in $O((1/\varepsilon)\log^2(1/\varepsilon))$ time, such that point location queries on F can be answered in $O(\log(1/\varepsilon))$ time (see [18]). Overall, this preprocessing takes $O((n/\varepsilon)\log^2(1/\varepsilon))$ time.

To answer a approximate shortest path query for a query point q, the algorithm must locate the face of $\mathcal{P}$ containing q. If $\mathcal{P}$ is a polyhedral terrain, we project the terrain into the xy-plane and preprocess it, in $O(n\log n)$ time, for planar point location. If $\mathcal{P}$ is a convex polytope, it can be preprocessed in linear time to answer point location queries in $O(\log n)$ time (see [9]). Otherwise, we preprocess $\mathcal{P}$ for spatial point-location in $O(n^2\log n)$ time, and $O(n^2)$ space, with $O(\log n)$ query time, using the algorithm of [23].

Given any query point q on $\mathcal{P}$, the algorithm computes the face F of $\mathcal{P}$ that contains q in $O(\log n)$ time. Locating the face of the subdivision Π that contains q takes an additional $O(\log(1/\varepsilon))$ time. Thus, ε-approximate shortest path queries for $\mathcal{P}$ can be answered in $O(\log(n/\varepsilon))$ time. (If the face containing q is already known, the query time reduces to $O(\log(1/\varepsilon))$.) ■

Definition Let P be a convex body in $\mathbb{R}^3$. An *outer path* of P is a curve γ connecting two points on ∂P and disjoint from the interior of P.

Remark 3.2 Let P be a convex polytope in $\mathbb{R}^3$, s a source point on P, T a set of points on P, and $0 < \varepsilon \leq 1$ a prescribed parameter. One can ε-approximate the length of the shortest path from s to all the points of T on P, in $O((n+|T|)/\varepsilon^3+((n+|T|)/\varepsilon^{1.5})\log(n+|T|))$ time, by adding the points of T as vertices to P and by using the algorithm of [1, Sec. 6].

The algorithm of [1] works by computing an approximation polytope for each point of T, and by computing the exact distance from s to the point on this polytope.

Moreover, suppose that T can be partitioned into m sets $T_1, \ldots, T_m$, such that $d_{P,s}(t) \leq c \cdot d_{P,s}(t')$, for all $t, t' \in T_i$ and for each $i = 1, \ldots, m$, where c is a prescribed constant, and all the points of T_i belong to the same edge of P, for any fixed $i = 1, \ldots, m$. Then it is possible to speed up the above algorithm, as follows. Instead of constructing an approximation polytope for each destination point separately, we construct an approximation polytope that can be used to approximate the distances from s to all the points of T_i, for $i = 1, \ldots, m$. This is done by ensuring that all the points of T_i lie on the boundary of the approximation polytope calculated by the algorithm, which can be enforced by intersecting it with a supporting plane of P passing through the edge containing the points of T_i (adding at most one new face to the approximation polytope). We also need to use a more refined approximation polytope, so as to achieve the claimed error bound, but since c is a constant this does not change the asymptotic complexity of the algorithm. See [1] for the technical details.

This improves the running time to

$$O\left(n + \frac{m}{\varepsilon^3} + \frac{m}{\varepsilon^{1.5}}\log n + \frac{|T|}{\varepsilon^{1.5}}\right),$$

by constructing an approximation polytope for the points of T_i (in $O(\frac{m}{\varepsilon^{1.5}} \log n)$ time), computing the exact distance map from the source point on the approximation polytope (in $O(1/\varepsilon^3)$ time), and extracting the shortest path to each point of T_i, repeating all this, for $T_1, \ldots, T_m$. Moreover, for each point $t \in T_i$, the algorithm computes a polygonal outer path of P, made out of $O(1/\varepsilon^{1.5})$ segments, that realizes the approximated distance.

Remark 3.3 The algorithm of [16] works for arbitrary polyhedral surfaces; in particular, it is not restricted to polyhedral terrains. Thus, the algorithm of Theorem 3.1 also works for general polyhedral surfaces.

Remark 3.4 For a convex polytope P with n edges in $\mathbb{R}^3$, one can compute an approximation map that can be used to compute an outer path that realizes the approximate distance. This is done by modifying the algorithm of Theorem 3.1, such that it stores an outer path from the source point to each of the constructed sites, where the outer path realizes its ε-approximate distance. Such a path is readily available from the procedure used to compute the approximate distance to the site, and the complexity of such a path is $O(1/\varepsilon^{1.5})$ (See [1]). The space needed to store the extended approximation map is $O(n/\varepsilon^{2.5} \log(1/\varepsilon))$, and the computation time remains $O((n/\varepsilon^3)\log(1/\varepsilon) + (n/\varepsilon^{1.5})\log(1/\varepsilon) \log n)$.

The new map can be used to answer approximate shortest path queries, in $O(\log(n/\varepsilon))$ time, and also compute, in additional $O(1/\varepsilon^{1.5})$ time, an outer path of the convex polytope realizing this distance. Such an outer path can be projected onto the boundary of the convex polytope, in additional $O(n \log(1/\varepsilon) + 1/\varepsilon^3)$ time, resulting in a path on ∂P which is an ε-approximation to the shortest path, see [1]. Note however that the performance of the enhanced data structure is poorer both in terms of storage and query time.

4 Constructing Spatial Approximate Shortest-Path Maps in $\mathbb{R}^3$

Let $\mathcal{O}$ be a collection of pairwise-disjoint polyhedral obstacles in $\mathbb{R}^3$, s a source point in $\mathbb{R}^3 \setminus int \bigcup \mathcal{O}$, and $0 < \varepsilon \leq 1$ a parameter. In this section, we present an algorithm for preprocessing $\mathcal{O}$ such that for any point in $\mathbb{R}^3$ (or, more precisely, for any 'free' point that can be reached from s without penetrating into an obstacle) one can compute, in $O(\log(n/\varepsilon))$ time, a distance $\Delta_{\mathcal{O},s}(t)$ satisfying $d_{\mathcal{O},s}(t) \leq \Delta_{\mathcal{O},s}(t) \leq (1+\varepsilon)d_{\mathcal{O},s}(t)$, where $d_{\mathcal{O},s}(t)$ is the length of a shortest path between s and t that avoids the interiors of the obstacles.

The preprocessing time of the algorithm is roughly $O(n^4/\varepsilon^6)$, which shows that the problem of approximating the distance from a single source to all the 'free' points in $\mathbb{R}^3$, is not much harder (computationally) than approximating the distance between any specific pair of points (which can be done in roughly $O(n^2/\varepsilon^4)$ time, see [7]). In fact, for a fixed source point and many destination points, our algorithm will actually be faster. The problem of computing the *exact* distance between two points in $\mathbb{R}^3$ among polyhedral obstacles is NP-hard, as shown by Canny and Reif [3], and the fastest available algorithms for this problem run in time that is exponential in the total number of obstacle vertices [20, 21, 22].

Definition Let $\mathcal{O}$ be a collection of pairwise-disjoint polyhedral obstacles with a total of n edges in $\mathbb{R}^3$, and s a source point in $FP(\mathcal{O}) = \mathbb{R}^3 \setminus int \bigcup_{O \in \mathcal{O}} O$. Let $FP(\mathcal{O}, s)$ denote the set of all points in $FP(\mathcal{O})$ that can be connected to s by a path that avoids the interiors of the obstacles of $\mathcal{O}$.

For any $t \in FP(\mathcal{O}, s)$, we denote, as above, by $d_{\mathcal{O},s}(t)$ the length of a shortest path between s and t, that avoids the interiors of the obstacles of $\mathcal{O}$.

As noted in Example 2.2 (ii), $d_{\mathcal{O},s}(\cdot)$ is a distance function over $FP(\mathcal{O}, s)$, and for any convex set $\mathcal{I} \subseteq FP(\mathcal{O}, s)$ such that $s \notin \mathcal{I}$, the function $d_{\mathcal{O},s}(\cdot)$ is boundary induced over $\mathcal{I}$.

Theorem 4.1 (Clarkson [7]) *Given a set $\mathcal{O}$ of polyhedral obstacles in $\mathbb{R}^3$, and points s and t, an ε-approximate path between s and t that does not penetrate into any obstacle in $\mathcal{O}$ can be computed in*

$$O\left(\frac{n^2}{\varepsilon^4}\beta(n)\log\frac{n}{\varepsilon} + n^2\log(n\rho)\log(n\log\rho)\right)$$

time, where n is the number of obstacle edges, and ρ is the ratio of the length of the longest edge in $\mathcal{O}$ to the Euclidean distance between s and t, $\beta(n) = \alpha(n)^{O(\alpha(n))^{O(1)}}$, and $\alpha(n)$ is the inverse of the Ackermann function.

The following theorem is the main result of this section.

Theorem 4.2 *Let $\mathcal{O}$ be a collection of pairwise-disjoint polyhedral obstacles with n edges in $\mathbb{R}^3$, s a source point in $FP(\mathcal{O})$, and $0 < \varepsilon < 1$ a parameter. Then a subdivision $\mathcal{M}$ of $FP(\mathcal{O}, s)$ of complexity $O(n^2/\varepsilon^{4+\delta})$, for any $1 > \delta > 0$, can be computed in*

$$O\left(\frac{n^4}{\varepsilon^2}\left(\frac{\beta(n)}{\varepsilon^4}\log\frac{n}{\varepsilon} + \log(n\rho)\log(n\log\rho)\right)\log\frac{1}{\varepsilon}\right)$$

time, where ρ, and $\beta(n)$ are as above.

For any query point $t \in FP(\mathcal{O}, s)$, one can compute in $O(\log(n/\varepsilon))$ time a distance $\Delta_{\mathcal{O},s}(t)$, such that $d_{\mathcal{O},s}(t) \leq \Delta_{\mathcal{O},s}(t) \leq (1+\varepsilon)d_{\mathcal{O},s}(t)$.

Proof: First, we partition $FP(\mathcal{O})$ into $O(n^2)$ vertical prisms. This can be easily done by erecting a vertical wall from each edge of the obstacles. For an edge e of the obstacles, such a wall is the set of all points in $FP(\mathcal{O})$ that lie on vertical rays emanating from the edge, and not intersecting the obstacles. Let $\mathcal{M}'''$ denote the resulting partition of $FP(\mathcal{O})$. It is easy to verify that the complexity of $\mathcal{M}'''$ is $O(n^2)$, and that it can be computed in $O(n^2 \log n)$ time.

We refine $\mathcal{M}'''$, by further partitioning each cell of $\mathcal{M}'''$ into vertical triangular prisms. This is done by projecting each cell of $\mathcal{M}'''$ into the xy-plane, and by triangulating the resulting polygon, in $O(m \log m)$ time (see [18]), where m is the number of vertices of the polygon. For each new edge created, we erect a corresponding vertical wall inside the cell. Let $\mathcal{M}''$ be the resulting subdivision of $FP(\mathcal{O})$. Clearly, the complexity of $\mathcal{M}''$ remains $O(n^2)$, and it can be computed in additional $O(n^2 \log n)$ time.

Let T_v be the vertical prism in $\mathcal{M}''$ that contains s. We construct an adjacency graph G on the vertical prisms of $\mathcal{M}''$. By computing the connected component of G that contains T_v, one obtains the subdivision $\mathcal{M}' = \mathcal{M}'' \cap FP(\mathcal{O}, s)$.

Each cell $\mathcal{I}$ of $\mathcal{M}'$ is a vertical prism, having at most 5 faces. We can approximate the distance function $d_{\mathcal{O},s}(t)$ inside $\mathcal{I}$ by computing a weighted set $\mathcal{S}_\mathcal{I} = (S_\mathcal{I}, w_\mathcal{I})$, as specified in the proofs of Lemma 2.7 and Lemma 2.14. To do so, it is necessary to $(\varepsilon/8)$-approximate the value of $d_{\mathcal{O},s}(\cdot)$

for $O((1/\varepsilon^2)\log(1/\varepsilon))$ points (i.e., the points of $\mathcal{S}_\mathcal{I}$). By Theorem 4.1, this takes

$$O\left(\left(\frac{\beta(n)}{\varepsilon^4}\log\frac{n}{\varepsilon}+\log(n\rho)\log(n\log\rho)\right)\frac{n^2}{\varepsilon^2}\log\frac{1}{\varepsilon}\right)$$

time. The weighted Voronoi diagram $\mathcal{V}_{\mathcal{S}_\mathcal{I}}$ induced by $\mathcal{S}_\mathcal{I}$ inside $\mathcal{I}$ approximates $d_{\mathcal{O},s}$ inside $\mathcal{I}$ up to a factor of $1+\varepsilon$.

Let $\mathcal{S} = \cup_{\mathcal{I}\in\mathcal{M}'}\mathcal{S}_\mathcal{I}$. Clearly, one can $(\varepsilon/8)$-approximate the distance between s and all the sites of $\mathcal{S}$ in

$$O\left(\frac{n^4}{\varepsilon^2}\left(\frac{\beta(n)}{\varepsilon^4}\log\frac{n}{\varepsilon}+\log(n\rho)\log(n\log\rho)\right)\log\frac{1}{\varepsilon}\right)$$

time.

Let $\mathcal{M}$ be the subdivision $\bigcup_{\mathcal{I}\in\mathcal{M}'}(\mathcal{V}_{\mathcal{S}_\mathcal{I}}\cap\mathcal{I})$. We preprocess $\mathcal{M}$ for spatial point location, by constructing a two-level spatial point location data structure. First, we preprocess $\mathcal{M}'$ for point location in $O(n^2\log n)$ time, using the algorithm of [23]. Next, we preprocess each cell $\mathcal{I}$ of $\mathcal{M}'$ for nearest neighbor queries for the weighted set $\mathcal{S}_\mathcal{I}$. By Lemma 4.3 below, performing this preprocessing for all the cells of $\mathcal{M}'$ takes a total of $O\left(n^2/\varepsilon^{4+\delta}\right)$ randomized expected time and space, for any $\delta > 0$.

For any query point $t \in FP(\mathcal{O}, s)$, we can compute in $O(\log n + \log(1/\varepsilon)) = O(\log(n/\varepsilon))$ time, the cell of $\mathcal{M}$ that contains t; that is, in $O(\log(n/\varepsilon))$ time, one can compute a distance $\Delta_{\mathcal{O},s}(t)$, such that $d_{\mathcal{O},s}(t) \le \Delta_{\mathcal{O},s}(t) \le (1+\varepsilon)d_{\mathcal{O},s}(t)$. ∎

To complete the description and analysis of the algorithm, we next show how to preprocess a weighted set in $\mathbb{R}^3$ so that one can perform efficient nearest neighbor queries in the additive weighted Voronoi diagram that it induces.

Definition Let $\mathcal{S} = (S, w)$ be a weighted set in $\mathbb{R}^3$. We decompose the weighted Voronoi diagram $\mathcal{V}_S$ into "simpler" cells in the following way: For each cell C in $\mathcal{V}_S$, we compute the spherical map S_C of the cell, by projecting the boundary of the cell onto the sphere of directions centered at p_C, where p_C is the site of C in S. (We use here the well-known property that C is star-shaped with respect to p_C.) We decompose S_C into pseudo-vertical subcells on the sphere of directions, by drawing a meridian arc upwards and downwards from each vertex S_C, and from each locally longitude-extremal point on any arc of S_C, and by extending each of these meridian arcs until it hits another arc of S_C or, failing this, all the way to the poles of the sphere of directions. Clearly, the complexity of S_C is linear in the complexity of the cell C.

We project each "vertical" trapezoid in S_C back into C, to obtain the portion within C of the cone with apex p_C spanned by the trapezoid. This defines a decomposition of C into simple subcells, such that each subcell is uniquely defined by at most 6 points of S. We decompose all the cells of $\mathcal{V}_S$ in a similar manner, and let $\mathcal{C}(\mathcal{S})$ denote the resulting subdivision. We call $\mathcal{C}(\mathcal{S})$ the *spherical decomposition* of $\mathcal{V}_S$.

For a weighted set $\mathcal{R} \subseteq \mathcal{S}$ and a subcell $\mathcal{T} \in \mathcal{C}(\mathcal{R})$, a weighted point $(p, w_p) \in \mathcal{S}$ *conflicts with* $\mathcal{T}$ if there exists a point $t \in \mathcal{T}$, such that $V_{(p,w_p)}(t) < V_\mathcal{R}(t)$. Let $K(\mathcal{S},\mathcal{T})$ denote the set of all the points of $\mathcal{S}$ that conflict with $\mathcal{T}$. The *conflict size* of $\mathcal{T}$ is $w(\mathcal{S},\mathcal{T}) = |K(\mathcal{S},\mathcal{T})|$.

Lemma 4.3 *Let $\mathcal{S} = (S, w)$ be a weighted set of m points in $\mathbb{R}^3$, and $\delta > 0$ be a parameter. One can compute, in $O(m^{2+\delta})$ randomized expected time, a data structure for nearest-neighbor queries, of size $O(m^{2+\delta})$, such that for any point $p \in \mathbb{R}^3$, one can compute, in $O(\log m)$ time, the cell of $\mathcal{V}_S$ that contains p; that is, the point in S realizing the distance $V_S(p)$.*

Proof: We construct the data-structure using a randomized divide and conquer algorithm. We randomly pick a subset R of S of size r, where r is a parameter to be specified later. One can compute the weighted Voronoi diagram of $\mathcal{R} = (R, w)$, in $O(r^2)$ time, by Remark 2.3, and construct the spherical decomposition $\mathcal{C}(\mathcal{R})$ in $O(r^2\log r)$ additional time, using plane sweeping techniques on the sphere of directions (see [18]).

For each subcell $\mathcal{T}$ in $\mathcal{C}(\mathcal{R})$, we compute its conflict size $w(\mathcal{S},\mathcal{T})$. Each subcell in $\mathcal{C}(\mathcal{R})$ is uniquely defined by at most 6 sites in $\mathcal{R}$, and if $K(\mathcal{S},\mathcal{T}) \cap \mathcal{R} \neq \emptyset$ then $\mathcal{T} \notin \mathcal{C}(\mathcal{R})$. We can thus apply the analysis of Clarkson and Shor. By [8, Corollary 3.8], $w(\mathcal{S},\mathcal{T}) \le c\cdot(n\log r)/r$, for all $\mathcal{T} \in \mathcal{C}(\mathcal{R})$, with probability at least 1/2, where $c > 0$ is an appropriate constant. We sample $\mathcal{R}$ from $\mathcal{S}$ repeatedly until we get a sample that fulfills this condition. Overall, this stage takes $O(mr^2 + r^2\log r)$ expected running time. For each cell $\mathcal{T} \in \mathcal{C}(\mathcal{R})$, we construct recursively a data-structure for point-location in the Voronoi diagram $\mathcal{V}_{K(\mathcal{S},\mathcal{T})}$.

For any query point p, locating the subcell $\mathcal{T}$ in $\mathcal{C}(\mathcal{R})$ that contains p is done by a brute force search inside $\mathcal{C}(\mathcal{R})$, in $O(r^2)$ time. Then, we compute the point realizing $V_S(p)$ by recursively performing a nearest neighbor query in the data-structure computed for $\mathcal{V}_{K(\mathcal{S},\mathcal{T})}$. Thus, a query takes $Q(m) = Q(c(m\log r)/r) + O(r^2)$ time, and the data-structure can be computed, in randomized expected time

$$T(m) = T(r) + O(r^2)T\left(\frac{cm\log r}{r}\right) + O\left(mr^2 + r^2\log r\right).$$

Choosing r to be a sufficiently large constant, we have $Q(m) = O(\log m)$, and $T(m) = O(m^{2+\delta})$ (where the constants of proportionality depend on δ). A similar bound holds for the space required by the algorithm. ∎

Remark 4.4 The only stage in the algorithm of Theorem 4.2 that uses randomization is the construction of the spatial point-location data described in Lemma 4.3. This can be replaced by a deterministic data-structure as follows.

We observe that each spherical cell, in the decomposition described above, can be parameterized by 24 parameters (6 sites and their respective weights). Thus, we define a range space $(\mathcal{S}, \mathfrak{R})$, where $\mathfrak{R}$ is the set of all possible subsets of $\mathcal{S}$ that are contained inside such a spherical cell. It is easy to verify that this is a range space having finite VC-dimension. By a result of Matoušek [15], we can compute, in $O(mr^{O(1)})$ time, a subset $\mathcal{R}$ of $\mathcal{S}$ having $O(r\log r)$ points, which is $(1/r)$-net of $(\mathcal{S}, \mathfrak{R})$. In particular, the set $\mathcal{R}$ can replace the random sample in the proof of Lemma 4.3, see [15]. This yields a deterministic algorithm with the same time/space complexity as in Lemma 4.3.

Alternatively, one can naively preprocess the Voronoi diagram $\mathcal{V}_S$ for spatial point-location directly, see [23]. However, this approach is considerably less efficient than the approach proposed above.

5 Conclusions

In this paper we have presented two results for computing approximate maps that facilitate shortest paths queries on the surface of a convex polytope or on a polyhedral surface in 3-space, or among polyhedral obstacles in 3-space. We conclude by mentioning the following open problems.

- Can an ε-approximate shortest path between two points on a polyhedral terrain, or on the surface of a nonconvex polyhedron, be computed in time that is near-linear in the number of edges? A recent subquadratic solution has been obtained by Varadarajan and Agarwal [24], but it only computes a constant-factor approximation to the shortest path.
- Can the exact shortest path between two points on a convex polyhedron be computed in near-linear time? in subquadratic time?
- Can the methods and techniques used in this paper be extended to handle shortest path queries for weighted surfaces (as in [13, 14])?

Acknowledgments

The author wishes to thank Pankaj Agarwal and Micha Sharir for helpful discussions concerning the problems studied in this paper and related problems. Micha Sharir has also suggested Lemma 4.3. The author also wishes to thank the referees for their comments and suggestions.

References

[1] AGARWAL, P., HAR-PELED, S., SHARIR, M., AND VARADARAJAN, K. R. Approximate shortest paths on a convex polytope in three dimensions. *J. Assoc. Comput. Mach. 44* (1997), 567–584.

[2] AURENHAMMER, F. Voronoi diagrams: A survey of a fundamental geometric data structure. *ACM Comput. Surv. 23* (1991), 345–405.

[3] CANNY, J., AND REIF, J. H. New lower bound techniques for robot motion planning problems. In *Proc. 28th Annu. IEEE Sympos. Found. Comput. Sci.* (1987), pp. 49–60.

[4] CHAZELLE, B. Triangulating a simple polygon in linear time. *Discrete Comput. Geom. 6* (1991), 485–524.

[5] CHEN, J., AND HAN, Y. Shortest paths on a polyhedron; Part I: computing shortest paths. *Int. J. Comput. Geom. & Appl. 6*, 2 (1996), 127–144.

[6] CHOI, J., SELLEN, J., AND YAP, C. K. Approximate Euclidean shortest path in 3-space. In *Proc. 10th Annu. ACM Sympos. Comput. Geom.* (1994), pp. 41–48.

[7] CLARKSON, K. L. Approximation algorithms for shortest path motion planning. In *Proc. 19th Annu. ACM Sympos. Theory Comput.* (1987), pp. 56–65.

[8] CLARKSON, K. L., AND SHOR, P. W. Applications of random sampling in computational geometry, II. *Discrete Comput. Geom. 4* (1989), 387–421.

[9] DOBKIN, D. P., AND KIRKPATRICK, D. G. A linear algorithm for determining the separation of convex polyhedra. *J. Algorithms 6* (1985), 381–392.

[10] FORTUNE, S. J. A sweepline algorithm for Voronoi diagrams. *Algorithmica 2* (1987), 153–174.

[11] HAR-PELED, S. Approximate shortest paths and Geodesic diameters on convex polytopes in three dimensions. In *Proc. 13th Annu. ACM Sympos. Comput. Geom.* (1997), pp. 359–365.

[12] HERSHBERGER, J., AND SURI, S. Practical methods for approximating shortest paths on a convex polytope in $\mathbb{R}^3$. In *Proc. 6th ACM-SIAM Sympos. Discrete Algorithms* (1995), pp. 447–456.

[13] LANTHIER, M., MAHESHWARI, A., AND SACK, J.-R. Approximating weighted shortest paths on polyhedral surfaces. In *Proc. 13th Annu. ACM Sympos. Comput. Geom.* (1997), pp. 274–283.

[14] MATA, C., AND MITCHELL, J. A new algorithm for computing the shortest paths in planar subdivisions. In *Proc. 13th Annu. ACM Sympos. Comput. Geom.* (1997), pp. 264–273.

[15] MATOUŠEK, J. Approximations and optimal geometric divide-and-conquer. In *Proc. 23rd Annu. ACM Sympos. Theory Comput.* (1991), pp. 505–511. Also to appear in *J. Comput. Syst. Sci.*

[16] MITCHELL, J. S., MOUNT, D. M., AND PAPADIMITRIOU, C. H. The discrete geodesic problem. *SIAM J. Comput. 16* (1987), 647–668.

[17] MOUNT, D. M. Storing the subdivision of a polyhedral surface. *Discrete Comput. Geom. 2* (1987), 153–174.

[18] O'ROURKE, J. *Computational Geometry in C.* Cambridge University Press, Cambridge, 1994.

[19] PAPADIMITRIOU, C. H. An algorithm for shortest-path motion in three dimensions. *Inform. Process. Lett. 20* (1985), 259–263.

[20] REIF, J. H., AND STORER, J. A. A single-exponential upper bound for finding shortest paths in three dimensions. *J. ACM 41*, 5 (1994), 1013–1019.

[21] SHARIR, M. On shortest paths amidst convex polyhedra. *SIAM J. Comput. 16* (1987), 561–572.

[22] SHARIR, M., AND SCHORR, A. On shortest paths in polyhedral spaces. *Siam J. Comput. 15* (1986), 193–215.

[23] TAN, X.-H., HIRATA, T., AND INAGAKI, Y. Spatial point location and its applications. In *Proc. 1st Annu. SIGAL Internat. Sympos. Algorithms* (1990), vol. 450 of *Lecture Notes in Computer Science*, Springer-Verlag, pp. 241–250.

[24] VARADARAJAN, K., AND AGARWAL, P. Approximating shortest paths on a nonconvex polyhedron. In *Proc. 38th Annu. IEEE Sympos. Found. Comput. Sci.* (1997), pp. 182–191.

Curvature-Constrained Shortest Paths in a Convex Polygon

(Extended Abstract)

Pankaj K. Agarwal* Therese Biedl† Sylvain Lazard‡

Steve Robbins§ Subhash Suri¶ Sue Whitesides||

Abstract

Let B be a point robot moving in the plane, whose path is constrained to have curvature at most 1, and let $\mathcal{P}$ be a convex polygon with n vertices. We study the collision-free, optimal path-planning problem for B moving between two *configurations* inside $\mathcal{P}$ (a configuration specifies both a location and a direction of travel). We present an $O(n^2 \log n)$ time algorithm for determining whether a collision-free path exists for B between two given configurations. If such a path exists, the algorithm returns a shortest one. We provide a detailed classification of curvature-constrained shortest paths inside a convex polygon and prove several properties of them, which are interesting in their own right. Some of the properties are quite general and shed some light on curvature-constrained shortest paths amid obstacles.

*Center for Geometric Computing, Computer Science Department, Duke University, Box 90129, Durham, NC 27708–0129, USA; pankaj@cs.duke.edu; http://www.cs.duke.edu/~pankaj/. Supported in part by National Science Foundation research grant CCR–93–01259, by Army Research Office MURI grant DAAH04–96–1–0013, by a Sloan fellowship, by a National Science Foundation NYI award and matching funds from Xerox Corporation, and by a grant from the U.S.-Israeli Binational Science Foundation.

†School of Computer Science, McGill University, 3480 University Street, Montreal, Qc, H3A 2A7, Canada; therese@cgm.cs.mcgill.ca.

‡School of Computer Science, McGill University; lazard@cgm.cs.mcgill.ca. Supported in part by an INRIA postdoctoral award.

§School of Computer Science, McGill University; stever@cs.mcgill.ca. Supported by an FCAR scholarship.

¶Department of Computer Science, Washington University Campus Box 1045, One Brookings Drive, St. Louis, MO 63130-4899, USA; suri@cs.wustl.edu;http://www.cs.wustl.edu/~suri/. Research partially supported by NSF Grant CCR-9501494.

||School of Computer Science, McGill University; sue@cs.mcgill.ca. Supported by NSERC and FCAR research grants.

SCG 98 Minneapolis Minnesota USA

1 Introduction

The *path-planning* problem, a central problem in robotics, involves planning a collision-free path for a robot moving amid obstacles, and has been widely studied (see, e.g., the book by Latombe [17] and the survey papers by Schwartz and Sharir [25] and Halperin, Kavraki and Latombe [12]). In the simplest form, given a moving point robot B, a set of obstacles, and a pair of configurations I and F specifying locations for B, we wish to find a continuous, collision-free path for B from I to F. This formulation, however, does not take into account the dynamic constraints (for instance, bounds on velocity, acceleration or curvature), the so-called *nonholonomic constraints*, imposed on a robot by its physical limitations (see [17] for a more detailed discussion). Although there has been considerable recent work in the robotics literature on nonholonomic motion-planning problems (see [3, 4, 14, 16, 18, 20, 26, 31, 32] and references therein), relatively little theoretical work has been done in this important area.

In this paper, we study the path-planning problem for a point robot whose configurations are specified by giving both a location and a direction of travel. This means that any solution to the path-planning problem for given initial and final configurations I and F must respect the directions of travel specified by I and F as well as the locations they specify. Furthermore, we require the path of the robot to have curvature at most 1. This curvature constraint arises naturally when the point robot models a real-world robot with a minimum turning radius; see for example [17]. Recently Reif and Wang [24] confirmed that the problem of deciding whether there exists a collision-free curvature-constrained path for B between two given configurations amid obstacles is NP-hard. This motivates interest in studying various special cases. In this paper we propose an efficient algorithm for computing a curvature-constrained shortest path inside a convex polygon.

We establish several new properties of shortest paths inside a convex polygon and use these properties to characterize shortest paths. Using these properties of shortest paths and some results in computational geometry [2, 8], we present an efficient algorithm that, given initial and desired final configurations I and F in the polygon, determines whether a curvature-constrained path from I to F exists, and if so, computes a shortest one.

1.1 Previous results

Dubins [10] was perhaps the first to study curvature constrained shortest paths. He proved that, in the absence of obstacles, a curvature-constrained shortest path from any start configuration to any final configuration consists of at most three segments, each of which is either a straight line or an arc of a circle of unit radius, assuming that the curvature of the path is upper bounded by 1. Reeds and Shepp [23] extended this obstacle-free characterization to robots that are allowed to make reversals, that is, to back up. Using ideas from control theory, Boissonnat, Cérézo and Leblond [4] gave an alternative proof for both cases, and recently Sussmann [29] was able to extend the characterization to the 3-dimensional case. In the presence of obstacles, Fortune and Wilfong [11] gave a $2^{poly(n,m)}$ time algorithm, where n is the total number of vertices in the polygons defining the obstacles and m is the number of bits of precision with which all points are specified; their algorithm only decides whether a path exists, without necessarily finding one. Jacobs and Canny [13], Wang and Agarwal [30], and Sellen [27, 28] gave approximation algorithms for computing an ε-*robust* path. (Informally, a path is ε-robust if ε-perturbations of certain points along the path do not violate the feasibility of the path.) For the restricted case of pairwise disjoint *moderate obstacles*, i.e., convex obstacles whose boundaries have curvature bounded by 1, Agarwal, Raghavan and Tamaki [1] gave efficient approximation algorithms. Boissonnat and Lazard [5] gave an $O(n^2 \log n)$ time algorithm for computing an exact shortest path for the case when the edges of the pairwise disjoint moderate obstacles are circular arcs of unit radius or line segments. Their algorithm can be used to compute an optimal curvature constrained path inside a convex polygon in time $O(n^7)$. Wilfong [31] studied a restricted problem in which the robot must stay on one of m line segments (thought of as "lanes"), except to turn between lanes. For a scene with n obstacle vertices, his algorithm preprocesses the scene in time $O(m^2(n^2 + \log m))$, following which queries are answered in time $O(m^2)$. There has also been work on computing curvature-constrained paths when B is allowed to make reversals [3, 19, 21]. Other, more general, dynamic constraints have been considered in [6, 7, 9, 22].

1.2 Our model and results

Let B be a point robot and $\mathcal{P}$ a closed convex polygon with n vertices. For simplicity we assume that the edges of $\mathcal{P}$ are in general position: no two edges are parallel and no unit-radius circle is tangent to three edges of $\mathcal{P}$. A *configuration* X for B is a pair $(\text{LOC}(X), \psi(X))$, where $\text{LOC}(X)$ is a point in the plane representing the location of the robot and $\psi(X)$ is an angle between 0 and 2π representing its orientation. When the meaning is clear, we often write X instead of $\text{LOC}(X)$.

The image of a differentiable function $\Pi : [0, l] \to \mathbb{R}^2$ is called a *path*. We denote both the function and the path it defines by Π. We regard a path Π as oriented from $\Pi(0)$ to $\Pi(l)$. We assume a path Π is parameterized by its arc length, and we let $||\Pi||$ denote its length. We say that Π is a path from a configuration X to another configuration Y if $\Pi(0) = \text{LOC}(X)$, $\Pi(l) = \text{LOC}(Y)$, and the oriented angles (with respect to the positive x-axis) of $\Pi'(0)$ and $\Pi'(l)$ are $\psi(X)$ and $\psi(Y)$, respectively. A path is called *moderate* if its average curvature is at most 1 in every positive-length interval.[1] This implies that the curvature is at most 1 whenever it is defined.

Any curve that lies entirely within the closed polygon $\mathcal{P}$ is called *free*. A path is *feasible* if it is moderate and free. A feasible path Π from a configuration X to another configuration Y is *optimal* if its length is minimum among all feasible paths from X to Y (it can be shown that whenever a feasible path from X to Y exists, then an optimal such path also exists [13]).

Main Results. Let $\mathcal{P}$ be an n-vertex convex polygon in the plane, and let I and F be two configurations inside $\mathcal{P}$.

(i) We prove that an optimal path from I to F consists of at most eight maximal segments, each of which is either a line segment or a circular arc of unit radius.

(ii) We give an $O(n^2 \log n)$ time algorithm to determine whether a feasible path from I to F exists. If such a path exists, then the algorithm returns an optimal path from I to F. If there are only k edges of $\mathcal{P}$ within distance 6 from both I and F, then the running time of our algorithm can be improved to $O((n + k^2) \log n)$,

Our algorithm is significantly faster than the algorithm implicit in the work of Boissonnat and Lazard [5], whose running time would be $O(n^7)$. Our paper is organized as follows. In Section 2, we present basic definitions, notation, and useful known results. In Section 3, we give a classification of the optimal path. In Sections 4 and 5, we describe our algorithms. Section 6 concludes.

2 Geometric Preliminaries

Given a configuration X, the oriented line passing through $\text{LOC}(X)$ with orientation $\psi(X)$ is denoted L_X. A configuration X *belongs* to an oriented path (or curve) Π if $\text{LOC}(X) \in \Pi$ and L_X is the oriented tangent line to Π at $\text{LOC}(X)$. Note that a configuration X belongs to two oriented unit-radius circles. We will use $\mathcal{C}_X^+$ (resp. $\mathcal{C}_X^-$) to denote the two circles of unit radius, oriented counterclockwise (resp. clockwise) to which the configuration X belongs.

If X and Y are two points on a simple closed curve γ, then $\gamma^+[X, Y]$ (resp. $\gamma^-[X, Y]$) denotes the portion of γ

[1] The *average curvature* of a path Π in the interval $[s_1, s_2]$ is defined by $||\Pi'(s_1) - \Pi'(s_2)||/|s_1 - s_2|$.

from X to Y in the counterclockwise (resp. clockwise) direction, including X and Y; we will use $\gamma^+(X,Y), \gamma^-(X,Y)$ to denote portions excluding X, Y. Similarly, for a path Π and two configurations $X, Y \in \Pi$, we will use $\Pi[X,Y]$ to denote the portion of Π from X to Y.

Segments and Dubins paths. Let Π be a feasible path. We call a nonempty subpath of Π a *C-segment* (resp. *S-segment*) if it is a circular arc of unit radius (resp. line segment) and maximal. A *segment* is either a C-segment or an S-segment. When referring to a C-segment on a path Π, we will call it a C^+-segment (resp. C^--segment) if Π induces a counterclockwise (resp. clockwise) orientation on it. Suppose Π consists of a C-segment, an S-segment, and a C-segment; then we will say that Π is of type CSC, or C_1SC_2 if we want to distinguish between the two C-segments; superscripts $+$ and $-$ will be used to specify the orientations of C-segments of Π. Abusing the notation slightly, we will also use C_1, C_2 to denote the C-segments and S to denote the S-segment of Π. The above notation can be generalized to an arbitrarily long sequence. Dubins [10] proved the following result.

Lemma 2.1 (Dubins [10]) *In an obstacle-free environment, an optimal path between any two configurations is of type CCC or CSC, or a substring thereof.*

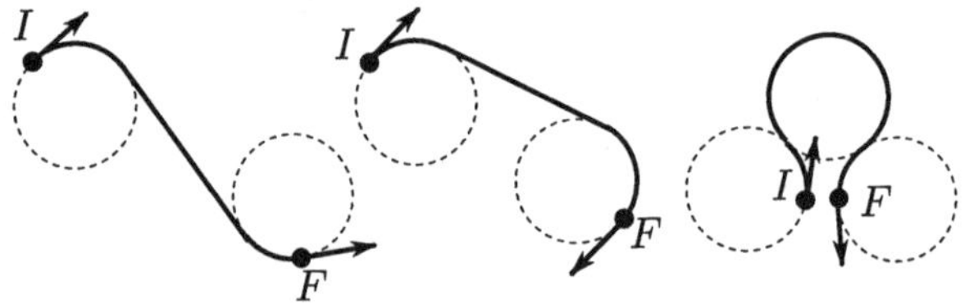

Figure 1. Different types of Dubins paths.

We will refer to paths of type CCC or CSC or substrings thereof as *Dubins paths*. In the presence of obstacles, Jacobs and Canny [13] observed that any subpath of an optimal path that does not touch any obstacle except at the endpoints is a Dubins path. In particular, they proved the following.

Lemma 2.2 (Jacobs and Canny [13]) *Let Ω be a closed polygonal environment, I an initial configuration, and F a final configuration. Then an optimal path from I to F in Ω consists of a sequence $\Pi_1 \cdots \Pi_k$ of feasible paths, where each Π_i is a Dubins path from a configuration X_{i-1} to a configuration X_i, such that $X_0 = I$, $X_k = F$, and, for $0 < i < k$, $\text{LOC}(X_i) \in \partial\Omega$.*

The above lemma implies that an optimal path in a closed polygonal environment consists of C- and S-segments. In the following, we will consider only those paths that are formed by S- and C-segments. We will refer to circles and circular arcs of unit radius simply as circles and circular arcs. Notationally, we differentiate between a C-segment and its supporting circle by using calligraphic font for the latter.

Terminal and nonterminal segments. A segment of a feasible path Π is called *terminal* if it is the first or the last segment of Π; otherwise it is called *nonterminal.* We apply the adjectives terminal and nonterminal to subpaths as well. If the first or last segment in Π is a C-segment, we will refer to it as a C_I-segment or a C_F-segment, respectively. $\mathcal{C}_I^+$, $\mathcal{C}_I^-$, $\mathcal{C}_F^+$, and $\mathcal{C}_F^-$ are called *terminal circles* (see Figure 3).

The following lemma states some basic known properties of optimal paths; see [1, 10, 13].

Lemma 2.3 *In an optimal path inside $\mathcal{P}$,*

(i) any nonterminal C-segment has length greater than π,
(ii) any nonterminal C-segment is tangent to $\partial\mathcal{P}$ or to a terminal circle in at least one point, and
(iii) no nonterminal subpath has type CCC.

Lemma 2.4 *Let Π be an optimal path of type C_1C_2S inside $\mathcal{P}$. Let X be the common endpoint of the C_1- and C_2-segments, and let Y be the last tangent point of the C_2-segment with $\partial\mathcal{P}$ along Π. Then the length of the C_2-segment between X and Y is greater than π, i.e., $\|\Pi[X,Y]\| > \pi$.*

Proof (Sketch): By Lemma 2.3(i), the C_2-segment is greater than π. Let X' be the antipodal point of X on C_2. Figure 2 shows how to shorten the path of type C_1C_2S [10]. This shortening can be done unless an obstacle obstructs the shortcut, i.e., unless an obstacle is tangent to the C_2-segment between X' and the last point of the C_2-segment. $\square$

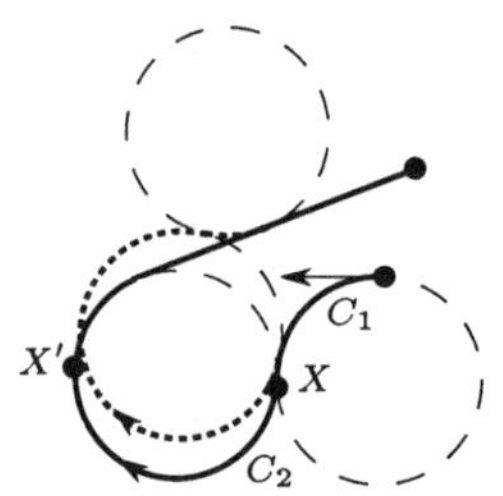

Figure 2. Shortcut.

Anchored segments. A C-segment or circle is called *anchored* if it is tangent to $\partial\mathcal{P}$ or to terminal circles at two points. The terminal circles are not considered anchored. An anchored C-segment is denoted by $\bar{\bar{C}}$. By our general-position assumption on $\mathcal{P}$, there are a finite number of anchored circles. A C-segment with at least one point of tangency with $\partial\mathcal{P}$ is denoted by $\bar{C}$.

An anchored C-segment or circle is *$\mathcal{PP}$-anchored* if it is tangent to $\partial\mathcal{P}$ at two points and *$\mathcal{PC}$-anchored* if it is tangent to $\partial\mathcal{P}$ at one point and tangent to a terminal circle at another point; see Figure 3.

A circular arc is called *long* if its length is greater than π; otherwise it is called *short.* A $\mathcal{PP}$-anchored C-segment is called *strongly $\mathcal{PP}$-anchored* if it contains the long arc defined by the tangent points of its supporting circle with $\partial\mathcal{P}$ (see Figure 4(b)). Similarly, a $\mathcal{PC}$-anchored C-segment is called *strongly $\mathcal{PC}$-anchored* if it contains the long arc between a tangency point of its supporting circle $\mathcal{C}$ with $\partial\mathcal{P}$ and a tangency point of $\mathcal{C}$ with a terminal circle (see Figure 5(a)).

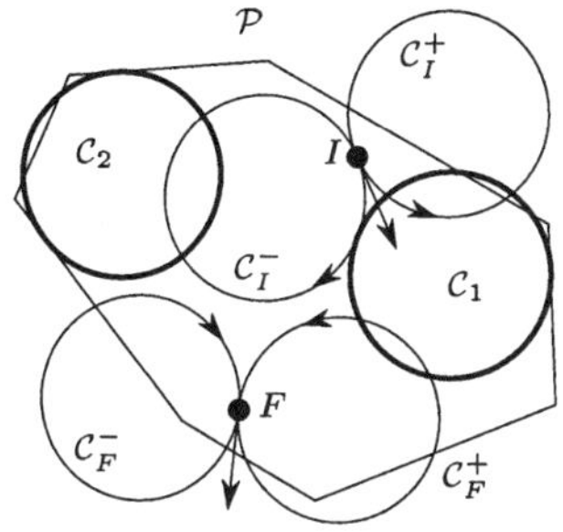

Figure 3. $\mathcal{PC}$-anchored (C_1) and $\mathcal{PP}$-anchored (C_2) circles.

Pockets. Let $\mathcal{C}$ be a circle intersecting $\partial\mathcal{P}$ at two or more points, and let X, Y be two consecutive intersection points of $\partial\mathcal{P}$ with $\mathcal{C}$ so that the short arc of $\mathcal{C}$ joining X and Y lies inside $\mathcal{P}$. If $\mathcal{C}^+[X,Y]$ is the short arc and the turning angle[2] of $\partial\mathcal{P}^+(X,Y)$ is less than π, then the closed region bounded by $\partial\mathcal{P}^+[X,Y]$ and $\mathcal{C}^+[X,Y]$ is called a *pocket* (see Figure 4) and is denoted by $\Lambda_\mathcal{C}[X,Y]$. Similarly we define a pocket $\Lambda_\mathcal{C}[X,Y]$ when $\mathcal{C}^-[X,Y]$ is the shorter arc. We will mostly be interested in pockets for which $\mathcal{C}$ is tangent to $\partial\mathcal{P}$ at X.

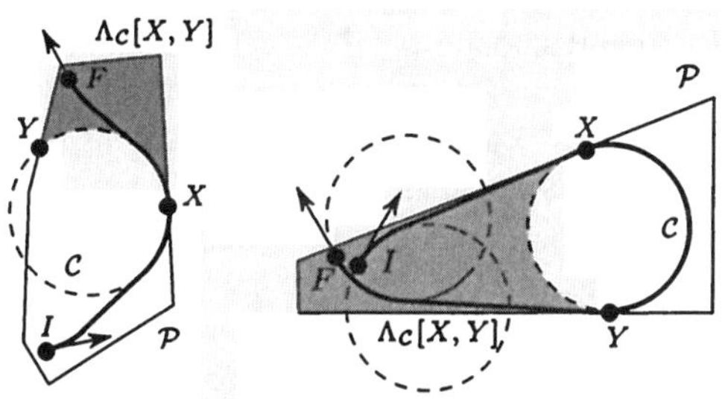

(a) (b)

Figure 4. Pockets.

It can be verified that the condition on the turning angle implies that a pocket does not have enough room to contain a unit circle. Using this simple observation, we can prove the following lemma, which will be crucial for characterizing the optimal paths containing a strongly anchored C-segment. In particular, the lemma implies that if a feasible path enters the interior of a pocket, then it cannot escape the pocket (see Figure 4).

Lemma 2.5 *Let $\mathcal{C}$ be a circle tangent to $\partial\mathcal{P}$ at X that defines a pocket $\Lambda_\mathcal{C}[X,Y]$. If a feasible path Π from I to F enters (resp. escapes) $\Lambda_\mathcal{C}[X,Y]$ at X, then either Π contains the small arc of $\mathcal{C}$ joining X and Y, or $\Pi[X,F] \subset \Lambda_\mathcal{C}[X,Y]$ (resp. $\Pi[I,X] \subset \Lambda_\mathcal{C}[X,Y]$).*

[2]The *turning angle* of a convex polygonal chain is $\sum_i(\pi - \theta_i)$, where θ_i is the interior angle at vertex i.

3 Classification of Optimal Paths

The goal of this section is to prove the first of our main results, namely a detailed characterization of optimal paths in convex polygons. We show that any optimal path is of type $C_I CSCCSCC_F$ or a subsequence of this form. However, not every subsequence of the above sequence can form an optimal path. The following theorem gives a more refined description of optimal path types. Recall that a segment has non-zero length by definition. In the following, we use $\cdot$ to denote a subpath of zero length.

Theorem 3.1 *An optimal path Π inside $\mathcal{P}$ either is a Dubins path or has one of the types listed below. Except in case (B.i), all the C-segments labeled $\bar{\bar{C}}$ are strongly anchored.*

(A) If Π has no nonterminal CC subpath, then Π has one of the following types:
(A.i) $\Pi_I S\bar{\bar{C}} S\Pi_F$ where $\Pi_I \in \{C_I, \cdot\}$ and $\Pi_F \in \{C_F, \cdot\}$ (see Figure 4(b))
(A.ii) $\Pi_I S\Pi_F$ where $\Pi_I \in \{C_I\bar{\bar{C}}, C_I, \cdot\}$ and $\Pi_F \in \{\bar{\bar{C}}C_F, C_F, \cdot\}$ (see Figure 5(a))

(B) If Π has a nonterminal CC subpath, then Π has one of the following types:
(B.i) $C_I C\bar{\bar{C}} C_F$ or $C_I \bar{\bar{C}} C C_F$
(B.ii) $\Pi_I S\bar{\bar{C}} C C_F$ or $C_I C\bar{\bar{C}} S\Pi_F$ where $\Pi_I \in \{C_I, \cdot\}$ and $\Pi_F \in \{C_F, \cdot\}$
(B.iii) $\Pi_I \bar{C}\bar{C}\Pi_F$ where $\Pi_I \in \{C_I\bar{\bar{C}}S, C_I S, C_I, S\}$ and $\Pi_F \in \{S\bar{\bar{C}}C_F, SC_F, C_F, S\}$ (see Figures 5(b), (c))

Proposition 3.2 *The type $C_I CS\bar{C}\bar{C}SCC_F$, having eight segments, does occur as an optimal path type.*

Proof (Sketch): Figure 5(c) shows an instance of $\mathcal{P}$ and initial and final configurations in which a feasible path has eight segments. We can argue that no paths of the other types described in Theorem 3.1 are feasible, which implies that the optimal path is of the given type. □

The proof of Theorem 3.1 is based on the following lemmas.

Lemma 3.3 (Agarwal, Raghavan and Tamaki [1]) *An optimal path has at most one nonterminal CC subpath. Moreover, any nonterminal C-segment that precedes (resp. follows) a C_1C_2 subpath is oriented the same way as C_1 (resp. C_2).*

Next, we state a lemma, which can be proved using geometric perturbations similar to the ones used in [1, 5].

Lemma 3.4 *(i) If an optimal path has a subpath of type SCS, then the C-segment in that subpath is strongly $\mathcal{PP}$-anchored.*
(ii) If an optimal path has a subpath of type $C_1C_2C_3S$ (or $SC_3C_2C_1$) so that the C-segment C_2 does not touch $\partial\mathcal{P}$, then C_3 is strongly $\mathcal{PP}$-anchored.

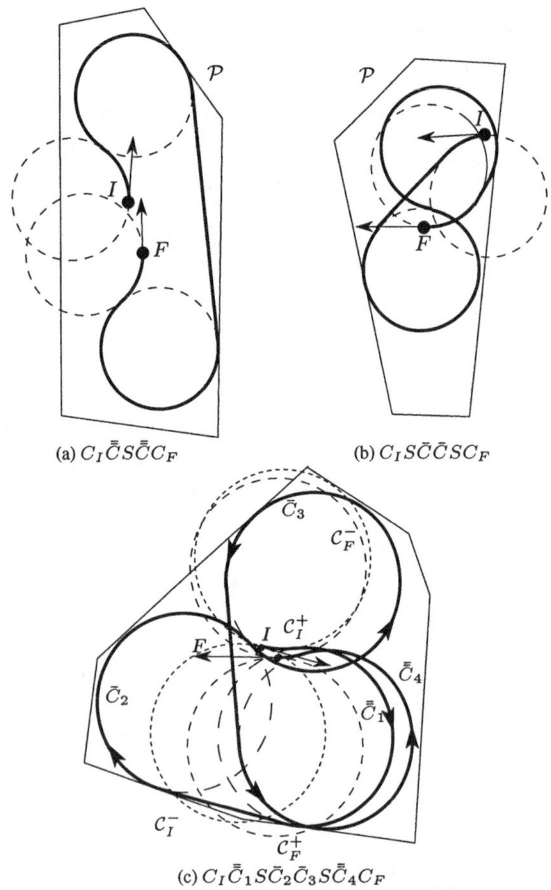

(a) $C_I\bar{\bar{C}}S\bar{\bar{C}}C_F$ (b) $C_IS\bar{\bar{C}}\bar{\bar{C}}SC_F$

(c) $C_I\bar{\bar{C}}_1S\bar{\bar{C}}_2\bar{\bar{C}}_3S\bar{\bar{C}}_4C_F$

Figure 5. Examples of shortest paths.

We next characterize the optimal paths that contain a strongly $\mathcal{PP}$-anchored C-segment.

Lemma 3.5 *If an optimal path* Π *contains a strongly* $\mathcal{PP}$*-anchored* C*-segment* $\bar{\bar{C}}$*, then* Π *is of type* $C_IS\bar{\bar{C}}SC_F$*,* $C_IC\bar{\bar{C}}SC_F$*,* $C_IS\bar{\bar{C}}CC_F$*, or a substring thereof (containing* $\bar{\bar{C}}$*).*

Proof (Sketch): By assumption, $\Pi = \Pi_I\bar{\bar{C}}\Pi_F$. $\bar{\bar{C}}$ is strongly $\mathcal{PP}$-anchored; hence its supporting circle, $\bar{\bar{\mathcal{C}}}$, has two or more intersections with $\partial\mathcal{P}$. Let X denote the first tangent point of $\bar{\bar{C}}$ with $\partial\mathcal{P}$ along Π. Let Y be the first point from X on $\bar{\bar{\mathcal{C}}}$ — moving in the opposite sense of $\bar{\bar{C}}$'s orientation — which intersects $\partial\mathcal{P}$ (see Figure 6). It is easy to prove that such a Y exists, and that $\Lambda_{\bar{\bar{\mathcal{C}}}}[X, Y]$ defines a pocket. Lemma 2.5 implies that the path up to X, i.e. Π_I and perhaps part of $\bar{\bar{C}}$, is contained in the pocket. We can also prove that Π_I consists of at most two segments, so Π_I is either C_IC, C_IS, or a substring thereof. Likewise, Π_F is CC_F, SC_F, or a substring thereof. The result follows by noting that paths of type $C_IC\bar{\bar{C}}CC_F$ are ruled out by Lemma 2.3(iii). □

We state now another lemma which will be useful for the algorithm. The proof is similar to that of Lemma 3.5.

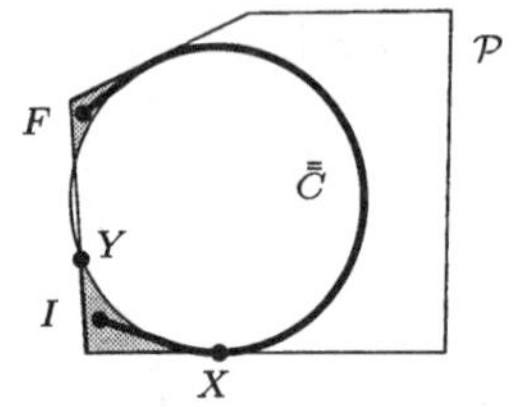

Figure 6. For the proof of Lemma 3.5. An optimal path containing a strongly $\mathcal{PP}$-anchored C-segment must start and end in a pocket.

Lemma 3.6 *If an optimal path* Π *contains a strongly* $\mathcal{PC}$*-anchored* C*-segment* $\bar{\bar{C}}$ *whose supporting circle is not free, then* Π *is of type* $C_I\bar{\bar{C}}SC_F$*,* $C_I\bar{\bar{C}}CC_F$*,* $C_IS\bar{\bar{C}}C_F$*,* $C_IC\bar{\bar{C}}C_F$*, or a substring thereof (containing* $\bar{\bar{C}}$*).*

We now prove Theorem 3.1.

Proof of Theorem 3.1: The proof proceeds by considering how a nonterminal C-segment may appear in Π. If there is no nonterminal C-segment in Π, then Π is of type C_ISC_F or a substring thereof, i.e., Π is a Dubins path.

Assume now that there is a nonterminal C-segment in Π. Then such a segment belongs to a subpath of Π of type either SCS or CC. Suppose Π contains a subpath of type SCS. By Lemma 3.4, the C-segment in SCS must be strongly $\mathcal{PP}$-anchored. Thus, by Lemma 3.5, Π is of type $C_IS\bar{\bar{C}}SC_F$, or substrings (containing $S\bar{\bar{C}}S$) thereof. In other words, Π is of type (A.i).

If Π contains a nonterminal C-segment but not a subpath of type SCS, we know it must contain a subpath of type CC. There are two cases to consider, depending on whether the CC subpath is terminal.

Case 1: Π does not contain any nonterminal subpath of type CC. Thus, one of the C-segments in any CC subpath must be a terminal segment. Either Π is of type C_IC_F, C_ICC_F (i.e., a Dubins path), or any nonterminal C-segment is also adjacent to an S-segment. Π must then be of type C_ICSCC_F, or any substring thereof containing S and a terminal CC. By Lemma 2.4, the nonterminal C-segments are strongly anchored. All these types of paths are covered by type (A.ii).

Case 2: Π contains a nonterminal subpath of type CC. By Lemma 3.3, it is the *only* nonterminal CC subpath in Π. Thus Π has the form $\Pi_ICC\Pi_F$. A nonterminal C-segment in Π_I must be followed by an S-segment, otherwise there will be a nonterminal CCC subpath in Π (Lemma 2.3(iii)). Furthermore, since we have no SCS subpath in Π, a nonterminal C segment must be preceded by a terminal C-segment. This means $\Pi_I = C_ICS$ or a subsequence of it. The subsequence cannot not be empty, for otherwise the middle CC subpath would be terminal; nor can it be simply CC, as noted above. Thus, $\Pi_I \in \{C_ICS, C_IS, C_I, S\}$. Similarly, $\Pi_F \in \{SCC_F, SC_F, C_F, S\}$.

If $\Pi_I = C_ICS$ or $\Pi_F = SCC_F$, then the nonterminal C-segment in Π_I or Π_F is strongly anchored by Lemma 2.4.

If both Π_I and Π_F contain an S-segment, then the nonterminal CC subpath in Π is preceded and followed by an S-segment. Thus, both C-segments of the nonterminal CC subpath in Π touch $\partial\mathcal{P}$. Indeed, otherwise Π contains a subpath of type SCC or CCS that does not touch $\partial\mathcal{P}$, which contradicts Lemma 2.2. Hence, if both Π_I and Π_F contain an S, Π is of type (B.iii).

Suppose that neither Π_I nor Π_F contains an S-segment. Then, the path is of type C_ICCC_F. One of the nonterminal C-segments must touch $\partial\mathcal{P}$ by Lemma 2.2. This C-segment is also tangent to a terminal circle and is therefore $\mathcal{PC}$-anchored. Thus the path is of type (B.i). Note that if both nonterminal C-segments touch $\partial\mathcal{P}$, then the path is of type $C_I\bar{\bar{C}}\bar{\bar{C}}C_F$ which can be considered as type (B.i) or (B.iii).

The last case to consider is when exactly one of Π_I or Π_F contains an S-segment. Say $\Pi_I = C_I$ and $\Pi_F \neq C_F$. The path has form $C_IC_1C_2\Pi_F$ where Π_F starts with an S-segment. We know that C_2 must touch $\partial\mathcal{P}$ by Lemma 2.3(ii). If C_1 also touches $\partial\mathcal{P}$, then the path Π is of type (B.iii). Otherwise, if C_1 does not touch $\partial\mathcal{P}$, then by Lemma 3.4(ii), C_2 must be strongly $\mathcal{PP}$-anchored. Lemma 3.5 then restricts the path Π to be of type (B.ii). Similarly, if $\Pi_I \neq C_I$ and $\Pi_F = C_F$, the path Π is of type (B.ii). □

4 A Simple Algorithm

Theorem 3.1 can be used to obtain the following simple algorithm for computing an optimal path inside $\mathcal{P}$. We enumerate candidate paths of types described in Theorem 3.1. Our candidate set is guaranteed to contain an optimal path, if any exist. For each such path, we check whether it is feasible, and if so compute the length. Finally, we either return the shortest feasible path, or report that no feasible path exists.

In order to determine whether a path is feasible, we rely on the circle-shooting data structure by Agarwal and Sharir [2] that preprocesses $\mathcal{P}$ in $O(n \log n)$ time into a data structure that makes it possible to determine in $O(\log^2 n)$ time whether a given circular arc intersects $\partial\mathcal{P}$. If the radius of all query circles is the same, then using fractional cascading [8], the data structure may be modified without affecting the preprocessing time, so that a query is answered in $O(\log n)$ time. This immediately implies the following lemma.

Lemma 4.1 *$\mathcal{P}$ can be preprocessed in $O(n \log n)$ time into a data structure that enables us to determine in $O(m \log n)$ time whether a given path consisting of m C- and S-segments is feasible.*

To bound the running time of this simple algorithm, we must count the number of candidate paths to check. We note that once a path type is given, and the supporting circles for C-segments are known, there are $O(1)$ candidate paths. These are determined by the choices of the orientations for the C-segments. Hence we are interested in the number of possible supporting circles for each path type. Note that there may be $\Omega(n^2)$ $\mathcal{PP}$-anchored circles and $\Omega(n)$ $\mathcal{PC}$-anchored circles.

There are $O(1)$ Dubins path candidates.

For paths of type (A.i) and (B.ii), once the $\mathcal{PP}$-anchored circle is chosen, there are $O(1)$ choices for other supporting circles, and hence $O(1)$ candidate paths. Since there are $O(n^2)$ $\mathcal{PP}$-anchored circles, there are $O(n^2)$ candidate paths for these two path types.

For type (A.ii), the path may have up to two $\mathcal{PC}$-anchored segments. Once their supporting circles are chosen, there are $O(1)$ path candidates. There are $O(n)$ potential $\mathcal{PC}$-anchored circles. If both anchored segments are present, we have $O(n^2)$ paths to check; otherwise, we have only $O(n)$.

Paths of type (B.i) are also determined by a $\mathcal{PC}$-anchored circle; hence there are $O(n)$ of them as well.

Paths of type (B.iii), i.e. of type $C_I\bar{\bar{C}}_1S\bar{C}_i\bar{C}_jS\bar{\bar{C}}_2SC_F$, present a special problem. If we know the supporting circles of the $\bar{C}\bar{C}$ subpath, the rest of the path is determined by a pair of $\mathcal{PC}$-anchored circles $\mathcal{C}_1, \mathcal{C}_2$, for which there are $O(n^2)$ possibilities. Unfortunately, there is an infinite family of supporting circles for the $\bar{C}\bar{C}$ subpath. The following result by Boissonnat and Lazard [5] allows us to consider only a finite set of $\bar{C}\bar{C}$ subpaths.

Lemma 4.2 (Boissonnat and Lazard [5]) *Given two configurations X and Y, and two edges e_i, e_j of $\mathcal{P}$, we can compute[3] in $O(1)$ time a finite set of paths from X to Y of type $C_1S\bar{C}_i\bar{C}_jSC_2$, where $\bar{C}_i$ and $\bar{C}_j$ are tangent to edges e_i and e_j, respectively. This set contains all optimal paths from X to Y of type $C_1S\bar{C}_i\bar{C}_jSC_2$.*

Given a pair of edges e_i, e_j and a pair of $\mathcal{PC}$-anchored circles $\mathcal{C}_1, \mathcal{C}_2$, tangent to C_I and C_F, respectively, we choose X to be the configuration determined by the intersection of C_I and $\mathcal{C}_1$ and Y to be the configuration determined by C_F and $\mathcal{C}_2$. Now by the above lemma, we can compute in $O(1)$ time a constant number of candidate paths for this pair of edges and anchored circles. Doing this for all possible pairs of edges (e_i, e_j), and pairs of $(\mathcal{C}_1, \mathcal{C}_2)$, we determine $O(n^4)$ path candidates of type (B.iii) in $O(n^4)$ time.

In summary, the simple algorithm examines $O(n^4)$ candidate paths, and for each, spends $O(\log n)$ time checking feasibility, by Lemma 4.1 with $m \leq 8$. Therefore, the overall running time is $O(n^4 \log n)$.

5 An Efficient Algorithm

In this section we prove additional properties of optimal paths that drastically reduce the number of candidates to examine. We have already shown that we need to consider only $O(1)$ Dubins paths and $O(n)$ candidates for paths of type (B.i). We will show that it suffices to consider only $O(1)$ candidate paths of type (A.i) and (B.ii), $O(n)$ candidate paths of type (A.ii), and $O(n^2)$ candidate paths of type (B.iii).

[3] The computation is performed by solving four algebraic systems of three equations in three indeterminates.

Computing paths of type (A.i) and (B.ii). The paths of types (A.i) and (B.ii) contain a strongly $\mathcal{PP}$-anchored C-segment $\bar{\bar{C}}$. The circle $\bar{\bar{\mathcal{C}}}$ supporting $\bar{\bar{C}}$ defines one or two pockets that contain a point of tangency of $\bar{\bar{\mathcal{C}}}$ with $\partial\mathcal{P}$ (see Figures 4(b) and 6). By Lemma 2.5, we know that I and F must belong to these pockets. The following lemma states that there exists at most one circle with these properties.

Lemma 5.1 *For a fixed pair of configurations I, F, there exists at most one $\mathcal{PP}$-anchored circle $\bar{\bar{\mathcal{C}}}$ so that the long arc defined by the tangent points of $\bar{\bar{\mathcal{C}}}$ with $\partial\mathcal{P}$ is free and so that I and F belong to the pocket(s) defined by $\bar{\bar{\mathcal{C}}}$ and its tangent points with $\partial\mathcal{P}$. This circle can be computed in $O(n)$ time.*

By the lemma, we can compute, in $O(n)$ time, a set of $O(1)$ candidate paths of types (A.i) and (B.ii). The candidate paths may be checked for feasibility in $O(\log n)$ time. Therefore, an optimal path of type (A.i) or (B.ii) can be computed in $O(n)$ time.

A monotonicity property of $CCSC$ paths. Subpaths of type $CCSC$ occur in both (A.ii) and (B.iii) path types. In this subsection, we ignore the polygon $\mathcal{P}$, and study paths from X to Y of type $C_1C_2SC_3$, with specified orientations on the C-segments. Then the circles $\mathcal{C}_1$ and $\mathcal{C}_3$ supporting C_1 and C_3, respectively, are fixed. Circle $\mathcal{C}_2$ is determined by M, its tangent point with $\mathcal{C}_1$. For each $M \in \mathcal{C}_1$, there is at most one path $\Pi(M)$ of type $C_1C_2SC_3$ with the specified orientations on C-segments. For certain positions of M, one of the segments may vanish. These positions of M are called *singular points*. The following lemma is proved by calculus.

Lemma 5.2 *As M moves along the oriented circle $\mathcal{C}_1$, $\|\Pi(M)\|$ increases monotonically, except at singular points.*

At singular points where a C-segment vanishes, the path length changes by $\pm 2\pi$. The S-segment vanishes when $\mathcal{C}_2$ and $\mathcal{C}_3$ have opposite orientation and are tangent.[4] Thus, there may be two singular points where the S-segment vanishes. If there are two, they split the circle $\mathcal{C}_1$ into two arcs. Along one of the arcs, circles $\mathcal{C}_2$ and $\mathcal{C}_3$ properly intersect, and so $\Pi(M)$ is not defined there. Thus, the singular points corresponding to a vanishing S-segment are the endpoints of the arc of $\mathcal{C}_1$ on which the path is defined. There may be up to six singular points. See Figure 7 for an illustration of six singular points in a path of type $C^+C^-SC^+$. All the singular points can be computed in $O(1)$ time.

Computing type (A.ii) paths. As mentioned in Section 4, we can compute in $O(n \log n)$ time the feasible candidates of type (A.ii) paths with at most one $\mathcal{PC}$-anchored segment. If the path is of type $C_I\bar{\bar{C}}S\bar{\bar{C}}C_F$, a simple analysis gives $O(n^2)$ candidates to check; we now use Lemma 5.2 to reduce the number of candidates and to compute them in $O(n \log n)$ time.

[4]The S-segment may vanish even if $\mathcal{C}_2$ and $\mathcal{C}_3$ have the same orientation and $\mathcal{C}_2 = \mathcal{C}_3$, but in this case C_2-segment also vanishes.

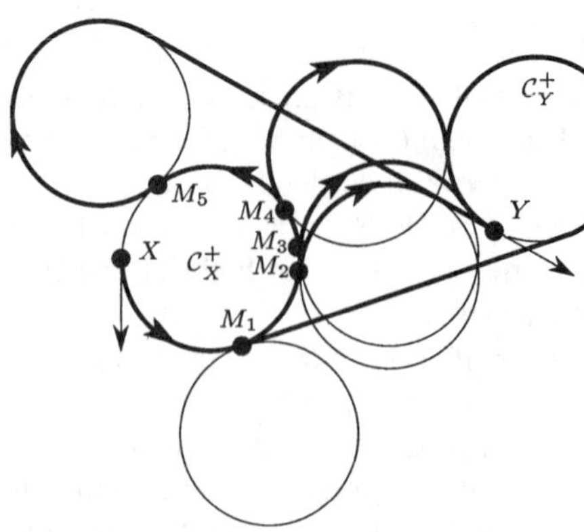

Figure 7. Paths of type $C^+C^-SC^+$ from X to Y and the six singular points X, M_1, M_2, M_3, M_4 and M_5 on $\mathcal{C}_X^+$.

Fix the orientations of the terminal C-segments, and let $\mathcal{C}_I$ and $\mathcal{C}_F$ denote the circles supporting C_I and C_F, respectively. Let $\mathcal{K}_I$ be the sequence of $\mathcal{PC}$-anchored circles that touch $\mathcal{C}_I$ and that are free, sorted by their tangent points with $\mathcal{C}_I$. The set $\mathcal{K}_F$ is defined analogously, for $\mathcal{PC}$-anchored circles tangent to $\mathcal{C}_F$. Note that $\mathcal{K}_I$ and $\mathcal{K}_F$ can be computed in $O(n \log n)$ time, and they have $O(n)$ elements.

By Lemma 3.6, circles supporting the $\bar{\bar{C}}$-segments in an optimal path Π of type $C_I\bar{\bar{C}}_1S\bar{\bar{C}}_2C_F$ are free. Therefore, the $\bar{\bar{C}}_2$-segment of Π lies on a circle of $\mathcal{K}_F$. Suppose $\mathcal{C}_2 \in \mathcal{K}_F$ supports the $\bar{\bar{C}}_2$-segment of Π. This fixes the terminal configuration of the subpath $C_I\bar{\bar{C}}_1S\bar{\bar{C}}_2$. The above subsection on monotonicity implies we have up to up to six singular points on $\mathcal{C}_I$.

Let $\delta \subseteq \mathcal{C}_I$ be an arc joining two singular points and let $\mathcal{K}_I(\delta) \subseteq \mathcal{K}_I$ be the subsequence of circles that touch $\mathcal{C}_I$ at a point in δ. By Lemma 5.2, only the first circle of $\mathcal{K}_I(\delta)$ is a candidate for $\mathcal{C}_1$. Hence, at most six circles in $\mathcal{K}_I$ are candidates for $\mathcal{C}_1$, and they can be computed in $O(\log n)$ time by performing a binary search. By examining each $\mathcal{C}_2 \in \mathcal{K}_F$ in turn, we compute $O(n)$ candidate paths in $O(n \log n)$ time. We can therefore conclude that an optimal path of type (A.ii) can be computed in $O(n \log n)$ time.

Computing type (B.iii) paths. Let Π be an optimal path of the form $\Pi_I\bar{C}_i\bar{C}_j\Pi_F$, i.e. of type (B.iii). Suppose we know the edges e_i, e_j that are tangent to $\bar{C}_i$ and $\bar{C}_j$, respectively.

If Π does not contain any $\bar{\bar{C}}$-segment in Π_I or Π_F, then we can compute Π in $O(\log n)$ time using Lemmas 4.2 and 4.1.

Consider now the case in which Π_I and Π_F each contains a $\bar{\bar{C}}$-segment, i.e. Π is of type $C_I\bar{\bar{C}}S\bar{C}_i\bar{C}_jS\bar{\bar{C}}C_F$. We show that, given e_i and e_j, we can compute, in $O(\log n)$ time, a set of $O(1)$ candidate circles that contains the $\bar{\bar{C}}$-segments of Π. Given this set, we can compute the shortest feasible path of the above type in $O(\log n)$ time, by Lemmas 4.1 and 4.2. Thus, by considering all $O(n^2)$ pairs of edges of $\mathcal{P}$, we can compute in $O(n^2 \log n)$ time a set of $O(n^2)$ candidate paths for this case. However, we will see later that in some cases we need not consider all $O(n^2)$ pairs of edges of $\mathcal{P}$.

We first establish some simple properties of an optimal path Π of type $C_I\bar{\bar{C}}_1S\bar{C}_i\bar{C}_jS\bar{\bar{C}}_2C_F$. Assume without loss of generality that $\bar{C}_i, \bar{C}_j$ are oriented clockwise and counterclockwise, respectively. By Lemma 3.3, the $\bar{\bar{C}}_1$-segment is oriented clockwise, and the $\bar{\bar{C}}_2$-segment is oriented counterclockwise, i.e., Π is of type $C_I^+\bar{\bar{C}}_1^-S\bar{C}_i^-\bar{C}_j^+S\bar{\bar{C}}_2^+C_F^-$. Let $\bar{\bar{\mathcal{C}}}_1, \bar{\mathcal{C}}_i, \bar{\mathcal{C}}_j$, and $\bar{\bar{\mathcal{C}}}_2$ denote the circles supporting the C-segments $\bar{\bar{C}}_1, \bar{C}_i, \bar{C}_j$, and $\bar{\bar{C}}_2$, respectively.

Lemma 5.3 *If an optimal path is of type $C_I\bar{\bar{C}}_1S\bar{C}_i\bar{C}_jS\bar{\bar{C}}_2C_F$, then the circles $\bar{\bar{\mathcal{C}}}_1$, $\bar{\mathcal{C}}_i$, $\bar{\mathcal{C}}_j$, and $\bar{\bar{\mathcal{C}}}_2$ are free.*

Proof: Lemma 3.6 directly yields that $\bar{\bar{\mathcal{C}}}_1$ and $\bar{\bar{\mathcal{C}}}_2$ are free. Suppose now for a contradiction that $\bar{\mathcal{C}}_j$ is not free. As before, we assume that the orientations are such that $\Pi = C_I^+\bar{\bar{C}}_1^-S\bar{C}_i^-\bar{C}_j^+S\bar{\bar{C}}_2^+C_F^-$. Let T be the tangent point between $\bar{C}_i$ and $\bar{C}_j$. Moving along $\bar{\mathcal{C}}_j^+$, let X be the last tangent point between $\bar{\mathcal{C}}_j$ and $\partial\mathcal{P}$. Starting at X and moving along $\bar{\mathcal{C}}_j^+$, let Y be the first proper intersection point between $\bar{\mathcal{C}}_j$ and $\partial\mathcal{P}$ (see Figure 8).

By Lemma 2.4, the length of $\bar{C}_j$ between T and X is greater than π, i.e. $\|\bar{\mathcal{C}}_j^+[T,X]\| > \pi$. It follows that $\bar{\mathcal{C}}_j$, X and Y define a pocket $\Lambda_{\bar{\mathcal{C}}_j}[X,Y]$ (see Figure 8). By Lemma 2.5, this pocket contains $\Pi[X,F]$ and therefore contains $\bar{\bar{C}}_2$. We know the free circle $\bar{\bar{\mathcal{C}}}_2$ cannot be entirely inside a pocket. The path $\bar{C}_jS\bar{\bar{C}}_2$ enters the pocket at X, and since $\bar{\bar{\mathcal{C}}}_2$ is free, it is possible to escape the pocket by extending segment $\bar{\bar{C}}_2$. This contradicts Lemma 2.5, establishing that $\bar{\mathcal{C}}_j$ is free. A symmetric argument shows that circle $\bar{\mathcal{C}}_i$ is free. □

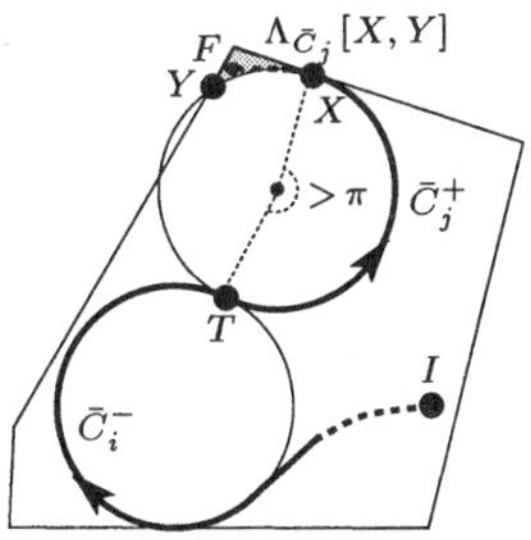

Figure 8. For the proof of Lemma 5.3.

We now introduce the following simple definition. Given a circle $\mathcal{C}$ and a point $X \in \mathcal{C}$, a point $M \in \mathcal{C}$ is called the *first free point after X along $\mathcal{C}^+$* if and only if the circle tangent to $\mathcal{C}$ at M is free and for any $M' \in \mathcal{C}^+[X,M)$, the circle tangent to $\mathcal{C}$ at M' is not free (in Figure 9, M^* is the first free point after M_L along $\mathcal{C}_I^+$). Note that M could be X. The circle tangent to $\mathcal{C}$ at the first free point after X is called the *first free circle after X along $\mathcal{C}^+$*.

We show that, given I, F, e_i and e_j, we can compute in $O(\log n)$ time a set of $O(1)$ candidate circles that contain the $\bar{\bar{C}}$-segments of any optimal path from I to F of type $C_I^+\bar{\bar{C}}_1^-S\bar{C}_i^-\bar{C}_j^+S\bar{\bar{C}}_2^+C_F^-$. We show how to compute candidate circles for $\bar{\bar{C}}_1$; computing candidate circles for $\bar{\bar{C}}_2$ is similar.

We identify two circles $\mathcal{C}'$ and $\mathcal{C}''$ that are the candidate circles for $\bar{\bar{C}}_1$. See Figure 9. Let $\mathcal{C}'$ be the first free circle after I along $\mathcal{C}_I^+$. If there is no free circle after I along $\mathcal{C}_I^+$, then $\mathcal{C}'$ and $\mathcal{C}''$ are not defined. Assume, after a possible rotation, that the line L through e_i is horizontal and $\mathcal{P}$ is above L. If the distance between L and the center of $\mathcal{C}_I^+$ is greater than 2, then $\mathcal{C}''$ is not defined. Otherwise, there exist two circles that are above L and tangent to both $\mathcal{C}_I^+$ and L. Let $\mathcal{C}_L$ be the leftmost of these two circles, and let M_L be its tangent point with $\mathcal{C}_I^+$. Let $\mathcal{C}''$ be the first free circle after M_L along $\mathcal{C}_I^+$. Note that $\mathcal{C}'$ and $\mathcal{C}''$ only depend on I, $\mathcal{C}_I^+$, and on the line L through e_i.

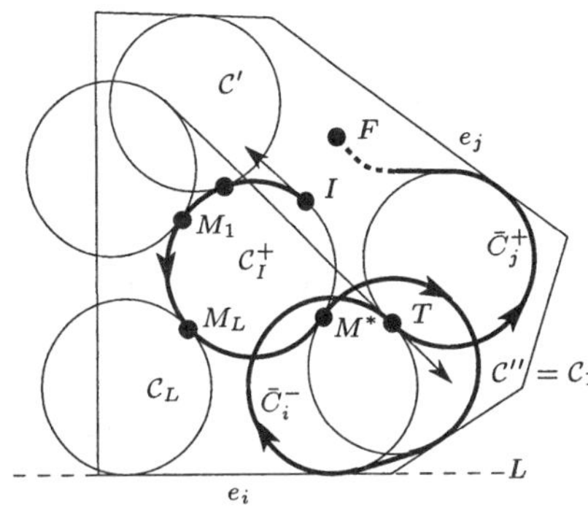

Figure 9. For the proof of Lemma 5.4.

Lemma 5.4 *Let Π be an optimal path of type $C_I^+\bar{\bar{C}}_1^-S\,\bar{C}_i^-\bar{C}_j^+$ $S\bar{\bar{C}}_2^+C_F^-$, and let L be the line through the edge tangent to $\bar{C}_i$. Then $\bar{\bar{C}}_1$ is supported by $\mathcal{C}'$ or $\mathcal{C}''$.*

Proof (Sketch): We prove the lemma only in the case where $\mathcal{C}_I^+$ and $\mathcal{C}_i$ properly intersect. Let $T \in \Pi$ be the configuration at the tangent point between $\bar{C}_i$ and $\bar{C}_j$. See Figure 9.

The circle $\mathcal{C}_1$ supporting the $\bar{\bar{C}}_1$-segment is tangent to $\mathcal{C}_I^+$. As before, any choice of a point $M \in \mathcal{C}_I^+$ defines at most one path $\Pi(M)$ of the form $C_I^+C_1^-SC_i^-$, which begins at I and ends at T, and where $\mathcal{C}_I^+$ and $\mathcal{C}_1^-$ are tangent at M. Let M^* be the intersection point of the C_I^+- and $\bar{\bar{C}}_1^-$ segments of the optimal path Π. Then $\Pi(M^*)$ is a subpath of Π and so it is an optimal path from I to T. By the monotonicity property (Lemma 5.2), and since $\mathcal{C}_1$ and $\mathcal{C}_i$ are free (Lemma 5.3), M^* must be the first free point along $\mathcal{C}_I^+$ after a singular point of $\Pi(M)$. Since $\mathcal{C}_I^+$ and $\mathcal{C}_i$ properly intersect, there are only two singular points I and M_1 of $\Pi(M)$, where M_1 corresponds to the vanishing of $\bar{\bar{C}}_i$.

If M^* is the first free point after I along $\mathcal{C}_I^+$, then $\bar{\bar{C}}_1$ is supported by $\mathcal{C}'$, the first free circle after I. If M^* is the first free point after M_1, then we show that $\bar{\bar{C}}_1$ is supported by $\mathcal{C}''$, the first free circle after M_L.

By Lemma 2.4, the arc length of $\bar{C}_i$ from its tangent point with L to T must be at least π. In other words, T must be in the right half of $\bar{C}_i$ (as L is horizontal and $\mathcal{P}$ is above L). Therefore by definition of M_L, the arc length of C_i in $\Pi(M_L)$ is less than π.

It follows that for any point $M \in C_I^+[M_1, M_L]$, the arc length of C_i in $\Pi(M)$ is less than π, so by Lemma 2.3, $\Pi(M)$ cannot be part of the optimal path. Thus, M^* does not belong to $C_I^+[M_1, M_L]$. So if M^* is the first free point after M_1, then it is the first free point after M_L. In other words, $\bar{\bar{C}}_1$ is supported by $\mathcal{C}''$. □

Lemma 5.5 *$\mathcal{C}'$ and $\mathcal{C}''$ can be computed in $O(\log n)$ time.*

Proof: Let Γ be the circle of radius 2 concentric with C_I^+. Let I_Γ (resp. M_Γ) be the intersection point between Γ and the ray emanating from the center of C_I^+ and going through I (resp. M_L) (see Figure 10). Let $\mathcal{R}$ be the *retracted polygon* of $\mathcal{P}$ with respect to a unit circle, i.e., $\mathcal{R}$ is the set of points p such that the unit circle centered at p lies inside $\mathcal{P}$; $\mathcal{R}$ is a convex polygonal region with at most n edges, and it can be computed in linear time. Let O' be the first intersection point between Γ and $\mathcal{R}$ starting at I_Γ and moving along Γ^+. The center of $\mathcal{C}'$ is O'. Indeed, by definition of $\mathcal{R}$, the circle centered at O' is free, and any circle (of unit radius) centered at a point on $\Gamma^+[I_\Gamma, O')$ is not free. Similarly, the center of $\mathcal{C}''$ is the first intersection point between Γ and $\mathcal{R}$ starting at M_Γ and moving along Γ^+. Using the circle-shooting data structure by Agarwal and Sharir [2], $\mathcal{R}$ can be preprocessed in $O(n \log n)$ time, so that $\mathcal{C}'$ and $\mathcal{C}''$ can be computed in $O(\log n)$ time. □

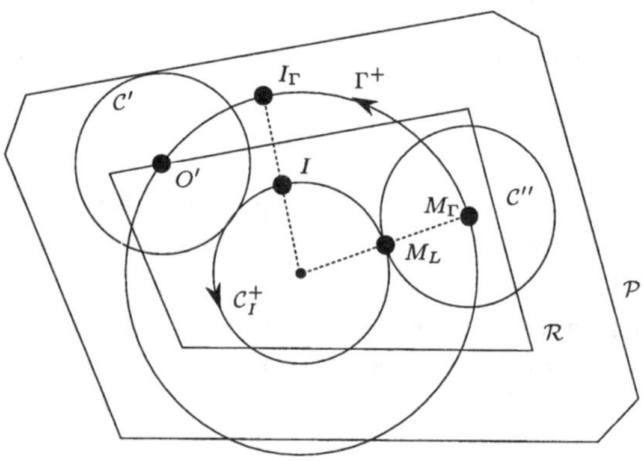

Figure 10. For the proof of Lemma 5.5.

By Lemmas 5.4 and 5.5, we can compute, in $O(\log n)$ time, two candidates for the circle supporting segment $\bar{\bar{C}}_1$. We can similarly compute two candidates for the circle supporting segment $\bar{\bar{C}}_2$. By Lemma 4.2, this gives us $O(1)$ candidate paths, for which we may check the feasibility in $O(\log n)$ time. Hence, given two edges e_i and e_j of $\mathcal{P}$, we can compute in $O(\log n)$ time, an optimal path of type $C_I\bar{\bar{C}}S\bar{C}_i\bar{C}_jS\bar{\bar{C}}C_F$, where $\bar{C}_i$ and $\bar{C}_j$ are tangent to e_i and e_j, respectively.

In the cases where the optimal path is of type (B.iii) with only one $\bar{\bar{C}}$-segment in Π_I or Π_F, we get similar results. For example, if an optimal path is of type $C_I\bar{\bar{C}}_1S\bar{C}_i\bar{C}_jSC_F$, then $\bar{\bar{C}}_1$ and $\bar{C}_i$ are free, and $\bar{\bar{C}}_1$ is supported by the $\mathcal{C}'$ or $\mathcal{C}''$ defined above. Thus we obtain the following lemma.

Lemma 5.6 *Let e_i, e_j be edges of $\mathcal{P}$. In $O(\log n)$ time, we can compute an optimal path of type $\Pi_I\bar{C}_i\bar{C}_j\Pi_F$ where $\Pi_I \in \{C_I\bar{\bar{C}}S, C_IS, C_I, S\}$, $\Pi_F \in \{S\bar{\bar{C}}C_F, SC_F, C_F, S\}$ and where $\bar{C}_i$ and $\bar{C}_j$ are tangent to e_i and e_j, respectively.*

Now we describe how to find a suitable set of pairs of edges $\mathcal{E}$ such that if an optimal path from I to F is of type (B.iii) (i.e., $\Pi_I\bar{C}_i\bar{C}_j\Pi_F$), then the pair of edges (e_i, e_j) tangent to $\bar{C}_i$ and $\bar{C}_j$ is in the set $\mathcal{E}$.

From [1], we know that if an optimal path from I to F is of type $\Pi_I\bar{C}_i^+\bar{C}_j^-\Pi_F$ such that $\bar{C}_i$ and $\bar{C}_j$ are nonterminal, then C_I^+ intersects $\bar{\mathcal{C}}_j$ (the circle supporting $\bar{C}_j$), and C_F^- intersects $\bar{\mathcal{C}}_i$ (the circle supporting $\bar{C}_i$). Thus, the center of $\bar{\mathcal{C}}_j$, which is at most distance 1 from the boundary of the polygon, is at most distance 3 from I. Since centers of $\bar{\mathcal{C}}_i$ and $\bar{\mathcal{C}}_j$ are distance 2 apart, they are each distance less than 5 from I. Thus, edges e_i and e_j are distance less than 6 from I. By symmetry, they are also distance less than 6 from F. Therefore, we can consider $\mathcal{E}$ to be the set of pairs of edges of $\mathcal{P}$ that are distance less than 6 from both I and F. Let k denote the number of edges of $\mathcal{P}$ distance less than 6 from both I and F. Then $|\mathcal{E}| = k^2$, and $\mathcal{E}$ can be computed in $O(k^2)$ time. Lemma 5.6 then gives:

Lemma 5.7 *An optimal path of type (B.iii) can be computed in $O(k^2 \log n)$ time.*

Putting everything together, we obtain the following.

Theorem 5.8 *Given a convex polygon $\mathcal{P}$, an initial configuration I, and a final configuration F, an optimal path from I to F inside $\mathcal{P}$ can be computed in time $O((n + k^2) \log n)$, where k is the number of edges of $\mathcal{P}$ at distance less than 6 from both I and F.*

Proof: We have shown in the previous subsections that the Dubins paths and the optimal paths of type (A.i), (A.ii), (B.i), and (B.ii) can be computed in $O(n \log n)$ time, while paths of type (B.iii) can be computed in $O(k^2 \log n)$ time. Choosing the shortest among all those paths yields an optimal path. □

6 Conclusion

Our classification of path types in a convex polygon yields a fast algorithm for computing an optimal path. An interesting question is whether the running time can be improved to $O(n \log n)$ by proving additional properties of paths of type (B.iii). Our results show that even for a convex polygon, optimal paths between two configurations can be rather complex. Such complex paths may be difficult to track by a mobile robot. Furthermore, they may arise as artifacts of a tightly constricted environment. A direction for future research is to seek a realistic notion of feasibility that rejects hard-to-follow paths, while admitting fast computation of optimal paths.

Acknowledgments

This paper arose from research begun at the International Workshop on Bounded Radius of Curvature, organized by S. Whitesides and held at the Bellairs Research Institute of McGill University, March 9-16, 1997. We would like to thank Hazel Everett, Micha Godau, and Steve Wismath for many useful discussions.

References

[1] P. K. Agarwal, P. Raghavan, and H. Tamaki. Motion planning for a steering-constrained robot through moderate obstacles. In *Proc. 27th Annu. ACM Sympos. Theory Comput.*, pages 343–352, 1995.

[2] P. K. Agarwal and M. Sharir. Circle shooting in a simple polygon. *J. Algorithms*, 14:69–87, 1993.

[3] J. Barraquand and J.-C. Latombe. Nonholonomic multi-body mobile robots: Controllability and motion planning in the presence of obstacles. *Algorithmica*, 10:121–155, 1993.

[4] J.-D. Boissonnat, A. Cérézo, and J. Leblond. Shortest paths of bounded curvature in the plane. *Internat. J. Intell. Syst.*, 10:1–16, 1994.

[5] J.-D. Boissonnat and S. Lazard. A polynomial-time algorithm for computing a shortest path of bounded curvature amidst moderate obstacles. In *Proc. 12th Annu. ACM Sympos. Comput. Geom.*, pages 242–251, 1996.

[6] J. Canny, B. R. Donald, J. Reif, and P. Xavier. On the complexity of kinodynamic planning. In *Proc. 29th Annu. IEEE Sympos. Found. Comput. Sci.*, pages 306–316, 1988.

[7] J. Canny, A. Rege, and J. Reif. An exact algorithm for kinodynamic planning in the plane. *Discrete Comput. Geom.*, 6:461–484, 1991.

[8] B. Chazelle and L. J. Guibas. Fractional cascading: I. A data structuring technique. *Algorithmica*, 1:133–162, 1986.

[9] B. R. Donald and P. Xavier. A provably good approximation algorithm for optimal-time trajectory planning. In *Proc. IEEE Internat. Conf. Robot. Autom.*, pages 958–963, 1989.

[10] L. E. Dubins. On curves of minimal length with a constraint on average curvature and with prescribed initial and terminal positions and tangents. *Amer. J. Math.*, 79:497–516, 1957.

[11] S. Fortune and G. Wilfong. Planning constrained motion. *Ann. Math. Artif. Intell.*, 3:21–82, 1991.

[12] D. Halperin, L. Kavraki, and J.-C. Latombe. Robotics. In *CRC Handbook of Computational Geometry* (J. Goodmand and J. O'Rourke, eds.), pages 755–778. CRC Press, Boca Raton, NY, 1997.

[13] P. Jacobs and J. Canny. Planning smooth paths for mobile robots. In *Nonholonomic Motion Planning* (Z. Li and J. Canny, eds.), pages 271–342. Kluwer Academic Publishers, Norwell, MA, 1992.

[14] P. Jacobs, J.-P. Laumond, and M. Taix. Efficient motion planners for nonholonomic mobile robots. *In Proc. of the IEEE/RSJ Internat. Workshop on Intell. Robots and Systems*, pages 1229–1235, 1991.

[15] K. Kedem, R. Livne, J. Pach, and M. Sharir. On the union of Jordan regions and collision-free translational motion amidst polygonal obstacles. *Discrete Comput. Geom.*, 1:59–71, 1986.

[16] J.-C. Latombe. A fast path-planner for a car-like indoor mobile robot. *In Proc. of the 9th National Conf. on Artif. Intell.*, pages 659–665, 1991.

[17] J.-C. Latombe. *Robot Motion Planning*. Kluwer Academic Publishers, Norwell, MA, 1991.

[18] J.-P. Laumond. Finding collision-free smooth trajectories for a non-holonomic mobile robot. *In Proc. of the Internat. Joint Conf. on Artif. Intell.*, pages 1120–1123, 1987.

[19] J.-P. Laumond, P. Jacobs, M. Taix, and R. Murray. A motion planner for nonholonomic mobile robots. *IEEE Trans. on Robot. and Autom.*, 10:577–593, 1994.

[20] J.-P. Laumond, M. Taix, and P. Jacobs. A motion planner for car-like robots based on global/local approach. *In Proc. of the IEEE/RSJ Internat. Workshop on Intell. Robots and Systems*, pages 765–773, 1990.

[21] B. Mirtich and J. Canny. Using skeletons for nonholonomic path planning among obstacles. In *Proc. 9th IEEE Internat. Conf. Robot. Autom.*, pages 2533–2540, 1992.

[22] C. Ó'Dúnlaing. Motion-planning with inertial constraints. *Algorithmica*, 2:431–475, 1987.

[23] J. A. Reeds and L. A. Shepp. Optimal paths for a car that goes both forwards and backwards. *Pacific J. Math.*, 145(2), 1990.

[24] J. Reif and H. Wang. The complexity of the two dimensional curvature-constrained shortest-path problem. In *Proc. 3rd Workshop on the Algo. Found. of Robotics* (P. K. Agarwal, L. E. Kavraki, and M. Mason, eds.). A. K. Peters, Wellesley, MA, 1998.

[25] J. T. Schwartz and M. Sharir. Algorithmic motion planning in robotics. In *Algorithms and Complexity*, volume A of *Handbook of Theoretical Computer Science* (J. van Leeuwen, ed.), pages 391–430. Elsevier, Amsterdam, 1990.

[26] S. Sekhavat, P. Švestka, J.-P. Laumond, and M. Overmars. Multi-level path planning for non-holonomic robots using semi-holonomic subsystems. In *Workshop on the Algo. Found. of Robotics* (J.-P. Laumond and M. Overmars, eds.), pages 79–96. A. K. Peters, Wellesley, MA, 1996.

[27] J. Sellen. Planning paths of minimal curvature, *In Proc. of the IEEE Internat. Conf. Robot. and Autom.*, 1995.

[28] J. Sellen. Approximation and decision algorithms for curvature constrained path planning: A state-space approach. In *Proc. 3rd Workshop on the Algorithmic Foundations of Robotics* (P. K. Agarwal, L. E. Kavraki, and M. Mason, eds.). A. K. Peters, Wellesley, MA, 1998.

[29] H. J. Sussmann. Shortest 3-dimentional paths with a prescribed curvature bound. In *Conf. on Decision & Control*, pages 3306–3311, 1995.

[30] H. Wang and P. K. Agarwal. Approximation algorithms for curvature constrained shortest paths. In *Proc. 7th ACM-SIAM Sympos. Discrete Algorithms*, pages 409–418, 1996.

[31] G. Wilfong. Motion planning for an autonomous vehicle. In *Proc. IEEE Internat. Conf. Robot. Autom.*, pages 529–533, 1988.

[32] G. Wilfong. Shortest paths for autonomous vehicles. *In Proc. of the IEEE Internat. Conf. Robot. and Autom.*, pages 15–20, 1989.

7th Annual Video Review of Computational Geometry

The following notes give brief descriptions of the video segments that make up the accompanying video proceedings. As in previous years, the videos use images to enhance our understanding of geometric ideas. In some cases this involves the illustration of algorithms in action. In others, we see applications of geometric algorithms in other disciplines. As the field matures, we see the videos serving as valuable adjuncts in each of these functions.

The video segments were selected by a video program committee meeting in Tel Aviv on January 30th, 1998. The segments were judged based on their clarity, quality, and contribution to the field. Some accepted videos have been revised to reflect the comments of the committee.

We thank the members of the Video Program Committee for their help in evaluating the entries. The members of the committee were Gershon Elber (Technion), Dan Halperin (Tel Aviv University), Leo Joskowicz (Hebrew University), Matthew Katz (Ben-Gurion University), Joseph S.B. Mitchell (SUNY Stony Brook), and Ayellet Tal (Technion). We also thank Gill Barequet (Johns Hopkins University) for assisting with the duplication of the final tape.

Dan Halperin
Video Review Chair

Voronoi Diagrams by Divergences with Additive Weights

K. Sadakane*, H. Imai*, K. Onishi*, M. Inaba*, F. Takeuchi* and K. Imai†

*Department of Information Science, University of Tokyo, Tokyo, Japan

†Department of Information and System Engineering, Chuo University, Tokyo, Japan

Abstract

We introduce the Voronoi diagram by the divergence determined by a convex function with additive weights. This class of Voronoi diagrams includes the Euclidean case and further the Voronoi diagram for normal distributions in a statistically meaningful setting. With the additive weights, the Voronoi diagram for circles is also included. These Voronoi diagrams can be handled in a unified way via appropriate potential functions and its tangent hypersurfaces.

1 Divergence Voronoi diagram

In this section, we define the Voronoi diagram by the divergence determined by a given convex function. The following preliminary results in sections 1.1 and 1.2 may be found in [9] and [1, 2], respectively.

1.1 Conjugacy of convex functions

Let S be an open convex set in $\mathbf{R}^d$. Let ψ be a twice differentiable and strictly convex function on S which diverges to $\pm\infty$ at the infinity/boundary. Define $\partial_i\psi = \partial_i\psi(\theta)$ for $\theta = [\theta^i] \in S$[1] by

$$\partial_i\psi = \frac{\partial\psi}{\partial\theta^i}$$

and denote $[\partial_i\psi]$ in $\mathbf{R}^d$ by $\partial\psi$. Define $S^\psi \subset \mathbf{R}^d$ by

$$S^\psi = \{\partial\psi(\theta) \mid \theta \in S\}$$

Lemma 1 *When ψ is twice differentiable and strictly convex, S^* is an open convex set.*

Define a function $\phi: S^\psi \to \mathbf{R}$ by

$$\phi(\eta) = \sup_{\theta\in S}\{\sum_{i=1}^{d}\theta^i\eta_i - \psi(\theta)\}$$

for $\eta \in S^\psi$. The supremum is attained for θ with $\partial\psi = \eta$, and so define $\eta(\theta)$ by

$$\eta(\theta) = \partial\psi(\theta).$$

[1]The tensor notation is used.

SCG 98 Minneapolis Minnesota USA

Then,

$$\phi(\eta(\theta)) = \sum_{i=1}^{d}\theta^i\eta(\theta)_i - \psi(\theta).$$

Lemma 2 *ϕ is a twice differentiable and strictly convex function on S^ψ.*

ϕ is called the conjugate function of ψ.

1.2 Divergence

For θ, $\tilde\theta \in S$, define the divergence $D^\psi(\theta\|\tilde\theta)$ determined by ψ as

$$D^\psi(\theta\|\tilde\theta) = \psi(\theta) + \phi(\tilde\eta) - \sum_{i=1}^{d}\theta^i\tilde\eta_i$$

for $\tilde\eta = \partial\psi(\tilde\theta)$ [1, 2]. By the definition of the conjugate function,

$$D^\psi(\theta\|\tilde\theta) \geq 0.$$

and the equality holds iff $\theta = \tilde\theta$.

1.3 Voronoi diagram by the divergence

Definition 1 (∇-Voronoi diagram by D^ψ) *For n points $\theta^{(j)}$ $(j = 1, \ldots, n)$ in S with additive weights $w^{(j)} \in \mathbf{R}$, the Voronoi region $V(p_j)$ of p_j is defined by*

$$V(p_j) = \bigcap_{k\neq j}\{\theta \mid D^\psi(\theta\|\theta^{(j)}) + w^{(j)} < D^\psi(\theta\|\theta^{(k)}) + w^{(k)}\}.$$

$V(p_j)$ $(j = 1, \ldots, n)$ partitions S, which is called the ∇-Voronoi diagram of $\theta^{(j)}$ with weight $w^{(j)}$.

Note that there is a bijection between θ on S and η on S^ψ and hence, corresponding to the Voronoi diagram on S, a Voronoi diagram on S^ψ having the same structure can be naturally defined. Therefore, we may handle the diagram in an appropriate side among S and S^ψ to theoretically analyze and algorithmically handle them.

1.4 Examples

1.4.1 Euclidean Voronoi diagram

Let $S = \mathbf{R}^d$, and $\psi(\theta) = \frac{1}{2}\sum_{i=1}^{d}(\theta^i)^2$. Then, the divergence is a half of the square of Euclidean distance between two points. The ∇-Voronoi diagram with $s^{(j)} = 0$ is the ordinary Euclidean diagram.

1.4.2 Voronoi diagram of normal distributions with respect to the Kullback-Leibler divergence

The probability density function of a normal distribution with mean μ and standard deviation σ is given by

$$p(x;\mu,\sigma) = \frac{1}{\sqrt{2\pi}\sigma}\exp\left(-\frac{(x-\mu)^2}{2\sigma^2}\right)$$

and

$$\tilde{S} = \{[\mu,\sigma] \mid -\infty < \mu < +\infty,\ 0 < \sigma < +\infty\}$$

is its parameter space. Instead of $\tilde{S}$, we consider another domain S defined by

$$S = \left\{[\theta^1,\theta^2] \mid \theta^1 = \frac{\mu}{\sigma^2},\ \theta^2 = -\frac{1}{2\sigma^2}\right\} \qquad (1)$$

Then,

$$\psi(\theta) = -\frac{(\theta^1)^2}{4\theta^2} + \frac{1}{2}\log\left(-\frac{\pi}{\theta^2}\right)$$

is a strictly convex function, and we can obtain the Voronoi diagram for this function ψ. This can can be generalized to the class of exponential family [1, 2]. In this case, the divergence corresponds to the Kullback-Leibler divergence $D(p(x;\theta)\|p(x;\widetilde{\theta}))$ which is well known in information theory [4] and statistics [1]:

$$D(p(x;\theta)\|p(x;\widetilde{\theta})) \equiv \int p(x;\theta)\log\frac{p(x;\theta)}{p(x;\widetilde{\theta})}dx = D^{\psi}(p(x;\widetilde{\theta})\|p(x;\theta))$$

2 Unified approach

The following theorem can be obtained. For details, see Onishi and Imai [7, 8] with a more solid background of differential geometry [1, 2].

Theorem 1 *Suppose all additive weights vanish. Introducing the $(d+1)$-th coordinate z, for given n points $\theta^{(j)}$ $(j = 1,\ldots,n)$ in S, consider a tangent hyperplane at $[\theta^{(j)},\psi(\theta^{(j)})]$ of $z = \psi(\theta)$ in the $[\theta,z]$-space for each $\theta^{(j)}$. Then, the ∇-Voronoi diagram is the projection of the upper envelope of these tangent hyperplanes to S.*

This extends the result for Euclidean diagrams [5]. Note that the arrangement of tangent hyperplanes represent all the higher-order Voronoi diagram. Also, note that the hyperplane arrangement may be defined in the whole $[\theta,z]$-space, but only part projected to S corresponds to Voronoi diagrams. A theorem for generalized Delaunay triangulations can be also derived.

To derive the Voronoi diagram with additive weights, we have only to lift each tangent hyperplane for $\theta^{(j)}$ by $w^{(j)}$ in the z-direction. The Euclidean case with additive weights is called the Laguerre-Voronoi diagram for circles $C^{(j)}$ [6] ($w^{(j)}$ is the square of radius for circle $C^{(j)}$) and the power diagram [3].

3 Video descriptions

The Euclidean case: We begin by demonstrating the Voronoi diagram and the potential function for the Euclidean case. Points in the $[x,y]$-plane are lifted up to the potential function, the parabola of $z = (1/2)(x^2+y^2)$. The arrangement of the tangent hyperplanes of the points is taken. The projection of the upper envelope of this arrangement is the Voronoi diagram. By lifting up each tangent hyperplane by the square of the weight of its corresponding point, the Voronoi diagram with additive weight is obtained. This is the Laguerre Voronoi diagram.
The case of normal distributions: We then show the Voronoi diagram and potential function for the case of normal distributions, to exhibit the statistical parametric space. A normal distribution is parameterized by its mean μ and deviation σ, and corresponds to a point (μ,σ) in the plane. We show the ∇-Voronoi diagram for these points in the plane. This diagram uses divergence mentioned in section 1.2 as "distance." We show a curve with its value the subtraction of the divergence from the potential function. This curve becomes a straight line for the Euclidean case, however, for the case here, it is not straight. Thus, we have to handle arrangements of surfaces (in Euclidean space) to compute the Voronoi diagram.
Transformation of coordinates: The coordinate transformation in formula (1) transforms the tangent hypersurfaces of the arrangement to tangent hyperplanes. Thus, the computation of Voronoi diagram, k-th order Voronoi diagram and furthest Voronoi diagram in statistical parametric space can be done similarly as the Euclidean case, handling arrangements of hyperplanes in Euclidean space. In the video, we show the potential function after coordinate transformation, the tangent hyperplanes for four points and their arrangement. The upper envelope of this arrangement, the ∇-Voronoi diagram, and the lower envelope, the furthest ∇-Voronoi diagram, are shown.
Dynamic Voronoi diagram: Finally, we animate the transition of Voronoi diagrams according to the movement of a generator point. The Voronoi diagram of the original (μ,σ) space is shown above, and the transformed (θ^1,θ^2) space is shown below. The abscissa is the boundary of S and $\tilde{S}$ in section 1.4.2, and special phenomena caused by this boundary are shown which do not happen in the Euclidean case.

Acknowledgment This research was supported in part by the Grant-in-Aid of the Ministry of Education, Japan.

References

[1] S. Amari: *Differential Geometrical Methods in Statistics.* Lecture Notes in Statistics, Vol.28, Springer-Verlag, New York, 1985.

[2] S. Amari and H. Nagaoka: *Methods of Information Geometry* (in Japanese). Iwanami Shoten, Tokyo, 1993.

[3] F. Aurenhammer: Power Diagrams: Properties, Algorithms and Applications. *SIAM Journal on Computing,* Vol.16 (1987), pp.87–96.

[4] T. M. Cover and J. A. Thomas: *Elements of information theory.* Wiley-Interscience, New York, 1991.

[5] H. Edelsbrunner: *Algorithms in Combinatorial Geometry.* Springer-Verlag, Berlin, 1987.

[6] H. Imai, M. Iri and K. Murota: Voronoi Diagram in the Laguerre Geometry and Its Applications. *SIAM Journal on Computing,* Vol.14, No.1 (1985), pp.93–105.

[7] K. Onishi and H. Imai: Voronoi Diagram in Statistical Parametric Space by Kullback-Leibler Divergence. *Proceedings of the 13th ACM Symposium on Computational Geometry,* Communication Section, 1997, pp.463–465.

[8] K. Onishi and H. Imai: Towards Riemannian Computational Geometry — Voronoi Diagrams, Convex Hulls and Delaunay-Type Triangulations. Preprint, 1997.

[9] R. T. Rockafellar: *Convex Analysis.* Princeton University Press, Princeton, 1970.

GASP-II – A Geometric Algorithm Animation System for an Electronic Classroom

Maria Shneerson
Department of Applied Mathematics
The Weizmann Institute of Science
Rehovot, Israel, 76100

Ayellet Tal
Department of Electrical Engineering
Technion – Israel Institute of Technology
Haifa, Israel, 32000

1 GASP-II

This video presents GASP-II, a geometric algorithm animation system for a distributed electronic classroom [2]. GASP-II allows the presentation and exploration of three dimensional geometric algorithms. Views of the same running algorithm may reside on different machines, that can be distributed in various geographic locations. While in most other electronic classroom systems, the students are completely passive and the system simply displays the same window on multiple machines, GASP-II allows the students to participate in the process of learning by interacting with the algorithm and collaborating with the instructor.

Each student can choose a different *style* for the animation which is suitable to the student's taste. This is done through a comfortable user interface, without any need to write or modify code. During a lecture, each student can view a different animation of the same running algorithm.

GASP-II provides every student with a control panel for starting, pausing, stopping, rewinding and controlling the speed of the animation. It is thus possible to rewind and rerun the difficult parts of the algorithm until they are fully understood.

GASP-II allows two modes of studying – *self-study* and *guided-study*. In the self-study mode, the lecture becomes similar to studying a chapter from an interactive electronic textbook. In the guided-study mode, the students must follow the instructor throughout the lecture. It is up to the instructor to decide when, during the lecture, a self-study mode is appropriate and when the students should be guided by the instructor. In the latter case, the system ensures that all the views are synchronized with the instructor's view.

The design of GASP-II allows each machine to be in charge of the graphical work displayed on it. One advantage of this architecture is that each student can have a full control over the animation. Another benefit is that potential network traffic bottlenecks are avoided since only a relatively small amount of short messages is sent over the network. Moreover, in contrast to the X approach, the least powerful machine does not determine the rate of the whole class.

GASP-II follows the framework of GASP [3] regarding the creation of algorithm animations. It allows programmers to create animations quickly and easily regardless of the complexity of the algorithm. The programmers need not be concerned with the visual aspects of the animation. However, while GASP has focused on facilitating the debugging process of geometric computing, GASP-II concentrates on supporting electronic classroom environments.

2 The video

The video shows a couple of screens of a running algorithm, in this case the BOXTREE algorithm [1]. One screen is the instructor's screen (Figure 1) while the other is a student's screen (Figure 2). Each screen consists of *an algorithm window, a control panel* and *text window.*

The video demonstrates some of the major features of GASP-II, as described below.

- **Study modes:** During the first segment of the video, the instructor chooses a self-study mode. In this mode, the student need not follow the instructor throughout the lecture. Rather, the student is free to learn the algorithm at his own pace. It is shown in the video how the student can run the animation backwards, and rerun certain parts of the algorithm.
- **Style:** Next, the video illustrates how easy it is for end-users to modify the default animation in

[0]This work has been supported in part by the Israeli Ministry of Science, Eshkol Grant 0554-2-96.

SCG 98 Minneapolis Minnesota USA

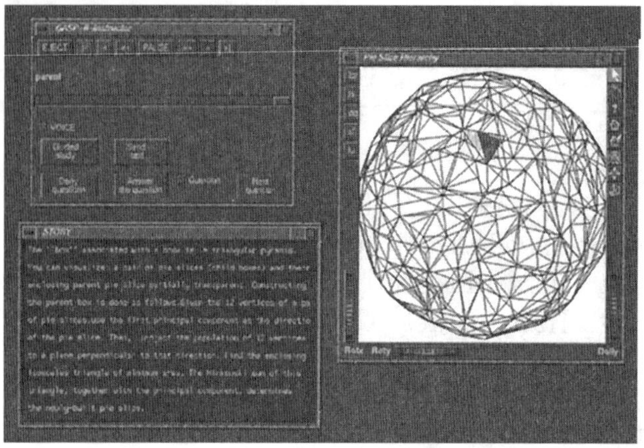

Figure 1: The instructor's Screen

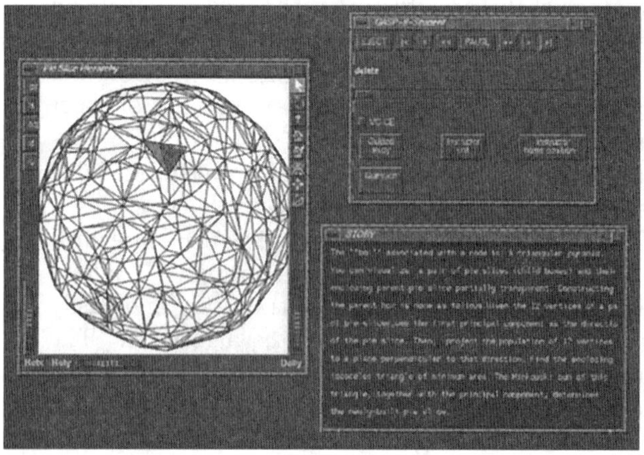

Figure 2: A Student's Screen

order to accommodate their personal taste. This is done by changing a few parameters in the style panel (Figure 3) without any need to write code. The video shows how the student modifies a couple of parameters in the style panel. In our case, the student chose different colors for the objects. The animation is then executed for the second time. However, now the student is viewing the modified animation while the instructor is viewing the original one.

- **Animation synchronization:** The student's animation can be synchronized at any time with the instructor's animation. The student can either ask for synchronization of the animation's phase, or for the exact position of the objects in the scene. The video illustrates these options.

- **Electronic discussions:** GASP-II provides the means for electronic discussions. An electronic discussion is more than a text-exchange, since it is accompanied by a three-dimensional animation of the algorithm, which illustrates the topic discussed. The video demonstrates a scenario in which the student wants to ask a question. GASP-II lets the student type the question in a text window, and send it to the instructor. An indicator on the instructor's control panel indicates that a student asked a question. The instructor asks to read the question and to watch the animation of the appropriate algorithm phase. The instructor can then edit an answer and send it to the student (as well as to the whole class). The student reads the answer in the corresponding window. Finally, the instructor returns to the phase of the animation which was observed before considering the student's question. At this point, the electronic lecture can be resumed.

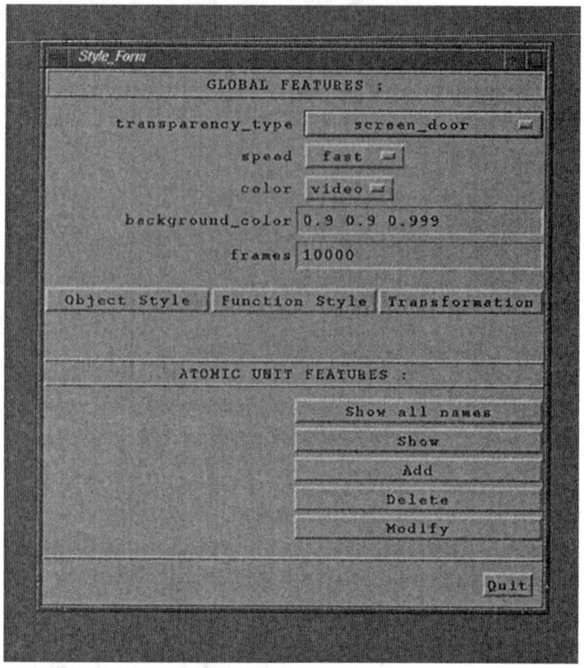

Figure 3: The Style Form

References

[1] G. Barequet, B. Chazelle, L. Guibas, J. Mitchell and A. Tal. BOXTREE: A Hierarchical Representation for Surfaces in 3D, *Graphics Forum,* Vol. 15, No. 3, August 1996, 387–396, 484.

[2] M. Shneerson and A. Tal. GASP-II: A Geometric Algorithm Animation System for an Electronic Classroom, *Thirteenth Annual ACM Symposium on Computational Geometry,* June 1997, 379–381.

[3] A. Tal and D. Dobkin. Visualization of Geometric Algorithms, *IEEE Transactions on Visualization and Computer Graphics,* Vol. 1, No. 2, June 1995, 194–204.

Visualization of Color Image Quantization using Pairwise Clustering

Luiz Velho[1], Jonas Gomes[1], Marcos Vinicius Rayol Sobreiro[1,2]

[1]IMPA–Instituto de Matemática Pura e Aplicada, Estrada Dona Castorina, 110, Rio de Janeiro, Brazil
{lvelho,jonas,mrayol}@visgraf.impa.br

[2]PUC-Rio, Pontifícia Universidade Católica, Rio de Janeiro, Brazil
mrayol@inf.puc-rio.br

1 Introduction

The quantization of a digital image consists in discretizing the image's color gamut, which implies in a reduction of the color information associated with each pixel in the image.

Color image quantization is an important problem in Computer Graphics and Image Processing. One practical reason for quantization is related to image display. In order to represent an image in a graphics device, the image's color gamut cannot be greater than the gamut of the device.

On the other hand, the discretization of color information reduces the image size and is closely related to compression.

The ultimate goal of quantization is to change the color resolution of an image (number of bits in the color representation) with minimum distortion. Quantization is naturally posed as an optimization problem in cluster analysis [1]: The optimal solution should minimize the quantization error over all possible N-element partitions of the color space.

In this video we describe the problem of color image quantization and present a new color image quantization method that iterates a local optimization technique to reduce the number of colors in the image's gamut, until the desired number of color quantization levels is achieved. The method is described in more detail in [2].

This video is divided into four parts. In the first part we introduce the concepts of image representation and color image quantization. The second part shows an execution of the pairwise clustering quantization algorithm in the 2-dimensional color space of an image projected to the Red-Green plane. Our algorithm running on a full 3-dimensional color space of an image with Red, Green an Blue components is showed in the third part of the video. Finally, a sequence of images quantized from 16 million to 16 colors is showed in the last part of the video.

SCG 98 Minneapolis Minnesota USA

2 A Brief Description of the Algorithm

The method consists of a relaxation process that computes a sequence of quantization levels by performing locally optimal pairwise cluster merging.

The input to quantization is the finite set C of M colors from the image gamut, $C = \{c_1, c_2, \ldots, c_M\}$. Each color c_i has a frequency $F_i = F(c_i)$. To each color c_i we associate an accumulated quantization error $E_A(c_i)$, which is initially set to 0.

The method is composed of the following steps:

1. Compute the image histogram.

2. Compute the quantization error $E(c_i, c_j)$ of every two color cluster $\{c_i, c_j\}$ represented by their quantization levels c_i, c_j from the color input color set.

$$E(c_i, c_j) = \frac{F_i F_j^2 + F_j F_i^2}{(F_i + F_j)^2} \, ||c_i - c_j||^2$$

3. Choose the two color cluster $K_0 = \{c_i, c_j\}$ from the input color set that minimizes the quantization error $E(c_i, c_j)$ computed in the previous step.

4. Compute the quantization level c of the cluster $K_0 = \{c_i, c_j\}$ chosen in the step 3.

5. Substitute the cluster $K_0 = \{c_i, c_j\}$ by its quantization level c_k. This results in a quantized color set C' with $M - 1$ colors. The frequency $F(c_k)$ of the color c_k is given by the sum of the frequencies of c_i and c_j: $F(c_k) = F_i + F_j$. The accumulated quantization error $E_A(c_k)$ of the color c_k is given by

$$E_A(c_k) = E(c_i, c_j) + E_A(c_i) + E_A(c_j).$$

6. Compute the quantization error for all two cluster colors $\{c_k, c_l\}$, from the quantized color set C'.

7. Use the quantized color set C' as input to step 3 of the algorithm. Repeat steps 3 to 7 until the desired number of quantization levels is obtained.

After the quantization levels are obtained using the pairwise clustering algorithm described above, we obtain the quantization cell using equation

$$q(c) = c'_i \iff d(c, c'_i) \leq d(c, c'_j),$$

for all $1 \leq j \leq M$ with $j \neq i$.

The final quantization is obtained by computing the optimal quantization level for the cell using

$$c = \frac{1}{\sum_i F_i} \sum_j F_j c_j,$$

where $F_i = F(c_i)$ is the frequency of the color.

3 Video Description

This video is divided into four parts. First, we introduce the concepts of image representation and color image quantization. In the second and third parts we make a visualization of the algorithm. We finalize the video showing some results of image quantized using our algorithm.

In this section we describe each part of the video in more detail.

3.1 Color Image Quantization Problem

We begin the video discussing the problem of representing an image in a computer. We talk about the discretization of the image plane, a process called sampling, and the discretization of the color space which is a process known as quantization.

We briefly discuss the main problem in image quantization: the choice of the colors in the quantized image. We make a comparison of the same image quantized with two different methods: one using an uniform quantization algorithm and the other one uses our adaptive method.

3.2 2-D Visualization of the Algorithm

For a better illustration of our quantization method, we present an execution of our algorithm in a 2-dimensional color space of an image projected to the Red-Green plane. The partition of the color space into quantized cells is given by the Voronoi diagram associated with the set of quantized colors (i.e. the centroids of the quantization cells). In this execution we can notice that our algorithm picks pair of similar colors and replace them by a third and intermediate one. This process is repeated until the desired number of colors is obtained. Then, the original colors in the image are replaced with the best representative colors.

Finalizing this second part, we show an image with no quantization and the same image quantized to 16 colors.

3.3 3-D Visualization of the Algorithm

In this third part of the video we show our algorithm running on an image that the color space is the RGB cube. By displaying the distribution of the colors in the RGB cube we can notice that the colors are clustered in some regions and our algorithm is able to identify those clusters. The size of the spheres representing the quantized colors is proportional to the frequency of colors in the associated quantization cell. We also notice that the algorithm reduces the number of colors in the image but preserves the structure of the original distribution of the colors in the color space.

3.4 Sequence of Examples

A sequence of images quantized from 16 million colors to 16 colors is showed in this fourth part. The distortion introduced by the quantization of that images are also showed. We can notice that the distortion produced by the method is minimal in the beginning of the process and keeps uniformly distributed as the numbers of colors in the image decreases.

4 Production of the Video

The images in the video were quantized using our implementation of the pairwise clustering quantization algorithm. This implementation also generates data for the visualization of the algorithm running.

All Voronoi diagrams were created using a program by Steve Fortune [3].

All images were rendered using Geomview. Then, they were captured to a Sony LVR 3000 Laser Videodisc Recorder and finalized using the Turbo Cube Video Editing System.

This research has been developed in the laboratory of the VISGRAF project at IMPA. This project is sponsored by research grants from CNPq, FAPERJ, FINEP, and IBM Brasil.

References

[1] H. Spath. *Cluster Analysis Algorithms*. Halsted Press, 1980.

[2] Luiz Velho, Jonas Gomes, and Marcos R. Sobreiro. Color image quantization by pairwise clustering. In L. H. de Figueiredo and M. L. Netto, editors, *X brazilian symposium of computer graphics and image processing*, pages 203–210, Campos do Jordao, SP, October 1997. IEEE Computer Society, Los Alamitos, CA.

[3] Steve J. Fortune. A sweepline algorithm for Voronoi diagrams. *Algorithmica*, 2:153–174, 1987.

Optimal Floodlight Illumination of Stages

Felipe Contreras
Dept. of Comp. Science
University of Ottawa
Ottawa, Ontario, Canada
K1N 9B4
fhca@csi.uottawa.ca

Jurek Czyzowicz
Dépt. d'Informatique
Université du Québec à Hull
Hull, Québec, Canada,
J8X 3X7
czyzowic@uqah.uquebec.ca

Eduardo Rivera-Campo
Dept. de Matemáticas,
Univ. Autonóma Metropolitana-I
México, D.F. 09340, México
erc@xanum.uam.mx

Jorge Urrutia
Dept. of Comp. Science
University of Ottawa
Ottawa, Ontario, Canada
K1N 9B4
jorge@csi.uottawa.ca

1. Abstract

In this video we present an algorithm finding an optimal illumination of a segment by a collection of floodlights. The input of the algorithm are the points at which the floodlights must originate, as well as the position of the segment. The illumination is said to be optimal if the sum of the floodlight angles is the smallest possible. The algorithm works in optimal O(*n log n*) time and O(*n*) space.

2. Introduction

Although interest in the Art Gallery problems started some time ago [1], many problems have been studied in the past two decades (see [3], [5]). Recently, growing interest has been concentrated around the area of floodlight illumination (c.f. [4]). One of the most difficult questions, open for more than ten years, is the stage illumination problem: given a line segment (the stage) and a collection of *n* floodlights, whose origins and angles are given, find an algorithm deciding whether the floodlights may be rotated to illuminate the stage (see Figure 1). Only an exponential decision procedure is known to answer this question ([4]).

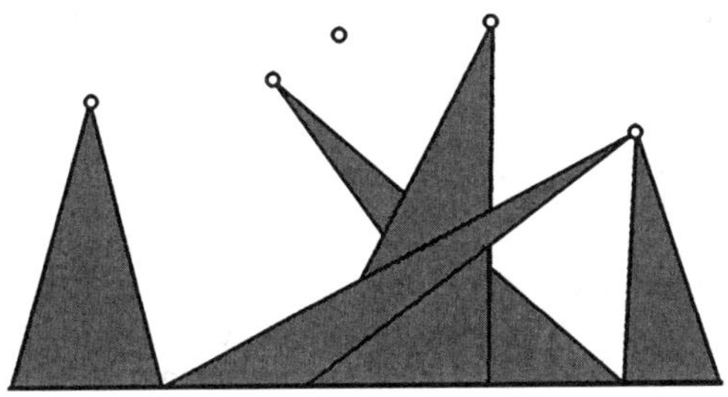

Figure 1: Stage illumination

3. The Problem

An efficient solution exists, however, for a different version of the stage illumination problem: given a stage and the origins of the floodlights, find an algorithm to illuminate the stage minimizing the sum of the floodlight angles. This solution has been presented in [2].

SCG 98 Minneapolis Minnesota USA
Copyright ACM 1998 0-89791-973-4/98/ 6...$5.00

More formally, suppose that we are given a segment *s* and a collection *F* of *n* points on the same side of *s*. Find a set of wedges with apexes at *F*, whose sum of apertures is the smallest possible, such that *s* belongs to their union. We allow some points of *F* to be used as sources of more than one floodlight, as well as some points to be not used at all. The output of the algorithm is to be produced as a partition of the stage into a sequence of subsegments (intervals), each one corresponding to a portion of *s* which will be illuminated from the same floodlight.

4. The Idea of the Algorithm

The idea of the algorithm is to decide for any point *X* of segment *s*, which is the best floodlight to illuminate it. In order to compare which point of the collection, *P* or *Q*, offers a better floodlight illuminating *X*, we take into consideration the smallest possible floodlights from *P* and *Q* illuminating $\varepsilon(X)$, some neighborhood of *X*. We say that *P* is a better floodlight

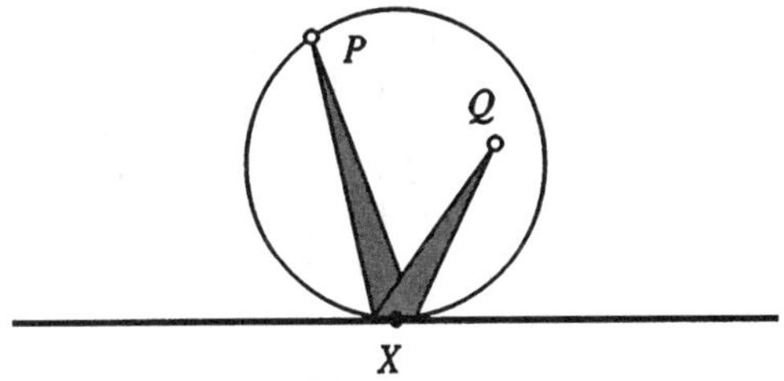

Figure 2: *P* is better than *Q* to illuminate *X*

than *Q* to illuminate *X*, if for any sufficiently small $\varepsilon(X)$ the angle of floodlight *P* illuminating $\varepsilon(X)$ is smaller than the angle of floodlight *Q* illuminating $\varepsilon(X)$ (see Figure 2). It is proven that the best floodlight originates on the boundary of the smallest circle containing *F* and tangent to *s* in point *X*. In the optimal illumination it is always possible to find a partition of segment *s* into at most *n*+1 subsegments, such that each subsegment is illuminated from the same floodlight. The algorithm computes a sequence of *critical circles*, such that each critical circle

1. encloses all the points of collection *F*,
2. passes through two points of *F*,
3. is tangent to segment *s*.

The points of tangency of critical circles to *s* (event-points) are output by the algorithm in order of, say, their right-to-left appearance on *s*. They are the endpoints of the subsegments of mutually disjoint interiors. Each subsegment is illuminated from the same point of *F* and their union is the entire stage. In order to report all the event-points, we walk clockwise on

$CH(F)$, the convex hull of F, starting from point p_1, the closest to s. At each step of this walk we consider a circle passing

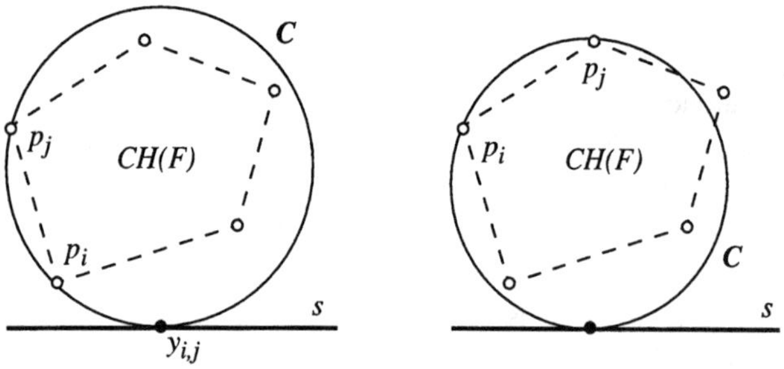

(a) A critical circle C and an event-point $y_{i,j}$

(b) C is not a critical circle: p_j is not used to illuminate s

Figure 3

through p_i and p_j, two consecutive points on $CH(F)$. If the circle encloses all remaining points, it is a critical circle and the corresponding event-point is output. Otherwise, point p_j is discarded from consideration and removed from $CH(F)$ (see Figure 3).

5. The Algorithm Complexity

At each step of the walk around $CH(F)$, either an event-point is determined or some point of $CH(F)$ is discarded. As the input set F, and the output set of event-points are both of linear size, this part of the algorithm requires a linear number of steps. During each step, we have to answer a query of whether a circle considered encloses all points of $CH(F)$. For this purpose, we will precompute $FSV(F)$, the furthest-site Voronoi diagram of F. The center of the circle will then be point-located in $FSV(F)$, to answer the query. As each of the linear number of steps requires logarithmic time for the point location, this part of the algorithm requires $O(n \log n)$ time.

$CH(F)$ and $FSV(F)$ may be both found also in $O(n \log n)$ time, hence this is the complexity of the algorithm.

6. The Lower Bound

We prove the $\Omega(n \log n)$ lower bound by $O(n)$ reduction from sorting.

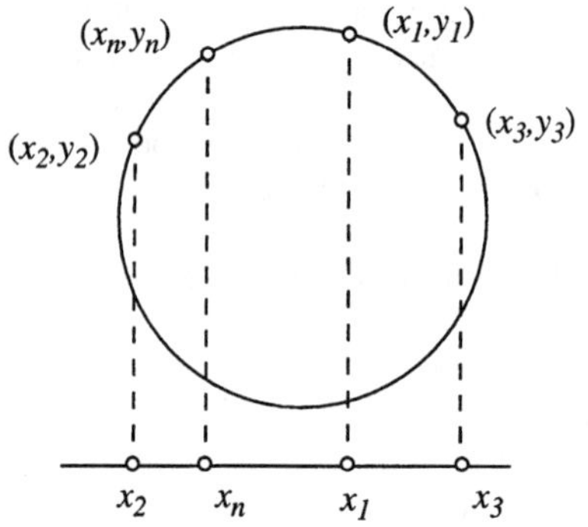

Figure 4: For the lower bound

Suppose that we are given a set of real numbers $\{x_1, x_2,...,x_n\}$. Let C be a sufficiently large circle lying above a given segment s. Map the numbers $\{x_1, x_2,...,x_n\}$ into a set of points $\{(x_1,y_1), (x_2,y_2),...,(x_n,y_n)\}$, such that each point (x_i,y_i) lies in the upper half of C (see Figure 4). According to the observation from section 4, it is easy to see that the order in which the event-points are generated (i.e. the order of illumination of the intervals of s) is the reverse sorted order of $\{x_1, x_2,...,x_n\}$.

7. Technical Data

This video was entirely produced with the aid of Microsoft's Power Point 7. The audio track was recorded and then digitalized using Sound Effects for MacIntosh. Digital manipulation for synchronization was possible using CoolEdit freeware.

8. Acknowledgments

The authors thank Prosenjit Bose, Jörg-Rudiger Sack and Mark Lanthier - professors at the School of Computer Science of Carleton University for many constructive comments and advice.

The narration was performed with the voice of Pierre McNiccoll.

The music by Chaz Guarino was obtained from The Soundroom (http://snhungar.kings.edu/Sndroom.html) created by Scott Hungarter.

Juan Lopez-Michelone from UNAM (México) and François Malric from Multimedia Communications Research Lab. of Univ. of Ottawa were responsible for the VHS recording. Luz Gasca from UNAM provided technical support.

This research was partially supported by NSERC grants OGP000977 and OGP0023992 as well as Mexican Council for Science and Technology (CONACyT) scholarship 110147/110159.

9. References

[1] Chvátal, V.; A combinatorial theorem in plane geometry, *Journal of Combinatorial. Theory, Ser. B*, 18 (1975), 39-41.

[2] Czyzowicz, J., E. Rivera-Campo and J. Urrutia, Optimal floodlight illumination of stages, *Proc. of the Fifth Canadian Conference in Computational Geometry*, (1993) 393-398.

[3] O'Rourke, J.; *Art Gallery Theorems and Algorithms*, Oxford University Press (1987).

[4] O'Rourke, J.; Combinatorics of visibility and illumination: Open problems, in *Discrete and Computational Geometry: Ten Years Later* (B. Chazelle, J.E. Goodman and R. Pollack eds.), to appear.

[5] Shermer, T; Recent Results in Art Galleries, *Proc. IEEE*, Vol. 80, no. 9, (1992) 1384-1399.

GAWAIN: Visualizing Geometric Algorithms with Web-based Animation

Alejo Hausner and David P. Dobkin *

Computer Science Department
Princeton University
Princeton, NJ 08544 U.S.A.
email: {ah,dpd}@cs.princeton.edu

Introduction

The video presents Gawain, a system for animating geometric algorithms on the WWW. Gawain handles all graphics, animation and communication with the user, freeing the programmer to concentrate on geometry.

Gawain can animate algorithms that have been implemented in any language. To animate an algorithm, a programmer inserts function calls that inform Gawain of an algorithm event. These calls indicate that an object has been created or destroyed, or is active in some sense. We provide geometric classes implemented in Java, to encourage geometers to implement their algorithms in this portable, robust, safe and easily learned language.

Features

In the video, we highlight the many features in Gawain that simplify the task of developing geometric algorithms and animating them. These include:

- Gawain can run as an applet inside a web page, providing a means to illustrate technical papers and other textual descriptions of algorithms.
- Algorithms can be visualized in two modes:

*This work supported in part by NSF Grant CCR-9643913 and by the US Army Research Office under Grant DAAH04-96-1-0181

SCG 98 Minneapolis Minnesota USA
Copyright ACM 1998 0-89791-973-4/98/ 6...$5.00

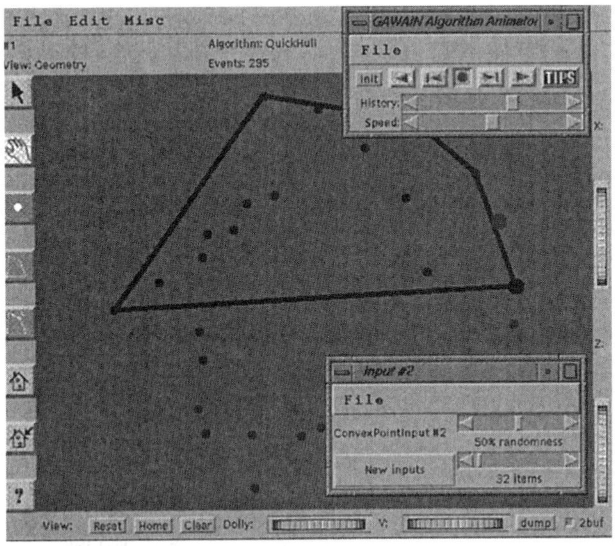

Figure 1: A screen shot showing Gawain in action. An animation of quick-hull is in progress

Immediate: the user changes some data with the mouse, and the program's final results are displayed. This provides an interactive illustration that demonstrates a geometric construct, such as a convex hull.

Animation: the algorithm's events are portrayed in the order they occurred, resulting in an animation.

- Animations can be viewed at variable speed, forwards or backwards, with either gradual or discrete scene changes. This is made possible by saving all the events produced by an algorithm as data in a history list. The events are animated using smooth transitions [7]. The user can interactively move the animation to any point in the algorithm's history.
- Gawain, like other systems [3] [4], provides many input generators, for random and ordered numerical and geometric data. This makes it easier to test algorithms on special cases. A standard interface is provided for adding new input generators.
- Algorithms can be viewed in more than one way. This means that multiple aspects of an algorithm can be

visualized. The user might see one view showing the algorithm's actions on the input data, or another showing the changes in an internal data structure.

- Simultaneous multiple views allow advanced interactions. For example, we can
  1. View two aspects of the same algorithm (see previous item).
  2. Race several algorithms on the same data.
  3. Compare the same algorithm's behavior on different inputs. This helps in debugging, as a user can make small changes in the inputs and make a side-by-side comparison of the original and altered behavior.
- To encourage the use of Java as a programming language for geometers, Gawain includes a set of geometric primitives, modelled on the LEDA [5] classes. This set includes points, line segments, infinite lines, rays, circles and polygons. All these primitives are three-dimensional. Although Gawain does not currently support polyhedral meshes, the next version will.
- Algorithms which have already been implemented in another language can be adapted to work with Gawain by inserting function calls to an interface library, at those points where "interesting events" happen. The interface library sends events and receives Gawain commands over the network, in a way similar to the Mocha system [1]. We provide this feature to avoid the time-consuming and error-prone process of re-implementing algorithms in a new language, and to ease the process of learning a new system.

  The video shows animations based on existing code for triangulation and polygon intersection from O'Rourke's book [6]. It also includes an animation of *qhull* [2].
- Gawain includes an interface to a standard debugger (not shown in the video), which lets a programmer visualize geometric data generated by a program. Debugging can benefit from animation, but it often needs a program that is close to being free of bugs, a requirement that cannot be satisfied in the early stages of implementation. Gawain's interface can help to bridge that gap.
- Gawain includes a three-dimensional environment which lets the user view and change the geometric data. Achieving responsiveness while implementing this in Java, an interpreted language, presented some interesting challenges.
- Gawain can act as a front-end to an SGI Inventor renderer. This provides for higher-quality animations, similar to GASP [8]. The user selects and configures the (possibly multiple) algorithms to be animated, then pulls down a menu. Gawain will then save the animation's scenes as Inventor data files, which can be rendered off-line later.

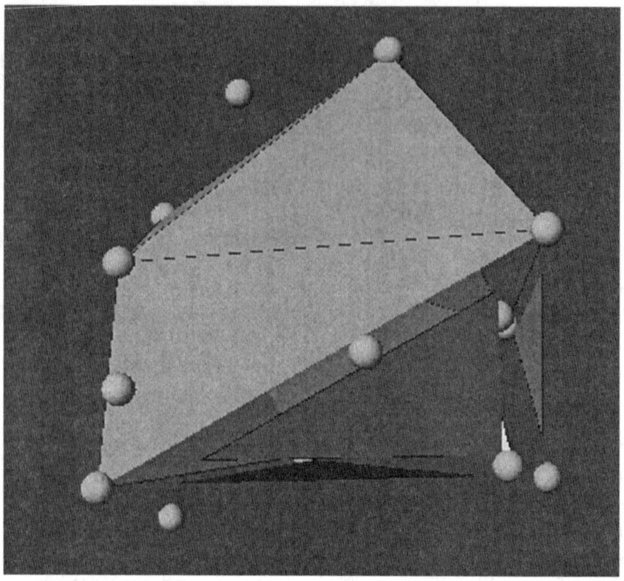

Figure 2: A frame from the animation of *qhull*

References

[1] J. E. Baker, I. F. Cruz, G. Liotta, and R. Tamassia. "A new model for algorithm animation over the WWW", *ACM Computing Surveys*, 27(4):568-572, 1995.

[2] C.B. Barber, D.P. Dobkin, H. Huhdanpaa, "The Quickhull algorithm for convex hulls" *ACM Transactions On Mathematical Software* 22: (4) 469-483 Dec 1996

[3] M.H. Brown, "Exploring Algorithms Using Balsa-II", *IEEE Computer* (May 1998) pp. 14-36.

[4] M.H. Brown, "Zeus: A System for Algorithm Animation and Multi-View Editing", *Proceedings, 1991 IEEE Workshop on Visual Languages* pp 4-9.

[5] K. Mehlhorn and S. Naher, "LEDA - A Platform For Combinatorial And Geometric Computing", *Communications Of The ACM* 38: (1) 96-102 Jan 1995

[6] O'Rourke, Joseph, *Computational geometry in C*, New York, Cambridge University Press, 1994.

[7] J.T. Stasko, "Tango: A Framework and System for Algorithm Animation", *IEEE Computer* (September 1990), pp. 27-39.

[8] A.Y. Tal and D.P. Dobkin, "Visualization of Geometric Algorithms", *IEEE Transactions on Visualization and Computer Graphics* Volume 1, Number 2.

[9] `http://www.cs.princeton.edu/~ah/alg_anim/index.html` (the current version of Gawain).

Viewspace Partitioning of Densely Occluded Scenes

Yiorgos Chrysanthou
Department of Computer Science
University College London, London, UK

Daniel Cohen-Or and Eyal Zadicario
Department of Computer Science
Tel-Aviv University, Ramat-Aviv 69978, Israel

1 Introduction

Computing the visibility of out-door scenes is often much harder than of in-door scenes. A typical urban scene, for example, is densely occluded, and it is effective to precompute its visibility space, since from a given point only a small fraction of the scene is visible. The difficulty is that although the majority of objects are hidden, some parts might be visible at a distance in an arbitrary location, and it is not clear how to detect them quickly. Visibility problems are notoriously complicated. This is reflected in the fact that a small change in the viewspace might cause large changes in the visibility. Having the visibility solved at one point does not help much in solving the visibility at a nearby view. Most previously developed methods for precomputing the visibility of in-door scenes take advantage of the inherent properties of such scenes [1, 6, 8].

In this video we show the results of our method to partition the viewspace into cells, where each cell contains a conservative superset of visible objects. For a given cell the method tests the visibility of all the objects in the scene. For each object it searches for a *strong occluder* which guarantees that the object is not visible from any point within the cell.

2 Outline of the Approach

Let's assume an urban model consisting of a large number of buildings. For simplicity we assume that each building is a single box, where each box is made of a number of triangles. The viewspace is partitioned into *cells* which are sometimes referred to as *viewspace cells*. Each cell defines a *set of visible triangles*, that is, a set of triangles, each of which is at least **partially** visible from some point in the cell. For a given cell our method computes a conservative superset of the visible triangles. Instead of looking for the visible triangles we look for triangles which are guaranteed to be hidden from any viewpoint within the cell.

Given a viewspace cell C and two objects S and T, we say that T is a *strong occluder* of S with respect to C, if T fully occludes S from every viewpoint in C. Since the cell C will be clear from the context we will simply say that T is a strong occluder for S. Given a polyhedral object S our algorithm looks for a convex strong occluder for S. If no strong occluder for S is found , we say that S is *potentially visible*. If a potentially visible object is found to be occluded inside our viewpoint cell, we call it a *weakly occluded* object. We call the set of potentially visible objects for a given viewspace cell, a *conservative visibility set.*

In [3] we show the effectiveness of our method for densely occluded sets by showing that: (i) the number of weakly occluded objects in each of our conservative visibility sets is relatively small, and (ii) we can construct the potentially visible sets efficiently (far more efficiently than it would have taken to compute the visibility set precisely).

Given a viewpoint s and a convex polyhedron P with vertices v_i, if all the rays connecting s and v_i intersect a single convex occluder O, then O is a strong occluder of P with respect to s. That is, for the viewpoint s, O is guaranteed to occlude any point v' in P [4]. Now, given a viewspace cell C defined by its vertices s_j, if O is a strong occluder of P with respect to all s_j, then O is easily verified to be a strong occluder of P with respect to C.

The definition of a strong occluder directly implies an algorithm to detect the strong occluder of a given polyhedron P with respect to a visibility cell [2]. A naive algorithm casts rays from every vertex of the cell towards each vertex of the polyhedron P. Each ray reports a list of potential occluders which is the set of convex objects that have been intersected by the ray. If the intersection of all these lists is not empty then P is reported as being strongly occluded. The conservative visibility set of a given cell is constructed by traversing all the objects of the scene. Each object that is strongly occluded is culled away and not included in the conservative visibility set. **Remark:** Even if O is non-convex, it may still be a strong occluder. However, determining that O is a strong occluder becomes rather complicated.

Assume for simplicity that the n objects of the scene are

SCG 98 Minneapolis Minnesota USA

boxes, and we are constructing the set of conservative visible bounding boxes, without refining them down to the triangle level. For each one of the n boxes the algorithm casts 8×8 rays, where each ray tests its intersection with all the $n - 1$ boxes. However, in a typical model, like an urban model, where the objects are distributed rather uniformly, it is common to use some space subdivision techniques by which a ray needs to test its intersection with only roughly $c\sqrt{n}$ objects. Thus, the cost of generating the conservative visible set of a given cell is roughly $64nc\sqrt{n}$. This estimate is for a naive implementation. In [3] we discuss some optimizations by which the cost in practice becomes linear in n.

3 The Video

The video shows the results of the algorithm tested on a city model consisting of 2500 textured buildings, 75% of them simple boxes, and the rest made up of three boxes each. The base of each building is a 100×100 square, its height is a random value between 200 and 900. In total, there are 3324 boxes, each of which consists of ten polygons, so the total number of polygons is 33240. In the center of the city we have removed four buildings and for this region we are computing the visibility space. The size of the viewspace region is 400x400 units.

In our measurements we use the number of ray/box intersections as the cost estimate for estimating the cost effectiveness of the technique. Our results show that the cost effectiveness gets higher as the cell size decreases. However, we need to account also for the size of the cell relative to the entire viewspace, since small cells require more computation to cover the entire viewspace. By dividing the cost effectiveness by the relative cell size, we got that cells smaller than 100 and larger than 50 are effective with a best preference for a cell size of around 80.

Only ground level cells have been computed (height of 50). The video shows the results by rendering the city from a birds eye view and from a ground level walkthrough. It shows the effect of removing the strongly occluded buildings, leaving only the potentially visible buildings. We first show the performance the walkthrough when it renders the entire model and then only the potentially visible polygon from the current cell (see the birds eye view in Figure 1).

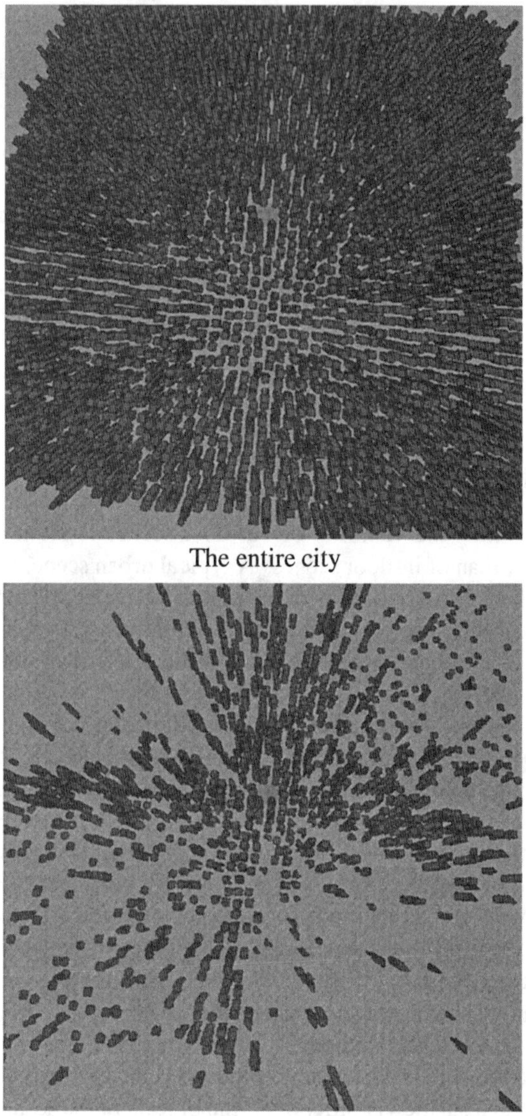

The entire city

The potentially visible polygons

Figure 1: Two birds-eye views from the video.

References

[1] J.M. Airey, J.H. Rohlf, and F.P. Brooks, Jr.. Towards image realism with interactive update rates in complex virtual building environments. *Computer Graphics (1990 Symposium on Interactive 3D Graphics)*, 24(2):41–50, March 1990.

[2] D. Cohen-Or and E. Zadicario, Visibility Streaming for Network-based Walkthroughs, to appear in Graphics Interface'98, June 1998.

[3] D. Cohen-Or, G. Fibich, D. Halperin and E. Zadicario. Conservative Visibility and Strong Occlusion for Viewspace Partitioning of Densely Occluded Scenes. to appear in EUROGRAPHICS'98, Lisbon, September 1998.

[4] S. Coorg and S. Teller. Temporally coherent conservative visibility. In the *Proceedings of the 12th ACM Symposium on Computational Geometry*, 1996.

[5] S. Coorg and S. Teller. Real-time occlusion culling for models with large occluders. In the *Proceedings of the 1997 Symposium on Interactive Graphics.*.

[6] S. Teller and C.H. Sequin. Visibility preprocessing for interactive walkthroughs. In *Proc. of ACM Siggraph*, Pages 61–69, 1991.

[7] H. Plantinga and R. Dyer. Visibility, occlusion, and aspect graphs. The international journal of computer vision, 5(2):137–160, 1990.

[8] H. Plantinga. Conservative visibility preprocessing for efficient walkthroughs of 3D scenes. *Graphics Interface'93*, 166–173, 1993.

A path router for graph drawing

David P. Dobkin *
Princeton University
Princeton, NJ 08544
dpd@cs.princeton.edu

Emden R. Gansner
Eleftherios Koutsofios
Stephen C. North
AT&T Laboratories
Florham Park, NJ 07932
{erg,ek,north}@research.att.com

1 Introduction

A good graph layout can clarify complex relationships among objects. Our goal is to generate layouts that are as readable as good hand-made drawings. We create layouts by first positioning the nodes of the graph, and then routing the edges between the nodes they join. A typical positioning of a node involves placing a label (such as text in a rectangle) that identifies the object represented by the node. Once the nodes are placed, we seek edge routings that are easy to read. This involves generating smooth curves that avoid touching unrelated nodes and edges (*i.e.* stay within an admissable region).

In this video we animate algorithms to solve 2 cases of the routing problem. The first case is routing within a simple polygon. The second case is routing around holes in a polygon. The video is a companion to the paper in which the algorithms are fully described [2].

2 Approach

It has been observed in practice that the most aesthetically pleasing curves are those that are both smooth and close to shortest paths. So, we compute edge routes by first finding shortest polyline paths connecting the source and destination vertices, then smoothing them.

For the case of routing inside a simple polygon case, we apply the path-finding algorithm of Hershberger and Snoeyink [3]. This algorithm builds a funnel delineating allowable paths from the current apex node. The apex is originally at the source vertex, and is moved as the partial shortest path grows, updating the funnel.

For routing around obstacles, we use the algorithm of Overmars and Welzl [4]. This involves first building a visibility graph, and then applying Dijkstra's algorithm [5] to find the shortest path.

Once a path has been found, we set about replacing it by a smooth curve. In this step, the input path may actually be any simple path that satisfies the constraints. We begin by filling in Bézier control points that assure continuity of value and slope across the vertices of the shortest path polyline. We then use these points to fit a smooth Bézier to the points. For smoothness and efficiency, we try to use as few cubic Bézier segments as possible, by testing when sections on fewer control points are possible. For each section, it is necessary to assure that the curve does not move out of the admissible region. This means checking against a set of forbidden edges. In the simple polygon case, the forbidden edges are the edges of the polygon. In the case of routing around obstacles, they are the edges of the holes.

The video illustrates both algorithms in action.

To make the video, we used the *GASP* algorithm animation system [1] on an SGI Onyx Infinite Reality 2. We taped video segments in D3 format; these were edited digitally to produce the final submission. The algorithms were implemented as part of ongoing work at AT&T on network visualization systems. The path router library shown in the video will be released to a broad user community. It comes with a Tcl/Tk user interface allowing interactive entry of routing configurations and capturing of output. We also developed a scripting tool, *tclgasp*, as an interface for GASP. This tool is of independent interest as the framework for a 3D algorithm an-

*Portions of the work of this author done while visiting AT&T Laboratories. This work supported in part by NSF Grant CCR-9643913 and by the US Army Research Office under Grant DAAH04-96-1-0181

SCG 98 Minneapolis Minnesota USA

imation language.

3 Acknowledgments

We thank Ayellet Tal for lots of guidance in using GASP, and we thank Keith Bauer for recording and editing the video. We also thank Joe Mitchell for sharing code that implements a version of the Hershberger-Snoeyink algorithm.

References

[1] D. Dobkin and A. Tal. Visualization of geometric algorithms. *IEEE Transactions on Visualization and Computer Graphics*, 1(2):194–204, 1995.

[2] David Dobkin, Emden Gansner, Eleftherios Koutsofios, and Stephen North. Implementing a general-purpose edge router. In G. Di Battista, editor, *Graph Drawing 97*, volume 1353 of *Lecture Notes in Computer Science*, Rome, Italy, 1998. Springer-Verlag.

[3] J. Hershberger and J. Snoeyink. Computing minimum length paths of a given homotopy class. *Comput. Geom. Theory Appl.*, 4:63–98, 1994.

[4] M. H. Overmars and E. Welzl. New methods for computing visibility graphs. In *Proc. 4th Annu. ACM Sympos. Comput. Geom.*, pages 164–171, 1988.

[5] R. Sedgewick. *Algorithms.* Addison-Wesley, Reading, 1985.

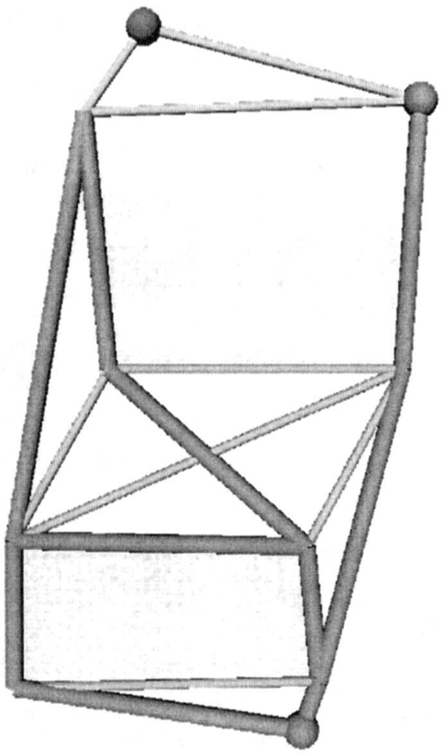

Figure 2: Finding a shortest path from the visibility graph.

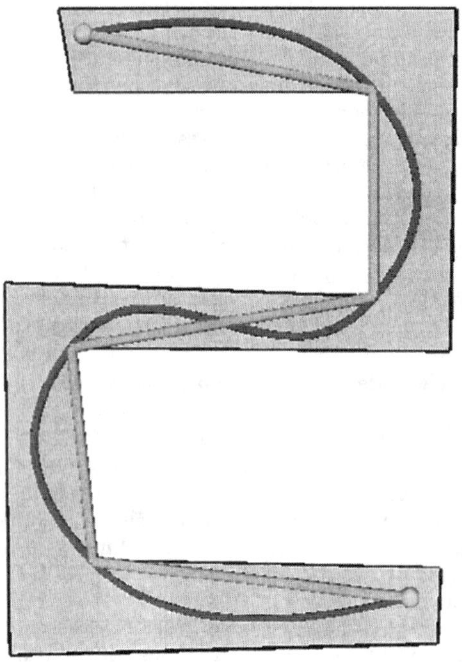

Figure 1: Routing inside a polygon

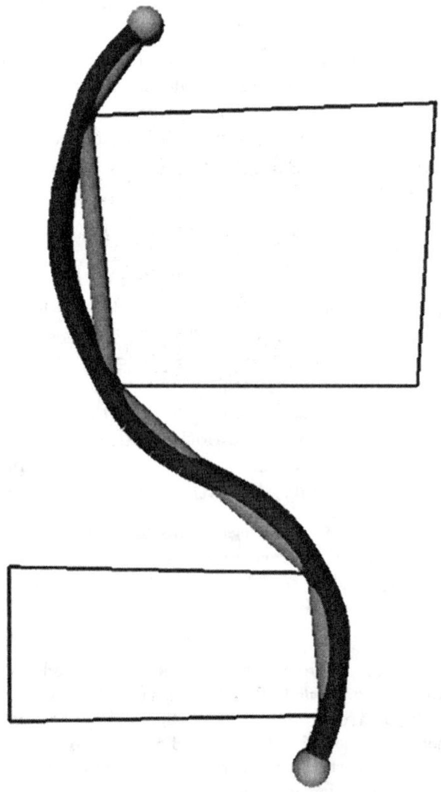

Figure 3: Routing around polygonal obstacles

AUTHOR INDEX

NOTES

NOTES

NOTES

NOTES